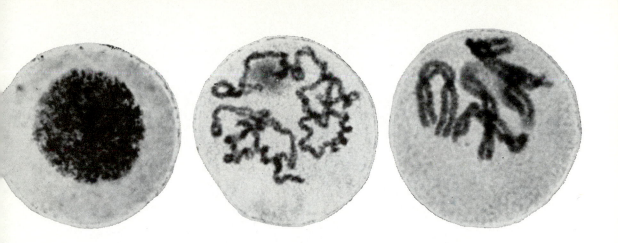

Life
An Introduction to Biology

SECOND EDITION

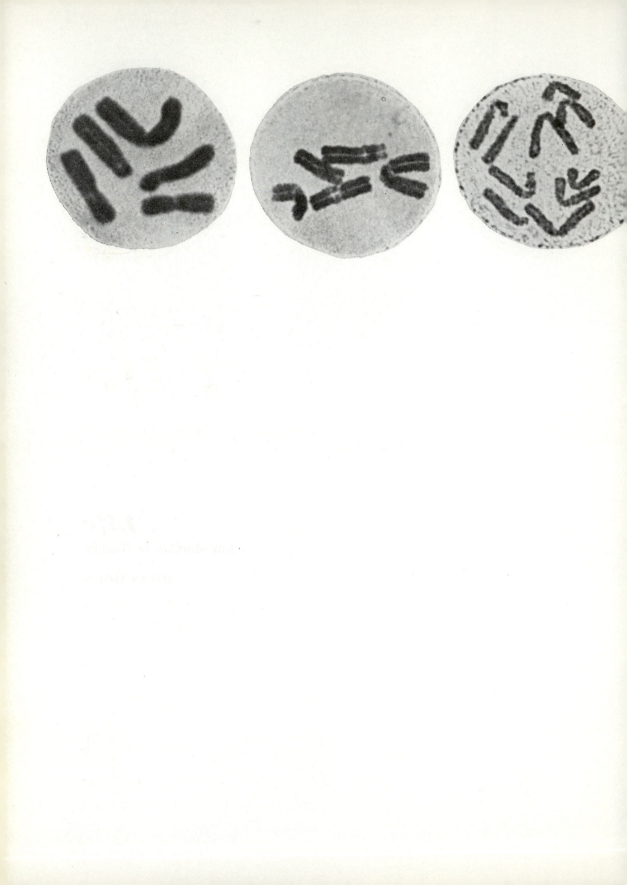

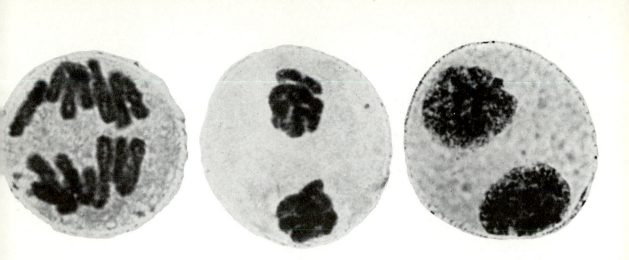

Life

An Introduction to Biology

SECOND EDITION

GEORGE GAYLORD SIMPSON
Harvard University

WILLIAM S. BECK
Harvard University

Harcourt, Brace & World, Inc.

New York · Chicago · Burlingame

COVER PHOTO: *Actinoptychus* (diatoms), from Carl Strüwe.

PHOTOS ON PAGE i–iii: Mitosis, from Dr. A. H. Sparrow, Brookhaven National Laboratory.

Library of Congress Catalog Card Number: 65–14384

Printed in the United States of America

FROM PREFACE TO FIRST EDITION

This book is based on strong convictions. We believe that there is a unified science of life, a general biology that is distinct from a shotgun marriage of botany and zoology, or any others of the special life sciences. We believe that this science has a body of established and working principles. We believe that literally nothing on earth is more important to a rational living being than basic acquaintance with those principles.

This, then, is a "principles approach" to general biology. We have tried to give more than lip service to that aim, to keep principles always foremost in mind and to organize the whole text around them. We have tried to avoid the common error of presenting merely summary or introductory descriptions as principles. We recognize, however, that principles are meaningless unless they arise from concrete data and can be applied to particular problems. We have tried to underpin the principles with supporting facts and to show how the principles arise from these facts.

We have discussed how scientists approach problems, which of course is all that is meant by the catchword "scientific method." Where appropriate and convenient we have introduced in sufficient detail a number of particular examples of important biological problems and their solutions, partial or complete. This has usually been done by a historical approach here and there in the text, additional and complementary to the more general historical summary in the last chapter.

We hope to discourage the idea that merely learning a specialized vocabulary is educational in a broader sense or has any useful relationship to wisdom, but we have not hesitated to use carefully defined technical terms when these really contribute to the easier communication and comprehension of principles. For selected examples we have introduced anatomical terminology in figures and their legends, where the user of the book may take it or leave it. These labeled drawings may also serve as a bridge between the text and the study of organisms in the laboratory, which is the place to learn such details of anatomy as may be useful or interesting to any particular reader.

The study of life is a lively science. Along with a body of well-established knowledge, it includes uncertainties, speculations to be tested, places where knowledge now ends and where the seeking of new knowledge should therefore begin. We have not concealed these dynamic aspects of the search. On the contrary, in appropriate places we have stressed them.

Like other human activities, biological teaching and research have fads that change from place to place and from time to time. We have tried to avoid them and to give as well-rounded a treatment as we could manage. We have, for instance, treated systematics and biogeography on the same level and to the same depth as, say, biochemistry or genetics. One of the fads we have avoided is centering the treatment on man, writing a human biology or using man as a "typical animal" (a horrid expression). We thoroughly agree that the best reason for studying biology is the most human reason: Know thyself! But too narrow a striving toward that objective is self-defeating, because a true understanding of man can come only from placing him in perspective in the whole realm of life. Man is used as an example when he is a good example, and the human implications of, for instance, population growth or the evolution of behavior are specified. The subject, however, is always considered to be broader than our own species.

The most general principle of all in biology is evolution. Most books on biology make such a statement, but many are lacking in conviction that it is really true. Some relegate evolution to a single chapter and treat all the rest of biology as if it really had nothing to do with evolution. Others adopt what they call an "evolutionary approach" and equate biology with the description of organisms in a "phylogenetic sequence." The sequence is never in fact phylogenetic, and this approach rarely teaches much about the real principles of evolution. We have tried to make evolution as pervasive as it really is in the world of life. Every topic has its evolutionary background and aspects.

Only those who have also tried to encompass our tremendous subject in one book know how difficult is the problem of organizing the material and how impossible is the achievement of a completely consistent and logical sequence. In general we have advanced by levels of inclusiveness: the cell as the basic unit in our subject; then the organism in itself; then like organisms in their reproductive capacity; the further reproductive processes and interactions that lead to evolutionary changes; the diversity wrought by those changes; the aggregations of the diverse organisms into populations and communities; the spread of such aggregations in the dimensions of space; and finally the history that embraces all previous levels and dimensions plus that of time.

This book is written neither for the nonprofessional nor for the student who is beginning a professional career, but for both at once. It is another of our convictions that grounding in the *principles* of *general* biology is equally useful to all. It is the most nearly indispensable (and the most interesting) view of the subject for the nonbiological student or the general public. It is at the same time the best basis from which to go on to specialization in botany, zoology, biochemistry, or any other of the life sciences.

We have received much help during the long years of preparing this book. Some aid cannot well be specifically acknowledged, but we can acknowledge with gratitude that the whole manuscript has been read and criticized in detail by W. H. Camp, A. D. Chiquoine, Waldo H. Furgason, Albert S. Gordon, Ella Thea Smith, and Kenneth V. Thimann. Anne Roe has also read most of the manuscript and has more particularly helped with the discussion of behavior. Other friends criticized our treatments of specific subjects as follows: Frank H. Johnson; (cell theory and physiology) A. D. Hershey and A. E. Mirsky (genes); D. L. Lindsley (reproduction and genetics); J. T. Bonner (development).

It is a pleasure to give thanks to Charles Halgren and his associates at the CARU Studios for the execution of the figures. In illustrating our book we have, of course, drawn freely on earlier works. In all but a few instances, however, figures have been either redrawn or newly drawn for our particular purpose.

GEORGE GAYLORD SIMPSON

COLIN S. PITTENDRIGH

LEWIS H. TIFFANY

PREFACE TO SECOND EDITION

In this revision, we have retained the general aims and approach announced in the Preface to the first edition. The revision has nevertheless been extensive, and much of the book has been rewritten, newly written, or reorganized. We have tried not only to take account of the many recent advances in knowledge, but also to improve the arrangement and precision of the text. Among the more obvious changes are these:

Purely introductory matter has been condensed and in part replaced by a more explicit and more extensive basis in biochemistry and molecular biology.

Biochemistry and a number of other topics have been given more sophisticated treatment throughout the text.

Reproduction has been treated in a more coherent sequence, being discussed first in terms of cellular processes and then through heredity to its fully organismal aspects. The study of reproduction thus precedes and introduces organismal physiology instead of following it.

The chapters on ecology (23–25) have been extensively reorganized.

Some repetition and redundancy have been eliminated; for example, protists are now discussed in essentially one and not several places.

In making this revision, we have profited from many helpful comments and constructive criticisms from teachers, students, and general readers. They are too numerous to mention individually, but we heartily thank them collectively. We must at least express our gratitude to Dr. Miklos D. F. Udvardy, who, in addition to other assistance, prepared a careful and detailed plan for reorganization of the whole subject of ecology. Although we have not followed his plan exactly, we have benefited from it.

Dr. Otto T. Solbrig critically reread the whole book with special reference to its botanical materials, on which he gave us extensive advice. Dr. Anne Roe again reviewed the chapter on behavior (now Chapter 14).

It has been impracticable for Drs. C. S. Pittendrigh and L. H. Tiffany, coauthors of the first edition, to take part in the revision. It is, however, thankfully recognized that they supplied substantial parts of the basis for the present version. No invidious comparison is involved in noting that this is particularly true of Dr. Pittendrigh, who as second of the original three coauthors was also second in the magnitude of his contribution.

The new coauthor has been especially but not exclusively concerned with the more biochemical and physiological aspects of text—notably Chapters 2–5, 7, and 10–12. Both of us will be grateful for renewed comments from readers and users of the book.

GEORGE GAYLORD SIMPSON
WILLIAM S. BECK

CONTENTS

4 The Maintenance and Integration of the Organism 281

6 *The Diversity of Life* 487

7 The Life of Populations and Communities 625

8 The Geography of Life 703

10 *The History of Biology* 819

Life
An Introduction to Biology

SECOND EDITION

Part 1

Introduction

The picture introducing this part, which is itself an introduction to the whole book, is a photograph of a spider's web. The web is not alive, but it serves as a beautiful symbol of many of the general themes in the study of living things. It reflects the complexity and organization of the animal that made it. It has a function, which is to help provide food, and therefore both materials and energy, for the maintenance of the spider. It is the product of its maker's behavior. It expresses the intricacy and precision of the spider's adaptation—its ability to live from and within its environment; an aspect of that adaptation—and of all adaptations whatever—is the ability to reproduce, and hence to maintain a population of spiders through the ages, a population that endures although the individuals in it perish. The web is finally but pre-eminently an outcome of a history of change, of organic evolution, that started long, long before there were any spiders and that continues to produce extraordinary results.

These are the major themes of life and of this text, and we shall examine them further in Chapter 1 as they are seen in a forest and on a coral reef. We shall then organize the inquiry to be pursued throughout the rest of the book in terms both of these main themes and of levels of organization in nature, from molecules through cells, whole organisms, specific populations, and multispecific communities to the entire world of living things in time and space.

Then we shall consider the nature of our inquiry and of science itself as a way of knowing, understanding, and coping with the world in which we live. We shall also briefly trace the expanding scope and impact of science to its culmination when Darwin at last brought the phenomena of life fully into the domain of naturalistic study.

1

The Living World
and Its Study

*Rabbits, a rich and perennial crop,
fall prey to hawks, coyotes, weasels, and
other hungry carnivores.*

"Know Thyself"

You are alive, and all around you are other living things. You would not yourself be alive if you were not part of the whole complex world of life. This is true not only in the sense that you depend on other forms of life for food but also in other and larger senses. You live in a community, a community of other humans and also of many other living things in greater diversity and of greater impact on your own life than you may have realized as yet. You share with them many processes of living. The study of these processes in other animals and in plants is necessary for an understanding of your own life. Moreover, you are literally related to all the other living things, just as truly as you are related to your sisters and your cousins and your aunts. You share a common ancestry with every other animal and every plant; you are a product of the same long, intricate history.

The real reason for studying biology is the old admonition "Know thyself." The better you know yourself, the happier, healthier, more comprehending, and richer will be your life. You cannot, however, really know yourself if that is *all* you know. True understanding can come only from knowledge of life in general. The meaning of biology is its human meaning, its significance to you as a person, but that meaning can be made clear only if human biology is seen as a part of the biology of all life.

Awareness of the world of life and knowledge of it begin when we look at living things and wonder about them. For a start let us look and wonder at two very different scenes in nature, both abounding in a multiplicity of things alive.

A Forest

THE SCENE AND THE ACTORS

Here (Fig. 1–1) is one of the beautiful forests of ponderosa pines in our Southwest. The stately old trees are 150 feet or more in height and 300 to 400 years old. They stand well apart, as if intolerant of each other's shadows, as indeed they are. They need full sunlight and, like all plants, have special requirements as to soil, slope, drainage, temperature, rainfall, and other conditions. We say that they have "preferences," but they make no choices, growing unconsciously when seeds fall where their needs are met.

The light that filters through the pines falls on a forest floor carpeted with herbaceous plants that make it green and flowery in summer. Along a nearby stream are other trees with different needs: alders, willows, narrowleaf cottonwoods. On exposed, rocky slopes are still others: piñon pines, tree junipers, and oak scrub. There are plants everywhere, from the cactus on a sandy, dry, southern slope to the iris on the marshy bank of a pool, each in the particular place that best supplies its particular needs.

On a calm summer day the plants are motionless and silent. They seem to lack the ceaseless activity that we associate with being alive, but their external appearance is merely an illusion. Within them tremendous activity is going on. Water and chemicals are being drawn from the soil by roots, and gases from the air by leaves. At the same time the leaves are capturing the energy of sunlight, energy released by fierce nuclear reactions in a star 93 million miles distant. From these materials and with this energy the plants are compounding special foods that, as we shall see, are the beginning point for all the vital processes of the forest and its denizens.

Throughout the scene are animals whose most evident activity is consumption of the plants. Squirrels tear apart the pine cones and eat the nutritious seeds. Deer wander daintily along, cropping leaves from the lower, more succulent vegetation. Wild turkeys flock in the oaks and piñon pines to gorge themselves on acorns and pine nuts. The plant-eaters are preyed on by other animals. The deer are killed and eaten by mountain lions. Wildcats stalk the turkeys. Rabbits, a rich and perennial crop, fall prey to hawks, coyotes, weasels, and other hungry carnivores. The methods are endlessly varied, but the business is the same for all from pine tree to weasel: the gathering of materials and energy.

Everywhere there are plants and animals less obvious than the few so far noticed. There are literally hundreds of kinds of insects, enough to occupy a whole congress of experts in their identification. They draw their materials and energy from every living thing in sight, from sources ranging from other insects to the nectar in flowers. Even the apparently lifeless soil is in fact swarming with life in forms often far too small to be detected with the unaided vision. Here, among many others, are mites and tiny worms, bacteria and molds. Even the clean mountain air is full of such living things as spores, bacteria, and pollen.

SOME THEMES OF THE DRAMA

The extremely diverse forms and the no less diverse activities of the plants and animals of the forest may seem at first sight to be an ill-organized jumble. More thoughtful consideration quickly reveals that all, from bacterium in the soil to hawk in the sky, make up an integrated cast of actors playing in a unified drama. That drama has certain major themes in which every actor has his part.

Organization

We may view the drama at many different levels, and at each level we shall find that it is in fact intricately and extremely well organized. The structure of each actor is no hodgepodge but amazingly the contrary. The most complex and highly organized things in the world are the living things, to such an extent that the word *organism* has become synonymous with *living thing*. There is nothing random in the anatomy of a bacterium or a hawk. We see that the bacterium is a cell and that a hawk is made up of millions of cells. There is nothing random within a cell, either. We are here looking at a level of organization below that of the individual organism, but it is no less organized. At a level above that of the organism, we find the individuals living together in groups of their own kind, reproducing, and maintaining populations that endure although the individuals perish. Here organization differs greatly in degree and kind among the various groups. It is comparatively loose and simple in bacteria and tight and complex in hawks, but both populations are organized. A level higher still comprises the whole community as we

1–1 A ponderosa pine forest. The ponderosas stand well apart as if intolerant of each other's shadows. The stately trees are 150 feet or more in height and 300 to 400 years old.

Wild turkeys flock in the oak and piñon pines to gorge themselves on acorns and pine nuts.

Deer wander daintily, cropping leaves from the succulent vegetation.

Lower right: Along the stream bank are different kinds of trees.

Below: Insects are everywhere and make up in numbers what they lack in bulk. Here one beetle larva preys upon another.

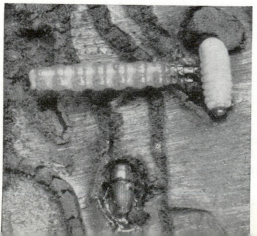

first saw it, with its myriads of different species. Organization at this level is most evident in relationship to a second major theme of the drama.

Traffic in materials and energy

Organisms are obviously built of materials, and a study of organisms must sooner or later examine what those materials are at a level still lower than that of the cell—the molecular level. In our forest a part of the drama already glimpsed is the individual acquisition of materials, by plants from soil and air, by animals from plants or from other animals. Within the individual, at an appropriately lower level of observation, the materials thus acquired are transformed so as to be useful in the new system of organization that they have entered. In the community as a whole there is a constant flow of materials from soil and air to plant, from plant to animal, from animal to other animal, perhaps through a long sequence of animals, eventually from plant and animal to bacterium, and back to soil and air. That flow is the plainest element in organization at the community level.

Along with the materials goes *energy*, the capacity to do work. Every organism ceaselessly does work and so must ceaselessly acquire energy. A hawk obviously is doing work when it flies, screeches, or dives and seizes its prey. It is less obvious but still more important that a hawk is also expending energy when it simply sits still, or that a bacterium or a tree constantly utilizes energy. The fact is that energy is required not only to achieve but also to maintain organization. Unless work is put into them for maintenance, all systems tend toward disorder and eventual loss of organization. This is bitter truth to the owner of a letter file, a house, or a factory. It is no less true of the systems that are organisms. They must spend energy or become disorganized, and disorganization is death.

With some exceptions of no importance at this point, all the energy used by all the organisms in the world (including industrial man) comes ultimately from the sun. In our forest we noted that the green leaves were capturing the radiant energy of sunlight. The plants transform that energy into a different form, into a successive series of chemical compounds, molecules that embody both materials and energy. The first step is the compounding of an energy-rich sugar, which later yields energy for other vital processes and materials for other organic compounds. All of the available energy and the most important of the materials in the entire forest community entered that organization through plants, which thus are the most essential actors in the whole drama. The flea that bites a mountain lion that ate a deer that browsed on leaves is acquiring energy from the sun, energy that would not be available for any other organisms if it had not been captured and transformed by plants.

Behavior

Behavior is not so much a theme of the drama as the way certain roles are played. As such it is a very necessary part of the drama, and we mention it here because it must also be one of the separate themes in our study of biology. In its overt forms—the browsing of the deer, the attack of the mountain lion, the hopping of the flea—it is practically confined to animals. It is essential to their acquisition of materials and energy, their reproduction, and many other of their life's activities. It depends on, and so its study becomes inseparable from, nonbehavioral activities at the lower levels of organization, for example, within the cells of muscles and nerves.

There are other major themes, but we shall bring them up in the scene to which we now turn.

A Coral Island

ANOTHER SCENE, ANOTHER CAST

The distribution of coral reefs (Fig. 1–2) is limited to shallow, warm, clear, and well-lit seas in tropical and subtropical regions. The restriction is due to the fact that the reefs are built by organisms that, like all living things, thrive only under particular conditions. The principal builders of the reefs are corals, sedentary relatives of the jellyfishes. Many different species of coral secrete calcium carbonate as a stony external skeleton (Fig. 1–3). Some corals are solitary, but the most important reef-builders divide repeat-

AUSTRALIA

GREAT BARRIER REEF

Between the clumps of living coral, brilliantly colored striped and spotted fishes dart along the channels.

1–2 The Great Barrier Reef. The world's greatest display of coral is the Great Barrier Reef, which lies from 15 to 100 miles off the northeastern coast of Australia, running for nearly 1200 miles roughly north and south.

The coral *Tubipora*, which secretes a red skeleton. Note the expanded tentacles of the animals to the right.

The male (*left*) and female (*right*) of the crab *Hapalocarcinus*, which lives in the coral.

A sea cucumber (bêche-de-mer).

Turtles come ashore to lay their eggs.

A tern nesting on a coral island in the Great Barrier Reef.

Much of the reef's life can be seen through a glass-bottomed boat.

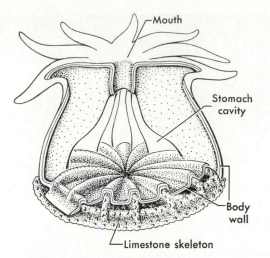

1–3 A coral animal and its limestone skeleton.

edly as they grow and eventually form whole colonies of interconnected individuals.

At certain times of the year the coral animals shed minute eggs into the surrounding water. The eggs develop into larvae, which drift until they encounter favorable conditions, settle down, grow into coral animals, and initiate new colonies. When colonies die, the stony skeletons remain. This cycle, repeated through many generations, leads to the accumulation of massive reefs.

The world's greatest display of coral is the Great Barrier Reef, running roughly north and south for nearly 1200 miles from 15 to 100 miles off the northeastern coast of Australia. Much of the reef is well below low tide, but in places oval, platformlike patches have been built up to the surface. On some of these platforms, barely awash at low tide, winds and waves have piled coral sand above the high-water mark, forming islands. The islands are soon invaded by pandanus palms and other vegetation. Here sea birds roost or burrow, and turtles come ashore to lay their eggs.

The living reef around an island is one of the most spectacular exhibits offered by the world of life. The corals themselves form dense, brilliantly variegated, forestlike growths in pools on the platform and down its outer slopes into the deep water. The skeletons are usually white, sometimes red, but the living animals are green, red, violet, yellow, and most of the other colors of the rainbow. Curiously enough, these colors do not belong to the corals' own tissues but to thousands of microscopic plantlike organisms that live in the stomachs and in the cells of the coral animals.

Embedded in the clumps of coral are giant clams and tubes made by worms—gorgeous creatures with colored tentacles unlike our drab garden worms. Soft corals (without skeletons), sea anemones (animals related to corals), and many seaweeds also find footing on the stony reef. On channel floors are starfishes, sea cucumbers (animals, despite the name), and many other animals. Brilliantly colored striped and spotted fishes dart in and out among the branches of the coral forest. The apparently clear water swarms with microscopic floating plants and animals. The diversity of species is incredible, even surpassing that of the pine forest.

When, as has repeatedly been done, a deep hole is drilled into a coral reef, we learn other things of supreme significance for the drama of life. Under the thin film of living corals and other abundant surface life, there are hundreds of feet of dead rock deposited long ago by once living corals and other reef organisms. A boring in the Great Barrier Reef found down to 506 feet the skeletons of corals that lived *several thousand years ago* but that nevertheless were practically identical with those living today. A boring on Bikini brought up from a depth of 2556 feet samples of coral skeletons laid down *20 to 25 million years ago*. Reef corals of today never grow at a depth below 250 feet. Moreover, the corals and other reef animals from the bottom of the Bikini boring are of species quite different from those now growing on the reef. The presence of coral skeletons at depths where they cannot now live, the persistence of their species through thousands of years, and the change of species through millions of years are three clues leading to other great themes in the drama of life.

THEMES OLD AND NEW

New versions of old themes

Here in this extraordinarily different setting, the themes we met in the forest are again exemplified. Certainly the life of the reef is no less organized at all the different levels from molecule to community. The traffic in materials and energy is no less constant and vital. In the forest the plants, which take the essential first steps in that traffic, were more conspicuous than the animals. Here on the reef animals are more conspicuous, but of course and necessarily plants are here, too; there are seaweeds but also and more especially immense populations of microscopically small plants in the waters around the reef. Many reef animals, unlike any in the forest, spend their whole lives in one spot after a brief larval stage. They can do so because every drop of water brings them food as it washes by them. Not a single species is common to forest and reef, but the pattern of the traffic is the same: energy from sun to plant to herbivorous animal to carnivorous animal and eventually to bacterium; materials from the nonliving surroundings to plant to animal and through varying cycles back to the surroundings again. The difference is mostly in the details of how the traffic is conducted in the two vastly different environments. Behavior on the reef is also directed toward just the same ends as in the forest, but it differs in detail as the environments differ.

Reproduction

Reproduction is a dramatic theme just as important and ubiquitous in the forest as on the reef. We have, however, more explicitly noticed it on the reef because it is responsible for the very existence of the reef itself. It was also by examination of the reef that we learned that species and communities like those of today have persisted for thousands of years. All living things reproduce, giving origin to others like themselves. This is the most fundamental point in any attempt to define life. The almost incredible accuracy of the process is attested by the persistence of whole communities for thousands of years. Through reproduction life continues although all organisms die.

Evolution

The Bikini boring introduced two other major themes: through millions of years the earth, the environment of life, has changed, and living things themselves have changed.

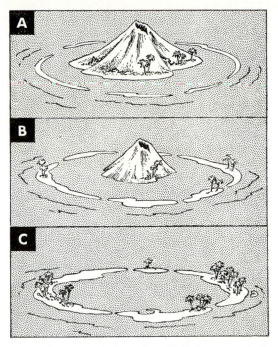

1–4 Darwin's theory of atoll formation. *A.* A coral reef fringes the coastline of an island. *B.* As the sea bottom (the earth's crust) sinks, the island slowly becomes submerged. The rate of sinking is slow enough for the growth of the coral reef to keep pace. *C.* Eventually the island is entirely submerged, but the characteristically circular atoll remains. The coral skeletons formed at *A* are now at depths below the growth of living coral.

Coral skeletons where now no coral can grow testify to changes in the relative levels of sea and land. Darwin, who was already a famous scientist before he wrote on organic evolution, was impressed by this fact as a young man. He developed a theory of the formation of atolls (circular reefs) that is still accepted in essence today (Fig. 1–4).

The change in organisms over long periods of time is, of course, organic evolution. Since Darwin, few biologists (and those few surely deluded) have doubted that all the millions of species of plants and animals have arisen from a remote, single common ancestry by a natural process acting through eons of time. The reasons for that conclusion are multiple and complex—it is, indeed, their multiplicity and complexity that makes them absolutely convincing. Evolution is the most general, pervasive theme in the whole drama of life. It is, on that account, also the central, recurrent theme of this book, and the reasons why evolution is an inescapable conclusion will emerge bit by bit as we take up its different aspects.

Two of the broader aspects of evolution do call for explicit mention here because they deserve to rate as major themes in themselves and because they are abundantly illustrated in our forest and our reef and indeed in any community. These are *adaptation* in organisms and *diversification* among them.

Every individual in a community is so organized as to achieve the greatest ends in life: the acquisition of energy and materials and the reproduction of the population. Each kind of organism is adapted—usefully organized—to survive and reproduce. It seems banal to assert that organisms are fitted to live where they do in fact live, but the statement has deep implications. To say that a hawk is adapted to feed on small birds and mammals means more than simply that it can and does feed on them. It means that the detailed organization of the hawk specifically serves this end and that the details, far from being obvious, require close study to be understood. Such considerations lead to one of the most profound questions a biologist can ask: "How did the hawk become adapted?"

The diversity of the actors in our two scenes has been sufficiently emphasized. Here we only make two additional points. First, it is a diversity in adaptation; each species of organism in the community has a way of life different in some respect from that of any of its neighbors, and each is specifically adapted to its own way of life. Second, the diversity is an outcome of evolution, which not only adapts each species to its own way but also somehow parcels out all the ways of life among different species.

Geography of life

Mountain forest and coral reef together illustrate another theme—that life has a geography. This geography was visible on a small scale within each scene. The species of forest plants growing on a sunny hillside differ from those on a shaded stream bank. The species of reef animals living on the sandy island differ from those beneath the surrounding waters. These differences relate to the special needs of each species and to differences of environment within the local scene. More broadly we observed that the American forest and the Australian reef do not have a single species in common (except such as may have been introduced by man). This situation, too, might be ascribed to differences in the environments, but a glance at another scene shows that this is not the whole story. There are forests in Australia growing in environments like that of our American forest. Yet the forests in the two countries do not share a single species. The geography of life does not depend only on its relationships with the environment, its *ecology*; it also depends on evolutionary history and especially those factors that have facilitated or impeded the geographic spread of species over the earth.

History of life

Mention of the historical element in the geography of life brings us to our final major theme, one also exemplified, although on a very small scale, by the boring on Bikini. This theme is the consideration of evolution not only, as heretofore, as a process and an explanation of vital phenomena but also as history, as the actual sequence of events that have occurred in the time since life began. Here we reach the highest level of all, one embracing the whole world of living things through the billions of years of our planet's existence. Here we may reasonably feel that our project of inquiring into life may be brought to an end.

Organizing the Inquiry

We have introduced the living world as anyone is most likely to approach it—seen with the unaided eye in the communities of forest, shore, field, or park. On that basis we have encountered the major themes of biology and also the concept of different levels of organization. There is no such thing as a wholly *separate* theme or level, in the sense of one that can be intelligently studied and understood without any reference to other themes and other levels. The hawk's behavior in seizing prey, for instance, becomes fully significant only when we follow the consequences in one direction down into the hawk's cells and the energetic molecular reactions there and in another direction to the origin, reproduction, and destiny of the population of hawks. All biological subjects are interconnected in so many ways that comprehension of any one demands a little knowledge, at least, of all. But one cannot study any broad field of science by learning something about every part of it all at once. To be effective and systematic, the procedure must be more like unraveling a tangled skein; one has to find an end and continue from there.

Among many possibilities, we have found it most practical and enlightening to orient our inquiry in large part by levels of organization. The level of communities was best for a first over-all glimpse of our themes, but it is not the best end for unraveling. The various levels in nature are not so distinct that a definitive and clear-cut sequence arises automatically, but for our purposes these levels can be listed:

> Atoms
> Molecules
> Cells
> Tissues
> Organs and functional systems in individuals
> Individual organisms as wholes
> Populations (reproducing groups of organisms)
> Phyletic lineages (evolving populations)
> Communities (associations of different populations and lineages)
> The world of life as a whole, in space and in time

Although our subject is life, our discussions must start at the top of this list. The first level at which life is present in the fullest sense is that of cells. Of course, what goes on within cells also involves the levels of atoms and molecules, and we must attend to those, too, insofar as they are basic for life in the cells. We shall start there at the level of the subliving and proceed in a general way upward through the scale of levels to its end.

At each level the various themes will appear as they are applicable and appropriate. All themes run through several if not through all levels. They thus supply a continuity or, in extension of the previous metaphor, a set of connecting threads as we weave the unraveled yarn into a fabric. Some themes distinctly belong to certain levels rather than to others; behavior, for example, emerges only when the level of the individual is reached. Other themes are highly pertinent at all levels but have different aspects at each level. The traffic in materials and energy, for instance, goes on at every level, but the forms it takes, the aspects of it that we examine, within a cell and, say, within a community are quite different. The most omnipresent of all themes, the thread running through the entire fabric, is evolution.

Pursuing the Inquiry

THE NATURE OF SCIENCE

A visitor to the Great Barrier Reef experiences many emotions. He is awed by the vastness and strangeness. He rejoices in the beauty. He is curious about all the details of the scene. Each emotion is a basis for human institutions and activities. Man's highest response to the sense of awe is religion. Joy in beauty leads to art. Science is the systematized human activity that responds to man's curiosity.

Motivation

Science, a human activity, has varied human motivations. Curiosity, the desire to know, the pursuit of knowledge for its own sake, is a basic motivation. But it is at least equally human to want knowledge for our own material ends: to supply growing needs, to increase comfort, to amuse us, and even (alas!) to kill our fellow men. It has been found that knowledge originally acquired for its own sake almost always, sooner or later, can be used to gain such material ends. It is, then, human to seek knowledge with those ends directly in view, motivated by gain rather than by curiosity. Whatever the motivation of the individual scientist may be, the tremendous increase in scientific activity that characterizes our civilization undoubtedly is due in largest measure to such material considerations. Science works. The methods of science yield knowledge that is both sound and materially useful. Certainly scientists do not scorn this outcome or consider material motivation as unworthy. On the contrary, most of them rejoice that science is useful and are happy themselves to share in the benefits. (Of the science that we are now studying, health is one of the benefits, only one among a great many.)

The interrelationships between the pursuit and the application of knowledge are so close and numerous that the two sometimes seem indistinguishable. It is true that they intergrade so that no sharp line can be drawn between them, but that does not make them identical. Besides intergrading, science and technology also depend each on the other, but they are not the same. Science is a response to curiosity, a way to acquire new knowledge, a body of knowledge acquired, a means of insight into relationships among facts, an explanation of those facts. Technology is a response to felt needs, the application of knowledge to material ends. Biology has many extremely valuable technological applications, in agriculture, in medicine, and elsewhere, but those are not our subjects. We are concerned with biology as a science, also as an indispensable basis for technologies, but we are not here concerned with the technologies themselves.

Question and answer; testability

Scientific method is refined common sense. In their examination of the world, the scientists' purpose is to ask sensible questions and to seek sensible answers. It is common sense that the questions should be about observable things, should arise from the phenomena of our world. What kind of coral is that? What does it eat? What happens to the food? How does it reproduce? Simple questions, but sensible ones, and scientific. How many invisible angels can dance on the point of a pin? We need not discuss whether this question is sensible or not, but we do insist that it is not scientific; it does not arise from an observed phenomenon. The first task of a scientist is learning to ask the right kinds of questions.

Scientific answers must also be sought in the right place, in the observation of the world. We consider this common sense, too, but it has not always been so considered. During the Middle Ages many highly intelligent men thought it sensible to seek answers about nature in the writings of ancient Greek philosophers rather than in nature itself. (And the ancient philosophers had rarely interrogated nature directly or asked the right questions.) Here the most important point is that to be scientific an answer must be testable, and must in fact be tested, by observation. The history of biology, more than that of most sciences, has been afflicted with untestable and therefore unscientific answers on various matters. For example, to the question "What causes evolution?" Henri Bergson, a French philosopher (1859–1941), proposed the answer *"Élan vital* [life drive]." But the *élan vital* was postulated as unobservable. The answer could, in some sense, be true for all we know, but that qualification is precisely what makes it unscientific; we cannot know, for there is no way to check it. Moreover—and this leads to our next point—even if true in some sense or other, the answer would not explain anything, any more than an *élan locomotif* could explain how a locomotive works.

Observation, generalization, and explanation

Observation of a particular object or occurrence is the fact, the datum, from which all scientific knowledge stems and to which all scientific theories must return. It is essential to the nature of science that an observation, against which the answers to scientific questions are tested, be repeatable. What a scientist has observed correctly, anyone else can observe and corroborate. This is the basis for one of many characterizations of the field of science: science deals with matters about which general agreement is possible. Nevertheless, an isolated fact, no matter how repeatable observation of it may be, is only a basis for science and is not the essence of scientific endeavor. It has little or no interest unless it can somehow be *generalized*, related to other facts, and ultimately be *explained*.

The isolated observation of a coral animal having a defined number of tentacles around a mouth and resting on a stony base of a certain form is generalized in one way when it is observed that the coral animal is one of a large population of individuals with closely similar characteristics. This kind of generalization is widespread and fundamental in biology. A similar kind of biological generalization is embodied in the statement that all normal mammals have sex (the concept of sex is itself a generalization) and in this respect fall into one of two categories, male and female. Once reached, such a generalization leads to a search for extensions of it; in this example the generalization as to sex was found to apply also to most (but not all) organisms other than mammals.

Such generalizations involve only the extension of observation to cover a large group of individual facts. The extension may even legitimately go beyond the range of observations actually made. If we have examined a million normal mammals of many species and found that all of them are either male or female, we run little risk of error if we conclude that those we have not examined follow the same pattern. Still all we have done is to establish some characteristics of a group rather than of an individual. We are still describing, not explaining.

Another kind or level of generalization deals with relationships between phenomena or between generalizations of different orders. Usually, although not quite invariably (few generalizations are absolutely invariable in biology), offspring of sexual organisms are produced after, and only after, some form of union of male and female elements has occurred. Here the nature and sequence of the phenomena are such that we feel that the generalization does have explanatory meaning. We feel entitled to say that offspring are produced *because* the two sexes come together. Given a particular observation of reproduction, we may cite this generalization as an explanation of the single incident.

Biology is characterized by three kinds of explanations. One kind is shared by the biological and the physical sciences and constitutes the connection between the two. It relates organic structures and activities to the chemical structures and physical activities that compose or underlie them and that in this special sense *cause* them. For example, as we shall see later in this book, essential features in reproduction are explicable by the properties of a remarkable substance called deoxyribonucleic acid, or DNA. This kind of explanation may be called *physiological,* with the word applied more broadly than usual.

The term *teleonomic* has been proposed for a second kind of explanation in biology. This relates organic structures and activities to the functions served in or for organisms. Physiologically digestion is explained as the action of enzymes and other chemicals on food. Teleonomically it is explained as a step in the process of providing the organism with needed materials and energy. Part of the teleonomic explanation of reproduction is that it leads to perpetuation of populations.

The third kind of explanation in biology is *historical*. This relates organic structures and activities to their evolutionary origins. The processes of digestion and reproduction in us are different from those in our most remote ancestors. Tracing their origins and changes through the evolutionary sequence results in an explanation different from the physiological and teleonomic explanations.

Most biological facts and generalizations can be adequately understood only when all three kinds of explanations are provided. There are still many phenomena in biology for which they are not yet available.

Hypothesis and theory

The goal of science is to establish generalizations and explanations for observed facts. The mere gathering of facts and their description is quite useless unless the observations are directed toward this goal and, indeed, dictated by it. Many of the most creative scientists have remarked that hunch and sudden inspiration have been instrumental in their discoveries of explanations. It must, however, be noted that the hunch or inspiration came only when they knew some pertinent facts and were seeking an explanation for them. And that is only the beginning of the process that produces an accepted scientific theory.

The first step toward explanation is taken when a *hypothesis* is formed. A hypothesis is a statement of an explanation that a scientist considers possible. It may be only a hunch or an informed guess, or it may seem inevitable as the only conceivable explanation of the facts at hand. We need not emphasize again that a hypothesis, as a suggested answer to a scientific question, must be testable by observation if it is to become a scientific explanation. The next step, then, is the testing. This involves seeking exceptions to generalizations embodied in the hypothesis. It also involves logical deduction of consequences from the hypothesis and observation of whether these consequences occur. The more consequences a hypothesis entails, the better it is, not only because it can be more thoroughly tested but also because it leads to new useful observations, to the formulation of still other good hypotheses, and to a general increase in knowledge. By way of example, we shall later in this book consider the hypothesis—now an accepted theory—of chromosomal inheritance.

If exceptions are found or if deductions fail to hold during the testing of a hypothesis, the hypothesis must be modified accordingly or rejected altogether. If the hypothesis does meet extended tests, our confidence in it grows. At some not precisely definable point, we feel justified in accepting it as the most probable explanation and in proceeding as if it were true. Then we begin to speak of the hypothesis as a *theory*. A theory is simply a hypothesis sufficiently tested that a scientist has confidence in its probability. We speak in terms of "acceptance," "confidence," and "probability," not of "proof." If by proof is meant the establishment of eternal and absolute truth, open to no possible exception or modification, then proof has no place in the natural sciences.[1] Alternatively, proof in a natural science, such as biology, must be defined as

the attainment of a high degree of confidence. A theory proved in this sense is still open to possible exception, which could amount to disproof, and even more often to modification. A good scientist must be prepared to change, sometimes even to discard, his most prized theories, sustained by the conviction that we thus advance in knowledge and toward truth.

The fact that theories are not subject to absolute and final proof has led to a serious vulgar misapprehension. Theory is contrasted with fact as if the two had no relationship or were antitheses: "Evolution is *only* a theory, not a fact." Of course, theories are not facts. They are generalizations about facts and explanations of facts, based on and tested by facts. As such they may be just as certain— merit just as much confidence—as what are popularly termed "facts." Belief that the sun will rise tomorrow is the confident application of a generalization. The theory that life has evolved is founded on much more evidence than supports the generalization that the sun rises every day. In the vernacular, we are justified in calling both "facts."

Biology and the Scientific Conceptual Scheme

ORIGINS OF SCIENCE

Men have surely been asking questions about the world and proposing answers ever since the human species began. What questions are asked and how answers are sought always depend on a conceptual scheme, an attitude toward the world and a set of postulates or beliefs about it. A conceptual scheme that sees the world as capricious and chaotic gives rise to no questions about order and natural law. A conceptual scheme that embraces magic and invisible spirits as causes of phenomena does not evoke answers testable by the phenomena themselves. Some such primitive conceptual schemes still have a lingering influence, but in the light of present knowledge, the result of science, we consider them superstitions. At a more sophisticated level are conceptual schemes that seek answers to questions about the material world not from that world but from dogmatic au-

---

[1] The natural sciences are those concerned with the phenomena of the material universe. In general, throughout this book when we say "science" we mean natural science. In mathematics, not a natural science, proof usually has a quite different meaning. It is the demonstration that a conclusion is tautologically inherent in whatever postulates or axioms were laid down to begin with. Such proof may be taken as absolutely and necessarily true, but only if the original postulates, which have *not* been proved, are accepted.

thority or by deduction from subjective philosophical premises. Such answers, not even regarded as subject to observational test, cannot be scientific.

The ancient world of Babylonia, Egypt, and Greece made great advances in knowledge of the world and laid the foundations for science and technology, but the ancients never developed a fully scientific conceptual scheme. Europe of the Middle Ages inherited from the ancient world a conceptual scheme that pictured the world as orderly but that sought answers about its orderliness in authority and in philosophical deduction more than in the world itself. It is an extraordinary fact that the scientific conceptual scheme arose so late in human history, within a single culture, that of western Europe, and over a comparatively brief span of time, roughly definable as from Nicolaus Copernicus (1473–1543) to Charles Darwin (1809–1882). Beyond the already general concept of order in the world, its whole basis was the strict relation of questions and answers to the observation of the world, the seeking of natural explanations for natural phenomena, the proposal of testable hypotheses, and the testing of them.

It seems quite simple and obvious to us that the only logical means of investigating the material world is by observing it. But we have grown up in a civilization in which the scientific conceptual scheme already existed and was generally (although not exclusively) accepted as the effective way of acquiring material knowledge. Copernicus' scientific theory that the earth circles the sun was firmly grounded in observation, but it was violently rejected by those whose conceptual scheme was still based on authority and philosophical deduction.

EARLY, INCOMPLETELY
SCIENTIFIC BIOLOGY

The scientific conceptual scheme arose first in the physical sciences. It brought about a revolution in human thought. Its insistence that natural phenomena obey natural, impersonal laws was a bitter and, at first, a deeply resented blow at age-old superstitions embedded in nonscientific conceptual schemes. Nevertheless, the scientific scheme responded to a refined concept of common sense, and it

worked. As regards the physical aspects of the world, its acceptance was soon general, if not quite universal. Yet well into the nineteenth century the great majority of people—even among the most intelligent and most learned—clung to a conceptual scheme in which essential phenomena of life, and most particularly of human life, were believed to transcend physical laws and not to be amenable to strictly scientific explanation. Biology is as old a science as any. It had its roots in antiquity, and in its physical or plainly material aspects it became a true science along with physics, chemistry, astronomy, and the rest when the scientific conceptual scheme was developed in the sixteenth and seventeenth centuries. Until 1859, however, it was impeded by the common view that some of its subject matter did not fit into that scheme.

BIOLOGY FULLY ENTERS
THE SCHEME

It was in 1859 that Darwin's book *The Origin of Species* was published. This book accomplished two main objectives. First, it established the *theory of evolution*, the broadest generalization ever made about the interrelationships of living things. This theory (which in common speech we are now justified in calling a fact) states that all organisms have arisen from common ancestors by a natural, historical process of change and diversification. Second, the book propounded a theory to explain the *causes and results of evolution*. The most important point that had to be explained was the apparent purposiveness of life, the observation that organisms seem to be designed precisely for the functions they carry out. It was this, more than anything else, that had supported the claim that vital structures and processes could not have entirely natural causes and hence did not fit the scientific conceptual scheme.

Darwin's complex explanation was only partially successful, but its most essential element, *natural selection*, has stood the test of time and is accepted today in somewhat modified form. We shall return to that subject, and also to the problem of purpose in nature, in Chapters 15 and 16, after sufficient basis for comprehension has been laid. What is significant here is that Darwin sufficiently demonstrated that *natural* explanations for *all* of the material phenomena of life should be sought and can be found. Thus *The Origin of Species* actually accomplished a third objective, most important of all: it finally brought biology as a whole, in all its aspects, within the conceptual scheme of science.[2]

CHAPTER SUMMARY

Life seen in communities: a forest and a coral reef as examples.

Major themes in the drama of life: organization; levels of organization and of study; traffic in materials and energy; behavior;

reproduction; evolution; geography of life; history of life.

Organization of this book: by levels and by themes within each level and between levels.

Nature of science: a human activity motivated by curiosity and by a desire to supply needs and wants; science and technology interdependent but distinct; science a system of questions and answers arising from observation and tested by observation; generalizations as extended observations and as explanations; three kinds of explanations in biology—physiological, teleonomic, and historical; hypothesis a proposed answer to a scientific question; theory a hypothesis in which we have confidence; the relationship of theory and fact.

Biology and the scientific conceptual scheme: the dependence of the pursuit of knowledge on conceptual schemes; superstitions and the nonscientific nature of old schemes; the rise of the scientific scheme from Copernicus to Darwin; the incomplete incorporation of biology in the scheme until 1859; Darwin's accomplishments—the theory of evolution, the theory of natural selection, and the inclusion of all biological phenomena in the scientific conceptual scheme.

[2] Of course, Darwin had forerunners, individual biologists who were, as far as possible to them, working within the conceptual scheme of science, who had formulated the generalization of evolution, and who had even proposed cruder forms of a hypothesis of natural selection. Darwin himself was so permeated by the scientific conceptual scheme that he took this phase of his work for granted and did not consciously draw attention to the third accomplishment of his great book. Nevertheless, it did have this effect in the history of biology. And it was Darwin who advanced evolution and natural selection to the status of accepted theories, tested by an irrefutable body of observational evidence.

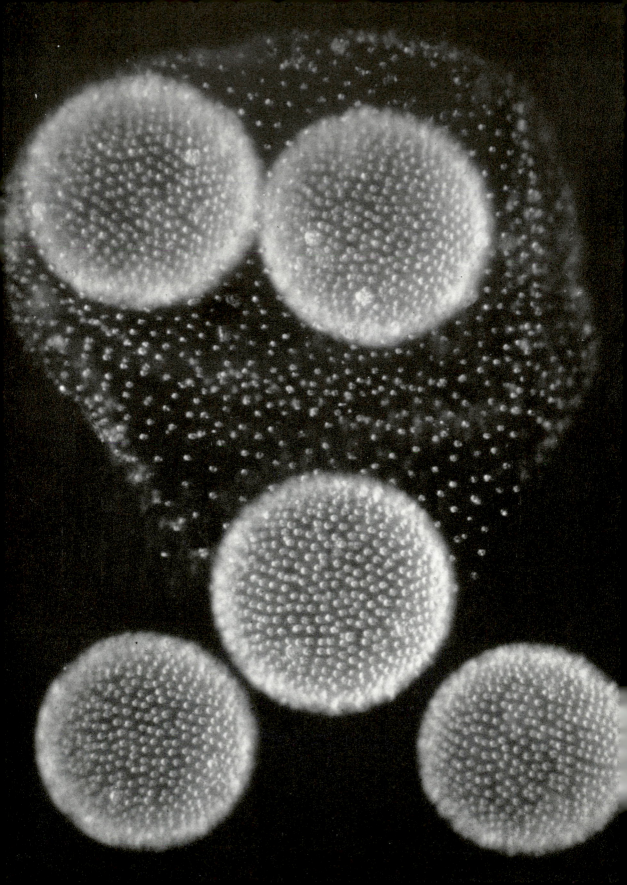

Part 2

The Basis of Life

The photograph introducing Part 2 is of a microscopically small organism called *Volvox*. It is breaking up and liberating five young *Volvox*. At the high magnification at which the photograph was taken, the structures of both parent and young resolve into hundreds of glistening round particles. These particles are cells; they are the basic structural units of all life.

Organisms are composed of cells. Plant or animal, large or small, *Volvox* or man— the generalization holds for *all* living things; they are either single cells or groups of cells. The discovery of this great truth, which is the heart of the cell theory, was one of the outstanding achievements of the nineteenth century and ranks with the theory of evolution as a cornerstone of modern biology.

Every cell is, so to speak, a world of life in miniature. Like the organism of which it is a part—or even the whole—the cell is complex and organized. It responds to stimuli and is capable of movement; it feeds and respires, expending energy to maintain its ordered state and execute other work such as secretion; and, like other larger living systems, it can reproduce itself. In large and structurally complex organisms like ourselves, all special parts like bones, muscles, and nerves are composed of special cells—bone cells, muscle cells, and nerve cells.

Although the cell is the basic unit of life, it is made up of units still lower in the hierarchic sequence. For full comprehension, the search for knowledge of life must start at those levels even below the cell, with a study of the atoms, molecules, and chemical reactions that enter into the structures and the activities of cells. These are the subjects of Chapter 2.

Chapter 3 describes the organization of cells, so intricate at microscopic and submicroscopic levels. It also discusses the differentiation of cells of different kinds and their organization into tissues of diverse functions in various organisms.

In Chapter 4 some of the most essential and widespread functions of cells are outlined. Here are the dynamic reactions that are the basis of the activities we call life.

2

Molecular Aspects
of Biology

This is a model of the glucose molecule,
a key constituent in the life of all cells.
The large, black atoms are carbon;
the medium-sized, stippled ones are oxygen;
and the small, white ones are hydrogen.

The preceding chapter spoke of the hierarchical arrangement of living organisms. The cell, it was noted, is the simplest structure in which life is unambiguously present, existing at a certain level of the organizational ladder. But cells are components of more complex structures, tissues and organs, and these in turn are parts of still higher organizations—and so on up the ladder. It is obvious that one can also begin with the cell and descend the ladder, for within the cell are substructures such as the nucleus, then its component parts, and then the parts within those parts. Eventually, one reaches the molecules and atoms.

One of the most fruitful and exciting chapters of contemporary biology has been the study of the molecules of living organisms and the recognition of their importance to those wishing to understand how organisms function. Early scientific study of living things consisted mainly of descriptive and rather static observations of *structure,* or *morphology.* But investigators were not content merely with anatomical data revealed by the eye, the knife, and the microscope, and there arose a second great movement which sought to comprehend the *function,* or *physiology,* of living things.

We shall see that both types of inquiry have been incisively extended in recent years into the molecular realm. By applying the techniques of chemistry and physics to the problems of life, scientists have developed the new and still actively growing disciplines of *biochemistry* and *biophysics.* So successfully have these fledgling sciences provided rational explanations for certain phenomena of life that it is no longer possible to discuss many areas of biology without reference to biochemical and biophysical principles.

One of the most stimulating results of the biochemical approach to biological problems was the discovery that despite the great diversity in the size, form, and structure of living organisms as a group, they have a great deal in common—much more, in fact, than was suspected following the discovery of the fundamental sameness of all cells. Although each type of living thing is unique in its particular combination of materials, processes, and relationships, all require just the same general sorts of chemical materials. They require them in amounts and proportions that vary considerably, to be sure, but that vary within prescribed limits. Yet what any two kinds of organism do with these materials is never precisely the same.

This unity-with-diversity can be strikingly illustrated by two exceptional cases. One would say—and, up to a point, quite correctly—that cellulose, the material of many plant cell walls and hence of wood, is a substance peculiar to plants as opposed to animals. Similarly, hemoglobin, the chemical that makes the blood red, seems completely characteristic of animals as opposed to plants. Yet cellulose occurs in a group of animals, the odd marine forms called *tunicates* (Chapter 21), and hemoglobin has been found in the root nodules of some leguminous plants.

These exceptions show that the diversity among organisms depends, after all, on still more fundamental similarities. Cellulose is a complex compound built up from sugar molecules. The glycogen (animal starch) involved in muscular energy is similarly built up from sugar molecules—and the analogy is further driven home by the fact that a few plants also make glycogen. Hemoglobin is a chemical combination of heme and globin. Compounds of the same general types are common in plants. For instance, chlorophyll, the green pigment of plants, is chemically very similar to heme. So the exceptional occurrence of cellulose in animals and of hemoglobin in plants points up a similarity in materials and in the way in which these are combined, even when the resulting combinations are different.

This chapter summarizes a number of elementary chemical and biochemical principles. You must know these fundamentals before you can expect to grasp certain basic principles of biology. It should be clearly understood, however, that the study of biology is not synonymous with the study of chemistry. What is given here in the way of chemical principles is merely an essential preliminary to discussions of the living organism in contemporary terms. You will quickly discover that many of the molecules described—particularly the proteins and nucleic acids—are uniquely associated with living organisms. So, though we speak here of molecules, we are off on an exploration of the molecular basis of *life*, and the living organism must stay in the forefront of our thoughts.

Chemical Substances

You surely know that the smallest particles in nature, electrons, protons, and the rest, are commonly organized into *atoms*. Atoms represent in most basic form the chemical *elements*, the building blocks for all larger chemical units. In view of the tremendous complexity of the matter composed of them, the number of elements is surprisingly small. Only 103 are known, and 11 of these have been made artificially and may not occur in nature. About 35 are common in nature and important for life.

You also doubtless know that two or more atoms (up to thousands) can combine with each other to form a *molecule*. If the molecule contains more than one kind of atom, it is a *compound*. As we shall see below, the atoms of a compound are held together by specific forces called *bonds*. Atoms or molecules or both can *react* with each other and produce different kinds of molecules. In ordinary chemical reactions, the number and kinds of atoms are the same before and after the reaction. Only their combinations in molecules change.

Each element has a name and a symbolic abbreviation that stands for the name. The elements most important in biology are hydrogen, abbreviation H; oxygen, O; phosphorus, P; carbon, C; and nitrogen, N. The simple (or empirical) formula for a molecule shows what atoms it contains. Water is H_2O because it contains two atoms of hydrogen for every one of oxygen.

ATOMS

So that you may understand how chemical bonds form between atoms, we must briefly consider the internal structures of atoms. Atoms, the smallest units capable of retaining their elemental identity when chemical reactions take place, are constructed of smaller diverse components known collectively as *elementary particles*. The elemen-

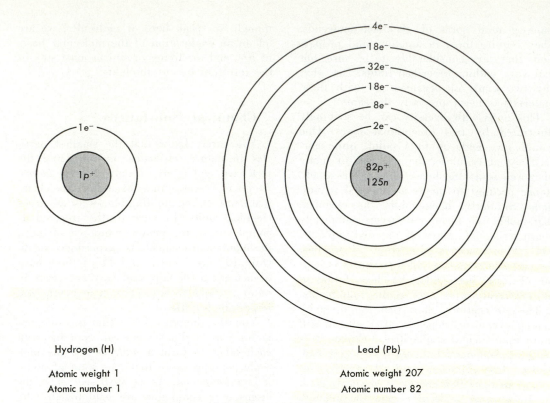

Hydrogen (H)

Atomic weight 1

Atomic number 1

Lead (Pb)

Atomic weight 207

Atomic number 82

2–1 The atomic structures of hydrogen and lead. The symbol e^- denotes an electron, p^+ a proton, and n a neutron. The nucleus of a hydrogen atom contains only 1 proton. The nucleus of a lead atom contains 82 protons and 125 neutrons.

tary particles that concern us are the *protons, electrons,* and *neutrons.* Every atom consists of a single *nucleus* and one or more electrons. The atomic nucleus contains the protons and neutrons. A proton has *mass,* or *weight.* The mass of every proton is the same, and it is given the arbitrary value of 1. A proton also has a positive electric charge. A neutron also has a mass of 1, but it is electrically neutral. The mass of an electron is very much less than 1—it is 1/1836th that of the lightest nucleus—hence its weight is almost negligible. An electron bears a negative electric charge equal in magnitude to the positive charge of the proton.

Although nuclei are extremely small indeed, the total mass of a whole atom is concentrated almost entirely in its nucleus. The total mass of the nucleus—that is, the number of protons and neutrons present—determines its *atomic weight.* For example, the simplest atom, that of hydrogen, has a nucleus con-

sisting of a single proton; neutrons are absent (Fig. 2–1). Since the nucleus has a mass of 1, the atomic weight of hydrogen is said to be 1. By contrast, the nucleus of an atom of lead contains 82 protons and 125 neutrons. Therefore, the atomic weight of lead is 207. Note that the number of electrons (or protons) defines the *atomic number* of an element. The atomic number of hydrogen is 1; that of lead is 82.

Ordinary matter is electrically neutral—that is, it contains equal amounts of positive and negative electric charges. Normally, the number of protons exactly equals the number of electrons. The electronegative electrons in an atom are attracted by the positively charged nucleus and move rapidly about it in variously located orbital pathways. In fact, electrons travel only at certain fixed distances from the nucleus, the paths of the electrons in these orbits marking out a series of specific concentric *shells.* Each shell can hold only a

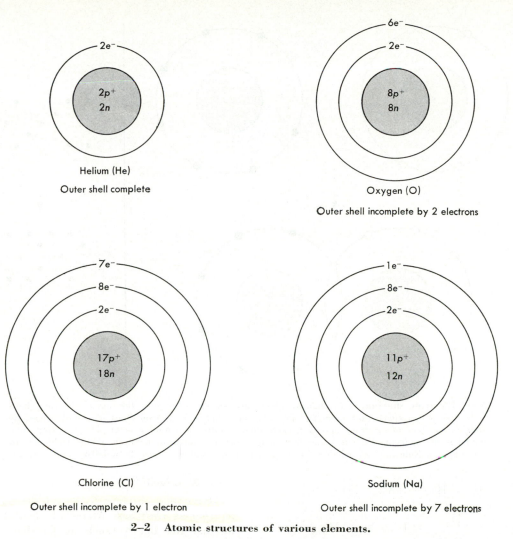

2–2 **Atomic structures of various elements.**

fixed maximum number of electrons. Thus the first shell, proximal to the atomic nucleus, can hold a maximum of 2 electrons, the second shell 8, the third 8, and so on.

The hydrogen atom has a single electron orbiting in its first shell. Since this shell can hold 2 electrons, hydrogen is said to have an incomplete shell. An atom of helium (He) does have 2 electrons in the first shell (Fig. 2–2). Since this is the maximum possible number for this shell, it is said to be complete. Oxygen has 8 orbital electrons, 2 in the first shell and 6 in the second. Since the second shell can hold a maximum of 8 electrons, it is incomplete in the oxygen atom.

Atoms whose electron shells are complete

are notably stable. This means that they are chemically inert and rarely react with other atoms. Helium with its 2 electrons is such a substance. So are the other inert "noble gases": neon, 10 electrons in 2 shells; argon, 18 electrons in 3 shells; krypton, 36 electrons in 4 shells; xenon, 54 electrons in 5 shells; and radon, 86 electrons in 6 shells. All contain just enough electrons to complete the orbital shells present.

The atoms of all other elements have incomplete outermost shells and are therefore unstable—that is, they are chemically reactive. Electron-dot symbols provide a useful shorthand notation for representing the outermost electron shell.

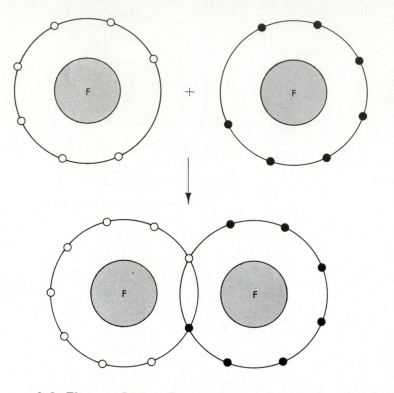

2–3 Electron sharing. Two fluorine atoms are shown with only the outer electron shells indicated. When apart, each atom has an outer shell that is 1 electron short of completeness. By sharing one pair of electrons, however, each atom acquires a complete octet. The resulting compound is the molecule F₂. The same reaction is shown in shorthand notation on the right side of the diagram. Note that shared-electron or covalent bonds can also form between dissimilar atoms.

H   He·   :A: :O: ·C·

:F·   :Cl·   Na   Mg·

Note that the inner electrons are not shown. When appropriate kinds of such atoms are brought together under favorable conditions, their incomplete outer electron shells may make them react with one another. A chemical bond forms between the atoms, and a chemical compound is produced.

CHEMICAL BONDS

Chemical reactions occur because every atom tends to complete its outer electron shell. An originally incomplete outer shell can become complete in several ways; therefore, the resulting bonds that hold the atoms together in a compound are of several distinct types.

Covalent bond

Atoms may become electronically stable by *sharing* electrons. Atoms of elements of the halogen family, such as fluorine (F), have outer shells with 1 less than the complete number of electrons. One way in which a fluorine atom can achieve completeness is by reacting with another fluorine atom (Fig. 2–3). The atoms share electrons in such a way that each behaves as if it actually possessed a complete outer shell. The F₂ combination is stable, and we say that a *covalent bond* has been formed. This type of bond may also form between atoms of different kinds. Indeed, the covalent bond is so commonplace that G. N. Lewis, the discoverer of its electronic basis, called it *the* chemical bond. Functionally, covalent bonds are notable for the large amount of energy needed to break them. This means that they hold atoms together very tightly.

Ionic bond

An *ion* is an atom or a group of atoms that has either acquired or lost 1 or more electrons. For example, consider the outer shells of the sodium and chlorine atoms:

$$\overset{.}{Na} \qquad \overset{..}{\underset{..}{\cdot Cl :}}$$

The chlorine atom is unstable because it has 1 electron too few in its outer shell. The sodium atom is unstable because it has 7 electrons too few—or, to put it another way, because it has 1 electron too many, since it is considerably easier for sodium to achieve a complete outer shell by losing 1 electron than by gaining 7. Under these circumstances, both atoms can attain stability simultaneously if a single electron is transferred from sodium to chlorine. Note that before the transfer of the electron the sodium atom is electrically neutral, its 11 orbiting electrons being counterbalanced exactly by the 11 positively charged protons in the nucleus. During the reaction, the sodium loses 1 negative charge when it loses an electron. Therefore, the sodium atom becomes positively charged, because the 11 protons are still present though there are only 10 electrons.

$$Na \rightleftharpoons e^- + Na^+$$

Conversely, chlorine loses its initial electrical neutrality by acquiring a single negative charge.

$$Cl + e^- \rightleftharpoons Cl^-$$

The charged forms, Na^+ and Cl^-, are ions. Negatively charged ions are called *anions*, and positively charged ions, *cations*. (A symbol such as Na^+, Mg^{++}, or Cl^- denotes an ion and indicates the size and electrical character of the charge.) Oppositely charged ions are attracted to one another, and the strong electrostatic force of attraction that binds them together is the *ionic bond*. A group of ions so bonded is an *ionic compound*. For example, in the reaction

$$Na^+ + Cl^- \rightarrow NaCl$$

the product is the ionic compound NaCl, or sodium chloride.[1] Such an anion-cation com-

[1] In writing ionic compounds symbolically, one usually does not include the electric charges of the ions. Thus, instead of writing Na^+Cl^-, one may write

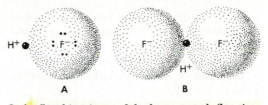

2–4 Combinations of hydrogen and fluorine. *A.* The hydrogen fluoride molecule. *B.* The hydrogen difluoride ion, containing a hydrogen bond. (From L. Pauling, *General Chemistry*, 2nd ed., Freeman, San Francisco, 1953.)

bination is called a *salt*; salts are noted for their solubility in water and the ability of their solutions to conduct electric currents.

The formation of NaCl from Na^+ and Cl^- is an *electron-transfer* or *ionic reaction*. In such a reaction, the total number of positive charges carried by the cation equals the total number of negative charges carried by the anion. Every electron transferred establishes one ionic bond. For example, the magnesium ion, Mg^{++}, resulting from the transfer of 2 electrons, forms two ionic bonds with other ions. The number of bonds an ion can form with other ions is its *ionic valence*. The sodium ion has a valence of +1, the magnesium ion has a valence of +2, and the fluoride or chloride ion has a valence of −1.

Hydrogen bond

Still another type of chemical bond, though weaker in strength than a covalent or ionic bond, holds electrically charged groups together under certain circumstances. A *hydrogen bond* is an interaction between a hydrogen atom attached to a negatively charged atom in one molecule and another negatively charged atom in the same or in a different molecule. To illustrate, a molecule of hydrogen fluoride (HF) is represented in Fig. 2–4. Note that the tiny hydrogen ion, having lost its electron, is a bare atomic nucleus resting on the surface of the larger and negatively charged fluoride ion. Since it carries a positive charge, it is able to attract to itself another negatively charged fluoride ion. When one is found, a new ionic grouping occurs with the structure $(F^-H^+F^-)^-$, or HF_2^-. Such

NaCl. Even though such notations do not specifically show the ionic nature of the components, ions are in fact present, and ionic or electrostatic bonds unite them.

an ion is stable. The bond holding it together is a hydrogen bond, indicated in a structural formula by a dashed line: $F^- —H^+ \cdots F^- —H^+$. The weakness of the hydrogen bond is evident in the relative ease with which it can be broken.

In the living organism, the hydrogen bond is as important as any covalent bond. We shall later see that hydrogen bonds play a fundamental role in determining the properties of proteins and nucleic acids. We shall also learn of the unusual tendency of water itself to form hydrogen bonds. This arises from the fact that each molecule has two attached hydrogen atoms and two unshared electron pairs. It is clear that this distinctive characteristic of water accounts for many of its other extraordinary properties, properties which are of profound biological significance.

Major Elements of the Living Organism

All the materials of life are *ultimately* derived from relatively simple elements and compounds. Once they are assimilated by a living organism, these ultimate materials often become elaborated into extremely complex forms, which are passed on from one organism to another. Most of these elaborated chemicals are eventually broken down into simple elemental forms again. There is, then, a cycle, with simple materials at the beginning and end; it thus seems logical to begin a consideration of life's materials with them. They include *carbon, oxygen, hydrogen, nitrogen,* and a great variety of what may be called, as a group, *minerals.* The relative abundance of these elements in living organisms is summarized in Table 2-1.

CARBON

During the early years of modern science, it was widely believed that living organisms possessed mysterious vital forces that made them fundamentally different from nonliving objects. When chemists turned to the study of living things, "vitalist" concepts seemed bolstered by the discovery of countless substances that were previously unknown— for example, uric acid in urine, lactic acid in sour milk, and citric acid in lemons. Such substances, it appeared, could be formed only by living organisms. Hence they were designated *organic* compounds in contrast to the *inorganic* materials of the lifeless rocks and minerals.

It is usually stated that the studies of the German chemist Friedrich Wöhler in 1826 revolutionized thought on this question. At first by accident and later by careful intent, Wöhler converted the familiar "inorganic" substance ammonium cyanate (NH_4OCN) to the "organic" substance urea (NH_2CONH_2).[2] This achievement and the many test-tube syntheses that followed gradually demolished the barrier between the two chemistries with the growing recognition that the fundamental laws of both are identical. The term "organic chemistry" remained in use when it was found that almost all of the important substances produced by animate nature contain carbon. Organic chemistry thus became the chemistry of the compounds of carbon. (The study of the remaining substances has been lumped to-

TABLE 2-1

Approximate relative amounts of the chemical elements in the human body

Element	Symbol	Per cent by weight
Oxygen	O	65
Carbon	C	18
Hydrogen	H	10
Nitrogen	N	3
Calcium	Ca	2*
Phosphorus	P	1.1*
Potassium	K	0.35
Sulfur	S	0.25
Chlorine	Cl	0.15
Sodium	Na	0.15
Magnesium	Mg	0.05
Iron	Fe	0.006
Iodine	I	0.00006
Cobalt, zinc, and other trace elements		Data incomplete

*Estimates differ widely.

[2] It is a fact, however, that the "inorganic" substance ammonium cyanate was usually prepared commercially from animal horns and hoofs. Furthermore, George Wald has pointed out that in Wöhler's experiments a living organism was still essential for urea synthesis, the living organism being Wöhler.

gether under the heading "inorganic chemistry.")

Once it has been incorporated into a living thing, a carbon atom may enter hundreds or thousands of different combinations within one organism and then, commonly, within others and yet still others as the materials are passed on. The beginning and end of the cycle in the inorganic environment involve carbon in the simple compound carbon dioxide, CO_2. CO_2 is produced when pure elemental carbon burns. It also forms within organisms. Under ordinary conditions, CO_2 is a gas. (At temperatures lower than those in nature it solidifies; solid CO_2 is called *dry ice*.) The gas makes up about 0.03 per cent of the atmosphere. This tiny fraction is the main inorganic reservoir of carbon as a material for life.

The gas CO_2 does not usually enter directly into chemical reactions. However, it dissolves in water (as everyone knows who has opened a bottle containing a carbonated drink) and then reacts readily in various ways. This activity is due in part to the fact that CO_2 in solution reacts with water itself to form a weak acid, carbonic acid, H_2CO_3, in accordance with the following equation:

$$CO_2 + H_2O \rightleftharpoons H_2CO_3$$

This equation, like many of those important in life processes, can readily go in either direction. Slight changes in pressure, heat, concentration, or presence of other chemicals can reverse the reaction

$$CO_2 + H_2O \rightarrow H_2CO_3$$

and produce the opposite:

$$H_2CO_3 \rightarrow CO_2 + H_2O$$

The beginning of the incorporation of atmospheric CO_2 into the materials of life is a more complex series of reactions with water. In green plants these two raw materials are combined into simple sugars in the process called *photosynthesis*. This is an extremely important and basic synthesis for the whole world of life. It is discussed in a later chapter. From the simple sugars, carbon is converted into many other substances.

The end of carbon's participation in the chemistry of life usually involves its withdrawal from the more complex compounds and its combination with oxygen. This yields CO_2 again, most of which finds its way more or less directly back into the atmosphere. Almost all organisms, both plants and animals, are constantly producing CO_2. The accumulation of this gas, or of the H_2CO_3 readily formed from it, in cells or in fluids such as blood soon becomes harmful. Its elimination is therefore necessary.

HYDROGEN AND OXYGEN

The great majority of the compounds involved in the substances and processes of life contain carbon, hydrogen, and oxygen. Some contain no other elements. Others contain quantities, usually much smaller quantities, of various additional elements.

Hydrogen, the lightest element, is unique in nature. Free H_2, a colorless, odorless gas, does not occur in living tissue. Instead, hydrogen atoms are extensively combined with carbon, oxygen, and the other elements. As we shall presently see, the ability to form hydrogen ions (H^+) is a characteristic property of a particular class of compounds.

The inorganic source of much oxygen in organic compounds is water. The reaction for the formation of water from its elements is

$$2H_2 + O_2 \rightarrow 2H_2O$$

Two molecules of hydrogen (each with two atoms) combine with one of oxygen (also with two atoms) and produce two molecules of water. There were four atoms of hydrogen and two of oxygen to begin with, and of course the same six atoms are there in the water after the reaction. There is considerably less energy in two molecules of water than in two of hydrogen and one of oxygen. In this reaction, therefore, energy is released from chemical form and appears in some other form, such as heat. The most meaningful and complete formula for the reaction is thus

$$2H_2 + O_2 \rightarrow 2H_2O + \text{Energy (heat)}$$

In addition to water and oxygen-containing organic foods, most organisms also require extra oxygen in the form of the element itself, O (or, as a molecule, O_2). Along with water and some salts, this is an inorganic material that can be utilized *directly* by animals as well as plants, in marked contrast to the carbon source, CO_2. (There are, however, some lower organisms, mainly bacteria and parasites, that can live without oxygen and may even be killed by it.)

Oxygen is a gas, and its great inorganic reservoir is the atmosphere, which contains about 20 per cent (by volume) of elemental oxygen (that is, not combined with other elements). Oxygen dissolves readily in water, remaining in the elemental form and thus available to aquatic organisms. In fact, oxygen must be in solution in water to enter organisms living in air. (See Chapter 10).

The principal role of elemental oxygen in cells is to combine with carbon and hydrogen from the breakdown of organic compounds, producing carbon dioxide and water. The water so formed may be further utilized by the organism, but the carbon dioxide is almost entirely eliminated. The input of oxygen and output of carbon dioxide depend on the chemical activity of the cells and of the organism as a whole. Oxygen consumption is a fairly good measure of total metabolism in most organisms. This is the principle involved in the clinical procedure known as the basal metabolism test.

If carbon (for instance, charcoal) is burned in air, it combines in a simple way with oxygen and produces heat, which is, of course, one form of energy.

$$C + O_2 \rightarrow CO_2 + Energy$$

Because of this reaction, so familiar to everyone, and because the body's oxygen consumption and carbon dioxide elimination do tend to be proportional to total activity, it used to be supposed that some such reaction accounted for the body's heat and other energy. The notion is held by many people that "fuel"—organic compounds such as sugars and fats—is "burned"—combined directly with oxygen—in the body and that this is how we obtain energy. It is now known, however, that this "burning" of "fuel" is rarely if ever a significant source of useful organic energy. Most such energy is released by chemical reactions that do *not* require oxygen. As a rule, oxygen is used and CO_2 is formed only *after* the important energy release, in a sort of sweeping up of the debris of earlier re-

actions and setting up for their repetition, as will be explained in Chapter 4.

Animal metabolism utilizes but does not produce elemental oxygen. The process of combining CO_2 and H_2O to form sugars in green plants, however, involves the release of O_2, and while this synthesis is going on, it usually produces more oxygen than the plants need. Green plants in light thus generally give off oxygen to the surrounding water or air. In the dark, where the active phase of sugar synthesis does not proceed, plants produce no spare oxygen but continue to utilize oxygen and give off CO_2. This is why fish in a pond with green plants may be suffocated during the night. The dissolved oxygen in the water is being used up, and none is being produced by the plants.

NITROGEN

After carbon, hydrogen, and oxygen, the most common element in the materials of life is nitrogen. It is, in particular, a constituent of all proteins, an important class of organic compounds to be discussed later in this chapter.

There is a tremendous store of nitrogen in the atmosphere, which is (by volume) almost 80 per cent elemental nitrogen. Yet no animals and few plants can make direct use of this nitrogen. Most plants can utilize nitrogen from the environment only if it is in the form of various inorganic compounds: *ammonia* (NH_3) or its compounds; *nitrates* (salts containing NO_3); or *nitrites* (salts containing NO_2). Animals acquire their nitrogen in the form of compounds, especially proteins, which they obtain from plants or from other animals. The withdrawal of nitrogen from the atmospheric reservoir and its incorporation into life thus depend entirely on processes that make ammonia, nitrates, and nitrites. Some inorganic processes do this, especially lightning, but formation of these compounds is due mainly to a few kinds of organisms that are able to convert elemental nitrogen into ammonia, nitrates, and nitrites through the process known as *nitrogen fixation*. Some bacteria and simple plants (some fungi and algae) can do this. Most noteworthy are the bacteria that live in nodules in the roots of beans and related plants (legumes).

On the other hand, few organisms decompose compounds of nitrogen completely and produce the element nitrogen. Some denitrifying bacteria do this, but the end products of nitrogen metabolism in most organisms are still organic compounds, such as urea or uric acid in animals, or, at the simplest, ammonia. Since even ammonia can be utilized by some plants and is readily turned into nitrates and nitrites by certain bacteria, the nitrogen remains available to life for long periods of time and passage through many different organisms.

MINERALS [3]

Besides carbon, oxygen, hydrogen, and nitrogen, many other elements are required in smaller amounts by all forms of life. The inorganic sources of these are, in most cases, salts dissolved in water—water in the soil or in lakes, streams, and seas. Plants and, to some degree, aquatic animals usually acquire these salts directly from the water of the environment. Land animals also acquire and utilize salts directly to some extent but obtain much of their mineral requirement more or less incidentally along with their organic food. This is one reason why physicians recommend a balanced diet—we have not only the obvious requirements for building materials and energy in our food but also the need for small amounts of many different minerals, not all of which are likely to be present in the essential forms and quantities in any one food. In exceptional cases extra quantities of iron, calcium, iodine, or sodium may be needed and supplied medicinally, and on occasion reduction of mineral intake (for instance, of table salt) may be indicated.

The large number of necessary mineral elements is often divided into two broad groups: the *macronutrients*, which are required in large quantities; and the *micronutrients*, which are needed only in small amounts. The latter are also known as *trace elements*. Although mineral needs vary from cell to cell,

[3] Strictly speaking, the inorganic sources of nitrogen just mentioned are mineral salts, which therefore belong under this heading. However, because of their special significance, they have been discussed separately.

and particularly from plant cell to animal cell, the essential macronutrients for most cells include calcium, phosphorus, chlorine, sulfur, potassium, sodium (plants do not need sodium), magnesium, iodine, and iron. Micronutrients that are widely required, but perhaps not by all forms of life, include manganese, copper, zinc, fluorine, cobalt, and possibly vanadium and selenium. In addition, plants require boron and molybdenum, and, for some species, vanadium.

It should be remembered that about 99 per cent of the mineral content of most organisms consists of the so-called macronutrient elements. Nevertheless, the micronutrient elements have great significance. For example, we shall see later that micronutrients have an indispensable catalytic function in many enzyme systems (p. 63). Much research is currently being done on these interesting trace components of the living organism.

Properties of Compounds

Most elements, as we have seen, are electronically unstable and therefore chemically reactive. For this reason, these elements almost without exception do not exist as free atoms on the earth's surface. Surely this is true of elements composing the living organism, for they are joined together in innumerable chemical compounds.

It is useful to separate chemical compounds into two general categories: *electrolytes* and *nonelectrolytes*. Electrolytes (or ionic compounds) are compounds containing ionic bonds (p. 27), that is, bonds formed through electron transfer. We can easily identify electrolytes in the laboratory by demonstrating that solutions in which they are dissolved permit the passage of electric currents. Nonelectrolytes are compounds that have bonds of the covalent type (p. 26), that is, their stability has been achieved through the sharing of electrons. Their solutions do not transmit electric currents.

We must consider these two groups of compounds because their properties (and particularly certain *differences* in their properties) are of great biological importance.

However, let us first consider that often neglected substance whose unusual behavior helps us distinguish electrolytes from non-electrolytes, *water*.

UNUSUAL PROPERTIES OF WATER

It is a notable fact that the major portion of the total weight of living tissue consists of water. In man, about 60 per cent of the red blood cells and 92 per cent of the blood plasma are water; muscle tissue is approximately 80 per cent water; and water makes up considerably more than half—often more than three-quarters—of most other tissues. The only exceptions are such relatively inert tissues as hair, nails, and the solid portion of bone. The spores of certain plants and bacteria have low water contents, but these are comparatively inactive cells. When they are transformed into cells that show active metabolism and growth, their water contents

2–5 Part of an ice crystal. At the top the relations between molecule size and interatomic distances are approximately correct. Note the hydrogen bonds and the open structure that gives ice its low density, so that it is lighter than water. In the molecules at the bottom the small spheres represent oxygen atoms, and the still smaller spheres hydrogen atoms. (From L. Pauling, *General Chemistry*, 2nd ed., Freeman, San Francisco, 1953.)

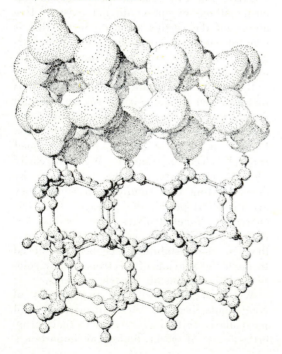

must increase. Thus water appears to be the indispensable matrix for the structural components and functioning of living cells. We are all familiar with the discomfort of animals (such as humans) and the wilting of plants when deprived of water. Indeed, all organisms die if their water supply continues to be inadequate. Water is also a principal constituent of the environment in which organisms live. Although much of the earth's water is liquid, large portions of it are in the vapor and solid phases. These forms of water are also of profound significance to the living world.

To the chemist, water is a strange and unusual substance with behavior full of apparent anomalies. It is this uniqueness that has given water a key role in shaping the character of the physical and biological worlds and in guiding the course of their evolutionary development. Detailed knowledge of the relationship between the structure and the function of the water molecule dates especially from a 1933 publication by J. D. Bernal and R. H. Fowler entitled "A Theory of Water and Ionic Solutions, with Particular Reference to Hydrogen and Hydrogen Ions." Today we recognize that a sample of pure water is considerably more complex than a mere collection of randomly distributed inert molecules having no particular connection with one another and that a number of unusual phenomena result from this fact. These include (1) its expansion on freezing, (2) its uniquely high surface tension, (3) its odd thermal properties, and (4) its exceedingly high dielectric constant. Let us briefly examine these phenomena.

Freezing

It is well known that most substances decrease in volume and therefore increase in density as their temperature decreases. Water is remarkable in that there is a temperature at which its density exceeds that at higher or lower temperatures. This temperature is 4°C. With further cooling, the volume of a sample of water *increases*. This extraordinary property of water is related to an unusual facility of formation of hydrogen bonds between water molecules. This causes ice to have an unusually open structure and hence a lower density than if the molecules were more closely packed (Fig. 2–5). As ice melts, the open crystal structure is partially destroyed, and the molecules move closer together. This explains why cool water is denser than ice. Only at 4°C. does the expansion from increasing molecular agitation start to overcome this effect and cause water to show the usual decrease of density with increasing temperature.

Very few molecular compounds have the property of expanding on freezing. The fact that water does has deep biological meaning. If ice were heavier than water, it would sink to the bottom on freezing. In fact, water just above the freezing point is heavier than water at the freezing point; therefore, it moves toward the bottom, freezing begins at the surface, and the bottom is last to freeze. Organisms living at the bottoms of fresh-water lakes are thereby protected from freezing. Organisms in the primitive ocean of the world, if they existed there before it was made salty by the erosion of the land surfaces, probably enjoyed the same protection.

Surface tension

The surface of a liquid, like a stretched membrane, tends to contract in area as much as possible. The phenomenon is called *surface tension*. With no important exceptions, water has the highest surface tension of any known liquid. One consequence of the high surface tension of water is that water rises to unusually high levels in narrow capillary tubes, a fact of great signficance in plant physiology (although surface tension alone does not account for the rise of water in tall plants and trees).

It is also of interest that most substances lower the surface tension of water when they dissolve in it. This means that such substances tend to collect at the interfaces between water's liquid phase and other phases. This tendency must be remembered in considering the passage of substances through a membrane and the formation and structure of the membrane itself.

Thermal properties

The thermal properties of water have helped to make it such an excellent medium for the origin, development, and maintenance of life. Because of them, water plays a major part in keeping the earth's surface at a relatively constant temperature. Since organisms can survive only within a restricted temperature range, this is a matter of the greatest possible importance.

The *heat capacity* of a substance is the amount of heat required to raise the temperature of a unit quantity of the substance —1 mole or 1 gram—by 1°C. without changing its phase (that is, without converting it from solid to liquid or liquid to gas).[4] The *heat of vaporization* of a substance is the amount of heat absorbed per mole or gram during the process of conversion from liquid to gas. The *heat of fusion* is the amount of heat needed to convert a mole or gram of a solid substance into a liquid at the melting point. When these values for water are compared to those for a large variety of other substances, they are found to be exceptionally high.

The biological implications of this fact are numerous. Because of the high heat capacity of water, much more heat is necessary to increase the temperature of a given amount of water by a given number of degrees than is necessary to heat most other substances. Stated another way, a given amount of heat produces a smaller temperature rise in water than in most other substances. Thus water acts as a temperature buffer, maintaining its temperature against the challenge of shifting environmental temperature more success-

[4] In this book we generally follow scientific practice and use the metric system of grams, liters, meters, and their multiples and fractions for weights and measures and the centigrade (C) system for temperatures. The equivalents in nonmetric units and Fahrenheit temperatures can be found in dictionaries and other reference works. A *mole*, frequently employed in chemical calculations, is the number of grams equal to the molecular weight of a given substance. The molecular weight of water is 18 (2 hydrogen atoms of atomic weight 1 each and 1 oxygen atom of atomic weight 16), and so a mole of water is 18 grams. A *molar solution* is one containing 1 mole of a given substance per liter. Thus a molar solution of common salt (sodium chloride, NaCl, molecular weight approximately 58.5) is one containing approximately 58.5 grams of salt in 1 liter. Note that a calorie is defined as the amount of heat required to raise 1 gram of water from 14.5 to 15.5°C.

fully than most other substances.[5] Water's high heats of vaporization and fusion are also important in regulating environmental temperature. We know that the air above a body of water takes up an increasing amount of water vapor as the temperature rises. Since water enters the atmosphere by evaporation from the sea and since water evaporation utilizes large amounts of heat because of the high heat of vaporization, the net result is another mechanism whereby water tends to prevent rapid rises in temperature at times of sudden heat input (such as sunrise). Likewise, the high heat transfers involved in the freezing of water explain why large bodies of water seldom develop temperatures below the freezing point.

Dielectric constant

The special property of water that, perhaps more than any other, accounts for its significance in biology is the abnormally high value of what is called its *dielectric constant*. The dielectric constant may be defined in several different ways, and its explanation, which involves water's power to form hydrogen bonds, is complex. In the present connection, the most important point is that in a fluid with a high dielectric constant the attraction between ions of opposite charge is weakened. In crystalline sodium chloride, for example, the Na and Cl atoms are strongly united by an ionic bond. Their attraction is only 1/80th as great in water, so that the Na^+ and Cl^- ions there readily separate and move off the crystal into the surrounding water. In other words, the salt dissolves. The dielectric constant of water is uniquely high, and water is a solvent for an extraordinarily large variety of molecules, which form aqueous (watery), ionized solutions. This property is essential for the chemical mechanisms that occur in living things.

[5] The high heat capacity of water also prevents sudden temperature shifts *within* organisms. The metabolism of a living organism, which provides the energy for its survival and activity, invariably involves the production of heat and, therefore, raises the temperature. Thus an average man, weighing 60 kilograms, may produce in the course of an average day 2500 kilocalories of heat. This heat in a closed system containing 60 kilograms of water would raise the temperature by more than 40°C. A much greater rise would occur with most other liquids.

IONIC DISSOCIATION

The electrostatic attraction existing between the sodium and chloride atoms in NaCl accounts for the regularities that manifest themselves in the characteristic structure of the NaCl crystal. But when solid NaCl dissolves in water, as we have just seen, Na^+ and Cl^- ions leave the crystal and roam randomly among the water molecules. This separation into free ions is termed *ionic dissociation*.

Dissociation of salts in water

In part, the solvent action of water is due to its high dielectric constant. Another phenomenon helps keep the Na^+ and Cl^- ions apart once they have left the NaCl crystal.

Let us consider what happens when the ions of a salt dissolve in water. We may picture the simple ions as if they were tiny charged spheres. The positive charge on the sodium ion tends to attract water molecules around it, the negative (oxygen) ends of the water molecules pointing toward the cation. The electric field around the ion is very intense, and so the orienting force is very great; a whole cluster of water molecules becomes arranged around the ion, and the intense electric field causes them to pack very closely. Similarly, water molecules cluster around the anion, in this case with their positive (hydrogen) ends pointing toward the ion. These shells of water molecules around the ions produce electric fields of their own, and these fields are so oriented as to oppose the fields arising from the ions themselves. Hence the forces of attraction between the ions are weakened; the ions are, as it were, *kept apart* from one another by the water molecules gathered around them. Thus water is an effective solvent for salts not only because of its high dielectric constant but also because water molecules tend to combine with ions to form *hydrated* ions.

Almost all chemical reactions of biological importance occur in aqueous solutions. Ionic compounds dissociate in water and yield free ions, as illustrated by NaCl and others.

NaCl	$\rightleftharpoons$	Na^+	$+$	Cl^-
Sodium chloride		Sodium ion		Chloride ion

$$CH_3COOH \rightleftharpoons CH_3COO^- + H^+$$

Acetic acid   Acetate ion   Hydrogen ion

$$NH_4OH \rightleftharpoons NH_4^+ + OH^-$$

Ammonium hydroxide   Ammonium ion   Hydroxyl ion

Since the dissociated ions permit the passage of electric currents, the compounds are electrolytes. Water is also an excellent solvent for many nonelectrolytes.

Acids and bases

The equation showing the dissociation of acetic acid reveals that a hydrogen ion, H^+, is liberated. Indeed, this is why the compound is called an *acid*. According to the modern theory of J. N. Brønsted, an acid is strictly defined as *any* compound that can liberate a hydrogen ion. Thus hydrochloric acid, HCl, is an acid because it dissociates in water to form H^+ and Cl^-. A *base* is characterized by the ability to accept and combine with the hydrogen ion furnished by an acid. Note in the following examples of acids and bases that water can serve as an acid *and* a base. Any substance performing in such a dual capacity is said to be *amphoteric*. Note also that acids and bases can be either neutral molecules (for example, HCl and NaOH) or charged ions (for example, NH_4^+ and HPO_4^{--}), since both can liberate or accept hydrogen ions.

Acids

$$HCl \rightleftharpoons H^+ + Cl^-$$
$$H_2CO_3 \rightleftharpoons H^+ + HCO_3^-$$
$$NH_4^+ \rightleftharpoons H^+ + NH_3$$
$$H_2O \rightleftharpoons H^+ + OH^-$$

Bases

$$NH_3 + H^+ \rightleftharpoons NH_4^+$$
$$NaOH + H^+ \rightleftharpoons Na^+ + H_2O$$
$$HPO_4^{--} + H^+ \rightleftharpoons H_2PO_4^-$$
$$H_2O + H^+ \rightleftharpoons H_3O^+$$

Acids and bases vary significantly in the *extent* to which they dissociate into free ions in aqueous solutions. Those that dissociate completely are called *strong* acids or bases. *Weak* acids and bases dissociate only to a limited and varying degree. For example, a strong acid such as HCl dissociates com-

pletely into H^+ and Cl^-. Virtually no HCl remains in the form of the intact ionic compound. The same is true of the strong base sodium hydroxide, NaOH. However, when acetic acid, which may be symbolized as HAc, dissolves in water, only a small fraction of the molecules dissociates to H^+ and Ac^-; the rest remain in the form of intact HAc. Most of the acids and bases of biological interest are weak. All salts, of both strong and weak acids, dissociate completely in water to free cations and anions.

The strength of an acid depends on the concentration of hydrogen ions in it. Since that concentration can vary enormously, from 1 mole of H^+ per liter to 0.00000000000001 mole per liter, it is convenient to represent it on a logarithmic scale. Because the value is generally a fraction and hence has a negative logarithm, a positive scale is obtained by using the negative of the logarithm. This is the meaning of the symbol pH, found in all scientific discussions of the acidity of solutions; the pH is the negative logarithm of the hydrogen ion concentration in moles per liter of solution. For a concentration of 1 mole per liter, the pH is 0, and for a concentration of 0.00000000000001 mole per liter, the pH is 14.

Pure water at ordinary temperatures dissociates to free H^+ and OH^-, but only to a slight extent. Its pH is 7, and this is defined as *neutrality*. Solutions with pH's below 7 are *acidic*, and they are more acidic the lower the figure. Solutions with pH's above 7 are *basic* or *alkaline*, and they are more basic the higher the figure. Since the scale is logarithmic, a change of 1 pH unit indicates a 10-fold change in absolute concentration of H^+, or in acidity versus alkalinity.

Buffers

The pH's of fluids in living organisms are usually near neutrality. For instance, the pH of human blood is 7.4, just slightly alkaline. There are some exceptions (such as some strongly acidic digestive fluids), but an organism ordinarily cannot tolerate significant alterations of internal pH's. The main devices for preserving nearly constant pH's in the face of varying chemical circumstances are the *buffers* of the organism.

Changes in pH in a dynamic chemical system depend on the ratios of the concentration of H^+ acceptors, which are bases, to the concentration of H^+ donors, which are acids. For example, in a solution of H_2CO_3 (carbonic acid), the pH depends on the ratio $[HCO_3^-]/[H_2CO_3]$, where the brackets denote concentrations. In order to produce a pH of 7.4, the ratio would have to be 20, but since H_2CO_3 dissociates only slightly, so high a proportion of HCO_3^- cannot arise from its dissociation alone. If, however, the salt $NaHCO_3$ (sodium bicarbonate) is added to the solution, it dissociates completely into Na^+ and HCO_3^-. The concentration of HCO_3^- ions from this source can readily be adjusted to make the ratio $[HCO_3^-]/[H_2CO_3]$ equal 20 and thus to produce a pH of 7.4. In the absence of such a salt, the addition of some other substance, such as the strong base $NaOH$ (sodium hydroxide), may cause a radical change in pH. In such a case, the further addition of a completely dissociated salt of the weak acid minimizes, that is, buffers, the effect on the pH.[6]

A biological buffering system, then, is one that achieves and maintains a pH adaptive for the organism in question. It consists of a solution containing a mixture of a slightly ionized weak acid with one of its completely ionized salts. Later chapters will show that the bicarbonate buffer (HCO_3^-/H_2CO_3) and a phosphate buffer ($HPO_4^{--}/H_2PO_4^-$) are particularly important in living organisms.

PROPERTIES
OF ORGANIC COMPOUNDS

A large percentage of the biologically important inorganic molecules are electrolytes, or ionic compounds. In contrast, organic molecular compounds are usually built up by covalent bonds. However, even though most organic molecules are considered to be nonelectrolytes, it would be incorrect to state that they do not undergo ionization at all. A compound held together by covalent bonds may possess certain groupings in one or more regions of the molecule that do undergo ionization. These groupings will be dealt with later. The discussion in this section concerns mainly the covalently bonded portions of an organic molecule.

There are, of course, as many kinds of substances as there are different kinds of molecules, and to whatever extent a substance is stable and enduring, to that extent its molecules are fixed and enduring in their individuality. Since many hundreds of thousands of kinds of pure substances (and hence of molecules) exist but only a hundred-odd kinds of atoms from which they can be constructed, it follows that the uniqueness of a molecule must depend upon the *number*, *type*, and *spatial arrangement* of its component atoms.

The ordinary chemical formula merely indicates the numbers and types of atoms present in each molecule; thus carbonic acid is H_2CO_3. Much can be learned from such a formula. Consider the following simple examples: carbon dioxide, CO_2; carbon tetrachloride, CCl_4; methane, CH_4; chloroform, $CHCl_3$; formaldehyde, CH_2O; and ethane, C_2H_6. Among other things, it can be seen that one carbon atom has four combining sites. It can combine with four hydrogens because hydrogen has only 1 electron to share.[7] Oxygen has a covalence of 2 (that is, it can form two covalent bonds), nitrogen 3, and carbon 4.

The three-dimensional structure of a molecule must be compatible with the covalences of its component atoms. The two hydrogens of water, for example, cannot be attached to one another, for although the valence of each would be satisfied, no means for attaching oxygen would remain. We would have H_2 and O, not H_2O. Furthermore, the molecule cannot have the structure H—H—O, because this would give a hydrogen atom a valence of 2, whereas its valence is actually only 1, and an oxygen atom a valence of 1, whereas its valence is really 2. The only possible structure for a water molecule is there-

[6] Just why this should be and how the effect ensues is a matter of physical chemistry. For our present biological purposes, we ask you to accept the fact without detailed consideration of its known explanation. This can be found in all elementary textbooks of chemistry.

[7] Note that hydrogen can contribute its single electron in an ionic electron-transfer reaction or it can share its electron in the formation of a covalent bond. In the former case, we say that hydrogen has an ionic valence, or *electrovalence*, of +1. In the latter, we say that it has a *covalence* of 1.

fore H—O—H (with each of the two hydrogens attached directly to the oxygen). Similarly, in methane, CH_4, each of the hydrogens must attach to the carbon and not to each other, so that the atoms necessarily are combined and distributed in space as follows:

$$
\begin{array}{c}
\text{H} \\
| \\
\text{H}-\text{C}-\text{H} \\
| \\
\text{H}
\end{array}
$$

Chloroform, $CHCl_3$, is comparable:

$$
\begin{array}{c}
\text{H} \\
| \\
\text{Cl}-\text{C}-\text{Cl} \\
| \\
\text{Cl}
\end{array}
$$

Likewise, the only possible arrangement of atoms in ethane, C_2H_6, is

$$
\begin{array}{c}
\text{H} \quad \text{H} \\
| \quad\; | \\
\text{H}-\text{C}-\text{C}-\text{H} \\
| \quad\; | \\
\text{H} \quad \text{H}
\end{array}
$$

When molecular structures include atoms of valences higher than 1, the possible arrangements become elaborate. Carbon dioxide, CO_2, is straightforward enough, the two oxygens being attached to the carbon by two double bonds, $O{=}C{=}O$. However, one of the bonds can open and still leave the oxygen and carbon firmly held together by the other. This happens when CO_2 is dissolved in water. The two molecules combine to form carbonic acid, H_2CO_3, or

The H and OH of the water molecule are added to the O and C ends, respectively, of the opened bond.

This depiction of carbonic acid constitutes a *structural formula*. The *empirical formula* merely gives the ratio of atoms in a molecule. The structural formula indicates the spatial distribution of the atoms or atomic groupings. Even more precise representations are possible. Actually the atoms within a molecule do not lie within a single plane as a printed structural formula might suggest. In

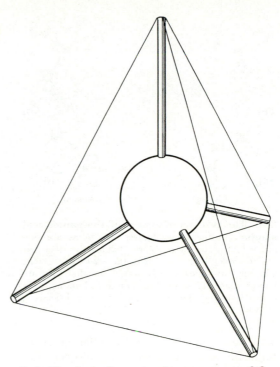

2–6 **The three-dimensional arrangement of the carbon atom.** The four potential bonds point to the vertexes of an imaginary symmetrical tetrahedron.

three-dimensional reality, the four potential bonds of carbon point to the vertexes of an imaginary symmetrical tetrahedron, a pyramid whose three sides and base are identical equilateral triangles (Fig. 2–6). To obtain a truer picture of the structure of a molecule, chemists build models in which bond lengths and angles are portrayed in accurate proportion. A three-dimensional model of the glucose molecule is illustrated on p. 22.

Structural formulas and molecular models make it readily apparent that a given empirical formula may represent more than one structural arrangement. Consider the formula C_2H_6O. The laws of valence indicate that all the hydrogens must be attached to oxygen or carbon and that the carbons must be bonded together, but two different structures are still possible.

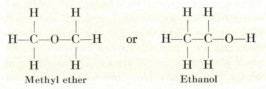

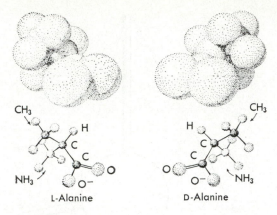

CH₃ ... H ... C ... C ... O ... NH₃ ... O⁻
L-Alanine

CH₃ ... H ... C ... C ... O ... NH₃ ... O⁻
D-Alanine

2–7 The two stereoisomers of the amino acid alanine. In the upper figures the atoms are diagramed to scale. In the lower figures the diagrams are in skeletal form, so that the structures of the molecules are more clearly seen. Note that they are mirror images of each other. (From L. Pauling, *General Chemistry*, 2nd ed., Freeman, San Francisco, 1953.)

Two such compounds with differing structural formulas and identical molecular formulas are *isomers* of one another. Given these two isomers, the only two known substances with the formula C_2H_6O, the chemist must use special chemical techniques to detect which is which.

In one form of isomerism, two molecules have the same structural formulas, but one is structurally the *mirror image* of the other (Fig. 2–7). This situation, known as *stereoisomerism*, can occur whenever a molecule contains a carbon atom attached to four different atoms or groups of atoms. Stereoisomers are sometimes called *epimers*.

The study of stereoisomerism has had a prominent place in the history of chemistry and biology, some of the classical work in this field having been performed by Pasteur. Stereoisomers have identical chemical properties and hence cannot be distinguished by chemical tests as can ordinary isomers like methyl ether and ethanol. They are recognizable in the laboratory chiefly on the basis of two other properties: (1) one of a pair of stereoisomers rotates a beam of polarized light to the right, whereas the other rotates it to the left; and (2) many enzymes, as we shall see, attack one of a pair of stereoisomers and not the other (for this reason, chemical

stereospecificity is of great biological significance). When compounds showing stereoisomerism are synthesized in the laboratory, equal numbers of right-handed and left-handed molecules are produced, and the mixtures are optically inactive. Such compounds synthesized in organisms by stereospecific enzymes are, on the contrary, always either right-handed or left-handed and not both, and they are therefore optically active.

Functional groups

Organic compounds would consist solely of carbon (and its associated hydrogens) in chains and rings of varying sizes were it not for certain distinctive groupings of atoms that recur frequently in organic structural formulas. In the compounds just discussed, for example, we encountered the hydroxyl (—OH) group, which replaces one hydrogen in ethane,

$$
\begin{array}{ccc}
 & H & H \\
 & | & | \\
H- & C- & C-H \\
 & | & | \\
 & H & H
\end{array}
$$

to make ethanol,

$$
\begin{array}{ccc}
 & H & H \\
 & | & | \\
H- & C- & C-OH \\
 & | & | \\
 & H & H
\end{array}
$$

The chemical properties of such groups play a large role in determining the chemical behavior of the molecules of which they are a part. Hence, they are called *functional groups* (or *radicals*).

The terminal functional group is essential in establishing the chemical behavior of the compounds in the following series:

$CH_3—CH_2—CH_3$	Propane
$CH_3—CH_2—CH_2—OH$	Propanol
$CH_3—CH_2—CHO$	Propionaldehyde
$CH_3—CH_2—COOH$	Propionic acid

Note that the last-named compound is an acid because it possesses a *carboxyl* group (—COOH), which dissociates to yield a hydrogen ion (here R denotes $CH_3—CH_2—$):

$$R—COOH \rightleftharpoons R—COO^- + H^+$$

The carboxyl group is the hallmark of the organic acid. (All organic acids, incidentally, are weak acids.) It was to compounds such as these that we referred earlier. Even though their carbon-to-carbon bonds are all covalent, they can possess one or more functional groups containing ionic bonds that dissociate in water. Indeed, in some of the most biologically important organic molecules (for example, amino acids), there is one acidic functional group (that liberates H^+ in water) *and* one basic functional group (that accepts H^+).

Table 2–2 lists some of the more common organic functional groups. Note that many are in series. The open bonds in the second and third columns of the table are a reminder that functional groups do not occur by themselves.

Organic Compounds in Living Organisms

It is now believed that the origin of life on earth was the result of a long series of chemical reactions of gradually increasing complexity. We have every reason to suppose that only simple organic molecules existed on the surface of the primitive earth—compounds such as methane (CH_4), carbon monoxide (CO), and perhaps some carbon dioxide (CO_2). There was a great deal of free hydrogen, nitrogen was mainly in the form of ammonia (NH_3), and oxygen was almost all in the form of water. We shall consider the origin of life in Chapter 5. But in anticipation of the following discussion of the major types of organic molecules within living organisms, we should note here that these complex organic chemical structures had themselves to evolve before the first living organisms came into being.

One might reasonably ask why the bulk of living organisms is fabricated of compounds of carbon. Why were the earliest living organisms not made primarily of compounds of silicon (Si), another element of valence 4, one that the earth possesses in abundance far beyond that of carbon? Why does organic chemistry deal with the compounds of carbon and not of silicon? In fact, writers of science

fiction have already reviewed the prospects of siliceous beings in which quartz, or SiO_2, is analogous to the CO_2 of real life.

Part of the answer is simply that, for a variety of reasons, carbon atoms bond together to form chains more readily than do silicon atoms. Hence larger and more varied molecular structures may develop from carbon. These include long and short chains, straight and branched chains, and ring structures of all descriptions. In addition, many (but not all) carbon compounds are readily soluble in water. The same is not true of silicon compounds. Most silicon has been continuously bound up in relatively insoluble minerals in the earth's crust. Carbon, though present in many rocks as carbonate, has remained in the earth's water and atmosphere as a constituent of labile and reactive small molecules capable of further combination into countless numbers of organic compounds.

Many of the details of early chemical evolution are not yet known. But the point is worth making that the relatively few compounds that we shall treat here are found in *all* living organisms. We must conclude, therefore, that these compounds were uniquely suited to form the coherent self-duplicating system, or organism, that emerged from a diffuse background, succeeded, and survived. These molecules are now synthesized only by living organisms, but in early prebiotic times they somehow arose without the aid of living creatures. We may assume that there existed some kind of process that caused these molecules destined for perpetuation to be selected from the welter of useless compounds.

Whatever the nature of this selective process, it centered on four major groups of simple organic molecules (and a number of minor ones). The simple molecules linked together in various ways to produce the large so-called *macromolecules* that are such prominent objects of contemporary biochemical research. In the cases of sugars, amino acids, and nucleotides, the formation of macromolecules consists in the production of long chains, or *polymers*, in which the simple compounds are the links of the chain, or *monomers*.

TABLE 2–2

Some organic functional groups

Name of group	Molecular formula	Structural formula
Amino	$-NH_2$	(structure)
Alkyl	$-C_nH_{2n+1}$	(structure)
Methyl	$-CH_3$	(structure)
Ethyl	$-C_2H_5$	(structure)
Propyl	$-C_3H_7$	(structure)
Carboxyl	$-COOH$	(structure)
Hydroxyl	$-OH$	$-O-H$
Aldehyde	$-CHO$	(structure)
Keto (carbonyl)	$=O$	$=O$
Sulfhydryl (thiol)	$-SH$	$-S-H$
Phenyl	$-C_6H_5$	(structure)

Simple organic compounds in living organisms	Macromolecules formed from these compounds
Monosaccharide sugars	Polysaccharides
Fatty acids	Simple and complex lipids
Amino acids	Proteins
Nucleotides	Nucleic acids

We shall attempt here simply to familiarize ourselves with the chemical natures of these substances. Later chapters will describe the metabolic activities in which they are involved. We shall also consider later the reasons why the gross structure of an organism persists while its individual chemical components are undergoing continuous synthesis, modification, and replacement. As you survey these compounds and classes of compounds, it will be helpful to bear in mind that most of the organic molecules of living organisms have four broad functions: (1) some are essential to body structure; (2) some serve primarily as energy-rich fuels; (3) some convey information controlling growth, differentiation, and biological specificity from one generation to another; and (4) some operate primarily as catalytic agents in the body's chemical processes.

SIMPLE AND COMPLEX CARBOHYDRATES

The *carbohydrates*, comprising the sugars and their many derivatives, consist of carbon, hydrogen, and oxygen, the last two elements ordinarily being present in the same proportions as in water. Most simple carbohydrates can therefore be represented by the empirical formula $C_x(H_2O)_y$. This makes it seem that they are *hydrates* of carbon, whence the name "carbohydrate." In fact, this is merely an illusion of the molecular formula; the water molecule as such does not appear in the structural formula of a carbohydrate. Rather, a carbohydrate is made up fundamentally of short carbon chains bearing many hydroxyl groups plus a single *aldehyde* or *keto* group.

Simple sugars are called *monosaccharides*. As noted earlier, these can join in long complex chains called *polysaccharides*. Monosaccharides function chiefly as immediate

sources of energy. Polysaccharides are storage forms of carbohydrate and, in plants, structural elements.

Monosaccharides are usually named according to the number of carbon atoms in each molecule, with the characteristic *-ose* ending of all carbohydrate nomenclature. Thus there are *trioses* ($C_3H_6O_3$), *tetroses* ($C_4H_8O_4$), *pentoses* ($C_5H_{10}O_5$), *hexoses* ($C_6H_{12}O_6$), *heptoses* ($C_7H_{14}O_7$), *octoses* ($C_8H_{16}O_8$), and *nonoses* ($C_9H_{18}O_9$). All of the sugars containing aldehyde groups are known collectively as *aldoses*, and those containing keto groups are known as *ketoses*. Since each category of monosaccharide includes aldoses and ketoses, one may speak of aldopentoses, ketopentoses, aldohexoses, and so on.

The monosaccharides of chief interest to us are the aldoses glyceraldehyde, ribose, 2-deoxyribose, and glucose (see p. 42) and the ketoses dihydroxyacetone, ribulose, and fructose. The formulas of these are shown in Fig. 2–8. Note that they are portrayed as ordinary straight-chain compounds. As it happens, many chemical reactions of the sugars can be explained only if the molecules are arranged as ring structures. Therefore, Fig. 2–8 gives the ring formulations of glucose and ribose. In the literature of biochemistry, both straight-chain and ring structures are employed, depending upon the circumstances.

The simplest polysaccharides are the *disaccharides*, each of which contains two molecules of the same monosaccharide or of two different monosaccharides. A disaccharide arises when two simple sugars are linked together. When such a linkage takes place, one hydrogen atom and one hydroxyl group are lost from the sites of attachment in the carbohydrate molecule and combined to form water. Thus sucrose (glucose-fructose) is produced as in Fig. 2–9. (The process as it actually occurs is far more complex than the simple one-step reaction shown. The equation gives only the net reaction, summing up what goes into the complex process and what comes out of it.) Other common disaccharides are lactose (glucose-galactose) and maltose (glucose-glucose).

A similar process can link together several

Straight-chain formulas

ALDOSES

Aldehyde group →

```
      CHO              CHO              CHO              CHO
   H—C—OH          H—C—OH          H—C—H           H—C—OH
   CH₂OH           H—C—OH          H—C—OH          HO—C—H
Glyceraldehyde     H—C—OH          H—C—OH           H—C—OH
   (triose)        CH₂OH           CH₂OH            H—C—OH
                   Ribose        2-Deoxyribose       CH₂OH
                  (pentose)        (pentose)        Glucose
                                                    (hexose)
```

CHO / $H-C-OH$ / CH_2OH — Glyceraldehyde (triose)

CHO / $H-C-OH$ / $H-C-OH$ / $H-C-OH$ / CH_2OH — Ribose (pentose)

CHO / $H-C-H$ / $H-C-OH$ / $H-C-OH$ / CH_2OH — 2-Deoxyribose (pentose)

CHO / $H-C-OH$ / $HO-C-H$ / $H-C-OH$ / $H-C-OH$ / CH_2OH — Glucose (hexose)

KETOSES

Keto group →

CH_2OH / $C=O$ / CH_2OH — Dihydroxyacetone (triose)

CH_2OH / $C=O$ / $H-C-OH$ / $H-C-OH$ / CH_2OH — Ribulose (pentose)

CH_2OH / $C=O$ / $HO-C-H$ / $H-C-OH$ / $H-C-OH$ / CH_2OH — Fructose (hexose)

Ring formulas

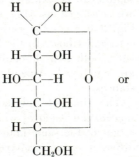

Glucose

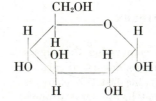

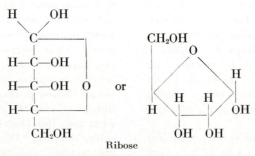

Ribose

2–8 Biologically important monosaccharides. The lower part of the diagram shows that some sugars can be formulated structurally as straight chains or as rings.

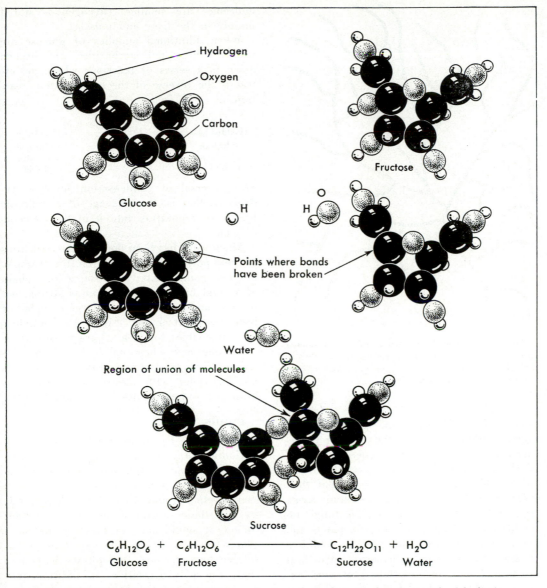

Labels in figure:
- Hydrogen
- Oxygen
- Carbon
- Glucose
- Fructose
- Points where bonds have been broken
- Water
- Region of union of molecules
- Sucrose

$$C_6H_{12}O_6 \ + \ C_6H_{12}O_6 \ \longrightarrow \ C_{12}H_{22}O_{11} \ + \ H_2O$$

| Glucose | Fructose | Sucrose | Water |

2-9 Carbohydrate molecules. Glucose, fructose, and their combination, with dehydration, to form sucrose.

monosaccharide molecules, one molecule of water being eliminated for each unit added.

$$n(C_6H_{12}O_6) \xrightarrow{\text{condensation}} (C_6H_{10}O_5)_n + (n-1)H_2O$$

n Glucose molecules ⟶ Starch ⟶ $(n-1)$ Water molecules

(This process as it actually occurs in cells also goes through many steps and is even more complicated than the last.) Here n indicates the number of hexose units involved,

which is large in this case. The result is a relatively large complex polysaccharide. The building up of a large molecule from simple sugar units is called *condensation*; it is a *dehydration* synthesis, that is, one involving the elimination of water. This is one of the principal ways in which the materials of life are compounded.

Like simple sugars, polysaccharides may have the same empirical formula and yet be

2–10 The general structure of amylopectin and glycogen. Each circle represents a glucose radical. Ernest Baldwin, *The Nature of Biochemistry*, Cambridge University Press.

quite different because the atoms are arranged differently. Thus the many polysaccharides with the same general formula $(C_6H_{10}O_5)_n$ include glycogen, cellulose, and the starches. All are polyglucoses, but they differ in molecular weight, branching structure, and solubility. Cellulose, the tough insoluble constituent of cell walls in plants and a very few animals (for example, tunicates or sea squirts) has a primarily structural function. It has no importance for most animals, except as a food for some herbivores. Starch, on the other hand, has no structural function in plants and animals but is a major form of storage carbohydrate in plants and a food source of carbohydrate for animals. The potato consists mainly of starch stores.[8] Glycogen is the insoluble glucose polymer that is produced in animals as a carbohydrate reservoir. It resembles starch in structure, but its

[8] In dahlias and the Jerusalem artichoke, starch, a polymer of glucose, is replaced by inulin, a polymer of fructose. Grasses contain other polymers of fructose called *levans*.

molecules are smaller. Glycogen is stored mainly in the liver and muscle.

When additional supplies of glucose are needed, these polysaccharides are broken down in a series of reactions whose net effect is the reverse of the condensation reaction.

$$(C_6H_{10}O_5)_n + (n-1)H_2O \xrightarrow{\text{hydrolysis}} n(C_6H_{12}O_6)$$

| Starch or glycogen (insoluble) | $(n-1)$ Water molecules | n Glucose molecules (soluble) |

This reversal of condensation involves *hydrolysis*, that is, the cleavage of a molecule by the incorporation into it of the elements of water.

Much work has been done on the structures of these polysaccharides. Starch, for example, has two components: *amylose*, a long chain of several hundred glucose units strung end to end; and *amylopectin*, a large complex of branched short chains (Fig. 2–10). Amylose is responsible for the characteristic blue color obtained when starch is mixed with iodine. Glycogen lacks amylose; thus it gives no blue color with iodine. Its structure is branched and resembles that of amylopectin.

LIPIDS

The *lipids*, or *fats*, are a heterogeneous collection of organic substances that are generally insoluble or sparingly soluble in water but freely soluble in organic solvents such as ethanol and ether. We shall presently see that these substances owe some of their biological significance to the fact that as a group they can bridge the gap from water-soluble to water-insoluble phases without a sharp line of demarcation. In addition, they are important fuels whose metabolic oxidation provides cells with a large proportion of their energy.

The group includes a variety of fats, oils, waxes, and related compounds. Many of these are extremely complex. However, the most abundant lipids, and the ones with which we are mainly concerned, are referred to as *simple lipids*. Simple lipids contain only carbon, hydrogen, and oxygen; compared to carbohydrates, they have less oxygen relative to carbon and hydrogen. Stearin, for example, has the empirical formula $C_{57}H_{110}O_6$.

Upon hydrolysis, a simple lipid yields *glycerol* and *fatty acids*. Glycerol contains three hydroxyl groups and thus is an alcohol. Conversely, a simple lipid, or *triglyceride*, results when the acidic —COOH groups of fatty acids react with the three alcoholic —OH groups of glycerol.

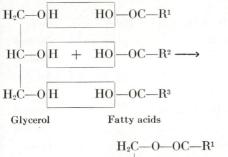

| Glycerol | Fatty acids |

$$H_2C—O—OC—R^1$$
$$HC—O—OC—R^2 + 3H_2O$$
$$H_2O—O—OC—R^3$$

| Triglyceride | Water |

The R^1, R^2, and R^3 compounds can be identical or different fatty acids.

The term "fatty acid" is not easy to define accurately. Usually it is taken to include the straight-chain members of the acetic acid series of organic acids. All possess *even* numbers of carbon atoms. In other words, if $CH_3(CH_2)_nCOOH$ is the general formula for a fatty acid, n is ordinarily an even number. Some branched-chain fatty acids occur as minor components of lipids. The more common fatty acids are palmitic acid, $CH_3(CH_2)_{14}$-COOH (or $C_{15}H_{31}COOH$), and stearic acid, $CH_3(CH_2)_{16}COOH$ (or $C_{17}H_{35}COOH$).

Fatty acids are divided into two classes depending on whether or not the carbon chain carries the maximum possible number of attached hydrogens. The *saturated* fatty acids have structures like

$$\begin{array}{ccccc} H & H & H & H & \\ | & | & | & | & \\ —C— & C— & C— & C— & COOH \\ | & | & | & | & \\ H & H & H & H & \end{array}$$

In the *unsaturated* fatty acids, double bonds join the carbon atoms that are not fully saturated with hydrogen, as in

$$\begin{array}{ccccc} H & H & H & H & \\ | & | & | & | & \\ —C— & C= & C— & C— & COOH \\ | & & & | & \\ H & & & H & \end{array}$$

A well-known unsaturated fatty acid is oleic acid, $CH_3(CH_2)_7CH{=}CH(CH_2)_7COOH$ (or $C_{17}H_{33}COOH$). If the double bond in oleic acid were eliminated by the addition of two hydrogens, the product would be stearic acid.

The higher members of the saturated class of fatty acids (lauric, myristic, palmitic, stearic, etc., acids) in combination with glycerol form the bulk of the body fat in most organisms. The short-chain fatty acids are not widely distributed in nature. We shall find in the next chapter that lipids are important components of cell membranes and membranous submicroscopic cellular particles.

One of the fatty acids making up a simple lipid may be replaced by a compound containing phosphorus and nitrogen. The result is a *phospholipid*. Members of this category also constitute an important part of the tissue lipids. The additional group makes them soluble in both water and organic solvents. For this reason, they serve to bind water-soluble compounds (such as proteins) to water-insoluble compounds.

The *steroids* are fat-soluble derivatives of the *cyclopentanoperhydrophenanthrene* nucleus, a complex organic molecule of the following structure:

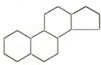

Compounds of this type are widely distributed in nature, and they are often associated with body fat. Hence they are usually discussed with the lipids. They include cholesterol, vitamin D, and a large number of animal hormones.

AMINO ACIDS AND PROTEINS

Proteins are large, complex, and fragile chainlike polymers of *amino acids*. When proteins are broken down into their low—molecular weight components, the amino acids, it is found that these links in the chain

TABLE 2–3 — *Major amino acids:*

$$R-\underset{\underset{NH_2}{|}}{\overset{\overset{H}{|}}{C}}-COOH$$

Name	Abbreviation	R	Category
Alanine	ALA	CH_3-	Neutral
Arginine	ARG	$NH_2-\overset{\overset{NH}{\|\|}}{C}-NH-CH_2-CH_2-CH_2-$	Basic
Asparagine	ASP-NH₂	$NH_2-CO-CH_2-$	Neutral
Aspartic acid	ASP	$COOH-CH_2-$	Acidic
Cysteine	CYS	$SH-CH_2-$	Acidic
Glutamic acid	GLU	$COOH-CH_2-CH_2-$	Acidic
Glutamine	GLU-NH₂	$NH_2-CO-CH_2-CH_2-$	Neutral
Glycine	GLY	$H-$	Neutral
Histidine	HIS	$H-C{=}C-CH_2-$ (imidazole ring: $H-N$, N, $\overset{C}{H}$)	Basic
Isoleucine	ILEU	$CH_3-CH_2-\underset{\underset{CH_3}{\|}}{CH}-$	Neutral
Leucine	LEU	$CH_3-\underset{\underset{CH_3}{\|}}{CH}-CH_2-$	Neutral
Lysine	LYS	$NH_2-CH_2-CH_2-CH_2-CH_2-$	Basic
Methionine	MET	$CH_3-S-CH_2-CH_2-$	Neutral
Phenylalanine	PHE	(phenyl)$-CH_2-$	Neutral
Proline	PRO	*	Neutral
Serine	SER	CH_2OH-	Neutral
Threonine	THR	$CH_3-CHOH-$	Neutral
Tryptophan	TRY	(indole ring) $-C(=)-CH_2-$, $\overset{C}{H}$, $\overset{N}{H}$	Neutral
Tyrosine	TYR	$HO-$(phenyl)$-CH_2-$	Acidic
Valine	VAL	$CH_3-\underset{\underset{CH_3}{\|}}{CH}-$	Neutral

* The structure of proline does not fit the prototype. It is as follows:

$$\begin{array}{c} CH_2-CH-COOH \\ | \qquad\quad | \\ CH_2 \qquad NH \\ \diagdown\;CH_2-\diagup \end{array}$$

E-VAL-ASP-GLU-HIS-LEU-CYS——GLY-SER-HIS-LEU-VAL-GLU-ALA-LEU-TYR-LEU-VAL——CYS-GLY-GLU-ARG-GLY-PHE-PHE-TYR-THR-PRO-LYS-ALA

$\quad$ NH₂ NH₂ $\qquad$ S $\qquad\qquad\qquad\qquad\qquad\qquad\qquad\qquad\qquad\qquad\qquad\qquad$ S

$\qquad\qquad\qquad\qquad\qquad$ S $\qquad\qquad\qquad\qquad\qquad\qquad\qquad\qquad\qquad\qquad\qquad\qquad\qquad\qquad\quad$ S

-ILEU-VAL-GLU-GLU-CYS-CYS-ALA-SER-VAL-CYS-SER-LEU-TYR-GLU-LEU-GLU-ASP-TYR-CYS-ASP

$\qquad$ NH₂ $\quad$ └——S————S——┘ $\qquad\qquad\qquad\qquad$ NH₂ $\qquad\qquad$ NH₂ $\qquad\qquad$ NH₂

2–11 The structure of insulin. The insulin molecule is a complex sequence of amino acids in two unequal chains joined by disulfide bridges.

all contain *nitrogen* as well as carbon and hydrogen.

Proteins have a unique role in biology. To a large extent, the complexity and diversity of life itself are dependent upon the complexity and diversity of proteins. In order to appreciate this singular fact, we must first learn how protein molecules are constructed.

The units that make up a protein molecule, the amino acids, are organic acids with the general formula

$$\begin{array}{c} H \\ | \\ R\!-\!C\!-\!COOH \\ | \\ NH_2 \end{array}$$

All the amino acids in proteins have this structure.[10] The individual amino acids differ only in the composition of the R group. This may have any of two dozen or so structures, according to current knowledge. The 20 most commonly encountered R groups are listed in Table 2–3. When R is simply a hydrogen atom, we have glycine,

$$\begin{array}{c} H \\ | \\ H\!-\!C\!-\!COOH \\ | \\ NH_2 \end{array}$$

The other R groups are more complex. Note that each amino acid has a standard abbreviation. The abbreviations are convenient for representing amino acid *sequence* (see, for example, Fig. 2–11). Note also that each amino acid has a basic —NH₂ group on the

[10] In the formal nomenclature of organic acids, the C immediately adjacent to the carboxyl C is designated the α carbon. The next C is β, and so on. The general formula given here is of an α-amino acid—that is, the amino group attaches to the α carbon. The amino acids making up protein are all α-amino acids; β-amino acids are found in nature but not in proteins.

carbon next to the acidic —COOH carbon. An amino acid containing no other acidic or basic functional groups is called *neutral* because it possesses one basic group and one acidic group.[11] An amino acid designated as *basic* or *acidic* possesses an additional basic (for example, —NH₂) or acidic (for example, —COOH) functional group elsewhere in the molecule.

The amino acids linked together in a protein molecule have been joined end to end. The carboxyl group of one amino acid combines with the amino group of the next to form a *peptide bond* (—CO—NH—), and a molecule of water is liberated.

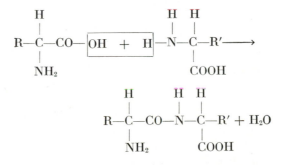

[11] We noted on p. 35 that any substance that can act as an acid *and* a base is called an "amphoteric" compound. Water is such a substance. So is any amino acid. For example, consider glycine, the simplest amino acid, NH₂CH₂COOH. Its dissociation behavior varies with the pH of the solution as follows:

$$^+NH_3CH_2COOH \underset{+H^+}{\overset{-H^+}{\rightleftharpoons}} {}^+NH_3CH_2COO^- \underset{+H^+}{\overset{-H^+}{\rightleftharpoons}}$$

$$\text{pH 1} \qquad\qquad \text{pH 6}$$

$$NH_2CH_2COO^-$$
$$\text{pH 11}$$

At pH 6, the doubly charged ion ⁺NH₃CH₂COO⁻ predominates. This dipolar ion can function as an acid or a base. When base is added and the pH is raised, the ion acts as an acid and gives up a hydrogen ion. When acid is added and the pH is lowered, it becomes a base and accepts a hydrogen ion. These shifts affect the solubilities of amino acids—and of proteins, which are therefore also amphoteric.

Short chains of amino acid residues are *peptides*; two amino acids linked together constitute a *dipeptide*, three a *tripeptide*, and a large number a *polypeptide*.[12] Polypeptide chains combine to form proteins.

An outstanding recent development was the discovery of methods for determining the amino acid sequence of an entire protein molecule. Because of the enormous amount of work involved, few proteins have yet been studied completely. *Insulin*, a hormone of the animal pancreas, was the first protein to be fully analyzed (Fig. 2–11). It was found that the insulin molecule consists of two chains of unequal length held together by disulfide (—S—S—) bonds formed between the —SH groups of opposed cysteine residues. One chain contains 30 amino acid units, and the other 21 units.

The amino acid sequence is one aspect of protein structure; the *shape* of the molecule is another. Thus we speak of *primary*, *secondary*, and *tertiary* protein structures. Knowledge of the primary structure of a protein molecule requires information on the number of polypeptide chains, the amino acid sequence in each, and the natures and positions of the covalent cross links between different chains or different portions of the same chain. As in insulin, such cross links are usually disulfide bonds forming between —SH groups projecting from cysteine units. It is important to remember that a protein may contain any or all of the approximately 24 known amino acids; that it may contain virtually any *number* of amino acids; [13] and that these may occur in any *sequence*.

At the secondary-structure level, we consider the spatial arrangements of the individual polypeptide chains. For example, the polypeptide chains of many proteins exist as *helical coils* (Fig. 2–12). These are stabilized by hydrogen bonds of the type —NH···OC—, which form between the CO— and NH—

groups at each turn. Other, more irregular, configurations in secondary structure also arise. The tertiary structure involves the foldings and loopings of the already coiled polypeptide chains. It too depends upon stabilizing bonds, which are of many types, including covalent disulfide bonds, hydrogen bonds between various groups, and others. Protein molecules whose stabilizing bonds have been irreversibly ruptured are said to be *denatured*.[14]

It would be difficult indeeed to overstate the biological importance of protein structure. A large, complex molecule that consists of long chains of simple amino acids arranged in diverse patterns is ideally suited to extensive and subtle variation. We shall later find that amino acid sequences are solely responsible for the enormous variety of living organisms. Their uniqueness, or *specificity*, is seen, for example, in the slight but characteristic differences in the amino acid sequences of horse, sheep, and pig insulins. Proteins are the basic structural components of most living tissues, but they are very much more. Certain protein molecules are *enzymes*, the *catalysts* of chemical reactions in organisms. Enzymes make possible chemical reactions that would otherwise virtually fail to occur. We shall say more about them presently, but the point to be emphasized here is the great specificity of enzyme action. In general, every chemical reaction in the living organism is catalyzed by a specific enzyme. It is now known that this specificity is a consequence of a precise amino acid sequence in the polypeptide chain and its three-dimensional configuration.

One other fact will become increasingly significant as we go along. Besides the *simple proteins* just discusssed—that is, those consisting entirely of amino acid chains in various arrangements—there are a large number of *conjugated proteins*. These have, in addition to the amino acids, nonprotein groups chemically bound onto their molecules. The nonprotein groups are called *prosthetic groups*. Among the important conjugated proteins are (1) *nucleoproteins*, compounds in which one or several molecules of protein are combined with a *nucleic acid*; (2) *glycopro-*

[12] Note that every peptide has a free —NH$_2$ group in one terminal amino acid, the *N-terminal* amino acid, and a free —COOH group in the other, the *C-terminal* amino acid.

[13] The great size of the average protein molecule is suggested by the following empirical formulas: casein of milk, $C_{708}H_{1130}O_{224}N_{180}S_4P_4$; gliadin of wheat, $C_{685}H_{1068}O_{211}N_{196}S_5$; human hemoglobin, $C_{3032}H_{4816}O_{872}N_{780}S_8Fe_4$.

[14] The action of heat upon egg white is a classic example of denaturation.

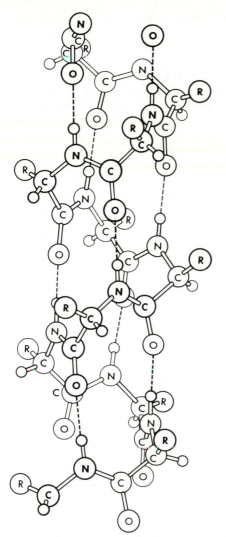
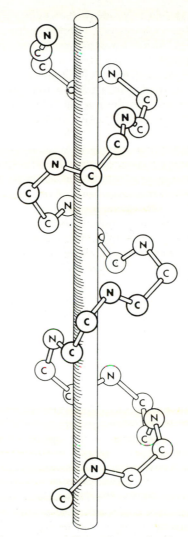

2–12 The α helix, a hydrogen-bonded helical configuration of the polypeptide chain present in many proteins. The polypeptide chain is coiled as a left-handed screw, with about 3.6 amino acid residues per turn of the helix. The circles labeled R represent the side chains of the various amino acid residues. The sketch at the right emphasizes the helical arrangement by picturing the carbon and nitrogen linkages circling a rod.

teins, compounds with polysaccharide prosthetic groups; (3) *lipoproteins,* proteins conjugated to triglycerides or other lipids; and (4) *chromoproteins,* compounds with chromophoric (that is, colored) prosthetic groups. The last category includes *hemoglobin,* the pigment upon which animal cells depend for oxygen.

NUCLEOTIDES AND NUCLEIC ACIDS

The nucleic acids are the largest of the "biomolecules." Like polysaccharides and pro-

teins, they are polymers made up of repeating small-molecule units (monomers). However, if proteins are viewed as long words composed from a 20-letter alphabet, the nucleic acids are longer words written from a 4-letter alphabet.[15] The four repeating monomers are called *nucleotides.* Thus nucleic acids may alternatively be termed *polynucleotides.* In the organism nucleic acids, or polynucleotides,

[15] The sense in which we here speak of words and letters and of their ability to convey information will be considered in Chapter 7.

are attached to proteins, the resulting nucleoproteins comprising conjugated proteins whose prosthetic groups are the nucleic acid polymers. It is important to understand the nature of nucleotides, for they have at least one major physiological function other than that of links in the nucleic acid chain—they are often essential participants in enzyme reactions.

A nucleotide is more complex than most of the other small molecules, since it consists of three simpler molecules in direct linkage: a *nitrogen-containing organic base*, a *sugar* residue, and *phosphoric acid*. Since the identities of the base and sugar may vary, there are a number of nucleotides. The bases all fall into two groups, one containing the *purine* ring and the other the *pyrimidine* ring.

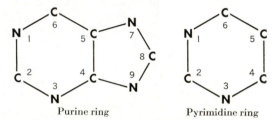

Purine ring   Pyrimidine ring

As shown in Table 2–4, the bases found in nucleic acid nucleotides are the purines adenine and guanine and the pyrimidines cytosine, thymine, and uracil. The two main nucleic acid sugars are deoxyribose and ribose (see Fig. 2–8), both five-carbon monosaccharides or pentoses, the former differing from the latter only in lacking one oxygen atom on carbon 2. The deletion of an oxygen atom is implied by the *deoxy*- prefix.

Nucleic acids fall into two major classes depending on whether their nucleotides con-

tain ribose (that is, are *ribonucleotides*) or deoxyribose (that is, are *deoxyribonucleotides*). Thus there are *deoxyribonucleic acid*, or *DNA*, and *ribonucleic acid*, or *RNA*. Interestingly, DNA and RNA have one other fundamental *chemical* difference (in addition to the physiological differences to be described later). Although both contain only four nitrogenous bases, of which three are adenine, guanine and cytosine, the fourth base is not the same in RNA and DNA. The second pyrimidine in RNA is uracil; in DNA it is thymine, a methyl derivative of uracil. Therefore, DNA may be represented as a chain of the following general structure, the base sequence being variable:

Deoxyribose—Adenine

Phosphate

Deoxyribose—Thymine

Phosphate

Deoxyribose—Guanine

Phosphate

Deoxyribose—Cytosine

Phosphate

Deoxyribose—Cytosine

And RNA may be represented similarly as follows:

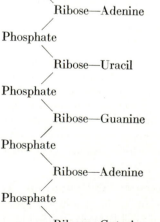

Ribose—Adenine

Phosphate

Ribose—Uracil

Phosphate

Ribose—Guanine

Phosphate

Ribose—Adenine

Phosphate

Ribose—Cytosine

TABLE 2–4
*Compositions of DNA and RNA**

		DNA	RNA
Bases			
	Purines	Adenine (A)	Adenine (A)
		Guanine (G)	Guanine (G)
	Pyrimidines	Thymine (T)	Uracil (U)
		Cytosine (C)	Cytosine (C)
Sugar		Deoxyribose (dR)	Ribose (R)
Phosphoric acid		Phosphoric acid (P)	Phosphoric acid (P)

* The letters in parentheses are frequently used abbreviations.

In 1953 J. D. Watson and F. H. C. Crick discovered that the DNA molecule consists of a *double-stranded helix* resembling a long twisted ladder (Fig. 2–13) with two parallel nucleotide chains winding around a cylindrical space. The sides of the ladder are alternating deoxyribose and phosphate groups. The rungs between the sugars are *paired* purines and pyrimidines. Base pairing is the key to the structure, for it has been found that *bases can be paired only in certain ways*—adenine with thymine only and guanine with cytosine only (Fig. 2–14). The explanation for the specificity of base pairing is seen in Fig. 2–14. The positions of the two sugar-phosphate chains are so fixed that only pairings involving a "large" purine and a "small' pyrimidine will fit in the confined space. Moreover, a base with a keto group on carbon 6 must stand opposite one with an amino group on carbon 6, since only such a pair will form the hydrogen bond required to hold the two strands together. The two DNA strands are *complementary*, not identical, and if one strand contains genetic information in its base sequence, then the other strand must contain the identical information in a complementary code. In Chapter 7 we shall learn how complementary double-strandedness makes it possible for the DNA molecule to perform its most notable functions: transmission of the genetic information handed down in the long processes of evolution, and accurate self-replication.

The structure of RNA is less well understood than that of DNA. Moreover, as we shall see later, the cell contains at least three kinds of RNA, which differ in function and structure. At least some of the RNA is a single-stranded nonhelical polymer. The principal function of RNA is in the control of *protein synthesis*. In Chapter 7 we shall encounter a most satisfying relationship among these isolated facts concerning DNA, RNA, and protein structure.

Properties of Chemical Reactions

We have remarked that enzymes catalyze biochemical reactions. So that we may under-

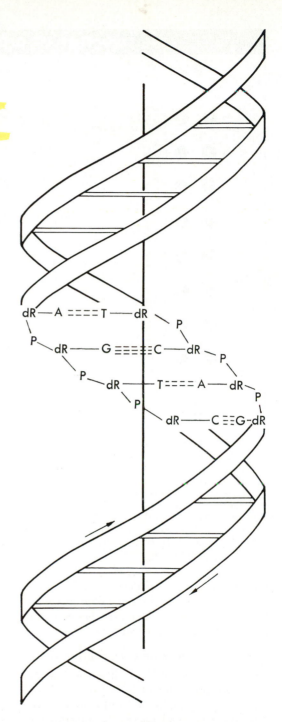

2–13 The double-stranded helical configuration of the DNA molecule. The two nucleotide strands are held together by hydrogen bonds forming between complementary purine–pyrimidine pairs.

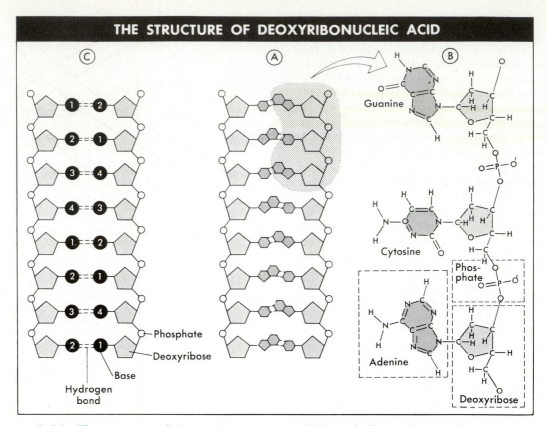

THE STRUCTURE OF DEOXYRIBONUCLEIC ACID

Guanine

Cytosine

Adenine

Phosphate

Deoxyribose

Phosphate

Deoxyribose

Base

Hydrogen bond

2–14 The structure of deoxyribonucleic acid (DNA). *A.* The DNA molecule is a long ladderlike structure that is twisted into a helix; in these figures the ladder is shown in untwisted form for simplicity. The "uprights" of the ladder consist of pentose sugars (light gray pentagons in the figure) and phosphate groups (open circles). The "cross rungs" of the ladder consist of pairs of nitrogenous bases (dark gray in *A* and *B*), of which there are four: thymine, guanine, cytosine, and adenine. *B.* The detailed chemical structure of part of the DNA ladder. *C.* The structure of the molecule is further simplified, the four bases being represented by black discs numbered as follows: *1*, cytosine; *2*, guanine; *3*, thymine; *4*, adenine. The important point is that the cross rungs of the DNA ladder always consist of one of these two pairs: *1–2*, cytosine-guanine; or *3-4*, thymine-adenine.

stand the principles of enzyme action, let us briefly consider certain general properties of chemical reactions. What are chemical reactions? Why do they occur?

ENERGY AND WORK IN GENERAL

Energy and *work* are well enough defined by our ordinary use of both words: work is something accomplished (ultimately, something moved); and energy is something required to perform work. The physicist's definition of energy is different only in wording: energy is the capacity to accomplish work. This definition holds whether or not the work is done. A boulder at the top of a hill has

potential energy while it is at rest—while not doing work; this energy becomes *kinetic* energy as the boulder does work in rolling down the hill. The work was performed at the expense of the boulder's potential energy inherent in its former position at the top of the hill.

Kinetic energy can be thought of as the motion of matter, and work as movement accomplished. Kinetic energy may take many forms. The movement of the boulder was *mechanical* energy. *Light* is kinetic energy and may be thought of as the movement of minute particles, *photons*. An *electric current* is another form of kinetic energy, the movement of electrons from atom to atom in the wire through which the current is flowing. The *chemical* form of kinetic energy is the movement of atoms within or between molecules that takes place in a chemical reaction. And, finally, *heat* is the ceaseless, random motion of atoms in all directions within a gas, liquid, or solid. Hot and cold water differ in the speed at which their molecules are moving; in the hot water the motion is faster.

All these forms of energy are interconvertible to some extent, and the conversions are governed by two general laws called the *first and second laws of thermodynamics*. The first law is all that we need at the moment, and it states that energy is neither lost nor gained when it is converted from one form to another. We may transform the mechanical energy of a waterfall into electricity by means of a turbine,[16] but we will never get more energy out of the transformation than entered it. Nor will we lose any.[17]

The form of energy with which the cell deals is principally chemical energy; its work is largely chemical work. Chemical work consists of the transformation of molecules in a chemical reaction such as [18]

$$A + B \rightarrow C + D$$
Reactants   Products

The formula sums up the fact that the two molecules *A* and *B* (the reactants) interact in such a way as to recombine their atomic parts, forming new molecules *C* and *D* (the products). Two groups of factors influence the reaction: (1) factors affecting the collision and contact of the molecules; and (2) factors relating to energy expenditures in the performance of the chemical work.

FACTORS GOVERNING REACTION VELOCITY

The molecules *A* and *B* are dispersed, un-

[16] Think out other forms of energy transformation commonly employed by man. How is the potential chemical energy in coal or oil eventually exploited in frying an egg on an electric stove? (Coal → Steam → Turbine→ Electricity→ Heat)

[17] Some of the energy will, however, be transformed into a condition in which it can no longer be used or perform work.

[18] Other types of chemical reactions are possible. Thus one molecule may be broken into two ($A \rightarrow B + C$); or two may combine into one ($A + B \rightarrow C$); and so on. What we say about $A + B \rightarrow C + D$ applies to all.

dergoing continual random movements in all directions. Only when they collide and make contact can they react with each other. The dependence of the reaction on collision explains several laws governing chemical reactions. All those factors that increase collision frequency increase the rate at which the reaction proceeds. The more *concentrated* the reactants are in a solution, the greater is their chance of colliding, and the faster is the reaction.[19] Accordingly, subjecting the solution to *pressure*, which effectively concentrates it, increases collision frequency and speeds up the reaction. An increase in *temperature* speeds up the movement of the molecules and thereby raises the probability that they will contact and react.

The contact necessary for a reaction between two molecules can be facilitated in another way—by the *adsorption* of both reactants onto a surface where they are brought close together. This is what happens in *surface catalysis*. Molecules of sulfur dioxide (SO_2) and oxygen (O_2) in air may contact each other so rarely that reaction between them is negligible. If, however, air is passed over finely ground platinum, SO_2 and O_2 molecules are adsorbed to the platinum surface, where they contact and react. The platinum is a catalyst, which catalyzes the reaction. The

[19] Consider a reaction in which molecule *A* decomposes spontaneously into smaller molecules *B* and *C*.

$$A \rightarrow B + C$$

The number of molecules of *A* decomposing in any unit of time is ordinarily proportional to the number present. If this concentration is represented as [*A*],

$$\text{Rate} = k[A]$$

If the rate constant *k* equals 0.001 and time is measured in seconds, this equation tells us that 1/1000th of the *remaining* molecules of *A* decompose each second. Reactions of this type are called *first-order reactions*.

Second-order reactions depend upon the collision and interaction of two molecules.

$$A + B \rightarrow C + D$$

Here the rate depends upon the frequency of collisions and is proportional to the product of the concentrations of *A* and *B*. Thus

$$\text{Rate} = k[A][B]$$

where *k* is the rate constant.

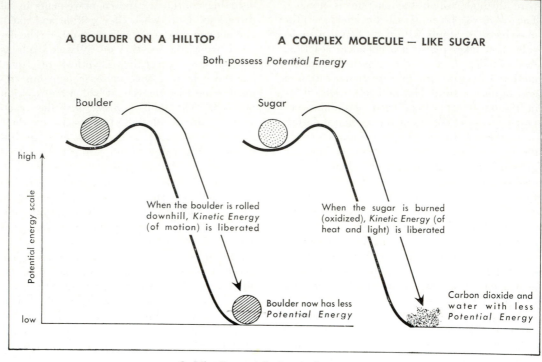

A BOULDER ON A HILLTOP   **A COMPLEX MOLECULE — LIKE SUGAR**

Both possess *Potential Energy*

Boulder

Sugar

high

Potential energy scale

When the boulder is rolled downhill, *Kinetic Energy* (of motion) is liberated

When the sugar is burned (oxidized), *Kinetic Energy* (of heat and light) is liberated

Carbon dioxide and water with less *Potential Energy*

Boulder now has less *Potential Energy*

low

2–15   Potential chemical energy.

adsorption to the catalyst surface has an effect similar to that of raising the pressure or temperature or concentrating the reactants. Usually, however, there is more to catalysis than this; and the effect of a catalyst is far greater than any other. It is a defining characteristic of a catalyst that it is never itself consumed in the course of a reaction. When the reaction is complete, the catalyst's surface is freed to act again. A minute amount of catalyst can thus do a great deal of work.

THE ENERGY OF CHEMICAL REACTION

Contact between molecules is necessary if they are to interact, but it is not sufficient. Energy considerations also play a dominant role in chemical reactions. Let us begin with the notion of the potential energy inherent in chemical structure.

Potential energy of chemical structure

It is a familiar fact that useful kinetic energy can be obtained from chemical compounds like sugar, coal, gas, or wood. When

any of these compounds is burned, its complex molecular structure is destroyed, and energy is liberated. In summary, the burning of sugar can be written as follows:

$$C_6H_{12}O_6 + 6O_2 \rightarrow 6CO_2 + 6H_2O + \text{Energy}$$

Sugar   Oxygen   Carbon   Water
                     dioxide

The large sugar molecule is broken down by an *oxidation* process. The products of the reaction are the smaller, simpler molecules CO_2 and H_2O. The energy liberated takes the obvious forms of light and heat. The important point is that potential energy is inherent in the structure of the large molecule. The complicated positions of its atoms might be compared to the position of the boulder at the top of a hill, in which potential energy is inherent. When sugar is disintegrated by oxidation into the simple constituents CO_2 and H_2O, the energy inherent in its complex chemical structure is freed, just as the boulder's energy is freed when it gives up its position on the hilltop. Fig. 2–15 compares the two processes.

Work can be done with the kinetic energy liberated from both the boulder and the sugar. And both the boulder at the bottom of the hill and the collection of small CO_2 and H_2O molecules can be used for work again, but only if they are first restored to their original positions. The boulder must be pushed up the hill, and the CO_2 and H_2O must be pushed up a "chemical hill." These processes are both work, and both demand new energy resources. Notice that in repeatedly doing work the *same matter* can be used over and over. There may be a cycle of materials used, but there is *no cycle of energy*. It is a continuous expenditure of capital funds. Where does all this capital lie? Ultimately all usable energy comes from the sun; this is a theme that we shall develop in later chapters. For the moment it suffices to note that the relocation of matter to a position enabling it to do work demands fresh energy.

Energy of activation

Let us return now to chemical reaction in general, keeping these energy ideas in mind. We must distinguish two kinds of energy function in the reaction. Once the boulder is started down the hill, it liberates huge quantities of energy; once sugar is started on its oxidation, it also liberates considerable energy. But both processes need a little energy to get them started. There is a little rise in front of the boulder that is, indeed, what keeps it in place. You must supply *activation energy* to push it over this rise and start it on its downward path. And so it is with a chemical reaction. There is always a need for activation energy to get the process under way.

Every molecule is in constant motion; this is true of the molecule as a whole and of the atomic parts within it. In terms of Fig. 2–16, this means that the sugar molecule is constantly dancing about inside the "energy valley" that keeps it from uniting with O_2 and then rolling downhill to CO_2 and H_2O. Sometimes—very rarely—a random movement takes it over the energy barrier and on to the reaction

$$C_6H_{12}O_6 + 6O_2 \rightarrow 6CO_2 + 6H_2O + \text{Energy}$$

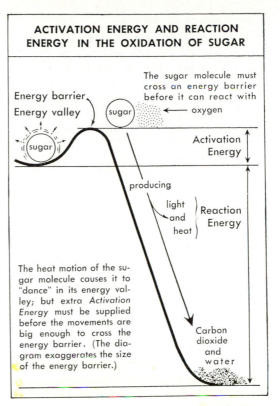

ACTIVATION ENERGY AND REACTION ENERGY IN THE OXIDATION OF SUGAR

Energy barrier
Energy valley

The sugar molecule must cross an energy barrier before it can react with oxygen

Activation Energy

producing
light and heat
} Reaction Energy

The heat motion of the sugar molecule causes it to "dance" in its energy valley; but extra *Activation Energy* must be supplied before the movements are big enough to cross the energy barrier. (The diagram exaggerates the size of the energy barrier.)

Carbon dioxide and water

2–16   **Activation and reaction energies.**

But this possibility is so extremely unlikely as to be utterly negligible. If left to themselves, sugar and oxygen will effectively *never* interact. The reaction is encouraged by the application of just a little activation energy. The great kinetic energy liberated from the first full downhill roll activates the other molecules, and off they all go. The minute kinetic energy of a single spark can liberate untold amounts of energy once it is applied to a gasoline dump.

In the last section we noted that a catalyst can have an effect similar to that of raising concentration, pressure, or temperature. It brings molecules together by adsorbing them to a surface, but its effect is always greater than could be accounted for by this property alone. Now we can go further. Catalysts speed up chemical reactions by lowering the activation-energy requirements. How they do this is not entirely clear. One might think of them picturesquely as tunneling through the activation-energy barrier.

Free energy of reaction

Another fundamental concept concerns the energy released once the molecule is over its activation hump. This is called the *free energy* of the reaction (see Fig. 2–16).

A reaction like the oxidation of sugar, in which the energy of reaction is given off or liberated, is called an *exergonic* reaction. We have already mentioned that it is possible to regenerate sugar (plus oxygen) from CO_2 and H_2O and that, like pushing the boulder up the hill, the process requires energy. A reaction like this—the combination of CO_2 and H_2O to form sugar—is called *endergonic* because energy must be paid into it. The amount of energy that must be supplied is exactly the same as that released when sugar was broken down.

As stated earlier, the laws of thermodynamics account for the energy changes in any physical or chemical process. These energy changes may be manifested either in the absorption or liberation of heat or in the performance of work. If a system does no work upon its surroundings, the total energy change in the system equals the change in heat energy.

Experience tells us that certain processes can occur spontaneously—for example, the diffusion of molecules from a region of high concentration to a region of low concentration, the running down of a clock, the descent of a boulder down a hill. In each of these processes, the system is changing its state in the direction of equilibrium. In order to reverse such a process, work must be introduced by some outside agency.

The second law of thermodynamics relates these facts in formal terms. The maximum useful work obtainable from a chemical reaction is usually signified by the symbol ΔF, where the Δ sign means "change in." According to the law,

$$\Delta F = \Delta H - T \Delta S$$

In this equation ΔF denotes the change in free energy that occurs when the system goes from one state to another. ΔH denotes the change in *total heat*. If heat is given off, ΔH is negative; if heat is absorbed, ΔH is positive. T is the absolute temperature; and ΔS is the change in a quantity called *entropy*. Entropy

is a measure of the degree of order in a system, high order always representing the most improbable condition. The more ordered or improbable the state of any system, the lower its entropy; the more randomly ordered or probable its state, the higher its entropy.[20] In simpler words, the entropy of a system indicates how much the system has run down.

When ΔF has a negative value (that is, when free usable energy is produced), the reaction is exergonic, and it usually will occur spontaneously (although initial activation may be necessary); ΔF is positive in endergonic reactions (that is, energy must be invested before the reaction will take place). If ΔF equals zero, the system is at equilibrium, neither liberating nor consuming energy.

One significant implication of this equation is that the change in free energy (ΔF) is more meaningful than ΔH as an index of the capacity of a chemical reaction to do work, since ΔF takes into account the energy $T \Delta S$ that is nonutilizable. Another is that the change in free energy (that is, the energy available for useful work) differs from the total energy change of a reaction. Thus

Change in usable energy = Change in total energy − Change in unusable energy

When the boulder in Fig. 2–15 rolls down the hill, energy is liberated and work is performed in amounts depending upon the size of the boulder and the height of the hill. In any such system, however, only a portion of the initial potential energy is available for the performance of work. The rolling boulder wastes some of its potential energy in overcoming frictional forces. Likewise, in chemical reactions larger or smaller parts of the total energy are unavailable. Hence we must distinguish between the *free* or *available energy* and the *total energy* of the system.

Living organisms are continuously engaged

[20] A sugar crystal dissolving in water provides an example of a system progressing from order to disorder. Initially the sugar molecules are in a highly ordered array. When solution has occurred and the molecules have diffused throughout the water, their arrangement is random and disordered. Statistically, the former is a far more improbable condition than the latter. Hence solution and diffusion increase the entropy of the system.

in the manufacture of large and complex molecules from smaller ones. They are, in short, always executing endergonic reactions, into which they must pay energy. It should be emphasized that enzymes do not in any way contribute energy to endergonic reactions. Their role is simply to accelerate reactions that are already thermodynamically possible.

OXIDATION—REDUCTION:
ENERGY EXCHANGE

The processes of burning sugar, coal, wood, and gasoline all consume oxygen and liberate energy. They are all special cases of the general process of oxidation, and all oxidations are energy-liberating chemical reactions. The reverse process is called *reduction*, and in it oxygen, instead of being consumed, is released, and energy, instead of being liberated, is bound into the chemical that is reduced. In the green leaves of plants, CO_2 and H_2O combine in a series of chemical reactions (photosynthesis) during which energy is consumed, CO_2 is reduced, and O_2 is liberated.

The oxidation or reduction of a compound can be accomplished in another way besides by the addition or removal of oxygen—by the removal or addition of hydrogen. The removal of hydrogen is an oxidation; its addition is a reduction. When these alternative methods of oxidizing and reducing are analyzed and compared, they are found to be special cases of an even more general process of energy exchange: the transfer of electrons from one atom to another. Energy is released by the *removal of electrons*; this is the general definition of oxidation. Either the addition of oxygen or the removal of hydrogen effectively removes an electron from an oxidized compound and thus liberates energy. Similarly, the general definition of reduction is the *addition of electrons*; either the removal of oxygen or the addition of hydrogen to a compound effectively adds an electron and thus binds energy.

In terms of the model in Fig. 2–16, oxidation occurs as the molecule rolls downhill, liberating energy; and when the molecules at the bottom of the hill are reduced, they are pushed uphill again. *Oxidations* are *exergonic, and reductions* are *endergonic.*

All these relations are summarized in Table 2–5. The reason for going into these matters here is that they are fundamental in the energy economy of the living organism. Most biological energy transfers involve oxi-

TABLE 2–5

Oxidation–reduction processes

Oxidation	Reduction
The *oxidation* of a compound (A or A^-) basically involves a *loss* of electrons.	The *reduction* of a compound (A^+) basically involves a *gain* of electrons.
$$A \rightarrow A^+ + e^-$$ $$A^- \rightarrow A + e^-$$	$$A + e^- \rightarrow A^-$$ $$A^+ + e^- \rightarrow A$$
The reaction releases energy from A. However, electron loss may be accompanied by 1. Addition of O	The reaction increases the energy content of A. However, electron gain may be accompanied by 1. Removal of O
$$A + \underset{\substack{\text{Oxidizing} \\ \text{agent} \\ \text{(O donor)}}}{BO} \rightarrow AO + B$$	$$AO + \underset{\substack{\text{Reducing} \\ \text{agent} \\ \text{(O acceptor)}}}{B} \rightarrow A + BO$$
A is oxidized; BO is reduced.	AO is reduced; B is oxidized.
2. Removal of H	2. Addition of H
$$AH + \underset{\substack{\text{Oxidizing} \\ \text{agent} \\ \text{(H acceptor)}}}{B} \rightarrow A + BH$$	$$A + \underset{\substack{\text{Reducing} \\ \text{agent} \\ \text{(H donor)}}}{BH} \rightarrow AH + B$$
AH is oxidized; B is reduced.	A is reduced; BH is oxidized.

dation-reduction processes, and most of them are accomplished by the removal of hydrogen (dehydrogenation) from a compound. Thus sugar is oxidized with a release of energy into the cell by the removal of hydrogen from the sugar molecule, not by the addition of oxygen to it. The oxygen consumed in degrading sugar in a cell acts as a hydrogen acceptor, not as an oxygen donor. (See Table 2–5. Oxygen acts as B, and sugar as AH, in equation 2 under *Oxidation*).

CHEMICAL EQUILIBRIUM

Three questions can be raised about every proposed chemical reaction. Will it occur? If so, at what rate? And to what extent? *Chemical thermodynamics*, as we have just seen, deals with the first question. *Chemical kinetics* is that branch of physical chemistry concerned with the rates of chemical reactions. It is important not to confuse the *rate* of a reaction and the *extent* of its progress toward completion. In reactions that can operate in both directions, the extent of the reaction depends upon the ratio between the rate constants of the forward and reverse reactions, k_1 and k_2. In a reversible second-order reaction,

$$A + B \underset{k_2}{\overset{k_1}{\rightleftharpoons}} C + D$$

the rate of the forward reaction is $k_1[A][B]$, and that of the backward reaction is $k_2[C][D]$. We say that the reaction has attained equilibrium when the forward rate equals the backward rate.

$$k_1[A][B] = k_2[C][D]$$

At this point, $[A]$, $[B]$, $[C]$, and $[D]$ have reached fixed values, and the ratio $[C][D] / [A][B]$ is constant.

$$\frac{[C][D]}{[A][B]} = \frac{k_1}{k_2} = K$$

The new constant K is the *equilibrium constant*. Clearly the final concentrations of the reactants would be the same regardless of whether we started with A and B or C and D. We now see why the reaction does not go to completion in one direction or the other. If it did in a forward direction, $[A]$ and $[B]$ would be zero, and K would be infinity. Since

$K = k_1/k_2$, it could not equal infinity unless k_1 equaled infinity or k_2 equaled zero. This is not the case, for both the forward and backward reactions occur at finite rates.

A number of circumstances may intervene, however, to make such a reaction go to completion in actuality. If one of the products, C or D, is an escaping gas or an insoluble precipitate, it is in effect removed as fast as it is formed, and its concentration drops to zero in the equilibrium equation. Thus the reaction continues until the initial reactants are consumed. This situation is known as "pulling a reaction to the right." Such circumstances arise in living organisms in the decomposition of H_2CO_3 to CO_2 and H_2O and in the precipitation of Ca^{++} as insoluble bone salts.

Enzymes

One of the distinguishing characteristics of living organisms is the presence within them of the special catalysts called enzymes. The existence of these catalysts in living tissues accounts for the rapid execution and control of an exceedingly large number of biochemical reactions. Enzymes possess all the properties we have noted for catalysts in general. They speed up reactions that are otherwise so slow that they would virtually never occur; they are not consumed in the reactions and in small amounts perform much work. In addition, they have many special properties that must be mentioned. Although enzymes are produced only by living organisms, they may be extracted from tissues, purified, crystallized, and studied in isolation. Through such studies, the knowledge of enzymes and their catalytic properties has expanded dramatically in the last two decades.

The names first given to enzymes, such as pepsin, trypsin, and ptyalin, offered no clues to their functions. The present convention of enzyme nomenclature often derives the name of an enzyme from that of the *substrate* (that is, the substance attacked by the enzyme), adding the suffix *-ase*. Thus we speak of the enzymes that split starch as *amylases* (*amylum* is Latin for "starch"), of the enzymes that split lipids as *lipases*, of the enzyme that acts on uric acid as *uricase*, and of the enzyme that acts on urea as *urease*. Groups of enzymes may therefore be designated as carbohydrases, proteases, glucosidases, and so forth. Enzymes may also be named according to the types of reactions they catalyze. One classification, for example, divides enzymes into five large functional groups: *hydrolases, transferases, synthetases, isomerases,* and *oxidoreductases.*

MECHANISMS OF ENZYME ACTION

We have seen that a catalyst, whether enzymatic or not, affects only the rate of a reaction. It has no effect whatever on the equilibrium of a reaction—that is, on the extent to which it proceeds to completion. Enzymes represent a special class of catalysts in their outstanding catalytic efficiency as well as in their specificity; molecule for molecule, no inorganic catalyst even approaches an enzyme in catalytic efficiency. As in the case of inorganic catalysts, the mechanisms of action of enzymes are not completely understood, although they constitute an area of active research. In part, an enzyme functions like an inorganic catalyst by bringing reactants into contact on a surface. Then, somehow, the enzyme makes the substrate molecules more reactive to the other molecules in the environment than before; the activation energy for the reaction is reduced. Perhaps this is due to the fact that the enzyme combines chemically with the substrate for a brief moment. Good evidence exists that the following reaction takes place:

$$E + S \rightleftharpoons ES \rightarrow E + \text{Products}$$

where E is the enzyme, S the substrate, and ES an unstable enzyme-substrate complex. Some theories of enzyme action hold that the formation of such a complex causes a slight deformation of substrate molecules that somehow "activates" them and facilitates their reaction.

The fact that enzymes are proteins is undoubtedly the key to any interpretation of their mode of action and specificity. Specificity remains one of the most striking properties of enzymes: individual enzymes catalyze specific reactions of specific substrates. Enzymes differ in the degree of their specificity. For instance, lipases, which catalyze the

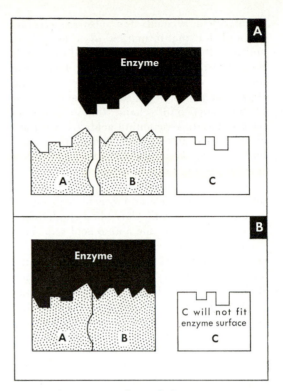

2–17 Enzyme catalysts bring reactants together in close contact and proper orientation for the reaction to proceed. The diagram merely illustrates the selective fitting of different complementary shapes; the real shapes of the molecules concerned are not like these.

breakdown of triglycerides to their constituents, glycerol and fatty acids, affect only the ester linkages in the triglycerides, the identity of the fatty acids being of no moment. Other enzymes have even more specificity. Many catalyze the reactions of only a single, unique substrate molecule.

Diagrams such as those in Fig. 2–17 are usually drawn to illustrate the concept of enzymatic specificity and to emphasize the importance of the enzyme surface. You should remember, however, that the shapes shown in the figure are purely imaginary. In reality the substrate consists of organic molecules with a variety of functional groups, and the enzyme is a large and complex protein molecule. Moreover, the active catalytic site on the enzyme is a unique three-dimensional configuration of the polypeptide chain whose various functional groups are precisely complementary to those of the substrate. This arrangement accounts for substrate specificity and for the formation of a temporary enzyme-substrate complex. In addition, it is now clear that the active center of the enzyme molecule, which chemically alters the substrate, comprises only a relatively small sector of the polypeptide chain, for investigators have removed some amino acids from certain enzyme molecules without altering their catalytic activity. Thus the entire molecule is not essential in the catalytic process.

Now that it has become possible to determine the complete amino acid sequence of a pure protein, studies have been undertaken for the first time of the amino acid sequences in the immediate vicinity of the active center in attempts to elucidate further the mechanisms of enzyme action. Although such studies have just begun, they have already yielded unexpected results. It appears that the active centers of several hydrolytic enzymes catalyzing a number of quite different reactions contain the amino acid residue serine flanked on one side by aspartic or glutamic acid (similar dicarboxylic acids—see Table 2–3) and on the other by glycine or alanine (both simple, neutral monocarboxylic acids). Though we still lack an understanding of the mechanistic significance of this brief amino acid sequence, we can scarcely imagine a more convincing example of nature's tendency to make repeated use of inventions that have operated successfully.

Temperature affects enzyme-catalyzed reactions in the same way that it affects ordinary chemical reactions. High temperatures, however, inactivate most enzymes by destroying (denaturing) their tertiary structures. Biological phenomena thus have optimal temperatures. This fact places serious limitations on organisms. Only a few have heat-resistant enzymes capable of surviving high temperatures. Many cells lose the ability to carry on metabolic processes above $40\,°C$. The "normal" temperature of the human body (that is, the average for healthy adults) is about $37\,°C$. So that body metabolism may proceed independently of environmental temperature variations, man and other warm-blooded organisms have evolved remarkable systems for precise body-temperature control. These will be described in later chapters.

ENZYME INHIBITORS

The complementary physical structures of enzyme and substrate ensure that they fit closely together. The phenomenon of *competitive inhibition* is an interesting consequence of this fit. (Fig. 2–18). Consider the reaction in which succinic dehydrogenase catalyzes the oxidation of succinic acid.

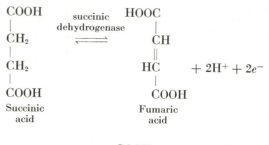

If malonic acid, $\begin{matrix} \text{COOH} \\ | \\ \text{CH}_2 \\ | \\ \text{COOH} \end{matrix}$ a substance whose structure closely resembles that of succinic acid, is added to the solution, the rate of the enzyme reaction is greatly diminished, although malonic acid itself undergoes no chemical transformation. It is now known that malonic acid attaches itself to succinic dehydrogenase in the location normally reserved for succinic acid. It in fact *competes* with succinic acid for a position on the enzyme molecule. The competition occurs because of the similarity of the structures of malonic and succinic acids. Presumably malonic acid combines with *part* of the active site, thereby occluding the normal substrate and preventing enzyme activity. Competitive inhibitors are known for many enzymes.[21]

Noncompetitive inhibition also occurs. This results from the reaction of certain agents with various functional groups of the enzyme protein (for example, —SH groups) that are essential components of the active center or are involved in preserving the necessary tertiary structure. The list of these agents includes many substances normally classed as

[21] Some competitive inhibitors are employed as drugs in the treatment of bacterial infection. One of these, sulfanilamide, acts by competitively preventing certain bacteria from utilizing *p*-aminobenzoic acid, a bacterial vitamin similar to it in molecular structure.

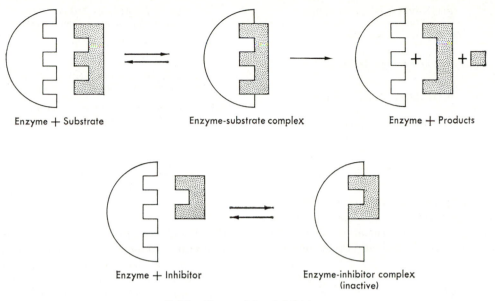

Enzyme + Substrate   Enzyme-substrate complex   Enzyme + Products

Enzyme + Inhibitor   Enzyme-inhibitor complex
                      (inactive)

2–18   Competitive inhibition.

poisons; by inhibiting normal reactions, they cause malfunctioning or death.

COENZYMES

In the well-known technique called *dialysis* (Fig. 2–19), a solution in which a mixture of large and small molecules is dissolved is placed in a bag made of cellophane or some other porous membrane. When the bag is immersed in water, the small molecules pass through the membrane into the surrounding water, leaving behind all the large molecules. In this way low–molecular weight organic molecules such as monosaccharides, amino

2–19   Separation of large (○) and small (·) molecules by dialysis.

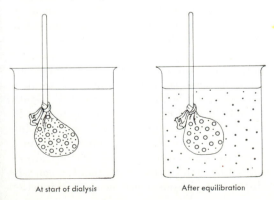

At start of dialysis   After equilibration

acids, and nucleotides are readily separated from macromolecules such as polysaccharides, proteins, and nucleic acids.

When early biochemists exhaustively dialyzed tissue extracts containing active oxidative enzymes, they observed a gradual loss of the catalytic power of the material remaining in the bag. They then found that activity was fully restored after the addition of a little boiled tissue juice. This effect could not have been due to the addition of new enzymes, since boiled enzymes are ordinarily denatured. The dialyzed enzymes in the bag could also be reactivated by the addition of the materials that had passed out of the bag—even though they had been boiled in the interim. It was concluded, therefore, that tissues contain molecules that are (1) essential for the catalytic activity of oxidative enzymes; (2) dialyzable and hence low in molecular weight; and (3) heat-stable and hence not proteins. The molecules were called *coenzymes.*[22]

The major coenzymes of the dehydrogenases are two dinucleotides:[23]

[22] The inactive enzymatic protein left in the bag is called the *apoenzyme;* the active combination of apoenzyme and coenzyme is called the *holoenzyme.*

[23] Until recently, these compounds were known as *diphosphopyridine nucleotide,* or *DPN,* and *triphosphopyridine nucleotide,* or *TPN,* since the base

Ⓟ—Ribose—Nicotinamide
|
Ⓟ—Ribose—Adenine

Nicotinamide-adenine
dinucleotide (NAD)

and

Ⓟ—Ribose—Nicotinamide
|
Ⓟ—Ribose—Adenine
|
Ⓟ

Nicotinamide-adenine
dinucleotide phosphate (NADP)

Some dehydrogenases operate with NAD, and others with NADP, but all employ the nucleotide coenzymes as intermediary donor-acceptors of hydrogen in oxidation-reduction reac-

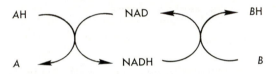

tions. Note that *A*H transfers its hydrogen to NAD to form NADH, which is converted back to NAD by the transfer of its hydrogen to *B* to form *B*H. NAD is not consumed in the reaction, its continuous regeneration requiring that it be present in only small quantities. Thus it functions as if it were a catalyst.

Tissues also contain other nucleotides that function as coenzymes. One outstanding type comprises the ribonucleotides of adenine, particularly the di- and triphosphates.

---

moiety, nicotinamide, is a *pyridine*. In the formulas, the symbol —Ⓟ designates the *orthophosphoric*

$$acid \text{ radical, } —OPO_3H_2, \text{ or } —O—\overset{\overset{O}{\|}}{\underset{\underset{OH}{|}}{P}}—OH. \quad Ⓟ—Ⓟ$$

and Ⓟ—Ⓟ—Ⓟ denote the *pyrophosphate* linkages

$$—O—\overset{\overset{O}{\|}}{\underset{\underset{OH}{|}}{P}}—O—\overset{\overset{O}{\|}}{\underset{\underset{OH}{|}}{P}}—OH$$

and

$$—O—\overset{\overset{O}{\|}}{\underset{\underset{OH}{|}}{P}}—O—\overset{\overset{O}{\|}}{\underset{\underset{OH}{|}}{P}}—O—\overset{\overset{O}{\|}}{\underset{\underset{OH}{|}}{P}}—OH,$$

respectively.

Adenine—Ribose—Ⓟ

Adenosine monophosphate (AMP)

Adenine—Ribose—Ⓟ—Ⓟ

Adenosine diphosphate (ADP)

Adenine—Ribose—Ⓟ—Ⓟ—Ⓟ

Adenosine triphosphate (ATP)

ADP serves as a Ⓟ acceptor in certain reactions, the result being ATP. Conversely, ATP donates Ⓟ to become ADP.

$$X—Ⓟ + ADP \rightleftharpoons X + ATP$$
$$ATP + Y \rightleftharpoons ADP + Y—Ⓟ$$
$$\overline{X—Ⓟ + Y \rightleftharpoons X + Y—Ⓟ}$$

Here *X* and *Y* represent any of a large group of organic compounds. It should be noted that again the coenzymes are donor-acceptors of some group (as NAD and NADP are donor-acceptors of hydrogen). This is a recurrent phenomenon of great significance in biochemistry. We shall presently see that ADP and ATP, the donor-acceptors of phosphate groups, are of profound biological importance. The functions and structures of other coenzymes will be discussed in detail in later chapters.

Although we shall consider *vitamins* later, it should be mentioned here that vitamins are converted in the body to coenzymes. Thus NAD and NADP contain the vitamin nicotinamide; FAD contains riboflavin; coenzyme A contains pantothenic acid; etc. The consequences of vitamin deficiency stem from deficiencies of specific coenzymes needed for enzyme function.

It is evident, then, that an enzyme reaction may involve more than the fitting of a substrate molecule, or of substrate molecules, onto the surface of an enzyme. The participation of a low–molecular weight nonprotein coenzyme may be necessary for the reaction to occur. Some coenzymes (for example, the porphyrin groups of the cytochromes) are more or less permanently attached to their apoenzymes, being extraordinarily tightly bound. These are more difficult to remove by dialysis than are the loosely bound pyridine nucleotides. In addition to vitamin-containing coenzymes, certain inorganic *cofactors* are required by many enzymes. So that they may be distinguished from coenzymes, which function as donor-acceptors in group-transfer reactions, these factors are referred to as *activators*. The best known of them are metals of the trace-element group (p. 31). Copper, for example, is needed by certain oxidative enzymes, magnesium is essential for the transfer of phosphate groups from ATP, and so on. The mechanism of action of metal activators is currently under investigation in many laboratories.

CHAPTER SUMMARY

Importance of chemical and biochemical principles for an understanding of life.

Nature of chemical substances: atoms and atomic structure; chemical bonds; the covalent bond and its origin in electron sharing; the ionic bond and its origin in electron transfer; the hydrogen bond and its special biological significance.

Major elements of the living organism: carbon, the fundamental constituent of organic molecules; oxygen and hydrogen, basic ingredients in oxidation and reduction reactions and components of water and many organic molecules; nitrogen, a key constituent of proteins; minerals, required in small amounts by all organisms.

Properties of compounds: the unusual properties of water—its biological importance, its expansion on freezing, its uniquely high surface tension, its unusual thermal properties, its high dielectric constant, which makes it an excellent solvent; ionic dissociation— dissociation of salts in water, acids and bases, the concept of pH, buffers; properties of organic compounds—compounds that contain carbon and are synthesized by living systems, properties of organic molecules, structural and empirical formulas, isomers, functional groups.

Major organic compounds in living organisms: how they arose prior to the origin of life; carbohydrates—sugars and polysaccharides, energy sources and structural materials; lipids—high hydrogen content, high energy yield; amino acids and proteins—the largeness and complexity of proteins, how amino acids form chains linked together through peptide bonds, the importance of protein structure; nucleotides and nucleic acids—properties of nucleic acid chains, RNA and DNA, purines and pyrimidines.

Properties of chemical reactions: energy and work in general; collision and contact; the energy of chemical reaction—potential energy of chemical structure, energy of activation, free energy of reaction; oxidation-reduction; chemical equilibrium.

Enzymes: classification and nomenclature; mechanisms of enzyme action; enzyme inhibitors; coenzymes—NAD, NADP, ATP, and ADP; activators.

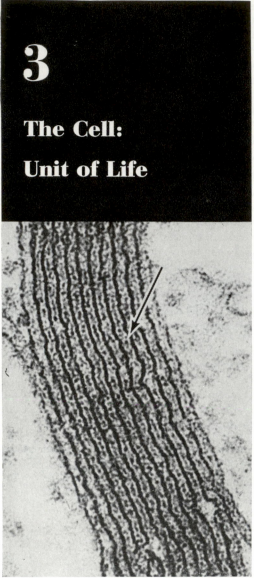

3

The Cell:

Unit of Life

H. Fernández-Morán

Transverse section of the myelin sheath from a frog nerve. Note the indications of regular fine structure (arrow) within the concentric array of dense and intermediate layers (×280,000).

The Cell Theory

DEVELOPMENT OF THE CELL THEORY

"We have seen that all organisms are composed of essentially like parts, namely, of cells." (Schwann, 1839)

"Where a cell exists there must have been a pre-existing cell, just as the animal arises only from an animal and the plant only from a plant. The principle is thus established, even though the strict proof has not yet been produced for every detail, that throughout the whole series of living forms, whether entire animal or plant organisms, or their component parts, there rules an eternal law of continuous development [or continuous reproduction]." (Virchow, 1858)

These two statements, one made in 1839 and the other in 1858, mark the emergence of the *cell theory* in its definitive form. The cell theory is one of the two great foundations of modern biology; the other is the theory of evolution (1859). The two theories have much in common: they reached their definitive forms almost simultaneously, and both had been developing over a long period of time. In their modern forms they merge almost inextricably.

A *cell* is the basic unit of life. It is the smallest unit of matter of which we can meaningfully say, "This is alive." Nearly all cells are small. Most of them, indeed, are not visible as distinct units without the aid of a lens or microscope.

It is impossible to be sure when living cells were first observed. Certainly it was long after the necessary instruments were available. Simple lenses were known many years before the Christian era began, and spectacles were commonly worn by the wealthy of the fourteenth century. But there is no record of the use of optical instruments to examine living specimens before the new curiosity of the Renaissance somehow suggested it. Two seventeenth-century biologists did see what we today call "cells." Robert Hooke (1635–1703) observed with a lens that wood charcoal, cork, and other plant tissues are made up of small cavities separated by walls;[1] in

[1] Hooke was in fact viewing only the thickened walls of dead cells whose bodies had been lost, and

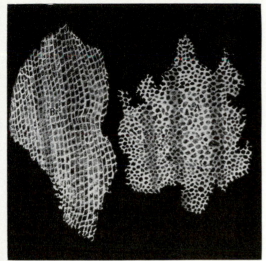

3–1 Hooke's drawings (1665) of the cellular organization in cork.

1665 he described cellular organization in plants (Fig. 3–1). A few years later Anton

though he called them "cells," he did not comprehend their nature. His choice of the term "cells" has often been lamented. "Cells" means "little rooms" which is what Hooke thought the spaces were. In the light of later developments, the term "corpuscles," meaning "little bodies," would have been more appropriate.

van Leeuwenhoek first saw minute single-celled organisms in a drop of pond water. But neither of these men, nor the many others who saw and drew cells throughout the eighteenth century, realized the significance of their observations.

It was not until 1839, when Theodor Schwann published the book quoted at the beginning of this chapter, that any *generalization* was made. Schwann clearly perceived that *all living organisms consist of cells; they are either single cells or groups of cells.* There is no life apart from the life of cells. Even in a large and complex organism like a man or a giant redwood tree, the whole body proves to be an aggregation of some billions of cells. Every specialized part of such an organism— skin, bone, muscle, nerve, and even blood; wood, bark, flower, and root—is composed of specialized cells plus a variety of intercellular materials that are largely produced by the cells (Fig. 3–2).

It is a long way from merely observing cells to reaching the great inductive generalization that Schwann announced. A generalization as broad as Schwann's surely indicates some underlying natural process of still greater consequence than the generalization itself. The underlying process is made clear in Rudolf

3–2 The cellular nature of plant and animal tissue. *Left.* A section of part of the growing tip of an onion root (the cells are all internal; surface cells are not shown). *Right.* A piece of salamander skin. Nuclei and dividing cells can be seen in both tissues. In those cells that are dividing, the membrane bounding the nucleus is not visible, but nuclear threadlike bodies called *chromosomes* are distinct.

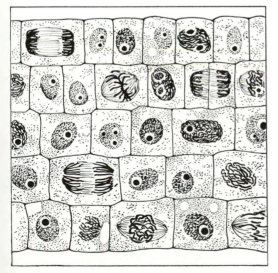
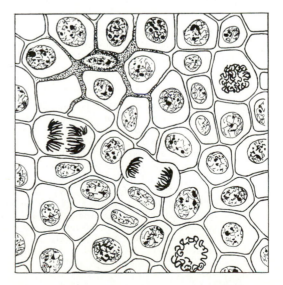

von Virchow's later statement that all cells arise only from preexisting cells. The full significance of this statement emerges when we consider how it relates the cell concept to two other major generalizations that appeared at about the same time. First, it conforms completely with the contemporary work of the French scientist Louis Pasteur (1822–1895). Second, it merges with the theory of evolution published by Darwin in 1859, the year after Virchow's book.

Between 1859 and 1861 Pasteur proved as conclusively as science can that in the modern world no living thing arises except from other living things. It was the general opinion for many centuries that lower organisms could originate from mud or, especially, from the flesh of dead animals. Aristotle (384–322 B.C.) taught that fleas and mosquitoes arise from putrefying matter, and no one ventured to contradict him during the long course of antiquity and the Middle Ages. An eminent seventeenth-century scientist was certain he had seen rats develop from bran and old rags. It was even easier to be certain that maggots arise spontaneously in rotten meat. Another seventeenth-century scientist, Francesco Redi (1627–1697), disproved this idea. By careful experiments he showed that maggots never appear unless flies have laid eggs in the meat.[2]

Perhaps Redi's work should have settled the matter, but it did not. A really popular error dies hard, if it ever dies completely. This one is still alive, and some people in the United

[2] These particular flies look somewhat like bees, so much so that they fooled some ancient writers, who recommended their use for the replenishment of beehives!

States still believe that a horsehair can turn into a worm if soaked long enough in water. As far as biologists are concerned, however, the problem was finally resolved by the work of Pasteur. In his day it was known that microorganisms appear in milk, wine, meat broth, and other substances even if they are protected from flies and other apparent sources. Pasteur examined every known example of such supposed spontaneous generation of living things. He showed that heating would kill any microorganisms present and that if the sterile substance were then protected from air, no other microorganisms would appear. (We still call this heating procedure *pasteurization*.) He demonstrated that microorganisms are carried through the air and that spontaneous generation does not occur in any known case.

Pasteur's critics were still inclined to the view that the living germs responsible for fermentation did not come from the air but arose spontaneously from the nonliving materials in the wine or milk. Pasteur won his point with a wonderfully simple and conclusive experiment. He took two flasks that contained a nutrient broth infected with germs capable of causing putrefaction. One of these flasks was subjected to prolonged heating. The other flask was not heated. The unheated broth putrefied immediately. The heated broth remained free of putrefaction for months until the neck of the flask was broken, exposing it to the entry of germs from the air. Once the neck was broken, the broth putrefied immediately, and Pasteur showed that its putrefaction was again due to *germs*. Pasteur's experiments marked the end of a belief in spontaneous generation and the establishment of the prin-

---

3–3   A diversity of cells. All ×700 except J, which is ×1500. A. Human muscle cells. Note their elongate form. B. A fat storage cell from human connective tissue. The nucleus is a small body near the cell membrane. The whole of the (unstippled) central part of the cell is occupied by a large globule of fat. C. Human red blood cells. D. Human white blood cells. E. Three epithelial cells from the intestine of an axolotl, a larval salamander. The flask-shaped central cell secretes mucus, which provides lubrication for the passage of food. F. An epithelial cell from the lining of the human vagina. Several bacteria lie beside it. G. Two cells from the liver of a mouse; the Golgi element can be seen in both of them. H. Two epithelial cells from the rat's intestine; they contain threadlike mitochondria lying above and below the nucleus. I. A human nerve cell. Note how it is drawn out into fiberlike processes. The nerve cell itself (black in the figure) is encased by a second cell containing many nuclei. J. A human sperm cell (×1500). K. A human egg cell, with a sperm entering it. L. Part of the human placenta; five separate cells are overlaid by a syncytium that bears cilia. A syncytium is a multinucleate mass of protoplasm that is not separated by membranes into distinct cells. M. Three cells from the human eye (retina) containing pigment granules.

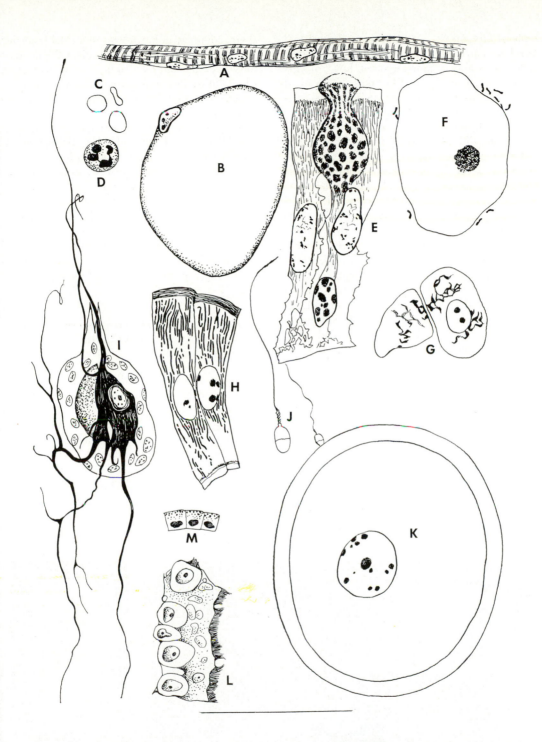

ciple of *biogenesis*—"all life comes from life." Virchow's principle that "all cells come from cells" is simply a more explicit form of this same truth, because all life takes the form of cells. Pasteur's minute "germs" were single-celled organisms, bacteria and yeasts.

The relationship of the cell theory to evolution follows from Virchow's sentence "The principle is thus established . . . that throughout the whole series of living forms, whether entire animal or plant organisms, or their component parts, there rules an eternal law of *continuous development.*" Here Virchow is glimpsing, in his mind's eye, an unbroken continuity of cell generations stretching back almost endlessly in time, back to the beginning of life. It is only a step from this picture of all cells arising from other cells to the insight that all cells have a common ancestry. Schwann's perception "that all organisms are composed of essentially like parts, namely of cells" recalls the perception of the imaginative Comte de Buffon (1707–1788) that the structures of all vertebrate animals have much in common. Both insights are explained by the theory of evolution. Vertebrates, no matter how diverse, share a common structural plan because they share a common ancestry. In similar fashion all living cells, no matter how diverse they may be (Fig. 3–3), show structural resemblances that are their common inheritance from still earlier forms of life. In cell and organism the novelty and diversity that have appeared in the course of evolution never obscure entirely those limitations imposed by an inheritance from ancient ancestors.

The transition from earlier investigators like Hooke and Leeuwenhoek to Schwann is a transition from a simple observation of fact to an inductive generalization of great breadth. And the transition from Schwann to Virchow is from a simple generalization to a *theory* of major importance that explains the generalization.

Before entering into a more detailed study of the cell, let us consider some of the principal implications of the cell theory.

THE CELL: THE MINIMUM MATERIAL ORGANIZATION POSSESSING LIFE

There are some particles that are smaller than any cells and that have been regarded by some biologists as alive. These are the *viruses* (Fig. 3–4), familiar as the agents responsible for such human diseases as influenza, poliomyelitis, and virus pneumonia. Many different kinds of viruses exist; however, all of them are associated with and utterly dependent up-

3–4 Various viruses. Photographed by electron microscopy (×50,000) at the Virus Laboratory, University of California, Berkeley. *Left to right.* Bacteriophage (by Dean Fraser and Robley C. Williams). Poliomyelitis (by C. E. Schmerdt and Robley C. Williams). Influenza (by Robley C. Williams). Vaccinia (by Robley C. Williams).

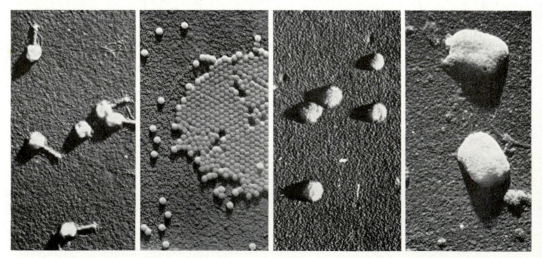

on living cells. When ordinary viruses enter the living cells of organisms as invaders, they cause the machinery of the host cell to cease its normal activities and to produce new viruses. In the process the host cell is destroyed. One could hardly term viral reproduction *self-reproduction* since it occurs only through the mechanisms of host cells.

A capacity for *self*-reproduction is surely one of the main characteristics of a living system. Another is a capacity for *self-regulation*. Neither of these attributes is possessed by the isolated virus. The cell, in fact, is the minimum organization of matter that, in the modern world, is capable of all those processes we refer to collectively as "life." [3] In this generalization we include the qualification "in the modern world" because we are sure that life first arose from the nonliving world billions of years ago in much simpler form than the cell as we know it now. For elementary as the cells appear in Fig. 3–3, they represent an enormous degree of complexity and organization, something much more than a single step away from nonliving matter. There are several reasons why precellular organizations of the kind that led to living organisms no longer exist. In the first place, the environment has changed radically since life began; the conditions that made the precellular precursors of life possible have long since passed. Secondly, progressively more efficient reproducing systems

[3] In the light of the important discoveries that began to emerge in the late 1950's and that are discussed on later pages, the boundary between the living and the nonliving is becoming increasingly indefinite. Today biochemists can place purified nucleic acids and other appropriate organic materials in a test-tube system and synthesize both duplicates of these nucleic acids and specific protein molecules. Many of the essential features of life can thus be made to arise experimentally without cellular organization. The various physical and chemical properties of living organisms did not originate all at once, and some, perhaps all, of the vital reactions can be made to occur separately and outside of cells. To this extent there is a continuous spectrum from nonliving to living, with a middle ground between the two where application of either term is more or less arbitrary. Nevertheless, the combination of *all* vital phenomena within a single and *natural unit* occurs only with organization of the degree and kind seen at a minimum in a cell. It is therefore meaningful and useful to consider the cell as the basic unit of life.

have developed as a universal aspect of organic evolution. The cell as we know it today, then, is a highly evolved organization of matter in which basic processes characteristic of life are performed with precision and assurance—indeed, often with double assurance, as we shall see. It has long since replaced (probably for well over a billion years) the cruder precellular transitions between it and the nonliving world.

The past existence of precellular life must not blind us to the significance of our generalization about life today. The cell is life's minimum unit. In the cell and its relations with other cells, we must seek the organization and mechanisms that underlie all life's processes: the intake, storage, and release of energy; the intake of materials and their metabolism, leading to growth; sense perception and the response to stimuli; movement; and, above all, life's most distinctive process, reproduction. [4]

Organization of the Cell

It is not our present purpose to describe in detail the countless differences among the many different kinds of cells. Rather, we propose to focus attention on those features of structural organization that are common to all cells and that are clearly related to the fundamental activities of life as a whole.

SIZES AND SHAPES OF CELLS

Although a few cells are visible to the naked eye, most are microscopic in size (Fig. 3–5). In a large, multicellular animal like man, the average diameter of a cell is about 10 microns, that is, 0.01 millimeter. A bacterial cell may be as small as 0.4 micron in diameter, which is near the limit of vision with an ordinary microscope. Some other cells are almost that

[4] The German physiologist Max Verworn wrote in 1895, as the cell theory was fully maturing: "It is to the cell that the study of every bodily function sooner or later drives us. In the muscle cell lies the problem of the heartbeat and that of muscular contractions; in the gland cell reside the causes of secretion; in the epithelial cell, in the white blood cell, lies the problem of the absorption of food; and the secrets of the mind are hidden in the ganglion cell."

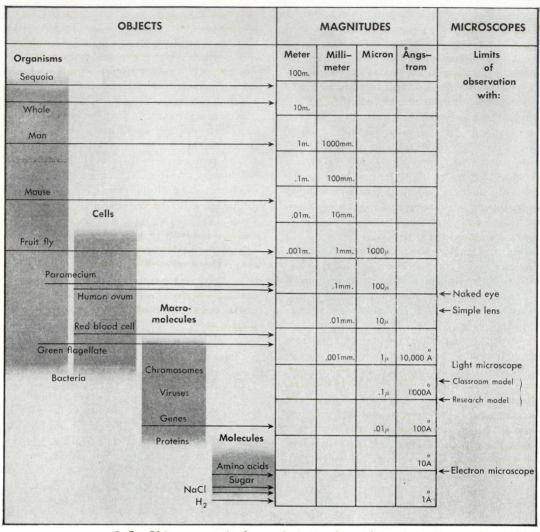

3–5 Objects, magnitudes, and appropriate microscopes.

small. There are, however, exceptions to the rule of small size. A single nerve cell in a large animal may reach a length of several feet, although its diameter is relatively small. Eggs of animals are single cells before development begins, and they are commonly visible to the naked eye. Human eggs, which are unusually small as eggs go, can be seen without a lens but only as specks smaller than the periods on this page. The eggs of an extinct bird (*Aepyornis* of Madagascar) had capacities of more than 2 gallons and were, in volume, the largest known cells.[5]

It is impossible to generalize about the shapes of cells. They assume an endless variety of shapes, largely in relation to the special functions they serve in the organism as a whole. Human skin cells are flattened and platelike; nerve cells are enormously elongated in accordance with their function of transmitting impulses over long distances; muscle cells are elongate; blood cells are biconcave and disclike. If there is any tendency to a general shape in cells, it is to the spherical; but even cells that are relatively unspecialized in func-

[5] Only the yolk of a bird's egg is a cell. It is usually greatly inflated with stored food material and surrounded by accessory materials familiar as the "white" and the shell of the egg.

tion and tend toward this shape are deformed into polyhedrons by close packing.

PROTOPLASM: A USEFUL IF MISLEADING WORD

It has been the custom to refer to all the constituents of the living cell collectively as *protoplasm*, which T. H. Huxley characterized as "the physical basis of life." Later biologists have given the special name *nucleoplasm* to the protoplasm in the nucleus and identified the protoplasm outside the nucleus as *cytoplasm*. All these terms are useful for convenient reference. At least, they have the merit of brevity. However, "protoplasm" is a misleading word for living matter if it is interpreted to mean a homogeneous substance of which we can have, like water, a representative drop or piece. Protoplasm is a complexly organized system of heterogeneous parts, the smallest representative piece of which is the cell. Strictly speaking, though, it is no more accurate or meaningful to say that a cell is composed of protoplasm than it is to say that a radio is composed of "radioplasm." How can there be a representative particle of radioplasm when the whole radio is a complex and heterogeneous organization of tubes, condensers, wires, and the like? Thus it is only as a convenient symbol for the still incompletely understood substance of the cell that we still use the term "protoplasm."

The Generalized Cell

Until the mid-1950's most textbooks of biology contained an old and familiar drawing that portrayed the "generalized" cell as it appeared through the ordinary *light microscope*. Unchanged for perhaps 30 years and many times reprinted, it presented an idealized composite picture of an object that does not exist in reality. As we have seen, cells exhibit an infinite variety in size and shape. But for all their variety, they possess certain features in common, and it is these that were depicted in the "generalized" cell. It showed, for example, a large central *nucleus* embedded in surrounding *cytoplasm*, the whole being encased in a *membrane*. And within the cytoplasm were

various *organelles*—small particulate bodies that were really nondescript.

In recent years new techniques of electron microscopy, giving much greater magnification than the old light-microscopic methods,[6] have disclosed exquisite fine structures in the subcellular particles, and it is now possible to present the contemporary diagram of the generalized cell shown in Fig. 3–6. This illustration should be compared with the actual electron micrographs reproduced in subsequent figures.

THE CELL MEMBRANE AND WALL

The main chemical constituent of the cell is water, which may account for over 90 per cent of the cell's weight. Accordingly, cell contents always have a fluid character; the fact that cells maintain any individuality and form is due to the universal presence of a *cell membrane*. The membrane retains the rest of the living system within it. The cell membrane, however, is not something outside the living system but an integral part of it. Indeed, the membrane plays a vital role in regulating what passes into and out of the cell. The cell membrane, also called the *plasma membrane*, can hardly be seen with the ordinary light microscope, and the new science of electron microscopy had to solve some difficult technical problems before this remarkable boundary structure could be visualized.

Even before it was clearly seen, however, a complex functioning membrane was known to be present simply because certain substances could be shown to enter and leave cells freely while others could not. Red blood cells and nerve cells, for example, distinguish between

[6] In place of visible light, which has a wavelength of 4000 to 7000 Ångstrom units, the *electron microscope* employs a beam of electrons, whose wavelength is 1/20th of an Ångstrom unit. In principle, the electron microscope is identical to the light microscope, but it uses magnetic fields to focus electron beams instead of light-focusing optical lenses and projects its image upon a fluorescent screen or photographic plate instead of the human eye. By this means, magnifications 100 times as great as those of the light microscope are easily attained. Photographs of electron-microscope images are called *electron micrographs*. The main drawbacks of electron microscopy are the elaborate procedures involved in the preparation of specimens and the expensive, bulky equipment.

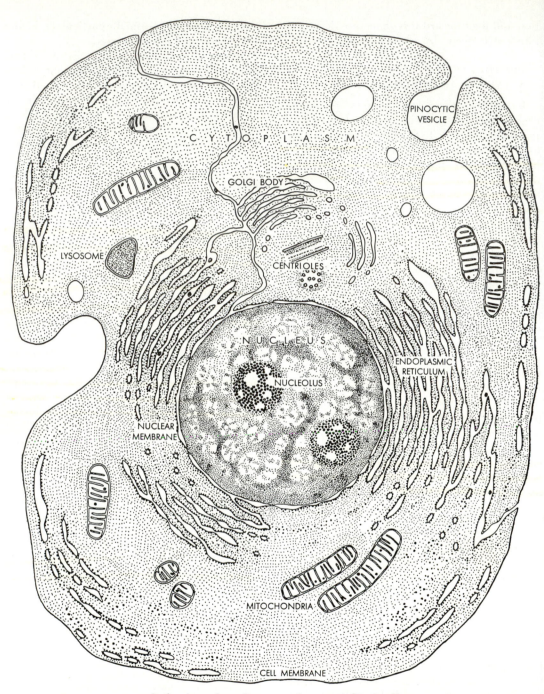

3–6 A modern diagram of a generalized cell.

sodium and potassium ions despite the similarities in size and charge of these ions. Potassium ions pass into the cells, and sodium ions are somehow forced out. This behavior implies the existence of both a selectively permeable barrier and an active mechanism analogous to a pump within the surface membrane. As we shall learn later, this distinction by the cell membrane between sodium and potassium is of great importance in the functioning of

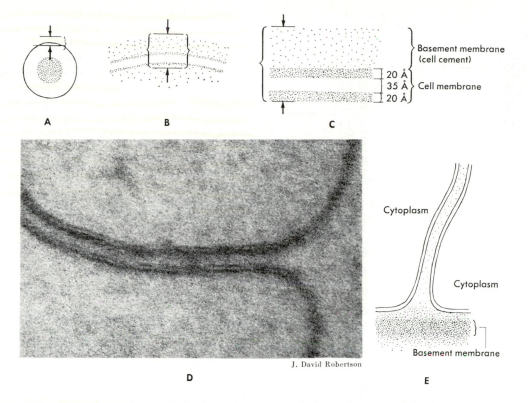

3–7   The cell membrane. *A.* A cell membrane as seen by light microscopy. *B.* Low-power electron microscopy of the region delimited by arrows in *A*. *C.* High-power electron microscopy of the delimited area in *B*. *D.* The appearance of adjacent cell membranes of two neighboring cells (×333,000,-000). The gap between cell membranes contains "cell cement." *E.* Demonstration of the fusion of cell cement into the basement membrane, which forms a supporting surface for a sheet of cells.

nerves, muscles, and many other cell types. The cell membrane is also capable of actively transporting certain large molecules and particles into the cell interior. The membrane is more than a mere boundary surface.[7]

Electron microscopy has revealed that the cell membrane consists of two extremely thin membranes separated by an intervening gap (Fig. 3–7). There is reason to believe that these layers are composed of fatty acid molecules lying side by side with their —COOH groups facing outward and an outside sheath of protein molecules (Fig. 3–8).[8] Presumably,

minute pores exist through which molecules are selectively passed. On the cell surface is a cementlike substance that appears to hold neighboring cells together. When cells are arranged in sheets of tissue, this substance fuses them together into an underlying, supporting basement membrane.

[7] Microsurgical experiments also indicate the presence of a cell membrane. For example, if a dye is injected directly into an individual cell with a small needle, the dye will diffuse freely throughout the cytoplasm but will be retained within the cell.

[8] The peculiar structure of fatty acids (Chapter 2) ideally equips them to serve as building blocks in membranes. A long molecule like oleic acid has in fact two functional groups: the carboxyl group

that is freely soluble in water and the long chain that resembles a simple oxygen-free (and hence water-insoluble) hydrocarbon. The latter component accounts for the solubility properties of lipids. As a result of the differences in solubility between the two ends of the molecules, fatty acids in an immiscible mixture of water and benzene orient themselves at the interface in a characteristic manner. The hydrocarbon chains project into the organic solvent while the polar carboxyl groups point toward the water layer. The affinity of the hydrocarbon residues for one another tends to spread the fatty acids into a monomolecular film in which the hydrocarbon chains are parallel to one another. Such configurations are frequently found in the functionally important surface layers of many tissue membranes.

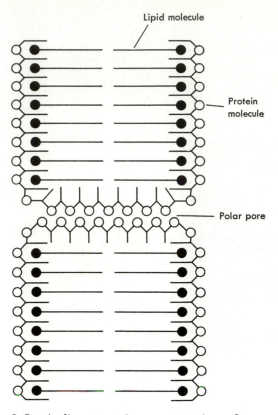

Lipid molecule

Protein molecule

Polar pore

3–8 A diagrammatic representation of one theory of cell membrane structure (Danielli, 1954).

Because of the fluid nature of the cell substance, nearly every organism is faced with mechanical problems in maintaining a definite form. When a single cell lives its own independent life, the cell membrane alone may suffice. But this is not true for a large multicellular organism, which would collapse under its own weight without a support of greater rigidity than the fluid protoplasm. In multicellular animals generally, this problem is met by the development of special *skeletons*, familiar instances of which are our own bones. In plants, however, mechanical support is not derived from wholly separate skeletal systems. Support comes from the *cell wall* that lies outside the membrane of each cell. Although some cells specialize in producing extra thick and strong walls benefiting the whole organism, *each* plant cell has a wall of its own. This wall is the most distinctive feature of a plant cell.

The cell wall should be distinguished from the cell membrane that lies within it. The wall is much thicker than the membrane. Moreover, it plays a smaller part than the membrane in controlling the passage of materials in and out of the cell; its function is mainly mechanical. The great functional differences between the wall and the membrane are reflected by the relatively simple structure of the wall (which is generally composed mostly of cellulose and related molecules) and by its durability, which contrast so sharply with the extremely complex and fragile nature of the membrane.[9]

THE NUCLEUS

Inside the membrane of every plant and animal cell, the nucleus is usually the largest distinct body. Despite its prominence, it is one of the least-understood components of the cell. It is a roughly spherical body, denser than the cytoplasm in which it is embedded, and surrounded by a thin *nuclear membrane*. Its position is commonly central, and its role in the life of the cell is certainly central in the metaphoric sense, for it is the governing headquarters of the cell. It contains within it a tangled network of material called *chromatin*, which is darkly stained by certain dyes in preparations for microscopic study. Chromatin strands coalesce during cell division to form a fixed number of distinct linear or threadlike bodies, the *chromosomes*, which are the carriers of the cell's heredity (see Fig. 3–2). The chromosomes' affinity for certain dyes is the property from which they derive their name—"chromosomes," from the Greek *chroma* (color) and *soma* (body). The chromosomes guide the development of the organism and ultimately maintain the order and organization of the entire living system. How they effect this control is a topic for later chapters. In essence, they manufacture special chemical compounds that eventually enter the cytoplasm as the agents of control, regulating the chemical processes that go on there.

In some cells the chromosomes can be rec-

[9] It was sheets of plant cell walls that Hooke observed in his early microscopic studies of cork (see Fig. 3–1). We may say, therefore, that plant cell walls were the first cell structures to be seen and animal cell membranes the last.

ognized as distinct bodies at all times, but usually they are readily seen only when the cell and its nucleus are dividing. When the nucleus is not dividing, a spherical body of uncertain function, the *nucleolus*, can be distinguished (see Fig. 3–6). It disappears during nuclear division. Special staining techniques have shown that the deoxyribonucleic acid (DNA) of the cell is in the chromosomes. The nucleolus also contains much ribonucleic acid (RNA) and is considered a site of RNA synthesis.

Electron microscopy has added some information to this picture. As shown in Figs. 3–6, 3–9, and 3–10, the nucleus is surrounded by an envelope consisting of an inner membrane and an outer membrane. Indeed, the nuclear membrane (as well as the other membranous structures to be described later) bears a striking similarity to the two-layered cell membrane (see Fig. 3–7). At certain sites the two membranes of the nucleus are joined together around small *pores*. Through these pores the nucleus communicates directly with the cytoplasm. The interior of the nucleus, in sharp contrast with the cytoplasm, is free of membranes.

THE CYTOPLASM

The electron microscope has also revealed that bacteria and certain protists are much more simply organized than the familiar cells of plants and animals. Although the former contain DNA, the carrier of genetic information, they lack nuclei, nuclear membranes, and chromosomes. Such cells are called *prokaryotes* to distinguish them from *eukaryotes*, cells containing characteristic nuclei and chromosomes.

The cell substance outside the nucleus is the complex, viscous, fluidlike material called "cytoplasm." Of the several cytological discoveries that can be credited to electron microscopy, surely the most significant is that of the cytoplasm's rich content of membranes and membrane-limited elements, unsuspected from light microscopy. The only membranes visible with the light microscope were the external cell membrane and the nuclear envelope; the cytoplasmic organelles were depicted as solid granules or filaments.

In disclosing the true *membranous* structure of each of the classic organelles, electron microscopy revealed the existence within the cytoplasm of a rich and diverse system of membrane-enclosed spaces. When cut across, these structures appear as narrow tubules, but when the dimension of depth is added, their walls are found to be sheetlike membranes extending into all parts of the cytoplasm.

Ribosomes and the endoplasmic reticulum

Almost every cell—and especially one engaged actively in the synthesis of protein—contains this labyrinthine system of membrane-limited channels that follow a meandering course throughout the cytoplasm (Figs. 3–6, 3–9, 3–10, 3–11, and 3–12). The system, a revelation wholly attributable to electron microscopy, has been called the *endoplasmic reticulum*. As shown diagrammatically in Fig. 3–6, the tubelike passageways of the endoplasmic reticulum open into the space between the two layers of the nuclear envelope. Conceivably the cytoplasmic part of the system is an outgrowth of the nuclear envelope. The connection seems to provide a channel between nucleus and cytoplasm. It is thought also that the tubules connect with the exterior of the cell through small pores in the cell membrane.

Two forms of endoplasmic reticulum are encountered: a *rough* or *granular* form and a *smooth* or *agranular* form. Typical granular endoplasmic reticulum (see Figs. 3–9 and 3–11) contains many parallel reticular membranes encrusted with minute dense particles visible at high magnifications. These are the *ribosomes*, the ribonucleoprotein particles whose role in protein synthesis has been so brilliantly elucidated in recent years (Chapter 7). Although located chiefly in the cytoplasm—indeed, most of the cell's RNA is within cytoplasmic ribosomes—similar particles are found in the nucleus (see Fig. 3–9), particularly in the nucleolus. These account in part for the high RNA content of the nucleolus. Whatever the function of the nucleolar ribosomes may be, the cytoplasmic ribosomes are the major sites of protein synthesis. Protein molecules synthesized on the ribosomes are moved through the supporting membrane and segregated from the rest of the cytoplasm within the membrane-limited spaces of the endoplasmic reticulum. We shall discuss pro-

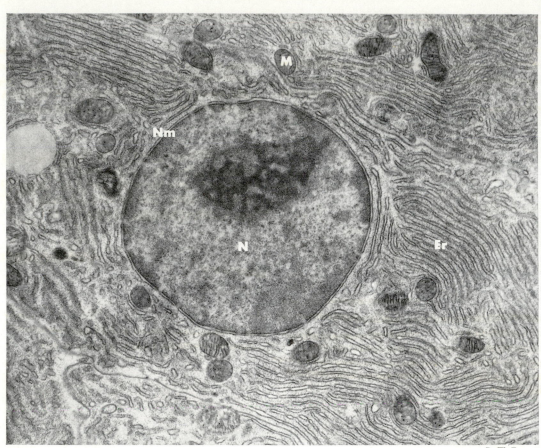

3–9 An electron micrograph of a liver cell. The nucleus (N), nuclear membrane (Nm), rough endoplasmic reticulum (Er), ribosomes, and mitochondria (M) are shown (×11,000).

tein synthesis in a later chapter, but it is useful to recognize at this point that cells synthesize two classes of materials: (1) those that remain within the cell; and (2) those destined to be extruded through the cell membrane. The extrusion process, called *secretion*, is characteristic of certain types of cells (for example, gland cells). These are the cells with the most profuse systems of granular endoplasmic reticulum.

Agranular endoplasmic reticulum has been less well studied than granular endoplasmic reticulum but is equally common and may appear in the same cell with the granular form (see Fig. 3–12). The agranular form displays smooth membranous surfaces that are devoid of ribosomal granules. Various theories implicating agranular endoplasmic reticulum in lipid, steroid, and polysaccharide metabolism

are as yet unsubstantiated. Finally, it should be mentioned that in some cells the endoplasmic reticulum is represented almost solely by the nuclear envelope, other parts having disappeared in the course of cellular development (see Fig. 3–10).

The Golgi complex

Another interesting cytoplasmic component is the *Golgi complex*, or *Golgi body*, first noted in 1898 in nerve cells of barn owls by the Italian cytologist Camille Golgi. Light microscopists suspected that it was involved in secretory mechanisms but were unable to delineate its ultrastructure. Electron microscopy revealed that the multilayered complex is made of smooth membranes that are probably continuous with the endoplasmic reticulum (see Figs. 3–6, 3–10, and 3–11).

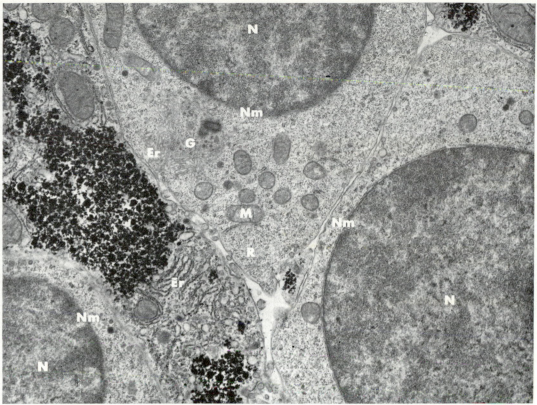

3–10 An electron micrograph of erythroblasts. Nuclei, nuclear membranes, scant endoplasmic reticulum, free ribosomes (R), mitochondria, and parts of Golgi complex (G) are visible (×10,000).

It is possible that the Golgi complex participates in the secretion of proteins destined to leave the cell—probably not as a site of synthesis but as a sort of collecting and packaging station, where particles of matter to be secreted are somehow enclosed in membranous envelopes capable of fusing with the external cell membrane and ultimately passing through it. These *secretory granules* occur only in specialized secretory cells (see Fig. 3–11). However, Golgi bodies themselves occur also in nonsecretory cells, and their function is disputed and unknown.

Mitochondria

The cytoplasm of nearly all cells contains small bodies called *mitochondria* (see Figs. 3–6, 3–9, 3–10, 3–11, and 3–12). Though barely visible with the light microscope, they exhibit highly intricate fine structures in electron micrographs. Their numbers vary from cell to cell, a fact of interest since recent studies of isolated mitochondria have identified them as centers of cellular respiration and energy-yielding metabolism. It seems logical, therefore, that mitochondria tend to aggregate in those regions of the cell that are most actively engaged in metabolic processes. The occasional cell types containing no mitochondria (such as red blood cells) are biological curiosities, and they must rely upon notably less efficient means of energy production.

Electron microscopy discloses a remarkable internal structure in a mitochondrion (Fig. 3–13). The individual mitochondrion is bounded by a double-walled surface membrane, flattened infoldings of the inner wall form perpendicular platelike *cristae*, and the internal space is filled with fluid. We shall presently encounter evidence indicating that the respiratory enzymes are anchored in or on these structures (Chapter 4). We should note, therefore, how much the *area* of the mitochondrial membrane is expanded by its architectural plan. Recent observations have shown that mitochondria swell and contract in the course of physiological activity. We shall later consider the possible meaning of these geometrical changes.

Lysosomes

The electron microscope clearly distinguishes between mitochondria and another newly identified group of bodies of similar size, the *lysosomes*. These curious particles contain many of the hydrolytic enzymes that cleave macromolecules into smaller molecules capable of being oxidized in the mitochondria—among them phosphatase, ribonuclease, and various proteases. The enzymes appear to be sheathed by lipoprotein envelopes, which isolate them from the rest of the cytoplasm. When rupture of the lysosomal membranes releases the enzymes, dissolution of the cell quickly follows. This phenomenon accounts for the breakdown of aging and dead cells.

Centrioles and kinetosomes

The *centrioles* are a pair of small, cylindrical bodies lying near the nucleus. Electron microscopy reveals that a centriole consists of a cluster of exactly nine groups of delicate tubule-like structures, each group containing three tubules (Figs. 3–6 and 3–14). Curiously, the elements of one member of a centriole pair always lie at right angles to those of the other. The significance of their orientation and their critical importance in cell division will be discussed in Chapter 5. Centrioles are rarely found in plant cells.

In cells equipped with *cilia* upon their surfaces (Chapter 14), at the base of each cilium is a *kinetosome*, whose structure precisely resembles that of the centriole. Indeed, cen-

3–11   An electron micrograph of pancreatic glandular cells of a bat. Nuclei, nuclear membranes, profuse rough endoplasmic reticulum, ribosomes, mitochondria, and Golgi complexes are shown (×3,260).

Keith R. Porter and Mary A. Bonneville

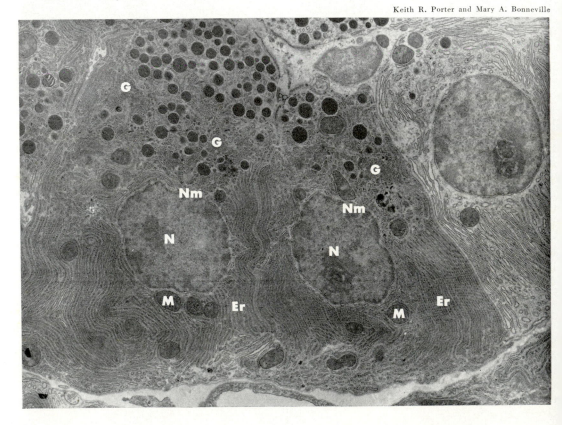

triole and kinetosome may be the same. Since both cilia and certain components arising in cell division are contractile, it is conceivable that centrioles and kinetosomes function similarly as contractile or locomotory units within a cell, analogous to the muscles of multicellular animals.

Plastids and vacuoles

In most plants and one-celled organisms, but not in higher animals, special bodies in the cytoplasm called *plastids* are often found. They are usually associated with the formation or storage or both of particular substances important in the metabolism of the organisms. They may be thought of as specialized factories and warehouses. The green pigment, *chlorophyll*, of green plants nearly always (invariably in higher plants) occurs in plastids, called *chloroplasts*, which are the sites of the fundamental process of photosynthesis (Chapter 4). There are other pigmented and unpigmented plastids in plants; starch

and lipids are formed and stored in some of them, and pigments other than chlorophyll in others.

Vacuoles are cavities in the cytoplasm bounded by definite membranes. They contain water with various substances in solution. They are absent from most animal cells and submicroscopic in young plant cells, but in a mature plant cell one or more large vacuoles develop within the cytoplasm. The viscous part of the cytoplasm outside the vacuole membrane then becomes restricted to a thin layer just inside the cell membrane and wall. The nucleus always lies outside the vacuole in the viscous cytoplasm.

Large vacuoles rarely occur in animal cells, but smaller ones of various sorts may be present. In some cells the cell membrane forms a pouch around a food particle, which later separates and becomes a temporary *food vacuole*. The food is broken down in the vacuole by juices secreted into it from the body of the cytoplasm. Another kind of vacuole is con-

3–12 An electron micrograph of a rat liver cell. Rough and smooth endoplasmic reticulum and mitochondria are present (×16,000).

Keith R. Porter

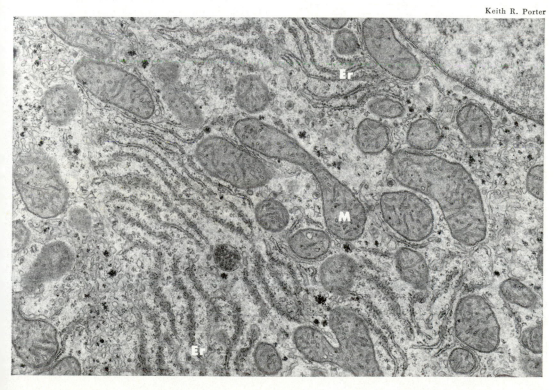

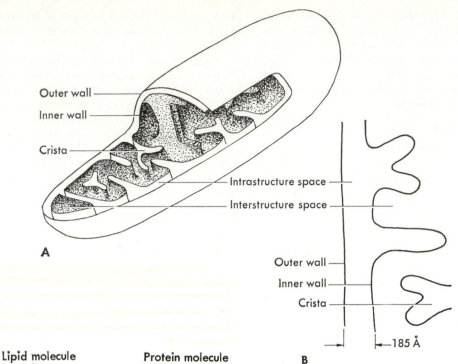

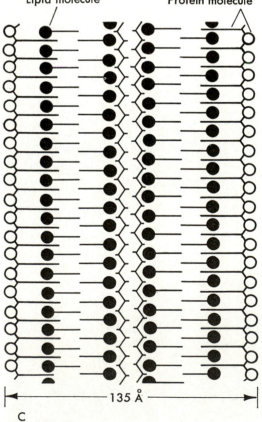

3–13 The structure of a mitochondrion at different levels of magnification. *A*. A cutaway drawing showing two membrane layers separated by a fluid-filled *intrastructure space* and invaginations of inner membrane, the *cristae*. *B*. The same structures at a higher magnification. *C*. The postulated molecular structure of each membrane. Note the ordered layers of protein and lipid molecules.

cerned with the maintenance of a proper water balance in the cell. Water continually passes into some single-celled organisms for physical reasons over which the cells have no control (cf. osmosis, p. 85). These cells meet the problem by continually bailing out excess water, which is first concentrated into *contractile vacuoles* whose rhythmic contractions eject the water.

PHYSICAL ORGANIZATION OF PROTOPLASM

Problem of invisible organization

Our survey of the visible structures within the cell strongly suggests that highly important features of the cell's organization exist beyond the limits of even the electron microscope. By indirect methods of analysis, we are

TABLE 3–1
Chemical constituents of protoplasm

	Sea urchin	Man *
Water	78.3	66.0
Protein	15.2	16.0
Lipids	4.8	12.4
Carbohydrates	1.4	0.6
Ash †	0.3	5.0

* Data are for a newborn child.

† Ash includes all the elements not included in the other categories (iron, potassium, and other metals).

beginning to understand aspects of the ultra-structure of protoplasm. The fact that it is not directly visible does not mean that it cannot be analyzed.

Table 3–1 gives the gross chemical composition of protoplasm from two different sources, a sea urchin and man. The predominance of water we have noted before. The proteins form the next largest protoplasmic constituent. Then follow carbohydrates and lipids, molecules of much smaller dimensions. The term "ash" includes a host of still smaller molecules, especially salts of various kinds. We are sure, then, to begin with, that protoplasm must be largely an aqueous solution of many kinds of molecules, but observation of its behavior indicates that there is more to it than that. For instance, even in its most fluid condition, protoplasm is thicker, that is, more viscous, than water or most solutions in water. Moreover, even within one and the same cell part, the viscosity varies; sometimes a piece of protoplasm flows freely, and at other times it acquires a thickly set jellylike consistency. Experiment shows that the viscosity is readily affected by acids and heat. This and other properties of protoplasm emphasize that it cannot be exclusively regarded as a simple solution of chemicals in water; it must be, in part, what is called a *colloidal system*. When we reach this conclusion, we see how a fine-scale structural organization may exist within the cell at the molecular level, far outside the range of direct vision. What is a *colloid*?

Colloidal systems

Gases, liquids, and solids are the three most obvious phases of matter. The same substance may exist in any of the three phases. Water is a familiar example: it is a gas in clear air (or steam *before* the steam condenses and becomes visible), a liquid in a drink, a solid in an ice cube. There are, however, other phases of matter. Suppose that a cube of sugar, a solid, is dissolved in water. It is no longer a solid, and yet it is not exactly a liquid, either. It is in *solution* in a liquid. As we saw in Chapter 2, dissolved molecules are separate from each other, and each moves more or less independently through the *solvent*, the medium in which it is dissolved.

If instead of sugar fine sand is stirred in a glass of water, the sand particles do not dissolve. For a time many of the unchanged, solid particles remain in *suspension* in the water. The system is not stable, however, and the particles eventually settle out of suspension and collect at the bottom of the glass. If extremely fine clay particles are stirred in the glass of water, some may settle out, but many

3–14 Sister centrioles in an animal cell (×90,000). They lie at right angles to each other. The cross section of the bottom one shows that a centriole is made up of nine groups, each containing three tubules.

Wilhelm Bernhard

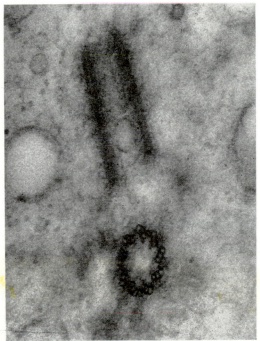

remain suspended indefinitely; in other words, the system is *stable*. If the particles are small enough, they never do settle out unless something is done to change the system—unless, that is, it is disturbed by an influence from outside. *A stable suspension is a colloid.*

A colloid may also be formed by two liquids instead of a liquid and a solid. If you stir kerosene into water, some small drops of kerosene remain suspended in the water for a while, but eventually they rise to the top. Such a system of one liquid dispersed in another is an *emulsion.* You will find it difficult or impossible to make drops of kerosene small enough to remain suspended in water indefinitely, so that a stable emulsion results. Stable emulsions—colloids of liquid in liquid—do exist, however. Homogenized milk is one; a special procedure makes the liquid fat globules so small that they remain dispersed in the milk and do not rise to the top as cream.

The differences among true solutions, colloids, and unstable suspensions or emulsions hinge on the sizes of the particles involved; and colloids are also dependent on the electric charges that the large particles carry. Consider the extreme cases first. Suspensions settle out eventually because of the pull of gravity on the large particles. Sugar molecules in water do not settle out because the pull of gravity is negligible on particles so small. All the molecules in a liquid and true solution are constantly in random dancing movement in all directions. This random movement of the liquid (water) molecules is, in part, the reason why colloidal particles, which are of intermediate size, are not settled out by the slight pull of gravity on them. The colloidal particles are constantly being bombarded and pushed in all directions, up as well as down, by the solvent molecules that surround them. They are also prevented from settling because their electric charges are all of the same sign (sometimes all positive and sometimes all negative); consequently, they mutually repel each other and stay uniformly dispersed. When chemicals added to a colloid neutralize the charges in the particles, they no longer repel each other, and then they precipitate.

There is no sharp line between the sizes of particles that form solutions, colloids, and un-stable suspensions or emulsions. A substance is usually considered to be in true solution if it separates into individual molecules (or their ions) that are smaller than 0.001 micron. Aggregates of unseparated molecules are rarely as small as 0.001 micron and usually, therefore, form colloids. Larger aggregates form unstable suspensions or emulsions.

Proteins as colloids and fibers

Some single molecules are larger than 0.001 micron, and these include proteins, important constituents of protoplasm. Proteins usually dissolve in water and form solutions in the sense that their molecules do separate. Yet these molecules are often as large as the particles in colloids, and they do in fact form colloids in water. The result may be said to be simultaneously a solution and a colloid, or *chemically a solution* and *physically a colloid.*

Gelatin and albumen are proteins with very large molecules that form colloids in water. The white of an egg is a colloidal system of albumen in water, and the familiar Jello is also an aqueous colloid of gelatin. Jello, of course, can exist either as a freely flowing fluid (when the gelatin is dissolved in warm water) or as a thickly set jelly. This is a general property of colloids: they can assume either a *sol* or *gel* state and moreover can be transformed from one state to the other.[10]

The colloidal properties of protoplasm account for some of its known organization and behavior. Its capacity to form semisolid gels helps us to understand how a system that is 90 per cent water can maintain any structure whatsoever. The integrity of the cell membrane as a retaining boundary depends on this ability of the protoplasmic proteins to form gels. And the variations in viscosity that are observable in different parts of a cell, and in the same part at different times, are reflections of reversible sol–gel changes.

[10] Basically, the change involves the withdrawal or addition of water in the system. In the sol state water is said to be the *continuous* phase. In the gel state, when water has been withdrawn, the protein molecules compact into a spongelike network that is the continuous phase; the water becomes the *disperse* phase. It is the spongelike network of the protein molecules, having various shapes including fibers, that gives the gel its semisolid consistency.

Although we are now part way toward an understanding of the invisible ultrastructure of protoplasm, we are still far from our goal. The properties of protoplasm leave no doubt that it is simultaneously a complex solution and a colloid. That is, it consists of water in which small molecules (like sugars and salts) and their dissociated parts are dissolved and in which there are also dispersed much larger molecules (proteins and lipids) and molecular aggregates of colloidal size. What we ultimately must look for, then, is some basis for maintaining order and organization within a semifluid system. It is difficult to imagine how any order can be preserved in a true solution, the essence of which is the random movement and uniform distribution of its constituent molecules. Yet various hypotheses, all unproven as yet, have been advanced. Most of them relate to the remarkably large sizes of proteins and the diversity of shapes, including fibers, that they can assume. The hypotheses envisage a semisolid framework of fibers and colloidal particles as the basis for spatial order and organization.

Diffusion

Molecules move about continuously in colloids as well as in true solutions. It is essential to recognize the principles governing their movements because the processes of cellular metabolism depend upon this molecular traffic. In turn, the movements of molecules in protoplasm are affected by its physical organization.

It is a physical property of matter, even of solids, that molecules are always moving. This is simply demonstrated when a bottle of perfume is uncorked in a closed, still room. You may stand several feet from the bottle, but soon you will smell the perfume. Molecules have moved out of the bottle and through the air of the room. Your nose, which is one of the most sensitive chemical detection systems known, has recognized the dispersed molecules.

Molecular movement in gases, such as that of the perfume molecules in air, is considerably faster than in liquids. It is faster in liquids than in solids, although it still goes on in the densest solid. It is faster in hot than in cold substances; in fact, the movement of molecules *is* the phenomenon that is known by the name "heat."

Molecules are much too small to see. Consequently, we cannot actually or directly see molecular motion, but there are simple ways to see it indirectly, by some of its visible results. Put a little face powder in a drop of water, and look at it under the high power of an ordinary microscope. The particles of face powder (which are composed of many molecules) will be seen darting about. They are being bombarded by the moving water molecules and are small enough to move when hit.

In a drop of water millions of molecules move about virtually at random. The *average* result, the net effect of these movements in every direction, is nil. In spite of all the activity within it, the drop does not go anywhere, nor do the molecules become more concentrated in one part of the drop than in another. This is the usual situation when molecules are evenly distributed through the space or substance being studied, even when there is a uniform mixture of different kinds of molecules.

When the perfume bottle was opened, however, something else happened. There was a change in net effect; the perfume molecules did go somewhere. When the bottle was first opened, there were no perfume molecules outside it. Soon there were such molecules at increasing distances away from the bottle. A short time later the molecules were still highly concentrated in the bottle but also were highly concentrated near it. Concentration away from the bottle was progressively lower, until it became zero in the farthest part of the room. The motion of the molecules was random. Some even moved back into the bottle, but more moved out. The net effect, the average over millions of molecules, was that more molecules moved from regions of high concentration to regions of low concentration than vice versa.

This tendency for molecules to spread from regions of higher to those of lower concentration is quite general and is called *diffusion*. If it continues, the concentration eventually becomes the same everywhere. The perfume molecules become evenly distributed throughout the room. There is a state of equilibrium.

What happens then? Does diffusion stop? Do the perfume molecules stop moving?

Now consider another simple experiment. If a crystal of copper sulfate (or any soluble, colored compound) is put in the bottom of a glass and the glass is filled with clear water, the water soon begins to turn colored around the crystal. The zone of colored water becomes larger from day to day. What is happening? At the surface of the crystal, copper sulfate is going, molecule by molecule, into *solution* in the water, the *solvent*. Naturally the dissolved molecules are more concentrated right at the surface of the crystal than elsewhere. They therefore diffuse into the surrounding water, and so the colored zone spreads.

It spreads very slowly, however, much more slowly than the perfume molecules spread through the air. The perfume can be smelled 3 feet from the bottle within a few minutes, but it takes more than a year for copper sulfate in recognizable amounts to diffuse this far through water. If you have ever tried to run in shallow water, you know the reason for this difference in the rates of diffusion. Water is a *denser* medium than air. This means that in air the molecules are farther apart than in water. The molecules are thus less likely to collide, and the diffusing molecules are less often slowed up or bounced back, than in water. In either air or water, however, the molecules are very small in proportion to the space between them. There is plenty of room between them for the oncoming molecules of perfume or of copper sulfate, and the total volume of air or water is not increased by the diffusion.

In these examples, then, diffusion does not push the molecules of gas or liquid farther apart. On the other hand, in a solid or a dense colloid, the molecules are so close together that diffusion among them may force them apart. If a thin piece of dry gelatin is placed in water, it swells. Water has diffused into the gelatin and separated its molecules. The resulting volume of the piece of gelatin is greater than that of the dry gelatin alone but less than that of the dry gelatin plus the original volume of the water diffused into it. This particular sort of diffusion is called *imbibition*. The word simply means "drinking in." Imbibition is a familiar part of daily life and

of life processes. It makes wooden doors stick in rainy weather, and it makes bean seeds swell and burst their seed coats just before they sprout. The pressures caused by imbibition may be enormous. Imbibition in starch can develop pressures up to 15 tons per square inch. Ships loaded with rice have been split asunder when water got into the holds and was imbibed in the rice.

Diffusion plays a leading role in the organization of the protoplasm. Frequently molecules within a cell are unevenly distributed. Diffusion then tends to equalize the distribution, providing one way for materials to move about within the cell.

Semipermeability

Now that we have discussed diffusion in general, we must consider diffusion through membranes, a special circumstance of peculiar importance in the lives of cells. Some substances diffuse quite readily through some membranes. If a membrane is no particular barrier to a given substance, it is said to be *permeable* to it. Of course, if a substance cannot pass through a membrane, the membrane is *impermeable* to it. If a membrane is permeable to some substances and less permeable or impermeable to others, it is called *semipermeable* (or *differentially permeable*, which may be more precise but is clumsier).

The relevance of these facts to our study of the cell is that a cell membrane is semipermeable. It is generally permeable to water and impermeable to colloids. The membrane holds in the colloidal elements of the cytoplasm and yet permits water to move rather freely into and out of the cell. The cell membrane is also more or less permeable to various materials dissolved in water. Thus needed materials from the outside can diffuse through the membrane. Once within the cell, they can be retained in the colloidal mesh or in compounds to which the membrane is not permeable. Waste products can be converted into forms to which the membrane is permeable and so can leave the cell.

Actually the situation is more complicated than this explanation indicates, and in a very interesting way. Like other parts of the cell, the membrane is in a continual state of flux. The degree of permeability to different substances is not constant, varying quite markedly from time to time. Thus even without any change in its own state or composition, a substance may pass through the membrane at some times and be held back by it at others. The variation depends on many things, such as sugar content, acidity, electrical properties, and conditions of colloids within the cell and temperature, acidity, heat, light, concentrations of materials, and other factors outside the cell. Alterations in membrane permeability greatly influence the kinds and amounts of materials diffusing into and out of the cell. These processes are essential in the regulation of basic life processes in the cell.

We have noted that a cell needs to obtain and to lose certain materials at various times. The membrane's permeability corresponds with the cell's needs and helps, when all goes normally, to ensure that they are met.[11]

Osmosis

Suppose that we make a container of a membrane permeable to water, such as a pig's bladder, a frog's skin, or collodion. If we fill it with water, close it tightly, and immerse it in water, nothing noticeable will happen. Although the container is permeable to water, the concentration of water is the same inside as outside. Therefore, as many molecules move out as move in, and equilibrium or the status quo is maintained. However, if we put a sugar solution in the container (preferably leaving it somewhat slack but without air inside) and immerse it in water (preferably distilled water) again, the container will swell up (Fig. 3–15).

What happens? Evidently more water moves in than out through the membrane. Why? What you already know of diffusion supplies the answer. Water molecules are less

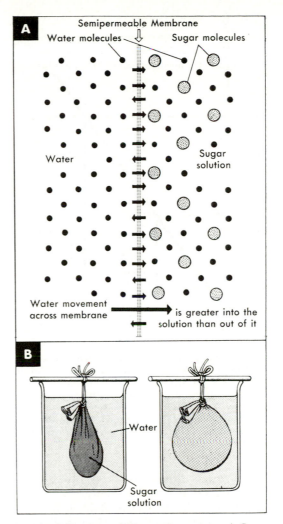

3–15 **Semipermeability and osmosis.** *A.* Sugar molecules cannot cross the membrane; they are too big. Water molecules are less concentrated on the sugar-solution side of the membrane than on the other side, and the chance that they will hit pores in the membrane, and thus cross it, is consequently smaller on the sugar-solution side. More water therefore enters the solution than leaves it. *B.* The net movement of water into a sugar solution enclosed within a membrane container creates a pressure (osmotic pressure) that causes the container to swell.

[11] You might be tempted to say that the membrane has its characteristics *because* of the cell's needs—but there you would part company with scientific thought. Such an assumption slips in a hidden metaphysical postulate that needs can *cause* their fulfillment; it is therefore to be rejected alike by science and by straightforward common sense. If you are at all inclined to balk at this conclusion, consider facts of the following sort. Cell membranes may admit materials of no possible use to the cell (such as nitrogen into green plant cells); they may let out so much water that the cell dries up and dies; they may let in poisons that kill the cell.

concentrated in the sugar solution than in pure water. Molecules move predominantly from the pure water into the solution. If the membrane were equally permeable to sugar and to water, sugar molecules would also move out. But the membrane is semipermeable, being much more permeable to water

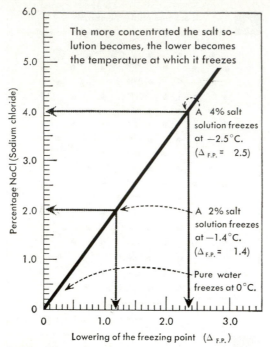

The more concentrated the salt solution becomes, the lower becomes the temperature at which it freezes

A 4% salt solution freezes at −2.5°C. ($\Delta_{F.P.} = 2.5$)

A 2% salt solution freezes at −1.4°C. ($\Delta_{F.P.} = 1.4$)

Pure water freezes at 0°C.

Percentage NaCl (Sodium chloride)

Lowering of the freezing point ($\Delta_{F.P.}$)

3–16 The relationship of freezing point to concentration of a sodium chloride solution. The higher the salt concentration, the lower the freezing point of the solution. The freezing-point depression ($\Delta_{F.P.}$) is therefore a convenient measure of the osmotic value of a solution because this, too, is dependent on the concentration of salt and other molecules.

than to sugar. So no (or very little) sugar moves out of the container, more water moves in than out, the amount of fluid in the container increases, and the container swells. This is the process of *osmosis.*

The influx of water into the container produces pressure, *osmotic pressure.* The flow through the membrane continues until the pressure inside forces water molecules out through the membrane as fast as osmosis brings them in. The pressure then ceases to rise, becoming steady. A state of equilibrium has been reached. The amount of osmotic pressure that can develop in a solution separated from distilled water by a semipermeable membrane depends on the concentration of the solution, the temperature, and other factors. Under given conditions, the pressure has a characteristic, constant value for any soluble substance. This is the *osmotic value* of the substance. Usually the value is not expressed

directly in terms of pressure but by some correlated figure. You will often find it expressed in terms of a change, symbolized as Δ, in the freezing point of a solution as compared with that of pure water. The relationship is rather complicated, and we shall not consider it in detail here (but see Fig. 3–16). But it may be convenient for you to know that $\Delta = 1°$, for instance, indicates a certain level of osmotic pressure, that $\Delta = 2°$ indicates a higher level, and so on.

In real life it would be unusual, indeed impossible, to find such a simple situation, with a solution of a single substance on one side of the membrane and pure water on the other side. There are always solutions of several or many different substances on both sides. Then each solution has its own total osmotic value, Δ, resulting from the combination of all the things dissolved in it. What happens in such a case? If Δ is the same on both sides of the membrane, nothing happens. If it is higher on one side than the other, water will diffuse (predominantly) from the side of *lower* Δ to that of *higher* Δ. In general, the greater the difference, the faster the diffusion, or, at any rate, the longer the diffusion continues before equilibrium is reached.

Osmosis goes on, to some degree, most of the time in most of the cells of any living organism. It is essential to the movement of materials into and out of cells. It usually pro-

3–17 Plasmolysis of a plant cell. In *A* the osmotic pressure inside the cell (Δ_i) is greater than that outside (Δ_o). Consequently, more water tends to move into the vacuole (*Vac*) than out of it. Increase of cell volume is resisted by the firm wall, and the cell thus becomes turgid. In *B* and *C* the osmotic pressures outside the cell (Δ_o) are experimentally increased. In *C*, because Δ_o is geater than Δ_i, more water leaves than enters the vacuole, which consequently shrinks away from the cell wall.

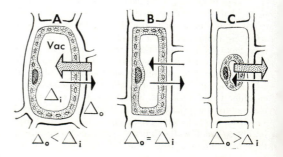

ceeds in a businesslike way, quietly and without any really striking results. Sometimes, however, its effects are clearly visible and even dramatic.

If a cell of almost any water plant—a filament of the little pond-scum alga *Spirogyra* will do nicely—is soaked in a 15 per cent solution of common sugar, the cell contents soon separate from the wall and form a mass in the center of the cell. The sugar solution soaks through the cell wall and comes in contact with the cell membrane. Remember that in plants the cell *wall* is different from the underlying *membrane*. The cell wall, unlike the membrane, is not semipermeable; the membrane is permeable to water and impermeable to sugar. The large vacuole in the center of the cell contains a dilute solution, with Δ definitely lower than that of the sugar solution. Osmosis therefore occurs; water diffuses out of the vacuole through the membrane into the space between membrane and wall. As the vacuole loses water and becomes smaller, this space enlarges. The membrane and its contained protoplasm are finally crowded into a ball at the center of the cell. This phenomenon is called *plasmolysis* (Fig. 3–17). Plasmolyzed cells ultimately die, but if you remove the cell from the sugar solution quickly enough and put it in pure water, it will recover and the protoplasm will return to its position near the wall. Why? (Fig. 3–18.)

You can demonstrate radical effects of osmosis even more simply in a kitchen. Put a leaf of lettuce in salt water. It will wilt, because the cells lose water by osmosis. Remove the leaf quickly, and put it in pure water. It will recover its crispness. The process is similar to the wilting of a plant on a dry, sunny day and its recovery when the plant is watered, to the extent that both involve the loss and gain of water by the cells; only the mechanism of water loss is different—it is osmosis between solutions in the lettuce and evaporation in the growing plant.

The Cell as an Organism

The plants and animals with which we are most familiar are composed of very large numbers of cells. Their individual cells do not have separate and independent lives of their own; in other words, the cells are not organisms themselves but are the parts of which organisms are constructed. However, even a multicellular organism does usually have a stage in its life history when it is a single cell. That is, it commonly develops from a single cell such as a fertilized egg. Sometimes, as in the swimming spores of certain relatively simple organisms, the single cell may lead an independent life for some time and behave like an active organism on its own, but it is still only a temporary stage in the life history.

There are numerous small organisms, separate and distinct individuals, in which the body is never at any stage in the life history partitioned into separate cells. The body is a single mass of protoplasm, a single cell, divided only into the usual cytoplasm and nucleus. Such organisms are *unicellular organisms,* single cells living wholly independent lives. Some are commonly called "plants," and some "animals," according to their resemblances to unquestionable, multicellular plants and animals. Among the unicellular organisms, however, the distinction is usually not clear. In fact, one and the same organism may be classified sometimes as a plant and sometimes as an animal.

In recent years many students have recognized that the terms "plant" and "animal" may be equivocal when applied to unicellular organisms and, further, that all unicellular organisms have much in common. They therefore propose to call them neither plants nor animals but Protista or *protists.* In a way this name only increases the difficulty of telling plants from animals, because some protists really are very like plants and almost surely closely related to them, and others are similarly allied to animals. An additional factor is also involved. Many protists have quite complex anatomies, even including specialized cytoplasmic parts (organelles) analogous to the mouths, eyes, fins, and so on of multicellular animals (Fig. 3–19). Some also have several nuclei, even though there are no cell membranes or walls between the nuclei. It is therefore quite possible to argue that a protist is not equivalent to a single cell of a multicellular organism but is really analogous to the whole of such an organism. From this point of

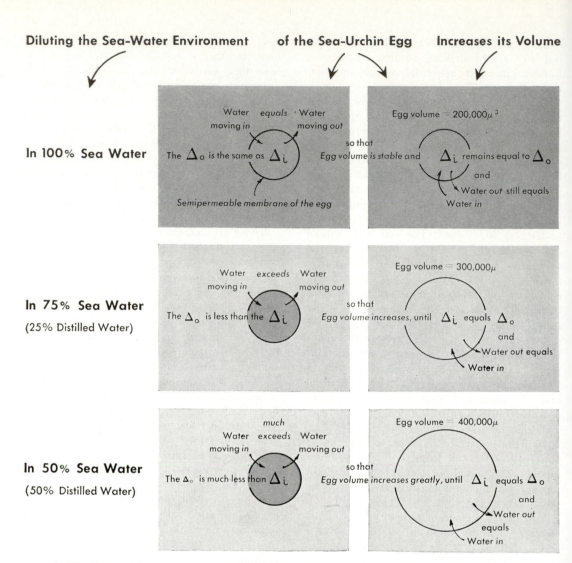

3–18 Sea-urchin egg as an osmometer. Dilution of the sea water surrounding the egg makes Δ_o (Δ outside the cell) less than Δ_i (Δ inside the cell). The osmotic pressure that develops is not resisted by a firm cell wall, as in a plant cell (cf. Fig. 3–17), and consequently the egg increases in volume until $\Delta_i = \Delta_o$.

view, protists are not unicellular plants or animals but are *acellular organisms*, that is, organisms not divided into separate cells. To sum up, the very same organisms, the protists, are considered unicellular when they are compared with single cells of higher organisms, and acellular when they are compared with whole higher organisms.

Note that there is no argument about the facts. The anatomical structures of the unicellular or, alternatively, acellular protists are well known and not disputed. No one denies

that some of them are plantlike in metabolism or physiology, others animal-like, and others intermediate between the two or different from either. All that is involved is two different ways of looking at the same facts. As a matter of convenience, we shall call these organisms "protists" unless we are referring to some particular group of them.

Some protists, such as amebas (see Fig. 3–19), are comparatively simple, with little specialization of parts. Others have high degrees of differentiation into varying kinds of

3-19   Cells as organisms. *A.* An ameba. *B. Epidinium.* The ameba shows very little visible specialization of parts, but in *Epidinium* such specialization is very extensive. All food enters through a mouth, whose location is fixed. The undigested remains of food are evacuated through a well-formed rectum and anus. The form of the animal is partly maintained by a definite skeletal structure. Retractile musclelike fibers move the mouth and esophagus, and the movement of these fibers is controlled by nervelike fibers. Similar "neurofibrils" control the movement of elaborate membranelles, which are organs of locomotion. All the neurofibrils connect with a central "motor mass," which may be likened to a brain.

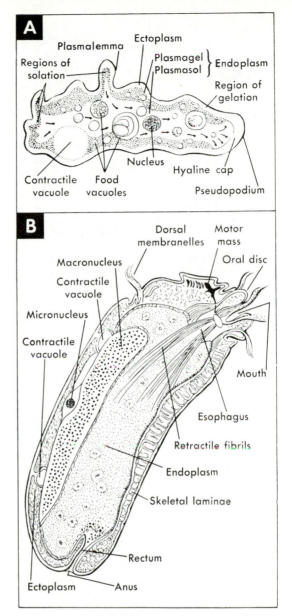

cytoplasm, corresponding to the tissues of multicellular organisms, and structures concerned with particular activities. Such structures are, in fact, organs just as much as are the structures with similar functions in multicellular organisms, although they are usually called "organelles" to distinguish them from organs composed of cells. Protists may have skeletons, excretory organs, light-sensitive organs (which may have lenses), conducting mechanisms analogous to nerves, contracting mechanisms analogous to muscles, explosive organs with which food is captured, mouths, anuses, a number of different sorts of organs of locomotion, and still other organs.

The Multicellular Organism

PROTISTAN COLONIES
AND CELLULAR DIFFERENTIATION

When a protist reproduces, it does so like any other single cell. The nucleus undergoes *mitosis* (see Chapter 5), and the cell mass constricts into two halves, each carrying a daughter nucleus. Sometimes, however, several or many organisms formed by division from what was originally one protist remain clumped together for a while. They may stick together at their outer surfaces or be caught in a gelatinous envelope that they have secreted. In a clump each cell (or protist) continues to be essentially independent in form and in activity. The clump is merely an aggregate in space, with no particular biological interaction among its units.

Some protists, especially certain of the green *flagellates*, carry the process of aggregation several steps further. The individual cells resulting from repeated divisions remain together not incidentally and in chance clumps

but in *colonies* of definite shape and characteristic size when mature. A common arrangement is for the cells to be held in the outer part of a gelatinous envelope forming a hollow sphere. The number of cells, usually a power of 2 (why?), may vary from a few to about 50,000. Reproduction occurs when one or more cells divide and form smaller daughter colonies inside the sphere. Eventually these colonies break out of the parental sphere and grow into separate, new, mature colonies.

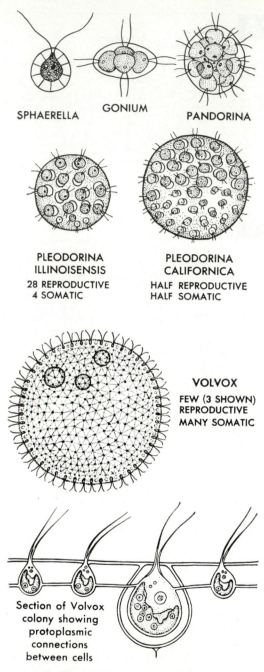

SPHAERELLA   GONIUM   PANDORINA

PLEODORINA
ILLINOISENSIS

28 REPRODUCTIVE
4 SOMATIC

PLEODORINA
CALIFORNICA

HALF REPRODUCTIVE
HALF SOMATIC

VOLVOX
FEW (3 SHOWN)
REPRODUCTIVE
MANY SOMATIC

Section of Volvox
colony showing
protoplasmic
connections
between cells

3–20 Colonial green flagellates.

In some colonial forms (Fig. 3–20) the cells are not connected with each other and are all exactly alike. Each can give rise to a daughter colony (although all do not necessarily do so). In such cases the cells are still essentially

distinct individuals, protists that live together. Other forms, of which some species of *Volvox* are examples, are more complex and of peculiar interest: the cells are connected to each other by strands of protoplasm running through the gelatinous envelope (see Fig. 3–20). Thus the cells are not fully independent; happenings in one cell may affect its neighbors. Moreover, the cells are not quite alike. The number of cells may run into the thousands, but only 4 to 20 larger cells in the back part of the colony are capable of reproducing, giving rise to daughter colonies. The cells in the front part of the colony cannot reproduce. They are smaller and have relatively larger light-sensitive organelles, which influence the orientation and movement of the colony.

Is this a colony of protists, or is it a multicellular organism? On the one hand, it is made up of similar cells and quite surely evolved from a unicellular or acellular protistan individual. On the other hand, the cells display some differences in form and activity and a certain amount of coordination throughout the whole colony, as do the cells of unquestionably multicellular organisms.

The colonies of *Volvox* and its relatives reveal a puzzling biological phenomenon that will appear repeatedly in our study of living organisms. Like the cells of multicellular organisms, although to a much lesser degree, the cells of a *Volvox* colony differ in size, shape, potentialities, and other characteristics. In a word, they *differentiate* as they mature; each specializes in a division of the colony's total labor. And yet their heredity is exactly the same. They all arose from one cell, and at each mitosis the new nuclei were endowed with identical sets of chromosomal controls. Finding an explanation for cellular differentiation in spite of identical heredity is one of biology's greatest and most interesting problems. It will be treated further in Chapter 8.

CELLULAR DIFFERENTIATION: CELLS, TISSUES, ORGANS, AND SYSTEMS

If one examines a complex multicellular organism such as a tree or a man, it is plain, first of all, that the organism has its own individuality. It is a unit and acts as such. Yet the unit is made up of visibly different parts. The basic units, the cells, which we have been discussing in this chapter, have many forms and functions within the individual. They are differentiated, and through differentiation they have become functionally *specialized*. A cell in a root tip is quite different from one in the surface of a leaf, and the leaf surface cell in turn differs from a cell within the leaf. In the human body a nerve cell, a muscle cell, and a blood cell are decisively different.

Each cell type usually occurs with many others of its kind and shares the same life processes. The whole leaf surface is covered with similar cells. A muscle is made up not of one cell but of thousands, all much the same in appearance and properties. An aggregation of cells is called a *tissue*. In *simple tissues* all cells are of the same type. *Composite tissues* contain two or more cell types in characteristic relationship with one another.

Tissues of several or many different kinds usually make up the complex parts of organisms. The hand, for instance, is a functionally distinct part, or *organ*, of the body, composed of skin, bones, muscles, nerves, and other tissues. Leaves and flowers or eyes and ears are other organs that are complexes of different tissues. Organs may, in turn, be cooperating and interacting parts of a larger complex, an anatomical and physiological *system*. The human digestive system, for example, is a sequence of organs from the mouth through the esophagus (the passageway from the mouth to the stomach), the stomach, the small intestine, and the large intestine to the anus. Each part or organ is different, but all interact one after the other in the processes of digestion. Each organ in the system is composed of tissues, and each tissue is composed of cells.

The striking differences that exist among tissues are related to the different functions that they perform; the structure of muscle is related to its function of movement, and the structure of nerve to its function of communication. This is in fact but one illustration, at the level of the cell itself, of a paramount principle that pervades all biology: *the structure of a living thing is intimately related to its functions*. Plants and animals are distinguished by great differences in their modes of life. Plants, nearly always sedentary, feed directly upon the materials of soil and air in a manner that requires a minimum of movement. Their immobile existence is clearly reflected in their tissue organization. Animals, on the other hand, are dependent upon living food, which they must actively ingest in solid form and which, moreover, most animals must actively pursue. Their constant movements and their searching activities are possible only because they possess certain organs and tissues—sensory, muscular, nervous, and connective—of which plants have no need. These basic differences between plants and animals should be borne in mind as we survey their characteristic tissues.

PLANT CELLS AND TISSUES

The bodies of the more highly evolved plants consist of two organ systems. The *root system* occurs below ground; and the *shoot system*, including *stem* and *leaves*, occurs above ground (Fig. 3–21). In these two organ systems four classes of tissue can be recognized. These are described briefly in the following paragraphs. Later sections will show how the four tissue classes are organized in leaves, stems, and roots.

Meristematic tissues

The two main aspects of plant growth are *primary* growth, the growth in length of shoots and roots, and *secondary* growth, the subsequent growth in thickness. Each kind of growth is accomplished by *meristematic tissue*. This tissue, called *meristem* for short, consists of thin-walled, unvacuolated, and unspecialized cells. When growth is active, cell divisions occur throughout a meristem, producing new cells, each of which differentiates into one of the three other classes of specialized (protective, fundamental, or conductive) tissue. Apical meristems, at the tips of roots and shoots, are responsible for primary growth. Meristematic tissue elsewhere causes secondary growth (Fig. 3–22).

Protective tissue

Protective tissue (*epidermis* or *cork*) comprises the outermost layers of cells that cover and protect underlying tissue in both organ systems.

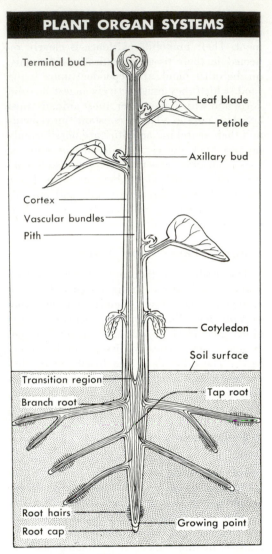

PLANT ORGAN SYSTEMS

Terminal bud

Leaf blade

Petiole

Axillary bud

Cortex

Vascular bundles

Pith

Cotyledon

Soil surface

Transition region

Branch root

Tap root

Root hairs

Root cap

Growing point

3–21 Plant organ systems.

Fundamental tissue

Various types of cells are rather loosely categorized collectively as *fundamental tissue*. Some of them (*sclerenchyma* and *collenchyma*) have thick walls, which add greatly to the mechanical strength of an organ—root, stem, or leaf (Fig. 3–23). Others (*parenchyma*) are thin-walled and highly vacuolated; some of them contain the green chloroplasts, which are the sites of photosynthesis.

Conductive tissue

Conductive (or *vascular*) *tissue* is concerned with transport in a plant, as the name

implies. There are two kinds of conductive tissue. The *xylem* carries water and dissolved minerals upward from the roots, where they are absorbed from the soil, to all other parts of the plant. Xylem cells, like all other specialized types, begin life as unspecialized cells, derived from a meristem. As they mature, their walls become heavily thickened, often with conspicuous spiral bands. In the more highly evolved plants, adjacent xylem cells, one above the other, eventually lose their end walls and fuse with each other to form a continuously open vessel running up the axis of the plant. They also lose their cytoplasm and nuclei, but though they are dead, their function as conducting vessels is unimpaired. As we shall see, each plant organ is supplied with an abundance of xylem vessels.

The second kind of conductive tissue is *phloem*. It carries the food materials manufactured in the leaves to all parts of the plant, both above and below the leaves. Mature phloem cells are elongate cylinders whose walls are not so heavily thickened as those of xylem cells. Nor do the end walls of the phloem cells break down. Instead, connection between phloem cells is effected through a series of pores in their end walls, which are appropriately called *sieve plates*. The nuclei of phloem cells disintegrate, but their cytoplasmic contents remain, fusing through the sieve-plate pores and forming a protoplasmic highway along which food is carried (Fig. 3–24).

Tissue organization in the leaf

Each leaf bud arises during the growing season from a small mound of meristematic tissue. Later on each opens, and, as the leaf grows to maturity, its constituent cells differentiate into specialized tissues. A simple leaf, as on a willow or lilac, has a flattened blade and a cylindrical piece (the *petiole*) by which it is joined to the stem of the plant. To examine the tissues, one may cut a thin slice across the blade and view it with a microscope. The tissues are shown in Fig. 3–25.

Protective tissue is present in the form of a single layer of cells, tightly fitted together and looking like a jigsaw puzzle in surface view. This is the epidermis. Its cells, especially on the leaf's upper surface, are glossy and waterproof because of a heavy deposit of a fatlike

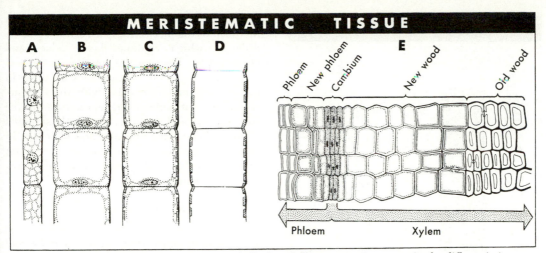

A B C D E

Phloem New phloem Cambium New wood Old wood

Phloem Xylem

3–22 Meristematic tissue in plants. *A, B, C,* and *D* are successive stages in the differentiation of a xylem vessel (*D*) from a sheet of cambial cells (*A*). Note how the cells enlarge, vacuolate, acquire thickened walls, and eventually lose their cytoplasm and nuclei. *E* is a cross section through part of a stem. Division of the cambial cells (stippled) produces new cells, which on the right differentiate into xylem and on the left differentiate into phloem.

3–23 Some plant fundamental tissue. Two types (sclerenchyma and collenchyma) of fundamental tissue are supportive. Their cell walls are especially thick, conferring strength and rigidity on the whole tissue. In sclerenchyma the wall is uniformly and heavily thickened at all points. In collenchyma the thickening is heaviest at the corners of the cell.

substance (*cutin*). Between the upper and lower epidermis there are parenchyma cells containing chloroplasts. The parenchyma cells, which are pulled apart during the growth of the leaf, have open, gas-filled spaces between them.

In the epidermis, especially on the lower side of the leaf, are numerous small openings (*stomata*; singular *stoma*) between adjacent specialized cells; these usually open in the light and close in the dark. The mechanism by which the stomata open and close exemplifies the biological importance of osmosis. As illustrated in Fig. 3–26, *guard cells* (usually a pair) surround the opening of a stoma. Light falling on the guard cells starts a series of enzymatic reactions that convert the osmotically inactive insoluble starch to a large number of soluble sugar molecules. These increase the osmotic pressure and thereby cause water to enter from adjacent cells. The guard cells swell, and the swelling pulls apart the walls of the stoma and opens it. In the dark the reactions are reversed, and the stoma closes. The

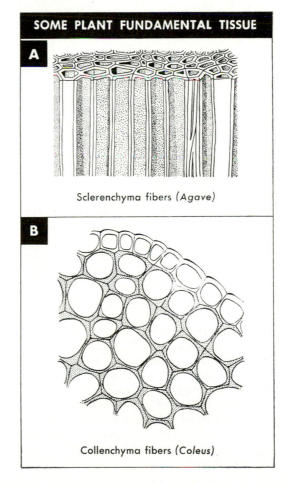

SOME PLANT FUNDAMENTAL TISSUE

A

Sclerenchyma fibers (*Agave*)

B

Collenchyma fibers (*Coleus*)

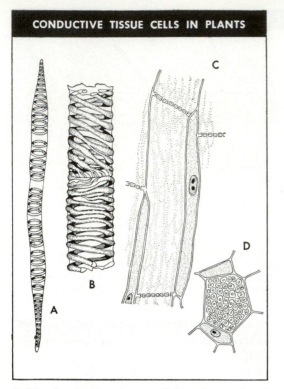

CONDUCTIVE TISSUE CELLS IN PLANTS

3–24 Conductive tissue cells in plants. *A* and *B* are xylem cells. Note the thickening of the cell walls. *C* and *D* are phloem cells. *C* shows one complete phloem cell whose end walls are sieve plates. The nucleated cell lying to the right of the phloem cell is a companion cell, a characteristic constituent of phloem tissue. *D* is an end view of a sieve plate; two companion cells lie beside it.

stomata open into the intercellular spaces of the parenchyma; gases involved in photosynthesis and other reactions in the parenchyma cells diffuse through them to the outer air.

The parenchyma cells contain large numbers of chloroplasts, in which is found the green pigment chlorophyll. Photosynthesis occurs in the chloroplasts, and it is now believed that the effective participation of chlorophyll in this process depends in part upon the way it is physically organized within the chloroplasts. We shall study photosynthesis in detail later. Here we merely note in passing the extremely elaborate internal organization of a chloroplast (Fig. 3–27). As in the mitochondria, this can be perceived only with the aid of an electron microscope. The chloroplast is built up of numerous layers, or *lamellae*, which contain the chlorophyll.

The conductive tissue in the leaf appears in the form of distinct veins, or *vascular bundles*, each of which contains both xylem and phloem cells. The vascular bundles in the leaf blade are, of course, continuous with those in the petiole and, ultimately, with the conductive tissue throughout the main axis of the root and shoot. The leaf is constantly supplied through the xylem with water and minerals. In turn, it supplies to the rest of the plant through its phloem the sugars it manufactures in its parenchyma cells.

Tissue organization in the stem

The word "stem" applies to all parts of the shoot system other than leaves—trunks and main branches as well as what are more commonly called "stems" in popular language. All these structures have similarities in their tissues and in their relationships to the life of the plant. In general, the differentiated stem tissues are involved in the support, growth, and protection of the plant and in the movement of solutions. In addition, some stems contain cells with chloroplasts and consequently manufacture sugars, but this is predominantly a leaf function. Stems also are often places of storage for food and water.

The tissues of the stem of a woody plant, such as ash or box elder (Fig. 3–28), are best studied in a section cut transversely across the stem. All the tissues are organized in a series of concentric cylinders surrounding a central mass of loose parenchyma cells, the *pith*. The first cylinder of tissue around the pith is the *wood*, composed of xylem cells. This is followed, in order, by cylinders of meristematic tissue (the *cambium*), phloem, parenchyma, meristematic tissue (the *cork cambium*), and finally cork.

The two cylinders of meristematic tissue permit the stem to grow in thickness as it gets older. The protective tissue of the cork consists of cells whose walls are all impregnated with a waterproofing substance (*suberin*). This effectively protects underlying tissues from water loss but prevents the cork cells themselves from obtaining water, so that they die. As they are sloughed off, they are replaced by new cells from the cork cambium.

The principal layer of meristematic tissue is the cambium, between the xylem and the

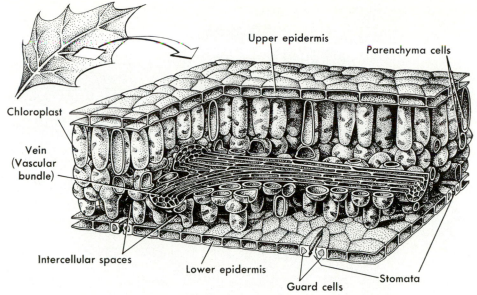

3–25   Tissue organization in a leaf. The upper and lower surfaces of the leaf are bounded by an epidermis (protective tissue), which is one cell in thickness. Stomata are present in the lower epidermis. The bulk of the internal tissue is parenchyma (fundamental tissue), whose cells contain chloroplasts. The parenchyma cells are loosely packed and are therefore surrounded by intercellular spaces, which, via the stomata, are continuous with the external atmosphere. The mass of parenchyma is penetrated by veins (or vascular bundles) that consist mainly of conductive tissue (xylem and phloem) and some supportive fibers (fundamental tissue).

3–26   The mechanism by which a stoma opens and closes.

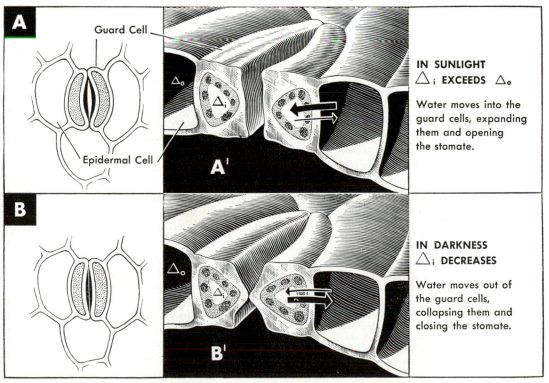

A'   IN SUNLIGHT $\triangle_i$ EXCEEDS $\triangle_o$

Water moves into the guard cells, expanding them and opening the stomate.

B'   IN DARKNESS $\triangle_i$ DECREASES

Water moves out of the guard cells, collapsing them and closing the stomate.

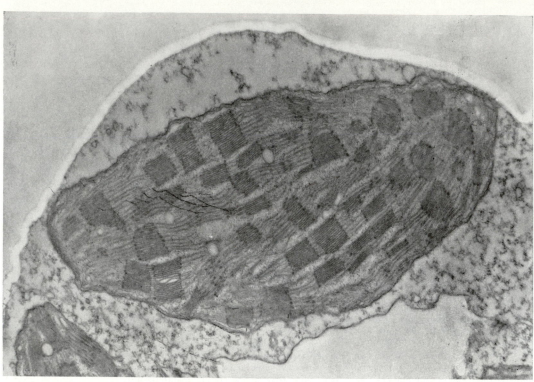

3–27 The chloroplast. Note the laminated structure as shown by electron microscopy (×22,000). The stacks of dense lamellae are the grana, the main loci of the chlorophyll molecules.

phloem. In each growing season the new cells on its inner surface differentiate into new xylem, and those on its outer surface into new phloem. Thus in each growing season a new layer of xylem is added inside the cambium, and a new layer of phloem is added outside it. The cells laid down during the most favorable part (spring) of the growing season are usually larger than the later cells, with the result that the phloem and especially the xylem show, in cross section, annual growth rings (Fig. 3–29).[12] In climates without well-marked seasons, these rings are faint or absent.

The great masses of xylem cells in the wood

[12] The formation of annual rings in the xylem allows us to determine the age of a woody stem and to recognize, by the relative widths of the rings, when good and bad growing seasons have occurred. The timbers of giant trees in western America are so old that their annual rings have revealed the history of climatic change over more than 1000 years. And the timbers in the abandoned cliff dwellings of the Mesa Verde in Colorado have permitted exact dating of their age.

of a tree like the oak are dead vessel elements. The oak, symbol of sturdy life, is in one sense, therefore, more dead than alive. However, the presence of the actively growing cambium between the xylem and the phloem guarantees a continual supply of living cells, whose number remains nearly constant throughout the life of the tree. A tree's increasing bulk with age signifies an increase in dead wood tissue, whose function is a combination of conduction and mechanical support.

There is much less wood (xylem cells) in the stems of herbaceous plants, which usually have large amounts of pith. Their protective tissues commonly are green, containing chloroplasts and carrying on photosynthesis.

The organization of tissues we have outlined here applies generally to the plants in one (the *dicotyledons*) of two principal groups of higher plants. In the other group (the *monocotyledons*), which includes plants like the lily, bamboo, palm, corn, and other grasses, the same kinds of tissues are present,

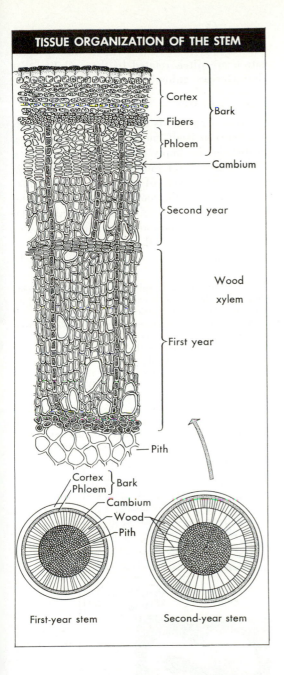

TISSUE ORGANIZATION OF THE STEM

Cortex
Bark
Fibers
Phloem
Cambium
Second year
Wood
xylem
First year
Pith

Cortex
Phloem } Bark
Cambium
Wood
Pith

First-year stem   Second-year stem

3–28 Tissue organization of a woody stem (box elder). The two lower figures show diagrammatically how the tissues are organized as concentric cylinders. In the left-hand figure there is only 1 year's growth of wood; in the right-hand figure a second year's growth forms a distinct annual ring in the wood. The upper figure shows the cellular detail of the various tissue layers in a segment of a 2-year-old stem.

Tissue organization in the root

Roots anchor plants, but their main role is the gathering in of water and dissolved salts from the soil. Among other specialized activities, they may also store materials, especially starch. Their tissues are quite similar to those of stems, but they have no pith, and the vas-

3–29 Annual rings in the wood of pine (×23).

Carl Strüwe

but they are not arranged in the series of concentric cylinders characteristic of the dicotyledons. The vascular bundles of xylem and phloem are scattered throughout the pith (Fig. 3–30). Nor do the monocotyledons have continuous cambium layers; consequently, most of these plants cannot maintain continuous growth in thickness.

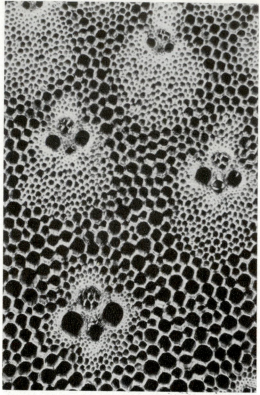

Carl Strüwe

3–30 Scattered vascular bundles in the stem of bamboo, a monocotyledonous plant (×75).

connective tissue. As we review these classes, bear in mind that their diversity and specialization are often related to the mobility that characterizes animals.

Surface tissue

Every organism is enclosed. It has a boundary between itself and the environment. In a protist this is simply the cell membrane. In a multicellular form, either plant or animal, it is one or more layers of epithelial cells. *Epithelium* is a tissue that covers a surface, and the external surface is only one of many in the animal body. Some internal surfaces are actually continuous with the outside; for example, the surfaces of the mouth, throat, stomach,

3–31 Plant root. Longitudinal section of a root, showing epidermal cells extending as root hairs into the soil between soil particles. The growing tip (region of cell multiplication) is protected by a root cap. The cells of the root cap are continuously abraded away by friction on the soil particles, as growth forces the tip downward; they are continuously regenerated by cell divisions in the growing tip. Immediately behind the region of cell multiplication, new cells elongate under the pressure of osmosis. These cells eventually mature (region of maturation) into specialized types such as xylem and phloem. The inset photograph shows root hairs on the young root of a radish plant germinating from its seed.

cular bundles are arranged in a somewhat different way. The surface cells just back of root tips usually have elongated extensions of their walls, forming root hairs (Fig. 3–31). The root hairs worm their way in among fine particles of soil. The development of root hairs is of functional significance, for they increase the surface through which water and dissolved minerals can be absorbed.

ANIMAL CELLS AND TISSUES

The higher animals have much more complex organs than do any plants, and the tissues and cells making up the organs are also more varied. Many tissues and tissue products are peculiar to certain groups of animals—for instance, bones to vertebrates and feathers to birds. There are, however, broad classes of tissues that occur in virtually all multicellular animals, differing only in detailed structure and origin. Among these are *surface* or *epithelial tissue, muscle tissue, nerve tissue,* and

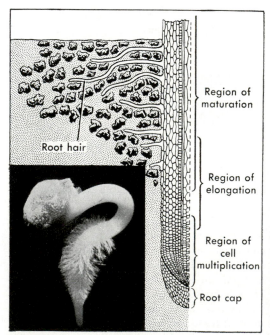

Root hair

Region of maturation

Region of elongation

Region of cell multiplication

Root cap

Hugh Spencer

intestines, and lungs are covered with epithelium. Other internal surfaces, like the linings of blood vessels and body cavities, have no connections with the outside but are nevertheless, since they are surfaces, covered with epithelium. The several types of epithelial cells are shown in Fig. 3–32. Note that *glands* are made of epithelial cells.

The functions of epithelium are determined by its surface position. They are to *protect* and to *repair*. In its exposed location, epithelium is subject to wear and tear, but there is continual replacement of its dead cells. The epithelial cells are lost either cell by cell or—as in human skin—in many-celled flakes,[13] so small that we are hardly aware that we are always shedding our skins. In some other animals—as in most snakes—the whole surface comes off in one piece from time to time.

As the organism's limiting boundary, the epithelium also controls what enters and leaves the body, just as a cell membrane controls traffic into and out of the individual cell. Thus the epithelial linings of the stomach and intestine absorb foodstuffs; oxygen enters the body through lung epithelium; urine is excreted by kidney epithelium; and so on. And many sense stimuli enter the body via epithelium, which, for this reason, is an important constituent of sense organs like the eye and nose.

The word "skin" is often used synonymously with epithelium, but in man, as in many other animals, it refers not to a single tissue but to a very complex organ. Human skin (Fig. 3–33) consists of many layers of cells, only the outermost of which are epithelial. Below are muscle cells, blood vessels, nerves, and much of the loose fibrous connective tissue described later.

Muscle tissue

Probably the protoplasm of all types of cells is capable of contraction to some extent. In muscle cells this capacity is fully developed; the muscle cell is a specialist in this protoplasmic function. Multicellular animals as a whole and the parts within them are generally moved by muscular tissue. Muscles occur in all but the lowest animals, and muscle tissue is

[13] As in the scaling associated with dandruff or the peeling that follows sunburn.

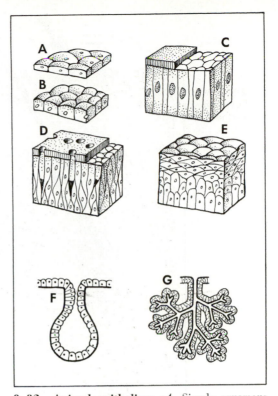

3–32 Animal epithelium. *A.* Simple squamous epithelium consists of flattened, tile-like cells; it lines cavities such as the mouth. *B.* Cuboidal epithelium occurs, for example, in kidney tubules. *C.* Columnar epithelium occurs, for example, in the lining of the stomach. *D.* Ciliated columnar epithelium lines the breathing duct (trachea) of a land vertebrate. *E.* Stratified epithelium occurs on surfaces subject to heavy wear, such as human skin. The lowest (germinative) layer of cells in the epithelium continually supplies, by mitosis, new cells to replace those worn away from the external layers. *F* and *G.* Simple and more complex glands, with the epithelium specialized to produce and secrete a particular substance or substances.

remarkably uniform throughout the animal kingdom.

In higher vertebrates, like man, three different sorts of muscle tissue are distinguished: *skeletal* (or *striated*) *muscle*; *smooth muscle*; and *heart muscle* (Fig. 3–34). Muscle cells generally are elongate, their contraction taking place along the long axes. The contraction of muscles is usually initiated by stimuli coming to them through *nerves*. Rapid transmission of nerve *impulses* (messages) and rapid response to them by muscle form the basis of the quick and beautifully coordinated move-

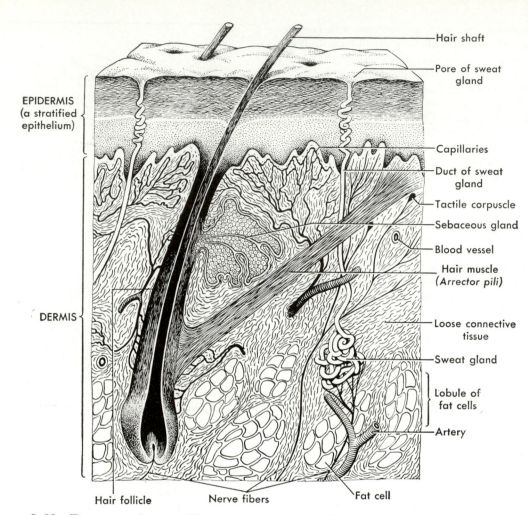

EPIDERMIS (a stratified epithelium)

DERMIS

Hair shaft

Pore of sweat gland

Capillaries

Duct of sweat gland

Tactile corpuscle

Sebaceous gland

Blood vessel

Hair muscle (Arrector pili)

Loose connective tissue

Sweat gland

Lobule of fat cells

Artery

Hair follicle

Nerve fibers

Fat cell

3–33 Tissue organization of human skin. On the outside is a many-layered epithelium, the outermost cells of which are continuously worn away. The tissue is maintained by the production of new cells from a generative layer (stippled in the figure) at the base of the epithelium. Below the epithelium lies the dermis, in which the following can be seen: capillary networks, arteries, and veins; nerve fibers and sense organs (tactile corpuscles); sweat glands; lobules of fat cells; hair follicles bearing hairs; muscles capable of raising the hairs; and loose connective tissue that serves to bind together the whole complex of other tissues.

ments of the higher animals. Skeletal muscle, which is conspicuously crossbanded when viewed under the microscope, brings about the movement of the whole animal by moving its skeleton; legs move because their bones are moved by muscles. The response of skeletal muscle to stimulation is rapid; and so is its relaxation, readying it for new messages. Smooth muscle, unstriated, is responsible for much of the movement of internal organs such as the stomach and intestines. Smooth muscle is slow to contract and capable of more pro-

longed contraction than skeletal muscle. Heart muscle exists only in the walls of the heart, as its name suggests. It is striated in much the same manner as skeletal tissue but is unique in other ways. For example, extensive branching of the muscle fibers is seen only in heart muscle.

Skeletal muscle is often referred to as "voluntary," in contrast to smooth and heart muscle, which are said to be "involuntary." This distinction arises from man's ability to exercise voluntary control over the contraction of skeletal muscle and his lack of conscious control over the others. The distinction is, of course, difficult to apply meaningfully to other animals.

Muscle tissue seems thoroughly amazing when one considers the work it performs in the heart, pumping decade after decade; its power in an individual accustomed to performing heavy labor; or its quick and precise responses in a trained athlete. The lightning-fast yet almost incredibly intricate chemical process by which muscular energy is released seems still more remarkable (Chapter 10).

Nerve tissue

A small boy stubs his toe and cries out. Unknown to him, he has demonstrated one of the most complex and extraordinary of all the phenomena of life—the transmission of impulses by nerve tissue. This tissue is as widespread among animals as is muscle tissue.[14] The association is significant, for nerves are involved in the stimulation and coordination of muscular contraction.

In spite of the expected variety in details, nerve tissue is even more uniform throughout the animal kingdom than is muscle tissue. Its basis is the nerve cell, or *neuron,* which, perhaps more clearly than any other cell, illustrates the relation of cellular structure to function (Fig. 3–35). The function of the nerve cell is the transmission of nerve impulses—or messages—often over long distances within the body. The nerve cell has a central cell body containing the nucleus and is drawn out into two or more long protoplasmic processes termed *nerve fibers.* These fibers are integral parts of the cell. (The fibers in connective tissue are *outside* the cells.) Some are simple, and others are greatly branched; some end very near the cell body, whereas others extend for some distance. In large animals this distance may be 3 to 4 feet. The fibers of some nerve cells are surrounded by accessory *Schwann cells,* which form a cellular sheath containing a lipid called *myelin.* A group of nerve fibers bound together by connective tissue constitutes a nerve.

All the complexities of perception, behavior, and conscious thought that we find in

[14] Sponges are the only major group of multicellular animals without nerves and muscles.

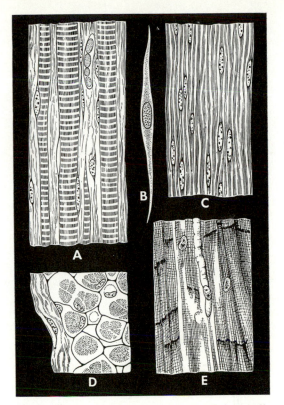

3–34   Muscle tissue. *A* and *D.* Skeletal (striated) muscle in longitudinal (*A*) and cross (*D*) sections. The nuclei can be seen lying at the periphery of the muscle fibers. The individual muscle fibers are packed together in bundles embedded in loose connective tissue. *B.* An individual smooth-muscle cell. *C.* A sheet of smooth muscle from the intestine of a cat. *E.* Heart, or cardiac, muscle.

ourselves and see, in varying degrees, in other animals are founded on a relatively simple basis. This is the nerve cell's capacity to transmit electrical impulses and join with other nerve cells in an intricately organized association—the nervous system. The importance of this system for understanding ourselves and other living things is obvious. It will be discussed at some length later (Chapters 13 and 14).

Connective tissue

A wide variety of tissue types is included under the heading of connective tissue; and they have a wide variety of functions (Fig. 3–36). All the tissues have one feature in common: the constituent cells lie in an extensive *matrix,* or bed, of extracellular material containing fibers. Both matrix and fibers are manufactured by the cells. The specialization of connective tissue is reflected in a characteristic specialization of the intercellular matrix and fibers. We shall distinguish two broad functional classes: supportive connective tissue and binding connective tissue.

SUPPORTIVE CONNECTIVE TISSUE. Most kinds of animals are built around or within a hard supporting *skeleton* that not only helps to hold them together but also serves as a mechanical framework for locomotion. It may also protect against attack by other animals and assist in waterproofing, among other roles. In most animals without backbones, like clams and insects, the skeleton is external to the body, secreted by—and lying on top of—an epithelial tissue. In vertebrates—animals with backbones—the skeleton consists of one or both of two supportive tissues, *cartilage* and *bone*, which we must treat more fully.

In cartilage (gristle) the intercellular matrix is extensive, consisting of organic compounds with a rubbery texture. This matrix gives the tissue an elasticity that resists compression. The fibers embedded in the matrix give cartilage added strength against pulling and stretching. Cartilage can be likened to the wall of a rubber tire; the rubber resists compression, and the internal nylon cords resist stretching. Most cartilage, like that in the nose and ear, is subject to stresses from all directions, and the fibers appropriately course in all directions within the matrix.

In some fishes (sharks and their relatives) cartilage forms the whole skeleton. In most vertebrates, however, the skeleton consists almost entirely of bone, cartilage being restricted largely to joints, where it forms resilient caps over the bone surfaces involved in the joint.

Bone, like other connective tissues, contains innumerable living cells. It owes its rigidity to

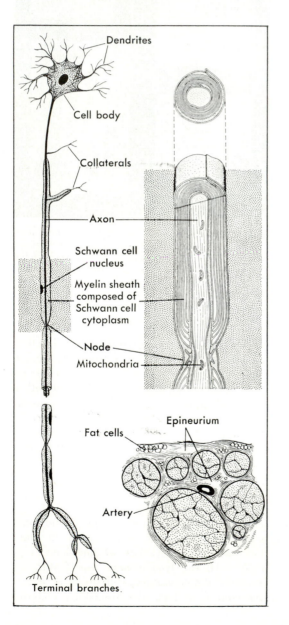

3–35 Nerve tissue. A single nerve cell, or neuron, is diagrammed at the left. Dendrites are incoming fibers that carry impulses to the cell body, in which the nucleus can be seen. The axon, which may branch, is the outgoing fiber along which the nerve impulse is transmitted. In some nerve cells the axon is surrounded by sheaths. In the one illustrated at left, there is a sheath partly composed of the fatty substance myelin. As shown in the upper right-hand figure (and in the remarkable electron microgram on p. 64), myelin is laid down in concentric lamellae. These arise from layers of Schwann cell cytoplasm that repeatedly wrap themselves around a new growing axon. In places the sheaths may be constricted, forming distinct nodes. Note the mitochrondria in the axon cytoplasm.

The lower right-hand figure is a partial cross section of a nerve. A nerve consists of many nerve fibers packed together by connective tissue (epineurium) into bundles. Blood vessels penetrate the nerve, supplying it with food and oxygen.

3–36   Connective tissue. The lower figure illustrates diagrammatically the structure and components of a long bone like the humerus of man. The head of the bone is capped by cartilage. Muscle is attached to the bone by a tendon whose cells (black in the figure) lie surrounded by inelastic fibers oriented along the axis of the tendon. The bone itself is essentially a cylinder. Within, the bone tissue has a spongy texture. A piece of the wall of the bone cylinder is shown at the lower left. It is perforated by blood vessels that supply food and oxygen to the living bone cells embedded in the hard ground substance. The bone (lower right) cells are organized in concentric cylinders at the center of which lies a blood vessel.

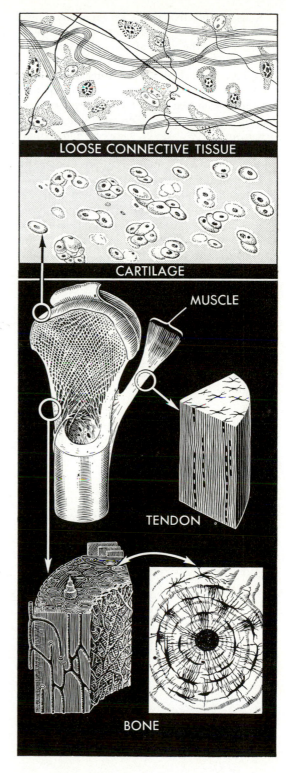

LOOSE CONNECTIVE TISSUE

CARTILAGE

MUSCLE

TENDON

BONE

the dense intercellular matrix secreted by the bone cells. The rigidity, lacking in cartilage, is what makes bone such excellent supportive tissue for large and heavy [15] animals. Bone is rigid because the bone matrix is impregnated with mineral salts (mainly a complex phosphate of calcium), which commonly make up about two-thirds of the weight of fresh bone. The matrix is maintained in proper condition by the bone cells that lie within it. These, in turn, remain alive only because the dense matrix is perforated by a system of canals carrying blood vessels that supply the cells with their needs.

BINDING CONNECTIVE TISSUE. The active movement characteristic of animals makes structural demands that are satisfied by many special tissues, including binding connective tissue. It includes *tendons, ligaments,* and loose connective tissue. In tendons and ligaments the intercellular matrix is packed with fibers that, unlike those of cartilage, are all oriented in one direction. Tendons bind muscles to bones and are subjected to strong stresses—always in one direction—when the muscles contract and the bones are moved. Tendons are appropriately inelastic—were they not, some muscle contraction would be wasted on their stretch. On the other hand, ligaments, which envelop bones at a joint, are appropriately elastic. They have the same resilient "give" characteristic of the knee bandage sometimes used to supplement them in supporting a weak joint.

[15] What is the significance of the fact that the only large animals with wholly cartilaginous skeletons are aquatic? While on this line of thought, consider the fact that the largest animal alive (a whale) is aquatic.

Loose connective tissue serves many functions. It binds constituent muscle cells together into the mass of the individual muscles (see Fig. 3–34) and, similarly, nerve fibers into individual nerves. Indeed, it binds all kinds of organs together in a loose but strong fashion that keeps them in proper place but allows them to move on each other as the organism moves. Again, between and within organs it serves as a loose elastic highway on which blood vessels and nerves are carried.

Internal fluids of the animal body, such as blood and lymph, contain cells and may be regarded as special connective tissues whose intercellular matrix is fluid. Lymph occurs in body cavities and between cells, although in many forms it also moves sluggishly through special vessels. Blood is pumped through a partly or completely closed system of vessels. We shall discuss these fluids in Chapter 11.

CHAPTER SUMMARY

The cell theory and its relation to the theory of evolution—the foundations of modern biology.

The cell as the minimum organization of matter that is alive, the basic unit of life.

Protoplasm, not the name of a single substance, but a name loosely applied to the living organization of matter.

Visible structure of a cell: membrane, wall, nucleus, cytoplasm—ribosomes, endoplasmic reticulum, the Golgi complex, mitochondria, lysosomes, centrioles, plastids, and vacuoles.

Physical organization of protoplasm, colloids, and fibers.

Molecular movements in solutions: diffusion; semipermeability and osmosis, the movement (and its control) of water into and out of the living cell.

The cell as a complete organism: protists; intracellular specialization.

Protistan colonies compared with multicellular organisms.

Tissues as aggregates of similar cells specialized in structure and function; organs as integrated aggregates of diverse tissues; systems as integrated aggregates of organs.

Plant cells and tissues: their organization into leaf, stem, and root.

Animal cells and tissues: their greater number and complexity; functional specializations related to the mobility of animals.

4

The Cell:

Its Metabolic

Machinery

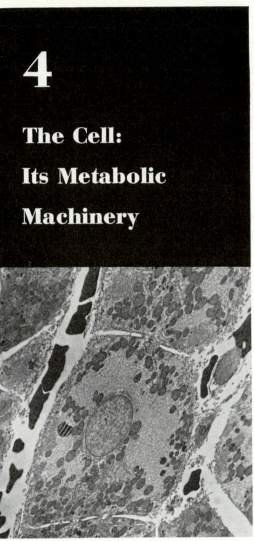

D. W. Fawcett

Part of a liver cell, as revealed by the electron microscope (×6500). The nucleus is the large oval body in the center. Lying in the cytoplasm are mitochondria, which play a central role in the cell's energy metabolism.

In Chapter 2 we were concerned with some of the basic chemical properties of living things, and in Chapter 3 with the structures of cells, units of life, and of tissues composed of cells. Both chapters were largely preliminary to a consideration of the dynamic processes that in a sense *are* life. Tissues and cells and their parts have *functions* in the lives of organisms. These functions include chemical processes that go on within cells, and this chapter is devoted to some of the most important of these processes.

Adaptation and Homeostasis

Every new organism is the product of reproduction. The individual then develops as did its parents and in the fashion characteristic of its species. Throughout its life it maintains and regulates the complex organization specific to it. There must obviously be some kind of program or information, a metaphorical blueprint, guiding development. This has been mentioned before and will be more fully dealt with in Part 3. Here we are more interested in the fact that throughout the whole history there must be a constant traffic in energy and materials if the organism is to exist, is to develop, and is to be maintained. We are now about to consider that traffic at one of its most significant levels: the events taking place within the cell. In approaching this subject, we first encounter two broad principles that will often reappear at other levels and in other contexts—those of *adaptation* and of *homeostasis*. Aspects of them occur in cells, in the tissues and organs of multicellular organisms, in whole individuals of all kinds, and in aggregations and communities of organisms. The word "adaptation" in this connection refers to the usefulness to an organism of its structures and functions. They tend, with more or less success, to enable the plant or animal to survive and maintain its normal form and activities in its given environment. They often have the peculiarity of seeming to provide for some *future* need, seeming literally providential in the old colloquial sense or teleological in the technical philosophical sense. As noted in Chapter 1, such apparently goal-directed structures and functions, which we call "teleonomic," can be explained in historical, evolutionary terms without the postulate of an actual goal. We shall have more to say on this matter, especially in Part 5.

Homeostasis, which is itself an adaptation and a teleonomic feature of life, is the tendency of an organism to maintain its internal composition and state with fair constancy and within a range suitable for its continued functioning. Processes such as respiration, digestion, metabolism, and excretion all plainly contribute to homeostasis. Autoregulation, the self-perpetuation of functional stability and structural integrity, or homeostasis in a broader sense, makes it possible for an organism to maintain itself in the face of a universe of destructive forces. Homeostasis is particularly evident in systems that involve feedback. Some biochemical examples of such systems within cells are discussed later in this chapter. Numerous more general aspects of maintenance, frequently involving an element of homeostasis or specifically of feedback, will be treated in Part 4.

Cell Metabolism: Some Basic Principles

ANABOLISM AND CATABOLISM

The molecular components of cells are in a constant state of flux in which they undergo continuous and rapid chemical transformation, replacement, and renewal. We speak of this dynamic state of the molecular constituents as *metabolism*. In its broadest sense, metabolism includes all of the enzyme-catalyzed chemical reactions taking place within an organism or its individual cells, whether they are involved in growth, reproduction, tissue repair, or energy production.

Metabolic processes are conveniently divided into two categories. Those reactions concerned with the synthesis of cell constituents and cell products from simpler substances are referred to collectively as *anabolism*. In general, these do not supply the energy needed by the cell. Indeed, they are usually endergonic reactions requiring an input of energy (Chapter 2). Energy production is the function of the second category of metabolic reactions, collectively called *catabolism*. In catabolism, complex molecules are broken down to simpler ones by exergonic processes that ultimately involve energy-liberating oxidative reactions.[1]

UNIVERSALITY OF FUNDAMENTAL METABOLIC PROCESSES

As we begin our survey of the main pathways of cellular metabolism, we should take note of the fact that most of these mechanisms were discovered quite recently. The brilliant explorations of biochemists in the years since World War II comprise one of the most distinguished chapters in the history of science. The new knowledge of cell biochemistry would surely have delighted Darwin, Schwann, and Virchow, because it permits an extension down to the molecular level of a combination of the cell theory and the theory of evolution. The theory of evolution sees a living thing as an evolving organization of materials that in themselves are not alive; the cell theory sees the cell as the unit of that living organization. Together, the two theories predict fundamental similarities in the chemical constitutions and organizations of all living cells. It is this broad prediction that modern biochemistry has now fulfilled.

The common ancestry of all cells—protists, plants, and animals—accounts for their marked similarity in constituent chemicals and processes. We glimpsed this similarity in Chapter 2. The lives of all cells are founded on the chemistry of four major classes of molecules and their derivatives—carbohydrates, lipids, proteins, and nucleic acids. Biochemistry now tells us that the basic processes of carbohydrate, lipid, protein, and nucleic acid synthesis are virtually identical in all living cells, whether of plants or animals. We know, too, that all cells synthesize their large organic molecules from simpler molecules that are available in their physical environments.[2]

---

[1] We shall later find that certain metabolic processes have both anabolic and catabolic functions. One authority has recently suggested that these be termed *amphibolic*, from the Greek *amphi*, meaning "both." (At the same time it was suggested that processes defying classification be called "diabolic.")

[2] We shall see in the following section, however, that cells do differ in the extent to which their organic molecules must be prefabricated by external agencies.

From these facts it is reasonable to conclude that long before the plant and animal kingdoms became separated in evolution, there existed in the ancient oceans various primitive forms of life that were endowed with most, if not all, of the principal metabolic processes occurring in contemporary living organisms.[3]

Obviously differences developed among the early organisms, for an enormous number of diverse species arose. Their diversity is reflected in metabolic differences among cells and species. Nevertheless, the metabolic similarities of all cells are striking and fundamental, and it is to them that we shall give our attention in this chapter. Thus we shall ask, "How does the cell manufacture its characteristic organic chemicals?" This question will lead us to recognize that the synthesis of complex molecules from simpler ones involves chemical work and, like other forms of work, requires the expenditure of energy. The key to understanding the pattern underlying the cell's metabolic framework is the concept of energy, and we shall need to review the remarkable mechanisms by which the cell obtains its energy and the way in which it uses its energy in the fabrication and organization of its chemical constituents.

CELLULAR ENVIRONMENT AND NUTRITION

All metabolism begins with the chemical raw materials that are referred to collectively as *food*. These material requirements of the organism are supplied in one form or another by the environment. The environment also, of course, affects the processes of metabolism by means of temperature, pressure, and other physical and chemical features of the background. But we are here concerned with the environment as the ultimate source of all raw materials.

It is fair to say that there are even more types of cellular environments than there are types of cells, for the environment of any one cell type is constantly changing. Moreover, the environment is determined not by physical,

[3] One interesting practical consequence of this "biochemical brotherhood" of all protoplasm has been the frequency with which scientists have discovered important new principles relevant to animal cells from experiments performed with bacteria or other lower forms.

geological, and meteorological factors alone. Every cell, every group of cells, every organism is dependent upon and influenced by other cells, other groups of cells, other organisms. Every organism lives in an environment of which it is itself a part, along with others of its kind, organisms of other kinds, and many nonliving substances such as air, water, and soil. No organism lives alone, and no living thing is sufficient unto itself. To consider a cell or organism apart from its environment is to consider the nonexistent, an abstraction that is not a real form of life. Relationships between life and environment are so pervasive a part of living that they are involved in all aspects of biology.

The fact that these relationships are so complex and that they do somehow affect every process of life makes it virtually impossible to study them as a distinct and separate topic. It is necessary to take them up repeatedly, from different points of view and at different levels. In this part of the book we are interested mostly in life at the cellular and subcellular levels. The relationships to be discussed here are thus mainly the exchanges and other interrelationships between cells and their environments, which are composed of other cells and the various media (mainly fluids) in or on which cells live.

The degree of interdependence varies according to the properties of cells, their positions with respect to other cells, the kinds of organisms, and the surroundings of cells and organisms. One very important sort of interdependence results from the fact that cells arise only from other cells; they are derived from other cells in sequence through time and related in growth, reproduction, and heredity. Another sort of interdependence is involved in the integration of cells in multicellular organism. Some cells secrete *hormones*, which modify the activities of other cells. Cells in organs and systems act to move or otherwise to affect the organism as a whole. Nerves (themselves cells) control and coordinate the activities of other cells.

Perhaps the most fundamental sort of interdependence of all is illustrated by the fact that organisms differ considerably in their ability to manufacture essential biomolecules with their own metabolic machinery. For example, *all* cells require certain sugars (and their derivatives) that are synthesized—or elaborated—*only* by the green cells of plants. These green plant cells synthesize the sugars from carbon dioxide (CO_2) molecules in the environment, provided there is adequate sunlight, and furnish them ready-made to animal cells. Similarly, some cells cannot make essential vitamins or amino acids from simple environmental precursors, whereas others can; again the nonsynthesizers must obtain the substances ready-made. We call the needed substances in such cases *essential nutrients or growth factors.*

The science of *nutrition* is the study of food requirements and of the methods by which foodstuffs are utilized.[4] As we have just learned, organisms are either *autotrophic* (from the Greek words meaning "self-feeder") or *heterotrophic* (meaning "other-feeder"). Autotrophs are organisms that grow and thrive in a purely inorganic medium. This means that they produce their own sugars, lipids, amino acids, etc., from CO_2 (as the carbon source), ammonia (NH_3) or nitrate (NO_3^-) (as the nitrogen source), and H_2O. Heterotrophs require preexisting carbohydrates and other organic molecules. In general, heterotrophs are animals, and autotrophs plants.

When put in such general terms, the difference between autotrophy and heterotrophy sounds clear-cut and absolute. So it is when we compare, say, a buttercup and a cow. If, however, we compare many different organisms, we find intergradations and a great diversity in powers of synthesis. We find also that autotrophs can be subdivided into two groups according to their methods of obtaining energy. *Photosynthetic* autotrophs convert light energy into chemical energy in the process of *photosynthesis.* Organisms in this group are the colored sulfur bacteria, blue-green,

[4] Like many everyday words, "food" is ambiguous. No doubt bread and meat are food. But plain water is the greatest material requirement of most organisms. Is it food? Vitamins are needed only in minute amounts. Are they food? It is probably best to reserve the term "food" for the organic substances that the cell must have in bulk quantities. It will be understood that the nutritional requirements also include water, vitamins, minerals, and a variety of trace elements.

red, brown, and green algae, and complex green plants. The colors in these organisms result from a number of special *pigments,* including chlorophyll, which plays a critical role in trapping light. By means of chlorophyll-mediated photosynthesis, green plants absorb the energy of sunlight and combine CO_2 with hydrogen to build the sugars that serve as the major energy sources for all living things:[5]

$$CO_2 + 2H_2O + Energy \rightarrow$$
$$(CH_2O) + H_2O + O_2$$

Interestingly, photosynthetic sulfur bacteria participate in a similar reaction, in which a special bacterial chlorophyll is involved and hydrogen sulfide (H_2S) is the hydrogen donor instead of H_2O.

$$CO_2 + 2H_2S + Energy \rightarrow$$
$$(CH_2O) + H_2O + 2S$$

In this case, elementary sulfur is formed instead of oxygen.

The second group of autotrophs comprises the *chemosynthetic* autotrophs. In these organisms, energy is derived not from light but from the oxidation of various inorganic compounds. It is then utilized in a basic organic synthesis of the type just shown. Energy may arise, for instance, from the oxidation of H_2S or NH_3, depending on the species. The summary oxidative reactions in these cases would be

$$2H_2S + O_2 \rightarrow 2H_2O + 2S + Energy$$

$$2NH_3 + 3O_2 \rightarrow 2H_2O + 2HNO_2 + Energy$$

Chemosynthetic autotrophs are not particularly numerous or important in the general economy of nature at the present time. They are also rather poorly understood. They are interesting because they illustrate the funda-

[5] In the equations that follow, (CH_2O) symbolizes the basic unit of a carbohydrate molecule (Chapter 2). Six of these units would be $C_6H_{12}O_6$, or glucose. It may seem unnecessary to put two water molecules on the left side of the equation since one remains on the right side, but, in fact, the H_2O on the right is not one of the two original molecules. The oxygen from the original $2H_2O$ is all released as O_2 in the reaction. O in the H_2O on the right comes from the CO_2 on the left. If we labeled the oxygen in the original $2H_2O$, which can be done with an oxygen isotope, and designated it with an asterisk, the equation would be $CO_2 + 2H_2O^* \rightarrow (CH_2O) + H_2O + O_2^*$.

mental diversity of biochemical systems and may throw light on the origins of such systems.

Heterotrophs are also classifiable into various subdivisions, and we shall mention only some of them. There are, for example, organisms (like the bacterium *Azotobacter*) that require preformed sugars but that can convert gaseous atmospheric nitrogen into inorganic or organic nitrogenous compounds. These *nitrogen-fixing* organisms are invaluable in agriculture because they are the natural fertilizers of soil. Other heterotrophs that do not require organic nitrogen sources can thrive on ammonia or nitrate (like the bacterium *Escherichia coli*). Still others require some of their nitrogen in the form of certain preformed amino acids, and others require certain vitamins or growth factors. It is generally assumed that if an organism requires a particular compound (such as a vitamin) in its food in order to survive and multiply, the organism cannot itself synthesize that compound. On the other hand, it is assumed that if a compound occurs and plays a vital role in an organism but is not required in its food, then the organism can and does synthesize it. Thus nutritional experiments reveal what syntheses are performed in an organism. Humans require ascorbic acid (vitamin C) in their diets and cannot synthesize it. Rats, although rather closely related to us and generally similar biochemically, can synthesize ascorbic acid, as can most plants and animals. Man is actually a very complex and exacting heterotroph who requires a carbohydrate energy source, certain fatty acids, at least 8 amino acids, and almost 20 vitamins.

There are many known examples of organisms that lack specific synthetic capacities. Some bacteria and other lowly organisms are especially revealing in this regard. Let us represent an anabolic pathway as a step-by-step sequence such as the following:

$$A \rightarrow B \rightarrow C \rightarrow D$$

Each arrow denotes a separate enzyme-catalyzed reaction. Some organisms can make all the steps; they synthesize D with no dietary requirement except A. Related forms can make steps $B \rightarrow C \rightarrow D$ but not $A \rightarrow B$. They therefore require B in their diets. Others can make the step $C \rightarrow D$ but not $A \rightarrow B$ or $B \rightarrow C$, and still others cannot make any of the steps. What do the last two types of organisms require in their diets?

The same sort of effect may appear in complex reactions of the type

$$\begin{aligned} A &\rightarrow C \\ B &\rightarrow D \\ C + D &\rightarrow E \end{aligned}$$

Some organisms carry through all three syntheses from the raw materials A and B. Others may require A and D or C and D or E in their diets. What syntheses are these organisms unable to perform?

There is some evidence that the ancestors of all living organisms went through autotrophic stages in which they could perform all necessary syntheses from elements and very simple inorganic compounds. The inability of present-day organisms to perform a particular synthesis may then demonstrate an evolutionary loss of a biosynthetic ability possessed by remote ancestors. We shall later learn (Chapter 7) that such a loss is due to a genetic mutation resulting in a heritable inability to synthesize a specific enzyme. Although new syntheses have certainly developed in some lines of evolution, on the whole the loss of syntheses has probably been a ruling factor in the evolution of biochemical systems. There can, for instance, be little doubt that ancestors of ours could synthesize ascorbic acid and that we need it in our food because the synthetic capacity was lost somewhere along the line.

In fact, it is likely that other ancestors of ours, very remote ancestors indeed, had the power of photosynthesis and then lost it. Among the living protists called *flagellates*, there are pairs in which the forms are almost exactly alike except that one has chlorophyll and performs photosynthesis and the other does not.[6] It is probable in each case that the nonphotosynthetic form was directly derived from the photosynthetic one by loss of the power to perform this synthesis. By experimental procedures it is possible to take some

---

[6] Here are the names of a few of these pairs: *Chlamydomonas–Polytoma, Cryptomonas–Chilomonas, Euglena–Astasia*. Among dinoflagellates there are several genera, such as *Gymnodinium*, in which some species are photosynthetic and others are not.

of the photosynthetic flagellates and to breed from them a strain that lacks chloroplasts and therefore cannot perform photosynthesis. (Why not?) This artificial strain lives and reproduces perfectly well as long as it is fed either the carbohydrates that it cannot synthesize or some other organic molecule as an energy source.

This experiment actually turns an autotrophic "plant" into a heterotrophic "animal" as far as nutrition is concerned. Some such change probably took place in the distant past when plants and animals became distinct and began to diverge along their separate paths of biochemical and bodily evolution. The experiment may actually be repeating the beginning of that epochal event. There is, indeed, reason to believe that the organisms among which the separation began may have been similar to the flagellates of today. As to whether the flagellates themselves, as a group, are really animals or plants, we may call them what we please. Among these organisms the distinction is arbitrary if not meaningless.

An organism that loses its capacity for photosynthesis immediately becomes dependent for food on organisms retaining a capacity for that synthesis. This is, of course, the basis of the dependence of the animal kingdom on the plant kingdom. Loss of the ability to perform any synthesis—and it has been mentioned that this is a frequent occurrence in the history of organisms—leads to greater dependence on other organisms. Such loss and increased dependence might be considered degeneration, and they have been so considered by some students of the subject. It is true that the real end of the trail (or is it the end of only one of the trails?) can hardly be deemed other than degenerate. This end is seen in some parasites that perform almost no syntheses and depend on the plants or animals within which they live for nearly all compounds. Nevertheless, the loss of synthetic capacity may be associated with increased activity, greater anatomical complexity, greater mobility, increased perceptiveness, and many other evolutionary changes that man, at least, considers progressive. Man himself, certainly the most intelligent if not in all respects the most advanced organism, has "degenerated" enormously by the loss of synthetic capacities and is as dependent on other organisms for food as most parasites.

HOW CELLS GET THEIR MATERIALS

In all cells some of the essential organic molecules are fabricated within the cell, and others come from outside. As we have noted, cells vary enormously as to what they make and what they take in. They vary also in what they release. In multicellular organisms there is great diversity in these respects among the cells of one organism. Its cells are chemical specialists. There is also great diversity among different organisms. We have seen that green plants can make sugars and most of the other essential foods, whereas animals cannot. These are matters of nutrition and of *food chains* [7] among organisms in communities, above the level of cells. At this point, we are most concerned with the fact that each cell must obtain materials from outside itself, regardless of what compounds it makes or what it does with materials taken in. The materials must also move within the cell.

Cells usually obtain their materials as molecules or as ions from watery solutions. There are some real and some merely apparent exceptions to this generalization. Cells in direct contact with air may acquire and lose gases in the absence of an external liquid solution. (The gases are usually in solution when *within* the cell.) Even in cases apparently of this sort an external solution may really be present. Your lungs do not extract oxygen directly from air, but from solution in a thin liquid film that covers the cells at the surfaces of the lungs. Cells may also take in undissolved material from their surroundings, small bits of solid food, globules of fat, or the like. You may see an ameba surrounding a food particle and taking it whole into its cell. There are cells in your own body that do the same thing. Among them are certain blood cells (phagocytes) that engulf and digest bacteria, thus providing resistance to bacterial disease. Their behavior is more of an apparent exception than a real one. The particles taken in

[7] A food chain is a series of organisms in a community, each of which devours the next in the series as its principal food source. The concept will be discussed in Chapter 24.

whole by an ameba or other cell are reduced to molecules in solution before they actually enter the protoplasm and take part in its chemical activities.

Thus even in these instances the rule holds that protoplasm usually obtains materials from a solution with which it is in contact. A cell, or rather a protist, living alone obtains its materials directly from the surrounding inorganic environment. The environment of such an organism is usually water and therefore a solution. (Chemically pure water does not occur in nature, and no organism can survive indefinitely in really pure water. Therefore, all water in which life exists is a weaker or stronger solution of some substance or substances.) Cells in multicellular organisms are usually also in constant contact with solutions, solutions in adjacent cells or, commonly, in liquids moving outside but among the cells. These liquids, of which the sap of a tree and blood plasma are special examples, are generally very elaborate solutions, intricately influenced by the chemistry of the whole organism.

The question of how cells get their materials thus reduces to the question of how molecules move about, sometimes in gases, but more often in the liquids in which they are in solution. Some movements of molecules are sufficiently obvious to need little discussion. If a molecule is in solution in a liquid or is part of a gas, it is moved along by movements of the liquid or the gas. It is less obvious that molecules also move under their own power, so to speak. This movement is governed by the principles of diffusion, the semipermeability of membranes, and osmosis, treated in Chapter 3.

RELATIONS OF CELL STRUCTURE AND CELL METABOLISM

The remainder of this chapter will deal principally with the major pathways of metabolism, both catabolic and anabolic, and the mechanisms controlling them. In effect, we shall be studying $A \rightarrow B \rightarrow C \rightarrow D$ sequences so that we may become acquainted with the compounds in each sequence and the modes of their chemical transformation.

This is an appropriate moment for an admonition. Even though we display the various metabolic pathways as abstract series of compounds and arrows, we must never forget that all of the changes occur in and on the many intricate subcellular structures described in Chapter 3 and so elegantly portrayed by electron micrography. The point to be stressed is that the cell is more than a mere bag of enzymes. It would be remarkable enough if it were. But, in fact, it is an intricate, highly compartmentalized structure consisting of many membranes and membrane-limited spaces. None of its exquisite structure is without importance in the conduct of metabolic processes. The outer cell membrane actively determines what substances may enter the cell. Other cytoplasmic membranes act as barriers to random molecular motion and thus tend to segregate certain reactions into certain cell regions. Perhaps the perfect example of such a segregating boundary is the membrane surrounding the lysosome. As long as it is intact, the cell is spared destruction by enzymes within the lysosome. When the membrane is broken, the cell is killed.

We should also remember that specific metabolic processes take place in or on specific intracellular structures. Protein synthesis, we have seen, occurs on the ribosomes, and the new protein molecules have particular relations to the endoplasmic reticulum. The main oxidative reactions of cell metabolism occur in the mitochondria, and we shall presently discover that the configuration of the mitochondrial membrane is such as to permit a particular spatial organization of the many interacting oxidative enzymes. We shall also have reason to appreciate that photosynthesis is enormously facilitated by the complicated organization of the chloroplast.

As we survey the various energy-producing and synthetic pathways of metabolism, we should keep in mind that all of the processes are taking place in a structural system whose complex arrangement of materials and high degree of order is an extraordinarily improbable event. Moreover, it is a system of great instability, one that can be maintained only by constant input of energy and synthetic activity. Thus metabolism depends for its continuance and efficiency on a complex physical structure, and in turn the structure depends for its continued existence on the transactions of metabolism.

Chemical Work of the Cell

We learned in Chapter 2 that energy is defined as the capacity to do work, that large quantities of potential energy are inherent in the chemical structures of sugars and other nutrient materials, that the potential energy is invested in these molecules in the process of photosynthesis, the ultimate energy source being the sun, and that exergonic oxidative reactions convert this potential energy to free kinetic energy, although such reactions frequently require initial investments of activating energy. We must now consider further the metabolic processes that generate energy and the forms in which this energy is transported, stored, and released.

The fact that the cell's complex organic molecules—sugars, lipids, and proteins—have high contents of potential chemical energy has a twofold significance. First, as has been said, it is by the chemical breakdown of large molecules that kinetic energy is released to power all the cell's and organism's many activities. You are expending kinetic energy when you run, when you merely lift a finger, when your chest muscles move in breathing, or when your blood courses in your veins. And in every activity the energy derives from the potential chemical energy in your own constituent molecules as they are oxidized. You literally consume yourself when you do work. It is, then, small wonder that you must constantly feed—and feed, moreover, by taking in more large and complex molecules.

But there is a second significance. *Only* living organisms possess these large molecules. It is true that after you have partly burned yourself up on hard work you can repair the damage by ingesting some other form of life. But it too did work and fed; where did its food come from? If the supply were limited, life would last just so long before it would stop because its chemical fuel resources had been expended. That life is not limited by a finite and expendable fuel source

4–1   **Sugar: crossroads of metabolism.**

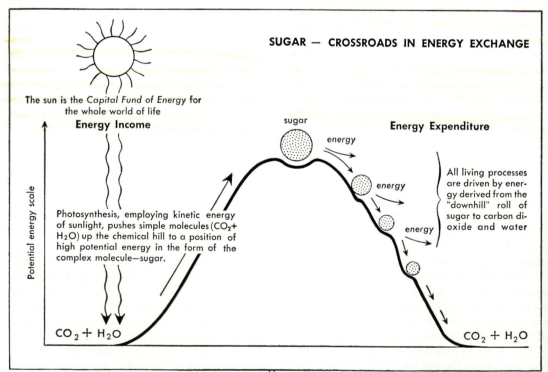

SUGAR — CROSSROADS IN ENERGY EXCHANGE

The sun is the *Capital Fund of Energy* for the whole world of life

Energy Income

sugar

energy

Energy Expenditure

Photosynthesis, employing kinetic energy of sunlight, pushes simple molecules (CO_2 + H_2O) up the chemical hill to a position of high potential energy in the form of the complex molecule—sugar.

energy

energy

All living processes are driven by energy derived from the "downhill" roll of sugar to carbon dioxide and water

Potential energy scale

$CO_2 + H_2O$

$CO_2 + H_2O$

TABLE 4–1

Cyclic interrelations of carbon dioxide, oxygen, sugar, water, and energy

Photosynthesis in plants (endergonic) is an energy-binding *reduction*.

$$6CO_2 + 6H_2O + \text{Kinetic energy of sunlight} \xrightarrow{\text{enzymes}} C_6H_{12}O_6 + 6O_2$$

$\underbrace{}_{\text{low potential energy}} \qquad \underbrace{\phantom{C_6H_{12}O_6 + 6O_2}}_{\text{high potential energy}}$

Respiration in plants and animals (exergonic) is an energy-releasing *oxidation*.

$$C_6H_{12}O_6 + 6O_2 \xrightarrow{\text{enzymes}} 6CO_2 + 6H_2O + \text{Kinetic energy of living processes}$$

$\underbrace{\phantom{C_6H_{12}O_6 + 6O_2}}_{\text{high potential energy}} \qquad \underbrace{}_{\text{low potential energy}}$

is due entirely to the existence of *some* living cells that can photosynthesize sugars. These cells can use the energy of sunlight to push CO_2 and H_2O uphill again into sugar (a reduction process). The chemical machinery in green plants is a turbine for the whole world of life. Just as the hydroelectric turbine exploits the resources of a waterfall, converting its energy into a form usable by other machines, so does the photosynthetic apparatus in a plant turn over the sun's energy to all other living things. It does so by transforming the kinetic energy of sunlight into the potential chemical energy that living systems can utilize.

There are many ways of looking at the metabolism of living cells and many details to fit together. But there is one feature that will always keep us oriented; like a major crossroads, it is the point to which we can always come back to find our way again. This feature is the central role of sugars in the economy of cells. In the balance sheet of life, all energy income leads to sugars, and all energy expenditures lead from them (Fig. 4–1).

The sugar fuels made by a plant can be broken down in an oxidative process that releases energy. This is so both in the plant itself and in the animal that eats the plant. The oxidative release of energy is *respiration*, and it is in one form or another coextensive with life. In the vast majority of cases respiration involves the consumption of oxygen; like the burning of coal or wood, respiration stops when oxygen is withheld.

In over-all view we can sum up the chemical processes underlying life's energy economy as in Table 4–1. Photosynthesis reduces CO_2 and binds energy with a release of O_2. Respiration

oxidizes a carbon molecule (sugar) and releases energy and CO_2. These two simple equations summarize the end results but omit the numerous complex intermediate reactions. Furthermore, they make no reference to the chemical mechanisms by which the energy released in respiration is put to use and the cell functions in which it is utilized.

The great similarities in the chemical properties of all cells derive largely from the fact that they all have to start from the sugar crossroads. They resemble each other most markedly in how they break sugar down piecemeal, in how they trap and transfer the energy released, and in how they use the energy to compound their other constituents, the fats and proteins. Before we can investigate the general patterns of energy expenditure and utilization shared by most cells, we must first examine the income side of the ledger in greater detail.

ENERGY CAPTURE: PHOTOSYNTHESIS

General plan of photosynthesis

It would be difficult indeed to overstate the fundamental biological importance of photosynthesis. The world of life has found no substitute for it as an energy transformer. As we have seen, some organisms (bacteria) derive energy from the oxidation of small compounds (for example, the oxidation of nitrites to nitrates). However, these creatures, the chemosynthetic autotrophs (p. 108), are quantitatively insignificant in the general picture of life, and even they incorporate energy into sugar.

Knowledge of the photosynthetic process

4–2 Oxygen cycle. In sealed chamber *A* the animal combines O_2 into CO_2 and dies. In sealed chamber *B* the plant releases O_2 from CO_2, and both plant and animal live.

was only slowly acquired through the seventeenth and eighteenth centuries. Jan Baptista van Helmont demonstrated in the seventeenth century that a potted plant gained weight that could not be accounted for by the withdrawal of food from the soil in the pot. Using a willow tree, he discovered that in 5 years it gained pounds while the soil lost only ounces. The gain in weight was evidently due to the uptake of gases from the atmosphere or of the water daily supplied to the pot or of both. The fact that a plant takes up CO_2 from the air and releases O_2 was discovered by Joseph Priestley in the eighteenth century.[8] He showed experimentally that a mouse soon dies if placed alone in a sealed chamber but survives if a green plant is also placed in the chamber. Photosynthesis takes up CO_2 from the mouse's respiration and releases O_2. The mouse utilizes

[8] Priestley (1733–1804) did not know CO_2 and O_2 by the modern terms. He spoke of "fixed" and "good" air, respectively.

the O_2 to continue to respire and live. (See Fig. 4–2.)

It was not, however, until the nineteenth century that these facts were understood. The weight gain Van Helmont noted was due to the manufacture of organic molecules from the CO_2 and H_2O. This manufacture was found to be dependent on light and on the green pigment chlorophyll in the leaf. Precisely *how* the sugar is manufactured in the green leaf was only recently resolved, through ingenious experiments involving the use of CO_2 containing radioactive carbon. From the biochemical pathway of the radioactive carbon, we can trace the various compounds formed in the course of sugar synthesis.

For purposes of discussion, it is helpful to divide the story of photosynthesis into two major topics. First, how is the energy of sunlight captured and made available for the performance of chemical work? And second, what metabolic pathway is followed in the conversion of CO_2 to sugar? Let us briefly consider these questions in turn.

Absorption of light energy

In order to be photochemically useful, light must first be absorbed. The particular molecules that are specially adapted to absorb visible light are the pigments. The color of any pigment is due to the fact that because of its molecular structure it absorbs some wavelengths of light more than others. Sunlight, as you doubtless recall, is a mixture of wavelengths ranging (as far as visible to our eyes) from the long waves that we perceive as red to the short waves that we perceive as violet. When sunlight strikes a colored object, part of this wavelength mixture is absorbed. The remaining wavelengths are reflected toward our eyes, which thus receive more of some wavelengths than of others. We perceive these dominant wavelengths as color. Chlorophyll absorbs mostly red and violet and adjacent wavelengths from sunlight and so transmits or reflects mainly the wavelengths around the middle of the visible spectrum, which we see as green. In many organic compounds color seems to be a purely incidental result of molecular structure and not, in itself, related to a biological function. In chlorophyll, however, the color is an indication that energy-rich radiation is being absorbed and used; the red and violet ends of the spectrum consist of rays with the richest energy content. (What color would trees and fields be if the middle portion of the spectrum had been the most energy-rich?)

Chlorophyll is commonly a mixture of two different compounds, chlorophyll a ($C_{55}H_{72}O_5N_4Mg$) and chlorophyll b ($C_{55}H_{70}O_6N_4Mg$). The chemical structure of chlorophyll a is given in Fig. 4–3. All green plants have chlorophyll a; many algae and a few other green plants lack chlorophyll b and may have other, related compounds. As we have learned (Chapter 3), except in the blue-green algae and bacteria chlorophyll occurs within cells only in the small bodies (plastids) called chloroplasts. Within the chloroplasts the pigment is found in numerous still smaller bodies, the *grana*. A single plant cell may contain up to 80 chloroplasts, and a single mature chloroplast —for instance, of spinach—may contain 40 to 60 grana. Within a granum the flat chlorophyll molecules are stacked in piles, rather like the plates of a battery, providing a relatively huge surface area. These molecules are the true loci of photosynthesis in the green plant.

In a chloroplast, chlorophyll is attached to protein, much as the similar compound, heme, is attached to the protein globin in animal hemoglobin. A chloroplast also contains other pigments, called *carotenoids*. Their molecules are essentially long carbon chains, parts of which may form the "tails" of the chlorophyll molecules. The carotenoids may help to transfer energy to chlorophyll. Light is generally necessary for the synthesis of chlorophyll. Iron and manganese are also necessary, although they are not present in chlorophyll.

In the first phase of photosynthesis, the radiant energy of sunlight activates or excites the chlorophyll molecule as follows:

Ch	+	Photon	→	Ch*
Chlorophyll		Light		Activated chlorophyll

The absorption of light energy by the chlorophyll molecule raises it to a short-lived *excited* state (designated by an asterisk), in which the pigment has the capacity to serve as an *electron donor*. A similar phenomenon

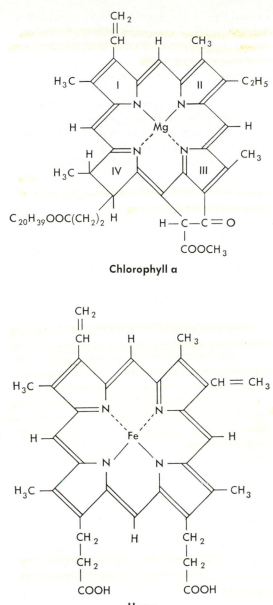

Chlorophyll a

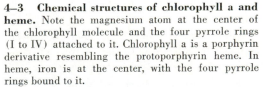

Heme

4–3 Chemical structures of chlorophyll a and heme. Note the magnesium atom at the center of the chlorophyll molecule and the four pyrrole rings (I to IV) attached to it. Chlorophyll a is a porphyrin derivative resembling the protoporphyrin heme. In heme, iron is at the center, with the four pyrrole rings bound to it.

occurs in photoelectric cells whose light-sensitive chemical substances generate electric currents. However, only chlorophyll is useful as an electron donor in photosynthesis. Moreover, it is useful only when it is in its natural crystal-like arrangement within a granum. ==Free chlorophyll, extracted from the granum, although chemically intact, is no longer active in photosynthesis.==

Organic *electron acceptors* within the granum accept the electrons (that is, they are *reduced*—see Chapter 2) from activated chlorophyll, and it then replaces its lost electrons by splitting water and taking some from the cleavage products. Water, it will be recalled, can dissociate into H^+ and OH^- ions. These ions can react in a variety of ways. ==The H^+ ion can indirectly reduce the coenzyme NADP to form $NADPH_2$== (Chapter 2), which in turn can serve as an essential reducing agent (that is, electron donor) in the conversion of $6CO_2$ to sugar ($C_6H_{12}O_6$). Four OH^- ions can interact to form two molecules of H_2O and one of gaseous O_2, which escapes. This is the free oxygen known to be liberated in photosynthesis.

==Many of the details of these reactions are unknown, but we may say in summary that (1) activated chlorophyll promotes the cleavage of water to H and OH; (2) the H of water provides the reducing power needed to transform the coenzyme NADP to $NADPH_2$, which then reduces CO_2 to sugar; and (3) free O_2 is a by-product, whose ultimate source is the water molecule.== Thus the net result of the first step in photosynthesis is

$$2H_2O + 2NADP \xrightarrow[\text{chlorophyll}]{\text{light}} 2NADPH_2 + O_2$$

Light energy, through its effect on chlorophyll, thereby converts the cleavage products of water into a metabolically useful compound. ==With the formation of $NADPH_2$, the light-dependent phase of photosynthesis is completed.==

Conversion of CO_2 to sugar

Present knowledge of the complex means whereby CO_2 is reduced to sugar has been achieved largely through the use of CO_2 containing the radioactive carbon isotope C^{14}.[9]

[9] Every chemical element has atoms of several forms, which vary in weight. These are the *isotopes* of the element. Though identical in their chemical behavior, isotopes are recognizable by their physical properties. They are either stable or unstable. Unstable isotopes, in the process of changing to a more stable condition, disintegrate, emitting subatomic

To the plant, $C^{14}O_2$ is chemically indistinguishable from ordinary CO_2 ($C^{12}O_2$); hence the metabolic fates of both types of CO_2 are identical. Thus when $C^{14}O_2$ is supplied to green cells in the light, all the intermediary compounds on the pathway to sugar become radioactive through its incorporation and thus traceable with Geiger counters and other detecting devices.

This approach to the problem of identifying the metabolic pathway of CO_2 was launched in the late 1940's by Melvin Calvin and his associates at the University of California. In experiments performed mainly with the unicellular algae *Chlorella* and *Scenedesmus*, they found that within less than 1 minute after the introduction of $C^{14}O_2$, radioactivity could be detected in many molecules within the cells, including sugars, sugar phosphates, amino acids, and organic acids. One of the *first* compounds to become radioactive was phosphoglyceric acid, or PGA, a well-known three-carbon acid,

Moreover, almost all of the radioactivity in the PGA formed during a brief exposure to

particles, or radiation; they are *radioactive*. Since the emitted particles cause molecules in their paths to ionize, radioactivity can be detected and measured. It is convenient to indicate the amount of time needed for half of an isotope to decay away as its *half-life*. Half-lives vary widely. The isotopes of major biological interest and their half-lives are as follows: carbon-14 (C^{14}), 5568 years; cobalt-60, 5.3 years; hydrogen-3 (tritium), 12.5 years; iodine-131, 8.1 days; iron-59, 45.1 days; phosphorus-32, 14.3 days; potassium-42, 12.5 hours; sodium-24, 15 hours; and sulfur-35, 87.1 days. The availability of these isotopes makes it possible for an investigator to incorporate them into certain molecules and follow their progress through the tissues of an organism and their metabolic transformations. Thus if CO_2 "labeled" with radioactive carbon (that is, $C^{14}O_2$) is administered to an organism and analysis then shows that certain sugars have become radioactive, it must be inferred that the labeled CO_2 was transformed into the sugars or their precursors. The rates of transformation can be determined precisely. Stable isotopes are also useful in such experiments. Since they are not radioactive, they can be detected and assayed only by means of a mass spectrometer.

$C^{14}O_2$ was located in its carboxyl (—COOH) carbon. We now know that a sequence of enzymatic reactions occurs in which the keto-pentose ribulose diphosphate accepts the $C^{14}O_2$ to form an unstable six-carbon compound, which splits to give two molecules of PGA.

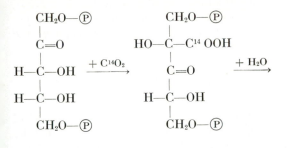

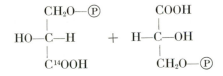

The photosynthetic process is completed by the conversion of PGA to glucose, through a series of reactions that is exactly the reverse of the pathway by which glucose is broken down to PGA in animal cells. An early step in the conversion of PGA to glucose requires the participation of the $NADPH_2$ formed in the first step of photosynthesis.

$$
\begin{array}{l}
\text{COOH} \\
| \\
\text{H—C—OH} \quad + \quad NADPH_2 \longrightarrow \\
| \\
\text{CH}_2\text{O—}\textcircled{P}
\end{array}
$$
Phosphoglyceric acid

$$
\begin{array}{l}
\text{CHO} \\
| \\
\text{H—C—OH} \quad + \quad NADP \;+\; H_2O \\
| \\
\text{CH}_2\text{O—}\textcircled{P}
\end{array}
$$
Phosphoglyceraldehyde

This reaction, the reduction of —COOH to —CHO, could not occur without the H atoms of $NADPH_2$.

ENERGY RELEASE AND EXPENDITURE

We have seen that the splitting of water gives rise to "reducing power" in the form of transferable H and to free O_2 as a by-product. It also happens to be a strongly exergonic process. The reducing power is harnessed for metabolic use with the synthesis of $NADPH_2$. We must now ask a critical question: How does the cell trap the released energy in a chemically useful form? The remarkable answer, it turns out, is equally valid for all cells, plant and animal—a striking example of the cell theory manifesting its universality at the biochemical level. In a sense, we have reached the great crossroads of metabolism.

Analogy with a business economy

It is useful to think about cellular energy expenditure in terms of a business community. Like a member of such a community, a cell does work—movement, heat production, manufacture of new complex chemicals (lipids, proteins), and, occasionally, even light production. Also, as in a business community, there is the costly task of maintaining order. For all this work, which includes both production and maintenance, payment must be made.

In a community of people, payment is made in currency, but currency is only a token for energy of some kind. It is simply more convenient to pay a man a dollar for some goods or services than to pay him by spading his garden for an hour. It is also convenient to carry only small change in one's pocket for immediate use and to keep reserves either in the bank for quick withdrawal or more deeply stored in bonds or real estate.

In the cell's economy we detect a similar over-all pattern. The cell keeps some of its energy-wealth in deep storage (especially starches and lipids), it keeps some on quick call (glucose and compounds called *phosphagens*), and it makes immediate payments in what proves to be small change (phosphate groups). Thus payments are made by the cell directly in energy—energy in the form of chemical structures.

Analogies are always dangerous when carried too far, but the present one does help us keep the details in perspective. Let us then look at this cellular economy, being careful to distinguish the following aspects: (1) the nature of work done; (2) methods of payment; (3) methods of currency (energy) conversion; (4) phosphate radicals as general currency; (5) glucose and phosphagens as current accounts; (6) starches and lipids as reserves.

Currency of energy exchange: energy-rich chemical bonds

When the cell executes any of the thousands upon thousands of endergonic reactions necessary for its organized chemical life, it must supply energy for two distinct operations: (1) energy of activation and (2) energy of reaction. *In principle* the chemist in his laboratory can duplicate, separately, the reactions that occur in the cell. In producing some of these reactions, he would heat the reactants both to activate them and to pay in the energy of reaction. But a cell would be killed if it were heated above a certain temperature (far less than 100°C.), a temperature that would be inadequate for the chemist's purpose. Most of the more complex molecules would be changed into less complex molecules. Specifically, the proteins, including enzymes, would be denatured.

There is another reason of more general importance and implication why heat cannot be used by the cell to do its organized chemical work. Heat would be nonspecific in its effects; it would affect all reactions at once. The orderly life of the cell hinges on the performance of the proper reaction in the proper place at the proper time. If you punched all the keys on a calculating machine, you would include all the motions necessary for a calculation, but your energy expenditure would produce chaos, not organized work. The cell's problem—how to pay for its endergonic reactions without resorting to heat energy—is met in two ways.

First, the cell employs enzymes to reduce the need for *activation energy*. As we have noted, enzymes catalyze specific reactions. You will now appreciate the significance of enzyme specificity; order would not be preserved within the cell if there were a "universal enzyme," simultaneously catalyzing all reactions.

Second, the cell pays in the *reaction energy* for endergonic reactions by avoiding them! We referred to this curious fact in Chapter 2. The explanation is quite simple. Let us expose

it in a generalized example. We shall assign arbitrary energy contents to the participating molecules in order to clarify the basic nature of the device.

Consider an endergonic or energy-binding reaction that forms a complex, high-energy molecule, D, from simpler, low-energy molecules, B and C—another molecule, E, incidentally resulting from the process. When it occurs *in a test tube*, a chemist supplies the needed energy as heat.

Reactants and products	Heat $+ B + C \rightarrow D + E$
Energies	$1 \ + 2 + 2 \rightarrow 3 + 2$

In the cell energy is provided by a *chemical source* in place of heat. Let us imagine that the metabolism of the cell has caused the synthesis from various raw materials of a certain energy-rich molecule, which may be symbolized as XR. If part of this molecule, say R, combines with B, it carries with it some of the energy of XR. Any such transfer is, of course, catalyzed by an enzyme. Thus

$$B + XR \xrightarrow{\text{enzyme}} BR + X + \text{Heat}$$
$$2 + 6 \longrightarrow 5 + 2 + 1$$

Now BR is an energy-rich molecule that can combine with C to form D and E.

$$BR + C \xrightarrow{\text{enzyme}} D + E + R + \text{Heat}$$
$$5 + 2 \longrightarrow 3 + 2 + 1 + 1$$

Not only is it now unecessary for heat to be applied from the outside—as in the test-tube synthesis of D and E—the two reactions just described turn out, oddly enough, to be energy-releasing, or exergonic, in spite of the fact that the products, D and E, together contain more energy than B and C. However, there has been no violation of the laws of thermodynamics. The extra energy came from XR, which had six energy units and was eventually split into X and R with a total of only three units. One of the three units thus freed went into the product D, and the other two were lost as heat.

This principle of *coupling* exergonic reactions is the heart of the cell's mechanism for meeting its energy payments. One of the great unifying themes in the chemical life of the cell comes from the use of the molecule we have symbolized as XR, which is nearly always the same in such reactions. The most important *XR* of the cell is the well-known nucleotide adenosine triphosphate, abbreviated ATP, which is adenine—ribose—℗~℗~℗ (see Chapter 2).

The ATP molecule has a high potential-energy content, largely concentrated in the bonds attaching the two terminal phosphates. An energy-rich bond is represented by a wavy line, ~ (in contrast to the conventional energy-poor bond, —).[10] The energy needed to "drive" an exergonic reaction such as $B + C \rightarrow D + E$ is provided in the following way:

$$B + ATP \xrightarrow{enzyme} B{\sim}℗ + ADP$$

The ATP gives one energy-rich phosphate, ~℗, to the molecule B. In doing so ATP is degraded to the energy-poorer form, ADP, which is adenine—ribose—℗~℗. The coupled exergonic reaction can then take place:

$$B{\sim}℗ + C \xrightarrow{enzyme} D + E + {-}℗$$

The phosphate group (—℗) that remains after the second reaction no longer is energy-rich. Thus the energy to drive $B + C \rightarrow D + E$ was provided by the reaction ATP → ADP + ~℗. Notice that the two equations involving ATP, ADP, and ~℗ are identical with the earlier ones using *XR*, *X*, and *R*. Clearly ATP is *XR*, ADP is *X*, and ~℗ is *R*.

---

[10] We cannot dwell here on the reasons why certain phosphate bonds are richer in energy than others. Suffice it to say that this situation is the consequence of certain structural features of the entire molecule. When an enzymatic reaction introduces these features into the molecular structure, low-energy phosphate bonds are converted to high-energy phosphate bonds. The two main types of reactions in which the bonds are so transformed are dehydrations and oxidations. An example of the former is the conversion of 2-phosphoglyceric acid to phosphoenolpyruvic acid:

$$\begin{array}{ccc} CH_2OH & & CH_3 \\ | & \xrightarrow{-H_2O} & | \\ CHO{-}℗ & & CO{\sim}℗ \\ | & & | \\ COOH & & COOH \end{array}$$

Removal of the elements of water from the glyceric acid molecule alters the molecular structure and redistributes intrinsic free energy in a manner that "concentrates" it in the region of the phosphate group. An example of the oxidative production of an energy-rich phosphate bond is deferred to p. 125.

In summary, the function of ATP is as follows: it pays out energy, in a currency of ~℗ radicals, to molecules unable themselves to react exergonically. It is as though ~℗ were the extra capital needed to allow a project to go ahead.

Let us now develop a more general view of cellular economy by returning to glucose. We have mentioned that the oxidative breakdown of glucose (respiration) is the source of cellular energy. How is energy released in the oxidation of sugar converted into the usable currency of ATP?

Respiration: releasing the energy of sugar

Fig. 4–4 shows in outline form how energy liberated in the respiration of glucose is ultimately converted to energy-rich phosphate groups. The oxidation of glucose proceeds stepwise, releasing small packets of energy adequate to regenerate ATP from ADP and —℗. The reconstituted ATP then can make further payments by donating ~℗ to other compounds that require energy.

In seeking to understand how the oxidation of sugar generates utilizable energy as energy-rich phosphate bonds in ATP, it is helpful to consider a few simple quantitative relationships. We learned earlier (Table 4–1) that glucose is oxidized according to the following reaction:

$$C_6H_{12}O_6 + 6O_2 \rightarrow 6CO_2 + 6H_2O + Energy$$

Thermodynamics tells us that the amount of energy liberated in the combustion of a substance is always the same, no matter what means is used to accomplish the combustion. When a mole of glucose (180 grams) is oxidized by actual burning, 690,000 calories are released. We may assume, therefore, that this much free energy is liberated in the stepwise metabolic oxidation of glucose.

In the late 1930's, it was found that when simple suspensions of ground muscle or kidney tissue are incubated with glucose in the presence of oxygen, glucose is oxidized, and inorganic phosphate is converted to ATP phosphate. Furthermore, it was found that the utilization of oxygen and the disappearance of inorganic phosphate are coupled; for each atom of oxygen utilized, 3 atoms of phosphate are incorporated into ATP. It was this dis-

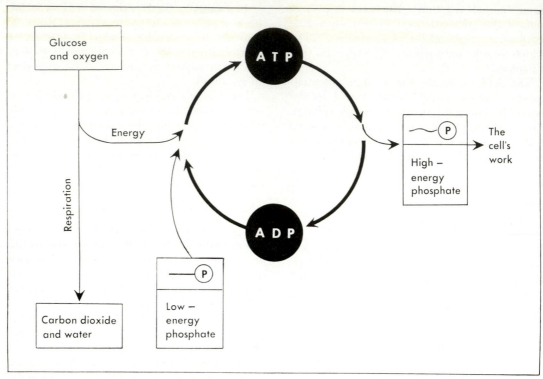

4–4 The ADP–ATP cycle.

covery of *oxidative phosphorylation* that led to the realization that ATP production is dependent upon oxidative metabolism. A P/O ratio of 3 means that the utilization of 12 atoms of oxygen in the total oxidation of glucose (see the equation) must be accompanied by the production of 36 molecules of ATP. Subsequent experiments showed that, in fact, 38 molecules of ADP are converted to ATP; the extra 2 will be explained later. These considerations indicate that the net equation of glucose oxidation should be expanded as follows:

$$C_6H_{12}O_6 + 6O_2 + 38ADP + 38P \rightarrow$$
$$6CO_2 + 6H_2O + 38ATP$$

These figures permit some interesting calculations. Since the terminal $\sim$ⓟ group of a mole of ATP contains about 10,000 calories, 38 moles of ATP represent an energy yield of 380,000 calories, or 55 per cent of the 690,000 calories originally present in the glucose. The remainder is dissipated as heat. Even the best modern steam-generating plants convert no more than 30 per cent of the invested energy to useful work!

The many individual transformations summarized in the net equation just given occur in four major stages (Fig. 4–5). In inspecting the figure, you should attempt to identify the stage at which each of the several components of the net reaction makes its entrance and exit. In particular, note what happens to the carbon skeleton of glucose and at what point electrons or hydrogen atoms are transferred [11] and $\sim$ⓟ groups created from —ⓟ groups.

In the first stage, the six-carbon glucose molecule is split into two molecules of three carbons each, and these are converted at

[11] In this discussion we shall speak of the transfer of electrons and the transfer of hydrogens in oxidative reactions as if they were synonymous (see Chapter 2).

4–5 The four stages of glucose metabolism. Carbon is released as CO_2, energy is released as $\sim$ⓟ groups, hydrogen is transferred in oxidative reactions, and molecular oxygen is consumed. Note that many intermediary compounds are omitted in this scheme.

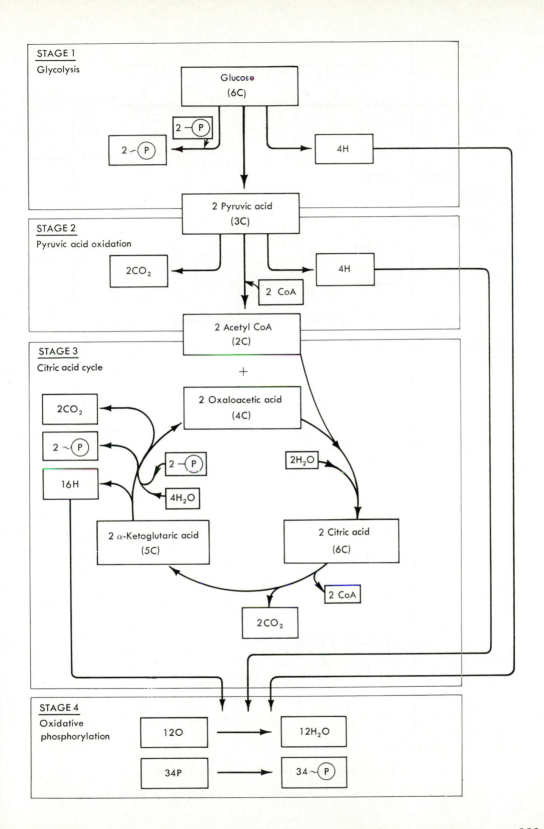

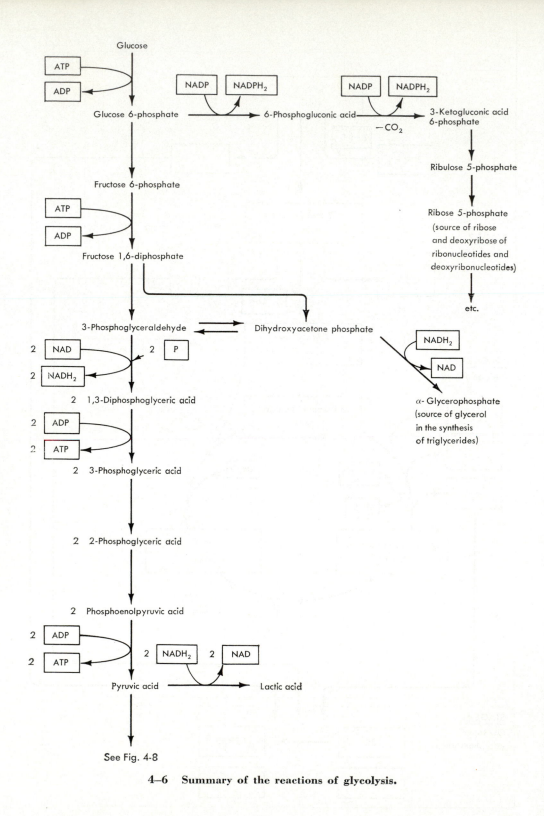

4–6 Summary of the reactions of glycolysis.

length into the three-carbon compound pyruvic acid. This process, involving no less than 10 sequential enzymatic steps, is called *glycolysis* (Fig. 4–6). Although some ATP is utilized early in the sequence, enough new ATP is produced in later reactions to yield a net gain of two ATP molecules per molecule of glucose cleaved. Note also that glycolysis involves the oxidative transfer of four hydrogens. In this preliminary stage, no CO_2 is produced, all the carbons being accounted for in pyruvic acid, and no O_2 is utilized. In sum, stage one involves the following net transformations:

Glucose + 2ADP + 2P →

2 Pyruvic acid + 2ATP + 4H

One of the three-carbon intermediates of glycolysis is phosphoglyceraldehyde. This compound is oxidized to phosphoglyceric acid (PGA) on its way to pyruvic acid. Note that this oxidative step is a reversal of the $NADPH_2$-dependent reduction occuring early in the conversion of CO_2 to glucose in photosynthesis (see p. 117). The reaction sequence from phosphoglyceraldehyde to glucose in photosynthesis is also exactly the reverse of the glucose-to-phosphoglyceraldehyde sequence in glycolysis.

In the second stage of sugar metabolism, each of the three-carbon pyruvic acid molecules is converted to a two-carbon derivative of acetic acid called *acetyl coenzyme A* (or acetyl CoA).[12] Other coenzymes participating in the reactions at this stage are NAD, thiamine pyrophosphate, and lipoic acid. For each pyruvic acid molecule, the reaction sequence liberates one carbon atom as CO_2, produces one molecule each of acetyl CoA and $NADH_2$, and oxidatively removes two hydrogen atoms per pyruvic acid molecule (meaning that four hydrogens are removed per original glucose molecule). As in the first stage, no oxygen is utilized, and relatively little energy is generated. In sum, stage two involves the following net transformations:

2 Pyruvic acid + 2CoA →

2 Acetyl CoA + 2CO_2 + 4H

Acetyl CoA is a "crossroads compound" of formidable versatility. Indeed, as shown in Fig. 4–7, it occupies the central position in intermediary metabolism. It is also produced in the breakdown of fatty acids and certain amino acids. We shall later see that some molecules of acetyl CoA—those that escape further oxidative metabolism—serve as key building blocks in biosynthetic reactions.

In the third stage of glucose metabolism, the two two-carbon molecules of acetyl CoA enter into a remarkable metabolic cycle in

[12] Coenzyme A (abbreviated CoA) is essential in a large number of group-transfer reactions, in which the group transferred is an *acyl* radical, that is, the R—CO— portion of an organic acid, R—COOH. The CoA molecule, like most coenzymes, contains a vitamin—in this instance, pantothenic acid.

4–7   The central position of acetyl CoA in intermediary metabolism.

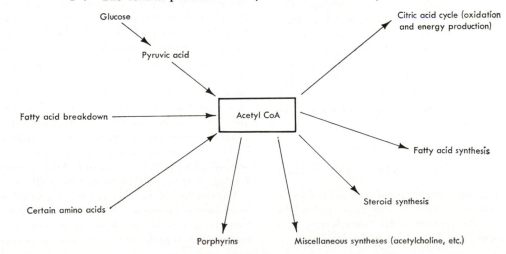

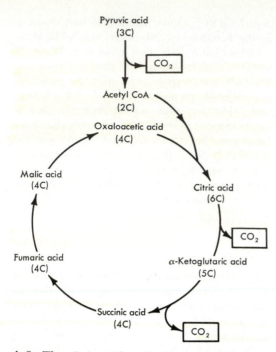

Pyruvic acid
(3C)

CO_2

Acetyl CoA
(2C)

Oxaloacetic acid
(4C)

Malic acid
(4C)

Citric acid
(6C)

CO_2

Fumaric acid
(4C)

α-Ketoglutaric acid
(5C)

Succinic acid
(4C)

CO_2

4–8 The citric acid cycle. The two-carbon compound acetyl CoA enters the cycle by condensing with oxaloacetic acid; at each turn of the cycle, two carbons are given off as CO_2, and one molecule of oxaloacetic acid is regenerated.

which their four carbons are ingeniously converted to CO_2 (Fig. 4–8). First, acetyl CoA combines with a four-carbon molecule, *oxaloacetic acid*, to produce a new six-carbon molecule, *citric acid* (and free coenzyme A). Citric acid is then systematically oxidized, first to a five-carbon and then to a four-carbon molecule. In the course of these reactions, two molecules of CO_2 are released, and four pairs of hydrogens are removed, for each molecule of acetyl CoA entering the cycle. The four-carbon molecule emerging from this sequence is oxaloacetic acid, which is free to combine with more acetyl CoA, so that the cycle begins again. The net transformations of stage three are as follows:

$$2 \text{ Acetyl CoA} + 6H_2O + 2ADP + 2P \rightarrow$$
$$4CO_2 + 2ATP + 16H + 2CoA$$

This stage has been variously termed the *citric acid cycle, tricarboxylic acid cycle*, and *Krebs cycle* (in recognition of Sir Hans Krebs, who first postulated its existence in 1937).

The reactions of the citric acid cycle convert all the available carbon of acetyl CoA to CO_2. But they (with one exception) produce no new ATP. Instead, there is an accumulation of hydrogens, or electrons, that have arisen in the oxidations (that is, dehydrogenations) of the first three stages—the hydrogens being in the form of reduced coenzymes. It is in the fourth stage that all these hydrogens are transferred through a complex chain of coenzymatic hydrogen donor-acceptors (including pyridine nucleotides, flavoproteins, and cytochromes) to oxygen, the final hydrogen acceptor, converting it to water and incidentally freeing the coenzymes to function again as hydrogen acceptors. Thus oxygen is utilized at last, and thus the fourth reaction sequence is appropriately designated as the *respiratory chain*. In the course of this final hydrogen transfer, the bulk of the ATP is generated. The net transformations of stage four are as follows:

$$24H + 6O_2 + 34ADP + 34P \rightarrow$$
$$12H_2O + 34ATP$$

If one adds up the equations describing the over-all changes in each of the four stages of glucose metabolism, the net result is identical to the summary equation on p. 120.

OXIDATIVE PHOSPHORYLATION. Let us examine the fourth stage more closely. At the completion of two turns of the citric acid cycle, the six original glucose carbons have been converted to CO_2. Six molecules of H_2O have been utilized, and 12 pairs of hydrogens have been transferred to coenzymatic hydrogen acceptors. Only four molecules of ATP have been produced, however, and the major remaining task is that of stage four, oxidative phosphorylation: (1) transferring the 24 accumulated coenzyme-bound hydrogens to 12 atoms of oxygen so that coenzyme can be regenerated for another cycle; (2) establishing a couple, or connection, between these oxidative hydrogen transfers and the conversion of inorganic phosphate to $\sim$ⓟ; and (3) then transferring $\sim$ⓟ to ADP to make new ATP.

Exactly how this takes place is an unsolved biochemical problem. It is well established that coenzyme-borne hydrogens (or electrons) are transferred from one molecular hydrogen carrier to another along the reaction sequence shown in Fig. 4–9. This respiratory chain is a series of interlocking cyclic processes. Each carrier molecule is reduced when it accepts hydrogen and becomes reoxidized by passing its hydrogen to the next link in the chain. In three of the cycles there occurs an intermediate step in which the oxidation causes the synthesis of a $\sim ⓟ$ group, which is then transferred to ADP to form ATP.[13]

What molecules serve as hydrogen carriers and phosphate carriers in the respiratory chain? The hydrogen carrier next in line after NAD is FAD (flavin adenine dinucleotide), a nucleotide containing the vitamin riboflavin.[14] FAD transfers its hydrogen to the acceptors called *cytochromes*. Cytochromes are chromoproteins (Chapter 2), found in all oxygen-utilizing cells. The colored prosthetic groups of these proteins are complex porphyrin structures similar to those that we have already encountered in chlorophyll and heme (see Fig. 4–3). The metal of the cytochromes is iron. The various cytochromes (known as cytochrome b, cytochrome c, cytochrome a, and cytochrome a_3) differ from one another in the fine structures of their proteins. But the iron-porphyrin groups of all four function similarly as electron carriers, the central iron atom undergoing reversible reduction and oxidation.

$$Fe^{+++} + e^- \rightleftharpoons Fe^{++}$$

The phosphate carriers indicated cryptically in Fig. 4–9 as X, Y, and Z have not yet been identified. Vitamin K and a new quinone compound, coenzyme Q, appear to function in the respiratory chain, but it is not known where, and it seems doubtful that either is one of the elusive X, Y, and Z compounds. It is also unclear just how free inorganic phosphate is

changed to $\sim ⓟ$ in concert with the oxidative transport of hydrogen.[15]

In summary, oxidative phosphorylation involves the transport of the 24 hydrogens (or electrons) derived from one molecule of glucose to oxygen via the following route: NAD to FAD to cytochromes to 12 oxygens. Simultaneously, 35 molecules of inorganic phosphate are converted to the $\sim ⓟ$ groups of ATP. These plus the one generated in the oxidation of α-ketoglutaric acid total 36.

We can now answer a question raised at the start of this discussion. In the photosynthetic splitting of water by chlorophyll, much energy is released (in addition to the hydrogens needed for CO_2 reduction). We inquired how the energy liberated in this exergonic process could be harnessed for the cell's purposes. The answer is: by being converted into $\sim ⓟ$ groups. Although plant and animal cells alike depend upon the process of oxidative phosphorylation just described as a principal source of ATP, plant cells also derive some ATP from the reactions associated with the photolysis of water.

$$H + OH^+ \underset{H_2O}{\overset{ADP + —ⓟ}{\rightleftharpoons}} ATP$$

Consequently, it is possible for green plants to produce ATP without having to oxidize stored carbohydrates. Hence they can manufacture more carbohydrates than they use, and the stored excess can be utilized by other organisms to supply their energy requirements.

ROLE OF THE MITOCHONDRIA. The enzymatic apparatus of the citric acid cycle and the respiratory chain is of such complexity that it could scarcely function efficiently if the individual participating enzymes were distributed randomly in the fluids of the cell. The precision of the bookkeeping is astonishing

---

[13] The picture of the respiratory chain has been modified somewhat by the discovery that one pair of hydrogen atoms enters the respiratory chain at the middle and thus yields only two molecules of ATP. The deficit is made up, however, by the conversion of ADP to ATP that accompanies the conversion of α-ketoglutaric acid to succinic acid in the citric acid cycle (see Fig. 4–8). The respiratory chain remains the primary site of $\sim ⓟ$ production.

[14] The molecular structure of this nucleotide is identical to that of NAD except that NAD contains the vitamin nicotinamide.

[15] Conceivably, the creation of one new $\sim ⓟ$ is coupled to an oxidative reaction by the following scheme, in which A and B represent hydrogen carriers and X represents a phosphate carrier:

$$AH_2—X + B \rightleftharpoons A{\sim}X + BH_2$$
$$A{\sim}X + —ⓟ \rightleftharpoons X{\sim}ⓟ + A$$
$$X{\sim}ⓟ + ADP \rightleftharpoons ATP + X$$

In any event, this is an instance in which oxidation results in a new high-energy bond. Dehydration, another mechanism of energy-rich bond production, was described on p. 119.

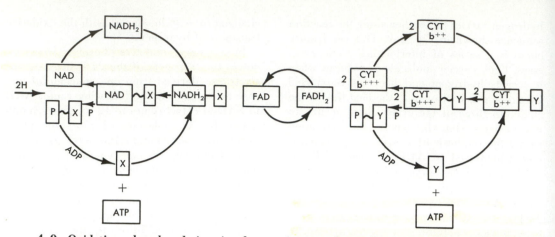

4–9 Oxidative phosphorylation in the respiratory chain. Respiratory enzymes transfer energy by a series of cyclic reactions, each set in motion by the one preceding it. A pair of hydrogen atoms released in the citric acid cycle reduces one coenzyme; this is oxidized again by reducing the next coenzyme, and so on. At the end of the chain the hydrogen combines with oxygen to form water. The known coenzymatic carriers in the chain are nicotinamide-adenine nucleotide (NAD), flavin adenine dinucleotide (FAD), and four cytochromes. Coupled with the reduction-oxidation cycles of three carriers (NAD, cytochrome b, and cytochrome a) are reactions with unidentified enzymes (designated X, Y, and Z) that transfer energy released in the cycles to ATP. As noted in the text, the transfer is not fully understood. When two hydrogen atoms are passed down the whole chain, they give rise to three molecules of ATP.

and is in fact the result of a highly ordered physical arrangement of the enzymes of the citric acid cycle and respiratory chain within the mitochondria. Since only the mitochondria are fully capable of converting pyruvic acid to CO_2 and H_2O when isolated from the cell, they are truly the power plants of the cell.

Investigators are now attempting to determine the exact locations of the enzymes within the mitochondrial membrane, whose structure we saw in Fig. 3–13. When a mitochondrial membrane is broken by intense sound waves or detergents, the internal matrix escapes, and insoluble membrane fragments can be separated by centrifugation. The matrix is found to contain the enzymes of the citric acid cycle; the enzymes of the respiratory sequence are exclusively in the membrane fragments. Recent theoretical estimates suggest that the respiratory-chain enzymes are organized in assemblies, or sets, each containing one molecule of each enzyme (Fig. 4–10). A complete assembly would contain 15 or more active protein molecules close together in a precise geometrical array. These assemblies are now thought to comprise 30 to 40 per cent of the mitochondrial substance. Significantly,

4–10 The localization of respiratory-chain enzymes in the mitochondrial membrane. The black spheres represent the enzymes arranged according to an early diagrammatic scheme of A. L. Lehninger. Such an assembly is believed to occur at regular intervals in the protein layers of the membrane and to contain the full complement of enzymes of the respiratory chain.

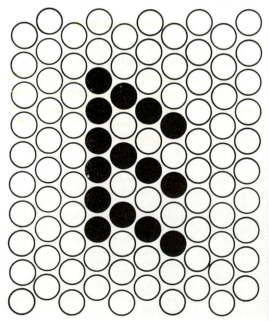

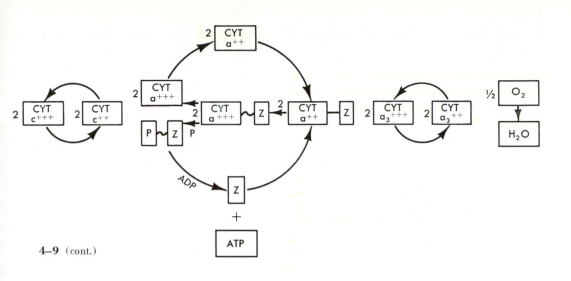

4–9 (cont.)

4–11 **Demonstration of the elementary particles of the mitochondrion.** *Top.* Longitudinal cross section of a mitochondrion as seen by electron micrography (×42,000). *Bottom.* Enlargement of a small area at the upper left of the upper photograph (×280,000). Note the particles attached to the membrane of the crista.

H. Fernández-Morán

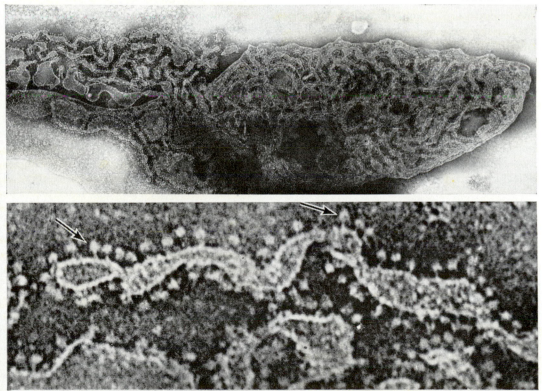

the prediction that such assemblies exist is supported by recent electron-microscopic studies using new high-resolution techniques. They show that the membranes of the external envelope and internal cristae of a mitochondrion are built from small paired ribosome-like particles (Fig. 4–11). These so-called *elementary particles* appear as regular subunits, 100 Ångstrom units in diameter, within the lipoprotein framework of the mitochondrial membrane. That each of these minute mitochondrial subunits contains all of the enzymes of the respiratory chain now seems likely. This means that the mitochondrial membrane is far more than an inert wall or container. Rather, it is a functioning metabolic machine, whose highly ordered pattern of enzyme molecules determines the organization and programing of the enzymatic activity of the living cell.

Mitochondria have been observed to swell and shrink, and the mitochondrial membrane itself changes its dimensions in the course of its activity. Like a sheet of muscle tissue, it can relax or contract, its behavior apparently being related to the local ATP concentration. The membrane contracts when the ATP concentration is high and relaxes when it is low. This suggests that the rate of oxidative metabolism is somehow regulated by shifts in the internal structure of the mitochondrion, which keep the rate of power production in accord with the needs of the cell. Clearly, the integration of chemical behavior and physical structure in the mitochondrion provides a fine and important example of the ultimate convergence of structure and function in biological systems.

**ENERGY UTILIZATION:
SYNTHESIS OF CELL
COMPONENTS**

We have now learned how remarkably intricate and efficient is the cell's machinery for deriving usable energy from the molecular fuel glucose. In essence, the cell achieves this efficiency by not liberating more energy from glucose at any one time than can be picked up and stored as ~$\textcircled{P}$. Second, the stepwise breakdown of sugar yields fragments with four-, three-, and two-carbon skeletons intermediate between the six-carbon glucose and the one-carbon CO_2. These intermediates play leading roles in the rest of the cell's economy, for they are the raw materials used in the manufacture of all lipids, proteins, and nucleic acids. Let us turn now to this aspect of cellular work.

Principles of biosynthesis

So far we have treated the cell's economy as though it were all a matter of energy—wealth and its expenditure. But an economy needs more than energy to do its work; it needs raw materials.[16] The four main constituents in the cell's raw materials are carbon, oxygen, hydrogen, and nitrogen. The *ultimate* source of all the carbon is atmospheric CO_2, which is fixed (reduced) in photosynthesis (and to an extent insignificant in most cells by nonphotosynthetic processes). Oxygen enters the cell in gaseous form from the air and also in water. Hydrogen and nitrogen never enter as gases despite their abundance in the atmosphere. Hydrogen enters ultimately in water. Nitrogen enters plants ultimately as nitrates (for example, potassium nitrate, KNO_3) absorbed from the soil by roots; and animals are ultimately dependent on plants for nitrogen. The point to be noted is that three of these four elements—carbon, oxygen, and hydrogen—enter the life of the cell by way of the glucose crossroads. We are back at the crossroads again in discussing materials, just as we were in discussing energy.

We shall try here to clarify only the basic outline of the cell's economy. Therefore, we shall deal with just the bare essentials of synthetic processes, as follows: (1) all synthetic steps are enzyme-controlled; (2) all syntheses consume energy; (3) generally speaking, all cells build complex molecules from the same relatively small group of molecular building blocks; (4) in the macromolecules formed from simpler organic molecules, the simpler units are frequently linked together chemically by *anhydride*, or *ester*, linkages (that is, bonds formed by the removal of water)—however, the metabolic pathways that ultimately create these linkages may be very complicated, involving many separate reactions,

---

[16] We need not treat the question of raw-material resources in any detail here. You were introduced to the problem earlier in the chapter, and diversity in the over-all picture is taken up in later chapters.

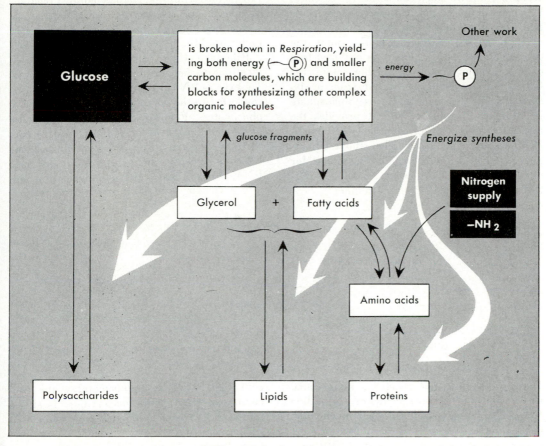

4–12   Synthetic pathways.

enzymes, and coenzymes; and (5) all syntheses obtain their energy and materials (or
both) from the catabolic processes described
earlier (Fig. 4–12).

Let us now consider briefly and selectively
the methods by which the cell converts its
small building blocks into recognizable cell
components.

*Ribose and the nucleotide precursors
of RNA and DNA*

As we have seen, a major function of the
sugars is energy production. Another is the
production of the versatile building block acetyl CoA. In addition, intact monosaccharide
molecules themselves can serve as building
blocks for a number of critically important
cell constituents. For example, they join together to form the polysaccharides—glycogen
in animals and starch in plants (Chapter 2);

and in certain cells unusual polysaccharides
such as heparin and hyaluronic acid are synthesized from various carbohydrate derivatives, such as acetylglucosamine and glucu-
ronic acid.

One of the most significant functions of the
sugars is the synthesis of ribose, the five-
carbon sugar of RNA, and its ribonucleotides.
Ribose synthesis begins with glucose 6-phos-
phate. As indicated in Fig. 4–6, this compound has an alternative route open to it. It
can be converted to fructose 6-phosphate and
thus enter the main pathway of glycolysis. Or
it can enter a series of oxidations involving
NADP that eventually produces ribose and
thus is known as the *oxidative*, or *pentose
phosphate*, *shunt*.[17] The NADPH$_2$ resulting

[17] Glucose 6-phosphate actually can traverse two
other pathways. In the third, the compound enters a
series of reactions that ultimately yields glycogen. In

from its first two steps can act as the essential hydrogen donor in the syntheses of fatty acids and other important molecules.[18]

After the first oxidation in the shunt, the product, 6-phosphogluconate, is oxidized to a compound that loses one carbon atom (as CO_2), thereby being converted from a six-carbon sugar to a pentose. Ribose 5-phosphate is ultimately produced, and it provides the ribose necessary for the synthesis of ribonucleotides and then RNA. The ribose 5-phosphate formed in the shunt is not available for nucleotide synthesis, however, until it is further acted upon by ATP in this manner:

Purine or pyrimidine ribonucleotide precursor then needs only to undergo several more transformations before it contains a finished purine or pyrimidine.

Ribonucleotides must undergo several modifications to become nucleic acid precursors. To serve as RNA precursors, they must first be converted to the corresponding nucleoside triphosphates.[19] For instance, the monophosphate of the ribonucleoside of guanine must be converted to the triphosphate before the enzyme RNA polymerase can incorporate it into new RNA. We now know that ribonucleotides are

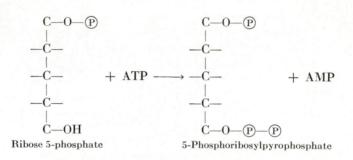

Ribose 5-phosphate   5-Phosphoribosylpyrophosphate

The product of this reaction, abbreviated PRPP, reacts with a free purine or pyrimidine (or its precursor) to form a ribonucleotide. For example,

Purine (or pyrimidine) precursor $+$ PRPP $\rightleftharpoons$
    Ribonucleotide precursor $+$ PP

converted to deoxyribonucleotides, the precursors of DNA. Since deoxyribose is ribose minus one oxygen atom, this conversion is a reduction. Thus one may visualize the production of deoxyribonucleotides from ribonucleotides according to the following scheme:[20]

$$
\begin{array}{lll}
A \cdot R \cdot PP \rightarrow A \cdot dR \cdot PP & \longrightarrow & A \cdot dR \cdot PPP \\
G \cdot R \cdot PP \rightarrow G \cdot dR \cdot PP & \longrightarrow & G \cdot dR \cdot PPP \\
C \cdot R \cdot PP \rightarrow C \cdot dR \cdot PP & \longrightarrow & C \cdot dR \cdot PPP \\
U \cdot R \cdot PP \rightarrow U \cdot dR \cdot PP \rightarrow U \cdot dR \cdot PPP & & T \cdot dR \cdot PPP \\
& U \cdot dR \cdot P \longrightarrow T \cdot dR \cdot P &
\end{array}
$$

$\xrightarrow{\text{polymerase}}$ DNA

the fourth, which occurs only in certain specialized body cells (namely, liver cells), glucose 6-phosphate is cleaved to free glucose. Thus when liver glycogen breaks down to form glucose 6-phosphate, liver cells can convert it to glucose. In this way the liver produces all the glucose present in blood and body fluids.

[18] If $NADPH_2$ is not used in such syntheses, it can be oxidized by the FAD-cytochrome respiratory chain with generation of ATP and terminal reduction of molecular oxygen.

[19] The word *nucleoside* denotes a nucleotide minus its phosphate group—that is, a purine or pyrimidine linked to ribose in the case of ribonucleosides or to deoxyribose in the case of deoxyribonucleosides. Thus

a nucleotide is the same as a nucleoside phosphate.

[20] The abbreviations are based on the symbols in Table 2–4. To emphasize the identity of the sugar in each nucleotide, the abbreviation $A \cdot R \cdot PP$ is used here instead of the more conventional ADP, which emphasizes the number of phosphate groups. Note that ribonucleotides are converted to deoxyribonucleotides only as *di*phosphates but that uracil is converted to thymine only as a *mono*phosphate. As in the case of RNA, the polymerase that synthesizes DNA from its nucleotide precursors functions only if they are in the form of *tri*phosphates. Thus a curious sequence of phosphate additions and deletions is necessary before DNA can be assembled.

Note that thymine, the unique pyrimidine of DNA (see Table 2–4), is not formed from uracil until uracil has been converted to its deoxyribosyl derivative.

Nucleic acids and proteins

We shall defer to Chapter 7 further discussion of the means by which the cell polymerizes the triphosphates of the ribonucleosides and deoxyribonucleosides into RNA and DNA. For these are much more than mere enzymatic reactions of the conventional type. We shall learn in the following chapters that the genetic pattern of the cell is coded in a specific sequence of nucleotides in DNA and RNA. Hence these unique molecules must be capable of somehow being "stamped" with genetic information. The same is true of proteins, whose synthesis will also be described in Chapter 7. It is one thing to speak of the biochemical arrangements in which amino acids are strung together as proteins. But we must also recognize that the specificity of each protein molecule depends wholly upon the exquisite precision of amino acid sequence and that amino acid sequence is in turn the final expression of a specific nucleotide sequence in DNA. This is the stuff of genetics, and we shall save it for the section on reproduction.

Despite this reservation, it is possible to remark briefly on proteins in the nongenetic context of cell metabolism. Protein synthesis requires a constant supply of amino acids. Many of these are present ready-made in the animal diet, and indeed some are essential in the diet since the animal cannot make them for itself. Others can be synthesized in the body cells.

A number of mechanisms exist by which a cell can synthesize a needed amino acid that is not provided by its environment. For example, there is the reversible process called *transamination*, in which an amino group is transferred from an amino acid to a keto acid—thereby converting the latter to an amino acid and the former to a keto acid.

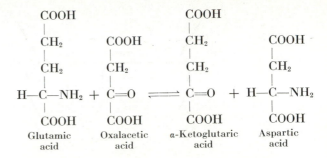

| Glutamic acid | Oxalacetic acid | α-Ketoglutaric acid | Aspartic acid |

Then there is *reductive amination*, in which nitrogen enters into the reaction in the form of ammonia and α-ketoglutaric acid is converted to glutamic acid.

$$
\begin{array}{c}
\text{COOH} \\
| \\
\text{CH}_2 \\
| \\
\text{CH}_2 \\
| \\
\text{C}=\text{O} \\
| \\
\text{COOH}
\end{array}
+ \text{NADH}_2 + \text{NH}_3 \rightleftharpoons
$$

α-Ketoglutaric acid

$$
\begin{array}{c}
\text{COOH} \\
| \\
\text{CH}_2 \\
| \\
\text{CH}_2 \\
| \\
\text{H}-\text{C}-\text{NH}_2 \\
| \\
\text{COOH}
\end{array}
+ \text{NAD} + \text{H}_2\text{O}
$$

Glutamic acid

It is apparent that NH_3 can be incorporated into many of the carbon skeletons arising in the citric acid cycle (see Fig. 4–8). Thus amino acid biosynthesis is largely dependent upon active carbohydrate metabolism. Different cell types vary in this regard, but Fig. 4–13 is a useful general summary of the pathways of amino acid synthesis. Note that the ammonia removed in *deamination* may be reutilized in amination. In animal metabolism, however, an excess of ammonia is highly toxic, and most of the ammonia is converted to the innocuous compound urea, $(NH_2)_2CO$, by the liver cells.[21]

---

[21] In Chapter 11 we shall learn that the kidney excretes urea, which is the principal by-product of amino acid and protein breakdown.

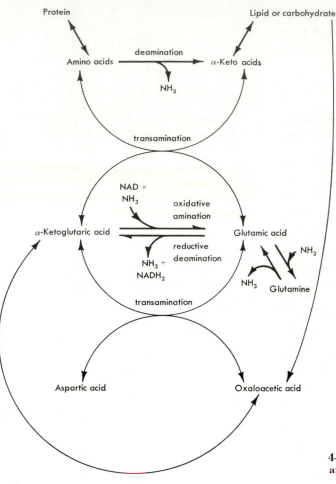

Protein   Lipid or carbohydrate

Amino acids — deamination → α-Keto acids

NH₃

transamination

NAD +
NH₃ oxidative amination

α-Ketoglutaric acid   Glutamic acid

NH₃ +   reductive deamination   NH₃

NADH₂

NH₃   Glutamine

transamination

Aspartic acid   Oxaloacetic acid

4–13 Some reactions in the synthesis and degradation of amino acids.

Lipids and fatty acids

Like carbohydrates, lipids (as well as other organic components of the cell) may contribute to the production of pyruvic acid and acetyl CoA (see Fig. 4–7). Indeed, they make up a most important storage form of substances capable of yielding large amounts of metabolic energy. As in the case of carbohydrates, the breakdown of lipids involves a long sequence of enzymatic reactions. We shall not attempt to review this pathway in detail. However, several of its interesting features must be noted.

We saw earlier (Chapter 2) that a simple lipid or fat is formed when the acidic —COOH groups of three fatty acids react with the three —OH groups of glycerol. When the resulting triglyceride is later used by the cell as a fuel, it is first split into fatty acids and glycerol. The glycerol is eventually converted into 3-phosphoglyceraldehyde (see Fig. 4–6), which then follows the usual glycolytic route to pyruvic acid and acetyl CoA. The remaining fatty acids have an equally interesting fate. A naturally occurring fatty acid always contains an even number of carbon atoms. Through ingenious biochemical investigations, it was found that a long-chain fatty acid is degraded entirely by successive removals of carbon atoms *in pairs*. This elegant series of reactions can occur only if the fatty acid is first attached to coenzyme A. The resulting two-carbon units then become two-carbon acetyl CoA molecules. Since carbohydrates (and many amino acids) also break down to acetyl CoA, it is evident that many independent lines of metabolic breakdown converge in a common compound. Thus lipids, carbohydrates, and proteins all are degraded along their own catabolic pathways, but all arrive in the end at a remarkable focal point of cell metabolism, acetyl CoA.

It was long believed necessary merely to reverse the pathway of degradation in order to synthesize a fatty acid. However, recently a major synthetic pathway has been discovered that begins with malonyl CoA, a three-carbon compound produced on the addition of CO_2 to acetyl CoA.

$$CH_3$$
$$|$$
$$CO—SCoA \quad + CO_2 + ATP \longrightarrow$$
Acetyl CoA

$$CH_2—COOH$$
$$| \qquad\qquad + ADP$$
$$CO—SCoA$$
Malonyl CoA

Malonyl CoA condenses with acetyl CoA or the CoA derivative of a longer acid, and the CO_2 is again liberated so that the chain becomes lengthened by two carbons in a sequence of reactions involving NADP. For the cell there is a distinct safety factor in having synthesis follow a pathway differing from that for degradation (thus the reactions synthesizing glycogen are also different from those breaking it down). Among other things, this situation permits rate-controlling agencies, such as hormones, to speed up one of the two processes without affecting the other.[22] For the advantages of being able to synthesize fatty acids from malonyl CoA, the cell must pay a price: one extra ATP molecule per two-carbon unit added to the chain.

Fatty acids are involved at some stage in the syntheses of all three main kinds of molecules —carbohydrates, lipids, and proteins. In the world of life as a whole, solar energy is first made available through the respiration of sugars. But much of the energy one expends each day may derive more immediately from a beefsteak. Lipids and proteins, initially synthesized at the expense of sugar energy (via ATP), may in their turn be broken down to release energy. Their breakdown products include fatty acids, which may enter the citric acid cycle and be oxidized just as though they had arisen directly from sugar. Thus the three major groups of compounds (carbohydrates, lipids, and proteins) are interchangeable as energy resources and also as raw-material reservoirs. The carbon-chain skeleton of a fatty acid today may be part of a protein, lipid, or carbohydrate tomorrow.[23]

Control of Cell Metabolism

Even the briefest survey of the cell's metabolic pathways must convince us of the necessity for regulatory systems capable of coordinating them. Complex traffic-controlling mechanisms must exist to allow cells to channel energy into the performance of specific tasks, to function uninterruptedly during periods of nutritional deprivation, and to adapt to adverse or injurious conditions.

Although the study of control systems— cellular and otherwise—has only just begun, it is evident that control of any activity in the cell can be effected only by the acceleration or deceleration of some particular metabolic process. Thus the maintenance of the so-called *steady state* within the cell becomes an exercise in the integrated control of metabolic rates.

When the nature of control systems is examined in abstract terms, it is seen that *all* operate in the same way; *effect acts back upon cause*, informing it of the consequences of its previous action and thereby permitting it to determine its future action. The most familiar example of such a system is the ordinary room thermostat, which turns the heat off when the rising air temperature exceeds the thermostat setting. Or consider the steersman of a boat: when he sees his vessel moving too far to

[22] For example, the hormone epinephrine (adrenalin) stimulates one specific enzyme (phosphorylase) in the pathway of glycogen breakdown. Thus epinephrine acts by supplying the body with needed glucose in emergencies. If the enzyme were also in the pathway of glycogen synthesis, epinephrine might simultaneously stimulate breakdown and synthesis, yielding no benefit to the organism.

[23] Organisms synthesize many other compounds, like pigments, alkaloids, essential oils, and steroids, that do not fall into the three major classes we have covered. It is not possible to discuss them here.

leeward, he swings the rudder to windward. The function of the steersman (or the thermostat) consists in holding the course (or the temperature) by swinging the rudder (or the heater) in a direction that will offset any deviation from that course (or temperature). This general mechanism is called *negative feedback*, since the response is opposite to the initiating stimulus, or negative with respect to it (for example, rising temperature decreases heat production).[24] Negative feedback systems demonstrate *oscillation* when for any reason there is a delay in response. The response then continues longer than it should to reach equilibrium, and the system overshoots. Feedback then occurs in the opposite direction. This accounts for the zig-zag course of the rudder-controlled boat. Oscillation of this type accounts for much normal physiological behavior, such as breathing and the heartbeat.

During the several decades that ended in the mid-1950's, biochemists were engaged chiefly in defining the individual steps of the metabolic pathways. Now that most of the principal pathways are known, it is possible to analyze the factors controlling the rates of their reactions. In general, these can be divided into two groups: (1) those that regulate traffic along a pathway by controlling the *level of activity* of individual enzymes; and (2) those that regulate traffic by controlling the *rate of formation* of individual enzymes.

FACTORS AFFECTING THE LEVEL OF ENZYME ACTIVITY

Although a specific reaction along a metabolic pathway depends upon the functioning of a specific enzyme, the activity level of that enzyme is a critical determinant of the rate of the entire pathway only when the reaction in

[24] *Positive feedback* can occur, but its undesirable characteristics are apparent. When an "enough" signal causes an effector to increase its output, instability is inevitable. Such a system is known as a *vicious cycle*. For example, the heart normally responds to blood loss by pumping more rapidly. If blood loss is so severe that the heart is weakened as a pump, there is a decrease in its output. This weakens it further, causing a further decrease in its pumping efficiency. When the stimulus of decreased output causes the output to decrease further, positive feedback exists. Unless the cycle is reversed, the result will be death.

question is the rate-limiting step, or bottleneck, of the whole sequence. In fact, many enzymes are present in cells in amounts far greater than traffic through the pathway demands. Small fluctuations in the activity levels of all but one of them have little consequence in terms of control. The one exception is the enzyme responsible for the reaction with the lowest maximum velocity, the rate-limiting step, which will be our primary concern here.

Substrate and coenzyme availability.

Up to a certain point, the rate of an enzyme reaction rises as the concentration of its substrate rises. Beyond that point, the enzyme is considered saturated with substrate molecules, and reaction velocity can be increased further only if the amount of enzyme present is increased.

It follows that the rate of a metabolic pathway varies with variations in the concentration of the initial substrate of the pathway. If the glucose supply were limited by starvation, glycolysis would be depressed (see Fig. 4–6). Conversely, a plethora of glucose would accelerate the pathway to the point at which the rate-limiting enzyme would be saturated with substrate. Reactions requiring the participation of a coenzyme are similarly affected by the availability of the coenzyme. Since most coenzymes are derivatives of vitamins—the only ones that do not contain vitamins are the coenzymatic nucleotides resembling the nucleotides of RNA (for example, ATP)—vitamin deficiency decreases coenzyme concentration and thereby depresses enzyme function.

The local availability of coenzyme and substrate is of particular regulatory significance in the case of substrates that stand at metabolic crossroads. A compound such as glucose 6-phosphate has at least four pathways open to it.[25] Ordinarily the glycolytic pathway is favored, in part because of the high affinity of glucose 6-phosphate for the isomerase. If, in addition, the supply of available NADP were low, the oxidation pathway would receive even less glucose 6-phosphate than usual.

[25] These are (1) isomerization to fructose 6-phosphate, (2) conversion to the glycogen precursor glucose 1-phosphate, (3) oxidation to 6-phosphogluconic acid, and (4) conversion to glucose (in liver cells).

Coupling

This rate-regulating mechanism, inherent in the design of a pathway, is best explained by the example of the controlled oxidations in the respiratory chain (see Fig. 4–9). The transfer of hydrogen (or electrons) along the carrier chain is possible only when simultaneously ADP and P are being converted into ATP within the mitochondrial structure. Thus demand controls supply in a simple interlocking manner. When ATP is utilized in the cell in the performance of work, ADP is produced, making possible the synthesis of new ATP. As a result, ATP is synthesized only when it is required, and glucose is oxidized only when it is necessary to make ATP.

Many other such couplings exist between metabolic systems.

Hormone action

Hormones are agents that are secreted into the body fluids by special glands called *endocrine* glands and that are transported to distant locations where they affect the metabolic rates of certain cells or groups of cells. In other words, hormones arise in special cells that are remote from the cells whose enzymes they control—and their chief importance is in the functioning of multicellular organisms. It may therefore be inappropriate to mention them in connection with the means by which a cell regulates its own metabolism. Nevertheless, hormonal control appears to involve principles that may well be valid in intracellular control systems.

Hormones were the first rate-determinants recognized. It was early found that epinephrine (adrenalin), the hormone secreted by the adrenal medulla (Chapter 12), stimulates the breakdown of glycogen. We now know that epinephrine activates phosphorylase, the enzyme that cleaves glycogen. Epinephrine functions by converting a catalytically inert enzyme precursor (dephosphorylase) to its active form. Other hormones have recently been shown to change the physical structures of specific enzyme proteins. Although it is likely that all hormones act by controlling the behavior of specific enzymes, we are only now beginning to understand the details of these relationships in molecular terms.

Feedback inhibition

In *feedback inhibition* an enzyme appearing early in a biosynthetic pathway is inhibited by the final product of the pathway. Let us picture a pathway that contains a branching point, thus:

$$E \to F \to G \to H \to I$$
$$A \to B \to C \to D$$
$$J \to K \to L \to M \to N$$

In many pathways of this type, a final product (for example, compound N) is directly inhibitory to the first enzyme in the sequence that is unambiguously committed to the synthesis of that product (here the enzyme converting compound D to compound J). The inhibition can be demonstrated in a test tube with purified preparations of the enzyme and the inhibitory end product (and in such experiments the inhibitory effect is a powerful and instantaneous one); hence the process of inhibition does not require a cell division or genetic intervention. Obviously, with such an arrangement the final product of the pathway will be fabricated only in the requisite amounts—and only when its concentration falls below the value at which it is inhibitory to a key biosynthetic enzyme.

In general, feedback inhibition serves the important purpose of preventing the wasteful synthesis of unneeded end products. It performs this function rapidly and with precision. If, for example, a cell capable of synthesizing an essential amino acid suddenly receives an outside supply of the amino acid, feedback inhibition instantaneously shuts off synthesis within the cell, thus promoting economy within the cell and avoiding possibly toxic accumulations of biosynthetic intermediates.

The occurrence of this phenomenon implies that an enzyme may have two kinds of specificity—the usual specificity with respect to substrate and an additional specificity with respect to inhibitor. There need be little chemical similarity between the inhibitory end product and the substrate for which the enzyme has affinity.

REGULATION OF THE RATE
OF ENZYME FORMATION

Induced enzyme synthesis

Induced enzyme synthesis is defined as the increase in the rate of synthesis of a single enzyme relative to the rates of synthesis of other cell proteins resulting from exposure of the cell to compounds (*inducers*) identical or closely related to the substrates of the enzyme. For instance, a culture of the bacterium *Escherichia coli* growing in a medium with a carbon source such as succinic acid contains only trace amounts of the enzyme β-galactosidase, which splits galactosides. The addition of a suitable galactoside—for example, lactose, a disaccharide of glucose-galactose (Chapter 2)—is immediately followed by a sharp increase of over 10,000-fold in the rate of synthesis of β-galactosidase (Fig. 4–14). The high rate of synthesis is maintained as long as the bacteria grow in the presence of the inducing galactoside. When it is removed, the rate of synthesis drops to its original low level. No compounds other than galactosides induce the synthesis of β-galactosidase.

Enzymes such as β-galactosidase have been called *inducible enzymes* in contrast to the *constitutive enzymes* normally within the cell. It should be noted that only certain enzymes are inducible. Moreover, enzymes that are inducible in one cell strain may be constitutive in another. This means that in cells of the latter group enzyme synthesis occurs normally, uninfluenced by the presence or absence of external inducers. A few years ago it was believed that some unknown fundamental difference existed between the synthetic machinery of constitutive enzymes and that of inducible enzymes. In the light of the more recent theories, it appears that a cell acquires the ability to synthesize constitutive enzymes (that is, those whose synthesis requires no external inducer) when an inducible system becomes altered through genetic mutation.

Feedback repression

Feedback repression has one feature in common with feedback inhibition. Both depend upon the concentration of the final product of a reaction sequence—that is, they are both feedback mechanisms. In feedback repression, however, the end product does not inhibit the activity of a biosynthetic enzyme; it represses its *synthesis*. Conversely, when the concentration of end product is low, synthesis of the enzyme is accelerated as a result of derepression.

Although the mechanisms of enzyme induction and feedback repression are still under study, it is evident that the two are closely related. We shall consider a promising hypothesis explaining their relationship when we discuss gene action in Chapter 7. Suffice it to say here that repression is exhibited in most of the major biosynthetic pathways. Functionally it resembles feedback inhibition in that both prevent the wasteful synthesis of an end product within the cell in the presence of an adequate supply from outside the cell. Feedback inhibition, however, is by far the more rapid and sensitive of the two mechanisms. It appears, therefore, that inhibition is the chief regulator of small-molecule production,

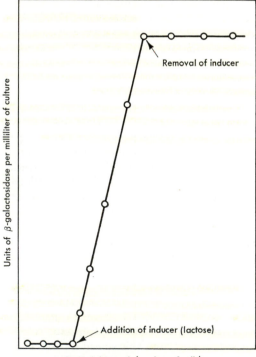

4–14 Induction of β-galactosidase in a culture of Escherichia coli.

whereas repression aims mainly at programing or orchestrating the processes of protein synthesis so as to provide an optimal combination of essential enzyme proteins with a maximum of economy. Repression represents the coarse adjustment of a metabolic machine, and inhibition the fine adjustment.

By harmonizing competing protein syntheses in the cell, repression prevents what otherwise would surely be a lethal overactivity of the mechanisms of protein synthesis. It has been shown, for example, that when the synthesis of an enzyme such as β-galactosidase or alkaline phosphatase is fully derepressed, the enzyme alone may constitute 8 per cent of the total cell protein. Obviously even a few such derepressed syntheses would wreck the economy of the cell. If the gene for each of the cell's several thousand enzymes were to attempt a synthesis of this magnitude, universal repressibility would be essential to survival.

Enzyme activation and molecular conversion

We have already seen that epinephrine converts an inactive dephosphorylase to active phosphorylase (p. 133). Other instances of enzyme activation have been observed, though they are few in number. The zymogens of the digestive tract are inactive precursors of the digestive enzymes that are secreted into the stomach and intestine. For example, we shall later learn (Chapter 10) that the pancreas produces a strong proteolytic enzyme, trypsin. In fact, the material secreted by the pancreas is the catalytically inert substance, trypsinogen, which is chemically converted in the intestine to the active protease trypsin. In the case of the powerful hydrolytic digestive enzymes, the existence of control mechanisms that delay "turning on" the activity until the enzymes are in the proper physical locations is of obvious value to the cell.

CHAPTER SUMMARY

Concept of homeostasis; the apparent purposiveness of organismic behavior.

Cell metabolism: anabolism and catabolism; the universality of fundamental metabolic processes; cellular environment and nutrition; autotrophism and heterotrophism; how cells obtain their materials; relations of cell structure and metabolism.

Review of the concepts of energy and work: energy as the capacity to accomplish work; the diverse forms of energy and their interchangeability—potential and kinetic energy; the cell's energy exchanges—sugar as the focal point of the economy, energy income and expenditure, energy storage and transport.

Photosynthesis: energy income leading to the formation of glucose; the capture of light energy by chlorophyll; chloroplasts; cleavage of water by energized chlorophyll, producing O_2 and H; the reduction of CO_2 to glucose by the released H.

Energy release and expenditure: analogy with a business economy; the use of phosphate as a convenient form of energy currency; the conversion of the released energy into the energy-rich phosphate bonds of ATP; the utilization of ATP energy in the coupling of exergonic reactions; coupled exergonic reactions as a substitute for endergonic reactions.

Four stages of glucose metabolism: (1) glycolysis, the cleavage of glucose to pyruvic or lactic acid; (2) the conversion of pyruvic acid to acetyl CoA plus CO_2; (3) the citric acid cycle; and (4) oxidative phosphorylation in the mitochondria.

Energy utilization and the synthesis of cell components: ribonucleotides and deoxyribonucleotides; nucleic acids and proteins; lipids and fatty acids.

Control of cell metabolism: principle of negative feedback; factors influencing the level of enzyme activity—substrate and coenzyme availability, coupling, hormone action, feedback inhibition; the regulation of the rate of the enzyme synthesis—induced enzyme synthesis, feedback repression, enzyme activation and molecular conversion.

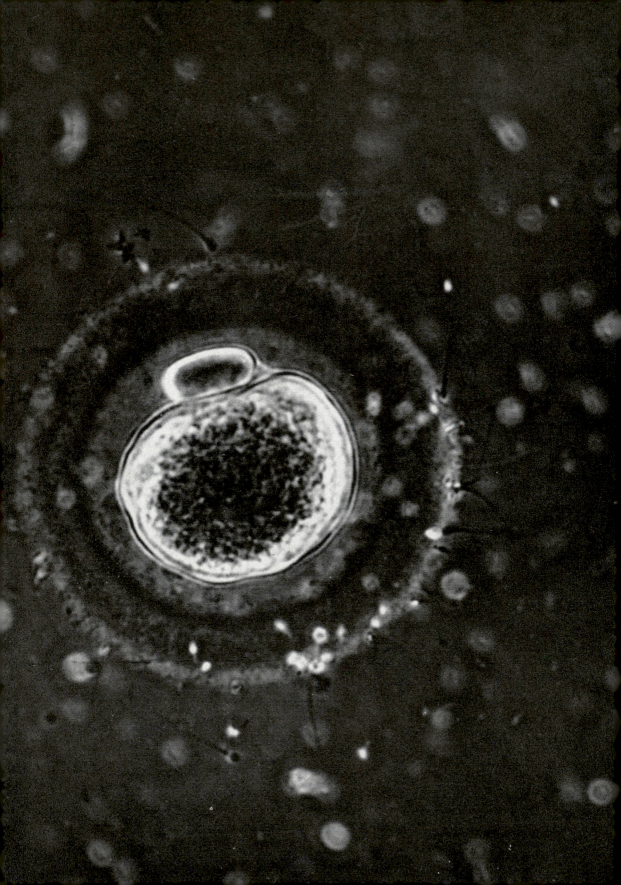

Part 3

Reproduction: The Continuity of Life

Self-reproduction, the most characteristic feature of all living systems, is the subject of Part 3. It is introduced by a remarkable photograph of a human egg cell with sperm cells attached. One of the sperms will fertilize the egg; in other words, its nucleus will migrate into the egg and there fuse with the egg nucleus. The fertilized egg cell—or zygote—contains nuclear material from both parents. It marks the beginning of the life of a new human being and is a useful focal point for presenting all the diverse aspects of organic reproduction.

In the first place, the nature of the zygote—a single cell—points up the fact that organic reproduction is basically a cellular process. Second, the origin of the zygote through fertilization illustrates the fact that organic reproduction nearly always involves the added complication of sex; the new individual organism originates with the union of two different cells, frequently, as in this example, from different kinds of parents, male and female, specialized for their roles in reproduction. Third, a zygote produced by human parents develops into a human being, not into a mouse or other species; thus one generalization about organic reproduction is that it involves heredity, the production of like by like. Fourth, an essential property of virtually all reproduction is elaborate development, whereby the parents' complex organization is created afresh in the offspring out of the simple beginnings afforded by a single cell.

Several of these aspects are to be considered in Part 3. We emphasize that the processes of development, which is essentially creative construction, cannot take their orderly course leading to a particular kind of adult without a set of specifications or instructions to guide and control them. Such specifications must be present in the zygote.

What an organism inherits from its parents—its heredity—is a body of information with specifications for proper development; and since the organism in turn transmits a copy of this same information to its own offspring, it is clear that reproduction is ultimately concerned with duplicating inherited information.

Chapter 5 shows that the inherited information is in the nucleus and mainly in the chromosomes. It describes how identical sets of chromosomes are transmitted to new cells in mitosis and meiosis.

Chapter 6 traces the growth of the major tenets of genetics, including the rigorous demonstration that the genes containing the inherited information are arranged in linear order as parts of chromosomes.

Chapter 7 discusses the chemical nature of the genes; it details the exciting recent discoveries illustrating in part how genes act to achieve their control over the life of the cell—in other words, how the inherited information is translated into the cellular machinery that determines how the cell is to be built and how it is to function.

Chapter 8 outlines the processes of development and suggests how they may be controlled by the inherited information in the chromosomes.

Chapter 9 is concerned in particular with adaptive specializations related to the sexuality of organisms.

5

Reproduction:
Cellular Aspects

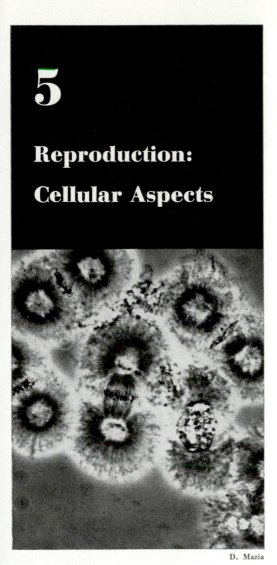

D. Mazia

A photomicrograph of isolated mitotic apparatuses from sea urchin eggs.

The individual organism—bacterium, rosebush, mouse, or man—is an elaborate and complex system whose structure and activities are highly organized. It can maintain its organization and its activities for varying lengths of time by capturing energy and expending it appropriately. The organism can adjust its structure and behavior, within limits, in such a way as to remain adapted to changing environmental conditions. Again within limits, it can repair damage due to accident and the inescapable ravages of wear and aging. We do not understand precisely how the machinery of the body deteriorates with age, but clearly it does; and ultimately the damage is beyond the organism's capacity to repair. Death comes to all living things—indeed, there is no surer or wider generalization that we can make in biology.

The persistence of life on earth in the face of death's certainty points up the universal ability of organisms to reproduce themselves as their most characteristic, and most defining, feature. The long-term endurance of life is not due to the capacity of the individual organism to repair and adjust itself. It is due to the fact that, in a sense, it can throw off its worn-out machinery and start again in the form of its offspring. These in turn can repair and adjust for only so long; ultimately they face the same fate as their parents. They live into posterity only insofar as they leave offspring. What survives on earth over the millennia is not the individual organism but the race, and its endurance depends on the act of reproduction as a vital bridge that spans successive mortal generations.

Major Features
of Organic Reproduction

HUMAN REPRODUCTION:
PROBLEMS AND PRINCIPLES

The phenomenon of reproduction is many-sided. It involves several different kinds of problems and several different features of the organism's structure and activities. We shall begin by sorting out the major features of reproduction as they are more or less familiar to us in humans, and in doing this we shall define the general problems to be given detailed attention in later sections.

In man, as in the vast majority of other species, the individual organisms fall into two categories, male and female, with respect to their roles in reproduction. The new human begins its life inside the body of its mother, and the role of the male is restricted to the act of copulation, during which he introduces into the female a fluid known as *semen*. It has been realized in Western culture for well over 2000 years that the act of copulation in man is causally related to pregnancy and the production of offspring. In spite of this knowledge, the real significance of copulation in man and other organisms remained obscure until the advent of the microscope and the clear formulation of the cell theory. The microscope revealed facts implying that reproduction, like so much else in biology, should be discussed and understood primarily in terms of cells.

The thick seminal fluid ejaculated by the male in copulation is a heavy suspension of single cells called *sperm* cells, or *spermatozoa*. Introduced from the penis of the male into the female vagina, these cells swim upward in special ducts, down one of which migrates a cell or cells contributed by the female. These are egg cells, or *ova*. The ultimate sexual event is the union of one egg cell with one sperm cell, an event known as *fertilization*. The product of fertilization is the *zygote*, or fertilized egg cell. In man, only one fertilized egg usually proceeds to grow further, although occasionally two or more may do so, leading to twins,[1] triplets, and so forth.

The zygote marks the real beginning of the new organism's life. In humans it lodges on the wall of the mother's uterus, where it remains for 9 months, undergoing growth and the initial development of the adult human's complex structure. The discovery of the cellular nature of fertilization was a crucial step forward. But the early microscopists did not grasp its full significance. Excited at the new world of life their microscopes revealed, they let their imaginations fill in the detail their imperfect instruments could not resolve. During the latter part of the seventeenth and early part of the eighteenth centuries, there raged

[1] This is not the only manner by which twins come into being (see Chapter 8).

one of the most absurd controversies that has ever marred the history of biology. One school of opinion (we should really say, "one school of imagination") asserted that in the egg cell there existed a preformed human being, minute but complete in every particular. These scholars, called *ovists*, viewed the sperm as serving only as a kind of trigger to initiate the growth of the minuscule creature contained within the egg. An opposing school—anxious perhaps to uphold the dignity of the male—claimed that the preformed adult lay in the sperm head. This group, known as the *spermists*, had as a member no less a figure than the great van Leeuwenhoek himself. Both schools were *preformist*—that is, they assumed that the adult was preformed in a germ cell (sperm or egg), and for them what we call "development" (*epigenesis*) today did not exist. The history of the young human in its mother's uterus was a simple one of expansion or unfolding of all the adult complexity present from the start. Each group acknowledged, but was not seriously embarrassed by, the astonishing logical implication of its position—that within a minute germ cell (sperm or egg) was a minute human with a reproductive organ containing eggs or sperm that in their turn contained still more minute humans. This meant that the eggs of Eve or the sperm of Adam (according to whether one was ovist or spermist) contained minute humans that contained minute humans that contained minute humans—and so on *ad infinitum*. One mathematically minded ovist came forward at the time with the "precise" conclusion as to how many million such human generations were packed one inside the next, all within an egg of Eve.

Ridiculous as a preformist view is, it is interesting for the following reason. The alternative, which turns out to be correct, is a truly astonishing fact. From a single cell (the zygote, or fertilized egg), which to the eye appears simple and unspecialized, the entire staggering complexity of an adult human is created or developed afresh in each generation.

The photograph used to open Part 3 is of the human egg cell as it normally exists in the mother during fertilization. The egg proper is the large cell (bright in the photograph) in

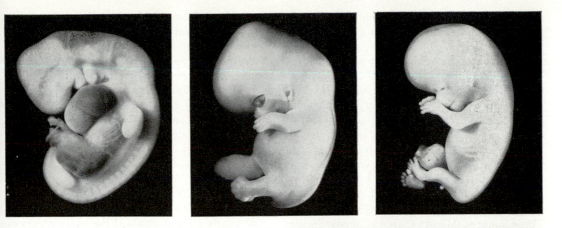

Top three photos, Chester F. Reather; bottom photo, Richard D. Grill

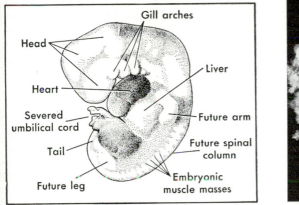

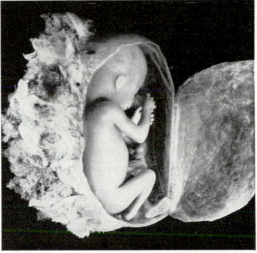

Gill arches

Head

Liver

Heart

Severed umbilical cord

Future arm

Tail

Future spinal column

Future leg

Embryonic muscle masses

5–1 Human embryos. *Top left.* A human embryo in its fourth week of development. *Bottom left.* The same embryo with some parts labeled. *Top center.* Embryo in its sixth week of development. The eye and ear are now recognizable, as well as the fore- and hind limbs. Note how, in the hind limb, the five toes are only roughly sketched out. *Top right.* Embryo in its eighth week of development. The mouth is clearly evident, and the digits (fingers and toes) on the limbs are fully formed. Note also the ribs. *Bottom right.* Fetus, in about the sixteenth week, lying in the fetal membranes.

the center of the mass. At its side lies a smaller cell (a polar body, p. 226), which will play no further role in development. The egg is surrounded by a wide membrane, which many spermatozoa are entering. Only one of these ultimately invades the egg proper to fertilize it. The single cell resulting from fertilization proceeds to divide into two replicas of itself. Each of these in turn divides, so that the embryo becomes four-celled. The multiplication of cells continues and is one of the basic elements in the increase in bulk of the embryo as a whole. Increase in bulk through cell reproduction is clearly not the whole story of development, however. Fig. 5–1 shows that the cell mass of the embryo soon takes on a definite form, although not, to begin with, a form that is recognizably human. The cell mass becomes elongate, corresponding with the future long axis (head to tail) of the body. At the side of the future spinal column, one can see regularly repeated bulges, which are destined to develop into the musculature of the adult. Only after about 5 weeks can the grotesque outline of a head be identified as such by the uninitiated; it is a head grossly out of proportion (by adult standards) with the rest of the embryo. Later the heart and limbs become evident, but each is only a rough expression of the fine structure of the adult, which is developed gradually in successive stages of embryonic life. There are two curious features of the early embryo's form that are of interest to us as evolutionary biologists: (1) a pronounced tail, extending well beyond the point where the hind-limb buds attach to the body; and (2) a set of so-called *gill arches* in the region of the pharynx just below the head. These features do not survive to appear in the adult; they reflect the early evolution of man and are relics of embryonic processes in early primates and fish, respectively.

We are not here concerned with the details of human development. It is our purpose only to emphasize the cardinal point that each reproductive cycle involves the fresh creation of adult form and that the development concerned is not a simple unfolding of detail present in the egg from the outset. The development of adult complexity in each generation is an integral part of the problem of reproduction as a whole.

Sexuality, development, and its cellular foundations are not the only major features of reproduction illustrated by humans. The simple fact that new humans arise only as offspring of preexisting humans implies two more generalizations. First, living things arise from other living things and do not appear spontaneously (without the intervention of life) from nonliving materials; this is the principle of biogenesis. Second, humans have human babies; flies reproduce flies; bacteria reproduce bacteria. This is the phenomenon of heredity—like begets like.

There are thus five general features of reproduction: *sexuality, development, cellularity, biogenesis,* and *heredity.* Of these, sexuality is the least general. Sexual processes occur in the great majority of organisms, but many reproduce themselves without sex. Asexual reproduction is common and familiar in horticulture, where many plants are propagated by cuttings, suckers, or bulbs; and most—probably all—bacteria reproduce asexually most of the time. Sexual and asexual reproduction

will be discussed further in Chapter 9.

We have learned that reproduction is basically a matter of cells rather than of whole, adult organisms. Therefore, biogenesis, development, and heredity are best approached as problems in cellular reproduction. Let us briefly examine these three great principles to see how they are related to each other.

BIOGENESIS: ARGUMENT FROM EXPERIMENT

In Chapter 3 we mentioned the old belief in the spontaneous generation of organisms from the nonliving and outlined the experiments of Redi and Pasteur to test that belief. As a matter of fact, their experiments did not disprove the possibility of spontaneous generation, or abiogenesis, but they did indicate that it is extremely improbable. All subsequent experimental results have supported their conclusion, and there is no serious doubt that biogenesis is the rule, that life comes only from other life, that a cell, the unit of life, is always and exclusively the product or offspring of another cell.

We take biogenesis as a fundamental principle of reproduction from the experimental evidence and also from theoretical considerations. It must, however, be remembered that life had to start sometime. Almost all biologists now think that the very earliest life originated from nonliving matter by natural processes (Chapter 29). It must have appeared at a very remote time, on the order of 2 billion years ago, when conditions on the earth were very different from what they are today. For the last few hundred million years, at least, the origin of new organisms has not been spontaneous but has been a matter of living cells producing other living cells. This fact accounts for the continuity of life as we know it.

BIOGENESIS: ARGUMENT FROM THEORY

Quite apart from the strong argument that experiment raises against the idea of the spontaneous generation of life under present world conditions, there are convincing theoretical reasons for rejecting the whole notion. An attempt to understand these theoretical reasons is worthwhile because the argument forces us to frame the problems of organic reproduction in terms of concepts that are most useful for further study.

We referred in Chapters 1 and 2 to the tendency of the physical world toward a state of uniform disorder—a tendency expressed technically in the second law of thermodynamics. The most probable state for any system of matter—the state it tends ultimately to assume—is one of simplicity and disorder. The idea of complexity embodies within it the idea of improbability; the more complex a thing is, the less probable it becomes that it arose by chance. This is also true of the ideas of order and organization. A complex organization is in itself improbable, and if left alone, it will decay. If no work is put into the system to maintain its organization, it will tend to assume its most probable state—simple disorder.

We first introduced these notions about simplicity and disorder versus complex organization in Chapter 1 in relation to the universal demand in organisms for an energy supply. In the face of the universal tendency for order to be lost, the complex organization of the living organism can be maintained only if work—involving the expenditure of energy—is performed to conserve the order. The organism is constantly adjusting, repairing, replacing, and this requires energy.

But the preservation of the complex, improbable organization of the living creature needs more than energy for the work. It calls for *information* or instructions *on how the energy should be expended* to maintain the improbable organization. The idea of information necessary for the maintenance and, as we shall see, creation of living systems is of great utility in approaching the biological problems of reproduction.

In the nonliving world of bricks, stones, and raw metals, the appearance of a skyscraper immediately indicates for us the presence of life, and not simply because we associate people with buildings. Nobody would ever assume that the complex organization of a modern building could come into existence spontaneously, of its own accord. The ordered nature of the network of bricks, girders, and so forth, is too improbable to arise by chance.

There is an enormous number of possible ways in which the constituents of a building *could* be arranged. The particular arrangement that is a functional building is only one out of the total ensemble or array of possible arrangements.

How is the element of chance removed in the construction of a skyscraper? It is removed by following instructions on how to put together the constituents in one particular way (out of the many possible ways). The information contained in the architect's plans is the agent that eliminates chance from the work of the engineer and leads to the production of a complex organization that is in itself too improbable to arise by chance.

A modern building is certainly a complex and highly ordered structure, but its complexity cannot begin to compare with that of the living system. And for precisely the same reasons that we reject the idea of a building's coming into existence spontaneously, we are forced to reject the idea that anything as complex as an organism could arise spontaneously from the materials of the nonliving world. The materials that go into the construction of an organism—chemical elements like carbon, nitrogen, hydrogen, oxygen, and the rest—could assume so many possible configurations that the chance origin of the particular arrangement which is a particular kind of living cell is utterly unlikely.

It is interesting to note in passing that a precise and quantitative way of talking about information has been developed in recent years and that with this mathematical technique we can estimate just how improbable the chance origin of a living organism is. In other words, we can predict how much information is needed to guarantee the creation of an organism. It is obvious that if someone has to perform a certain task and there are only two possible courses of action open to him, one of which is correct, it takes less information to control and guarantee his correct performance than would be necessary in a situation in which he had to choose among a hundred possible courses of action. The idea of information, in the sense in which it can be treated as an exact and useful scientific concept, is the idea of how much specification is essential to exclude all but the correct or desired possibility out of an array of many possibilities. The more possibilities that exist, the more information that must be given to specify one in particular. Since we know roughly the number of atoms of each chemical element present in the human body, for example, we can estimate the number of possible combinations of elements in the human body. By estimating how many configurations are possible, we are also estimating how much information is needed to specify the particular configuration that is man. Although the numbers involved are only estimates, they are staggering. They indicate an amount of information that, if translated into human language, would fill all the books of several major libraries. That anything requiring as much specification as a human being could arise by chance is out of the question. Theory and experiment together refute the ancient myth of spontaneous generation; and the theoretical argument helps us recognize what we must learn about reproduction as a whole.

That a complex organism cannot develop by chance but must proceed according to some set program raises the question as to how the information for the program is acquired. This is the fundamental problem of evolution, and it will be discussed in Part 5. Our present concern is with the fact that information does exist, with its location and nature, and with the way in which it acts.

HEREDITY AS INFORMATION FOR THE CONTROL OF DEVELOPMENT

The hallmark of living systems is their complexity and their organization, and the focal point of the reproduction problem is how complexity and organization can be reproduced. You cannot reproduce skyscrapers by cutting them in two. This process yields half-skyscrapers, not skyscrapers, because the essential complexity and organization of a building is three-dimensional in space; it is destroyed by being cloven in two. To duplicate a skyscraper, you must use again the information in the initial blueprint—or at any rate a copy of it—and build or develop from scratch.

Precisely the same considerations apply to a living organism, whose organization is also three-dimensional in space.[2] Simple cleavage of a living system like man into two parts does not lead to reproduction; it leads to the destruction of his essential complexity of structure. The reproduction of man, as of all other living systems, involves the development or building from a simple start of the complexity that is to be duplicated. The phenomena of embryonic development are an integral part of the process of reproduction.

Our understanding of the need for the initial blueprint (or a copy of it) as a prerequisite for the reproduction of a building forms the proper starting point for defining the problems involved in organic reproduction. The redevelopment of complexity in each successive act of reproduction demands information for its control. The same construction is undertaken generation after generation in living systems. Embryologically, we and you were constructed in the same fashion, and so were our parents and grandparents before us. In each generation there is a supply of information that regulates development, and in each generation the "blueprint" containing this information is the same. What we inherited from our parents at the outset of our lives as single cells was the information that controlled our development and that determines our fundamental behavior as adults even today.

In its immediate aspect, heredity is the phenomenon of like begetting like in successive acts of reproduction. Like begets like *because* parent and offspring both develop by processes controlled by the same kind of information. Clearly there must be a sense in which it is true to say that what is reproduced and transmitted from generation to generation of living organisms is the information needed for their creation. We shall direct our study of development and heredity along these lines.

We must discover (1) the nature of the information that controls the development of an organism; (2) how the information is reproduced; (3) how copies of it are transmitted from generation to generation; and (4) how the information achieves control.

[2] Actually, the organization of a living system is four-dimensional in that organized sequences of events in time are an essential aspect of it.

Cellular Basis of Reproduction

VIRCHOW'S DOCTRINE: "ALL CELLS FROM CELLS"

We learned in Chapter 3 that the cell theory in its explicit form, the doctrine that all living systems are built of cells, was announced in 1838–1839 by Matthias Schleiden and by Theodor Schwann but that it was not until about 1860 that its full implications began to be evident. The decade from 1850 to 1860 still saw much discussion about what was called "free cell formation," or the "spontaneous origin" of cells from noncellular materials. As the whole concept of spontaneous generation began to fall, it became increasingly clear that the aphorism, "All life from life," could be translated, as it was by the German physiologist Virchow in 1855, into a more precise formulation, *"Omnis cellula e cellula"*—"All cells from cells."

Cells are the elementary structural units of living systems; if new life arises from old, it must take the form of new cells from old. In most organisms, life begins, as we have seen in man, as a single cell, the zygote. The growth and development of one cell to the massive complexity of adult man is, in one major respect, due to the activity of the original cell and its progeny of cells in reproducing themselves. When the multicellular organism is mature and reproduces itself, the fundamental event is again one of producing a new cell —egg or sperm.

Obviously the single-celled zygote from which the adult develops contains within its minute structure the controlling information we have argued must be present—the real inheritance of the child from its parents. We shall now proceed to determine where in the cell the control center must be and how copies of it are transmitted to new cells as they are produced.

THE NUCLEUS AS CONTROL CENTER

Several kinds of observation indicate that the nucleus is the control center for the whole cell's activities. In the first place, nuclei can be removed from some cells that are sufficiently big to permit operation on them with micromanipulating equipment. A cell devoid of its nucleus soon dies, and so we know at least that the nucleus is indispensable for the enduring welfare of the cell. But by other means it can be demonstrated more directly to be the actual site of the cell's controls, the store of information for the regulation of its organization and behavior.

There is a group of marine protists that is especially suitable for experiments bearing on the role of the nucleus. In spite of their single-celled nature, these protists attain a surprisingly large size and complexity of structure. Fig. 5–2 shows the two species *Acetabularia mediterranea* and *Acetabularia crenulata*. For simplicity we shall refer to them as *med* and *cren*, respectively. In both species three different parts can be distinguished: a cap or hat, a "stem," and a base that branches in a way suggestive of roots. The nucleus of the single cell lies in the base. In *med* the cap is disc-shaped, and in *cren* it is branched. The organisms, although single-celled, are big enough for operations to be performed on them with ease. If the stem and cap are cut away, the cell *redevelops* a new stem and cap; and under the control of the normal nucleus, the development repeats the original and produces the typical cap. Another kind of operation is possible. The stem and cap can be cut away, and the stem (lacking the cap) of the other species can be transplanted or grafted onto the cut base. This is illustrated in Fig. 5–2C, where a capless stem of *cren* is grafted on the base of *med*. The *med* base contains a *med* nucleus, and when a new cap is regenerated from the *cren* stem material, the cap assumes the form of a *med* cap. Clearly the course of development that the *cren* stem material follows is controlled by information in the *med* nucleus. The converse experiment gives similar results: a *med* stem transplanted to a *cren* base regenerates a *cren* cap.

The experiments of Theodor Boveri on sea-urchin eggs illustrated in Fig. 5–3 lead to the same conclusion. If the nucleus is removed from an egg of *Sphaerechinus* before it is fertilized, and if it is then fertilized by the sperm of a different sea urchin, *Echinus*, then the egg develops solely under the control of the *Echinus* nucleus. Minute as the *Echinus* nucleus is in relation to the mass of *Sphaerechinus* cytoplasm that it enters, it neverthe-

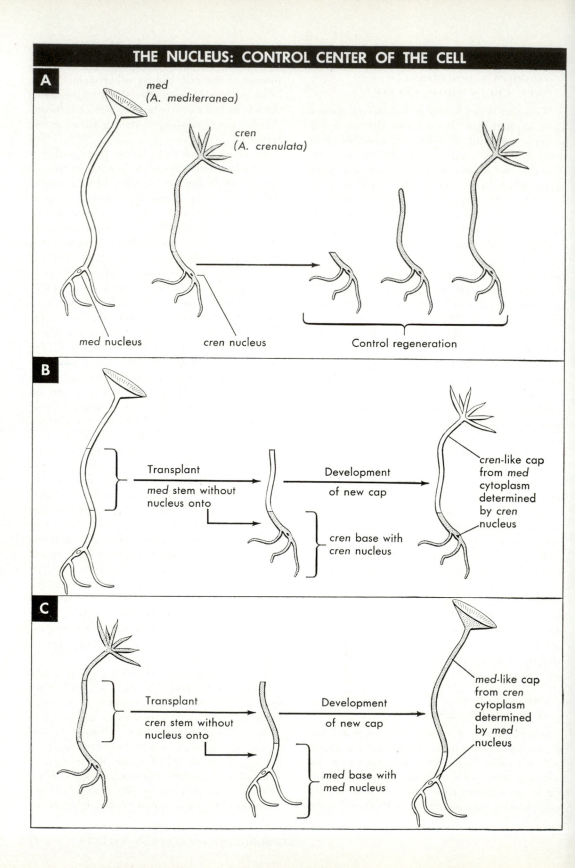

A

med
(A. mediterranea)

cren
(A. crenulata)

med nucleus

cren nucleus

Control regeneration

B

Transplant

med stem without
nucleus onto

Development
of new cap

cren base with
cren nucleus

cren-like cap
from med
cytoplasm
determined
by cren
nucleus

C

Transplant

cren stem without
nucleus onto

Development
of new cap

med base with
med nucleus

med-like cap
from cren
cytoplasm
determined
by med
nucleus

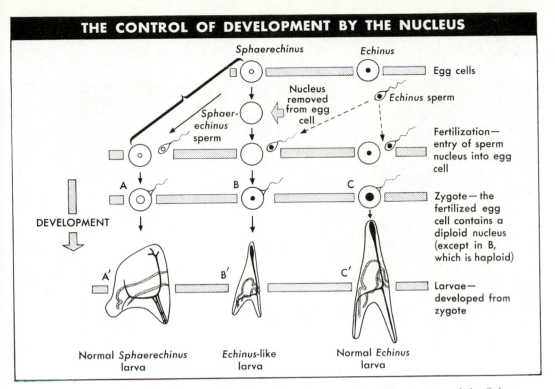

5–3   The control of development by the nucleus in sea urchins. The structure of the *Sphae-rechinus* larva is markedly different from that of the *Echinus* larva. Entry of the haploid *Echinus* sperm nucleus (shown in black) into the enucleated *Sphaerechinus* egg produces a larva that, in spite of its small size (because it is haploid), is clearly *Echinus* in its morphology.

less exhibits its controlling function. It carries into the *Sphaerechinus* form the specifications for development of the *Echinus* form, which the larva assumes.

The experiments with *Acetabularia* and the sea urchins indicate that we are probably on the right track in our search for the controlling information in the organism if we pursue our study of the nucleus. To do so, we must consider the mechanisms of cell reproduction.

MITOSIS

Cells multiply by dividing. This arithmetic oddity means that the number of cells increases by the splitting in two of single cells. This splitting of a cell is preceded by complex events involved in the division of the nucleus,

a process called *mitosis* [3] (Fig. 5–4, 5–5, and 5–6).

When it is not reproducing, the nucleus gives little evidence of its intimate structure. It lies in the cytoplasm, revealing only its nucleolus and a fine granular appearance. Outside the nucleus in animal cells (although not in most plants) lies the centrosome, or aster, which is a star-shaped body, and at the very center of the centrosome is the minute centriole. Details of nuclear structure become abundantly apparent, however, as soon as the cell begins to reproduce.

The onset of reproductive activity is marked by the division of the aster and its

[3] Strictly speaking, the division of the cell itself should be called *cytokinesis*, and the term "mitosis" should be reserved for the division of the nucleus alone. It follows that mitosis takes place only in eukaryotes (see Chapter 3). Bacteria and other prokaryotes lack nuclei and chromosomes and do not undergo mitosis.

5–2   The nucleus as control center of the cell. The plants being used in this experiment are species of the marine protist *Acetabularia.*

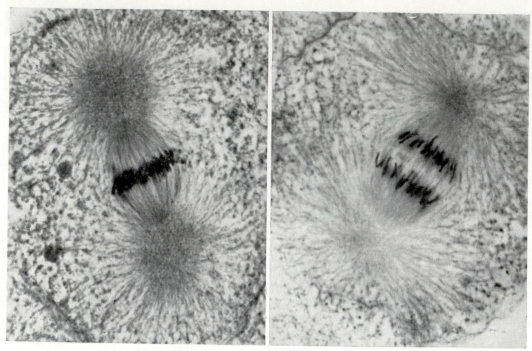

5-4 Mitosis in cells of the whitefish. The spindle lying between the two star-shaped centrosomes is very clear. The chromosomes lie on the central part, or equator, of the spindle. In the right-hand figure the two duplicate sets of chromosomes can be seen on their way to opposite ends of the spindle.

centriole into two parts and the separation of these toward opposite sides of the nucleus. Within the nuclear membrane, visible changes begin as elongate threads replace the previous fine granules. These threadlike structures are the chromosomes. As we shall see, they are the carriers of the cell's inherited controls, and, consequently, the details of their structure and behavior are what we want to understand.

The diagrams in Fig. 5–6 represent the successive events that occur during the mitotic division of a cell into two new daughter cells. The word "mitosis" derives from the Greek root *mitos* (thread) and refers to the threadlike nature of the chromosomes at mitosis. The nucleus in the figure belongs to a purely hypothetical organism in which the number of chromosomes is kept small for diagrammatic purposes. It is convenient to treat the long [4] and continuous process of mitosis by recognizing in it a sequence of more or less distinct

[4] The duration of mitosis varies in different cell types; it may last for minutes or for several hours.

stages or phases. The stages are *prophase*, *metaphase*, *anaphase*, and *telophase*.

Prophase

In prophase the chromosome threads are elongate to begin with and progressively shorten, becoming at the same time apparently thicker and more heavily stainable when treated with dyes in laboratory preparations. The shortened and thickened appearance is due to the fact that the elongate thread of the earliest prophase is thrown into a helix (Fig. 5–7), which becomes coated with a matrix of material possibly derived from the nucleolus. The nucleolus becomes progressively smaller during prophase.

Even from the earliest prophase, we can recognize two characteristics of the chromosomes that are fundamental and shared by almost all organisms: (1) They occur in pairs. In our hypothetical form there are two pairs (A^f and A^m, B^f and B^m). The members of each pair are similar and are said to be *ho-*

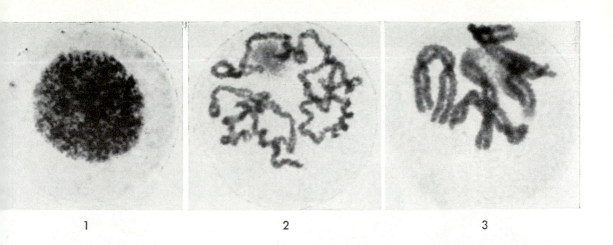

1   2   3

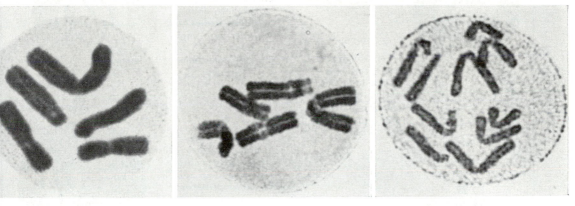

4   5   6

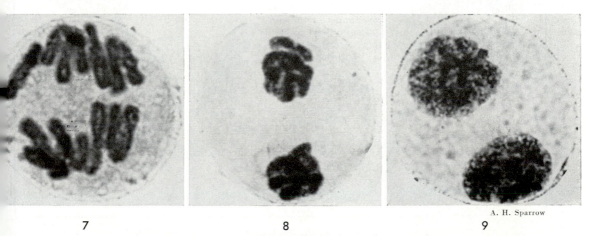

A. H. Sparrow

7   8   9

5–5 Mitosis in *Trillium*. The centrosome is missing in plant cells, and the spindle, which is so clear in animal cells (see Fig. 5–4), is not easily stained and seen, although it is present. In *1* the nucleus has not yet entered mitosis; the chromosomal material appears diffuse and structureless. In *2, 3,* and *4* the chromosomes are progressively clearer as they shorten and thicken. There are five chromosomes, and each is present in duplicate. In *5* they are lying on the equator of the spindle, which is unstained and therefore invisible in this preparation. In *6* and *7* the duplicate sets of chromosomes move to opposite ends of the spindle. In *8* and *9* new nuclei are formed from the two sets of chromosomes, and the process of mitosis is completed. A wall eventually develops between the nuclei, and two new cells are thus established.

MITOSIS

8. INTERPHASE
The mitotic cycle is completed; the two daughter cells have identical sets of chromosomes.

1. INTERPHASE
Nucleolus visible; chromosomes uncoiled and appearing as scattered chromatin.

2. PROPHASE
Chromosomes appear as distinct, double threadlike bodies held together by a single centromere (A^m, A^f, B^m, B^f); centrosome divides.

7. TELOPHASE
Separation complete; chromosomes elongate; new nuclear membranes form; cell divides into two daughter cells.

3. LATE PROPHASE
Chromosomes condense; nuclear membrane disintegrates; spindle develops between centrosomes.

6. LATE ANAPHASE
Sister centromeres continue their separation toward opposite poles of the spindle; cell begins to divide.

4. METAPHASE
Chromosomes short; nuclear membrane gone; centromeres orient on equator of spindle; centromeres still undivided.

5. ANAPHASE
Centromeres divide; daughter centromeres move apart, dragging daughter chromosomes with them.

5–6   The behavior of the nucleus during its mitotic cycle.

mologous. (2) Each chromosome is itself double-stranded. The two strands of each chromosome are held together by a small body called the *centromere*. Each chromosome has one centromere.

As the prophase of mitosis progresses, the two strands of each chromosome coil into helixes independently of each other; when prophase is complete, each chromosome has the appearance shown in Fig. 5–7.

Metaphase

Metaphase is the stage of mitosis that follows prophase. At the end of prophase the nuclear membrane disappears,[5] and in the space between the two centrioles there develops a remarkable structure called a *spindle* (see Fig. 5–4). The spindle consists of fibers that radiate from the two centrioles, producing a biconical structure. At metaphase the chromosomes migrate to the central or equatorial portion of this spindle. The arms of the chromosomes may lie loosely off the equator of the spindle itself, but the centromeres lie precisely on it. Some fibers of the spindle somehow become attached to the centromere of each chromosome. The centromeres then split in two in a definite plane (see Fig. 5–6), as the anaphase of mitosis commences.

Anaphase

Each of the two daughter centromeres of a chromosome is attached to a spindle fiber. The centromeres begin to move apart. As they move, they carry with them the daughter chromosomes. Thus at each of the two ends, or poles, of the spindle there accumulates a complete set of chromosomes: A^m, A^f, B^m, and B^f.

Telophase

Telophase, which follows anaphase, is in a way the reverse of prophase. A new nuclear membrane develops around the chromosomes, which uncoil and resume their original appear-

[5] The nuclear membrane, like that on the outside of the cell, owes its firmness and individuality to what is basically its gel-colloid nature. Its disappearance and later reappearance illustrate the great importance of the capacity of colloids to shift from gel to sol and back to gel. The cell can build and tear down boundary membranes where and when necessary and thus maintain order within its semifluid, semisolid state.

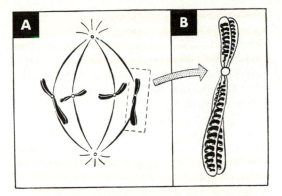

5–7 The helical structure of the metaphase chromosomes. *A*. The four chromosomes of Fig. 5–6 seen on the metaphase spindle. *B*. One of the chromosomes enlarged to show how its thread is thrown into a helix and embedded in a matrix.

ance as elongate, poorly stainable threads, and the nucleolus re-forms.

Interphase

Interphase is the stage during which the nucleus is not undergoing mitosis. During interphase the chromosomes remain virtually unstainable, although we know, from a series of observations too involved to describe here, that they retain their identities as distinct bodies.

Movements of the chromosomes

The highly organized movements of the chromosomes, first in their orientation on the spindle equator at metaphase and later in their separation and progression to the spindle poles, is one of the great wonders of biology. We still do not understand fully what causes these movements. Does the anaphase movement result from the spindle fibers' contracting, like muscles, so that they drag the daughter centromeres to the poles? Or does the middle region of the spindle expand and push the centromeres toward the poles? Perhaps the movement of the chromosomes is in part autonomous; that is, the forces responsible for their movement may be internal, so that they *go* to the poles rather than being *forced* to the poles.

The problem of explaining the chromosomes' movements during mitosis has attracted and baffled many biologists. The most widely held view at present is that the spindle fibers do exert a pulling effect on the centromeres to which they are attached, so that the chromosomes are drawn to the poles rather than moving there independently. One thing is certain. Both the centromere and the spindle fiber are essential for the movement. Chromosomes without centromeres do not proceed poleward; and chromosomes that have centromeres but fail to get onto the spindle and attach to fibers also fail to reach the poles. The centromere and the spindle fiber are therefore indispensable for orderly chromosome movement, and in some way responsible for it.

In animals an additional element is necessary for this movement—the centriole. Oddly enough, centrioles are not known to exist in higher plants, and so they are not universally necessary. Where centrioles do occur, it is not surprising to find them involved in the movements of chromosomes. In flagellated single cells like some protists and sperm cells, the flagellum, which is the organelle [6] of movement, is organized and controlled by the centriole. Nor are we surprised by another related point: the centromere and centriole are evidently basically the same structure. There is a group of snails in which the centromeres seem to be poorly anchored to the chromosomes, sometimes getting loose from them. The chromosomes losing their centromeres never get onto the spindle; anl the free centromeres aggregate around the centrioles. Later each centromere reveals its essential identity with a centriole by, like it, organizing a flagellum. The sperms from these atypical snails come, in this way, to have not one but several sperm tails; and the number of extra tails corresponds, as might be expected, with the number of chromosomes that are lost from the spindle —stranded without their organelles of movement.

These relations of centriole and centromere are the kind that we, as evolutionists, expect to find in living organisms. As we shall see later in greater detail (Chapter 17), living systems evolve in an opportunistic manner, using what is already available and suitable for the solution of new problems as they arise. Clearly the

[6] The word "organelle" is used to denote a functionally distinct part of a cell, just as the word "organ" is used to denote a functionally distinct part of a whole organism.

cell in its long evolutionary history has made use of essentially one and the same body, centromere-centriole, for quite distinct tasks whose only common denominator is involvement in organized movement. The properties of this kinetic center (or movement organizer) must surely have evolved first in response to one of the several functions it now serves, but which one we do not know. At a later time its presence and properties were put to use in solving other problems.

Ordered separation of duplicates

We digressed briefly to the subject of the chromosome movements not only because it is one of the main biological problems studied today but for the more general reason that the organized and controlled movements of the chromosomes at mitosis merit emphasis as a focal point in organic reproduction in general.

If we survey mitosis throughout the plant and animal kingdoms, we can discover fascinating differences in detail, but we study these largely because they are so rare. It is the other side of the picture that we want to stress: mitosis is virtually coextensive with life, and its major features are amazingly constant. The universality and constancy of mitosis bespeak something fundamental, and what is fundamental is clear enough in the light of our earlier discussion of heredity as information. The mitotic mechanism is the basic mechanism of hereditary transmission. As a result of mitosis, two cells are developed from one, and to each of the daughter cells is transmitted—by virtue of the orderly movements of the chromosomes on the spindle—a copy of all the chromosomes.

Duplication

At the beginning of each prophase, each chromosome is already duplicated, and the duplicates are tied together by a single centromere. Mitosis involves only the orderly separation of the two duplicate sets and their transmission to different cells.

When and how does the actual duplication take place? At anaphase and telophase each chromosome consists of only one of the two strands seen at the previous prophase. It follows that the single strand is duplicated some-

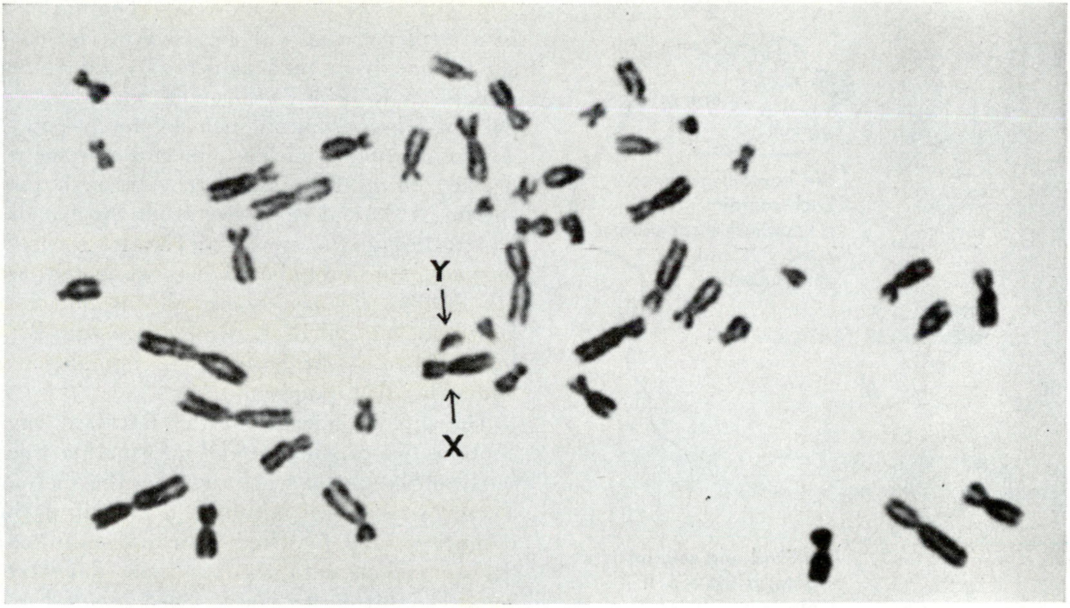

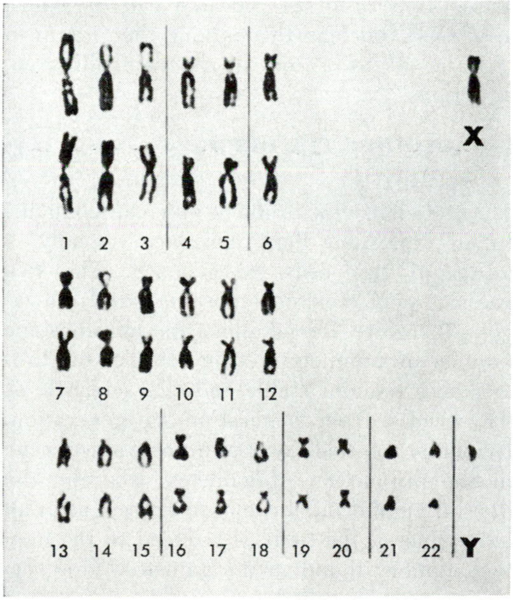

J. H. Tjio

5–8 Chromosomes from a human male cell and their karyotype. Cells to be studied are cultivated briefly in a tissue culture. Then a drug (for example, colchicine) is added to stop mitosis in metaphase. The cells are then literally squashed. This procedure disperses the chromosomes so that they can be photographed. Such a presentation reveals the general morphology of the chromosomes, or their karyotype.

time between telophase, when it is single, and the following prophase, when it is double. Exactly *when* during the interphase duplication, or replication, occurs and *how* it is effected we do not know. However, one point must be emphasized. Sometimes biologists speak of a chromosome's splitting in two, but this is at best an unfortunate phrasing. If the duplex nature of the prophase chromosome resulted only from a longitudinal split of the earlier telophase chromosome, it could be repeated only so often—a cake can be cut into smaller and smaller pieces, but eventually none is left! Obviously the duplication of a chromosome is a real reproduction of a copy of the already existing one, and not simply a splitting in two of the existing one. In the next chapter we shall return to the problem of how the chromosome is reproduced.

DIPLOIDY: ITS ORIGIN
IN FERTILIZATION

One important point that demands explanation is that the chromosomes in the nucleus occur in pairs. Biologists usually represent the number of pairs as *n*. In our simplified hypothetical organism in Fig. 5–6, there are 2 pairs (A^m and A^f, B^m and B^f); therefore, $n = 2$. In man there are 23 pairs (46 chromosomes) (Fig. 5–8). In some relatives of the lobster,

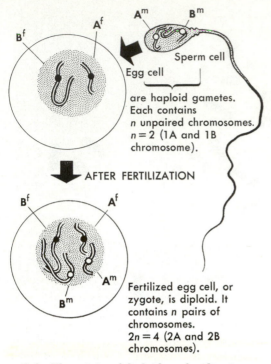

are haploid gametes.
Each contains
n unpaired chromosomes.
n = 2 (1A and 1B
chromosome).

AFTER FERTILIZATION

Fertilized egg cell, or
zygote, is diploid. It
contains *n* pairs of
chromosomes.
2*n* = 4 (2A and 2B
chromosomes).

5–9   The origin of diploidy at fertilization.

there are 100 pairs. In the fruit fly *Drosophila*, to which we give much attention in the next chapter, there are 4 pairs. In some ferns there are over 500 pairs, the highest numbers known in any organisms, but even at this high level, the normal number is fixed for any given species. What is the significance of this regularity—the fact that the nucleus of each species contains a definite number, *n*, of chromosome pairs?

The answer is found in the nature of sexual reproduction. The sexual act leads to the union of two cells, one contributed by each parent. These cells (sperm and egg), called *gametes*, prove to be special with respect to their chromosome contents. Each contains half the number of chromosomes present in the nucleus of the zygote and in the nuclei of body cells in adult multicellular organisms. Thus in the organism of Fig. 5–6, with four chromosomes (two A's and two B's), each gamete contains only two (one A and one B); the union of the egg and sperm is followed by a pooling of their chromosomes so that the fertilized egg contains four (Fig. 5–9).

The nucleus of a gamete is said to be *hap-*

loid, or to have *n* chromosomes. The nucleus of a fertilized egg, with two sets of chromosomes, one from each gamete, is said to be *diploid,* or to have 2*n* chromosomes. In each pair of chromosomes (such as the A pair) within the diploid nucleus, one chromosome is derived from the male parent through the sperm (A^m), and the other from the female parent through the egg (A^f). The two A-chromosomes are said to be a homologous pair, as are the two B-chromosomes. A^m is homologous with A^f, and B^m is homologous with B^f; but neither A-chromosome is homologous with either B-chromosome.

The diploid nucleus of the fertilized egg contains two complete sets of information, one set from the egg and one set from the sperm. These two sets are faithfully copied in each interphase, and the duplicates are separated during mitosis into the two new daughter cells. Thus, from the single-celled zygote produced by fertilization, the multicellular adult arises as a result of repeated cell divisions, and every nucleus throughout the organism contains its own copy of the controlling instructions.

HAPLOIDY: ITS ORIGIN IN MEIOSIS

An obvious new problem now confronts us. In all organisms that reproduce sexually, a union of two cells takes place. The two gametes pool their chromosomes, and the zygote therefore has double the chromosome number of a gamete. Yet the number of chromosomes remains stable and characteristic of the species from generation to generation. Evidently a special form of mitosis must occur in the production of gametes, whereby the diploid number of chromosomes present in all other cells of the body is reduced to the haploid number found in the gametes. This special mitosis is called *meiosis.*

Where meiosis takes place

Let us suppose that the hypothetical organism of Fig. 5–6 is some kind of fly whose adult form is outlined in Fig. 5–10. This figure shows that, after the fusion of haploid gametes, the diploid zygote undergoes successive cell divisions during which the duplicated chromosomes are transmitted faithfully by

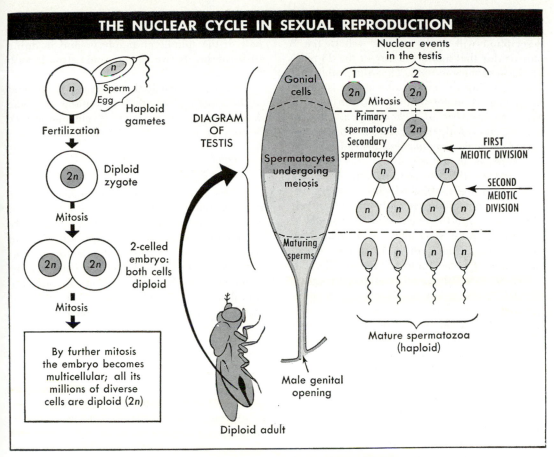

THE NUCLEAR CYCLE IN SEXUAL REPRODUCTION

5–10 The nuclear cycle in sexual reproduction.

mitosis to all new cells. All the tissues of the adult fly are composed of diploid cells, including those of the reproductive organ. Let us suppose further that this individual fly is male; the reproductive organ is therefore a testis. The sperms produced by this testis are, we know, haploid, and so meiosis—the special nuclear division yielding haploid cells—must occur in cells in the testis.

The cellular organization of the testis is represented in highly schematized fashion in Fig. 5–10. At the head of the testis are *spermatogonia* (or *gonial* cells), which divide mitotically. One of the daughter cells (labeled 1) produced by a spermatogonial mitosis remains as a gonial cell, and the other (labeled 2) becomes a *spermatocyte*. A spermatocyte is a cell in which meiosis takes place. The entire meiotic process comprises two cell divisions, conveniently designated as meiosis I and mei-

osis II. The cell in which meiosis I takes place is diploid, being derived by mitosis from a gonial cell; it is the *primary* spermatocyte. Meiosis I produces two *secondary* spermatocytes, which then undergo meiosis II, thus yielding four cells, all of which are haploid and are gametes.

Here our task is to investigate the special chromosome movements in meiosis that are responsible for the transition from the diploid condition of the primary spermatocyte to the haploid condition of the four sperms derived from each primary spermatocyte.

Pairing, or synapsis, at prophase I

The complexities of meiosis are best understood if we focus our attention on the behavior of the centromeres and confine our discussion initially to one of the pairs of homologous centromeres with their attached chromosomes. We shall follow the A pair (A^m and A^f) in our hypothetical fly, although our description will apply equally well to the B pair or, for that matter, to any of the 23 pairs of chromosomes in man. Thus the problem in understanding the events of meiosis is essentially one of understanding how in meiosis *only one member of a homologous pair of centromeres* (with its chromosome) *is transmitted to each new nucleus* instead of both members, as in mitosis.

In mitosis the two homologous A centromeres with their attached chromosomes behave absolutely independently of each other. They move onto the equator of the spindle at metaphase quite separately, and then each splits in two so that both an A^m and A^f centromere move to each pole of the spindle. The behavior of the centromeres during the first meiotic division is different in two respects. First, the homologous centromeres (A^m and A^f) do not behave independently of each other; and second, they do not split as they do in mitosis but move instead one to each pole. We shall examine these differences now in more detail.

The specialized nature of the first meiotic division (Fig. 5–11) is indicated from its very beginning by the fact that the individual chromosomes are still single-stranded, as they were at the previous telophase. The duplication of the single strands, which normally occurs during interphase, has been delayed. (You will recall that in mitosis the prophase chromosomes are already duplicated.) These single-stranded chromosomes now begin to pair up; A^m pairs with A^f, and B^m with B^f. Strictly corresponding, or homologous, regions of the chromosome pairs are brought next to each other. Only *after* this pairing, or *synapsis*, do the individual chromosomes undergo the normal process of duplication. As a result, the chromosome pair becomes four-stranded and includes two centromeres.

Separation, or segregation, at anaphase I

The pairing persists throughout prophase. When the spindle is formed, the chromosomes move onto its equator at metaphase, still in their paired condition. The homologous centromeres (A^m, A^f, etc.) do *not* now split. Instead, one intact centromere moves to each pole of the spindle at anaphase, and in doing so it carries its two chromosome strands with it.

Thus each nucleus that re-forms at the end of the spindle at telophase I contains only one A centromere and only one B centromere (see bottom of Fig. 5–11).

Second meiotic division (meiosis II)

In the interphase following meiosis I, the nucleus contains only one of each kind (A, B, etc.) of centromere. Each centromere already carries two chromosome strands with it, and accordingly no further duplication takes place. When the second division (meiosis II) commences, each centromere moves onto the spindle at metaphase and splits, so that one of the two strands goes to each pole.

Arithmetic of meiosis

The four strands present in the paired chromosome at metaphase I finish up separated, one in each of four gametes. In the whole process of meiosis, there are two chromosome separations on a spindle—one at the first division and another at the second division. But chromosome duplication precedes only one of these divisions, the first; and the centromere divides only once, at the second division. The chromosome separation at the first division is based not on the *division* of individual centromeres but on the *separation* of paired homologous centromeres.

5–11 Mitosis and meiosis compared. The top row compares mitosis and meiosis from the previous interphase through metaphase. Homologous chromosomes, like A^f and A^m, are single-stranded at the beginning of the meiotic prophase (cf. mitosis) and pair up before duplication. Note how A^m and A^f, for example, are paired at meiotic metaphase, whereas they are separate entities at mitotic metaphase. (Space does not permit representation of the true proportions of the B-chromosomes on the mitotic spindle.) The two bottom rows continue the comparison of mitosis and meiosis from metaphase through two divisions. Chromosome duplication (suppressed in the interphase *between* the first and second meiotic divisions) occurs normally during interphase *after* the second meiotic division.

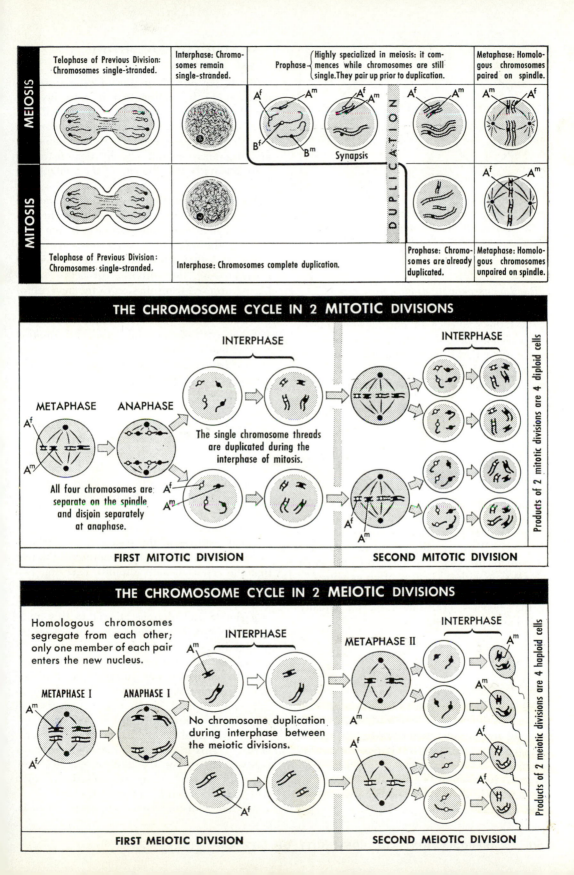

MEIOSIS	Telophase of Previous Division: Chromosomes single-stranded.	Interphase: Chromosomes remain single-stranded.	Prophase: Highly specialized in meiosis: it commences while chromosomes are still single. They pair up prior to duplication.		Metaphase: Homologous chromosomes paired on spindle.

Synapsis

MITOSIS	

Telophase of Previous Division: Chromosomes single-stranded.

Interphase: Chromosomes complete duplication.

Prophase: Chromosomes are already duplicated.

Metaphase: Homologous chromosomes unpaired on spindle.

THE CHROMOSOME CYCLE IN 2 MITOTIC DIVISIONS

INTERPHASE

INTERPHASE

METAPHASE ANAPHASE

The single chromosome threads are duplicated during the interphase of mitosis.

A^f

A^m

All four chromosomes are separate on the spindle and disjoin separately at anaphase.

A^f

A^m

A^f

A^m

Products of 2 mitotic divisions are 4 diploid cells

FIRST MITOTIC DIVISION

SECOND MITOTIC DIVISION

THE CHROMOSOME CYCLE IN 2 MEIOTIC DIVISIONS

Homologous chromosomes segregate from each other; only one member of each pair enters the new nucleus.

INTERPHASE

METAPHASE II

INTERPHASE

A^m

A^m

METAPHASE I ANAPHASE I

A^m

No chromosome duplication during interphase between the meiotic divisions.

A^m

A^f

A^m

A^f

A^f

A^f

A^f

Products of 2 meiotic divisions are 4 haploid cells

FIRST MEIOTIC DIVISION

SECOND MEIOTIC DIVISION

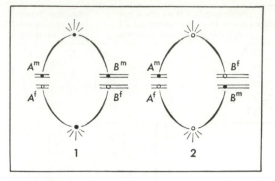

5–12 Independent chromosome assortment.
The orientations of the chromosomes shown in *1* and *2* are equally probable at metaphase I.

Random assortment of chromosomes

In our account of meiosis so far, we have stressed the sequence of chromosome movements responsible for separating the members of a single pair of homologous chromosomes. An additional item of great importance is necessary to make the picture complete. Fig. 5–12 (*1*) shows that A^f, the A-chromosome derived originally from the mother of the organism we are studying, goes to one pole of the spindle in meiosis I and A^m to the other. The figure also shows B^f going to the same pole as A^f. This is not the only possible way that the chromosomes could behave. The orientation of the B pair on the spindle is quite independent of that of the A pair. The metaphase arrangements shown on the spindles in Fig. 5–12(*2*) are equally frequent, since they are equally probable. B^m may go to the same spindle pole with A^f just as often as with A^m. The following combinations of chromosomes in the gametes are, therefore, all equally probable and frequent.

$$A^m \text{ and } B^m$$
$$A^f \text{ and } B^f$$
$$A^m \text{ and } B^f$$
$$A^f \text{ and } B^m$$

All that the mechanism of meiosis guarantees is that the gametes produced by an organism will contain one member of each pair of homologous chromosomes. It does not guarantee that the two groups of maternal and paternal representatives remain together in the gametes in the combinations in which they were inherited.

SIGNIFICANCE OF MITOSIS AND MEIOSIS

We began our discussion of cellular reproduction (1) by noting that it must involve, somehow, the transmission, from one cell generation to the next, of the information controlling the maintenance of the cell's complexity and organization and (2) by demonstrating experimentally that the site of the control was the nucleus.

We have found that the nucleus contains a number of elongate threads, the chromosomes. The chromosomes prove to be the vehicles of the control mechanisms. We have found also that when the cell reproduces, it provides—in the mitotic process—for a highly ordered transmission to the new cell of the information (the chromosomes) it requires.

When mitosis begins, the information needed has already been copied. The two copies are held together by the centromere. The centromere does not divide until it is properly oriented on the spindle. The spindle's fibers guide the duplicates of each chromosome to two opposite poles. These poles are foci for the gathering together of all necessary information into two strictly equivalent packets promptly enclosed in new nuclear membranes at telophase. The significance of the spindle is apparent: it is a device for the orderly separation of the two copies, or blueprints, of the chromosomes.

Other features of the system need clarification, however. What is the meaning of the fact that the cell is usually diploid, carrying two basic and equivalent sets of information, one of which seems, in a sense, redundant? There is a simple and obvious answer to this question: diploidy is a consequence of sexual reproduction, since it results from the fusion of representative nuclei from two parents. Each parent contributes one complete copy of the total information necessary. To understand fully why diploidy is so nearly universal in organisms, we must answer a second question: "Why is sexual reproduction so nearly universal?" This is a question that we are not yet ready to treat, for the answer emerges only from an understanding of evolution, which will be taken up in Chapters 15 through 17.

Similar considerations apply to meiosis.

There are simple features of meiosis that are easily explained in terms of sexual reproduction. Since this does involve nuclear fusion in each generation, the meiotic process must occur if the number of chromosomes is to be held constant. Again, however, we must emphasize that, like diploidy, meiosis can be understood in all its detail only in relation to evolution, and so we must return to it later.

CHAPTER SUMMARY

Continuity of life dependent on the universal ability of organisms to reproduce themselves.

Major features of reproduction illustrated by reference to the familiar case of man: sexuality; embryonic development—preformationism versus epigenesis (the creation of adult complexity); biogenesis; and heredity.

Biogenesis: experimental evidence against spontaneous generation in the modern world; theoretical argument against modern spontaneous generation.

Complexity and information: heredity as information for the control of development.

What the study of reproduction must reveal: the nature of the information that controls the development of the organism; how the information is reproduced; how copies of it are transmitted from generation to generation; how the information acts to achieve control.

Cellular basis of reproduction: "all cells from cells"; the nucleus as the control center.

Mitosis: nuclear division; chromosomes at mitosis; the sequence of stages—prophase, metaphase, anaphase, telophase, interphase; chromosome movements; duplication of chromosomes; ordered separation of duplicates.

Diploidy: its origin in fertilization.

Haploidy: its origin in meiosis; where meiosis takes place; chromosome pairing at prophase I; separation at anaphase I; the second meiotic division; the arithmetic of meiosis; the random assortment of chromosomes; the significance of mitosis and meiosis.

6

The Chromosome
Theory of Heredity

Fruit fly after Sturtevant and Beadle; chromosomes after Dobzhansky; both from A. M. Srb and R. D. Owen, *General Genetics*, Freeman, San Francisco, 1952

The development of the chromosome theory of heredity is one of the great achievements of twentieth-century science. The little fruit fly Drosophila *(2 mm.), illustrated here with its eight chromosomes, has played a major role in the growth of genetics as a principal object of study.*

"We hold these truths to be self-evident: that all men are created equal. . . ." All men are equal before the law, and all are equal in dignity as human beings: that is what the writers of the Declaration of Independence meant. It is, however, one of the profound lessons of genetics that only identical twins are born equal biologically. Unless you are an identical twin (and only about 0.4 per cent of births yield identical twins), you are not equal to anyone else on earth in the sense of being biologically the same. No judgment is involved here as to who is better and who is worse, or as to whether "better" and "worse" have valid meanings in this connection. Nevertheless, the fact remains that the mechanisms of heredity make it virtually impossible that any two nontwins ever have inherited just the same genes.

Centuries of human striving—political, moral, and physical—have gone into establishing the truth of the political equality embodied in the Declaration of Independence. There is an equally long history of human endeavor behind the establishment of the truth of the biological dissimilarity and inequality of men, even though the deliberately straightforward account of the physical basis of heredity that we developed in the last chapter may have left some impression that the problem was simple.

In the last chapter we showed that heredity involves the transmission of controlling information, not only from generation to generation but also from cell to cell. We showed that the controls are in the nucleus. Their physical basis is in linear "tapes" of information (the chromosomes) that are accurately copied and transmitted to new cells in orderly fashion like so many blueprints for the government of the cells' activities. The logical sequence we followed was dictated by our present understanding of the problem. We wish to emphasize now that the logical sequence bears little or no relation to the historical sequence of discovery and development of understanding. Cell division and the details of mitosis were studied well before there was rigorous proof that the nucleus was the cell's control center. Indeed, the facts of cell and nuclear division were being studied at the same time that many able students were still arguing for a kind of spontaneous generation called *free cell formation*.

Whenever a great advance is made in science, it is in the form of a theory—a scheme of explanation in terms of which all the facts, previously scattered and confusing, seem clearly and simply to fall into a pattern. T. H. Huxley is said to have remarked after reading Charles Darwin's great book *The Origin of Species*, "Why didn't I think of it?" Until a theory is found, the meaning of the facts is obscure.

The historical development of a theory nearly always involves several distinct lines of investigation. In the development of the chromosome theory of heredity, there were two distinct lines that ultimately fused. One of these we have looked at already, although not in a historical way: the study of the cell's visible structures and their behavior in cell reproduction. The other line of inquiry was the study of differences and similarities between parents and offspring, with a view to defining general rules of inheritance. This direct study of the regularities of heredity is very much older than the study of the cell. Democritus and Aristotle, among other ancients, had discussed the problems of heredity, and eighteenth- and nineteenth-century scientists were much preoccupied with the same problems before the cell theory gave biology its firm start about 1840. We shall shortly see that the two lines of study yielded conclusions that supplemented and required each other. The theory that began to emerge from their fusion in 1902 was like the opening of a floodgate. It produced a flow of biological investigation and new insight that is still at its height over 60 years later. This, as we said in Chapter 1, is characteristic of the best theories; they not only explain old facts but also point ways to new knowledge.

Pre-Mendelian Ideas on Heredity

We may well ask, "If a theory, once discovered, is the main guide to inquiry and research after it is discovered, what guided inquiry before?" There are really two answers to this question.

First, much of the earliest inquiry was in effect random, helter-skelter, and unguided. Consequently, it produced relatively little except oddities such as notions about strange hybrids issuing from the mating of camels and leopards or myths concerning the lingering effect of a woman's first husband on her children by a later husband. Other erroneous ideas more important to the history of the subject will be treated later.

However, there is a second answer to our question. Men are really never at a loss for some kind of scheme in terms of which they can talk about their problems. The natural human tendency to seek order in the world leads them to find analogies or models of some kind that offer frameworks for discussing things. There is no doubt that the search for analogies or models for comparison is a basic tool of human thought and the pursuit of understanding. Somehow satisfaction is derived from comparisons that reveal similarities. The search for analogies is essentially an attempt to find something familiar, something already known, a model for the explanation of the new and unfamiliar; and this search goes on continuously, sometimes consciously and often subconsciously.

It is not surprising that the most common source of analogies or models to which men have turned in the absence of exact theories is the realm best known to them: human nature and human society. Before the scientific development of the modern Western world, the universal tendency was for men to try to explain nature by talking about it in the familiar terms of human attributes; the forces of nature took on human will and motivations. The conscious effort to avoid such analogies and to depersonalize the explanation of the nonhuman world is a real hallmark of the Western scientific movement. But the influence of human models has nevertheless remained, often subconsciously. For instance, the cytologist Virchow used a model from human experience when he sought to interpret how discrete cellular units work in subservience to the welfare of the whole organism; he called the organism a *cell-state*.

BLENDING OF INHERITANCE

It used to be believed that blood was particularly involved in inheritance. Even those of us who know better still speak of "blood lines," "bad blood," "blue-blooded," and the like. It was supposed that the blood of our ancestors mingled and was finally poured into us. According to this concept, your inheritance is a half-and-half blend of blood from your father and mother, and hence a quarter each from your grandparents, an eighth each from your great-grandparents, and so on.

This erroneous idea about heredity springs from two sources. First, there is the notion that since blood is fundamental to life, it must be what we inherit. Second, there is the intuitive but erroneous notion that all our ancestors must necessarily contribute their due share to our make-up. If we have a great man somewhere in our lineage, it is nice to consider precisely how much he has contributed to our inheritance. The nearly universal tendency in human cultures to revere ancestors must have added somewhat to the plausibility of a blending of inheritance on the one hand —and on the other hand have received some justification from it.

We learned in the last chapter that it is the information carried in chromosomes that we inherit rather than blood; and the briefest consideration of the rules of chromosome inheritance reveals a fact that must be disquieting to people unduly concerned with their ancestry. Families proud of their descent from some famous Revolutionary or Pilgrim ancestor have a good chance of lacking any chromosomes at all from that famous ancestor.

The familiar idea of blending—of mixing and getting intermediates as we do with paints—must have been at the root of those other quaint and amusing speculations on the origins of strange hybrids. A giraffe was supposed to have issued from the mating of a camel and a leopard—blending the leopard's spots and the camel's long neck. Camels were indeed favorite and versatile hybridizers. One authority stated that an ostrich was a cross between a camel and a sparrow—a curious blend to say the least. Arabian scholars thought that sea cows were crosses between humans and fishes, and Greek mythology is full of half-human, half-animal hybrids. Many a visitor to the zoo still explains the queer animals as crosses between the most diverse parents.

Of course, only animals of the same or very closely related species can cross and produce offspring. Even when two closely related species cross, the offspring produced are usually not fertile; witness the mule. It is entirely impossible for animals as distinct as man and ape, cat and dog, or horse and cow (let alone camel and sparrow) to engender offspring.

INHERITANCE OF ACQUIRED CHARACTERISTICS

The persistence of other errors about heredity is as nothing compared with the persistence of the belief in inheritance of acquired characteristics. A man who exercised and developed large muscles would, of course, pass on his muscular development to his children. An animal that stretched its neck reaching for leaves would have offspring with longer necks than if he had been content to browse near the ground. Hence, in time, a giraffe would develop.

At one time this theory was especially in vogue in relation to problems of evolution. Although it has gone under the name of *Lamarckism*, after Jean Baptiste de Lamarck, the French evolutionist of the early 1800's, the idea is at least 2000 years older than Lamarck. It was discussed by many Greek scholars, and its origin was independent from evolutionary thought. It doubtless arose because it has a common-sense plausibility about it, but for a notion to have common-sense plausibility, it must be familiar in some respect. A familiar analogy to the inheritance of acquired characteristics is not hard to find. The everyday experience of man has acquainted him with the inheritance of acquired characteristics in the social and legal realm. A man inherits his father's estate—what his father acquired by dint of work, good fortune, and, in his turn, inheritance. One generation inherits all the cultural advances and setbacks acquired by the last. Such social, legal, or cultural Lamarckism was doubtless the model initially predisposing human minds to such a view of biological heredity. The word "heredity" itself, which now has a strictly biological meaning, is a derivative of "inheritance," which refers to the social and legal phenomenon. Nothing illustrates better than this example the dangers of the use of analogies.

The catch in the whole situation is that the mechanism of social or legal heredity is utterly different from the mechanism of biological heredity. The mechanism of biological heredity renders the inheritance of acquired characteristics impossible. The idea is historically important only because it was part and parcel of the first full and consistent theory of evolution—Lamarck's; and it is politically important only because the Communist Party has professed it.[1]

PANGENESIS

Pangenesis, like the idea of the inheritance of acquired characteristics, which demands it, was in vogue in the late nineteenth century before the development of the chromosome theory of heredity. Again, the Greeks, Democritus in particular, had discussed pangenesis in only slightly cruder form 2000 years earlier.

If a blacksmith's enlarged muscles, acquired by virtue of his work, are inherited by his son, there must be some mechanism whereby the condition of his muscles can be represented in what heredity transmits to his son. Democritus spoke of representative particles, *pangens*, coming from all parts of the bodily organization and entering the semen introduced into the female in copulation.[2]

In his later years Darwin resorted to a theory of the inheritance of acquired characteristics and to a revived form of Democritus's pangenesis. He spoke of *gemmules*, representative particles again, entering the germinal material.

It has been fashionable in learned circles to heap endless blame on Aristotle for the confusion and fallacy in medieval biology. Certainly much of later myth had its origin in his writings. But in a curious and somewhat ironi-

cal way, it is also true that Aristotle had some remarkable biological insights, and among his other real achievements must be counted the rigorous demolition of the Democritean theory of pangenesis, as well as its Darwinian form, proposed some 2000 years later. How, asks Aristotle, can we believe that the hereditary material consists of representative particles derived from all over the body when a man who has lost an arm nevertheless has a child with the usual complement of two? Where did the pangens for the missing arm come from? His argument and his illustrations are more extensive, but this one alone is final—for Democritus, for Darwin, and for all other versions of the same line of human thought.

The Aristotelian conclusion about heredity has an astonishingly modern ring to it. Aristotle concluded that what was inherited was a *potentiality to develop.*

WEISMANN'S ONE-WAY RELATIONSHIP BETWEEN GERM CELLS AND SOMA

The advent of the cell theory in the mid-nineteenth century brought with it the seeds [3] of many advances, including especially a clarification of ideas about heredity. We saw some detailed fruits of the cell theory in the last chapter. Once it became clear that all organisms were derived from single cells, it became difficult to subscribe to pangenesis and associated ideas.

We may diagram the relations between the germ cells and the body cells as in Fig. 6–1. All the diverse body cells (muscle, nerve, bone, etc.) are descendants of the single zygote cell, as the arrows indicate. From which type of cell is the egg or sperm derived? It cannot be a descendant of all these differentiated types; except for the zygotes produced by fertilization, every cell is a descendant of a single cell. August Weismann provided the solution to this problem by pointing out that the germ cells of each generation descend directly through a lineage of unspecialized cells from the germ cells of the previous genera-

[1] In 1948 the Party decided that all good Communists must teach and say that acquired characteristics are inherited. More recently this political dogma has stumbled against some cold biological facts, but it has still not been wholly abandoned in the Soviet Union.

[2] The length to which a speculative theorist may be driven in his search for facts supporting his theory—rather than facts critically testing it—is shown by one of the arguments Democritus offered in support of pangenesis. The intensity of orgasm, he claimed, was explained by the simultaneous rush of pangens into the semen from all over the body.

[3] Notice again how models or analogies from the realm of practical everyday human experiences affect our references to other things. We use such a model when we say that further scientific advances grew from "seeds" in the cell theory.

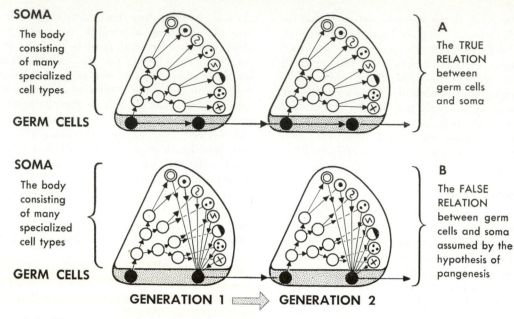

SOMA
The body consisting of many specialized cell types

GERM CELLS

A
The TRUE RELATION between germ cells and soma

SOMA
The body consisting of many specialized cell types

GERM CELLS

B
The FALSE RELATION between germ cells and soma assumed by the hypothesis of pangenesis

GENERATION 1 ⟹ GENERATION 2

6–1 The one-way relation between germ cells and the differentiated cells of the soma.

tion. That is, the specialized body cells of each generation are related to germ cells in a one-way fashion: they are derived from germ cells but do not give rise to them. This insight dealt a death blow to pangenesis in all its forms. The very essence of pangenesis is the assumption that the hereditary material transmitted from one generation to the next is derived from and represents—like members of a parliament from various districts—the various specialized regions of the "cell-state" that is the organism. As Fig. 6–1 indicates, such a two-way relationship between *soma* (body) and *germinal* (or hereditary) cells does not, in fact, exist.

The great clarification that Weismann's insight brings is due again to the study of the cell itself. We shall now turn to the study of heredity, of resemblances between parent and offspring, the other main line of investigation leading to the chromosome theory.

Mendel's Principles of Heredity

Gregor Mendel (1822–1884) was a monk in the Augustinian monastery of Brünn, Austria (now Brno, Czechoslovakia). He taught natural science in the monastery school and thus

became interested in the problems of heredity. He devised ingenious and careful experimental techniques, and, by crossing different strains of peas, he discovered the fundamental principles of *genetics*, the science of heredity.

Mendel's results were published in 1866, but they were long neglected by other students. Mendel himself did not follow up his discoveries, and their importance was not recognized by other biologists for 35 years. Finally, about 1900, three other experimenters independently rediscovered the Mendelian principles: Karl Correns in Germany, Hugo De Vries in the Netherlands, and Erich von Tschermak in Austria. This situation is a striking example of the fact that a theory, even though it is correct, may not be accepted and bear fruit until the general progress of science creates an atmosphere receptive to it.

MENDEL'S FIRST EXPERIMENTAL RESULTS: A SINGLE CHARACTER DIFFERENCE

Most of Mendel's work was done with a sweet pea (*Lathyrus*). Among other advantages, this pea offers an abundance of variant types that can be crossed or hybridized. Many garden varieties differ in a clean-cut,

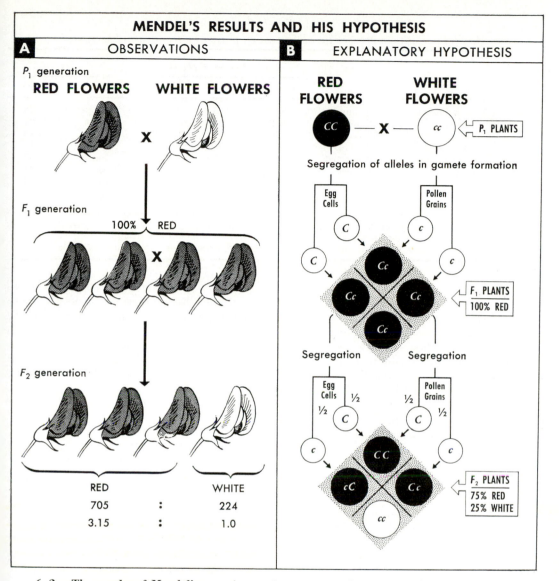

MENDEL'S RESULTS AND HIS HYPOTHESIS

A   OBSERVATIONS   **B**   EXPLANATORY HYPOTHESIS

P_1 generation
RED FLOWERS   **WHITE FLOWERS**

X

F_1 generation
100%   RED

X

F_2 generation

RED		WHITE
705	:	224
3.15	:	1.0

RED FLOWERS   **WHITE FLOWERS**

CC — X — *cc* ⟵ P_1 PLANTS

Segregation of alleles in gamete formation

Egg Cells   Pollen Grains

C   *c*

C   *c*

Cc   *Cc*   *Cc*   *Cc* ⟵ F_1 PLANTS 100% RED

Segregation   Segregation

Egg Cells ½   ½ Pollen Grains

C   *C* ½

c   *c*

CC   *cC*   *Cc*   *cc* ⟵ F_2 PLANTS 75% RED 25% WHITE

6–2   **The results of Mendel's experiments in crossing red- and white-flowered peas and his interpretation of the results.** P_1 refers to the parental generation with which an experiment is started; F_1 refers to the offspring of that generation, or the first filial generation; F_2 refers to the next, or second, filial generation; and so on.

either—or way. Some have colored flowers, others white; some have yellow seeds, others green; some are tall, others dwarf; some have flowers clustered at the apex of the stem, and others spread out. The most valuable feature is that these differences in character are clean-cut. Mendel seems to have been aware of this advantage; he sensed that he was correctly attempting to study and analyze the simplest

possible kind of heredity. This was a major reason for the success of his work. The complexity involved in the inheritance of most commonly studied human characters had been a block to their successful analysis. Mendel's first experiments were crosses between varieties that differed in only one visible respect. He crossed tall plants with dwarf plants, colored flowers with white flowers, and

TABLE 6–1

Mendel's results in crosses involving seven pairs of alternative characters

Characters	F_1	F_2: Number of plants			F_2: Per cent	
		Dominant *	Recessive *	Total	Dominant	Recessive
Seeds: round vs. wrinkled	All round	5474	1850	7324	74.74	25.26
Seeds: yellow vs. green	All yellow	6022	2001	8023	75.06	24.94
Flowers: red vs. white	All red	705	224	929	75.90	24.10
Flowers: axial vs. terminal †	All axial	651	207	858	75.87	24.13
Pods: inflated vs. constricted	All inflated	882	299	1181	74.68	25.32
Pods: green vs. yellow	All green	428	152	580	73.79	26.21
Stem length: tall vs. dwarf	All tall	787	277	1064	73.96	26.04
Totals		14,949	5010	19,959	74.90	25.10

* The dominant character is the one found in F_1, and the recessive character is the one reappearing in smaller numbers in F_2. Dominance and recessiveness are discussed further in this chapter and in Chapter 7.

† Axial flowers occur all along the stem (axis) of a plant; terminal flowers cluster at the end of the stem.

so on. He had seven pairs of alternative characters for study, and in all seven cases he found the same result as he did, for example, in the cross between red-flowered and white-flowered types.[4] When red-flowered plants were crossed with white-flowered plants, the hybrid offspring were all red-flowered. This result was obtained whether the red form was used as the male (pollen) parent or as the female (ovule) parent. However, when two hybrid reds were crossed, the offspring surprisingly comprised both reds and whites in a ratio of about three reds to every white (Fig. 6–2). These results can be summarized schematically as follows:

P_1 plants Red-flowered × White-flowered
F_1 plants   100% Red-flowered × Red-flowered
                           ↓

F_2 plants	Red : White
Number	705 : 224
Per cent	75.9 : 24.1
Ratio	3.15 : 1.00

P_1 designates the initial parents; F_1 the first hybrid (or filial) generation; and F_2 the second hybrid generation, arising from the crossing of F_1 plants.

The two important results are these: (1) F_1 consists entirely of plants resembling only one of the parents; and (2) F_2 consists of plants resembling both parents in the P_1 generation. The parental character missing in F_1 appears in about one-fourth of the individuals in F_2.

[4] The flower color is actually a violet-red, simplified here as red.

Table 6–1 lists the actual counts Mendel obtained from his other experiments. In every instance the general result was the same: F_1 was all of one type, and F_2 contained both types, the type missing in F_1 making up about 25 per cent of F_2. The generality of the result immediately suggests its significance. Mendel perceived this and proceeded to seek an explanation, which he found and then tested.

MENDEL'S HYPOTHESIS OF PAIRED FACTORS: SEGREGATION AND DOMINANCE

The most remarkable aspect of Mendel's results is the reappearance of the white-flowered plants in the F_2 generation. It is obvious that, although the F_1 plants do not show white flowers, they nevertheless must carry some hereditary factor for them, because when they are crossed among themselves, white flowers appear in approximately one-fourth of the offspring. Indeed, each F_1 plant has a factor for white flowers that it passes on to half of its offspring. This line of argument suggests two other conclusions: first, that each plant carries at least two hereditary factors for each flower color; and second, that the factor for white is completely dominated by the factor for color when both are present in the same plant.

Mendel saw that he could explain his results if he made the following assumptions: (1) there is in each plant a pair of hereditary factors controlling flower color; (2) the two

TABLE 6–2
Theoretical results of breeding heterozygotes

There are two kinds of gametes from the female parent (½ are C; ½ are c)	and they may be fertilized by	either one of two kinds of gametes from the male parent (½ are C; ½ are c)	to give the following zygotes in F_2
1. C	×	C	CC (¼ of total possible)
2. C	×	c	Cc ⎫ (½ of total possible)
3. c	×	C	cC ⎭
4. c	×	c	cc (¼ of total possible)

factors in each pair are derived from the plant's parents—one member of the pair from each parent; (3) the two factors in each pair separate, or segregate, during the formation of germ cells, so that each germ cell receives only one factor; (4) the factors for red flowers and white flowers are alternative forms of the same factor, the red being dominant over the white.

For discussion and understanding of Mendel's scheme and the whole science of genetics that has developed from it, we need to define several terms.

1. The paired hereditary factors are called *genes*.[5]

2. The alternative forms of the same gene are called *alleles*.[6] Thus the genes for red flowers and for white flowers are alleles of each other. Again, the gene for plant size occurs in two allelic forms; there is an allele for tallness and an allele for dwarfness. Tallness, however, is not an allele of redness.

3. One allele is *dominant* over the other, *recessive,* allele. The alleles of a gene are symbolized by the same single letters or brief combinations of letters. The dominant allele is often written as a capital letter (here C represents the red allele), and the recessive allele as a small letter (here c represents the white allele).

4. When both members of the pair of alleles in a plant are the same (for example, cc or CC), the plant is said to be a *homozygote* ("like joined"). When the two alleles differ (for example, Cc), the plant is said to be a *heterozygote* ("differently joined").[7]

[5] This term was introduced long after Mendel's analysis by a Danish biologist, Wilhelm Johannsen.

[6] The original term, also proposed after Mendel, was *allelomorphs* ("alternative forms").

[7] The adjectives are *homozygous* and *heterozygous.*

5. Although all the offspring in the first hybrid generation (F_1) are alike and indistinguishable in appearance from their parents (P_1), they nevertheless have different hereditary constitutions. We shall constantly have to make this distinction between appearance and hereditary constitution. In doing so, we shall speak of the *phenotype* ("visible type") and the *genotype* ("hereditary type") of the organism. The phenotype of the F_1 plants in our example has red flowers in every case, but each plant has a factor for white flowers in its genotype.

Now we can follow Mendel's interpretation, as given in Fig. 6–2B. In the P_1 generation both the red-flowered plant and the white-flowered plant are homozygotes. The genotype of the red-flowered plant is CC, and the genotype of the white-flowered plant is cc. When they produce gametes, the genes segregate, so that the red-flowered plant produces only C gametes and the white-flowered plant produces only c gametes. Union of these gametes in the zygote at fertilization yields only a heterozygous F_1 plant with a Cc genotype (we can equally well write cC; the order of the symbols is meaningless). The phenotype of this plant is, however, red-flowered and indistinguishable from the red-flowered parent.

The F_2 arises from crossing two F_1 reds, each genotypically Cc. What gametes will these F_1 plants produce? In both parents the Cc pair segregates so that each parent produces two kinds of gametes, C and c. Mendel correctly assumed, moreover, that these two kinds of gametes (C and c) must be produced in equal numbers by each parent. Thus, if the F_1 plant used as male parent produces 1000 pollen grains, 500 will be C, and 500 will be c.

Only one kind of fertilization was possible in the formation of F_1 plants—the union of C and c, giving Cc as the F_1 genotype. However, in the mating of the two F_1 plants, more than one kind of fertilization is possible. In fact there are four types, and all four are equally probable or frequent, as shown in Table 6–2.

Fig. 6–2B shows a checkerboard system making it easy to visualize fertilizations listed in Table 6–2. The second and third types of fertilization in the table produce F_2 plants with the genotypes Cc and cC: these two heterozygotes are identical. Thus we conclude that, on the basis of the Mendelian scheme, the F_2 generation should contain three kinds of genotypes in the following proportion:

$$CC \quad : \quad Cc \quad : \quad cc$$

CC	Cc	cc	
¼	½	¼	genotypic ratios
¾		¼	phenotypic ratios
Red		White	

Because the CC homozygotes and Cc heterozygotes are phenotypically the same, only two classes of phenotypes will appear in F_2, and they will tend to appear in the ratio of ¾ red : ¼ white. The class of plants that are cc homozygotes are those with white flowers that Mendel had found were missing in F_1. They constitute about one-fourth of the F_2.

MENDEL'S TESTS OF THE HYPOTHESIS

Let us, in spite of our present greater knowledge, assume ourselves to be in Mendel's position for a moment. We have no knowledge of chromosomes and their role in heredity. Indeed, we have no knowledge at all of the real physical basis of heredity. We have just performed some crosses with garden plants differing in flower color and obtained some quantitative results that are summarized in Table 6–1. Then to explain these results we have *created a hypothesis* that assumes the existence of hereditary factors which we will call genes, although at this stage we have no knowledge of what they are physically or how they are related to cell structure. We did not observe the hypothesis; it was a pure invention, as given in Fig. 6–2B.

The fact that the hypothesis will explain the data is not in itself a sufficient basis to accept it as true; there may be other hypotheses that could explain the facts equally well. What we now seek, therefore, is some further basis in fact for the acceptance or rejection of our scheme: we must *test the hypothesis*.

In Chapter 1 we stated that a proper scientific hypothesis is testable; Mendel's hypothesis is scientific because it is testable. In this case and many others the test takes the form of finding out whether or not certain predictions arising out of the scheme hold good. We must look in the hypothesis for predictions as to the outcome of new crosses yet to be performed.

The most obvious tests of the scheme hinge around the fact that red-flowered plants are of two kinds genotypically. The red in the P_1 generation is, by hypothesis, homozygous CC and when crossed with a white-flowered plant (cc) can produce only red-flowered offspring that are Cc heterozygotes. But the superficially similar mating of F_1 red with white will have a different outcome. The prediction (Fig. 6–3) is that the progeny will be one-half red and one-half white because, although the white parent produces only c gametes, the red parent produces one-half C and one-half c, so that Cc and cc fertilizations will be equally frequent. Mendel performed such a cross and obtained 166 plants, of which 85 were red-flowered and 81 were white-flowered. The result is as near to equality as could be expected in an actual experiment with that many plants.

It is very important to notice that the prediction was not just a *qualitative* one. The scheme predicted not only that red-flowered plants would arise from this cross of F_1 red with white plants but also that the offspring would tend toward a particular *quantitative* relationship—½ red : ½ white. When specific forecasts are thus fulfilled both qualitatively and quantitatively, we have that much more reason to believe in the validity of the hypothesis that produced them.

It is clear from these and other experiments that the Mendelian hypothesis is formally or mathematically valid. Mendel did not know at the time what the paired factors or genes were, but he recognized that they did exist, and in pairs that segregate when gametes are formed. His hypothesis became a sound theory that

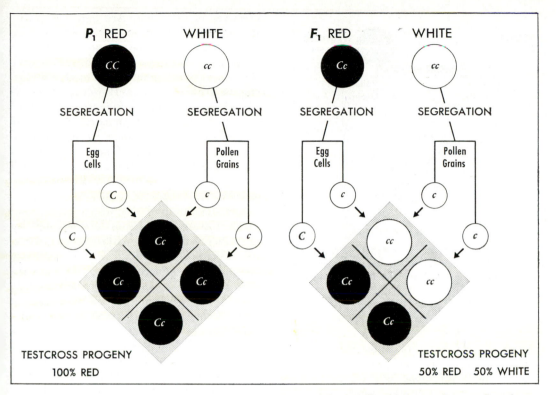

6–3 A testcross that distinguishes the genotypes of P_1 and F_1 red-flowered peas. P_1 red crossed with white gives a progeny that is 100 per cent red. F_1 red crossed with white gives a progeny that is 50 per cent red and 50 per cent white.

has since been supplemented but never supplanted.

TWO CHARACTER DIFFERENCES: INDEPENDENT SEGREGATION

Having discovered the rules governing the hereditary transmission of genes controlling one character difference (such as flower color), Mendel proceeded to a more complex situation. He traced the inheritance of two distinct character differences simultaneously: seed shape and seed color in peas. Each of these character differences he had studied separately, finding them to obey the same rules as flower color (see Table 6–1).

We will see in Chapter 9 that a seed contains a young or embryonic plant in which root, stem, and two embryonic leaves, *cotyledons*, can be distinguished. Seed color and shape both relate to the cotyledons. First, the cotyledons may be either yellow (controlled by the dominant gene Y) or green (y, reces-

sive). Second, they may be fully swollen, making the whole pea seed round (controlled by the dominant gene R), or shrunken, making the seed wrinkled (r, recessive). For each pair of characters (yellow versus green, round versus wrinkled), there are, as with flower color, three possible genotypes.

	Homozygous dominant	*Hetero-zygote*	*Homozygous recessive*
Seed shape	RR	Rr	rr
	Round phenotype		Wrinkled phenotype
Seed color	YY	Yy	yy
	Yellow phenotype		Green phenotype

Mendel began by crossing plants raised from round, yellow seeds ($RRYY$) with plants raised from wrinkled, green seeds ($rryy$). In meiosis, when gametes are formed, each pair of alleles (for example, RR or YY) segregates independently of each other. Thus the round

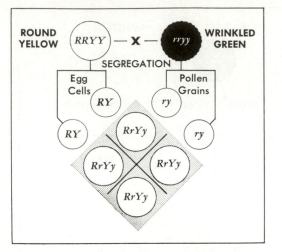

6–4 All F_1 plants from the cross round yellow × wrinkled green have the genotype $RrYy$.

yellow plants ($RRYY$) produce only RY gametes, never RR or YY. In other words, *each pair of alleles is always represented in the gametes by one of its members.* The wrinkled green plants ($rryy$) produce only ry gametes. The F_1 consists of seeds that are round and yellow, with the genotype $RrYy$ (Fig. 6–4). The F_1 is heterozygous for both pairs of alleles and is called *dihybrid*. (Why?)

What are we to expect when two such dihybrid F_1 plants are crossed to produce the F_2 generation? We know from the study of each character separately that in the formation of gametes R will segregate from r, and Y will segregate from y. However, we are left with an uncertainty as to whether there will be two or four kinds of gametes. Thus

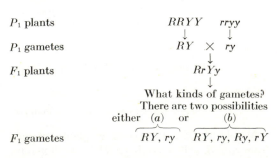

When the F_1, $RrYy$, was formed, the gametes from the parental plants were of two types, RY and ry. The two dominant alleles were associated in one gamete, and the two recessives in the other. Are they necessarily always associated? If so, we expect only two classes of gametes to be produced by the F_1: RY and ry. These two types of gametes (*a* above) are said to contain *parental* combinations of alleles. (Why?) However, if the Rr pair of alleles segregates independently of the Yy pair, and the original combinations are not necessarily maintained, then (as in *b* above) two new kinds of gametes (Ry and rY), in addition to the two original ones, will be produced by F_1. Ry and rY are said to be *recombination* types of gametes. (Why?)

It is easy to find out which of the two possibilities (*a* or *b*) is in fact realized, because they will lead to different F_2 generations, as indicated in Fig. 6–5: (1) If the only combinations produced by the gametes are the parental ones, RY and ry, there will be only three classes of genotypes in F_2 and two classes of phenotypes—round yellow $\frac{3}{4}$: wrinkled green $\frac{1}{4}$ (see Fig. 6–5*A*). (2) If the two recombination types of gametes are produced in addition to the parental types, and all four occur in equal frequencies, there will be nine classes of genotypes in F_2 and four classes of phenotypes. The four phenotypes and their expected frequencies are round yellow $\frac{9}{16}$: round green $\frac{3}{16}$: wrinkled yellow $\frac{3}{16}$: wrinkled green $\frac{1}{16}$ (see Fig. 6–5*B*).

At the bottom of Fig. 6–5 are the results from Mendel's own experiment. It is clear that the four kinds of genotypes are produced; the parental combination RY and ry can be recombined in the F_1 to yield gametes that are Ry and rY. Moreover, all four are produced with the same frequency, as shown by the close agreement between expected and observed proportions.

Mendel's demonstration of the independence of the Rr and Yr pairs of alleles in their segregation is often referred to as the rule of *independent assortment.* We shall shortly see that it is not universal. Exceptions were found in the outburst of genetic work that followed the rediscovery of Mendel's principles in 1900, and it was by understanding the causes both of the rule and of its exceptions (which, as it has turned out, outnumber the cases in which the so-called rule applies) that still further advances were made. The explanation of these exceptions is discussed on p. 185.

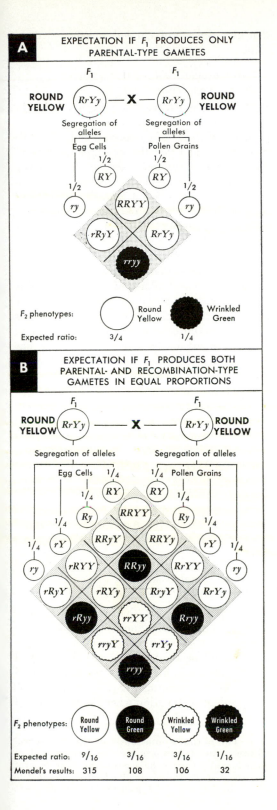

A EXPECTATION IF F₁ PRODUCES ONLY PARENTAL-TYPE GAMETES

F₂ phenotypes: ○ Round Yellow ● Wrinkled Green

Expected ratio: 3/4 1/4

B EXPECTATION IF F₁ PRODUCES BOTH PARENTAL- AND RECOMBINATION-TYPE GAMETES IN EQUAL PROPORTIONS

F₂ phenotypes:	Round Yellow	Round Green	Wrinkled Yellow	Wrinkled Green
Expected ratio:	9/16	3/16	3/16	1/16
Mendel's results:	315	108	106	32

Physical Basis of Heredity

THE CHROMOSOME THEORY

The reader, armed with the information from the last chapter, will long ago have perceived the significance of the main Mendelian results—proofs of the paired nature of the genes, one derived from each parent, and their segregation in gamete formation. Clearly these characteristics of genes must be related to the like characteristics of chromosomes discussed in Chapter 5. We have deliberately avoided using what we know of chromosome behavior here to emphasize the brilliance and adequacy of Mendel's analysis, for Mendel was actually ignorant of chromosome behavior when he performed and analyzed his experiments in 1865. He lived until 1882, by which time much, although far from all, of our basic knowledge of chromosomes had been amassed. Mendel himself did not follow up his brilliant first discoveries. In his later years he was drawn more and more away from science into monastery affairs and became embroiled in some wrangles within the church.

The people who did pioneer research on chromosomes overlooked the importance of Mendel's work. It was not until after De Vries, Correns, and von Tschermak had rediscovered the Mendelian phenomena that biologists put the two lines of study together.

Sutton's formulation of the chromosome theory

In 1902 an American cytologist, W. S. Sutton,[8] saw the implications of Mendel's analysis in relation to chromosome behavior. Sutton maintained that the hereditary factors —or genes—are carried on the chromosomes or are parts of chromosomes. The theory that genes are chromosome parts explains (1) why they occur in pairs (because chromosomes do); (2) why the two members of a pair are derived one from each parent (because chromosomes are so derived); and (3) why genes

[8] Theodor Boveri and De Vries made the same analysis independently, but Sutton's was the first and most complete.

6–5 **Alternative expectations from the cross** *RrYy × RrYy,* **and the fit of Mendel's actual results to the second alternative.**

segregate at meiosis (because chromosomes do). Other features of Mendel's results are immediately explainable by the assumption that the genes are carried on the chromosomes.

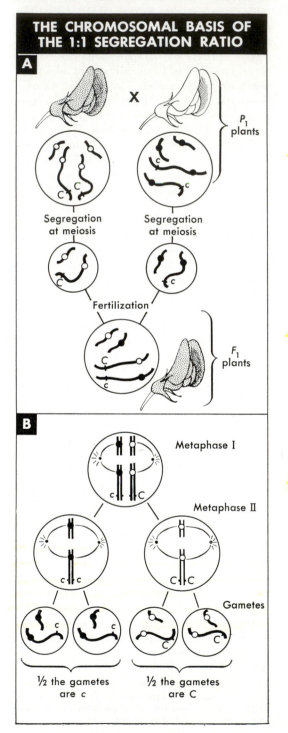

THE CHROMOSOMAL BASIS OF THE 1:1 SEGREGATION RATIO

A

Segregation at meiosis

Segregation at meiosis

Fertilization

P_1 plants

F_1 plants

B

Metaphase I

Metaphase II

Gametes

½ the gametes are c

½ the gametes are C

Chromosomal basis of segregation: the 1:1 segregation ratio

Mendel assumed that the Cc genes in his F_1 (red-flowered) heterozygotes segregated to produce two types of gametes, C and c; and he further assumed that the two types were produced in equal numbers (see Fig. 6–2B). The chromosome theory shows his assumption to be correct and explains it. The F_1 red-flowered plant must contain a pair of chromosomes carrying the alleles C and c. Fig. 6–6A shows schematically the origin of the F_1 red and its chromosome constitution, with the same hypothetical nucleus that illustrated meiosis in Chapter 5. At metaphase of meiosis I in this F_1 plant, the chromosomes are at the equator of the spindle (Fig. 5–6B). The larger pair of homologous chromosomes carries C and c. Each member of the chromosome pair has duplicated, and the two members are going to different poles. Each telophase I nucleus contains a large chromosome carrying either C or c. The two strands of each chromosome are separated at meiosis II, and so eventually each cell produces *precisely* two gametes with C and two gametes with c. If a pea plant produces 4000 pollen grains, these must have arisen from 1000 diploid cells in which meiosis occurred. Each of the 1000 cells yielded four gametes, two of which are C and two of which are c. No matter how many gametes are produced, the segregation ratio will always be 1:1.

Chromosomal basis of independent segregation

Sutton pointed out that the chromosome theory could explain another feature of the Mendelian results—the independent segregation (or independent assortment) that Mendel had discovered in his experiment on the inheritance of two characters. Fig. 6–7 depicts independent segregation when the two pairs of

6–6 The chromosomal basis of the 1:1 segregation ratio. *A.* The origin of the F_1 plant and its nucleus, which carries the genes C and c on its large chromosomes. *B.* Meiosis in the F_1 plant, leading to the production of c and C gametes in equal numbers.

genes (Rr and Yy) are carried on different chromosomes. Suppose that the Yy alleles are carried on the larger (B) pair of chromosomes and that the Rr alleles are carried on the smaller (A) pair of chromosomes. The orientation of the large pair of chromosomes on the equator of the spindle is independent of the orientation of the small pair. Thus the arrangements given in Fig. 6–7A and Fig. 6–7B are equally likely, and, accordingly, the allele Y is just as likely to enter the same nucleus with r as it is with R. Four types of gametes (YR, yr, Yr, and rR), then, are produced with equal frequency (check with Fig. 6–5B).

Statistical nature of Mendelian heredity

The ratios of different genotypes and phenotypes in F_2 generations are often referred to as *Mendelian ratios*. For instance, when two F_1 red-flowered plants are crossed, the Mendelian ratios expected in F_2 are as follows: phenotypes, $3/4$ red : $1/4$ white; genotypes, $1/4 CC : 1/2 Cc : 1/4 cc$. In actual experiments the ratios are realized only as statistical approximations. Why is this so? Why, for example, did Mendel obtain 75.9 per cent red and 24.1 per cent white, not exactly 75 per cent and 25 per cent, respectively? The answer to this question emerges from the understanding that any *sample* is only an *approximate representation* of the population of events or things from which it is drawn.

Consider, first, what happens in tossing a coin. We expect heads half the time and tails half the time. If we were to throw a coin a million times, the ratio of heads to tails would be very close to 500,000 heads : 500,000 tails ($1/2 : 1/2$). But if we throw it only four times, there is a good chance (actually 1 in 16) that we may get four heads instead of the expected ratio of 2 heads : 2 tails. The more times we toss the coin, the less likely is the observed ratio of heads to tails to depart far from the expected ratio of $1/2 : 1/2$.

When a geneticist crosses two F_1 red-flowered pea plants, he obtains a relatively small number (say, 100) of F_2 plants. The 100 ovules and 100 pollen grains that gave rise to the F_2 plants were only a sample of the entire population of gametes produced by the F_1 plants. Like a small sample of coin throws, the sample of gametes only approximates the exact ratio of $1/2 C : 1/2 c$ that applies to the gamete population as a whole. Consequently, the sample of zygotes obtained merely approximates the $1/4 CC : 1/2 Cc : 1/4 cc$ ratio to be expected in an infinitely large progeny.

In any such situation we do not in fact expect our sample to have exactly the ratio inferred from the hypothesis being tested. By statistical methods, however, we can calculate how likely we are to find the observed ratio in our sample if the hypothesis is correct. If the observed ratio is unlikely, we suspect that the hypothesis is wrong, and perhaps we discard it altogether. If it is likely, the hypothesis is supported to some extent, although not proved. In Mendel's experiment summarized in Fig. 6–5B, the theoretically expected and actually observed numbers compare as follows:

Expected from theory: 315.5 105.2 105.2 35.1
Observed by Mendel: 315 108 106 32

The agreement is remarkably close. The experiment gives us no reason to doubt the hypothesis. On the contrary, it greatly strengthens our confidence in it.

TESTS OF THE CHROMOSOME THEORY

With the announcement of the chromosome theory of heredity in 1902, the science of genetics in its modern form was born. This theory, unlike earlier views on heredity, was precise and quantitative and had its foundations in cellular structures (chromosomes) whose behavior and properties were open to direct observation and interpretation. The earlier theory of pangens (which nobody could see) made virtually no specific predictions [9] by which its merits could be judged. On the other hand, the chromosome theory offered abundant predictions, the testing of which led to the rapid growth of genetics as an exact science. From 1906 onward, much of the experimental work in genetics was carried out with the common fruit fly *Drosophila melanogaster*, the small yellowish fly that hovers around garbage cans and fruit in the late summer and fall.

[9] Except the incorrect ones that Aristotle saw and pounced upon 2000 years ago.

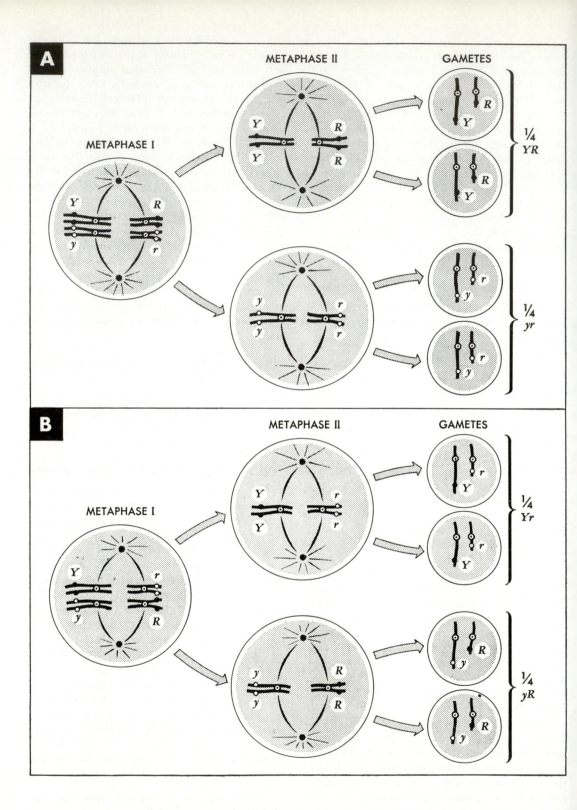

Drosophila melanogaster

Someone should erect a much-larger-than-life monument to this tiny (2 millimeter) insect.[10] It is an ideal laboratory animal and has done more than any other to enlarge our knowledge of genetics. It is big enough to work with but small enough to raise by the hundreds in containers without crowding a laboratory. It breeds readily in captivity, it is prolific, and it has short generations (as little as 10 days). All these properties make experimentation with *Drosophila* relatively easy, quick, and cheap. The experiments that took Mendel 7 years with peas and required a large garden can be repeated with *Drosophila* in a few months in a dozen or so flasks tucked away on a shelf. *Drosophila* has two other advantages not realized when experiments with it were begun. It has a small number of chromosomes (four pairs in the most-used species), and its salivary glands contain chromosomes enormously larger than those of its other parts or of most other organisms. The small number of chromosomes simplifies the study of the grouping of genes in chromosomes. The giant chromosomes greatly aid in correlating heredity with the anatomy of the chromosomes.

Genetic experiments have been carried out with many organisms besides *Drosophila*, from bacteria and the viruses that infect them to corn, pine trees, chicks, and mice. Many interesting peculiarities of heredity occur in one species or another. Thousands of experiments on hundreds of species have, neverthe-

[10] Rapid progress in genetics began in 1906, when T. H. Morgan (1866–1945) started experiments with *Drosophila* at Columbia University. Morgan got the idea of using *Drosophila* from another pioneer geneticist, W. E. Castle (1867–1962), who had been conducting other sorts of experiments with fruit flies at Harvard since 1901. A *Drosophila melanogaster* female and its chromosomes are shown on the chapter opening page.

6–7   The chromosomal basis of independent segregation. The chromosomes are shown at meiosis in the dihybrid plants *YyRr*, with the *Yy* genes on the larger pair of chromosomes and the *Rr* genes on the smaller pair. *A* and *B* are equally probable ways in which the chromosomes can orient themselves at metaphase I, as explained in Chapter 5. The consequence is that the four kinds of gametes—*YR, yr, Yr,* and *yR*—are produced in equal numbers.

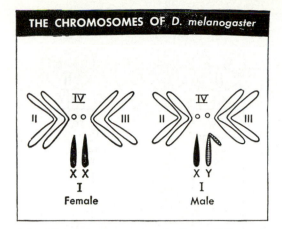

6–8   The chromosomes of *Drosophila melanogaster*.

less, confirmed that the *principles* of heredity are basically the same in all organisms. Most of the principles learned from *Drosophila* apply in the widest way to all living things.

Sex determination

In brief, Sutton's theory said that the correlation between the behavior of genes, which were inferred but not visible, and the behavior of chromosomes, which are visible with a microscope, is so great that we must conclude that the genes are in the chromosomes.

Clearly a further step in developing this theory would be the attempt to associate a specific gene and a specific chromosome, to find where particular genes are located in the chromosomes.

For this task it would be convenient if there were in the nucleus one pair of chromosomes different in appearance and behavior from all the others. If its behavior were sufficiently aberrant, but understood, we ought to be able to predict how the genes it carries will be inherited. There actually is such a pair of special chromosomes in the nucleus of nearly every higher animal—the sex chromosomes. It is no coincidence that the attention of modern geneticists for a time focused strongly on the behavior of sex chromosomes in *Drosophila*.

Fig. 6–8 is a simplified diagram of the chromosomes at metaphase of mitosis in a male and female *Drosophila*. There are only four pairs of chromosomes. One pair is dotlike and very small; it is called pair IV. Two larger

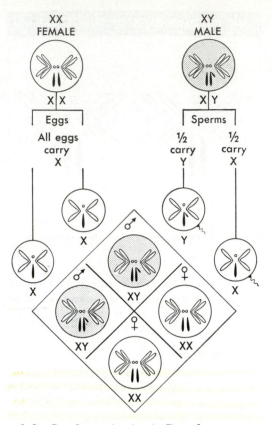

6–9 Sex determination in *D. melanogaster.*

V-shaped pairs with the centromere in the middle of the chromosome, are called pairs II and III. These three chromosome pairs are collectively designated *autosomes*,[11] to distinguish them from pair I, designated *sex chromosomes.* The sex chromosomes are different in the two sexes. In the female they are both rod-shaped, with the centromere near the end. In the male one is rod-shaped, but its partner (unique to the male) is J-shaped, with the centromere at the bend. The two sex chromosomes in the female are known as X-chromosomes; in the male there is one X-chromosome and one special Y-chromosome, which is the J-shaped member. It is obvious that although the female will produce only one kind of gamete so far as the sex chromosomes are concerned, the male will produce two kinds. All the female's eggs will carry an X-chromosome, but of the sperms produced by

[11] In organisms with other numbers of chromosomes, all except the sex chromosomes are called autosomes.

6–10 Some types of sex determination. In man ($2n = 46$), as in *Drosophila*, the male is the heterogametic sex; that is, the male produces two kinds of gametes (X and Y). The female is homogametic; all her gametes are of one kind (X). In the grasshopper ($2n = 22$ in the male) and many other insects, the male is heterogametic again, but the two kinds of gametes are X and no X, rather than X and Y. In the fowl ($2n = 18$), as in all birds and also butterflies and moths, the female is heterogametic. The sex chromosomes are designated Z and W to distinguish female heterogamety from male heterogamety (XY). In some butterflies and moths a condition analogous to XO has evolved: there are ZO females. The honeybee ($2n = 32$) and many other members of the insect order Hymenoptera (bees, wasps, ants) have a remarkable sex-determining mechanism: unfertilized eggs develop into (haploid) males; fertilized eggs develop into (diploid) females.

the male, half will carry an X-chromosome, and the other half a Y-chromosome.

Fig. 6–9 indicates the outcome of the gamete constitution at fertilization. One-half the fertilizations yield zygotes with two X-chromosomes; the other half yield zygotes with an X-chromosome and a Y-chromosome. The former become females, and the latter males. Thus sex determination, which like other aspects of heredity had been discussed in vague and mythical terms for over 2000 years, is clarified by the chromosome theory. In most cases, sex in animals is determined at the moment of fertilization, and the decisive factor is whether the sperm carries an X-chromosome or a Y-chromosome.[12]

Offhand, it looks as if X-chromosomes carry genes for femaleness and Y-chromosomes those for maleness, but, as often happens in biology, things turn out to be not so simple.[13] The Y-chromosome has little to do

[12] There are, however, numerous exceptions in animals, and sex determination in plants, in many of which a single individual has both male and female organs (or gamete producers), usually follows other principles. Some of these variations are treated later.

[13] In recent years, new techniques of cell study have been widely used in human cases of abnormal sexual development. Many have shown abnormalities in the number of X- and Y-chromosomes (or both). We shall presently consider the mechanisms by which some of these abnormalities develop. Here we simply point out one important conclusion of such studies. It appears that in man the addition of a Y-chromosome to an X-chromosome (or to a group of them) convert a potential female into a male, the Y-chromosome tending to mask the expression of the genes on the X-chromosome.

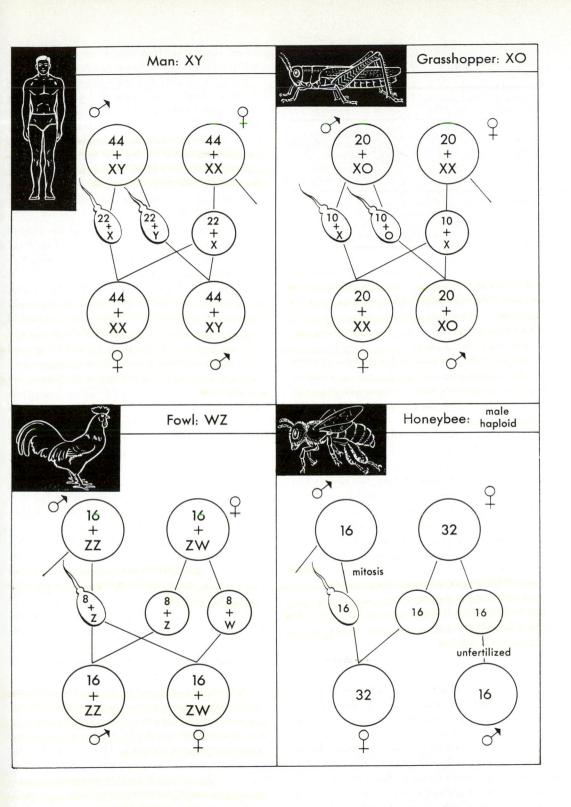

Man: XY

♂   ♀

44   44
+   +
XY   XX

22   22   22
+   +   +
X   Y   X

44   44
+   +
XX   XY

♀   ♂

Grasshopper: XO

♂   ♀

20   20
+   +
XO   XX

10   10   10
+   +   +
X   O   X

20   20
+   +
XX   XO

♀   ♂

Fowl: WZ

♂   ♀

16   16
+   +
ZZ   ZW

8   8   8
+   +   +
Z   Z   W

16   16
+   +
ZZ   ZW

♂   ♀

Honeybee: male haploid

♂   ♀

16   32

mitosis

16   16   16

32   16

unfertilized

♀   ♂

TABLE 6–3
Effect of X/A ratio

Number of X-chromosomes	Sets of autosomes (A) (three in a set)	Ratio X/A	Sex phenotype
3	2	1.5	Superfemale *
2	2	1.0	Normal female
2	3	0.67	Intersex
1	2	0.5	Normal male
1	3	0.33	Supermale *

* Both supermales and superfemales are weaker flies than their normal counterparts. Indeed, there is nothing "super" about them except that their chromosomal balance exceeds that characteristic of their sex.

with sex determination. Its genetic effect is slight in man and apparently nil in some other animals. In some, indeed, there is no Y-chromosome (Fig. 6–10). It has been established in some animals, and probably is a valid generalization, that the development of sex is determined by an interaction between genes on the X-chromosomes and genes on various autosomes. As usual, the best evidence comes from *Drosophila*.

In experiments with *Drosophila* it has been possible to obtain flies differing in the ratio of X-chromosomes to autosomes. Table 6–3 illustrates the different ratios obtained and shows how the scale of X/A ratios from 0.3 to 1.5 is a scale of increasing femaleness. The normal male has an X/A ratio of 0.5 (1 X-chromosome : 2 sets of autosomes); the normal female has an X/A ratio of 1.0 (2 X-chromosomes : 2 sets of autosomes). Evidently the genes of the autosomes tend to produce males, and genes concentrated on the X-chromosomes interact with those, producing females when they overbalance the autosomal genes.

Sex linkage

Among the characters in *Drosophila* that Morgan and his students first investigated were many, like vestigial wings, which followed precisely the simple pattern of inheritance Mendel had discovered in peas. There were, however, other characters, like white eye color or the trait known as *Bar* eye (in which the eye is abnormally small in size), that showed a markedly aberrant pattern of inher-

itance suggesting that they were controlled by genes on the sex chromosomes.

First let us note the inheritance of vestigial versus normal wings in the fly. As in Mendel's pea experiments, it does not matter which parent—male or female—carries a particular character. Whether the vestigial-wing character is in the male or female parent, the F_1 is always all normal-winged, and the F_2 contains three normals to one vestigial. Moreover, the 3:1 ratio applies to both the male and female members of F_2.

The situation is very different in the case of white eye color versus red eye color. Fig. 6–11 shows that the constitution of the F_1 depends on whether the white-eyed parent is father or mother. And the F_2 differs in the two cases. Hopelessly aberrant as this inheritance looks at first sight, it is exactly what we would expect if (1) the genes for white and red eye color were carried by the X-chromosome and (2) the Y-chromosome were an empty shell carrying no genes at all or at any rate not these genes. Note Fig. 6–11A, which interprets the cross between a white female and a red male. All the male offspring are white-eyed because a male's X-chromosome is always derived from its mother; from its father it receives only the Y-chromosome, which evidently is empty of genes, at least as far as this character is concerned. On the other hand, a female always receives one X from its father and one from its mother. Since red (R) is dominant over white (r), all daughters in this cross must be red-eyed because the father was red-eyed; all sons must be white-eyed because both the mother's X-chromosomes carry the recessive gene r. Follow the segregations and fertilizations involved in both parts of the figure to see how the assumption that the genes are on the X-chromosome and missing from the Y-chromosome explain all the results.

The conclusion that the Y-chromosome carries few or no genes fits in with some observations on other insects, such as the grasshopper, in which the Y-chromosome has been dispensed with altogether. Here again the female has two X-chromosomes, and the male has one. As in the XY system, the male produces two kinds of gametes, but they are X and no X rather than X and Y (see Fig.

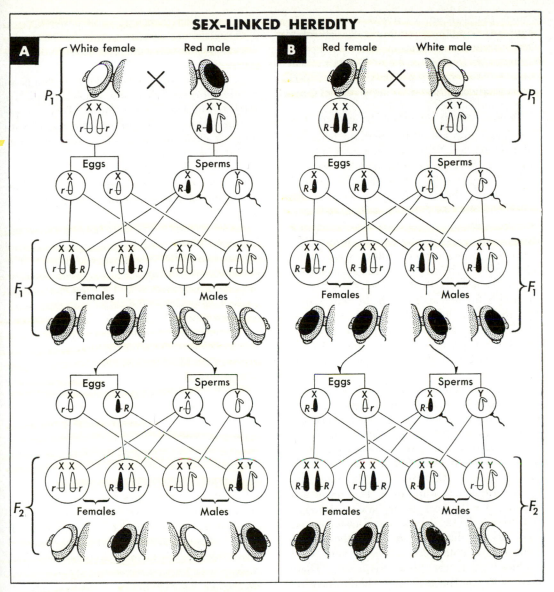

6–11   Sex-linked heredity.

6–10). Notwithstanding these findings, there is now conclusive evidence that genes are present in the Y-chromosomes of some animals.

A number of recent observations in human and mouse cells have raised the interesting possibility that in these species (and probably in others) differences exist between the two X-chromosomes of the female. Mary Lyon suggested in 1961 that one of the two X-chromosomes is somehow inactivated. Indeed, one does observe that one X-chromo-

some becomes smaller and more aggregated in appearance and synthesizes its DNA later than its partner. This "Lyonization" of one X-chromosome is believed to occur early in embryonic life in each cell of the developing organism. As a result, the cells in the body of a woman can be divided into two groups, in which one or the other X-chromosome remains active. Since the active X-chromosome may derive from the woman's mother or from her father, the eye of a woman heterozygous for such an X-carried gene as color-blindness contains patches of colorblind light receptors alternating with patches of normal ones. This probably explains why some females carrying the colorblind gene are not affected by it, whereas others are.

Nondisjunction (or failure to segregate)

The assumption that the gene for white eye color is on an X-chromosome accounts for unusual patterns of eye-color inheritance so precisely that we have the strongest grounds for believing the assumption to be true. We have correlated the behavior not just of genes and chromosomes in general, but of a particular gene and a particular chromosome. This particular gene—for eye color—was studied intensively by the early American geneticists. One of them, C. B. Bridges (1889–1938), discovered a further anomaly in its inheritance. The discovery led to final and convincing proof that the gene was carried from cell to cell, from generation to generation, by the X-chromosome in the fashion we have outlined.

Fig. 6–11A demonstrated that when a white-eyed female is crossed with a red-eyed male, all the daughters will be red-eyed (heterozygotes Rr), and all the sons will be white-eyed (r alone with no companion gene from the Y chromosome). Bridges found that very rarely (about once in 3000 times) a daughter was produced that had white eyes. This result was startling, for the normal mechanism demands that the daughter have one X-chromosome carrying the dominant R from the father. Bridges also noted that red-eyed sons were produced at a very low frequency. These again were unexpected—they should have received their X-chromosomes from their mothers (rr) and so have been white-eyed.

Bridges perceived that the exceptional offspring could be explained by the hypothesis that the two X-chromosomes in the white-eyed mother sometimes fail to segregate (or disjoin) during the first meiotic division. Fig. 6–12 contrasts the pattern according to Bridges' hypothesis with the normal course of events in the first meiotic division. In the case envisaged by Bridges, the X-chromosomes do not segregate at anaphase but go instead to the same pole of the spindle. Their behavior leads to the production of eggs that either lack an X-chromosome altogether or else have two X-chromosomes (both of which carry the gene r).

The figure also shows the four kinds of zygotes formed when the unusual eggs are fertilized by the red-eyed father. Two of these fertilizations we may ignore: the zygote with only a Y-chromosome dies because at least one X-chromosome is essential for development—it carries indispensable parts of the total blueprint; the zygote with three X-chromosomes is female, and red-eyed like her normal sisters.

It is the other two fertilizations that are important here. One will produce a red-eyed son, and the other a white-eyed daughter. The red-eyed son will arise from the egg that received no X-chromosome from the mother but an X-chromosome (and therefore an R gene) from the father. XO, like XY, is male. The real difference between male and female is not the Y-chromosome but rather the single X-chromosome in the male and thus an X/A ratio of 0.5 (see Table 6–3). XO is the same as XY as far as sex determination is concerned. Each has one X. The female condition is determined by two X's. Therefore, the presence of a Y-chromosome does not cause an XXY fly to be male; it will be female because it has two X's, and it will be white-eyed because both X's came from the white-eyed mother.

Fig. 6–12 summarizes Bridges' hypothesis as to how a breakdown in the usual mechanism causes the appearance of exceptional offspring. This hypothesis, based on the theory of chromosomal inheritance, has great importance because it yields two specific and testable predictions, as follows: (1) An exceptional white-eyed daughter will contain a Y-chromosome, unlike all its much commoner sisters. It is white and not red because it received its father's Y-chromosome, not his R-carrying X-chromosome. (2) An exceptional red-eyed son will entirely lack a Y-chromosome, unlike all its much commoner brothers. It received its father's X-chromosome (with its R gene), not his Y-chromosome. From its mother it received no sex chromosome at all.

Bridges tested and confirmed these predictions by looking through a microscope at the chromosomes of the exceptional offspring. His

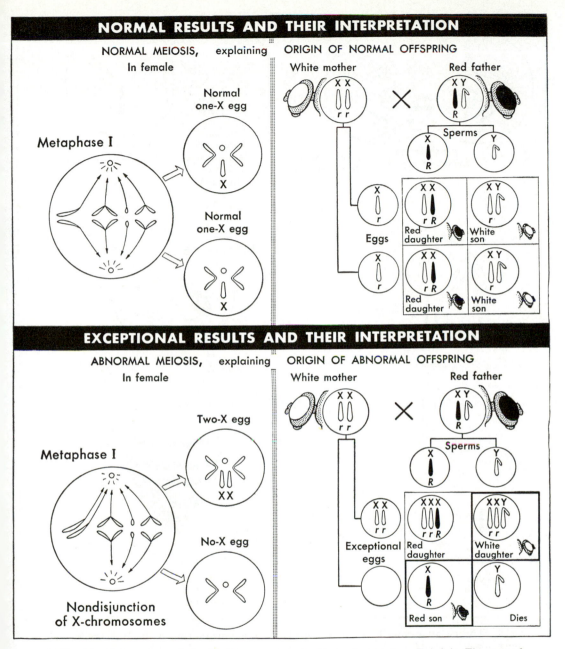

6–12 Nondisjunction of the X-chromosomes in *D. melanogaster*. *Top left.* The normal course of meiosis I leading to eggs that have one X-chromosome each. *Top right.* The normal genetic results of a cross between a white female and a red male. The usual form of the genetic checkerboard has been modified to include a summary of the chromosome content, as well as the gene content, of each fertilized egg. *Bottom left.* The abnormal course of meiosis I, which Bridges guessed was the cause of the exceptional offspring he had found. *Bottom right.* The genetic results predicted by this hypothesis in the form of a genetic checkerboard. The two squares in heavy outline explain the exceptional offspring and lead to the crucial prediction that made this study of Bridges' a landmark in genetics. What is the prediction? For simplicity we have represented the oöcyte (at metaphase I) as giving rise to two eggs. In reality, an oöcyte gives rise to only one egg. This does not, however, affect the implications of our diagram. In the lower half of the figure, we show the abnormal oöcyte giving rise to both a two-X egg and a no-X egg, whereas in nature it gives rise to either one or the other. But since a female lays large numbers of eggs, she will produce in the long run both two-X and no-X eggs as a result of nondisjunction.

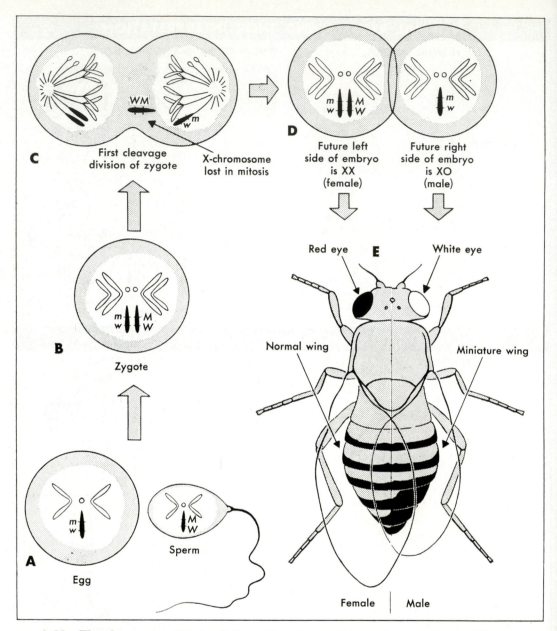

6–13 The chromosomal basis of the origin of gynandromorphism in *Drosophila melanogaster.*

analysis of the nondisjunction of X-chromosomes (1913–1918) was a landmark in the history of genetics. It also exemplified the fact that observations which may at first seem to contradict or even disprove a theory may actually confirm it and broaden the range of phenomena explicable by it. The chromosome theory was found to explain not only the usual results but also rare, apparently anomalous, ones. Such unusual occurrences are now being extensively studied in man by the new techniques of chromosomal analysis, and it is evident that nondisjunction is the main explanation for the known abnormalities of chromosome number and sex chromosome pattern (such as XXY, XXXX, and many others).

Gynandromorphs

Gynandromorphs (literally "female-male forms") are rare monsters, part female and part male. They were known long before the development of the general chromosome theory of heredity and its derivative, the special theory of sex determination by sex chromosomes. Obviously, whatever theory of sex determination ultimately proved valid, it would have to account for these exceptional forms.

The sex-chromosome theory explains the origin of gynandromorphs. Fig. 6–13 indicates how an X-chromosome can be lost from the spindle at mitosis of an early cell division in the embryonic life of a female. All the tissues that later derive from the right-hand cell in the diagram carry only one X-chromosome. Consequently, the mature fly is male on the right side and female on the left. In the case illustrated, the fly has a white eye and miniature wing on the right (male) side of the body. This is exactly what is expected if the chromosome theory is correct and gynandromorphs originate because of the loss of an X-chromosome. The zygote from which this fly arose was heterozygous for the genes controlling eye color and wing length, both of which are on the X-chromosome. Moreover, the recessive genes for miniature (versus long) wings and white (versus red) eye color are together on one of the X-chromosomes. The fly would normally be long-winged and red-eyed because of the dominance of these characters—as on the left side of the body, where both X-chromosomes are retained, causing femaleness. The X-chromosome lost early in the life of the fly was the X carrying the dominants long and red. The single surviving X-chromosome is responsible for maleness, miniature wing, and white eye on the right side.

LINKAGE AND CROSSING OVER

There remains one striking aspect of the chromosome theory of heredity that may have occurred to the reader before now: the independent segregation that Mendel discovered when he followed the inheritance of two pairs

of genes simultaneously must be the exception rather than the rule. Sutton explained independent segregation by assuming that the two gene pairs must be on separate chromosomes; the physical basis of gene segregation is chromosome segregation. In *Drosophila* there are only four pairs of chromosomes. It follows that the independent segregation of genes could only be a valid generalization for *Drosophila* if there were only four genes in its entire hereditary blueprint—one gene pair on each chromosome pair.

We must surely anticipate far more than four gene pairs in the hereditary blueprint of an organism as complex as *Drosophila*. And we must therefore anticipate that many gene pairs will not segregate independently at meiosis because they will be linked together, carried by the same chromosome. Such a linkage —or nonindependence—between genes was detected in the early growth of genetics after 1900. It was found, in fact, that in *Drosophila* all the genes (several hundred) fall into four groups called *linkage groups*. All the genes within a linkage group tend to segregate as a unit. The fact that there are four such groups in *Drosophila* is obviously significant, since there are four pairs of chromosomes. One of the four groups is very small and clearly belongs to the dotlike chromosome pair, IV. Another group is sex-linked and belongs, as we have seen, to the X-chromosomes. The other two groups are very large and are associated with the two big V-shaped pairs of chromosomes (II and III; see Fig. 6–8).

Let us take an example of linkage between two genes in corn, which is one of the plants best known genetically. In corn there are two genes, each with two alleles, that affect the color and texture of kernels as follows:

	Alleles	Effect on kernel
1st gene pair	$\begin{cases} C \\ c \end{cases}$	Colored Colorless
2nd gene pair	$\begin{cases} S \\ s \end{cases}$	Smooth (or full) Shrunken

We begin with the following cross:

P generation	CCSS	×	ccss
P gametes	CS	↓	cs
F_1 generation		CcSs	

If these two genes follow Mendel's rule of independent assortment, we should expect

TABLE 6–4

A testcross to determine what kinds and proportions of gametes the F_1 corn plant (CcSs) produces

	F_1 plant CcSs			$\times$	Tester plant ccss
	Possible F_1 gametes				
(1) Genotypes of F_1 gametes	CS	cs	Cs	cS	cs — There is *only one* *possible* kind of gamete from the tester plant
(2) Genotypes of F_1 testcross progeny	CcSs	ccss	Ccss	ccsS	
(3) Phenotypes (characters of seeds) of testcross progeny	CS colored smooth	cs colorless shrunken	Cs colored shrunken	cS colorless smooth	
(4) Ratios indicating *no linkage* between C and S	25%	25%	25%	25%	50% recombination gametes
(5) Ratios indicating *complete linkage* between C and S	50%	50%	0%	0%	0% recombination gametes
(6) Observed ratios indicating *incomplete linkage* between C and S	[4030] 48.25%	[4030] 48.25%	[150] 1.75%	[150] 1.75%	3.5% recombination gametes

(1) that the F_1 would produce four types of gametes—CS, cs, Cs, cS—and (2) that it would produce them in equal numbers, with each type constituting one-fourth of the total pool of gametes produced. Furthermore, if all four classes of gametes were produced in these frequencies, we should expect the F_2 generation ($F_1 \times F_1 \to F_2$) to contain kernels in the ratio $\frac{9}{16}$ colored smooth : $\frac{3}{16}$ colored shrunken : $\frac{3}{16}$ colorless smooth : $\frac{1}{16}$ colorless shrunken. (Why? Cf. Fig. 6–5B.) We would test whether or not the F_1 did produce all four kinds of gametes expected on the basis of independent segregation by making an F_2. There is, however, a more direct and generally useful way—known appropriately as a *testcross* [14]—of finding out what kinds of gametes the F_1 produces. It is performed by crossing the F_1 (CcSs) to a plant that is homozygous for both recessive alleles (ccss). The double recessive (ccss) plant is known as the *tester*; clearly it can produce only cs gametes.

Table 6–4 shows that the phenotypes of the

testcross offspring indicate immediately what kinds of gametes came from the tested plant (CcSs) because it alone contributed dominant alleles. Scoring the frequencies of phenotypes in testcross progeny amounts to scoring the kinds and frequencies of gametes produced by the tested plant (CcSs in the present case). If the Cc and Ss gene pairs segregate independently, we can expect the four possible phenotypes in the testcross progeny to occur in equal numbers, $\frac{1}{4}$ colored smooth : $\frac{1}{4}$ colored shrunken : $\frac{1}{4}$ colorless smooth : $\frac{1}{4}$ colorless shrunken (row 4 in the table—no linkage). If, however, the two kinds of genes are on the same chromosome, and cannot segregate independently, we can expect the tested plant to produce only CS and cs gametes in equal numbers, leading to a testcross progeny of $\frac{1}{2}$ colored smooth : $\frac{1}{2}$ colorless shrunken (row 5 in the table—complete linkage).

Row 6 shows what kinds of testcross progeny were *actually* obtained in this experiment.[15] It is apparent that the two genes are

[14] Compare with our earlier testcross in Fig. 6–3.

[15] Carried out in 1922 at Cornell Agricultural Experiment Station by C. B. Hutchinson.

not independent in their segregation; the ratio among the *CS, cs, Cs,* and *cS* gametes is far from ¼ : ¼ : ¼ : ¼. It is much closer to ½*CS* : ½*cs*, which is what we would expect if linkage were complete. Therefore, we must be on the right track in attributing the results to linkage. The commonest allele combinations produced by the *CcSs* plant are the combinations (*CS* and *cs*) that existed in the gametes producing the plant. These combinations are appropriately called the "parental combinations"; the new types (*cS* and *Cs*) are called the "recombinations." It is clear that we cannot ignore the few recombination gametes. Yet our notions about linkage so far demand that we anticipate progeny in the ratios given either by row 4 or by row 5. We cannot have our cake and eat it, so to speak; either the genes are linked, or they are not.

This is another excellent illustration of the importance of exceptions in scientific progress. By focusing attention on the exceptions in the present case—the recombination gametes—genetics made one of its most significant advances.

A close study of the behavior of chromosomes in the prophase of the first meiotic division reveals details that explain the appearance of a small number of recombination gametes. When the homologous chromosomes, such as those carrying *CS* and *cs* in our tested plant, pair up in meiosis, the two homologous partners often undergo an exchange of parts. This exchange of chromosome parts between homologous chromosomes at meiosis is called *crossing over.* The process is diagramed in Fig. 6–14, where the *C* and *S* genes and their alleles *c* and *s* are situated on one pair of homologous chromosomes. We have used again, for the sake of simplicity, our nucleus with only two chromosome pairs. Actually in corn there are 10 pairs. The genes for seed color and texture are on the B-chromosomes. *C* and *S* are on the B^f-chromosome, which came from the female parent (*CCSS*). Both chromosomes from the male parent (A^m and B^m) are drawn in black to contrast with those (A^f and B^f) from the female parent.

In Fig. 6–14*A* the exchange, or *crossover,* has taken place *between* the genes *C* and *S.* The diagram entails a discovery that was one of the great achievements of genetics; each

gene (such as *C* or *S*) has a definite and fixed place, or *locus,* on the chromosome. The crossover involves only two of the four strands. It arises because the B^m and B^f strands broke at strictly homologous points. When the broken ends reunited, they did so in such a way that the B^f strand joined with the B^m strand. As a consequence, one-half of the four gametes produced contain B-chromosomes that are mixtures of B^m and B^f parts and are thus recombination types (*Cs* and *cS*).

In this particular meiosis the crossover occurred between *C* and *S,* but it does not always occur in this region of the chromosome. Sometimes it occurs outside the region bounded by *C* and *S,* and when this happens, no recombination gametes for these genes will be formed (Fig. 6–14*B*). A moment's thought discloses that, once we have the data given in row 6 of Table 6–4, we actually know how often a crossover occurs between the genes *C* and *S.* Altogether, there were 8067 parental-combination gametes produced by the *CcSs* plant, and there were 301 recombination gametes. Let us round these numbers out to 8060 (96.5 per cent) and 300 (3.5 per cent). All told, we have 8360 gametes derived from 2090 cells by meiosis (2090 × 4 = 8360). Remember that in every cell in which a crossover occurs between *C* and *S* half the gametes will be crossovers and half will not. It follows that a crossover occurred between *C* and *S* in only 7 per cent of all the cells undergoing meiosis.

What is the significance of the fact that a crossover occurred between *C* and *S* in only 7 per cent of all the cells that underwent meiosis in the corn plant? Why is it 7 per cent and not, for example, 20 per cent? The answer to this question was perceived by Morgan and his collaborators, especially A. H. Sturtevant, at Columbia University between 1910 and 1915. It provided geneticists with a means of making maps of chromosomes and pinpointing on the maps the positions of individual genes.

Here is their answer: Each chromosome is a long thread on which the genes are located at fixed places in a definite sequence. The frequency of crossing over between two genes is approximately proportional to the distance between them. Thus, if genes are arranged in linear order on a chromosome,

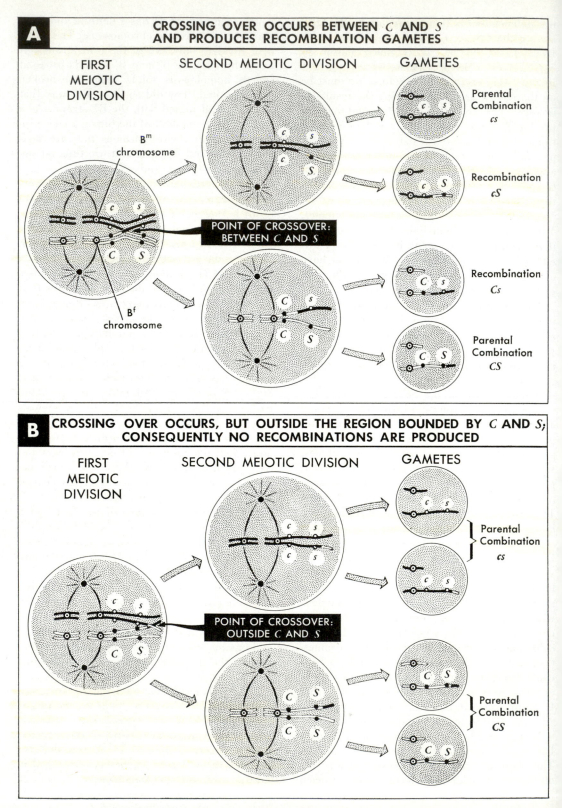

6–14   The chromosomal events involved in crossing over.

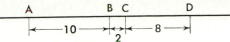

the frequency of crossing over between *A* and *B* (10 per cent) is greater than between *B* and *C* (2 per cent), which are closer together. By measuring crossover frequencies between genes, we can determine not only their relative spacing on the chromosome but also the sequence, or order, in which they are arranged. The procedure is as follows. Suppose that we study another gene, *E*, which is on the same chromosome as *A*, *B*, *C*, and *D*. First we measure the frequency of crossing over between *D* and *E*. It turns out to be 6 per cent. Where does *E* lie? It could be in either of two places.

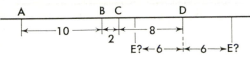

To settle the issue, we measure the frequency of crossing over between *C* and *E*. It proves to be 2 per cent. We now know that the following map is correct.

Where would *E* lie if the *C-E* value had been 14 per cent?

Many of the hundreds of genes so far discovered in *Drosophila* have been mapped (Fig. 6–15). A start, although only the barest, has even been made in mapping human genes. With his long generations and his aversion to controlled breeding, man is the worst of all animals for genetic experiments. Nevertheless, there are exceptionally full nonexperimental data on human pedigrees and on the distribution of many hereditary characters in large human populations. By applying to these data the principles learned from experiments in other species, we have gained a great deal of knowledge about human genetics, and this is a flourishing field of research at present.

Quantitative Inheritance

THE THEORY OF MULTIPLE FACTORS

The new science of genetics, which Mendel founded and later workers developed as the chromosome theory, is based on the assump-

tion of distinct genes arranged in linear file on the chromosome. The early success of genetic analysis depended on the distinctness and recognizability of some individual genes. Their transmission from one generation to the next could now be followed as if it were the transmission of particles. Genetics, like physics and chemistry, became quantitative largely because of its particulate—or "atomic"—foundation, for, in a sense, the gene is an atom of heredity.

During the first decades of this century, an important aspect of genetic phenomena seemed inconsistent with the new theoretical scheme. Experimentation in the classic or Mendelian way centered almost entirely on characters with only a few easily distinguishable alternatives. A flower is red or white; a seed is full or shrunken; a chicken is black, blue, or spotty white. Most variations among organisms are not so simple. Instead of falling into a few sharply distinct groupings, they consist of numerous intergrading conditions. Men are not either tall or short, either black or white; cows do not produce either much or little milk; field mice are not either pale or dark; ears of corn are not either small or large. In all these characters, and indeed in a majority of the characters of plants and animals, there is a broad spectrum between extremes. Individuals cannot be classified merely as one thing or another but can only be placed somewhere along a scale.

The widespread occurrence of continuous, quantitative variation was a serious problem for genetics. It seemed at first that the Mendelian principles did not apply to it. If they did not, they would have decidedly limited value, and a significant part of heredity would remain unexplained. It was, however, possible to frame and test a hypothesis attributing continuous variation to genes inherited in accordance with the Mendelian principles. The hypothesis is that (apparently) continuous variation is controlled by numerous different genes, whose effects tend to add up *without distinct dominance*. This is a reasonable hypothesis, for it has been established that two or more genes can affect one character and that gene effects can add up without dominance. Let us see how the hypothesis interprets continuous variation.

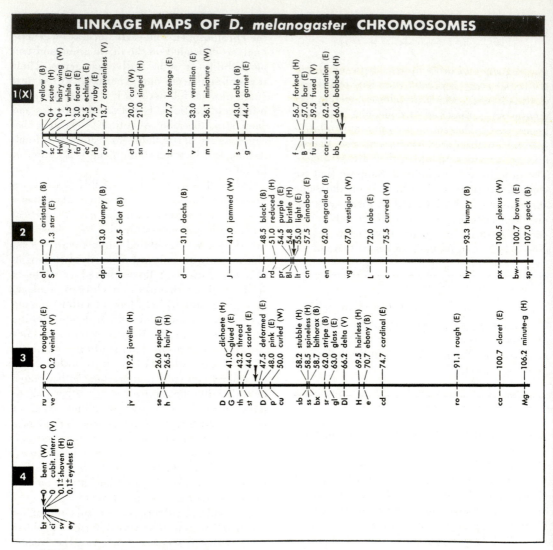

6–15 Linkage maps of *D. melanogaster* chromosomes. The map is to be read by turning the page on its side. The chromosomes are numbered 1 through 4, corresponding with the usage of Fig. 6–8: chromosome 1 is the X-chromosome, 2 and 3 are the large V-shaped autosomes, and 4 is the small dot-shaped autosome. On the map of each chromosome given here, the centromere is marked by an arrow. The descriptive name of each gene (for example, "bent") is given to the right of its position, and the standard symbol for the gene (for example, "bt") is given to the left. The capital letters in parentheses following the name designate the part of the fly most affected by the gene: W = wing, V = veins of wing, H = hairs, E = eyes, B = body. On each map one gene is taken as an arbitrary starting point; its map distance is taken as zero. The positions of the other genes are then mapped in relation to this. The number given for each gene (for example, 13.0 for "dumpy" on the chromosome 2 map) is the standard map distance for that gene; it is discovered experimentally through a study of crossing over. The map distance for "dumpy" implies that there is essentially 13 per cent crossing over between "aristaless" and "dumpy"; there is 3.5 per cent crossing over between "dumpy" and "clot."

Suppose that a character such as height in a plant depends on only two genes, A and B, each with two alleles, A_1 and A_2 and B_1 and B_2. Suppose that A_1 and B_1 make for tallness and A_2 and B_2 for lowness. Then the tallest plants would have the genotype $A_1A_1B_1B_1$, and the lowest would have the genotype $A_2A_2B_2B_2$. If these two were used as parents, all the offspring in F_1 would have the genotype $A_1A_2B_1B_2$. Since we have postulated that no dominance is involved, these offspring with half tall and half low alleles would be intermediate in height. Now suppose that the intermediate F_1 is interbred. The checkerboard (Fig. 6–16) predicts that five height classes would appear in F_2 in the proportions indicated in Table 6–5.

If tallest and lowest were crossed as parents, and if, instead of just two genes, a great many were involved, the hypothesis would lead to these predictions:

F_1 would be intermediate between the parents, with few size classes, and no individuals as tall or as low as the extreme parents.

F_2 would have many size classes, so close together as to intergrade in practice.

The extreme-size classes of F_2 would be about equal to the tallest and lowest parents.

There would be very few individuals in the extreme-size classes of F_2 and increasing numbers of individuals in classes nearer intermediate size.

Do you see how these predictions follow by extension from the simple two-gene situation?

TABLE 6–5

F_2 *progeny from the cross*
$A_1A_2B_1B_2 \times A_1A_2B_1B_2$

Genotypes	Expected ratio	Phenotypes	Expected ratio
$A_1A_1B_1B_1$	1	Tallest	1
$A_1A_1B_1B_2$	2	Medium tall	4
$A_1A_2B_1B_1$	2		
$A_1A_1B_2B_2$	1	Intermediate	6
$A_1A_2B_1B_2$	4		
$A_2A_2B_1B_1$	1		
$A_1A_2B_2B_2$	2	Medium low	4
$A_2A_2B_1B_2$	2		
$A_2A_2B_2B_2$	1	Lowest	1

TABLE 6–6

The inheritance of ear size in corn

| Length of ears in centimeters | Number of individuals | | |
	P_1	F_1	F_2
21	2		1
20	7		2
19	10		1
18	15		12
17	26		11
16	15		10
15	12	4	13
14	11	9	21
13	3	17	27
12		14	33
11		12	33
10		12	33
9		1	17
8	8		5
7	24		2
6	21		
5	4		

They can be checked experimentally. The results in Table 6–6 and Fig. 6–17 were obtained by E. M. East (1879–1938) in a now-classic experiment. Corn with short and long ears was used for the parents.

The predictions from the hypothesis are well confirmed. Many similar tests have been made, and other sorts of tests have been devised and carried out. They leave little doubt that the hypothesis is correct and that continuous variation is due to *multiple factors*, which are genes inherited according to Mendelian principles.

There is another interesting agreement with the theory of multiple factors. Continuous variation in natural populations almost always has the sort of distribution seen in F_2 of experiments like that of East. Most individuals are near the intermediate or average condition, and the number of individuals becomes smaller the greater the deviation from the average.[16] This is one of the most important generalizations on variation in nature (Chapter

[16] As an exercise, account for the following facts: The offspring of a Negro without white ancestors and a white without Negro ancestors are always intermediate in skin color, with little variation. The offspring of couples of mixed Negro and white ancestry and of intermediate color also tend to be intermediate in color more often than not, but they vary greatly, all the way from pure black to pure white.

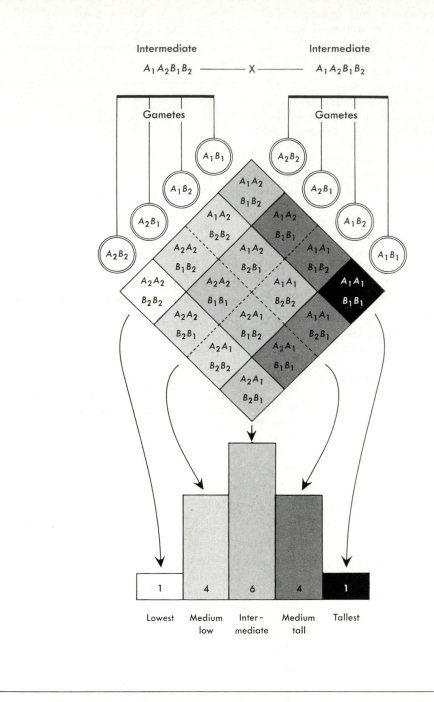

6–16 Multiple-factor inheritance. For an explanation see the text and Table 6–5.

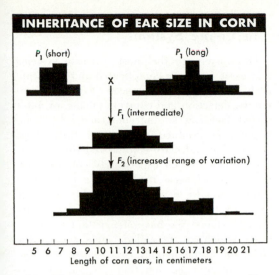

INHERITANCE OF EAR SIZE IN CORN

P₁ (short)   X   P₁ (long)

↓ F₁ (intermediate)

↓ F₂ (increased range of variation)

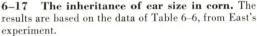

5 6 7 8 9 10 11 12 13 14 15 16 17 18 19 20 21
Length of corn ears, in centimeters

6–17 The inheritance of ear size in corn. The results are based on the data of Table 6–6, from East's experiment.

17). Although variation is also affected by nonhereditary, environmental factors, to the extent (usually large) that it is hereditary, it is explained by the theory of multiple factors.

C H A P T E R   S U M M A R Y

The chromosome theory of heredity: two lines of investigation leading to it and fused by it; discovery of the theory opening up the flood of work that is modern genetics.

Pre-Mendelian ideas on heredity: their derivation from analogies with human nature and human affairs; the blending of inheritance, related to the idea of blood as a vehicle of heredity; the inheritance of acquired characters, an ancient idea derived from legal and cultural backgrounds; pangenesis, and its dismissal by Aristotle and by Weismann, who showed the one-way relationship between germ line and soma.

Mendel's principles of heredity (1866): results of a single-character cross; Mendel's hypothesis of paired factors to explain his results; his tests of the hypothesis; the results and interpretation of a two-character cross: independent segregation (or assortment).

Physical basis of heredity: Sutton's (1902) formulation of the chromosome theory: genes as parts of chromosomes; the chromosomal basis of the 1:1 segregation ratio, of independent assortment, and of the statistical nature of Mendelian heredity.

Tests of the chromosome theory: the role of *Drosophila melanogaster* in the history of genetics; sex linkage; nondisjunction; gynandromorphs.

Linkage and crossing over: linkage groups— their number equal to the number of chromosome pairs; crossing over and its basis in chromosome behavior—crossover maps.

Quantitative inheritance: an early difficulty of the Mendelian theory, explained by the theory of multiple factors and illustrated by East's experiments with ear size in corn.

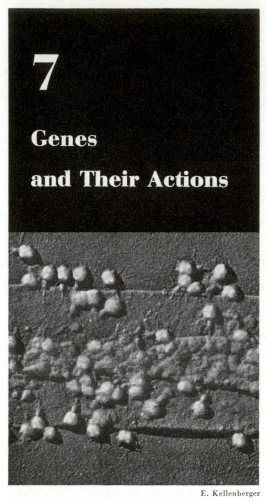

E. Kellenberger

What is a gene? How does it act to control the life of the cell? Partial answers to these questions have come from many experiments, including studies on the heredity of viruses. This is a photograph (×36,000) of viruses attached to a bacterial cell they are about to parasitize.

7

Genes

and Their Actions

The Inherited Message in Living Systems

Organisms are the most intricate systems that exist, far more complicated and elaborate than even the most advanced electronic computers, detection and control systems, or automated factories so far constructed by man. Organisms produce replicas of themselves, and we have already emphasized that the reproduction of such exceedingly complex systems must of necessity involve a transfer of information by devices comparable with the plans or blueprints used in organizing man-made systems. We have shown that these devices, which can be called "messages" in the strict and in the metaphorical senses, are present in organisms, that they are passed from parents to offspring, and that they control the development of the offspring into the likeness of the parents. We have also shown that a major part, at least, of these hereditary specifications is located in the chromosomes. Further, we have noted how genetic experimentation led to the theoretical recognition of genes as the operational units in the processes of heredity and revealed the basic facts about how genes are transmitted and reassorted in the usual circumstances of reproduction.

That something to which the name "gene" was given *does* exist in chromosomes and that a gene *does* produce specifiable results could be and was thoroughly established before anything was really known about the nature of the gene and its behavior. This latter knowledge, still incomplete, has been gained largely in the past few years. The investigations of the last decade on the chemical nature and activity of genes have been especially productive and now constitute a brilliant chapter in the history of biology. In surveying it, we should not merely assimilate facts and conclusions. Here is an unusual opportunity to witness the vitality and pace of modern biological science and to observe how its various subdivisions can all shed light on fundamental problems.

The exploration of heredity must continue in two directions, corresponding with different aspects of scientific explanation (see Chapter 1). The direction just mentioned is concerned with physiological explanation, with how information, figuratively speaking, is coded in the genes and then put into effect. That is the subject of this chapter. It is, however, at least equally important to know what the message means, and why and how it arose in the first place. That is the direction of teleonomic and historical explanation, to be discussed later. In the meantime, let us remember that the significance of the processes considered in this section will be fully appreciated only when they are brought into the broader evolutionary picture.

Nature of the Genetic Material

We shall begin with the question, "What are the physical and chemical characteristics of the genetic material?"

SMALL SIZE OF THE INHERITED MESSAGE

The number and the size of the chromosomes are fixed characteristics[1] of any particular species. All normal humans have a diploid number of 46, and all normal *Drosophila melanogaster* flies have a diploid number of 8. The number and size of chromosomes differ considerably between species, as is illustrated by the comparison of fruit flies and men. The commonest diploid number is 12 or thereabout, although the known range extends from 2 to hundreds. There is no correlation at all between the chromosome number and the degree of complexity of the organism. It is true that man (46) has more chromosomes than the fruit fly (8), but there are ferns and some crabs with chromosome numbers in the hundreds.

The most remarkable fact about the number and size of the chromosomes is the extremely small bulk of material that carries the inherited message. It is not more and probably much less than 1/1000th of the material in the fertilized egg. Moreover, it is by no means certain—it is even unlikely—that *all* the material of the chromosomes is directly concerned with carrying hereditary information.

[1] There are cases in which the number and size of the chromosomes differ in different tissues of the same organism. But these are exceptions that do not invalidate the generalization.

Indeed, we now know that chromosomes are complex chemically as well as structurally. They consist of several ingredients, some of which serve only as skeletal supports for the rest. There is some analogy between a chromosome and the cards or tape that feed information into a computer, but there is a radical difference in that a chromosome is part of a *living* system. As such, it is metabolically active, containing enzymes and metabolic machinery that are not necessarily involved in message carrying.

GENES: SIZE AND NUMBER

It has been possible, with ingenious and indirect approaches, to obtain estimates of both the size and the number of genes in a nucleus. One method for gene size employs a principle that could be used to determine the size of a target at the far end of a dark room. We could calculate the size of such a target from the volleys of small shot fired by a shotgun. If we knew, for example, that there is one pellet per square inch within the area covered by a volley, and if we hit the target, on an average, twice every time we fire a volley, then we could conclude that the target has an approximate area of 2 square inches; but if we hit it, on the average, only once every two times we fire a volley, we could conclude that the target is only about 1/2 square inch in area.

To determine the size of a gene in *Drosophila*, we use volleys of x-ray particles instead of lead pellets. The density of x-ray showers can be controlled. We can tell when a gene has been hit by an x-ray particle, for its structure changes—it undergoes *mutation*—and it produces a recognizably different fly in the next generation. Thus we have a technique for finding out how often we hit the gene with an x-ray beam whose particle density we know. We can then calculate the target size. The results of these and other studies show that a gene is an astonishingly small part of the chromosome, about 10 to 100 millimicrons in diameter. (A millimicron is 0.000001 millimeter.) Similarly indirect approaches tell us that in the four chromosomes of *Drosophila* there are altogether about 10,000 genes (not less than 5000 and not more than 15,000).

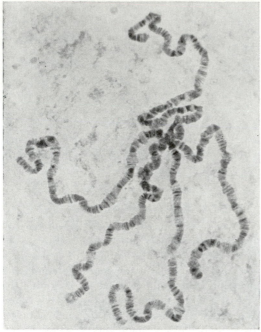

7–1 Salivary-gland chromosomes in *Drosophila melanogaster*. In the salivary glands of many flies, including *Drosophila*, the chromosomes are enormously enlarged and clearly visible even during interphase. Moreover, they are strikingly crossbanded. The bands (local concentrations of nucleic acids) are constant in position from nucleus to nucleus. Thus it is possible to construct topographic maps of chromosomes. Particular genes have been localized in some bands.

WHAT IS A GENE?

We cannot see genes as distinct parts of the chromosome even with the most powerful electron microscopes. Most chromosomes give no visible evidence at all of linear differentiation. Some chromosomes, however, show a linear sequence of small swellings (*chromomeres*) at prophase of mitosis and meiosis; others, in special tissues like the salivary glands of *Drosophila*, are enormously enlarged, with clearly defined crossbands (Fig. 7–1). Some geneticists have suggested that these swellings and bands represent individual genes. However, this is unlikely to be true in general because it has been discovered that at least *some* chromomeres and bands contain more than one gene.

If we cannot see the separate genes on the

chromosome, how do we know that they exist? Let us briefly recount why geneticists have concluded that there are two distinct and separable genes controlling eye color (red versus white) and wing length (normal versus miniature) on the *Drosophila* X-chromosome.

If we cross a "white, miniature" fly to a normal one, the result will be a normal F_1 because the alleles for red eyes and long wings are dominant over those for white eyes and miniature wings (Fig. 7–2). Crossing this F_1 normal fly to a "white, miniature" fly will give us a progeny that will tell us what gametes the F_1 produced. (The principles involved were explained in Chapter 6.)

Suppose that we obtain from this cross only two classes of flies in equal numbers: 50 per cent normal and 50 per cent "white, miniature." What would be the simplest hypothesis about the genetic control of eye color and wing length? For simplicity, we would have to assume that they were controlled by one factor, or one gene [2] (Fig. 7–3).

Entirely different conclusions are forced on us if we find four classes of flies in the testcross progeny. In this case the two new types of fly—white eyes, normal wings, and red eyes, miniature wings—tell us that eye color and wing length must be controlled by quite separate pieces of the chromosome (separate genes) that recombine by crossing over.

Here is a most important point: in this kind of experimental study of inheritance, we can be sure that two genes, rather than one, are involved *only if* crossing over occurs between them and so proves that they are distinct parts of the chromosome. On the other hand, even though there actually are two genes present, if crossing over does not occur between them, we shall be forced to conclude erroneously that there is only one gene.

Thus a possible answer to the question "What is a gene?" is as follows: Genes are the smallest segments of chromosomes known to be separable by crossing over. We shall discover later in this chapter that this answer, although valid in its own context, is not complete and is not the only possible answer.

[2] As we shall see later, it is usual for a gene to affect more than one character and for a character to be affected by more than one gene.

CHEMISTRY OF THE GENETIC MATERIAL

When we ask what chromosomes are made of, difficulties immediately confront us. How indeed can we put a chromosome—much less a gene—into a test tube for the usual type of chemical analysis? We cannot, and the study of chromosome chemistry has necessitated the development of special indirect methods of microchemical analysis. If we have a colorless solution that reacts with substance x to yield a specific color, then we have a tool for detecting the location of substance x in the cell. Thin sections of cells can be treated with the test solution and studied under the microscope; the local appearance of color within cell structures signifies the presence of substance x. Test-staining techniques have revealed that the chromosomes contain, principally, two classes of molecules: protein and nucleic acid. The two kinds of molecules occur in the chromosomes in a saltlike chemical combination, nucleoprotein.

We have already noted the special importance of the two forms of nucleic acid, ribonucleic acid (RNA) and deoxyribonucleic acid (DNA).[3] It will be recalled that DNA is found almost exclusively in the chromosomes of the nucleus, whereas RNA is found mainly in the cytoplasm, although, as we shall see, it is probably synthesized in the nucleus. Let us now consider the evidence that DNA is the primary message-bearing substance of the gene.

GENETIC SIGNIFICANCE OF DNA

The very presence of DNA in the chromosomes strongly suggests that it is the message-carrying fraction of the genetic material,[4] and a recent series of important scientific discoveries greatly strengthened this presumption. An early indication came in the 1930's from studies on the pneumonia-causing bacterium,

[3] It would be helpful at this point for the reader to review the section on nucleic acids and nucleotide chemistry in Chapter 2.

[4] All living nuclei contain DNA, and the amount of it in a nucleus is mathematically related to the number of chromosomes, the DNA level of haploid cells (for example, spermatozoa) being half that of diploid cells. The quantity of DNA doubles during mitotic interphase, just prior to cell division.

Pneumococcus. In *Pneumococcus* there are many different hereditary cell types, and it was learned that a particular type (for example, type III) could be changed into one of the other known types (type II) by a chemical extract taken from type II. The chemical extracted from type II was called the *transforming principle;* it transformed the heredity of type III to that of type II.

The transforming principle was subjected to intensive chemical investigation and found to consist of DNA; the most careful scrutiny revealed only a minute trace of protein. This was an historic experiment in biology because for the first time part of the hereditary message of one genetic strain was extracted as an inert chemical and introduced into and accepted by another genetically different strain.

Equally strong evidence that DNA is the actual carrier of hereditary information has come from recent research on the heredity of viruses. Viruses are extremely small particles, many of them no bigger than some protein molecules (see Fig. 3–4). We have seen (Chapter 3) that viruses are incapable of self-reproduction. Rather, they must enter a living host cell, which is then caused to start producing new viruses. The life cycle of a typical virus (bacteriophage) with bacteria as the host is outlined in Fig. 7–4.

Note that the virus shown consists of little more than a small bit of DNA with a protein coat. Its reproduction, therefore, involves the synthesis of new virus DNA and new virus protein and their assembly into new virus particles. The new viruses are like the old ones, and virus reproduction has the basic characteristics of reproduction by living systems; we may therefore speak of *virus heredity*.

The particular virus in Fig. 7–4 is a widely studied one that infects, and ultimately destroys, the common colon bacillus *Escherichia coli*. This bacterial virus attacks the host cell tail first (see the illustration on p. 194). The tail penetrates the cell membrane. The contents of the main body of the virus then enter the cell as though squeezed from a syringe through the tail, which can be likened to a hypodermic needle. The empty shell of the virus is left outside. What has entered the bacterium is the carrier of virus heredity, and

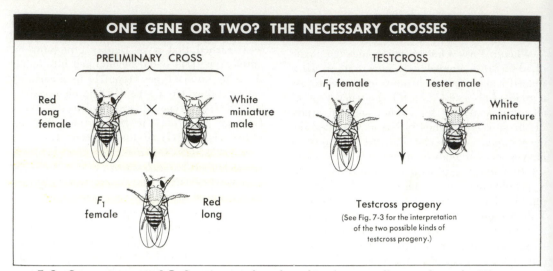

ONE GENE OR TWO? THE NECESSARY CROSSES

PRELIMINARY CROSS

Red long female × White miniature male

F_1 female Red long

TESTCROSS

F_1 female × Tester male

White miniature

Testcross progeny
(See Fig. 7-3 for the interpretation
of the two possible kinds of
testcross progeny.)

7–2 One gene or two? I. Genetic tests show that white (versus red) eye color and miniature (versus long) wings are controlled by the genetic material in the X-chromosomes; they are sex-linked characters. Are these two characters controlled by two separate genes or by one? This question is attacked by making the following crosses: (1) Preliminary cross: red, long female with white, miniature male. The F_1 is red and long since these characters are dominant. (2) Testcross: F_1 female with "tester" male, who is white and miniature. Since white and miniature are recessive, the testcross progeny immediately reveal what kinds of gametes the F_1 female produces.

once inside the cell, it is copied again and again until the cell is destroyed and breaks open, yielding hundreds of new bacteriophage particles complete with protein coat, DNA, and tail.

That bacteriophage protein does not enter the host cell during the original infection was demonstrated in a classic experiment in 1952. *E. coli* cells were grown on a nutrient medium containing both radioactive phosphorus and radioactive sulfur. They incorporated both radioactive elements into their protoplasm. When these cells were infected with virus, the new virus particles produced in them also incorporated both radioactive elements.

The reason for using radioactive phosphorus and sulfur was simple. All DNA contains phosphorus in large amounts, but it contains no sulfur; on the other hand, proteins contain almost no phosphorus, but they do commonly contain sulfur. When, therefore, we have viruses that contain radioactive phosphorus and sulfur, we know that the phosphorus is in the nucleic acid and the sulfur is in the proteins.

Such radioactive viruses were used to infect fresh bacterial cells free of all radioactiv-ity. After the virus material entered the cell, the empty virus cases on the outside were shaken off. It was then discovered that the empty cases contained radioactive sulfur but no radioactive phosphorus; it followed that they contained protein but no nucleic acid. On the other hand, the infected cells contained radioactive phosphorus but no radioactive sulfur. Thus the hereditary material that entered the cell was DNA, not protein. This experiment confirmed the inference, from the earlier study of *Pneumococcus,* that DNA is the principal genetic material.[5]

We must now consider a question of over-

[5] This conclusion introduced a problem. Although bacterial viruses consist of protein and DNA, many viruses that reproduce in plant and animal cells consist of protein and RNA. For example, tobacco mosaic virus (TMV), a well-known virus of tobacco plants, is an RNA-containing virus, as is human poliomyelitis virus. Some interesting experiments have been performed with TMV. It is possible by chemical means to separate its protein from its RNA and to "infect" a tobacco leaf with the RNA alone. Application of pure RNA to the leaf causes new whole TMV viruses to be produced within the leaf cells. TMV protein, on the other hand, causes no new virus production. Therefore, TMV RNA carries all the information for the synthesis of both TMV protein and RNA.

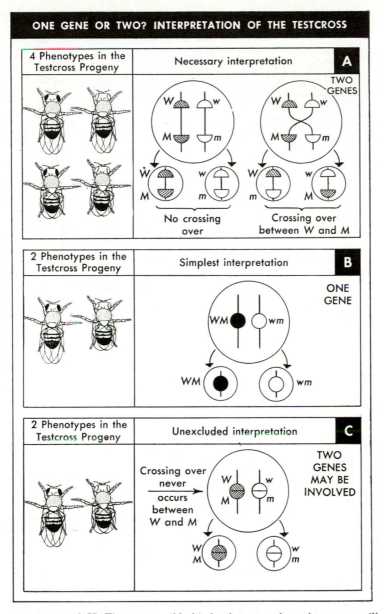

ONE GENE OR TWO? INTERPRETATION OF THE TESTCROSS

4 Phenotypes in the Testcross Progeny

Necessary interpretation

A

TWO GENES

No crossing over

Crossing over between W and M

2 Phenotypes in the Testcross Progeny

Simplest interpretation

B

ONE GENE

2 Phenotypes in the Testcross Progeny

Unexcluded interpretation

C

TWO GENES MAY BE INVOLVED

Crossing over never occurs between W and M

7–3   One gene or two? II. The two possible kinds of progeny from the testcross illustrated in Fig. 7–2 are as follows: (1) Four phenotypes present: white miniature, red long, white long, and red miniature; and (2) only two phenotypes present: red long and white miniature. The interpretation of these alternatives is as follows: *A.* Four phenotypes present: the X-chromosome carries two physically distinct genes. One of these (*W*) affects the eye, and the other (*M*) affects the wing. These genes are separated by crossing over in some cells that undergo meiosis. The F_1 female, therefore, produces four kinds of eggs: *WM*, *Wm*, *wM*, and *wm*. *B.* Two phenotypes present: the simplest interpretation. A single gene (call it *WM*) on the X-chromosome affects both eye color and wing length. Thus the heterozygous F_1 female is *WMwm* and produces only two kinds of eggs: *WM* and *wm*. *C.* Two phenotypes present: an unexcluded possibility. The X-chromosome carries two physically distinct genes: *W* and *M*. They lie close together, and crossing over never separates them. The heterozygous F_1 female is *WwMm* and produces only two kinds of eggs: *Wm* and *wm*.

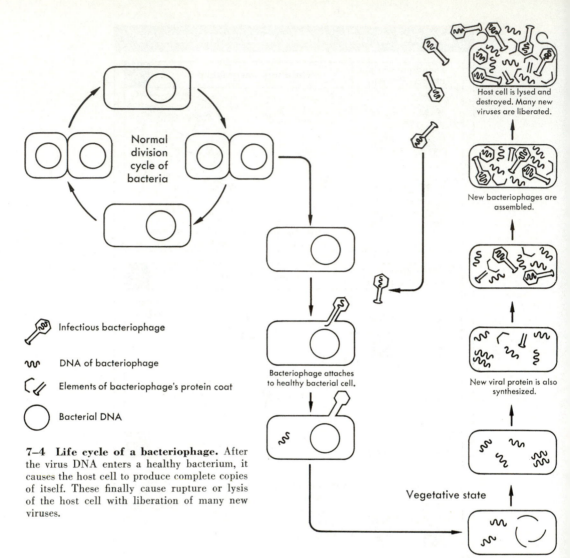

Infectious bacteriophage

DNA of bacteriophage

Elements of bacteriophage's protein coat

Bacterial DNA

7–4 Life cycle of a bacteriophage. After the virus DNA enters a healthy bacterium, it causes the host cell to produce complete copies of itself. These finally cause rupture or lysis of the host cell with liberation of many new viruses.

Normal division cycle of bacteria

Bacteriophage attaches to healthy bacterial cell.

Vegetative state

Bacteriophage takes over metabolic machinery of host cell. Host cell DNA disappears. New viral DNA is synthesized.

New viral protein is also synthesized.

New bacteriophages are assembled.

Host cell is lysed and destroyed. Many new viruses are liberated.

riding importance. We have referred to DNA as the message-carrying molecule in *Pneumococcus* bacteria, in viruses, and presumably in all organisms. How can one kind of molecule carry all the detailed and different specifications for such diverse cases? The answer is that all DNA is not the same.

Chemical analysis of DNA (see Table 2–4) shows that it contains four different nucleotide bases—thymine (T), adenine (A), guanine (G), and cytosine (C). However, to say that it consists only of four nucleotides is like saying that this book consists only of 26 letters and a few numbers, or that a long telegraph message consists only of dots and dashes. In fact, the DNA molecule is immensely long. Monotonous though its constituent parts are, the possibilities for long unique sequences are nevertheless almost numberless. The message in the DNA molecule must reside in the sequence of its parts.

As we saw in Chapter 2, the usual form of DNA is a double-stranded helix with a ladder-like structure whose cross rungs are A-T pairs and G-C pairs, each pair being held together by hydrogen bonds (see Fig. 2–13). Thus the two nucleotide chains are complementary to one another; T is always opposite A, and G is always opposite C. If the hereditary code resides in a certain base sequence in one nucleotide chain, it is obvious that the other strand must contain the identical information in a complementary sequence.

Reproduction of the Genetic Material

So far we have examined the hereditary material by asking: What is its chemical nature? What is it made of? We must now ask: What does it do? How does it do it? Genetic material does two things: (1) it reproduces itself (or it somehow governs its own reproduction by the cell); and (2) it effects general control over the activity of the cell and organism as a whole. In this section we shall turn our attention to the process of DNA replication.

The discovery that DNA is a double-stranded molecule containing complementary base pairings held together by hydrogen bonds suggested a plausible mechanism for the transfer of genetic information from generation to generation. Prior to a cell division, the two DNA strands would separate, but each would maintain its integrity and serve as a template [6] for the formation of its partner strand. Since each base in the chain could accommodate only the correct complementary base, the parent strand would force the synthesis of a complementary image of itself in the new complementary strand. The daughter DNA then would be similarly double-stranded, and the correct nucleotide sequences, identical in every way to those in the parent DNA, would be carried over into the new generation.

Although this view of the replication of DNA was first offered speculatively following

[6] The template function of a DNA strand in determining the character of its partner strand is understandable by analogy with one of the commonplaces of everyday life. Complex surfaces, such as a sculptor's work, are copied by casting in molds or templates. The use of templates in industry for the easy mass production of all kinds of items is well known.

the discovery of the double-stranded structure of DNA, critical evidence soon confirmed its correctness. In one brilliant experiment, M. Meselson and F. Stahl used heavy nitrogen, N^{15}, a nonradioactive isotope of nitrogen, as a label for DNA and ascertained its presence in the daughter DNA molecule.[7] When *E. coli* were grown for several generations in a medium containing N^{15} as the only nitrogen source, all the nitrogen in the cells, including that of the DNA, came to be N^{15}. When heavy DNA (heavy because it contains N^{15}) was isolated from the cells, its increased density made it readily distinguishable from ordinary light DNA in the ultracentrifuge (Fig. 7–5). In the next step of their experiment, Meselson and Stahl replaced the N^{15} in the medium by ordinary N^{14} and allowed the bacteria to undergo several divisions. By sampling the culture after each division and determining the density of the DNA, they obtained a clear picture of the process of DNA replication. After the first division, when the total amount of DNA had doubled, isolated DNA was found to have an intermediate density—that is, the new DNA was not heavy N^{15}-DNA or light N^{14}-DNA but a hybrid, half of whose nitrogen was N^{15} and half N^{14}. After a second cell division and a second doubling of DNA in the N^{14} medium, the centrifuge revealed two kinds of DNA, normal N^{14}-DNA and hybrid DNA, in equal amounts. These results are clearly accounted for by the diagram in Fig. 7–5. If DNA were to duplicate by first undergoing a breakdown into small fragments, the DNA isolated after the first division would have a variety of densities, ranging from that of pure N^{14}-DNA to that of pure N^{15}-DNA. Instead, before duplication the two DNA strands separate (presumably by unwinding from their coiled arrangement), and then, *remaining intact*, each acts as a template for the formation of a new companion chain. Thus the new

[7] The periodic table indicates that nitrogen has an atomic weight close to 14. Most atoms of nitrogen have this weight. However, there is a nonradioactive isotope with an additional neutron in its nucleus and thus an atomic weight of 15. This is heavy nitrogen, or N^{15}. Bacteria will grow equally well in media containing either pure $N^{14}O_3^-$ or pure $N^{15}O_3^-$ as the sole nitrogen source. In the latter case, all N-containing molecules of the cell (including DNA) will contain N^{15} instead of N^{14}.

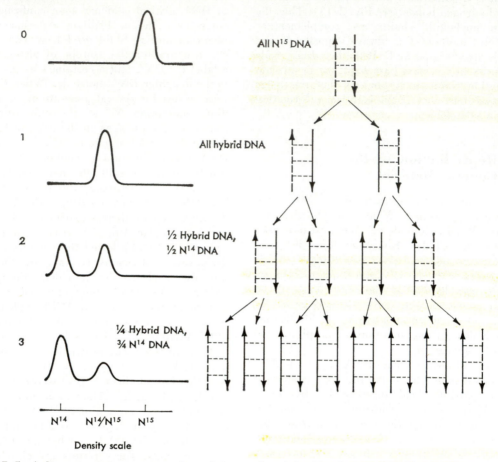

Generations

0   All N¹⁵ DNA

1   All hybrid DNA

2   ½ Hybrid DNA, ½ N¹⁴ DNA

3   ¼ Hybrid DNA, ¾ N¹⁴ DNA

N¹⁴   N¹⁴/N¹⁵   N¹⁵

Density scale

7–5 A diagrammatic representation of the results of the Meselson and Stahl experiment on the mechanism of DNA replication. - - -, N^{15}-containing strand; ——, N^{14}-containing strand. Details are given in the text.

DNA contains an old N^{15}-strand and a new N^{14}-strand. In the next division, the N^{15}-strands form new hybrids, and the N^{14}-strands form normal N^{14}-DNA.

In another ingenious experiment, J. H. Taylor, P. S. Woods, and W. L. Hughes made the DNA in the root tip of the broad bean, *Vicia faba*, radioactive by growing the plant in the presence of radioactive thymidine (a thymine-containing DNA precursor). Tritium, a radioactive isotope of hydrogen (having an atomic weight of 3 instead of the 1 of ordinary hydrogen) was used as the radioactive label. When a minute object containing tritium is placed on a photographic plate, only the molecules of sensitive emulsion that directly touch the tritium-labeled molecules are exposed. When

the plate is developed, the exposed areas accurately identify those fine structures within a cell that are radioactive. This technique is called *radioautography*. The investigators' radioautographs showed that radioactivity was equally distributed throughout all of the chromosomes. If the cells were then permitted to divide once in the *absence* of radioactive thymidine, the radioautographs showed that a chromosome contained one labeled strand and one unlabeled strand (Fig. 7–6). Note that this experiment, in contrast to the N^{15} experiment, reveals the doubleness of the replicating genetic structure at the chromosomal, not the molecular, level. The two experiments, however, complement one another. The tritium results demonstrate that the chromosome con-

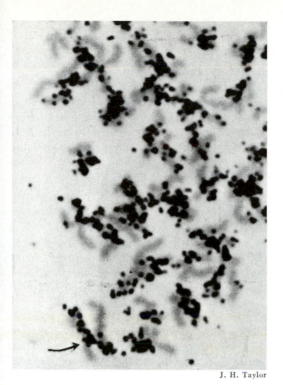

J. H. Taylor

A

7–6 The chromosome-duplication experiment of Taylor, Woods, and Hughes. *A.* A photograph, superimposed by exposed grains of photographic emulsion, of several metaphase chromosomes at the second division after labeling by H³-thymidine (× approximately 3000). Note that only one of each of two chromatids is labeled. Chromatid exchanges sometimes occur as shown at the arrow. *B.* A diagrammatic interpretation of the observation in *A. 1.* Chromosome before duplication; seen as one strand under the microscope, it is revealed to be double-stranded in this experiment. *2.* Chromosome after duplication in the presence of H³-thymidine. *3.* Appearance of metaphase chromosomes, showing uniform labeling. *4.* Separation of the uniformly labeled chromosomes. *5.* Duplication in the absence of H³-thymidine. *6.* Duplicated chromosomes at the next metaphase; one is radioactive, and the other is not. Uniform labeling at the first metaphase but not at the second metaphase results from the double-strandedness of the chromosomal DNA. The solid lines indicate unlabeled strands, and the dashed lines indicate radioactive strands; shadowing indicates grains seen in the radioautographs, resulting from the presence of H³-thymidine.

B

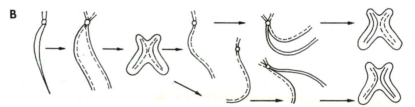

Duplication with labeled thymidine   First metaphase after labeling   Duplication without labeled thymidine   Second metaphase after labeling

tains two DNA subunits (strands), both of which are duplicated and then segregated from one another in cell division.

A final and decisive bit of evidence on the mechanism of DNA replication is found in the enzymatic studies of Arthur Kornberg, who successfully synthesized DNA in a test tube. Soon after the discovery of the double-strandedness of DNA, Kornberg discovered DNA polymerase, the enzyme that catalyzes the formation of the DNA chain from its four component deoxyribonucleotides (in the form of their triphosphate derivatives). Interestingly, the synthesis requires not only the four nucleotides, magnesium ions and DNA poly-

merase but also a small amount of single-stranded fully formed DNA to act as an information-containing and sequence-ordering primer or template for the reaction. The reaction can be formulated as follows: [8]

[8] As noted in Table 2–4, the letters A, G, C, U, and T denote the purines and pyrimidines adenine, guanine, cytosine, uracil, and thymine, respectively. The symbols R and dR refer to ribose and deoxyribose, so that $A \cdot dR \cdot P$ is deoxyadenosine *mono*phosphate and $A \cdot dR \cdot PPP$ the corresponding *tri*phosphate. Likewise, $A \cdot R \cdot P$ is adenosine monophosphate, and $A \cdot R \cdot PPP$ adenosine triphosphate. Note that DNA and RNA are polymers consisting of x molecules and that the true substrates of both DNA and RNA polymerase are nucleoside triphosphates.

$$\begin{array}{l} x \text{ molecules of A} \cdot \text{dR} \cdot \text{PPP} \\ x \text{ molecules of T} \cdot \text{dR} \cdot \text{PPP} \\ y \text{ molecules of G} \cdot \text{dR} \cdot \text{PPP} \\ y \text{ molecules of C} \cdot \text{dR} \cdot \text{PPP} \end{array} + \text{DNA (primer)} \rightleftharpoons \begin{pmatrix} x\text{A} \cdot \text{dR} \cdot \text{P} \\ x\text{T} \cdot \text{dR} \cdot \text{P} \\ y\text{G} \cdot \text{dR} \cdot \text{P} \\ y\text{C} \cdot \text{dR} \cdot \text{P} \end{pmatrix}_x \cdot \text{DNA} + (2x + 2y)\text{PP}$$

It is significant that polymerization does not occur unless single-stranded DNA[9] is present. When the primer has a measurable biological property, such as transforming activity (p. 197), the newly synthesized DNA also has it. Biochemists have not yet devised a method for determining nucleotide sequences, and so the replication of genetic information leaves us with the presumption, if not the proof, that the base sequence of the primer is present in the new DNA strands.

Mechanism of Gene Action

We learned in Chapter 6 how genes are transmitted to the offspring, how they are recombined, and how certain regularities observed in breeding experiments can thereby be explained. We have also seen that the primary genetic material is DNA, a double-stranded structure that duplicates by separating its two strands, each of which then serves as a template in the formation of a new partner strand. In this way the unique nucleotide sequence of each DNA molecule is handed down to daughter cells. We must now inquire how genes determine the existence of given traits. In recent years it has become apparent that genes act primarily by controlling the structures and rates of synthesis of specific proteins (notably enzymes). In late 1961 discoveries were made of the form in which the relevant information is coded in the nucleic acid molecule, the method of communicating the coded message to the sites of protein synthesis, and the mechanisms governing the amino acid sequences upon which specificity depends. Let us consider some of the evidence underlying these new and epoch-making concepts of gene action.

THE ONE GENE—ONE ENZYME THEORY

The preceding chapters have indicated that once the zygote of a future organism has been

formed and its future traits determined, it becomes a self-regulating metabolic system engaged primarily in extracting energy from nutrient materials and converting them to building blocks for the compounds essential in tissue synthesis. Metabolism thus may be illustrated quite simply.

$$A \xrightarrow{e_1} B \xrightarrow{e_2} C \xrightarrow{e_3} D \xrightarrow{e_4} E \xrightarrow{e_5} F$$

A is a nutrient (or its derivative); F is a compound that the organism must have; B, C, D, and E are products and precursors in an orderly sequence of stepwise chemical reactions; and e_1, e_2, etc., are enzymes, each of which catalyzes one of the reactions.

We may regard the traits of an organism as visible consequences of sequences of enzyme reactions. For example, there are genes that cause hair to be black; but the direct cause of black hair is a black pigment synthesized from simpler materials by a sequence of enzyme reactions in skin cells. Other phenotypic traits may similarly be attributed to specific sequences of enzyme reactions, although the relationships may be much less direct than in the example.

The work of George Beadle and Edward Tatum, published in 1945, led to one of the first great simplifications in our ideas of gene action—the one gene—one enzyme theory. These investigators performed their experiments with a common red bread mold, *Neurospora crassa*. A typical experiment is diagramed in Fig. 7–7. When organisms that were fully capable of synthesizing the vitamin thiamine were exposed to x-irradiation, they produced offspring that would not grow unless thiamine was supplied. Radiation had apparently eliminated the capacity of the offspring to produce thiamine for themselves from ordinary dietary carbon sources such as glucose. It was found that this loss of synthetic ability was associated with the absence (or nonfunction) of a specific enzyme. If we represent this enzyme as e_3 in our sequence, assuming it to

[9] Single-stranded DNA is easily prepared in the laboratory from purified DNA heated to separate the strands and then quickly cooled to prevent reassociation.

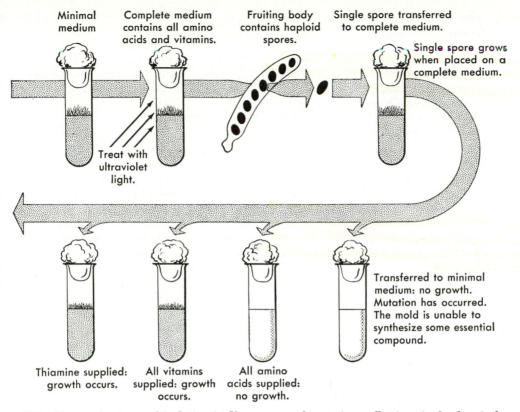

Minimal medium

Complete medium contains all amino acids and vitamins.

Fruiting body contains haploid spores.

Single spore transferred to complete medium.

Single spore grows when placed on a complete medium.

Treat with ultraviolet light.

Transferred to minimal medium: no growth. Mutation has occurred. The mold is unable to synthesize some essential compound.

Thiamine supplied: growth occurs.

All vitamins supplied: growth occurs.

All amino acids supplied: no growth.

7–7   The production and isolation in *Neurospora* of mutations affecting single chemical reactions. The sequence of steps illustrated demonstrates that a mutation has been induced that affects the synthesis of the vitamin thiamine.

be the pathway of thiamine synthesis, the organism had lost the capacity to convert *C* to *D*. When growth was sustained with an artificial supply of thiamine, reproduction continued normally, and all subsequent generations displayed the same enzyme deletion. Hence they were in perpetual need of the product of the missing enzyme.

The implications of this classic experiment are obvious. A mutation, induced by radiation, produced strains lacking one enzyme. Such strains are called *auxotrophic* mutants (from Greek roots meaning increased nutrition) because, through single gene mutations, they have acquired one more nutritional requirement than the *prototrophic* (from roots meaning first or basic nutrition) parent strain. As long as the environment provides an ample supply of the compound the organism can no longer make for itself, the heritable loss of an enzyme does not jeopardize survival.[10] On the basis of many similar data, Beadle and Tatum postulated that a single gene governs the synthesis of a single specific enzyme.

[10] This suggests an explanation for certain of the nutritional needs of man and other higher organisms. For example, thiamine is just as necessary in human metabolism as in *Neurospora* metabolism, but the normal (wild) *Neurospora* can make it from simpler compounds whereas man cannot. Hence human beings are thiamine-requiring auxotrophs. Vitamins are defined as essential substances that an organism cannot make for itself; thus thiamine is a vitamin for man and for the *Neurospora* mutant but not for wild *Neurospora*. We may reasonably speculate that our remote evolutionary ancestors could make thiamine but that, through mutation, the enzymatic machinery for its synthesis was lost. Had there been no organisms in the environment (for example, *Neurospora*, wheat plants, and other foods), that could make thiamine for the mutants, they would have been eliminated, and we should either retain the ability to synthesize thiamine—or be extinct.

HOW DNA CONTROLS PROTEIN SYNTHESIS

It is now evident that genes act by controlling amino acid sequences in the synthesis of proteins (including enzymes). Proteins contain 20 amino acids. DNA contains only four bases. How, then, can DNA be coded so that its genetic information can be read off as an amino acid sequence?

The answer to this question emerged in late 1961 and 1962. DNA was first found to direct the synthesis of RNA in much the same way that it directs its own replication. In 1960 an enzyme was discovered that catalyzes the synthesis of RNA—an RNA polymerase that, like DNA polymerase, demands the presence of a DNA primer. Moreover, the base ratio of the newly formed RNA was fundamentally identical to that of the DNA primer (if the U in RNA is taken as the equivalent of T in DNA).

$$
\left.\begin{array}{l}
a \text{ molecules of A} \cdot \text{R} \cdot \text{PPP} \\
b \text{ molecules of U} \cdot \text{R} \cdot \text{PPP} \\
c \text{ molecules of G} \cdot \text{R} \cdot \text{PPP} \\
d \text{ molecules of C} \cdot \text{R} \cdot \text{PPP}
\end{array}\right\} + \text{DNA (primer)} \rightleftharpoons \left(\begin{array}{l}
a\text{A} \cdot \text{R} \cdot \text{P} \\
b\text{U} \cdot \text{R} \cdot \text{P} \\
c\text{G} \cdot \text{R} \cdot \text{P} \\
d\text{C} \cdot \text{R} \cdot \text{P}
\end{array}\right) + \text{PP} + \text{DNA}
$$

The requirement for DNA in RNA synthesis suggested the following means for transmitting information from the nuclear gene to the cytoplasmic ribosome, the site of protein synthesis (Chapter 3): the coded instructions for protein synthesis contained in the DNA nucleotide sequence are duplicated in the arrangement of nucleotides in the DNA-determined RNA strand; the RNA then carries these instructions from the nucleus to the ribosome. The existence of this *messenger RNA* and its template function were postulated as a theoretical necessity in early 1961 and confirmed experimentally a few months later.

Protein synthesis begins only after another series of events has taken place (Fig. 7–8). First the individual amino acids must be carried to the ribosome. It was learned in 1957 that another type of RNA, much smaller than messenger RNA, serves as an adaptor molecule and performs this transport function. Because of its role, it was termed *transfer RNA*. Special enzymes (plus ATP) activate the amino acids, and then a specific transfer-RNA

molecule, a different one for each of the 20 amino acids, moves in, attaches itself to an amino acid, and carries it to the ribosome. Transfer RNA thus provides the mechanism by which the amino acids are brought into confrontation with the messenger RNA.

The genetic code

How is specificity stamped into the amino acid sequence? How is the amino acid chain faultlessly arranged in a given order? How can a four-letter alphabet (the nucleotides) give a 20-word dictionary (the amino acid determinants)?

It was clear fairly early in 1961 that the amino acid sequence depends on the nucleotide sequence of the messenger RNA that has coated the ribosome surface (Fig. 7–9). It seemed likely that the correct amino acid sequence could be achieved if a particular group of bases in each strand of transfer RNA would attach to a complementary group of bases in the messenger RNA on the ribosome. For example, the base sequence uracil-adenine-guanine (UAG) in the transfer RNA would attach to the complementary sequence adenine-uracil-cytosine (AUC) in the messenger RNA.[11] The amino acid would be left dangling in a manner that would allow it to form peptide bonds with the amino acids dangling on either side of it. (The chemical mechanisms that link amino acids were discussed in Chapter 2.) When a full chain of amino acids had joined in the sequence dictated by the gene's messenger, the resulting protein would peel away from the ribosome and enter the cell sap to serve as an enzyme, hormone, or structural element.

The spectacular breakthrough in deciphering the amino acid–ordering code was made possible by the production of synthetic or unnatural messenger RNA's containing combinations of nucleotides chosen by the investigator. When a synthetic messenger RNA containing only uracil nucleotides was added to a

[11] Bear in mind that A is always complementary to U (or T) and C to G.

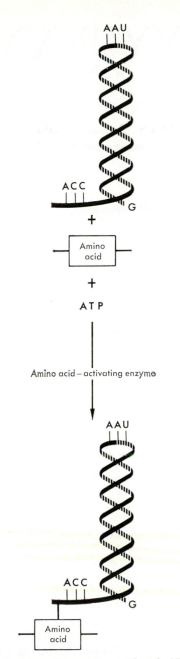

7–8 Activation of an amino acid with ATP and amino acid–activating enzyme and attachment of the activated amino acid to a molecule of transfer RNA specific for the amino acid. Note that transfer RNA is a helical form of RNA containing one unpaired triplet. It is this triplet (AAU in the example) that provides the means by which transfer RNA "recognizes" the messenger RNA triplet specific for the amino acid. Transfer RNA carries ACC at the point where the amino acid attaches and G at the opposite end.

cell-free protein-synthesizing system (complete with amino acids, ATP, isolated ribosomes, and transfer RNA's), it was discovered that the artificial protein synthesized contained only one amino acid, phenylalanine. It was apparent that the base uracil (U or some combination of U's, such as UU, UUU, or UUUU) was the code word that instructed the ribosome to insert phenylalanine in the chain.

When the synthetic messenger RNA contained one molecule of adenine for every two molecules of uracil,[12] it was found that the artificial protein synthesized consisted mostly of phenylalanine with small amounts of isoleucine. The 2:1 proportion of uracil to adenine in the synthetic messenger RNA indicated that the code word for isoleucine must contain two U's for one A (that is, UUA, UAU, or AUU). The incorporation of phenylalanine into the protein presumably reflected the many stretches of UUU that would be expected in the messenger RNA. When the ratio of adenine to uracil was 2:1, asparagine was present in the artificial protein—in addition to phenylalanine and isoleucine. The three amino acids were incorporated in direct proportion to the statistical frequency with which 3U, 2U:1A, and 1U:2A would be expected in the messenger RNA. Thus the code designation for asparagine is 2A:1U (sequence unknown).

By such studies the code has been elucidated for all 20 amino acids (Table 7–1). Although some amino acids have more than one code word—the significance of this fact is unknown—it is believed that each nucleotide triplet is specific for a particular amino acid.[13]

[12] Messenger RNA containing bases in any proportions (for example, 2U:1A, 2A:1U) can be made, but the sequence of the bases cannot yet be ordered. It is easily calculated, however, that 2A:1U messenger RNA would contain 3A, 2A:1U, 2U:1A, and 3U sequences in a frequency ratio of 1:2:4:8. Such an artificial messenger would, for example, stimulate the incorporation of twice as much phenylalanine (UUU) as isoleucine or leucine (both UUA or 2U:1A).

[13] A code with multiple words for each term coded is called "degenerate." Table 7–1 shows that the genetic code is indeed degenerate. However, degeneracy of this kind does not imply lack of specificity in protein synthesis. It simply means that a specific amino acid can be directed to the proper site in a protein chain by more than one code word. Presumably this flexibility in coding benefits the cell.

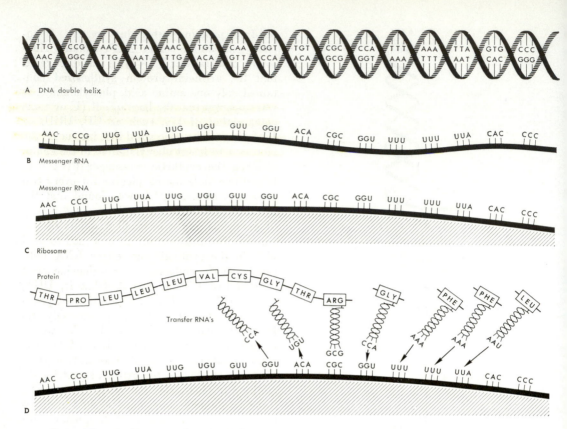

7–9   The mechanism of action of messenger RNA. *A.* The DNA molecule is a series of nucleotide triplets. Note the A-T and G-C pairings in complementary strands. *B.* Messenger RNA is a series of triplets complementary to one of the two DNA strands. *C.* Messenger RNA finds its way to a ribosome, the site of protein synthesis. *D.* Amino acids are carried to the proper sites on messenger RNA by molecules of transfer RNA. The amino acids are then linked by peptide bonds, and the resulting protein "peels off."

Therefore, if the sequence of bases in messenger RNA were UUU-UUA-AAU-UUU, the sequence of amino acids would be phenylalanine-isoleucine-asparagine-phenylalanine—or, using the standard amino acid symbols (see Table 2–3), PHE-ILEU-ASP(NH)₂-PHE. In summary, it appears that nonoverlapping nucleotide triplets make up the code that determines the sequence of amino acids in a protein.

Elegant indirect support for the genetic code is furnished by a number of recent experiments. For example, the genetically determined human disease called *sickle-cell anemia* is associated with the presence of an abnormal hemoglobin molecule. Vernon Ingram conclusively demonstrated that this abnormality is due entirely to the insertion of *one erroneous amino acid* in the chain of over 300 amino acids that comprises the hemoglobin molecule. Instead of the sequence

-HIS-LEU-THR-PRO-GLU-GLU-LYS-

the abnormal molecule contains

-HIS-LEU-THR-PRO-VAL-GLU-LYS-

Through mutation, valine has replaced glutamic acid in the amino acid chain. Ingram's discovery, made before the elucidation of the genetic code, was significant because it showed a specific *biochemical* consequence of mutation and indicated the localized nature of the mutational event, for surely this localized error in the gene product must reflect an equally localized change in the messenger RNA and its parent DNA. When the genetic code was discovered, it was found that one code word for glutamic acid (UAG) differs

TABLE 7–1

The genetic code

Amino acid	Code word *
Phenylalanine	UUU
Alanine	CCG, UCG
Arginine	CGC, AGA, UCG
Asparagine	ACA, UAA
Aspartic acid	UAG
Cysteine	UUG
Glutamic acid	GAA, UAG
Glutamine	ACA, AGA, UCA
Glycine	AGG, UGG
Histidine	ACC
Isoleucine	UAA, UUA
Leucine	UUC, UUG, UUA
Lysine	AAA, AAG, UAA
Methionine	UAG
Proline	CCA, CCC, UCC
Serine	UCG, UCC, UUC
Threonine	CAA, CAC
Tryptophan	UGG
Tyrosine	UUA
Valine	UUG

* The precise sequence of bases in each triplet has not yet been determined. This may explain why some triplets represent more than one amino acid. Furthermore, some amino acids are represented by more than one triplet. Presumably possible triplets not listed here are "nonsense words."

from that for valine (UUG) by only *one* base. Thus it may be concluded that the mutation causing the synthesis of an abnormal hemoglobin—possibly many hundreds of generations ago—in fact altered no more than one nucleotide in the DNA sequence.

Structural, regulator, and operator genes

The classical concept of genes used up to this point considers them as independent molecular units that determine the structures of individual cellular constituents. It has been suggested that they be called *structural genes*, since they act by forming a cytoplasmic transcript of themselves, the messenger RNA, which in turn governs protein structure.

We learned in Chapter 4 that several cellular mechanisms control the synthesis of enzymes; in enzyme induction, enzyme synthesis is greatly accelerated, and in feedback repression, it is diminished. These are obviously key devices utilized by the cell to regulate its metabolic machinery. The capacity to form inducible or repressible enzymes is a genetic

endowment. We know this because we can bring about its loss through genetic mutation (which affects only the gene). Therefore, we infer that genes other than the structural genes exist. The function of these other genes is to initiate, operate, or regulate the activities of associated structural genes. The function of the structural genes is to determine the amino acid sequence of the enzyme. When the structural genes are mutated, the resulting protein is defective (as in sickle-cell hemoglobin).

Although these ideas are still in a developmental stage, it has been confirmed that the synthesis of messenger RNA occurs at certain regions of the DNA strand. These are the structural genes. The initiation of messenger-RNA synthesis is under the control of neighboring regions of the DNA strand. These are the *operator genes,* or *operators.* They may be regarded as switches that turn messenger-RNA synthesis on and off. In most cases, a single operator governs the activities of a cluster of structural genes, a cluster containing patterns of all the enzymes of a metabolic pathway. The combination of an operator gene and the structural genes under its control is called an *operon* (Fig. 7–10).

In addition, *regulator genes* are thought to exist. They are believed to produce a macromolecular substance (which may also be a form of RNA) known as *aporepressor.* It is probably a transcript of a regulator gene, just as messenger RNA is a transcript of a structural gene. However, it does not provide structural information for protein synthesis. One current view holds that aporepressor formed by a specific regulator gene has two important combining properties. First, it has an affinity for a specific operator gene. When the two combine (reversibly, presumably by complementary base pairing), the operator is prevented (or largely prevented) from initiating the synthesis of messenger RNA by the structural genes.

Second, it has an affinity for certain small molecules in the cytoplasm. The combinations are of two types. In some systems (those called *inducible*), the small molecule is an inducer. Combination with an inducer inactivates aporepressor, and enzyme synthesis takes place without limitation. In other systems (those called *repressible*), the small

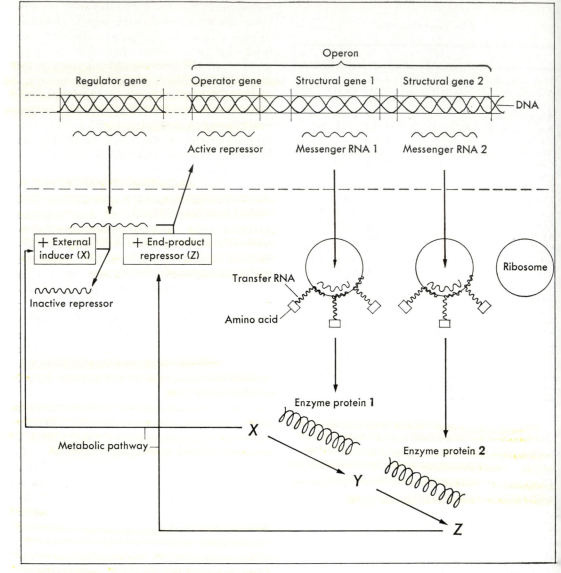

7–10   The general model of the regulation of enzyme synthesis according to F. Jacob and J. Monod.

molecule is the end product of a metabolic pathway, or a *repressor*. Combination of aporepressor with a repressor blocks operator function. In the absence of a repressor, aporepressor is ineffective, and enzyme synthesis is unlimited (derepressed). In the presence of a repressor, aporepressor is active, and enzyme synthesis is repressed. Under ordinary conditions, the synthesis of most enzymes proceeds at a rate substantially below the maximum possible rate. This situation is attributable to the fact that some repressor is usually available to combine with aporepressor.

The repression or induction of enzyme formation is of great importance, and the ability of an organism to perform such adaptive maneuvers is part of its genetic inheritance. Hence it is included in a genetic constellation that must be viewed as a series of blueprints for protein structure *and* a coordinated program for protein synthesis.

MUTATIONS

We have now considered the gene from the viewpoint of the classical geneticists, whose methodology is the breeding experiment and the test of progeny in intact organisms, and from the viewpoint of biochemistry, which seeks to understand molecular mechanisms of replication and sequence coding. The problems of mutation—the alteration of the gene in an irreversible manner—may also be considered from these two viewpoints. Over the short period of human experience, the most obvious aspect of heredity is its highly conservative nature. It assures that each normal individual will develop just those characteristics that define its species. All members of *Homo sapiens,* or of any other species, resemble all other members in essential ways and, as a rule, tend to resemble their parents more than more distant relatives within the species. This conservativism is evidently explained by the replication of DNA and the distribution of chromosomes in mitosis and meiosis.

Nevertheless, heredity can and does change. If it did not, evolution could not occur, and man and all the other species of today would not exist. Therefore, change is just as normal, characteristic, and important in heredity as is conservatism. Initial changes take place as sudden, discrete events in individual cells. They are mutations in a broad sense.[14]

Plant and animal breeders have long known that an individual sharply different from its parents may suddenly appear in the most uniform, carefully bred strain. Such "sports," as the early breeders called them, or *mutants,* as they are now called, have been a source of numerous distinctive true-breeding varieties. The ancon mutation (Fig. 7–11), producing animals too short-legged to jump fences, has occurred at least twice in sheep. The bulldog-

[14] The word "mutation," from the Latin for change, was given more or less its present genetic meaning by De Vries. Unknown to him, the term had already been used in several different senses by zoologists and paleontologists over a century or more. The resulting ambiguity led to misunderstanding, but the genetic usage is now practically universal.

Courtesy LIFE Magazine © TIME, Inc., 1947

7–11 Ancon, a dominant mutation in sheep. The short-legged ram on the right carries a mutant gene ("ancon") that causes a shortened leg. Bred to a normal ewe (*left*), it produces an ancon F_1. This particular mutant is agriculturally useful, for it cannot jump fences.

faced (locally called *ñata*) breed of Argentine cattle represents a mutation that has also occurred in swine and other animals as well as in dogs. Mutations for hornlessness have occurred in many horned animals and have been used to develop hornless breeds. Horses that pace by heredity rather than by training have been bred from mutants. Many popular horticultural varieties—dwarf, double-flowered, variegated—have appeared spontaneously as mutants. Many mutations have also been observed in humans. Queen Victoria's tragic hemophilia arose as a mutation in the germ cells of either herself or her parents, although this was not known at the time.

Mutations result in heritable alterations in the genotype. They may have all sorts of visible effects, from barely perceptible to great and obvious. Mutations are, indeed, the ultimate sources of all *new* genetic materials, which then are endlessly shuffled in the processes of sexual reproduction. All evolutionary change depends in the final analysis on mutations. Of course, not all mutations lead to evolutionary change. Most of them are disadvantageous and are eventually eliminated. What happens to mutations after they occur—whether they vanish, remain as rare variants, or spread and cause significant change—will be dealt with in Chapter 15.

Mutations have been seen in all organisms that have been carefully studied for any length of time, from bacteria to the highest plants and animals. That they appear in all organisms is a very fundamental generalization; a capacity for mutation is one of the universal and definitive characteristics of life, and all organic evolution is contingent on it.[15]

Breeding experiments together with microscopic observations of chromosome sets and structures have shown that mutations may occur in several different ways, in fact, in as many ways as the copying and transmission mechanisms may vary. All mutations fall into one of the following categories:

Chromosome mutations

Change in chromosome number
Change in chromosome structure

Gene mutations

Change in an individual gene

The gene mutations are believed to have the most far-reaching significance. Indeed, when the word "mutation" is used without specification as to whether gene or chromosome mutation is meant, gene mutation is generally implied. Chromosomal mutations do, however, play important roles in some phases of variation and evolution. Let us briefly consider them before we discuss gene mutation.

Changes in chromosome number: polyploidy

We have learned that an organism normally has a basic set of chromosomes, the haploid (simple) set, in which the number of chromosomes may be symbolized as n. A biparental organism usually receives one full set from each parent and therefore has two sets, each with n chromosomes, for a total of $2n$ chromosomes; $2n$ is the diploid (double) number. In both mitosis and meiosis there is a stage at which the numbers are doubled, so that for a brief time $4n$ chromosomes are present in each cell. In mitosis a single division then

produces two diploid cells with $2n$ chromosomes each. In meiosis two divisions produce four haploid cells with n chromosomes each.

Now suppose that mitosis and meiosis do not proceed normally. If the mitotic-spindle mechanism fails (in any one of several ways) to operate properly, the two duplicate diploid sets will not separate into the nuclei of the two new daughter cells. Instead they will remain clumped together and reconstitute a nucleus with double the usual chromosome number—$4n$ instead of $2n$. Other irregularities in division may eliminate some chromosomes or duplicate some. For instance, if the spindle mechanism fails on only one of the metaphase chromosomes, one daughter cell will have $2n + 1$, and the other will have $2n - 1$. Such irregularities in chromosome distribution can, but rarely do, occur in both mitosis and meiosis.

Individuals can therefore originate with not only $4n$ but also $3n$, $5n$, $6n$, and so on, chromosomes, as well as with odd numbers. Such numbers are *polyploid* (multiple), the new individuals are *polyploids*, and the conditions giving rise to them are summed up as *polyploidy* (Fig. 7–12).

Polyploids differ in various ways from their parents. Their cells, containing more chromosomes, are usually larger. Often the organisms themselves are larger. Polyploids derived from the crossing or hybridization of different races and species may be more or less intermediate but often also have some new characters of their own. Thus polyploidy results in the sudden appearance of new kinds of organisms. It is a form of mutation. Polyploidy has been important in the systematic improvement of agricultural crops—both by the uninformed selection of favorable strains that were later found to be polyploid and by scientifically planned hybridization programs.[16]

Chromosome rearrangements

Genes are arranged in single file along a chromosome. In any chromosome they have a definite sequence, so that if they are designated as A, B, C, D, and so on, a chromosome can be represented as A-B-C-D-E-F-. . . . (Actually, there are hundreds or thousands of

[15] As we shall see later, a mutation is usually defined as an "error" in reproduction. The designation is accurate in a sense, because the occurrence of a mutation indicates that the usually conservative mechanisms of cell heredity failed to make an exact copy. However, since mutations are normal, frequent, and absolutely necessary for the development of the whole world of life, to call them errors may seem misleading. Moreover, the word "error" as it has been used in other contexts is one of those emotion-laden, anthropomorphic concepts that are unjustifiable in scientific discussion.

[16] Hybridization may be successful without giving rise to polyploids.

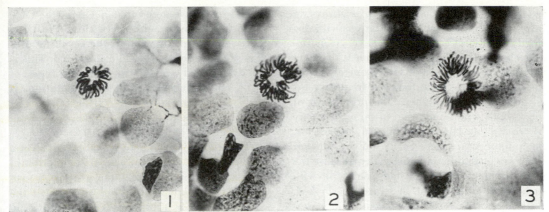

7–12 Polyploid nuclei in salamanders. *1.* A normal diploid nucleus $(2n = 22)$. *2.* A triploid nucleus $(3n = 33)$. *3.* A pentaploid nucleus $(5n = 55)$. Interestingly, polyploid salamanders have been shown to be more intelligent than their diploid fellows.

genes in most chromosomes.) During meiosis, paired chromosomes often exchange segments in the normal process of crossing over (Chapter 6). This process does not disturb the gene sequences. Sometimes, however, crossing over, too, is irregular, and the gene sequences are altered. The following types of changes are possible:

1. A gene is dropped out. *A-B-C-D-E-F* becomes *A-B-D-E-F* (*deficiency*).

2. A gene is duplicated. *A-B-C-D-E-F* becomes *A-B-B-C-D-E-F* (*duplication*).

3. Part of the gene sequence is reversed. *A-B-C-D-E-F* becomes *A-B-E-D-C-F* (*inversion*).

4. A segment of one chromosome changes places with a segment belonging to a chromosome of another pair (not its own mate). Two nonpaired chromosomes, *A-B-C-D-E-F* and *G-H-I-J-K-L*, become *A-B-C-J-K-L* and *G-H-I-D-E-F* (*translocation*).

Since the genes control development, it is easy to see that a deficiency (the absence of a gene) or a duplication (a double dose of a gene) would be likely to affect development adversely. This probability is confirmed in *Drosophila*, where, for example, an apparent gene deficiency in a short X-chromosome produces a notch in the wings in the phenotype. One might suppose, however, that an inversion or a translocation would have no effect. The same genes are present; only their positions are changed. Sometimes, indeed, no

difference in the phenotype is apparent even though inversion or translocation is visible by microscopic study of the chromosomes. Still, when chromosomes in only one parent have undergone inversion or translocation, the pairing of the chromosomes in meiosis may result in a decrease in fertility or various peculiarities in inheritance. Sometimes the action of a gene is modified simply because it has new neighbors. In *Drosophila* it has been found that the size of the eye is influenced by the identities of the genes adjacent to the gene specific for eye size. This *position effect* is evidence that genes interact.

Gene mutations

We shall now consider mutations that are associated *not* with gross changes in the structure or number of the chromosomes but rather with changes in the DNA molecule. For many years it was entirely unclear how or why such genic alterations occurred, but that they did occur was manifested by their effects. Since mutation is the only known mechanism whereby new genes, the principal raw materials of evolution, can arise, we may assume that the various alleles of genes, such as those extensively treated in Chapter 6, have all arisen through mutation. This was certainly true of the alleles used in Mendel's experiments, although the mutations occurred at some unknown time before he began his work.

The frequency of mutation varies greatly

from gene to gene, but mutation is always an infrequent event. Tests for mutations with distinctly visible effects in corn have given frequencies for single genes of from 0 to about 500 mutations per 1,000,000 gametes. There are so many genes that the total frequency for all genes is much higher. The proportion of individuals with a mutation in some gene may be around 5 to 10 per cent in many organisms. Such figures are necessarily only estimates. They are probably underestimates because the most frequent mutations may be those with such slight effects that they are overlooked. With ingenious statistical procedures it is possible to estimate the rates of some mutations with strong effects in man. Mutations producing hemophilia are found in 1 to 5 X-chromosomes in every 100,000. Fortunately this is a small figure, but it is large enough to keep hemophilia present in human populations even if no one ever passed the gene on to descendants. It is possible that at least 1 per cent of the babies born have some mutation. Some students place the figure at 10 per cent or even higher.

The frequency of mutation is subject to a certain amount of control by mechanisms within the organism itself. For instance, it was found that a particular population of *Drosophila* in Florida possessed a gene that increased the mutation frequency of the other genes in the chromosomes to a value 10 times as high as that in an Ohio population of the same species.

Research on mutations (chiefly in *Drosophila*) proceeded very slowly when geneticists had to wait for them to turn up spontaneously. The field was revolutionized in 1927, however, when H. J. Muller discovered that mutations could be induced by the application of x-rays. This method is illustrated in Fig. 7–7. Since 1927, several other means of increasing the frequency of mutations have been found. These *mutagenic agents* have consisted mainly of various types of radiation and chemicals. For the most part they merely speed things up. The mutations produced are those that would occur naturally less often. Also, there is no way to predict what mutation will occur in any given experiment.

MOLECULAR MECHANISMS OF GENE MUTATIONS. In our earlier discussion of the genetic code, we noted that if UAG (the code word for glutamic acid) were altered to UUG (the code word for valine), the resulting protein (hemoglobin) would be abnormal. This is an example of one of two possible types of DNA change occurring in gene mutation. It yields a protein altered by virtue of the fact that it contains amino acid Y instead of amino acid X.

The second type of gene mutation involves a failure to code an amino acid at some point in the sequence. Then either the protein is synthesized with one amino acid missing, or the synthesis of the protein is stopped. The absence of an amino acid is easy to establish experimentally, since the altered protein may be isolated and identified. It is more difficult to prove that protein synthesis has been stopped, for the experimenter may not know what was present before the mutation.

In any event, we may conclude that a mutation is generally a change of a single nucleotide in the DNA chain. The change can be brought about in several ways. Oddly enough, we do not yet know how x-rays, the most potent mutagenic agents known and the first to be discovered, cause base substitutions. The chemical mutagenic agents are much better understood. They are of various kinds. For example, the inorganic acid nitrous acid (HNO_2) is a powerful oxidizing agent that converts amino groups to hydroxyl groups. It therefore converts cytosine to uracil.

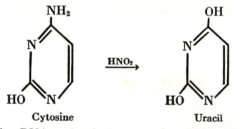

Cytosine   Uracil

If a DNA molecule is exposed to HNO_2, the following events may occur. First, cytosine is converted to uracil.

Then the chains separate, complementary chains are synthesized, and the new chains separate.

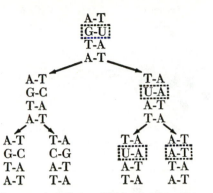

The change (or "error") in base pairing is now perpetuated, and the mutant DNA contains an A-T pair in place of a G-C pair.

Some of the many mutagenic agents now utilized are chemical analogues of natural purines and pyrimidines that "trick" the DNA into making a pairing "error." Others, by some unknown means, cause the synthesis of DNA molecules containing one more or one less nucleotide. All, it appears, act by producing a localized alteration in the nucleotide sequence of DNA. In light of the ease with which experimental mutations can be induced, the rarity of natural mutations is astonishing.

DOMINANCE AND MULTIPLE ALLELES

When the one gene–one enzyme theory was first advanced, a simple interpretation was also given to the familiar genetic phenomenon of dominance. It was supposed that, although the dominant allele of a gene was capable of making its specific enzyme, the recessive allele was not. This situation was summed up as the *presence and absence theory* of dominance and recessiveness. It is certainly attractive in its simplicity and nicely explains the similarity of the heterozygote (Aa) to the homozygous dominant (AA); both of them have the enzyme necessary for a particular reaction that fails in the homozygous recessive from complete lack of the enzyme. Unfortunately, the hypothesis has proved too simple.

Dominance has turned out to be relative and seldom complete. It is only sometimes true that the heterozygote (Aa) has precisely the same phenotype as the homozygous dominant (AA). Often the heterozygote is different from both homozygotes (AA and aa). This is

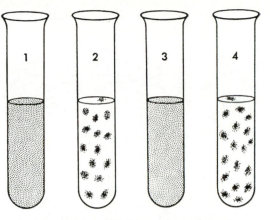

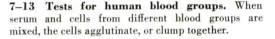

1. Serum of A, cells of A → No agglutination
2. Serum of A, cells of B → Agglutination
3. Serum of B, cells of B → No agglutination
4. Serum of B, cells of A → Agglutination

7–13 Tests for human blood groups. When serum and cells from different blood groups are mixed, the cells agglutinate, or clump together.

obvious, for instance, in Andalusian chickens. When homozygous, one allele for feather color produces black, and another produces white with small spots. When both alleles are present in heterozygous combination, the feathers are blue. It is hardly accurate to say that either allele is dominant. Their effects blend. But note that the *genes* do not blend. They segregate as usual when gametes are formed. When blue fowls are interbred, the characteristic ratio appears: 1 black : 2 blues : 1 spotty white.

It is impossible to reconcile such cases with the presence and absence theory of the enzyme. Two distinct forms of the gene must be present, both of them active. This is even more clearly demonstrated in the inheritance of the human blood groups.

All people fall into one of four blood groups—A, B, AB, and O. Each blood group is characterized by a particular protein antigen [17] on the surfaces of the red blood cells. Thus group A people have antigen A on their red cells; group B people have antigen B

[17] Antigens are substances (usually proteins) that induce certain cells in the animal body to form *antibodies*. These are proteins that react specifically with antigen molecules, inactivating them. Most antigens are foreign substances. Thus antibody synthesis, the phenomenon called *immunity*, is an important defense mechanism.

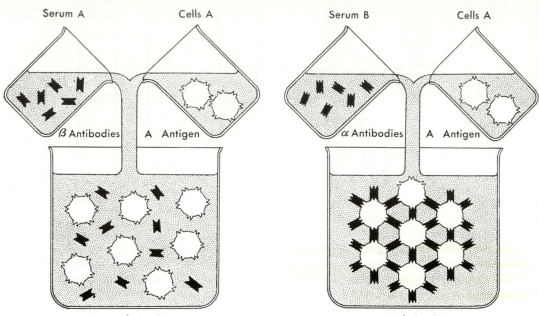

Serum A	Cells A	Serum B	Cells A
β Antibodies	A Antigen	α Antibodies	A Antigen

No agglutination   Agglutination

7–14 The mechanism of agglutination. α antibodies have surfaces complementary to those of A antigens; β antibodies do not. α antibodies serve to bind together A antigens and hence to bind together those red blood cells that carry A antigens. (Adapted from A. M. Srb and R. D. Owen, *General Genetics*, Freeman, San Francisco, 1952.)

on their red cells; group AB people have both antigens A and B; and group O people have neither antigen A nor antigen B. It is a remarkable fact that group A people also have an antibody in their blood *serum* (Chapter 11) that, given an opportunity, specifically combines with the red cell antigen of group B people. This causes clumping or *agglutination* of the red cells (Figs. 7–13 and 7–14); thus the antibodies are known as *agglutinins*. For clarity, the anti-B antibody in the serum of group A people is called the β antibody; the serum anti-A antibody in group B people is the α antibody.

Of immediate interest to us here is the inheritance of the blood groups. They are absolutely determined by three alleles at a single gene locus,[18] L^A, L^B, and L^O. L^A produces A antigen, L^B produces B antigen, and L^O produces neither. The genotypes $L^A L^A$ and $L^A L^O$ produce type A blood; $L^B L^B$ and $L^B L^O$ produce type B blood; $L^A L^B$ produces type AB blood; and $L^O L^O$ produces type O blood.

There are three noteworthy points about the

[18] The gene is labeled L for Karl Landsteiner, who first discovered the A, B, AB, and O blood groups.

heredity of the blood groups. First, the gene concerned acts directly, causing the synthesis of proteins with specific configurations. Second, no dominance is involved. In the $L^A L^B$ heterozygote, each allele independently does its own job in synthesizing its own protein product. Third, one gene can have more than two alleles. The situation illustrated by human blood groups, with three alleles, is by no means exceptional. Geneticists believe that most, if not all, genes can occur in many allelic forms. In *Drosophila* there are known to be at least 12 alleles of one gene controlling eye color.

The existence of multiple alleles and the common absence of dominance force us to reject the idea that the total range of possible actions for a particular gene is a simple pair of alternatives—such as making or not making the enzyme. The relations between the gene and the reaction it normally controls can be far more varied than this. In a heterozygous organism, all gradations between complete homozygosity and complete heterozygosity are present. Thus allelic interaction is basic to all of diploid genetics.

ACTION OF THE GENOTYPE AS A WHOLE

Genes and development

Our discussion of how an individual gene ultimately effects its control may seem remote from the more familiar aspects of an organism. Are not size, shape, pattern, and color the more obvious features whose development we want to understand? How does the molecular mechanism of gene action relate to them?

With respect to color the connection is clear enough. Red petals or white, the observable difference between fully developed organisms is a chemical difference in the pigments of their flowers. Indeed, in some cases the chemical difference is fully known, as well as the particular reaction that did or did not take place, leading to the development, or lack of development, of the pigment concerned.

With respect to size and shape the connection between cell chemistry and the final organism is less clear. But here too there is no need to alter our concept of gene action. There is less difficulty in envisaging the connection between the primary biochemical effect of a gene and its eventual morphological (size, shape) effects when we realize that *all* the genes in the total inherited message tend to act together as an integrated whole in the regulation of development. We speak of a single gene for red flowers in peas, for wrinkled seeds in corn, or for baldness in man. This practice reflects a scientific truth, for in each case—redness, wrinkledness, baldness—the development of the character is dependent on the presence of one particular gene allele that we have been able to study. But it is also dependent on many other genes. We should not fall into the habit of thinking that an organism has a set number of characters with one gene governing each character. The experimental evidence indicates that genes never work altogether separately. Organisms are not patchworks with one gene in charge of each of the patches. They are integrated wholes, whose development is controlled by the entire set of genes acting cooperatively. The rules are (1) that *each character is affected by many genes* and (2) that *a single gene may affect many characters.*

7–15 The action of modifying genes in mice. All six mice are heterozygous (*Ww*) for a gene producing white spots, but they differ in the number of modifying genes they carry for bolstering the effect of the primary (white-spotting) gene. *1.* Fewest modifiers present. *6.* Most modifiers present.

Many genes affect one character

In the blue Andalusian fowl two different alleles of the same gene work together, and the result is a phenotype distinct from that produced by either allele alone (that is, when homozygous). It is also common for different genes to interact, so that phenotypic characters depend on two, three, or many genes and not on one alone. In mice there is a gene that tends to produce white spotting, but its effect is modified by a whole series of other genes, called *modifiers* in genetic terminology. If numerous modifiers are present, the animal is almost pure white. Practically all intermediate conditions also occur, with various numbers of modifiers (Fig. 7–15). Another example is a Norwegian family in which for over four generations about half the children had short index fingers. In some children the finger was very short, and in others it was just barely shorter than usual. Study of the pedigree strongly suggests that two genes are involved, one determining that the finger will be short and the other determining the degree of shortness.

TABLE 7–2

Gene interaction in Drosophila

Gene	Length of life in days *
"Purple" (eye color) alone	24.5
"Arc" (wing shape) alone	26.8
"Purple" and "arc" together	33.7
Nonmutant	39.7

* The mutants "purple" and "arc" each reduce length of life when compared with a nonmutant; when both mutants occur in the same fly, they interact, producing a new effect.

We have already seen experimental evidence (Chapter 6) demonstrating the multiple-gene control of an organism's more general aspects like size and shape. Sometimes the many gene steps seem to be simply additive, as in the inheritance of size in corn and the determination of sex in *Drosophila*, where sex depends on the net balance of autosomal and X-chromosomal genes. Often, however, the various genes controlling a character—like that of the Andalusian fowl—interact in a way that is not simply additive. In *Drosophila* the two genes "purple" (eye color) and "arc" (wing shape) reduce the vigor and longevity of flies carrying either of them separately in homozygous condition. However, when both genes are homozygous in the same fly (Table 7–2), the individual has a life expectancy almost as great as that of a normal fly and much greater than that of either a purple or an arc homozygote. Again the effects are complexly interactive rather than simply additive.

When multiple-factor effects can be traced to a few genes, the genes can sometimes be identified and counted. More often, as in corn size, so many genes are involved that their separate identification is a practical impossibility. When a single gene has an obvious effect on a mature organism, we can recognize Mendelian inheritance, but even then we are merely picking out one element from an integrated system.

One gene affects many characters: pleiotropy

In the development of an organism, a single constructional step has trivial or far-reaching effects depending on when it occurs and how basic it is. The majority of clean-cut inherited differences involve trivial and terminal steps in development; the formation of a pigment in a particular organ, eye, or petal may fail, but little else is affected. But failure or abnormal execution of earlier or more basic single steps has far-reaching, ramifying results. In mice a particular gene mutation is known that improperly specifies a single step in bone formation. Its consequences are seen everywhere in the body where bone formation takes place. In *Drosophila* the gene "vestigial" is so called because its most obvious effect is to reduce the single pair of wings to mere vestiges of the normal (Fig. 7–16). Careful observation shows that it also affects the balancers behind the wings, certain of the bristles, the reproductive organs, and other parts. More significantly, it also shortens life and lowers egg-producing capacity. Clearly the gene directly or indirectly influences the development of a wide range of the organism's characters.

In man there is a recessive mutant allele that, when homozygous, produces imbeciles of pale skin coloring. They excrete phenylpyruvic acid, a compound absent from the urine of normal people, who possess the nonmutant form of the gene. The disorder, known as *phenylketonuria*, has widely diverse symptoms—abnormal skin color, urine composition, and brain function—and all are effects of one gene. It is now known that individuals with phenylketonuria lack the enzyme that normally converts phenylalanine to tyrosine (a precursor of the skin pigment melanin). Phenylalanine is alternatively converted to phenylpyruvic acid, which is excreted.

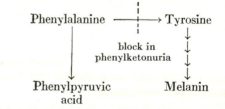

The missing enzyme normally controls one step in a complex metabolic cycle—involving phenyl compounds and producing at various points in the organism not only pigments but also substances necessary (in some unknown way) for proper brain development.

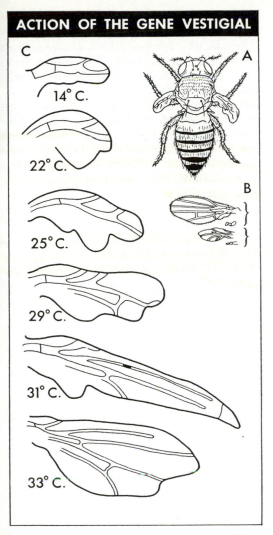

ACTION OF THE GENE VESTIGIAL

C

14° C.

22° C.

25° C.

29° C.

31° C.

33° C.

A

B

7–16 The action of the gene "vestigial" in Drosophila. *A.* An adult female fly homozygous (*vg-vg*) for the gene "vestigial." Note the rudimentary nature of the wings. *B.* The normal (*upper*) and vestigial (*lower*) wings and balancers compared. *C.* The wings of *vgvg* flies reared at different temperatures. The action of the gene is dependent on the temperature: at lower temperatures its effect (in causing departure from the normal) is greatest; at 33°C. the wing is nearly normal.

The phenomenon of genes affecting many characters is called *pleiotropy* ("more changing") and is a genetic generalization of great importance. Some students point out that the facts we currently possess concerning pleiotropy do not exclude the possibility that a gene has many primary actions, that is, that

one gene produces more than one enzyme, and so on. We need not pursue this difficult question here. It is only necessary to note that the multiple effects of a single gene *can* be understood in terms of its having a single primary action that has diverse consequences all over the finished structure. Not only could a single primary action occur over and over again in different places (all cells have a complete genotype), but the product of the primary action could be used as a building part in various subsequent developmental processes. Most geneticists studying the developmental effects of genes favor the one gene–one primary product (enzyme) interpretation of pleiotropy. Although it may turn out to be too simple, for the present it is a useful hypothesis.

Environment affects gene action: nature and nurture

We have just shown that the action of a gene may be influenced by other genes. We could say that a gene's action is affected by its *genetic* environment. And it is similarly affected by the *physical* environment in which it acts. A fly homozygous for the recessive gene "vestigial" (*vgvg*) served earlier for illustration; the gene *vg* affects many characters besides the wings, and its action is affected by many other genes. Its action is also strongly dependent on the temperature at which the fly develops. At high temperatures (30 to 31°C.) the wings grow nearly as long as in a fly carrying the normal gene (see Fig. 7–16).

A tendency to develop *diabetes mellitus* (manifested as an elevation of the blood sugar level) is hereditary in man. The manner of gene control is unknown, but the facts are consistent with the belief that an enzyme or other compound essential for the formation of insulin is produced by one allele of a gene and not by another allele. Insulin, in turn, affects the metabolism of carbohydrates, and hence the level of sugar in the blood; high levels are reflected by the presence of sugar in the urine.

It has been found that persons who have the genotype for diabetes do not necessarily have the disease; that is, it does not always appear phenotypically. An investigator who studied

63 pairs of identical twins, one or the other of whom had diabetes, found that in 10 of the pairs only one twin was afflicted. Identical twins are so called because they have identical genotypes (Chapter 8). Therefore, in these 10 pairs, the twin who did not have overt diabetes nevertheless had a diabetic genotype. It was learned that in some instances the twin who developed diabetes had a different diet, one that made larger demands for carbohydrate metabolism. Thus the appearance of diabetes does not depend on the genotype alone but on the genotype plus the diet.

We may then ask "Is diabetes mellitus caused by inheritance or by environment?" The disease is caused neither by inheritance alone nor by environment alone but by the interaction of the two. This is true of the great majority of the characters of organisms. It is one of the most important principles of biology that development is an *interaction* between heredity and environment.

Human blood types are genetically determined in the zygote, even before blood itself has developed; and as far as is known, no environmental influence thereafter changes the blood type in the least. This is one extreme. At the other extreme, human behavior seems to be almost endlessly modifiable. Yet it only seems that way, as a moment's thought will confirm. No human being really "crawls on his belly like a reptile," as some sideshow barkers would have us believe. No one swims literally like a fish, and no airman flies truly like a bird. Genetic limitations on human behavior are extraordinarily far apart, but they exist.

There is every gradation between characters that are not (as far as we know) modifiable by the environment and those that are greatly modifiable. Height is strongly affected by heredity, but it may be equally influenced by diet and even more strongly influenced by glandular disease or by the administration of hormones. Skin color is primarily determined by heredity, but it is also much modified by exposure to radiation, as well as by some diseases and chemicals. It is common knowledge that plants of the same varieties and races (hence closely similar in genotypes) have different sizes and shapes (different phenotypes) when grown on different soils,

with different fertilizers, with different water supplies, or in different climates.

In sum, what is inherited is a developmental mechanism. The mechanism determines how the organism will develop under given environmental conditions. It sets narrower or wider limits to the developing organism's reactions to different environments. This is the principle of the reaction range. All phenotypes that actually occur, including those that we consider abnormal or pathological, are necessarily within the reaction ranges of the underlying genotypes. The meaningful question about any trait is not "Is it hereditary or environmental—due to nature or to nurture?" but "What is the reaction range of its genotype, and what are the environmental factors correlated with this particular position in the range?" The processes of development will be treated in more detail in the next chapter.

AGAIN, WHAT IS A GENE?

So far we have not given a really clear and complete answer to the question "What is a gene?" posed near the beginning of this chapter. In fact, a clear and complete answer is not forthcoming, not only because, in spite of all recent breakthroughs, knowledge is still incomplete but also because the term "gene" has become ambiguous.

The gene was originally defined as something that produced a clean-cut heritable effect on the phenotype of an organism. Later it was learned that genes existed within chromosomes and that they could be identified and placed in sequence if (1) they had alleles producing distinct phenotypic effects and (2) they were separable by crossing over in meiosis. So we came to an operational definition, as in the first pages of this chapter, identifying a gene not by what it is but by what it does and by how we recognize it. At this stage it was possible to think of genes as if each one were a perfectly definite and distinguishable bead strung on a chromosome string.

Now we know that, insofar as a gene can be considered a material object and not just an operational concept, it is somehow contained within a molecule of DNA. It must, in fact, be a segment of such a molecule. But how large a segment and how definable in concrete, molecular terms? While this relationship was being discovered and explored, it was found by elaborate and refined methods that many of the operational genes of classic Mendelian experimentation were compound, containing "subgenes" or "pseudoalleles" not apparent earlier. It also was found, as we have outlined, that a gene mutation is often, at least, a change in a single nucleotide in the DNA chain. Evidently the unit of gene mutation is *not* the whole gene. As a functional unit, a gene, or at any rate a structural gene, specifies a protein—usually an enzyme—or a subdivision of protein and must therefore have considerable length, probably involving hundreds and perhaps thousands of base pairings. Thus it is certainly not a mutational unit, or a single base pairing. It is not really physically discrete at all but rather a long, still not delimited section of a DNA molecule.

Because of these ambiguities, some molecular biologists prefer not to use the term "gene." They call a functional subdivision of the DNA molecule (defined by position effects) a *cistron*. A cistron is in many and perhaps most cases at least approximately equivalent to the geneticists' gene. The unit of mutation within a DNA molecule they call a *muton*. This is much smaller than a cistron or a gene and is believed in most cases to be a single base pair. The term *recon* is applied to the smallest unit known to recombine genetically. It is smaller than a cistron or a gene, approaching a muton in size.

CYTOPLASMIC INHERITANCE

We have spoken thus far of genes and information-bearing DNA as though there were no reason for invoking other mechanisms of hereditary transmission. Are we justified in such an inference? Are there mechanisms of heredity that do not involve the nuclear genes?

In fact, there are modes of inheritance that are attributable solely to self-duplicating cytoplasmic particles. Since they are transmitted independently from one cell generation to the next, they are endowed with genetic continuity and can be regarded as hereditary units. Certain familiar cytoplasmic structures such as chloroplasts and mitochondria appear to

have a capacity for self-duplication.[19] However, one must distinguish between the possible independence with which such structures duplicate and their possible independence of control by nuclear genes through the mediation of messenger RNA.

Let us consider two examples that are believed to illustrate true cytoplasmic inheritance. One is the phenomenon of *plant variegation*—that is, the appearance in otherwse normal green leaves of pale areas and blotches. We may assume that in diploid genetics cytoplasmic hereditary determinants are more likely to come from the maternal gamete than from the paternal gamete since in gametic fusion the large egg contributes the bulk of the cytoplasm and the sperm contributes relatively little. Our assumption is borne out by the pattern of plant variegation, to which the Mendelian laws do not apply. If the maternal line is pale, the progeny are pale; if the maternal line is green, the progeny are green. Such breeding data suggested that a nonnuclear element derived from the maternal plant determines the trait, and this has been found to be the chloroplast (see Chapter 3). If a leaf loses some of its chloroplasts (and thus is pale in spots), no nuclear gene is capable of initiating their formation *de novo*. Moreover, chloroplasts, like genes, can undergo mutation, and mutant chloroplasts reproduce themselves just as mutant genes do. An altered or mutant chloroplast is responsible for variegation. Since variegation is heritable, we must conclude that it is an instance of extranuclear inheritance.

Nuclear genes nevertheless have some effect upon chloroplasts and similar semiautonomous cytoplasmic particles, as is revealed by our second example, the famous *kappa* particles of *Paramecium aurelia*, demonstrated by Tracy Sonneborn and his collaborators. *P. aurelia* is a single-celled protozoan with a

[19] The precise manner in which these and other complex cytoplasmic structures reproduce during cell division has not yet been established. A theory advanced early in this century held that mitochondria—visible then only as small specks in the field of the light microscope—are autonomous bacteria living within the cytoplasm and dividing on their own! We now know that this is erroneous. Recent discovery that mitochondria and chloroplasts contain traces of DNA that differs from nuclear DNA increases the possibility of its participation in their duplication.

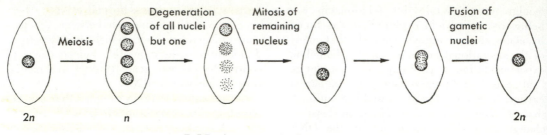

7–17 Autogamy in *Paramecium*.

normal mitotic cycle. It has three nuclei, one macronucleus and two diploid micronuclei. The macronucleus governs the cell during ordinary asexual (mitotic) reproduction, whereas the micronuclei undergo meiosis and participate in sexual reproduction (the conjugation of two cells of opposite mating types; see Fig. 19–11. In this process, haploid micronuclei are exchanged by the mating partners, which then separate. They are now genetically identical heterozygotes, but their cytoplasms may differ. *P. aurelia* also exhibits self-fertilization, or *autogamy*, in which any genotype produces homozygous progeny (Fig. 7–17); for example, the heterozygote *Aa* produces either of two homozygotes, *AA* or *aa.*

Certain strains of *P. aurelia*, designated "killers," secrete a substance into the surrounding medium that kills strains sensitive to it. To ascertain the genetic basis of the killer trait, one can cross a killer animal with a sensitive animal and examine the phenotypes of the two resulting cell lines. If the killer trait is carried by a nuclear gene, both of the two exconjugants should be either killers or sensitives, since conjugation yields identical heterozygotes. Surprisingly, however, the animals that were killers remain killers after conjugation, and those that were sensitives remain sensitives. Thus the killer trait, which persists despite the nuclear identity, must be determined by the parental cytoplasm. We now know that the secretion of the killer substance is due to the cytoplasmic kappa particles, which contain DNA and are 0.2 to 0.8 micron in diameter (that is, smaller than some bacteria and larger than viruses).

When killers undergo autogamy, 50 per cent of the progeny are killers, and 50 per cent are sensitives. When sensitives undergo autog-amy, all the progeny are sensitives. During conjugation the kappa particles are usually retained in the parental cell (although in prolonged conjugation some cytoplasm may be exchanged and some sensitives may thereby be transformed into killers). However, the fact that in autogamy there is *segregation* of killers and sensitives indicates the involvement of a specific nuclear gene, *K*, in the replication of the kappa particles. After autogamy, a killer has the genotype *KK*, and a sensitive has the genotype *kk*. After conjugation, both killer and sensitive are genetically *Kk* heterozygotes, even though only one contains kappa particles. Autogamy of *Kk* killer produces *KK* killer and *kk* sensitive. The latter contains kappa particles but lacks the necessary *K* gene. Autogamy of *Kk* sensitive yields *KK* and *kk*, but neither is a killer because neither has kappa particles.

The difficulties in deciding whether kappa particles are normal inclusions within the cytoplasm of *P. aurelia* or parasites recall the earlier views of the origin of mitochondria (see footnote, p. 221). In any event, they are self-duplicating, DNA-containing particles that are transmitted from generation to generation, and they exemplify well a type of extranuclear inheritance that is being discovered from time to time in other species.

There is at present much doubt and debate as to how widespread nonnuclear inheritance is and how important it may be. Some students speak of nonnuclear or nonchromosomal genes as if they constituted a true, general genetic system on a par with that in the chromosomes. Others believe that such phenomena are sporadic, under the ultimate control of the chromosomal system, and of little or no special significance for long-range genetic studies. Perhaps, as has so often happened, the truth will be found to lie partly between and partly outside of these alternatives.

One more point relevant to this discussion is that every new individual produced by organic reproduction starts as a cell that is already highly organized. This circumstance is most obvious in protists reproducing by fission. Here each daughter cell possesses from the beginning all the structures of the mother cell. Since there is little development in the usual sense, the statement that chromosomes primarily control development does not here seem particularly appropriate. Therefore, let us counteract an impression that may have arisen from our concentration on heredity. Chromosomes and their contained genes form an essential, operating part of almost every living cell whether or not reproduction or development is occurring. However, we have come to realize that the expression of certain hereditary traits can be influenced by the cytoplasm in which the genetic material acts —a cytoplasmic effect to be distinguished from the actions of autonomous particles that reproduce and act in the cytoplasm.

Interestingly, such cytoplasmic effects are not limited to protists. Many examples are now known of substances in the cytoplasm of the egg cell of a multicellular organism that influence the phenotypic expression of certain genetic characters during the early development of the offspring. Egg cells, as we all know, are often notable for their huge cytoplasmic masses. If this large cytoplasm contains a chemical substance that is a product of a maternal gene but is capable of affecting the phenotype of the progeny irrespective of its genotype, the so-called *maternal effect* may not be diluted out until many cell divisions have taken place in the embryo. In some cases (such as the inheritance of coiling direction in the shells of the snail *Limnaea*), the influence of the maternal cytoplasm is a permanent one.

CHAPTER SUMMARY

The inherited message in living systems: information; chromosomes compared to the information tape that controls the work of automatic machines; the study of gene action.

Nature of the genetic material: the small size of the inherited message; gene size and number; the difficulty in defining a gene; genes as the smallest units of chromosome separable by crossing over.

Chemistry of the genetic material: proteins and nucleic acids; evidence that DNA is the message-carrying molecule: transforming activity in *Pneumococci* and other organisms; the demonstration that bacteriophage DNA alone carries the instructions for bacteriophage replication.

Reproduction of the genetic material: the significance of the double-strandedness of the DNA molecule and of complementary base pairing; proof that one strand orders the nucleotide sequence of its new partner strand, thus serving as a template that preserves the sequence from generation to generation.

Mechanism of gene action: the "one gene–one enzyme" theory; mutation as the ultimate cause of loss or alteration of an enzyme; the control of protein synthesis by DNA; evidence that DNA orders the sequence of "messenger RNA," which then travels to the ribosome and orders the amino acid sequence during protein synthesis; the genetic code; structural, regulator, and operator genes.

Mutations as errors in gene reproduction: meanings of "mutation"; chromosome and gene mutations; the experimental production of mutations by radiation and chemicals; the molecular mechanism of gene mutation.

Dominance and multiple alleles: the argument against the simple presence-absence hypothesis of dominance-recessiveness.

Action of the genotype as a whole: genes and the control of development; many genes affecting one character—modifier genes; one gene affecting many characters—pleiotropy; the environment affecting gene action.

Cytoplasmic inheritance: the examples of plant variegation and of kappa particles in *P. aurelia*; the maternal effect.

8

Development

© Douglas P. Wilson

This is a developmental stage (the veliger larva) of a marine snail. Its subsequent development will be so accurately controlled that its mature form will duplicate almost exactly that of its parents.

In the whole realm of biology there is nothing more remarkable or more baffling than the development of a mature organism from a zygote. Here is a single cell, often of microscopic size and always very small in comparison with the developed organism. It divides; the two cells divide again; and the resulting cells divide again and again. Unerringly from this process there emerge all the tissues and organs of a rosebush or a man, an immensely complex, patterned organism whose essential characteristics were somehow determined from the start by the specifications in the zygote's chromosomes. In the previous chapter we discussed the nature of the inherited specifications that control development. We saw that those specifications involve molecular contents and structures and that their primary action is chemical. That knowledge is a substantial step toward the goal of understanding heredity in general, but it is only a beginning. We still cannot explain precisely how one egg becomes a rosebush and another a man. Now we shall turn directly to the processes of development. By knowing in more detail what has to be controlled, we may get hints as to how it is controlled, although we must admit at the outset that many unsolved problems remain in this realm of biology.

Processes of Development

In a study of the development of a multicellular organism from a zygote, it is useful, for purposes of description and analysis, to distinguish the three constituent processes: (1) growth, with cell division; (2) morphogenesis; (3) differentiation. These processes are neither successive nor independent. At various times in development one or the other may predominate, but as a rule all proceed together, and each may influence the others.

GROWTH AND CELL DIVISION

At first, offspring are always smaller than their parents. When a protist divides, the two offspring are necessarily about half the size of the parent. In higher plants and animals parents are commonly thousands or millions of times as large as the zygote that begins a new generation. The process of development, then, always involves growth. Cell division and a consequent increase in the number of cells is one aspect of the over-all growth process. But, of course, growth also necessarily means that some individual cells increase in size.

MORPHOGENESIS: THE CREATION OF PATTERN AND SHAPE

Compare yourself with the tiny, spherical zygote from which you developed. You are enormously larger than the zygote and contain a trillion times as many cells, or more. In a word, you have grown a great deal. But there is clearly more to your development than growth alone. You differ from the zygote even more strikingly in having, as a whole organism, a definite, complex structure and functional pattern and shape created by the foldings and mass movements of groups of cells.

DIFFERENTIATION: CELL SPECIALIZATION

You also differ strikingly from the zygote in that your constituent parts (cells and tissues) are visibly differentiated into several highly specialized types. Cellular differentiation is a distinct developmental process to which we shall pay detailed attention later; but it is also inextricably a part of morphogenesis. Much of the characteristic pattern of the adult organism derives from the intricate interrelationships that develop between specialized tissues.

It is characteristic for differentiation to affect most of the cells and to be irreversible. Specific kinds and arrangements of tissues arise; after the cells are well differentiated, they cannot revert to an undifferentiated form. By the time a human infant is born, its organic pattern is fully established. Subsequent changes are almost entirely in size and shape. Such development is called *closed* development. The least differentiated cells in animals and those that longest retain the embryonic capacity for repeated division are the *germ cells*. Concentrated in the primary sexual organs, these continue to divide and to produce gametes through most of the life of the organism.

In plants differentiation may also be irreversible once a plant is fully developed. However, even the higher plants, in which differentiation is most elaborate and complete, retain throughout life extensive tracts of undifferentiated cells. These constitute the meristematic tissue, a persistently embryonic tissue (Chapter 3). From it specialized tissues, such as those of wood and bark, and organs, such as leaves and flowers, continue to be differentiated periodically throughout life. Such development is called *open* development.

Like so many distinctions between plants and animals, that between open and closed development is not absolute. Even here there is a unity in the diversity. The capacities of the germ tissue of animals are more restricted than those of meristematic tissue, but both types of tissue are composed of persistent, relatively unspecialized, more or less embryonic cells, and both give rise to the reproductive cells from which the next generation originates. The extent to which the commitment of other tissues to specialization is irrevocable varies greatly from one tissue to another and one organism to another, as we shall learn shortly.

The actual continuity between generations is not established through the differentiated cells that comprise most of the developed organism. Continuity passes by way of embryonic tissues and the germ cells derived from them. The following mode of continuity applies to the majority of multicellular organisms, both animals and plants:

Differentiated cells, P　　　　*Differentiated cells, F_1*

Embryonic tissue → Germ cells　→　Embryonic tissue → Germ cells　→　to F_2, etc.

P generation　　　　F_1 generation

This is in another form Weismann's principle of continuity of the germ plasm, which we discussed in Chapter 6.

Patterns of Morphogenesis and Differentiation in Animals

MATURATION OF THE EGG CELL

The process of meiosis is the same in both sexes as far as chromosome behavior is con-

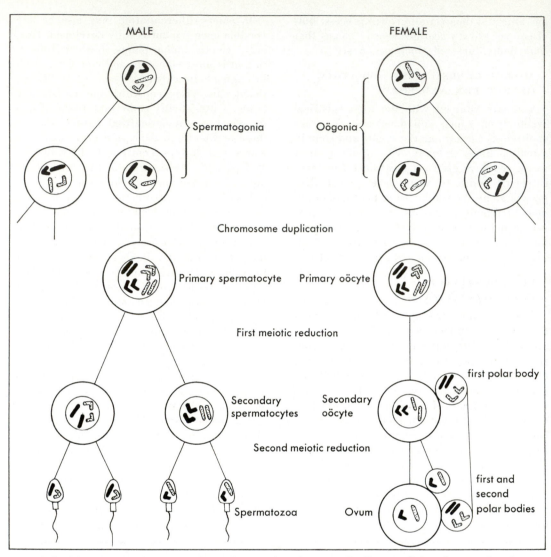

MALE   FEMALE

Spermatogonia   Oögonia

Chromosome duplication

Primary spermatocyte   Primary oöcyte

First meiotic reduction

first polar body

Secondary   Secondary
spermatocytes   oöcyte

Second meiotic reduction

first and
second
polar bodies

Spermatozoa   Ovum

8–1 Maturation of gametes in male and female animals. For simplicity, the organism is represented as having four chromosomes. The ovum is actually relatively much larger than indicated in the diagram. The polar bodies degenerate after the ovum is fertilized and have no part in subsequent development.

cerned. It follows the course outlined in Chapter 6, whereby the chromosome number is reduced from the diploid to the haploid number. In the female, however, nonchromosomal aspects of gamete formation are modified in a way directly bearing on the general problem of development (Fig. 8–1).

In sperm production, or *spermatogenesis,* the cells that give rise to spermatozoa divide equally twice in the course of meiosis. Each original diploid cell (spermatogonium) thus

becomes four haploid cells (spermatocytes), which in turn mature into four spermatozoa. The size of the cells decreases during these processes, and the cytoplasm is greatly reduced, so that a spermatozoan consists of little more than a small nucleus, essentially just a set of chromosomes, and a motile tail. In egg production, or *oögenesis,* on the other hand, each original diploid cell (oögonium) gives rise to only one mature haploid ovum, which is larger than the original cell. The other nuclei produced in the course of meiosis form small separate bodies (called *polar bodies*), which simply degenerate. The cell that matures into an ovum, called an *oöcyte* during its meiotic development, is enclosed and nourished by a *follicle* of other cells in the ovary. There it grows substantially, partly by the addition of food materials, or yolk. The result is that a mature egg is a relatively large cell, much larger than a sperm, with much cytoplasm and generally also food that provides energy and material for the growth of the embryo.

The large cytoplasmic content of the egg is not uniform throughout. It has a high degree of organization, such as we have found in other active, living cells. It also has a characteristic spatial arrangement and other properties which, as we have seen, can profoundly influence development, morphogenesis, and differentiation.

FERTILIZATION

Cytoplasmic organization is complete by the time meiosis occurs in the egg. Meiosis commonly is arrested at the metaphase of the second meiotic division and is completed only when the entry of the fertilizing sperm cell acts as a signal to proceed. Indeed the response of the egg cell to the sperm's entry is not only to complete meiosis but also to proceed with the rest of development.

In some species artificial stimuli such as the prick of a fine needle can cause the egg to complete meiosis and to start development. This shows that the sperm's nucleus is not always essential for development. A haploid egg cell can go ahead in some instances and complete all or nearly all of its development. Haploid salamanders have been raised in this way. Haploid embryos can also be raised by removing the egg nucleus before fertilization. The haploid nucleus of the fertilizing sperm proves adequate. Indeed, in a few instances some early development can be achieved by an egg cell artificially deprived of any nuclear controls. To be sure, such eggs do not develop far along the normal path and soon die, but the fact that they develop at all tells us that the cytoplasm itself carries some information on how to proceed. This reflects the early differentiation of the egg cytoplasm that took place

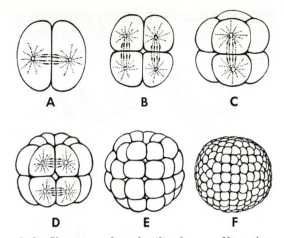

8–2 Cleavage of a fertilized egg. Note that successive cell divisions proceed without cell growth; at the end of cleavage, the embryo, though many-celled, is no larger than the zygote after the first cleavage division (*A*). (Although some animals, such as the starfishes, follow this simple sequence rather closely, in others it may be greatly altered.)

in the egg's maturation (which was nuclear-controlled) before fertilization.

CLEAVAGE

In normal development the fertilized egg proceeds to divide immediately following the formation of a diploid nucleus by union of egg and sperm chromosomes. Cell division continues for some time in the absence of any protoplasmic growth. Thus the cleaving embryo remains constant in over-all size, and cell dimensions steadily decrease with successive divisions until the growth accomplished by the egg's maturation is canceled out (Fig. 8–2). A usual result of egg cleavage is a raspberry- or mulberrylike cluster of cells called a *morula* (Latin for "mulberry"). With further divisions the number of cells continues to increase, and a cavity may develop in among them. In this stage, which does not occur in all animals, the whole embryo has the form of a hollow sphere and is called a *blastula*.

Experiments with cleavage-stage embryos are very instructive. With many species it is possible to shake apart the individual cells and observe their subsequent behavior. In some animals, these separate cells are able to start all over again and complete development; in other animals this is not possible.

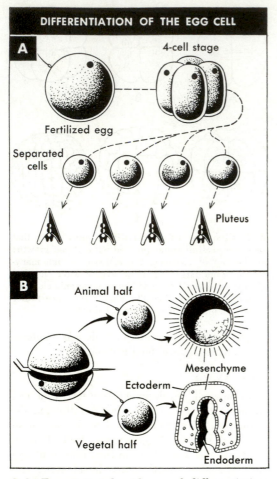

DIFFERENTIATION OF THE EGG CELL

A

4-cell stage

Fertilized egg

Separated cells

Pluteus

B

Animal half

Mesenchyme

Ectoderm

Vegetal half

Endoderm

8–3 Experimental evidence of differentiation in the zygote's cytoplasm. *A*. The fertilized egg of a sea urchin cleaves along the axis of differentiation, producing four cells, all of which are the same and contain a complete sample of the field of differentiation; shaken apart, each of the four cells can produce a complete embryo. *B*. When the fertilized egg is cleaved with a needle across the axis of differentiation, a different result is obtained. Neither of the two halves ("animal" and "vegetal") contains a complete sample of the field of differentiation; consequently, neither develops into a complete embryo. The vegetal half of the egg reaches only an advanced gastrula stage, and the animal half stops development while a blastula.

When renewed development is possible, as with sea urchins and frogs, we must conclude that the cells were not irreversibly differentiated; that is, they all were still essentially alike, and each was capable of complete development, just like the original egg. Lack of differentiation in the cells of young embryos is

surprising, for we noted earlier that the mature egg cell is regionally differentiated. How is it that the cleavage of the egg cytoplasm into distinct cellular compartments does not produce differentiated cells? For instance, the sea urchin, whose egg is visibly differentiated, is one of those animals whose early cleavage stages (up to the four-cell stage) can be shaken apart to produce cells that are still capable of further development. The explanation lies in the plane of cleavage in the egg. The first divisions of the zygote cleave the egg *along* the axis of differentiation (Fig. 8–3*A*). If, however, another egg cell is experimentally cleaved *across* the axis of visible differentiation, the separated halves fail to develop into normal embryos (Fig. 8–3*B*). In those animals like snails and worms, in which cell differentiation starts with the first cleavage, the first cleavage plane cuts across the field of differentiation in the egg cytoplasm.

Embryos that develop from separate cells derived from the same morula do, of course, possess exactly the same heredity. Their nuclei were produced by the mitotic division of a single zygote nucleus. This is the way in which identical twins arise. The potentiality for separate development may persist later than the two-cell stage. Identical quadruplets can arise by separation of cells in the four-cell stage, after two cleavages. (How could identical triplets arise?) In man this potentiality for total development persists even after three cleavages. This is one way the Dionne quintuplets may have arisen. There are other possible ways in which identical twins, triplets, and so on, can arise from one zygote, but all depend on the absence of irreversible differentiation among cells in the young embryo.[1]

[1] Twins are not necessarily identical. Sometimes two eggs are liberated from the ovaries at the same time, are fertilized by separate sperms, and develop simultaneously. Twins developed in this way are neither more nor less similar in heredity than brothers and sisters born separately. They are called *fraternal* twins (even if both are girls). *Identical* twins are necessarily of the same sex. Fraternal twins may or may not be. Triplets may be all fraternal, all identical, or two identical and one fraternal.

Multiple births are relatively rare in humans. The numbers vary considerably in different groups, but in the United States twins occur in about 1 birth in 90 among whites and in about 1 birth in 70 among Negroes. Of these, about a third are identical twins, and

Sooner or later there is some differentiation of cells within the developing embryo. Thereafter a single cell removed from the rest will not develop into a whole organism, and development of the remaining cells does not produce a normal organism. Parts that would have developed from the removed cell are missing.

The cleavage of the zygote is often profoundly modified by the amount of food materials (yolk) and other factors in the embryo. Fig. 8–4 compares cleavages in Amphioxus, frog, and bird—a series of embryos with increasing amounts of yolk material. In the hen's egg the zygote cell is the whole yellow mass we designate as yolk. The nucleus and cytoplasm from which the embryo develops are localized on a part of the surface of the yolk. Much of the original zygote cell is never involved in cleavage. Cleavage of the cytoplasm occurs first in a platelike area on one side, and the growing embryo gradually consumes the yolk proper, finally occupying the whole available space.

In general, species whose eggs have little yolk develop in one of two ways. The immature organism, still midway in its development, (1) becomes self-feeding or (2) is nourished directly by its mother, as in the mammals. Self-feeding immature stages are *larvae*. They are common in many invertebrate groups of animals and are familiar to us in caterpillars (which develop further into moths and butterflies) and maggots (imma-

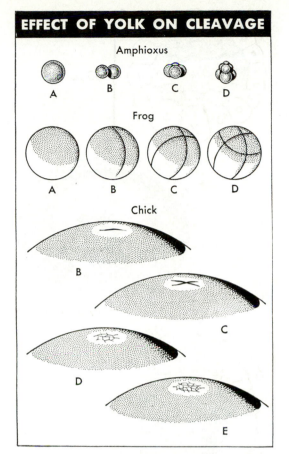

EFFECT OF YOLK ON CLEAVAGE

8–4 **The effect of yolk on the cleavage of an egg.** In the Amphioxus egg, which contains virtually no yolk, cleavage is complete. In the frog's egg there is considerable yolk on one side (unstippled). The yolk material is an inert mass, and the transverse cleavage plane is displaced above it. The hen's egg is almost entirely yolk, and consequently cleavage is restricted to a small area. *B, C,* and *D* are corresponding stages in all three animals.

about two-thirds fraternal twins. Triplets occur about once in 8000 or 9000 births, and quadruplets only once in 600,000 or 700,000. A tendency to produce twins or other multiple births is sometimes hereditary, and so the proportions may be much higher than quoted in twin-prone families. Oddly enough, this applies only to fraternal twins. There is no clear evidence that production of identical twins runs in families, that is to say, is influenced by heredity.

Most of the larger mammals, like man, usually have one young at a time, with occasional twins and still rarer triplets. Small mammals, especially the rodents, usually have multiple births. In members of the mouse family quintuplets are common, and 10, 15, or even 20 young may occasionally be born at once. Multiple offspring are also the rule among other vertebrates and especially among invertebrates, some of which produce thousands and even millions of eggs in one breeding season. Such multiple births are usually fraternal.

ture flies). They will be discussed more fully in Chapter 9.

GASTRULATION: A PRINCIPAL MORPHOGENETIC PHASE

Even in the earliest stages, development follows different lines in various groups of animals. As the process goes further, there is tremendous diversity in detail. However, some features can be seen in more or less modified form in most animals. Among such features is the formation of separate layers of cells, which we shall describe in its simplest form—

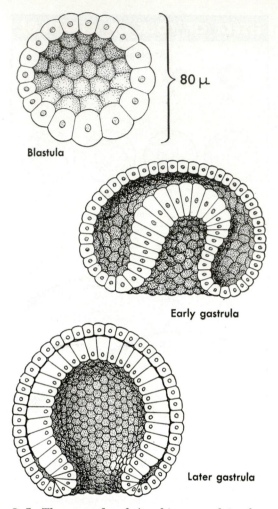

Blastula

80 μ

Early gastrula

Later gastrula

8–5 The gastrula of Amphioxus and its formation by invagination. The endoderm of the gastrula has larger cells than the ectoderm.

a form so simple, indeed, as to be rare in actual occurrence. Almost always the process is profoundly modified, sometimes almost unrecognizably so. Nevertheless, even the most complex forms of development do involve layer formation analogous to that in the simple cases we describe.

The process of *gastrulation,* whereby the embryo becomes a two-layered structure, follows the morula and blastula stages. The simplest, though not the most common, form of gastrulation is for one side of the blastula to cave in, much as does the side of a soft rubber ball that is poked with a finger. The result is a cup- or bowl-shaped *gastrula* with two

layers of cells forming its walls (Fig. 8–5). The outer layer is the *ectoderm* ("outside skin"); the layer lining the cavity of the gastrula is the *endoderm* ("inside skin"). Figs. 8–5 and 8–6 compare the simplest mode of gastrulation, as found in Amphioxus, for instance, with gastrulation in the frog's embryo. In the frog one region of the blastular wall grows more rapidly than the rest, and a lip of cells is formed where the more rapidly growing tissue folds inside the cavity of the blastula.

The cavity inside the double-walled gastrula becomes the alimentary canal of the adult animal. Here in the gastrula we can recognize the gross outlines of the adult's future pattern. Gastrulation is indeed principally a morphogenetic phase of development, producing the main outlines of organic pattern rather than differentiation of cells into special types.

In coelenterates (corals, *Hydra*, etc.) gastrulation is more than the main outline of morphogenesis. It is almost as far as the whole process goes because these animals are essentially just pouches with two well-defined layers of cells, although a few other cells may

8–6 Early development of the frog. *1.* The sperm enters the egg. The arrow indicates the future head-to-tail axis of the embryo. *2.* The sperm nucleus migrates to the egg nucleus—fertilization. *3.* The blastula in cross section. Note the large yolk cells in the lower ("vegetal") half of the egg. *4, 5.* Early gastrulas. The sheet of cells forming the upper ("animal") half of the blastula is beginning to grow actively and roll inward at the blastopore. The actual point of involution of cells is the dorsal lip of the blastopore. *6–10.* Later stages of gastrulation. The blastocoele (original cavity of the blastula) is nearly obliterated (*7*). The new cavity formed by the ingrowth of cells at the blastopore is the archenteron, or primitive digestive tract. The blastopore (at this stage closed by a plug of yolk cells) is destined to become the anus of the adult. The mouth remains to be developed at the other end of the archenteron (cf. *16*). The sheet of cells that has grown in from the blastopore is the future mesoderm (*8, 9*). Endoderm on the floor of the archenteron grows upward and arches over to form the future digestive tract (*9, 10*); the sheet of mesoderm then lies between the roof of the digestive tract and the ectoderm (*10*). The notochord arises fom the mesoderm immediately above the archenteron (*10*). The coelomic cavity arises as a split within the mass of mesoderm. *10–15.* On the dorsal surface of the embryo, two folds of ectoderm grow upward as the neural folds; the flat

THE EARLY DEVELOPMENT OF THE FROG

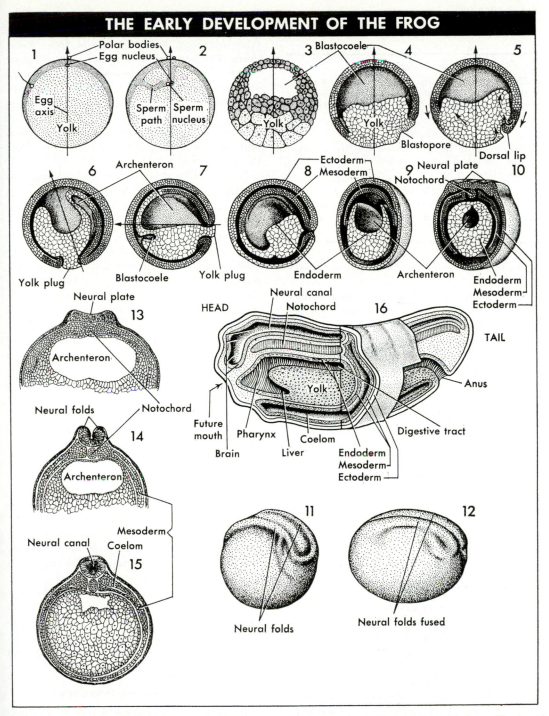

layer of ectoderm between them is the neural plate; the tube resulting from closure of the folds is the neural canal. In this manner the tubular dorsal nerve cord—typical of all vertebrates—is formed. In the anterior portion of the neural canal, three swellings correspond with the future fore-, mid-, and hindbrain portions of the adult brain. *16.* A late frog embryo,

showing major outlines of the adult form. The head-to-tail axis is clear. The neural canal and brain lie above the notochord. The digestive tract still lacks a mouth opening, which will later develop as an invagination of ectoderm; an evagination from the digestive tract marks the beginning of the liver; the mesoderm and coelom are clear.

TABLE 8–1
Tissues derived from germ layers

Ectoderm

Epidermis of the skin, nails, and hair; sweat glands in the skin; all nervous tissue; receptor cells in the sense organs; epidermis of the mouth, nostrils, and anus.

Endoderm

Epidermis lining the gut, trachea, bronchi, lungs, urinary bladder, and uretha; liver; pancreas; thyroid gland.

Mesoderm

All muscles; blood; connective tissue (including bone); kidneys; testes and ovaries; epithelia lining the body cavities.

---

migrate between the bases of the distinctly layered cells.

In most other animals a more complex third layer of cells (the *mesoderm*) develops between the ectoderm and endoderm. The mesoderm may arise by pouchlike foldings from the endoderm region or as individual cells that migrate into the space between ectoderm and endoderm and proliferate there. In either case the mesoderm soon becomes quite complex and loses whatever clear resemblance to a layer it may have had. Usually at a rather early stage a cavity, the *coelom*, appears in it, lined with mesodermal layers or masses. A distinction can sometimes be made between a coelom that develops from endodermal pouches and one that appears later within a mesodermal mass. The distinction is important for classification and evolution and will be further mentioned in Chapter 21.

DIFFERENTIATION AND ORGAN FORMATION

The three germ layers represent an early sorting out and arrangement of cells as they increase in number in the embryo. The cells in the layers still are not strongly differentiated. When the cells do later form distinct tissues and organs, the layer arrangement is no longer evident. Many of the tissues can, however, be traced back more or less clearly to derivation from one layer or another. The alimentary canal, and the glands and other organs developed from it, and the respiratory system are lined with tissue from the endoderm surrounded by other tissues of mesodermal origin. Skin has outer parts that arise from ectoderm and inner parts from mesoderm. The nervous system and parts of sense organs closely associated with it arise from ectoderm by a remarkable process to be illustrated below. Most connective tissue, muscles, bones, blood vessels and blood, and many internal organs such as the heart, kidneys, and gonads (testes and ovaries) arise from the mesoderm (Table 8–1).

It used to be thought that every tissue was clearly and necessarily derived from one germ layer and always from the same germ layer for any one sort of tissue. The inference, then, was that the germ layers, as such, represent early and irreversible differentiation of the cells in them. For instance, it was believed that ectoderm cells were, by the mere fact that they are in the ectoderm, partially differentiated so that they could become nerve or skin cells, but could not become muscle or bone cells. This has been proved incorrect. Gastrulation and germ-layer formation are processes of morphogenesis rather than differentiation; they are the processes whereby adult form is roughly sketched out and masses of potentially multipurpose cells are stockpiled ready for later cell differentiation and development of detailed organs. The later differentiation of a cell depends not so much on whether it is broadly ecto-, meso-, or endoderm, as on precisely *where* it is in the over-all pattern of the embryo. Muscles and bones usually develop from mesoderm only because mesoderm normally happens to occur in those parts of the embryo where muscle and bone are differentiated. But if by some variation of development in a particular species ectoderm happens to come into a region of muscle or bone development, muscle and bone develop from ectoderm. In other words, final and definitive differentiation depends on regional relationships and on later anatomical controls, not basically on the germ layers.

INDUCTION AND ORGANIZERS

Discussion of the germ layers emphasized the fact that later differentiation of a cell depends not only on what but especially on where it is—that is, on interactions among adjacent cells and tissues as well as on processes within single cells. Experiments have revealed something of the nature of the interactions among developing cells, although as a whole they are not yet understood. In typical experiments bits of tissue removed from a living organism are kept in a solution of suitable temperature, osmotic pressure, and composition such that they there continue to develop. If kidney cells from a mouse are so cultured, they grow in a formless way, like undifferentiated embryonic cells. If, however, kidney connective tissue is added to the culture, the cells differentiate and tend to form kidney tubules. The connective tissue somehow *induces* differentiation in the kidney cells.

Experiments on embryos show that *induction* likewise and even more elaborately occurs in normal development. The German experimental embryologist Hans Spemann (1869–1941) first showed that there is a special inducing region at the upper edge of the infolding lip of the gastrula in frogs and salamanders (see stages *4* and *5* in Fig. 8–6). He called this region the *organizer*. Its removal prevents normal differentiation. Transplantation of the organizer into another embryo at a similar stage of development causes double differentiation and development of a sort of Siamese twins (Fig. 8–7).

The organizer of frogs and salamanders is ultimately responsible for much of their differentiation. Specifically, the tissue [2] just inside the organizer lip induces formation of the nervous system in the overlying ectoderm. In these animals, as in all vertebrates, the nervous system arises first as a plate or thickening (the *neural plate*) in the ectoderm along what is to become the back of the animal. Later the sides of the neural plate are elevated, forming the *neural folds*, which fuse together to form a tube that eventually becomes the brain and spinal cord (see stages *10–16* in Fig. 8–6). Experiments prove that this process is induced by the organizer tissue, which nevertheless does not itself form any of the nervous system. If the organizer tissue is removed just before the neural plate would have been formed in the ectoderm, no neural plate develops. If the organizer is transplanted and

[2] This tissue, destined to become mesoderm, is technically known as *chordamesoderm*.

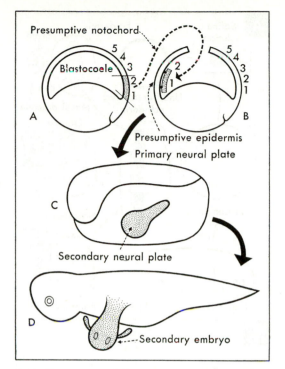

8–7 Experimental demonstration of induction by the organizer. *A.* The piece of the organizer (dorsal lip of blastopore) that induces the formation of neural plate is presumptive notochord; it will become notochord eventually. *B.* This presumptive notochord is removed from one embryo and transplanted into the blastocoele of another, where it attaches at a point corresponding with the future left side of the embryo. *C.* It here induces a secondary neural plate. *D.* The result is a two-headed monster.

placed beneath another area of ectoderm, a neural plate will develop there instead of in the normal position.

Further experiments indicate that the induction of the neural plate, at least, is due to a chemical diffused from the inducing organizer tissue into the differentiating ectoderm.

Other experiments involving the transplantation of pieces of the embryo to new locations elucidate further aspects of the process of normal development. In the gastrula stage of the amphibian embryo, careful observation and experiment have enabled embryologists to draw a map showing what organs various parts of the gastrula will eventually produce. Fig. 8–8*A* shows the location of a mass of cells that in normal development would give rise to an eye. These gastrula cells can be trans-

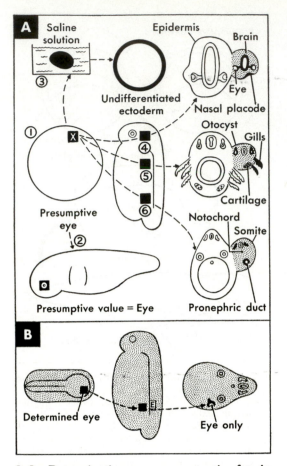

8–8 Determination versus presumptive fate in embryonic development. *A. 1, 2.* A piece of embryonic tissue (marked *X*) in the amphibian gastrula normally develops into an eye; its presumptive fate is eye. *3.* If, however, it is removed from the gastrula and allowed to develop in saline solution, it does not differentiate into an eye. *4, 5, 6.* If it is transplanted to unusual locations in older embryos, it differentiates, not into an eye, but into a structure characteristic of the location to which it has been moved. In the gastrula the fate of *X* as an eye is only presumptive and is contingent on its embryonic location. *B.* If the same tissue is transplanted from a much older embryo than a gastrula, its fate is found to have become determined; no matter where it is transplanted, it differentiates into an eye.

planted to various locations on an older embryo. When this is done, their future development is again controlled by the particular environment of cells into which they have been transplanted. Transplanted to the head region, they develop into the brain and eye, characteristic of that region; transplanted be-

hind the mouth, they develop into gills; and when placed in the tail end of the embryo, they develop into, among other things, the kidney duct characteristic of the region.

All the embryo's cells, arising by mitosis from the single-celled zygote, arrange themselves into a definite form and differentiate in a gradual process. What they differentiate into is initially controlled by their location in the embryo. Their differentiation cannot be primarily controlled by differences in heredity because all the cells receive one and the same set of chromosomes from the zygote by the process of mitosis. However, their differentiation, initially guided by embryonic location, eventually does reach a point of no return. In the normal embryo a stage of differentiation is ultimately reached by the cells we have discussed when their ultimate fate to be eye cells is no longer *presumptive* but *determined*. After differentiation has reached this state of determination, the cells can be transplanted and will go ahead and become an eye independent of their new location (Fig. 8–8*B*).

THE PROBLEM OF GENETIC CONTROL

After a brief survey of the kinds of processes that go on in the normal development of organisms, we must now return to an earlier problem. How does the inherited set of chromosome specifications control these processes? To begin with, how can we reconcile the two following facts?

1. Development involves the organization of a multitude of cells into a complex but definite pattern and the differentiation of these cells into many diversified and specialized cell types.

2. All these cells [3] with their widely different fates in the organism have identical heredity. They arose by mitosis from the single-celled zygote and have the same chromosomes.

There is no evidence that, in successive cell divisions in the growth of the embryo, neatly packaged fractions of the total blueprint are partitioned out—one set to presumptive leg

[3] With, in some organisms, a few exceptions hardly pertinent to the general problem.

tissue, one set to presumptive eye tissue, and so on. Indeed, there is evidence that this is not the case. The chromosome set remains visibly complete in different tissues. And, as we have seen, cells can proceed along one path of differentiation for some time and still retain their ability to differentiate into something entirely different if they are transplanted.

In plants many somatic cells can give rise to whole new plants without requiring the participation of the normally reproductive cells —hence the possibility of propagation by cuttings and vegetative reproduction in general. The cells involved are usually poorly differentiated (especially meristematic cells, as previously noted), but it has recently been found that even differentiated cells, notably almost any in a carrot, may act in this manner. Vegetative reproduction also occurs in a few lowly animals, and all animals retain some capacity for the regeneration of cells and organs, that is, for newly differentiating lost specialized cells and structures. Even animals as complex as salamanders readily regenerate whole legs.

These phenomena testify that the genetic endowment is usually the same in all cells. They thus make still more profound the mystery of how different cells assume such radically different forms and functions, chemically and otherwise, in spite of having the same nuclear controls.

Our earlier survey of gene action provides some leads toward the eventual solution of this problem. A gene apparently acts through its control of unitary chemical processes. This control, at least in most cases, is effected through the intermediary of an enzyme. But, as we have noted, the gene's action is not an absolutely fixed affair. There is not even a flat alternative of producing or not producing an enzyme—of a reaction's taking place or not. There are instances in which a reaction (for example, the synthesis of pantothenic acid in some bacteria) fails to occur even when the necessary enzyme is present. Any reaction, such as the conversion of a substance A to a substance B, depends on many conditions, only one of which is the presence of the enzyme. For one thing, it depends on the rate at which A is supplied and the rate at which B is removed by being converted into other substances in other reactions (and the supply of

A and the removal of B are influenced by the actions of other genes). It is also sensitive to physical conditions, such as acidity and temperature. A gene's action is, in short, subject to both the genetic and physical environment in which it finds itself. Consequently, the diverse effects of the same genotype must be due to the different cellular environments existing in the embryo.

There are many illustrations of the reality of this principle. A simple one involves rabbits with a gene that when homozygous produces a characteristic color pattern: the coat is white except for the nose, ears, paws, and tail, which are black. The blackness is due to an enzyme-controlled synthesis of pigment. Our basic problem can here be stated in clear-cut terms: we are sure that the enzyme is gene-controlled and that each cell of the body contains the gene. How is it that the gene acts, ultimately producing pigment, in only some cells (in the nose and ears, for instance)? Experiment reveals that the pigment is produced only—that is, the reaction proceeds only—below a certain temperature (92° F.). When the temperature is higher, the reaction fails in spite of the presence of the gene. The restriction of pigment to the paws, tail, and nose is due simply to the fact that the high surface/volume ratio of these extremities causes them to be slightly colder than the rest of the body. If the back of the rabbit is shaved and cooled with an ice pack while new hair is grown, it comes in black.

The general results of transplant experiments point in the same direction. What a cell becomes depends at first entirely on where it is in the embryo; later it also depends on both where it is and what it is; and what it is depends on where it was earlier. The direction of differentiation seems then to be determined by differences in strictly local physical and chemical conditions in the embryo. All the embryo's cells have identical controls, but which genes act and how they act in any particular cell depend on the strictly local environment in which the gene finds itself. This principle involves a kind of chain reaction. Gene Z's action may be dependent on products of X and Y; and gene X's on the products of W. A slight peculiarity in local conditions of, say, one part of the mesoderm might inhibit the action of gene W only; but this in turn will affect genes X and Z. Evidently minute environmental differences within the embryo can have profound, snowballing effects on the genotype as a whole.

We studied in the preceding chapter the mechanisms of action of the so-called structural genes, which control the structures or amino acid sequences of proteins (that is, enzymes), and of the recently discovered regulator and operator genes, which control the rate at which structural genes instruct the cell to synthesize certain proteins. We saw also how the regulator and operator genes provide a reasonable explanation for the remarkable phenomena whereby enzyme synthesis is *induced* by the presence of substrate molecules and *repressed* by the presence of metabolic end products (Fig. 7–10). These important discoveries may also lead to a more acceptable explanation for differentiation, a process that involves the systematic acquisition of capacities for the synthesis of new proteins. It is likely that differentiation depends upon intricate sequences of enzyme inductions and repressions, the basic information for which resides in the genes of the fertilized egg cell. Thus, when a product of one metabolic pathway serves as the substrate for an enzyme of another pathway, one could predict an increase in the amount of new enzyme. Many such schemes could be imagined and interconnected into regulatory circuits endowed with virtually any property.

These explanations support the conviction that the genes do in fact control development, and they hold out hope that the control processes will eventually be explained in physical and chemical terms. At present, however, there are still gaps in our knowledge of, on the one hand, the DNA-segment gene and its determination of an enzyme and, on the other hand, the systematic, structural development of an embryo. How and why should one collection of proteins produce the structure and function of a human brain and another set, different only in amino acid sequences, those of a rose bloom? We know enough to have faith that such questions are answerable, but as of now we do not have the final answers. Here, in all probability, is the greatest opportunity for research in biology.

Patterns of Growth

In most multicellular organisms the anatomical pattern of differentiated cells, tissues, and organs is established rather early in the life cycle. As the organism matures, however, there are changes in structure and function. For instance, in vertebrates the integration and coordination of the nervous system may continue long after the anatomical pattern is complete, and the sex organs are not effectively in operation for a long time after they are present and differentiated. The changes during maturation actually involve size more than structure and function. A period of growth usually intervenes between the period of embryonic differentiation and the relatively static period of adult life. Although plants have more open development than animals and proliferate new specialized organs throughout life, they have growth patterns not fundamentally different from those of animals.

THE GROWTH CURVE

An over-all pattern of growth is obtained if an organism is measured at suitable intervals and the measurements are plotted against time. Measurements of individuals are, of course, likely to show a good many irregularities, especially if environmental conditions are not absolutely uniform. (Even in a laboratory, it is extremely difficult to maintain a constant environment.) Nevertheless, the pattern of growth frequently looks more or less like Fig. 8–9*A*. Marked deviations from this pattern can usually be attributed to special circumstances.

The curve of Fig. 8–9*B* shows that in most cases the increase in size is at first rather slow but soon becomes rapid and finally tapers off as adult size is reached. This sequence is familiar in ourselves. The slow beginning occurs in the embryo. By the time of birth, growth is going on very rapidly, and it continues so into adolescence. Then growth slows down and finally stops altogether. The timing differs considerably for different individuals, sexes, and measurements. In humans increase in height usually stops in the early twenties, or in almost all healthy individuals sometime between 18 and 30, at least. Increase in weight continues longer, frequently into the forties, but also usually stops eventually. When growth stops, an actual decrease in size soon sets in. Man is likely to reach his greatest height in the early twenties, and then to start shrinking in height, even though his waistline may increase for another 20 or 30 years. The decrease in height will be slow and not noticeable for a long time, but it is likely to speed up in the fifties. If he reaches a ripe old age, he will probably be visibly shorter and lighter than he was in his prime.

When other animals and plants are considered, it is not usual for growth to reach a definite stopping point, and it is still less usual for growth to be followed by shrinking. Definite cessation of growth is especially characteristic of mammals. The reason or, at least, the mechanism involves the way the vertebrae and limb bones grow in man and other mammals. While growing, these bones have separate plates or pieces at their ends, and growth occurs by addition of bone tissue between these pieces and the main body of the bone. Eventually fusion occurs, and normal bone growth is subsequently impossible. However, birds also tend to grow to a definite size and then stop, although they do not have this particular mechanism of bone growth. In other vertebrates, in many invertebrates, and in most plants growth may continue throughout life. Growth becomes very slow in the adult, however, and such species do have a maximum size that is never exceeded.

In insects and their relatives with hard external skeletons (most of the arthropods), another variation in growth pattern occurs. The skeleton does not grow, and it confines the other tissues and prevents growth of the animal as a whole. Growth therefore occurs in spurts, with the shedding or molting of the old skeleton and the growth of a new, larger one (Fig. 8–10). Growth ceases after the last molt. Insects do not molt after the adult stage is reached, and adult insects are therefore fixed in size. (In us, too, the outer layer of the skin is dead and does not grow. How is it, then, that we can grow continuously without outgrowing our skins?)

It is obvious that growth varies with variations in environment and food. All organisms have a temperature range, for instance, in

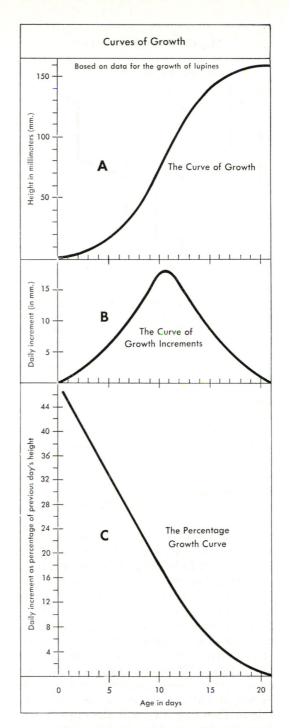

8–9 **Curves of growth.** The data from which the curves were plotted relate to the growth of lupine plants. Similar patterns covering widely different ranges of time are widespread among both plants and animals.

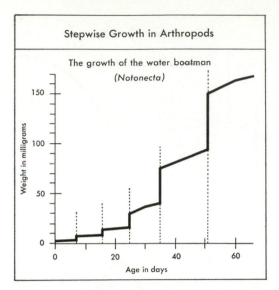

Stepwise Growth in Arthropods

The growth of the water boatman (Notonecta)

8–10 Stepwise growth in arthropods. The hard exoskeleton of an arthropod must be shed periodically to permit growth, which therefore takes on a stepwise character. Each abrupt step in the growth curve occurs at a time when the old hard skeleton has been discarded and the new one below is still soft enough to permit growth.

which growth is most rapid, and all have an optimum (best) combination of foods and other materials for growth. Growth may also be strongly influenced by normal changes in the internal environment, especially those directly or indirectly influencing the growth hormones. This is well illustrated by changes in human growth rates at puberty, when the sex glands begin to function and the general hormonal balance of the body changes. There is then a quite marked and relatively rapid increase both in height and in weight. On an average this occurs about the age of 12 or 13 in girls, who then shoot up and become gawky, in some groups becoming taller than boys of the same age. When they are 14 to 16, the boys shoot up in their turn and become (on an average) decidedly taller than the girls (Fig. 8–11). When they are 15 to 17 girls often have a second period of rapid increase in weight but not, or not so evidently, in height. Then they fill out in the magic transformation from gawkiness to femininity.

PERCENTAGE GROWTH

It is one of the principles of growth and, indeed, of life that the products of growth are

themselves living and capable of growth.[4] The process is not like building a wall of bricks, which stay put. In growth, as soon as a "brick" is added, the brick itself adds others around it. The activity of growth thus depends on how much material is already there—in other words, on how much growth has already occurred. In judging how active the growth process is, it is therefore more enlightening to consider the *percentage* increase rather than the absolute increase. If a baby weighing 10 kilos (22 pounds) gains 2 kilos in a year, it is growing more actively than an adolescent weighing 50 kilos and gaining 5 kilos in a year. This essential fact is brought out by saying not that the baby gained 2 and the adolescent 5 kilos, but that they gained 20 per cent and 10 per cent, respectively.

If growth curves are plotted by percentage, they tend to take the form of Fig. 8–9C instead of that of Fig. 8–9A. In terms of percentage increase there is no early period of slow growth that later increases to a maximum.

[4] This is a valid generalization for growth as a whole, but it does not always apply to all the separate parts of organisms. Notably, it is not true of the non-cellular parts of supportive and connective tissues (Chapter 3).

8–11 The percentage growth of boys (Minot's data). Note at *a* the slight but characteristic increase in growth rate around 14 to 16.

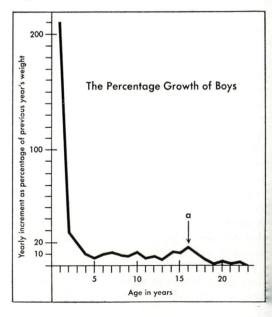

The Percentage Growth of Boys

Once growth starts at all,[5] the percentage increase is fastest right at the beginning and then tends to slow down steadily throughout life or until growth stops. Perhaps you will agree with C. Minot, a leading student of growth, that the capacity for rapid growth is a youthful characteristic and that its loss is a sign of aging. If so, you must also agree with Minot that you started to grow old before you were born, and that you were aging most rapidly during your first days in your mother's womb. The young grow older rapidly, but the aged age slowly (see Fig. 8–11).

This situation may be related to something you have probably noticed, or if not, will before you are much older. Time seems to pass more slowly when you are young than when you are older. This is perhaps because the young grow and live faster. An extension of this idea is that a dog, for instance, has used up about as much of its life span and is physiologically about as aged at 10 as a man at 70. This is the concept of *physiological time*, which, according to another student of growth (P. B. Medawar), is "biology's claim to be considered at least as obscure to the lay mind as theoretical physics." Obviously an animal with a short life span *ages* more rapidly than one with a long span. Whether, however, it *lives* more rapidly is indeed obscure.

RELATIVE GROWTH

So far we have talked as if growth were just a matter of becoming larger. If you look at a calf and a cow or at a baby and yourself, you will see that there is more to growth than that. There are changes in shape as well as in size. Frequently, the changed shape results from the fact that something new has been added, such as horns on a cow or breasts on a human female. Even more frequently, shapes change because parts grow at different rates. A calf or a baby has a relatively larger head than a grown cow or you. Evidently in cattle or men (and this is true as a generalization about mammals) the head either grows more slowly or stops growing earlier than the body as a whole.

Fig. 8–12*A* compares growth rates for the human body and some of its organs. Body

[5] You recall that growth usually does not occur in the very earliest phase of development.

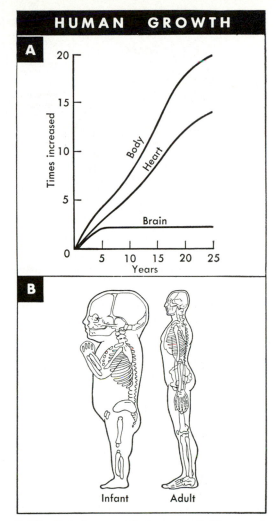

8–12 Relative growth in man. *A*. The growth rates of body, heart, and brain compared; the heart and brain grow at a relatively slower rate than the body, especially in later years. *B*. The effects of differences in the relative growth rates of parts of the body are strikingly brought out by a comparison of the proportions of infant and adult.

weight continues to increase fairly steadily up to the age of 25 (and beyond). Heart and brain grow at nearly the same rate initially but they grow more slowly than the whole body. Their growth also levels off long before the body's weight stops increasing. The weight of the brain increases about as rapidly as that of the heart up to about 4 years, but then it stops growing. Of course, this agrees with and in one way explains the fact that the head is relatively larger in infants than in adults. Fig.

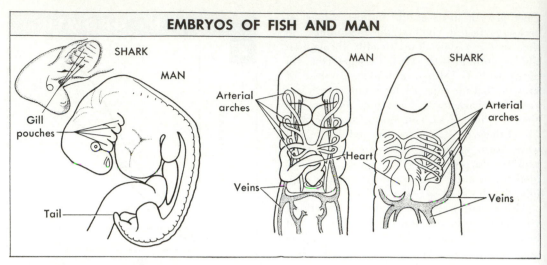

EMBRYOS OF FISH AND MAN

SHARK

MAN

Gill pouches

Tail

MAN

SHARK

Arterial arches

Heart

Veins

Arterial arches

Veins

8–13 An example of the biogenetic principle. *Left.* Embryos of shark and man are compared, showing the presence in both of gill pouches in the neck region. (See Fig. 18–5 for the subsequent developmental history in man of the skeletal arches in these gill pouches.) *Right.* Embryos of shark and man are again compared, showing the great similarity in their circulatory systems. Adult sharks and men differ much more in these respects.

8–12*B* shows how strikingly external and skeletal proportions change in man as a result of different growth rates of different parts.

Evolution of Development

In some respects the development of embryos may be singularly indirect. Human embryos (Fig. 8–13) develop tails, which later disappear. They also develop gill-like pouches in the neck region, which disappear as such and are in part transformed into quite nongill-like structures, including the ear canal. At this stage the embryo looks a little like a fish, although the resemblance is not so close as is sometimes suggested.[6] Still it seems very roundabout for a developing human to go through a stage even remotely fishlike and to have structures so little related to adult anatomy. Until the reality of evolution was recognized such facts were inexplicable. Then it was realized that they have a historical, evolutionary basis.

SUPPOSED RECAPITULATION

The fact is that the tailed, gill-pouched stage of the human embryo is not much like

an adult fish but is very like an *embryonic* fish. K. E. von Baer (1792–1876) noted this fact even before the fact of evolution was recognized. He considered it an example of the so-called biogenetic law, which is not really a law but a descriptive generalization with many exceptions. According to this generalization the earlier stages of embryos resemble those of other animals lower in the scale of nature, or, as we would now say, more like those of related or ancestral groups. As development proceeds, the embryos of different animals become more and more dissimilar. In its very earliest cleavage stages, a human embryo is rather like that of a starfish. In somewhat later stages it is still very similar to that of a (true) fish, amphibian, or reptile. Even later, it is quite like the embryos of other mammals. Finally, well before birth, it becomes clearly human and unmistakably distinct from any other species.

Early evolutionists, especially E. H. Haeckel (1834–1919), rephrased that generalization as the *principle of recapitulation:* "ontogeny repeats phylogeny,"[7] that is, successive stages

---

[6] The human embryo does not have any differentiated gill tissue, and the gill-like pouches do not have open gill slits as in fishes. Fins are lacking. The tail is

not at all like any fish's tail. Indeed, the resemblance to an adult fish is vague and superficial.

[7] *Ontogeny* is the "development of being" (the individual organism), and *phylogeny* the "development (or descent) of races."

of individual development correspond with successive adult ancestors of the line of evolutionary descent. The vaguely fishlike stage of the human embryo was believed to represent the stage when our adult ancestors were fishes. Von Baer had more correctly generalized the facts, but at a time when the principles underlying those facts could not be understood. Haeckel correctly pointed out that the observed facts must result from evolution, as Darwin had already done, but Haeckel misstated the evolutionary principle involved.

It is now firmly established that ontogeny does *not* repeat phylogeny.[8] Ontogeny repeats ontogeny, with variations. Phylogeny is a series of ontogenies. Evolution is not manifested by a sequence of adults giving rise to later, modified adults. An individual organism has a time dimension. It is the same organism from zygote to death and is to be understood only as a dynamically developing living system. What is passed on from one generation to the next is a developmental mechanism. It is also the developmental mechanism that evolves. Heritable change in the adult can occur only on the basis of change in that developmental mechanism.

The developmental mechanism that produced a fish in our ancestors of about 300 million years ago has been inherited by us. In the meantime, however, it has undergone many and profound evolutionary changes and it produces quite a different kind of adult organism. The changes are more evident in later than in earlier developmental stages, and that is why an early human embryo is still rather like a correspondingly early fish embryo. There have, however, been some important evolutionary changes even in the earliest stages.

[8] You may well ask why we bother with principles that turned out to be wrong. There are two reasons. In the first place, belief in recapitulation became so widespread that it is still evident in some writings about biology and evolution. You should therefore know what recapitulation is supposed to be, and you should know that it does not really occur. In the second place, this is a good example of how scientific knowledge is gained. Von Baer and Haeckel were not flatly or wholly wrong. They made successive approximations to truth, and our present closer approximation is based on their accumulation of facts and attempts at explanation.

CHANGES IN ONTOGENY

Haeckel's principle of recapitulation would be correct if changes in the developmental mechanism were usually additions of new stages at the end of development. For instance, a sequence of stages *a-b-c-d* might produce an adult fish, and addition of new stages after this, making the ontogeny *a-b-c-d-e-f-g*, might produce a reptile or a mammal. Students who followed up Haeckel's lead did at first think that this is what commonly happens. Something more or less like it may, indeed, sometimes happen, but simple addition of new stages at the end of ontogeny is certainly not the usual course of evolution. An adult fish has developed about as far from the zygote as has an adult man, but it has developed in a *different direction.*

The changes in ontogeny that produce evolutionary change are usually changes in direction of development. Often, in accordance with von Baer's generalization, they are more evident in later stages. Usually, however, they have some effect throughout the whole course of development, and they may be most apparent at any one stage. As would be expected, the earlier the effects of a change in direction, a *deviation,* become evident, the greater the usual difference in adult structure. Up to a point, gill-like structures develop similarly in fishes and in mammals, but deviation occurs early, and the adult differences in this region are profound.

In some instances, which are exceptions to von Baer's generalization, the young of related species differ more than the adults. This is especially likely among animals that have feeding larvae quite different in form from the reproducing adults; caterpillars and butterflies are a familiar example. The larval forms have special adaptations for their own way of life, usually strikingly unlike the way of life of the same animal when adult. Larval adaptations may evolve without any great influence on adult structure. Thus it happens among some flies or worms that the larvae differ greatly and the adults are closely similar.

Even in forms without larvae, there are usually special adaptations to embryonic life that never occurred in any adult form and that may be quite different in different groups. In man and other high mammals, the umbilical cord, the placenta, and the membranes surrounding the fetus are embryonic adaptations of essential evolutionary importance. They differ in several respects from adaptations in marsupial mammals and are profoundly different from the embryonic adaptations of, for instance, fishes or birds.

Apparent oddities and indirections in development may give helpful clues as to the way structures evolved. For instance, the way jaws develop in some fishes clearly indicates that they arose from gill arches of earlier, jawless vertebrates (see Chapter 16). Of course this is not because the jawless stage is "recapitulated." It is because the developmental mechanism for jaws and for a particular pair of gill arches started out as the same and has not deviated unrecognizably. In other instances almost unrecognizable deviation has occurred. For example, there is little question that the liver in vertebrates originated in the evolutionary sequence (phylogenetically) as a pouchlike protrusion from the intestine. In mammals, however, it does not arise embryologically (ontogenetically) in that way, but as a compact mass of cells that differentiates directly into liver tissue. Stages present in the ancestral developmental sequence, but no longer functionally significant, have been dropped out of ontogeny.

There is another mode of evolution of ontogeny that may have had exceptional importance in some instances. This is the retention in adults of features that occurred in earlier stages in the ancestors.[9]

It does happen, although not commonly, that a final stage of development is simply dropped off the life history. Salamanders normally have a larval stage in which they are aquatic and have gills, and an adult stage in which the gills are lost and they breathe with lungs. However, some salamanders become sexually mature and reproduce without ever losing the gills or acquiring the fully adult form.

There are other ways in which ontogenies have changed in the course of the history of life. The preceding few examples suffice to

[9] There are complicated and sometimes conflicting technical terminologies for the many kinds of changes that can occur in ontogeny. The term most often applied to this particular process is *neoteny*.

emphasize and illustrate the fundamental principle that phylogeny, evolutionary descent, is a sequence of ontogenies and that the course of evolution is through changes in ontogeny. Some other striking examples among both plants and animals will appear in the discussion of life histories in the next chapter.

RELATIVE GROWTH AND EVOLUTION

Changes in the relative sizes of parts because they grow at different rates and stop growing at different times are extremely common among animals. Often they are even more striking than in man. Since it is really the growth mechanism that produces changes of form in the course of evolution, this phenomenon has significance in the history of life.

Striking changes can occur as a result of relative (or differential) growth persisting for longer or shorter periods. A famous example is the extinct Irish elk, which had tremendous antlers, the largest known. In the smaller living allies of this animal, the antlers increase in size at a rate faster than the body as a whole. The larger the animal, the larger its antlers, not only in absolute size but also in proportion to the rest of the body. Now if the living elk, which are much smaller, grew to be as large as the extinct Irish elk, they would have antlers as disproportionately large as in that animal. This strongly suggests that the Irish elk had the same sort of growth mechanism as its living relatives. The evolutionary change that produced that species was, then, just an increase in total body size. The established growth mechanism automatically resulted in the enormous antlers. Change in body size is one of the commonest evolutionary events, and usually results in differences of shape through the mechanism of relative growth.

In the example of the Irish elk, the pattern of relative growth was not changed. It only continued further in a larger animal. In some other cases changes in the pattern are important. Fig. 8–14 shows a comparison of two related recent fishes by a method developed by D'Arcy Thompson. The porcupine fish has more usual proportions and probably resembles the ancestor of both fishes. The extraordinary form of the sunfish results mainly

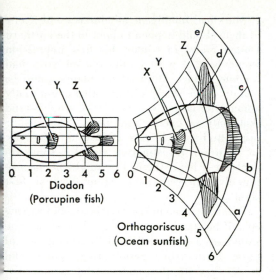

8–14 Evolutionary change in the pattern of growth. A grid of coordinates is drawn over both the porcupine fish and the sunfish in such a way that homologous points (*X, Y,* and *Z*) in the two fishes fall on corresponding points on the overlying coordinate grid. Comparison of the grids shows that the sunfish grows much faster from top to bottom in the tail region than does the porcupine fish.

from a change of growth pattern such that the nearer a part is to the hind end, the faster it grows in the top-to-bottom direction.

CHAPTER SUMMARY

Developmental processes controlled by the inherited message in the chromosomes: growth and cell division; morphogenesis, the creation of pattern and shape; differentiation, the specialization of cell types; closed development in animals; open development in plants; continuity of germ cells; phases of development in which one or another of the developmental processes predominates.

Patterns of morphogenesis and differentiation in animals: maturation of the egg cell includes differentiation of egg cytoplasm; cleavage—cell division without growth; the blastula; twinning in relation to the plane of cleavage and the axis of the egg's differentiation; the effect of yolk in cleavage.

Gastrulation and morphogenesis: ectoderm and endoderm; gastrulation in amphioxus and frog compared; origin of mesoderm and of coelom; differentiation of tissue types; their derivation from germ layers; organ formation.

Induction and organizers; the organizer in frogs; experimental demonstration of its inducing action; presumptive fate, dependent on embryonic environment; determination.

The problem of genetic control of development—in essence the problem of how the same inherited message (identical chromosome sets) is consistent with the differentiation of diverse tissue types: its solution to be sought in knowledge of how gene action is affected by the chemical and physical environment; known to be different in different cells after cleavage of the egg, the gene controlling pigment formation in rabbits as an example.

Patterns of growth: the general form of the growth curve; variations in vertebrates; stepwise growth in arthropods; endocrine effects on growth in man; percentage growth; relative growth.

Evolution of development: similarity of human and fish embryos; the biogenetic principle; Haeckel's misstatement of the principle; changes in ontogeny—deviation, larval specialization; relative growth and evolution (the case of the Irish elk).

9

Reproduction:
Organismic Aspects

American Museum of Natural History

These young chickadees are awaiting the return of their parent with food. Parental care is one of the many aspects of reproduction related to the whole organism.

If Mars had intelligent inhabitants and if one of them should happen to land in the northern United States in winter, his first impression would be that most of the life on earth had become extinct. Many of the plants would be plainly dead. Most of the others, especially among the larger woody plants, would stand skeletonlike, gaunt and bare. The visitor might find a few insects, all dead or seemingly so. If wider investigation revealed some rodents or a bear, these, too, might be so profoundly quiescent as to seem dead at first sight.

What amazement the Martian would experience if he stayed until spring! Everywhere he would see a resurrection of life. Plants would burst forth from germinating spores and seeds. Trees would break out into vividly living greenery. From tiny eggs and mummy cases millions of insects would emerge. Birds would appear, seemingly from nowhere, and bustle about the business of nesting and reproducing. The hibernators would awaken, among them the she-bear, trailed by her cubs. The visitor would thus conclude that life on earth is cyclic. Each longer life has a recurrent pattern through the cycle of the seasons. All lives, long or short, are links in reproductive cycles in which one life follows another, each life repeating the pattern of its forebears within its species.

The seasonal cycle is not everywhere so apparent as in the so-called Temperate Zone, with its intemperate alternation of summer and winter. Yet the reproductive cycle is completely universal. It exists for all living things wherever they may be on earth and whether their life spans be measured in minutes or in centuries. All kinds of organisms are born and differ only in the manner of their origin. "Birth" may be the fission of a parent, the germination of a seed, the emergence from an egg or from the womb; but the resemblances are more fundamental than the differences. In every kind of organism, an individual in its turn gives birth in its own fashion to others, and so begins another cycle of life.

The Generalized Reproductive Cycle

The individual life cycle begins with reproduction from a parent or two parents. Development of the individual follows (Chapter 8) and then reproduction again, starting another cycle. There has thus been in all organisms a continuous sequence: reproduction–development–developed organism–reproduction–development–developed organism– . . . and so on through the centuries and millennia, ever since life began. It would be foolish to say that one phase is more important than another in a process that is continuous, with each phase dependent on the others. However, in some sorts of studies (for example, those of transfers of energy and materials in communities) attention is naturally focused on the developed organism. Then reproduction may be viewed merely as part of the background, as the process that keeps up a continuous supply of organisms. In consideration of the life cycle, on the other hand, reproduction is a particularly crucial phase. It begins the individual cycle and is the connection between the generations that are links in the long chain. From this point of view, the developing and developed individual is the medium of reproduction, or what intervenes between the crucial episodes of reproduction. The subject of life cycles may, then, be considered first of all in terms of reproductive cycles.

ASEXUAL AND SEXUAL REPRODUCTION

In Chapter 5 the subject of cellular reproduction introduced us to the study of heredity, and the distinction was there made between sexual and asexual reproduction. In *sexual reproduction* there is a fusion of two nuclei from different cells (gametes) into one cell (zygote); and in *asexual reproduction* there is not. Asexual reproduction is necessarily uniparental. Sexual reproduction is ordinarily biparental but may also be uniparental, as, for instance, in the many self-fertilizing flowering plants.

One of the two usual forms of asexual reproduction (Fig. 9–1) is *vegetative reproduction,* which occurs in both animals and plants but is much more common among plants. Both protists and multicellular organisms may re-

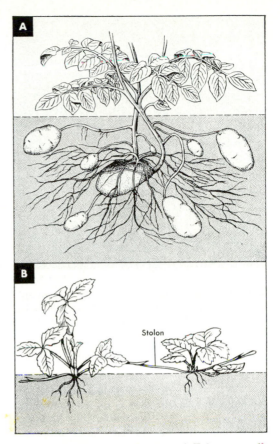

9–1 Vegetative reproduction. *A.* Tuberous swellings of the stems in the potato plant serve to propagate the species vegetatively. The bulk of tissue is a food reserve (starch), but embryonic tissue is present in the "eyes" of the potato, and from these new shoot and root systems can develop. *B.* In the strawberry the shoot system effects vegetative reproduction. Stolons, shoots growing along the surface of the ground, may develop new root systems and establish a new plant.

produce vegetatively by fission of the body, each part developing into a separate organism. In many plants and a few animals, almost any reasonably large part will grow into a new oragnism under proper conditions. Thus *cuttings* are often used to propagate cultivated plants;[1] thus a planarian (p. 327) develops

[1] A special sort of propagation by cuttings is grafting, the attachment of a cutting from one plant to the growing stem or root of another. The graft and its host become part of the same physiological system and look and grow like a single plant, but there is no exchange or mixture of chromosomes and each part retains its own heredity. A peach may be grafted to a plum, and then the same tree will bear both peaches and plums on different branches.

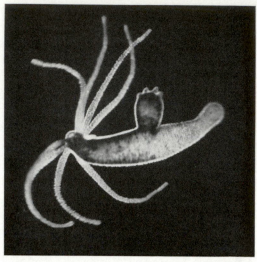

Hugh Spencer

9–2 Vegetative reproduction in *Hydra*, an animal. A bud in *Hydra* is an evagination of the polyp wall containing both ectoderm and endoderm tissue layers. It develops tentacles (barely visible in the photo) and a mouth. Eventually, severing itself completely from the parent, it becomes a new individual.

from a portion of its parent. Quite a few animals, of which *Hydra* is an example (Fig. 9–2), reproduce vegetatively by *budding.* Many plants develop special *organs of vegetative reproduction.* Some of these are surely familiar to you: the tubers of potatoes, the bulbs of onions or tulips, the bulblike organs of gladioli, the runners of strawberries, and the runnerlike underground structures of many grasses. Many different parts and processes may be involved in vegetative reproduction, but the principle is always the same and is simple: a part of the parental organism, genetically identical with its other parts, develops into a separate organism.

The other usual form of asexual reproduction is by *spores,* plant reproductive cells each of which develops into a separate organism without fertilization.

The essential feature of sexual reproduction is the fusion of two nuclei within a single cell, which develops into a separate organism. You will recall from Chapter 5 that the specialized sexual reproductive cells of plants and animals are gametes and that they usually are of two kinds: smaller, more mobile male gametes, or sperms, and larger, more passive fe-

male gametes, or eggs. The union of male and female gametes is fertilization, and the resulting single cell is a zygote. In some protists and lower plants, the two cells whose nuclei fuse are not visibly different from each other or form ordinary cells (or, in protists, whole organisms) of the species. The male and female are not distinguishable, and the lack of apparent specialization for reproduction may make the term "gamete" seem inappropriate. Nevertheless, such cells have gone through meiosis and are fully comparable with gametes in this preparation for fusion by reduction of chromosome number.[2]

Modes of achieving sexual reproduction have become amazingly diverse in plants and animals. The gametes have acquired numerous different characteristics in different groups. Organs associated with their production, with fertilization, and with subsequent development of the offspring have become extremely specialized and have also taken innumerable different forms. Nevertheless, sexual reproduction is essentially the same and has the same evolutionary and biological significance whether it occurs in paramecia, roses, or humans. The basic phenomenon is that half the chromosomes from each of two individuals have been united in a single new individual.

DISTRIBUTION AND SIGNIFICANCE OF SEX IN ORGANISMS

It seems probable that the first organisms, perhaps 2 billion years ago, when life was young, had simple, asexual reproductive cycles. They must have reproduced solely by fission—division of protoplasm and genetic materials—with each fission product becoming an individual of the same kind. Even to-

[2] Some biologists object to applying the term "fertilization" to fusion of nuclei in protists and do not consider this as sexual reproduction. They call passage of the nucleus, only, from one individual to another "conjugation," and fusion of two individuals, both nuclei and cytoplasm, "syngamy." Nevertheless, in essence and real biological significance conjugation or syngamy in protists and fertilization in multicellular organisms are fundamentally the same. In both, haploid cells (or their nuclei) fuse and produce a new diploid individual. It is not confusing two different things but recognizing their essential equivalence when the same terms are applied to the process in protists and in other organisms.

day, simple, asexual cell division and the mitotic transmission of identical nuclear controls to each new daughter cell is the fundamental mode of cellular reproduction.

Sex clearly is not essential to the reproductive process, even though it is almost universal in organisms. In fact, at the cellular level and (as a rule) in protists, it generally results in no increase in the number of cells and may cause a decrease (through the fusion of hitherto separate gametes). In multicellular forms, on the other hand, sex is directly associated with reproduction. The initial reason for this association is evident. The basic sexual process is the combination in one cell of genetic materials from two sources. In organisms multicellular as adults, the one-celled stage at which any sexual process must occur is necessarily the reproductive stage at which a new generation begins. In most multicellular organisms development from the one-celled stage does not normally proceed unless there has been sexual fertilization. This association, however, does not explain why sex is so widespread.

The significance of sex in organisms is that it is a device for promoting genetic variability and adaptation. We saw in Chapter 6 that the gametes produced in meiosis contain a recombination, or reshuffling, of gene combinations present in the parents' gametes. Consider two organisms that are homozygous for two gene pairs as follows: *AABB* and *aabb*. As long as each reproduces only by asexual means, the offspring can be only *AABB* and *aabb*, respectively, generation after generation until the comparatively rare process of mutation changes one of the alleles. But were these organisms to reproduce sexually—were they to mate—producing hybrids (*AaBb*), the very next generation of sexually produced offspring would contain the following recombination genotypes: *AABB*, *AABb*, *AAbb*, *AaBB*, *AaBb*, *Aabb*, *aaBB*, *aaBb*, and *aabb*. Of course, normal sexual reproduction yields an even richer array of variants than this. Since an organism contains thousands of gene pairs, not just two, as we noted in Chapter 7, the number of genotypes possible with recombination among offspring is enormous.

The basis for evolution is genetic variation in a population (see Chapter 15). Asexual reproduction keeps such variation at a minimum, whereas sexual reproduction maximizes it. There is, therefore, no mystery attaching to the nearly universal occurrence of sex in organisms. Those populations of organisms most able to vary have been those most able to survive changing conditions in the environment and those most able to evolve new ways of life as the opportunities arose. Sex is widespread because, like any other adaptation, it has promoted the long-term survival of the populations having it.

The term "sex," as ordinarily used in biology, refers to meiosis followed by nuclear fusion. This implies a fully developed nuclear, chromosomal, and mitotic system. As we learned in Chapter 3, such systems are not found in all cells. Those that do possess them are the eukaryotes, the group that includes the cells of all multicellular organisms and the majority of protists. Thus the structures and systems necessary for sexual reproduction are present in multicellular organisms and most protists, and sex occurs almost universally in this group. Prokaryotes, on the other hand, are the bacteria and blue-green algae, lowly organisms possessing no nuclei, no true chromosomes, and no mitotic or meiotic apparatuses. It was long believed that sex was absent from these groups, but it is now known that processes with the same result as sex—that is, the combination within one organism of genetic material from separate ancestors—occur in them. To distinguish these processes from sexual processes with meiosis and nuclear fusion, they are sometimes called *parasexual*.

Although bacteria are prokaryotes lacking true chromosomes and meiosis, certain mechanisms do exist for the transmission of genetic material from one bacterial cell to another. It can be transferred in the form of soluble DNA liberated when a cell breaks up or by the hands of a biochemist. This is the phenomenon called *transformation* (Chapter 7). In *transduction*, DNA is carried from one cell to another by an infecting virus. There is still another process called *conjugation* in which a donor cell makes direct contact with a recipient cell and inserts into it the bacterial equivalent of a chromosome. In structure and other properties it is not a true

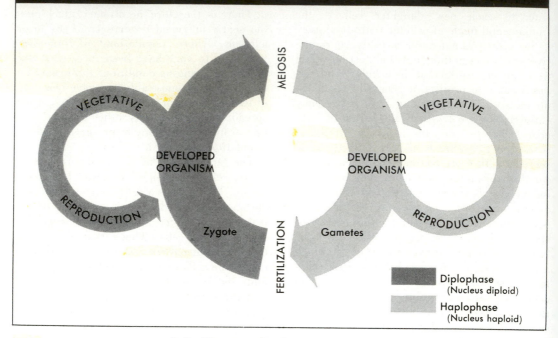

GENERALIZED REPRODUCTIVE CYCLE

MEIOSIS

VEGETATIVE

VEGETATIVE

DEVELOPED ORGANISM

DEVELOPED ORGANISM

REPRODUCTION

REPRODUCTION

Zygote

FERTILIZATION

Gametes

Diplophase
(Nucleus diploid)

Haplophase
(Nucleus haploid)

9–3   The generalized reproductive cycle.

chromosome, however, but a linear physical structure containing DNA and a linearly arrayed "linkage group" of genetic determinants. Some have regarded conjugation as a sexual process because it involves the transfer of genetic material from a "male" to a "female" cell.[3] However, it is not true sexual union because it does not involve meiosis and nuclear fusion. Nevertheless, it comes close to sexuality and does produce genetic recombination analogous to that occurring when chromosomal recombination takes place.

Neither sexual nor parasexual processes have yet been observed in blue-green algae. With that doubtful exception, however, we can generalize that all now-living organisms either exhibit some process of genetic recombination or are modified descendants of ancestors that exhibited such a process.

[3] Many authors loosely refer to the object transferred in bacterial conjugation as a "chromosome" because it behaves as if it were a chromosome. One might also point out that when conjugation takes place in a population of bacteria, it is performed only by certain specially endowed organisms within the population. Conjugation will be discussed further in Chapter 19.

THE HAPLOPHASE–DIPLOPHASE REPRODUCTIVE CYCLE

Most protists, a great many plants, and quite a number of animals have both asexual and sexual reproduction. In simplest form possible combinations of reproductive cycles are generalized in Fig. 9–3. The numbers and sequences of asexual and sexual cycles vary greatly. Most protists go through an asexual cycle (by fission) repeatedly and then once in a while go through a single sexual cycle, followed by more asexual cycles. In animals and plants the sequences may be quite irregular. Not infrequently asexual and sexual cycles are simultaneous (in different offspring). For instance *Hydra* often reproduces sexually and by budding at the same time.

Fertilization, which is nuclear fusion at the cellular level, is the definitive process of sexual reproduction. You have seen that fertilization must be preceded by meiosis, cell division with halving of chromosome number. Otherwise the number of chromosomes would double in each repetition of the cycle and would soon become impossibly large. On the other hand, if meiosis recurred in a cycle and fertilization did not, the chromosome number would be cut in half in each repetition of the cycle. This is also impossible because normal life demands the presence of at least one full *set* of chromosomes in each cell. Therefore the sexual cycle must include both of two key processes: meiosis and fertilization.

Meiosis and fertilization divide the whole life cycle of sexual organisms into two parts. After meiosis and before fertilization cells and organisms have a reduced number of chromosomes; they usually have *n* chromosomes and are haploid; this phase of the reproductive cycle is the *haplophase*. After fertilization and before meiosis they have the full complement of chromosomes: the number is usually $2n$, and they are diploid; this is the *diplophase* of the reproductive cycle.

Among most organisms, especially those most familiar to us, the developed organism occurs in the diplophase of the sexual cycle. The more conspicuous and longer part of the cycle, the one in which the organism matures and lives an independent life, acquiring its own materials and energy from the environment, involves development from the (diploid) zygote and subsequent maturity. The developed organism of the generalized diagram (see Fig. 9–3) is then on the left side of the diagram. Then it may be that only gametes, sperms and eggs, occur on the right side of the diagram, in the haplophase of the cycle. This is the situation in human beings and most other animals.

Gametes are, in fact, haplophase organisms themselves. They have only a single set of chromosomes, but they do have one complete set, which is sufficient for an organism to live and carry out all necessary metabolism. Specialized gametes usually are short-lived, acquiring materials only from the parent and having a brief period of independent life during which they acquire little or nothing on their own. Then they unite into a zygote or die. Nevertheless it is possible for a cell produced by meiosis, hence haploid, to lead a longer life, and even for it to acquire materials on its own and to have all the characteristics of an independent organism. It may go so far as to develop and become multicellular. The occurrence of a developed organism (some development in meiosis) in the haplophase of the sexual cycle (to the right in our diagram) is rare, but not absent, among animals.[4] It is usual in plants.

There are three possibilities for the occurrence of developed organisms in the sexual cycle. (1) They may occur in the diplophase only. This is usual in animals, as has been noted, and common in protists. (2) They may occur in *both* the diplophase and the haplophase. This is a rare, unusual modification in animals, but it is almost universal in lower plants. There is some development of both phases in all higher plants, too, but the haplophase ordinarily consists only of a few parasitic cells. (3) Finally, they may occur only in the haplophase. This is the rarest of the three possibilities, but it does characterize some protists and lowly plants, and it may be the primitive condition from which the others evolved. Note also that, wherever they occur in the sexual cycle, whether diplophase or haplophase, many developed organisms can reproduce vegetatively as well as sexually. The descriptive facts reviewed up to this point are summarized in Fig. 9–3. No one organism is known to go through *all* the specific processes indicated. However, all the reproductive cycles occurring in organisms are shown in the diagram.

The valuable thing about the diagram is that it shows the biological relationships between the numerous kinds of reproductive cycles that occur among organisms. For instance, the reproductive cycles of most animals may seem to be fundamentally different from those of most plants, and they are usually so represented. Yet if, as we go along, the special cycles in plants and animals are compared with the generalized cycle, they will be seen to be fundamentally similar. The only essential difference is that in animals the cells (the haplophase) resulting from meiosis usually develop no further before fertilization (they are simply gametes), while in plants the haplophase usually does develop somewhat further, becoming multicellular and, in some cases, may even develop into an independent organism.[5]

---

[4] Male bees, for instance, are haploid.

[5] Some exceptions are diatoms and a few other algae.

MEIOSIS

FERTILIZATION

Diplophase

Haplophase

9–4   The reproductive cycle in the protist *Chlamydomonas*. The diplophase is of minor significance; most of the life cycle is passed in the haplophase, which alone undergoes vegetative reproduction. *1.* Young zygote. *2.* Mature zygote, in which meiosis occurs. *3.* Liberation of four haploid cells (products of meiosis) from the old zygote case. *4.* Mature haplophase cell. *5, 6.* Vegetative reproduction. *7, 8.* Two haplophase cells acting as gametes. *9.* Fertilization by union of the gametes.

Reproductive Cycles in Plants

GAMETOPHYTE AND SPOROPHYTE

The occurrence of development in the haplophase of the reproductive cycle is so nearly universal in plants and so rare in animals[6] that this difference is even more definitive than the fact that most plants are photosynthetic. Since developed organisms commonly occur in both the haplophase and diplophase of plants (both to the right and to the left in our diagram), it is convenient to distinguish these by their technical names. The developed haplophase organism (between meiosis and fertilization and therefore haploid or with reduced

[6] You will recall that the animal egg cell does undergo some internal differentiation—which is development of a sort—but it never becomes either multicellular or an independently feeding organism before fertilization.

chromosome number) is called the *gametophyte* ("gamete plant") because it produces gametes. The developed diplophase organism (between fertilization and meiosis and therefore diploid) is called the *sporophyte* ("spore plant") because it produces spores. One of the striking things about reproductive cycles in plants is their great diversity in relative development of gametophyte and sporophyte.

PLANTLIKE PROTISTS

Chlamydomonas (Fig. 9–4) is a single-celled photosynthetic organism. The cell, which swims around actively photosynthesizing for most of the life cycle, is haploid. It is capable of asexual reproduction in the haplophase. Occasionally two cells fuse to form a diploid zygote which immediately undergoes meiosis, yielding four new haplophase cells. Thus in *Chlamydomonas*, as in all other flagellated protists (which are believed to be the

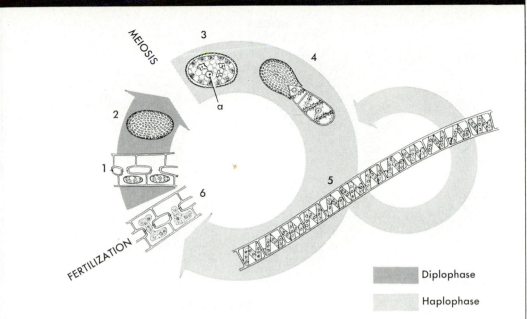

MEIOSIS

FERTILIZATION

Diplophase

Haplophase

9–5 The reproductive cycle in the green alga Spirogyra. As in *Chlamydomonas*, the diplophase is of minor significance; most of the life cycle is passed in the haplophase, which alone undergoes vegetative reproduction. *1*. Two young zygotes in the cells of a former haplophase filament. *2*. Mature zygote, which has developed its own cell wall (or case). *3*. Four haploid nuclei (products of meiosis) within the old zygote case. Three of these disintegrate; one (*a*) develops further. *4*. Young haploid filament, developed from the surviving cell in *3*, germinating from the old zygote case. *5*. Mature haplophase filament, which can vegetatively reproduce. *6*. Conjugation of two haplophase filaments. The cytoplasm and nuclei of the cells in one conjugating filament migrate through specially developed conjugation tubes to the cells of the other conjugating filament; zygotes are formed in each cell of the latter filament.

most primitive of cellular-chromosomal organisms), the life cycle is spent mostly in the haplophase. The diplophase is not extended beyond the time necessary to execute meiosis leading to another generation of haploid cells.

ALGAE

Algae is a name applied to numerous plants, relatively simple in structure, almost all aquatic. Various groups of algae probably arose independently from protists.[7] In many algae like *Spirogyra* (Fig. 9–5) the primitive form of the haplophase-diplophase cycle persists; that is, the diplophase undergoes no

[7] Indeed, photosynthetic protists and their allies are classified as algae by many botanists. This is true of *Chlamydomonas*, for instance. Some other botanists classify all algae as protists rather than as plants.

special development but immediately enters meiosis to yield new haplophase organisms that grow and propagate asexually before renewed sexual reproduction. There is, then, no sporophyte[8] in these plants; the actively growing and dividing plant is a gametophyte.

Within the algae themselves—most primitive of all multicellular plants—there has been an evolutionary tendency to switch emphasis from the haplophase to the diplophase part of the cycle. We will find that this trend applies broadly to plants as a whole. In *Ectocarpus*, a common brown seaweed that often grows on other algae, a definite sporophyte is present. Indeed, it is as prominent as the gametophyte,

[8] Recall that the terms "gametophyte" and "sporophyte" are to be applied to developed (some nuclear division) organisms in the haplophase and diplophase, respectively.

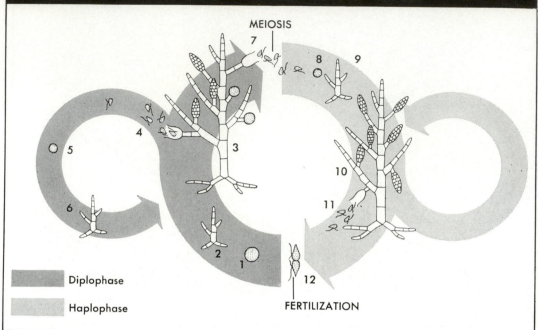

MEIOSIS

7

8   9

4

3

5

10

6

11

2   1

12

FERTILIZATION

Diplophase

Haplophase

9–6   The reproductive cycle in the brown alga *Ectocarpus*. Haplophase (gametophyte) and diplophase (sporophyte) are equally prominent and very similar. Vegetative reproduction occurs in both. *1.* Zygote. *2.* Young sporophyte. *3.* Mature sporophyte. *4, 5.* Diploid zoospores liberated from a sporangium. *6.* Young sporophyte. *7, 8.* Haploid zoospores (products of meiosis) liberated from a sporangium. *9.* Young gametophyte. *10.* Mature gametophyte. *11.* Gametes being liberated from a gametophyte. *12.* Fertilization.

and the two generations [9] cannot be distinguished without microscopic study. The cycle in *Ectocarpus* is shown in Fig. 9–6. The cycle in the somewhat more familiar sea lettuce, *Ulva,* is the same.

The sporophyte in *Laminaria,* a brown alga commonly known as kelp, is far more conspicuous than the gametophyte (Fig. 9–7). The sporophyte is a flat, crinkly blade, often 2 meters or so in length, single or fork-like, attached to rocks by what look like roots. (The "roots" are, however, anatomically and physiologically unlike true roots, which do not oc-

cur in algae.) In some of the surface cells in the blade the nuclei and cytoplasm divide repeatedly and give rise to tiny spores. Meiosis occurs in the course of these divisions, and the spores are therefore haploid. The spores are released from the parent plant into the surrounding water, where they swim by means of flagella. The spores develop into separate male and female [10] gametophytes, each microscopic in size and consisting of only a few cells. The male plants produce and release tiny sperms,[11] which also have flagella and swim about. The female plants produce somewhat

[9] In plants in which distinctly developed multinucleate organisms (gametophyte and sporophyte) occur in both haplophase and diplophase parts of the reproductive cycle, it has been customary to speak of an "alternation of generations." Strictly speaking, this is a confusing usage, because both gametophyte and sporophyte are essential constituents of one full generation (a full reproductive cycle).

[10] It is useful here to introduce in the diagrams the biological symbols for the sexes: ♀ for female and ♂ for male. ♀ is Venus's mirror (a round hand mirror with a handle below), and ♂ is Mars's round shield with his spear sticking out from behind. Astronomers use these same symbols for the planets Venus and Mars.

[11] Another botanical name for them is "antherozoids."

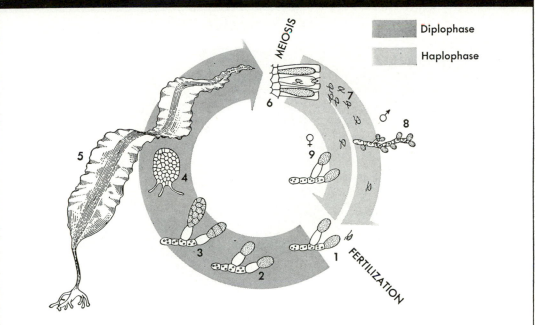

Diplophase

Haplophase

9–7   The reproductive cycle in the brown alga *Laminaria*. The diplophase is the dominant part of the cycle; a large sporophyte is developed. The gametophyte generation is a small filament of cells and occurs as two distinct sexes. *1.* Fertilization: a sperm approaching a female gametophyte on which there are two egg cells, one of which has extruded from its cell wall and is ready for fertilization. *2.* Female gametophyte carrying two fertilized eggs or zygotes, each of which marks the beginning of the diplophase. *3.* Young sporophytes, each still consisting of only a few cells, attached to the old female gametophyte. *4.* Older sporophyte. *5.* Mature sporophyte, about 6 feet long. *6.* Meiosis, in some surface cells of the sporophyte, liberating haploid motile spores (7) that mark the beginning of the haplophase. *8, 9.* Male and female gametophytes, respectively, which have developed from the haploid spores.

larger gametes—eggs—which are pushed out of the cell walls but remain attached to the gametophyte. If a sperm in its wanderings encounters an egg, fertilization occurs. The zygote begins to divide while still attached to the female gametophyte, and there forms an embryonic sporophyte. After just a few cell divisions, the sporophyte becomes detached from the parent gametophyte, attaches itself by holdfasts, and develops into the large, familiar seaweed.

Fucus is another brown alga (Fig. 9–8). In it the emphasis on the diplophase has reached a limit: there is no gametophyte generation.

MOSSES

Mosses have evolved or have inherited from ancestral algae a reproductive cycle opposite in tendency from those of higher plants (ferns and seed plants). In mosses the conspicuous developed organism, the one that has leaflike organs and that carries on most of the vital syntheses for the whole cycle, occurs in the haplophase; it is a gametophyte. The gametophytes develop sperms and eggs, on the same plant in some species and in others on different male and female plants. The sperms are swimming, flagellated cells. Most mosses are nonaquatic but, if fertilization is to occur, there must be a film, at least, of water through which the sperms can swim to the eggs.[12] The

[12] The way in which the sperms find the eggs is amusing. The egg-bearing organs release sugar that diffuses into the surrounding water. The sperms are attracted by sugar (a forced movement and not a preference) and move toward a higher sugar concentration. They thus end up at an egg.

MEIOSIS

FERTILIZATION

Diplophase

Haplophase

Hugh Spencer

9–8 The reproductive cycle in the brown alga *Fucus*. The dominance of the diplophase (sporophyte) is here complete. The haplophase never produces a fully developed (gametophyte) organism; it is restricted to the gametes, which are of two distinct sexes. *1.* Zygote. *2–4.* Successively older stages in the development of the sporophyte. *5.* Mature sporophyte, photographed below. *6.* Swollen end of a sporophyte branch in which there are cavities containing tissue that undergoes meiosis, producing gametes. *6'.* Diagram of one such cavity. For simplicity it is shown producing both male (*7*) and female (*8*) gametes. In fact, each individual cavity produces only eggs or sperms. *9.* Fertilization.

eggs are retained in the special organs (archegonia) in which they develop. After fertilization the zygote also remains there and develops into a sporophyte that is attached to the parental (maternal) gametophyte. The usual form of the sporophyte is simply a long, slender stalk with a capsule at its upper end. Within this capsule meiosis occurs, producing spores. Liberated spores are widely distributed through the air, and under favorable conditions they settle down and develop into gametophytes. The cycle is shown in Fig. 9–9.

Mosses also have remarkable powers of veg-

etative reproduction, which occurs only in the haploid phase (in our diagram, the right side) of the reproductive cycle. In a few species sporophytes are unknown and sexual reproduction has apparently been lost.

FERNS

In ferns and all higher plants the conspicuous leafy organism is the sporophyte (Fig. 9–10). The undersides of fern leaves have brownish dots on them. These are the spore-producing organs, or *sporangia*. If a spore falls on a favorable spot, it develops into a

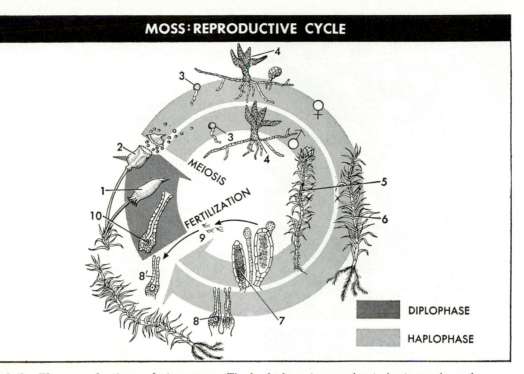

DIPLOPHASE

HAPLOPHASE

9–9   The reproductive cycle in a moss. The haplophase (gametophyte) dominates the cycle; the sporophyte is a small structure, consisting of only a stalk and a sporangium and completely parasitic on the gametophyte. *1.* Sporangium (capsule) of the sporophyte generation. *2.* Liberation of haploid spores that have been produced (through meiosis) in the sporangium. *3.* Germination of the spores to produce young gametophytes, which are of two distinct sexes. *4.* Older gametophytes. *5.* Mature male gametophyte. *6.* Mature female gametophyte. *7.* Male sex organ, or antheridium, which produces motile sperm (*9*). *8.* Female sex organ, or archegonium, containing a single egg cell. *8′.* Fertilization of the egg cell at the base of the flask-shaped archegonium by the sperm (*9*), which swims over the surface of the female plant and eventually down the neck of the archegonium to reach the egg. *10.* Zygote, marking the beginning of the diplophase, at the base of the archegonium.

small, flat, more or less heart-shaped multicellular organism, the gametophyte or *prothallus* of the fern. The gametophytes are seldom more than 5 or 6 millimeters in diameter, so small and inconspicuous that they are almost never noticed except by botanists, who know what to look for and where to look. The same gametophyte produces both sperms and eggs in different organs and at different times, the sperms first. The sperms swim to the eggs, which are retained in the egg-producing organs (archegonia, Fig. 9–11), much as in mosses. As in mosses, too, the sporophyte embryo starts developing on the gametophyte and is at first parasitic. Unlike the mosses, however, the sporophyte soon develops its own roots, stems, and leaves and grows into a

relatively large, independent plant, while its parent gametophyte dies and disappears. The parallel with the alga *Laminaria* (see Fig. 9–7) is obvious. In fact, the reproductive processes in these two plants are practically identical, even though the anatomy of the sporophytes is very different.

SEED PLANTS

The highest plants, that is to say, those most complex and belonging to groups of most recent evolutionary origin, were derived from fernlike ancestors. In them, as in ferns and the brown alga *Fucus*, the conspicuous, vegetative plants are the sporophytes. The gametophytes are greatly reduced in number in comparison with the ferns and are microscopic in size. The

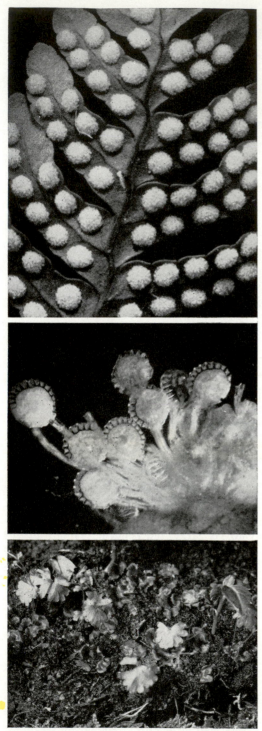

Hugh Spencer and American Museum of Natural History

9–10 Ferns. *Top to bottom.* Clusters of sporangia on the back of the polypody fern's leaf. A single cluster of (about 10) sporangia from the Christmas fern. Gametophytes.

female gametophyte is parasitic in the tissues of its parent sporophyte, and the free existence of the male gametophyte is only during the period when it is being transported as pollen and when little or no development occurs. These characteristics are associated with the production of seeds, in which the sporophyte embryo is enclosed with nutritive material inside a protective coat. Seeds are an adaptation to plant life in the open air and have made possible the great diversity and abundance of land plants.

Flower

The sporophyte in flowering plants, like that of ferns, bears special spore-producing organs, or sporangia. They are of two kinds, male and female. The male sporangia are *anthers*, and the female sporangia are *ovules*. These sporangia are not uniformly distributed over leaf surfaces, as they are in ferns; they are clustered together in a *flower*. Thus the familiar flower is essentially an aggregate of sporangia surrounded by modified, often colored, leaves called *petals* and *sepals* (Fig. 9–12). The significance of the (often) colored petals is something we return to later; it is related to the problem of getting the male gametophyte from one flower to the female gametophyte of another.

Gametophytes

The *microspores* are those destined to develop into male gametophytes. They originate as products of meiosis in the anthers (male sporangia). Development of the male gametophyte, which starts while the spore is still in the anther, consists simply of division of the nucleus into two and then of one of these into two more, which become the male gametes (see Fig. 9–12). Sometimes there is also a division of the cytoplasm by cell membranes, but often there is not. The partly developed male gametophyte when it leaves the anther (it usually has two nuclei at this stage) is a *pollen grain*. It is carried, generally by wind

FERN: REPRODUCTIVE CYCLE

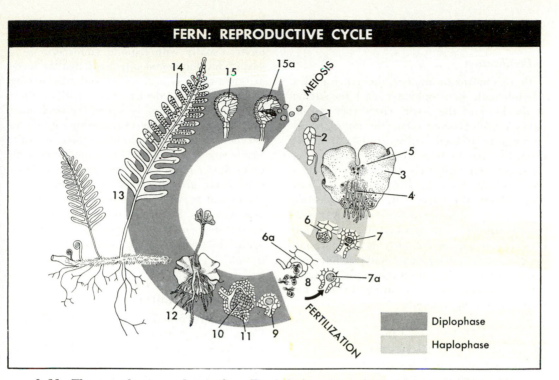

9–11   The reproductive cycle in a fern. The diplophase (sporophyte) dominates the life cycle; the gametophyte is a small, short-lived structure bearing male and female organs. *1.* Haploid spore (meiotic product) shed by the sporophyte. *2.* Young gametophyte. *3.* Mature gametophyte, or prothallus. *4.* Antheridia (male sex organs) among rootlike structures on the underside of the gametophyte. *5.* Archegonia (female sex organs). *6.* Individual antheridium, liberating sperms (*6a*). *7.* Individual archegonium with an egg cell (*7a*) at its base being fertilized (*8*) by sperm. *9.* Diploid zygote at the base of the archegonium. *10.* Embryo sporophyte still in the archegonium (*11*). *12.* Young sporophyte still attached to the parent gametophyte. *13.* Mature sporophyte. *14.* Clusters of sporangia, on the back of the sporophyte leaf. *15.* Individual sporangium bursting (*15a*) to yield haploid spores produced by meiosis in the sporangial cells.

or insects, to the sticky upper end (stigma) of the pistil of the same or another flower.

The female organs of the flower are collectively designated the *pistil*. This comprises stigma, style, and the ovary, inside of which are several ovules. Each ovule is destined ultimately to become a *seed*. Anatomically it is basically a sporangium, for within it meiosis takes place, producing *megaspores*. Meiosis occurs within only one cell of the sporangium, producing four megaspores. Three of these megaspores degenerate [13] in a manner that strikingly recalls the degeneration of polar bodies in egg production in animals. The single remaining megaspore in each ovule under-

goes three subsequent mitotic divisions, becoming an eight-nucleate female gametophyte.[14] Only one of the eight nuclei becomes the female gamete (egg). The remaining seven represent much-reduced gametophyte tissue. The eight nuclei characteristically assume the positions shown in the figure. The egg nucleus is associated with two others that are thought to be evolutionary vestiges of the flask-shaped archegonium that surrounds the egg in gametophytes of mosses and ferns (cf. Fig. 9–10). The remainder are thought to be vestiges of the general photosynthetic tissues of these earlier gametophytes. Two nuclei are

[13] There are (we are tempted to add "of course") exceptions among lilies and other plants.

[14] The eight-nucleate female gametophyte is technically known as an embryo sac. There may be as few as four or as many as 16 nuclei in the embryo sac of other seed plants.

of special interest; they fuse, giving rise to a diploid *fusion nucleus.*

Fertilization

By one means or another, the pollen grains (mobile male gametophytes) reach the stigma at the head of the pistil. There they germinate. A protoplasmic tube (the *pollen tube*) emerges from each pollen grain and grows down the style to enter the ovary, penetrating an ovule and ultimately the female gametophyte within the wall. The two sperm nuclei move down the tube, entering the female gametophyte. One sperm nucleus unites with the egg nucleus to form a zygote: fertilization occurs. The zygote then begins development and forms an embryo sporophyte. Note that this embryo is still within the ovule (sporangium), and the ovule is in turn still within the ovary of the sporophyte.

The second sperm nucleus unites with the fusion nucleus. You recall that the fusion nucleus is already diploid, so when another haploid nucleus joins it, it becomes triploid, with three sets of chromosomes. The triploid cell then divides and develops into a multicellular mass called the *endosperm.* This is a most extraordinary structure, which really has no parallel in reproductive processes outside the flowering plants, not even in the other (nonflowering) seed plants.[15] It looks almost like the start of a third kind of developed organism in addition to the haploid gametophyte and the diploid sporophyte. There is, however, no evidence that such is its evolutionary significance. In any event, no real differentiation occurs in it. It becomes merely a uniform tissue, rich in foods usable by the growing sporophyte embryo. It is the starchy endosperm that forms the bulk of our food grains, corn, wheat, and so on. Peas and beans, however, have no endosperm when mature because it has been absorbed by the growing embryo, which thus acquires its food value for humans.

Seeds and fruit

The ovule (sporangial) tissues around the young embryo harden to form a tough *seed*

[15] In the seeds of pines and some other nonflowering plants, there is a tissue that looks like endosperm and is sometimes so called. It is, however, really the nonreproductive part of the female gametophyte.

coat. The final product, the seed, is thus a sporangium containing an embryonic sporophyte, usually with triploid endosperm as a sort of developmental sideline. Often the seeds are shed free (as in the pea) and may or may not have various devices that facilitate their dispersal. Sometimes, however, the seeds may be retained within the ovary, which itself undergoes further growth to become a fruit. Fruits are modified ovaries, sometimes with other flower parts, in which seeds are embedded. Their growth sometimes produces succulent and sweet tissues attractive to animals, although dry fruits are more common. The significance of such modified ovaries is, again, in their relation to problems of dispersal.

THE PROBLEM OF PLANT IMMOBILITY

Plants are usually anchored or rooted in place. This is one of their striking differences from most animals (although there are anchored animals, too). Their immobility is related, through the processes of evolution, to

9–12 The reproductive cycle of a flowering plant. *Top.* Broad features of the cycle. The diplophase (sporophyte) is the dominant part of the cycle. The male and female gametophytes are minute structures with a transient existence associated with fertilization. *1.* Mature sporophyte with a flower: *a,* sepal; *b,* petal; *c,* stamen, consisting of filament (*d*) and anther (*e*); *f,* pistil, consisting of stigma (*g*), style (*h*), and ovary (*i*). *2.* Pollen tube (male gametophyte) growing out of the pollen grain. *3.* Embryo sac (female gametophyte). *4.* Seed, containing an embryo sporophyte (*a*) and nutritive endosperm (*b*). *5.* Seedling sporophyte. *Bottom.* Detail of the gametophytes and fertilization. *1.* Mature sporophyte with flower. *2.* Young pistil. The ovary (black in the figure) is, historically speaking, a modified leaf (or leaves) folded over to enclose the sporangium (or sporangia) that it bears. This female sporangium is the ovule (gray in the figure); within the ovule is a single cell, which will undergo meiosis, producing four haploid megaspores. *3.* Ovule containing the four megaspores. *4.* Ovule with the one surviving megaspore, which marks the beginning of the female gametophyte (embryo sac). *5–7.* Successive mitotic divisions in the development of the female gametophyte, leading to its eight-nucleate state. *8.* Mature embryo sac with eight nuclei. Three of these (*a*) are the antipodal nuclei; two others (*b*) are the future fusion nucleus; of the other three (*c*), the largest is the female gamete (egg cell), and the other two

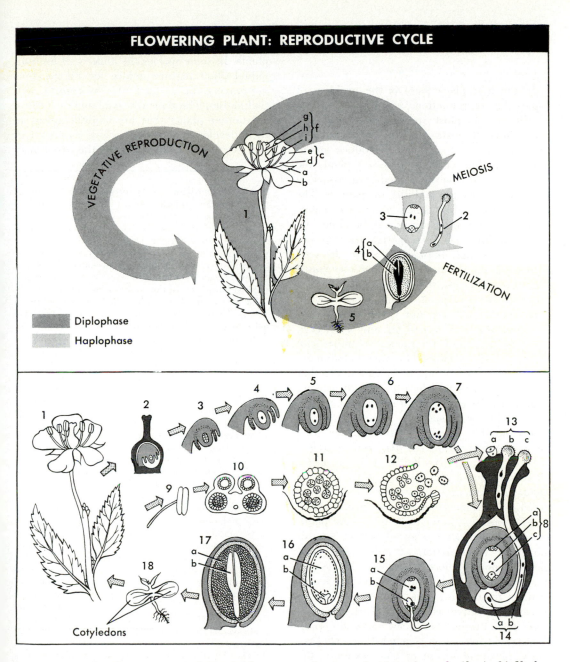

Diplophase

Haplophase

Cotyledons

are sometimes thought to represent the vestigial remains of an archegonium. *9.* Stamen. *10.* Cross section through the anther (sporangium), which has four separate chambers; in two of these can be seen cells that will undergo meiosis, yielding microspores (pollen grains). *11.* Section through a single anther chamber containing cells in which meiosis has occurred. *12.* Same anther chamber at a later stage, bursting and yielding pollen grains that have already developed into young male gametophytes; their nuclei have divided. *13.* Development of a pollen grain on the stigma (*a*), with germination of the pollen grain and growth of the male gametophyte (pollen tube) down the style to the embryo sac (female gametophyte) in the ovule (*b*, *c*). *14.* Nuclear constitution of the male gametophyte: *a*, tube nucleus, which is not a gamete; *b*, two male gametes. *15.* Fertilization: *a*, one of the male gametes fuses with the diploid fusion nucleus of the embryo sac, producing a triploid nucleus that will later develop into the endosperm of the seed (*16a*); *b*, the other male gamete nucleus fuses with the egg cell, producing a diploid zygote that marks the beginning of the new sporophyte. *16, 17.* Younger and older seeds including (*a*) endosperm and (*b*) embryo sporophyte. *18.* Seedling sporophyte still showing its embryonic leaves (cotyledons), of which there are two in this particular (dicotyledonous) flowering plant.

many special developments in the reproductive structures and cycles in plants.

Getting the sexes together

In the first place there is the problem of biparental reproduction, which has been a condition of most progressive evolution throughout the history of life. Two immobile plants cannot get together to reproduce, as can any two mobile animals. In lower plants the immobility of the developed organisms is counteracted by the mobility of gametes. The male gametes of attached algae are waterborne and are often motile (self-mobile) by means of *flagella*.[16] Fern sperms are also motile, even though limited in range. In seed plants it is the pollen grains, partly developed male gametophytes, that are mobile. In most nonflowering seed plants pollen is produced in enormous quantities, and the grains are so light that they are spread far and wide, even for hundreds and thousands of miles, by winds and air currents. This was undoubtedly the primitive condition for the first plants that were able to rise well into the air, dispensing with a requirement for water in which gametes could be dispersed. Many flowering plants, notably among the grasses, have secondarily returned to wind dispersal of pollen.

It is probably significant that so many plants, including the vast majority of seed plants, are *hermaphroditic,* or in more strictly botanical parlance *monoecious,* that is, both sexes are present in the same individual plant. In spite of devices (reviewed below) to promote safe transmission of pollen from one plant to another, there is always a chance that this transmission will fail. Many plants can fertilize themselves; this may be viewed in broad perspective as a last resort, assuring some seed production.[17] But self-pollination, if sustained over several generations, is effectively an abandonment of true sexuality, the essence of which is the reshuffling of different genotypes. Many plants, accordingly, have

evolved quite elaborate mechanisms to assure *cross-pollination* between different individual plants. In some species the anthers of an individual plant mature before its stigma, thus minimizing the amount of (last-resort) self-pollination. The pollen is available to fertilize only other plants that are slightly ahead in their development and therefore possessed of ripe stigmas. In other species the order may

---

9–13 Insect-flower relationships. *A.* The *Yucca* flower and the *Pronuba* moth. *1.* The moth collects a ball of pollen from the stamens. *2.* It packs the pollen onto the head of the stigma. *B.* The flowers of *Aristolochia* are among those in which the sex organs mature at different times; in this particular plant the female organs ripen first. Small insects enter the young flower (*1*) and push their way past the downward-pointed hairs (*a*) on the corolla tube to reach the dilated base of the flower where the ripe young stigma (*b*) is. Their exit is prevented, however, by the same hairs they pushed by on entry. Thus they are trapped in the flower for some days until the stamens mature (*c*) and shed their pollen, which dusts the insects. Then the hairs wither (*d*), and the insects leave, only to enter another, younger flower, whose ripe stigma is inevitably dusted by the pollen the insects bring with them. *C.* The flowers of *Salvia* (sage) are among those in which male organs ripen first. The flowers have, moreover, evolved very elaborate morphological adaptations to exploit bee visitors as cross-pollinating agents. The essential feature is best seen in *3:* the filament of the stamen (*c*) is short; the anther is highly asymmetrical, one part (*d*) being extremely short and shaped like a shield, so that it gets in the way of the bee's proboscis when it seeks the nectar (*1*). This short arm acts as a lever: pushed by the bee's proboscis, it is lifted, and the other, long arm of the anther (*e*) is pushed down and touches the bee's abdomen. *2.* The anther in its "rest" position. *3.* The anther in the position to which it is forced by the bee. In young *Salvias* (*1*) the style has not completed its growth, and so the stigma (*a*) does not reach to the bee's abdomen; the anther (*b*), on the other hand, does. In older flowers (*4*), however, the style has completed its curved growth, and the stigma is always brushed onto the bee, receiving pollen from other, younger flowers the bee has recently visited. *D.* The Australian orchid *Cryptostylis leptochila* is one of the flowers that mimics a female insect and thus is cross-pollinated by seducing males of the insect species concerned. In this case the insect is the ichneumon wasp *Lissopimpla semipunctata* (*1*). *2.* Part of the flower (*a*) is strikingly like the female's abdomen; other parts (*b, c*) simulate legs and antenna. *3.* The male ichneumon attempts to copulate with the flower; when he leaves, he carries away pollen, which he transmits to the next plant he mounts. *E.* The Mediterranean orchid *Ophrys fusca* and the male bee *Andrena trimmerana* that copulates with it.

---

[16] Flagella are small hairlike projections that propel the organism about by their whiplike or lashing movement. They are discussed in Chapter 14.

[17] We will note later that hermaphroditism in animals is common in *sessile* (immobile) forms and in others (like specialized parasites) in which there is great risk of never encountering a mate.

INSECT - FLOWER RELATIONSHIPS

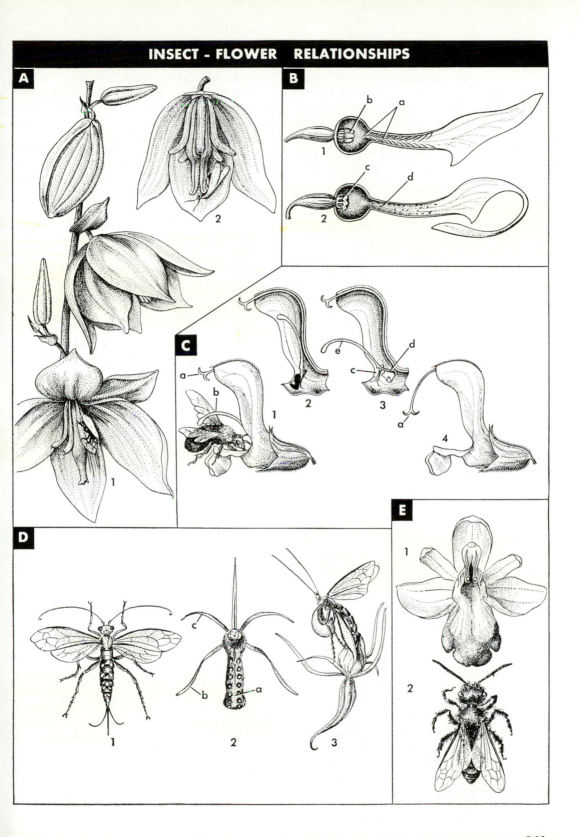

be reversed; the stigma may ripen first. The effect is the same in either case.

It is especially characteristic of flowering plants that so many of them have the pollen carried by animals, usually insects but often birds or bats and occasionally other animals. It is clear that many features in the evolution of flowers are related to this complex process of pollen dissemination. Here is the biological significance of color, scent, and nectar in flowers. Nectar is a rich sugary secretion much prized as food by a host of insects—flies, butterflies, moths, bees, and so on—and also by hummingbirds and bats. Many flower structures have evolved as guarantees that the visiting nectar-seeker will pick up and transmit pollen from one flower to another in the same species (Fig. 9–13). *Salvia* (Fig. 9–13C), a kind of wild sage, has a flower with a trigger that pulls down a stamen and dusts a nectar-seeking visitor with pollen. Scent and flower color act as advertisements or recognition signals that guide the visiting insects to food, and hence serve as means of cross-pollination in the plants. Flowers visited by bees are often white, yellow, violet, or ultraviolet, which can be seen by bees, although not by us. They are rarely red, which looks black to a bee. Hummingbirds, however, can see red as a distinct color. Many flowers, especially in the tropics, where hummingbirds abound, are brilliantly red, advertising a nectar supply to which hummingbirds are partial. The structure of hummingbird-flowers is often long and tubular, preventing nearly all visitors other than the long-tongued hummingbird from getting the nectar.

There are many amazing and more elaborate relationships between flowering plants and the animals they attract as pollinating agents. Three of these we give here as samples. Some orchids have evolved flowers that look remarkably like the females of a particular bee that lives in the neighborhood (Fig. 9–13E). Once a year this orchid flowers, and does so when male bees have emerged and are flying but the females are still in pupal cases. So good is the flower's mimicry of the female that even the male bee is deceived. He mounts the flower as he would a female bee and copulates with it. The stamens of the orchid are so arranged that the male bee's genital organs become dusted with pollen. The bee then carries the pollen and transmits it to the next orchid that seduces him.

When Smyrna fig trees were first grown outside the regions where they had long been cultivated, the trees failed to produce any fruit. Eventually it was found that the reason for the failure was the absence of a particular kind of small wasp. The female wasp lays her eggs in fig flowers, which are the only places where the larval wasps develop normally, and in turn the wasp is the only means of pollinating the figs so that they bear fruit. The figs have separate male (staminate) and female (pistillate) flowers. Wasps can develop only in the male flowers. When the female wasps emerge from the flowers in which they were born, they are dusted with pollen. They then enter some other flower, lay their eggs, and die. The greatest peculiarity is this: if the female lays her eggs in a male flower, her eggs develop but no pollination occurs; if she lays them in a female flower, her eggs do not develop, but pollination occurs and fig seeds and fruit are produced. If all the wasps laid eggs in male flowers, there would soon be no more figs, and then no more wasps. But if all eggs were laid in female flowers, there would soon be no more wasps—and then no more figs.

Some yuccas bear seeds only if yucca moths are present. The moth collects balls of pollen and stuffs them into the tubular end of the pistil, which assures fertilization of the yucca and development of seeds and fruits. The moth then lays her eggs in the ovary of the yucca, and when the larvae hatch, they eat some of the developing seeds. There is more than enough for their appetites, so some yucca seeds mature (Fig. 9–13A).[18]

The most primitive seed plants did not have flowers, and their pollination was entirely by wind. It still is for living nonflowering seed plants (gymnosperms), as was mentioned before. The evolution of flowers undoubtedly was connected with increasing animal pollination, which as a rule is more efficient and precise than wind pollination. Nevertheless, as

[18] Do you think that fig wasps and yucca moths know what they are doing and have decided on these ways of assuring the future of their species? If not, how can such intricate relationships and behavior have arisen?

9–14   A bee gathering pollen and nectar from a *Lobelia* flower. Its traffic from flower to flower serves to transport pollen and cross-fertilize the *Lobelia* population.

we have noted, some whole groups of highly effective flowering plants have reverted to wind pollination—an evolutionary development for which there is as yet no obvious explanation.

A final noteworthy point relates further to the problem of getting the sexes together. If fertilization is to occur (and therefore if the species is to persist), it is necessary that availability of pollen and access to a mature stigma be simultaneous. This would be impossible if liberation of pollen and opening of flowers with stigmas occurred at different times of the year or, in some cases, even at different hours of the day. Plants have evolved the necessary means to synchronize flowering time among individual plants in the same species population. It is commonly known that all crocuses bloom at the same time early in spring, and chrysanthemums only in the fall. But how is this *timing* accomplished? Many factors seem to be involved, like temperature and light intensity, but probably the most important of all is the relative length of day and night. No

matter how fickle other conditions may be, the relative length of day to night[19] is a rigorously constant signal of the season. It is the signal that usually synchronizes flowering, although in some instances flowering is affected or may be primarily controlled by temperature or, in arid regions, rainfall.

Many plants control flowering time not only as to season but also as to time of day. Their flowers open and close rhythmically with a 24-hour frequency controlled by some internal timing device that establishes opening time at a fixed hour relative to dawn. Bees make use of the fact and economize on time and effort by visiting the right flowers at the right time of day (Fig. 9–14).

Dispersal

So much for the problem of getting sexes together in biparental reproduction. There is

[19] What most plants apparently measure is night length.

Hugh Spencer

9–15   Seed dispersal. The cocklebur's seeds (*left*) are armed with hooks that attach them to passing animals, who unwittingly aid dispersal of the plant. The milkweed's seeds (*right*) are supplied with delicate parachutes of many fibers. They are dispersed by wind.

the further question of how plants spread geographically. In lower plants dispersal usually occurs in the spore phase of the cycle, and in those plants spores can develop into independent organisms. Like their motile gametes, the spores of algae are water-borne and are often motile, having flagella. Fungi and ferns produce clouds of light spores that are carried by air currents far and wide, a few of them even for thousands of miles. The air you breathe is seldom free of spores. In seed plants the spores as such do not leave the parent organism (only pollen, which cannot by itself produce a new plant as spores can). In them dispersal is by means of seeds. The great success of the seed plants is clearly related to two characteristics of seeds: they protect the embryo through a dormant period and during times of drought or cold, and they are readily dispersed.

Adaptations for dispersal in seeds (Fig. 9–15) are as varied as those for pollination in flowers. Dandelions and many other plants have parachute seeds scattered by winds. Burrs stick to passing animals. Tempting fruits have hard-coated seeds that pass unharmed through the digestive systems of fruit-eaters. Coconuts

and mangroves have floating seeds, resistant to long voyages in the currents of the sea. Vetches and a number of other plants have pods that open suddenly and scatter seeds like the fragments of a grenade. In Russian thistles—the tumbleweeds of Western song—the globular plant breaks off from its roots and rolls across the countryside, scattering seeds as it goes. Can you think of other ways in which seeds are dispersed?

Most individual plants are immobile as developed organisms, but seed dispersal is so effective that *populations* of plants may spread more widely and rapidly than populations of mobile animals.

Reproductive Cycles in Animals

DOMINANCE OF THE DIPLOPHASE

We have seen that both within the algae (most lowly of plants) and in the plant kingdom as a whole there has been a strong evolutionary tendency to emphasize the diplophase (sporophyte). In the most elaborately evolved algae, like *Fucus*, and in the seed plants, the haplophase has been greatly reduced.[20] What is the significance of this trend? The answer is not entirely clear, but it probably involves two factors. First, an organism with a double set of chromosomes has a more complex and probably as a rule a more adaptively flexible biochemical system. Duplication of genes may not only be a sort of "fail safe" provision but may also, when there are different alleles, provide alternative means of maintaining homeostasis under fluctuating conditions. Second, populations of diploid organisms can store more genetic variability than populations of haploid organisms. Diploid organisms have therefore probably been favored for the same general reasons that sex itself was favored. This point will become clearer when we discuss population genetics in Part 5. In the meantime, we note that these are at least plausible reasons for the facts that the diplophase in plants has tended to be emphasized and that virtually all animals are diplophase-developed organisms.

Indeed, the history of animals reveals only slight traces of any original emphasis on the haplophase. Biologists suspect that multicellular animals may have evolved from animal-like (nonphotosynthetic) flagellated protists, much as algae probably evolved from photosynthetic flagellates like *Chlamydomonas*. So far as is known, the animal-like flagellates are also haplophase organisms, the diplophase being restricted to the zygote itself. In any case, once animal life evolved above the protistan level, it became exclusively diplophase,[21] the haplophase being represented only by the gametes. It is not clear why some plants have been able to survive as haplophase organisms while animals have not.

The usual reproductive cycle in animals is, then, that familiar to you in man; there is no analogue of a gametophyte. Modifications of that cycle relate mainly to secondary regression in sexuality and to addition of vegetative (asexual) reproduction in the diplophase of the basic cycle (Fig. 9–16). Both of these processes are less common in animals than in plants, but they also occur rather widely in animals. Other evolutionary changes in reproduction and life histories in animals have involved not so much the over-all reproductive *cycle* as the structures and processes of reproduction and development. In these respects animals are extremely diverse.

SEXUAL PLUS VEGETATIVE REPRODUCTION

Vegetative reproduction is lacking in human beings and in the animals most familiar to us, the other vertebrates, but some animals do reproduce vegetatively. Such reproduction is not the rule in animals and there are few—perhaps no—animal species in which it is the *sole* means of reproduction. Nevertheless, it does occur in many different kinds of invertebrates and even in tunicates, animals related to the vertebrates. Vegetative reproduction in animals is usually by budding (as in *Hydra*) or by fission (as in planarians), processes not always clearly distinguishable.

In some insects a mass of cells formed early in embryonic development may bud off into many smaller clumps, each of which develops into a separate organism. When the process occurs before tissue differentiation, it does not differ in principle from multiple identical twinning. Oddly enough, a closely similar process occurs in one mammal, an armadillo, which generally produces identical quadruplets. In some insects division into several individuals may occur later in the life history, in larval or pupal stages.

In many coelenterates, relatives of the jellyfishes, sexual and vegetative reproduction alternate regularly. In *Obelia* (Fig. 9–17), a classic example, the zygote develops through a larval stage into an attached individual that grows into a branched colony of polyps by budding without separation. Parts of the colony then produce further buds that do become detached and that swim off as medusas,[22]

[20] *Fucus* is highly evolved, for an alga, in another significant respect: it has vascular tissue not unlike that of true vascular plants. Both in its possession of vascular tissue and its elimination of the gametophyte, *Fucus* evolution has converged with that of seed plants.

[21] There are the usual exceptions, such as the male bees noted earlier in this chapter. But they are clearly secondary specializations and outstanding by virtue of their rarity.

[22] After the girl in Greek mythology whose hair was turned into snakes. The animals have numerous tentacles hanging down all around, and someone with

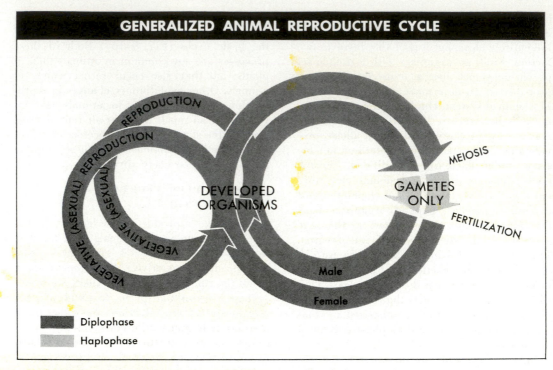

GENERALIZED ANIMAL REPRODUCTIVE CYCLE

REPRODUCTION

REPRODUCTION (ASEXUAL)

VEGETATIVE (ASEXUAL)

VEGETATIVE (ASEXUAL)

DEVELOPED ORGANISMS

GAMETES ONLY

MEIOSIS

FERTILIZATION

Male

Female

■ Diplophase

▨ Haplophase

9–16   Generalized animal reproductive cycle. No analogue of the gametophyte generation in plants exists in animals; the haplophase is represented by the gametes only. (There is one remarkable exception to the generalization in this figure: male bees and other Hymenoptera are haploid. See Fig. 9–27.)

which look like small jellyfishes. The medusas are male or female (in separate individuals), and they produce eggs and sperms that unite in fertilization and give rise to new attached colonies. The complex cycle, summarized in Fig. 9–17, thus includes developed organisms of two strikingly distinct forms: the attached, sexless, colonial polyps and the free-swimming, sexed, noncolonial medusas. Both forms are diploid and belong on the left side of our reproductive-cycle diagrams—the diplophase. Their anatomical structure is fundamentally the same beneath the superficial differences of attachment and proportions. Although it is most common in coelenterates, alternation of sexual and vegetative reproduction also occurs in several other groups of invertebrates. You already know that this same alternation also occurs (usually with less regularity) in plants. It is, however, funda-

mentally different from the other alternation of sporophytes (diplophase) and gametophytes (haplophase), with which it is often confused under the misnomer of "alternation of generations."

FERTILIZATION

Almost all animals produce specialized and sharply distinct female and male gametes: relatively large, approximately spherical eggs, immobile except as they may be passively moved by surrounding fluids, and relatively small, tailed or flagellated sperms, actively motile. The egg may remain attached within the maternal tissues until it is fertilized and starts to develop, as is usual in plants and also in sponges among animals. Much more often in animals the egg is detached from the maternal tissues before fertilization, although it may remain in cavities or passages in the maternal body and the developing organism may even, as in man, become re-attached.

In the majority of aquatic invertebrates—

a lively imagination thought that the unfortunate girl must have looked like this.

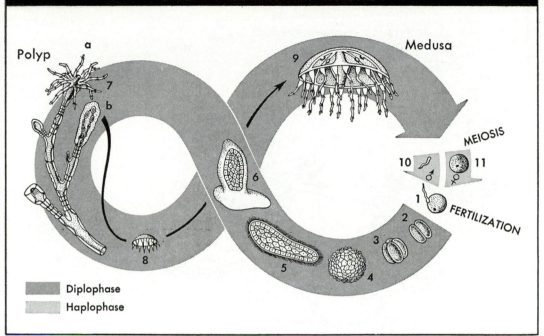

Polyp

Medusa

MEIOSIS

FERTILIZATION

Diplophase

Haplophase

9–17 **The reproductive cycle in coelenterates like *Obelia*.** There is a so-called "alternation of generations" (medusa–polyp–medusa–polyp, etc.) in certain coelenterates like *Obelia*. However, this is not at all comparable to the alternation of sporophyte and gametophyte in plants; in coelenterates both polyp and medusa are diplophase. As in other animals, the haplophase is represented by the gametes only. *1.* Fertilization—union of egg and sperm. *2–4.* Successive stages in the cleavage of the egg and growth of the embryo. *5.* Ciliated swimming larva. *6.* Larva metamorphosing into a polyp adult. *7.* Adult colony of polyps: *a*, feeding polyp; *b*, reproductive polyp, which asexually buds off medusas. *8.* Young, free-swimming medusa. *9.* Adult medusa, which produces sperms (*10*) and eggs (*11*).

American oysters will serve as a concrete example—both eggs and sperm are simply shed into the water. There some of the swimming sperms, so enormously numerous that the water may be milky with them, eventually encounter eggs. Fertilization and development follow, with no special relationship to the parents. There are, however, innumerable modifications and complications of this simple process. In European oysters, for example, the eggs are retained in cavities in the mother, and are there fertilized and develop into larvae. In some fresh-water clams the eggs, after leaving the ovary in which they are formed, are retained under the gills. Water currents passed through the gills draw in sperms, which fertilize the eggs. Development of small larvae proceeds, still within the interior of the gills. After being expelled, the larvae die un-

less they encounter a fish on the fins or gills of which they live as parasites until they have developed far enough to take up independent life.

In many fishes, also—indeed, in an extremely wide range of aquatic animals—eggs and sperms are shed into the surrounding water, where fertilization occurs. This seems, and sometimes is, an extremely haphazard process (Fig. 9–18). A small pond, not to mention an ocean, is a vast volume in which a sperm can wander and never contact an egg. Animals have evolved, as have plants, a variety of devices that synchronize their sexual activity and so reduce the chance of wasting gametes. Most have seasonal periods of sexual activity, just as flowers are seasonal. And as in the flowering plants, many animals achieve synchronized seasonal reproductive activity

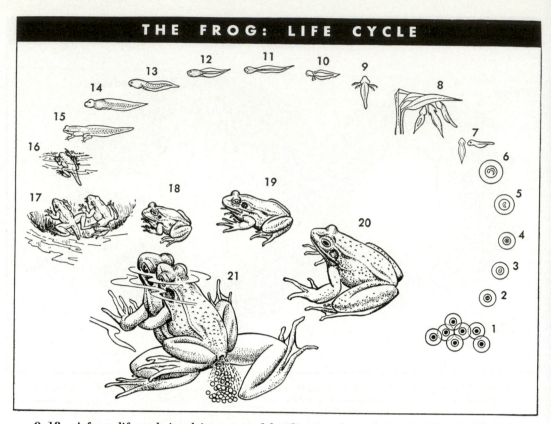

9–18   A frog: life cycle involving external fertilization. Stages *1* through *6* show development of the fertilized egg, which floats freely and unprotected in pond water. *7, 8.* Freshly hatched larvae —tadpoles. *9–15.* Progressively older larvae (*15,* larva with limbs). *16, 17.* Metamorphosis from a tadpole larva to an adult frog. *18.* One-year-old adult. *19.* Two-year-old adult. *20.* Three-year-old adult. *21.* Adults spawning in water: eggs are shed by the female; the male ejects sperms onto them when they are outside the female and free in the water.

by responding to day–night length as a reliable season marker. Of course, this is not a conscious recognition; the relative lengths of day and night act as purely physiological triggers to initiate sexual activity. Many animals, again like plants, also use internal timing devices—they have been metaphorically called "clocks"—to synchronize activity more finely to time of month and day. Thus many marine animals release their gametes on a 28-day cycle in relation to a particular moon phase. The palolo worm in Pacific coral reefs makes the sea milky with eggs and sperm, but only at full moon.[23]

In addition to these broad physiological processes which tend to synchronize sexual activity, motile animals have evolved elaborate behavioral adaptations that synchronize their release of gametes. Males and females may be stimulated to expel gametes only in the presence of each other. There may even be a more or less elaborate *courtship* that has the result that eggs and sperms are discharged at the same time and place (Fig. 9–19; see also Chapter 14 and its example of stickleback courtship).

Fertilization is possible only in watery surroundings and with eggs that do not have a complete tough coating. Such a coating excludes sperms. An egg without a coating, a

[23] Lunar cycles of reproductive activity are also known in some mammals; the 28-day periodicity of the estrus cycle in women may be regulated by a timing system inherited from earlier mammals whose sexual activity occurred in phase with the moon. The

lunar clock in women, however, has lost capacity to be synchronized (brought into phase) with the moon.

9–19   The courtship of the great crested grebe.
The behavioral device of courtship, which serves to synchronize the release of gametes (and also to prevent interspecific matings), is often an elaborate ritual. The figure shows several incidents in the grebe's courtship display. *1.* Mutual head shaking. *2.* The female displaying before the male, who has dived and shoots out of the water in front of her. *3.* Two views of the male rising from his dive and displaying his "collar" of feathers. *4.* Both sexes have dived and brought weeds, which they display to one another.

shell of some kind, must be surrounded by fluid or it dries and dies. Fluid is also a necessary medium through which sperms may swim and reach eggs. Fertilization cannot take place in the open air, where an egg without a shell rapidly dies and where a living sperm cannot reach the egg. The same limitations apply for the same reasons to fertilization in plants, and the limitations are really overcome in the same way in land plants and land animals even though the organs and activities are so different in the two: in both, fertilization takes place within the maternal organism. In the immobile higher plants, as you know, the male gamete is brought to the egg (somewhat indirectly) by the mobility of the pollen. Land animals are mobile and, in simpler fashion, the male takes the sperms to the female and injects them into her: copulation occurs. In all animals fully adapted to life on dry land, notably insects, reptiles, birds, and mammals, fertilization is internal following copulation (Fig. 9–20).[24] In insects the shell is formed before fertilization, but there is a small hole in the shell through which a sperm enters. In reptiles and birds the shell forms after fertilization. In (true) mammals no shell forms, the zygote and resulting embryo being retained within the mother.

Copulation entails anatomical specializations. The male must have an organ (a penis) that can be inserted into the female and through which sperms are ejected. The female must have a receptacle from which the sperms can move to the eggs through a fluid internal

[24] Copulation also occurs in some aquatic animals, such as sharks, in which the egg either is surrounded with a protective shell before being expelled or is retained and develops inside the mother.

COURTSHIP OF GREAT CRESTED GREBE

1

2

3

4

Hugh Spencer

9–20   The copulation of female (*left*) and male (*right*) *Cecropia* moths.

medium. In insects but not in vertebrates [25] there is an additional specialization. The females have sacs within which sperms are retained and kept alive for considerable periods of time. Copulation occurs only once, and thereafter sperms are released from the sacs whenever fertilized eggs are to be laid. The queen bee is an extreme case; she copulates only once and stores the sperms received for the rest of her life, sometimes as long as 17 years.

LARVAE

Development uses energy and materials. If all the necessary foods are supplied in some

way or other within the egg, by a parent, or both, development is usually more or less direct. That is to say, when the developing individual reaches the stage of being an independent, self-maintaining organism it is already adapted to a life like that of its adult parents and is recognizably similar to them. Familiar examples are reptiles, birds, and mammals. Reptile and bird eggs contain large stores of food (in the yolk), and this suffices until the young animal hatches. Reptiles hatch as essentially small adults, able to fend for themselves. Most birds after hatching are fed for some time by their parents. Mammal eggs contain almost no food, but the needs of the develop-

[25] There are, however, some mammals (not including man) in which live sperm may be stored in the female sex organs for long periods and there may even be successive fertilizations from a single copulation.

ing embryo are provided through the mother. For a time after birth all needed food is supplied by the mother in the form of milk.

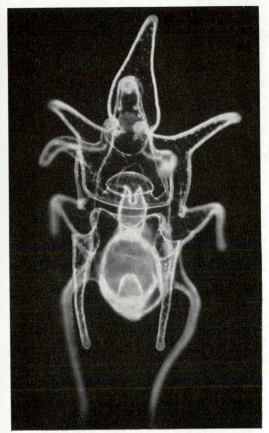

Top, © Douglas P. Wilson; bottom, V. B. Schaeffer from U.S. Fish and Wildlife Service

9–21   The brachiolaria larva and the adult of the starfish, *Asterias*. The photograph of the larva is highly magnified.

In most animals, including the overwhelming majority of species of invertebrates, many fishes, and most amphibians, the embryo does not develop directly into an adultlike organism. Free, postembryonic life, no longer dependent on food stored in the egg, is begun in a form unlike the adult. The organism in this phase differs more or less from its adult parents in environment, food, locomotion, and other particulars of the way of life. It is a larva. A stranger to the earth would never guess that a butterfly is an adult caterpillar, a frog an adult tadpole, or a starfish an adult brachiolaria. Caterpillar, tadpole, and brachiolaria are larvae (Figs. 9–22, 9–18, and 9–21).

Larvae have several different roles in the life histories of animals. In aquatic invertebrates (also notably in parasites) the larva is a means of dispersal. To that extent it is analogous to a plant seed, and indeed it occupies a similar position in the life cycle as a distinctive phase preceding the mature diploid organism. Many animals are attached and immobile (sessile) as adults: sponges, corals, most sea "lilies" (which are animals), oysters, and numerous others. Without exception they have unattached larvae that are scattered by currents in the water in which the larvae float or swim. Even animals that are mobile as adults—starfishes, for instance—may be transported much farther as larvae than they could travel in adult form.

Larvae also often represent a special feeding phase in the life cycle. This is particularly true of insects. Most of the feeding and growth may occur in the larval stage.[26] In fact, some adult insects do not eat at all. They are reproductive entirely and in their brief span use up materials and energy acquired while they were larvae. In most insects both larvae and adults eat, but they usually eat different food. Each may rely on a food more available at a particular time of year; for instance, caterpillars may feed on leaves, and butterflies on nectar. In this example, too, all the materials for growth are obtained by the caterpillar. A

[26] This, too, has its analogy in the life cycle of plants, but the analogy is more remote. The gametophyte of a moss is a "feeding" phase, and the sporophyte more exclusively reproductive.

Hugh Spencer

9–22   Stages in the life cycle of the swallowtail butterfly, *Papilio*. *Left to right.* Caterpillar larva. Pupa (or chrysalis). Adult (imago) freshly emerged from pupal case. *Below.* Adult with fully expanded wings.

butterfly does not grow, but it does acquire energy by eating.

Change from larval to adult form and way of life is *metamorphosis* ("form changing") (see Fig. 9–22). Metamorphosis may be gradual or it may be dramatically abrupt. In grasshoppers the larva is quite like the adult except in proportions and absence of wings. Adult form is acquired gradually through a succession of molts of the external skeleton. In dragonflies the larva is aquatic and markedly different from the aerial adult. Metamorphosis occurs during a single molt, involving mostly a change in the relative developments of tissues and organs already present. In most in-

sects, with butterflies as a common example, there is a special, outwardly inactive phase, the *pupa*, between the larva and the adult. The pupa is enclosed in a case or cocoon, where it undergoes radical metamorphosis. Almost all organs and tissues of the larva are destroyed, and what is essentially a new organism grows from small cell clumps or buds. (Do you see any parallel between this metamorphosis and the phases of the life cycle of *Obelia?*) In insects that metamorphose, the adult is called an *imago,* so the full life sequence is larva→ pupa→imago.

EVOLUTION OF VERTEBRATE REPRODUCTION

There have been many different lines of evolution, and they have led to many and highly diverse specialized kinds of reproductive structures and processes. Even among the vertebrates there are myriad different lines of descent leading to markedly different reproductive systems among recent animals. Of course, too, no species now living are in the lines ancestral to man, nor are they likely to have retained precisely the conditions that did occur in the human ancestry. Nevertheless, comparison of them in the light of phylogeny as revealed by fossils permits highly probable inference as to the course followed in the evolution of human reproduction. In this history there has been a trend toward greater protection and care of the embryo and young, and a major transformation connected with the change from water to land life.

In our remote fish ancestors, eggs without shells but with gelatinous envelopes were expelled and fertilized externally. Similar eggs, sperms, and fertilization were retained in the ancestral amphibians, and so was the early, larval phase of the fish life history, but there was a more radical metamorphosis leading to four-legged, semiterrestrial adults. Pairing (but not copulation) [27] probably arose in this evolutionary stage, a male remaining near or on a female while the eggs were being laid and extruding sperm under this stimulus (see Fig. 9–18). Transition to completely terrestrial life in early reptiles involved many changes. Fertilization became internal, following copulation. The fertilized egg (a zygote soon becoming an embryo) was enclosed in a protective shell before being laid. Within the shell a membrane (the amnion, Fig. 9–23) developed from the embryonic tissues and formed a fluid-filled sac enclosing the embryo proper, which thus continued to develop in a self-contained aquatic environment. Other membranes (allantois and chorion) formed sacs and surfaces aiding in respiration, in absorption of food within the egg, in storage of waste materials, and in further protection of the whole complex inside of the shell. That is the reptilian condition, retained with variations in the reptiles still living.

In the gradual transition from early reptiles to advanced mammals, the eggs were retained during development in the lower (or posterior) parts of the tubes leading from the ovaries to the exterior. These parts of the tubes became enlarged and thickened to form *uteri* (singular, uterus). (In man and some other mammals the two paired uteri have fused into one medial uterus.) The egg shell was reduced and eventually lost. Then the membranes of the embryo came in direct contact with the wall of the uterus and began to exchange substances with that wall by diffusion. Oxygen and dissolved foodstuffs were absorbed from the uterus, and carbon dioxide and wastes such as urea were given off to it. Finally, this exchange was made most effective

[27] As previously mentioned, some fishes pair, and some copulate, but in them these are later, independent specializations. The fishes ancestral to land vertebrates almost certainly had not yet evolved close pairing or copulation.

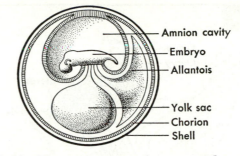

9–23 The amniotic egg of a terrestrial vertebrate.

by a complex intergrowth of tissues from the embryo and those in the wall of the uterus, forming a *placenta*. In the placenta, capillaries from the maternal and the embryonic circulatory system come into close contact, although normally there is no exchange of blood but only of dissolved substances by diffusion. Special vessels, which become inoperative at birth, connect the embryo with the placenta.

In the meantime special provision for the young after birth was also evolving. Even before the eggs were retained in the uterus, *mammary glands* yielding milk arose in females and provided a rich, balanced liquid diet for the newborn. Finally, in man, even after the child is weaned, there is a long period of mental and social training by the parents.

Parental care is an extremely useful adaptation because it promotes survival of the next generation at the most dangerous period in life. It is therefore not surprising that parental care, extreme in man, is by no means confined to man but has evolved over and over again. The meticulous care given to larvae in the hives of bees and the nests of ants is well known. Although parental care cannot be considered characteristic of lower vertebrates, it does occur among numerous fishes, amphibians, and reptiles. In some form or other it is characteristic and indeed universal among birds and mammals.

Mammalian reproduction

Man is a fairly typical higher mammal as regards reproduction and may be used as our example. The organs involved are shown in Fig. 9–24. Sperms develop in large numbers in the testicles and are stored in the coiled

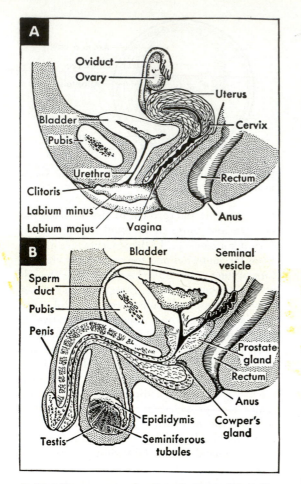

9–24 Human reproduction: I. Genitalia. *A*. Female. *B*. Male. These are drawn in longitudinal section, with the anterior to the left.

tubes of the epididymis and the connecting ducts. The seminal vesicles and prostate gland secrete fluid in which the sperms are transported. Prior to copulation, the penis becomes stiff and erect by the pressure of blood temporarily trapped in its internal tissues. It is then inserted into the female vagina. Frictional stimulation leads to a sudden reflex ejaculation of semen (sperms and fluid) through the urethra in the penis and out into the vagina. The sperms are motile and tend to move from the vagina through the uterus and thence up the uterine tubes. If an egg is encountered in one of the tubes, fertilization is likely to occur.

Eggs develop and ripen, usually one at a time in humans, in small, bubble-like struc-

tures, follicles (Fig. 9–25), in the ovaries. When an egg is mature, the follicle bursts, releasing the egg, which normally passes into the funnel-like end of the uterine tube. If there are no live sperms there at the time, the egg disintegrates or passes on down the tube, through uterus and vagina to the exterior. The egg is so small that its passage is usually wholly unnoticeable. If the egg encounters live sperms in the uterine tube, fertilization is likely to occur (Fig. 9–26). Development begins in the tube and continues as the embryo slowly passes on down the tube into the uterus. There the embryo becomes embedded in the wall of the uterus, in due course a placenta is formed, and development of the young (gestation) proceeds. After some time (10 days in mice, 9 months in humans) intra-uterine development is complete; then the muscular walls of the uterus contract rhythmically, expelling the young through the vagina and then tearing the placenta loose and expelling it.

The estrus cycle

In most male mammals mature sperms are produced and sex hormone secretion is intense only at certain times during the year. Sexual desire and behavior are strongly influenced by hormone concentration, and it is only during these periods that the male is ready and willing to copulate. Similarly in most female mammals, eggs mature in ripening follicles once or a few times a year. Simultaneously sex hormone secretion increases and is greatest at about the time when eggs are released from the ovary, the only time when most female mammals are willing, or anxious, to copulate. They are then said to be in *heat*. The whole process of buildup and drop in hormone concentration, ripening of follicles and release of eggs, and simultaneous changes in the uterus is called the *estrus* [28] *cycle*.

As regards that cycle humans are aberrant mammals. In men maturation of sperms and secretion of sex hormones is approximately continuous from early adolescence with decreasing vigor into old age. Women have an estrus cycle, as do other female mammals, but it is unusual in that it has constant periods of about 28 days, instead of occurring a lim-

---

[28] The word originally meant "gadfly" in Greek.

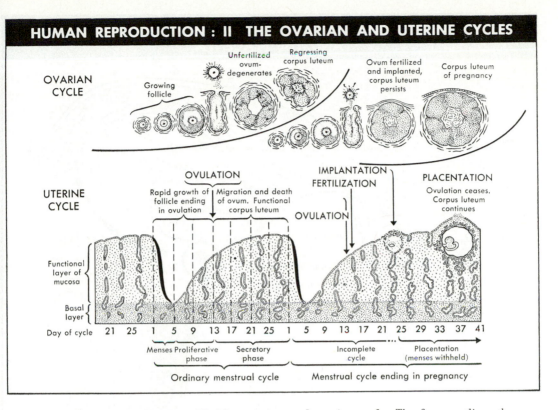

9–25 Human reproduction: II. The ovarian and uterine cycle. The figure outlines the changes that occur in the ovary and in the wall of the uterus (*1*) during an ordinary menstrual cycle and (*2*) during a menstrual cycle that includes fertilization and ends in pregnancy. *1*. The ordinary cycle in the ovary involves the growth of a follicle with its contained egg cell or ovum. When its growth is complete, the follicle bursts, discharging the ovum from the ovary; this event of ovulation occurs, on the average, on the fourteenth day of the cycle. If the ovum fails to be fertilized in the oviduct, it degenerates. After ovulation the old follicle undergoes a special development into a corpus luteum (yellow body). There is a parallel cycle of change in the wall of the uterus. The uterine cycle is indeed controlled, through hormones, by the ovarian cycle. While the follicle is growing in the ovary, the female sex hormone estrogen is secreted into the blood stream; it stimulates development of a special functional layer (the endometrium) of the uterine wall. This layer is richly supplied with blood vessels and built, so to speak, in anticipation of fertilization and the implantation on it of an embryo to be nourished. After ovulation the corpus luteum that develops from the old follicle secretes another hormone, progesterone, which maintains the growth of the endometrium. In the ordinary menstrual cycle (no fertilization), the corpus luteum eventually degenerates, and the endometrium is shed with a loss of blood during a 4- or 5-day period, the menses. The whole cycle is renewed as another follicle develops in the ovary and estrogen is again secreted, causing the endometrium to regrow. *2*. If ovulation is followed by fertilization, the sequence of events is different. The fertilized egg becomes implanted in the endometrial layer of the uterus, and a placenta develops (Fig. 9–26). Following placentation, ovulation ceases because the high progesterone level in the blood (maintained by the corpus luteum, which does not, this time, degenerate) causes the pituitary to stop secreting the gonadotropic hormone that stimulates follicle growth. Menstruation does not occur, owing to the high progesterone level, and the endometrium is retained as a functional part of the placenta.

ited number of times at definite seasons. The cycle in women is continuous from early adolescence until the late forties or early fifties except during pregnancies. The main features

of the human female cycle are summarized in Fig. 9–25. Fertilization can occur only shortly after release of an egg from its follicle and, although highly variable, that usually happens

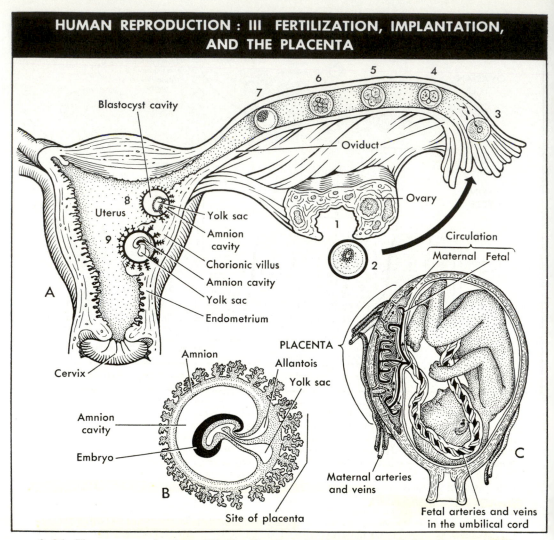

9–26 Human reproduction: III. Fertilization, implantation, and the placenta (schematic). *A. 1, 2.* Liberation of an egg from a follicle in the ovary, which shows corpora lutea. *3.* Fertilization of the egg in the oviduct shortly after it has entered (during ovulation the ovary is pressed closely to the mouth of the oviduct). *4–7.* Cleavage stages. *8, 9.* Embryo, with its amnion cavity already clear, implanting on the endometrium of the uterine wall. The chorion of the embryo has developed fingerlike villi that burrow deep into the endometrium and serve as the embryo's agents of exchange of nutrients and wastes with the maternal tissues of the placenta. *B.* The implanted embryo. *C.* An advanced fetus in the uterus, showing the fetal circulation in the umbilical cord leading to and from the placenta.

in women about midway between menstruations.

Regression of Sexuality

In this chapter we have been concerned mostly with main trends in evolution and with the more usual kinds of reproduction and life cycles. As we have noted, there is reason to believe that some device for genetic recombination has occurred in the ancestry of all recent organisms, including bacteria (and viruses, if these are considered organisms). In all but bacteria and blue-green algae (and viruses) recombination is achieved through a

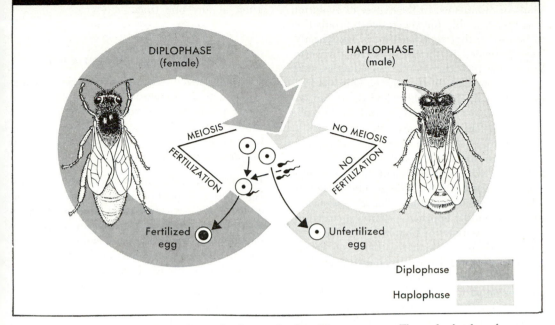

THE BEE AND OTHER HYMENOPTERA: REPRODUCTIVE CYCLE

DIPLOPHASE
(female)

HAPLOPHASE
(male)

MEIOSIS

FERTILIZATION

NO MEIOSIS

NO FERTILIZATION

Fertilized egg

Unfertilized egg

Diplophase

Haplophase

9–27   The reproductive cycle in the bee and other Hymenoptera. The male develops from an unfertilized egg and is, therefore, haploid. All fertilized eggs give rise to females (diploid). Only a few of these ever become functional females, however; the majority become sterile workers. Since the males themselves are haploid, they produce normal haploid sperm without meiosis.

sexual process. Therefore we have stressed reproductive cycles in which fully sexual reproduction, including meiosis and fertilization, is present. Nevertheless, we have noted from time to time that the sexual cycle is modified or absent in a number of highly diverse organisms. In ancestors of most of these organisms, if not of all, some form of sexuality once occurred, so that they represent a *regression* or loss of sexuality. This phenomenon is now to be briefly considered.

Self-fertilization is still sexual, but from an evolutionary point of view, at least, it must be considered as a partial regression. Biparental sexual reproduction must be older than uniparental self-fertilization. The latter process necessarily leads to marked reduction of genetic variation, which is the important evolutionary consequence of all kinds of regression of sexuality. In *parthenogenesis*, which occurs sporadically in both the higher plants and the higher animals, the sexual apparatus is present and eggs, specialized sex cells, are produced, but these develop without fertilization.

Some organisms are parthenogenetic and fully sexual (biparental) in reproduction at different times, sometimes with a fairly regular alternation, as in many aphids or plant lice. In bees and many of their relatives males are produced from unfertilized eggs and are haploid, while females arise from fertilized eggs and are diploid, an extraordinary modification of the sexual reproductive cycle that can be diagramed as in Fig. 9–27. Sometimes parthenogenesis is the only method of reproduction, as also exemplified among aphids. Then the cycle resembles the sexual cycle in passing through a gamete (or what was a true gamete in some ancestral form), the egg, but both meiosis and fertilization are lacking. (Why cannot meiosis occur in such a reproductive cycle if fertilization does not?) The two key processes are, so to speak, short-circuited. The effect is the same as in vegetative reproduction: variation is reduced to a minimum because until a mutation occurs all the descendants of an individual have exactly the same heredity.

You have seen how frequently both sexual and vegetative reproduction occur in the same kinds of organisms. In the course of evolution it is then possible for vegetative reproduction to be lost so that reproduction is sexual only. This happened in the ancestors of the vertebrates. On the other hand, reproduction may also become entirely vegetative. This has happened to many species here and there, but not to the whole of any large and progressive group,[29] a fact that is certainly significant in relationship to the role of sex in evolution.

When organisms are well adapted to a uniform and stable environment, it is advantageous for them to have little genetic variation. (Any mutation or other change increasing variation will be opposed by natural selection, a process discussed in Chapter 15). These are the conditions in which regression of sexuality, which always decreases variation, is likely to occur. When, on the contrary, environments change markedly in the course of time and when a varied environment offers opportunity for diverse new ways of life, groups with high genetic variation are more favorable for survival and progressive change. Then evolution puts a premium on sexual reproduction. In the history of life there has been a balance between the two processes. Sexual regression has frequently prevailed in particular circumstances, but it leads to evolutionary dead ends and strongly limits any further change.

Finally we may inquire why sexual regression is so much more common in protists and algae than in higher plants or animals, and why it is more common in higher plants than in higher animals. A complete explanation is not yet possible and will not be simple, but it seems that these factors are among those involved: shortness of life cycle, relative simplicity of organism, open as opposed to closed development, and relatively greater breadth of reaction ranges. Such characteristics may make adaptation to environmental change possible without a high degree of genetic variation at any one time.

CHAPTER SUMMARY

The generalized reproductive cycle: asexual reproduction—vegetative propagation and spore production; sexual reproduction—the production and union (fertilization) of haploid gametes, forming a diploid zygote that initiates a new generation.

Distribution and significance of sex in organisms: sex as a phenomenon distinct from and added to the basic (asexual) process of reproduction; its evolutionary function in producing adaptive flexibility and genetic recombination; other mechanisms that produce genetic recombination in bacteria.

The haplophase-diplophase reproductive cycle: haplophase initiated by meiosis and concluded by fertilization; diplophase initiated by fertilization and concluded by meiosis; major variations on the haplophase–diplophase reproductive cycle.

Reproductive cycles in plants generally: gametophyte and sporophyte as the developed organisms of the haplophase and diplophase, respectively; the reproductive cycles of green protists, algae, mosses, and ferns, showing an increasing evolutionary emphasis on the sporophyte (diplophase).

Reproduction of seed plants: the flower as a reproductive organ; sepals and petals, modified leaves, functionally related to effecting pollination; anthers, male sporangia; ovules, female sporangia; the ovary, formed by leaflike structures bearing the ovules; the embryo sac and the pollen tube as much reduced (vestigial) female and male gametophytes; seeds as ovules containing the embryonic sporophytes and food reserves (endosperm and embryonic leaves); fruits as modified ovaries or other flower parts containing seeds.

Reproductive problems inherent in the immobility of plants: (1) getting the sexes together; insects as agents for carrying pollen; devices that ensure cross-pollination;

---

[29] The largest group that may be wholly asexual is that of the blue-green algae, some 2500 living species. Their evolution has been retrogressive rather than progressive in many respects, and they owe survival and abundance to their high resistance to what are highly adverse conditions for most organisms. As noted before, the absence of sexuality in them has not been established with certainty.

(2) the dispersal of seeds; adaptations with wind, animals, and insects as dispersal agents.

Reproductive cycles in animals: the complete dominance of the diplophase, with rare exceptions like male bees; asexual reproduction in the diplophase of animals; the alternation of generations in coelenterates entirely *within* the diplophase.

Fertilization in animals: external fertilization in aquatic animals; devices that ensure the simultaneous release of male and female gametes; synchronization with the moon; synchronization by mutual stimulation in courtship; internal fertilization as mainly an adaptation to land life.

Larval forms: their function as feeding stages and, especially, as agents of the species' dispersal.

Evolution of vertebrate reproduction: external fertilization in primitive aquatic vertebrates (fishes, amphibians); the reptilian evolution of internal fertilization and egg adaptations (membranes, shell) suited to land life; the reproductive specialization of mammals—live birth.

Mammalian reproduction: gametes; copulation; fertilization in the oviduct; implantation of the embryo in the uterine wall; formation of a placenta—its function in nourishing the fetus; the estrus cycle in women and other female mammals.

Regression of sexuality: self-fertilization and parthenogenesis; its interpretation as an adaptive abandonment of genetic recombination in organisms already well adapted to a uniform and stable environment.

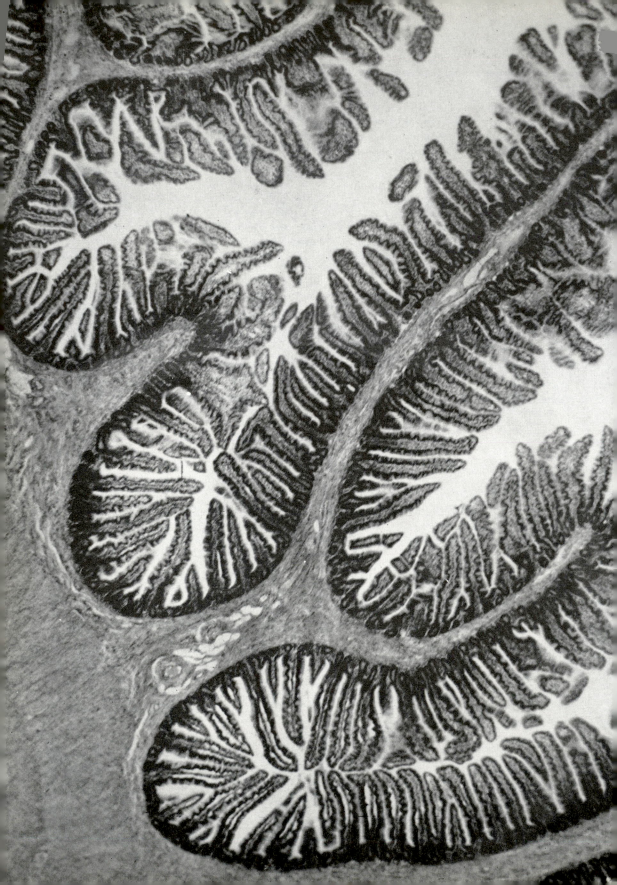

Part 4

The Maintenance and Integration of the Organism

Whatever complexity of organization we may infer from electron micrographs or genetic studies or metabolic data of a single cell becomes insignificant when we consider the whole multicellular organism.

The photograph introducing Part 4 shows a greatly enlarged section through a piece of the wall of a human intestine. It contains a host of specialized cells—of epithelial, glandular, muscular, connective, and nerve tissue. To the trained eye each cell's specialization of structure is obvious and can be related to the role it plays in the organized activity not only of the gut wall but also of the entire organism. Multicellular organisms are large enough for us to observe and experiment on directly. Thus it is from their study that we have gained much of our present knowledge on how living systems integrate and maintain their organization. Part 4 is devoted to these topics.

Chapters 10 and 11 discuss the procurement and processing of needed materials and energy. One theme running through these chapters has to do with the size of a multicellular organism. Although a large size has certain advantages, it also creates difficulties that are overcome only by special features in the organism. One such feature is a rapid-transport system, like that of blood.

Chapter 12 focuses attention on the idea of organization itself—the integration of constituent parts and processes into an ordered whole. Coordination among isolated parts demands communication within the organism. The chapter describes the system of chemical messengers (hormones), which are transported from place to place in animals by the movement of the blood and from cell to cell in plants by a mechanism that is still obscure.

Chapter 13 is devoted entirely to the second main kind of communications network in organisms, the nervous system of an animal. In addition to effecting communication among internal parts, the nervous system is concerned with obtaining and handling information about the external environment and with regulating the organism's behavior.

Chapter 14 deals with the behavior of an organism—its ordered movements in relation to other organisms and the environment in which it must survive. This is an essential aspect of organic maintenance.

10

Organic
Maintenance: I

Carl Strüwe

The insect tracheae illustrated here (×145) are one of many kinds of adaptive adjustments to increased size found in large organisms.

The metabolism of energy and materials surveyed in Chapter 4 is the economy not only of cells but also of whole organisms and communities of organisms. It is the economy of life. Living systems differ only in their methods of fulfilling the fundamental energetic and synthetic requirements that they have in common. There are two broad, and never completely separate, categories of differences. The first, most fully exemplified by the difference between a green plant and an animal (that is, between an autotroph and a heterotroph), has to do with the degree of dependence on prefabricated sources of energy and materials. The second is related to the size of the living system. In the individual cell the procurement, transportation, and processing of raw materials is manifestly a different problem from that in a large multicellular organism like man or an oak tree. The total economy—the living system—is confronted with new technical problems as it grows in size. For example, the internal distances to be traversed become so great that transportation can no longer be entrusted to simple diffusion and protoplasmic streaming. Specialized systems for the mass movement of products, parts, and raw materials play a novel and major role in the economy of the multicellular organism.

These differences among living systems— plant or animal, large or small—do not obscure a fundamental similarity in their ultimate material needs. In this chapter and the next, on the maintenance of the whole organism, you will recognize that, when we discuss diverse modes of digestion, assimilation, respiration, transportation, and excretion, we are discussing diverse ways of achieving ends already familiar from our study of the individual cell.

Procurement Processes

INTAKE

The passage of materials through an organic metabolic system begins when they are taken into the organism. The most widespread method of intake of animals is the one we have ourselves: solids and liquids are taken into a tube, the *alimentary canal*, running through the body. Fig. 10–1 outlines the alimentary

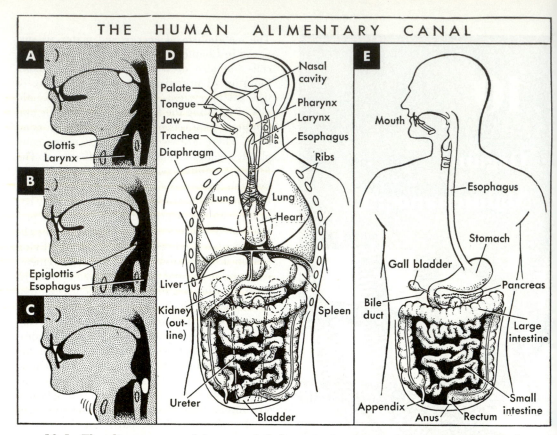

THE HUMAN ALIMENTARY CANAL

A

Glottis
Larynx

B

Epiglottis
Esophagus

C

D

Palate
Tongue
Jaw
Trachea
Diaphragm

Nasal cavity
Pharynx
Larynx
Esophagus
Ribs

Lung
Lung
Heart

Liver
Kidney (outline)
Spleen

Ureter
Bladder

E

Mouth

Esophagus

Stomach

Gall bladder

Bile duct

Pancreas

Large intestine

Appendix
Anus
Rectum
Small intestine

10–1 The alimentary canal in man. *A–C.* Stages in the passage of food through the mouth cavity and pharynx; the epiglottis prevents the passage of food into the larynx and, hence, into the lungs. *D.* The alimentary canal in relation to other internal organs in man. *E.* Part of the alimentary canal. (From G. Hardin, *Biology: Its Human* Implications, 2nd ed., Freeman, San Francisco, 1952; after Sturtevant and Beadle.)

canal in man and identifies its parts. Within the alimentary canal parts of the solids go into solution, and liquids and substances in solution are absorbed through the wall of the tube.

Among the animals that take solid food into an alimentary canal there are many different systems for the actual acquisition of the food. Without attempting to consider all the details at this point, we can distinguish three main groups. Some eat sizable plants and animals or chunks of them. That is true of man, of course, and also of the animals most familiar to us. Usually such animals have a method of seizing food, a mouth, and some way of breaking up the chunks. Many animals, however, eat food in particles very small in comparison with the animals themselves. These animals have filters to catch food, which is then swal-

lowed whole in filtered masses (Fig. 10–2). The variety of filter feeders is surprising. They range from oysters to whalebone whales. Finally, there are some animals, fewer than those with other feeding habits, that simply take in samples of their whole environment, digest any food in it, and evacuate the rest. Earthworms are of this sort: they pass bits of whole earth, the medium in which they live, through the alimentary canal. Clearly this system will not work unless the environment is fairly nutritious; there are no worms in sand dunes.

Protists absorb water and dissolved materials directly through the body membrane. The body membrane is analogous to a cell membrane, if not exactly the same thing, and of course the absorption involves diffusion and osmosis. It is really the same sort of process as

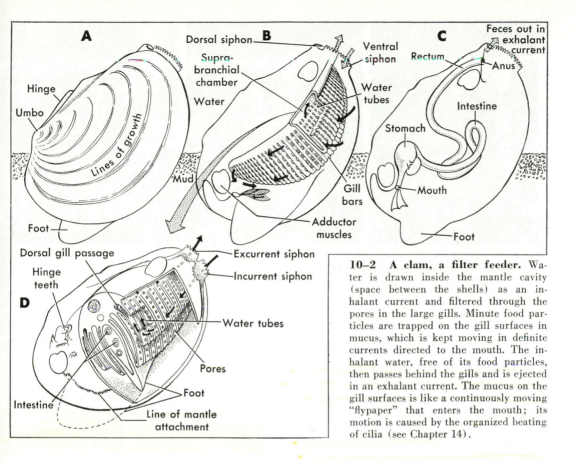

10–2 A clam, a filter feeder. Water is drawn inside the mantle cavity (space between the shells) as an inhalant current and filtered through the pores in the large gills. Minute food particles are trapped on the gill surfaces in mucus, which is kept moving in definite currents directed to the mouth. The inhalant water, free of its food particles, then passes behind the gills and is ejected in an exhalant current. The mucus on the gill surfaces is like a continuously moving "flypaper" that enters the mouth; its motion is caused by the organized beating of cilia (see Chapter 14).

absorption by cells lining the tube or sac—the alimentary canal—in multicellular animals. Many protists, especially some of those that are often called animals, also take in solid food. They eat other protists and even small multicellular organisms, just as man eats plants and animals—although they often do not bother to kill their food before eating it. The solid food of protists is simply surrounded or captured by various devices. It is taken in through the body membrane and becomes enclosed in a food vacuole, in which digestion occurs. This method looks very different from our way of eating and digesting solid food, but the food vacuole corresponds with our stomach and intestine, and the same sort of processes go on in it. The only essential difference is not functional but structural: the food vacuole is not a permanent part of the protist's anatomy but is improvised anew for each meal or "mouthful."

Some protists utilize solid food without ac-

tually taking it into the body for digestion. They secrete digestive fluid onto food outside the body wall and then absorb the dissolved products. Many starfishes and some other multicellular animals do essentially the same thing: they hold their prey outside the body and surround it with digestive fluids in various ways. It could be said that they digest their food before they eat it. But from the point of view of food intake this is not fundamentally different from digestion within the alimentary canal.

A few plants take in solid food in a way that is much the same as in carnivorous animals. These plants, including the pitcher plant, sundew, and Venus's flytrap, have modified leaves that trap insects (Fig. 10–3). The traps also are cavities, analogous to an animal's alimentary canal, in which digestion occurs and from which the digested food is absorbed. These extraordinary plants have a special appeal to the imagination, and they are striking exam-

10–3 Carnivorous plants. *Left to right.* Venus's flytrap, *Dionaea* (natural size). Pitcher plant, *Sarracenia* (one-third natural size). Sundew, *Drosera* (slightly reduced in size).

ples of both the unity and the diversity of life and its processes. They are, however, exceptional plants.

The overwhelming majority of plants take in materials in only one way: by diffusion through the cell walls and membranes. In aquatic plants all materials, including water itself, are acquired from the water solution surrounding them. In terrestrial plants water and dissolved salts are acquired mainly by diffusion into root hairs and other cells on the surface of the roots. The other needed materials, oxygen and carbon dioxide, diffuse from their gaseous condition in the atmosphere into solution within cells of aerial parts, especially leaves. The intake of oxygen is part of the process of respiration, which will be discussed as a separate topic. Carbon dioxide, however, is a raw material for food.

Separate problems are involved in the diffusion into plant roots of water and its dissolved inorganic materials (mineral salts). Water is pulled into the roots by osmosis. Fig. 10–4 is a schematic section across a plant root from its hair cells to its xylem vessels. Suppose, first, that all the cells along the path from hair to xylem have the same osmotic pressure (Chapter 3). The pressure in the cells is much higher than that of the soil water because of the high concentration of sugars and salts in the cell. Water is consequently drawn by osmotic pressure into the root hair cell, slightly lowering

its osmotic value. The hair cell then yields water to the adjacent cell inside the root which has a higher osmotic value. This continues across the whole cortex (outer layer) of the plant to the xylem, into which the soil's water is pumped by the gradient of osmotic pressure. The pressure developed by this gradient of osmotic values across the root is often very large. It is called *root pressure;* we shall have more to say about it in the next chapter.

The movement into root hairs of dissolved substances in the soil water raises quite different problems. Where the incoming salt is immediately consumed in some synthesis or is bound to protein colloids in the cytoplasm, its concentration as a free molecule (or ion) in the cell may be kept very low. The cell is then assured of continued simple diffusion of the salt from higher concentrations in the soil to the lower concentrations in the cytoplasm. In some plants, however, the cell requires concentrations of salts above those found in soil water. It has then to expend energy—in a manner still unknown—to pull salts into the cell against a concentration gradient.

In general, then, substances taken in as food or as materials for food may be solid, liquid, or gaseous. Intake of a gas, in addition to oxygen for respiration, is essential in almost all green plants but is rarely important in other organisms. Solids are not actually absorbed and do not really enter into the life

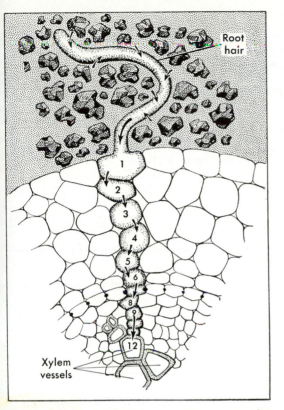

10–4 Water uptake by roots. Water enters the root hair by osmosis from the soil and then follows an osmotic gradient through successive cells (*1–11*) until it reaches a vessel (*12*).

processes of the organism until they have been transformed into solutions. The ultimate, most common method of intake is thus absorption from aqueous solutions. Small globules of liquid fats may, however, be absorbed without going into solution.

DIGESTION

The principal feature of digestion is that it transforms foods into forms that can readily be absorbed. Substances insoluble in water or with very large molecules or in large globules cannot as a rule be absorbed; they usually cannot pass through the membranes that surround cells and vacuoles. Most foods fall into these classes of substances difficult or impossible to absorb. Simple carbohydrates (sugars) are generally soluble in water and are readily absorbed, but polysaccharides such as starch are water insoluble. So are lipids.

Proteins are large molecules to which cell membranes are ordinarily impermeable.

It appears, therefore, that most of the more complex compounds built up in cells cannot be passed on to other cells unless they are torn down again. Life would be impossible without elaborate organic foods. Yet in many cases these foods must be changed back into the substances from which they were derived before they can serve as food for other parts of the same organism or for other organisms. However, this process is not as inefficient as it may appear. It is true that organisms are less than 100 per cent efficient, but further consideration shows that these apparently inefficient features of food utilization have an essential function. If large molecules were absorbed through cell membranes, so that cells could take in complex proteins as such, then colloidal protoplasm could not always be retained within the cell. If cellulose were soluble, then fluids moving in plants would dissolve the vessels through which they flow. If starch and glycogen were soluble, they could not be retained in cells in extra quantities available as storage against future needs. Moreover, the diversity of life and the actual characteristics of its different kinds depend on the fact that each kind has its own proteins. Each kind, therefore, cannot utilize the proteins of other organisms directly as they are taken in but must tear them down and rebuild them.

Most of the chemical reactions of digestion have the net effect of reversing the synthetic processes described in Chapter 4. The most fundamental of these are summarized in Table 10–1. All of the syntheses ultimately involve a linking together of molecules with the elimination of water. Digestion in general is the opposite process—a hydrolysis, or separation of molecules through the addition of the elements of water to them.

We normally think of digestion as the first operation after food is taken into the organism from outside. Exactly the same reactions, however, occur within almost all body cells whenever insoluble or colloidal materials pass into them from other cells. In biological terms the reactions are digestion whether they occur at the first intake of food or later on. For example, the first digestion of a polysaccha-

TABLE 10–1

Comparison of the net reactions of biosynthesis and digestion

Starch and other polysaccharides

$$n \text{ Monosaccharides} \underset{\text{digestion}}{\overset{\text{synthesis}}{\rightleftarrows}} \text{Polysaccharide} + (n-1)H_2O$$

Lipids

$$\text{Glycerol} + 3 \text{ Fatty acids} \underset{\text{digestion}}{\overset{\text{synthesis}}{\rightleftarrows}} \text{Triglyceride} + 3H_2O$$

Proteins

$$n \text{ Amino acids} \underset{\text{digestion}}{\overset{\text{synthesis}}{\rightleftarrows}} \text{Protein} + (n-1)H_2O$$

ride (usually starch) converts it to simple sugars, which are carried to the liver and there resynthesized into another polysaccharide (glycogen), which in turn is later digested in the liver cells to simple sugars and passed on to other parts of the body.

This cyclic relationship between digestion and synthesis suggests answers to many questions about the phenomena of metabolism. How does starch get into the underground tubers of a potato plant? How do stem and root tips obtain the amino acids necessary for their growth? What takes place before a bite of meat affects the cells in the fingers or toes? How does sugar in the diet become a source of energy in the leg muscles during active exercise? Complete answers to all such questions cannot be given here, but we have covered most of the essential underlying principles.

Digestive enzymes

Enzymes catalyze all the hydrolytic reactions of digestion. Usually they are not the same enzymes as those involved in the corresponding synthesis. Since digestion actually proceeds as a series of small steps, what is summarized as a single reaction may require a number of different enzymes. Unlike ordinary enzymes, which are found only within cells in minute quantities, the digestive enzymes of the animal alimentary canal are poured into the gut from surrounding cells in relatively large amounts—and in fairly simple mixtures. Thus they have been extraordinarily well

studied. Table 10–2 lists some of the enzymes participating in human digestion.

It should be noted that evolution has introduced an additional digestive device that is unique to human beings—cooking. Though many foods can be digested without cooking, the cooking process performs an initial digestive function without which certain foods would be relatively indigestible. Cooking denatures proteins and bursts the granules of natural starch, facilitating their attack by digestive enzymes. The only highly starchy food commonly eaten without cooking is the banana. This explains why it is often the first solid food given to babies.

Two solutions important in human digestion are *not* enzymes. First, the gastric juice in the stomach is a strong solution (approximately 0.1 molar) of *hydrochloric acid.* Pepsin, the principal enzyme of the stomach, is not effective unless it is in acid solution. (Some other enzymes, such as trypsin, require an alkaline environment; acid from the stomach is neutralized by alkaline secretions when the food passes into the duodenum.) Second, bile from the liver is, in part, a solution of *bile salts.* These are poured into the small intestine, where they help to break up fats into small globules, forming an emulsion. This action does not change the fats chemically, but it puts them in a condition in which chemical reaction (catalyzed by lipase) occurs more readily.

Since the alimentary canal digests meat in

TABLE 10–2

Some digestive enzymes in man

Source of enzyme	Enzyme	Substances acted upon	Products of digestion
Salivary glands	Salivary amylase	Polysaccharides (cooked starch and glycogen)	Disaccharide (maltose)
Stomach glands	Pepsin	Proteins	Peptides
	Rennin	Milk casein	Clotted casein (curds)
Pancreas	Trypsin Chymotrypsin Carboxypeptidase	Proteins and peptides	Amino acids
	Pancreatic amylase	Starch	Maltose
	Lipase	Triglycerides	Fatty acids, glycerol, and monoglycerides
Small intestine	Peptidase	Peptides	Amino acids
	Sucrase Maltase Lactase	Disaccharides	Monosaccharides

the stomach and in the upper part of the small intestine (the duodenum), one might ask why it does not digest itself. Obviously it must have a lining that is not normally susceptible to the action of the digestive juices, but just why it is not susceptible is far from clear. In fact the juices of the stomach and duodenum do sometimes digest parts of their walls: the result is an ulcer. That ulcers and, more commonly, indigestion may be brought on by anxiety or nervousness illustrates the great complexity of interactions in an integrated organism.

Enzymes like those in the human digestive system occur widely among other animals. Similar, though not necessarily identical, enzymes, catalyzing the digestion of similar substrates, occur in both plants and animals.

Some organic compounds widespread in nature are indigestible to most organisms. Probably the most striking example is the polysaccharide cellulose, a usual constituent of plant cell walls. Cellulose is entirely indigestible by man and most other animals, although some crabs, snails, and insects have an enzyme that hydrolyzes cellulose. In cows and termites, along with some other animals, cellulose is digested in an interesting way: these animals cannot themselves digest it, but their alimentary canals contain large numbers of bacteria or other protists that can. The cow absorbs the sugars produced by the protists' digestion of the cellulose. Keratin, a protein in wool and other types of hair, is indigestible to practically all animals except the larva of the clothes moth. In that larva the keratin is first reduced and then hydrolyzed by a special enzyme—a good trick, but one we humans cannot wholly admire.

Digestive systems

Since green plants are autotrophs—that is, they do not take in complex organic foods but make their own—they have no special systems or organs for digestion.[1] Nevertheless, digestion does occur in green plants. It is necessary if food manufactured or stored in one cell is to be passed along to another cell, and it may occur in any plant cell when food, enzymes, and the other necessary conditions exist. Such digestion within a cell, also common in animals, is called *intracellular* digestion.

In nongreen plants, which obtain food from the outside, the actual reactions and processes of digestion are essentially the same, but the procedure is different. For example, *Rhizopus*, the common bread mold (Fig. 10–5), is a fungus that grows on bread, cake, and other foods at room temperature. Its food is in the

---

[1] Exceptions are the few insect-eating plants, mentioned above (pp. 285).

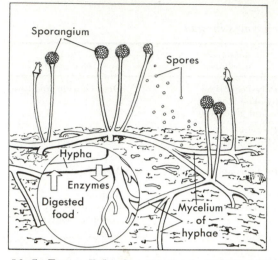

10–5 Extracellular digestion by the bread mold *Rhizopus.* The body of the mold consists of a branching system (or mycelium) of threadlike structures, the hyphae. Enzymes produced by the hyphal protoplasm are secreted to the outside, where they digest the complex carbohydrates in the bread. The sugars resulting from this extracellular digestion are then absorbed by the hyphae.

bread or other material on which it grows. Since the organism cannot absorb the food until it is digested, enzymes diffuse from its cells and catalyze the digestion of starches and other foods outside it, and the resulting compounds in solution diffuse into the cells. This method resembles the one by which humans and other animals digest solid food: digestion is *extracellular,* or outside the cells. In all such instances digestion occurs outside the organism, strictly speaking. (The inside of your stomach is topologically outside your body.) The only real difference is in the anatomical organs involved. In man and bread mold—indeed, in practically all organisms— intracellular digestion also occurs, facilitating the further movement and utilization of foods after they have been absorbed into the organism.

Some protists have permanent mouths, and a few have permanent anuses, but most have no special digestive organs at all. If solid food is taken in through the body membrane, temporary food vacuoles form. Food manufactured in the cell or entering through the membrane is digested diffusely throughout the cytoplasm.

In the group of animals to which the corals and jellyfishes belong, and also in the flatworms, there is a cavity surrounded by the body and freely open to the outside through a single opening (Fig. 10–6). Most of the animal's food is first brought into this cavity, which is filled with the water in which the animal lives but which also contains some enzymes secreted by the cells around it. Digestion begins in the cavity, but in most of the animals it is not completed there. Usually small bits of food are taken bodily into the cells lining the cavity and are finally digested in food vacuoles in the cells, just as solid food is usually digested in protists. Thus digestion in these animals is intermediate between the food-vacuole system of protists and the alimentary-canal system of higher animals.

With innumerable modifications of form and of the organs into which it is subdivided, an alimentary canal occurs in practically all the animals above the level of the coral or the flatworm. It is present and is even quite complex in the earthworm (see Fig. 10–6), which is much more advanced than a flatworm and belongs to a different branch of evolution.

With such supplementary information as you now have about the processes of digestion, the human digestive system (see Fig. 10–1) will serve as an example of the system in higher animals. It should be understood

---

10–6 Digestive systems in animals. *A.* Intracellular digestion occurs in protists. *Trichonympha,* which inhabits the alimentary canal of termites, engulfs solid wood particles eaten by the termite. Digestion of the wood takes place in a food vacuole inside the cell. In *Paramecium* a food vacuole forms under pressure of the water driven down the gullet by the beating of cilia. Bacteria and other microorganisms are engulfed in the food vacuole, which moves along a fixed path through the cytoplasm. *B, C.* Digestion in *Hydra* (and other coelenterates as well as in flatworms) is partly extracellular and partly intracellular. The alimentary canal is a blind sac (one opening only) and lacks any specialization of parts. Partly digested (but still solid) food in the gut cavity is engulfed by cells into food vacuoles, as in protists. Digestion is completed as an intracellular process. *D.* Extracellular digestion occurs in an earthworm (*Lumbricus*) and all higher animals. Here the alimentary canal has two openings, and food moves in one direction (mouth to anus). This situation permits specialization of the alimentary canal along its length.

DIGESTIVE SYSTEMS

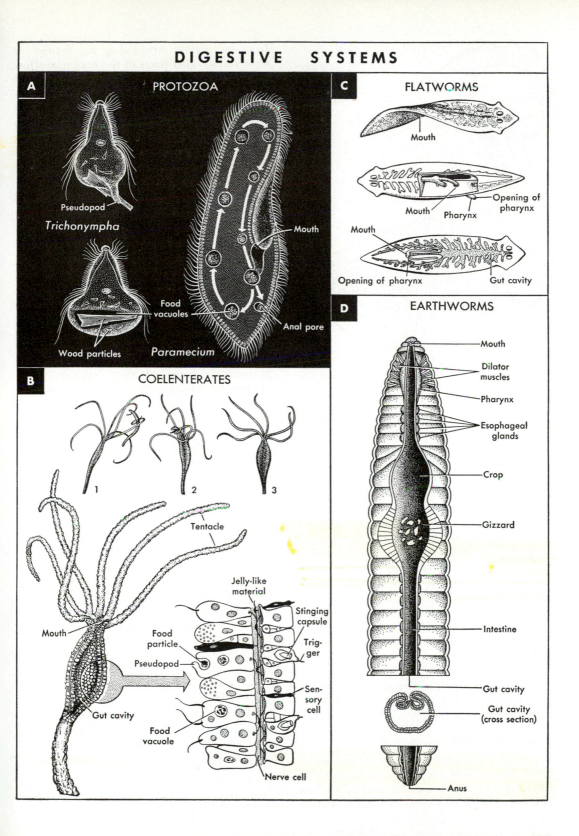

A PROTOZOA

Pseudopod

Trichonympha

Food vacuoles

Wood particles

Paramecium

Mouth

Anal pore

B COELENTERATES

1 2 3

Tentacle

Mouth

Food particle

Pseudopod

Gut cavity

Food vacuole

Jelly-like material

Stinging capsule

Trigger

Sensory cell

Nerve cell

C FLATWORMS

Mouth

Mouth Pharynx

Opening of pharynx

Mouth

Opening of pharynx Gut cavity

D EARTHWORMS

Mouth

Dilator muscles

Pharynx

Esophageal glands

Crop

Gizzard

Intestine

Gut cavity

Gut cavity (cross section)

Anus

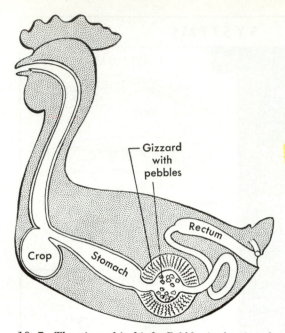

10–7 The gizzard in birds. Pebbles in the gizzard functionally replace teeth by grinding up food (cf. earthworms, Fig. 10–6 *D*).

that this or any other example is not typical. Every variety of animal has its own peculiarities of structure, enzymes, and so on that are adapted to its particular way of life. Cows have an extraordinary series of stomachlike cavities and rechew their food after digestion has started. (That is what they are doing when they chew their cuds.) Birds have no teeth and grind food not in the mouth but in a specialized part of the alimentary canal called the *gizzard* (Fig. 10–7). Some insects do not swallow solid food but digest it in the mouth (or, more strictly, the pharynx), swallowing the solutions and spitting out the undigested parts. Many animals live on solutions such as nectar from flowers or blood from other animals, and their digestive systems and processes are correspondingly modified. In this, as in so many respects, the diversity of life seems almost endless. And yet, once more, there is unity in basic principles.

ASSIMILATION

Digestion changes food into soluble molecules able to enter cells and thus become incorporated into the body of the organism. The next event is the movement of these molecules from the regions where they are formed or absorbed to other parts of the organism. How this movement takes place is a special topic for discussion in the next chapter. Eventually the products of digestion pass into tissues and cells where they are immediately utilized in metabolism or temporarily stored. As we have already noted, storage usually requires the resynthesis of insoluble molecules, especially polysaccharides and lipids. The stored foods are redigested and commonly redistributed by the same sorts of metabolic processes as the rest of the food materials.

Once in the cells where they are utilized, the soluble nutrients participate in all the metabolic reactions that occur in protoplasm. Since digestion converts them into simple forms, they reach the cells mainly as simple sugars, fatty acids, glycerol, and amino acids. Sometimes these compounds are utilized directly, but often they are rapidly converted into the complex compounds making up much of the protoplasm. Some of these complex compounds are subsequently incorporated into the actual cell structure, and some supply energy for cell activities. As we have seen, the distinction is not clear-cut. The cell-building materials are in a constant state of flux, as are the materials more largely utilized as energy sources, and the same materials may be used for both purposes.

Cells form some compounds intended for export (that is, secretion) and not for their own consumption. Such compounds may be poured out from glands, as are the enzymes in the various digestive juices; they may be carried in the blood to other parts of the organism, where they influence other activities, as are the hormones; or they may be final products of metabolism no longer useful anywhere in the organism. The processes of secretion and excretion will be surveyed in the next chapter.

NUTRITION

Adequate nutrition supplies the organism with all the materials it needs for synthesizing the compounds that make up its structure and provide its energy.[2] Autotrophic organisms,

[2] This discussion of nutrition in a large multicellular organism parallels the discussion of cellular nutrition in Chapter 4.

especially the green plants, can start their syntheses with simple inorganic compounds and elements. Heterotrophic forms must start their syntheses with complex organic compounds, such as sugars and amino acids. Man belongs to the latter group; he has low synthetic ability and requires a wide variety of elaborated foods. He thrives best on a diet containing, besides water and a considerable number of mineral salts, all three major classes of foods—carbohydrates, fats, and proteins—with some diversity in each class. Any digestible food of any of these classes can furnish him with energy. In this respect the classes are largely interchangeable. They are not interchangeable as building materials or as sources for the synthesis of enzymes and other compounds. For these needs, proteins (or a number of amino acids) are absolutely indispensable. A man who ate enough carbohydrates to satisfy his hunger and fulfill all his energy demands would, if he ate nothing else, be fatally undernourished.[3]

There is another nutritional requirement, mentioned only in passing heretofore—vitamins. A man who has an apparently sufficient diet of carbohydrates, fats, and proteins, with adequate material and energy intake, may nevertheless develop serious deficiency diseases. A deficiency disease is caused by a lack of substances essential for normal metabolism. One of the earliest such diseases to be studied was scurvy, a disorder producing bleeding gums, painfully swollen joints, and eventually death. It frequently appeared among sailors on long voyages in the days when fresh foods were not available on shipboard. Long before there was any knowledge of the cause of scurvy, it was discovered by chance that lime juice prevents it.

Later experiments with animals and clinical investigations with humans revealed that scurvy and a long list of other deficiency diseases result from a lack of vitamins. Vitamins are organic compounds required in small amounts for normal metabolism but not synthesized by the organism and therefore neces-

sarily present in the diet.[4] Vitamins do not belong to any one chemical family. The same substance may be synthesized by one organism, for which it is thus not a vitamin, and be required as a vitamin in the food of another.[5] Some of the vitamins essential to man are listed in Table 10–3.

Although commercial advertisements of vitamin preparations give the opposite impression, it is a fact that the varied diet customarily eaten by all but the poorest Americans usually provides sufficient amounts of all the needed vitamins—particularly if the diet includes a considerable proportion of vegetables and fruit in addition to meat. Heavy drinkers are likely to develop deficiency diseases, especially polyneuritis (from a deficiency of thiamine), because alcohol supplies a large part of their energy requirements and there is a tendency in alcoholism to omit other nutritious foods that are sources of vitamins. A specific vitamin deficiency should be given specific treatment. Overdoses of vitamin D are harmful, and the possibility of harm from the overuse of some other vitamins has not been ruled out.

Most vitamins can be synthesized by one plant or another. Although the compounds are therefore not vitamins for plants, the fact that plants synthesize them suggests that they are necessary in plant metabolism, and their chemical roles are generally similar in plants and animals.

In animals these roles are well understood for all but a few vitamins. As we have seen (Chapter 2), almost all vitamins are converted in the body to coenzymes, in which form they participate in cell metabolism. Thus, for example, nicotinic acid is converted to the familiar coenzymes NAD and NADP. In nicotinic acid deficiency, these coenzymes are lacking,

---

[3] As a general rule, foods rich in carbohydrates are cheap, and those rich in lipids and proteins are expensive. What social implications does this have? Is deficiency in energy the major nutritional problem of underprivileged peoples?

[4] Other organic substances, such as particular lipids, may be required in small amounts. Yet they are not usually called vitamins, even though they correspond to this technical definition of vitamins. Some other materials, such as iron and iodine salts, may also be required in small amounts and be similar to vitamins in the effects of deficiency; they are not called vitamins because they are not organic compounds.

[5] We have already mentioned the genetic mechanisms that led in evolution to the loss of synthetic ability and hence to vitamin requirements (Chapter 7).

TABLE 10–3

Some vitamins important in human nutrition

Vitamin	Some deficiency symptoms	Some sources
FAT SOLUBLE		
A	Dry, scaly skin; night blindness	Milk, butter, liver oils, yellow and green vegetables *
D	Rickets (defective growth of bones and other hard tissues)	Milk, egg yolk, liver oils †
E	Degeneration of muscles (also sterility in rats and possibly in other organisms)	Green leafy vegetables, oils from seeds, egg yolk, meat
K	Delayed clotting of blood	Green leafy vegetables †
WATER SOLUBLE		
Thiamine (B_1)	Beriberi, polyneuritis (inflammation and degeneration of nerves)	Yeast, whole-grain cereals, lean meat
Riboflavin (B_2, G)	Soreness around mouth, inflammation of eyes	Vegetables, yeast, milk, liver, eggs
B_{12}	Anemia due to the inability of blood-cell precursors to synthesize DNA; nerve damage	Liver and lean meats
Ascorbic acid (C)	Scurvy	Citrus fruit, tomatoes, green peppers
Nicotinic acid or niacin	Pellagra (a disease affecting skin, alimentary canal, and nerves)	Yeast, meat
Folic acid	Anemia similar to that of vitamin B_{12} deficiency	Leafy vegetables

* Vitamin A is not required as such, since the human body can synthesize it from yellow pigments, carotenes. These are required if vitamin A itself is deficient.

† The human body can synthesize vitamins D and K, but the amounts synthesized may be inadequate.

and the metabolic pathways upon which they depend function subnormally. This is the ultimate cause of the complex group of disorders constituting pellagra.

Respiration

THE MEANING OF RESPIRATION

In the language of the physiologists, breathing in is *inspiration*, breathing out is *expiration*, and the whole process is *respiration*. Inspiration draws in oxygen, and expiration expels carbon dioxide. Oxygen is not used solely in the lungs, any more than food is used solely in the stomach and intestine. In both cases the organs named are involved in the intake of materials used throughout the whole body. Thus, respiration is a process that takes place not in the lungs alone but in cells and tissues through the organism.

It is unfortunate when the same word comes to be used in more than one way, and this has happened to the word "respiration": it means breathing in and out of the lungs, and it also means the consumption of oxygen and production of carbon dioxide by metabolic processes in an organism. Popular writers to the contrary notwithstanding, plants and a great many animals never breathe. Practically all organisms and individual cells within organisms do respire, in the broader sense of the word. In this book we apply the term "cellular respiration" to the processes in cells and tissues comparable with chemical oxidation.

CELLULAR RESPIRATION

As we have noted, energy from solar radiation is bound into chemical compounds by photosynthesis. Among the first and simplest energy-rich compounds formed are simple sugars; the other, more complex, organic compounds also contain bound energy. Respiration, or oxidation-reduction, in cells includes the various processes by which this bound energy is released. We covered the fundamental chemical aspects of this respiratory-energy release in Chapter 4. Here we must draw attention to variations on the basic pattern. One of these variations—the cellular respiration of muscles—is related to transport problems arising from the rapid energy utilization and large size of some multicellular animals.

Carbohydrates enter the animal as glucose from the alimentary canal. Picked up in the blood, the glucose is transported to the liver through a special short circuit in the blood system (the hepatic portal vein). It is converted in the liver into glycogen (animal starch), an insoluble polysaccharide suitable for storage. Carbohydrates are withdrawn from liver storage as glucose produced by digestion of the glycogen. The glucose is transported all over the body in the blood stream. In muscles it is again converted to glycogen for temporary storage. Fig. 10–8 summarizes these relations.

The respiratory breakdown of glycogen in muscle takes the usual pathway from glucose through a series of intermediates to pyruvic acid. This is the pathway outlined in Chapter 4 and diagrammed in Figs. 4–5 and 4–6. In the presence of abundant oxygen it continues along the usual pathway, ending as CO_2 and H_2O. The energy is released as $\sim\textcircled{P}$ and trapped immediately by ADP to form ATP. The ATP ultimately supplies the energy (again as $\sim\textcircled{P}$) consumed in the contraction of muscle.

When a large animal, like man, performs rapid movements, as in running, it consumes a great deal of energy. This implies that it demands a continuous supply of ATP, which it utilizes at a great rate. In extreme exercise the rate of ATP consumption exceeds the rate at which it can be supplied by the normal aerobic (oxygen-consuming) respiration of glucose. The inadequacy of aerobic respiration is due to the relatively slow rate at which oxygen is transported from the lungs to the muscles through the circulation. Limitation in the oxygen supply does not, however, set an upper limit to the rate at which energy can be released in the muscle. In the last analysis, respiratory energy is released in oxidative

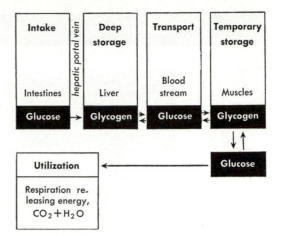

10–8 The absorption, transportation, storage, and utilization of glucose.

phosphorylation by the removal of hydrogen, not by the addition of oxygen. Oxygen, as we saw in Chapter 4, acts, in a sense, as a sponge to clean up the hydrogens stripped from sugar. In the respiratory chain (see Fig. 4–9) it is the final hydrogen acceptor.

During violent muscular exercise the amount of hydrogen removed from sugar is much too great to be taken care of by the inadequate oxygen supply. Instead, the hydrogens are picked up by pyruvic acid itself (see Fig. 4–6), which is thereby transformed into lactic acid. Lactic acid can accumulate in large quantities, and as long as the demand for energy is maintained, it continues to accumulate. The muscle does its work with the ATP formed by the partial, and anaerobic, respiration of glycogen to lactic acid—that is, by anaerobic glycolysis.

The accumulation of lactic acid eventually fatigues the muscles, and the runner gets cramps. When he stops, he pants deeply for a long time until his system returns to normal. The rapid panting serves to bring in oxygen as rapidly as possible and pay off the oxygen debt he has accumulated. Actually the oxygen completes the final respiration of only a portion of the accumulated lactic acid. This is oxidized to CO_2 and H_2O, liberating, of course, some energy as ATP (energy 2 in Fig. 10–9). The energy released by this portion of lactic acid is utilized to resynthesize the remainder back into glycogen, ready again for heavy withdrawals.

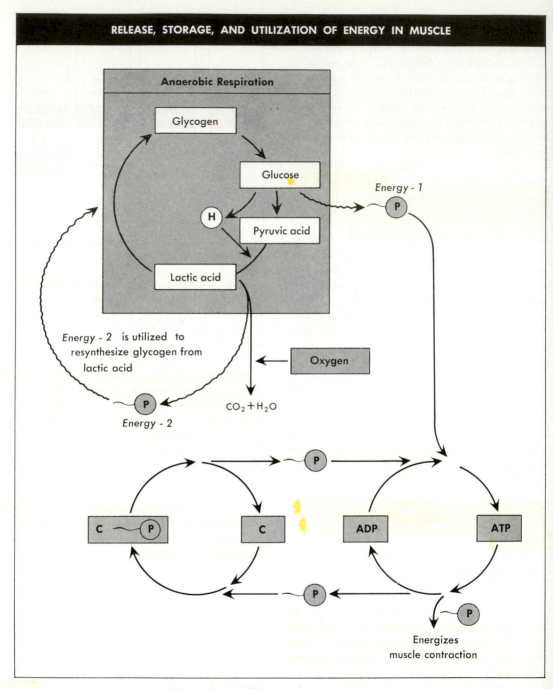

10–9   Intracellular respiration of muscle.

One other aspect of energy release in muscle is noteworthy; like glycolysis, it relates to the very heavy demands that a muscle is likely to make at any moment. No cells carry a very large supply of ADP and ATP. *Quickly* with-drawable reserves of $\sim\!\text{P}$ are stored by animals in compounds called *phosphagens*. In man and other vertebrates[6] the phosphagen

[6] See p. 580 for discussion of other phosphagens in

is creatine phosphate. In Fig. 10–9 the creatine molecule is designated simply as C, creatine phosphate as $C\sim\text{\textcircled{P}}$. Of course, even when energy demands are low, muscle respiration (aerobic) continues, and the $\sim\text{\textcircled{P}}$ radicals generated are passed, as usual, first to ADP to form ATP. As the ADP–ATP cycle becomes saturated (no more ADP), some ATP passes off $\sim\text{\textcircled{P}}$ to creatine, which is present in large amounts, for storage as $C\sim\text{\textcircled{P}}$. Later, when rapid exercise makes high demands for $\sim\text{\textcircled{P}}$, these are first paid from $C\sim\text{\textcircled{P}}$ reserves, but *through the agency of the ADP–ATP cycle*. All these relations are summarized in Fig. 10–9.

Anaerobic respiration also occurs in some plant cells, where it takes a slightly different course from that in animals: it proceeds to pyruvic acid, as in animals, but this compound, instead of becoming lactic acid, is transformed into ethyl alcohol and carbon dioxide (see Fig. 4–6). Anaerobic respiration in plants is, indeed, the process of *fermentation*, which, carried out by yeast cells, supplies us with alcohol.

$$C_6H_{12}O_6 \xrightarrow{\text{enzymes}} 2C_2H_5OH + 2CO_2 + \text{Energy}$$

Sugar Alcohol

For comparison, we summarize the anaerobic respiration in muscle as follows:

$$C_6H_{12}O_6 \xrightarrow{\text{enzymes}} 2C_3H_6O_3 + \text{Energy}$$

Sugar Lactic acid

Although most organisms require aerobic respiration to complete the cycle, as we do, they can often get along on anaerobic respiration alone for shorter or longer periods. In man, the period is very short, indeed, because our nervous system requires rapid aerobic respiration. If no oxygen is delivered to the brain, unconsciousness ensues in a few seconds. (Since this is true, how does it happen that you can hold your breath for 2 minutes or so without losing consciousness?) Some lower animals and plants can live by anaerobic respiration without oxygen for hours, days, or weeks, even though they do require aerobic respiration eventually. Some organisms, too, are normally anaerobic throughout their lives.

___

animals and the evolutionary significance of their distribution in different animal groups.

This fact has great importance for medicine and health, for these anaerobic organisms are mostly bacteria and intestinal parasites. The ability of tetanus germs to thrive in deep closed cuts and punctures, and of some worms to spend long periods in the intestines of men or other animals, is due to their anaerobic respiration.

Part of the energy released in cells by respiration appears as heat and is eventually lost. Even this energy cannot be considered wasted. It helps to maintain body temperature in warm-blooded animals (birds and mammals), and even in other animals and in plants it influences the rates of other physiological processes. Much of the released energy is used in syntheses. The building up of complex from simpler compounds, for instance of proteins from amino acids, requires the expenditure of energy, and the energy comes from respiration. This is the reason why every cell must respire. It needs energy for building its own materials, even if it does not seem to be doing anything energetic. The released energy may also bring about physical changes in the protoplasm and in cell membranes. It may produce radiation and electrical phenomena. More familiarly, it may be expended as motion.

RESPIRATORY RATES

Even though anaerobic respiration also occurs, most plants and animals have respiratory cycles that are aerobic at one stage or another. This means that their oxygen consumption is at least roughly in proportion to their total energy consumption. This, in turn, is proportional to the total amount of activity, or work, of all kinds going on throughout the organism as a whole. Thus oxygen consumption is a valuable and interesting measure of organic activity. The amount of oxygen consumed per hour or day and in proportion to the weight of the organism is the respiratory rate.[7]

___

[7] This is the principle of the basal metabolism test used by physicians for diagnostic purposes. The basal metabolic rate, or B.M.R., is usually expressed in terms of heat production per unit of body surface area per day. What is actually measured is oxygen

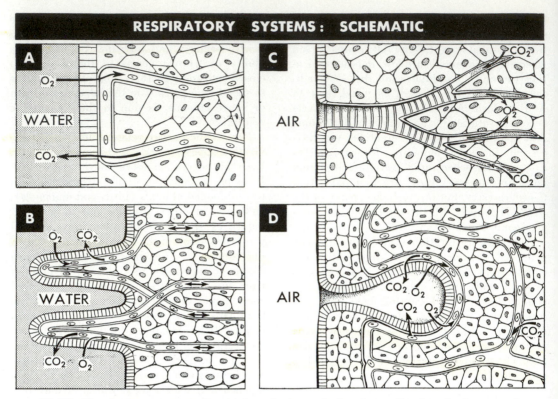

10–10 Respiratory systems. In aquatic animals respiration is usually through the external body surface, either (*A*) generally or (*B*) in localized areas where the surface for O_2–CO_2 exchange is greatly increased by the development of gills. In land animals the respiration surfaces are internal because they must be kept moist. The two commonest systems are tracheae (*C*) and lungs (*D*). Tracheae (in insects and some other arthropods) are tubular ducts through which air passes directly to and from tissue cells. The fine ultimate branches (tracheoles) are filled with water and supply individual cells. Lungs are internal cavities whose moist surfaces are richly supplied with blood vessels that transport O_2 and CO_2 to and from tissue cells. Animals like the earthworm, which respire all over their body surfaces, are not fully terrestrial; they are restricted to moist soil. Why?

Many measurements of respiratory rates in plants and animals have been made, and some of the results are surprising. The rate in a man at rest is comparable to that in a carrot. It is much lower than in bacteria, some of which have the highest rates known. Protists also tend to have high rates, but rates are generally low in the lower multicellular animals. Cold-blooded animals often have rates about as high as in man and other warm-blooded animals. Warm-bloodedness is not a matter of greater heat production but of more even production and better maintenance of heat level.

consumption, and heat production is calculated from this. The surface area likewise is not actually measured but is calculated on the basis of height, weight, age, and sex.

It is not surprising to find that respiratory rate increases greatly with muscular activity in animals. A butterfly's respiratory rate was found to be more than 150 times as high while flying as while resting. The rate in a running mouse was eight times that of a resting mouse. In man, other mammals, and many other animals, the rate tends to be higher in small than in large individuals. The rate is usually higher in young than in old animals and also higher in males than in females. It will be no news to parents that a small active boy is spending energy at a rate that is tops for mankind.

In plants most of the useful energy is expended in chemical synthesis and other metabolic activities. These activities tend to vary with the temperature and with light, water supply, and other changing environmental conditions. Green plants produce oxygen in the course of photosynthesis and also use it in respiration. Conversely, they use carbon dioxide in photosynthesis and produce it in respiration. These relationships involve an interesting balance. When photosynthesis is most active, more food is being made and more oxygen produced than are used in respiration. As photosynthesis decreases, which it does as the light grows dimmer, a *compensation point* is reached where the two processes balance. Below that point respiration takes the lead: more food and oxygen are consumed than are produced and more carbon dioxide is produced than is used.

RESPIRATORY SYSTEMS

As was seen to be true of digestion, respiration usually involves no special organs in protists or in aquatic plants and many small and relatively simple aquatic animals. The required oxygen simply diffuses from solution in the surrounding water into the organism through cell membranes. Carbon dioxide similarly diffuses out from the cell into the water of the environment. In land plants a similar process of diffusion occurs between cells and the surrounding air. Many of these forms also develop air-filled spaces among internal cells, and these spaces often have special openings—stomata—which open and close through the mechanism of guard cells. You have met this arrangement before (Chapter 3). More complex anatomical arrangements for O_2–CO_2 exchange do not occur among plants.

In higher animals, on the other hand, elaborate anatomical organs and systems have evolved (Fig. 10–10). Animals living in water or in a moist environment often receive much of their oxygen by simple diffusion through the cells of the skin. This is still true in the frog, for instance, even though it has gills as a tadpole and lungs as an adult. Worms may have such diffusion not only through the outer skin but also through cells lining the gut. Some worms have an enlarged, thin-walled hind-gut and rhythmically take in and expel water through the anus. This part of the gut is thus a true respiratory organ, in addition to serving its more usual function. In sea cucum-

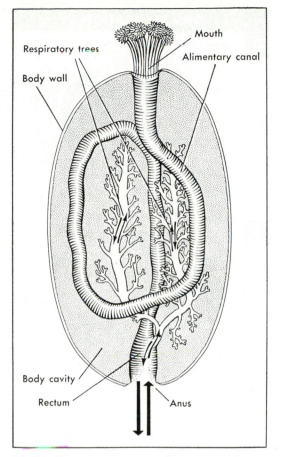

10–11 **Respiratory trees in the sea cucumber, a starfish relative.**

bers (Fig. 10–11) this type of respiration is further developed. Water, pumped in and out through the anus, circulates through much-branched *respiratory trees*, which extend throughout the body.

Most relatively large and complex animals require more oxygen than can be readily taken in by diffusion through the unmodified body surface. Moreover, in many of these forms, both aquatic and terrestrial, there is a protective skin through which diffusion is slight or absent, a necessity for regulation of their water content. It is among such animals that the most specialized respiratory organs have evolved. In aquatic forms these are usually *gills*, which have arisen independently in various groups and are highly diverse. Most of them have in common a filamentous or

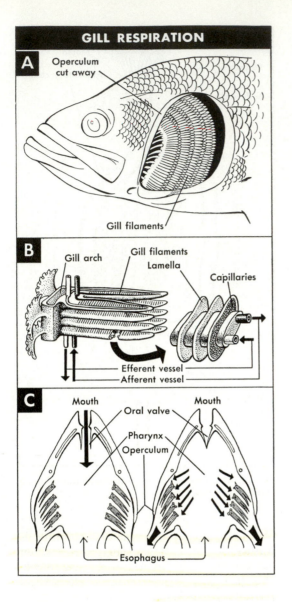

GILL RESPIRATION

A Operculum cut away

Gill filaments

B Gill arch — Gill filaments — Lamella — Capillaries

Efferent vessel
Afferent vessel

C Mouth — Oral valve — Pharynx — Operculum — Mouth

Esophagus

10–12 Gill respiration in fishes. The principal structural feature of gills is the way they are organized to offer as great a surface as possible to the water flowing over them; the greater their surface, the greater their capacity to pick up oxygen and give up carbon dioxide to the external water. *A.* The gill cover, or operculum, of the fish has been cut away, and four gills can be seen. Each gill carries a double row of gill filaments. *B.* Each gill filament is arranged as a series of lamellae. An afferent blood vessel carries blood to the gills, where it breaks up into a fine capillary bed at the surface of each lamella. Here gaseous exchange occurs. An efferent vessel carries the oxygenated blood away from the gills to the rest of the body. *C.* The water that carries oxygen into the gills enters through the mouth into the pharynx. As it is compressed by contraction of the pharynx, it is prevented from escaping through the mouth by the closure of oral valves. The water is thus forced through slits between the gills, finally leaving under the operculum.

insects. In them there is a system of tubes, the *tracheae* (Fig. 10–14), with paired openings, *spiracles*, through the body wall. The tracheae branch repeatedly, and their smallest terminal extensions deliver air to the cells throughout the body. In many insects rhythmic movements of the abdomen and synchronized opening and closing of the spiracles produce a definite circulation of air through the tracheal system.

In animals with gills or lungs these organs facilitate O_2–CO_2 exchange between the blood and the external water or air. It is the blood that exchanges O_2 and CO_2 directly with the cells in which respiration occurs. In the insects, however, the body fluids are not involved in respiration (with unimportant exceptions), and O_2–CO_2 exchange is directly between air in the tracheae and the cells of the body.

platelike structure, which crowds a large surface area into a small bulk, and some means of circulating water past or through the gills (Fig. 10–12). Gill-like structures have become adapted to O_2–CO_2 exchange with the air in some land-living forms, such as the spiders and scorpions. Other air-breathing forms have developed lunglike pouches, as in certain snails. True *lungs*, like man's (Fig. 10–13), occur in some fishes, most amphibians, and all reptiles, birds, and mammals. A different and unique system for air breathing has evolved in the

CHAPTER SUMMARY

Effects of large size on multicellular organisms.

Modes of food intake: in solid form by animals; by diffusion in protists and plants.

Digestion, as the converse of synthesis: extracellular and intracellular digestion; the neces-

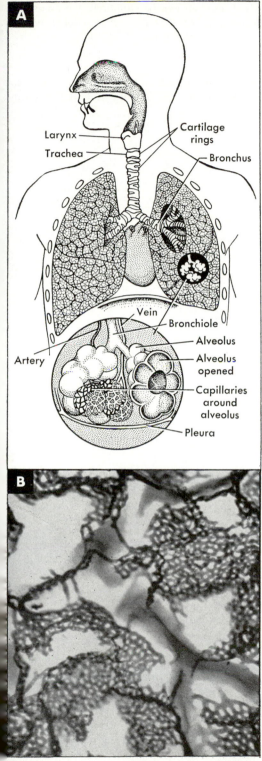

A

Larynx
Trachea
Cartilage rings
Bronchus

Vein
Bronchiole
Artery
Alveolus
Alveolus opened
Capillaries around alveolus
Pleura

B

Carl Strüwe

sity of intracellular digestion for the movement of materials across cell membranes; digestive enzymes; digestive systems.

Assimilation, the incorporation of digested food into the organic system.

10–13 Lung respiration in man. *A.* Air, drawn in through the mouth or nose, reaches the lungs through the trachea and its branches, the bronchi and bronchioles. The cartilaginous rings around the trachea give it sufficient rigidity to prevent its collapse and thus guarantee a continuously open air passage to the lungs. The bronchioles lead into small cavities, alveoli, which are richly supplied with blood capillaries (p. 315). Gaseous exchange occurs in the alveoli. The increased internal surface of the lung achieved by its organization into alveoli may be compared with the increased external surface of the gill achieved by its arrangement into filaments and lamellae (Fig. 10–12). What is the significance in both cases? All the alveoli in the lung are bound tightly together by a surrounding epithelium, the pleura. *B.* The capillaries around the alveoli (× 250).

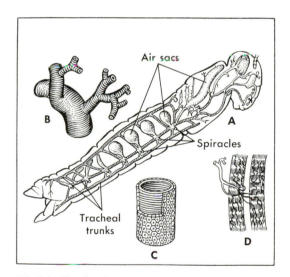

Air sacs
B
A
Spiracles
Tracheal trunks
C
D

10–14 Tracheal respiration in a grasshopper. *A.* The distribution of spiracles, main tracheal trunks, and air sacs on one side of the insect. The air sacs are reservoirs. *B.* A portion of a main tracheal trunk and its branches. *C.* A portion of a tracheal tube, showing its single layer of cells and the internal support of chitin (a protein substance) they secrete (cf. the cartilage rings around the human trachea, Fig. 10–13). *D.* Ultimate tracheal branches supplying muscles.

Nutrition and vitamins.

The meaning of the term "respiration": (1) gas exchange, as in breathing; (2) intracellular processes of oxidation that liberate energy.

Cellular respiration: aerobic and anaerobic respiration, the former limited by the rate at which oxygen can be supplied; the anaerobic respiration of muscle; oxygen debt; energy reserves in the form of phosphagens.

Respiratory rates.

Respiratory systems: their diversity of form related to the peculiarities of the environment.

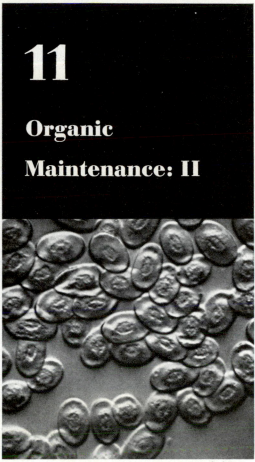

Internal transportation systems are indispensable for the maintenance of higher organisms. Outstanding among such transportation systems is the blood. This photomicrograph shows the red blood cells of a frog.

11

Organic

Maintenance: II

Transportation and Circulation

An organism may be as small as a bacterium or as large as a whale or a sequoia. Whatever its size and structure, substances have to move about in it. Each cell requires material from outside, and these materials must move to where they are needed within the cell. The cells produce wastes that must move out. In a multicellular organism substances formed in one cell are required by other cells. Products of digestion have to be distributed through the body. In aerobic respiration O_2 must be delivered and CO_2 removed. Transportation within organisms is as important for their maintenance as any other process we have discussed. The mechanisms of internal transportation—widely different in protists, plants, and higher animals—occupy us through most of this chapter.

TRANSPORTATION IN PROTISTS AND PLANTS

In protists and within single cells the movement of materials is largely by diffusion, atom by atom and molecule by molecule. In many of these organisms there is also some movement of the cytoplasm that helps to distribute materials around the cell. In the plants that have no transport vessels and that are called *nonvascular* [1] on this account, movement of materials is also almost entirely by diffusion and cytoplasmic motion. There is some specialization of roles among the cells of such plants—particularly for reproduction—but the majority of cells may still be relatively independent, making their foods with materials diffused directly from the environment. Such transportation of materials as must occur between one cell and another is across the cell walls and membranes or along cytoplasmic threads joining the cells, not in distinct transport vessels. With more differentiation of parts and processes, however, and with development of organisms more complex and

[1] Few biological generalizations are without exception. The so-called nonvascular plants are, as a group, relatively small and characteristically lack special transport vessels. A few seaweeds, giant kelps, have attained considerable size and have developed phloemlike tissue similar to the true phloem that forms the vessels of vascular plants (ferns and seed plants).

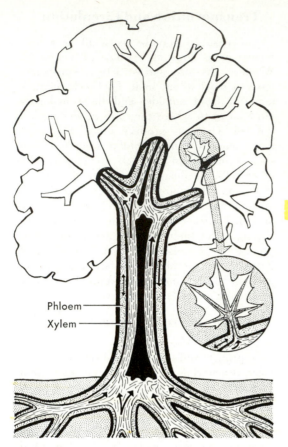

11–1 Direction of transportation in the xylem and phloem of a tree.

Phloem

Xylem

ganism. Frequently the spaces develop as long tubes, blood vessels in animals and various tubular structures or ducts in plants, in which there is a definite and directional flow of fluid.

The ducts of the vascular plants are of two sorts, xylem and phloem; they occur in bundles in roots, stems, and leaves. We have already seen examples of them (Chapter 3). Flow in the xylem is mainly upward from roots to stem and leaves, whereas products from leaves and other organs can move in both directions and commonly move downward through the phloem (Fig. 11–1). An enormous amount of water enters the roots of a plant, moves through the vessels, and eventually leaves the plant by evaporation. Of the 2000 tons of water absorbed by corn roots during one growing season on 1 acre in Illinois, all but 5 tons were lost by evaporation, or *transpiration*, as this type of water loss by plants is called.

We speak of transpired water as lost, and one may be inclined to think of it as wasted. But is it really lost? Where did it go? And where, before it entered the plants, did it come from? Study shows that the "loss" is essential to the life of the plant and that transpired water may be as necessary and useful as water retained in the plant's cells because transpiration is part of the mechanism that generates the energy for raising sap. Plants have no hearts or other organs for this purpose. Yet sap does rise in them, sometimes to an elevation of hundreds of feet in a tall tree, and, as in the manual operation of a water pump, much energy is expended in the process. Transpiration contributes some of it. More comes from the root pressure described earlier (Chapter 10).

The pressure developed by the gradient of osmotic pressures across the root and into the xylem serves to drive the water up the xylem for a considerable height. The power of this pressure is easily demonstrated in a laboratory experiment like that shown in Fig. 11–2. The effect of root pressure is also readily observed on the cut surface of a plant stem whose roots remain in the soil; sap continues to be forced up by the osmotic pumps in the roots.

It is doubtful, however, if this root pressure

more fully integrated, more effective methods of transportation are necessary and have evolved.[2]

The more advanced systems of transportation all involve the movement of fluids *outside* the living [3] cells but *inside* the organism. In plants this fluid is sap. In animals it is often blood, although other animal fluids are involved in internal transportation. Such fluids necessarily are contained in and move through otherwise open spaces within the or-

[2] They did not evolve because they are necessary; the point is that if they had not happened to evolve no such complex organisms would exist.

[3] It is debatable whether the phloem cells are dead. They are without nuclei but do contain cytoplasm, whose "life"—if it can be said to exist—may be rendered possible by the nuclei of the companion cells that are immediately adjacent to each phloem cell. (See Fig. 3–24.)

can account entirely for the movement of water up to the tops of very high trees. The theory currently favored by most plant physiologists is the *transpiration–cohesion–tension theory*. According to this theory, the transpiration of water is not a "loss" but a functionally valuable process. Visualize columns of water extending through the xylem from roots to leaves. Now water has a high degree of *cohesion*; adjacent molecules tend to stick together and strongly resist separation. In the leaves especially, water molecules diffuse outward from the cells and evaporate into the air. As they go, other molecules move in behind them and exert a pull that affects the whole column of water. This *transpiration pull* is a principal factor in the supply of sap to the upper parts of vascular plants (Fig. 11–3). The energy expended comes mainly from the heat necessary to produce evaporation. The heat comes directly or indirectly from the sun, and so in this process, as in photosynthesis, plants use solar energy. The total upward movement is effected by the combined power of a solar engine and an osmotic pump.[4]

Most of the liquid in plants moves in one direction only, from soil into roots, up through the vessels, and out into the air through the leaves and other surfaces. Solutions also frequently move in the opposite direction (in the phloem), but generally with less regularity. There is no directed *circular* movement of liquids, and to this extent plants do not have circulation or circulatory systems.

TRANSPORTATION IN ANIMALS

The simpler multicellular animals—sponges, corals, jellyfishes, and their relatives—resemble the nonvascular plants in the absence of vessels or other special anatomical arrangements for the movement of internal fluids (Fig. 11–4). They do, however, have passages or cavities in which there is move-

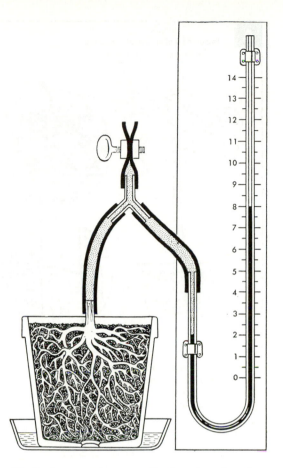

11–2 An osmotic pump. Sap forced up the xylem by root pressure exudes from the cut stem into the closed rubber tube, forcing in turn the mercury up the open arm of the manometer. This device permits the quantitative determination of root pressure.

ment of *external* fluids—the sea or fresh water of their environment.

There is one group of animals whose whole form dramatically emphasizes the profound importance of a vascular system to a multicellular organism. This is the group of flatworms. The point of interest about these worms is their flatness itself and its relation to the absence of a vascular system. Their innermost cells can obtain oxygen from outside and eliminate carbon dioxide fast enough only if the cells are kept close to the epithelium. This is accomplished by the flattening of the whole animal; as in a pancake, nothing in the inside is far from the outside. Again, a problem arises in nourishing the whole cell mass in

---

[4] The relative importance of the solar engine (transpiration pull) and osmotic pump (root pressure) is an unsettled and debated issue among plant physiologists. It should be added that although transpiration is generally necessary to land plants, its results can be catastrophic if the lost water cannot be replaced from the soil.

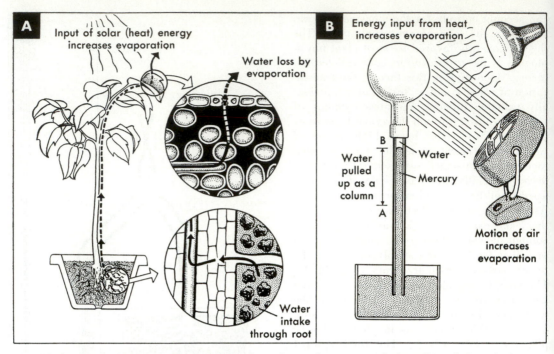

11-3 Transpiration pull. *A.* An unbroken column of water extends from the soil through the root hair and root cortex (cf. Fig. 10–4) into the xylem and thence to the leaves and leaf surfaces. The water column is maintained in an upward movement because its apex is drawn off as transpired water vapor. *B.* The efficiency of evaporation in raising a continuous water column can be demonstrated experimentally with a porous bulb at the head of a water and mercury column. As heat and wind from the lamp and fan increase evaporation from the bulb, the upward movement of the liquid column can be watched. (Adapted from J. Bonner and A. W. Galston, *Principles of Plant Physiology,* Freeman, San Francisco, 1952.)

the absence of a circulatory system to distribute food products from the gut. This is offset by a branching of the gut so elaborate that no internal cell is far from part of it.

Most other multicellular animals do have *internal* spaces or cavities between the outer body wall and the alimentary canal, and some of these spaces are devoted to transport systems. Internal-transport cavities usually take the form of definite vessels through which body fluids move. Local regions of these vessels are enlarged and have strongly muscular walls that contract rhythmically. The contractions force fluid through the vessels, and a system of valves keeps the motion always in the same direction. These organs may properly be called *hearts* in all animals, although they have evolved independently in different groups and differ markedly in number, arrangement, and structure. A common earthworm, for instance, has 10 hearts, arranged in

pairs around a forward part of the alimentary canal and pumping blood from an upper (*dorsal*) to a lower (*ventral*) longitudinal vessel. A squid has one heart that pumps blood to the body as a whole and two separate gill hearts that pump blood through the gills, from which it passes through vessels to the body heart. These are just two examples of the many arrangements of hearts in animals (Fig. 11–5).

In most animals one or more hearts constantly pump blood in the same direction, and there is no steady loss of body fluid analogous to transpiration in plants.[5] It follows that the blood must eventually return to the heart; it

[5] There is a continuous water loss from the lungs—the breath is moist—but this makes no important contribution to the movement of body fluids. Nor does the more occasional or sporadic loss of water in perspiration effect any significant internal movement of body water.

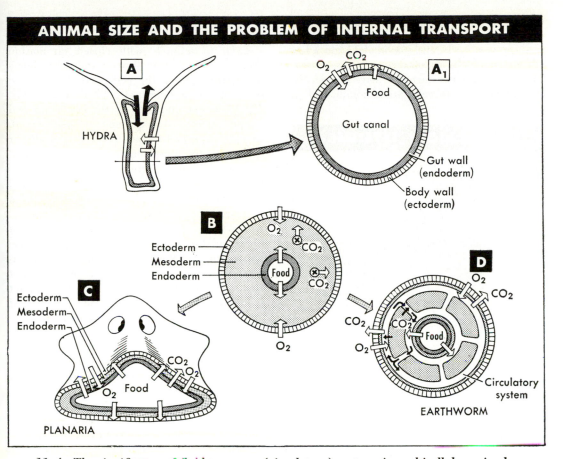

11–4 The significance of fluid-transport (circulatory) systems in multicellular animals.
A. In a coelenterate, such as *Hydra*, the body wall is essentially only two cells thick. Water circulates in and out of the body cavity (black arrows), and diffusion is sufficient for the transport of food substances, O_2, and CO_2 to and from all cells (white arrows). A_1. A cross section through a coelenterate. *B.* A cross section through a hypothetical animal with extensive mesoderm tissues between the gut and the body wall. The cells (marked X) in the middle of these tissues are too far removed from the gut and from the outside to rely on diffusion for transport of their food, O_2, and CO_2. *C.* In a flatworm, natural selection has produced a flattened body so that no cell is too far from the gut or the body wall to be served efficiently by diffusion. *D.* All animals higher than flatworms have evolved a more effective solution to the problem: a circulatory system that overcomes the limitations imposed by diffusion and permits animals to attain relatively enormous sizes. Food, O_2, and CO_2 are transported by the circulating blood over far greater distances than diffusion could cover.

must somehow make a circuit of the body. There is, in short, true *circulation*. In some animals, including some worms and many mollusks (snails, clams, and their relatives) and insects, vessels from the heart pour blood into irregular spaces or channels, *sinuses*,[6] among the tissues. Here the blood moves rather sluggishly, exchanging materials with the cells in the various tissues and eventually seeping back into collecting vessels that return it to the heart. This is an *open* circulatory system, because the blood is not retained in

[6] "Sinus" is another word applied in a somewhat confusing way to different things. The internal sinuses to which we are now referring have nothing to do with the nasal or frontal sinuses that may become inflamed and cause headaches and other discomforts. The latter sinuses are not, strictly speaking, internal. They normally open to the outside and do not contain blood.

distinct vessels through the whole circuit. In most worms, all vertebrates (fishes, amphibians, reptiles, birds, and mammals), and a few other animals the circulatory system is *closed*. The blood does not normally leave the vessels. Its exchanges with cells are accomplished in networks of very tiny, thin-walled vessels called *capillaries*.

Even in animals with closed circulation there are in addition sinuses and other internal spaces among the cells and tissues, and these spaces are normally filled with fluids. Thus two (or more) body fluids—blood and lymph—freely exchange many substances although separately contained in vessels or spaces. In the most complex types of animal transport systems, there are two separate systems of vessels, one transporting blood and the other transporting lymph. There may even be a separate heart for the lymphatic system, as in the frog. Man has no lymph heart, but his circulatory system is otherwise one of the most complex. We shall consider the mechanism of man's circulation in a moment, but first it is advisable to learn a little more about blood, the fluid that is the main means of transportation in the animal body.

Blood plasma

An examination of human blood under a microscope reveals cell-like objects floating in a colorless fluid. Since some of these objects lack nuclei, there is a question as to whether or not they should be called cells. Therefore, they are collectively referred to by the evasive term *formed elements*. The clear liquid is *plasma*, a solution, partly colloidal, of an extraordinarily large number of different substances. Some of the formed elements and the substances dissolved in plasma are related to the maintenance of the blood itself, and to its several activities. Others are merely in the process of being transported, transportation being one of the main functions of blood.

Blood donation and the use of serum and other blood derivatives in transfusions have become so commonplace that it is of interest to know the following relationships:

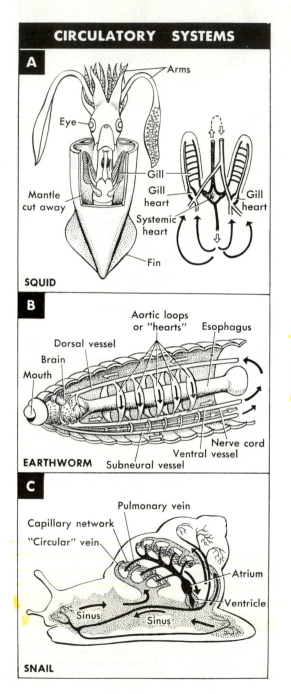

CIRCULATORY SYSTEMS

A

Arms
Eye
Gill
Gill heart
Mantle cut away
Gill heart
Systemic heart
Fin

SQUID

B

Aortic loops or "hearts"
Esophagus
Dorsal vessel
Brain
Mouth
Nerve cord
Ventral vessel
Subneural vessel

EARTHWORM

C

Pulmonary vein
Capillary network
"Circular" vein
Atrium
Ventricle
Sinus
Sinus

SNAIL

11–5 Circulatory systems. *A.* A squid: a closed circulatory system with three hearts. *B.* An earthworm: a closed circulatory system with 10 hearts. *C.* A snail: an open circulatory system with a single heart. Blood from the heart reaches the muscular foot via a distinct artery; in the foot the blood enters an open sinus, from which it then drains once again into a distinct vessel (the circular vein) in the surface of the lung. Capillaries connect this vessel to other veins that return the blood to the heart.

Whole blood	= formed elements plus plasma (including fibrinogen and other clotting factors)
Plasma	= whole blood minus formed elements
Serum	= plasma minus fibrinogen and other clotting factors
Defibrinated blood	= whole blood minus fibrinogen and other clotting factors, or = serum plus formed elements

Except for the respiratory gases, most of the blood's substances are transported in the plasma. The distinction between materials being transported and substances involved in properties and processes of the plasma itself is not an absolute one. The presence of dissolved materials being transported of course does affect the properties of the solution. Also, the same substances may play roles both in the plasma and in cells and tissues to which they are transported. Nevertheless, we can make a useful rough division. First, we note that virtually every substance used by cells is found dissolved in plasma at one time or another. With few exceptions, every substance secreted or discarded by cells is also present in plasma. We shall not try to list all these substances, but some of the more important ones are listed in Table 11–1. These substances are dissolved or suspended in water, which makes up about 90 per cent of the plasma.

Plasma (minus its proteins that are too large to pass through the capillary walls) is in contact with most of the cells inside the body. It is, in effect, the environment in which the internal cells live. It is an environment that provides all necessary materials—this is the function of transportation—and that also maintains the chemical and physical conditions most favorable for the cells. This maintenance depends not only on the inorganic salts (as noted in the table) but also on the blood sugar—indeed, on the whole complex of dissolved materials as well as the rate of flow and the temperature.

Plasma and whole blood are pumped under pressure through a closed system of vessels. Even a small opening in this system as a consequence of injury or disease would lead to the loss of most of the blood and rapid death if it were not for a defensive activity of plasma, *clotting*. The reactions involved in clotting are complex, but in summary the process is approximately as follows. When an injury occurs, the injured cells and certain of the formed elements in the blood (the platelets) release substances that react together to produce *thromboplastin*, which catalyzes the conversion of a plasma protein called *prothrombin* to *thrombin*. Thrombin in turn reacts with *fibrinogen*, another protein dissolved in the plasma, to produce *fibrin*. Fibrin comes out of solution as a tangled network of fibers, which shrink and form a hard clot that plugs the opening in the circulatory system. There are a few unfortunate people, "bleeders," whose blood does not clot readily and who are in danger of death from the slightest injury. Their condition, known as hemophilia, is hereditary and has become famous because of its prevalence among the male descendants of Queen Victoria. The scheme of blood clotting as outlined omits several of the clotting factors necessary for the formation of thromboplastin. When any one of these is missing owing to gene mutation, a hemophilia-like disease results. The hemophilia of Victoria's descendants was due to a lack of only one, antihemophilic globulin.

Clotting depends on the coagulating property of proteins. With most proteins, including some closely related to fibrinogen, this property is not useful and may be harmful in unusual circumstances. The clotting reaction itself has some serious dangers, too, and often leads to accidents with which the organism cannot cope. A clot may form within a blood vessel and thus block the blood supply of a vital organ; when it remains in place, it is called a *thrombus*. Or a clot fragment (*embolus*) may break loose into the blood stream and finally lodge in a vessel essential for life. Both conditions, known as *thrombosis* and *embolism*, respectively, are serious. Most of us know people who have suffered one of these accidents and have died or become invalids as a result.

Formed elements of the blood

The formed elements (Fig. 11–6) include red blood cells (*erythrocytes*), white blood cells (*leucocytes*), and platelets (*thrombocytes*). Of these, the red blood cells are the principal vehicles of transportation. The

TABLE 11–1

Some materials in the plasma of human blood

Materials being transported	From	To	Significance
Sugars (chiefly glucose)	Intestine, liver	Whole body	Food for cells, especially energy source. Also transported to liver for storage.
Lipids and related compounds	Intestine, storage deposits in body	Whole body	Food for cells, energy source. Includes cholesterol (essential to nerve cells) and other special compounds. Fats are also transported *to* storage deposits.
Amino acids	Intestine	Whole body	Food, precursors for synthesis of proteins.
Inorganic salts	Intestine, storage, many tissues	Whole body	Essential in building protoplasm, enzymes, and other compounds and hard tissues such as bones and teeth.
Hormones	Endocrine glands	Whole body and special "target" tissues	Regulate and coordinate organic activity (see Chapter 12).
Urea, other nitrogenous compounds, acids, and salts	Whole body	Kidneys	Waste products being removed.
Gases			
Oxygen	(Mainly in red cells rather than plasma)		
Nitrogen	Lungs	Lungs	Inert. Taken in and given out incidentally in gas exchange in lungs.
Carbon dioxide	Whole body	Lungs	Waste product being removed.

Materials active in plasma	From	Activity
Blood proteins		
Fibrinogen	Liver	Clotting of blood.
Albumin and others	Liver	Metabolism, viscosity, and osmotic pressure of blood; also, indirectly, CO_2 transportation.
Antibody globulins	Lymphoid tissues	Act against antigenic bacteria, toxins, foreign proteins.
Inorganic salts	Intestine, storage, other tissues	Maintain osmotic level, degree of acidity, internal environment favorable for cells. Calcium necessary for clotting of blood.

white blood cells, of which five different types are commonly distinguished, are active mainly in combating bacterial infections. The platelets break down in the vicinity of an injury and liberate substances involved in the early stages of the clotting reaction.

In mammals (man included) the mature red blood cells are curious biconcave discs with no nuclei; therefore they are not organized like ordinary cells.[7] For one thing, they cannot divide or reproduce. Young red cells, which do have nuclei and do reproduce, are not present in circulating blood. They form in connective tissue in the marrow of bones. As they lose their nuclei and mature, they pass into the blood stream, where they live and function for about 120 days (in man). Eventually they are devoured and digested by cells in the spleen and liver. The number of red cells involved in this ceaseless activity is very large. In an average adult approximately 27 million million—27,000,000,000,000—mature red cells are present in the blood at one time. Every second about 26 million new cells pass from the marrow into the blood stream, and

[7] The red cells of birds, reptiles, and fishes *do* have nuclei. See, for example, the frog red cells in the illustration on p. 303.

about 26 million old cells are devoured in the liver and spleen. If the replenishment is inadequate or if destruction occurs too rapidly, *anemia* ensues. (Anemia may also result from a deficiency of hemoglobin in the red cells even when their number is not greatly diminished.)

The red substance in a red blood cell is *hemoglobin*, which we have mentioned before (see p. 49). It is a conjugated protein, globin, with heme as a prosthetic group (p. 115). This compound has the important property of combining loosely and reversibly with oxygen. When it is in an oxygen-rich environment, as in the capillaries of the lungs, it combines with oxygen. In an oxygen-poor environment, as in capillaries among cells in need of oxygen, it gives up the oxygen again. This is how oxygen is transferred from the lungs to the cells. Hemoglobin is somewhat lighter and brighter in color when combined with oxygen. This is why the blood in arteries, on its way to the tissues, is bright red, and the blood in veins, returning from the tissues, is "blue" (actually a darker and more purplish red).

Hemoglobin has another property, one that happens to be dangerous in our civilization. Hemoglobin combines with carbon monoxide, CO, even more readily than with oxygen, and when it has done so, it is incapable of picking up oxygen. Carbon monoxide is produced in large quantities in gasoline motors and in fumes from gas ranges and heaters or charcoal grills. Carbon monoxide poisoning, quickly fatal, is a suffocation of the cells of the body because of loss of capacity of the red blood cells to carry oxygen.

Some carbon dioxide, CO_2, is transported in the plasma, as we have already noted, but most of it is carried in the red cells (erythrocytes). About a fifth of the total CO_2 combines chemically with amino acids in the hemoglobin, but the bulk is carried within the red cells as carbonic acid formed from CO_2 and water by a specific enzyme in the erythrocytes. The resulting acidity in the cell, which would rapidly become harmful if not counteracted, is neutralized by the buffering action of hemoglobin (Chapter 2). On a lesser scale, the same sort of reaction occurs with some of the proteins in the plasma.

The white blood cells are nucleated cells

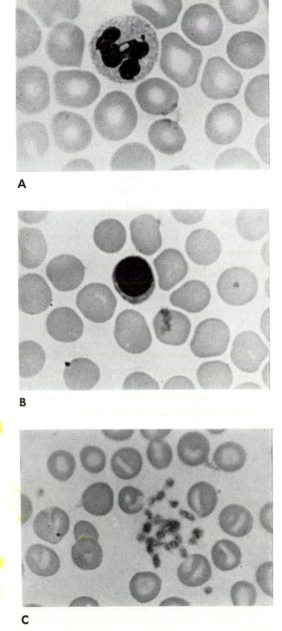

A

B

C

11–6 **Photomicrographs of a stained film of human blood showing the major formed elements.** *A.* (×1600) The large nucleated cell is a neutrophilic polymorphonuclear leukocyte. The little "drumstick" extending upward from the right side of the nucleus is the sex chromatin that indicates that the cell comes from a female. The non-nucleated cells are erythrocytes or red cells. *B.* (×1600) The nucleated cell is a lymphocyte. *C.* (×1700) A clump of platelets surrounded by red cells.

that live in the blood (and tissues) almost as if they were separate, parasitic protists. In fact, some of them look and act much like an ameba, which *is* an independent protist and sometimes a very harmful parasite. The white cells, however, are not harmful; on the contrary, they are an indispensable part of the body's defenses. Their main role is in combating bacterial infections. When an infection occurs, white cells move out from the blood vessels (from the capillaries) in great numbers. They surround and digest bacteria, and some of them are killed by bacterial poisons. Soon the area is strewn with living and dead bacteria and white blood cells, dead tissue cells, digestive enzymes, cell fluids, and cell fragments in all stages of disintegration. The resulting semifluid mass is *pus*. An *abscess* is a deeply embedded, walled-off accumulation of pus. Marked increase of white cells in the blood is nearly always a sign of infection somewhere in the body, and a count of the cells is a useful diagnostic technique much employed by the physician.

Occasionally, for unknown reasons, the production of white cells becomes uncontrolled, and their numbers tremendously increase, even though there is no infection. This abnormal condition, known as *leukemia*, is usually fatal.

Although we have spoken only of human blood, its properties and activities are generally similar to those of the blood of other animals, despite diversity in detail. The most striking peculiarity of any one group distinguishes insects, whose blood does not transport oxygen and carbon dioxide.[8] The pigment involved in oxygen transportation is usually hemoglobin, or rather *a* hemoglobin, since this compound is not exactly the same in different kinds of animals. There are several other types of pigments, however, such as hemocyanin, which occur in a number of snails and crustaceans. Hemocyanin contains

[8] It is hard to find any generalization without exceptions, and a sweeping statement like this sometimes ignores facts that simply are not essential for present purposes. Some oxygen and carbon dioxide do go into solution in the blood of insects, but since there is no oxygen-carrying pigment, the amounts are practically negligible. Even this last statement has an exception, for hemoglobin is present in the blood in a few larvae of one family of insects.

copper rather than iron (as in hemoglobin), and it is blue rather than red. In these forms and some others, including a few with hemoglobin, the pigment is not in cells or formed elements but is dissolved in the plasma. All true blood in animals contains some cells that correspond more or less to man's white cells.

The mechanism of circulation

We shall use man to illustrate the circulatory mechanism. The arrangement is practically the same in all mammals, and the principles involved are similar in most animals having a definite circulation, even though the differences in details are legion. Examples of some differences have already been mentioned (p. 306).

The center of the mechanism is the heart. Maintenance of a constant flow of blood through the whole system requires the application of pumping pressure at some point, and this is the function of the heart. In man (and other mammals) the heart is really two separate organs (Fig. 11–7), existing side by side, beating in unison, and yet quite distinct. The reason for this oddity, as might be anticipated, is historical. Our two hearts evolved by the gradual separation of a single heart. The heart is still single in fishes. Our right heart, or the right half of our double heart, pumps blood to the lungs, and the left heart pumps it to the body tissues as a whole.

Blood coming to either side of the heart first enters a smaller and weaker upper chamber, the right or left *atrium* (Fig. 11–8). When the heart relaxes—the part of the beat called *diastole*—blood surges through the atrium into the larger and stronger chamber below it, the *ventricle*. This blood surge into the ventricle is initiated by the ventricular relaxation (diastole), but it is completed by the contraction of the atrium (atrial *systole*). The filled ventricle itself contracts (ventricular systole) immediately thereafter, pumping blood upward and out of the heart through the main artery, the *aorta*. Flaplike valves between atrium and ventricle automatically close as the ventricular systole begins and prevent the blood from moving back into the atrium. If you listen with a doctor's stethoscope, at this point you will hear a dull "boom," which is partly the sound of the

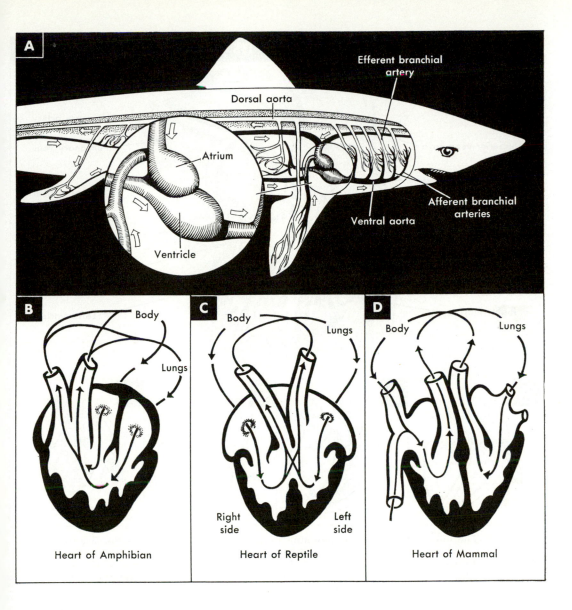

11–7 Vertebrate hearts. *A.* In the shark and other fishes, the heart consists of a single atrium and a single ventricle. It drives blood forward through a ventral aorta to the gills. The blood enters the gills through afferent branchial arteries (cf. Fig. 10–12) and leaves them via efferent branchial arteries. It then passes to the rest of the body through a dorsal aorta. Thus in fishes there is only a single stream of blood (deoxygenated) passing through the heart. In terrestrial vertebrates, however, there are two streams of blood surging through the heart—deoxygenated blood and blood freshly oxygenated in the lungs, which returns to the heart to be pumped to the rest of the body. *B.* In the frog and other amphibians, the freshly oxygenated blood enters the heart through a distinct atrium (the left) but becomes partly mixed with deoxygenated blood when it passes on into the single ventricle. This system is clearly inefficient because some of the blood pumped to the body through the dorsal aorta is already exhausted of its oxygen. *C.* In reptiles the inefficiency is largely overcome by an incomplete partition in the ventricle. *D.* In mammals the partition is complete; two distinct ventricles are present. Freshly oxygenated blood returning from the lungs via the left atrium is pumped by the left ventricle to the rest of the body without any contamination with deoxygenated blood.

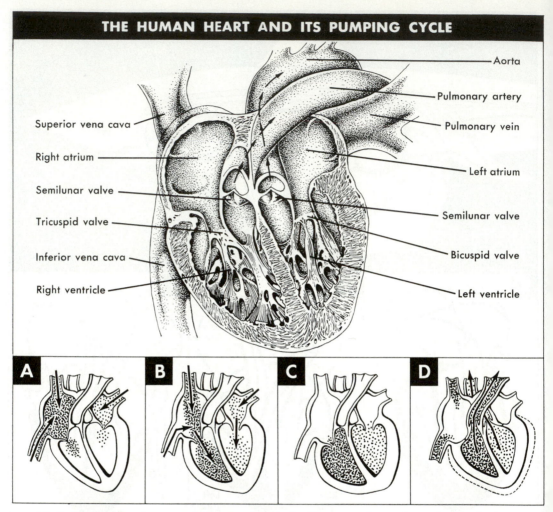

Aorta

Pulmonary artery

Pulmonary vein

Superior vena cava

Right atrium

Left atrium

Semilunar valve

Tricuspid valve

Semilunar valve

Inferior vena cava

Bicuspid valve

Right ventricle

Left ventricle

A   B   C   D

11–8 The human heart and its pumping cycle. The upper figure shows the detail of the heart's four chambers and its valves. Note that the wall is thicker in the left ventricle than elsewhere. What is the significance of this? *A, B, C,* and *D* are successive stages in the pumping cycle. *A.* The atria fill with blood as their walls relax (atrial diastole). *B.* The relaxation (diastole) of the ventricles causes blood to flow into them from the atria. *C.* Contraction (systole) of the atria completes the filling of the ventricles. *D.* Contraction of the ventricles (ventricular systole) drives blood from the ventricles into the aorta and pulmonary artery. The blood is prevented from returning to the atria by the bicuspid and tricuspid valves. During diastole it is prevented from returning to the ventricles from the arteries by the semilunar valves.

valves closing and partly the sound of the ventricle's muscles contracting. When the ventricles relax in their next diastole, three smaller, crescent-shaped valves in each of the blood vessels close and prevent flow of the expelled blood back into the heart. The sound of these valves as they close is also distinctly audible with a stethoscope—a shorter, higher, almost clicking sound. If the valves do not close perfectly, some blood does leak back and the sound becomes softer, somewhat hissing. This sound is a heart *murmur.* It is a sign that the valves are defective.

The blood that leaves the left ventricle goes into a very large artery, the aorta. From this, repeated branching distributes the blood throughout the body in smaller and smaller arteries. (Vessels carrying blood from the heart are *arteries;* those carrying it to the heart are *veins.*) Eventually the blood goes through a capillary network, where the vessels are so extremely small that a red blood cell may be bent double in being forced through. Every organ in the body has capillary networks, and this is where the blood makes its deliveries and pickups (Fig. 11–9). From the capillary network the blood moves into small veins, thence into progressively larger ones, and finally into two *venae cavae,*[9] which deliver it to the right atrium.

At this point the blood has not yet finished a complete circuit, because we started with it from the left heart and now it is back to the right, not the left, heart. From the right atrium it goes into the right ventricle and then is pumped to the lungs through the pulmonary artery and its two branches. In the capillary network of the lungs, it drops carbon dioxide and picks up oxygen. From there it is collected into the two pulmonary veins, which join and deliver the blood into the left atrium. Then it goes into the left ventricle, and so the circuit is completed and the blood is ready to go around again. A complete round trip takes about 25 seconds.

There is just one complication in the arrangement of the blood circulation that we must mention briefly because it is an unusually interesting example of effective correlation of mechanism and process. When digestion occurs, large amounts of sugar are absorbed by the capillaries of the intestines. If this sugar went into the general circulation, it would cause a sharp rise in blood sugar. On the other hand, when digestion was not occurring, the sugar level would drop. Body cells, especially nerve cells, are very sensitive to the sugar concentration in the blood. Sustained high sugar content produces the symptoms of diabetes, and low sugar content produces convulsions and unconsciousness. The sugar-rich blood from the intestinal capillaries does not go into the general circulation but into the *portal vein.* This vein carries it to another capillary network in the liver, where sugar is removed and stored. When sugar is not being absorbed from the intestines, the liver feeds a steady supply of sugar from storage through its capillaries into the general circulation.

Pulse and blood pressure

Every time the left ventricle contracts, it sends a column of blood surging through the arteries under strong pressure. As the ventricle relaxes, the pressure drops. It is these alternations of pressure that we feel as the pulse. There is, of course, a pulse in all arteries and not only in the wrist, where it is usually taken for convenience. Feeling the pulse is simply a way of checking how fast and regularly the heart is working. Even when the ventricle is relaxing, in diastole, there is some pressure in the arteries because they are elastic. When the systolic surge reaches them they expand. Then they contract again, and the contraction keeps the blood under some, but less, pressure when the systolic wave has passed.

The familiar procedure used to measure blood pressure involves compression of the arteries in the arm until the top of the systolic wave can no longer force blood through them. The pressure is then relaxed until even the low pressure of diastole forces blood through. Thus the highest (systolic) and lowest (diastolic) pressures of the whole pulse sequence are determined. As with many physical variables, each individual has his own average blood pressure, which is healthy and normal for him even though it may not be average for all people of his age. The blood pressure also can and indeed should vary considerably under the influence of emotion or physical factors. Extremes of high or low pressure may, of course, be danger signals.

There is friction in the blood vessels, as there is in any system of tubes through which fluid is flowing. This friction reduces the pressure steadily as the blood moves progressively farther from the heart. In the capillaries friction is greatest. Here the pressure is radically reduced, and the pulse surge disappears. Blood returning in the veins has no pulse and is under little or no pressure from the heart. Blood flow in the veins depends in good part on breathing and other muscular motion. The motion alternately compresses the veins and lets them expand, and blood is kept moving

[9] "Vena cava," singular; "venae cavae," plural. The superior vena cava delivers blood from the upper part of the body, the inferior vena cava from the lower part.

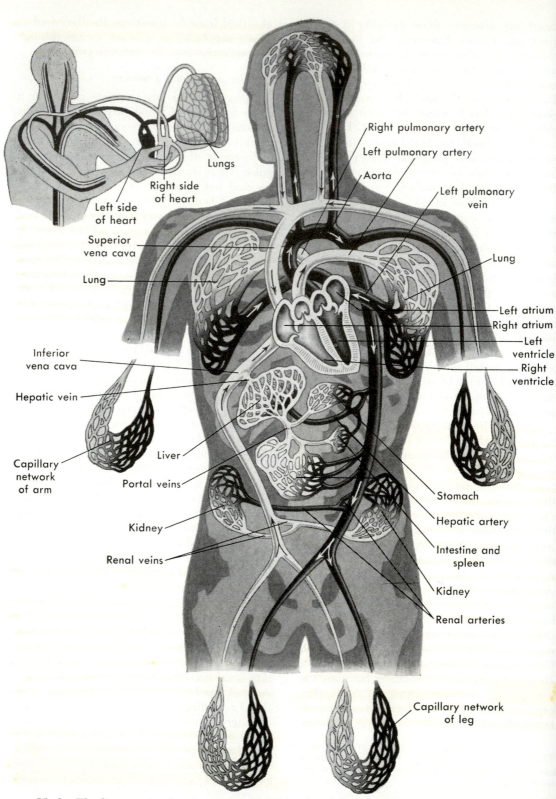

Right pulmonary artery

Left pulmonary artery

Aorta

Left pulmonary vein

Lungs

Right side of heart

Left side of heart

Superior vena cava

Lung

Lung

Left atrium

Right atrium

Left ventricle

Right ventricle

Inferior vena cava

Hepatic vein

Capillary network of arm

Liver

Portal veins

Stomach

Hepatic artery

Intestine and spleen

Kidney

Kidney

Renal veins

Renal arteries

Capillary network of leg

11–9   The human circulatory system. (Top left adapted from G. Hardin, *Biology: Its Human Implications*, 2nd ed., Freeman, San Francisco, 1952.)

toward the heart by a series of valves permitting flow in that direction only. Thus your feet are likely to swell if you sit or stand for a long time without moving. The swelling is caused by accumulation of blood in the veins in the absence of muscular movement around them.

The lymphatic system

We have noted that there is fluid in various internal spaces of the body and, indeed, around almost all the cells. Much of this fluid, with the materials dissolved in it, has diffused from the blood in the capillaries and may diffuse back into it. It is, in fact, blood plasma minus the proteins held back by the capillary walls. Other materials have diffused from the cells bathed by the fluid, and these materials, too, may diffuse into the blood in the capillaries. The fluid is thus intimately involved in the delivery and pickup activities of the transport system of the blood. However, not all the fluid finds its way back into the blood directly.

In most parts of the body there is another set of capillaries quite separate from those through which the blood flows. These are the *lymphatic capillaries*. Much of the fluid outside the blood vessels and tissue cells finds its way by diffusion into the lymphatic capillaries (Fig. 11–10). From them lymph passes into successively larger vessels, much as blood collects in veins after passing through the blood capillaries. The lymphatic system of vessels, however, runs in one direction only. Its capillaries end blindly and pick up materials only by diffusion through their walls. There is no delivery to them corresponding with the arterial part of the blood circulation.

The lymph vessels have internal valves, and flow through them depends on pressure and motion of surrounding tissues, as in veins, with no pressure from the heart. Eventually all the lymph in vessels reaches the upper part of the trunk and is emptied into veins near the shoulders (most of it on the left side). Thus the lymph, too, does finally enter the blood.

As would be expected from its relationship to the blood, the composition of lymph is similar to that of plasma, but it contains much less of the blood proteins. Since it flows only from and not to the body tissues, it also contains less dissolved food materials in transit, but it does transport waste products from the tissues. There is one important exception to the statement that lymph is not involved in food transportation: it picks up small globules of fat from the intestines and delivers them to the blood (Figs. 11–10B and 11–11). Undissolved fats seem to enter the lymphatic capillaries more readily than those of the blood. Red blood cells cannot normally leave the blood vessels, and they occur in lymph only as a result of injury. Some kinds of white blood cells can move out of or into vessels by an amebalike movement through the walls, and these cells are abundant in lymph. Many are also formed in the lymphatic system and poured into the blood.

At points along the lymph vessels are lumpy enlargements, the *lymph nodes* (see Fig. 11–10). *Lymphocytes,* one form of leukocyte (see Fig. 11–6B), originate in lymph nodes and other lymphoid tissues. Nodes are also filters which remove and destroy bacteria and solid particles. Lymph nodes near the lungs of city dwellers are often black with soot and dust particles. During infection, the nodes may become swollen and sore.

The lymphatic system, then, returns body fluid to the blood, assists with the intake of fats, filters out foreign particles, and helps to combat infection. It would seem that all these operations could be performed just as well by the blood circulatory system, and the need for a second system of vessels is not obvious. The reason is certainly historical—evolutionary— but is still obscure, although not beyond conjecture. Lymph in the sense of body fluid outside the cells and blood vessels occurs in all multicellular animals. A lymphatic vascular system is characteristic of the vertebrates. It seems odd that lymph nodes, which in man are involved in the more special activities of the lymphatic system, occur only in mammals and some birds. Many of the lower vertebrates, without lymph nodes, have lymph hearts that actively pump the lymph along, but mammals lack them.

Secretion and Excretion

We are now approaching the end of our discussion of the processes that are metabolic,

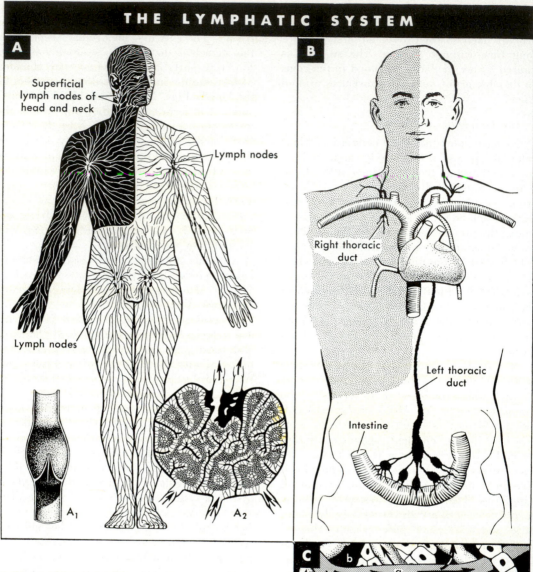

A. Superficial lymph nodes of head and neck

Lymph nodes

Lymph nodes

A_1

A_2

B. Right thoracic duct

Left thoracic duct

Intestine

11–10   The lymphatic system. *A.* The distribution of superficial lymphatic vessels and lymph nodes in man. *A₁.* A valve in a lymph vessel. *A₂.* A lymph node. *B.* The left and right thoracic ducts emptying into the veins (subclavian) that drain the arms. The left thoracic duct is the larger one, draining the whole of the unshaded area in *A* and *B*. It carries fats absorbed by the lacteals in intestinal villi (Fig. 11–11). *C.* Detail of blood and lymph capillaries. Plasma flows from the blood capillaries (*a*) into intercellular spaces (*b*) and then into the lymph capillaries (*c*), which end blindly.

C

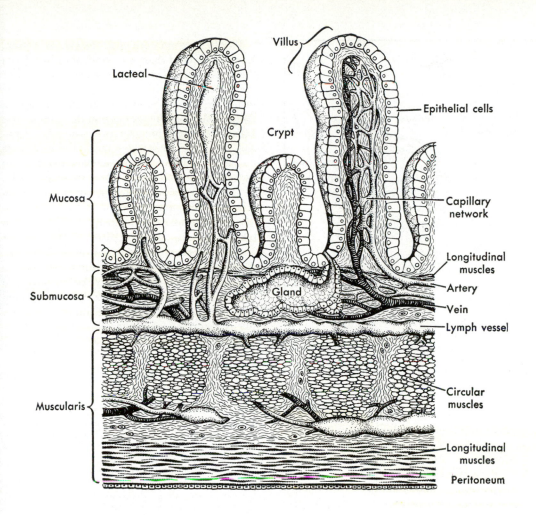

11–11 Lacteals and capillaries in the intestinal villi. The lining of the intestine in verte-brates, including man, is molded into many villi, which serve to increase the effective absorptive area in the gut (cf. the earthworm, Fig. 10–6). Each villus contains blind-ending lymphatic vessels, the lacteals, into which fats enter. Each villus also is richly supplied with capillaries, which absorb other foodstuffs. In the diagram each villus is shown, for simplicity only, with either a lacteal or a capillary network.

that is, concerned primarily with the feeding and maintenance of the organism. But before we turn to other activities of the organism, we must consider a final aspect of metabolism, the disposition of its end products.

Reactions in cells and tissues may yield products that are not needed or useful at the sites where they are formed. Some of these products are necessary elsewhere in the organism, and some are not. Thus CO_2 produced by respiration in plant cells is useless, even harmful, as far as respiration is concerned, but it is necessary for photosynthesis at other loci in the plant. This is not the usual situation, how-ever, and to distinguish the two phenomena, we call the production of substances useful elsewhere in the organism *secretion,* and the elimination of those of no further use *excretion.*

SECRETION

Secretion occurs widely, indeed universally, in multicellular organisms. Specialized secre-tions include those responsible for odors and tastes and the poisons formed in mushrooms and many flowering plants. Numerous other organisms, from protists to snakes, also se-crete poisons that are useful for defense or

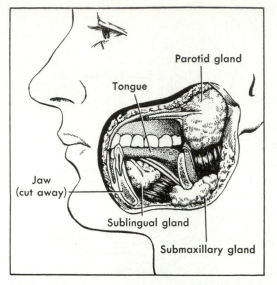

11–12 The human salivary glands. The three pairs of glands pour saliva containing digestive enzymes into the mouth cavity. As the names imply, the sublingual gland lies under the tongue, the submaxillary under the jaw, and the parotid near the ear.

for the capture of food. All kinds of synthesized molecules in both plants and animals are called secretions if they are used by the organisms elsewhere than in the cells where they are elaborated.

Groups of cells active mainly in producing secretions are called *glands* (see Fig. 3–22). They are more common and varied in animals than in plants. In some flowers the nectaries secrete a sweet fluid. In man the salivary glands (Fig. 11–12) and others secrete digestive juices. The lachrymal glands secrete a fluid that moistens the eyes and has a peculiar tendency to overflow in the act of weeping. The secretions of the endocrine glands are the hormones. These have unique importance, and we shall discuss them separately in the next chapter.

EXCRETION IN PLANTS AND LOWER ANIMALS

The primary end products of metabolism are not secretions but true waste products, no longer of any use to the organism. They are usually poisonous if they accumulate in any considerable quantities; thus they must be eliminated.

Plants have no special organs of excretion. End products of their metabolism diffuse from the cells into the surrounding water or air, or they accumulate in vacuoles or in harmless, insoluble [10] form elsewhere in the organism. In protists and the simpler multicellular animals also, excretion occurs mainly by diffusion from the individual cells. Some protists do have an organ (or organelle) involved in excretion, the contractile vacuole (Fig. 11–13). The principal activity of this organ is to control the water content of the cell, counteracting osmosis, and most of the excretion takes place by diffusion through the cell or body membrane. There is, however, some evidence that waste products dissolved in the water are also expelled by the contractile vacuole. This is the simplest level of a phenomenon that is widespread among animals: the excretory organs generally help to regulate water content in addition to eliminating waste

[10] Why is an insoluble substance likely to be harmless to the life of the cell?

11–13 The contractile vacuoles in *Paramecium*. Each vacuole goes through the cycle illustrated. *A.* The central vacuole is full, and the radial collecting vessels that penetrate the cytoplasm are empty. *B.* With systole (contraction) of the central vacuole, water is ejected through the pore. The radiating canals are filling (diastole). *C.* The central vacuole's systole and the radiating canals' diastole are complete. *D.* Systole of the radiating canals fills the central vacuole.

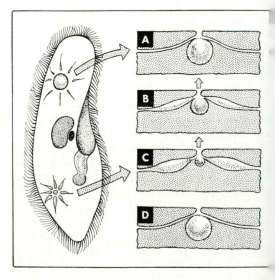

products. Such organs are, however, absent in most protists.

Above the protists, three of the major groups of animals have no special organs for water regulation or for excretion: Porifera (sponges), Coelenterata (corals, jellyfishes, etc.), and Echinodermata (starfishes, sea urchins, etc.). It is significant that all these animals are entirely aquatic and that the great majority are marine. They include one widespread family of fresh-water sponges and a few fresh-water coelenterates such as *Hydra*, often used as a laboratory animal, but in both groups marine forms greatly predominate. All echinoderms are marine. What bearing do you think their marine environment has on their lack of water-regulating and excretory organs? The osmotic pressure (Δ) of the sea water is so nearly the same as that of their body fluids that there is no serious tendency for excess water to move into the organisms by osmosis. Moreover, since they are completely bathed in water, they can simply excrete wastes from all their surfaces.

WATER PROBLEMS

The unusual fact that animals living in the sea may have great difficulty in protecting their cells from dehydration has had profound effects on the history of life. Simpler sea animals and most seaweeds have cells and internal fluids with an osmotic pressure similar to that of sea water. They all live in a good water equilibrium. If their osmotic pressure falls, they lose water by osmosis to the sea water, and the osmotic pressure comes right back up to normal. If their osmotic pressure rises, they gain water by osmosis from the sea water, and the osmotic pressure drops to normal by dilution. But in many higher animals, including most fishes, the osmotic pressure of the body fluids is well below that of sea water. Wherever there is a semipermeable membrane in contact with the sea water, in the gills, for instance, these fishes therefore tend to lose water continuously by osmosis. If there were no compensatory mechanisms, their cells would quickly lose so much water they would die of extreme dehydration although completely surrounded by water.

Obviously there are compensatory mechanisms, because fishes do live quite successfully

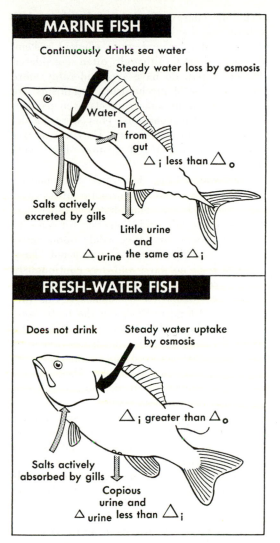

MARINE FISH

Continuously drinks sea water

Steady water loss by osmosis

Water in from gut

Δ_i less than Δ_o

Salts actively excreted by gills

Little urine and Δ urine the same as Δ_i

FRESH-WATER FISH

Does not drink

Steady water uptake by osmosis

Δ_i greater than Δ_o

Salts actively absorbed by gills

Copious urine and Δ urine less than Δ_i

11–14 The maintenance of water balance in fishes.

in the sea (Fig. 11–14). In most marine fishes these are the essentials: they drink sea water continually and in large quantities in replacement of osmotic water loss, and they secrete very little urine, avoiding loss of water in that way. But this results in an enormous intake of salt, which would upset their metabolism (based on fluids with low osmotic pressures) if it were not quickly gotten rid of. It is rapidly and continuously excreted from the gills. This excretion goes *against* the direction of osmotic pressure, and it therefore requires work to be done in the cells. Much of the metabolic energy of marine fishes is used in getting rid of salt. Almost any engineer could design a simpler and more efficient system. In fact, sharks, although often considered "lower" types, do have a considerably more efficient osmotic mechanism. Organisms are not perfect or in some respects even particularly efficient mechanisms. They were not designed in a purposeful way. They are what a long, blind, unplanned history has made them.

Fresh-water fishes have just the opposite problem. Their internal osmotic pressure is considerably higher than that of fresh water. The osmotic tendency is therefore for water to flow into them, and this would soon cause death by swelling if not counteracted. They drink little or no water and they continuously secrete great quantities of urine, getting rid of extra water acquired by osmosis. But this alone would tend to flush out the body salts, which would quickly kill the fishes. Before the urine leaves the body the fishes reabsorb much of their dissolved salts, and they also absorb salt through the gills. This, too, is work done against osmosis, and it takes a great deal of metabolic energy.

In most environments there are special difficulties in maintaining enough but not too much water in the cells. In the myriads of organisms there are many, often quite extraordinary, mechanisms that accomplish this in diverse ways and thus make life possible.

EXCRETION IN HIGHER ANIMALS

What we have already learned of metabolism should suggest to us the major substances that have to be excreted by animals. The end product of respiration is CO_2. This usually leaves the body through the same organs that acquire O_2, that is, through the respiratory organs—gills or lungs in most animals. Respiration and numerous other metabolic processes also produce water,[11] in addition to the water that is taken in as such. Water formed by metabolism is as useful as that drunk and is commonly retained for longer or shorter periods.

[11] Recall that when oxygen acts as the final hydrogen acceptor in oxidative phosphorylation, the product of the reduction is water (Chapter 4).

All animals do lose some water. In some situations this is an unavoidable loss of a needed substance rather than excretion of an unneeded end product; but water is often in excess and actively excreted. Water is a necessary part of urinary excretion, as we shall see in a moment. Water may also be lost in larger or smaller amounts from any surface of the organism. In man large quantities are lost from the lungs. This is not active excretion, but evaporation as an incidental result of the lung mechanism. We also lose much water through our sweat glands. Even when we are not actively sweating (when no visible drops of water appear on our skins), we lose a pint or so of water per day in this way. Several quarts may be lost on a hot day. This water is secreted by glands, but its significance is in heat regulation, not excretion. The loss of heat in the evaporation of sweat serves as a cooling mechanism.

Regulation of salt concentration and the production of salts by the metabolic breakdown of some compounds require that there be some mechanism for excretion of inorganic salts. In fishes that drink sea water and thus acquire excess salts, the gills are excretory organs for salt as well as for CO_2 (see Fig. 11–14). Some salts are always excreted in fishes' urine, and this is usually the most important means of salt excretion, as it is in man. Because sweat is salty, it is widely believed that perspiration is an important route of salt excretion. Such is not the case. Loss of salt in sweat is quantitatively insignificant in comparison with loss of salt in urine and usually plays no essential part in body salt regulation. Exceptions may, however, occur in situations in which men sweat copiously for prolonged periods.

Contrary to another common impression, the intestine is primarily an organ of absorption, not of excretion. The feces consist almost entirely of undigested food (or indigestible matter taken in with the food) and masses of bacteria from the intestines—materials that have not really been absorbed into the body and therefore cannot have been excreted. There is, however, some important excretion into the intestines. Dark brown bile pigments are excreted by the liver and flow into the intestines. These pigments are mainly end

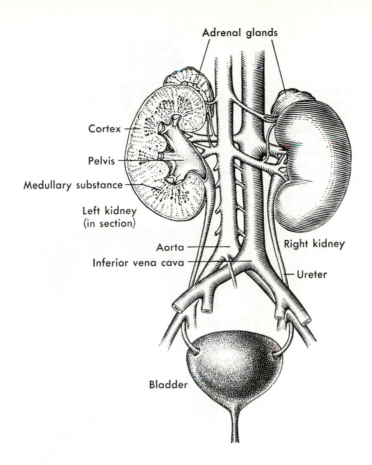

11–15 The urinary system in man, viewed from behind.

Labels: Adrenal glands · Cortex · Pelvis · Medullary substance · Left kidney (in section) · Aorta · Inferior vena cava · Right kidney · Ureter · Bladder

products of the destruction of hemoglobin in the liver and spleen.

All these modes and avenues of excretion in animals still fail to account for a large, important, and often toxic class of end products—those containing nitrogen. They are the end products of protein metabolism; you recall that proteins are the major category of body materials that contain nitrogen. Among the common end products are urea, $CO(NH_2)_2$; uric acid $(C_5H_4N_4O_3)$; and ammonia (NH_3) (see Chapter 4). All three occur in human urine, largely as the result of the breakdown of different proteins.[12] Urea is,

[12] Here is a curious fact. In almost all mammals uric acid is converted into a different compound before it is excreted. The exceptions are men, apes, and Dalmatian dogs. Except for that one breed, dogs follow the general rule and differ from man in the composition of their urine.

however, about 20 times as abundant as the other two put together.

Animals differ greatly in their most abundant nitrogenous end products, and some interesting relationships are involved. Almost all aquatic animals, both invertebrates and vertebrates, excrete more ammonia than anything else. Ammonia is highly soluble, but it is also extremely poisonous. Its excretion involves solution in large amounts of water and copious, continuous or frequent flow of urine. Land animals excrete predominantly urea or uric acid. Almost all animals that lay eggs outside of water excrete more uric acid: land snails, insects, reptiles, and birds. Land animals that lay eggs in water or that do not lay eggs at all excrete more urea: adult amphibians and mammals. Uric acid is almost insoluble and leaves the body in solid crystals. It may also be safely stored for long periods

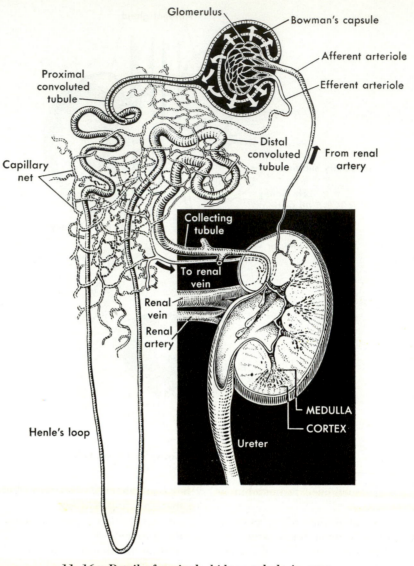

11–16   Detail of a single kidney tubule in man.

because its insolubility prevents absorption. Urea is highly soluble and must be excreted in solution, but it is not seriously toxic except in unusually high concentrations. Can you think of any relationships between the ways of life of these various animals and their nitrogen excretion?

The kidney

All the animals we have just mentioned have special organs that excrete nitrogenous wastes. These organs also largely affect the amount of water that leaves the body. In vertebrates the organ involved is the kidney, and we shall study the human kidney as an example (Fig. 11–15).

The work of the kidney is done in minute tubules, over a million of them in each kidney. At one end of the tubule is a filter (called *Bowman's capsule*) closely applied to a ball of blood capillaries (the *glomerulus*) (Fig. 11–16). Here a filtrate of plasma passes into the tubule. Its composition is essentially unchanged at this point, but the formed elements of the blood and colloidal proteins are filtered out and do not enter the tubule. As the plasma filtrate passes down the tubule, which is also surrounded by capillaries, most of the water is reabsorbed and returned to the blood.[13] Other substances such as glucose are also returned. Oddly enough, so is about half of the urea. Some additional substances also here move from the blood into the tubule. At the end of the tubule the fluid, which must now be called urine and is greatly altered in composition from the plasma, flows into branched collecting tubes and finally through the ureter into the bladder. The bladder is only a storage reservoir and does not further alter the urine.

This process does more than excrete waste products. Selective absorption in the tubule and selective return of constituents of the plasma do much to regulate the composition of the blood. Substances in less than normal concentration in the blood are usually not retained in the tubule, and substances in more than normal concentration are usually retained in part. For instance, blood sugar is usually wholly returned to the blood, but if its concentration becomes abnormally high some is excreted in the urine. The tubules also re-

[13] One of the more remarkable aspects of kidney function is the quantitative relationship between the amount of plasma filtrate that forms in the glomerulus and the amount of urine that eventually leaves the kidney. In an average-sized man, for example, the total volume of blood plasma is about 3200 milliliters. It has been shown that 130 milliliters of plasma filtrate are produced in the glomeruli each minute. This means that the whole plasma volume is reworked by glomerular filtration and tubular reabsorption 58 times a day or once every 25 minutes. It also means that about 187 liters (48 gallons) of glomerular filtrate are formed each day. Since only 1 to 2 liters of urine emerge from the kidneys each day, the kidney tubules must reabsorb 185 to 186 liters of water and selected solutes each day. Thus the kidney operates by first excreting everything and then taking back those items (in whole or in part) that the body needs to retain.

tain and pass on to the bladder more water when the water content of the body is high and less when it is low. Thus control by the kidneys of the composition and volume of the blood, and hence of the internal environment of all body cells, is most effective.

CHAPTER SUMMARY

Transportation within the organism: its simplicity in protists; one of the major functions demanding special adaptations in large organisms.

Transportation in the xylem and phloem of vascular plants: the role of transpiration and root pressure in moving water up the xylem.

Circulatory systems in animals: hearts; open and closed circulations.

Composition and function of blood: plasma, serum, and the formed elements of the blood; the clotting reaction; hemoglobin and the transportation of oxygen; the transportation of carbon dioxide.

The mechanism of circulation: diastole and systole of the heart; blood pressure.

The lymphatic system: the movement of plasma from blood capillaries through intercellular spaces to lymphatic capillaries; the mechanism of one-way lymph movement; drainage into the subclavian veins; lymph nodes; fat transportation.

Glands and secretion.

Excretion: its simplicity in protists and plants; excretory problems and systems in multicellular animals; water problems in fishes; ammonia, urea, and uric acid; the organization and functions of the kidney in man.

12

Organization and Integration

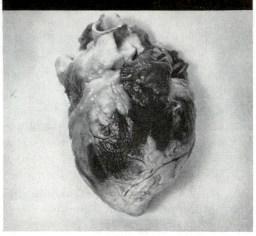

Arthur F. Jacques

The ceaseless beating of the human heart, and its automatic regulation, exemplify that "wisdom of the body" that is the subject of this chapter.

Organization

The essence of organism is stated in the name: an organism is an *organization* of materials and functions that work together for their own preservation and for that of the species. We have already learned of the organism's intricate system of levels of organization, a pattern that characterizes all living things. Only this rising hierarchy—molecules, macromolecules, cells, tissues, organs, systems—makes possible the remarkable self-regulatory behavior that maintains within an organism the steady state, or homeostasis (Chapter 4). Homeostasis could never exist without an enormous variety of control processes. Since control, as we have seen, can come only from a system that includes negative-feedback, control requires organization.

Earlier (in Chapter 4) we mentioned the apparent purposiveness or goal-seeking quality of the various vital activities of the organism. We concluded that this teleonomic (not teleologic) serving of future needs in development, self-maintenance, and reproduction is in fact a necessary feature of living systems that has evolved under the direction of natural selection. In the last analysis it is the end-serving activities that unite the organism into an ordered, self-regulating, and single whole and, by enabling it to make internal adjustments, impart to it a degree of independence from the vicissitudes of the environment.

In protists the organism is a single cell, and it obviously has all the properties and processes of a living system at its level. In multicellular organisms, composed essentially of cells and their products, the total system is in a sense the sum of the properties and activities of the individual constituent cells. In another sense, however, the system is more than the sum of its parts, for the parts themselves, the individual cells, interact and have different properties within the system from those they would have if isolated. Thus they act together with results that would be unobtainable by single cells. For example, the specialized conductive properties of a nerve cell could not possibly evolve in a cell living alone. Moreover, the intricate structure that is a brain has over-all properties of perception and reaction that could not possibly occur in any one cell within that organ. It is some of these over-all properties of organismal integration, function, and homeostasis that we shall study in this chapter and the next two.

Some of the most important aspects of organismal integration were discussed in Chapter 8. The striking phenomena of regeneration in plants and animals, mentioned in that chapter, are especially pertinent here and are well illustrated by experiments on planarians, flatworms usually about 10 to 25 millimeters in length. In spite of their small size, they are rather complex, with muscles, nerves, a primitive sort of brain, eyes, digestive organs, and specialized excretory and reproductive systems. If a piece of reasonable size is cut from front, middle, or hind end, it will regrow all the missing parts and become a whole, normal animal (Fig. 12–1). Any piece cut off will have a front and a rear. A new head will grow only at the end that was toward the head to begin with, and a new tail only at the end that was toward the tail.

The planarian as a whole has a field of organization within which the individual cells are arrayed and which gives them singly and in unison properties they would not have if isolated. In this field, in a way not true of each single cell, there is a front, a rear, a right side, a left side, a top, a bottom, and so on. Similar fields occur in most multicellular organisms, sometimes in simple and sometimes in more complex form.

INTERNAL ENVIRONMENT

We have seen that how a cell is organized and how it reacts depend upon where it is. Not only the individual cell but also its relationships to adjacent cells and to the organism as a whole are involved. This is another aspect of the principle that many activities of life are interactions between cells and their environments, a principle discussed earlier in a different context.

Protists and simpler plants and animals

In protists the organization of the whole body is that of a single cell which is, necessarily, in direct contact with the external environment. For active protists, that environment

REGENERATION IN PLANARIA

12–1 Regeneration in planaria. The animal at the left is cut crosswise into two halves. The anterior (head) half regenerates a posterior (tail) half, and the posterior half regenerates an anterior half. Note in the lower sequence that the head itself is regenerated first and the pharyngeal region later. As the regeneration of the anterior half proceeds, the posterior portion decreases in size until a whole organism of normal proportions is achieved. A similar reduction of the anterior half occurs when it regenerates a posterior half, as shown in the upper sequence.

is always a watery solution of various substances. The solution may be sea water or the water of ponds and streams. For the many parasitic protists it is blood, protoplasm, or some other solution in a multicellular organism. In simpler multicellular water plants or animals most individual cells are also in direct contact with the external environment. In an alga (plant) or in a sponge or coral (animals), the environment of a single cell includes the adjacent cells and the water in which the animal lives. Organization is brought about and maintained by contacts with other cells. Metabolism is maintained mostly by exchanges with the water of the external environment, which in these lowly organisms is partly contained in pouches and canals where the solution is modified by activities of the surrounding cells. There are internal cells in many of these organisms, but they are not far removed from the environmental solution. The internal cells also make metabolic exchanges with the external environment but do so through intervening cells and with fluids seeping between the cells.

The point is that in these organisms (protists and the simpler aquatic plants and animals) the environment of each cell is essentially that of the organism as a whole. It is the general external environment, which is only slightly modified in the immediate vicinity of the organism by its own activities. The cells are thus affected by any change in the external environment. Any cell has a narrow range of conditions under which it can remain alive and a still narrower optimum range in which its activities are most effective. In the organisms that we are now considering, life depends on the environment, which must not fluctuate beyond the range in which all the cells of the organism can live. The outside environment must also be near the optimum during the most active periods of the organism's life. Put in another way, the evolutionary possibilities for such organisms are strictly limited. They can occupy only those environments that do not fluctuate too greatly; and the new metabolic processes they can evolve are restricted by the conditions existing in available external environments. This is true of all their cells.

The water of the sea is the most stable and in some other respects, as well, the most favorable environment for organisms. Changes in composition, temperature, and other vital conditions are slower and less extreme in the sea. The inorganic material dissolved in sea water is similar to that in many cells, and the osmotic pressure also often is similar. It is not surprising that life arose in the sea and that almost all organisms with cells dependent on the external environment still live in the sea.

Conditions are less favorable in fresh water. Fluctuations are greater, dissolved salts may differ more from those in cells, and osmotic pressure is lower. Many protists and simpler (nonvascular) plants and a few of the simplest animals do live in fresh water, but they are here much less abundant than in the sea. They have special adaptations that enable them to survive the fluctuations as well as the constant features of fresh water. For instance, the low osmotic pressure of fresh water may cause too much water to diffuse into cells. This may be countered by lower osmotic pressure in cells, by reduction of diffusion in various ways, by greater elimination of water, or by a combination of such protective adaptations.[1]

Land is the most difficult environment of all. Fluctuations in temperature and other physical conditions are far greater than in any body of water, even a small pool of fresh water. The air contains few of the materials required by cells. Worst of all, it contains very little water, which is absolutely essential for cells. No protists or simpler animals truly live on land. Some live in damp soil, but there they are really in an aquatic environment. Many nonvascular plants (among them fungi, lichens, mosses, and liverworts) do succeed in living on land by a variety of special adaptations to this unfavorable environment. Most of them are small, and many grow in damp or downright wet situations. Others have special means of obtaining and retaining water. Among them are lichens, which may attain considerable size and which grow in drier, less favorable circumstances than any other plants.

Vascular plants and animals

Evolution early proceeded beyond the level of the forms of life that we have been talking about. Although this is a point for later discussion, it is of interest to note here in passing that such evolution happened earlier in animals than in plants. Organisms evolved with more layers of cells and with greater differentiation and complication of internal tissues and organs. In them, the deeper cells are so far removed from the external environment that no effective interchange between deep cells and the external environment occurs. In fact, the environment in which these cells live is quite different from the environment outside the organism. It is an *internal environment*. With further evolution along these lines, many of the external cells have become relatively inactive. Some still serve for absorption

[1] See the discussion of contractile vacuoles, Chapter 13.

and excretion, but many are only protective, like cork cells in the bark of a tree. Most of the activities in higher organisms, especially animals but also plants, depend largely upon internal cells.

The evolution of more abundant and complex internal tissues and organs was accompanied by the elaboration of internal fluids. This elaboration also involved the development of vessels in which the fluids move. All higher organisms, both plants and animals, are vascular. Life in such vast aggregations of internal cells, far removed from an external medium, would be quite impossible if there were not an internal medium and a means for moving it. The complication of internal organs did not ensue because the vascular system made it possible, nor did the vascular system arise because it is necessary in such complex organisms. The two necessarily developed together. If they had not, neither one would have evolved.

The evolution of internal fluids and vascular systems provided internal cells with an environment separated from that outside the organism. This carried the possibility that the internal environment could be more stable than the external environment and more specifically adapted to the particular needs of each kind of organism. It resulted in a new and closer integration, bringing all or most of the internal cells in contact with a fluid environment specific to the organism. These cells became relatively independent of the environment of the organism as a whole, although, of course, the materials for construction and maintenance of the internal environment still had to come ultimately from outside.

Evolution of vascular systems made possible the full conquest of the land. The difficulties of this most difficult of environments are in considerable part removed for vascular organisms. Most of their cells are not really living in the inhospitable open air but in a highly favorable internal liquid environment of their own. All the truly terrestrial animals and an overwhelming majority of the terrestrial plants are vascular.

The sea within us?

The course of evolution has been a progression from organisms without a truly separate

TABLE 12–1

Relative amounts of some elements in sea water and some plasmas *

	Sodium	Potassium	Calcium	Magnesium	Chlorine
Sea water	100	3.61	3.91	12.1	181
Plasma of					
King crab	100	5.62	4.06	11.2	187
Lobster	100	3.73	4.85	1.7	171
Man	100	6.75	3.10	0.7	129

* These data are from A. B. Macallum, a proponent of the view that plasma and sea water are substantially the same. In each case the amounts are relative to the amount of sodium present.

internal environment to those with one. Most organisms without such clear separation live in the sea, and there are resemblances between sea water and the blood plasma of vascular animals. Some writers have suggested that when animals moved out of the sea they preserved a sea-water environment for their cells by retaining an equivalent of sea water in their vascular systems. Popularizers have gone so far as to say that the cells of our bodies are lapped by waves of the seas in which our ancestors swam hundreds of millions of years ago. That is poetic, but unfortunately untrue.

Let us consider first whether the compositions of sea water and plasma are in fact the same. Some data are given in Table 12–1. There is considerable similarity between the composition of sea water and the three plasmas listed, but there are also some notable differences. In relative terms, the plasma of man has almost twice the potassium but less than a fifteenth the magnesium of sea water. Even the lobster, which lives in sea water as its ancestors have always done, has more calcium and much less magnesium. An attempt has been made to explain these differences by postulating that the composition of sea water has changed and that the plasmas of different lines of descent were shut off in circulatory systems at different times. However, this theory is contradicted by the data. To account for the amounts of magnesium would demand that the system arose first in the ancestry of man, somewhat later in that of the lobster, and much later (almost recently) in the king crab. The evidence of fossils and the relationships of these forms show that this certainly is not true. A circulatory system arose at the same time in the ancestors of king crab and lobster and probably later (surely not any earlier) in the ancestry of man.

There are distinct differences among these plasmas and between any one of them and sea water. These differences cannot be explained by any changes that may have occurred in sea water, but they can be explained in two other ways. In the first place, plasma is not really derived from sea water. Even in primitive marine forms plasma is a fluid that develops in organisms; it is not sea water somehow trapped in them. The fact that plasma is in a closed system separated from the water of the environment makes possible its *differences* from sea water. In the second place, plasma has evolved. It is different in different organisms and has specific characteristics developed in each, just as have the other materials and structures of various organisms.

It is still true that all plasmas have similarities and that they somewhat resemble sea water. The common-sense explanation of these facts is quite simple. The conditions under which cells can live actively are not exactly the same for all cells, but they are similar for all. Cells can live in both sea water and plasma. If sea water and plasma were not rather similar, this would not be possible. If sea water had a decidedly different composition, primitive life of the sort that did arise in the sea would have been impossible. As animals with plasma evolved, their plasmas were necessarily similar to each other and to sea water because markedly different solutions would not have been compatible with the activities of cells.

REGULATION OF THE INTERNAL ENVIRONMENT

The internal environment must be suitable for the life of all the cells it touches. It must also be adaptable to changes in organic activities. Further, it becomes involved in integration of the whole organism and coordination of various processes within the organism. All these facts require that there be precise control of the internal environment. It must on one hand be kept constant within certain limits and on the other hand change within those limits as the activities of the organism change.

All vascular organisms do have mechanisms of control of the internal environment, but the mechanisms and the degree and nature of control differ greatly in different organisms. Control in plants is less rigid, less localized or more diffuse, and less independent of the outer environment than in most animals. The difference may be related to a more definite and more complex differentiation of organs in animals, to the mobility of many animals, and, above all, to the fact that animals have nervous systems and plants do not.

The composition of the ascending sap in plants is necessarily limited by the materials that happen to be in solution in the soil. The initial composition of solutions entering the plant is further determined by the roots, which are to some degree selective and absorb some substances more readily than others. The composition of the sap is further modified by absorption and secretion from cells throughout the plant as the sap moves along. The rate of flow is largely, although not entirely, determined by the rate of transpiration, which in turn depends on both external and internal factors. The humidity, heat, and light of the surrounding air play a part, and so do opening and closing of stomates and other conditions in the leaves. In some plants leaves may turn toward or away from the sunlight, or become reduced in area by rolling up, and these influence rate of transpiration as well as other processes.

In plants there is no mechanism for control of internal temperature, which is usually very near the external temperature.

Chemical coordination by secretion of hormones (substances that affect processes elsewhere in the plant) is important in vascular plants. It is not fully understood but is certainly both extensive and complex. It will be discussed more fully later in this chapter.

Man belongs to a group of organisms (the mammals) in which the internal environment is most fully controlled. We may therefore draw more specific examples of such control mostly from human physiology, with some comparisons with other forms of life.

Control of body-fluid composition

The importance of the chemical constitution of the blood plasma and lymph has already been made clear. The main ways in which this is controlled need brief review in this new context. New materials are picked up in the intestine, where the lymph acquires mainly fats and the plasma takes on a large variety of absorbed materials. The plasma and, to a smaller extent, the lymph exchange materials with all the active cells of the body. Excess foods and salts tend to diffuse from the plasma into tissues where they are stored. If the plasma becomes deficient in any of these materials, diffusion tends to be reversed and substances pass from storage into the plasma. These more or less automatic diffusion processes help to maintain a balance between materials in tissues and in their internal environment. Such processes are not, however, sufficient in themselves to stabilize composition and concentration because they do not change the total amount of materials in the system.

The liver plays a special role in the regulation of plasma solutions. By means of the portal vein the liver removes much of the glucose formed by digestion before it circulates through the body as a whole. Glucose in temporary excess is stored as glycogen, which is reconverted to glucose and poured back into the plasma as the blood-sugar concentration drops between meals. Some of the plasma proteins are formed in the liver. Most of the other plasma proteins (for example, the globulins) come from lymphoid tissues. The liver also has an essential excretory function. Excess amino acids are here converted into urea and other end products, and toxic products from other tissues or from infections may here be converted into less harmful substances. The end products thus formed are not immediately excreted from the body, but pass back into the plasma and are eventually excreted through the kidneys. Besides these processes that influence the composition of plasma, you will remember that the liver produces substances necessary for digestion of fats and that it and the spleen destroy old red blood cells. About a dozen other known processes occur in the liver. No wonder that it is one of the largest organs of the body and that serious disturbance of liver function may be quickly fatal.

The final and most delicate regulation of plasma composition is in the kidney, the action of which was discussed in Chapter 11. This regulation is purely negative. The kidney does not add anything to plasma but it alters and controls the solution by selecting what substances it removes and how much of each. The concentration of the solution and, indirectly, the water content of tissues throughout the body are also controlled largely by the kidney. The mechanism for this is the balance between the amount of water filtered into the kidney and the amount reabsorbed and passed back to the plasma by the tubules.

Control of plasma composition, mainly by the liver and kidneys, is achieved in all vertebrates, from jawless "fishes" upward, by mechanisms identical to those in man. Many invertebrates have glandular pouches along the intestine which resemble the vertebrate liver in the way they develop. These are sometimes called "livers" and may sometimes store glycogen, but they do not necessarily correspond with the true vertebrate liver either in structure or in function.

Practically all vascular animals have well-developed organs with essentially the same regulatory functions as our kidneys. These organs selectively remove end products from the plasma and help to control the water content of the body. In all these animals the basic mechanism is the same: internal fluid flows or is filtered into tubules and eventually is passed outside the animal. In all but the simplest animals, there is selective excretion along the tubule, much as in man, although the organs may look quite different. In worms and some other invertebrates, separate tubules occur throughout the body. In some more advanced invertebrates, such as crustaceans, the tubules combine into more complex and localized organs. Localization of such organs is related to the development of effective and complex circulatory systems. It might be said that in worms the excretory organs go to the plasma but in lobster or man the plasma is brought to them.

We have been speaking thus far of plasma and lymph. It is these solutions that bathe the cells and form the internal environment in the strictest sense. The formed elements of the blood are in the internal environment and are also a part of the mechanism of its regulation. The red blood cells carry O_2 and CO_2. The withdrawal of CO_2 from the plasma into the red blood cells, where it is not part of the internal environment, is an especially striking part of environment regulation. The removal of foreign invaders, such as disease-producing bacteria, by white cells is also an evident part of regulation of the environment. Finally, the prevention of loss of fluid due to injury is also regulatory, and the whole clotting mechanism may be mentioned again in this connection.

Flow: the heartbeat

There is no more impressive example of the dynamic power of life than the ceaseless beat of the heart, day and night, week after week, year after year. This incessant activity is evidence that the internal environment cannot be a static system. It must move constantly if it is to remain a livable environment for the cells of the body. Depleted materials must be restored, and waste products must be removed. It is this motion that makes the blood a powerful force of integration of the whole organism, for its motion can carry substances from any part of the body to all other parts.

The necessity of blood flow for life is dramatically established by the fact that cessation of the heartbeat is the symbol and the definition of death. That death in man seems to occur immediately when the heart stops beating is due to the extreme demands of our nervous system, especially the brain. If the blood flow stops, the brain is without oxygen or energy reserves, and it ceases to function adequately within a few seconds. In animals with less demanding nervous systems, activity may continue for considerable periods after the heart stops. Even in man other tissues retain vital activity for variable periods after the nervous system has died; death is not instantaneous throughout the organism.

William Harvey (1578–1657), the discoverer of the circulation, rather despairingly concluded that "The motion of the heart was to be comprehended only by God." Biologists cannot yet say that they have answered the last "why" about this or anything else, but the motion of the heart has now been described in fullest detail and a great deal is known about its causes and controls. In man and other vertebrates, the motion is inherent in the heart itself. A turtle's heart, for instance, entirely removed from the body of the turtle, can keep on beating for a long time. Even small bits of heart tissue, kept alive in solutions resembling lymph, may continue to contract rhythmically. The property of contracting, then relaxing, then contracting again, and so on, is, so to speak, built into vertebrate heart muscle, which is visibly different in structure from muscles that contract only when stimulated through nerves (see Fig. 3–34).

The heartbeat starts in a small region in the wall of the right atrium, picturesquely called the *pacemaker*. From this point, an impulse spreads to the left atrium, which thus starts to contract a little later, and then to both ventricles, which then contract together. Thus the contractions of the four chambers of the heart are kept in effective rhythm and sequence (Fig. 12–2).

In a man at rest the heart rate or pulse is usually about 70 beats per minute. In a hummingbird the rate may be as much as 1000, and in an adult elephant it is only about 30. Considerably slower rates occur in some lower animals. Pulse rate, like other activities, depends on the size and other characteristics of each particular sort of organism. Your own heart beats much faster, as much as twice as fast, at some times than at others. The tissues need more supplies from the blood, especially oxygen, at certain times. It is important to organic maintenance that the blood flow be variable to meet these varying needs. The amount of blood pumped by the heart can be increased in two ways: by increase in the volume pumped at each stroke, and by increase in the number of strokes per minute.

The volume of blood per stroke depends largely on the amount of blood reaching the heart between contractions. Flow of blood through the veins to the heart is greatly increased by muscular action. These relationships automatically ensure that the heart will pump more blood per stroke during periods of muscular activity.

The rate of beating can be modified by a large number of factors. It is increased by heat, as from exercise or by fever. It is markedly increased by epinephrine (a hormone

THE CONTROL OF THE HEARTBEAT

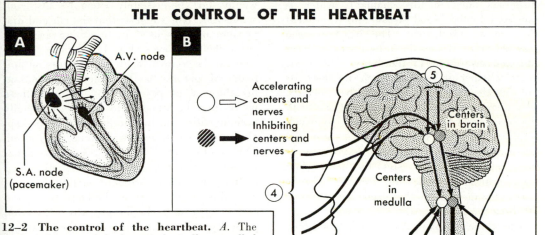

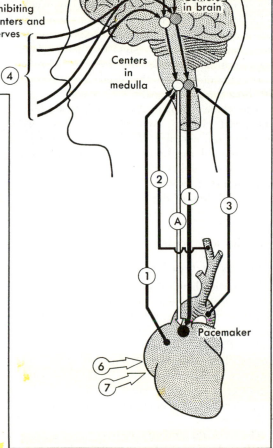

12–2 The control of the heartbeat. *A.* The rhythmic beating of the heart is basically controlled by the pacemaker, or sinoatrial (S.A.) node, in the right atrium. Its impulses spread to the left atrium, which contracts a little later than the right atrium. The impulses from the pacemaker later stimulate the atrioventricular (A.V.) node, which in turn initiates impulses that travel through the ventricle walls, causing them to contract. The impulses from the pacemaker thus underlie the rhythmic contraction of the heart's four chambers. *B.* The pacemaker's activity is, however, subject to modification. Centers in the medulla of the brain can transmit nerve impulses to it. Impulses from one of these centers inhibit (*I*), or slow down, the pacemaker; impulses from the other center accelerate (*A*), or speed up, the pacemaker. These inhibiting or accelerating impulses have several causes. 1. The increase of muscle movement during exercise forces more blood than usual through the veins and thence into the right atrium, which is stretched by the resulting high pressure. The stretching of the right atrium's wall initiates nerve impulses that are transmitted through channel *1* to the accelerating center in the medulla; this then sends further impulses back through channel *A* to the pacemaker, with a resulting increase in the rate of heartbeat. (Channel *A* is part of the sympathetic nervous system.) 2. The increase of respiration during exercise causes an increase in the CO_2 concentration in the blood stream. This is detected by special sense organs located at one site in the carotid artery. Nerve impulses initiated in the carotid artery by the high CO_2 concentration are transmitted through channel *2* to the accelerating center in the medulla, thus ultimately causing the increase in pacemaker activity appropriate to exercise. 3. The heart is protected against beating at too fast a rate by the presence of pressure-sensitive nerve endings in the dorsal aorta. As the rate of the heartbeat increases, blood pressure in the aorta builds up and stimulates these nerves, which transmit impulses through channel *3* to the inhibitory (or decelerating) center in the medulla. This in turn initiates impulses that pass back through channel *I* and slow down the pacemaker. 4. Wholly external stimuli, received through the olfactory, auditory, or visual sense organs, can affect the rate of the heartbeat; these stimuli are relayed through higher centers in the brain to the centers in the medulla, finally accelerating or decelerating the pacemaker. What perceptions will stimulate the accelerating centers? What perceptions will stimulate the inhibitory centers? 5. Purely internal stimuli in the form of ideas arising in the cerebral cortex can be similarly relayed to the centers in the medulla. Can you think of examples of ideas that affect your heartbeat? 6, 7. The pacemaker's activity can also be influenced through non-nervous channels. An increase of body heat during exercise or fever increases pacemaker activity, as does an increase in the flow of the hormone epinephrine into the blood.

discussed later in this chapter), which is poured into the blood stream during violent exercise or excitement. The heart is also speeded up or slowed down by neural reflexes initiated elsewhere in the body. The heart beats whether it receives such impulses or not, but the rate is modified by the impulses. The nerves involved have connections with the brain, and this is another way in which emotions influence heartbeat. There are also nerve connections responding to pressure in the arteries, tending to slow the heart when rapid beating increases the pressure. This is part of the mechanism that keeps the rate near 70 when the body is calm and at rest.

The whole mechanism by which blood flow and heart rate are controlled is remarkably intricate and complex. We have by no means gone into all the known complications, and there are still unknown factors involved. Even in summary, however, this is a good example of "the wisdom of the body," of how internal conditions are kept nearly constant under normal conditions but also respond to varying activities and needs.

Control of breathing

The basic chemical processes of respiration occur in cells; in man and most other organisms the cells sooner or later use O_2 and produce CO_2. These substances are carried from and to the lungs in the blood (Chapter 11). For the body as a whole, rate of respiration depends on how fast O_2 is taken up and CO_2 given off in the lungs. These exchanges in the lungs depend, in turn, on the rate and depth of breathing.

The lungs themselves have no muscles. They are forced to expand and permitted to contract again by movements of the ribs, which form a box around the lungs, and of the *diaphragm*, a muscular partition that closes the lower side of the box. Since this motion continues rhythmically even when a person is paying no attention to it or is asleep, the muscles might seem to have an inherent rhythm like the heart muscles. It is, however, easy to see that this is not so. A person can stop breathing immediately if he wishes to, but he cannot commit suicide by holding his breath; even if he had will power enough to hold it as long as he were conscious, breathing

would start immediately when he lost consciousness. This suggests that the rib and diaphragm muscles used in breathing are fully controlled by nerves (Fig. 12–3). The conclusion can be confirmed by experiments on animals; if nerves to these muscles are cut, breathing stops permanently. In man, these nerves may be attacked by disease, especially poliomyelitis, and the victim can then be kept alive only by continuous artificial respiration. This can be provided by the respiratory apparatus known as the "iron lung." By producing a slight decrease in pressure outside the chest wall, this machine causes the chest to expand and air to enter the lungs, which function perfectly. It is only the nerves to the breathing muscles that are damaged.

The rhythm of breathing is maintained not in the lungs themselves nor in muscles of ribs and diaphragm, but in a nerve center. This breathing center is located in the *medulla oblongata*, which is the lower part of the brain at the top of the spinal cord (Fig. 12–4). From this center, rhythmic impulses are sent to the muscles of breathing, controlling their alternate contraction and expansion. The breathing center is situated in the central nervous system, with connections to the brain and to other parts of the body. Thus nerve impulses to the breathing center can modify the rate of breathing and coordinate it with other bodily activities. When a person purposely holds his breath, an impulse travels from the brain and stops the rhythmic impulses from the breathing center to the muscles of breathing. Breathing also stops when a person swallows or quickly catches his breath when suddenly smelling irritating fumes. The mechanism is the same: a nerve impulse from throat or nose to the breathing center.

In usual circumstances the rate of breathing and even the fact that breathing occurs at all is controlled by the amount of CO_2 in the blood. CO_2 acts on the respiratory center and causes it to send out its impulses to the muscles of breathing. The more CO_2, the faster the rhythm of the impulses. If there is extremely little CO_2 in the blood (a condition not normal but one that can be produced experimentally), the impulses stop and so, of course, breathing stops. The effect of CO_2 on the respiratory center is a powerful one. When a

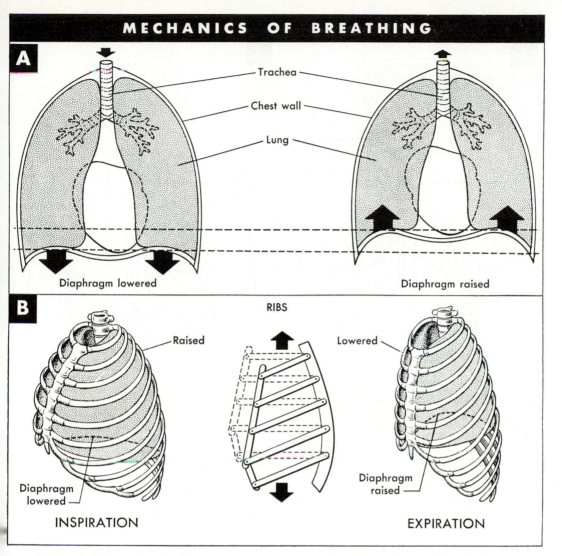

MECHANICS OF BREATHING

A

Trachea

Chest wall

Lung

Diaphragm lowered

Diaphragm raised

B

RIBS

Raised

Lowered

Diaphragm lowered

Diaphragm raised

INSPIRATION

EXPIRATION

12–3 Mechanics of breathing. Ventilation of the lungs (movement of air in and out) is due to the bellowslike action of the thoracic cavity (bounded by the chest wall and the diaphragm), in which the lungs lie. Air is forced into the lungs when the cavity is enlarged by (1) the elevation of the ribs and (2) the depression of the diaphragm. Air is forced out of the lungs when the thoracic cavity is decreased in volume by (1) the depression of the ribs and (2) the elevation of the diaphragm. *A* shows the extent to which the diaphragm moves during inspiration (*left*) and expiration (*right*). *B* shows the extent to which the ribs move during inspiration and expiration. The central figure schematizes the way in which the ribs are loosely joined to the spinal column and the sternum (breastbone); it shows how elevation of the ribs enlarges the thoracic cavity.

person tries to hold his breath for a long time, CO_2 accumulates until its influence is so strong that he starts breathing again in spite of himself. Production of CO_2 by the cells and its diffusion into the blood are approximately proportional to the consumption of O_2. Thus there is an automatic regulation that is unusu-ally direct and simple as organic processes go. Consumption of O_2 in cells leads to increase of CO_2 in blood, which leads to increase in depth and rate of breathing, which leads to increase of O_2 in the blood available for use by the cells. Another interesting point is involved here. CO_2 is a waste product, harmful in any

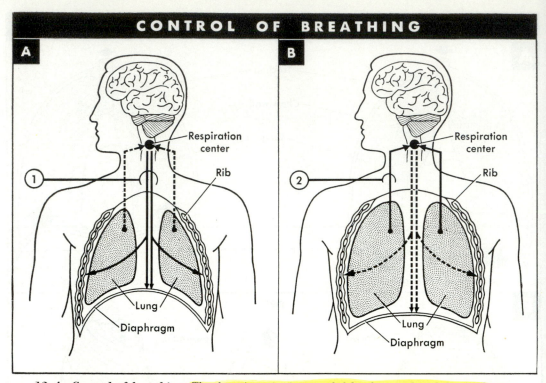

A

B

Respiration center

Rib

1

2

Respiration center

Rib

Lung

Diaphragm

Lung

Diaphragm

12–4   Control of breathing. The thoracic cavity is expanded by the muscles of the diaphragm and the intercostal muscles, which elevate the ribs. *A.* These muscles are stimulated into action, thus enlarging the thoracic cavity, by nerve impulses that reach them (via channel *1*) from a respiratory nerve center in the medulla. *B.* The expansion of the thoracic cavity enlarges the lungs and thus draws air in. In the lungs are nerves sensitive to the tension developed when the lungs are fully stretched. Stimulated by the stretch of the lung walls, these nerves initiate impulses that are transmitted (via channel 2) to the medulla's respiratory center, whose activity they inhibit. This inhibition of the respiratory center causes the diaphragm and the intercostal muscles to relax and, consequently, the thoracic cavity to collapse, resulting in expiration. When expiration is complete, impulses from the "stretch receptors" in the lung cease, and so, consequently, does inhibition of the respiratory center. Its renewed activity again stimulates (via channel *1*) the appropriate muscles to expand the chest cavity.

considerable concentration, and rapidly discarded by the organism. Yet it is also an essential part of the bodily mechanism, and on its way out it acts as a messenger from the body as a whole to the respiratory center. (Compare this picture of CO_2 as a "poison" and as an essential agent with what we had to say about blood clotting and water loss in plants [Chapter 11]. The general point arises again and again throughout biology.)

Control of temperature

Life can exist only within a relatively narrow range of temperatures. Particular types of organisms are at their best in an even narrower range than that of life as a whole. The

rates and even the kinds of reactions in cells are strongly modified by changes in temperature.[2] The physical properties of protoplasm, such as fluidity and elasticity, are also affected by temperature.

It is not surprising that temperature regulation should have evolved as part of the mecha-

[2] Other things being equal, there is a tendency for average rates to double for every 10°C. rise in temperature. The approximation is rough, and there is great variation. Some processes reach a maximum rate at a certain temperature and become slower when this point is passed. Thus potato plants synthesize starch more rapidly at 20 than at 30°C., even though they utilize sugar much more rapidly at the latter temperature; the sugar that is not utilized is stored as starch.

nism stabilizing the internal environment. It *is* surprising that such regulation is poor or absent in the great majority of organisms. Man happens to be among the relatively few forms in which the temperature of the internal environment is closely controlled. This is so exceptional that use of man as an example should be preceded by a brief review of the very different conditions in other organisms.

The ranges of environmental temperatures in the sea are much narrower than those in other environments. The whole range of surface temperatures is about −2 to 40°C. (about 28 to 104°F.), and these are very local extremes in different places. The upper limit, especially, is most exceptional and can occur only in shallow rock pools in very hot climates. At any one place and depth, the range is usually only a few degrees. Marine organisms are markedly affected in their distribution by water temperature. Those living in warm seas are quite different from those of colder water. In each place, however, the temperature changes are slight. Occasional incidents such as the upwelling of cold water from the depths may cause death to millions of animals adapted to warmer surface waters. Yet the usual fluctuations pose no severe problems, and temperature regulation has never evolved in sea animals. (Penguins, seals, and whales have highly developed temperature regulation, but they inherited this from ancestors that lived on land.)

Fresh water has more variable temperatures than sea water, but still the variation is limited. Its temperature cannot drop much below freezing or rise above the temperature of the adjacent air. Organisms in waters with relatively great seasonal variation in temperature often pass the winter in an inactive state.

It is the land that is the most difficult environment. Here temperature ranges are extreme, not only over a continent but also in single localities through the year. In the tropics temperatures change little from one season to another, but they have a daily rise and fall that may be much greater than any seasonal change. Tropical animals, from mosquitoes to monkeys, commonly have a daily rhythm. They are active at the times of day or night when the temperatures suit them; they stay in shelter when temperatures are much higher or lower.

The most extreme seasonal changes occur in the so-called Temperate Zone, where the climates are most intemperate. If you live anywhere in the central region of the United States, you probably have some summer days hotter than any in the tropics and some winter nights as cold as those in the far north. In this zone, land animals frequently either hibernate (as many mammals do) or migrate to warmer climes (as many birds do) in winter. There are many annual animals and plants. These die when cold weather comes, and their offspring survive the winter as inactive, well-protected eggs or seeds. Many insects and most leafy plants meet the problem of temperature range in this way. Perennial plants and trees become quiescent in winter, and most of them shed all their leaves in the fall. Some shed all their aerial parts and grow them again from the subterranean parts when warm weather returns. In one way or another all plants suspend activity in the coldest weather. It is noteworthy that those most successful in meeting low winter temperatures may be most susceptible to a late spring frost that comes when they have started active new growth.

Practically the only organisms fully active throughout the year in the Temperate and Frigid Zones are among those having internal temperature regulation—birds and mammals. As far as has been detected, no protists, plants, or invertebrate animals have any means of temperature regulation.[3] They live in environments with relatively slight fluctuations in temperature, or they have adaptations (a great variety of them) permitting them to be physiologically inactive when temperatures are too low or too high for their normal activities. It is a striking anomaly that the two groups of land organisms that are the most abundant and diverse—flowering plants and insects—have no internal temperature regulation. There are, to be sure, a few important exceptions among the insects. Although no insects maintain constant body temperatures, some do regulate their temperatures to a certain extent. Locusts move about so as to ab-

---

[3] Some invertebrates, however, have evolved regulatory devices that maintain some vital processes at a constant rate in spite of variations in temperature.

sorb more or less heat from the sun if they are too cool or warm. Ants move their larvae to warmer or cooler places in the nest. Some insects warm themselves up for action by exercise, much as athletes do. Honeybees keep the temperature in their hives within livable limits. If it drops to 13°C. (about 55°F.), the bees in the hive become very active and by release of body heat raise it to about 25°C. (77°F.). In summer they ventilate the hive with their wings.

Among the vertebrates, some are "cold-blooded" (fishes, amphibians, and reptiles) and some are "warm-blooded" (birds and mammals). A "cold-blooded" animal may very well have a higher temperature than a "warm-blooded" one. The difference is not in coldness or warmth but in the fact that "warm-blooded" animals have mechanisms for keeping the internal temperature constant and "cold-blooded" animals do not. Since the terms "warm-blooded" and "cold-blooded" are therefore flatly wrong, it is better to call those with internal temperature-regulating mechanisms *homeothermous*, and those without, *poikilothermous*.[4]

Poikilothermous animals tend to take on the temperature of their environments. This is only a tendency: poikilotherms do have some heat-regulating ability. A swiftly moving animal, even a fish, raises the body temperature well above that of its surroundings, and this warming in turn helps to speed up the reactions that produce it. Reptiles often bask in the sun on cool days, and absorption of radiation can raise their temperatures above that of the air. On hot days they avoid the sun and may burrow into cooler earth.

Only homeotherms tend to maintain a constant temperature by entirely internal mechanisms. The efficiency of the mechanisms varies. Variations of as much as 15°C. in internal temperature may not be fatal in some birds, but under usual conditions the temperature range is kept within a degree, more or less. In man the body temperature is normally

[4] Homeothermous means same temperature; poikilothermous means variable temperature. Still better in principle and preferred by many zoologists are *endothermic*, for internal regulation of temperature, and *exothermic*, for environmental regulation. Those terms, however, are used in a completely different sense in physical chemistry.

close to 98.6°F. (37.0°C.) ; a rise of less than 2°F. may mean trouble. (Although 98.6°F. is marked as "normal" on clinical thermometers, this is not necessarily true for all individuals. Perfectly healthy individuals may have somewhat lower or higher temperatures normal for *them*, and, in the same individual, a slight temperature change need not signify illness.) The temperature of a normal, healthy man remains nearly the same whether the air around him is at 115°F. or at −40°F. Obviously we have some means of producing heat, some means of losing it, and some means of striking a balance between the two.

Heat is produced by oxidative processes in the cells. The most important source is the muscles, which produce much heat when exercised or even when tensed. We tend voluntarily to move about more when it is cold than when it is hot around us. Even if we do not, our muscles automatically become tense in the cold. If the tenseness is extreme, the muscles begin to quiver, and we say we are shivering. When we warm up, the tenseness disappears and we feel relaxed.

Heat is lost by radiation and evaporation. The blood flows through a network of capillaries in the skin, where it is so near the surface that it loses heat by radiation. The more blood flows here, the more heat is lost. The amount of flow is regulated by constriction or enlargement (dilation) of the capillaries. In cold air the capillaries constrict; less blood flows near the skin, and less heat is radiated. In warm air the capillaries dilate; more blood flows, more heat is lost. Evaporation causes heat loss and hence is cooling, as everyone knows who has ever put a wet cloth on his brow. Cooling evaporation occurs all over our skin, mostly from sweat glands which secrete more liquid when the air is warm, less when it is cool. There is also much evaporation from our lungs.

In summary, these mechanisms work as follows:

When it is cool
{
skin capillaries are constricted—and less heat is radiated.
sweat glands are inactive—and less heat is lost by evaporation.
muscles are tenser—and more heat is produced.
}

When it is warm
- skin capillaries are dilated—and more **heat** is radiated.
- **sweat glands** are active—and more **heat** is lost by evaporation.
- muscles relax—and less heat is produced.

These diverse mechanisms are put into action by nervous impulses in nerves outside the sphere of conscious control. We cannot stop sweating when we want to, even to the extent that we can control breathing. The nerve impulses involved originate and are coordinated in a temperature-regulating center at the base of the brain.[5] The center reacts automatically to signals from sensory nerves in the skin where the sensations "hot" and "cold" originate. It is also possible that the center reacts directly to the temperature of the blood around it.[6]

Human beings differ in two ways from most other homeotherms. First, few other animals have as many sweat glands. Birds have none, and dogs and other carnivores very few. Heat loss is increased by more rapid breathing and by secretion of more fluid in the mouth. Everyone has seen dogs panting, with tongues hanging out and dripping, in hot weather. The second difference is that humans have no really useful natural insulation. Practically all other homeotherms are covered with thick fur or feathers, highly effective insulation both for keeping heat out when the sun is hot and for keeping it in when the air is cool. There is another reflex that raises hairs and feathers, increasing the thickness of the insulating layer in very cold air.[7] Oddly enough, man still has this reflex although we have lost the fur that makes it effective. In us it just produces goose flesh. We have made good our loss of fur, as we have many of our other deficiencies, by artificial means. We wear clothes, in cold

[5] In the part called the thalamus.

[6] Since in man cooling mechanisms operate when the body temperature is above 37°C. and heating mechanisms operate only when it is below 37°C., it follows that the thalamus must contain something equivalent to a thermometer capable of an absolute measurement of temperature. Or it may be thought of as a thermostat with a fixed temperature (37°C.) setting.

[7] This is brought about by contraction of the individual hair muscles, one of which is clearly shown in Fig. 3–33.

weather at least. Do you suppose there is any relationship between our loss of fur and exaggerated development of sweat glands?

Chemical Coordination

We have repeatedly emphasized that the whole multicellular organism is a unit, whose component parts interact with one another in a highly coordinated manner. Now let us think a little more about how the organism is unified, what ties it together. In the first place, it is bound together mechanically. It is enclosed in a skin of some kind. The various internal cells and tissues are in contact with each other and fastened together in various ways and at many points. There are skeletons, a bony framework, an external shell, connective tissue, the united cell walls of plants, or other arrangements that enclose part or all of the organism and to which tissues are fastened.

Another kind of unification is achieved through the nervous system. A nervous system is absent in protists,[8] plants, and a few animals but is present in the great majority of animals. We know that it is of supreme importance in man, so much so that it will be considered at some length in the next chapter.

Still another mechanism of coordination exists, and this one is present in *all* organisms: it is the movement of chemical substances from one part of the organism to another. Chemicals diffuse and flow in the protoplasm among the parts of a protist within its single, cell-like mass. In nonvascular multicellular plants and animals, substances also diffuse from one cell to another and may be carried by water of the external environment in cavities or otherwise in the immediate vicinity of the organism. In vascular organisms, movements within and between cells occur, too, and another feature has been added, especially apt for carrying chemical substances through the organism. There are internal fluids which, sooner or later, may move from any active cell of the body to any other.

As multicellular organization has been

[8] In Chapter 19 we shall describe the merest beginnings of a nervous system analogue in the most highly evolved protists, the ciliates.

more intensively studied, it has become evident that all living cells must have some influence on adjacent cells. This influence is usually if not always chemical in large part, although it may also have mechanical, electrical, and other elements. It has also become clear that any cell may produce substances that influence parts of the body quite distant from it. In other words, chemical coordination occurs everywhere in the organism, from and to all its cells, and not just between particular organs. A good but relatively simple example is provided by CO_2. Produced by all actively respiring cells, CO_2 passes into the blood stream and produces a specific effect at one point, the respiratory center, through which it initiates changes in the whole body.

Chemical coordination seems to be a general phenomenon of life, occurring in and between all the living parts of all organisms. However, certain organs produce specific chemical compounds that diffuse into the blood and have special, sometimes spectacular, results in the organism. These compounds, the hormones, have been intensively studied in man and other vertebrates.

The organs that produce hormones are the endocrine glands. It is interesting to realize that the two systems bearing the major responsibility for the control and integration of body functions—the endocrine system and the nervous system—have much in common. In both, the process being controlled is located some distance from a control center. And, to a great extent, both operate according to classical feedback principles (Chapter 4). There are also important differences between the two systems. The controlling messages of one system are nerve impulses, which travel along anatomically defined fibers and activate a limited number of cells within the influence of their peripheral endings. The messages of the endocrine system are the hormones, which enter the blood and, spreading widely, may influence cells and tissues in every part of the body. In general (as we shall see in Chapter 13), nerve impulses control rapidly changing activities such as skeletal-muscle movements. Hormones, in contrast, act by altering the rates of cellular metabolic processes (Chapter 4). Thus it appears that hormone action is only a special instance of a much more general

phenomenon. With this in mind, we may consider the hormones of man as our first examples of chemical coordination.

HORMONES OF THE HUMAN BODY

The hormones of man and the organs that produce them seem, in the light of present knowledge, to be more elaborated and advanced than they are in most other organisms. In such a situation there are two equally logical methods for beginning a study. We can start with the more complex end product of evolution and then try to trace its relationships to simpler or earlier products (the simpler are not necessarily the earlier), or we can start with the simpler or earlier and work up. In this book we have used both methods on different topics. Since the hormones of man are fairly familiar and those of other organisms far less so, let us proceed here from the better to the less well known.

The human hormones are produced by endocrine glands or by endocrine tissues in organs that also have other activities. The word "endocrine" derives from the Greek and means "inside separation"—that is, internal secretion.[9] It was Claude Bernard (1813–1878) who coined the term "internal secretion" and started biologists on the way to understanding chemical coordination. The secretion is called internal because the secreted substances, the hormones, diffuse directly into the blood and do not leave the tissue—go outside of it—through tubes, as do the secretions of the so-called exocrine glands. In man, but not in all other organisms, each separate endocrine gland is definitely localized in the body, whether it constitutes a whole organ or not. The various local endocrine glands are scattered through the head, neck, and trunk without any particular anatomical relationship to each other or to their activities (Fig. 12–5). In fact, their activities characteristically affect many or all parts of the body.

Since the secretions of endocrine glands are rapidly carried everywhere by the circulatory system, no particular placement is necessary

[9] The word "hormone" (from the Greek "excite") was first used by the famous English physiologist E. H. Starling in 1905 in referring to the newly discovered intestinal hormone, secretin.

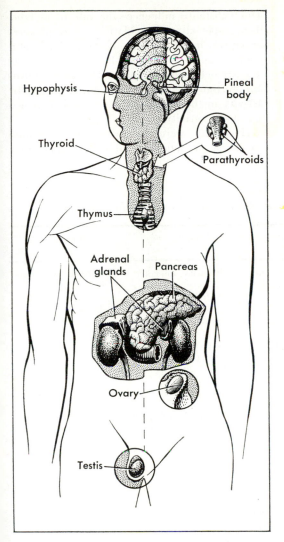

12–5 The locations of the human endocrine glands. The left side of the illustration includes one of the pair of glands (testes) characteristic of the male, and the right side one of the pair of glands (ovaries) characteristic of the female. Since the parathyroids are not visible in the over-all view, they are shown from the back in the separate drawing.

for their effectiveness. It seems logical, somehow, that the male hormone should be produced in the same organs that produce the male sex cells, sperms. Undoubtedly there is a relationship, but it is historical rather than functional. The male sex hormone also affects processes of the body far distant from the testis, for instance, growth of hair on the face and changes in the vocal cords. There is no obvious reason why the endocrine gland most involved in general metabolic level, the thyroid gland, should be in the neck. The reason is, again, historical: the thyroid has evolved from an organ that *did* have a functional relationship to this region of the anatomy (see Fig. 21–21).

The hormones resemble each other only in being secreted into the blood and influencing processes at various other points in the body, or throughout the body. They are all chemical messengers, mechanisms of coordination. They are not members of any one family of chemical compounds. At least one (insulin) is a protein, but others belong to quite diverse groups of simpler compounds. Interestingly, the hormones of the hypophysis are now known to be peptides containing relatively small numbers of amino acids. The two hormones of the neurohypophysis contain only nine amino acids. These have been artificially synthesized in the laboratory.

Table 12–2 lists some of the better-known endocrine glands, their hormones, and their effects. The list is not exhaustive. For example, the duodenum produces secretin. The stomach also produces a hormone, and the liver is thought to, also. Other organs, such as the pineal body in the brain and the thymus gland at the base of the neck, may be endocrine glands, but their activities are not yet clearly understood.

Note the kinds of functions that hormones control. Some influence chemical balances, especially in the blood. In other words, they help importantly to stabilize the internal environment. Others affect the rates of reactions concerned with growth and general bodily activity. Still others operate at particular times or periods of life to coordinate special developments and processes. Try making a classification of hormones along these lines. You will find that the distinctions are not clear-cut. The hormones, too, interact and are parts of an integrated system in the unified organism.

Thyroxin

We do not propose to discuss each endocrine gland and hormone in detail, but we shall make additional remarks about some of them to exemplify their action or to bring out a few points of special interest. We have already mentioned the thyroid and might consider it in most detail as an example. Its secretion, which has been precisely identified and synthesized chemically, is a combination of an amino acid with iodine; it is called *thyroxin*.[10]

[10] As actually found in the thyroid gland, the iodine-containing amino acid is part of the larger molecule of a protein, but the amino acid alone has all the hormonal effects.

This substance influences the rate of metabolism, especially of oxidative or respiratory processes, in all the cells of the body. It normally is delivered to the blood and circulates in extremely small amounts. Decreases in this minute but essential amount lower metabolic activity in the whole organism, and increases speed up activity.

Hormones were first discovered because an excess or deficiency of one of them may

TABLE 12–2

Some human endocrine glands and hormones and some of their effects *

Endocrine gland	Hormone	Processes affected or controlled	Symptoms of excess	Symptoms of deficiency
Thyroid	Thyroxin	Level of metabolism, oxidation rate, etc.	Irritability, nervous activity, exophthalmos [1]	Cretinism [2] when severe in infancy; lethargy, myxedema [3]
Parathyroids	Parathyroid hormone	Calcium balance	Bone deformation	Spasms; death if severe [4]
Adrenal medulla [5]	Epinephrine	Stimulation similar to that of sympathetic nervous system [6]	Increased blood pressure, pulse rate, blood glucose [7]	None
Adrenal cortex [8]	Aldosterone	Salt balance	Accumulation of body fluid [9]	Addison's disease [10]
	Corticosterone and cortisol	Carbohydrate metabolism, etc.	Abnormality of sugar and protein metabolism	
Pancreas [11]	Insulin	Glucose metabolism	Shock, coma	Diabetes [12]
Ovary [13]	Estrogen	Female sex development and menstrual cycle		Interference with menstrual cycle and sexual activity
	Progesterone	Control of ovary and uterus during pregnancy		Sterility or miscarriage
Testis [14]	Testosterone	Male sex development and activity		Lessened development of male characters and lessened sexual activity
Hypophysis [15] Adenohypophysis	Adrenocorticotropic hormone (ACTH)	Control of adrenal cortex	Symptoms related to glands controlled.	
	Thyrotropic hormone	Control of thyroid		
	Gonadotropic hormones	Control of sex glands, etc.		
	Growth hormone	Growth	Gigantism	Dwarfism
Neurohypophysis	Antidiuretic hormone (vasopressin)	Kidney action	Excessive water in body	Excessive loss of water
	Oxytocin	Milk production and contraction of uterine muscle		Lessened or no milk production

produce striking abnormalities or diseases. This is still the basis of the widespread interest in them. The association of thyroid degeneration with a disease was noted in 1874, one of the earliest definite recognitions of endocrine action. Much later thyroxin was isolated, and finally in 1927 it was artificially synthesized. The most striking and tragic effect of thyroxin deficiency occurs if severe deficiency starts in early infancy. The victim fails to grow physically and mentally, and becomes a mentally subnormal dwarf, a cretin. Thyroxin deficiency developing after normal growth has been completed has less drastic but still serious effects. If severe, it results in myxedema, a disease characterized by decreased bodily vigor, sluggish mentality, puffing of the skin, and often anemia.

In both cretinism and myxedema the symptoms reflect lowered metabolic rate, the difference depending on whether this affects early growth. Excess thyroxin production, which rarely occurs before early adulthood, has results that would be expected from heightened metabolism: loss of weight, nervousness and excitability, fast pulse, excessive sweating, and sometimes damage to the overworked heart and digestive upsets. It also usually has an oddly unexpected result, and one that may immediately identify a victim of this disorder: the eyes may bulge out markedly (Fig. 12–6).

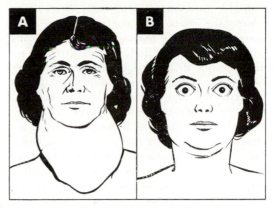

12–6　Effects of thyroid gland malfunction. *A.* Goiter. *B.* Exophthalmos.

Excess thyroxin production can be corrected by removal of part of the thyroid gland. Deficient production may be caused either by deficiency of the thyroid gland itself or by inadequate supply of iodine, an essential element in the thyroxin molecule. It can be corrected by continual doses of thyroxin or, if only the iodine supply is at fault, by adding iodine to food. Once cretinism has developed, it can be ameliorated but not cured.

The fact that iodine in very small amounts is necessary for thyroxin production has some interesting ramifications. Iodine must be ac-

TABLE 12–2 (Cont.)

* Do not memorize this list. Read it through to get a general idea of endocrine activity, and then use it for reference on particular questions that may arise.
[1] Exophthalmos is excessive protrusion of the eyes.
[2] Cretins are mentally deficient dwarfs.
[3] Myxedema is a disease whose symptoms include loss of vigor, falling hair, and puffy skin.
[4] Parathyroid deficiency used to follow the surgical removal of thyroid tissue when part or all of the parathyroids were accidentally removed at the same time. Surgeons are now aware of this danger and avoid it.
[5] The medulla is the inner part of the adrenal gland. The hormone epinephrine is also known as adrenalin.
[6] See Chapter 13
[7] The symptoms noted are apparently normal results of increased adrenal secretion, but they may result from other causes not involving excessive epinephrine secretion.
[8] The cortex is the outer part of the adrenal gland.
[9] Aldosterone causes the kidney to retain sodium. Excessive sodium retention causes the retention of an osmotically equivalent amount of water.
[10] Addison's disease, fortunately rare, has among its symptoms coloration (bronzing) of the skin, low blood pressure, general weakness, loss of water from the body, and upset carbohydrate metabolism.
[11] The pancreas as a whole is not an endocrine gland, but it contains clusters of cells, the islets of Langerhans, that have an endocrine function and secrete insulin.
[12] The principal symptom of diabetes is a marked increase of sugar in the blood.
[13] The ovary as a whole is not an endocrine gland, but some of the cells around the developing eggs produce estrogen. After an egg leaves the ovary a "yellow body," the *corpus luteum*, develops, and this secretes progesterone.
[14] Like the ovary, the testis is not exclusively an endocrine gland but does contain hormone-producing tissue.
[15] This double gland is also known as the *pituitary gland*, and its two divisions, the *adenohypophysis* and *neurohypophysis*, as the anterior pituitary and the posterior pituitary, respectively. Although these terms are in common use hypophysis is the favored designation.

quired in foods or drinking water. With either source, the iodine must come ultimately from the soil. There are regions in which the soils contain little or no iodine, and in these regions people drinking local water and eating local foods develop thyroxin deficiency. It was noticed long ago that cretinism, myxedema, and less severe cases of thyroxin deficiency are relatively common in certain areas, including parts of Switzerland and of the Great Lakes region of North America. Students were very slow to make the connection between this distribution of disease and the lack of iodine in the soil. Once they did, the remedy was obvious. Iodine is supplied in the diet, most conveniently by addition of small amounts to table salt. As you know, iodized salt is now sold everywhere in the United States.

Goiters were common in many people in regions of iodine lack. In fact, such regions are often called "goiter belts." A goiter is an enlarged thyroid gland. With a low iodine supply, there is a compensatory enlargement of the thyroid, which increases thyroxin output. Sometimes this suffices, and the individual is entirely normal except for having a goiter—which, incidentally, was considered a desirable female adornment by some of our forebears. Thyroxin deficiency may, however, persist even after great enlargement of the thyroid.

The cure seems obvious enough. Since a goiter indicates iodine deficiency, add iodine to the diet. But once again things are not so simple. People with an adequate iodine supply and an excess production of thyroxin also frequently have goiters. Any enlargement of the thyroid not compensatory for scanty iodine brings on an excess of thyroxin. Thus a goiter may mean either too much or too little thyroxin, and it is necessary to have a physician find out which.

Insulin

The insulin molecule was the first protein to have its amino acid sequence determined (see Fig. 2–11). Therefore, its study led to one of biochemistry's greatest triumphs. It also represents a historic medical triumph that has alleviated a great deal of human misery. The disease of diabetes (more strictly "sugar diabetes" or *diabetes mellitus*) is a disturbance of carbohydrate metabolism that results in abnormally large amounts of sugar in the blood plasma. It is diagnosed by the constant presence of considerable sugar in the urine.[11] The disease is rather common and had been well known for centuries before any way of treating it was found.

In 1884 experiments on the function of the pancreas revealed that dogs lacking the pancreas develop all the symptoms of diabetes. There are two kinds of glandular tissue in the pancreas. One secretes digestive enzymes, as the experimenters knew. The other, forming what are called the islets of Langerhans (after their discoverer, Paul Langerhans), was of unknown function. Further experiments showed that diabetes probably resulted from the lack of some substance secreted by this tissue. Years were spent by numerous investigators trying to extract a substance from the pancreas that would prevent diabetes. All efforts failed until F. G. Banting, collaborating with C. H. Best, J. J. R. McLeod, and J. B. Collip, developed a method of extracting secretions not from the whole pancreas but from the islets only. The experimental difficulties had arisen from the odd fact that the other parts of the pancreas produce proteolytic enzymes that destroy the protein hormone secreted by the islets in the pancreas. The substance finally extracted in 1922 was insulin.

Since 1922 methods have been developed for producing insulin in the amounts needed for treatment of diabetes. The hormone must be injected; it cannot be taken by mouth because insulin is destroyed by digestive fluids. Even now there is no known *cure* for diabetes, but periodic injection of insulin counteracts the symptoms so effectively that diabetics can now look forward to long and nearly normal lives. One merely artificially supplies the body with a hormone not sufficiently supplied by an abnormal endocrine tissue.

Epinephrine

It is worthwhile to comment briefly on epinephrine as a hormone of quite a different kind from thyroxin or insulin. The adrenal glands (Fig. 12–7) consist of two separate

[11] Traces of sugar may occasionally be present in normal urine, as in times of stress, anger, or fear.

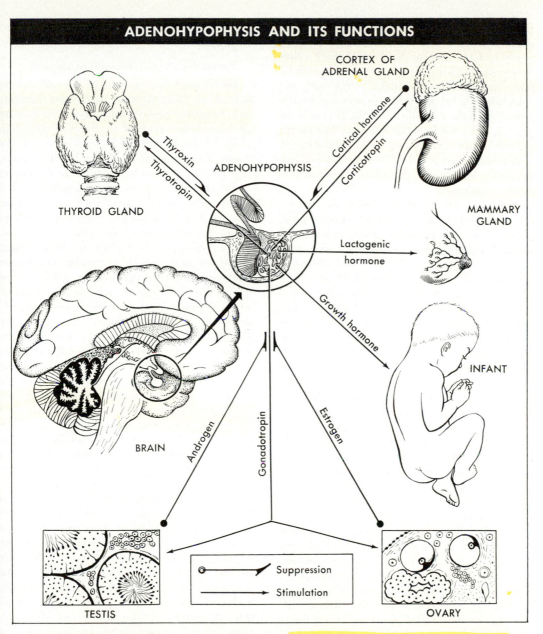

CORTEX OF
ADRENAL GLAND

Thyroxin

Thyrotropin

ADENOHYPOPHYSIS

Cortical hormone

Corticotropin

MAMMARY
GLAND

THYROID GLAND

Lactogenic
hormone

Growth hormone

INFANT

Androgen

Gonadotropin

Estrogen

BRAIN

⊙ ➤ Suppression

───➤ Stimulation

TESTIS

OVARY

12–7 The adenohypophysis and its functions. The hypophysis lies at the base of the brain. Most of its important hormones are secreted by its anterior lobe. These include: (1) a thyrotropic hormone, which stimulates the thyroid gland to secrete thyroxin; (2) an adrenocorticotropic hormone, or ACTH, which stimulates the adrenal cortex to secrete its hormones; (3) a hormone promoting growth; and (4) three gonadotropic hormones, which stimulate the secretions of the ovaries and testes and prepare the mammary glands for milk secretion. The thyrotropic, adrenocorticotropic, and gonadotropic hormones control the secretions of the endocrine glands that they affect, and in turn the hypophysis is regulated by the concentrations of the hormones it causes the other glands to secrete. Thus, when thyroxin (a thyroid hormone) reaches a certain concentration in the blood, it causes the hypophysis to stop secreting thyrotropic hormone, and the adrenocortical and sex hormones act similarly to regulate the secretion by the hypophysis of ACTH and gonadotropins. These are excellent examples of the way the body is organized to maintain automatically an optimum condition.

parts, inner and outer, which are of different origin, produce different hormones, and apparently have nothing to do with each other except that their history has brought them into nonfunctional contact.[12] *Epinephrine,* sometimes called adrenalin, is secreted by the inner part (the medulla). Unlike thyroxin or insulin, it is not essential for life. The entire inner (but not outer) portions of the adrenal glands can be removed without producing any illness or apparently changing any reactions in the organism.

Although the gland producing it can be removed without any apparent effect, epinephrine is a very powerful drug. It has been isolated and identified chemically and is also made synthetically. Injection of either the natural or synthetic compound results in a number of strong reactions: the heart beats faster, blood pressure rises, the glucose content of the plasma increases, the pupil of the eye is dilated, and hair stands on end or goose flesh appears. There is a relationship between these apparently unrelated reactions: they all happen in man and in many other mammals, such as dogs, under the stress of fright, surprise, or great excitement. But, and this is itself very surprising, they also happen under these circumstances whether epinephrine is released into the blood or not. They are reactions produced not only by epinephrine but also and equally strongly by stimulation of part of the nervous system, in the autonomic or sympathetic system (Chapter 13).

This functional parallelism is a puzzling situation and one that should not be glossed over by the usual statement that epinephrine mobilizes us for action in moments of sudden stress. If this is really a mobilization (and anyone who has been, as they say, frozen with fear may doubt it), then there are two systems of mobilization, one chemical and one nervous, which seem to do exactly the same thing. Perhaps there is a clue in the fact that the nerves producing these reactions also produce a substance very like epinephrine. Perhaps we are in an evolutionary stage of changeover from one system to the other. But the clue raises new questions. Which way is the changeover going, and above all, why? It is a

[12] Incidentally, the same is true of the pancreas and the islets of Langerhans.

question that at present cannot be answered with certainty.

Hormones of the adenohypophysis

The *hypophysis* of man is a small nubbin only about 15 millimeters in greatest diameter and weighing about half a gram. It is situated inside the skull at the base of the brain (Fig. 12–7). In spite of its small size, it is the most complicated of all the endocrine glands. It produces at least nine different hormones and perhaps several more. It has been called "the master gland," and with some reason because it is the principal chemical coordinator of the activities of the other endocrines. The coordination is mainly by hypophyseal hormones that act on particular endocrines and stimulate their growth or activity. For instance, one hormone acts on the thyroid, which degenerates if the hormone is deficient; at least three act in different ways on the sex glands; and others act on the islets of Langerhans in the pancreas and the outer part (or cortex) of the adrenal gland. In at least one case the control is by a balance of opposite reactions: one hypophyseal hormone acts in a way opposite to that of insulin; that is, it increases blood sugar. Still other hormones balance similar reactions. Hypophyseal hormones affect metabolism of carbohydrates, for instance, and this metabolism is also affected by hormones from other endocrines.

Besides coordinating the whole system of hormones, the hypophysis produces hormones with special reactions of their own. The most spectacular of these is the *growth hormone,* which promotes and controls normal increase in size of the body. Overproduction of the hormone during early life produces giants, and underproduction, midgets. Robert Wadlow, an American who was 8 feet 10 inches high and still growing when he died at the age of 22, was such a giant; Martina de la Cruz, who lived to the age of 74 without topping 1 foot 9 inches, was such a midget.[13] Such giants

[13] Most dwarfism is not of hypophyseal origin. We have seen that cretinism is due to thyroxin deficiency. The dwarfs usually seen in circuses, with large heads and trunks but short arms and legs, do not owe their deformity to endocrine disturbances but to the inheritance of a factor producing short limbs during prenatal development.

and midgets are usually of fairly normal proportions and structure. If secretion of the growth hormone is excessive *after* growth is completed, however, serious deformity may result from the enlargement of hands, feet, jaw, and facial bones (Fig. 12–8).

We must now ask, "If the hypophysis coordinates the other endocrine glands what coordinates it?" There is some evidence that the other endocrine glands reciprocally help to regulate the hypophysis by a feedback mechanism that works like the governor on an engine. When the thyroxin concentration rises in the blood, the production of thyrotropic hormone in the adenohypophysis is slowed down, and a fall of the thyroxin concentration speeds up the production of the thyrotropic hormone. Good evidence also exists that the adjacent portion of the brain, the hypothalamus, is itself an endocrine gland, which secretes a substance that travels the short distance to the adenohypophysis through special blood vessels (the hypothalamo-hypophyseal portal system) and causes its cells to release hormones into the blood. This presumably explains how external environmental conditions such as cold and stress can influence the hypophysis through the nervous system. The details of this mechanism are questions a future biologist will have to answer.

HORMONES OF OTHER ANIMALS

Other vertebrates

As would be expected, animals closely related to man resemble him in endocrine glands and hormones. Most or all mammals seem to have endocrine systems similar to that of man. This makes it possible to obtain hormones used in clinical medicine as a by-product from our food animals. It has also made possible most of our knowledge of human hormones by experimentation on other mammals, especially dogs.[14] The other verte-

[14] Let us point out here that most of our intimate knowledge of human and animal physiology has involved experimentation with animals. It has resulted in the saving of many lives and the alleviation of endless misery for both men and animals. It is unfortunate that we cannot explain things to the animals and call for volunteers, but anyone who understands the situation must conclude that experimentation, humanely performed, is justified.

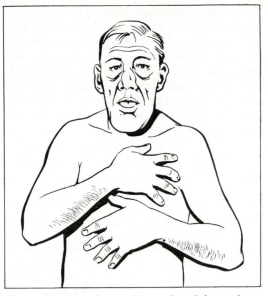

12–8 Acromegaly, an example of hypophyseal malfunction. Note the abnormal enlargement of hands, jaws, and facial bones caused by excessive secretion of growth hormone.

brates have similar endocrine systems, but differences begin to appear.

Investigations of recent years on the chemistry and biology of hormones have led to a number of interesting generalizations. It is clear, for example, that hormones having the same biological function in two different species are not necessarily identical chemically. Hormones that have different chemical structures may possess similar biological activities. As noted above, hormones obtained from one species may be active in another—but the converse is also true: certain hormones are species-specific with respect to biological activity. And it has been found that a hormone derived from one species may elicit a different type of response in the same target organ of a different species.

It is also a fact that the same hormone (that is, the corresponding hormone) in different species may have different functions. The human hormone *lactogenic hormone* stimulates the secretion of milk when a child is born. Lactogenic hormone has the same role in all mammals, as would be expected. But the corresponding hormone in pigeons stimulates the secretion of "pigeon's milk," which is not milk at all, but a secretion in the crop—

part of the alimentary canal—that is regurgitated and fed to the young. In hens the hormone produces broodiness. It is even present in some fishes. The evidence suggests that lactogenic hormone evolved long before milk did, that it came to have various effects on maternal behavior, and that in mammals it somehow took over regulation of a new, related process.

In many fishes, amphibians, and reptiles hypophyseal hormones control variable coloration of the skin. The color is controlled by nerves in some species, but by hormones in most. It is a most peculiar fact that these hormones also occur in mammals, including man, although no mammals have the special cells in the skin that produce color variation. Man has half the mechanism for matching himself to his surroundings, but he lacks the effective half. Whether these hormones are only useless baggage inherited from our remote ancestors or whether they do have some other effects in us is entirely unknown.

It is less puzzling to find that different but related processes may be regulated by the same hormone in different groups of animals. Change from a tadpole to a frog can be brought on prematurely by injection of thyroxin or can be prevented altogether by removal of the thyroid. This metamorphosis is quite different from anything in our life histories, but it involves heightened metabolism, and thyroxin raises our metabolism, too. Going still further back, there is convincing evidence that the thyroid gland in the earliest vertebrates evolved from a grooved structure that carried food particles into the front end of the digestive tract. This explains why the thyroid gland is in the neck. (We have said that there was an historical explanation.) This raises a bigger question: why and how did a feeding mechanism turn into an endocrine gland? It is possible to frame a hypothesis, at least, and instructive to try to invent one.

When the more ancient groups of vertebrates are compared with the younger, another tendency appears. The older groups seem to have simpler endocrine systems. At least some of the hormones present in mammals are absent, as far as has been determined, in fishes. Hypophyseal extract from fishes lacks the thyroid- and adrenal-regulating hormones and even the growth hormone. Fishes do grow, and they do so in a very well-regulated and coordinated way. It is impossible to believe that chemical regulation is not part of the process. The sensible conclusion is not that chemical growth regulators are absent, but that when fishes evolved these had not yet become localized as the product of one particular gland.

Invertebrates

Chemical coordination has been less studied in the invertebrate animals. This is natural enough, since the strongest motive for such study in vertebrates, and especially in mammals, has been immediate applicability to problems of human health and medicine. Even from this point of view, however, it is desirable and truly practical to learn all we can about other organisms. This knowledge can be expected to throw light on the most basic aspects of chemical control. It is from such wide comparisons that new points of attack and really fundamental advances are likely to arise.

Endocrine glands and hormones secreted by them are definitely known to occur among some of the more active mollusks (octopuses, cuttlefishes) and in crustaceans and insects. These are all relatively advanced animals with well-developed circulatory and nervous systems. In several sorts of insects it has been demonstrated that metamorphosis from the larval ("worm" or caterpillar) stage to the adult involves hormones. In at least some of these insects hormones active in the process are secreted by endocrine glands in two different locations. One gland, closely associated with the brain, secretes a hormone that stimulates metabolism and secretion in the other, which is situated farther back in the thorax, the three body segments to which legs and wings are attached. The hormone from the thoracic endocrine then regulates the metamorphosis. In some insects there is also evidence that hormones from the glands in the head have an influence, probably indirect, on sexual processes, on molting, and on more general metabolic processes.

The known insect endocrine glands and hormones strongly invite comparison with some basic features in the vertebrate system.

In both there are glands in the head (the hypophysis in the vertebrates) closely associated with coordinating centers of the nervous system. In both, hormones from these master glands affect widespread metabolic activities, including the stimulation of endocrine glands elsewhere in the body. The two systems certainly evolved entirely separately. The ancestries of insects and vertebrates separated well over 500 million years ago and at a stage when there cannot have been a definite endocrine system. The similarity is adaptive. It bears witness to the fact that all animal tissues have fundamentally similar reactions. It also suggests that the development of local endocrine glands secreting specific hormones was based on a more primitive, more diffuse system of chemical coordination.

Some insects, mollusks, and crustaceans also have hormones that effect temporary changes in color. These hormones, too, have been shown to be secretions of definite endocrine glands in the head region—different glands in the different groups. You recall that vertebrates also have hormones that influence variable skin colors. Those vertebrate hormones injected into crustaceans sometimes produce color changes. There must actually be a close chemical similarity in some of the hormones in spite of the very distant relationships of the animals producing them. This, again, is good evidence of similar reactions in all animals and is suggestive of an extremely ancient basis for chemical coordination. There is more evidence to back up this suggestion. For instance, epinephrine has been identified in a sea snail and even in a protist, and the eggs of sea urchins contain a compound related to thyroxin and with similar (though not quite identical) effects on tadpoles.

Even in insects, with the most complex endocrine systems yet found among invertebrates, the systems are much simpler than in the vertebrates. In invertebrates other than insects and perhaps crustaceans (which are especially related to insects) the systems are still simpler. In fact, in the vast majority no endocrine glands or tissues have been identified. In this majority of lower animals it is thus unjustified to speak of endocrine systems at all. This does not mean that chemical coordination is absent.

It is, as you can readily imagine, extraordinarily difficult to isolate and identify a coordinating chemical unless this is produced by a definite gland or is involved in some distinct event such as the metamorphosis of an insect. A substance secreted diffusely or in small groups of cells scattered through the body and constantly involved in maintenance of a stable condition is especially hard to identify experimentally. Yet we do know that compounds similar to and even identical with hormones occur in animals that have no endocrine glands. This and other evidence is sufficient for the highly probable conclusion that chemical coordination is present in all animals and even in protists.

Let us now look at the subject from the opposite direction. Chemical coordination appeared very early in the history of life. In protists and the simpler multicellular animals the coordinating chemicals, hormones by the definition of producing effects at a distance from their point of origin, moved and still move mainly by diffusion. The development of vascular circulatory systems provided a much more effective way in which the hormones could move throughout the body. In the earlier vascular animals and in their less modified descendants, the hormones were and are still produced throughout the body or in scattered cells. With further evolution came a tendency for complication in the system of hormones, for more hormones, and for each to be more specific in its action. The cells producing some of these also became more specific and came to be aggregated in local tissues and organs. This tendency has gone furthest in man and the other higher vertebrates. The organs formed are the endocrine glands. Even in man, however, there are activators or hormones that are not produced in endocrine tissues or glands, although they circulate in the blood, and there are still others that spread in the most primitive way, by diffusion.

Evolution of the nervous system was going on at the same time as evolution of the endocrine system. As we have seen, both are coordinating and integrating mechanisms for the organism as a whole. They could hardly be independent of each other, and they are not. Part of the evolutionary trend has been toward closer coordination between the two sys-

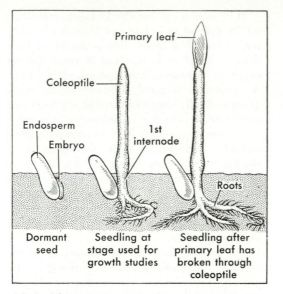

12–9 The germination of an oat seed. *Left.* Before germination. *Center.* After germination. The young shoot system consists of a tubular sheath, the coleoptile, which contains within it the primary leaf. *Right.* After the primary leaf has emerged. The coleoptile itself is the subject of growth experiments undertaken to study plant hormones. (Adapted from J. Bonner and A. W. Galston, *Principles of Plant Physiology*, Freeman, San Francisco, 1952.)

tems. In insects and vertebrates, independently, this trend has led to the development of a master endocrine gland actually attached to the main coordinating mass of the nervous system. Perhaps the two clearest examples of the interaction are the hypothalamo-hypophyseal portal system and the curious relationship between epinephrine, the hormone of the adrenal medulla, and the sympathetic nervous system.

HORMONES OF PLANTS

Everyone has heard of "plant hormones." They are widely advertised both as stimulants for the growth of desirable plants and as killers for weeds. Plants differ from the more highly evolved animals in that their hormones are secreted by unspecialized tissues. We concluded, above, that this was the primitive method of chemical regulation in animals also, and it still persists to some extent even in animals that also have well-developed endocrine systems. It is therefore a universal mechanism of coordination in organisms. Indeed, in plants no other special mechanism

has evolved, neither a more specialized chemical system nor a nervous system. The hormone system itself has certainly undergone evolutionary changes in plants, but exactly what the changes have been is still not known.

Since plants have no other active system of organic coordination but are nevertheless well-coordinated organisms, their hormones are particularly important. Each plant grows at a particular time of the year; each grows in its own characteristic way, each flowers at a definite time, and the fruits and leaves of some perennials also drop at definite times. There is nothing haphazard about these activities. They are coordinated processes of the organism, and the coordination must be largely if not altogether chemical in nature. It is, in fact, well established that hormones are involved in these and other processes. Yet only a few of the hormones have been definitely identified, so that this is a pioneering field of research at present. A special difficulty is the fact that the hormones are not secreted in endocrine tissues, where their isolation would be less difficult. The most fully known plant hormones are those called *auxins*. They are involved primarily in growth processes.

Auxins

"Auxin" is another invented word, coined from a Greek root meaning "to increase" or "grow." The term is still fairly precise in application to one definite class of plant hormones that promote elongation in growing cells, although they also have other activities. Their existence and some of their properties can be shown by simple experiments.

If oat seeds are germinated (Fig. 12–9), they first send upward a bluntly pointed, leafless shoot.[15] If you cut off the tip, the shoot stops growing. Now place the tip, cut side down, on a little cube of gelatin or, better, agar jelly (a seaweed product much used in biological work) and leave it there for about 2 hours. Put the cube of agar on the cut end of a decapitated shoot. The shoot starts growing again. Something diffused from the tip into the agar, and the something—an auxin—promotes growth in the shoot (Fig. 12–10).

[15] Actually the shoot has a leaf inside it. Botanists call the shoot a sheath or coleoptile because it surrounds the first leaf of the plant.

Does the experiment as we have outlined it really prove this? As a matter of fact, it does not. Before you read on, try to think of alternative explanations and of ways to test them as hypotheses.

There are at least two other possible explanations: (1) perhaps growth of the shoot was only temporarily stopped when it was decapitated and would have started again in a couple of hours anyway; or (2) perhaps the agar alone would have caused growth. To test these hypotheses the experiment requires *controls*. Decapitate three shoots at the same time. On one, put an agar cube that has had an opportunity to receive diffused material from a tip. On another put a cube exactly the same except that it has not been in contact with a tip. On the third do not put anything. Only the first shoot will grow, and the presence of a growth substance in the tip is substantially proved. It is still better to run multiple tests, giving each of the three treatments to three or four shoots.

Now try another experiment. Put a cube of agar that contains diffused auxin on a cut shoot in such a way as to cover only one side of the cut end. As the shoot grows, it will bend away from the side in contact with the jelly. That side receives more auxin and therefore grows faster, and the shoot is forced into a curve. This important reaction explains many of the motions and growth patterns of plants.

Three distinct but chemically related auxins have been isolated from plants and identified. The most widespread is a simple derivative of the amino acid tryptophan, indole-3-acetic acid:

$$\text{—CH}_2\text{—COOH}$$

Oddly enough, the same substances are common in animals, and one of them is present in human urine. Here is further evidence of the unity of life, although the relationship is not wholly clear.

In extremely small concentrations, auxins cause the elongation of cells and hence growth in length, but with an increase in concentration, they inhibit growth. The effect is different on different plant tissues. A concentration producing the most rapid elongation of stems

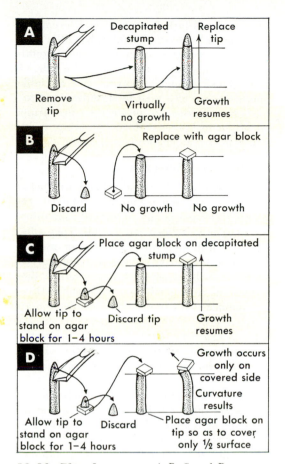

12–10 Plant hormones. *A*, *B*, *C*, and *D* are a series of experiments demonstrating the existence of a diffusible growth hormone (auxin) in the tip of the oat coleoptile.

stops the growth of roots. Thus auxins are not merely growth promoters. They are growth *regulators*, either stimulating or retarding growth depending on the circumstances. In addition, they may directly initiate or promote mitosis. The mechanism of this effect is unknown.

A number of chemical compounds not normally present in plants have auxinlike effects due to resemblances to the natural auxins in molecular composition and structure. Such chemicals can be synthesized in commercial quantities and have many increasingly practical applications. Probably the most familiar at present is 2,4-D.[16] This compound produces

[16] 2, 4-D is a merciful abbreviation of 2, 4-dichlorophenoxyacetic acid.

abnormal, distorted growth and eventually death in wide-leaved plants but has little effect on the narrow leaves of grass. It can therefore be used as a spray to free a lawn of weeds. (It kills clover, though.) Auxins and auxinlike synthetics are also used to stimulate root formation on cuttings, to produce seedless fruits, to keep stored potatoes from sprouting, and to delay or force blooming in flowers.

Other plant hormones

The presence of several other hormonelike substances in plants is known or inferred. The *giberellins* are a family of at least four different but closely related compounds. They affect growth but in ways different from auxins. For example, they do not cause a shoot (or coleoptile) to bend, as auxins do; rather, they induce elongation of the whole stem. Other compounds that can promote cell division are derived from purines and nucleic acids and known as *kinins*. Although they are effective in experiments, it is still uncertain whether they have an essential role in natural development.

Among the most important activities of plants are the induction of flowering and the consequent fruiting. These are influenced by the environment. The environmental control is usually the length of darkness, although it is commonly and confusingly referred to as the day length. "Short-day" (really "long-night") plants, such as ragweed and cocklebur, blossom as the ratio of light to dark decreases seasonally below a characteristic value. "Long-day" (or "short-night") plants, such as plantain and coneflower, bloom when the ratio increases to a fixed value. Elaborate experiments indicate that the ratio of light to darkness influences the formation of some substance in the leaves, perhaps only by the modification of the rate of formation. After its formation the substance moves only through living tissues, especially phloem, to the parts most concerned with flowering. The substance has not yet been surely identified. Part of the difficulty may be that it is not *a* hormone but a mixture or succession of different compounds. Also, it (or they) may not directly induce flowering but may overcome an inhibitor, which could be an auxin. This is one of the most significant of the many unsolved problems of plant physiology.

CHAPTER SUMMARY

Condition of free life: stability of the organism; capacity of the organism to regenerate; fields of organization.

Internal environment: its stabilization as part of evolutionary progress; vascular systems —the sea within us; the roles of the liver and kidneys in regulating plasma composition; control of blood flow and heartbeat—pacemaker and nervous regulation; control of breathing, especially the role of CO_2 as chemical messenger; control of body temperature —devices that cool the body and devices that retain body heat.

Coordination through chemical communication—the endocrine glands; the hormones of man—thyroxin, insulin, epinephrine, the hypophyseal hormones; hormones of other animals; plant hormones.

13

Responsiveness

J. Z. Young

These cells (×188) are from the cerebral cortex of the cat's brain. Their special functions have evolved from an ancient and universal property of cells to receive and react—in a word, to respond—to stimulation.

A sunflower turns to the sun. A hawk dives from the sky. There is more to life than vegetating, even for a vegetable. Most of the life processes that we have considered up to this point are of the sort sometimes called "vegetative." Whether they occur in plants or in animals (and we have seen many basic resemblances between the two), they involve primarily the maintenance of the organism, its metabolism, keeping it going, perhaps in the face of environmental difficulties and changes. While these processes of maintenance were being considered, it was obvious that they were a background for other activities. Organisms not only maintain themselves through environmental changes. They also respond to these changes by characteristic activities. Because everything cannot be studied at once, the existence of such responses has been taken for granted or mentioned incidentally in earlier chapters. Now we are ready to consider them more specifically and systematically.

Reactivity of Protoplasm

If you touch an ameba with a finely pointed needle or glass rod, it moves away (Fig. 13–1). An ameba has no organs (or organelles) for perception of touch—no nerves and no muscles. Nevertheless, in some sense of the word it felt being touched. Its reaction involved putting out an extension on the opposite side of the animal, so that a result of the touch somehow was conducted from one side to the other. The end result was movement away from the foreign, inedible, and possibly harmful object. We have no right to conclude that the ameba really felt the stimulus in the conscious way that we feel the prick of a needle, and we have every reason to believe that it did not. We also know that the conduction of an effect from this contact did not occur, as in us, along nerves. Nor did movement of the ameba involve muscles. Nevertheless, the whole sequence was similar to what happens if you stick your finger with a needle. There was a *stimulus*, there was *conduction* of a signal of some sort set off by the stimulus, and there was an appropriate *response*. The response was appropriate in the sense that, by and large, movement away from inedible ob-

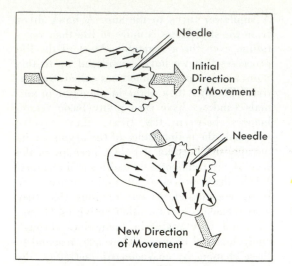

13–1 An ameba's response to being touched with a needle.

jects is likely to take the ameba into territory more propitious for its activity and survival.

The incident shows that an ameba can and does respond as a whole organism. It also shows, and this is our immediate concern here, that protoplasm reacts to environmental changes. It was the cell as a whole, or more particularly the cytoplasm, that reacted in the ameba. This is a general property of living protoplasm wherever it is found, in a protist or in any cell of a higher plant or animal. The response comes from inside and may have little relationship to the nature or strength of the stimulus. The ameba did not move away because it was pushed, and (within limits) its reaction does not depend on how hard you prick it.

Even in the simple ameba we saw that the whole reaction involved three factors or phases. First, the stimulus was received and started some sort of process in the protoplasm. Second, reaction spread from the point where the stimulus was received; there was conduction of a signal. Third, there was a definite response to the signal; in this example the response was a movement. In the ameba the three phases are practically inseparable and certainly are not localized. The same reaction occurs regardless of where the stimulus touches the organism. Conduction is through its cytoplasm as a whole, and the response may also occur anywhere.

The evolution of multicellular animals has been accompanied by clearer separation of the three processes of reaction to a stimulus. In all but the simplest of animals special cells, tissues, and organs are involved in each phase. There are *receptors,* and these are practically always specialized to receive particular sorts of stimuli. They are sense organs. Then there are *conductors,* usually nerves, which carry signals from the receptors. There are also special organs of response, which are called *effectors* because they carry out, or effect (not affect), a response characteristic for the given organism. The conductors, of course, carry signals not only from the receptors but also to the effectors.

The specialization of receptor-conductor-effector organs is highly characteristic of animals. Plants are reactive or responsive to many environmental stimuli, but the mechanisms of the reactions are simpler and less specific. Their reactions, usually slow and not complex or varied, occur mostly through the relatively undifferentiated tissues of growth and maintenance. There are no special conductor cells comparable to the nerves of animals, and receptor or effector organs are few and simple. Here is another of the really basic differences between plants and animals.

Stimulus and Response

KINDS OF STIMULI

A stimulus is a change in the environment capable of producing a response in an organism. Stimuli convey the information that something has happened around the organism. (Of course we do not mean that organisms as a rule are conscious of the information as such.) Any organism, in order to survive, must be sensitive to stimuli from the usual types of environmental changes. Certain broad kinds of stimuli affect any protoplasm. Particular, narrower kinds affect some organisms and not others.

Let us be more specific. All cells and therefore all organisms are affected by any considerable change in pressure, in chemical composition, in temperature, in some kinds of radiation, and in some electrical properties of the environment. An ameba reacts to some stimuli of all these sorts, and so does man. So, too, do plants, although in them the reactions may not be apparent in a short time. In the ameba reaction to pressure may be seen when the organism is touched. Reactions also occur if a bit of weak acid or salt solution (changing osmotic values) is added to the water around the ameba; if the water is heated; if a light (radiation) is shone on it from one side; or if a weak electric shock is sent through it.

Within these broad classes, discrimination of particular stimuli and reactions to them differ greatly among organisms. Both plants and animals usually react to the pressure of gravity. In most plants, the main stem turns upward, away from the direction of gravity. Many higher animals, including humans, have special organs that signal the position of the body or, especially, of the head in relationship to the pull of gravity. Even with his eyes closed a human being knows whether his head is upright or not. Fishes have (humans do not) special organs for reception of pressure changes in currents of water. The rhythmic pressure waves in air that we call "sound" are stimuli for many animals, and humans and numerous other animals can distinguish the frequency with which such waves reach them as well as their intensity.

Discrimination of different chemical stimuli may also be highly developed in animals. Even an ameba moves toward some chemical stimuli, those associated with food, and away from others. Humans have the special chemical senses of taste and smell. Their sense of sight discriminates among radiations by direction, intensity, pattern, and frequency, the last of which they perceive as color.

No organism reacts to any and all possible stimuli from the environment. For instance, all organisms react to *some* wavelengths of radiation but not to *all* wavelengths (Fig. 13–2). Our special receptors for radiation, the eyes, react only to a limited range, from violet (shorter waves) to red (longer waves). Some animals, such as bees, can see radiation that we cannot. Bees see ultraviolet light, invisible to us, but bees do not see red rays, visible to us. Other cells in our bodies do react to radiation that is not visible to us: we are tanned by ultraviolet, and we feel warm in infrared radiation.

No organism is known to have receptors for (or any reaction to) the radiations that are the basis of radio communication, which are much longer waves than in light. Man perceives them only by using devices that turn them into other sorts of stimuli: sound waves or (in television) visible light waves. It is improbable that magnetism, even when very intense, is a stimulus for any organisms.[1]

It is a striking fact that many possible stimuli capable of conveying information about the environment do not actually produce any reaction in organisms. As a rule, however, it seems that such information would not really be useful (and certainly is not necessary) to the organisms incapable of receiving it. The only radio waves in nature come from the stars or from lightning, and this kind of information about the stars or about lightning after it has struck could hardly be useful to a plant or animal. It certainly is not true that something or someone has neatly arranged for each organism to receive precisely all the information of any use to it. That would require purposeful, efficient planning, which is conspicuous in nature only by its absence. It is true that all organisms receive the information *necessary* to their ways of life. Few if any receive more.

EFFECTIVENESS OF STIMULI

A stimulus can produce a reaction only if an organism is capable of receiving it. That is a truism. There are, however, some less obvious points involved in the effectiveness of a stimulus. The first point is that a very weak stimulus may produce no reaction at all. For each sort of stimulus and for each organism there tends to be a strength or intensity below which no reaction occurs. This is the *threshold* of stimulation. If you take a glass of pure water and add one drop of dilute salt solution, the water will still be tasteless. Keep on adding and tasting, drop by drop, and finally at a definite concentration the water will begin to taste salty. This threshold varies greatly in different people and in the same person at

[1] There have been hypotheses that some birds detect the earth's magnetism and also that some of them react to radio waves, but at present these hypotheses are unproved and seem improbable.

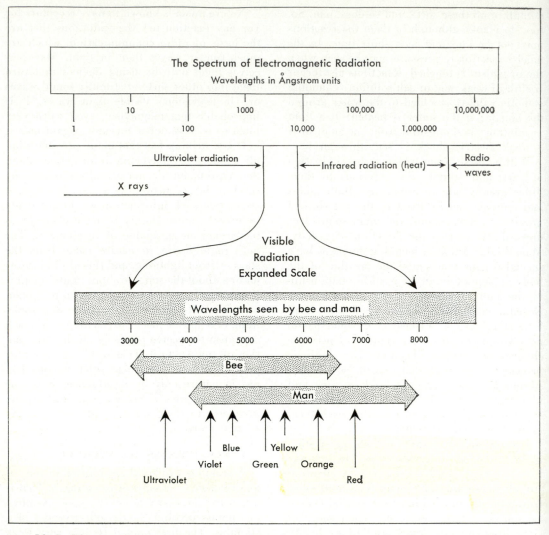

13–2 The sensitivity of man and the honeybee to radiation of different wavelengths.
Living organisms perceive as visible light only a small fraction of the total spectrum of electromagnetic radiation. The radiation visible to them falls within the wavelengths of 3000 to 8000 Ångstrom units (an Ångstrom, named after the Swedish physicist Anders Ångstrom, is 100,000,000th of a centimeter.) The figure indicates, on an expanded scale, the wavelengths seen by the honeybee and by man. Note that bees see ultraviolet light but do not see red light.

different times, but it is often about a 0.1 per cent concentration of salt in the solution.

For any stimulus there is generally an intensity below which a reaction is not produced. Above this intensity, whether a reaction is really produced, whether in fact the threshold is crossed, depends mainly on how long the stimulus continues and how rapidly it changes. In other words, there may be thresholds for duration and rate of change of stimulus as well as for intensity, and the actual threshold in an organism depends on all three.

RESPONSES

A stimulus may affect any process of change in a cell or organism. A response, then, may be anything that an organism can do. The common response to increase in heat, for instance, is simply a speeding up of metabolism. More complexly (Chapter 12) there may be additional responses that set cooling mechanisms in operation. Such reactions are part of organic maintenance, which we have already reviewed in previous chapters. Now we are especially interested in more specific responses to immediate environmental changes, to particular stimuli.

The nature of the response depends more on the organism than on the stimulus. This is a special characteristic of responsiveness in living things. How a football responds to "stimulation" depends precisely on the stimulus. If you kick it or throw it, the result is always a motion fully determined by the speed, strength, and direction of the force applied to it. If you stick a pin into it, the response is quite different, and if you set fire to it, the response is different still. In organisms, the responses are built in, so to speak, and have no such simple relationship to the stimulus. The response built into the nerve of a tooth is pain, and an exposed nerve gives a pain reaction whether you poke it, eat a lemon, drink cold water, or give it an electric shock. (Some dentists have a disagreeable way of determining whether a nerve is alive by an electrical test.) If someone hits you in the eye, you "see stars"—an apparent flash of light. A hard blow stimulates the optic nerve, the built-in response to which is the sensation of light. This specificity of response determined by the organism rather than by the stimulus is not confined to the nervous system. Some kinds and intensities of pressure, chemicals, radiation, and electricity may all produce the same response in an ameba—rounding into a ball. Muscular tissue responds to a variety of mechanical, chemical, and electrical stimuli, and always in the same way, by contracting.

There are some other special characteristics of responses in living matter. As a rule the response does not begin immediately when a stimulus, even a strong stimulus, is applied. There is a *latent period* before the response begins. This period varies greatly in different tissues, but is usually very short, often about 1/1000th of a second. Nerve and muscle tissue, in which these reactions have been most studied, also have the property of not responding again immediately after a response. For a short period, the *refractory period*, no stimulus is sufficient to produce a new response. In heart muscle this period may be relatively long, up to $1/5$ second or so. However this property may have evolved, it has important and useful consequences. No matter how often or continuously the heart is stimulated, the muscle cannot remain contracted (which of course would stop its pumping and lead to death). In ordinary skeletal muscle (see Chapter 3) the refractory period is much shorter, often about 1/1000th of a second. One can flex his finger faster than the heart can beat, and one can keep the finger muscles contracted indefinitely. There is some question whether a refractory period is a universal characteristic of protoplasm, but it does occur in amebas. An ameba that has rounded into a ball as response to a stimulus does not, and presumably cannot, do so again for several seconds.

Nerves and Nerve Action

In all but the lowest animals the most essential and characteristic part of the mechanism for specific responses to stimuli is the nervous system. All special receptor organs in animals transform stimuli into nerve impulses. A nerve impulse is an electrical signal transmitted along fiberlike extensions of nerve cells. The nerves are the conductors from receptors to effectors. Their pathways and connections determine what, in the end, the response will be. Increasing complications along these pathways are mechanisms for coordination of actions, for memory, for perception, for association, and finally for all the richness of our own mental lives. The nervous system, too, has come to share with chemical integration the wisdom of the body, automatic maintenance of the internal environment. (See Chapter 12.) The nervous system of a clam or of a worm is truly remarkable. What can we say of the nervous system of man, so vastly more complex? It is this that makes us mankind, organisms beyond the potentialities of any others that have ever existed.

NERVE CELLS

One of the most remarkable facts about the nervous system is that, with all its complication and flexibility of reaction, it is made up of cells all of which perform essentially the same action. The action is produced by the extremely complex, interlocking processes so characteristic of life, but the action itself is rather simple; it is merely the transmission of impulses from one end of the cell to another. The impulses are of the same sort in all nerves, although they may be involved in seeing red (either literally or figuratively), in composing a poem, in telling us that we are hot or cold, or in the coordinated contracting of muscles that eventuates in a song or a swift kick. It is not the separate nerve cells but their arrangements and connections that determine the qualitatively different outcomes. Again we see that properties of the organism are different from and more complex than those of its separate parts.

The unit of the nervous system is the nerve cell, or neuron (see Fig. 3–35). For all their differences in size and shape, neurons are fundamentally alike in all animals that have them. The nucleus lies in a central cell body that is drawn out into two or more fibrous projections. The projections may be short, but usually are long fibers, much thinner than hairs. Nerve impulses can pass in either direction along these fibers. In some lower animals an impulse may come into the cell body along any fiber and then move away through all the others, out to their tips. Most neurons in higher animals, however, are so arranged anatomically that impulses always travel through them in the same direction. A frequent, although not the only, arrangement in these cases is with numerous, shorter branching fibers, the *dendrites*, on one side, through which impulses come into the cell body, and one longer fiber, the *axon*, on the other side, carrying impulses away from the cell body. The fibers are often enclosed in one or more sheaths. The sheaths are not essential to conduction, for unsheathed fibers also occur and transmit impulses of the same sort. The sheaths look rather like insulation on electric wires, and they do include partial insulators that increase the speed of impulse transmission. One of the sheaths, when present, is also known to play a crucial role in the regeneration of injured fibers.

A nerve such as you are likely to see when you dissect an animal in the laboratory is not a single fiber but a whole bundle of fibers belonging to different neurons. In vertebrate animals most of the cell bodies of nerves are in or near the brain and spinal cord. Some of the fibers may run through nerves for long distances. In man some extend down to the toes from cell bodies in the small of the back. It is exceedingly difficult to follow a single fiber through its whole length, and earlier students could not believe that a fiber in a toe was actually part of a cell with its nucleus several feet away. Experimentation and more delicate dissection have proved that this is true.

NERVE IMPULSES

The nature of the nerve impulse has been the subject of long, ingenious investigation and of much dispute. The main features now seem to be well established, although certain details remain unknown. How would one go about finding what chemical and electrical changes are going on in a fiber much smaller than a hair? Just to make it more difficult, the changes take place in a small fraction of a second, and everything is back as it was before the impulse passed. Despite these difficulties, so much has been learned of the nerve impulse that we can only indicate its general nature here. The impulse is electrical, although associated with and set up by chemical changes. It is not simple conduction of an electrical current, as in the transmission of a message over a wire. The impulse accompanies a zone of electrical charges in the nerve fiber, a zone that moves along with the flux of small, purely local currents and ions between the fiber and the fluids immediately around it.

The mechanism of the nerve impulse has a bearing on the properties that, in turn, help in understanding the whole process of reaction to stimuli in animals. One consequence of the mechanism is that the impulse moves much more slowly than a current in a wire. The speed varies greatly from one nerve to another and one animal to another but is always slow enough that conduction takes appreciable time. In the fastest fibers of mammals, including man, the rate is about 100 meters per second. This is a little less than 225 miles per hour, which does not seem very fast to us in this age of much faster planes. In some of the slow invertebrate fibers, the rate is as low as 5 centimeters per second, or about 0.1 mile per hour. A tortoise can walk faster than this. Even in some human fibers the rate is lower than 2 meters per second, which is no faster than we can walk.

Another point about the mechanism of nerve conduction is that the strength of the impulse is standardized. It is not like putting an impulse of variable strength into a wire. The reaction is local in each part of the fiber. If an impulse is started at all, it starts at full strength and the strength is not affected by the length of the fiber. A usual way to make this sort of process understandable is to compare the fiber with a sprinkled line of gunpowder. If the powder is lit at one end, the flash travels to the other end. The rate of travel and the strength of the flash do not depend at all on the heat of the match with which it was lit or the length of the line. A nerve fiber similarly transmits impulses by local power at each point and has an *all-or-none reaction*. Comparison with the gunpowder cannot be carried further, because the reactions involved are really quite different.

No matter how strong the stimulus is, the impulse in a given nerve has a fixed strength. How, then, does it happen that the impulses do have varied intensities in their effects? The answer is that a stimulus seldom starts a single impulse. Unless the stimulus is extremely brief, repeated impulses pass along the fiber one after the other. A stronger stimulus results in impulses that are closer together, hence more frequent. More of them arrive at the other end, and so they can have a stronger effect even though each has the same intensity.

All protoplasm reacts to stimulation and conducts impulses. Nerve cells do not have unique properties, but are only specialized in the sense of heightening the particular property of conductivity and directing it anatomically. Nerve cells also carry on respiration and other metabolic processes of protoplasm in general. In fact nerve cells have particularly high oxidation (nerve cell mitochondria, the loci of this oxidative activity, can be seen in Fig. 3–35) rates and are especially sensi-

tive to variations in concentration of sugar, the principal source of energy. Thinking really is work, not just because some of us are reluctant to indulge in it, but because it does use energy even though in amounts smaller than for muscular activity.

NERVE CONNECTIONS

In some lowly animals, receptor cells are in direct contact with an effector. The receptor or sensory cell receives a stimulus and transmits an impulse (as if it were a neuron) direct to a muscle cell (Fig. 13–3A), which contracts when the receptor is stimulated. In the next

13–3   Receptor–conductor–effector systems.

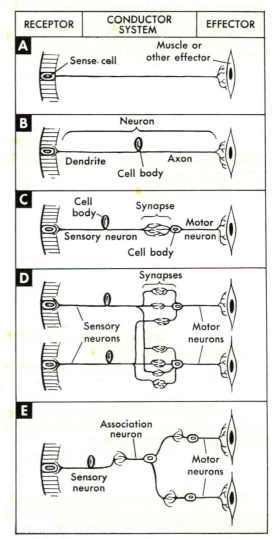

stage of complication, the sensory cell starts an impulse in a separate nerve cell, through which the impulse travels to a muscle fiber or other effector (Fig. 13–3B). In either of these systems, the reaction is necessarily simple and entirely inflexible. A sufficient stimulus, one above threshold, on the sensory cell invariably and necessarily results in an impulse to one particular effector, which responds in a way that usually varies only in duration.

In higher animals sensory cells are rarely in direct contact with effectors, and they never are in vertebrates. Almost always there is more than one nerve cell between receptor and effector, with branches in the possible lines traveled by nerve impulses. A sensory neuron carries impulses from the receptor and passes them on to a motor neuron. The motor neuron in turn conducts the impulse to an effector, characteristically a muscle fiber, which reacts. This is still a simple chain from receptor to effector. It is a reflex arc (Figs. 13–3C and 14–11). Even this relatively simple arrangement brings in the possibility of more effective and flexible response than direct connection between a receptor and an effector. When there are separate sensory and motor neurons, one sensory neuron can stimulate more than one motor neuron (Fig. 13–3D). Response to stimulation can be more extensive, and continued stimulation can spread so as to involve more and more effectors. Yet the reaction must still be a simple reflex, a connection from one particular receptor to one particular set of effectors.

The next complication, which has become practically universal among animals such as insects or vertebrates, is the occurrence of still other nerve cells, one or many, between the sensory and the motor neurons (Fig. 13–3E). These additional cells are association neurons. Through them, an impulse may be passed on selectively to any of a number of different effectors. Impulses from different receptors may also be brought together and routed to the same or different effectors. Simple reflexes can still occur, but the possibilities for more complex and flexible reactions are tremendously increased.

The fibers through which an impulse is passed from one neuron to another are not continuous or fused to each other. They may

be pressed close together, but there is always a tiny gap between them, a separation by cell membranes, at least. The point of transfer, with its tiny gap, is called a synapse. The actual mechanism of transfer is still incompletely known. When an impulse reaches a synapse, a minute quantity of a chemical compound appears there. The compounds that have been detected (there are at least two of them) have a stimulating effect on nerves. One of them is very like epinephrine (Chapter 12), and the other is acetylcholine (see also p. 385). These chemicals bring about the transmission of the nerve signal, an electrochemical impulse, from one nerve fiber to another.

The fact that impulses are not conducted through fibers from one neuron to another has some important consequences. One or a few impulses or widely spaced impulses may arrive at a synapse and not cross it. If numerous impulses arrive at close intervals, they add up (summation occurs, technically speaking), and finally they are transmitted across the synapse. A brief, weak stimulus produces few and widely spaced impulses. These may cross no synapses, so that no response occurs, or few synapses, so that response is weak and local. A long, strong stimulus produces many and closely spaced impulses. These readily cross synapses, so that response definitely occurs, and can cross many synapses, so that response may become strong and widespread, especially as association neurons become involved.

Another important property of synapses is that once an impulse has been transmitted across them, subsequent impulses pass more readily. It may take a long volley of closely spaced impulses to cross a synapse, but once this has occurred, a short sequence of impulses or even a single impulse may cross. The fact that impulses have already been transmitted facilitates transmission of later impulses; this phenomenon is called facilitation. If no further impulses do come along, the effect fades out, often in a matter of seconds or minutes. If impulses keep coming along before facilitation has entirely faded out, the facilitation is maintained and increased. Continual crossing of a particular synapse at appropriate intervals can maintain facilitation there for a lifetime. Thus there are established pathways in the nervous system, routes across facilitated synapses, along which impulses move more readily and rapidly.

Once a response or an association has occurred, it occurs more readily soon thereafter. If it occurs often, its readiness increases and it may be maintained indefinitely. Clearly this is a mechanism that goes far toward explaining habit, learning, and memory.

The termination of a nerve fiber on an effector, such as a muscle (see Fig. 13–3), is anatomically unlike a synapse, but it has similar properties of summation and facilitation.

NERVE PATTERNS AND RESPONSES

In animals with well-developed nervous systems, specific reactions to stimuli depend mainly on how nerves act and how they are arranged. The kinds of reactions that may occur depend on the nerve pattern: how many nerves there are, what receptors stimulate them, how they run to synapses and to effectors, what associative paths are present, and so on. This arrangement is determined mostly by heredity, although it may also be somewhat affected by the conditions of early development. The reactions that actually *do* occur depend first of all on the stimuli that an individual happens to receive, next on the mainly inherited nerve pattern, and finally on the past experiences and current condition of the individual. Facilitation, in particular, depends largely on what responses the individual organism has already made.

Thus in higher animals what an animal does, its whole pattern of behavior, and indeed what sort of creature it is are more closely related to its nervous system than to any other one factor. As neurons become more abundant and patterns more complex, culminating in man, responses and associations become more varied, too, and so do the experiences imprinted on the nervous system. Herein is the basis not only of the high and numerous capacities of mankind, but also of individuality and of personality.

Receptors

The nervous system is largely involved in handling information about the environment.

(It also handles information about the organism itself, but that is a different point for later consideration.) It is involved in responses to changes in the environment and also, with increasing complexity, with sorting out and associating information, storing impressions for future responses, varying responses according to current situations, and other activities all of which go back to the receipt of information about the environment. (Can you imagine what your mental life would be like if you could have no knowledge about things outside your own body?)

We have inherited from antiquity the popular notion that we have just five senses: sight, hearing, taste, smell, and touch. A little thought or simple experimentation suffices to show that this is false. We have five large, complex, and definitely localized sets of sense organs: eyes, the hearing mechanism of the ears, an organ of equilibration connected with the inner ear but producing very different sensations, the taste organs in the mouth, and the smelling organs in the nose. It is significant that all these localized organs are in the head. Elsewhere throughout most parts of the body we have an extremely large number of small, anatomically simple, scattered receptors. These vary and intergrade so much in structure and in the sensations produced that it is really impossible to say how many senses they represent.

In the skin, especially well provided with scattered receptors, there are at least four distinct senses, that is, there are are at least four kinds of receptors each of which on stimulation produces a different sort of sensation. The sensations are: warmth, cold, pain, and touch or pressure. A little exploration with a pin and small warm and cold rods will convince you that these sensations have receptors definitely localized at certain spots and distinct from each other. Many other receptors of diverse sorts occur within the body. We feel the tenseness of muscles, the motion of joints, hunger, thirst, internal pain, nausea, sexual orgasm, and other distinct sensations that originate inside ourselves.

An ameba or almost any single, undifferentiated cell reacts to stimuli of many different kinds. Now we are interested in the development of special receptors that react to particular sorts of stimuli. Such specialization clearly depends on protoplasm and has evolved from its general reactivity. In all but the simplest animals it has resulted in the development of organs in which stimulated receptor cells start impulses in nerve fibers. The most conspicuous, at least, of these organs can be classified as follows:

Receptors of	Names of organs
Light	Photoreceptors
Sound	Phonoreceptors
Touch and pressure	Tangoreceptors
Gravity and motion	Statoreceptors
Taste and smell	Chemoreceptors
Temperature	Thermoreceptors

LIGHT RECEPTORS

All protoplasm is sensitive to some kinds of radiation, including the radiation that is visible to us and that we therefore call *light*. (But sensitivity in most protoplasm is usually greater to ultraviolet, which is invisible to us.) As a stimulus giving information about the environment, light is in a class by itself in the amount of information it can give and in giving information about things at a distance from the organism. Comparisons of simple and more complex light receptors, or photoreceptors, are especially interesting because they show how more and more information can be gained from the same stimulus.

The very simplest sorts of light receptors— one could hardly call them "eyes" at this stage—occur even in some protists as well as in simple multicellular animals (Fig. 13–4). They are sensitive spots of light-absorbing pigment. They really give no information from a distance but only indicate whether light is present or absent where the animal is. Some or all of these simple animals also react to increase or decrease in intensity of light.

More complex eyes, such as occur in most invertebrates, have evolved with great diversity of details in form and structure. They almost always involve one or both of two features: the presence of a lens and of several to many separate light-sensitive cells, each capable of starting impulses in a nerve fiber. A lens concentrates light on the sensitive cells. It therefore permits reaction to weaker intensities of light and also finer discrimination between different intensities. An eye with a lens also gives a new sort of information: the direction from which light is coming. This clearly facilitates and directs the response, widespread in animals, of moving toward or away from light. Increase in number of sensitive cells also makes response to light and to changes in intensity more delicate. Moreover, if there are many sensory cells back of the lens, different cells will be stimulated in succession when light or dark objects move in front of the lens. In the simple kinds of such eyes no true image is formed. The organism cannot tell *what* is moving in front of its eye, but it can distinguish and respond to the fact that motion has occurred. This is a great advance in amount of information received, and it can include information (even though of a vague sort) about happenings at some distance from the animal.

There is no doubt that lenses first evolved not as image-formers but merely as mechanisms for concentrating light from particular directions. It is, however, a fact (you can call it a peculiarly fortunate coincidence) that lenses can form images by focusing light on a surface. If a lens is of appropriate shape, if the number of sensitive cells increases greatly, and if all the cells are arranged as a *retina* on the curved surface where the lens focuses an image, then it is possible for the eye to receive an image and to translate it into a pattern of nerve impulses (Fig. 13–5).

These evolutionary developments have occurred more than once. There need not have been, and quite surely was not, a definite point or sudden change when eyes began to receive images. Increased discrimination of light intensity, direction, and movement would gradually begin to produce a vague image. Variation such as occurs in all groups of animals would mean that the image was a little less vague for some than for others. It is definitely an advantage for a sufficiently complex animal to be able to discriminate what is approaching it; this becomes possible as even vague images are formed. On an average, animals with clearer images would have better chances to survive and to pass their characteristics down to posterity. Thus extremely slow but steady improvement in the image would occur. What do you think of the claim, sometimes made, that eyes like ours must have

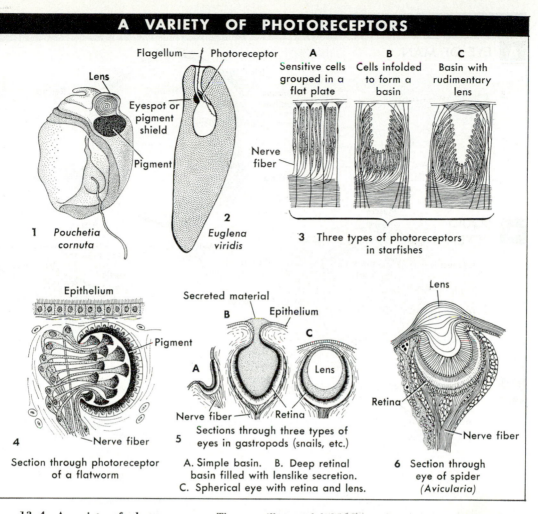

1 *Pouchetia cornuta*

2 *Euglena viridis*

Flagellum — Photoreceptor
Lens
Eyespot or pigment shield
Pigment
Nerve fiber

A Sensitive cells grouped in a flat plate
B Cells infolded to form a basin
C Basin with rudimentary lens

3 Three types of photoreceptors in starfishes

Epithelium
Pigment
Nerve fiber
4 Section through photoreceptor of a flatworm

Secreted material
Epithelium
B
C
A
Lens
Nerve fiber — Retina
5 Sections through three types of eyes in gastropods (snails, etc.)
A. Simple basin.   B. Deep retinal basin filled with lenslike secretion.
C. Spherical eye with retina and lens.

Lens
Retina
Nerve fiber
6 Section through eye of spider (*Avicularia*)

13–4 A variety of photoreceptors. The eyes illustrated belong to a diverse array of organisms. In groups as different as snails, starfishes, and spiders, there has been an evolutionary tendency to develop special light-absorbing cells and a lens that concentrates or focuses the light on these cells. The frequency with which this system has evolved in unrelated organisms attests to the importance of photoreceptors. Light energy is actually absorbed by pigment molecules in the receptor cells of the eye. The absorbed light energy initiates in these cells chemical reactions that ultimately stimulate associated nerve cells. *1* and *2* are examples of very simple photoreceptors in unicellular organisms. In some protists, like *Pouchetia* (*1*), a simple lens concentrates the light on the absorbing pigment molecules in the cell behind the lens. *3*. A simple kind (*A*) of photoreceptor in starfishes consists of a flat plate of nerve cells that contain the light-absorbing pigment. Other starfish photoreceptors (*B, C*) illustrate the evolution of a more efficient system with infolded nerve cells and eventually a simple lens formed by a thickening of the epithelium. *4*. In the flatworm's eye light-absorbing pigment is concentrated in a separate layer of cells lying in front of the nerve cells they ultimately stimulate. *5*. Snails (gastropods) have eyes of varying degrees of complexity. The simplest (*A*) is a basin of pigment-carrying cells supplied with nerves. The most complex (*C*), which has surely evolved from a simple beginning like *A*, has a spherical lens. *6*. The eye of the spider *Avicularia* is similar to that of the snail but arose entirely independently.

IMAGE-FORMING EYES - I

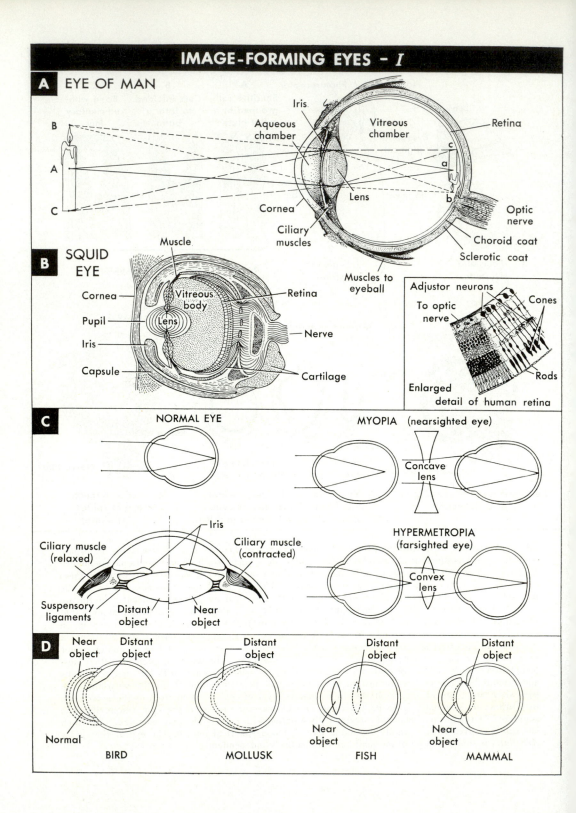

A EYE OF MAN

Iris
Aqueous chamber
Vitreous chamber
Retina
Cornea
Lens
Optic nerve
Ciliary muscles
Choroid coat
Sclerotic coat
Muscles to eyeball

B SQUID EYE

Muscle
Vitreous body
Retina
Cornea
Pupil
Lens
Nerve
Iris
Capsule
Cartilage

Adjustor neurons
To optic nerve
Cones
Rods
Enlarged detail of human retina

C

NORMAL EYE

MYOPIA (nearsighted eye)
Concave lens

Iris
Ciliary muscle (relaxed)
Ciliary muscle (contracted)
Suspensory ligaments
Distant object
Near object

HYPERMETROPIA (farsighted eye)
Convex lens

D

Near object
Distant object
Normal
BIRD

Distant object
MOLLUSK

Distant object
Near object
FISH

Distant object
Near object
MAMMAL

appeared all at once because they would be of no use at all until they were perfect?

Eyes with single lenses forming useful images have evolved, entirely independently, at least three times: in some of the more active and complex mollusks (such as the octopus), in some spiders, and in the early vertebrates, from which we and all other vertebrates have inherited them.

There are several other ways besides single lenses by which images can be formed mechanically, or rather, optically. Insects and some of their relatives evolved image-forming eyes that work on one of these other principles. Their *compound eyes* have a large number of tubes, *ommatidia,* each of which is an independent light-sensitive unit. Since each tube points in a slightly different direction, each receives light from a different part of the insect's surroundings. Each ommatidium therefore responds to more or less light from one area of the surroundings, and the sum of all ommatidial reactions is a rather crude but often recognizable image (Fig. 13–6).

There is another sort of information that can be conveyed by light and that may be useful in identifying objects and discriminating among them; this is *color,* which depends on the wavelengths of light transmitted or reflected. Animals may, and many do, see very well by light of various wavelengths without discriminating *differences* between the wavelengths. Such animals do not have color vision. (It really is not correct to call these species colorblind, because most of them simply never had color vision.) Color vision, the ability to distinguish different wavelengths of light, has evolved several times. Some crustaceans have it, and so do many insects. Some insects, including honeybees, not only see ultraviolet but also see it as a color, as different from, say, green. Of course we have not the slightest idea what ultraviolet color looks like to a bee, any more than a bee could imagine (if it had an imagination) what red looks like to us. Red is black, absence of light, to a bee as ultraviolet is to us.

Strangely enough, color vision may have been lost and regained in our ancestry. Many fishes and reptiles and most birds have color vision, but most mammals do not. Primitive mammals probably lacked color vision. At

---

13–5   Image-forming eyes of the camera type. Image-forming eyes comparable to cameras have evolved in at least three different groups of animals. The lens system has been transformed in the course of evolution from a mere light-concentrator to a precise optical device that focuses an image on a light-sensitive retina. *A.* The eye of man, a vertebrate. *B.* The eye of the squid, a mollusk. Essentially both eyes consist of a light-sensitive retina; a light-focusing system, cornea and lens; an iris that controls the amount of light entering the eye by adjusting the diameter of the pupil; and a protective coating. In the squid the protective coating is a complex and heavy cartilaginous casing; in man it is the sclerotic coat, a strong elastic connective tissue. (The cornea is the transparent and strongly curved anterior portion of this sclerotic coat.) The detail of the inset shows how the rods and cones (p. 367) are arranged in the human retina. Note how the nerve fibers from the retina run over its surface on their way to the optic nerve. *C.* Accommodation (or focusing) of the human eye. Incoming light rays are focused on the retina mainly by the strongly curved surface of the cornea. Final adjustment is made by the lens, which is suspended by ligaments immediately behind the iris. The curvature of the lens surface is regulated by the tension of the ciliary muscles. When the ciliary muscles contract, ligamentous tension on the lens decreases, and the elasticity of the lens causes it to bulge, increasing its surface curvature and bringing near objects into focus on the retina. Myopia (top right in *C*) is an abnormal condition in which the eye cannot focus on distant objects. Its usual cause is an eyeball too long for its lens system, so that, even when the ciliary muscles are fully relaxed, the focal plane lies in front of the retina. The condition can be compensated for by the use of concave lenses in eyeglasses. Hypermetropia (bottom right in *C*) is the reverse condition, in which the eyeball is too short. Convex lenses are used to compensate for this condition. *D.* Diverse methods of accommodation in camera eyes. Mammals generally, like man, accommodate by varying the curvature of the lens surface (far right in *D*). Other ways of adjusting focus have, however, been evolved in various animal groups. Accommodation is of unusually great importance in a bird of prey, which must focus on its victim not only when at a great distance above ground but also, following its swoop to earth, when its victim is immediately in front of it. Predatory birds accommodate by a radical change in the curvature of the cornea rather than of the lens. Some mollusks accommodate, not by changing the curvature of the refractive surfaces, but by shortening the eye itself. With the lens thus closer to the retina, the eye accommodates to distant objects. Fishes lack ciliary muscles and hence cannot change the curvature of the lens; instead muscles within the eye change the position of the lens, pulling it back closer to the retina to accommodate to distant objects.

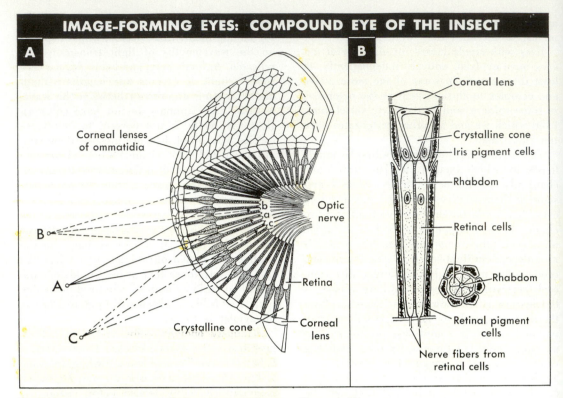

A. Corneal lenses of ommatidia · Optic nerve · Retina · Crystalline cone · Corneal lens

B. Corneal lens · Crystalline cone · Iris pigment cells · Rhabdom · Retinal cells · Rhabdom · Retinal pigment cells · Nerve fibers from retinal cells

13–6 Image-forming eyes of the compound type. *Above. A.* The compound eye of an insect cut away to show some of the hundreds of individual ommatidia. *B.* A single ommatidium. The corneal lens and crystalline cone focus incoming light rays onto the rhabdom, a clear, rod-shaped structure. Light passes from the rhabdom into eight retinal cells that surround it. Each retinal cell contributes a nerve fiber to the optic nerve. The whole ommatidium is surrounded by pigment cells, which prevent the leakage of light from one ommatidium to another. Compound eyes of this type form erect images (cf. Fig. 13–6*A* with Fig. 13–5*A*). In a camera-type eye the image is optically inverted when it falls on the retina and has to be inverted again in the nervous system. No such optical inversion is involved in the mechanism of the compound eye.

Edwin Way Teale

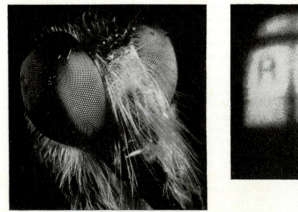

Left. The face of a robber fly. The individual ommatidia in the compound eye are readily seen.

Right. A photograph made through the eye of a firefly, showing the erect image transmitted to the insect's brain. The picture is of a church steeple seen through a window, on one of whose panes a capital *R* has been written.

least this is slight or absent among most present-day mammals except man and his nearest relatives, the apes and some monkeys. Dogs, cats, horses, and most other mammals may in some cases have very feeble discrimination of colors, but more probably they have no color vision.

The evolution of image-forming eyes has involved other refinements and extensions of the capacity to obtain information from light. Many of these refinements can be seen in our own eyes, which are about as highly developed as any. The whole apparatus is enclosed in a ball, which can be turned by its own muscles. Both eyes habitually focus on the same point, and the stereoscopic effect is a clue to distance. (This is uncommon in animals other than man; man also uses other clues to distance, and many other animals rely wholly on other clues than the stereoscopic effect.) The lens is elastic and can be focused. A variable diaphragm, the *iris*, automatically adjusts to light intensity. The sensory, retinal, cells are of two sorts, one set (the *rods*) especially sensitive and active in dim light, the other set (the *cones*) active in bright light. (Only the cones are involved in color vision, but precisely how is still uncertain.) Altogether, the eye in the higher vertebrates—perhaps even more in birds than in man or other mammals—is the most complex receptor that has ever evolved.

Before leaving the subject of light reception, we must mention another point of great importance here. There is much more to the sense of vision, or any other sense, than the receptor. Complex as it is, the eye does no more than send a series of signals, all of the same sort and intensity but varying in frequency, along a large number of nerve paths. If these signals were diverted to a muscle, the muscle would contract, and that is all. The fact that an optical image was involved in their formation would have no meaning. The image becomes meaningful—it is really information to the organism—only if all the separate signals are somehow associated into a whole pattern simultaneously grasped. This process does not occur mainly if at all in the eye. It takes place through an incredibly complex arrangement of extremely numerous associative and conductive neurons in the brain. Thus the image formed in the eye is transformed into a *perception* in the brain.[2]

It follows that an eye that is optically image-forming is not really image-forming for the organism unless it is accompanied by a complex associative system precisely related to it. The advanced sort of receptor could not evolve without this evolution in the brain. Organisms without brains or with simple brains can have only simple light receptors or none. At each level of complexity there is only so much information that can become meaningful for the organism and that can, therefore, really be information for it.

The effector system is also involved in the interrelationship that starts with the receptor. The more complex receptors make finer discriminations. A simple eye receives information adequate to conclude, "Something is moving." But a complex eye provides enough information for the discriminating observation, "A yellow house cat a foot long is coming toward me slowly from 10 feet away a little to my left." Such varied and detailed discrimination is of no use to the organism unless the possible responses are also varied and detailed. An ameba's responses are just about exhausted when it rolls into a ball, moves one way or another, or puts out a projection and engulfs a food particle. Discriminating precisely what was touching it would have no significance in the life of an ameba because the limited repertory of possible reactions would be precisely the same in any case. Discrimination in receptors simply does not evolve unless it is accompanied both by a correspondingly complex association system and by appropriately varied effector responses. Does this have a bearing on the near lack of receptors in plants?

TOUCH, PRESSURE, AND SOUND RECEPTORS

Reviewing the animal kingdom, we will find it difficult to distinguish sound receptors from touch and pressure detectors. Protoplasm as a

[2] There is a speculative current theory that perception (and not merely sensation) involves the sense organ in addition to or even in place of the brain. It is not generally accepted and seems improbable; even if it were true, appropriate reaction to a complex perception would still require a complex brain.

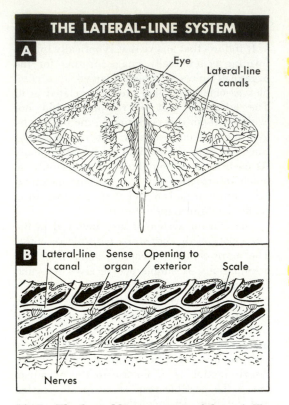

THE LATERAL-LINE SYSTEM

A

Eye

Lateral-line canals

B

Lateral-line canal Sense organ Opening to exterior Scale

Nerves

13–7 The lateral-line system in fishes. *A.* The distribution of the lateral-line canals on the surface of a ray. *B.* A section through the surface of a perch. It shows the scales embedded in the skin; a lateral-line canal with its openings to the exterior; and pressure-sensitive sense organs supplied with nerves.

whole—in protists, plants, and lower animals —is sensitive to touch and pressure, but not to sound as such. Sound receptors, however, clearly evolved in higher animals on the basis of this primitive sensitivity to pressure. These receptors are more discriminatory organs for pressure reception, specialized for reaction to vibratory changes of pressure (that is, a sound wave) in water or air. Besides these special organs, higher animals, including man, also have simpler receptors for other sorts of touch and pressure.

In most aquatic invertebrates touch and sound receptors, if present, are the same. Frequently they are sensitive hairs that respond to a touch and also vibrate when affected by sound. Insects also receive sound stimuli with vibratory hairs, but in addition they have small, special sound receptors that may be scattered in the body or grouped in rather

simple special organs. These organs discriminate intensities (loudness) and modulations (change of frequency) in sounds. They do not, however, permit the organism to recognize the absolute pitch or frequency of a sound.

Fishes have a special sensory system lacking in terrestrial vertebrates. This consists of a series of grooves or canals with clusters of sensory cells on head and body, the *lateral-line organs* (Fig. 13–7). They are sensitive to changes in pressure or currents in the surrounding water. These organs also occur in the aquatic, larval stages of amphibians, as in the tadpoles of frogs, but are usually lost in the adults. They are absent in reptiles, birds, and mammals, even those such as whales that have become secondarily aquatic.

Aside from their evident importance to most aquatic vertebrates, the lateral-line organs are of special interest because they seem to be the primitive, simple pressure receptors from which organs of hearing and equilibrium have evolved. In fishes the ear is entirely internal and is mainly an organ of equilibrium. There has been some argument as to whether fishes really hear, in the sense of discriminating the sort of vibrations that we detect as sound. The answer is that some do and some do not. Many fishes do not respond at all to sound vibrations, but some quite specialized forms do respond and also discriminate between different frequencies or pitches of sound waves.

The land vertebrates have specialized organs of hearing that seem to have evolved from parts of the pressure and equilibrium receptors of fishes and are still closely associated with those receptors in the ear. The anatomical details of sound receptors differ markedly in diverse groups and have evolved in an interesting way, but the same principles are involved in amphibians,[3] reptiles, birds, and mammals. Air vibrations (sound waves) hit an eardrum (technically, the tympanic membrane) and cause it to vibrate at the same rate as the sound wave. The membrane is attached to a small bone that vibrates with the membrane. Either directly or through one or

[3] In most living amphibians the hearing organs are aberrant or degenerate.

two other bones, this bone passes the vibrations on to a fluid-filled cavity of the inner ear. Here vibrations in the fluid stimulate sensory cells on another membrane in a special sac or tube, the cochlea.

The hearing organ is most highly developed in man and other mammals. Here there is an outer ear, with a tube leading to the eardrum. In the middle ear there is a chain of three small bones. The final receptor in the inner ear is a coiled tube with an elaborate arrangement of membranes and sensory cells. Such an arrangement is highly sensitive and discriminates delicately among all the changes that occur in sound, those of intensity (loudness), frequency (pitch), and timbre (pattern or quality).

THE SENSE OF EQUILIBRIUM

For an animal leading a free life, swimming in water, walking on land, or flying in the air, it is important to know which way is up. In other words, such animals usually need to orient themselves in the field of the earth's gravity. Most of them have special organs sensitive to gravity. In invertebrates the organ is usually a sac in which there is a small stony ball. The ball is acted on by gravity and stimulates sensory hairs or cells (Fig. 13–8). It is curious that insects, although usually well oriented, only exceptionally have a special gravity receptor.

Vertebrates have the same kind of receptor as that described for invertebrates. In the inner ear are two sacs, each with a usually stony secreted mass. Vertebrates also have a more specialized receptor, connected with this one, which responds to changes in rate and direction of motion. This consists of two (in the most primitive forms) or three (in almost all recent vertebrates) semicircular canals (Fig. 13–9). The canals, which are connected, are arranged approximately at right angles to each other and are full of fluid. Any increase or decrease of motion causes the fluid to flow in the canals, and the pattern of flow in the three canals depends on the direction of motion. The flow is detected by sensory cells. You can readily recognize the sensations from this apparatus if you close your eyes and nod or shake your head. In some way not wholly clear, there is a connection between this

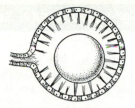

13–8   A statocyst from the mollusk *Pecten*. The central body is a statolith, a stony concretion of calcium carbonate. Gravity causes it to press on sensory hairs that line the hollow sphere of cells. Which sensory hairs are being stimulated is the animal's cue to its orientation to gravity.

mechanism and the abdominal region. If you go up or down in a fast elevator, the motion is sensed mainly in the semicircular canals, but it is most clearly felt in the pit of the stomach. This connection is instrumental in sea or air sickness. Do you suppose that there is any relationship between the fact that up or down motion is disturbing to us and the fact that extensive motion in these directions seldom occurs to land animals or primitive men and may be dangerous when it does occur?

CHEMICAL RECEPTORS

The chemistry of the environment is extremely important to any organism. It is not surprising that all protoplasm is sensitive to chemical changes around it and that in many animals special chemical receptors have evolved. Even without special receptors many protists and lower animals detect their food through a general chemical sensitivity. Molecules or ions from the food diffuse through water and act as stimuli. Harmful chemicals are also detected and produce defensive responses such as motion away from the stimulus or contraction into a less penetrable mass. Most aquatic invertebrates are sensitive to chemicals over their whole bodies, and many have developed rather simple local receptors. For instance, clams and some other marine mollusks have little patches of yellow cells that taste the water drawn in and circulated through the gills. The most specialized chemical receptors among invertebrates occur in insects. They have scattered simple receptors, rather like short, blunt hairs, and also have special taste receptors in or near the mouth and special smell receptors on the antennae. Some insects can taste things with their legs.

THE HUMAN EAR: SENSE OF SOUND; SENSE OF EQUILIBRIUM

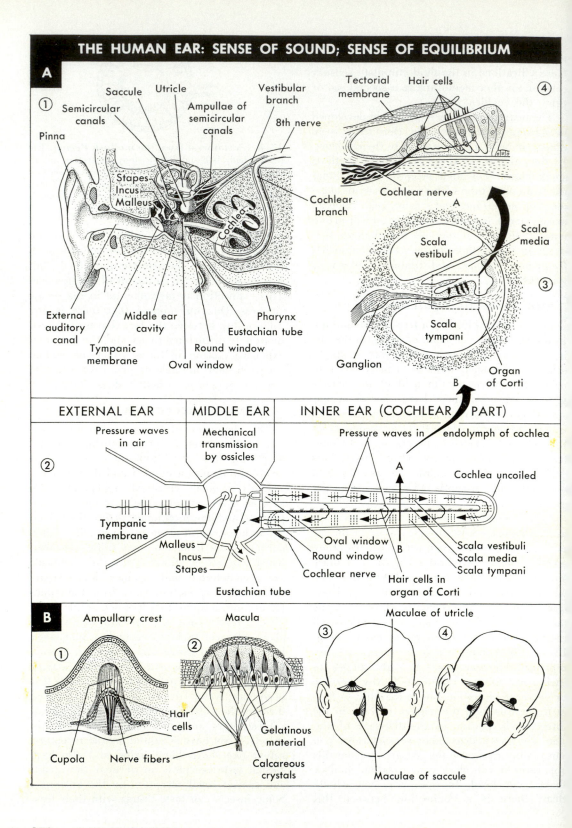

A

① Pinna · Semicircular canals · Saccule · Utricle · Ampullae of semicircular canals · Vestibular branch · 8th nerve · Stapes · Incus · Malleus · Cochlea · Cochlear branch · External auditory canal · Middle ear cavity · Tympanic membrane · Oval window · Round window · Pharynx · Eustachian tube

④ Tectorial membrane · Hair cells · Cochlear nerve

③ Scala vestibuli · Scala media · Scala tympani · Ganglion · Organ of Corti · A · B

EXTERNAL EAR	MIDDLE EAR	INNER EAR (COCHLEAR PART)

② Pressure waves in air · Tympanic membrane · Mechanical transmission by ossicles · Malleus · Incus · Stapes · Eustachian tube · Pressure waves in endolymph of cochlea · Oval window · Round window · Cochlear nerve · Hair cells in organ of Corti · A · B · Cochlea uncoiled · Scala vestibuli · Scala media · Scala tympani

B

① Ampullary crest · Hair cells · Cupola · Nerve fibers

② Macula · Gelatinous material · Calcareous crystals

③ Maculae of utricle · Maculae of saccule

④

Our own sense of smell plays so small a part in our lives that we are inclined to underestimate the importance of this sense in other vertebrates. We rely on our eyes for most of our information about the world around us, and derive relatively little useful or (to us) interesting information from the nose. Smell does greatly increase our esthetic pleasure in food, but aside from that it is more likely to be a useless annoyance. Yet this sense is of supreme importance to most fishes, amphibians, and reptiles, and even to most of our fellow mammals. The last point is familiar to anyone who has ever owned a dog. Dogs can see, hear, and feel very well, but for them smell is obviously the main source of really reliable and meaningful information. They agree with most other vertebrates in this, with man and most birds among the conspicuous exceptions. It is clear that dogs smell things that we do not.

Nostrils first evolved among early fishlike vertebrates as inlets for water to specialized chemical receptors. When air-breathing animals arose, this chemical sense was retained in the same place and became sensitive to molecules diffused in the air. Use of the nose for breathing was incidental and is still usually unnecessary (Fig. 13–10). Except for some secondarily aquatic forms, all air-breathers can and on occasion do breathe just as well through the mouth.

Even man, with his comparatively poor sense of smell, can detect extraordinarily small concentrations of some molecules in air and can discriminate among a very large number of smells. This ability to discriminate smells is baffling. Anatomically the odor-sensitive cells are simple and all look alike. The nervous impulses from them are all alike. Discrimination can be explained only on the basis that different cells respond to different stimuli and send impulses to different associative connections in the brain. But how can

---

13–9   The human ear. *A. Sense of sound. 1.* The components of the ear of man. The external ear includes the pinna (the ear in common language) and the external auditory canal. The middle ear is the cavity between the tympanic membrane and the two "windows" (oval and round) to the inner ear. The three ear ossicles (malleus, incus, and stapes) lie in the cavity; the malleus attaches to the tympanic membrane, and the stapes to the membrane covering the oval window. The middle ear vents into the throat, or pharynx, via the Eustachian tube. The Eustachian tube thus serves to maintain the air pressure of the middle ear cavity at equilibrium with atmospheric pressure. (Temporary closure of the Eustachian tube during a rapid change of altitude in an airplane or elevator causes unpleasant sensations in the ear, because the air pressure in the middle ear cavity is higher or lower than the outside pressure exerted on the tympanic membrane. Chewing motions open the Eustachian tube and equalize the pressure in the middle ear with that outside.) The inner ear comprises two major elements: (*a*) the vestibular apparatus, consisting of the semicircular canals, saccule, and utricle, which are all concerned with the sense of equilibrium; and (*b*) the cochlea, concerned with sound reception. The cochlea is a long coiled tube (shown in *2* as though uncoiled) partitioned lengthwise into three distinct chambers—the scala vestibuli, the scala media, and the scala tympani. The auditory sense cells of the cochlea are hair-bearing cells in the organ of Corti, which separates the scala media from the scala tympani (*2, 3, 4.*) *2*. The mechanism of sound reception. Pressure waves in the air pass down the external auditory canal and cause the tympanic membrane to vibrate. The vibration is transmitted mechanically by the three ear ossicles (malleus, incus, and stapes) to the membrane covering the oval window of the inner ear. The vibration of this membrane produces pressure waves in the fluid of the inner ear's canals. Pressure waves in the fluid of the scala media strike the organ of Corti and stimulate its hair cells. The hair cells then initiate impulses in the fibers of the cochlear nerve that supply the hair cells. *3*. A cross section through the cochlea showing the three cavities and the organ of Corti. *4*. The organ of Corti.

B. Sense of equilibrium. The position of the head is detected by means of sensory devices in the vestibular part of the inner ear. Each semicircular canal, filled with fluid, terminates in an ampulla (see *A1*), which contains a sense organ, the ampullary crest. Each crest (*1*) is a group of hair cells whose hairs are embedded in a gelatinous mass. When the head moves in a given direction, the fluid in the semicircular canals tends, because of its inertia, to move in the opposite direction. In so doing, it strikes the crest and thus stimulates its hair cells, initiating impulses in the fibers of the vestibular nerve that supply the hair cells. The saccule and utricle (see *A1*) of the inner ear contain similar structures, the maculae (*2*). Calcareous crystals deposited in the gelatinous material surrounding the hairs of the hair cells in each macula. This added mass makes the cells sensitive to gravity. Any change in the position of the head displaces the crystals relative to the hairs that suspend them, and the "push" or "pull" exerted on the hairs (*3, 4*) is registered in impulses initiated in the nerve supply to the hair cells.

cells that are apparently exactly alike have different specific reactions to particular molecules? At present the best-supported theory is that the cells contain different enzymes that somehow fit against or are otherwise activated by molecules of particular shapes.

The sense of taste is somewhat less puzzling because it makes much less complex discriminations, and those it does make are anatomically localized. The receptors are clusters of cells on the tongue and on the roof and back of the mouth. Only four tastes are discriminated: salt, sweet, sour, and bitter. Salt and sweet are tasted mainly at the tip of the tongue, sour along the sides, and bitter at the base (Fig. 13–11). You may object that you detect far more different flavors than these four in your food. This is true, but the complexity is due to the fact that you also smell the food you eat and do not wholly distinguish taste and smell in the blend of flavor. This is why food tastes odd and flat when you have a bad head cold. What you sense then really is the *taste* of food. The rest is smell, plus impressions of texture and temperature.

OTHER RECEPTORS

We have mentioned the considerable and indefinable number of other sensations that we receive from various parts of our bodies. Complex sensory reactions are evidently present in all higher animals, but they are seldom associated with well-defined and specialized organs except those we have mentioned. Of course we have not listed every single type of organ related to these senses. For instance, snakes and lizards have two little pouches in the roof of the mouth into which they run the tips of their forked tongues. The pouches are a special chemical receptor organ. When a reptile flicks its tongue in and out rapidly, it is not sticking its tongue out at you or expressing emotion: it is tasting the air.

There are also exceptions to the generalization that other senses, such as that of warmth,

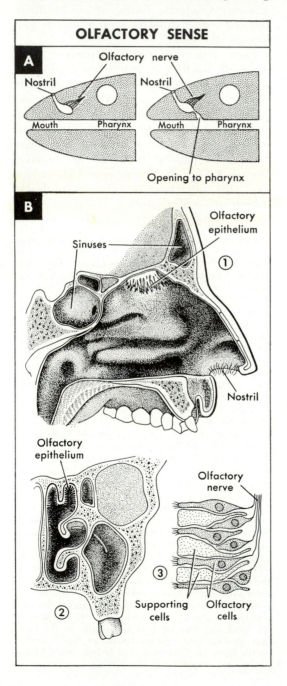

OLFACTORY SENSE

A

Olfactory nerve

Nostril · Nostril

Mouth · Pharynx · Mouth · Pharynx

Opening to pharynx

B

Sinuses

Olfactory epithelium

①

Nostril

Olfactory epithelium

Olfactory nerve

③

②

Supporting cells · Olfactory cells

13–10 **The vertebrate nostril and olfactory sense.** *A.* The position of the nostril in primitive aquatic vertebrates (*left*) and in terrestrial vertebrates (*right*). The nostril is primarily an opening and duct leading to a sensory epithelium devoted to smell. In terrestrial vertebrates the nostril's respiratory function is secondary; the duct leading to the olfactory epithelium has, so to speak, been extended to enter the pharynx. *B.* The nostril and olfactory epithelium in man. *1, 2.* The position of the olfactory epithelium in the nasal cavity. *3.* A somewhat schematic cross section of part of the olfactory epithelium, in which the ciliated olfactory sense cells lie between supporting cells.

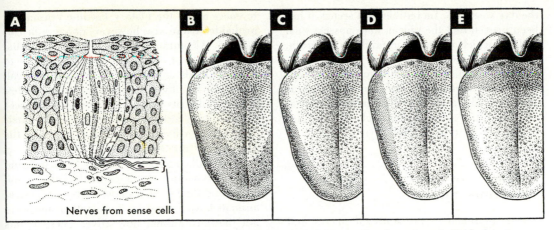

Nerves from sense cells

13–11 The taste receptors on the human tongue. *A* is a section through an individual taste receptor on the surface of the tongue. It is a pit in which lie sensory cells supplied with nerve fibers. Each such "taste bud" is sensitive to only one of four basic tastes: salt, sweet, sour, and bitter. The shaded areas of *B, C, D,* and *E* show the distribution on the tongue of salt receptors, sweet receptors, sour receptors, and bitter receptors, respectively. (Adapted from G. Hardin, *Biology: Its Implications* 2nd ed., Freeman, San Francisco, 1959.)

usually do not have complex and local receptor organs. Some snakes, called "pit vipers" on this account, have sensory pits on each side of the head between eyes and nostrils. The pits are extremely sensitive warmth receptors. They can detect the presence of a warm-blooded animal, such as a mouse, up to several feet away. Pit vipers, of which rattlesnakes are the most familiar examples in this country, feed on small warm-blooded animals, and the pits help to detect their prey.

Many insects sense and respond to humidity differences, but it is doubtful whether this represents a special sense. In the insects studied, the humidity sense organs either occur on antennae or are bristles on the back. It is still unknown exactly how these sense organs operate, but it is likely that they exploit one or both of the known sensitivities to temperature and pressure. The rate of water evaporation from the antennae varies inversely with the air's moisture content, and could well be registered by temperature receptors cooled by the evaporation. The large bristles on an insect's back bend as their moisture content varies, and this bending could well flex the pressure-sensitive skin where they are attached.

There is no entirely convincing evidence that any animal can sense stimuli altogether different in *kind* from any of ours. As for dogs that "know by a sixth sense" that a man is honest, or when their masters die—surely we do not have to point out that these are old wives' tales. If a dog does something you do not immediately understand, it is highly likely that he is simply reacting to a smell you have not detected. In scientific circles, discussion of a wholly different sense has often centered around homing and similar phenomena. Birds fly thousands of miles to a more or less precise destination. Bees and ants return unerringly (sometimes) to their homes. Salmon (also sometimes) find their way back from the open sea to the mountain stream where they were born. Male moths fly long distances to an unseen and unheard female. Such phenomena suggest a homing sense, or the use of clues, such as the earth's magnetism, undetectable by us. Strict study, however, has largely negated those ideas. Male moths can smell females from miles away, which is wonderful, certainly, but does not involve a sense absent in ourselves. Ants lay and follow scent trails. Even the homing of bees and birds does not seem to demand any new sense, although recent studies have revealed something almost as spectacular. Some brilliant experiments in Germany have shown that homing pigeons keep a true course by constantly referring to the position of the sun, like some ancient mariner. But, like the mariner, the bird needs a chronometer when it navigates, because it must compensate for the steady movement of the sun across the sky as the day wears on. Incredible as it seems, it has been proved that the pigeon does possess such a clock, which is almost certainly in its brain, but in exactly what form is still unknown.

Many organisms, including man, have built-in "clocks" that control the timing of various activities. These are, however, internal mechanisms and not sense organs such as are under discussion in this chapter.[4]

History of the Nervous System

Our approach to the study of nerves and receptors has been partly historical. We have pointed out a number of changes and trends that have occurred in the course of evolution. Now we are going to consider briefly some major features of the history of the nervous system as a whole. The only *direct* approach to such a history is by means of *fossils*, remains of the organisms of the past. For the nervous system, that approach is quite inadequate. Nerve tissue is practically never preserved in fossils. Fossils of invertebrate animals give no good evidence of what their nervous systems were like. Fossil vertebrates do give valuable evidence about part of the nervous system, especially the brain, because the brain in these animals was surrounded more or less closely by bone. The bone is preserved, and its shape gives good clues as to the nervous structures once contained in it. Still the evolution of the brain from early fish to modern man is only half, or considerably less, of the story.

In such a situation the historian of life must fall back on the study of living animals. Judicious comparison of their nervous systems gives evidence as to the history. Some animals have evolved farther and more rapidly than others. It is fair to assume that a clam of a few hundred million years ago had a nervous system somewhat like that of a clam today. By comparing animals that have changed less with those that have changed more, some in-

ferences about the history become reasonably probable. There are, however, dangers in this method. All animals have changed somewhat in the course of evolution, and comparison of living animals is fundamentally nonhistorical. It should be obvious (although even professional biologists sometimes forget this fact) that no living animal is ancestral to any other. Moreover, no living animal is exactly like the ancestor of another.

Above all, it is necessary to keep in mind the relationships of the animals being compared and the fact that many sorts of ancestral animals have disappeared altogether and were not like any animal now alive. These points can be illustrated by consideration of the nervous systems of vertebrates and of insects. Vertebrates have much more complex nervous systems than invertebrates, and they evolved from invertebrates. Insects have about the most complex nervous systems among living invertebrates. It is therefore tempting to assume that the insect nervous system represents the next stage below the vertebrates, that it tells us something about the transition from invertebrate to vertebrate nervous systems. But the assumption is not warranted.

The ancestry of insects and of vertebrates separated well over 500 million years ago, long before insects or vertebrates themselves had arisen. The very remote forms ancestral to both insects and vertebrates probably had nervous systems, but if so these were very simple and we can learn less about them from insects than we can from other and simpler animals. Unfortunately no living animal is likely to resemble that common ancestry closely, and we can infer what its nervous system was like only in general terms. The insect nervous system of today certainly is not like any stage in the ancestry of the vertebrate system, and it can tell us little about the history of the latter. Since nerve action is much the same anywhere, we can learn things about vertebrate or other nerves from insects or from any animals with nerves. We can also detect evolutionary tendencies that occurred separately in both insects and vertebrates. This must, however, be done cautiously and in the light of the long separate histories of the two groups.

We have made this rather long introduction

[4] There is also a hitherto unmentioned kind of information that some organisms other than man can obtain; this is the polarization of light. We refer the interested reader to Karl von Frisch's *The Dancing Bees*, Harcourt, Brace & World, New York, 1955, pp. 81 and 95.

to the present subject because it applies more widely than to this subject alone. In any discussion of anatomy, processes, and their histories, remember that the comparative study of living animals is not directly historical. It can give evidence on history, but that evidence is likely to be misleading if it is not interpreted cautiously and in the light of evidence that *is* directly historical.

ORIGIN

We have seen that in protists the whole body may conduct impulses of some sort. Stimulation at any point may be followed by response at any other point. In sponges, the only major group of multicellular animals without nerves, the protoplasm of each individual cell shows similar capacity to conduct impulses within that cell. In most ways the various cells operate quite independently of each other, but it is possible for a stimulus on one cell to be transmitted across another cell and to evoke a response in more distant cells. Sponges have evolved along a line all their own, but it is almost a logical necessity that some such stage as this existed in the earliest multicellular ancestors of other animals. Increasing cell specialization and the beginning of differentiation of tissues would result in the more regular activity of some intermediate cells as conductors between others. These intermediate cells, mainly involved in the conduction of impulses, would be primitive nerve cells.

Next simplest after the sponges among major groups of living animals are the coelenterates: the corals, jellyfishes, and their relatives, including the little fresh-water *Hydra,* which you may be able to study at first hand. The living coelenterates, too, are off the main lines of further evolution but they are nearer than sponges, and again there is some inherent probability that higher nervous systems passed through a stage like theirs. They have a *nerve net* composed of neurons, all nearly alike, each with several fibers of about equal length (Fig. 13–12). The neurons are spread rather evenly and thinly through outer layers of the body. The fibers are short and have synapses with fibers of adjacent neurons, roughly in a circle around each neuron. This is just the system one would expect if interme-

diate cells, such as we spoke of in the last paragraph, became specialized for conduction.

In a nerve net the main response to stimulation is local contraction. A stimulus anywhere can eventually spread to the whole net, and conduction through fibers is in either direction, depending on what point is stimulated. A stimulus must, however, be strong and long to spread far because the fibers are short and many synapses have to be passed. For the same reason, conduction is slow. Such a nervous system can coordinate and spread simple responses, mostly contractions of the whole body or parts of it, but that is about all it can do. It has little or no associative activity, and it cannot control or coordinate complex reactions. It does permit simple reflexes, for there are scattered sensory cells, and the nerves conduct impulses from them to effectors, muscular cells. The barest essentials of nervous reaction are present.

TRENDS IN HIGHER INVERTEBRATES

Free-living flatworms, such as the planarians, are very instructive animals because they show in simple and probably primitive form many of the most fundamental features of all higher animals. Such features as are shared by most higher forms are almost sure to be primitive. Complications peculiar to one group, such as the special characteristics of the insect brain, probably are not primitive for other groups. Such considerations suggest that the *general* characteristics of the planarians' nervous system (see Fig. 13–12) probably are primitive. The detailed anatomy of the system is undoubtedly specialized in living planarians, but its basic structural and functional characteristics were probably present in the common ancestry of all the more complex animals.

There is still an outer nerve net in planarians. Indeed, this simple type of nervous structure may persist or reappear in all sorts of higher animals: there is a nerve net in the wall of the human intestine. Planarians also have the rudiments of a *central nervous system.* This includes a series of *nerve cords* and an enlargement of them that may be called a *brain,* although so rudimentary as

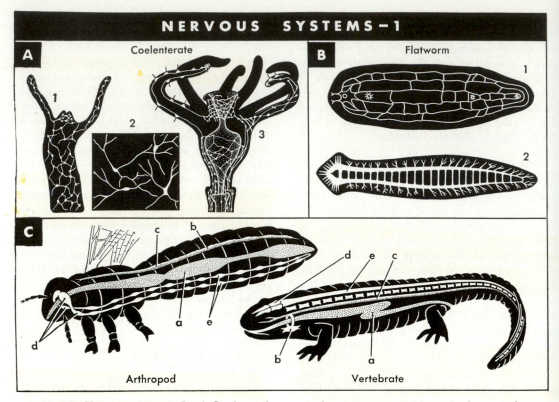

A Coelenterate

B Flatworm

C

Arthropod   Vertebrate

13–12 Nervous systems–1. *A.* In the coelenterates the nervous system is a simple network formed by the connected branches of individual neurons. *1* is the nerve network in *Hydra*, part of which is enlarged in *2* to show individual neurons. *3* is the nerve network in *Obelia*, showing a concentration of neurons around the mouth region. *B.* In some flatworms also (*1*), the nervous system is basically a simple net of neurons. In others (*2*) the beginning of a central nervous system is discernible in nerve cords running lengthwise down the animal and a brain anterior to the sense organs of the head. *C.* Fully evolved central nervous systems are present in the two most complex animal groups, the arthropods and the vertebrates. The schematized arthropod is an insect, and the vertebrate a salamander. The arthropod nervous system is typical of most invertebrates in being a double nerve cord (*e*) *ventral* to (that is, in front of) the alimentary canal (*a*); the brain (*d*) is a concentration of ganglia forming a ring around the esophagus. [Note, too, that in the arthropod the heart (*b*) and main blood vessels (*c*) are *dorsal* to (that is, in back of) the alimentary canal.] The vertebrate nervous system is a single tubular nerve cord (*e*) that, along with the brain (*d*), is completely *dorsal* to the alimentary canal (*a*). [The vertebrate heart (*b*) is *ventral* to the alimentary canal.]

barely to deserve the name. The nerve cords are main lines of conduction, each containing many neurons. The neurons have long fibers along the cord and are so arranged that impulses pass along the fibers in one direction only. From cell bodies in the nerve cords, fibers also connect with the nerve net and with other cells of the body.

Planarians have a definite front end in the direction of usual motion. Here sensory cells (which also occur all over the body) are especially numerous, and here are the eyes. The eyes are definite and well-developed receptor organs, sensitive to light intensity and direction, although they have no lenses and do not form an image. The nerve cords converge at the front end of the body and merge with an enlarged mass of nerve tissue which for courtesy we have called a brain. The eye and many sensory cells connect directly with the brain, which contains associative cells and a fairly complex arrangement of synapses between the sensory nerve fibers and other fibers from the nerve cords.

Simple as it is, the planarian nervous system permits control and coordination of special responses throughout the body. It mediates between sense organs and effectors, and it involves association and some variability of responses in addition to simple reflexes. All the rest of the evolution of the nervous system can be viewed as elaboration of characteristics already present in a planarian—as really tremendous elaboration. Among the important trends already suggested, at least, in planarians and carried further in the ancestors of the vertebrates are the following:

1. Formation and concentration of a *central nervous system*. Most cell bodies of neurons come to be concentrated in one or a few nerve cords or in masses (ganglia) near the cords. Connections from here to all other parts of the body are by a *peripheral nervous system*. The peripheral system does have some ganglia with cell bodies, but it is mainly composed of nerves, bundles of long nerve fibers from the cell bodies in the central nervous system.

2. Differentiation of *afferent* and *efferent* fibers and nerves. Most of the nerve impulses occurring in the body (although never all of them) become routed through the central nervous system, and a particular nerve carries impulses in one direction only. Those bringing impulses to the central nervous system are afferent ("carrying to"), and those conducting impulses away from it are efferent ("carrying away"). Sensory nerves are of course afferent, and motor nerves are efferent. A result of this arrangement is that even the simplest reflex passes through at least two neurons and is more flexible than a one-neuron reflex, which does not occur in higher animals. Usually many neurons are involved in a stimulus–response reaction.

3. Increased complexity of *association*. The afferent–efferent system accompanies increase in associative neurons throughout the central nervous system and in the number and complexity of nerve routes and connections.

4. Development and complication of a *brain*. The front end of the nerve cord (or cords) becomes enlarged, principally by the development here of large numbers of associative neurons and tracts. Eventually a large proportion of nervous impulses in the body are routed through this associative mass. Fully central coordination for complex responses is thus provided.

5. Increase in number, complexity, and sensitivity of *special sense organs*. Scattered and simple sensory cells and organs persist, but the progressive trend is for the development of complex organs at the front end of the organism, in the head. Here these organs are directly connected to the brain. As we have already noted, this development is necessarily accompanied by that of complex associative mechanisms in the brain, where volleys of nerve impulses from the receptors become transformed into perceptions (recall the discussion of the brain and the eye, p. 367).

STRUCTURE OF THE VERTEBRATE BRAIN

The precise number and pattern of nerves of course differ greatly from one species to another and even among individuals of the same species. Those anatomical details do not concern us here. In the vertebrates the tendency for concentration of central control reaches a peak, and the evolution of the brain is the most important single factor in the success of this group.

The central nervous system of all vertebrates consists of a *single, hollow* nerve cord that runs along the *back* (dorsal part) of the body. These basic anatomical characteristics of the nervous system do not occur in any living invertebrates—reason enough to infer that none of them preserve just the same structure that was present in those ancient invertebrates from which vertebrates evolved. The brain of the great majority of modern higher invertebrates is a ring of nerve tissue surrounding the esophageal part of the alimentary canal (see Fig. 13–12). The ring consists of two ganglia (masses of nerve-cell bodies) lying above the esophagus and connected to two others lying below. Two solid nerve cords leave the *sub*esophageal ganglia and extend backward along the length of the body under (or *ventral* to) the gut. The fact that the brain is an anterior inflation of the nerve cord is the principal similarity in the gross anatomy of the central nervous system of the more ad-

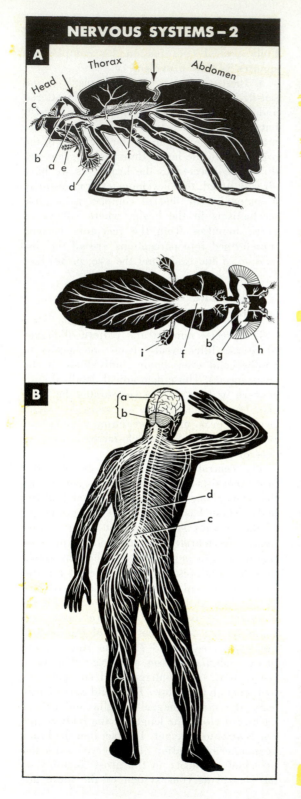

vanced living invertebrates and vertebrates (Figs. 13–12 and 13–13).[5]

The most primitive vertebrate brain consisted mainly of three irregular swellings of the hollow nerve cord, each with various thickenings of the walls. The three enlargements are the *forebrain, midbrain,* and *hindbrain,* which can still be distinguished in the human brain, with its vastly greater complications.[6] Very early, even in primitive fishes, further complications occurred. The forebrain became divided into three parts: (1) the *thalamus* and associated structures; (2) a pair of swellings farther forward and higher, the *cerebral hemispheres* (or, taken together, the *cerebrum*); and (3) the *olfactory bulbs,* which project as swellings from the lower front of each cerebral hemisphere (Fig. 13–14). The midbrain developed various swellings, especially an upper pair, the *optic lobes.* A large swelling developed on the forward, upper part of the hindbrain and became

---

[5] Some ingenious but completely unconvincing arguments have been made to the effect that invertebrates like those still living turned into vertebrates by rolling over on their backs.

[6] There are of course more technical terms for these parts, derived from Greek roots. Full description of any vertebrate brain, an extremely complex organ, requires an exceptionally large number of technical terms. We have here used only the few terms absolutely necessary to understand the general functions and significance of the brain.

13–13 Nervous systems–2. *A.* An arthropod nervous system, that of the fruit fly *Drosophila.* Note in the side view how the central nervous system (white) lies ventrally to the gut (stippled), which is shown only as it passes through the head and thorax, except in the head, where the brain encircles the esophagus: *a,* ventral portion of the brain; *b,* dorsal portion of the brain; *c,* antenna carrying diverse kinds of sense organs; *d,* sucking structure leading to the mouth; *e,* maxillary palp (*d* and *e* are also endowed with sense organs); *f,* large concentration of nerve cells in the thorax. The dorsal view shows also *g,* optic nerve from the compound eye (*h*), and *i,* balancer, a remarkable sense organ that assists the fly to maintain its orientation in flight. *B.* A vertebrate nervous system, that of man, showing the organization into brain (*a,* cerebrum; *b,* cerebellum) and dorsal spinal cord (*c*), from which individual nerves supply all parts of the body. The two sympathetic nerve trunks (*d*) lie on either side of the spinal cord (cf. Fig. 13–19).

PARTS OF THE VERTEBRATE BRAIN

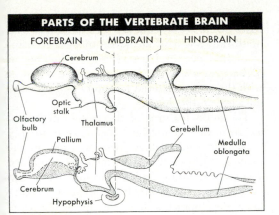

13–14   **Parts of the vertebrate brain, in generalized and schematized representation.** The lower figure is a longitudinal section showing local differences in thickness of the brain wall.

the *cerebellum*. The much thickened lower wall of the hindbrain is the *medulla oblongata*.

There are dozens of other distinguishable parts even in fairly primitive vertebrate brains, but the *main* parts from front to back are as shown in Figs. 13–14 and 13–15.

Forebrain — {
Olfactory bulbs
Cerebral hemispheres (the cerebrum)
Thalamus (with an upper epithalamus, lower hypothalamus, etc.)
}

Midbrain — Optic lobes, in mammals four (two pairs) swellings called the *corpora quadrigemina,* "quadruplet bodies"

Hindbrain — {
Cerebellum (forward and above)
Medulla oblongata (below)
}

13–15   **The expansion of the forebrain in vertebrate evolution.** Four vertebrate brains are shown in side view (odd numbers) and from above (even numbers). The proportionate sizes of fore-, mid-, and hindbrains may be judged from the width of the stippled pathway, which approximates the midbrain in each case. The huge proportionate increase in size of the forebrain is obvious. Although the forebrain was originally principally concerned with olfaction, its expansion is due to the growth of the cerebrum, which is concerned with general association and control. The most important development in the later stages of forebrain evolution involves the cerebrum's neopallium (Fig. 13–16). This is relatively small and unconvoluted in the shrew, a primitive mammal, but it makes up the entire convoluted outer surface of the horse cerebrum.

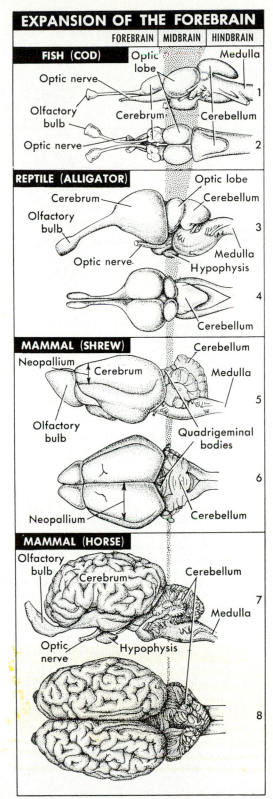

EXPANSION OF THE FOREBRAIN

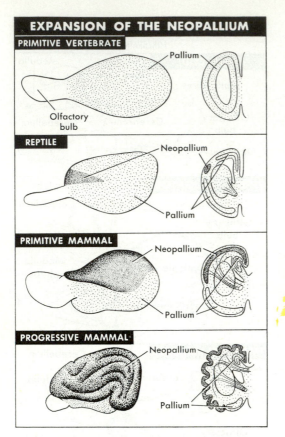

EXPANSION OF THE NEOPALLIUM

PRIMITIVE VERTEBRATE

Pallium

Olfactory bulb

REPTILE

Neopallium

Pallium

PRIMITIVE MAMMAL

Neopallium

Pallium

PROGRESSIVE MAMMAL

Neopallium

Pallium

13–16 The expansion of the neopallium in vertebrate evolution. *Top to bottom.* A generalized primitive vertebrate forebrain seen in side view (*left*) and in cross section (*right*): the cerebral hemisphere, lying behind the olfactory bulb, is stippled; its outer wall is a pallium of neurons. Comparable views of a generalized reptilian forebrain, with the neopallium (dark stipple) a small area on the front side of the cerebrum. Comparable views of a generalized brain in a primitive mammal, with the neopallium more extensive. Comparable views of a generalized brain in an advanced, or progressive, mammal, with the neopallium covering virtually the entire cerebrum; indeed, the growth of the neopallium is such that it has become folded, or convoluted, on the cerebral surface.

All parts of the brain connect directly or through chains of neurons with the spinal cord through the medulla oblongata, which grades into the spinal cord without sudden change. The brain also has a series of paired nerves of its own that connect it directly with some sense organs and muscles. The forebrain is connected with the smell receptors in the nose. The midbrain is connected with the light receptors in the eyes and, by two separate pairs of nerves, with the eye muscles. The hindbrain has a whole series of nerves—usually six pairs in lower forms and eight in higher forms, including man—that connect with the more scattered receptors and with the muscles of the head.

The main features of the vertebrate brain, as we have just outlined them, were already present in rudimentary form, at least, in the jawless "fishes" that are the earliest known vertebrates. Many and important changes in details have occurred, but the most striking later developments have involved the forebrain, especially the cerebrum. At first the cerebrum was greatly outweighed in bulk by the rest of the brain. It was only a pair of small, smooth swellings involved mainly in association of sensations of smell. Even the most progressive fishes and the amphibians have small, smooth cerebra that are dominantly olfactory; they are "smell brains."

In early reptiles significant changes began to appear. The cerebrum, although still forming less than half of the whole brain, became definitely larger. Most of it was still concerned with smells, but at the forward, upper part a new sort of nervous tract began to appear. This was and is involved in association and coordination of all kinds of impulses from various other receptors and brain centers, and not primarily with smell. This new part of the brain is the *neopallium,* which means "new cloak" (Fig. 13–16). It is a new sort of covering of gray matter on the cerebrum. This covering in general is the *cerebral cortex,* and the neopallial part can also be called *neocortex.*

Birds and mammals arose separately from reptiles, and each group evolved in its own way. In both the brain became larger and the cerebrum became the largest part of the brain, but the expansion of the cerebrum was very different in the two groups. In birds the smell areas were reduced and the neopallium or cortex as a whole did not expand. The expansion of the cerebrum in birds is almost entirely in a basal region which remains relatively small in all other vertebrates. This peculiarity of the brain is certainly related to the special behavior of birds, which is also unlike that of any other animals, although its significance is not fully understood. It is of interest, at any rate, that the part of the brain where we form more complex associative patterns—where, indeed, we think—is practically absent in birds. The epithet "bird-brain" therefore has some justification. On the other hand, birds perform very complicated *unlearned* procedures in courtship, nest-building, and the like, and we are deficient in a part of the brain well developed in them. So a bird would be justified in calling a particularly clumsy mate a "man-brain."

In primitive mammals the smell brain was not reduced. It remained large, as is reptiles, or even increased in size. The most significant evolutionary development was that the neopallium became separated from the rest of the cortex by a furrow and expanded greatly. In the most ancient and primitive known mammalian brain, which is some 150 million years old, the neopallium already forms the whole upper part of the cerebral cortex. A stage almost as primitive survives in some living mammals, notably the opossum. In most mammals, however, there has been considerably further increase in the area of the neopallium, and consequently in its number of associative neurons. This increase has occurred in two ways: by expansion of the whole upper part of the cerebrum and by folding or *convolution* of its surface. A convoluted hemisphere has more surface than a smooth one of the same diameter, and it is the surface area that determines the functional extent of the cortex of the brain.

Increase in size of the cerebrum, increase in relative extent of the neopallium, and increase in its surface by convolution have all occurred in varying degree in the evolution of most groups of mammals. All three are carried to an extreme in man. The cerebrum has expanded right over the other parts of the brain so that nothing but cerebrum is visible from above. The surface of the cerebrum is almost entirely neopallium. Only a small bit of smell brain is still visible in the middle of the bottom side of the cerebrum.

FUNCTION OF THE BRAIN IN MAMMALS

In all animals that have this organ, the brain is primarily an associative and coordi-

nating center for nerve impulses. It receives impulses from sensory receptors, organizes them, and transfers or initiates impulses to various effectors. In the vertebrate brain the number of associative neurons is enormous and their arrangement is extremely complex. In one group of vertebrates after another there has been a tendency for intensification of these characteristics, and the tendency culminates in mammals, especially in man.

In lower vertebrates the hindbrain is a sort of message center. Impulses in both directions between most of the body and the brain pass through here, and here preliminary associations are made. Some regulatory responses are started here: automatic or reflex adjustments of posture, changes in rate of heartbeat or breathing, and the like. Messages involved in more complex or modifiable responses are, however, passed on to the parts of the brain farther forward. The functional relationships of the hindbrain are essentially the same in mammals. This is the part of the brain that has changed the least in vertebrate evolution.

In lower vetebrates the forebrain is also a message center, but a simpler one than the hindbrain. It receives information from a single sense, that of smell, makes preliminary associations, and passes on the organized information to the midbrain. In these animals the midbrain is the main center of control and coordination. Here final associations are made with data from forebrain and hindbrain, and here the more complex and diversified responses are started.

In the rise of mammals from reptiles and further evolution among mammals, a great change has occurred. Central control has passed almost entirely from the midbrain to new centers in the neopallium of the forebrain. The midbrain has been much reduced in relative size. It has become a comparatively unimportant reflex center and a secondary message center. It relays messages between the forebrain and hindbrain (and thence to the spinal cord) and also between the forebrain and the eyes. Some secondary associations occur here, and most of the information is passed on to the neopallium.

Thus, in mammals, final coordination and control of most information and responses are concentrated in the neopallium, which makes

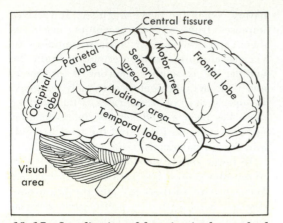

13–17 Localization of function in the cerebral cortex of man. The brain is viewed from the right side, with only the major convolutions of the cortex represented. These are remarkably constant from individual to individual and provide landmarks in the mapping of the distribution of special functions in different parts of the cortex. Note especially the sensory area and the motor area, which are posterior and anterior, respectively, to the central fissure. Details of the localization of function within these areas are given in Fig. 13–18. (Adapted from G. Hardin, *Biology: Its Human Implications,* 2nd ed., Freeman, San Francisco, 1952.)

up most of the cortex of the cerebrum. This development reaches its height in man. It is, indeed, what makes us human. It is the mechanism chiefly responsible for the extreme complexity and variability of our behavior and mental activity. To accomplish such results, even in mammals less complicated than man, the neopallium itself must be highly complex and must have some differentiation of structure, connections, and functions.

Perhaps you have seen in an old book or in a fortuneteller's display a phrenological chart. This is a picture of a human head marked off neatly into sections labeled "combativeness," "amativeness," and the like. It was an old idea that each such personality trait had a center in the brain. If a person was particularly amorous, say, his or her "center of amativeness" would be large, so large as to cause a bump on the skull. Few people take phrenology seriously now; yet it has been learned, little by little, that there is localization of functions in the brain. Phrenology has about the same relationship to our present topic that astrology has to astronomy.

Particular receptors and effectors have regions of the cortex with which their connec-tions are most direct and where messages from and to them are normally received and sent (Fig. 13–17). Sensory messages from the nose still go to the smell brain at the bottom of the cerebrum. Messages from the eyes go to an upper hind part of the neopallium. It may seem a little odd that messages from the eyes in front of the brain go to the *back* of the cerebrum, but remember that in our reptilian and earlier ancestors they went to the midbrain. They are still routed through the midbrain. Messages from the ears go to an area below or in front of the visual area. Farther forward on the upper part of the neopallium is an area which receives messages from scattered receptors all over the body and is therefore called the somatic sensory area (*somatic* means "of or relating to the body"). Just in front of this is a motor area that sends messages to muscles all over the body. It has even been possible to show where the various parts of the body are represented in these areas. The *right* side of the body is represented on the *left* side of the brain. (No one has yet discovered why the nerves cross over to the other side in this aspect of the nervous system.) In general the lower part of the body connects with areas along the midline of the brain. Areas for the upper part of the body are farther out, and those for the head are on the sides of the cerebrum (Fig. 13–18).

In mammals that have a relatively small neopallium, most of the region is occupied by fairly definite receptor (sensory) and effector (motor) areas like those we have just mentioned. Nevertheless there are regions that have no specific connection with receptors and effectors, and these apparently blank areas are proportionately larger when the neopallium as a whole is larger. Man has an enormous neopallium, and most of it is not specifically either sensory or motor. It is certain that these areas are not really blank or functionless. Some, at least, are involved in associations of a higher and less localized sort, one or many steps further removed from specific sensations and muscular actions. This is confirmed by such relationships as those between the front end of the brain and emotions and between lower lateral parts of the brain and speech. Conscious emotions and control of reactions to them are on a high level of mental activity

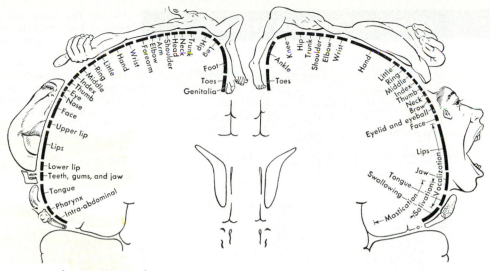

Sensory Homunculus   Motor Homunculus

13–18 Localization of function in the somatic sensory and motor areas of the cerebral cortex of man. The figure represents a cross section through the cerebrum, with the cortex indicated by a heavy black line broken into fragments labeled as to function. The left side of the figure is a section just posterior to the central fissure and thus is in the sensory area. The right side of the figure is a section just anterior to the central fissure and thus is in the motor area. Outside the heavy line is a drawing of the corresponding body (somatic) areas serviced by the cortex. The distorted representation of the body obtained in this way has been called a "homunculus." Compare the sensory and motor functions of various parts of the body. Note, for example, that more cortex is given over to the motor function of the hands than to their sensory function, whereas more sensory area than motor area is devoted to the lips.

and are localized, in part, in the front of the brain. Similarly speech, or language in general, represents a supreme sort of generalization and symbol formation and is definitely localized in an area that is not *directly* either sensory or motor. Functions of the non-sensory–nonmotor areas in man and other mammals are only now slowly being learned. Some of them are a sort of general-purpose associative mechanism, ready to take on such work as turns up, so to speak. Experiments on rats and other mammals have shown that if part of the cortex is removed its functions may in some instances be taken over by other parts.

The Nervous System and the Internal Environment

When we discussed the internal environment and its maintenance in Chapter 12 we were especially concerned with chemical controls. Even so, we noted many relationships

with the nervous system. It was apparent that the controls are partly chemical and partly nervous, and that the two are not independent but interact closely.

In the vertebrates and in some of the more complex invertebrates (including insects) nervous control of the internal environment is largely through a *visceral*[7] *nervous system.* This visceral system is closely connected with the central nervous system but can act with some independence. The arrangement is most highly developed and best known in mammals, but the principles of its operation are about the same everywhere. Afferent nerves from the various internal organs run to the central nervous system, to the spinal cord or the medulla oblongata in mammals. Here they form reflex arcs with efferent visceral nerves and also have associative connections with the central nervous system itself, including the brain. Much of the activity remains within the

[7] Internal organs, such as stomach, intestine, bladder, and kidney, are collectively designated as viscera.

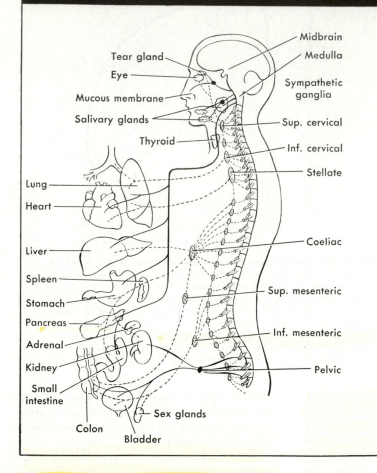

AUTONOMIC NERVOUS SYSTEM

Tear gland
Eye
Mucous membrane
Salivary glands
Thyroid

Midbrain
Medulla
Sympathetic ganglia
Sup. cervical
Inf. cervical
Stellate

Lung
Heart

Liver
Spleen
Stomach
Pancreas
Adrenal
Kidney
Small intestine
Colon
Bladder
Sex glands

Coeliac
Sup. mesenteric
Inf. mesenteric
Pelvic

13–19 The autonomic nervous system, in relation to the central nervous system (brain and spinal cord) and to the viscera it controls. The sympathetic and parasympathetic divisions of the autonomic system are distinguished by broken and solid lines, respectively. Note how the sympathetic system includes a distinct chain of ganglia (superior and inferior cervical, stellate, coeliac, and superior and inferior mesenteric) lying to the side of the spinal column (cf. Fig. 13–13). The parasympathetic system comprises two main elements, one arising from the medulla and the other from the lower (sacral) region of the spinal cord.

visceral reflexes: most of the processes of digestion, glandular secretion, and other internal regulations go on without our being aware of them or having any conscious control over them.

The efferent visceral nerves, which carry regulatory messages to the internal organs, form the *autonomic nervous system* (Fig. 13–19). There are two complete sets of these nerves, which run to most of the same organs but which have different connections with the central nervous system. The effects of the two sets of nerves are generally opposite. For instance, one set, called *sympathetic*, speeds up the heart and slows down digestion. The other set, *parasympathetic*, slows the heart and speeds digestion. The balance between the two normally keeps bodily processes near a con-

stant rate. Comparison can be made with keeping an automobile at a steady speed by using the accelerator on upgrades and the brake on downgrades. The balance is an important part of the maintenance of stability in the internal environment.

Like the chemical regulators, the autonomic nervous system is involved not only with maintenance of a steady state of activity but also with altering this state in exceptional circumstances. Stimulation through the central nervous system can increase the activity of the sympathetic nerves so that they overbalance the parasympathetic nerves. Then the heartbeat increases, blood pressure rises, and in general the effects of the hormone epinephrine (Chapter 12) appear. In fact, the nerve fibers secrete epinephrine (or something very like it) at their terminations, and sympathetic nerves also stimulate the adrenal endocrine to pour epinephrine into the blood. The reaction occurs when one is frightened or excited. Its net effect is to make one keyed up and ready for instant action.

We may now again consider the relationship between nervous control and hormonal control of reactions to fright and excitement. We see that they are not really alternative or duplicate systems but parts of the same system. The nerves deliver immediate stimulation at particular points in organs involved in a reaction. Acting through endocrines, the same nerves also produce the somewhat slower but more widespread and lasting delivery of epinephrine to the whole body through the blood stream. There is, indeed, some evidence that the endocrine organ concerned, the adrenal medulla, evolved later than the sympathetic nerves, as a secondary mechanism increasing and spreading the epinephrine output normal for some of these nerves.[8] If this is so, there is no question of a hormonal system having replaced a nervous one, or the other way around. There has been only a growth in the capacity and extent of the action of a combined nervous–chemical system.

We noted in Chapter 12 that most endocrines are not under *direct* nervous control. Their action is controlled, in one way or another, by hypophyseal hormones. The hypophysis, in turn, is closely connected with the midbrain, and its activities are largely controlled by autonomic nervous reflexes[9] through the hindbrain and midbrain. It has, moreover, recently been found that many neurons secrete a variety of *neurohumors*, which circulate through the body and may affect nerve and other reactions far from their sites of origin. They are, in fact, true hormones. In this respect the nervous system itself acts (in part) as an elaborate endocrine organ. Thus,

strictly speaking, the whole hormonal regulatory system is intricately interrelated with the nervous system and the endocrine glands.

In all but the simplest animals, the role of the nervous system in bodily regulation is still more extensive and complex. When water becomes deficient in the body, drying of throat tissues and reduction of saliva flow signal "thirst" to brain centers. Those centers make appropriate associations and send coordinated volleys of signals to the muscles; one then takes a drink of water. Water balance is restored. Here is a stabilizing mechanism, just as much as the activity of a hormone or of an autonomic reflex, but clearly much more complicated. Even animals with quite simple nervous systems often have this sort of stabilization. They seek food when they are hungry and in other ways respond with the whole organism to meet internal needs. Such responses are on a different level of activity. They are behavior, which is the subject of the next chapter.

CHAPTER SUMMARY

Responsiveness of protoplasm; stimulus–conduction–response; receptors–conductors–effectors.

Stimuli—their diversity, their significance in terms of survival, their effectiveness, thresholds; responses—determined by the organism (not the stimulus), latent and refractory periods.

Nerves: the structure of a neuron; the nature and speed of the nerve impulse; all-or-none reaction; sensory, motor, and association neurons; synapses and their properties; summation; facilitation; nerve patterns and behavior.

Receptors: light receptors—their diversity, lenses as light concentrators and their subsequent utilization in image-forming eyes, compound eyes, color vision, complex eyes related to complex brains; touch, pressure, and sound receptors—lateral-line organs, the vertebrate ear, the sense of equilibrium; chemical receptors—taste and smell; other receptors—

---

[8] Many nerves of the autonomic nervous system and most of those of the central nervous system produce another substance, acetylcholine, which also mediates the passage of a nerve impulse but whose action is more short-lived than that of epinephrine in synapses.

[9] There is a possibility that the hypophysis is also controlled by some hormonal influence through a blood vessel which reaches it from the hypothalamus of the brain.

warmth receptors, navigation by birds, chronometers in many organisms.

History of the nervous system: dangers attending historical conclusions based on comparisons of living forms; origin of the nervous system; nerve nets; a central nervous system (cords and brain); relation of the central nervous system to the peripheral nervous system; the differentiation of afferent (sensory) and efferent (motor) nerves; association fibers; the evolution of special sense organs; structure of the vertebrate brain—its complexity, comparison of vertebrate and invertebrate, central nervous systems, the three parts of the vertebrate brain (forebrain, midbrain, and hindbrain) and their subparts and functions; the forebrain—its initial association with the sense of smell, its evolution in higher vertebrates, the importance of the cerebrum and the development of its neopallium in mammals, the contrast between bird and mammal forebrains; function of the brain in mammals—the hindbrain, the message center and the seat of much automatic or reflex control, the midbrain, the secondary message center, and the forebrain, the site of important associations and the final control of coordination; the localization of function in the neopallium.

The nervous system and the internal environment; sympathetic and parasympathetic fibers in the autonomic (efferent) system; hormone secretion by nerves; nervous control of endocrine glands.

Australian News and Information Bureau

The Australian lyrebird, named for the shape of his showy tail, exhibits an elaborate behavioral pattern. As he paces in a carefully cleared space in the forest, he sings, mimicking sounds he has heard during the day.

14

Behavior

What Is Behavior?

An eagle perching quietly at the top of a tall tree is a wonderful sight. But how much more wonderful when it spreads its wings, soars overhead, and dives at a scuttling rabbit! You have learned that the motionless eagle and also the tree on which it perched were not inactive. Ceaseless, intense activity was going on at tremendous rates in every living cell of both organisms. They were also responding to stimuli. The rays of the sun, for instance, evoked responses in the heat-regulating mechanism of the eagle and in the photosynthetic cells of the tree. When the eagle flew, however, there was a different level of response and a quite different order of activity. We all feel this difference strongly, even after we have learned about the incessant internal activities of the organism. The bird sat, "not doing anything"; then it flew away, soared, and dived. That was really doing something, in the usual conceptions of our speech and thought. It is their doing things, in this sense, that makes animals so interesting, that makes them impressive as whole and individual creatures. This different order of activity is *behavior*.

Behavior is a term difficult or impossible to define with absolute precision. This is true of almost all terms that involve abstraction and do not apply to a separate, concrete object; you have encountered several previous examples. Perhaps as good a definition as any is that behavior is simply doing things, in the popular sense of the words. We still cannot define "doing things," but everyone has a reasonably accurate idea of what the words mean. A useful definition in more scientific terms is that behavior is externally directed activity. Activity is behavior if it is related to the surroundings of the organism and if it brings about some external change in the relationship between organism and environment.

Although we define behavior as involving an external change, it arises from within the organism and obviously also involves all the sorts of internal mechanisms discussed in previous chapters. It may be a direct response to an external stimulus—to a change in the environment—and it almost always is this in plants and lower animals. The mechanism of response is nevertheless internal. In many organisms, especially in higher animals, behavior may not have any immediate outer stimulus. The eagle may fly off even though it does not see a rabbit and nothing has changed in the environment. It just was hungry, or it was tired of sitting. Even so, the behavior involved previous experience with external stimuli. Such behavior requires complex associative processes and also memory in some sense of the word. It is usually dependent on a well-developed nervous system.

All behavior directly or indirectly involves the sequence of stimulus–conduction–response, discussed in the preceding chapter. Behavior therefore is strongly influenced and in large part determined by receptors and conductors and further by associative mechanisms when these are present in the conductor system. It is also necessarily influenced and in part determined by the mechanisms of response, the effectors. We have not hitherto paid much attention to effectors. Since they are so important in behavior, we had better begin this chapter on behavior with some discussion of them.

Effectors

MOVEMENTS

When the eagle flew away, its behavior was a sequence of motions, and the active effectors were the bird's muscles. It is possible to think of responses that involve no motion—or at least motion of nothing larger than molecules —but that could be called behavior. Some simple organisms, especially bacteria, release poisons without any visible motion of the organism. The action does lead to an external change of the environmental relationship, but is it behavior? Then, too, if we are thinking of the whole life pattern of activity of the eagle, we are likely to say that one of the things it does—part of its behavior—is to perch motionless on trees. Such points underline the difficulties of definition and the complexity of the concept labeled "behavior." It is nevertheless clear that the one common factor in practically every activity clearly recognized as behavior is motion.

Behavior certainly depends on a great deal more than motion, but we almost always recognize that behavior is occurring by the fact that there is motion. Behavior is also described almost entirely in terms of motion and of the relationship of motions to the circumstances in which they occur. The motion may involve movement of the entire organism from one place to another; then it is *locomotion*. It may, however, involve movement of only a small part of the organism. A wink is behavior just as much as is dashing a hundred yards or driving an automobile. In behavior, then, the effectors of primary importance are those that produce motion either of a part or of the whole of an organism. Can you think of any other sorts of effectors?

EFFECTORS IN PLANTS

Plants typically have few, simple, and poorly differentiated receptors, conductors, and effectors. Most plant motions of a sort that can be called behavior depend not on special effector cells or organs but on activities of the ordinary plant cells. The experiment outlined in Chapter 12 showed that more rapid elongation of cells in one place than in another can cause a plant to bend, a distinct motion. This mechanism is involved in most of the natural movements of plants, and in nature it is usually if not always effected by auxins, as it was in the experiment.

The only other mechanisms commonly involved in motions of parts of plants are changes in the turgidity (stiffness or degree of inflation) of cells, mainly controlled by their water content. Leaves and stems droop or stiffen according to whether their supportive cells are less or more turgid. Among the very few definite effector organs in plants is one that operates by the same sort of changes in water content and turgidity. In sensitive plants (Fig. 14–1) there are little swellings (called *pulvini*) at the base of each leaf attachment (petiole) and also at the base of each branch within the compound leaf. The swellings are loosely packed with thin-walled cells. When the leaf is touched, the upper or lower cells of the pulvinus lose water and become less turgid. The leaf then bends toward the side of less turgidity. That is about as far as plants have gone in developing effectors. Indeed it is not customary to speak of

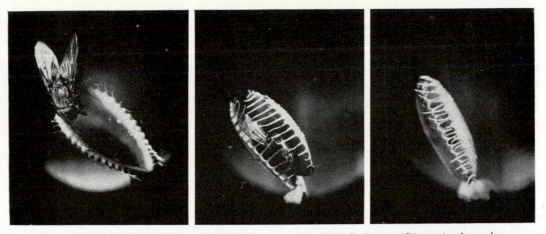

14–1 The behavior of plants. *Above.* The carnivorous Venus's flytrap (*Dionaea*), shown here capturing a housefly, displays relatively complex plant behavior. The leaf, shaped like a bear trap, is sensitive to touch and responds by rapidly closing, thus trapping whatever touches it. *Below.* Sensitive plants (some species of *Mimosa*) also exhibit complex plant behavior; their leaves close rapidly on being touched.

Courtesy, General Biological Supply House, Inc., Chicago

plants as having behavior in the usual sense of the word.

EFFECTORS IN PROTISTS

Behavioral motion occurs in almost all protists and animals and is usually produced by special effectors. Protoplasm has among its characteristics an ability to expand and contract. Motion in protists and animals involves various specializations of this general property of protoplasm.

Most behavioral motion is ameboid, ciliary, flagellary, or muscular. *Ameboid motion* is so called because it is characteristic of amebas, but it also occurs in many other protists. Through processes not yet understood, the

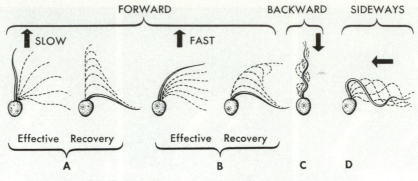

14–2 Flagellary motion in the protist *Monas*.

cell develops a bulge (or *pseudopod*, "false foot") into which cytoplasm flows, with a coincident withdrawal of cytoplasm and retraction of the cell membrane at the opposite end of the organism (see Fig. 13–1). A slow, oozing kind of locomotion results from repetition of this process. *Ciliary motion*, characteristic of protists such as *Paramecium*, which are for this reason called ciliates, involves numerous short hairlike projections, *cilia*, usually covering most of the exposed surface of the cell. The cilia row the protist along by a beating motion. Normally the movements of the numerous cilia are coordinated and rhythmic, again illustrating the complex interactions that are somehow packed into single cells without differentiated cellular organs. Flagella, especially characteristic of flagellate protists such as *Monas* (Fig. 14–2), are similar to cilia but longer and more whiplike; usually there are only a few or a single one on each cell. In *flagellary motion* they may push the cell by a gentle rippling motion or pull it along by sometimes very vigorous lashing. The flagellary movements executed by the same cell may be quite diverse.

Ameboid, ciliary, and flagellate motions are the principal (but not the only) means of movement in protists. All of them also occur in higher or true plants and animals but (with a few exceptions) not for moving the organism as a whole. In fact, all occur in humans—ameboid motion in the white cells of the blood, ciliary motion in the respiratory tracts (in cells that do not themselves move but sweep dust particles out), and flagellary motion in sperm cells. Although these motions and similar activities, which are widespread

in animals and also occur in many plants, are certainly important, they are rarely classed as behavior in the sense that we are using the word. Most behavioral motion in multicellular animals is *muscular*, an evolutionary development of the primitive ability of cytoplasm to contract.

Even in some protists there are contractile fibers, which of course in these acellular organisms are not made of muscle cells. They are localized strands within the cytoplasm that have the capacity of contracting more rapidly and pulling in a more definite direction than the rest of the protoplasm. In almost all multicellular animals there are specialized cells in which part, or usually all, of the cytoplasm has these properties. These are the muscle cells, usually united into strands or layers of muscular tissue and clustered in bundles forming separate muscles. Most animals above the level of sponges have muscles, and it is in these organisms that behavior is most elaborate and most significant, from our human point of view at least. Muscular effectors are therefore worthy of special consideration.

MUSCULAR EFFECTORS

In practically all animals above the level of sponges a muscle is the usual effector in a stimulus–conductor–effector sequence, whether this occurs in just that form, as a simple reflex, or with associative complexities. In all but the simplest forms, a fairly clear distinction can be made between muscles that produce the sorts of motions that we have defined as behavior and those involved only in internal processes. In higher animals the distinction may be so clear that even a small segment of a muscle can be identified as *skeletal* (*voluntary* in man) or *visceral* (*involuntary* in man).

A well-developed but still quite simple muscular system can be exemplified by that of an earthworm (Fig. 14–3). As in man, a cylindrical sheet of muscular tissue around the intestine moves food along by rhythmical contractions. There are also muscles in the walls of some blood vessels, especially in the five pairs of hearts. The intestinal and circulatory muscles are of course visceral muscles. The behavioral muscles form two nearly complete cylinders, one inside the other, throughout the body just inside the outer wall. In the outer layer the fibers are arranged in circles, and in the inner layer the fibers run lengthwise. A section of the body with the outer layer contracted and the inner layer relaxed becomes long and thin. Conversely, when the outer layer relaxes and the inner contracts, the section of the body becomes short and fat. There are also paired bristles down the sides of the body, each with its own small muscle fibers that erect it or pull it back.

An earthworm in its burrow moves along by bracing itself with its bristles and then sending waves of long-thin and short-fat along the body. The contractions forming the waves are controlled by nerves. The nerve cords (a fused pair) exercise central control, but what is by courtesy called the brain has little to do with the process. An earthworm with its front end cut off moves along just about as well as before. This simple reaction is nearly, although not quite, all there is to an earthworm's behavior. You are likely also to have noticed that an earthworm can contract the whole body at once and snap back into its burrow.

Development of really complicated motions and of the most complex sorts of behavior depends on a system of rigid braces, attachments, levers, and joints worked by muscles. Such an arrangement is a *skeletal system*. The skeleton often has other functions as well, particularly that of protection, but in its relationship to the behavioral muscles it is an integral part of the effectors of the organisms that have it. In animals with skeletons almost all the behavioral muscles are attached to skeletal parts and move them, the other tissues

naturally coming along. There are exceptions. For instance, the muscle most involved in kissing is not directly attached to the skeleton, and kissing is definitely behavior.

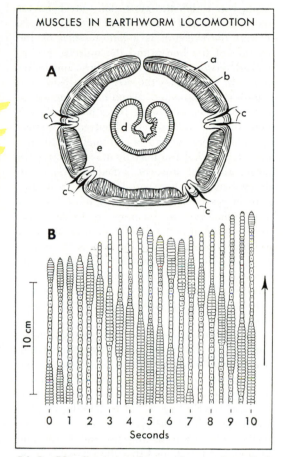

14–3 **Muscles in earthworm locomotion.** *A.* A cross section through the body of an earthworm: *a*, outer layer of circular muscles, whose contraction narrows the animal; *b*, inner layer of longitudinal muscles, whose contraction fattens the animal; *c*, four pairs of bristles, which serve to hold the animal against its burrow; *d*, alimentary canal; *e*, body cavity, or coelom. *B*. The shape and progress of an earthworm moving forward through 10 successive seconds. At the beginning, note that the anterior segments are fat (from the local contraction of the longitudinal muscles), gripping the burrow. Subsequently, the contraction of the longitudinal muscles moves backward as a wave, while the contraction of the circular muscles begins at the anterior end, thrusting the anterior segments forward. Part of the forward advance is conserved (from 5 to 7 seconds) when a new wave of longitudinal-muscle contraction begins in the anterior segments. This contraction fattens the animal anteriorly, slowing down the initial advance as the bristles engage the burrow wall.

There are two possible relationships between muscular and skeletal systems, and both arrangements have evolved more than once. They contrast characteristically in the two groups in which the most elaborate behavior has also evolved: insects and vertebrates. In insects the skeleton is a hollow shell on the *outside* of the body, and the muscles are attached *inside* the skeletal tubes and other parts. In vertebrates the skeleton is inside and muscles are attached around it, fastened to outer surfaces of bones. In rather small animals the two systems are about equally effective mechanically, and in extremely small ones an external skeleton may be more effective. In larger animals, however, sufficient strength in an external skeleton would require an inordinate increase in its bulk and almost insuperable difficulties in forming workable joints and lodging muscles strong enough to work the heavy mechanism. This is one of the reasons why insects are smaller than most vertebrates. As usual, there are other factors, too, but this alone would make an insect as big as a man impossible. Could a man as small as an insect exist?

The study of bones, or of insect skeletons, and muscles has great fascination for anyone who is at all mechanically minded. Most of the mechanical principles of simple and compound levers, for instance, may be illustrated in animals (Fig. 14–4). It is outside the plan and approach of this book to name and describe the vertebrate muscles and bones one by one, but it is hoped that the reader may be able to study them by dissections in the laboratory.

We must make just two other points about muscles in general: The first is that they work in only one direction. Force is exerted only when a muscle contracts. When it relaxes again, it may slowly resume its former shape spontaneously or it may be pulled back into shape, but in any event it releases no energy by its return. A consequence of this fact is that in a workable system there must be something pulling in the opposite direction from each muscle. This opposing force may be an elastic tissue of some sort, but with the skeletal muscles it is usually another muscle, or group of them. That system of opposing forces makes possible the delicacy and precision of behav-

ioral motions. You can illustrate it and feel it by bending your arm into any position. The arm can be moved and then held, with precision in small fractions of an inch. While in the set position you can flex the arm muscles strongly without moving the arm at all. Obviously, then, you are putting in play opposing sets of muscles and are applying exactly equal force in opposite directions.

The second point is that muscle fibers, like nerves, normally respond by an all-or-none reaction. If they contract at all, they go as far as they can. This is an effect in part of their own character and in part of the fact that the nerve impulses stimulating them are all-or-none phenomena. You will be right to question this statement from your certain knowledge that you can exert greater or less force with the same muscle, but perhaps you can think of a hypothesis to account for this fact.

The answer is that each muscle consists of a large number of fibers divided into many groups stimulated by different nerve fibers. Each group contracts fully or not at all, but the total force in the muscle depends on how many groups contract. Thus more impulses through more nerve fibers produce greater contraction of the whole muscle.[1] Even when we think we are not using our nerves and muscles, they are usually slightly active. It is usual for a few fiber groups in any skeletal muscle to be contracted all the time. This is muscle *tonus*. It does not tire the muscles, because the same groups do not remain contracted for long; one relaxes and rests while another contracts.

OTHER EFFECTORS

There are kinds of effectors that we have not discussed, but they are less important for an understanding of general biology than those we have dealt with, and we shall not devote much space to them. Two are mentioned as examples. Coelenterates (corals, jellyfishes, and so on) have special cells scattered over the surface that explode when touched (cf. Fig. 21–3). They shoot out threads or darts, some of which help the ani-

[1] In some animals nerve impulses of certain frequencies or along certain channels prevent contraction instead of stimulating it. This is another way of balancing the amount of contraction.

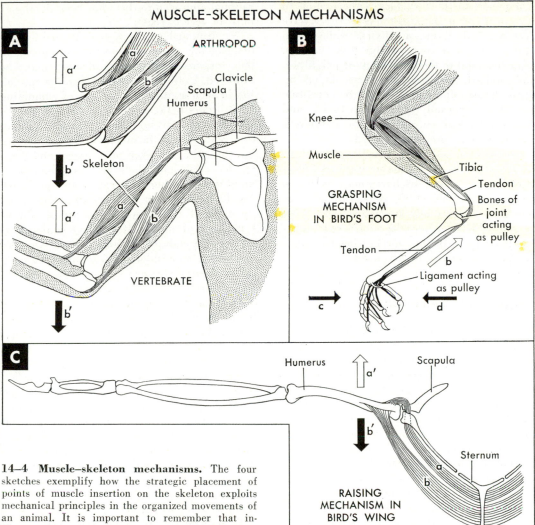

MUSCLE-SKELETON MECHANISMS

A ARTHROPOD

a'

b'

b

a

Clavicle

Scapula

Humerus

Skeleton

a'

a

b

b'

VERTEBRATE

B

Knee

Muscle

GRASPING
MECHANISM
IN BIRD'S FOOT

Tibia

Tendon

Bones of
joint
acting
as pulley

Tendon

b

Ligament acting
as pulley

c

d

C

Humerus

a'

Scapula

b'

Sternum

a

b

RAISING
MECHANISM IN
BIRD'S WING

Keel

14–4 Muscle–skeleton mechanisms. The four sketches exemplify how the strategic placement of points of muscle insertion on the skeleton exploits mechanical principles in the organized movements of an animal. It is important to remember that individual muscles exert force in one direction only; complex movements of organs depend on the interaction and placement of several different muscles as well as on the geometry of the skeleton. *A.* The movement of a limb joint in an arthropod (external skeleton) is compared with that in a vertebrate (internal skeleton). In each, contraction of muscle *a* causes limb movement in the direction *a'*; contraction of muscle *b* causes limb movement in the direction *b'*. The controlled motion of the limb in any one direction depends on the balance of the opposing contractions of muscles *a* and *b*. *B.* When a bird squats on a branch, its claw executes a grasping motion caused by the contraction of a single muscle in the middle part of the leg; there are no muscles in the foot region. The simple muscular contraction is translated into the complex claw motion by the geometry of the whole limb. The muscle responsible originates on the distal end of the thighbone (femur) and proximal end of the tibia; it ends in the lower

part of the leg as a long tendon that passes over the joint between leg and foot and through a pulley formed by a ligament at the base of the toes. Here it branches, sending parts to the tip of each toe. Contraction of the leg muscle pulls the long tendon upward (*b*) and drags its separate parts through the ligament pulley, causing the claws to move inward in the two directions *c* and *d*. *C.* The up and down beats of a bird's wing are caused by muscles (*a* and *b*, respectively) lying side by side. Both originate on the keel of the sternum, or breastbone, and both are inserted on the humerus. When *b* contracts, it pulls the humerus directly down (*b'*); when *a* contracts, it raises the humerus (*a'*), in spite of its position, simply because it passes over a pulley formed by the scapula or shoulder blade.

mal to adhere to a surface. Others cling to prey or pierce it and inject poison. If you have ever brushed against a jellyfish while swimming, you know how painful these can be. They are fatal to the usual prey of jellyfishes, and, indeed, humans have been killed by them. These discharges certainly constitute behavior, externally directed activity, and they involve motion, although of a different sort from any previously noted.

Our second example is the discharge of an electric shock by any of the several varieties of electrical fishes. The discharge may involve as much as 2000 watts at 200 volts, a tremendous jolt. This is "externally directed activity" with a vengeance, so it must be considered behavior. It is one of the few sorts of behavior that do not necessarily involve motion.[2]

Study of Behavior

We have now entered a field that is at the same time both biology and psychology. Since psychology is often defined as the science of behavior, it may, indeed, appear that the present subject is psychology rather than biology. But behavior is obviously a living activity, and it is impossible to pursue the study of life, which is biology, and to ignore behavior. There is a large area in which biology and psychology not only overlap but are actually the same subject. We shall not attempt to follow the psychologists far into their most complex and distinctive subject matter, human mental life and personality, but even that is inseparably grounded in all of biology.

Behavior is extremely complex in the manifestations of most interest and importance to us. It depends on a multiplicity of factors, all hard to isolate and identify—many of them internal processes such as we have already studied plus others that almost seem to defy observation. Men have always been interested in the behavior of other men and of all the animals around them. Study of behavior is one of the oldest occupations, but it is one of the youngest sciences. The problems of interpreting, of understanding, the behavior of other organisms are so great that a really

rational approach to them has been developed only in recent years.

We ordinarily see only the external aspects of behavior. An environmental stimulus and the response of some effector may be observed. Until very recently even the specialists in such studies had no way of examining what occurs between the stimulus and the response, and even now the information about this is extremely incomplete. We cannot see into the minds of even our nearest and dearest, let alone into whatever corresponds to a mind in a rat, a bug, a worm, or an ameba. We usually have to infer the internal process from its external result, which is almost always some form of motion. If our nearest and dearest goes through the motions we call "kissing" or "hitting," we judge the accompanying mental processes by the processes we would have if we performed the same actions.

Judging others by ourselves is all very well to the extent that others are really like ourselves. We can be quite mistaken even about our nearest and dearest. We are more likely to be mistaken about an Eskimo, although there is little biological difference in the mechanisms that determine his behavior. How much more likely we are to be mistaken about a dog, a fish, or a fly, in which there is a great difference in mechanisms. Yet from time immemorial most people have assumed that if some nonhuman animal performs an action its reasons for doing so (that is, the accompanying processes inside the organism) are the same as ours would be.

Interpretation of nonhuman actions in terms of human motives was one of the two greatest impediments in development of a *science* of animal behavior. The other was that animals were observed in surroundings so complicated that the stimuli really involved could not be accurately identified. Observations were *anecdotes* as to what an animal was seen to do in complex natural situations, in which the stimuli important to the particular animal were often wholly unknown to the observer. The anecdote was invariably interpreted *anthropomorphically*,[3] as if the animal were human. You know that this entirely non-scientific and non–common-sense procedure is still usual. Some popular magazines make a

[2] Note the interesting point that electric-organ effectors in fish are modified muscles.

[3] From the Greek, meaning "man form."

special point of gathering anecdotes about pets and other nonhumans and giving anthropomorphic interpretations of their behavior.

There is an old principle of logic and science that has been called Occam's razor, after a scholastic, William of Occam, who expressed it in the fourteenth century in an effort to eliminate nonsense. The principle is that if several different explanations are possible, the simplest is to be considered the most probable. The science of behavior can be dated, more than by any other one event, from the application of this principle by Lloyd Morgan (1852–1936). Morgan's canon (stated by him in somewhat different words) is that the actions of an animal should be interpreted in terms of the simplest possible mental process.

This application of the razor has certainly cut out a great deal of nonsense. When an ameba puts out a projection, surrounds a food particle, and digests this in a food vacuole, we do not say that it smelled food, liked the smell, decided to eat, and therefore seized the particle. We attribute to the ameba no mental activity at all. Even before Morgan, few would have been quite so anthropomorphic about an ameba. Yet many bird lovers still assume that a mother bird sits on eggs because she wants to have babies and that she cares for the hatched young because she loves them and knows they need food and warmth. Application of the razor would suggest that the bird reacted instinctively to physiological needs in herself, without the slightest idea that young would hatch from the eggs, with no conscious knowledge of the needs of the young, and no tender emotion toward them. In fact, control of the stimuli confirms this view. Birds will react to pebbles or to cuckoos' eggs in the same way as to their own eggs. They may feed a cuckoo hatchling and viciously attack their own young if the cuckoo presents the maternal stimulus and their young do not.

The razor can cut both ways, and most students of behavior now recognize that its previous use was sometimes in the wrong direction. On the face of the facts, the simplest hypothesis about what happens between stimulus and response is that nothing happens except conduction from receptor to effector—in other words, a simple reflex. Some researchers did go so far as to conclude that all reactions should be interpreted as reflexes and that the study of behavior reduces to correlation between stimuli and responses. This *behavioristic* approach, as it was called, can be applied to man, and it leads to the remarkable conclusion that our actions have nothing to do with our thoughts; nothing, at least, beyond the fact that we may notice our own actions. Now, we cannot get into the minds of others, but we are in our own minds. We know beyond any common-sense doubt that our actions—not all of them, but the great majority —*are* determined by complex mental processes and are not simple reflexes. The question arises whether, after all, it is a correct use of the razor to insist that nothing of the sort happens in other animals.

A dog sees its master and wags its tail. According to Morgan's canon we must conclude that this is no more than an established reflex: stimulus (sight of master)—conduction (from eyes to tail muscles)—response (wag). Is this the right application of the razor? We know the dog and know that recognition of his master has been built up by learning and is equally possible by different cues, not only sight but also sound and, above all, smell. We know that the dog responds to the master in different ways, not only by tail wagging but also by barking, running, cringing, picking up its leash, or otherwise, depending on the situation. We also know that the dog has a very complex brain, with all the parts that occur in ours although they are different in proportions and shapes. We can also determine that there is activity in the cortex of the brain, not only in reflex conductors, when the dog sees its master. This last is a recent technical development that makes it no longer true that behavior can be studied only through responses. By use of suitable electrical equipment it is possible to determine that nervous activity is occurring, and where. Surgical interference with the nervous system can also determine what parts are necessary for perception of a stimulus and production of a given response.

In the light of *all* the facts about the dog, should we infer that the dog recognizes its master, likes him, and responds by tail wagging, or that nothing but a reflex is involved?

The reflex does not explain what is known about the whole of the dog's behavior in the presence of its master and about the dog's associational equipment and its activities. To begin to explain all this would require an extremely complicated set of reflexes. It is actually simpler, and more consistent with the whole body of data, to conclude that the dog thinks. That is the way to apply the razor. In this case it cuts out Morgan's canon.

We still are not justified in concluding that the dog thinks like a man. Of course it does not. Its receptors, conductors, associators, and effectors are characteristic of its species, with variations peculiar to this individual. It thinks like a dog, and like this particular dog. To find out what that thinking is like we have to steer clear of anthropomorphism, of course, but we must steer equally clear of the unjustified application of Morgan's canon.

Modern study of behavior follows principles of good experimentation applicable in any field. Tests are made in a standardized situation kept as uniform as possible except in one respect. One environmental factor is varied, and this is a controlled stimulus. Responses, mainly motions, are observed. Stimuli are varied in kind and intensity in successive experiments. Finally, the mechanisms between stimulus and response are also studied by their anatomical relationships, by their activities, and by the effects of interference with them at various points.

Suppose you want to find out what sounds a dog can hear, what tones it can discriminate, whether it has an established reflex to any tones, whether it can learn a response to a tone, whether it can learn different responses to different tones, and whether it can learn to make two successive responses to a single tone. (You can readily think of still other things you might want to know about relationships between hearing and behavior in dogs.) Can you set up a series of experiments to find out these things?

Tropisms and Similar Movements

TROPISMS

The first movements to be well understood were those of plants. Such movements usually take a very simple form: a plant stem moves away from, or toward, a stimulus such as light. The same stimulus invariably produces the same response; there is no evidence that memory or learning is involved. The word *tropism* was originally coined for plant responses of this kind. Tropism derives from the Greek word meaning "turning," and in its original use it had a very precise meaning. Indeed, botanists carefully reserved the term "tropism" for the bending movements of stems and roots that are caused by differences of stimulation (for example, by light or gravity) on the two sides of a growing organ. They employed *taxis* for the movements of freely swimming unicellular plants. However, the word "tropism" was subsequently carried over by zoologists into the description of animal movements. We shall see that its meaning has been seriously blurred as a consequence.

Two elements are involved in the tropism concept. (1) The movement in a tropistic response has a definite direction, which is determined by differences in the intensity of stimulation on the two sides of the moving organ. (2) The mechanism of the turning response is *innate*—it is an inflexible part of the organism's inherited physiological apparatus. It is not subject to modification or control by the organism. The organism that executes a tropism has no choice about the matter. A man may turn his head when a pretty girl passes by, but he does so on his own responsibility. That is *not* a tropism.

Tropisms in plants

Much of the so-called behavior of plants consists of tropisms, the most obvious of which are responses to gravity and light. Both reactions in garden plants can easily be shown in experiments.

When you plant seeds, you do not bother to put them in right end up. No matter how the seed is oriented, the roots grow down and the stem grows up. There is a gravity (or earth) tropism, which is positive in the root (it grows toward the center of gravity) and negative in the stem (it grows away from the center). If you sprout some beans, a little cylinder, the developing root, appears. Turn a seedling so that the root is horizontal. Next day the tip will have turned and will point downward.

Only the growing tip turns downward; response to gravity is confined to this region. Tropism of the stem can be demonstrated equally simply: turn a pot containing a growing, tall-stemmed plant on its side. The growing end of the stem (and again *only* this part) will turn upward in a day or so.

Differences in growth rates are the mechanism of tropistic responses in plants: there is greater elongation of cells on one side of the root or stem than on the other. Auxins (Chapter 13) become more concentrated on one side. In the case of gravity, this is believed to form a concentration of auxins so great that it *inhibits* growth on the lower side.

Tropistic response to light is equally familiar and easy to demonstrate. Most green plants grow toward light and they may, like the sunflower, keep parts turned to the sun in its daily movement. If an oat shoot (like those used in the experiment in Chapter 13) is lit from one side, it bends toward that side. Or if one side is covered with carbon black, the shoot bends toward the uncovered side. (Charles Darwin carried out many experiments of this kind and determined most of the phenomena of light tropism.) In these tropisms differential growth stimulated by auxins is again involved, although here as influenced by the effect of light on auxin formation and loss. The crucial point is clear enough: the direction of the stem's movement is ultimately determined by the difference in intensity of light stimulus on its two sides.

Many other stimuli determine direction of growth in parts of plants and therefore produce tropistic responses. Roots grow toward water. Tendrils twine around what they touch.

Loeb and animal "tropisms"

The general nature of plant tropisms as due to differential growth stimulation on the two sides of an organ was well understood by the end of the nineteenth century. However, the description and analysis of animal behavioral movements was still encumbered by anecdotal techniques and anthropomorphic interpretation. As we noted earlier, Lloyd Morgan led a rebellion against this approach; and he was joined in his revolt by one of the most dynamic personalities in biology of the early twentieth century, Jacques Loeb (1859–1924). Loeb set out to place the study of all animal behavior on as sound a footing as that of plants by developing a theory of *animal tropisms* or *forced movements*. His main aim was to show that the pattern of animal movement—like that of a plant—was caused by differences of stimulation on its two sides.

We can illustrate Loeb's ideas with the behavior of a pill bug (or wood louse) in the presence of a light source such as a lamp. The animal moves directly toward a single light source, following a path like that in Fig. 14–5A. If it is placed near two lights, it moves along one of the paths shown in Fig. 14–5B. Loeb explained movements like these simply by substituting amount of muscular activity for amount of growth. Just as the two sides of a plant stem differ in the amount of light stimulus they receive, so do the two sides (eyes) of the pill bug; just as the difference in stimulation automatically causes a difference in growth in the stem, so does it automatically cause differences in muscle movement in the pill bug. The less illuminated side moves faster, and the animal curves in towards the light. As soon as the pill bug is oriented head-on to the light, its two eyes are equally stimulated and a straight path is maintained. The animal moves along the line where stimulation from the two sources is equal.

Loeb's theory of animal tropisms as forced movements is a zealous application of Morgan's canon, and it is attractive in its simplicity. It tempted many biologists into believing that a fully mechanical and simple explanation of animal movements (and therefore behavior, if one follows Morgan's canon to its limit) was at hand. Loeb devoted much of his life to pressing this extreme mechanical view of animal behavior. The story of his campaign is fascinating but complex enough to prevent our telling it fully here. It is to Loeb's credit, as well as to Morgan's, that behavioral study was freed from anthropomorphism; but on the debit side of his campaign we must place the ruination of tropism as a term with precise meaning and usefulness. Loeb stubbornly attempted to explain all animal movements as tropisms; yet the fact is that comparatively few of them are like those of the pill bug in the presence of light. And for that matter, many plant movements are not tropisms in the true meaning of the word. Let us look into these nontropistic movements before finally evaluating the meaning and utility of the tropism idea in general.

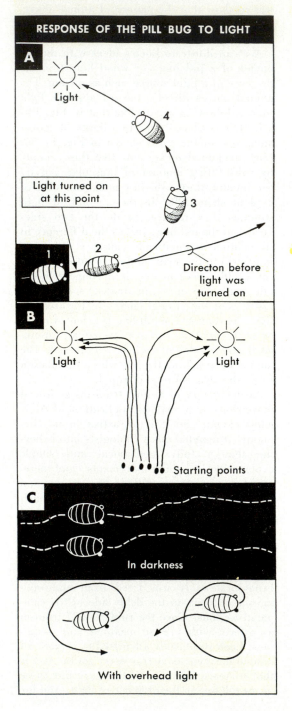

RESPONSE OF THE PILL BUG TO LIGHT

A

Light

4

Light turned on at this point

3

1

2

Direction before light was turned on

B

Light

Light

Starting points

C

In darkness

With overhead light

NONTROPISTIC MOVEMENTS

The direction of many plant movements is not controlled by differences in the intensity of the stimulus on the two sides of the moving organ. Such movements are, therefore, not tropisms. When the sun rises many leaves move upward from a nighttime position; when darkness arrives the leaves droop. The direction of either petal or leaf movement is not determined by differences in intensity of light on their two surfaces; it is determined by the anatomical features described on p. 388.

In protists and multicellular animals there is much behavior that looks tropistic—movements apparently away from or toward a source of stimulation. *Paramecium* is a complex ciliated protist much used in laboratory studies. It performs many seemingly tropistic movements: it moves upward away from the center of gravity, and when confronted with an obstacle it moves away from it. Many flagellated protists can photosynthesize, and they react markedly to light. They move toward a light that is not too strong but away from a light above a certain intensity. Many protists are sensitive to chemical stimuli and, like the ameba in Fig. 14–6, they will move directly toward food, guided by chemical

---

14–5 Forced movements of a pill bug. *A. 1, 2, 3,* and *4* are successive positions of the pill bug. Before the light is turned on, it is in position *1,* headed in the direction shown. When the light is turned on, it is illuminated differentially on its two sides, as the shading on the body and eyes indicates. Subsequently it turns until, in position *4,* it is illuminated equally on its two sides and eyes. It then maintains a straight path to the light. *B.* The tracks of six animals exposed to two lights follow a path on which the animals experience equal light intensities on their two sides. Ultimately they swing toward one light or the other. *C.* Two animals have had their right eyes covered with black paint. In darkness they maintain an essentially straight path. When exposed to an overhead light, they both start a "circus movement"—the pathway is circular and directed to the left, that is, toward the more illuminated eye. (How might you have improved this experiment? Would it have been better to paint the left eye in one of the animals? Why is the experiment in darkness necessary?) Loeb considered these reactions to be tropisms, but it is preferable to think of them as examples of phototaxis, innate reactions to the stimulus of light.

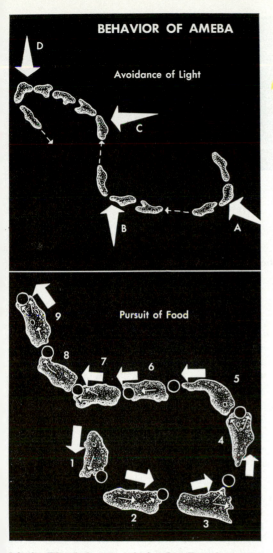

BEHAVIOR OF AMEBA

Avoidance of Light

Pursuit of Food

14–6 The behavior of *Ameba*. *Top.* Successive positions of an ameba after being subjected to successive lights coming from the directions indicated at *A, B, C,* and *D.* Each time it turns away from the light. *Bottom.* Successive positions of an ameba in pursuit of a food particle (circle). Even when the ameba loses actual contact with the particle, it can still detect its position, presumably by chemical stimuli.

stimulation. <mark>Even bacteria, which have about as little behavior as any organisms, move away from acids.</mark> Planarians, the little flatworms we have discussed before, move against a current, and they aggregate in the darker side of a dish. Fruit flies congregate in the drier side of a dish. <mark>Everywhere we find</mark>

<mark>animal movements that are guided somehow by external stimulation: gravity, light, chemicals, temperature, moisture, and so on. But by far the majority of these movements are not tropisms in the original and more exact meaning of that term.</mark>

The paramecium that avoids an obstacle does not turn tropistically, forced by stimulation differences on its two sides. When it blindly bumps into an obstacle, it has a shock reaction; it stops. It then reverses its ciliary beating and, backing up a way, makes a turn of about 30 degrees and goes forward again (Fig. 14–7). The angle of turn is not controlled by the pattern of stimulation; it is fixed. If the paramecium hits the same obstacle on its re-try it repeats the backup and 30-degree change of course. Sooner or later the obstacle is avoided. But in 12 repetitions the animal might be back where it started after missing several escape channels in the meantime. Similarly, <mark>flagellates do not, as a rule, move *directly* toward or away from a light. They have shock reactions to strong</mark>

14–7 The trial-and-error behavior of *Paramecium*. On encountering an obstacle, such as an alga, the paramecium backs away (*1–2*), makes a turn of approximately 30 degrees (*2–3*), and goes ahead again (*3–4*). If the obstacle is still present, the maneuver is repeated (*4–7*). (Based on Jennings.)

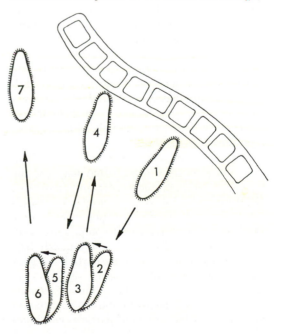

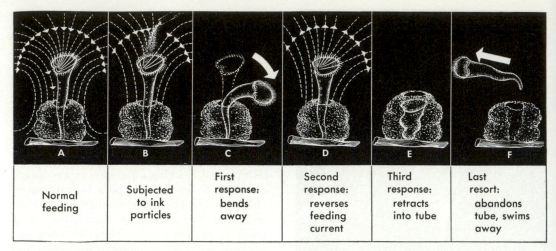

| Normal feeding | Subjected to ink particles | First response: bends away | Second response: reverses feeding current | Third response: retracts into tube | Last resort: abandons tube, swims away |
| A | B | C | D | E | F |

14–8 The behavior of the protist *Stentor* when irritated with ink particles. The pipette used to apply the ink is shown only in *B*.

light and darkness. Sooner or later trial and error lead them into light of intermediate intensity. Fruit flies in a dish that is drier on one side do not move directly to the dryness. They move about at random and eventually congregate in the dryness simply because they slow up there. This is no more a tropism (a directed movement) than the congregation of automobiles in the congested, slow traffic at intersections and the approaches to big cities.

Indeed in none of the examples we have mentioned (other than the pill bug) is the *net direction* of movement controlled by differences in stimulation on the two sides of the organism. The only resemblance they have to Loeb's tropisms is that many of them are forced movements—they are *fixed* or invariant responses that are released (forced as a mousetrap is forced) by a set stimulation. They are therefore best thought of as reflex or automatic movements.

Unfortunately, we cannot even leave it at that. Close scrutiny reveals that many of the responses are not wholly reflex and automatic. What the animal does is often open to modification in the light of past experience. *Stentor* (Fig. 14–8) is a relatively complex, stalked, tube-building protist. If it is continuously irritated by a stream of India ink particles, it goes through a series of changing responses in spite of the fact that the stimulus remains constant. First it bends to and fro in different directions; if the stream of particles continues

it *then* reverses its cilia so as to "blow" the particles away; when this fails, it pulls itself inside its tube. Whenever it ventures out, only to find the stimulus still present, it goes straight back into the tube. It does not make the responses (bending, "blowing") it earlier made to the same stimulus. Finally (Morgan and Loeb forbid us to say "in exasperation") it breaks away, moving off to build a new tube elsewhere. Such is genius among the protists.

ADAPTIVE NATURE OF BEHAVIOR

The net impact of Morgan's canon and the personality of Loeb was to free behavioral studies from silly and obstructive human comparisons. This was indeed a great step forward, and for the most part the only price paid for this advance was loss of useful meaning to the word "tropism." However, for a while there were other consequences: the insistence that conscious manlike purpose played no part in animal behavior tended to obscure the fact that "purpose" in another, quite different, sense is indeed involved. For it is a reasonable generalization about behavior in all organisms that it is *adaptive*. With some exceptions behavior in any species has *average* results beneficial to the species. Behavior *serves a purpose* when by means of it an animal gets food, avoids enemies (and India ink!), and not only finds but wins a mate. Behavior serves purposes, and it benefits the species not because animals are sensible or because their behavior is planned to help them. The reason is quite simple; behavior is adaptive because it has to be. A species with inadaptive behavior would not last long. This does not explain how particular sorts of adaptive behavior arose. That is another question, a much harder one that we postpone until Chapter 17.

Finally we note that Loeb's campaign for "forced movements" as a general explanation of all animal behavior founders on another hard fact. Most behavior is never wholly automatic and forced, whether by this we mean a true tropism or some other simple reflex movement. It is open to some degree of control and modifiability by the animal in the light of its experience, as even *Stentor* shows. We will, however, certainly do well to follow Morgan and Loeb in avoiding "wisdom," "conscious purpose," and such like in explaining this fact.

Behavior and Nervous Systems

In animals that have nervous systems, which means practically all multicellular animals except sponges, the nervous system is always involved in behavior. It at least acts as the conductor that sets off the effector response. Usually it does much more. It is not surprising, then, that the broad features of behavior in these animals approximately parallel the organization of their nervous systems. Complex behavior requires a complex nervous system. Alternative responses require alternative routes in the conductors. Coordination of different effectors in a single response requires associative mechanisms. Assembling of complex stimuli into units of information producing selective responses to different elements in the total situation requires still more complex associative mechanisms. *Centralized control* requires a centralized message center composed of conducting and associating neurons. Delayed response to stimulus requires a storage mechanism, a *memory* of some sort.

In considering the evolution of the nervous system we saw certain broad trends beyond the stage of a diffuse nerve net. Conducting paths were centralized, lengthened, and speeded up in nerve cords. Associative neurons multiplied in these cords. Coordination of information from receptors and action by effectors began to be centralized in nerve masses near the front end of the body. These masses—brains—tended to take over more and more of the coordination and control of effector activities, and hence of behavior. In the vertebrates this tendency was particularly evident, with development of extremely complex secondary and higher-level association centers in the brain. Finally, in mammals new centers of still higher level and greater complexity evolved.

All these major changes in nervous structure are paralleled by changes in sorts and complexities of behavior. That the nervous system became more complex and that behavior did also is broadly true, but it leaves out some extremely important points. In the first place, animals comparable in complexity may and usually do differ in their kinds of behavior. Still more striking is the changing balance between innate—built-in or instinctive—behavior and modifiable, or learned, behavior.

NERVE-NET BEHAVIOR

Human behavior may serve as a handy example of one extreme in kind and complexity of behavior and associated nervous system. Insects represent another extreme, of quite a different kind and along a widely separate line of evolution. Among living animals (as distinct from protists and except for sponges) the simplest nervous system and behavior [4] occur in the coelenterates, with their undifferentiated nerve nets. If a tentacle of a sea anemone touches a small bit of food, that tentacle alone bends toward the mouth. If the food is large or struggling, the stronger stimulus may spread and other tentacles may join in the reaction. There is no centralization or evident coordination. The reaction simply spreads out from wherever the stimulus occurs. In a swimming jellyfish, on the other hand, the whole

[4] Exceptions occur among various parasites with secondarily simplified nervous systems and behavior. Parasitism is a turning of evolutionary direction into a one-way bypath. It takes the animals out of any main line of evolutionary change. It is, nevertheless, highly adaptive, and whether it is degenerative depends on the point of view.

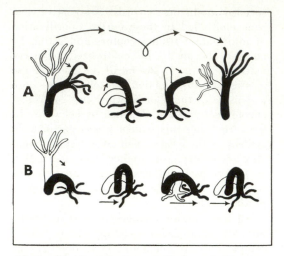

14–9   The locomotion of *Hydra*. *A.* Movement by somersaulting. *B.* Movement by looping, like a caterpillar.

bell rhythmically contracts at the same time. This certainly is coordination, but of a simple sort and without local control. The nerve net is circular and has practically simultaneous rhythmic impulses throughout.

Perhaps the most complex and certainly the most amusing behavior in a coelenterate is the locomotion in *Hydra*. *Hydra* often creeps along very slowly by ameboid movements in the basal cells by which it usually clings to a plant or to the bottom of the pool or stream in which it lives. Once in a while, however, it moves more rapidly. Then it bends double, clings with its tentacles, lets go at the basal cells, bends double again, clings with the basal cells in a new position, and straightens up (Fig. 14–9). It thus turns a complete somersault and comes up in a different place. This is not really a very elaborate maneuver, but still it is rather surprising that it can be done without central control or developed associative mechanisms. It looks like a chain reaction, one in which each step is the stimulus for the next step. (Some more clear-cut chain reactions are mentioned later in this chapter.) The locomotion of *Hydra* does not seem to be directive. It is not going after anything, but is only moving to new and possibly better feeding grounds by trial and error. In spite of the odd method of locomotion, its behavior is not much if any more advanced than that of *Paramecium*.

Echinoderms (starfishes, sea urchins, and their relatives) have a nerve ring and radial nerve cords in addition to a nerve net. There are associative neurons and true nervous reflex arcs from receptors and effectors. As you would expect, their reactions are more highly coordinated and varied than those in coelenterates, but the difference is not very striking. Most of their behavior is still of the trial-and-error variety, although once a stimulus is encountered, such as a succulent clam, it is dealt with by directive and unified actions of the whole body. There is even some evidence that a starfish can be trained to avoid unfavorable stimuli or to use one arm more than others, but the learning ability is slight at best. A disturbed starfish reacts very sluggishly and in a seemingly aimless way. The reactions of its cousins, the brittle stars, seem equally aimless but are far from sluggish. Violent lashing of the arms follows any disturbance, an example at a simple level of marked differences of behavior in animals with similar nervous systems.

BEHAVIOR WITH THE BEGINNINGS OF BRAINS

We have noted (p. 391) that an earthworm has rudiments of brains, anterior enlargements or *ganglia* of the nerve cords. We also there noted something of its behavior. Reflexes through the central nervous system are highly developed, but these occur locally in the segments of the nerve cords and involve the brain scarcely or not all. Locomotion and most other behavior is not seriously affected by removal of the brain. Nevertheless, the brain has some control over the tonus of the muscles and general sensitivity. Removal of the upper of the two anterior ganglia causes the front end to bend upward (because tonus is reduced in upper muscles), and removal of the lower ganglion results in downward bending. Removal of the upper ganglion also makes the worm more sensitive and restless. Almost everyone has noticed that a worm with its head cut off squirms more than one that is intact, but this is due simply to loss of inhibitory messages from the brain. There is extremely little central control in an earthworm, but the beginnings are there.

A flatworm such as a planarian (which is

not at all closely related to an earthworm) has slightly more central control. A planarian moves in several different ways: by beating cilia on the lower surface, by rippling motions of the body, and by a looping crawl a little like an inchworm caterpillar. Removal of the brain does not interfere at all with ciliary locomotion, interferes seriously with rippling but does not wholly prevent this motion, and stops crawling altogether. Thus the more complex behavioral reactions, or those requiring more extended cooperation of muscles, are coordinated and largely controlled in the brain.

Both flatworms and earthworms can be trained in simple ways. Planarians can be taught not to cross a rough surface in a dish by shaking the dish every time they come to the rough part. After a great many repetitions, they turn away from the rough area even when the dish is not shaken. Earthworms can be trained to turn to the right or left in a Y- or T-shaped passage by giving them an electric shock when they take the wrong turn. Some training is possible when the brain has been removed, but it usually takes longer. In other words, earthworms can learn with their nerve cords alone, but the brain helps.

Innate and Learned Behavior

INSTINCT AND LEARNING

In plants beyond any doubt, and in protists and coelenterates with only a whisper of a doubt, we were dealing with completely innate behavior. The whole mechanism of behavior is built into these organisms and is rigidly fixed. A certain stimulus always produces exactly the same response, and the organism can do nothing about it. Other individuals of the same species (if they are in the same physiological condition) have exactly the same responses to the same stimuli, whatever their past history has been. In the echinoderms dimly and questionably, and in the flatworms and earthworms simply but definitely, something else has appeared. Most of the behavior here still seems to be innate, but not all. Reaction to a given stimulus, a roughened surface, a right or left turn (or a dark or light passage in other, similar experiments) is not rigidly fixed. It is changed by training. After the training, the response is not the same as in

other individuals of the same species. There are differences due to the past experience of the animals.

Innate behavior is what is commonly called *instinct*. Instinct has usually been understood as something quite different from learned behavior. An action was *either* instinctive *or* learned. The task of the investigator was to find out which. There used to be a tremendous amount of discussion as to whether a particular item of behavior was learned or instinctive, or as to the relative amounts of the two sorts of behavior in a given animal. It was found out that many supposed instincts are not *purely* instinctive. Researchers began to shy away from the word "instinct" and to talk about "innate behavior" instead. Changing the name does not change the problem, but it does help a little in getting away from some of the wrong ideas that had become attached to the old name.

Scientists have had to learn, rather slowly and painfully in some subjects, to be suspicious of "either–or" questions. Such questions offer only two choices, and it is entirely possible that neither choice is correct. This is notably true of questions about whether a thing is inherited or acquired. It usually turns out, not only in the field of behavior but also in intelligence, size, blue eyes, or almost anything else, that many things are neither solely inherited nor solely acquired. The wrong question was being asked. We smile now at all the sound and fury of the "nature or nurture" battles that were raging up to a few years ago.

Some behavior is probably *completely* innate, although this is open to question except, perhaps, for true tropisms in plants. It is practically impossible to say that any behavior is *completely* learned in that it depends in no way on inherited mechanisms. To take an extreme example, nothing is more obviously learned, one might say, than human language. Infants born deaf and blind never talk unless they can be taught through the sense of touch as Helen Keller was.[5] Language is always

[5] Helen Keller became deaf and blind at 19 months as a result of illness. Some of the important aspects of her education in relation to language are described and discussed in Ernst Cassirer's *Essay on Man*, Yale University Press, 1944.

learned from those around us,[6] and we learn totally different languages, depending on who raises us. Nevertheless, there is an essential innate element in language. We are born with the inherited nervous mechanism and effectors for talking. Without them, as when the left temporal area of the brain is damaged, we cannot talk. Other animals are innately unable to talk.

The real point about linguistic behavior is not that it is not innate, but that it is highly *modifiable*. Its mechanism is innate, but it can be used in many different ways. The ability to talk is innate, but the particular language we talk is not.

At the opposite extreme, when a doctor taps just below your knee with a rubber hammer, you normally give a little kick. This is a highly innate response, a simple reflex. Yet you can learn to prevent the response. You can also learn to make the same response, or what looks exactly the same from outside you, whether the doctor taps you or not and without using the reflex arc. This behavior, too, is modifiable, although it is much less modifiable than linguistic behavior.

From this point of view, the significant thing about behavior is not so much whether it is innate or not but whether it is less or more modifiable. Every gradation exists from practically unmodifiable to almost endlessly modifiable behavior. The songs of birds illustrate this point very well. Individuals of some species of birds, if raised without ever hearing another bird, will nevertheless produce a song completely characteristic of their species. Not only the ability to sing but also a particular song is innate. If raised with another species, such birds still sing the song of their parents, which they never heard. Other birds, if raised with a different species, produce the song of that species and not of their own. Sometimes the song will have characteristics of both species. Some birds have one, fully stereotyped song, and some have wide variation of song. Some never alter their songs, and some can

[6] The Greek historian Herodotus wrote that Psammetichus, king of Egypt, had two children raised without ever hearing anyone speak. He claimed that the children started talking Phrygian to each other, which proves—or does it?—that language is innate and that Phrygian is the oldest language.

imitate the songs of almost any other birds and even the sounds of human voices. There is, in short, almost every degree of modifiability, and the question whether bird song is instinctive is not a "yes or no" one.

IMPRINTING, RELEASERS, AND ACTION PATTERNS

Relationships between an innate behavioral mechanism and the influence of environment are simply and strikingly illustrated by the peculiar phenomenon of imprinting. In the greylag goose (on which the first observations were made) and in a number of other birds, the young normally begin shortly after hatching to follow the mother about and in general to exhibit a somewhat flexible but quite characteristic pattern of behavior of offspring toward parent in the particular species. An ingenious series of experiments showed that there is a brief period soon after hatching when the young bird becomes "attached," so to speak, to the first moving object it sees. Moving objects seen before or after that period do not cause attachment, nor do any that are seen after the initial attachment has occurred. A particular object at a particular time becomes indelibly *imprinted* in an innate mechanism, and the bird thereafter responds characteristically to that one stimulus and not (in the same way) to any other. In the natural circumstances of wild birds, that object will almost always be the mother bird, and the behavior is thus normally adaptive. In experimental situations, however, a man, an animal of another species, or even an inanimate but moving object may be imprinted. The young bird will thereafter react to such an inappropriate object as if that object were its mother.

In other and more widespread instances, an innate behavioral mechanism may be set off not by one specific object but by a particular *kind* of stimulus. The kind of stimulus required may also be innate, and it is then called an *innate releasing mechanism*, or, for short, a *releaser*. It is so frequent as to be characteristic that the response is not to the whole of the apparent stimulus but to some particular aspect of it, a specific sign or signal. One of the best-studied examples concerns the European red-breasted robin (*Erithacus rubecula*, it only superficially resembles the American "robin," which is really a thrush). At certain seasons male robins establish territories from which they try to drive out any other males of their own species. Experiments show that the actual releaser is not a whole invading robin but solely its red breast. The defending bird will threaten or attack a tuft of red feathers just as fiercely as if it were a whole male bird, and it does not react so strongly to a young male that has not yet developed a red breast. Similar phenomena are very common. For instance, there is a fish (a stickleback, *Gasterosteus aculeatus*) that attacks red spots such as occur on the bellies of males of its species. Of course, releasers are not always colored red. Different releasers may produce different reactions within a single species, and similar reactions may have quite different releasers in different species.

Such innate mechanisms tend to lead to *fixed action patterns*, which may be very elaborate. They do not form a very clear-cut category because, as with bird songs, the precision or detail of a fixation varies greatly from one activity or one species to the next. In some instances the pattern is so firmly fixed that it must go on to completion once it has been released, even if the stimulus is no longer present. For example, the greylag goose has a definite action pattern for rolling a stray egg back into its nest, and it proceeds through the whole pattern even if the egg is taken away by the experimenter, continuing to roll an egg that is not there. Some of the most elaborate action patterns involve *chain reactions*, in which each step is the stimulus for the next. Many courtship patterns have longer or shorter chain reactions (Fig. 14–10). Chain reactions involving only one individual of a species are also common and are illustrated by an insect later in this chapter (p. 413).

Evidently the existence of action patterns and the interactions between innate mechanisms and environmental stimuli must depend in some way on the structure and function of the nervous system. The precise relationships are by no means fully understood and constitute one of the most interesting current fields of research. Much is, however, already known about some of the more general and basic relationships such as are involved in reflexes and their modifications.

REFLEXES

Any individual nervous system develops with a distinct and definite pattern. The nerve pattern is mainly dependent on the inherited pattern of growth in general. Nerves are not moving parts within the body. They grow into their pattern, which does not change its arrangement in accordance with behavioral reactions. For instance, an efferent nerve does not move from one effector to another in the production of different responses. Each effector must have its separate, fixed nerve, and difference in response must involve impulses through different nerves.

What is most innate and least modifiable in connection with behavior is the pattern of the nervous system. In animals with nervous control of behavior, a particular sort of behavior cannot occur unless there is an established nerve channel for it. Modifiability of behavior implies the presence of alternative channels. Even in the maze of associative neurons in the vertebrate brain, with millions of alternative channels across the various synapses, the synapses are fixed. Impulses can only move through an existing and fixed, even though extremely complex, pattern. The modifiability of behavior cannot, therefore, depend on changing the pattern of neuron connections. It consists of making some alternatives through a fixed maze of *possible routes* more easily followed than others. In the last analysis this must depend on something like the facilitation of synapses we described in Chapter 13. (There is some recent evidence that RNA information patterns may be involved, but this remains uncertain.)

The simplest and the only inherently unmodifiable behavior involves direct nerve connection from a receptor to an effector. Such connections occur in a few very simple nervous systems, but never (as far as known) in higher invertebrates or vertebrates. The simplest connection in them involves at least two and usually (always in vertebrates, at least) more neurons between receptor and effector. The connection is usually through a nerve cord, and the arrangement is a reflex arc (Fig. 14–11). Even at its simplest, the reflex arc involves the *possibility of modification*, because there are possible channels other than the one direct from receptor to effector. This is, however, the mechanism of the most strongly innate and least modifiable behavior of higher animals.

If you accidentally put your hand on a hot stove, the hand jerks back very rapidly. This is a reflex, and its retention in man is adaptive: speed minimizes injury, and a reflex is a lot faster than thinking. At the same time a message does go to the brain, where further behavior (quite possibly linguistic behavior) is started. We have a number of such emergency reflexes, as well as maintenance reflexes like those that narrow the pupil of the eye in strong light and widen it in dim light, or those that flex our muscles and keep us standing erect without our thinking about it. Reflexes are pretty well at a minimum in us, however. Most of our behavior is much more modifiable.

As a generalization, open to many exceptions in detail, lower animals have more of the simpler, least modifiable reflexes than higher animals have.[7] Most of an earthworm's behavior is controlled by quite simple reflexes through the nerve cords, probably even simpler than your heat–hand jerk reflex. If an

[7] This is one of the reasons why we call them "lower" and "higher." The terms are vague and can be quite misleading, but it is a convenience to use them here and should not confuse you seriously. You have been warned against the common mistaken belief that so-called lower living animals represent ancestral stages of higher animals.

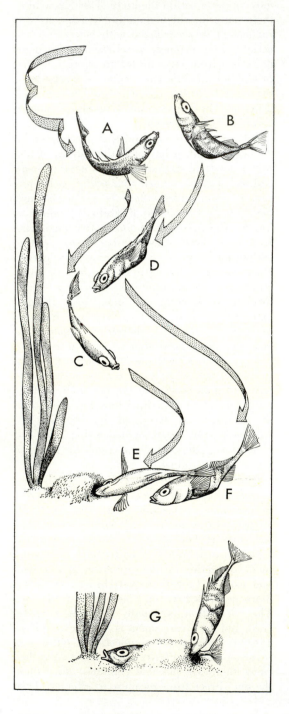

14–10 Chain-reaction behavior in a stickleback. In the three-spined stickleback the male is responsible for all the family chores: he builds the nest, induces a female to lay eggs in it, and attends the eggs after they have been laid and fertilized. The female's only contribution is to lay the eggs, and she even has to be led to this task in an elaborate behavioral dance initiated by the male. This courtship dance illustrates well the principles of chain-reaction behavior. Each movement made by the male is a stimulus that elicits the next, fixed movement from the female; in turn, each female response stimulates the male to the next movement. *A.* The male performs a zigzag dance directed at the female; this stimulates the female to *B. B.* The female turns toward the male and adopts an upright posture; this stimulates the male to *C. C.* The male swims toward the nest; this stimulates the female to *D. D.* The female follows; this stimulates the male to *E. E.* The male shows the female the nest by putting his snout in the entrance and rolling on his side; this stimulates the female to *F. F.* The female enters the nest; this stimulates the male to *G. G.* The male nuzzles the female at the base of the tail; this stimulates the female to lay eggs. She then leaves the nest. The male then enters the nest and fertilizes the eggs.

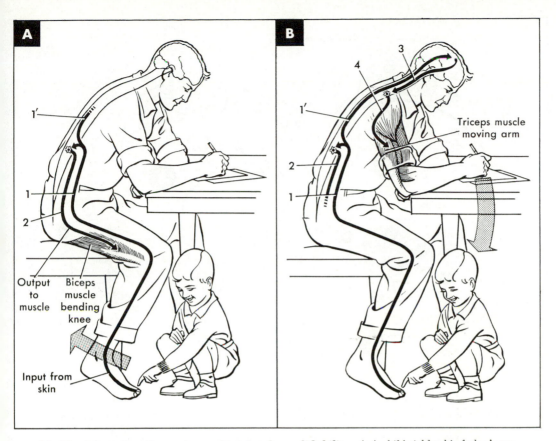

14–11 A human reflex action and its simple modifiability. *A.* A child tickles his father's toe. The stimulus is conducted along a very long sensory neuron (*1*), which in turn stimulates a motor neuron (*2*). The motor neuron, originating in the spinal cord, in turn stimulates the biceps muscle in the thigh, causing the knee to bend back reflexively from the stimulus. The stimulus, traveling along the sensory neuron (*1*), also reaches the brain via the path *1'*. *B.* When the child continues to "stimulate" and the simple reflex of knee bending proves inadequate, a second response (a modification of the initial reflex) results. The renewed stimulus follows the pathway *1–3–4*, eventually activating the triceps muscle of the arm and causing a more effective response to the stimulus. Modification (arm swipe) of the initial reflex follows only after many stimulations because the pathway *1–3–4* involves synapses less easily traversed, requiring facilitation by several stimulations.

earthworm gets near something too hot, its reflexes move it away without troubling the brain to act at all in the matter.

CONDITIONED REFLEXES

A *conditioned reflex* is one in which the response has been transferred from one stimulus to another. The classical experiment, by the Russian physiologist Ivan Pavlov (1849–1936) was based on the reflex that causes a dog to secrete more saliva when it smells food. Dogs were *conditioned* by ringing a bell every time food was placed before them. In time, ringing a bell produced the flow of saliva even

though food was absent. The response, saliva flow, was transferred from one stimulus, the smell of food, to another, the sound of a bell (Fig. 14–12). Many other experiments have since been performed, and it has been found that responses of all sorts can be attached by conditioning to almost any sort of stimulus that the animal can receive.

The conditioning of a reflex is clearly a modification of behavior, and in that sense it is learning: the dog learned to salivate at the sound of the bell. For a time after publication of Pavlov's work (his major works appeared in English translation from 1902 to 1929),

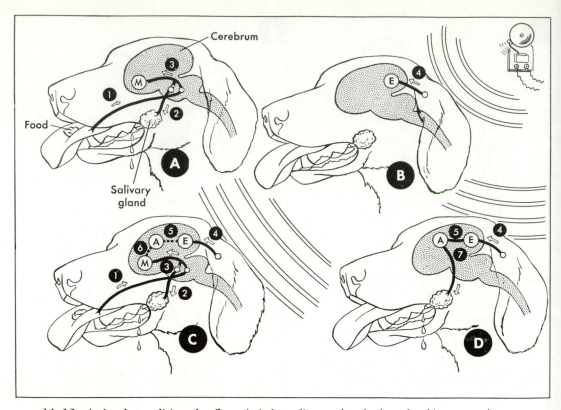

14–12 A simple conditioned reflex. *A.* A dog salivates when food reaches his tongue. A nerve impulse initiated at the tongue's taste buds passes along a sensory neuron (*1*), and then an afferent neuron (*2*) stimulates the salivary gland into activity. The nerve impulse also reaches the part of the cerebrum (*M*) concerned with mouth sensations such as taste; in other words, the dog has some conciousness of the taste. *B.* When a bell is rung, impulses pass from the ear to the auditory region (*E*) of the cerebrum. These impulses have, however, no way of passing into the afferent channels leading to the salivary gland, and so no salivation occurs. *C.* When food reaches the tongue at the same time that the bell is rung, the dog salivates because impulses reach the salivary gland via the old routes (*1, 2*). Impulses also reach a cerebral *association center* (*A*) from the mouth (*M*) and auditory (*E*) regions of the cerebrum via routes *5* and *6*. The dog thus learns to associate sound with food. *D.* When the association between sound and food has been well established, impulses initiating in the ear can reach the salivary gland via the routes *4, 5,* and *7*. (Adapted from G. Hardin, *Biology: Its Human Implications,* 2nd ed., Freeman, San Francisco, 1952.)

some enthusiasts believed that *all* behavior could be put in terms of reflexes. Innate behavior involved a simple reflex, learned behavior a conditioned reflex. This apparent simplification fitted in well with then current devotion to Morgan's canon. What could be simpler, and therefore more probable, than the hypothesis that all behavior, including human behavior, is reflex behavior?

Gradually it has been realized that instead of simplifying the concept of learned behavior, ascribing it to conditioned reflexes only complicates the concept of a reflex. A conditioned reflex is no longer a reflex in the strict meaning of the term. Association has occurred; a different path through the nervous system has been made habitual by constant use. It is, then, largely a matter of degree when the pathway becomes more and more complicated, leading through increasingly larger parts of the brain. If all these reactions are reflexes, then a reflex is any impulse following any route in the nervous system. It is simply nerve conduction. If conditioning is any change in such a route, then conditioning necessarily accompanies any modification of behavior. "Conditioned reflexes" became synonymous with "learning" and explained learning only by saying, in effect, that learning is due to learning.

The failure of the conditioned reflex to reduce learning to simple (truly) reflex processes does not mean that conditioned reflexes are of no significance. They cast a great deal of light on associational processes and the facilitation of nerve impulses along alternative routes—in other words, on what are accepted as the basic processes of learning. The point is that so-called conditioned reflexes are learned responses, whether or not they are reflexes strictly speaking. Recent students use the Pavlov technique, in varied forms, under the name of "conditioned response." It is an extremely useful technique not only for the study of learning but also for such investigations as those of discrimination. An animal may, for instance, be conditioned to respond differently to stimuli such as two tones. By making the tones more and more alike, the point at which the animal can no longer distinguish them is determined.

If you were instructed to find out whether bees recognize different colors, the task might appear hopeless because you cannot ask the bee—at least you cannot ask it outright and in English. But you can ask it indirectly by using the technique of conditioned responses. The problem is to determine whether or not the bee can distinguish a color from that intensity of grayness to which the color would correspond if the bee were colorblind and recognized therefore only degrees of brightness in a series of grays. The problem was solved by Karl von Frisch, as shown in Fig. 14–13.

CONDITIONED RESPONSES, HABITS, AND LEARNING

The learning seen in experiments on conditioned responses is the association of a particular, usually quite simple, response with a particular stimulus. A sheep is given an electric shock in one leg a few seconds after a metronome is started. After many repetitions the sheep learns to lift the leg as soon as it hears the metronome and before the shock is given. A stimulus (the sound of the metronome) that at first produced no response now produces a specific response that is invariable for a time. This invariability is not innate, however. The animal may forget the response

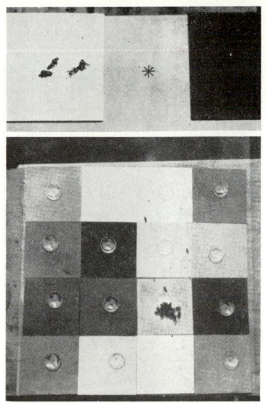

Karl von Frisch

14–13 Using a conditioned response to demonstrate color vision in the honeybee. *Top.* Bees that have been fed regularly on a blue paper are offered blank sheets of paper, one blue, the other red. They all settle on the blue paper, ignoring the red in their search for food. They have become conditioned to the blue paper. It now remains to be shown that they distinguish blue because of its color and not because of its apparent grayness, which clearly differentiates it from red. *Bottom.* A table is covered with squares of paper, each carrying an empty dish. The squares are different shades of gray except for one, which is blue. The bees concentrate on this square, proving that they recognize its color as such, distinguishing it from the equivalent shade of gray, which lies just above it and to the right.

from lack of practice, and the experimenter can purposely change it. The experiment is more similar to habit formation than to what we ordinarily think of as learning in ourselves. Habits are responses which, through long repetition, we come to (really *learn* to) make in a given situation, which is the stimulus. The habitual response becomes automatic in time, but it is not innate and it remains modifiable. We can change our habits, and

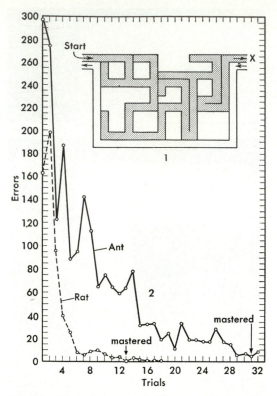

14–14   Maze learning. *1.* The ground plan of a maze offered to rats and ants. *2.* Graphs showing the progress of a rat and an ant in mastering the maze. The rat learned the maze more rapidly than the ant, mastering it by the thirteenth trial, whereas the ant required 31 trials.

they die out if they are not repeated more or less frequently. We can purposely break a habit by preventing or changing the response or by avoiding the stimulus.

Unless an animal is kept in controlled, experimental conditions all its life, it is extremely hard to distinguish reflexes from habits. In fact there is no sharp distinction, because both do depend on the innate nervous connections and both may be developed and are strengthened by repetition. A habit usually involves a more complex nerve path; it is more modifiable; and it does not become fixed without repetition. The difference between habit (or conditioned response) and other or, as we feel, higher types of learning is even less clear; it is one of degree only. The nerve paths become still more complex; they often bring in more evident control; they are more modifiable as a rule; and the greater central con-

trol may also reduce the requirement for repetition. Not much, even in ourselves, but some learning can occur at one trial, without repetition.

Most experiments on learning in nonhuman animals have involved, besides response conditioning, mazes or problem apparatus. A maze is a series of pathways with one or more points where the animal must choose which way to go. With the wrong choice the animal comes up against a blind end, is punished, or fails to achieve a reward. A Y- or T-shaped passage is a one-choice maze, and even an earthworm can learn this. Animals with more complex nervous systems learn more complicated mazes. One often used with rats is illustrated in Fig. 14–14. The rate of learning is scored by the number of mistakes made on successive trials. The graph shows that the error rate decreases (the rat learns) on successive trials. Ants tested in the same maze learned less rapidly. By varying rewards or other factors, the influence of different conditions on rate of learning can be determined.

Problem apparatuses of many different kinds have been used. One frequently used device is to place food in a compartment that can be opened by operating one or more pedals or catches.

MOTIVATION

Learning experiments always involve some system of rewards or punishments. The animals are rewarded, usually with food,[8] for a "right" choice, that is, the choice the experimenter wants them to learn, or punished, usually with an electric shock, for a "wrong" choice. As you would expect, most experiments show faster learning if both reward and punishment are used. This doubtless applies to humans, too. Many mothers offer a piece of candy or a whack depending on whether little Johnny does "right" (that is, what mama wants) or "wrong" (what mama does not want, but Johnny probably does). The method generally works.

Psychological experiments on learning may not seem very realistic when compared with

[8] At Harvard they tell a story about a mouse who boasted to another, "I've got that psychologist conditioned; every time I push this button he feeds me." *Somebody* had been learning.

learned behavior in animals (including man) not undergoing experimentation. Learning may be rewarded by food or in some other concrete way, but often it is not. Punishment in the form of an electric shock rarely occurs and ordinarily there is no punishment. Nevertheless, it is clear that animal behavior, including learned behavior, depends on something similar to reward and punishment. Animals do things because they "want" to. They learn new behavior because they "like" the results, and they learn to avoid behavior with results that they "dislike." "Want," "like," and "dislike" are anthropomorphic words and were put in quotation marks for that reason. They perhaps apply correctly to some animals rather like ourselves, but they are misleading if applied, say, to an ameba or a worm. A scientifically better way to put the matter is to say that animal behavior is *motivated*. Motivation need not be conscious or involve an emotion.

The most widespread motivation for behavior is the maintenance of stability in the organism. We have seen that there are elaborate chemical and nervous internal stabilizing mechanisms. Behavior is another powerful stabilizing mechanism, closely interrelated with the other two, as they are between themselves. Lack of energy or materials in an organism upsets its stability. The upset is translated into a felt (but not necessarily consciously felt) need: hunger. The organism, whether it is *Hydra* or *Homo*, responds by behavior that is likely to bring in food and restore stability. The behavior may be mostly innate, as in *Hydra* or a human infant, or mostly learned, as in more mature humans. It is very different in form in different kinds of organisms. But in all it has this in common: as a rule or on an average it tends to satisfy the organic need.

Motivation arising from a need to stabilize the organism is called a *biological* [9] *drive*. Hunger and thirst are obviously such drives. Sex is also so considered, but here there are other and more complex factors. At its simplest, the sex drive corresponds with an insta-

[9] Or biogenic, "born of life," as opposed to psychogenic, "born of the mind." There is much question as to whether this is a real distinction. What do you think?

bility arising not from deprivation but from the internal release of chemical substances, hormones. In higher animals, including most vertebrates, sex behavior also involves more or less complex patterns not related simply and directly to satisfaction of the drive.

Learned behavior motivated by biological drives does involve rewards and punishments. Restoration of internal stability is definitely a reward. In fact the food usually used as a reward in experiments is a reward precisely because it restores stability, and the same reward is frequent for learning in nature. Continuation and increase of internal instability is similarly a punishment.

Much behavior is not obviously related to internal instability. When a bee builds a honeycomb, gathers honey, and stores it in the comb, it is not satisfying any hunger it feels at the time. It seems nevertheless to be satisfying a need built into its nervous system. An inherited behavior pattern exists, and it is probably not going too far to conclude that there is an internal instability unless this pattern is carried out in action.

When you study this chapter and learn something about learning, you do not feel an organic drive or a restoration of your internal stability. Nor do you have an innate behavior pattern that is carried out when you read what we have written. Yet you do have motives for studying, and if you search far enough through the intricacies of your mind you may find that your motivation is not wholly divorced from biological drives.

As to the general relationships of motivation and learning, no learning occurs unless it is motivated in some way. That is a redundant statement, because motivation is definable as the cause of behavior. It is not so redundant to say that learning is faster if motivation is stronger. Degree of hunger can be measured, and learning with a food reward is faster in hungrier animals. (An animal that is not hungry at all usually has sense enough not to bother with the experiment.) Much study has been devoted to the relative effectiveness of different drives. For instance, some experiments with rats suggest that hunger is a stronger incentive than sex. One trouble with this conclusion, however, is that no one can measure equal intensities of hunger and sex drives. And rats, at least, do not die for lack of love, although they do for lack of food.

OTHER INFLUENCES ON LEARNING

Many factors other than motivation influence rate and kind of learning. We shall leave these to the psychologists on the whole, but perhaps we should mention a few of them. With rats learning a maze, a little help from teacher early in the learning period speeds things up. A great deal of help or help late in the learning period does not. Up to an age that corresponds with about 60 in man, old rats learn about as fast as young rats. Undernourished rats sometimes learn more slowly and sometimes faster than well-nourished rats. The results depend on the particular dietary deficiency and on motivation rather than on the general level of nutrition.

These are facts about learning in rats. To a biologist, it is extremely interesting to compare learning and other behavior in all sorts of animals, and we only wish that as much were known about others as about rats. The psychologists who have done most of these experiments were not (with certain exceptions) really interested in rat behavior. They really wanted to find out about behavior and learning in humans. Rats are used because they are convenient and cheap laboratory animals, and because they are simpler. An old rat is not likely to have habits and motivations that interfere with learning. (At least, rats can be raised so as to avoid such interference.) But an old man is almost certain to have interfering habits and is likely to lack motivation for learning. Basic biological factors, including those of the nervous system, are closely similar in rats and men, but due allowance must be made for some quite radical additional factors in man.

Many behavioral processes are similar in kind throughout much of the animal kingdom. This applies in essence to all the processes we have considered up to this point. Man's peculiarity depends largely on having more modifiable behavior and more extensive learning. Man also has mental faculties that are rudimentary or absent in most other animals. Before alluding further to these, we shall consider a group of animals in which both innate and learned behavior are well developed, but developed in quite different forms and degrees from man.

Behavior in Insects

Anyone who has ever spent much time watching ants or bees will agree that the behavior of insects has a peculiar fascination. Some of it looks so clever and farsighted! Bees build their elaborate, mathematically aligned combs and fill them with honey for the winter's needs. Some insect behavior looks so abysmally stupid. An ant, lugging a burden larger than itself, completely exhausts itself trying to pull the load over a stone or stick, when it could have gone around in a few steps. Some insect behavior seems so baffling as to require senses other than ours. Bees react as if they were following guiding patterns on flowers that are plain white to us. To us, insects have truly alien minds and have the puzzling attraction of something completely foreign.

The puzzle of insect behavior results in part from their different sensory equipment, to which we have already referred several times. The bees *are* following patterns on the "plain white flowers"; the patterns are in ultraviolet, which they can see but we cannot. More of the puzzle, however, arises from the peculiarities (as we consider them) of the insect nervous system. Insects have the most complex behavior, and correspondingly the most complex nervous systems, of any invertebrates. They also show more evidence of modifiable behavior and learning than other invertebrates, although one reason why they baffle us is that they perform complicated maneuvers that are *not* learned. We tend to think of animals by degrees of intelligence in comparison with ours. The outstanding point about insects is not that they are less intelligent than we are (although of course this is true), but that their intelligence is so very different in kind. They are at the end of a long evolutionary progression that has been diverging from ours for half a billion years and more.

Although it is often overemphasized, undoubtedly the most striking fact about insect behavior is that a large proportion of it is innate. Most insects live quite alone, never seeing their parents or their offspring, and seldom with any relationship to other members of their species except for brief mating. A few insects do not even mate; the females carry on the race without benefit of males. Yet each species has a distinct, frequently a very complex, behavior pattern carried on without training and with little variation from one generation to the next. The constancy of these behavior patterns can be explained only by innate characteristics of the nervous system. Social insects, in which learning is most evident, also have much innate behavior. Honeybee workers make their precise, six-sided cells without any learning.[10]

A solitary wasp, which has never seen its parent and need not have seen any other wasp doing the same thing, goes through a sequence of actions. (Fig. 14–15). It hunts for a nest site and digs a nest. It then hunts prey, usually a caterpillar, paralyzes the victim, drags it to the nest, and shoves it in. This is repeated several times. Then the wasp lays an egg in the nest, seals the opening, and smoothes it over so that the location is invisible. Later the egg hatches, and the larva eats the food provided, but the wasp that made the provision knows nothing of this. In the course of this chain of actions, there are immediate reflexes to separate stimuli; the stinging of the caterpillar has this appearance, at least. There is also learned behavior; the wasp's unerring return to the nest cannot be explained otherwise. But the sequence of acts as a whole is innate. Each act is the stimulus for the next, and thus this is another example of chain behavior (see p. 405). The wasp cannot vary the sequence, omit a step, or go back and repeat one. For instance, if caterpillars are removed from the nest when enough have been provided, the wasp lays an egg and seals it in anyway. It is perhaps in such complex combinations of innate chain behavior, simple reflexes, and learning that insects differ most from us, even though we also have innate behavior, reflexes, and learning.

Bees have no innate sense of direction that

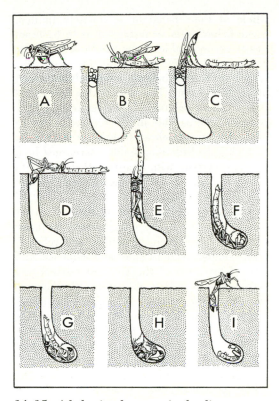

14–15   A behavioral pattern in the digger wasp.
A. Having dug its nest, the wasp captures and stings a caterpillar, paralyzing it. *B.* It transports its prey to the nest. *C.* After depositing its prey nearby, it opens the nest. *D–G.* It packs its victim into the nest. *H.* It lays an egg on its still-paralyzed victim. *I.* It crawls from the nest, reclosing it and leaving it permanently.

enables them to return to the nest. Part of their homing is learned behavior; young bees have to learn their way around just as much as we do, although of course they do so in their own way. The bee learns landmarks around the nest and recognizes them by sight. On longer flights the direction of the sun is noticed and used as a guide. But if the bee is moved to another, unfamiliar spot, it will fly in what *was* the right sun direction but here is not, and it will miss the hive altogether. The behavior is modifiable, but not very. The behavior, even the most modifiable of it, is innately limited. Modification can take place only as between a few simple alternatives.

Clearly it is not true that insects operate by instinct only. Their behavior can be modified. It is, however, reasonable to conclude that the

[10] This is not quite so much of a feat as it may seem. A series of equal cylinders crowded into a confined space automatically takes on the hexagonal precision of a honeycomb. It would be harder to build the cells in any other pattern.

whole of their behavior is innately limited. Some behavior is so restricted as to be practically unmodifiable. Some has broader limits, but none is highly modifiable. In itself this fact is not so striking; the same might be said of an earthworm. The remarkable specialization in insect behavior is that is may be very complex, comparable in this respect to that in many higher vertebrates, even in mammals, and yet is so little modifiable. Here must be a nervous system with many pathways established at birth, but with few connections and alternatives between them.

A final important point is that all insect behavior seems to be interpretable in terms (1) of innate structural pattern and reflexes, quite complex as to number and sequence, and (2) of learning of the simplest conditioned responses—limited modifications of responses to surrounding stimuli. There is little or no evidence of what we would call in ourselves sizing up a situation, thinking things over, solving a problem, deciding on a course of action, and the like. There is no evidence of foresight, in spite of storage for the winter. An insect does not plan; it acts. The apparent foresight is built into the species, not decided on by the individual. There is very little evidence of emotion. Certainly there are in our own behavior many factors weak or absent in insects. These factors have arisen in our own line of evolution, among the vertebrates.

Concepts and Symbols

When we come to the higher mental faculties, those just alluded to in most unpsychological terms, we are leaving the field of the biologist and are more definitely entering that of the psychologist. Thus far a background knowledge of general biology carries us. As we approach the most special aspects of human behavior and thought, a more specifically psychological background is required. We shall therefore be particularly brief in this section. Something must be said, however, because the biological status of man—that is, his place in the world of life and his special characteristics as an organism—depends largely on his psychology. It is also important to mention that even man's highest faculties

have biological bases that can be seen in other animals and down to the lowly protists.

The usual maze experiment prevents learning by any process except trial and error. This is not the characteristic human way of learning or solving problems. Trial and error cannot be wholly eliminated, but we usually try to size up the whole of any new situation as well as we can. Then when we act, our behavior is often right the first time. If it does prove to be wrong, it was still based on a great deal more than blundering along a path taken at random until we were punished or thwarted.

Biologically speaking, "right" behavior in any situation is behavior more likely than not to benefit the organism or its species. All organisms do tend to have right behavior in this sense. In animals, the usual primitive basis of such behavior is *perception*, the dim beginnings of which occur also in protists and in plants. "Perception" is another word hard to define, but roughly it is the process of organizing sensations in such a way as to give them meaning for the animal concerned.

If a *Paramecium* hits the stem of a water plant, it experiences a touch stimulus. A *Paramecium* has no mind in the human sense and probably is not conscious or aware of anything, but its reactions show that there was in a broader sense a sort of meaning derived from the sensation produced by this stimulus within the organism. The meaning was "obstacle," and the response was to back up. "Meaning" here merely labels a relationship between stimulus and response. This is an extremely dim and primitive perception, but it shows the beginnings of the process. At the other end of the scale, a botanist might come along and perceive the same water plant. He would have much more extensive sensations from stimuli, mostly light stimuli in his case, and would form a fully conscious, meaning-packed, very complex perception. He would, further, attach to this perception a symbol which would relate it to all his other perceptions of the same sort. He would call it *Spirogyra*.

The word *Spirogyra*, spoken or written, does not have the slightest resemblance to the little, filamentous water plant of which it is the name. It is a symbol that stands in our minds for the plant. It does not stand, however, just for the particular plant that the botanist perceives at a given moment. It stands for *all* plants that are like this one in certain well-defined ways. Strictly speaking, then, it does not really stand for the plant as a concrete object at all. It stands for the idea of a whole group of plants, of the things they have in common, and of their relationships with each other. Such an idea is a *concept*.

Aside from intensification of abilities common to many other animals, man's most distinctive characteristic is the ability to form abstract concepts and by this means to grasp complex relationships between things. This is the ability concerned in, among many other things, the human way of sizing up a problem and solving it without trial and error. The ability to form concepts has surely evolved from increasingly complex processes of perception, and, as we have seen, the rudiments of perception occur even in protists. Nevertheless, there is little evidence that the full quantity and quality of mental activity that we call "concept formation" occurs in animals other than man. Its rudiments probably are present in our nearest living relatives, monkeys and apes, so that even in this capacity we are not absolutely different from other animals. Rhesus monkeys and chimpanzees have been taught to match objects by *either* color *or* form depending on still another cue presented at the same time. Some degree of abstract thinking must be involved. Some experiments have shown that parrots can abstract numbers up to 7. No other animals, however, not even our useful friends the laboratory rats,[11] are known to do anything requiring that particular ability.

The further step of attaching a definite symbol to a concept seems a natural outgrowth of concept formation, but it is even more peculiarly human. Again, there are rudiments in a few higher animals, but fully developed symbolization occurs only in man. Our most complete and systematic set of symbols is language, which has had tremendous biological and social consequences. It has done as much as any other one factor to make us altogether unique organisms.[12] And it has done this not just because it is language in some broader sense, but because it is language employing symbols for abstract concepts like numbers and relationships. We can best show what this means by contrasting our own language with the so-called "language" of bees.

When a worker bee returns to the hive after finding a rich supply of honey (the nectar in flowers), it can promote the efficient use of its fellow workers' time and effort by "telling" them the location of the rich harvest. To do this, it dances on the vertical face of the honeycomb; and in the details of its movements is a message to fellow workers "saying" how far and in what direction the honey can be found. The dance (Fig. 14–16) is made in a circle, which the bee transects on a line whose angle to the vertical (defined by the sense of gravity) corresponds with the angle between the sun and the honey source. As it makes its dancing transect across the circle, it waggles its body, and the number of waggles is roughly proportional to the distance of the honey source from the hive. This is a nearly incredible degree of communication. It seems to be a language—but is it really? Does the bee *tell* its fellow workers how far and in what direction to fly? Or is it simply forced by its own recent experience to make a set of motions that act as a stimulus eliciting a predetermined, but appropriate, response from its fellows? In short, is it using a *language* or exhibiting innate, instinctive behavior of the most elaborate kind? Here we are caught in the crossfire of Morgan's canon, which confronts us wherever we turn for an explanation. For neither alternative is really simple; common sense demands a little of both views, and much uncertainty. At any rate there is surely no evidence that the dancing movements that constitute the bee's language are abstract symbols like the sounds or written words of our own language. They have a fixed denotation; they are *signs* for one concrete situation and refer only to distance and direction of flight.

[11] But rats and other animals have been trained in ways that suggest their formation of still more rudimentary concepts. Rats apparently can form a concept "triangle." Even bees readily react to a color, such as blue, regardless of form or other characteristics, but it is extremely unlikely that an abstract concept of blueness is involved.

[12] Many readers will greatly enjoy the chapters in Cassirer's *Essay on Man* that treat the use of symbols by man as his truly unique ability.

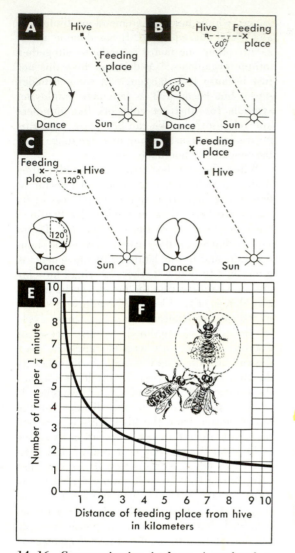

14–16 Communication in bees. A worker bee, returning to the hive after finding a food supply, imparts information to its fellow workers about the distance and direction of the food. The information is transmitted by signs illustrated in *F*. The bee dances along a line, wagging its abdomen; it turns a full half-circle and again dances along the line; it then makes a full half-circle in the other direction. All the while, it is followed closely by other workers. The speed of the dance, plotted in *E* as the number of runs down the diameter of the circle per ¼ minute, is related to the distance: the farther away the feeding place, the slower the dance. The direction in which food will be found corresponds with the orientation of the dance relative to gravity, as indicated in *A*, *B*, *C*, and *D*. The dance is performed on the dark vertical face of the honeycomb inside the hive. (Based on observations and experiments by Karl von Frisch.)

The bee has not freed these signs from their unique concrete meanings and used them to develop, for instance, a simple arithmetic.

Other Aspects of Behavior

In this chapter we have now touched on many of the aspects of behavior that are biologically important. Other aspects are treated briefly in other chapters. Still others may now be no more than just mentioned, so that you may have a more rounded idea of the contents of the science of behavior.

In Chapter 12 it was noted in passing that hormones may influence behavior. For instance, excess or deficient thyroxin is accompanied by more nervous or more phlegmatic behavior. When the importance of hormones was just becoming generally understood, it was a popular fad to describe behavior patterns and personality in terms of endocrines. "You are what your glands make you." Now we know that was a gross exaggeration. Any bodily function or disturbance can influence behavior. So can drugs and many other things. But in man and other higher animals the essential control is in the nervous system, through which these other influences secondarily affect behavior.

We have discussed behavior mostly, although not entirely, in terms of responses to external stimuli. It should be added that the signal for action may come from the central nervous system without any immediate environmental change. This is particularly true in higher animals and in man, because of processes of association, memory, and concept formation. We know that we sometimes decide to do a thing when there is no present stimulus for the action. There is, however, generally an association with a past stimulus. The initiation of behavior in the central nervous system of nonhumans is extremely difficult to study. It is, however, clear that this process is common only in man, and it is probably absent altogether in many lower animals.

The subject of *emotions* is another that is extremely difficult to study in any animal but man (and rather difficult in man, too). Emotions are internal phenomena and, unlike behavioral movements, not open to simple observation. They influence animal behavior in ways often erratic, but they cannot be directly observed. We almost have to judge emotions anthropomorphically. We have no way of telling whether a worm has emotions or what they are like, any more than we can tell whether a worm is conscious.

Personality is a subject that can be studied in other animals to the extent that it involves consistent individual differences in behavior. It is a fascinating subject, too. Personality depends on modifiable behavior. Invertebrates and lower vertebrates have little personality. Every bee behaves almost exactly like any other of the same species and caste. Personality is evident in most mammals, and culminates in man. We must leave these complexities to the psychologists.

In our brief consideration of innate behavior we have not referred to *maturation*. Changes of behavior in the course of life depend not only on experience, that is, on learning, but also on development of mechanisms. A butterfly's behavior is quite different from that of a caterpillar, but it is innate in equal degree. In almost all organisms one of the factors of changing behavior is growth, which was discussed in Chapter 8.

Social behavior is another extremely complex subject that must be left mostly to the specialists: social psychologists, sociologists, ethnologists, and politicians. As far as it is directly biological, it will be considered in Chapter 24.

CHAPTER SUMMARY

What behavior is: "doing"; externally directed activity; its intimate dependence on stimulus–conduction–response systems, especially on effectors.

Effectors: behavior as movement; effectors in plants; effectors in protists (ameboid motion, cilia, flagella); effectors in multicellular animals (muscles).

Muscular effectors: types of muscles (behavioral and visceral); muscles and skeletal systems; endo- and exoskeletons; control of movement by opposing muscles; all-or-none response of individual fibers; a muscle as a group of fibers.

Other effectors: stinging organs in coelenterates; electric organs.

Study of behavior: its relationship to psychology; its difficulties; dangers of anthropomorphism; Morgan's canon as a special case of Occam's razor relevant to study of behavior; dangers in Morgan's canon.

Tropisms and other reflex movements: initial botanical meaning of tropism; Loeb's extension of the concept to animals; behavior of the pill bug and the notion of forced movements—zealous application of Morgan's canon; nontropistic movements in plants and simple animals.

Adaptive nature of behavior:—simple cases of learning.

Behavior in relation to the structure of the nervous system: simplicity of behavior in animals with nerve nets—*Hydra*, echinoderms, and planarians; increased complexity and learning capacity in animals with brains.

Innate and learned behavior: instinct or learning as a falsely simple problem; language as learned behavior dependent on innate human capacities; learning of song in birds and its modifiability in some species.

Imprinting, releasers, and action patterns.

Reflexes and conditioned responses: the modifiability of reflexes; conditioned responses as simple cases of learning; habits; maze learning; motivation; maintenance of the organism's stability the most general source of motivation; age and other influences on learning.

Behavior in insects: its great complexity and relative lack of modifiability, exemplified by the digger wasp.

Concepts, symbols, and language.

Other aspects of behavior: endocrine effects; personality; social behavior.

Part 5

The Mechanism of Evolution

The photograph introducing Part 5 shows a population of seals lying on a rocky coast where, each year, they come ashore for several weeks to mate and rear their young. The scene points up two major features of our present topic, the mechanism of organic evolution: (1) The evolutionary process can be understood only in terms of populations; what evolves, as we shall see, is the pooled hereditary constitution of a population of individuals. (2) The mechanism of natural selection—key feature in the mesh of evolutionary causes—focuses on reproductive and genetic process we have previously outlined.

Seals are descendants of early carnivores that were terrestrial running creatures. Their evolution has been a history of change of habitat and way of life with a corresponding change in bodily structure and function. But in another, more fundamental, view, their evolution has been a history of change in hereditary constitution. In the last analysis the evolution of organisms is change in their genetic make-up—a gradual rewriting, so to speak, of their inherited message.

The first task in explaining evolution has already been performed, in preliminary fashion, by Part 3: to discover the factors that introduce variations, or innovations, into the hereditary make-up of organisms. The mechanisms of heredity do not guarantee a perfect similarity of offspring to parent; mutation and genetic recombination ensure that innovations will continually appear in successive versions of the chromosomal blueprint. Each seal in our photograph, for example, surely differs to some minor extent from the others in its private version of the basic seal genotype they all possess. There is, however, more to organic evolution than the variations that mutation and recombination constantly produce. For these processes, being completely mechanical and blind to the organism's needs, result in all manner of little changes, few of which improve the adaptive organization of the living thing. And one of the most conspicuous features of evolution is the way it has maintained and increased the adaptation of organisms to their environment.

The second task in explaining evolution is therefore the search for the processes that, given the random hereditary changes of mutation and recombination, mold from them the organized, nonrandom change that fits the organism to new environments or, just as importantly, enhances its fitness for the accustomed habitat. Those processes—complex and subtle in their detail—giving this direction and order to evolutionary change are what we refer to collectively as natural selection.

Chapter 15 is concerned with the fundamentals of population genetics; it discusses the nature of natural selection as basically a process of differential or nonrandom reproduction.

Chapter 16 deals with the fundamental concept of adaptation; it gives a fuller meaning to natural selection as an adaptive historical process.

Chapter 17 treats those factors in the evolution of populations that cause an increase in life's diversity through the formation of new species.

15

Elementary Processes of Evolution

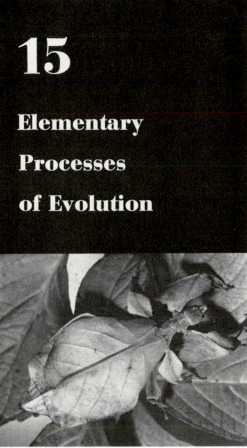

Charles Halgren

This leaf insect is a not-too-distant relative of the cockroach. Its remarkable resemblance to the leaves that form its background is a product of evolutionary processes, the elementary forms of which are the subject of this chapter.

Individuals, Populations, and Species

THE FOREST AND THE TREES

Not to see the forest for the trees has become a common metaphor for taking a short-sighted view of things, for being tangled up in details and failing to grasp the whole situation. If you look, literally, at a forest and try to understand it, you are likely, and quite rightly, to look first at trees. First questions are, "What is the structure of this particular tree? What is it doing? And how does it do it?" But even as you look at a single tree, you become conscious that it is not living alone. It is affected by surrounding trees and other plants with which it must share the necessities of life: space, water, other materials, and sunlight. Still more important, it reproduces. It is a temporary link in the continuity of its kind. If (as is usual) its reproduction is biparental, the process of continuity brings together substance and characteristics of different trees in the forest. In a matter of years, or of centuries at most, the one tree you are looking at will be gone, but the forest may endure without significant change.

To understand life it is necessary to see *both* the trees and the forest. So far we have been looking mostly at trees. Preceding chapters have reviewed the biology of individual organisms. Here and there relationships in groups of organisms have been touched on, for there are no really hard and fast divisions of knowledge. Yet even when we discussed continuity and genetics (Part 3), this was mainly in terms of basic processes as seen in individuals and their offspring.

Now the time has come to look at the forest. We have started to consider diversity, the different kinds or species of organisms. The species are groups. They cannot be understood in terms of individuals only. From here on, most of the rest of this book is devoted to the biology of populations and of communities, which are composed of populations.

UNITS OF LIFE

Organisms and populations as biological units

An obvious unit of life in nature is the individual. We have seen that the cell is a basic unit at a different level and that it is the pertinent unit when some, especially biochemical, processes of life are under consideration. In many protists cell and individual are identical. In multicellular organisms they are not identical. The difference in organizational level is a crucial element in our study. Whether of one cell or of trillions, the individual is the pertinent unit as regards physiological organization, total metabolism, responsiveness, development, and some other essential phenomena of living things. The individual is the separate and objectively concrete biological unit.[1] Intracellular processes take on full biological significance only when they are considered in relationship to the whole individual within which they occur.

Now, however, that we are turning more explicitly to the subject of evolution, the individual organism is no longer the most pertinent unit. Just as processes within cells must be related to individuals composed of cells, so must processes in individuals be related to populations composed of individuals. Evolution is something that takes place through long sequences of generations in which the individuals are evanescent and only the populations are continuous. In this continuity the discreteness of the individual tends to be lost. The continuity occurs by the passing on of hereditary codes of information, and as a rule the lineages of the transmittal are not discrete. They anastomose, separating when numerous descendants arise from one ancestral individual or pair and uniting when, as is usual, an individual derives parts of its genetic code from multiple ancestors.

The deme: a local population of similar individuals

Consider the individuals of a protist, say *Paramecium*, in a pond. All are much alike, they may reproduce for long periods without interbreeding,[2] and they have no social organization. They are about as independent from each other as individual organisms ever are. Nevertheless, the whole population of paramecia in the pond does constitute a naturally defined unit. That population as a whole fills a certain unitary role in the life of the pond. Descendants of any individual in it may spread anywhere in the pond, and, as far as the role of the population is concerned, it does not matter what particular individuals or progeny are carrying forward the activity at any time or place within the pond. Moreover, this population in this pond is a unit separable from other, similar populations of paramecia in other ponds in the vicinity.

Coming out on dry land, we may observe a grove of pine trees or a population of squirrels living in those trees. Here, too, there are clearly units of population, definable like the populations of paramecia because the group as a whole has a role that continues, regardless of what particular individuals happen to be there at any given time. Here, however, there is still another factor that helps to define the group: the individuals composing it are all related and they are interbreeding. The future populations of the unit may be derived from any or all individuals now present in this local group. They are less likely to be derived from any individuals of other pine groves or their squirrel occupants.

As another example, we may find among the pine trees a large anthill. Here is a still more sharply defined unit of population. All the ants swarming in the hill are closely related. Usually they include one resident queen and her daughters. Future populations, as long as the colony persists here, will be of the same descent. The population is also more closely knit than those previously exemplified

---

[1] There are some marginal or dubious cases. More than one stem and crown may grow from one root system. Coral colonies have distinct polyps organically connected with each other. In the Portuguese man-of-war and its relatives, the polyps are differentiated and seem to function as the organs of an individual of a different order. Such examples complicate but do not contradict the ordinary concept of an individual as a discrete unit. Colonies of the kinds mentioned arise from one zygote and hence are still genetically single individuals, and although organic continuity persists in them they are also still functionally single individuals.

[2] In fact, paramecia and most other protists do occasionally interchange nuclear material, which has the same effect as interbreeding. They may, however, serve as an example of the fact that the individuals within a unit of population in nature do not necessarily interbreed. What we say here applies equally well to the few populations that are strictly and always uniparental in reproduction.

because it has a social organization. The whole population works together as a cooperative unit, with division of labor among the several castes.

A general term for any definable local unit of population, like those of paramecia, pine trees, or ants, is *deme.* The examples have shown that demes can be several sorts or, at least, that they can be defined by different characteristics. When a deme is defined in part by interbreeding among its individuals, as in demes of pine trees or of squirrels, it can also be called a *genetic population.* (Some authorities use the synonymous term *Mendelian population.*) Social demes are usually also breeding units so that they tend to correspond with genetic populations. This need not, however, be strictly true, as the example of the anthill shows. Few individuals in a social deme of ants ever breed. When they do, they usually crossbreed with individuals from other social demes (anthills) and set up new ones. In the long run, the genetic population thus includes members of a large number of social units. All the anthills over a considerable area may represent the real genetic population, or a genetic deme of a larger and more permanent sort than the social demes.

Do you think the biological concept of demes is applicable to human populations? If so, how would you define various sorts of demes among mankind?

The species: a group of similar demes

Adjacent demes often intergrade, and the distinction of any one deme may be quite temporary. It is, in fact, characteristic of demes that they are commonly vague in definition and fluctuating in numbers. A grove of pine trees is distinguished as a deme and a genetical population because its continuity depends for the most part on interbreeding of individuals within the unit. It is, however, more likely than not that some pollen from other demes will reach this one and affect its reproduction, and also that some pollen from this deme will spread to others. There is some genetical unity in the deme, but it is not absolute. If pines grow up between this grove and the next, the two former demes may become quite indistinguishable. In a large pine forest, trees in one area are more likely to interbreed

among themselves than with distant trees in the same forest. There are vaguely separate genetic populations in the forest, but they intergrade so continuously that it would be entirely arbitrary to divide the forest into definitely bounded demes (Fig. 15–1).

Similarly, the squirrels in any one deme are likely on occasion to interbreed with those of surrounding demes. Heredity does pass from one deme to another. Adjacent demes may interbreed so freely as to become essentially one deme: the demes fuse. If all the squirrels in one grove or one part of a forest die out, a deme ceases to exist, but squirrels of surrounding demes may soon occupy the territory, or migrants may give rise to a new deme of the same sort. No essential or permanent change in the squirrel population has occurred. Demes fluctuate and intergrade, but there are larger units of population in nature which tend to be both more permanent and more clear-cut. These are species.

A *species* is a group of organisms so similar in structure and heredity that their demes intergrade, may fuse, and may take the place of each other without essential change in the nature and role of the group as a whole. All biologists agree that species are important units in nature and that they correspond more or less with that definition. It is, however, difficult or impossible to frame a really precise, fully meaningful definition which applies without question to all organisms in nature. This is the *species problem*, which will be discussed in Chapter 17. In order to understand the problem and the processes by which species arise, it is necessary first to know more about heredity in populations.

Genetics of Populations

A deme or any sort of population as it occurs in nature tends to persist for years, centuries, or millennia. It has *continuity*, as we have already emphasized. We have also emphasized that an essential element in that continuity is the passing on of chromosomes, with their genes, from one generation to the next. The persistence of a population without significant change in its characteristics implies that there has been little or no change in

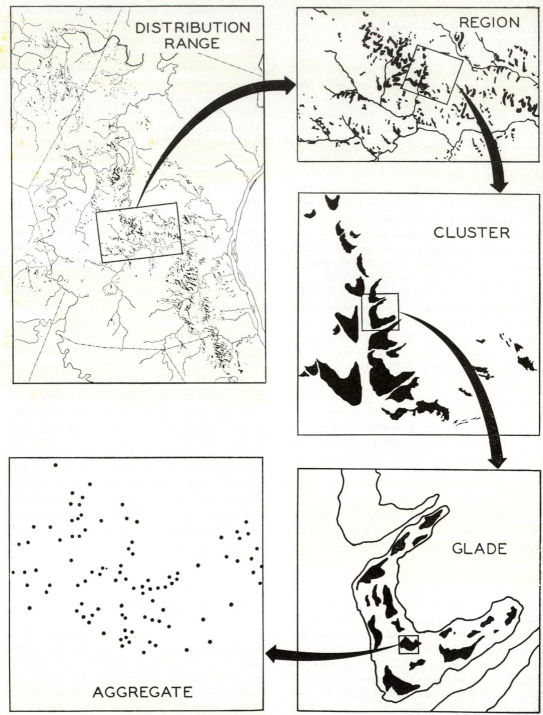

Ralph O. Erickson

15–1 Population structure in the plant *Clematis fremontii*. The arrows indicate the details of distribution in hierarchical fashion down to a deme in a particular glade.

its genetic or environmental factors. Change in a population over the generations—in other words, its evolution—must indicate change in genetic or environmental factors, or both. Since changes *directly* due to environmental factors are not heritable, we may for the moment ignore these. Changes in heredity are in the strictest sense the basis of evolution. To understand life and its history it is therefore necessary to know something about the heredity not only of individuals but also of continuously reproducing groups. This is the subject of *population genetics*.

The definition of a species given in the preceding section applies as well to uniparental as to biparental populations. In uniparental populations the genetic situation is comparatively simple and requires little discussion here.[3] Because uniparental offspring characteristically have just the same genetic make-up as their parents, genetic changes in the population as a whole arise only by the appearance of new mutations or the elimination of old ones. Biparental populations are both more usual and more important in the history of life. Their population genetics is also more complicated and more interesting.

VARIATION WITHIN BIPARENTAL POPULATIONS

There is always variation among the individual organisms within a population, and the variation of prime importance for evolution is genetic. Every biparental population that has ever been studied has been found to exhibit genetic differences among its individuals. *Drosophila* populations in local orchards and woodlands have been intensively investigated as examples. All have proved to contain genetic variation, including many of the mutant alleles studied in the laboratory. This is really

[3] Remember, from Chapter 9, that strict and long-continuing uniparentalism is rare. Most organisms, even including the viruses of doubtful organismic status, do at least exchange genetic materials (such as fragments of DNA) from time to time. Although this may not be biparental reproduction in the ordinary sense, it has an analogous effect in producing genetic recombination and makes applicable the main principles of evolution in biparental populations. Some higher plants and animals have secondarily and definitely abandoned biparentalism, but they are uncommon.

not surprising in view of what we know about mutation. A big enough population of any organism is bound to include different alleles at many gene loci in the whole set of chromosomes. The evidence indicates, moreover, that each gene locus may mutate to many more than just two allelic forms. With just two alleles present (A_1 and A_2) three diploid genotypes are possible among the individuals in a population: A_1A_1, A_1A_2, and A_2A_2. With three alleles (A_1, A_2, A_3) there are six genotypes possible. (What are they?) The number of possible genotypes as regards a single locus increases rapidly as the number of alleles increases. Then, too, because of crossing over and chromosome reshuffling at meiosis and recombination in zygote formation, each of the genotypes possible at one locus can be combined with any of the several possible at all other loci. The known processes of mutation and recombination in biparental (sexual) populations guarantee enormous variation within the population as a whole.

It will also appear in the course of our further discussion that variation in a population is commonly adaptive in itself and then is maintained by natural selection. The conclusion is that, with the trivial exception of identical (monozygotic) twins, genetic identity between any two individuals in even the largest populations is extremely rare or probably nonexistent. Most, if not all, of the individuals in natural populations are genetically unique.

EVOLUTION AS A CHANGE IN GENETIC POOLS

The idea of a population's genetic pool

One way, and for the purposes of this chapter the best way, to define evolution is as a change in the genetic constitution of a population. As individuals are almost always genetically unique, there is of course constant change from one generation to the next in the genetic constitutions of individuals within the population, but we do not consider that evolutionary change has occurred unless it involves the hereditary characteristics of the population as a whole, the pertinent evolutionary unit.

If we were to count all the alleles of the genes in a population, a tabulation of their ratios or percentages would be a measure of the genetic constitution of the population. It is this pooling of all the individual counts that characterizes what has been picturesquely called a *gene pool*. Chromosomal and, when present, extrachromosomal factors of heredity should also be taken into account. To encompass total heredity, we thus speak of a *genetic*, and not simply a gene, *pool*. It is, however, convenient at first to explain some of the more important processes of evolution in terms of alleles alone.

Processes within genetic pools

The processes that go on within genetic pools and that are therefore potential causes of evolutionary change are in part already familiar to you from Chapters 6 and 7. Others that emerge clearly only in the new context are as follows:

1. Mutation in a broad sense, including both gene and chromosome mutations.
2. Recombination in a broad sense, including crossing over and chromosome assortment in meiosis as well as sexual recombination of chromosomes of gametes.[4]
3. Gene migration or hybridization, that is, the introduction into a population of chromosomes (with their included genes) from some other population.
4. Natural selection.

Each of these processes can and in fact does on occasion produce changes in a genetic pool. They are, indeed, the elementary processes of evolution. Of course, the actual outcome in any given case depends on a multitude of other factors, including the conditions under which the processes take place and the way in which they interact. In some circumstances mutation and recombination lead not to change but to equilibrium. Gene migration may be classed as a special type of recombination. Each of the first three processes is normally random with respect to natural selection, which is the nonrandom, oriented, and adaptive process in evolution. But all three also interact with natural selection, so that

[4] There are some other mechanisms of recombination, especially in protists, that complicate but do not essentially alter the principles about to be discussed, but they need not be considered here.

their direct effects are modified and they, in turn, partly determine the effects of selection.

GENETIC EQUILIBRIUM

Mutational equilibrium

By definition, mutation in its various guises is the process that produces entirely new hereditary variations and that hence is the ultimate source of evolutionary change. Mutations are, however, repetitive; the same ones (in different individuals) occur over and over again. Furthermore, they are to some extent reversible; a mutant allele can mutate back again to the original allele.

As a simplified example, let us consider a gene with two alleles, A_1 and A_2. Suppose that we start with an experimental population having only the allele A_1 in all individuals. In the course of many generations, this gene will mutate to A_2 in a few gametes. Rates of mutation vary greatly, but they are usually low, often on the order of one mutation per million gametes. In spite of the low rate, if we have a large enough population and maintain it through enough generations, A_2 will accumulate through repeated mutation and subsequent reproduction, and evolutionary change will result. Some of the A_2 alleles may, however, mutate back to A_1, so that the situation at this one locus for the two alleles can be represented thus:

$$A_1 \underset{\text{back mutation}}{\overset{\text{forward mutation}}{\rightleftarrows}} A_2$$

The absolute number of mutations from A_1 to A_2 will at first be much larger than from A_2 to A_1, because we started with all A_1 and this allele will long remain abundant. As the process continues, however, the number of A_1 alleles and hence of mutations to A_2 will decrease while the number of A_2 alleles and of mutations back to A_1 will increase. A time will come, therefore, when the absolute numbers of mutations in the two directions will be equal. Change in the proportionate numbers of the two alleles will then cease, genetic equilibrium will have been reached, and evolution by mutation alone will stop. The exact point of this occurrence, in terms of the proportions of the two alleles, will depend on the relative rates of forward and back mutations.[5]

Although these considerations are logically impeccable and although such an equilibrium may occasionally be reached in nature, it is probably extremely rare and of quite minor importance in evolution. In the first place, back mutations are not so common as was once believed. Another forward mutation, say from A_2 to A_3, is likely to intervene, greatly reducing the chances that descendants of the particular gene will ever mutate back to A_1. It has further been found that many supposed back mutations are not so in fact. They do not really produce an exact duplicate of A_1 but still another allele that happens to have an identifying phenotypic effect similar to that of A_1. (But mutations away from A_2, whether true back mutations or not, will eventually be in equilibrium with mutations to A_2.) In the second place, the attainment of mutational equilibrium is normally so slow that some other process intervenes before it is reached. The most likely intervention is by natural selection favoring the spread of one allele or another in the population regardless of the mutational equilibrium point.

Combinational equilibrium and the Hardy-Weinberg law

A tendency toward genetic equilibrium more important than that arising from mutation is involved in recombination in biparental populations breeding at random.

Let us set up an imaginary experimental population of *Drosophila* and follow the history of its genetic variation from one generation to the next. Let A_1 symbolize one of two alleles of a particular gene and A_2 the other. We begin our experiment by introducing 200 *Drosophila* (100 males and 100 females) into a breeding cage, where one generation follows another without break so long as we continue to supply food. We cannot wait for mutation to produce variation in our experimental population, and so we deliberately include both alleles (A_1 and A_2) among the initial flies. Of the 100 flies in each sex, 49 have the

[5] Let the forward rate ($A_1 \rightarrow A_2$) equal u and the back rate ($A_2 \rightarrow A_1$) equal v. At equilibrium the frequency of $A_1 = \dfrac{v}{u+v}$.

genotype A_1A_1, 42 A_1A_2, and 9 A_2A_2. Thus we have an initial population as follows:

100 females				100 males		
A_1A_1	A_1A_2	A_2A_2	$\times$	A_1A_1	A_1A_2	A_2A_2
49	42	9		49	42	9

We allow the males and females to mate at random. Can we predict the relative frequencies of the genotypes that will appear in successive generations? It is common for people first confronted with this problem to have two impressions: (1) that the less common allele, A_2, will gradually be lost from the population; and (2) that the task of predicting exactly what genotypes will result from indiscriminate matings among so many flies is quite hopeless. Neither impression is correct.

The prediction problem would certainly be formidable if we had to figure out, one at a time, all the matings that could possibly occur and summate their outcomes. Fortunately the problem is much simpler than that. We can treat all the females as though they were one female and ask: What kinds and frequencies of eggs will be produced by the population's "composite female"? Treating males similarly, we can determine what sperms the "composite male" will produce.

Let us begin with the males. Every male produces enormous numbers of sperms, but we can simplify our arithmetic by assuming the number to be 10. This reduction in total numbers does not affect *ratios*, which are our real concern. The 100 males will produce gametes as shown in Table 15–1. The table

TABLE 15–1

Gene frequencies in an experimental population

Flies	Gametes		
	A_1	A_2	Totals
49 males are A_1A_1 and produce	490	0	(490)
42 males are A_1A_2 and produce	210	210	(420)
9 males are A_2A_2 and produce	0	90	(90)
The 100 males as a group produce	700	300	(1000)
Ratio of different gametes in the pooled population of sperms as a decimal fraction	0.7	0.3	(1.0)

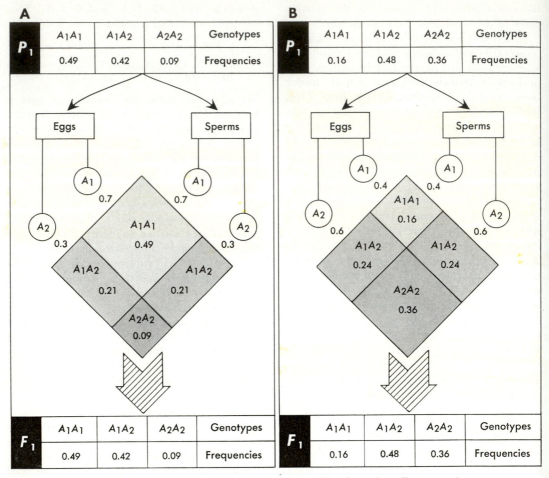

A				
P₁	A₁A₁	A₁A₂	A₂A₂	Genotypes
	0.49	0.42	0.09	Frequencies

	A₁A₁	A₁A₂	A₂A₂	Genotypes
F₁	0.49	0.42	0.09	Frequencies

B				
P₁	A₁A₁	A₁A₂	A₂A₂	Genotypes
	0.16	0.48	0.36	Frequencies

	A₁A₁	A₁A₂	A₂A₂	Genotypes
F₁	0.16	0.48	0.36	Frequencies

15–2 Genetic checkerboards illustrating the Hardy-Weinberg law. For an explanation, see the text.

shows that 3 out of every 10 sperms produced by the total population of males are A_2 and that 7 out of every 10 are A_1. The ratio between alleles is often conveniently called the gene frequency. The frequency of the gene (allele) A_1 in the present population is 0.7. Since the initial females in our experiment had the same genetic constitutions as the males, it follows that in their pooled gametes (eggs) the frequency of A_1 will again be 0.7 and that of A_2 0.3.

The next step in calculating what genotypes will appear in the next generation is to set up the genetic checkerboard shown in Fig. 15–2A. You are familiar with the checkerboard method from Chapter 6. There we used it to predict what genotypes would result from

fertilizations made from gamete pools taken from *one* female and *one* male. The method is just as applicable here where we have simply pooled all the gametes from 100 females and 100 males rather than from one each.

When we crossed two single heterozygote flies, the gene frequencies in the gamete pool of each parent were of course $0.5A_1 : 0.5A_2$ (cf. Fig. 6–2). In our experimental population, however, gametes are produced in the ratio $0.7A_1 : 0.3A_2$. The checkerboard in Fig. 15–2A shows that the zygotes produced by random matings in our population will have genotypes as follows: $0.49A_1A_1 : 0.42A_1A_2 : 0.09A_2A_2$. Comparing this ratio with that of the initial parental population, we notice that the two are identical.

Parental population
with genotypes

A_1A_1	A_1A_2	A_2A_2	$\left\{\begin{array}{l}\text{Gene frequency} \\ \text{is } 0.7A_1:0.3A_2.\end{array}\right.$
0.49	0.42	0.09	

↓

F_1 population
with genotypes

A_1A_1	A_1A_2	A_2A_2	$\left\{\begin{array}{l}\text{Gene frequency} \\ \text{is } 0.7A_1:0.3A_2.\end{array}\right.$
0.49	0.42	0.09	

We must of course conclude that the F_2, F_3, and all subsequent generations would continue to have the same gene frequencies and genotype ratios among individuals.

This result may well surprise you. It was certainly not self-evident to the early geneticists. The rule, discovered independently by two men and called after them the Hardy-Weinberg law, is one of the most fundamental laws of genetics. It states that under certain conditions gene frequencies and genotype ratios remain constant from one generation to the next in biparental (sexual) populations. This is true no matter how many alleles there are at each gene locus [6] or what their relative frequencies may be in the initial population. Of course, if the proportion of alleles differs in different populations, the genotype ratios will also differ. Fig. 15–2B gives, as further illustration, a checkerboard for a population in which the gene frequencies are $0.4A_1 : 0.6A_2$.[7]

[6] We have taken only the simplest case of two alleles, A_1 and A_2.

[7] Mathematically inclined students will have recognized before now that our genetic checkerboards are only graphic ways of expanding the binomial expression $(p + q)^2$, where $(p + q) = 1.0$. If we let p = the frequency of the allele A_1 in a population and q = the frequency of the allele A_2, then the algebraic expansion of $(p + q)^2$ describes the frequency of the three genotypes (A_1A_1, A_1A_2, and A_2A_2) in the population. Taking our experimental fly population (Table 15–1 and Fig. 15–2A) as an example, we have $p = 0.7$ = frequency of A_1; $q = 0.3$ = frequency of A_2. Then it follows that

$$\begin{aligned}
(1) \quad (p + q)^2 &= p^2 + 2pq + q^2 \\
(2) \quad &= A_1A_1 + A_1A_2 + A_2A_2 \\
(3) \quad (0.7 + 0.3)^2 &= 0.49 + 0.42 + 0.09
\end{aligned}$$

Line 1 gives the expansion of the binomial; line 2 shows the equivalence of zygote genotypes to the three terms in the expanded binomial; and line 3 shows how the frequencies of zygote genotypes are computed from the binomial. The familiar Mendelian F_2 ratio for a single pair of alleles is only a special case of this same general rule, in which $p = 0.5$.

The Hardy-Weinberg law explains one of the most striking and fundamental facts of nature: that while biparental populations always contain a variety of genotypes, the population as a whole, including its variations, may continue for generation after generation without significant change. Variation, which makes evolutionary change possible, is maintained even when evolutionary change is not occurring.

Conditions for Hardy-Weinberg equilibrium

The Hardy-Weinberg law can be restated in this form: the relative frequencies of alleles ($A_1 : A_2$ in our simplified example) and genotypes ($A_1A_1 : A_1A_2 : A_2A_2$ in the example) in a population tend to be constant through successive sexually reproduced generations. This tendency in nature partly explains the usual stability of populations and species over limited numbers of generations. Nevertheless, evolutionary change does eventually occur. It is therefore obvious that the Hardy-Weinberg law is not followed absolutely or indefinitely. It would be strictly applicable only under the following conditions:

1. Mutation either must not occur or must have reached its own equilibrium.
2. Chance changes in gene frequencies must be insignificant.
3. Gene migration either must not occur or must be a precisely balanced interchange between populations.
4. Reproduction must be random.

These conditions are rarely, indeed probably never, exactly met. Practically, then, Hardy-Weinberg equilibrium is only a tendency and not the actual course of the genetics of real populations.

EVOLUTION AS A DEVIATION FROM GENETIC EQUILIBRIUM

Basic causes of genetic change in a population

The four conditions just listed are necessary for genetic equilibrium, or lack of evolutionary change. It follows that their opposites are the basic causes of evolutionary change, thus:

1. Mutation must occur but must not have reached mutational equilibrium.

2. Chance changes in sexual recombination and reproduction must be significant.

3. Gene migration must not be in precise balance.

4. Reproduction must be nonrandom.

These conditions for evolution are clearly related to the four processes of population genetics given on p. 426. The first three will be briefly discussed in sequence, and the remainder of the chapter will be devoted to the last, most important, and most complex of them.

Mutations and evolutionary change

Mutations are the ultimate raw materials for evolution. Obviously a mutant allele, or chromosome, or set of chromosomes cannot be incorporated into the genetics of a population if it has never appeared in that population. It is, then, equally obvious that the possibilities for evolutionary change are dependent on the mutations that occur. This is, however, far from being all or even the greater part of the story. Mutations are of innumerable different kinds. Not all of them but, indeed, only a relatively small proportion do in fact eventually become characteristic of whole populations. Further, a mutant gene or chromosome does not act independently or in isolation. Its actual effect depends in considerable measure on the already established genetic system in which it appears and on the new combinations into which it is introduced in meiosis and sexual recombination.

Although mutation is necessary for continued evolutionary change, the fate of a mutation within a population depends largely on external factors. In a few exceptional cases— significant only in reference to some kinds of polyploidy, especially in plants (Chapter 17) —mutation may produce a variant individual distinct from its parents and forthwith capable of expanding reproductively into a new, genetically different population. In most cases, however, the individual mutant breeds back into the population in which it arose. The spread of the mutation in that population, if it occurs, is a gradual process over the course of generations. The actual outcome rests on factors that impede or promote the spread, much the most important of which is natural selection. Contrary to some early geneticists of 40 or 50 years ago,[8] mutation alone cannot determine a sustained direction of evolution and rarely determines the nature of any evolutionary change. It is a well-established observation that the most frequent mutations are not likely to be in the direction of past or apparently present evolution in the population in question.

A final point pertinent here arises from the previously mentioned fact that all natural populations have tremendous stores of genetic variation at any given time. Furthermore, the potentially possible combinations of alleles generally vastly outnumber the individuals in the population. This variation, both realized in individuals and potential through recombination, is the material on which natural selection can act. The variation is derived from earlier, ancestral mutations, and it can finally be exhausted. It does, however, permit extensive evolutionary change without the occurrence of any new mutations.

Chance changes in gene frequencies

When Mendel crossed two pea plants heterozygous for the flower-color gene (Cc), he obtained 23.1 per cent (224 out of 929) cc (white) homozygotes in his F_2 instead of the theoretical or ideal proportion, 25 per cent. We noted in Chapter 6 that such departures from the ideal genetic ratios are always experienced in practice. The gametes employed in producing any family are only an *approximate representation*—or *sample*—of the total population of gametes from which they are drawn. This is true whether we are considering the progeny from mating two individual organisms or the total progeny raised by a whole population of organisms. One hundred sperms taken at random from a population of males in which the gene frequency is 50 per cent A and 50 per cent a will contain *approximately* 50 a gametes, but the actual number is quite likely to be 53 or 48 or some other number in the general vicinity of 50. These chance departures from the ideal ratios are called *sampling errors*, and they become in-

[8] And contrary to one of the favorite themes of science fiction today.

creasingly serious the smaller the sample becomes. Conversely, they are less serious in larger samples; you will get closer to a 50 per cent incidence of heads the oftener you toss a coin.

In the reproduction of large populations of organisms, sampling errors are usually negligible; the initial ratios of alleles are fairly accurately represented in the large sample of gametes that initiates each new generation. But in small populations a considerable number of errors may accumulate because the sample of gametes that initiates each new generation is small. Thus the equilibrium of the population's genetic pool can be changed (can evolve) by purely chance processes. Such evolution is said to be *indeterminate* because, since the changes are due to chance, the genetic equilibrium is as likely to drift one way (for example, toward loss of the allele *A*) as the other (toward loss of the allele *a*).

There is debate among biologists as to the extent and importance of such indeterminate evolution in nature, but there is little doubt it plays a role, perhaps a minor one.

Another cause of chance change in genetic frequencies is embodied in what has been called the *founder principle*. When a species colonizes a new area, the individuals moving into that area are usually few—possibly only one, generally more, but probably never comparable in number to the main body of the species. These founding individuals, from whom new populations develop, rarely if ever include the whole genetic repertory of the species. The new populations thus represent only a sampling of the genetic pool. Especially if the sample is small, they may right from the start differ markedly and in a more or less random way from the parental population. The effect is most apparent on islands, which tend to isolate new populations, and it helps to account for the differentiation of island populations from their mainland relatives. Even in the continuous expansion of a species, the marginal populations invading new areas are likely to differ genetically from the populations near the center of the species' whole region of distribution. The founder principle probably has had more total influence on evolution than indeterminate evolution within a single small population.

Gene migration

Gene migration from one population to another, generally called *hybridization* if the populations are markedly different, has analogies with both mutation and recombination. It may, indeed, be imperfectly distinguishable from them as an evolutionary factor. Hybridization may introduce a genetic variation quite new to a population, as mutation does. The chances of its survival and spread may, however, be greater than for a mutation, because it has, as a rule, already been integrated into a genetic system similar to that of the population to which it has migrated. (Why is the genetic system of the donor population necessarily similar to that of the recipient population?) If the two populations involved are closely related—are, for instance, adjacent demes of the same species—then the phenomenon is hardly distinguishable from sexual recombination within a single deme.

An intermediate situation may be particularly important for adaptive evolution under the influence of selection. In ways that will be made clearer later on, a local population tends to become selectively adapted to its locality (environment). Achievement and maintenance of such adaptation require some degree of genetic isolation from populations adapted to other localities. However, complete isolation and perfect local adaptation may restrict variation and reduce the possibility of adaptive change, or of survival, if the local environment changes. A local population that is partly isolated but still has some genetic exchange (migration) can be adequately adapted to current conditions and yet have genetic access to further variation when those conditions change. An abundant, widespread species split up into many incompletely isolated local populations is in the best position to achieve and maintain adaptation throughout its range.

Hybridization may also, although not necessarily, produce polyploidy.

Natural Selection

NONRANDOM REPRODUCTION

Seen in broad perspective, the historical course of evolution has two major features: it produces *diversity* among living things; and it gives rise to their *adaptation*, their ability to survive and to reproduce in the environments they inhabit. The evolutionary factors that we have just reviewed are, in the main, random with respect to adaptation. That is, they have no direct and causal connection with adaptation, so that, in themselves and if other things are equal, they are as likely to be inadaptive as adaptive; indeed, under the usual natural conditions they are more likely to be inadaptive. The universality and intricacy of adaptation, as we shall find in the next chapter, are such that evolution by these random processs alone is entirely incredible. There must also be an evolutionary process that is nonrandom with respect to adaptation, that tends specifically and directionally toward the adaptedness of populations. That process is *nonrandom reproduction*, which in modern usage is synonymous with *natural selection.*

An example of nonrandom reproduction

The effect of natural selection on the frequency of mutant alleles in a population can be illustrated by hemophilia. You recall that hemophilia is an inherited disease in man affecting the ability of the blood to clot. Hemophiliacs are liable to die from loss of blood because a defective clot may fail to halt bleeding from even a small cut. This disability of the blood is caused by a mutant allele that we shall designate H_2. Blood in subjects with the allele H_1 clots normally.[9]

The mutation $H_1 \rightarrow H_2$ occurs at what is, as mutations go, a fairly high rate, about once in every 50,000 gametes. The rate of the back mutation ($H_2 \rightarrow H_1$) is not known exactly, but it is much lower than the forward rate. (Indeed, it is not certain that this back mutation occurs.) Consequently, if the mutational processes at the H gene locus were to reach an equilibrium, hemophilia (H_2) ought to be a common inherited disease in man. In fact, however, the disease is comparatively rare; the ratio of H_2 to H_1 alleles in human populations is about 1 : 10,000. Clearly H_2 is much

rarer than it would be if mutation and a possible mutational equilibrium were the only factors involved.

The cause of this discrepancy is not hard to find. Hemophiliacs commonly die while still too young to have reproduced. Even when they survive to sexual maturity, they are less likely to be accepted for marriage or (if married) are likely to raise fewer children than people with normal blood-clotting mechanisms. What does this difference in reproductive competence between normals and hemophiliacs imply in terms of the human population's gene pool?

We can clarify this problem by supposing that we have for study a population in which H_1 and H_2 *have* reached a mutational equilibrium. Since we do not know what the back-mutation ($H_2 \rightarrow H_1$) rate is, we do not know what its equilibrium is, but for purposes of illustration we may assume it to be $0.3H_1 : 0.7H_2$. The mutant allele is commoner than the normal one because the forward-mutation rate exceeds the back-mutation rate; thus hemophiliacs are common in the hypothetical population.

We would normally set up a genetic checkerboard (Fig. 15–3) in order to compute what zygote genotypes would appear in this equilibrium population. A checkerboard that assumed that all zygotes were equally competent reproducers would take the values $0.3H_1$ and $0.7H_2$ as the frequencies of alleles in the population's pool of gametes (Fig. 15–3*A*). On this assumption we would expect zygotes to appear generation after generation with the frequencies $H_1H_1(0.09) : H_1H_2(0.42) : H_2H_2(0.49)$. This assumption of *random* (or equally successful) reproduction on the part of the two alleles in our hypothetical population however, would be quite unjustified. We have seen that hemophiliacs (carriers of the allele H_2) differ from nonhemophiliacs (carriers of H_1) in being (1) more likely to die before sexual maturity, (2) less likely to marry, and (3) if married, likely to have less than average-sized families. Thus in the sample of gametes that actually initiates the next generation (those that actually reach the genetic checkerboard so to speak) the proportion of H_2 alleles is substantially reduced from the initial value of 70 per cent to something more nearly like 20

[9] The precise mode of inheritance and expression of the hemophilia allele involves other complications, but they do not materially affect the points we are making here.

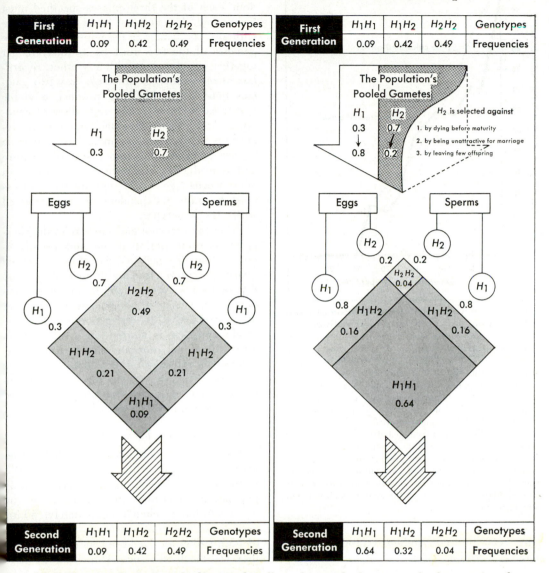

15–3 Genetic checkerboards showing the effect of natural selection on the frequencies of hypothetical alleles in a human population.

per cent (Fig. 15–3B). Zygote frequencies in the next generation will accordingly be different from those in the first generation: hemophilia will be rarer.

This can be summed up in a different way by saying that if the gametes that go to make up a new generation were drawn at random—were a fair sample of the parental population—the ratio of H_1 to H_2 would not change. But because a gamete containing H_2 is less likely

to be passed on to the next generation the sample is not a fair (or random) one; it is *biased*, as the statisticians say. Therefore, the proportion of H_2 in the population tends to decrease. H_2 would, indeed, eventually be eliminated entirely if it were not continually replenished by mutation. The actual ratio $H_1 : H_2$ is determined neither by mutational equilibrium nor by the Hardy-Weinberg equilibrium but by the relationship between muta-

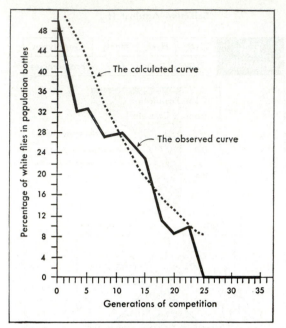

15-4 **An experimental demonstration of natural selection due to nonrandom mating.** *Drosophila* females discriminate against white-eyed males in favor of red-eyed males. By measuring experimentally a coefficient of their preference, it is possible to predict the rate at which the gene *w* (white eyes) will be naturally selected out of a population. Experiments to test the prediction are performed in population bottles—a population of *Drosophila*, in which the initial proportion of W and *w* genes is known, is allowed to breed in bottles, and the entire population is transferred to fresh bottles with new food at fixed intervals. The proportion of red-eyed to white-eyed flies is followed generation after generation. The graph shows the observed rate, as well as the calculated rate, at which the gene *w* is eliminated. The agreement is close.

tion rate and nonrandom reproduction or natural selection.

Elements in nonrandom reproduction

As it affects populations over successive generations, biparental reproduction involves (1) the bringing together of male and female gametes by mating or otherwise, (2) the production of offspring in the form of viable zygotes, and (3) the development and survival of the offspring until they are capable of producing offspring in their turn. In the present context, reproduction means a continuous sequence of life cycles and not just the production of offspring at one stage in the life of one generation. If the whole process is random, each of the three phases specified must be random. A male must be as likely to mate with any one female in the population (or the gametes of the two must be as likely to get together somehow) as with any other, *regardless of the genotypes involved*. Any two gametes must be as likely to produce a viable zygote as any others, also regardless of genotypes. And every individual zygote, again regardless of its genotype, must have the same chance of developing into a sexually mature and reproducing organism as any other. In other words, the whole reproductive process must be quite independent of—uncorrelated with—the genotypes.

From the opposite and more relevant viewpoint, reproduction may be and usually is *non*random in each of the three specified phases of reproduction and in several or many different ways. Nonrandom, in the sense pertinent here, means in some degree dependent on—correlated with—the genotypes. Thus reproduction that is nonrandom tends to produce directional, nonrandom changes in genetic pools and hence directional, nonrandom evolution. The three specified phases of reproduction will now be separately considered from this point of view.[10]

Nonrandom mating

There is a mutant allele (*w*) that causes white eyes instead of the usual red eyes (W) in *Drosophila*. If we set up an experimental fly population containing these alleles, we run into evolutionary change immediately. White-eyed males are unsuccessful in the courtship of both white- and red-eyed females; at least they are much less successful (literally have less sex appeal) than red-eyed males. As a result the white-eyed allele is quite rapidly lost by the population as a whole. The relative attractiveness of red- and white-eyed males to females can be measured, and the rate of elimination of the gene from the population can therefore be predicted. Fig. 15-4 shows that the experimentally observed evolution of

[10] Each of the three phases is, of course, dependent on the others and not a fully separate thing. The whole cycle could be divided into a greater number of phases, but for present purposes this analysis is adequate.

the population closely follows the predicted course. In one sense the evolution of this population is only a change in frequency of the gene w from 70 per cent to nearly 0. But in another and more important sense it is an evolutionary improvement in the mating (hence reproductive) efficiency of the population as a whole. Nonrandom mating has inevitably caused the evolution of adaptive improvement in the population.

A moment's thought reveals that mating customs, the accepted pattern of courtship, within a species constitute one of the most powerful and direct sources of nonrandom reproduction. Gene mutations that cause deviations from the common, accepted pattern of courtship will be immediately selected against; they will have little chance of entering the next (and therefore *any* subsequent) generation. On the other hand, accepted courtship patterns can act not only as a conservative agent, eliminating deviants (like white eyes in *Drosophila*), but also as agents of new evolution. In fishes, birds, and some other animals, bright-colored or showy parts may act as a stimulus to the opposite sex, the inherited "password" required before copulation is begun. When, as in some fishes and birds, the courtship password is a color spot or a showy pattern, almost inevitably larger spots and showier patterns evolve. Why? Any new mutations that make the courtship password more readily perceived or more effective are more likely to succeed in courtship than their longer-established alleles. The new mutant alleles have a favored entree into subsequent generations.

In the examples given there is sexual acceptance or refusal of one animal by another. This is what Darwin called *sexual selection.*[11] A nonrandom element can enter into this phase of the reproductive cycle, when male and female gametes are brought into propinquity with each other, even without sexual selection or mating in the sense of copulation. This phase is often nonrandom even in plants. Fertilization of an ovum is more likely to be by pollen from a nearby source than from a distant source. If, as commonly happens, the

[11] Darwin distinguished it from natural selection. We now see it as only one special case under the broad category of natural selection.

genetic pool of the local population differs from the average for the species as a whole, we have the equivalent of nonrandom mating.

Nonrandom fecundity

Fecundity is the actual number of offspring produced as viable zygotes—those capable of developing in a normal way. Two factors must be distinguished here: the number of gametes, hence of potential zygotes, produced and the proportion of these that do actually unite into viable zygotes. Both are involved in natural selection.

The sheer production of more offspring must plainly affect natural selection. If individuals with, say, an allele X_1 regularly produce 10 offspring while those with X_2 produce only 1, the proportion of X_1 in the population will surely tend to increase. This is, indeed, the most obvious form of natural selection as differential reproduction. Offhand, it might appear that selection would always favor fecundity and that the rise of organisms, such as man, that have comparatively few young would be an anomaly. There is, however, a balance between the factor of fecundity and the factor of survival, next to be considered. If individual chances of survival are low, then selection will favor fecundity. This situation applies, for example, to parasites with the hazards of complex life cycles and to many fishes with tremendous mortality among the young. Such organisms have become extremely fecund, with one female often producing literally millions of eggs. If, however, chances of individual survival are high, there may be no selection for fecundity. In fact, there may be selection against fecundity, because the production of many offspring may reduce the chances of their survival—for instance, by reducing the effectiveness of maternal care, as in many birds and mammals. If one female produces 3 offspring, all of which survive to breed in their turn, while another female produces 20 young, only one of which survives to breed, then selection clearly is favoring the genetic characters of the first female. Not only is there an interplay of different factors but also, as in so many aspects of evolution, there are alternative solutions to the same problem: high fecundity, low survival; low fecundity, high survival.

Specific differences in fecundity depend largely on the number of eggs (or ova) produced, which varies enormously but under the influence of natural selection has a characteristic average for each species. Another strong selective factor that enters into this phase of reproduction is inherent in the evident requirement that the zygotes be fully viable. If, for instance, as may happen in hybridization, gametes with incompatible genetic systems are united, the resulting zygote will not develop at all or will do so abnormally. The overt effect may be deferred to later generations. A hybrid may itself develop into a vigorous individual but have reduced or, as in the mule, virtually no fertility. The genetic characters involved are selected against in the longer run. There are, moreover, many mutant alleles (lethals) that do little or no harm when heterozygous but prevent normal development when homozygous. A zygote formed by the union of gametes that both carry such an allele does not survive to reproduce in its turn, and selection against the allele thus occurs.

Survival to reproduce

We have seen how mating (or, more generally, union of gametes) and fecundity are sources of nonrandom reproduction and of natural selection, therefore leaving their mark on the evolutionary process. They are not the whole story; most organisms have to go through a longer or shorter period of development, growth, and sexual maturation before they ever enter the final reproductive contests, so to speak, which mating and fecundity (the actual reproductive processes) set up. And it is nonrandom success in surviving to and through reproductive age that has given rise to many of the more obvious features of organic adaptation. The ultimate significance of the lion's speed, strength, and cunning is not that these adaptive features promote his survival in the sense of merely staying alive; it is that they promote his survival up to and throughout the period of his sexual maturity when he can make a contribution to the next generation's gene pool.

The crucial issue in natural selection is the leaving of offspring. In terms of our genetic checkerboards, the crucial issue is whether or not an organism succeeds in getting his gametes into those listed on the checkerboard as the source of the next generation. From the point of view of natural selection, what counts about a hen or a lioness is its capacity to produce offspring. Natural selection will never lead to improved survival capacity at the expense of reproductive efficiency. It could well be that a lion might gain in strength, cunning, and longevity from genetic changes that lower his fertility or his attractiveness to lionesses, but improved survival at such a cost will never endure in the evolutionary process revealed by population genetics. Such improved survival capacity does not get the genes causing it onto the genetic checkerboard.

Survival beyond the reproductive age is rare in nature. What happens to an organism after it has exhausted its possibility of contributing genes to the new generation rarely matters to natural selection. It is true that in some social groups, among insects or men, nonreproducers may help the population as a whole to raise the next generation and so promote significant survival, but that is an exceptional situation. Even in man the ailments that strike increasingly after the age of 45 or 50 testify to the fact that natural selection in our ancestors did not tend to promote survival after reproduction was completed. Medical science is more and more devoted to combating this unpleasant but quite natural consequence of the way we originated.

PHYSICAL ENVIRONMENT. It is obvious that survival of the individual to sexual maturity demands competence to withstand the rigors of the physical environments.

Insects have never been able to become big because of their external skeleton and their mode of respiration. Their small size has been a source of danger to them as land animals because it means that their surface is large in relation to their volume. Land animals tend to dry out by evaporation of their water into the unsaturated air which they inhabit, and the danger of death by water loss (a function of surface area) is proportionately greater the smaller the animal.

In insect populations those individuals whose genotypes render the outer coating less permeable to water are more likely to survive and therefore contribute to the next generation than are their brothers and sisters whose genotypes have opposite effects. Insects possess a host of diverse adaptations that are directed at conserving water, and the origin of those adaptations by natural selection is clear enough. Similarly all physical environments make demands that must be met for survival and that consequently are determinants of natural selection.

BIOTIC ENVIRONMENT. The environment in which an organism must make a living, survive, and raise a family contains important biological as well as physical elements. The organism has to live amid other organisms and the dangers and opportunities they afford. The major feature pervading any community of organisms is its traffic in energy and materials—the scramble to eat and avoid being eaten. This pattern of community life universally imposes natural selection on the community's members.

At the bottom of the heap are the plants, confronted with two sources of natural selection due to other organisms. First there is competition for soil room and sunlight arising from the abundance of other plants constantly showering the ground with seeds or thrusting in roots and runners from adjacent locations. Secondly, there is the constant threat of being consumed by herbivorous animals of all kinds before the season's crop of spores or seeds can be produced. These two *selection pressures* will elicit, in the long run, their appropriate adaptive adjustments.

In the desert, plant life is in a most difficult environment, and it comes under special threat of extinction largely because, being scarce to begin with, it is the more likely to be totally destroyed by hungry animal life; here as nowhere else adaptations are developed that discourage the hungry consumer. Thorns abound, as on cactus and euphorbia, and acrid juices, as in sagebrush.

Animals are under a whole complex of pressures in the web of relationships involved in the community's food economy. Most are under pressure to avoid being eaten as well as to find food themselves. Most birds *are* early birds because they have descended from ancestors who caught the worm. Keen vision in the hawk is always at a premium, and for this reason so are all genetic factors that improve

vision. Good eyesight is no less important in the mouse or rabbit; alleles in mouse populations that help to promote quick recognition of the hawk's sinister silhouette floating above make a greater contribution to the next generation's genetic pool than alleles that impair this recognition—by no matter how little.

NATURE OF NATURAL SELECTION

Selection and adaptation

Evolutionary changes resulting from nonrandom reproduction can have only one general direction. They are always of such a kind as to maintain or improve the average ability of populations to reproduce in the environments they inhabit. Effective reproduction is the only purpose that natural selection has or can possibly have. There are many different ways—literally millions of them—by which effectiveness of reproduction can be promoted, so that the specific effects of natural selection are various and innumerable. All, however, tend toward the same biological end.

We have mentioned that natural selection is the directive and adaptive factor in evolution. Adaptation is the subject of the next chapter, where it is defined and extensively exemplified. Here we must make just one point about it. If adaptation were defined as success in reproduction, then it would be a mere tautology to say that natural selection is adaptive. If, however, we define adaptation (as, with some further qualifications, we shall) as any feature of an organism that promotes its welfare, then it does not follow that natural selection is invariably or directly adaptive. Natural selection directly and always promotes the effective reproduction of populations. It promotes individual welfare only indirectly and only to the extent that individual welfare itself promotes population reproduction, as it does much more often than not.

Natural selection, then, does not always promote the welfare of individuals. It commonly (not necessarily or invariably) leads to the death of individuals before maturity; their individual welfare is obviously not promoted. Since it requires variation in populations, it also entails the probability that some individuals will be less well adapted than others and will, in a sense, suffer so that others may survive.

Changing concepts of natural selection

Old ideas about natural selection have become so deeply ingrained that they reappear even in some present-day discussions of the subject. To many of Darwin's contemporaries natural selection seemed to be a brutal struggle for survival in the universal carnage of "nature red in tooth and claw." Such catchwords as "the struggle for existence" and "the survival of the fittest" appear repeatedly even in Darwin's own works, although his views were less extreme than those of many of his followers.

From such concepts there developed a doctrine called "social Darwinism," although it was not supported by Darwin himself. Natural selection was believed to warrant as "right" all kinds of cutthroat competition, including wars between classes and nations, on the grounds that thus the "fittest" would survive and progress would ensue. The doctrine was completely unjustified, for two reasons. First, it is not a true picture of the way natural selection actually operates. Second, natural selection is not an ethical or moral principle that indicates what is right in human behavior. It is like the law of gravitation, a fact about nature which we must recognize as existing and affecting our lives but which is neither good nor bad in itself.

The question whether natural selection is "right" or "wrong" is not a scientific one, and there is no reason to discuss it here. Questions about what natural selection is and how it operates are scientific. The modern concept of natural selection, set forth in earlier pages of this chapter and now familiar to you, has developed from Darwin's, and yet it differs in some essential respects from nineteenth-century ideas on the subject. Natural selection is not struggle, competition, or survival; it is simply nonrandom reproduction.

Of course, animals do sometimes fight, and plants do compete for space, water, and sunlight. The competition has no bearing on populations and their genetic, evolutionary changes *unless* it leads to nonrandom reproduction. That is the real point of natural selection, and not the winning or losing of a struggle by one individual or another. Competition, struggle, and red-tooth killing often result in nonrandom reproduction, but not always. The concept of competition as a combat or literal struggle is also usually inapplicable to competition as it really occurs in nature. Moreover, the competitive aspects of nature are not the only ones that result in nonrandom breeding. An animal that gets along best with its neighbors may be precisely the one that has the most offspring. Then selection by nonrandom breeding favors absence of competition. Well-integrated plant and animal communities and, finally, animal (including human) social organizations have arisen under the directive influence of natural selection.

Creative selection

Some nineteenth-century critics of natural selection objected that its effects would be only negative and noncreative. It could, they admitted, account for the elimination of the unfit but not for the origin of the fit, a more important problem. Their objection had considerable force with the older concept of natural selection, which centered on the idea of survival or failure to survive. The objection is, however, completely answered by the modern concept, based on population genetics and centered on (indeed, identical with) nonrandom reproduction.

In the light of modern theory, it is easy to see that natural selection does have a positive and creative role in evolution. In the first place, the elimination of one allele from a population by selection does not occur unless there is an alternative allele that is, under the existing conditions, superior in terms of effective reproduction. The negative effect of elimination of the "unfit" allele and the positive increase in frequency of the "fit" allele are two sides of the same coin. You cannot have one without the other. In the example to be discussed in Chapter 16, selection for darker color literally *created* a moth population better fitted to survive and reproduce in the pine woods.

There is a second, more complex, and still more important way in which natural selection is creative. The various characteristics of an organism are not determined separately and independently by individual genes. As a rule, each gene affects numerous different characteristics, and each characteristic is affected by numerous different genes. Genes also commonly interact, so that a given allele may produce different effects when associated with different alleles of other genes. As will be further emphasized in a moment, natural selection acts not only on each allele but also on the genetic system as a whole.[12] It tends to produce gene associations and integrated genetic systems that would have little or no chance to arise and to spread through populations by any random process. This is a truly creative action.

Natural selection, variation, and the genetic system

For the sake of simplicity and clarity, we have so far discussed natural selection for the most part in terms of simple alternatives of alleles at a given gene locus. In fact, it is the whole genetic system, taken all together as a unit, that determines what the organism will be, and effective selection acts not only on each allele but also on the whole system in all its aspects. An important outcome of these facts is that selection does not tend simply to increase the frequency of the "best" or "fittest" allele of each gene and so eventually to produce a population completely homozygous for all those alleles.

One basic consideration here, which we have already mentioned in passing, is that variation is in itself adaptive for a population. A completely homozygous population would have nothing on which natural selection could act and no possibility of evolutionary change. If—or we can say "when," for it is inevitable in the long run—the environment changed, the population would no longer be well adapted and could not become so. The most rigorous of all the sanctions of natural selection would be applied: the population would become extinct. Populations that have survived have necessarily always maintained enough variation to permit adaptive change. Observations on natural populations of flies have shown that their genetic pools change

[12] And also, indeed, on the interaction of the chromosomal mechanism with the cytoplasm of the egg and in some instances with other hereditary factors outside the chromosomes (Chapter 7).

cyclically in adaptive response even to the brief alternations of the seasons. Such adaptation requires continuous preservation of the appropriate variations.

Another consideration pertinent here is that in many known instances a heterozygote is distinctly fitter for survival and reproduction than either of the corresponding homozygotes. But the highly fit heterozygotes cannot be maintained in the population unless comparatively unfit homozygous individuals are also regularly produced. (Why is this so?) Depending on the relative fitnesses, or selective values, of the heterozygotes and the two homozygotes, there is a definite ratio of alleles (or of different forms of homologous chromosomes) in the genetic pool that results in the optimum proportions of heterozygotes and homozygotes—optimum in the sense of being most effective for continued reproduction of the population. This kind of preservation of variation by selection is called *balanced polymorphism*. There is increasing evidence that the fittest populations, those that are reproductively most effective, maintain a high degree of heterozygosity.

Still another way in which natural selection acts on the genetic system and the genetic pool as a whole is by tending to bring and to keep together genes that act in unison, as a group, to produce an adaptive characteristic. Such genes, sometimes called *polygenes*, have additive or complementary effects and are less effective singly. They are more likely to be inherited together if they occur on the same chromosome than if they occur on different ones, and the closer together they are on the chromosome, the less likely they are to be separated by crossing over, as you will recall from Chapter 6. There are also genetic mechanisms that prevent crossing over so that blocks of cooperating genes can be kept together indefinitely. Once such an association has arisen by mutation, natural selection preserves it and spreads it through the population. The probabilities of its occurrence are increased by the selection of alleles that are near each other and that interact favorably.

Stabilizing selection

We have so far been concerned with natural selection as producing evolutionary change.

We have also seen that change could occur without selection but that such change would ordinarily be inadaptive. Another important role of natural selection is to limit or prevent inadaptive change. The simplest example is selection against harmful mutations, preventing their spread in populations and maintaining a steadily high ratio of more adaptive alleles at the same loci. This stabilizing effect undoubtedly occurs in all populations (including human). In a more general way, stabilizing selection often favors the reproduction of individuals near the mean, which are generally best adapted, at the expense of less well-adapted extreme deviants. One of the earliest students of natural selection in wild populations found after a storm that killed many birds that mortality had been relatively highest among the largest and smallest birds and lowest among those of average size. In a more subtle way, stabilizing selection seems to favor genetic and hence developmental systems that tend to produce normal organisms in spite of varying environmental influences during development. This flexibility in the production of what one might call standard items in the face of environmental modification apparently requires a considerable degree of heterozygosity in a population.

Limitations of selection

Natural selection obviously is not all-powerful. It can operate only on genetic variations actually present in a population, and so the changes that can occur ultimately depend on the mutations that have previously occurred. As mutations are most often inadaptive, the material for adaptive selection may be comparatively little, although selection usually assures that mutations are efficiently used. Within any one population over a limited span of time, selection can generally produce only relatively slight modifications of what was already there. Mutations with really large effects are almost always so poorly integrated with the rest of the developmental system that they yield monstrosities and are immediately eliminated by stabilizing selection. A dog cannot mutate into a viable cat. Changes of comparable and far greater scope indeed result from natural selection, but only by the slow accumulation of lesser changes over great

numbers of generations. The time scale of evolution is in millions of years. Moreover, mutations and recombinations of them that might be highly adaptive may simply not occur, although once a potentially adaptive mutation has occurred selection tends to bring it into adaptive combinations and to integrate it with the whole genetic system.

We have seen that selection can act at any point in the whole reproductive cycle from generation to generation. It can act directly on the genotype, for instance in requiring that the mitosis of the zygote be normal. Usually, however, it acts on the phenotype and hence at a point one or many steps removed from the actual genetic system, changes in which control the course of evolution. A fully recessive allele in a heterozygotic individual has no phenotypic expression and hence is shielded from natural selection. (It will, nevertheless, be exposed to comparatively slow natural selection in the whole population. How?) The many environmental modifications or acquired characters that contribute to variation in populations do not correspond with variations in genotypes and therefore selection on them has no direct evolutionary effect. Further, an adaptive phenotypic character may be produced by an allele that also has other and inadaptive effects. Selection for that character may therefore be inadaptive in other respects. The outcome will be a compromise, so to speak, balancing help against harm.

The limitations of natural selection have had a remarkable result in the history of life, one that probably will surprise you: the great majority of populations have failed to maintain continuous adaptation. Failure of adaptation is the general cause of extinction, and the fact is that most of the millions or probably billions of species that have ever lived became extinct without issue. Conditions changed. For all these extinct species, either variations suitable for the adaptive action of natural selection were lacking, or selection could not act rapidly enough.

Natural selection is a limited and blind process that cannot produce perfect or even adequate adaptation in any and all circumstances. The whole world of living things nevertheless attests that its over-all action is adaptive and that in the long run it is tremendously powerful and exceedingly delicate in that action. The next chapter will provide some striking evidence of its significance.

CHAPTER SUMMARY

Individuals, populations, and species: populations as the appropriate units for the study of evolution; the deme, a local population of similar individuals and a genetic population of interbreeding individuals; the species, a group of similar demes.

Genetics of populations: the universality of genetic variation in populations; the concept of a genetic pool—processes of possible change in genetic pools; mutation, recombination, gene migration, natural selection.

Genetic equilibrium: mutational equilibrium between forward and back mutations, rarely if ever attained; the Hardy-Weinberg law in biparental populations; necessary conditions for genetic equilibrium, involving mutation, chance changes in gene frequencies, gene migration, and random reproduction.

Evolution as a deviation from genetic equilibrium: changes due to mutation, chance changes in gene frequencies, and gene migration random with respect to adaptation.

Natural selection: equivalence to nonrandom reproduction; cause of basic evolutionary phenomena of diversity and adaptation; example of selective departure from Hardy-Weinberg equilibrium in hemophilia; elements in nonrandom reproduction; nonrandom mating; nonrandom fecundity; nonrandom survival—the effects of physical and biotic environments on survival and selection.

Nature of natural selection: an indirect cause of adaptation; a promoter of individual adaptation to the extent that adaptation promotes effective reproduction; changing concepts of natural selection; natural selection as a creative process; the relation of selection to variations and to the whole genetic system; stabilizing selection; the limitation of selection by materials available and modes of action.

16

The Evolution
of Adaptation

A woodpecker's adaptations equip it both for its narrow specialization of probing insects and for its broad specialization as a flying terrestrial animal. The origin of adaptations is in natural selection, the historical process that has directed life to its present condition.

Evolution and the Problem of Purpose

In ancient times and in primitive societies man's concept of his place in the universe was centered around himself and the place where he lives. The rise of modern science was accompanied by a physical reorientation. It was learned that the earth is not the center of our solar system, then that our sun is a relatively mediocre one among myriads in our galaxy, and finally that our galaxy is only one of millions in the observable cosmos. Those discoveries dealt heavy blows to human self-importance and the age-old ingrained idea that man is the center, in some sense the purpose, of creation. Still, even among well-educated men, until 1859 the vast majority could cling to the belief that although our habitat may be only a speck in an inconceivable vastness, the world of living things was created to be under the dominion of man. In 1859 Darwin shattered this last basic prejudice. No other truth revealed by science compares with that of organic evolution in its impact on human thought.

Many people growing up today go through a similar sequence of reorientations and suffer the same shocks. The child's world centers on himself, and as he develops further it expands, still around that center, to include family, community, nation, earth, and physical universe. He may be well along in his education before evolutionary biology invades his comfortably self-centered cosmos and carries him into unfamiliar and emotionally uncertain realms. He learns that the earth existed for billions of years before man entered the scene. He learns that man is only one among millions of species all akin to each other. He learns that all living things are products of the same natural processes by which life arose on the sterile planet Earth in forms simpler than the simplest cells of today and expanded into its present awesome complexity. It is impossible to contemplate the grand sweep of evolution and our own true place in nature without experiencing emotions of wonder and humility. It was with these emotions that Darwin closed *The Origin of Species*, beginning the final sentence, "There is a grandeur in this view of life . . . ."

The grandeur of which Darwin spoke so eloquently elicits many questions from us: "What is the meaning of this vastly greater world of which man is only a recent and minute fragment?" "What has caused evolution?" "What is its purpose?" These questions span both science and philosophy, for science rests on a philosophical foundation and cannot be absolutely separated from it. That our perceptions can be reliably related to objective phenomena, that the universe is orderly, that we can approach a true picture of its orderliness by observation—these and other propositions necessary for science are philosophical postulates that cannot be proved by scientific methods.

Yet, as we have already seen in Chapter 1, there is a touchstone by which many questions can be designated clearly as either scientific or philosophical. A question is scientific if it elicits answers testable by observation. "What has caused evolution?" is a scientific question because it has elicited answers that are testable and that have been reliably tested. "What is the meaning of the universe?" is a philosophical question because in an ultimate sense no observations seem to permit a choice among the many answers that have been given. Like any other man, a scientist seeks answers to the philosophical questions as well, and with his store of tested observations he is in a better position than most to find them. The point to stress here is that he should as far as possible distinguish between the two types of questions and that when he is speaking purely as a scientist he should stick to scientific questions and answers.

The greatest problems arise from questions that elicit both scientific and philosophical answers. Here our role as scientists, as biologists in the present instance, demands that we first consider and test the strictly scientific answers. Only if all those answers fail or when we push inquiry beyond the scope of testability should we enter the realm of philosophy, and we shall not enter that realm but only attempt to delimit it in this book devoted not to philosophy but to science. "What is the purpose of evolution?" is one of the questions that almost imperatively elicits both scientific and philosophical answers. A possible answer, one that still gives comfort to some people, is

that man is the goal toward which all evolution has been purposefully directed. Thus stated, this answer is untestable. It is strictly philosophical, and the scientist, as scientist, cannot argue with it. It has, however, been asserted that the facts demand this answer, that it can be tested, and that it is in accordance with observation. This assertion brings the answer into the field of science, and our conclusion as scientists is that in this field, on this level of inquiry, it is wrong.

Observation indicates unmistakably that there is no single direction of evolution. As far as evolution has yet gone, man is one end point or, if you like, pinnacle, but so is a tapeworm or a mosquito or any of the millions of existing species. There is a purposeful aspect in the world of life. That world is, in fact, permeated with what could in some sense be called "purpose." The highest purpose that is *scientifically* discernible is the universal one of survival to reproduce. More immediate purposes, already quite familiar to you, are diverse: food-getting, escape from enemies, and so on. The specific means of achieving these purposes are incomparably more diverse still, numbering in the millions. They are adaptations.

"How are adaptations achieved?" is the most important of all scientific questions about evolution—if not, indeed, about biology in general. Any theory of evolution must be judged by its success in answering that question. You already know the answer that most biologists now consider most successful: adaptation is the result of natural selection acting in and with the whole complex of biological factors—genetic, populational, ecological, etc. The rest of this chapter is devoted to the scientific examination of adaptation and of its relationship to natural selection.

Nature of Adaptations

We have encountered the phenomenon of adaptation in previous chapters and restrict ourselves here to a brief systematic review and definition of it.

A DEFINITION

Like so much else in biology, adaptation proves far from easy to define in a thoroughly

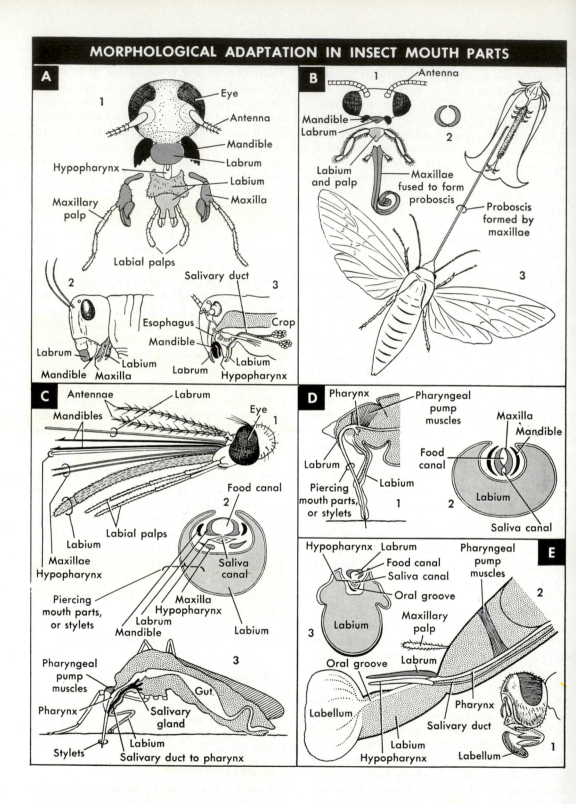

MORPHOLOGICAL ADAPTATION IN INSECT MOUTH PARTS

A

1

Eye
Antenna
Mandible
Labrum
Labium
Maxilla

Hypopharynx
Maxillary palp
Labial palps

2
Labrum
Mandible Maxilla
Labium

Salivary duct
3
Esophagus
Mandible
Labrum
Labium
Hypopharynx
Crop

B

1
Antenna
Mandible
Labrum
Labium and palp

2
Maxillae fused to form proboscis
Proboscis formed by maxillae

3

C

Antennae
Mandibles
Labrum
Eye 1

Labium
Maxillae
Hypopharynx

Labial palps

Piercing mouth parts, or stylets

Food canal
2
Saliva canal

Maxilla
Hypopharynx
Labrum
Mandible
Labium

Pharyngeal pump muscles
Pharynx
Stylets
Labium
Salivary duct to pharynx
Salivary gland
Gut
3

D

Pharynx
Pharyngeal pump muscles

Labrum
Labium
Piercing mouth parts, or stylets
1

Maxilla
Mandible
Food canal
Labium
Saliva canal
2

E

Hypopharynx Labrum
Food canal
Saliva canal
Oral groove
Maxillary palp
Labrum
Labium
3
Oral groove
Labellum
Labium
Hypopharynx
Pharynx
Salivary duct
Labellum
1

Pharyngeal pump muscles
2

exact manner. As in the case of life itself, adaptation is better defined by the whole discussion of it here and elsewhere in the book than by any statement condensed within the limits of a sentence. Nevertheless, its essentials are covered in the statement that *an adaptation is any aspect of an organism that promotes its welfare, or the general welfare of the species to which it belongs, in the environment it usually inhabits.* Individual welfare here means simply the organism's success in obtaining food, avoiding predators, and generally surviving and satisfying its whole range of biological needs. The welfare of the species is not only that of its individual members but also that of the group—in the maintenance or increase of the population. Adaptations are thus the apparently goal-directed features of living things that constantly impress us with the notion that organisms do have purposes, even though we cannot assume for any organisms other than ourselves that these purposes are conscious or that they are predetermined beyond the universal goals of survival and reproduction.

DIVERSITY OF ADAPTATIONS

Adaptations take any form—morphological, physiological, or behavioral—by which the welfare of the organism or the species is enhanced.

Morphological adaptations are among the most obvious and well documented. Examples are provided by the great structural diversity in insect mouth parts (Fig. 16–1) and the feet and beaks of birds (Fig. 16–2), all of which

16–1 Morphological adaptation in insect mouth parts. The basic insect mouth parts are as follows: a labrum, or upper lip; a pair of mandibles; a hypopharynx; a pair of maxillae (technically the first maxillae); and a labium, or lower lip (technically a fusion of the second pair of maxillae). *A.* The mouth parts of a primitive biting and chewing insect like a cockroach. The labium and maxillae manipulate the food, keeping it in the mouth cavity while it is being chewed by the mandibles. The food is salivated through the hypopharynx. *B.* The mouth parts of a moth and a butterfly, adapted to· sucking nectar from flowers. Superficially these mouth parts bear no resemblance to those of a biting and chewing insect, since they have been evolved to suit a very different activity. The mandibles are reduced and functionless, as are the labrum and labium, and the only functionally important element is the proboscis, a long tube formed by the close apposition of the two maxillae. Nectar is sucked from flowers through this proboscis. *1.* The mouth parts of a butterfly. *2.* A cross section through the proboscis, showing how it is formed by the two maxillae. *3.* A sphinx moth sucking nectar from a flower.

Many insects have evolved complex modifications of their mouth parts enabling them to pierce the surfaces of other animals or plants and then to suck nutritive body juices from their prey. Mosquitoes pierce the skins of animals and suck their blood, whereas the true bugs (Hemiptera) pierce plants and suck juices from their tissues. (Some, however, like the human bedbug, have turned to animals as a source of food.) *C.* The mosquito's mouth parts. The labium serves as a scabbard in which other mouth parts are housed. The labrum, mandibles, maxillae, and hypopharynx are all elongate, needle-like structures, modified to form collectively an elaborate "hypodermic needle" in which there are two distinct channels. One of these is the hypopharynx, down which saliva flows, facilitating the insertion of the mouth parts by providing lubrication. The saliva contains a chemical agent that prevents the victim's blood from coagulating and thus clogging the other delicate channel—in the labrum—through which the blood is sucked up into the pharynx. The sucking is done by a simple structural adaptation of the pharynx itself (*3*): muscles anchored to the exoskeleton of the head are also inserted on the wall of the pharynx; their rhythmic contraction and relaxation causes the pharynx to act as a bellows. This mechanical adaptation is found in many other sucking insects and is shown for the bug (*D1*) and a dipterous fly (*E2*). *D.* A plant bug's mouth parts, simpler than those of the mosquito. The labium again functions as a sort of protective scabbard for the long needle-like stylets, formed this time only by the maxillae and mandibles. A comparison of the mosquito and the bug brings out an interesting generalization about adaptations and their evolution: different (even though related) organisms are likely to solve a common adaptive problem in different ways. This may be called the principle of multiple solutions. The bug, like the mosquito, pumps a lubricating saliva down through one channel and sucks food up through another channel; but the morphological bases of the food and saliva channels differ for the two types of insects. In the bug the maxillae are complexly sculptured so that when closely apposed they form both channels, whereas in the mosquito the food canal is in the labrum and the saliva canal in the hypopharynx. *E.* The mouth parts of a housefly or other dipterous fly, adapted for sucking juices from free surfaces like those of fermenting fruits and decaying foods. The mouth parts collectively comprise the proboscis and are normally withdrawn under the head (*3*). Functionally they may be likened to a vacuum cleaner. At the end of the proboscis is the labellum, the tip of the labium. Flattened out and covered with capillary grooves, it is well fitted to sponge up fluids from wide surfaces. The necessary suction for the vacuum is supplied by the pharyngeal pump mechanism.

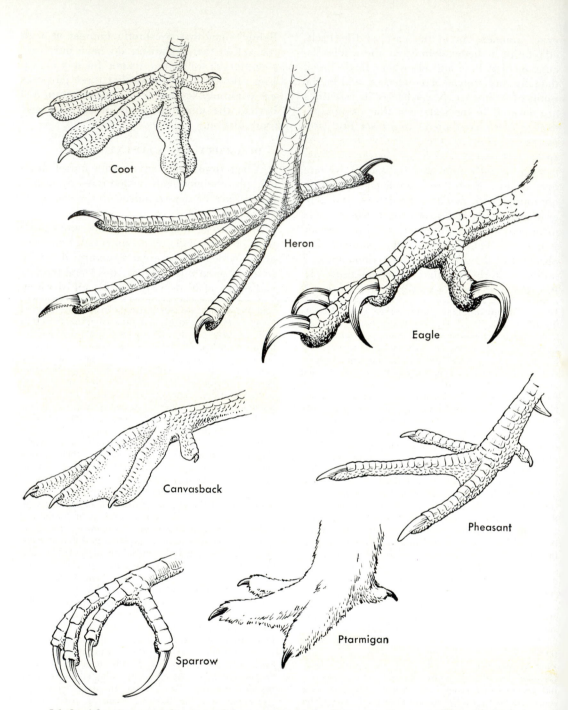

16–2 Adaptive specialization of feet among birds. The coot swims or paddles with its feet, which have lobed toes. The elongate toes in front and in back of the feet of the heron, a tall and large bird, give it a firm base for walking. The eagle is typical of birds of prey in having long talons on each toe with which to grasp its prey. The canvasback duck is a swimmer with fully webbed feet. The pheasant has feet suited to walking and scratching the ground for food. The sparrow is a typical perching bird, with feet equipped for grasping a branch. The ptarmigan, inhabitant of very cold regions, has feet stockinged by feathers.

16–3   The Indian leaf butterfly, *Kallima*.

relate to the efficient functioning of these species in their special environments (compare cockroach, mosquito, and butterfly mouth parts, and the feet of duck and eagle). Other clear morphological adaptations are the shapes and colors (discussed later in this chapter) by which animals are protectively concealed. A famous example is the leaf butterfly (*Kallima*) of Asia, which at rest is hard indeed to recognize in foliage (Fig. 16–3). Desert plants, possessing water-storage tissues, reduced or otherwise specialized leaves, and still other features, are spectacularly

adapted to the problem of conserving water in arid conditions.

Physiological adaptation, if less obvious, is always present in organisms. The shipworm *Teredo,* which causes so much trouble in the wooden pilings of wharfs and in the hulls of wooden ships, is able to exploit its remarkable habitat because it possesses special enzymes that digest wood. Shrimps that inhabit salt lakes like the Great Salt Lake in Utah have highly specialized powers of regulating their internal osmotic pressures.

Behavior in animals is almost as conspicu-

16–4   Adaptive behavior. The male of the (marine) 15-spined stickleback (*Spinachia vulgaris*) ensures good aeration of the eggs developing in his nest by fanning a current of water over them with his pectoral fins.

ously adaptive as structure. Figs. 16–4 and 16–5 provide examples, and the termite provides one still more remarkable. A termite, which like *Teredo* is able to live on wood, does so not because it can manufacture the enzymes necessary to digest wood but because its gut is always inhabited by flagellated protists that can digest the wood for it. A termite, like any other insect, must shed its skin periodically in order to grow. In the process it sheds the lining of its hindgut and with it loses the wood-digesting flagellates. The problem this raises—loss of a digestive system!—is overcome by a behavioral adaptation: as soon as it has shed its skin, the termite eats it and thus reinfects itself with the flagellates it needs. A young termite freshly hatched from the egg case acquires the protists it needs by the appropriate if indelicate action of licking the anus of an adult termite.

BROAD AND NARROW ADAPTATIONS

Adaptations so far mentioned are extreme and obvious. Too often in biology there has been a tendency to think of all adaptation in this way. The phenomenon has been regarded as a collection of bizarre curiosities whereby creatures succeed in unusual habitats or in usual habitats in unusual ways. Such a view is in fundamental error. Properly viewed, any entire organism is a bundle of adaptations; it is all adaptation insofar as it is appropriately organized to survive and reproduce in its habitual environment.

We may get a more useful perspective on kinds of adaptation if we classify them as broad and narrow rather than as morphologi-

16–5   Morphological and behavioral adaptation in the moth *Venusia*. The moth always rests by day on the leaf of a tree lily, with the striations of its wings carefully aligned with the vein striations of the leaf. The moth is thus inconspicuous and safe from predators.

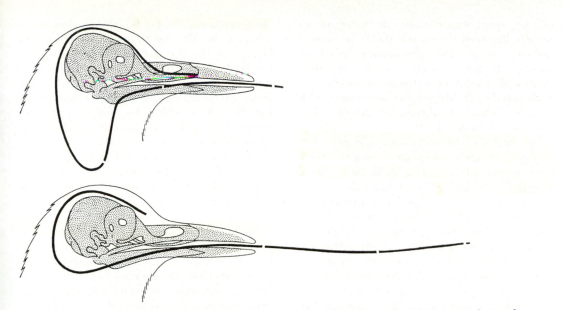

16–6 Adaptive specialization in a woodpecker's hyoid cartilage. In all higher vertebrates the tongue skeleton is the hyoid cartilage. In woodpeckers it is enormously long, adapted for probing insects from deep crevices in trees. The tip of a woodpecker's tongue is equipped with barbs.

cal, physiological, and behavioral. Let us consider the adaptations of a woodpecker on this basis. It shows several obvious *narrow* adaptations to its special way of life; these include its posture, its ability to hop up the vertical face of a tree trunk, its powerful neck muscles, which operate the head like a hammer, its large and chisel-like beak, and its long tongue, used to probe insects from the crevices it cuts (Fig. 16–6). In all these respects it is adapted to the narrowly defined occupation of extracting insects from tree trunks. But it is also more *broadly* adapted to the bird's way of life generally. Both its wings and its bones, which combine lightness with strength, are adaptations to flying shared by thousands of bird species. Its respiratory system, while similarly specialized to the bird's way of life, nevertheless is to be seen primarily as one of its broadest adaptations. It is shared by all birds, reptiles, and mammals as a fundamental and essential part of the vertebrate organization adapted to life on land. Of still wider adaptive significance is the woodpecker's possession of muscles and nerves intricately organized in relation to each other and to bones in effecting the precisely coordinated movements on which its whole life depends. Again, the fact that muscles and nerves are possessed by

every animal group of higher organization than sponges must not blind us to their adaptive nature. Broadest of all adaptations is the woodpecker's capacity to reproduce itself or, more strictly, to copy and pass on the hereditary message that it inherited from its parents. The mechanisms that copy chromosomes are as surely adaptive as the chiseling beak and the long probing tongue. But this ability to pass on genetic information is shared by all organisms. It is an adaptation of life in general.

INDIVIDUAL ADAPTABILITY

A nearly universal kind of adaptation, developed to a greater or lesser degree in different organisms, is exemplified by the response of *Paramecium* to increasing salt concentrations of the water in which it lives. The adaptation concerned is a capacity to adjust or habituate to changed conditions. *Paramecium* is a fresh-water creature whose environment usually contains little salt; in fact, the animal can readily be killed by a sudden and large increase of salt in its water. If, however, the salt is added gradually, *Paramecium* adapts or habituates to the salt; the concentration necessary to kill it becomes greater. Similar phenomena are known in other organisms, including man, whose ability to habituate to arsenic has been the central theme in many a mystery story. The prospective murderer slowly habituates himself to arsenic and then invites his victim to a dinner of which they both partake; the dinner has been spiced with arsenic adequate to dispose of the guest but not the habituated host.

Habituation to specific toxic compounds is a special example of a much more general phenomenon of somatic adjustment or *individual adaptability*. Mice or men raised in the lowlands have severe respiratory difficulties at high altitudes where oxygen is scarce. In the course of time their performance improves, however, because the body adjusts to the new stress placed on it by increasing the number of red cells and, thus, the oxygen-carrying capacity of the circulating blood. Many other instances of somatic adaptability are familiar in man: well-exercised muscles respond to the special and prolonged work load by appropriately increasing their size; exposure to a new infective agent elicits manufacture of an appropriate antibody; and so on.

Seen in broad perspective, even the capacity of animals to *learn* is itself just another example of the same phenomenon: behavior is appropriately adjusted in the light of past experience in such a way as to heighten the efficiency of living. Thus, whether earlier experience takes the form of unfavorable osmotic conditions, drug intake, low oxygen content, extra muscle work, or a behavioral problem, the organism possesses the ability to improve its organization for survival efficiency. In all these instances the improvement in adaptation occurs within an individual's life span.

We must therefore distinguish carefully two uses of the verb "to adapt." Looking back over the history of man, we might comment that some of his ancestors *adapted* to life in trees by evolving modifications of forelimb structure. That is a radically different process from the one referred to in a sentence like "Mr. Smith eventually *adapted* to high altitudes and decided to stay in Peru." The adaptation in man's ancestors involved an overhaul of their genetic make-up; it was an adaptive modification of the inherited chromosomal information that specified relatively rigidly the structure of limbs. But the process of adaptation in Mr. Smith took place with no change in the chromosomal specifications he received from his parents and transmitted later to his children. The physiological flexibility of Smith's body is guaranteed by his heredity; it is within his inherited reaction range. It is as though his chromosomes had given these instructions for his development: "Build a blood system that includes the following special physiological equipment that will permit automatic adjustment of the red cell content to a level appropriate to local conditions."

Individual adaptability is in itself an adaptation of the highest order.[1] Were a population to inhabit a rigorously stable environment, inherited information could be simplified by specifying a quite inflexible bodily organization appropriate to the enduring conditions. But no environments are completely stable; the biological environment is constantly changing, and other ecological changes are inevitable. The time eventually arrives for all populations when, in order to survive, they must move into new environments or must meet changing conditions where they are. In such circumstances those populations with the capacity to adjust somatically (that is, bodily or phenotypically) are at an advantage. Mutational modifications of inherited information causing development of an *adjustable* body in species meeting changing conditions are favored by natural selection as surely as the mutations that increase simpler fixed adaptive features like protective coloration.

Inheritance of Acquired Characters

LAMARCKISM OR NEO-LAMARCKISM

Two main scientific (as opposed to philosophical) hypotheses have been proposed to account for adaptation. One already familiar to you, natural selection, has successfully undergone such thorough testing that it has ad-

[1] We should recall from Chapter 4 the several mechanisms by which the individual organism can adjust or control its enzymes and their rates of activity in response to changing environmental conditions.

vanced to the status of an established theory.[2] The other, the inheritance of acquired characters, failed to meet the tests of observation and has been almost universally discarded by biologists.[3] We might, therefore, ignore it here were it not for the fact that it played an important part in the history of evolutionary thought. Moreover, it sometimes has so strong an appeal to beginning students of evolution that reasons for its rejection still need to be spelled out.

The hypothesis of the inheritance of acquired characters states that the results of individual, somatic adaptability are passed on by heredity to the offspring of the individual. If it were true, some adaptations, at least, could be explained by the accumulation of such modifications in the genetic system through the generations. A classic but grotesque example is the view that the giraffe's long neck is an evolutionary consequence of long-continued exercise in stretching as generation after generation of giraffes strove to reach leaves at the tops of trees. Frequently included in the hypothesis was the further idea that modifications from strictly environmental sources, including inadaptive changes due to injuries or malnutrition, also become heritable.

It is customary to call the hypothesis of the inheritance of acquired characters *Lamarckian* and the acceptance of it *Lamarckism*, after Lamarck, the most thoroughgoing of the pre-Darwinian evolutionists. It is important that you know these terms because of their frequent use in discussions of the history of evolutionary thought. They are, however, both inaccurate and unfair and would be abandoned if scientists were as meticulous in historical study as in science itself, for the following reasons: the hypothesis of the inheritance

of acquired characters was only a minor and subsidiary feature of Lamarck's theory of evolution;[4] Lamarck was not, by many centuries, and never claimed to be the originator of the hypothesis; Lamarck himself did not accept the hypothesis in just the same form or to the same extent now labeled "Lamarckian"; along with many other nineteenth-century naturalists, Darwin did accept the hypothesis in the way now labeled "Lamarckian" and misleadingly contrasted with "Darwinian." Although also subject to possibly erroneous interpretations, the now commonly used terms *Neo-Lamarckism* and *Neo-Darwinism* are less objectionable in reference to the debate that developed after Darwin about the inheritance of acquired characters versus natural selection as explanations of adaptation.[5]

When Lamarck wrote his *Zoological Philosophy* (1809) and still when Darwin wrote *The Origin of Species* (1859), the inheritance of acquired characters was a legitimate and indeed a good scientific hypothesis. It could be tested, and it was necessary for the advance of biology that it should be thoroughly tested. It was an age-old belief, and it also had at least two seemingly common-sense features that suggested its application to the newer concept of evolution. One of these is its apparent analogy with social or cultural evolution, in which the new acquisitions of one generation are indeed inherited by the next. The other is that, as we have seen, individual organisms do to a limited extent become somatically adapted. Why should not such adaptation lead to or even be identical with organic evolution? The same verb, "to adapt," is applied to both processes. In fact, here as in the comparison with social evolution, this version of "common sense" turns out to be misplaced trust in a false analogy.

FAILURE OF THE HYPOTHESIS

Intensive testing in the latter part of the nineteenth and early part of the twentieth

[2] Remember (from Chapter 1) that in proper scientific usage designation as a theory need not imply any serious doubt. In looser, vernacular language a scientific theory as well substantiated as natural selection can as easily be called "a fact" as "a theory."

[3] Michurinism, one of several nineteenth-century forms of this hypothesis, still has political support in the Soviet Union, but (to say the least) political sanctions are not criteria for scientific acceptability. That there are still also a few non-Soviet supporters of the hypothesis is hardly more significant than that there are still a few people who maintain that the earth is flat.

[4] Its main feature was the Aristotelian "ladder of nature" (Chapter 21), more philosophically than scientifically expressed.

[5] Although so considered by many of the debaters, this was not really an "either–or" question. Darwin himself accepted *both* of what later evolutionists called "Neo-Lamarckism" and "Neo-Darwinism."

centuries revealed many and insuperable weaknesses in the Neo-Lamarckian hypothesis. Most of these relate to five main points.

1. The hypothesis demands that hereditary information be passed backward, so to speak, from body cells to germ cells. Our present vastly expanded knowledge of heredity not only discloses no mechanism by which such a transfer could occur but also demonstrates that it is virtually impossible.

2. Enormous effort has repeatedly been put into experiments designed to demonstrate the inheritance of acquired characters. All these experiments have failed.

3. It seems quite impossible that many adaptations could be achieved either by efforts of the organisms themselves or by the direct effects of their environment. For example, the green coloration of insects that escape predation because they look like leaves is an evident adaptation. But how can an insect practice becoming green? And how can green leaves directly cause insects to become green with completely different pigments?

4. There are other adaptations that, for a different reason, cannot possibly result from the inheritance of acquired characters. The most noteworthy examples are among the neuter worker or soldier classes of insects, which do not breed and therefore cannot pass on any characters that they may acquire.[6] (Each individual must in fact inherit the capacity to develop its caste characters from parents that did not have those characters.)

5. If, as the usual Neo-Lamarckian form of the hypothesis demands, effects of the environment were directly heritable as such, inadaptive effects such as are caused by injury or malnutrition would inevitably accumulate along with any adaptive effects. For example, by now we must all have had many ancestors who lost a finger or toe before producing offspring, but possession of the normal five on each limb continues to be just as much a part of our inheritance as it was 300 million years ago.

[6] Although Darwin believed in the inheritance of acquired characters, he noted this objection. He correctly concluded (in other words) that these examples show that such inheritance is not a sufficient cause of adaptation but leave the question open whether it may not be in other instances.

At each point where the hypothesis fails or is inadequate, natural selection agrees with the observations and provides a sufficient explanation for them. The occurrence of natural selection in nature is abundantly attested, and its explanation of adaptation is consistent with everything we know about genetics and the history of life.

We do not mean by all this that acquired characters have nothing to do with adaptation. We have already stressed that individual *ability* to acquire adaptive characters is itself an adaptation *developed by natural selection.*

Natural Selection and Adaptive Coloration

Colors and color patterns, some brilliant and some subtle, are present in organisms of many different kinds. They are adaptive in highly diverse ways, and they provide some of the most striking illustrations of adaptation. Some of them have also provided crucial evidence in controversies about the role of natural selection in producing adaptations. From this point of view we shall briefly consider two kinds of adaptive coloration—cryptic coloration and warning coloration.

CRYPTIC COLORATION

The theory of cryptic coloration

It is a common observation that many animals are colored or shaped in such a way as to be hard to recognize in their natural surroundings (Fig. 16–7). *Cryptic, or concealing, coloration* is widespread among mammals, birds, and some other groups, but insects are perhaps the most obvious examples. Grasshopper species in the lush grass of meadows and stream banks are green; other species in dry prairie grasslands match the drab straw color of their backgrounds. Their relatives the stick insects can change their color to suit different backgrounds, and they add to the deception by being elongate and twiglike in shape. The common moths of birch and pine trees differ in wing coloration, each appropriately matching the surface upon which it habitually rests. Pale moths from birch trees are nearly invisible in their usual

16–7 Protective form and coloration and warning coloration. *Above left*. Cryptic form as a protective device in the pipefish, *Entelurus aequoreus*, which rests amidst eelgrass, *Zostera marina*. *Below*. Cryptic coloration in the flatfish topknot, *Zeugopterus punctatus*, which rests on a shell-gravel bottom. *Above right*. Warning coloration in the larvae of the cinnabar moth. Note the conspicuous crossbanding and the clustering together.

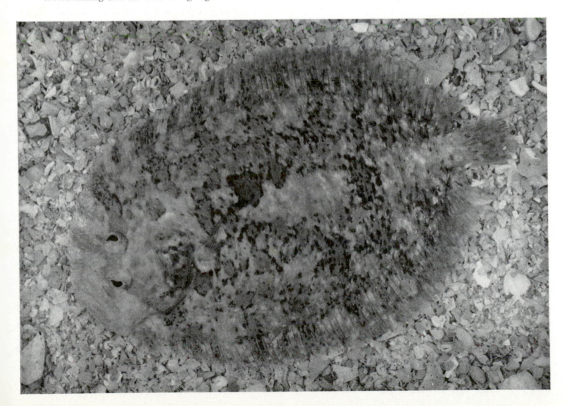

habitat, but if they land on the dark bark of a pine, they stand out clearly—fair game for any bird hunting insect food.

A more subtle kind of cryptic coloration is exhibited by most protectively colored mammals and birds, which are darker on their backs than underneath. Under ordinary conditions, with light coming mostly from above, this color arrangement tends to conceal the body by eliminating the comparative brightness of the upper, more illuminated parts and darkness of the lower, more shadowed parts. Some bold color patterns of stripes, irregular blotches, and the like, highly conspicuous when seen in isolation, are in fact protective in the natural habitat because they break up the outline of the body and make it unrecognizable as prey; the same device is used in military camouflage.

The theory of the origin of cryptic coloration by natural selection is as follows. In a population of animals subject to predation, as almost all are, there is hereditary variation in color and pattern. Some variations are more likely to deceive predators than others. On an average, the predators will find and kill more of the less well-protected variants than of the better-protected variants. This lethal form of natural selection will therefore produce, through the course of generations, a population as a whole more adequately protected.

A criticism

There have been severe critics of the theory that cryptic coloration is protective. One such critic made an extremely laborious count of allegedly protected and unprotected insects taken from the stomachs of birds that prey on insects. He found large numbers of cryptically colored individuals and concluded that their coloration does not protect them and therefore could not be favored by natural selection. His seemingly logical argument is, in fact, fallacious, and the fallacy has broad implications for the scientific method.

In spite of the very extensive data gathered, the observations did not really test the theory in question. In order to have any relevance, they would have had to be compared with the relative numbers of protected and unprotected insects *before* they were caught. We might find two protected insects for every one unprotected insect in a bird's stomach; but if before they were caught, there were, say, four protected to one unprotected, then the protection would still have been highly effective—in proportion to the whole population, twice as many unprotected as protected ones would have been caught. The observations actually made proved only that *some* protected insects are caught. To argue from this that they are not really protected is to imply that an adaptation must be 100 per cent effective. We could then only conclude that no organism is adapted because all eventually die—and that is nonsense.

The refrigerator fallacy

Another argument against the theory of protective coloration has it that since some animal populations survive without such coloration, protective coloration is not essential and is not adaptive. This has been called the "refrigerator fallacy" because it is analogous to saying that refrigerators are not an advance in efficient living, as our ancestors got along well without them and so do some people today. Once refrigerators evolved (were invented), they spread in the population because they are *useful*. The fact that they are not *necessary* has no particular bearing. Likewise, protective coloration and indeed all organic adaptations must be useful if natural selection is to spread them through populations, and it is not at all essential to the theory that they be indispensable.

Let us consider two similar but separate light-winged moth populations on trees with dark bark. Both are conspicuous and are heavily preyed on by birds. Suppose now that in population 1 but not in population 2 mutations occur that slightly darken the wings. It is a common error to assume that because the darker wings are only *slightly* less easily seen they are not an adaptive improvement. Even if the mutants are overlooked only 1 per cent more often than the individuals with normal alleles, natural selection will result. The proportion of individuals with darker wings will increase in succeeding generations, and eventually the mutant allele will entirely replace the original allele. It will do so because it is more effective and more adaptive, and the process of replacement is natural selection. This phenomenon of progressive darkening of the wings of moths more nearly to match a dark background has been observed in nature, and careful study has proved conclusively that it is indeed caused by the selective pressure of predation.

In the meantime, population 2 continues to flourish in spite of its lack of darkening mutations and its consequent retention of unprotective coloration. This circumstance does not invalidate the observation that darker wings are protective when they do occur. Population 2 meets the predation pressure in some other way, perhaps by a fecundity so high that a large population is maintained regardless of heavy losses. This is an adaptation, too, and one that commonly arises through natural selection when losses of individuals are high. It illustrates another important situation—the existence of alternative means of adaptation. There are ways of adapting to predation other than by protective coloration, just as there are ways of preserving food other than in a refrigerator.

WARNING COLORATION

The existence of alternative adaptations is pointed up by the fact that flashy color patterns, just the opposite of cryptic coloration, may also (but in quite a different fashion) protect their bearers against predation. This answers another argument against the theory of cryptic coloration, the argument that concealment is not adaptive because its opposite, advertisement by color pattern, also occurs in thriving populations. We have explained that some populations can get along in spite of the absence of cryptic coloration, but it may still seem baffling that some get along not in spite of but because of their conspicuousness.

There are many insects among beetles, bugs, bees, butterflies, and others that have unpleasant tastes, bristles, or stings making them disagreeable food for most predators. These features profit the insect nothing if they are noted only after it has been killed. If, however, it advertises its unacceptability by *warning coloration*, predators learn to leave it alone (see Fig. 16–7). The advertising message is "This is the color pattern that gave you trouble before."

The theory of warning coloration predicts

TABLE 16–1

The adaptive value of warning coloration *

	Accepted as food by monkey	Rejected as food by monkey	Totals
Warningly colored insect species	23	120	143
Cryptically colored insect species	83	18	101

* The monkey accepted as food 83 per cent (83/101) of all cryptically colored species but only 16 percent of all warningly colored species; 87 per cent (120/138) of all species rejected were warningly colored.

that predators learn to recognize and not to molest distasteful, warningly colored insects. Repeated experiments have borne out the prediction. In the example summarized in Table 16–1, G. D. H. Carpenter offered to a monkey over 200 species of insects, some warningly colored and some cryptically colored. Again we do not expect, and the theory does not demand, 100 per cent efficiency,[7] but it is clear that the warning coloration was, on an average, highly protective.

OTHER KINDS OF ADAPTIVE COLORATION

Obviously there are more ways of avoiding being eaten than by evolving cryptic coloration. Another, also involving coloration and pattern, is by *mimicry*. In one of the several kinds of mimicry, the mimics are animals (especially insects) that are quite acceptable as food but that look and act like others, usually with warning coloration, that are obnoxious to some predators. If a mimic is mistaken for an unpalatable species, its chances of survival are increased, and natural selection will tend to favor and to reproduce the mimicry, which is, of course, quite unconscious. (Can you think of still other ways of escaping predation?)

Coloration may also be adaptive without having anything to do with predation. Conspicuous colors may be recognition marks, helping birds of a feather to flock together. They may also be releasers or, so to speak,

[7] For one thing, the predator usually has to learn by experience that the gaudily colored insects are distasteful.

passwords that elicit sexual acceptance and appropriate mating behavior (Chapter 14). Colors may be adaptive to the physical environment. For example, dark integuments may facilitate heat radiation from within and screen off shorter frequencies (light, ultraviolet) from without, whereas light integuments may have the reverse effects. The colors and patterns of flowers commonly serve to attract and guide pollinators such as insects and birds. The omnipresent green of chlorophyll in fields and forests is adaptive in yet another way: it is incidental to the absorption of energy from light.

We have by no means exhausted the ways in which color is adaptive in organisms, but have mentioned only a few of them.

Natural Selection as a Historical Process

We have stressed that natural selection is not necessarily a process of competitive combat and strife. Nevertheless, competition, which is usually unconsciously entered, does play an important role in the direction of evolution. Competence to find food and make a living up to and throughout reproductive maturity is an essential part of that more general competence to leave offspring which is *the* feature of organisms that selection constantly maintains or improves.

The natural selection that results from competition for food and a place to live leads to a variety of evolutionary consequences. Broadly speaking, these may be (1) a narrowing and further specialization in competence, (2) an increased diversification and broadening in competence, or (3) a change to a different mode of life and kind of competence. Which of these evolutionary effects is realized depends on a variety of circumstances, some of which will now be considered.

OCCUPATION OF NEW ENVIRONMENTS

Pressure to diversify

All species of organisms are adapted to a particular environment that is limited, to a greater or lesser extent, in size and resources

(food, a place to live, etc.). Sooner or later the reproductive capacity of the species raises the population to a limit determined in large part by the availability of resources. This situation creates a selective process in which those individuals able to make use of otherwise unexploited environments and resources are at an advantage; their probability of successfully leaving progeny is heightened by the low competition in the new environment. Thus the theory of natural selection leads to an explanation of the diversity that is one of life's most striking features. When we survey the nearly incredible range of habitats, or environments, exploited by organisms—from hot springs to arctic waters, from ocean floor to mountain streams, deserts, rain forests, and the air above, from the intestines of other animals to the pages of library books—we are viewing the diverse habitats that have constituted opportunities for organisms to escape from competition in other, well-occupied environments.

It is surely unnecessary at this point to labor the fact that, when we speak of "escape from competition," we are not envisaging a deliberate attempt on the part of a squirrel, bird, or bacterium to find a new way of life where the going is easier. The universally present structural devices and behavior patterns that ensure random dispersal of a species cause it constantly (and for the most part quite unwillingly) to sample new environments.

Conditions for entry into new environments

To enter a new mode of life a species must be given the *opportunity* to do so in three distinct senses:

First it must have the *physical opportunity* to enter. This is the most obvious of the three conditions and needs little amplification. Conceivably a butterfly species now limited in distribution to South American forests might be competent to exploit some new and so far unused environment in African savannahs, but its evolutionary potential in this respect will remain unfulfilled so long as it lacks physical access to South Africa. The intestines of other animals have been exploited as new environments only by those groups (protists and worms with aquatic larvae) that have had prolonged physical access to animal intestinal tracts through their presence in drinking water and food.

Prolonged physical access to a new environment does not, however, guarantee its successful evolutionary invasion by a species. A *constitutional opportunity* must develop. We can put this another way by saying that a species has *constitutional access* to a new environment only when it already possesses some minimal adaptation adequate to sustain survival and reproduction while it gains a footing. Once entry is established, selection will steadily raise the level of adaptation to the new conditions. Examples given later in the chapter show that we are not begging the question of how new modes of life evolve when we say that a minimal degree of adaptation must exist before new habitats are entered. It is a commonplace of life's history that the adaptations which permit exploitation of *new* modes of life are only temporary makeshifts initially acquired as adaptation to *old* modes of life, and subsequently improved or sometimes completely replaced.

Physical and constitutional access are both necessary, but even together they are not in themselves sufficient to ensure invasion of new habitats. The species concerned must also have *ecological opportunity*; the ecological conditions prevailing must be appropriate. Ecological access always means that the competition encountered in the new habitat must be slight enough to permit survival of the new invader during its initial phase, when its adaptation may still be poor.

Much of life's history becomes understandable as exploitation of constitutional and ecological opportunities to enter new environments and thus escape the heavy competition in those that are fully occupied.

Darwin's finches: an adaptive radiation

The role that ecological opportunity plays in determining evolution is well illustrated by the history of a group of small land birds, the Geospizinae or Darwin's finches on the Galápagos Islands (Fig. 16–8). These islands are a compact group lying about 600 miles off the coast of Ecuador. Darwin's visit to them when he was serving as a naturalist aboard

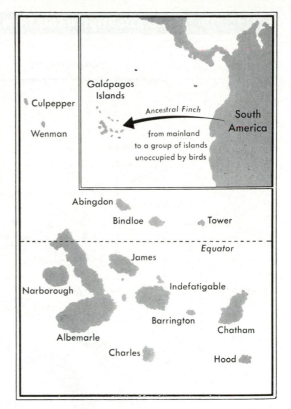

16–8   The Galápagos Islands, home of Darwin's finches.

the exploration ship H.M.S. *Beagle* in 1832 strongly influenced his later thought about evolution.

The biologist's interest in the Galápagos stems from the fact that they are oceanic islands thrust up from the ocean floor. They have had no connection with the mainland at any time in their history. Coming into existence late in the history of life, they initially constituted a completely unoccupied environment, and a remarkable ecological opportunity for such land organisms as could reach them. The nature of the present flora and fauna gives away completely the story of how the islands were initially colonized by life. The groups now present on the Galápagos are an extremely spotty sampling of those present on the South American mainland. In the absence of a land connection with the islands, only a few kinds of organisms have ever reached them. Their successful immigration was a rare event brought about by the chance

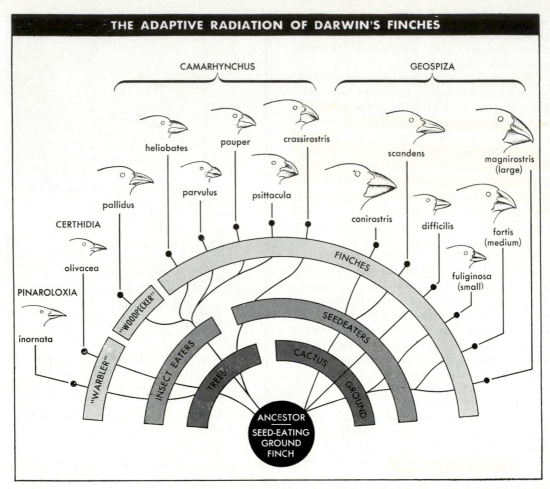

16–9   The adaptive radiation of Darwin's finches.

movement of winds and of debris floating along where currents drove them. As soon as vegetation became established on the islands, immigrant animals were free to enter any of the several new environments they encountered and to which they had constitutional access.

All the small land birds of the Galápagos today are descendants of a small finch from the South American mainland. Since its arrival this finch has evolved into at least 14 distinct species, each of which specializes to some extent in exploiting the resources of the islands (Fig. 16–9). The evidence indicates that the ancestral finches were ground birds feeding mainly on seeds and other vegetation. Of the 14 species that evolved from this stock, three, of different sizes, are still ground

finches feeding on seeds, two are mainly cactus finches, and one combines ground and cactus feeding. All the others have become tree finches, the majority of which are insectivorous. Within these broad categories (ground, cactus, and tree finches, some vegetarian and others insectivorous) still further specialization has developed. The species differ markedly in beak size and structure; this relates to the size of the food they capture and eat. One of the tree finches has become essentially a woodpecker. It lacks the long tongue that is an adaptation of the tree woodpeckers but substitutes a remarkable piece of behavior. After chiseling with its beak, it snaps a cactus spine and uses it to probe its insect prey from the crevice it has chiseled (Fig. 16–10). Another of the tree finches has

16–10 *Camarhynchus pallidus.* The adaptive radiation of Darwin's finches has produced one species that is essentially a woodpecker. It has mastered this way of life by evolving, not the morphological specializations of the familiar woodpecker, but a behavioral substitute. It uses cactus spines instead of a long tongue to probe insects from trees.

become to all intents and purposes a warbler.

The difference in evolutionary future between the initial finch immigrants to the Galápagos and their brothers and sisters on the mainland is striking. All the Galápagos finches of today have departed so far from the original ancestor that we can no longer identify it among the mainland finches. But we are sure of this—not one of the mainland finches (including the ancestor of the Galápagos birds) has been able to undergo the extensive adaptive diversification that occurred on the Galápagos, in spite of having identical physical and constitutional opportunities to do so; the mainland birds lacked the ecological opportunity created by the vacant habitats of the Galápagos (Fig. 16–11).

The evolution of new species on the Galápagos Islands was further enhanced by the island nature of the new territory. Island-to-island movement of small birds like the finches is extremely limited, though adequate to ensure the population of all the islands in time. The geographic isolation of each new island population promoted the initial breaking up of the finch population into new species. The various species specialized to some

extent to local conditions but ultimately spread all over the archipelago. The result has been that most of the islands now support a number of species. Roughly the same ecological opportunities existed on most of the islands. On all of them openings existed, for example, for large, medium, and small vegetarian ground finches. These openings were filled on all the islands, although not always by the same species. Even among the Galápagos finches themselves, we can discern how local differences in ecological opportunity from island to island have affected the history of individual species (Fig. 16–12).

The evolutionary phenomenon exemplified by the Galápagos finches is a general one called *adaptive radiation.* The descendants of an ancestral species that was itself adapted to a typically restricted way of life have radiated out into a diversity of new habitats. Adaptive radiations have characterized the evolution of life throughout its long history. Whenever for one reason or another a group of organisms has been confronted with a diversity of new ecological opportunities to which it has physical and constitutional access, radiations have occurred. Some radiations, like those of the Galápagos finches, are trivial in extent even if beautifully clear in understood detail. Other radiations have taken place on a more massive scale with far-reaching importance for the history of life in general.

MAJOR ADAPTIVE RADIATIONS

The total evolutionary opportunity afforded the Geospizinae on the Galápagos Islands was limited in two ways. First, the diversity of open habitats (ecological opportunities) was limited; the variety of vegetation types was restricted. Second, vegetarian ground finches have limited constitutional opportunities; many ways of life, for instance those of cats or of large rodents, remained open on the Galápagos, but the finches had no access to these ways of life because they lacked teeth and unspecialized forelimbs. Cats and rodents themselves did not exploit the opportunities on the Galápagos because they lacked physical access; they did not cross the sea barrier between South America and the islands. Evolutionary opportunity of far greater scope and significance has been created when organisms

A major adaptive radiation of great importance in our own history followed the first conquest of land by vertebrates. The earliest vertebrates that could emerge even temporarily on the land were fishes that (1) could breathe air to a limited extent with lungs that were extremely crude by modern standards and (2) could walk or wriggle in still cruder fashion with the aid of slightly modified fins.

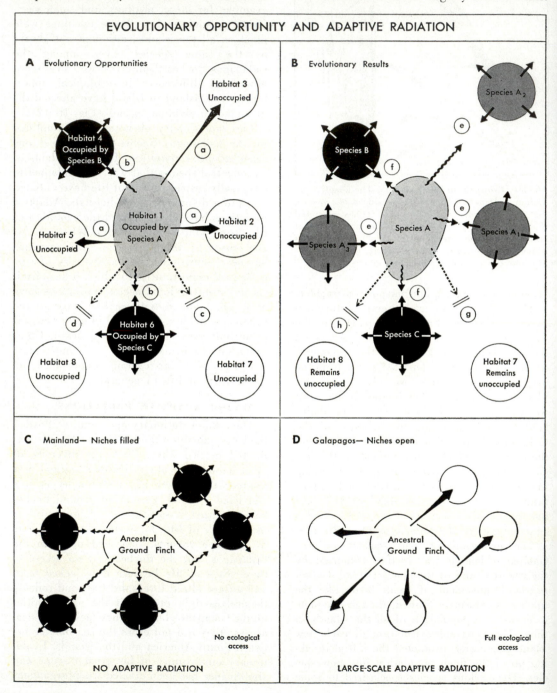

EVOLUTIONARY OPPORTUNITY AND ADAPTIVE RADIATION

A Evolutionary Opportunities

Habitat 3 Unoccupied

Habitat 4 Occupied by Species B

Habitat 1 Occupied by Species A

Habitat 2 Unoccupied

Habitat 5 Unoccupied

Habitat 6 Occupied by Species C

Habitat 8 Unoccupied

Habitat 7 Unoccupied

B Evolutionary Results

Species A₂

Species B

Species A

Species A₁

Species A₃

Species C

Habitat 8 Remains unoccupied

Habitat 7 Remains unoccupied

C Mainland— Niches filled

Ancestral Ground Finch

No ecological access

NO ADAPTIVE RADIATION

D Galapagos— Niches open

Ancestral Ground Finch

Full ecological access

LARGE-SCALE ADAPTIVE RADIATION

The evolution of these minimal constitutional prerequisites took place as part of a strictly fish radiation. What is important here is the fact that, once able to colonize the land to any extent, the vertebrates had clear sailing from then on; being the first large terrestrial animals, they met with no competition. Amphibians, reptiles, and mammals successively evolved through subsequent geological periods. The reptiles were the first fully competent land vertebrates. Equipped with improved reproductive apparatus, they could move inland away from water and exploit the rich array of wholly unoccupied land habitats. The reptile radiation was a grand one; it produced herbivores in rich assortment, a diversity of carnivores, a host of flying forms, and even forms that successfully returned to water, making a good living in spite of fish competition.

Two special points should be noted in connection with the reptilian radiation. First, it removed the ecological opportunity earlier open to the fishes that were becoming amphibious. The fact that fishes once evolved into land vertebrates but can no longer do so has proved puzzling to many people. How could fishes have accomplished the feat 300 million years ago if they cannot do so now? The answer is simple enough: the evolutionary step from water to land involves a transitional stage, neither fully aquatic nor fully terrestrial, that was extremely unlikely to be successful once efficient competitors were established on the land. No fishes today would, to put it plainly, stand a chance.

The second special point concerns the history of mammals. It is clear from the fossil record that one branch of the early reptilian radiation produced animals that later gave rise to mammals. Indeed, mammals of a sort were in evidence throughout the prolonged heyday of the reptilian dinosaurs, but they were of restricted variety and abundance. For reasons still obscure, many reptiles became extinct about 70 million years ago. With their passing, all the ways of life they had formerly filled became available. Into these now open environments the mammals, which had survived the great dying among the reptiles, were free to radiate.

As in the earlier great radiations, the ensuing adaptive radiation of mammals was rapid. Forms appeared that converged toward most of the earlier reptilian adaptive types: herbivores, carnivores, flying forms, swimmers, and so forth. The history of mammals has involved a whole series of still later, more restricted radiations, some of which will be discussed in later chapters.

THE EVOLUTIONARY LOTTERY

Multiplicity of evolutionary directions

In any given situation, a combination of physical, constitutional, and ecological opportunities controls the directions of evolution. It does not limit evolution to a single direction

16–11 **Evolutionary opportunity, or access: prequisite to adaptive radiation.** *A.* Habitat (or niche) 1 is filled by species A (consider it to be a bird like the ancestral ground finch that colonized the Galápagos Islands). Competition within the species constitutes a perpetual pressure on the population to diversify, to exploit new and unoccupied niches. The species can diversify only when it has physical, constitutional, and ecological access to (opportunity to enter) a new habitat. (*a*) Such is the case for species A with respect to habitats 2, 3, and 5. (*b*) Although species A has physical and constitutional access to habitats 4 and 6, it lacks ecological access, for these niches are already occupied by the well-adapted species B and C. (*c*) Species A (a bird) has ecological and physical access to habitat 7 (say, that of a cat), but it lacks constitutional access. (*d*) Species A could, constitutionally and ecologically, enter habitat 8, but it lacks physical access, for habitat 8 is on another island. *B.* The consequences of the conditions in *A* are as follows: (*e*) Three new species (A$_1$, A$_2$, and A$_3$) evolve as the original species A exploits the combined evolutionary opportunities afforded to it. (*f*) Habitats 4 and 6 are never entered because, in its transitional stage of incomplete adaptation, A cannot compete with the well-adapted occupants B and C. (*g, h*) Habitats 7 and 8 remain unoccupied. *C, D.* The Galápagos Islands versus the mainland. The figures compare schematically the evolutionary opportunities afforded the same species of finch on the South American mainland and in the Galápagos. For the most part, the adaptive radiation of the finches in the Galápagos has been limited only by the constitutional opportunities of the initial ground finches; having gained physical access to the islands, they had virtually unlimited ecological opportunities.

	CENTRAL ISLANDS	SMALL OUTLYING ISLANDS		
		TOWER	HOOD	CULPEPPER
LARGE GROUND FINCH	*magnirostris*	*magnirostris*		*conirostris*
CACTUS GROUND FINCH	*scandens*	*conirostris*	*conirostris*	*difficilis*
SMALL GROUND FINCH (ARID ZONES)	*fuliginosa*	*difficilis*	*fuliginosa*	*difficilis*
SMALL GROUND FINCH (HUMID WOODS)	*difficilis*	The outlying islands lack the moist woodland habitat occupied by *difficilis* in the central islands		

16–12   The exploitation of local evolutionary opportunities by *Geospiza* species in the Galápagos Islands. Several of the small outlying islands (Fig. 16–8) have never been colonized by some species from the central islands. *G. magnirostris*, for example, has failed to reach Hood and Culpepper. Its absence from these islands left open the large-ground-finch niche to which *G. conirostris* had physical and constitutional access. On Hood *conirostris* has a larger beak than in the central islands and occupies both the large-ground- and cactus-ground-finch niches; on Culpepper it has a still larger beak and occupies the large-ground-finch niche. In the central islands *G. difficilis* is a small ground finch of humid woodlands, a habitat not found in the arid outlying islands. In these outer islands *difficilis* has succeeded only where *fuliginosa* (the common arid-zone small ground finch) is absent and where it has had, therefore, the ecological opportunity to enter the new niche.

but repeatedly produces adaptive radiations like that of the reptiles, the first vertebrates to emerge fully onto the land. This is decidedly not a random process. It is strictly determined by the opportunities presented and by the resultant force of natural selection. These factors, however, depend on the past history and present status of any given population. They do not and cannot take account of the future, as is evident from the fact that the eventual fate of most populations is extinction. With reference to the ultimate outcome, then, the process may be likened to a lottery in which most of the ticket-holders are losers (extinction), some just happen to make limited gains, and a small minority become big winners.

Small winners

Two examples will sufficiently illustrate how features highly adaptive in their own way

may nevertheless limit future opportunities for further or still broader evolution.

The ancestral arthropods arose in the first great radiation of animals. They themselves have radiated and reradiated repeatedly, giving rise among other things to the exceedingly diverse insects. One reason that they prospered was that they acquired external skeletons. Nevertheless, that very characteristic forever debarred them from the further and in some sense higher opportunities open to the vertebrates with their internal skeletons.

Parasitic flatworms make up another highly successful but even more limited group. Their adaptation to parasitism entailed the loss of sense organs and other structures adaptive to nonparasitic life. Parasitism forever closes all other opportunities of evolution. No parasitic group known has escaped into a nonparasitic existence.

Big winners

Man shares a common ancestry with other mammals, with birds, with reptiles and fishes, with arthropods, and even with parasitic flatworms. Like all of them, he is descended from animals that underwent the first great adaptive radiation resulting in the animal phyla. His history traces through the later radiation of fishes (first vertebrates), through the still later radiation of reptiles, and so on.

The evolutionary advance that man represents when compared with all his relatives, close and distant, is the product of the same lottery process in which so many organisms lost future opportunities. Man simply happens to be the descendant of a long line of organisms that drew winning tickets in every successive adaptive radiation. The basic adaptations of his ancestors have proved, in hindsight, not to have closed out the evolutionary future.

An essential part of our present adaptive organization is our respiratory system. We owe our lungs to the happy accident that some Devonian fishes developed them, under pressure of selection, as an adaptation to strictly fish problems. They turned out to be more than valuable adaptations in the radiation of fishes; they helped to confer upon some early fishes constitutional access to environments then unoccupied. The sequence of causes here is clear: fishes did not evolve lungs *in order to* become land vertebrates; their descendants became land vertebrates because the ticket the earlier fishes drew in the lottery of fish radiation was a winning one.

So it is with all our adaptations. All arose in response to the generally quite different conditions under which our ancestors, near and remote, lived. All just happened to be winning tickets that opened opportunities for further evolution eventually culminating in man. The many millions of populations that drew other tickets—evolved different adaptations—therefore either became extinct or evolved into species other than *Homo sapiens.*

HISTORICAL OPPORTUNISM VERSUS DESIGN

The complexity of organic adaptations sometimes strains our credulity as to the adequacy of natural selection to explain them.

The task of molding something as complex as the human ear seems too much for natural selection if we look upon the ear as an invented and completed thing. But the adequacy of selection to bring about this result is evident if we follow the actual historical development. Selection accounts for this development at every step, but in a most devious way, which no pure conjecture could conceive.

The present-day mammalian ear is a patchwork of ducts, bones, and membranes that in earlier vertebrate history were variously employed in surprisingly diverse functions, many of which had nothing to do with sound reception (Fig. 16–13). The Eustachian tube was originally part of a gill slit serving the respiratory needs of our fish ancestors. The three bones that transmit vibrations from the eardrum to our inner ear also had a history just as incidental to their present function. One of them (the stapes) was at different times in its career a skeletal element in the throat and later a convenient prop to the jaw apparatus of the first biting vertebrates. The other two bones (incus and malleus) were part of the reptilian jaw of 200 million years ago. The present function of all three ear bones is due to a fortuitous combination of circumstances. First, their earlier roles happened, quite incidentally, to bring them into close proximity to a pressure-sensitive organ (the future inner ear) in the brain case. While functioning in their former capacity of jaw components, the bones happened, willy-nilly, to transmit sound (pressure) waves to this organ. Second, successive adaptive improvements in the vertebrate jaw structure made all three bones nonessential in their initial functions. Released in this way from one kind of selection, their future was determined by the ever present premium on efficient sense organs and the natural selection this creates; they became exclusively devoted to a function that was previously incidental and of minor importance.

Natural selection is always restricted in what it can accomplish by the opportunities with which it is confronted. Earlier in this chapter we discussed opportunities for selection arising from ecological accidents—encounters with unoccupied environments. Now in the history of the ear we have exempli-

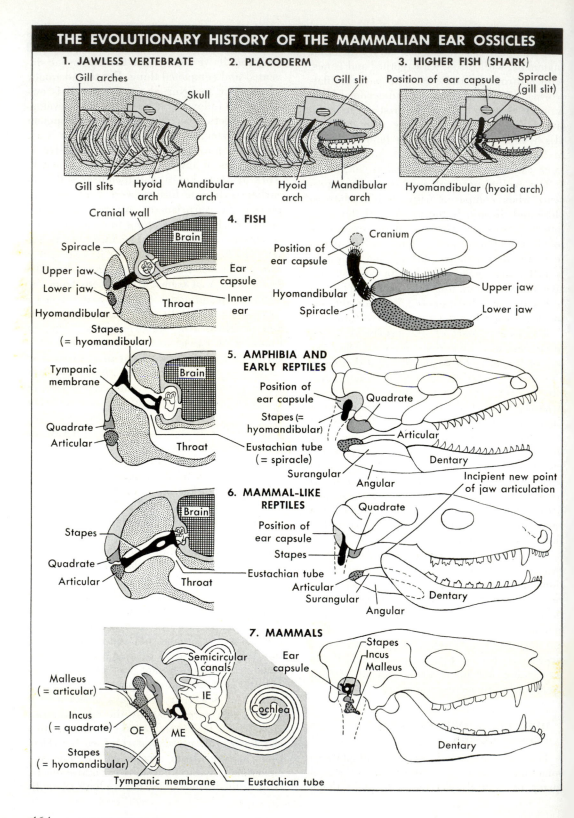

THE EVOLUTIONARY HISTORY OF THE MAMMALIAN EAR OSSICLES

1. JAWLESS VERTEBRATE

Gill arches
Skull
Gill slits
Hyoid arch
Mandibular arch

2. PLACODERM

Gill slit
Hyoid arch
Mandibular arch

3. HIGHER FISH (SHARK)

Position of ear capsule
Spiracle (gill slit)
Hyomandibular (hyoid arch)

4. FISH

Cranial wall
Brain
Spiracle
Upper jaw
Lower jaw
Ear capsule
Inner ear
Hyomandibular
Throat
Stapes (= hyomandibular)

Position of ear capsule
Cranium
Hyomandibular
Upper jaw
Spiracle
Lower jaw

5. AMPHIBIA AND EARLY REPTILES

Tympanic membrane
Brain
Quadrate
Articular
Throat

Position of ear capsule
Quadrate
Stapes (= hyomandibular)
Articular
Dentary
Eustachian tube (= spiracle)
Surangular
Angular
Incipient new point of jaw articulation

6. MAMMAL-LIKE REPTILES

Stapes
Brain
Quadrate
Articular
Throat

Position of ear capsule
Quadrate
Stapes
Eustachian tube
Articular
Surangular
Dentary
Angular

7. MAMMALS

Semicircular canals
Ear capsule
Malleus (= articular)
IE
Stapes
Incus
Malleus
Incus (= quadrate)
OE
ME
Cochlea
Stapes (= hyomandibular)
Tympanic membrane
Eustachian tube
Dentary

fied a distinct kind of evolutionary opportunity that arises from anatomical accidents like the physical proximity of jawbones and ear.

The history of the ear offers a general warning. In seeking a scientific explanation of the apparent design in organisms, we waste time if we look for an agent conscious of its goal or oriented toward the future. Our task is rather to unravel a historical succession of transient opportunities to meet needs. By the time we have traced an adaptation like the human ear, our feeling that it must have been purposely created is gone. Formerly regarded as an architectural masterpiece, it now seems more like one of those improvisations with pulleys, tilting buckets, and string-tied joints that imaginative cartoonists delight in drawing. But *this* improvisation works, and under the control of natural selection it works better in each successive stage of its evolution.

16–13 The evolutionary history of the mammalian ear ossicles. *1*. In the earliest jawless vertebrates (Agnatha, Chapter 22), the wall of the throat, or pharynx, is perforated by gill slits and supported, between the slits, by a series of skeletal elements, the gill arches. The first two of these are the mandibular and hyoid arches, respectively. The jaws of all later vertebrates are evolutionary transformations of the first (mandibular) arch. *2*. In the Placoderms (Chapter 22) the jaws are quite separate from the cranium, or brain case, and are attached to it only by ligaments. *3*. In later fishes, like the shark, the second (hyoid) arch has been moved forward, its upper element (the hyomandibular) being used as a brace for the jaw apparatus. At one end the hyomandibular attaches to the cranium at a point close to the ear capsule, the bony cavity housing the inner ear; at its other end the hyomandibular connects with the point of jaw articulation, firmly bracing it to the skull. In the hyoid arch's evolutionary movement forward, the first gill slit has been forced into a position above the jaws, near the ear capsule, and has been reduced. It is now the spiracle and is clearly visible behind the eye in the ray (close relative of the shark) pictured in Fig. 22–16. A highly schematic cross section through one side of the skull of a fish shows how the jaw articulation is braced to the skull by the hyomandibular at the ear capsule and how the spiracle opens to the exterior above the jaws. A right-side view further details the structure. *5*. In Amphibia and early reptiles the principal innovation is that the upper jaw has fused with the cranium; it is not a separate element, as it is in fishes. The main evolutionary consequence of this reorganization has been the liberation of the hyomandibular from its earlier function as a brace to the jaws. Its proximity to both the spiracle and the ear capsule has been exploited in a radical of function. It has come to lie in a middle ear cavity, derived from the early gill slit and the later spiracle. The part of the spiracle communicating with the throat has become the Eustachian tube. The middle ear cavity is closed off at the exterior by a tight membrane, the tympanic membrane or eardrum. Sound waves falling on the membrane cause it to vibrate, and the vibrations are transmitted to the inner ear by the hyo-mandibular, now the stapes. The stapes lies in the middle ear, attaching at one end to the tympanic membrane and at the other end to a membrane that covers an opening (the oval window, Fig. 13–9) into the inner ear; it is an efficient mechanical device to communicate sound vibrations from the surface of the skull to the auditory sense organ located deep in the head. The jaws themselves comprise many bones. The lower jaw, for instance, includes the dentary (bearing the teeth) and articular (articulating with the upper jaw) bones. The articulating bone of the upper jaw is the quadrate bone. (In *5*, *6*, and *7*, for simplicity, only the articular and quadrate bones are given the characteristic shading that shows their relationship with the whole of the lower and upper jaws in *1*, *2*, *3*, and *4*.) *6*. The condition of the mammal-like reptiles (actual ancestors of the mammals) is substantially the same as that of the Amphibia and earlier reptiles. However, two changes point the way to the ultimate condition of the mammals. First, the middle ear and tympanic membrane have shifted, the membrane now being very close to the point of quadrate-articular articulation. Second, the jaw itself is developing a new articulation with the skull: the end of the dentary bone is curved upward and is beginning to make contact with the cranium near the original quadrate-articular joint. *7*. In mammals the new articulation is complete, and the lower jaw consists solely of the dentary bone; the other former jawbones have assumed new functions or been lost. The articular bone has become the malleus. The quadrate bone has become the incus. Sound is now transmitted from the tympanic membrane to the oval window of the ear capsule by a chain of three bones —malleus, incus, and stapes. Thus, like the hyomandibular, the articular and quadrate bones are freed from one function to perform a new function. In their history there was no sharp point of functional switchover. Instead there was a period when, while still involved in jaw movement, they also helped to transmit sound vibrations to the nearby ear mechanism. Their transformation to an exclusively auditory role was an exploitation of an anatomical opportunity—the accident of their location near the middle ear.

Natural Selection and Inherited Information

We have repeatedly emphasized the fundamental problems posed for the biologist by the fact of life's complex organization. We have seen that organization requires work for its maintenance and that the universal quest for food is in part to provide the energy needed for this work. But the simple expenditure of energy is not sufficient to develop and maintain order. A bull in a china shop performs work, but he neither creates nor maintains organization. The work needed is *particular* work; it must follow specifications; it requires information on how to proceed.

Our treatment of the subject of genetics was presented in this light. We envisaged embryological development as leading to complex adult organization, and the study of heredity proper as the search for the inherited information that specified how the work of development must proceed. This search led us to the nucleus and its chromosomes, which proved to be the carriers of the inherited specifications ultimately responsible for the organization of the living system.

In showing that the organization of living matter is controlled by information in the chromosomes, genetics provides only the beginnings of a full explanation. We need to know not only where the information is and how it is decoded by the organism, but *how it got there.* The answer to this final question is given by the historical processes of evolution that we have reviewed.

The processes of mutation introduce new modifications into the inherited instructions; genetic recombination always reshuffles the variations of the inherited information that exist among the individuals of a population. Many of the variants thus produced in the chromosomal instructions are disadvantageous; they distort an otherwise clear and appropriate set of inherited specifications. In that case, however, they never persist long in the succession of generations, for their very inappropriateness guarantees their reproductive inefficiency. Natural selection keeps the population's inherited patterns in good repair. But it does more than that. *Some* of the novelties in the message *happen* to specify a more appropriate organization—a better-adapted organism—than did the original instructions. As a result of natural selection this more appropriate information ultimately becomes the prevalent pattern throughout the population. Thus natural selection is the agent that *created* the coded information pattern in the first place as the appropriate set of specifications which guarantee the adaptive organization of living things.

CHAPTER SUMMARY

Evolution and the problem of purpose: the impact of the knowledge of evolution on human orientation; the interdependence of science and philosophy; testability as the touchstone for scientific questions and answers; the purposive aspect of adaptation open to scientific examination.

Nature of adaptations: a definition; the diversity of adaptations—morphological, physiological, behavioral; broad and narrow adaptations; individual adaptability.

Inheritance of acquired characters: an alternative to natural selection as a hypothesis to explain adaptation; the misleading label "Lamarckism"; Neo-Lamarckism versus Neo-Darwinism; failure of the hypothesis—incompatible with principles of heredity, not supported by experiments, incapable of explaining many kinds of adaptations (insect coloration, neuter insects), indicative of inadaptive evolution; the success of natural selection where inheritance of acquired characters fails.

Natural selection and adaptive coloration: the cryptic coloration evolved by selective predation; the criticism that supposedly protected species are also preyed on; the refrigerator fallacy; the example of cryptic coloration versus fecundity in moths; warning coloration, the opposite of cryptic coloration but also caused by selective predation; mimicry and other kinds of adaptive coloration.

Natural selection as a historical process: the pressure to diversify generated by intraspecific competition; physical, constitutional, and ecological opportunities as conditions for entry into new environments; adaptive radiation exemplified by Darwin's finches on the Galápagos Islands—the ecological opportunities not previously exploited on the islands and radiation into 14 species in different niches; major radiations—reptiles, mammals.

The evolutionary lottery: evolution in many directions, determinate but not oriented toward the future, with ultimate extinction as the usual outcome; small winners—arthropods, parasites; big winners—the human ancestry; adaptations to ancestral conditions opening new opportunities for descendants.

Historical opportunism versus design: the evolution of the vertebrate ear as an example of the lack of design anticipatory of future needs.

Natural selection as the process that creates the information content of the chromosomes.

17

Variation, Species, and Speciation

Bob Plunkett

The Kaibab squirrel is a species distinct from its close relative, the Abert squirrel. They are now separated by the chasm of the Grand Canyon. This chapter discusses the nature of species and the processes, such as geographic isolation, that lead to new species.

The first reaction of a newcomer to a coral reef or a tropical rain forest is one of confusion. In such places life is most obviously abundant and diverse. The lavishness of nature is overwhelming, and no meaningful pattern is evident at first sight. If perception is not too dulled by familiarity, the same emotion may arise nearer home. A meadow particolored with wild flowers or a swarm of insects around a light has in its own degree the same massing of life and bewildering variety. The diversity of life is a product of the evolutionary process, a product arising from the principles of adaptation. It is basically a diversity of adaptations to different ways of life—a topic we have repeatedly touched upon in preceding chapters. One of the fundamental problems in the study of evolution is how diversity arises. That is the topic of this chapter.

Species: Unit of Population Diversity

Were we to look closely at all the individual organisms on a reef, in a forest, or in a meadow we would find every individual different in at least some minor respect. Armed with the genetic facts we have already discussed, this would not surprise us. On the other hand, we would be more impressed with the fact that the individual organisms seemed to fall into natural groups of *nearly* similar forms. We might notice the differences among buttercups in the meadow, but we would more certainly recognize their similarity. We would perceive that buttercups form a distinct kind of organism, a natural group of similar individuals—a *species*. The differences that would impress us would be those between species, as between buttercups and daisies. The diversity of life is a diversity of populations, and the species is a significant unit of population.

Intraspecific Variation

Diversity begins at the lowest level, with the fact already repeatedly mentioned that no two individuals are ever exactly alike. No matter how small a unit of population we study, there is always variation among its members. That very fundamental generalization lies at the heart of the diversity of life.

SOURCES OF VARIATION

The sources of variation have all been mentioned in previous chapters. Let us briefly review them as background for the present subject.

The characteristics of organisms arise in the course of their development by the interaction of heredity and environment. Heredity determines a reaction range. The circumstances of development (through the entire life span) determine just where in that reaction range an individual's characteristics will actually be. In the long run the differences that count the most are differences in reaction ranges and therefore in heredity. If one man or one tree is taller than another solely because of better diet or soil, this may be an important difference between *individuals* but it is not likely to have much significance for the *populations* to which they belong. Being nonhereditary, such variation cannot be passed on to the next generation; it has no continuity in the population. This does not mean that nonhereditary variations within reaction ranges have no biological or evolutionary significance, especially when reaction ranges are wide, as in man. For instance, religion affects attitudes toward reproduction which in turn affect natural selection and human evolution. The fact remains that there is no long-range evolutionary significance unless hereditary differences also become involved. Can you think of ways in which nonhereditary differences could affect the evolution of nonhuman organisms?

Variation due to differences between reaction ranges is of major biological importance and is the source material for evolutionary change and for the diversity of life. Within a population, the variation arises mainly from the shuffling of genes and chromosomes, especially in sexual reproduction. The different kinds of genes and chromosomes involved in this shuffling originated by mutation.

BELL-SHAPED DISTRIBUTION OF VARIATION IN DEMES

In order to convince yourself of the reality of variation and to learn something of its nature, you should now examine 50 or 100 specimens from one deme (see Chapter 15) of some one sort of organism. (If the specimens were collected in a limited area and over a brief period of time they are probably from a single deme.) These may be available in museum or college collections, but they will be more interesting and instructive if you gather them yourself. Among the innumerable possibilities are flowers, leaves (full-grown, each from a different plant of one species); seeds; shells; ants, grasshoppers, beetles, butterflies, or other insects; or field mice (skins, skulls, or both). Whatever you collect, some measurements of size will be possible. Probably also there are characters that can be counted (petals on a flower, scales of a pine cone, ribs on a shell, etc.). Other characters, perhaps colors and color patterns, may best be noted in words. These are the principal sorts of observations used in studying variation and classifying diversity.

In your sample some features will be the same in all the individuals. After all, it is characteristic of a deme that the organisms in it are similar. For instance, if you collect simple flowers, all will probably have the same number of petals and other flower parts, although they may not. Other characters are sure to vary in your sample. This is especially true of measurements of size and weight or of counts of such multiple parts as ribs on a shell or scales on a snake or lizard. Observations of such characters may be grouped and tabulated in the form of a *frequency distribution*. Examples of frequency distributions of a common sort are given in Table 17–1.

These observations can also be presented pictorially in graphs, as in Fig. 17–1. You see that these graphs have definite patterns, which are similar but not identical in the two examples. In both there is a particular range of values, a *class*, that is most frequent. You might say that this class is the fashion among these animals, and it is called by a name for a fashion; it is the *mode*. On each side of the mode, the frequencies fall off, with fewer and fewer (and finally no) individuals in each class.

Most variable characters that can take any one of a considerable number of values in the individuals of a deme tend to have frequency

TABLE 17–1

Frequency distributions

a. A measurement of size. Tail lengths of individuals from a deme of deer mice.

Measurements in millimeters	Numbers of individuals (frequencies)
52–53	1
54–55	3
56–57	11
58–59	18
60–61	21
62–63	20
64–65	9
66–67	2
68–69	1

b. A count of multiple parts. Numbers of scales (scutes) along the tails of individuals from a deme of king snakes.

Number of scales	Numbers of individuals (frequencies)
38–40	3
41–43	10
44–46	17
47–49	15
50–52	8
53–55	5
56–58	2

distributions similar to those of Fig. 17–1. This is the most important single generalization regarding variation in demes. The pattern approximates a bell-shaped mathematical curve, one which is a member of a particular family of curves that is known collectively as the normal curve of probability. Some differences from the precise mathematical curve always occur, because nature is not as tidy and regular as mathematicians, but the correspondence of a nearly symmetrical distribution (like Fig. 17–1*A*) may be quite close. This is the basis for some of the essential statistical methods in the technical study of variation. Some frequency distributions are distinctly lopsided (like Fig. 17–1*B*) or may otherwise differ from a normal curve. In statistical analysis the differences are also studied and may reveal important facts about variation in a deme.

It is easy enough to demonstrate that variation in demes often has a bell-shaped

frequency distribution. In your own sample you can almost certainly find some variation that has this sort of pattern. The fact is interesting, but it does not have much real scientific meaning unless we can link it with biological principles. What does the bell-shaped pattern mean? We know that variation may result either from hereditary differences in reaction ranges under similar conditions or from different environmental conditions interacting with similar reaction ranges. Either or both of these principles may be involved in the bell-shaped distributions.

If organisms with similar or identical genotypes (Fig. 17–2*B*), and hence similar or identical reaction ranges, develop in similar (but not identical) conditions, the phenotypes

17–1 Frequency distributions.

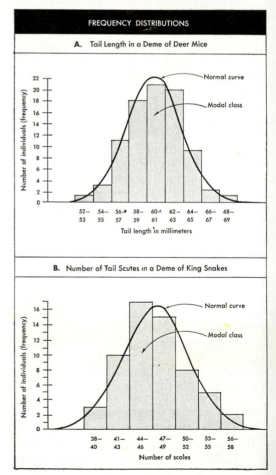

will tend to be similar. Most of them will fall into a modal class representing the usual interaction of genotype and conditions of development. Conditions of development producing phenotypes far from the modal class will be comparatively rare. Thus the effects of environment often tend to produce a bell-shaped distribution even apart from hereditary variations. This can be checked by comparing mature leaves from a single plant, actually parts of one individual and identical in genotype. Their variation in size usually has a bell-shaped distribution.

On the other hand, we have seen that characters influenced more or less equally by several or many genes also tend to produce a distribution that you now recognize as like a bell-shaped distribution in a deme. In the experiment previously studied (Chapter 6), crossing of large and small parents produced a broad, bell-shaped distribution in F_2. In natural populations with the same genes, the same sort of distribution results from interbreeding in a deme and tends to persist through the generations. When a distribution of this pattern (Fig. 17–2A) has a genetic basis, the most probable inference is that the genetic system is one with several or many genes affecting the character being studied.

Thus either environmental or hereditary factors may produce a bell-shaped distribution. In most instances both are involved simultaneously. It is difficult to disentangle the two, but this can be done, approximately, at least. One way is to conduct an experiment like that just mentioned, crossbreeding extreme variant individuals from a deme. If the results are similar to those of that experiment, hereditary variation is present, and further analysis can determine its nature and extent. Another way is to perform selection experiments of the type shown in Fig. 17–3. Of course, it is impractical to conduct breeding experiments on many of the millions of demes present in nature, and such experiments are wholly impossible for the even more numerous demes now extinct. Still it is possible to get an approximate idea of the amount of truly hereditary variation by the statistical analysis of variation within demes and among demes living under similar and different conditions.

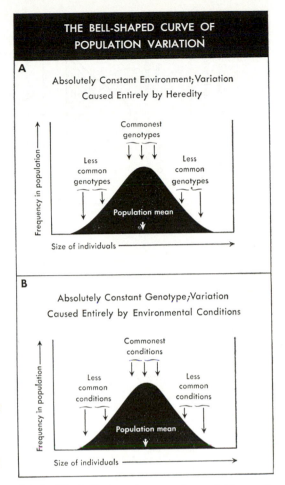

17–2   The bell-shaped curve of population variation. A bell-shaped distribution of phenotypic characteristics (size, weight, color, etc.) in a population may be due theoretically to either (A) variation among the genotypes of the population's members or (B) variation in the environmental conditions encountered by the population's members. The bell-shaped distributions of variation encountered in natural populations are nearly always the result of variation in both environmental conditions and individual genotypes.

POLYMORPHISM

There are white blackbirds, and they exemplify another aspect of variation in demes. We have just considered variation that seems to intergrade continuously or that has a sequence of numerous classes, usually with a bell-shaped distribution. There is also variation with few classes, often strikingly different forms which may occur in varying pro-

THE EFFECT OF SELECTION ON POPULATION VARIATION

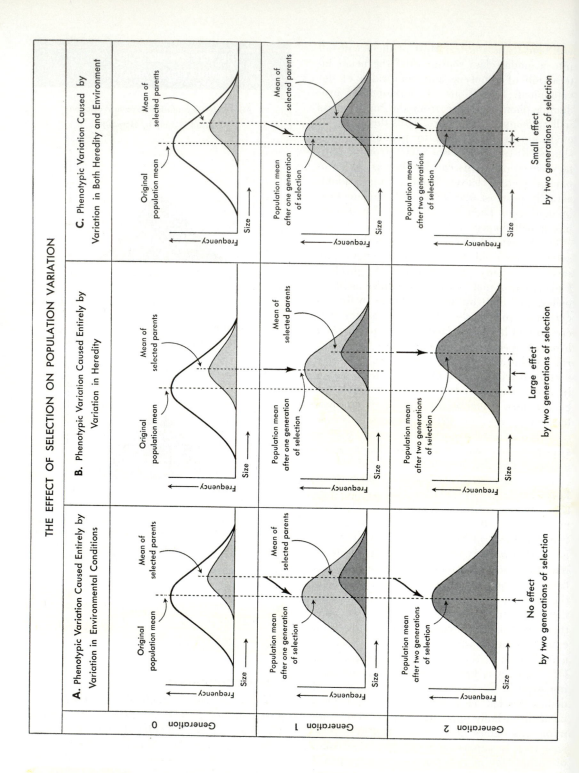

portions in different demes. White and black blackbirds are such decidedly distinct forms, and this particular variation has only two classes. Black blackbirds are the rule, and white blackbirds are rare everywhere in nature. Among some other birds a mixture of dark and white forms is usual in demes. There are herons among which demes normally include both gray and white birds, which interbreed. Sometimes there are about three white to one gray, a suggestive approach to the Mendelian ratio for combinations of equally numerous dominant and recessive alleles. In some bitterns (relatives of the herons), brown, white, and black (or unusually dark brown) forms commonly occur within single demes. On the other hand, some species of herons are always white and some always colored in the same way.

The natural occurrence of two or more sharply distinct forms in a single deme is *polymorphism* ("many forms"). Polymorphism is a widespread sort of variation and may be seen, in some respect and to some extent, in most demes. It involves not only color but many other characters. For instance, a deme of snails may be polymorphic in its

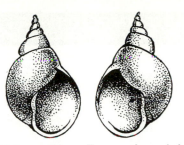

17–4 Polymorphism. Two members of the same snail population differ in the direction of shell coiling; one shell has a right-handed coil (the common condition), and the other has a left-handed coil.

direction of coiling, some individuals having shells coiled left-handedly and some shells coiled right-handedly (Fig. 17–4). Most human populations are polymorphic in blood types, which do not show but which are sharply distinct biochemically. The number of forms in a deme is usually small: two (colors of some herons, direction of coiling in some snails), three (colors of some bitterns), or four (the human A, B, AB, and O blood types). Less commonly many forms may occur; more than 120 distinct patterns have been found in a single species of platyfish (small tropical fish often kept in home aquariums).

What underlies polymorphism? You would not expect it to depend usually or mainly on environmental differences alone. That would mean either that a single reaction range included two or more sharply different phenotypes, or that members of one deme lived in two or more likewise sharply different environments. Either could happen, but both are unusual in nature. In fact, many breeding experiments have been made with the different forms of polymorphic demes, and it has been found that this sort of variation is usually hereditary.

In Chapter 6 you learned that alleles of single genes may accompany striking and sharply distinct differences in particular characters. Such were the characters studied by Mendel, and these are the genes of classical Mendelian experimentation. Polymorphic characters in demes are like the flower colors and seed characteristics of Mendel's peas. They are determined by one or a few genes, each with two or a small number of alleles in various members of the population. They con-

17–3 The effect of selection on population variation. *A.* When population variation is caused entirely by variation in environmental conditions, selective breeding from one generation to another will fail to change the distribution of population variation. Generation 1, the progeny bred from a select group of large individuals of generation 0, will show (if raised in the same range of environmental conditions) the same distribution of phenotypes that characterized generation 0 as a whole. *B.* The result is different when environmental conditions are constant and the population variation reflects an array of different genotypes in the population. In this case the mean size of generation 1, bred from selected parents, is considerably larger than the mean size of generation 0 as a whole. In fact, it is the same as the mean size of the selected parents. *C.* When both hereditary and environmental variations contribute to the variation of a population's phenotypes, selection leads to a generation 1 whose members are larger than those of generation 0, but the effect is relatively slight. The mean size of generation 1 is smaller than that of the selected parents. The reason is clear: although he carefully breeds from large individuals, the selector cannot avoid picking some individuals that are large because of favorable environmental factors and in spite of relatively poor genotypes. (Adapted from A. M. Srb and R. D. Owen, *General Genetics*, Freeman, San Francisco, 1952.)

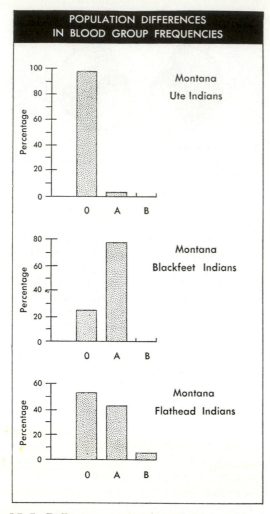

POPULATION DIFFERENCES IN BLOOD GROUP FREQUENCIES

Montana Ute Indians

Montana Blackfeet Indians

Montana Flathead Indians

17–5 Differences among populations in blood group frequencies. Each of three distinct populations of Montana Indians is characterized by the relative frequency (as percentage) within it of the three blood group alleles O, A, and B.

trast with characters determined by the interaction of larger numbers of genes, which usually have a bell-shaped distribution in demes. This is not an absolute distinction. There is a continuous scale of intensity of gene action and of single to multiple gene effects on phenotypic characteristics. Similarly, polymorphic and bell-shaped distributions intergrade in populations. They are, however, distinct in their more extreme or characteristic forms as well as in convenient methods for studying them and in some of their implications for evolution.

DIFFERENCES BETWEEN DEMES

Within a species divided into more or less distinct demes, as most species are, there are two kinds of variation. One is variation within any one deme, some features of which have been mentioned. The other is variation of the species in the form of differences between demes, for two demes are not likely to be precisely similar.

In species with biparental reproduction, there is occasionally and often regularly interbreeding between adjacent demes. Offspring from one deme may move into the other, a factor present in both biparental and uniparental species. For these reasons, adjacent demes almost always intergrade and are seldom completely distinct from each other in any characteristic. One deme may have members with larger *average* size than a neighboring deme. Probably some individuals in the first will be larger than any in the second, and some in the second smaller than any in the first. Nevertheless, there will be a considerable range of size represented in both demes. If size has a bell-shaped distribution, as it generally does, variation and intergradation of this sort between demes will be shown by an overlap of the frequency distributions.

Adjacent demes often differ but still usually intergrade in polymorphic characters. It is rare for the individuals of one deme to be all of one form and those of another, adjacent deme all of a different form. If a character takes two forms, adjacent demes will usually have both forms. If the demes differ in this character, the difference is, as a rule, in the percentages of individuals of each form. Variation of this sort between demes is exemplified in Fig. 17–5. The basis for these relationships is that the same genes and alleles are present in the adjacent demes but that some alleles are more frequent in one deme than in another.

Two demes of the same species do not occur in just the same place at the same time.[1] If they did, the demes would quickly fuse and become one. Variation between demes therefore has geographic patterns. The patterns are of great interest because they throw light on many problems such as the genetics of popula-

---

[1] There are a few exceptions, but even in them the occurrence is temporary.

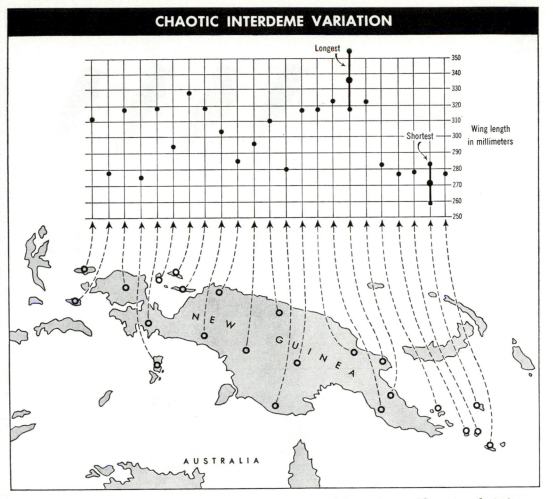

17–6 Chaotic variation in wing length among demes of the cockatoo (*Cacatua galerita*) on the island of New Guinea. The mean wing lengths in 23 demes of the cockatoo are plotted as solid dots on the graph above the map of New Guinea, which locates each deme. The variation is significant, as is indicated by the relatively small ranges of variation (solid vertical bars) in the longest-winged and shortest-winged populations. There is no obvious trend or pattern in the variation among the demes.

tions, the relationships of populations with their environments, and the origin of species and other units of classification.

Some geographic patterns of variation are quite irregular. Differences among demes of some species seem to be scattered about in checkerboard or kaleidoscopic fashion, without particular rhyme or reason, as in Fig. 17–6. Of course, there are reasons for this, although they are often difficult to determine. Such irregular patterns seem to have one of two causes, or a combination of the two. They may represent a scattering of small demes with little interbreeding between adjacent demes. Then each deme may have established its own characteristic mutations and genetic systems, kept distinct by the reproductive isolation of the deme. Or the patterns may represent local adaptations of demes to irregularly distributed environmental conditions. Then an irregular pattern reflects an irregularly heterogeneous environment. Highly irregular patterns are seldom found unless both factors apply to some extent, that is, unless the environment lacks uniformity and there is little interbreeding between demes or groups of

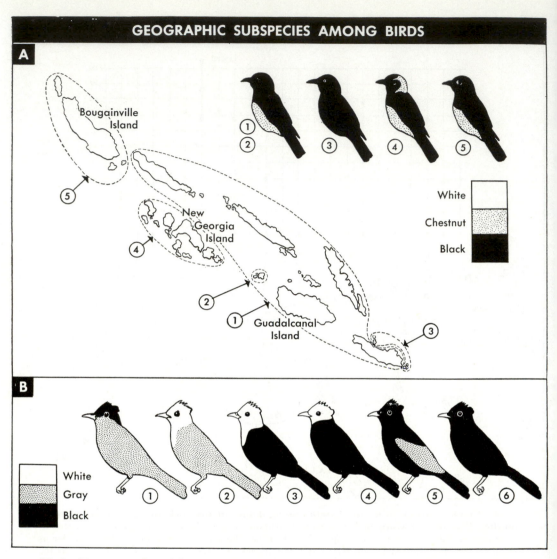

17–7 Two examples of bird species subdivided into distinct subspecies. *A.* Geographic subspecies, and their distribution, of the flycatcher *Monarcha castaneoventris* in the Solomon Islands. Four color patterns (involving white, chestnut, and black) are found among the five subspecies. *B.* Six subspecies of the Asiatic bulbul (*Microscelis leucocephalus*), distributed from India through China.

demes. This is another instance of interaction between the environment and genetic factors, the genetics in this case being that of populations.

It is unusual for each deme to be sharply distinctive or for the differences among demes to be highly irregular. Usually all the demes over a considerable area are closely similar. Such similarity is maintained by interbreeding between demes. Interbreeding over the whole area may, indeed, be so common that in effect the whole regional population is a large unit, a single deme or a complex without distinct division into demes. The large deme or complex of similar demes may constitute the whole of a species. More commonly there are two, several, or many such complexes, their demes also similar among themselves but different from those of other complexes. In classification, the large demes or deme com-

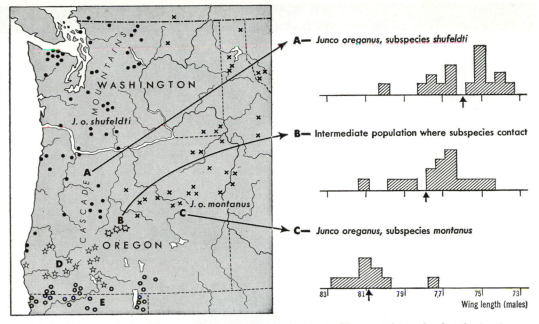

17–8 Intergradation where subspecies come in contact. The map shows the distribution in Oregon and Washington of three subspecies of the small snowbird *Junco oreganus* and of intergrading populations between subspecies. Each symbol identifies a locality from which samples have been studied. The three subspecies are as follows: **A** (• symbol), *J. o. shufeldti*; **C** (x symbol), *J. o. montanus*; **E**, (o symbol), *J. o. thurberi*. **B** (✿ symbol) represents intermediate populations between the two neighboring subspecies *shufeldti* and *montanus*. **D** (☆ symbol) represents intermediate populations between *shufeldti* and *thurberi*. Frequency distributions for male wing length are given for *shufeldti*, *montanus*, and the intergrading population (**B**), where these two subspecies meet. Population means are indicated by the arrows on the segmented line.

plexes in such a situation are usually designated *subspecies* (Fig. 17–7). Where two subspecies of one species are in contact, there is usually a zone of intergradation, due to interbreeding, but away from this zone the subspecies may be quite distinct in their characteristics. Fig. 17–8 illustrates this frequent sort of distribution of variation within a species.

Zones of intergradation between demes (or groups of similar demes) are not always narrow and well-defined as in the last example. In fact, a whole regional population may intergrade from one end of its distribution area to another. For instance, southern and northern individuals of a species may be quite different, and yet when the population is followed from south to north there may be no definite line or zone where the change is localized. Or differences may appear gradually as populations are followed from lower to higher elevations (Fig. 17–9) or from wetter to drier situations.

Continuity and complete intergradation imply that the interbreeding of adjacent demes is common, and that the gradual change is usually correlated with gradients of environmental conditions.[2]

Environmental gradients are common in nature. Every mountain range has a gradient from bottom to top, from warmer to colder, and usually simultaneously from drier to wetter, with gradients also in other climatic conditions and in soils. Plains or lowlands have similar gradients from south to north. Gradients of temperature, of salinity, and of light occur in lakes and seas. These gradients in physical environments are often accompanied by gradients in characteristics of the organisms inhabiting them. The gradients may be similar in a number of different species. There are, for example, three famous rules (gener-

[2] The technical term for a sequence of poorly separable demes with gradual, regular change from one area to another is *cline*.

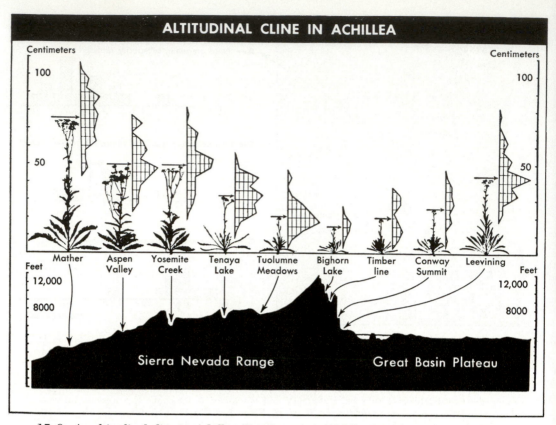

ALTITUDINAL CLINE IN ACHILLEA

17–9 An altitudinal cline in *Achillea*. The plant species *Achillea lanulosa* occurs at all altitudes in the Sierra Nevada Mountains of California, exhibiting considerable adaptive intraspecific variation in relation to the different environments at various altitudes. Populations at lower altitudes are taller and those at higher altitudes are shorter. Beside the plant representative of each altitudinal population is a graph showing the distribution of height variation within the population.

alizations with some exceptions) that apply to many mammals and birds:

1. Within any one species, the average size of the individuals tends to be smaller in warmer climates and larger in colder climates.[3]

2. Within any one species, protruding parts such as tails, ears, or bills tend to be shorter in colder climates than in warmer climates.[4]

3. Within any one species, colors tend to be darker in warm, moist climates and lighter in cold, dry climates.[5]

There are numerous other rules of this kind that apply to various groups of animals and plants. There is nothing especially mysterious

[3] This is commonly known as Bergmann's rule.
[4] Allen's rule.
[5] Gloger's rule.

about such rules. They are only special examples of a broader and profoundly important generalization: *Differences among demes tend to be correlated with differences in their environments.*

This is not true, or at least it is not definitely known to be true, of all differences among demes, but it is much more often true than not. The reason is that the environment-correlated distinctive characteristics of a deme help to fit it for more successful living in its particular environment. Among homeo-thermous animals heat production and conservation are more efficient in larger than in smaller individuals. That established fact underlies the greater fitness of larger animals in colder climates and explains the first of the three rules given above. The reason for greater fitness of dark-colored animals in warm, wet climates is more complex and is not surely established, although some likely hypotheses have been proposed. Can you think of one? What is a possible adaptive meaning of the second rule given above?

Another way to state the generalization about differences between demes is this: variation among subspecies or local populations of a species is largely adaptive. Most students of the subject agree that some non-adaptive variation also occurs in species and may even be common *within* demes, where variation involves *particular* characteristics of *individuals*. It is, however, probable that most of the variation represented by differences *between* demes is adaptive, for there variation involves *average* characteristics of natural *populations* of individuals.

Species and Speciation

THE SPECIES PROBLEM

It has already been noted (Chapter 15) that the basic unit of population in nature is the deme. There are still smaller units, such as a family consisting of parents and their offspring. Such units are obviously very temporary. They are distinguishable only for a generation or so. Their distinction and naming are not practical or significant for the broader study of the evolution of populations and the diversity of life. The same situation often applies to demes, which are not likely to persist indefinitely as distinct units. The smallest persistent unit in nature is the species.

All organisms are classified into species. Biological specialists from field naturalists to geneticists or biochemists recognize that a species is a very special kind of thing and one with fundamental significance for the study of life. All biologists think that they have a pretty good idea of what a species is, at least among the particular organisms on which they work. Yet for centuries biologists have been battling and baffling each other about "the species problem." The problem is simply to produce a clear, fully satisfactory answer to the question, "What is a species?"

Millions of words have been written in discussion of the species problem, and the subject is still just as lively today as it ever was.

This sounds very distressing and perhaps stupid: all agree that something is absolutely fundamental in their science, but they cannot agree as to exactly what it is. The situation is not really as bad as that, however. Two competent biologists may disagree heartily on the precise definition of a species, but if they look at the same populations in nature they will probably agree nine times out of ten on which groups are and which are not species. They may wrangle over the tenth group, but even then they usually find that they are saying much the same thing in different words. Everyone agrees on some of the characteristics usually defining species, and there is a modern consensus as to the kind of thing a species is.

It is one of the facts of life that an exact definition of a species applicable without question to all sorts of organisms is inherently impossible. It is no waste of time to discuss what species are; on the contrary, that is one of the most important subjects in the whole science of biology. It is, however, a waste of time to try to agree on one, infallible definition of species. In the first place, there are many ways, all valuable, to approach the subject. A geneticist naturally thinks in terms of breeding and an anatomist in terms of structure. In the second place, population units tend to be of different sorts among the tremendously varied products of evolution. Why should we expect to find precisely the same sorts of units among, say, yeasts, tapeworms, bees, and primroses? Third, natural populations are not static things that stay put neatly within the confines of rigid definitions. They are constantly changing, splitting up, reuniting, becoming more or less similar to each other, expanding, contracting, acquiring new habits, discarding old structures, and, in a word, *evolving*.

We have already (Chapter 15) characterized a species as a group of organisms so similar in structure and heredity that their demes intergrade, may fuse, and may take the place of one another without essential change in the nature and role of the group as a whole. A species may not be clearly subdivided into smaller units that can be called demes. Usually it is so subdivided, and then the point is that the demes do not necessarily have evolutionary continuity as distinct units. A species does tend to have long evolutionary continuity and distinctiveness. That, really, is what makes it such a significant unit. The species may then be defined *as a sequence of ancestral and descendent populations evolving independently of others and with its own separate and unitary evolutionary role and tendencies.* It is an essential part of the concept that a species is a sort of population, a group, the individuals within which are rather closely related to each other and therefore more or less similar in essential characteristics, although there is always variation among them. The kind and degree of variation are important features of a species. A species is not a group of individuals all of which have the same pattern; that is a non-evolutionary and old-fashioned idea, as will be evident when we discuss classification in Chapter 18.

The thing that maintains a species as a unit among biparental organisms is interbreeding. As long as the individuals of a group can interbreed, producing fertile offspring, and as long as they do so with some frequency, the whole group shares in a genetic pool and tends to have the unity and continuity that we have noted as characteristic of species. Thus we have the basis for another definition: *a species is a group of actually or potentially interbreeding natural populations that are reproductively isolated from other such groups.* This definition, stressing the genetic structure of a species, does not conflict with the previous definition, which stresses its evolutionary role. Provided that it is biparental (as most species are), the evolutionary species as seen *at any one time* almost always corresponds with the genetic species.

Neither definition demands that two distinct species never interbreed. In some groups of organisms hybrids between two closely related species are rather common. The genetic distinction of the species and their integral evolutionary roles are nevertheless maintained if breeding between them is decidedly less common than breeding within each one separately, or if the hybrids are distinctly less fertile than the offspring of parents both of the same species.

Of course there is no absolute distinction

between separate demes of one species that interbreed more freely and produce more fertile offspring, and separate species that interbreed less freely and produce less fertile offspring. The two sorts of groups intergrade in nature. The intergradation is part of the process of evolution. That is a reason why a species is not really an absolutely delimited unit and why it is futile to try to define it as such. But if the process goes a little further, until all the hybrid offspring are completely infertile or interbreeding can never occur, then the delimitation has become absolute and no one would seriously question that the two groups are different species.

SPECIATION

Chapter 15 dealt with the genetics of populations and how populations change. Such changes are part of the basis for the diversity of life. Obviously two species, necessarily derived from a common ancestry some time in the past, will not be different unless one or both have changed. It is also obvious that there will not be two species rather than one unless there has been another sort of change, a splitting of one group into two. Changes in a continuous population make for differences between earlier and later forms, but they do not increase the number of different populations or species. Another basis for the diversity of life is thus necessarily the splitting of a population into two or more. An eventual outcome of this process is adaptive radiation, discussed in the preceding chapter. We did not there consider just how the process goes on at the basic level where one species divides into two or more, and that is the topic to which we now turn.

There has been a great deal of argument as to whether speciation is a sudden or gradual process. Some early evolutionists thought that new species might normally arise as individual "sports" by what they called "saltation," literally, "jumping." The pioneer geneticist De Vries decided that this concept fitted in with his studies of mutations. He and some other early geneticists decided that species normally arise by "big" mutations—those with rather striking phenotypic effects. But a species is not a mutant or otherwise new form or type of organism. It is a group of organisms, a population. The population geneticists, developing the subject after the days of De Vries and his contemporaries, found that a single mutation very rarely leads to a new population. Indeed, it is usually impossible for it to do so. The mutant form may not breed at all, or it may breed with other members of the population in which the mutation occurred, in which case the mutation may or may not spread in the population in accordance with the principles treated in Chapter 15. In any event, the population continues to be the *same* population even if the mutation does eventually change some of its characteristics. Therefore, speciation cannot occur.

There is only one important class of exceptions to the rule that speciation does not occur by saltation or mutation. It has already been mentioned (Chapter 6) that chromosome mutations and hybridization resulting in polyploidy can produce individuals that do not interbreed with their parental species but may produce populations among themselves. This has happened often enough among some groups of plants to have had considerable effect on their diversification. It has happened only rarely among animals, and even among plants as a whole it is the exception rather than the rule. As far as is known, gene mutations never give rise directly to new species among biparental organisms, although they might possibly do so in asexual groups.

Geneticists and other biologists have now returned to the view that was held by Darwin: speciation is usually a gradual process. Demes or groups of similar demes (such as the subspecies of classifiers) often develop somewhat distinctive genetic characters. This may be just a matter of having different allele ratios in their genetic pools. There is somewhat freer interbreeding within these groups than between them. Usually that is all that happens. The groups continue to be subdivisions of a species and do not become so distinctive as to be considered separate species. Sometimes, however, interbreeding between groups becomes less and less frequent, and its resulting hybrids less and less fertile, or both. If this process continues, the group eventually become different species, although there is no sharply definable point at which speciation can be said to have occurred.

The key processes in evolution down at the level of the populations are thus two: genetic change within populations; and the splitting up of populations by sudden or, more commonly, gradual decrease of interbreeding between demes or other units of populations.

ISOLATING MECHANISMS

Anything that decreases interbreeding between groups or organisms is called an *isolating mechanism.* Isolating mechanisms are of many different kinds, and some of them are very curious.

The most generally effective isolating mechanism is simply space or geographic separation. Some students believe that speciation rarely or never occurs (always excepting new polyploid species of plants) unless the groups becoming isolated are in different areas. There may be rare exceptions, but certainly spatial separation is usually involved. Sooner or later other isolating mechanisms also arise, even if the effective isolation is purely geographic at first. Two nonmigratory animals a hundred or a thousand miles apart obviously cannot interbreed, and two plants at such distances are most unlikely to do so.[6] If plants or animals of the same species occur fairly continuously in the intervening region, there is still little isolation as far as the whole population is concerned. Individuals at opposite ends of the occupied region cannot interbreed directly, but they can pass on genes to any part of the population in the course of reproduction over a few generations. Often, however, there are spatial breaks in the distribution. Then there is less interbreeding across the gap, and in time the populations on the two sides of it may become different species.

There are rather similar populations of tuft-eared squirrels on the north and south rims of the Grand Canyon (Fig. 17–10). They must originally have come from one population, but in their present positions they seldom or never interbreed because they do not cross the canyon. They have become visibly different: the northern squirrels have darker underparts and whiter tails. The difference is not yet very great, but the two populations are usually considered distinct species, Abert squirrels on

[6] The bare possibility does exist for plants with wind-blown pollen.

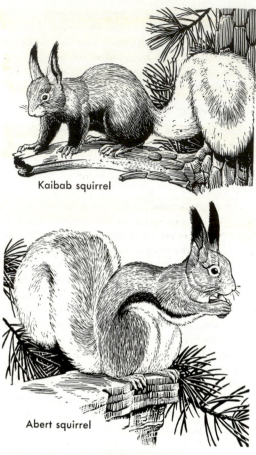

Kaibab squirrel

Abert squirrel

17–10 The Kaibab and Abert squirrels.

the south rim and Kaibab squirrels on the north rim. The distributional break involved in geographic speciation need not be so spectacular as the Grand Canyon. A stretch of grassland between two forests may represent a break between plant and animal populations of the forest. Narrow or wide discontinuities in distribution are abundant, and if they long persist, they frequently lead to specific separation between the populations on each side of them.

Geographic isolation alone is seldom permanently effective in decreasing interbreeding. Even the Grand Canyon might not entirely prevent the squirrels from interbreeding; some might cross, and populations can spread around the ends of the main barrier. (Also in broad geological view, even so tremendous a barrier as the Grand Canyon is only temporary.) In almost all instances of

speciation, biological as well as geographic factors eventually decrease interbreeding. It is the biological isolating mechanisms that are strangest and most varied.

Interbreeding between otherwise similar populations in the same region is often reduced simply because they have slightly different habitats in that region. In Florida there are two groups of turtles that do interbreed to some extent and are considered as belonging to the same species, but interbreeding is reduced by the fact that one group prefers to live in lakes and ditches and the other in running rivers. Anyone who has ever climbed a mountain knows that both plants and animals tend to live at characteristic elevations. Interbreeding between similar populations may be reduced by their preferences for different altitudes.

Interbreeding may also be reduced or even eliminated if animals have different breeding seasons or plants produce pollen at different times. In eastern and central United States there are two groups of common toads, considered distinct species. They can hybridize and occasionally do, producing fertile offspring. The populations as a whole are, however, kept quite distinct by the fact that one breeds early in the season and the other late.

In higher animals, especially insects and vertebrates, sexual isolation is frequently effective. Males and females of different populations simply do not like each other, or perhaps it would be more strictly scientific to say that they do not effectively stimulate each other sexually. Choice of partners may be exercised by males, females, or both. Some experiments with flies (*Drosophila* again) suggest that the males do not care but that the females rebuff "foreigners."[7] In some tropical fishes, however, the males do the selecting. The females have distinctive color patterns, and the males pick out the patterns of their own species and even show preferences for certain particular variants of pattern within their species.

Many animals have elaborate courtship procedures. In these species a female does not

[7] As applied to mutant individuals, we have seen an example of this (Chapter 15). Differences between species have been shown sometimes to have the same selective effect as such mutational differences.

breed unless she is properly stimulated by courtship characteristic of the male of her own species. A relatively simple instance familiar to most of us is the strutting display of male turkeys. Some birds go through much more elaborate performances (and so do some men). Even among fishes there may be complex courtship. There are, for example, specific differences in their breeding patterns (Chapter 14). Male sticklebacks build nests and then induce females to lay eggs in them. Differences in the procedure in different species are effective isolating mechanisms, for a female is persuaded only by the performance of her own sort of male. The following are some of the many differences between two species:

The male of one species	The male of the other species
Builds a nest with separate entrance and exit, hanging on water plants.	Builds a nest with entrance only, on the bottom.
Leads the female to the nest in a series of zig-zags and gets her to enter with a little prodding.	Puts on a special mating play in front of the nest and then forces the female in.

Interbreeding in plants is often restricted by the tendency of pollinating insects to visit only one species of plant on a single foraging trip. This sort of isolation is carried to a high degree in orchids, for instance, which commonly have flowers that attract one species of bee or fly to one species of orchid.

The mechanisms so far discussed reduce interbreeding even if the organisms concerned are completely fertile with each other. They are thus of particular importance in early stages of speciation, when the populations are still quite similar and may still be interfertile. Sooner or later another factor enters the picture: genetic isolation, reduction of the genetic capacity for reproduction between two populations. Genetic isolation may set in early in the process or may not become significant until long after the species are fully separated, but it does always occur in the long run. The main reason why most of the diverse sorts of organisms do not interbreed is simply that they cannot produce fertile offspring. Distantly related species usually cannot produce hybrid offspring at all. They are so different genetically that the chromosomes cannot get together and produce a zygote that will develop. More closely related species may produce hybrid offspring, but continued reproduction between them may be reduced or impossible. The hybrids may be inferior in survival capacity, or may have reduced fertility or be entirely sterile. Cats and dogs are both carnivores, but they belong to families that have been distinct for scores of millions of years. Their gametes do not produce hybrid zygotes. Goats and sheep belong to the same family but to long-distinct lines in the family. Their gametes form a hybrid zygote that begins to develop but dies before birth. Horses and donkeys are distinct but rather closely related species. Hybrid zygotes develop normally and produce vigorous, long-lived animals, mules, but the hybrids cannot reproduce. Some crosses of closely related kinds of cotton result in fertile hybrids, but reproduction among the hybrids produces a majority of abnormal, short-lived offspring.

Once genetic isolation is fully established, the evolutionary destinies of the two populations are forever separate. They are unquestionably and irreversibly established as distinct species, and the diversity of living things has been increased. The process has been repeated countless millions of times during the history of life, and it has produced the millions of separate species that fill the world of life.

Interspecific Selection

We previously (Chapter 15) discussed natural selection primarily in terms of single populations. Now we should mention a kind of selection that depends on principles familiar to you but the genetic aspect of which has hitherto been mostly implicit: *interspecific selection*, which is selection not within but between populations or especially species. If species are in contact with each other and if they must share some necessity of life in limited supply, competition will ensue. This generalization, earlier made in other contexts, also has a strong bearing on natural selection. Two different outcomes are possible. In the first case, the competitively more effective species may be more successful in reproduction, and its population may increase while that of the less effective species decreases, usually to local or total extinction. This is natural selection not by the increase of some and the decrease of other particular genetic characters within a population but by the increase of one whole isolated genetic pool at the expense of another.

In the other case, the intensity and direction of natural selection change within each of the two separate genetic pools. Competition intensifies selection, and in each population selection favors variants ecologically least similar to members of the other population and hence least competitive with them. As a result, the two species diverge, evolving until they occupy ecological niches sufficiently distinct so that competition is no longer significant. This is plainly one of the basic factors in adaptive radiation, although it applies only after speciation has occurred and the species in question are genetically isolated. (Why?)

CHAPTER SUMMARY

Species and diversity: the diversity of life as a diversity of populations adapted to different environments; the species (contrasted to the deme) as the significant unit of population diversity.

Intraspecific variation—its sources: variation within a deme—its bell-shaped distribution exemplified, the modal and the less frequent classes, the approximation of empirical bell-shaped distributions of variation to the normal curve of probability, heredity and environment as causes of bell-shaped distributions, polymorphism; variation between demes—continuously varying characters like size and weight, irregular patterns between demes, the usual existence of geographic pattern in the variation between demes, geographic subspecies, interbreeding and the intergradation of subspecies, interdeme variation correlated with different environments (Bergmann's, Allen's, and Gloger's rules).

The species problem: the problem of defining a species—the variety of approaches to a definition, the difficulty in giving a truly general definition because all species are always evolving; the species as the smallest *persistent unit* of population in nature; the general similarity of species members as a result of free interbreeding within the species, whose members therefore share a common genetic pool (a common heredity).

Speciation—the evolutionary process whereby one species population splits into two populations that no longer interbreed: speciation not a single step (saltation) except in the special case of new polyploid, usually plant, species; Darwin's correct view that speciation is gradual; the genetic divergence of isolated segments of a species population (genetic pools gradually changing) until interbreeding is permanently prevented.

Isolating mechanisms preventing interbreeding between populations: the importance of geographic isolation, especially in initiating speciation (the Abert and Kaibab squirrels of the Grand Canyon); the evolution of biological isolating mechanisms—ecological isolation (as in Florida turtles), sexual isolation because of periodism or courtship patterns in animals (*Drosophila* and sticklebacks) and habits of the pollinating insects in plants, hybrid sterility as the ultimate barrier separating species.

Interspecific competition: the outcome either expansion of one species at the expense of the other or divergent evolution until the two occupy different niches.

Part 6

The Diversity of Life

The photograph introducing Part 6 displays a rich assortment of organisms in a drop of sea water. Greatly magnified, it shows several diatoms—photosynthetic protists—as well as crustaceans and the larvae of echinoderms. Diverse as it is, the life in this drop of the sea can only suggest the over-all variety that confronts the biologist when he turns—as he must—to study the total array of different organisms that evolution has brought into existence. The truly immense diversity of life, attested to by the existence of well over a million species today, is bewildering. But bewilderment gives way to understanding as we perceive two general themes running through the vast variety of life.

The first theme is that the diversity of species is essentially one of mode of life: species differ in the adaptations they have evolved to diversified environmental opportunities.

The second theme is that all the different organisms are related to each other, some to greater, others to lesser, degree. Organic reproduction, under the control of heredity, is basically a conservative process. The adaptive innovations and diversity that have come about in life's evolution have rarely obscured, or transformed beyond our recognition, a fundamental and ancient pattern of organization. The trained observer can detect this organizational pattern underlying the more special adaptations wrought by natural selection. And to this extent he can trace the line of evolutionary descent by which an organism has come to its present condition. The elucidation of evolutionary relationships is the basis of the natural classification of organisms by which life's diversity is most effectively rendered understandable.

Chapter 18 deals with the evolutionary principles that underlie the process of classification. Chapters 19, 20, and 21 survey the major groups of organisms with particular reference to the two themes of mode of life and evolutionary affinity. Chapter 22 is entirely devoted to the three groups—mollusks, arthropods, and vertebrates—that represent the culmination of evolutionary change in three distinct directions in the animal kingdom.

18

Principles
of Classification

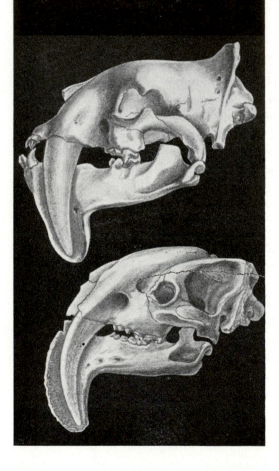

The skulls of these two extinct mammals look alike, but the two species were not closely related. The upper skull is that of a placental mammal (Eusmilus sicarius) *from the Oligocene of North America, and the lower one is that of a marsupial* (Thylacosmilus atrox) *from the Pliocene of South America. Both had enormous, saberlike canine teeth and flanges on the lower jaw that protected the canines when the mouth was closed. These peculiarities were adaptations to a particular way of life; both animals were predacious carnivores. The similar characteristics were not, however, inherited from a common ancestor but evolved separately in the ancestry of each. A principal task in evolutionary classification is to distinguish similarities reflecting only adaptation to like ways of life from those reflecting a common ancestry.*

Variation within demes and species and the nature and origin of species, as discussed in Chapter 17, are fundamental aspects of the diversity of life. Still they leave us far from real comprehension of that diversity. Even if all species were clearly delineated, we would end up with over a million units among living organisms alone. We could not possibly grasp so many different items or see meaningful relationships among them. We need to arrange the units in a more comprehensive, orderly way. In short, we must *classify* them. The most meaningful classification must rest on principles that are biologically significant. The next step, then, is to consider what principles best apply to the problems of classification.

Resemblance and Difference

Probably the most apparent thing about the diversity of life is that some organisms are more and some less alike. Two daisies are so much alike as to be hard to distinguish. A daisy and an aster are still a good deal alike but are easily distinguished. They are both very different from a pine tree. Red and gray squirrels are rather similar to each other, but they are increasingly different from cats, frogs, fishes, or earthworms. The most obvious way to make a classification of organisms would be to group together those that are alike and put those that are different into other groups. It sounds simple, but when you try to do it in a really clear and systematic way, you begin to run into serious difficulties.

The first problem is that there are different degrees of resemblance, and these degrees evidently should be taken into account. A red squirrel is more like a gray squirrel than like a cat, more like a cat than like a frog, more like a frog than like a fish, more like a fish than like an earthworm, and more like an earthworm than like a daisy. It is not too hard to meet this difficulty. The solution is simply to establish groups of different scope, like this:

ORGANISMS:
everything that is alive, including daisy, earthworm, fish, frog, cat, gray squirrel, and red squirrel.

ANIMALS:
no daisies, but including earthworm, fish, frog, cat, gray squirrel, and red squirrel.

VERTEBRATES:
no earthworms, but including fish, frog, cat, gray squirrel, and red squirrel.

TETRAPODS (four-limbed vertebrates):
no fishes, but including frog, cat, gray squirrel, and red squirrel.

MAMMALS:
no frogs, but including cat, gray squirrel, and red squirrel.

RODENTS:
no cats, but including all kinds of squirrels (also rats, porcupines, and so on).

SQUIRRELS:
divided into red and gray squirrels, among others.

This is a *hierarchy*,[1] an arrangement of groups of decreasing scope, one within another. A great many different hierarchies are possible. The example we have given happens to correspond with one now in general use, but there is nothing absolute and fixed about it. Many other arrangements could be made that would be just as natural and just as useful as long as students agreed on them and understood what was meant by them.

[1] From the Greek for "sacred rule." Obviously the word has greatly changed in meaning since ancient times. It came to be applied to the various ranks or orders of church officials and then to any series or arrangement in which each step has authority over or includes all those below it in the sequence.

The next and more serious problems arise from the fact that organisms resemble each other in different ways. A possible sort of classification, one that dates from antiquity if not from prehistory, is according to ways of life. Plants, for instance, can be arranged as aquatic, herbaceous, shrubby, or arborescent; or animals as swimming, walking, or flying. (These groups are very far from exhausting the possibilities, but they will do as examples.) Such a classification might also bring in resemblances and differences of habitat: swamp or alpine plants, marine or desert animals, and so on. Classifications of this sort, according to habit and habitat, are meaningful and valuable. They are, in fact, in wide use today, especially in the study of ecology, or communities and their relationships with environments (see Chapters 23 to 25). Nevertheless, they have rarely been used for the general, basic classification and naming of organisms.

Even in antiquity it was apparent to some students that classification by habit and habitat often brings together organisms that are different and separates those that are alike in some fundamental way. A yucca is shrubby, and a Joshua tree is arborescent, but even a casual observer feels that somehow they resemble each other more than a yucca resembles most other shrubs or a Joshua tree most other trees (Fig. 18–1). Alligators and otters have many similarities in habits and habitats that lizards and weasels do not have, yet in some more basic way an alligator is surely more like a lizard, and an otter more like a weasel.

Doubtless you already know that organisms are ordinarily classified on evidence derived, in large part, from their anatomy and physiology. Perhaps you are a little impatient with us for not coming right out and saying so without preamble. However, classification by anatomy and physiology is not a simple and clear-cut matter. It is not really meaningful to classify by some such procedure as this: a gray squirrel has 10,001 anatomical and physiological resemblances to a red squirrel, 8346 to a cat, 3921 to a frog, 2754 to a fish, and 172 to a daisy. The figures used certainly are not correct. We have no idea what the correct figures are or how they could be estab-

Left, National Park Service; right, Edward Lewis from Black Star

18–1 A yucca and a Joshua tree. The superficial similarity between a yucca (*left*) and a Joshua tree (*right*) is less than that between the skulls of marsupial and placental sabertooths (p. 488). Nevertheless, they are closely related, as the morphology of their flowers shows.

lished; that is one of several reasons why the most meaningful classifications are not really based solely on anatomical and physiological resemblances.

Cacti and many South African euphorbias are succulent, spiny, flowering plants physiologically adapted to very arid conditions. Cactus fanciers call both groups "cacti." This is unconscious use of classification by habit and habitat, but anatomical and physiological resemblances between the two groups are also real and numerous. Botanists ignore those resemblances in classifying the plants, separating the euphorbias sharply from true cacti by placing them in a different family along with many other plants, such as the poinsettias, which no one would think of calling cacti. Similarly, the Tasmanian wolf (now possibly extinct) was called a "wolf" because it is very like one in appearance and habits. It does have many anatomical and physiological resemblances to wolves. It has other, much less obvious resemblances to kangaroos. Nevertheless, in zoological classification it is grouped with the kangaroos and widely separated from the true wolves.

Classification by anatomy and physiology requires that characters be selected and interpreted. The fact that the ovary is superior in spiny euphorbias and inferior [2] in cacti is considered more important than the fact that both have spines. The fact that Tasmanian "wolves" have pouches and marsupial bones while real wolves do not is emphasized in classification, and the fact that both are four-footed running animals with flesh-cutting teeth is given secondary importance. What characters are selected, what meaning is assigned to them, and finally what classification is devised depend on the way in which the characteristics of organisms are interpreted. They depend on a decision as to one of the most profound questions of biology and of philosophy: What is the nature of a systematic unit among organisms, and how do the char-

[2] These technical botanical terms refer to the higher or lower position of the ovary in the flower, not to its better or worse qualities.

acteristics of such units originate? The principles on which biological classification is based arise from answers to that question.

Nature of Systematic Units

The form of classification of organisms still in use today is conventionally dated from the middle of the eighteenth century. It is dated especially from the book *Systema Naturae* by the Swedish botanist Carolus Linnaeus (Karl von Linné, 1707–1778), which appeared in 1735 and went through many editions. This dating applies, however, only to the *form*, the terms and names used. There have been two revolutionary changes in the principles of classification since Linnaeus. The result is that although a classification of plants or animals today *looks* almost the same as one of two centuries ago, it *means* something altogether different.

In Linnaeus' day classification was based on the philosophical doctrine that species are fixed, unchanging units. That doctrine was heavily reinforced by the reigning theological dogma (which was not then very old and does not date from Biblical times), that all the units were created as such by God at the beginning of the world. Some slight change within a species was admitted as possible, on the example of the various races of domesticated animals. There were also a few biologists who considered the evolution of new species possible. These ideas, however, had not yet had any real influence on systematics. The systematists' task then was supposed to be simply to recognize the units of divine creation. The units were presumed to be sharply distinct, like cats, dogs, sheep, pines, or maples, and to pose no "species problem." Linnaeus had little doubt that the 4235 species of animals listed by him were actually the "kinds" of the Creator and subject to no further revision by human systematists.

For most systematists of Linnaeus' day a species had its characteristics because it was created just so. This implies a pattern, one for each species. Deviations from that pattern, so evident in nature as variation within species, were considered accidental and irrelevant. In fact, the varying individuals belonging to a species were held to be of no particular importance for systematics. The essential thing, the ultimate or transcendental reality of a species, was believed not to be material and tangible but a pattern, a divine idea, a *type* or, as then often designated, an *archetype* ("primeval pattern"). The way for a systematist to look at organisms, then, was supposed to be to ignore individuals, to brush aside variation and all characteristics of populations as such, but to abstract an idea of what the individuals have in common. That abstraction was the type or archetype of the species.

The same concepts were applied to groups of wider scope than species. (Remember that we are still talking not about modern systematics, but about the eighteenth century, a necessary preliminary to understanding the present different principles.) Carnivorous mammals included numerous species. The Carnivora, as a group, were therefore not a "kind" as a unit of creation. Nevertheless, all Carnivora were thought to have certain characters common to all included species. Those common characters were also supposed to reveal an archetype, one broader than the specific archetypes, involving fewer but (in some way not too clear) more fundamental characters. Thus there was a whole series of "ideas," of divine patterns, becoming increasingly detailed and explicit.

These were the classical principles of systematics, the ones with which the science was established. The first revolution occurred when it was learned that species are not separate and unchanging creations but that they have evolved, one from another, in the long course of the history of life. This concept of species, and of all systematic groups made up of species, is profoundly different from the special-creationist and type concepts. Species are now known not to be fixed units, but to be changing things, in continual flux over immense periods of time. Hence the "species problem," because you cannot give a fixed definition to units that change and grade one into another.

Even more important than the new concept of evolving species was the discovery that relationships among species are not abstractions, reflections of metaphysical archetypes. Species are related to each other in the fully material sense that one descends from another and several or many have descended by the processes of organic reproduction from the same ancestral species. Conformance with a broader archetype was no longer the principle for grouping species into larger units of classification. Instead, the principle became, and still is today, that all species in any systematic grouping are of common ancestry.

Revolutionary change in the principles underlying classification made no difference in the form and, at first, little difference in the practice of classification. Preevolutionary classification grouped organisms according to the anatomical and physiological characters common to all within the group, interpreting these characters as physical manifestations of a metaphysical archetype. Evolutionary classification established groups in just the same way and, indeed, took over most of the groups already established by Linnaeus and other nonevolutionary biologists. The only real difference was that the characters in common were now considered as physically inherited from the common ancestry of the whole group.

Classification as it was practiced by the evolutionary systematists of the later nineteenth century did not fully do away with the preevolutionary archetype. As far as it really affected classification, the archetype was simply relabeled and differently interpreted. Each species had a type, now an individual specimen, supposed to be a sort of standard, a model, in a sense really an embodied archetype. Other individuals were compared one by one with type specimens and placed in the species the type of which they most nearly resembled. The fact that they were never *exactly* like the type was simply a nuisance. Variation in species was not an integral part of the species concept, but an imperfection of nature that had to be put up with by the unfortunate systematist.

Higher groups, including several or many species, also continued in practice to have archetypes in more or less veiled form. The groups were defined by characters in common, abstracted patterns quite like archetypes even though considered as somehow representing the characteristics of an ancestor. Thus although classification soon became evo-

lutionary in principle after publication of Darwin's *The Origin of Species* (1859), it remained largely preevolutionary in practice. It was still *typological*, as in Linnaeus' day. It continued in practice, regardless of stated principles, to picture the diversity of life as a series of idealized types, with individuals merely embodying the type patterns more or less adequately.

The second major revolution in principles of systematics since Linnaeus is the change from a typological to a *population* concept of the nature of systematic units. The turning point cannot be associated with one man and date, as that of evolutionary systematics can with Darwin and 1859. Some earlier students, including Darwin, did grasp the rudiments of a systematics of populations rather than of types. Full comprehension, however, required the discovery of Mendelian genetics and then the development of population genetics on that basis. Systematics clearly based on populations and explicitly nontypological is mainly an achievement of the second quarter of the twentieth century. In fact, this revolution in systematics is still going on.

The following, then, is the modern, fully evolutionary concept of systematic units:

A systematic unit of organisms in nature is a population or a group of related populations. Its anatomical and physiological characteristics are simply the total of those characteristics in the individuals making up the population. The pattern of characteristics is neither a real individual nor an idealized abstraction of the characters of an individual. It is a frequency distribution of the different variants of each character actually present at any given time. Species are populations of individuals of common descent, living together in similar environments in a particular region, with similar ecological relationships and tending to have a unified and continuing evolutionary role distinct from that of other species. In biparental species, the distinctiveness and continuation of the group are maintained by extensive interbreeding within it, and less or no interbreeding with members of other species. Demes and subspecies are subdivisions of species, more local in distribution, less isolated genetically, and without established and distinct continuing evolutionary roles for each such unit. Systematic units more inclusive than a species are groups of one or more species of common descent.

The best and the only *direct* evidence that an individual belongs to a particular species is not the anatomy or physiology of the individual as such, but the observation in nature that the individual is living with the specific population and functioning as a member of that population. Such direct evidence is not available in practice for organisms that have been removed from the context of nature—that are specimens in collections rather than parts of living populations. Fully direct evidence is also lacking for fossils. It is also usually impossible to obtain entirely direct evidence that a group of species is of common descent and hence is properly classified as a higher systematic unit. In all these cases pertinence to a systematic unit must be judged by indirect evidence. There are many kinds of indirect evidence, anatomy and physiology among them.

The one other point to make here is that use of anatomy, for instance, as *evidence* that an organism belongs to a particular species does not mean that a species as a systematic unit is *definable* in anatomical terms. A species or other systematic unit is defined in terms of populations and their biological, evolutionary relationships. These relationships have anatomical consequences, among others, from which they may be inferred. If two people look exactly alike, that is evidence that they are, or may be, identical twins. But they are not twins because they look alike; they look alike because they are twins. Similarly, typological systematics maintained that organisms belong to the same systematic unit because they have the same anatomical pattern. Modern systematics has learned that they have the same anatomical pattern (to the extent that they really do) because they belong to the same biological, evolutionary population or group of populations. An up-to-date systematist is not now engaged in classifying anatomy or any other sort of evidence. He is engaged in using the evidence to classify populations of organisms according to their evolutionary relationships.[3]

[3] It is true that some groups just are not well enough known as yet to achieve evolutionary classi-

Interpretation of Form and Descent

The resemblances between a yucca and a Joshua tree are somehow more important than the difference that one is a shrub and one a tree. On the other hand, the differences between a Tasmanian "wolf" and a true wolf are likewise in some way more important than the resemblance that they are four-footed carnivores similar in appearance and habits. "More important" means, in this connection, more reliable indications of evolutionary relationships and hence primary bases for modern classification of these organisms. The resemblances of yucca and Joshua tree are old and fundamental characteristics inherited from the common ancestry of the two sorts of plants. The differences between them are of relatively recent evolution and do not contradict close relationship. Botanists classify these plants as different species of the same

fication, even though what is known is interpreted by evolutionary principles. There is some doubt whether truly evolutionary classification is really possible in a few groups, mostly among the protists and notably including the bacteria, and the old typological method there continues in use mainly because no one has come up with a more practicable method.

One interesting new approach to systematics that has developed in the last few years is based on studies of the nucleotide composition of the DNA of each species. We will recall from Chapter 7 that the patterns of information determining *all* of an organism's hereditary characteristics reside in the sequences of purine and pyrimidine bases in the double-stranded DNA molecule. Since it is not yet technically possible to determine base *sequences* in the laboratory, investigators have been doing the next best thing: determining the *ratios* (on a molar basis) of the four bases to one another. This can be done conveniently by expressing the ratio of the G-C pair (guanosine–cytosine) to the A-T pair (adenine–thymine)—or even more simply by determining the "percentage G-C." (The rest is necessarily A-T.) It turns out that there are wide variations in the G-C content of DNA's from different species but there appear to be similarities within certain related groupings. For example, vertebrate DNA's range from 40 to 44 per cent G-C. Some higher plants have higher percentages; some have lower. Bacterial and viral DNA range very widely, from 25 to 75 per cent G-C. These differences are clearly compatible with the idea that the all-important base sequences of DNA *are* shaped by evolution and that their analysis may one day provide another useful approach to systematics.

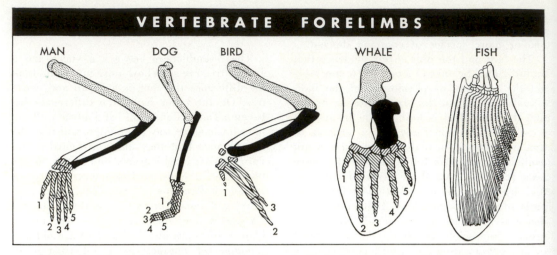

VERTEBRATE FORELIMBS

MAN　　　DOG　　　BIRD　　　WHALE　　　FISH

18–2　Vertebrate forelimbs. Homologous bones in the limbs of man, dog, bird, and whale are indicated as follows: stippled, humerus; white, radius; black, ulna; upper crosshatching, carpals; lower crosshatching, metacarpals and phalanges (finger bones). Numbers refer to digits (or fingers).

genus. In the example of the Tasmanian and true wolves, it is the differences that are phylogenetically old and the resemblances that are more recent and less fundamental. Zoologists place these animals in widely different major groups of mammals (among the marsupials and the placentals, respectively).

In such examples the decisive factor is whether resemblances between organisms have or have not been inherited from a common ancestry. Interpretation and application in classification depend on some historical, evolutionary principles and processes, which must now be reviewed.

HOMOLOGY

Homology is correspondence between structures of different organisms due to their inheritance of these structures from the same ancestry. Such structures are called *homologues* and are said to be *homologous*.[4]

[4] From the Greek for "agreement." Biological use of these terms stems mainly from a publication in 1843 by the eminent British anatomist Richard Owen (1804–1892). Like all biologists of those and earlier times, Owen recognized that some similarities between parts of organisms are more fundamental than others, and he proposed to call the more fundamentally similar parts "homologues." The then current explanation as to why the similarity is more fundamental was that the parts correspond with an archetype of a higher order (p. 491). Even Owen, who never explicitly accepted the truth of evolu-

If the bones of a man's arm and a dog's foreleg are compared (Fig. 18–2), the number and arrangement are remarkably similar. The resemblance also extends to the way the limbs arise embryologically and, in greater or less degree, to the arrangement in them of muscles, blood vessels, and nerves. The only reasonable explanation is that the limbs are homologous. Although the forelimbs are now used differently, in man mainly for manipulation and in the dog mainly for locomotion, the limbs were inherited from the same ancestry. The differences between them have arisen since the lines of descent leading to dogs and to men separated. From fossil (paleontological) evidence, we know that the separation occurred about 70 to 75 million years ago. It happens that evolution of the forelimb has not been particularly rapid either in men or in dogs, and in spite of the long lapse of time it is still clearly evident that not only the limbs as a whole but also the various bones in them are homologous.

If, now, comparison is made with the wing of a bird, the similarity is less striking. Nevertheless, it can be established beyond any doubt that man's arm, dog's foreleg, and bird's wing are all homologous, and even that

tion, later (in 1866, after publication of *The Origin of Species*) admitted that inheritance is the "most intelligible" explanation of homology.

some homologous bones are present in all three. The evidence involves anatomical and embryological comparisons and, still more convincingly, fossils of early mammals (from which men and dogs later evolved), of ancient birds, and of still older reptiles from which both mammals and birds arose. Separation of the reptilian ancestries of mammals and birds occurred not less than 200 million years ago, but a fundamental genetic resemblance has not been obliterated.

Let us now add the front (pectoral) fin of a fish to the comparison. The resemblance is even slighter, and this is not surprising because we are now comparing animals whose nearest common ancestry is more than 300 million years in the past. The fish's fin as a whole is still certainly homologous with the forelimb of bird, dog, or man, but it is no longer possible to designate homologous individual bones in the limb with any assurance. Extremely ancient fish fins are known from which both modern fish fins and the limbs of air-breathers have evolved, but the separate lines of evolution have undergone changes so profound that the homologies of the individual bones have been practically obliterated.

The degree of homology in the examples given is the basis for phylogenetic inferences which, in turn, are involved in the classification of these organisms. Man and dog are more nearly related to each other than either is to a bird. Man, dog, and bird are more nearly related than any is to a (modern) fish. Homology is the anatomical evidence for degrees of relationship among organisms. A main problem, then, is to distinguish between resemblances that are homologous and those that are not.

HOMOPLASY AND ANALOGY

Anatomical features that resemble each other but that are not inherited from the same feature in a common ancestry are called *homoplastic*, and the phenomenon is *homoplasy*.[5] Thus correspondences in anatomy between two plants or between two animals are either homologous or homoplastic. If they are homologous, they are evidence of relationship

or genetic affinity. If they are homoplastic, they are not such evidence. The terms are, of course, interpretive. They express an opinion or a deduction from the available evidence. The use of comparative anatomy in classification depends on this decision whether anatomical resemblances are homologous or homoplastic.

If we add the wings and legs of insects to the comparisons previously made, they provide clear-cut, sure examples of homoplasy (Fig. 18–3). The front leg of an insect has some sort of anatomical correspondence with the front limb of a dog or the arm of a man. It is, however, certain that the limbs of insects and mammals evolved wholly independently and were not inherited from an ancestry common to the two. Insects and mammals did have a common ancestry, but this was extremely remote, more than 600 million years ago, and the limbs of the two groups evolved after their ancestries had separated. The limbs are not homologous, but they are to some extent homoplastic.

The homoplastic forelegs of insects and arms of men do not function in the same way. If an insect and a dog are compared, however, there is a resemblance in function, for in both the limbs are used mostly for walking. When structures that are not homologous have a functional resemblance, they are *analogous* and are called *analogues*.[6] The wing of an insect is to some extent homoplastic and is completely analogous with the wing of a bird (Fig. 18–3A). It is certainly not homologous.

Homoplastic structures are usually also analogous. It is a rule of evolution that resemblance in structure not due to inheritance from a common ancestry is generally correlated with similarity of function. Homoplasy usually results from similar adaptations of organisms of different ancestry. Like almost all biological rules, this one has exceptions but is true in the majority of instances. Structures may, indeed, be analogous without having any noticeable degree of anatomical resemblance. The gills of a fish and the lungs of a mammal are anatomically so different that they would

---

[5] "Same-forming" (but not same inheritance).

[6] "Things proportionate to each other." The technical meaning in biology is different from and more precise than the Greek word or the meaning of "analogy" in everyday speech.

HOMOLOGY, HOMOPLASY, and ANALOGY

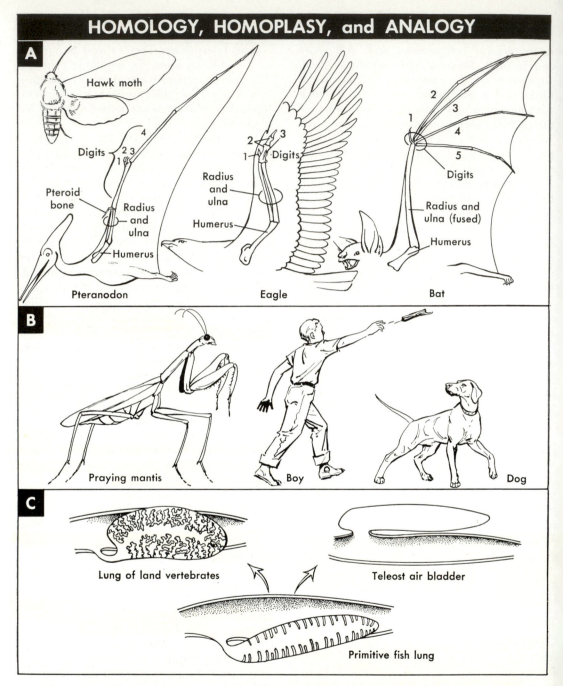

18–3 Homology, homoplasy, and analogy. *A, B.* The wings and legs of insects bear a homoplastic relation to those of vertebrates. The wings in *Pteranodon* (an extinct flying reptile), eagle, and bat evolved separately from walking forelimbs and so are homoplastic as wings, but they contain homologous bones. The wings of insects and of these vertebrates are analogous: they perform similar functions. The forelimbs of man and praying mantis are not only homoplastic but also analogous, for both are used to manipulate. The forelimbs of man and dog, which are homologous, are not fully analogous; the dog's forelimb is nearly exclusively for locomotion. *C.* Lungs of land vertebrates and air (or swim) bladders of fishes are homologous but not analogous.

hardly be called homoplastic, but both are organs of respiration and are in that respect analogous. The homologue of the lungs in fishes is the swim bladder (Fig. 18–3C).

TRANSFORMATION

Homologous structures may differ markedly both in anatomy and in function. In most living fishes the swim bladder is a simple closed sac filled with gases that decrease the specific gravity of the fish and help it to maintain a favorable depth in the water—a sort of internal water wings. It has nothing to do with breathing. In one group of fishes it has become even less lunglike and acts as a sort of sounding board or resonating chamber, the vibrations of which are communicated to the brain through a series of bones. It is, in fact, analogous to an eardrum (fishes do not have real eardrums), although still homologous with the lungs. This extraordinary development can be seen in goldfishes, as well as many other fresh-water fishes (Fig. 18–4).

It is odd enough that fishes should hear by means of a structure homologous with part of our breathing apparatus. It is at least as peculiar that *we* hear by means (in part) of structures homologous with parts of the jaw apparatus of reptiles, including our own reptilian ancestors. Two of the three little bones that transmit vibrations from our eardrums to our inner ears are homologous with the bones that form the joint between upper and lower jaws in reptiles (Figs. 16–13 and 18–5).

One of the most fascinating things about comparative anatomy, phylogeny, and classification is that homologous structures can be so very different both in form and in function. From the point of view of their evolution, this means that structures can and often do change radically both in appearance and in the way they work. Such radical changes are known, logically enough, as *transformations*, and they have been rather common in the history of life. They are especially likely to be involved in the origin of major new groups of organisms, such as the rise of mammals from reptiles or of flowering from nonflowering plants.

The widespread occurrence of transformations emphasizes that new sorts of organisms,

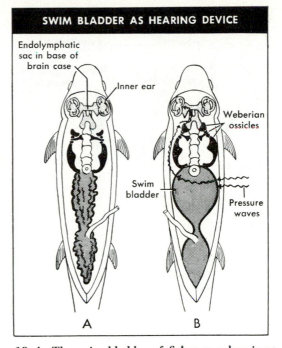

SWIM BLADDER AS HEARING DEVICE

18–4   The swim bladder of fishes as a hearing device. The swim bladder in modern bony fishes evolved first as a lung. In some later fishes it has been transformed into a hydrostatic organ used to adjust the net specific gravity of the fish. In some of these later fishes it has been exploited to serve still another function—hearing. Bony processes of the vertebrae closest to the skull have evolved into a chain of Weberian ossicles. They function analogously to the ear ossicles in land vertebrates: pressure waves striking the taut membrane of the swim bladder are transmitted mechanically by the Weberian ossicles to an endolymphatic sac in the base of the brain case, which communicates with the endolymph of the inner ear. *A.* The swim bladder collapsed to illustrate how the chain of Weberian ossicles is attached. *B.* The swim bladder inflated and the ossicles pressing on the endolymphatic sac.

new organs, or new adaptations evolve from what is already there. A good engineer should be able to design a better reproductive apparatus than a magnolia flower, a better means of walking than a salamander's leg, or a better sound receptor than an opossum's ear. The point is that these structures were not designed. They evolved on the basis of mutations affecting earlier structures that functioned differently. And even though they are not perfect, they do work. They have been molded by and have stood the test of millions of years of natural selection.

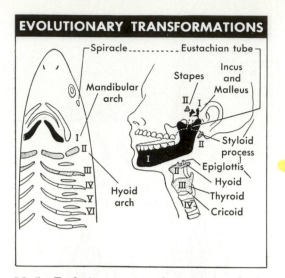

EVOLUTIONARY TRANSFORMATIONS

18–5 Evolutionary transformations of the primitive vertebrate gill arches. The evolutionary transformations of two of the primitive vertebrate gill arches into the mammalian ear ossicles have already been illustrated (Fig. 16–13). This figure shows the ultimate transformations of the remaining gill arches into several other features of the mammalian skull and pharyngeal region. For simplicity a mammal (man) is compared with a contemporary shark, which retains ancestral features of the gill arches but is not, of course, itself an ancestor of mammals. Homologous arches have the same roman numbers. They have been transformed in the mammal into the following structures: (I) the lower jaw and two of the ear ossicles (incus and malleus); (II) the third ear ossicle (stapes), the hyoid cartilage (tongue skeleton), the styloid process of the skull, and the styloid ligament; (III) the thyroid cartilage, which is part of the skeleton of the larynx; and (IV) the cricoid cartilage and epiglottal cartilages of the larynx.

IRREVOCABILITY OF EVOLUTION

The fact that evolutionary change occurs on the basis of what is already there—that is, of the results of all previous evolution—has extreme and widespread consequences. It makes evolution *irrevocable*. What has already happened before any given time necessarily affects and limits what can and does happen next. The future cannot change the past; what is past is irrevocable. Evolutionary change never wholly eradicates the effects of previous evolution. The most radical transformation does not wipe out influences of the previous condition of the structure transformed. If the little mammalian ear bones had not been part

of the reptilian jaws they would certainly not have the relationships that they do, in fact, have. If any of our ancestors, back to 2 billion B.C. or earlier, had been different from what they were, we too would be different in some respect and to some degree.

There is also another side to the principle of irrevocability. If the past cannot be wholly lost, neither can it be fully regained. Nothing quite like an earlier form of life ever evolves again, for the simple reason that time does not double back on itself. If there is a sequence of ancestral and descendent organisms $a \rightarrow b \rightarrow c$, then b evolved from a and would have been different if a had been different in any respect. Evolution cannot be reversed; b cannot evolve again from c, because c is different from a and cannot possibly give rise to quite the sort of organisms that arose from a. This aspect of the irrevocability of evolution is usually called the "irreversibility of evolution,"[7] and it is not always recognized that it is just a special case of the more general principle. Failure to grasp the broader principle has also led to misunderstanding of the meaning of the irreversibility of evolution. Some biologists formerly supposed, for instance, that if animals had grown larger in the course of their evolution they could not have smaller descendants. In fact, an evolutionary trend like that toward larger size can be and frequently has been reversed. The real point is that, although the descendants come to be of the same size as remote ancestors, they still are very different organisms. Even the genetic mechanism determining size is almost certain to be different.

The principle of irrevocability is of great importance in systematics. An example should make its relevance clear. Land vertebrates such as reptiles and mammals, arose remotely from aquatic forms, fishes. Some reptiles and mammals, such as the whales, became aquatic and fishlike in habits. Their evolution reversed the way of life, but of course this did not make them fishes again. The effect of

[7] You will also find it called "Dollo's law" in most books on evolution. Actually it is not a law, properly speaking, but a principle. Also, Louis Dollo (1857–1931), an eminent Belgian paleontologist, was not the first to state or discuss the principle, and we now know that his statement of it was inadequate

irrevocability can be seen throughout their anatomy, as clearly in their flippers as elsewhere (see Fig. 18–2). The flipper functions like a fish's fin, but it is quite different because, unlike the fin, it has passed through a stage when it was a leg. This is one aspect of irrevocability. However, even though the flipper has become finlike, it has not lost the effects of its land-living ancestry, and the bones in a whale flipper are still plainly homologous with those in the leg of a land mammal. This is the other aspect of irrevocability. The application to classification is that even though a whale is more fishlike in habitat, habits, and general appearance, it is more closely related to land animals than to fishes. It is classified as a mammal.

CONVERGENCE AND PARALLELISM

The example of whales and fishes shows that organisms may be alike in living conditions and appearance even though they are of quite different ancestry and relationships. Such resemblances may arise between organisms even more distantly related. Spending an evening in a southwestern garden, a visitor was astonished to see what were apparently large numbers of hummingbirds gathering nectar in the dusk. A closer look showed that they were not hummingbirds but hummingmoths, insects almost identical with the birds in actions, size, and superficial appearance (Fig. 18–6).

The evolutionary development of resemblance between organisms whose ancestors were less alike is called *convergence*. (What relationship does this have to homology and homoplasy?) Convergence is a common phenomenon in nature. The previously noted resemblance between cacti and some euphorbias is convergent, and so is that between Tasmanian and true wolves. The wolves are part of a large-scale convergence. In Australia, an island continent, the isolated evolution of marsupials has produced kinds convergent toward many different nonmarsupial (placental) mammals of the rest of the world. There are not only native "wolves" but also native "mice," "cats," "anteaters," "moles," and "sloths," as well as the squirrel- and flying squirrel-like phalangers and the groundhoglike wombats. These marsupials are only dis-

18–6 Hummingbird and hummingmoth, examples of convergent evolution. The bird and the moth have converged in form, flying habit, and feeding procedure in their common exploitation of the nectar in flowers as a food source. None of these resemblances existed in their extremely remote common ancestry.

tantly related to the placental true wolves, mice, cats, anteaters, moles, sloths, squirrels, and groundhogs. The resemblances result from convergent evolution (Fig. 18–7).

It is well for the systematists that evolution is irrevocable, for otherwise convergence would commonly be mistaken for community of ancestry. Such mistakes have been made in the past and probably a few are still being made, but by and large modern systematics has sorted out truly related from merely convergent organisms. The fact that convergent forms did have ancestors less similar means that they cannot become identical. Traces of the ancestral dissimilarities persist and can be recognized for what they are. No one studying specimens with a view to classifying them would really mistake a hummingmoth for a hummingbird or a euphorbia for a cactus.

In such examples the separate ancestries of convergent forms were so different that the convergent nature of the resemblance is obvious with a little study. Things become more difficult if the ancestries were not very different, and still more so if the ancestries were related and evolution in the descendent lines has simply followed more or less the same course. This sort of evolution intergrades with convergence and is not always clearly distinguishable, but it is given a different name:

CONVERGENT EVOLUTION OF PLACENTAL AND MARSUPIAL MAMMALS

PLACENTALS

MARSUPIALS

Wolf
(Canis)

Tasmanian
wolf
(Thylacinus)

Ocelot
(Felis)

Native cat
(Dasyurus)

Flying
squirrel
(Glaucomys)

Flying
phalanger
(Petaurus)

Ground
hog
(Marmota)

Wombat
(Phascolomys)

Anteater
(Myrmecophaga)

Anteater
(Myrmecobius)

Mole
(Talpa)

Mole
(Notoryctes)

Mouse
(Mus)

Mouse
(Dasycercus)

18–7 The convergent evolution of some placental and Australian marsupial mammals.

parallelism. Parallelism is even more common than convergence. Moreover, close parallelism may be practically impossible to distinguish from close community of ancestry unless the actual ancestors have been found as fossils. It was, for example, long assumed that the American and Old World porcupines are closely related and that their spininess is homologous and part of the evidence for that relationship. Later work raises a strong possibility that this is a case of parallelism, that the common ancestor was really very remote and was spineless. If this is so, the spines (and a number of other resemblances) have evolved independently in the two groups and are homoplastic, not homologous. The fossil evidence is still insufficient and the question remains open, but most systematists now think that parallelism is involved in the history of porcupines. In recent years it has become apparent that parallelism has occured far more widely among all sorts of organisms than was formerly recognized. This is one of the major problems of systematics today.

The cause of convergence is known; indeed you have probably already thought of it in the light of what you have learned about the factors of evolution. Plants or animals of different origin become adapted to similar habitats and habits or, in general, ways of life. Cacti and cactuslike euphorbias are both water-storing desert shrubs. Tasmanian and true wolves are both running predators preying on other animals of similar habits and small size. Adaptive similarity involves similarity also of structure and function. The mechanism of such evolution is natural selection: in fact, of all evolutionary phenomena, convergence seems to be most rigidly controlled by natural selection. Differences persist in convergent forms because they arise from different bases. Structures adapted to the new activities and conditions are likely to be different to start with and may not even be homologous. Mutations, materials for natural selection, are also likely to be different in the two groups.

Parallelism is also largely a matter of similar adaptation under the control of natural selection. The ill-defined difference between convergence and parallelism, however, involves factors that make parallel forms more similar than convergent ones. The term "parallelism" is applied to evolving organisms that were rather similar to begin with and that were related closely enough for some of the same mutations to arise in them. Close parallelism occurs only between groups that are, in fact, related. The resulting resemblances may then be misleading only as to the degree and not the fact of relationship.

DIVERGENCE

If you try to predict the future of two evolving groups of organisms there are plainly three possibilities: the groups can become more similar (convergence); they can evolve in much the same way, so as to remain about equally similar (parallelism); or they can become less similar (*divergence*). It is obvious that divergence is extremely common. Indeed, it is the universal rule for directions of evolution and is necessarily the basis for the whole of the diversity of nature. Within a species there is fluctuating, reversible divergence as demes or other subdivisions of the species develop differences from each other. As soon as speciation occurs, two (or more) species arise from one, and divergence between the two becomes essentially irreversible and tends to increase as time goes on.[8]

Special examples of divergence are unnecessary because it is illustrated by the manifest differences between any two species or organisms. The pattern of divergence recurs in all groups of organisms. It is the basic element reflected in classification. When we list the species of a genus, we are (if our classification is successful and correct by modern principles) listing populations that have diverged from a single ancestral population. The genera of a family, the families of an order, and so on, are also the representatives and results of

[8] It is conceivable that two parts of a specific population might permanently cease to interbreed without developing any readily visible differences. Some such instances are known; they are called "sibling" or "cryptic" species. They are comparatively rare, and in every example some differences have been found by detailed new study, even though the differences may be so slight as to have been missed by earlier systematists. Sibling species evidently do not persist very long without developing more clear-cut differences and hence ceasing to be sibling species.

divergence of increasingly long standing and on an increasingly large scale.

The frequent occurrence of convergence and parallelism do not at all contradict the fact that divergence is universal in evolution. There can be neither convergence nor parallelism unless there has been previous divergence. Furthermore, in comparing any two groups of organisms of near or remote common ancestry it never has been found that they are completely, in every respect, convergent, parallel, or divergent. In some respects the two have always retained some ancestral features unchanged; in some features they are always divergent; in others they may or may not be convergent or parallel.

The Practice of Classification

Classification consists mainly of three operations: (1) recognizing and describing related smaller and larger groups of organisms according to the principles of populations and of phylogeny; (2) fitting these groups into a formal hierarchy; and (3) providing names for the various groups. Only the first of these operations involves the direct observation and interpretation of nature. The second and third put the results of the first into meaningful form and supply the names by which we can think and talk about those results. The second and third operations are, however, necessarily subjective and more or less arbitrary. The third, nomenclature, involves completely artificial, legalistic symbolic devices.

THE SYSTEMATIC HIERARCHY

The hierarchic principle has been introduced on p. 489. By general consent, a particular form of hierarchy, modified from that of Linnaeus (p. 491), is now in general use, and a special term is applied to each recognized level or category.

Kingdom
 Phylum (plural, phyla)
  Class
   Order
    Family
     Genus (plural, genera)
      Species (identical in singular and plural; "specie" means "coin" and has no application in biology)

There is no reason in nature why the hierarchy should have seven steps. Groups of almost any inclusiveness could be recognized, for in nature there is no fixed size inherent in the facts. A hierarchy might have five basic steps or 50, and fixing on seven is only a matter of usage that has grown up among systematists. In fact, with the tremendous increase in the number of known kinds of organisms, most specialists have found that seven steps are not enough for their purposes and have supplied additional steps by prefixing "super-," "sub-," and "infra-." [9]

A full classification of a human subspecies, with the names applied to the groups at the various levels, shows how the system works.

Kingdom Animalia
 Phylum Chordata
  Subphylum Vertebrata
   Superclass Tetrapoda
    Class Mammalia
     Subclass Theria
      Infraclass Eutheria
       [Cohort Unguiculata—optional]
        Order Primates
         Suborder Anthropoidea
          Superfamily Hominoidea
           Family Hominidae
            Subfamily Homininae
             Genus *Homo*
              Subgenus *Homo* (*Homo*)
               Species *Homo sapiens*
                Subspecies *Homo sapiens sapiens* [10]

For reasons that will be apparent from the following discussion, no two authorities agree exactly as to how all organisms should be classified. There is, however, a reasonably good consensus on the most important

[9] "Sub-" categories are in general use at all levels. The others are less generally used, especially "infra-." Some botanists use "division" in place (approximately) of "phylum." A few other categories and terms are used by some students or in certain groups of organisms but are not fully standardized. Most common are "cohort," sometimes inserted between "class" and "order," and "tribe," sometimes inserted between "family" and "genus."

[10] Whites of European descent belong in this subspecies, by the accident that Linnaeus happened to belong in it. All living humans belong in the same species, but there are other subspecies.

features. A brief, modern, but conservative classification is provided in this book. It is mainly for reference purposes, and it has been placed at the end (Appendix) where it can be easily turned to. It should be consulted throughout the study of the rest of the book, whenever it is desirable to see how any one group fits in among its relatives or among organisms as a whole.

Sample and population

A species is a special sort of category in nature and in evolution, and therefore also in classification. It is a population that may be subdivided in various ways but that has an essential internal unity and continuity in evolutionary role, in geographic and ecological distribution, in genetic relationships, and in physical, phenotypic characteristics. It has some degree of external discontinuity, sometimes relative but tending to become absolute, from any other species. Groups higher in the hierarchy than species are less unified and continuous because they generally include more than one species. Groups lower than species are less clearly bounded because there is less or no discontinuity between them within the species. The species, then, is the fundamental population unit of classification.

Classifying a species involves determination of the characteristics of a population. The procedure can be compared with what you would do if, for example, you were a wholesaler who planned to buy the output of a large apple orchard. You would need to know what the whole crop is like, and you would be sure that the apples would vary considerably in size, color, ripeness, and other characteristics important to you. Surely you would not pick one apple and base your price and plans on the idea that the apple (a type) is sufficiently representative of the whole lot. Obviously, too, it would not be practical for you to examine every one of tens of thousands of apples in the orchard. (In systematics, examination of all individuals of a species is usually not merely impractical but downright impossible.) What you would do would be to take a sample of perhaps a hundred apples, being careful that the selection was at random so that it covered the range of variation fairly well and was not loaded with the better or poorer apples. If you

really knew your business, you would know methods (derived from the science of statistics) for deciding how many apples you need as an efficient sample for your purposes. By related methods you would estimate from the sample the average quality of the crop and its variation, and you would calculate how close your estimates are likely to be to the actual characteristics of the whole crop (the population).

That, in principle, is the modern procedure for determining (or more strictly speaking, for estimating) the characteristics of a species or smaller natural population unit in systematics. To be sure, some systematists still use the older and now clearly inadequate one-apple (typological) method, but their number is decreasing. In systematics, too, you often cannot obtain the most efficient sample. Sometimes you may even have to be content with one apple (specimen). Even so, you think of it as a sample of a varied population, not as a type to which other apples will or should conform. You still can tell from it something about the population, although you know that your estimates are rougher, less reliable, than if you had a better sample.

Higher categories

Placing species into genera, genera into families, and so on upward in the hierarchy is not based on inferences from samples about populations. The basic population units, the species, are successively combined in increasingly larger groups according to interpretation of their evolutionary relationships. In principle and with certain exceptions, all the species of one genus have evolved from one ancestral species, all the genera of one family from one ancestral genus, and so on. Of course, we do not yet know the natural relationships of all organisms with high probability and in sufficient detail to produce a definitive, final classification on which everyone can agree. Classification is constantly changing in some respects as we learn more about the course of evolution, and experts often disagree about details of classification.[11] Even when evolutionary affinities are ade-

[11] Experts who have tried to produce "objective" classifications, based wholly on the resemblances among organisms and not on their evolutionary re-

quately known and there is no disagreement about them, classification in higher categories does not automatically follow. It is impractical to express *all* the intricacies of relationships in a classification simple enough to be understandable and usable. All that can be done is to assure that the classification is *consistent* with a reasonable theory of evolutionary affinities. Different arrangements may be equally consistent with the same theory. Choice is a matter of usage and personal taste. Since stability is desirable, conservative systematists hold that a classification in general use should not be changed unless new study shows that it is probably inconsistent with the affinities of the organisms concerned.

NOMENCLATURE

Systematists come in for a good deal of derision because they call a rose by another name or stutter *Rattus rattus rattus* when they mean plain "rat." It is true that there are people who like to make things sound mysterious and difficult, but systematists have been forced into a complex and stuffy nomenclature whether they like it or not. There are millions of species that must be named, not to mention all the other groups up and down the hierarchy. No everyday language has enough names to go around. Innumerable sorts of plants and animals simply do not have common names. This situation applies, naturally, to all extinct organisms, but also to many still living. Worse yet, the common names that are available are usually vague in application, often misleading, and almost always are applied indiscriminately to different groups that the systematist must distinguish. In New Mexico, alone, there are 13 species of wild roses. In the Old World there are more than 560 species and subspecies of the genus *Rattus* and hundreds more called "rats" that do not even belong to this genus.

A less impelling but still important reason for not using common names is that they are different in every language. What is a *squirrel* in England becomes an *écureuil* across the

English Channel, an *ardilla* across the Pyrenees, an *Eichhörnchen* across the Rhine, and hundreds of other names here and there around the world.

The systematists had to invent an artificial system for naming organisms, and fortunately they agreed to use the same names everywhere, regardless of native language. This is one useful holdover from the days when all scholars wrote in Latin. Linnaeus wrote in Latin, and naturally he also named plants and animals in Latin. Innumerable scientific names are also derived from Greek, and in modern usage they may be derived from any language or from none, just made up. However, they are always latinized and treated as if they were Latin words. This is not too barbarous, for the Romans did the same thing, especially with Greek words. Biologists now write in English, German, Russian, Japanese, or whatever language they prefer, but they still use Latin or latinized names for groups of organisms. The name of any group from a kingdom down to a genus is a single capitalized word. Names of genera are usually printed in italics, but those of higher groups are not. The name of a species is two italicized words: the name of the genus followed by a name (not capitalized)[12] peculiar to the species. For example, the name of the human species is the two words *Homo sapiens*. This is why the nomenclature is called *binomial* ("two-named"). For subspecies, a third italicized word (not capitalized) is added to the name of the species.

It frequently happened that different workers applied different names to the same group or the same name to different groups. Frequently, also, some systematist has decided that two or more groups that had different names should be combined into one, or that a group with one name should be split into two. Even after systematists had agreed in principle on a system of nomenclature, such duplications and changes tended toward chaos. Finally international congresses and biological unions drew up codes of nomenclature de-

---

lationships, have disagreed still more. Resemblances still have to be selected and interpreted, and if evolutionary criteria for such evaluation are rejected, there are no acceptable criteria.

[12] Botanists formerly capitalized specific names derived from proper nouns and some others, but this capitalization is no longer required by the rules of botanical nomenclature, and it is banned by the zoological rules.

signed to solve these problems and to settle on one name for each group and a different name for each. The codes are complex and appeal more to the legalist than to the biologist, who is likely to regard them as a necessary evil. The codes depend heavily on the *rule of priority*. The valid name of a group is the first published name applied to it (if it was published after a fixed date and if it complied with certain requirements). If the same name has been given to two groups, the name belongs to the group to which it was first applied, and the other group must have a different name.[13] Priority sounds both simple and sensible, and it has worked out in most instances. However, in a good many cases the working of the rules is neither simple nor sensible. An international commission considers such cases when they are brought to its attention. The commission is empowered to suspend the rules if necessary to preserve a stable, widely known name that might become invalid under the rule of priority.

The nomenclatural codes require that there be designated type specimens for species, type species for genera, and type genera for families. The use of the term "type" in this connection is unfortunate because it recalls the old typological systematics from which it was inherited. Confusion results because the terminology suggests that the nomenclatural type is somehow typical or a standard of comparison. Modern biology and systematics recognize no "types" in that sense. The types required by the codes of nomenclature are only legalistic devices, of no real biological significance, used in order to specify what groups of real organisms the names are tied to. In that usage nomenclatural types are simply aids in stabilizing nomenclature.

CHAPTER SUMMARY

Resemblance and difference between species; their recognition the elementary act in classification; the hierarchy of groups in classification; the possibility of different hierarchies; the accepted system based on anatomical and physiological characters; the problem of selecting significant characters in classification; its relation to the nature of systematic units.

Modern classification, still Linnaean (eighteenth century) in form but justified by principles radically different: the twofold revolution in post-Linnaean systematics—(1) the advent of evolutionary thought and rejection of the concept of archetypes or divine patterns; the recognition that species are related in the material sense of common ancestry; but early evolutionary systematics still typological—concept of morphotype; (2) the replacement of a typological by a population concept of systematic units; the systematic unit as a varying population or group of populations related genetically; the nature of evidence in systematics.

Interpretation of form and descent: homology—structural correspondence between organisms due to inheritance from a common ancestor (exemplified by vertebrate forelimbs), homoplasy and analogy—homoplasy a structural correspondence not caused by common ancestry, and analogy a functional correspondence (exemplified by wings and legs in insects and vertebrates); transformation—a radical evolutionary change in structure and function (exemplified by the histories of fish lungs and swim bladders and of the mammalian ear mechanism); irrevocability of evolution—past structural evolution never wholly eradicable, old structure never wholly regainable, past evolution as a commitment setting conditions for future evolution (exemplified by aquatic mammals); convergence and parallelism—convergence as the evolution of resemblance between organisms with dissimilar ancestors (exemplified by hummingbirds and hummingmoths and by the history of Australian mammals) and parallelism as a comparable and commoner phenomenon involving ancestors less remotely related than in convergence, the evolution of similar adaptations as the cause of convergence and parallelism; divergence, the opposite of convergence and parallelism, and its universality.

[13] Except that a group of plants and one of animals can have the same name; botanical and zoological nomenclatures are entirely separate, although they follow the same general system.

The practice of classification: (1) treatment of species—sample and population, the special status of the species in systematics (more clearly bounded than either higher or lower categories in the hierarchy), the use of population samples for characterization of species; (2) treatment of higher categories—inference of evolutionary relationships, existence of alternative acceptable bases, consistency with evolutionary affinity as the main criterion of acceptability; (3) nomenclature —the practical need for technical nomenclature, the binomial system, rules of nomenclature, the priority rule, nomenclatural types (not to be confused with archetypes).

19

Protists and Simple Plants

These organisms are radiolarians, microscopically small protists with siliceous skeletons, related to amebas. Other protists and simple plants are the subject of this chapter.

In an appendix to this book there is a summary outline of one way in which living things can be classified. It is relegated to an appendix, not because it is unimportant, but because it is reference material. It cannot be read with pleasure, and memorizing it, although possibly useful, would not make you particularly wise. It is, however, important as a guidebook or a road map to country that you will be exploring in the next four chapters. We have now covered some of the principles underlying the fact that living things are and have long been of such an amazingly large number of different kinds. Next to be considered is what those different kinds are, how they resemble each other and how they differ, and what roles they play in the drama of life.

We shall be dealing mainly with the major subdivisions of each kingdom of living things, the large groups called *phyla* (singular, phylum).[1] Some individual examples will of course be mentioned, but it is not necessary (or possible) in a study of the basic principles of biology as a whole to enter into much detail as to the anatomy, physiology, and other peculiarities of many of the extremely numerous kinds of organisms. In some of the more important and more familiar phyla some of the lesser included groups, especially the *classes*, will be separately discussed. Further detail in this more descriptive and taxonomic part of the broad science of biology is the special province of the systematic biological subsciences: protistology,[2] botany, and zoology.

Few people have ever seen even one representative of the most abundant organisms in the world. The great majority of them are too small to be seen clearly, if at all, with the naked eye. They include a few giants as much as 2 centimeters in diameter, or even a bit more, but almost all of those present in our usual environments are between about 0.1 mi-

---

[1] Botanists frequently call the major groups "divisions" instead of "phyla," but it is also justified and is better in general biology to use the same terms for the classification of all forms of life.

[2] The kingdom Protista is not universally recognized as distinct from the plant and animal kingdoms, and its study, protistology, is commonly divided between botany and zoology, in elementary systematics courses, at least.

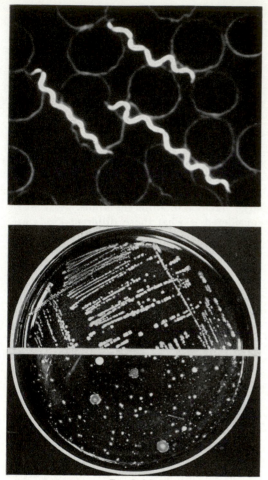

Top, Carl Strüwe; bottom, C. E. Clifton

19–1 Bacteria. *Top.* Spirochetes, the pathogenic (disease-causing) agents of relapsing fever in blood (×1100). *Bottom.* Bacterial colonies growing on plates of agar jelly. The figure is compounded from photographs of two agar plates. In the top half-plate the bacterial cells used to make the inoculation—that is, to start growth of the large colonies now visible—were streaked across the jelly with a needle. Individual cells left on the surface started the hundreds of round colonies that now lie along the paths of the needle. In the bottom half-plate the cells used in the inoculation were first suspended in water, which was then spread uniformly over the plate. Each of the colonies now visible again grew from a single cell.

cron and 200 microns in diameter.[3] They are therefore nearly or completely invisible as individuals, as you can readily understand if

[3] A micron, symbolized as μ, is one millionth of a meter.

you convert those sizes into more familiar terms (see Fig. 3–5). Sometimes they pile up in such numbers that it is obvious that *something* is there, but only a microscope can reveal the fact that the something consists of uncounted millions or billions of tiny organisms. In nature such a situation occurs, for instance, in the red tides that sometimes sweep an ocean shore. The red tides are caused by concentrations of up to 40 million red protists (generally species of *Gonyaulax* and of *Gymnodinium*) per cubic meter of water. Such red tides may be dangerously destructive because the microscopic organisms are poisonous to many fishes and also to man. In the laboratory, bacteria entirely invisible as individuals become easily visible as groups when they are cultured in large colonies (Fig. 19–1).

The fact that there is a whole world of life smaller than the eye can see has had a profound influence on human thought and history. Most of the organisms that cause human diseases belong in this microscopic realm. In the long millennia before their existence was suspected it was impossible to have a rational idea of the causes of disease, and irrational ideas became deeply embedded in human thought. They are not yet fully eradicated. This gap in the possibility of observation was also partly responsible for the fallacious notion of spontaneous generation. More widely still, the invisible protists have indispensable parts in the operation of nature, in the intricately intermeshed cycle of life. Life itself could not be understood, even in a superficial way, until these tiniest of its manifestations were discovered.[4] The fortunate man who made the discovery of protists was Anton van Leeuwenhoek, a minor Dutch official and amateur biologist, who lived from 1632 to 1723. His must have been one of the most exciting

[4] It is true that it is not always necessary to see an object in order to know that it exists. No one has ever seen an electron, or is likely to, but electrons surely exist, and a good deal is known about them. It is similarly true that we would have to postulate the existence of microorganisms even if we could not see them, and the existence of some of them was indeed postulated before they were seen. (How would you go about determining their existence and characteristics without using a microscope?) Nevertheless, general acceptance and understanding followed the visual demonstration.

experiences anyone ever had! He saw a whole new world more truly than did Columbus or any other explorer. Think of him the next time you look at a drop of pond water under a microscope.

General Characteristics

We have had several occasions before to mention that protists are, in general, one-celled or, in another sense, noncellular organisms. Most of them are organized like single cells; they are not highly developed systems of many cells specialized for different functions within the over-all organism. This description seems at first sight to be a clear definition, but nature (or evolution) rarely deals in absolute distinctions. The boundaries of the kingdom Protista are in fact vague and correspondingly disputed.

On the one hand, there are objects having some of the attributes of life, and therefore sometimes considered protists, that do not have all the vital attributes and that lack cellular organization. Prominent among these are viruses (see Fig. 3–4). When complete, they are comparatively simple combinations of nucleic acids and proteins. They are not self-reproducing, replicating only through the action of their nucleic acids within a living cell of a true organism. In spite of their biological importance, they are not themselves true organisms, and we need not deal with them further here.

The *rickettsias* are a group still more puzzling and more difficult to place in the spectrum from nonliving to living. They are in the size range of large viruses, and thus much smaller than bacteria. Under a microscope they look somewhat like bacteria, but they do not have demonstrably cell-like organizations. Like viruses, they reproduce only in living cells, where they are commonly pathogenic— they cause spotted fever and typhus in man. Pending future knowledge, we exclude them too from consideration as true organisms.

On the other hand, it is impossible to draw a hard and fast line between protists and higher multicellular animals and plants. Even among the bacteria there are some cells that join in strings and other aggregations, al-though the cells are all alike, without any differentiation. The so-called blue-green algae resemble bacteria in having both unicellular and aggregate forms and also in being prokaryotes lacking true nuclei. They are therefore often classed with the bacteria rather than with the true algae, and we follow this lead by including them among the protists. Some organisms still classed as algae or fungi are unicellular or undifferentiated aggregates apparently closely related to multicellular differentiated algae and fungi. As a result, some authorities include algae and fungi in the Protista, but we follow the more conservative course of classifying them as (true) plants. The distinction is really arbitrary. The so-called slime molds seem to be animal-like protists during part of their life cycle but later become multicellular aggregations within which there is some differentiation of cellular functions. We classify them as protists.

The fact that the extremely diverse protists do not make up a really clear-cut category is a consequence of evolution and is one of the evidences of evolution. Similar difficulties of categorization occur within the kingdom Protista. As previously noted (Chapter 4), some protists are autotrophic and some are heterotrophic, both in various ways and to varying degrees. It has been common to call the autotrophs among them "plants" and the heterotrophs "animals." However, some of the forms thus placed in different kingdoms are obviously very closely related, and some do not fit well in either kingdom. In some instances, exactly the same organisms (especially among the flagellates) appear as plants in textbooks of botany and as animals in textbooks of zoology.

The protists apparently represent a variety of organisms not now very closely related to each other or to anything else. These are age-old branches from the tree of life that split off before true or typically multicellular plants or animals had arisen. They have not followed any of the main paths of complication and differentiation seen in the usually larger true plants and animals. Nevertheless, many of them have become highly specialized within their own sphere. They include the bacteria, the blue-green algae, the flagellates and their

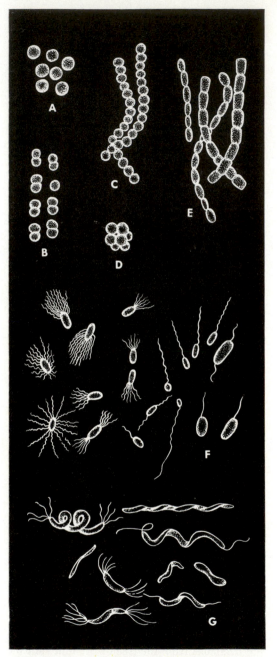

19–2 A diversity of bacterial cell types. *A–D.* Cocci (singular, coccus) are round cells that may grow singly, in pairs, or as long chains. *E.* Bacilli (singular, bacillus) are rod-shaped. They, too, may grow singly as well as in chains. *F.* Many bacteria are flagellated. *G.* Spirilla (singular, spirillum) are helical cells.

allies, the slime molds, and three more animallike phyla sometimes grouped together as Protozoa.

Bacteria

GENERAL CHARACTERISTICS

The most striking physical characteristic of bacteria (Fig. 19–2) is their extremely small size. They are the smallest indisputably living things; only some viruses (which do not meet usual definitions of living organisms) are smaller. The smallest bacteria are about 0.1 micron in diameter. There are viruses larger than that. A giant among bacteria may be as much as 60 microns long by about 6 microns in transverse diameter. With bacteria of usual size, between these extremes, it takes about a trillion (1,000,000,000,000, more simply written as 10^{12}) to weigh 1 gram.

With ordinary (light) microscopes, organisms of such almost inconceivably small size are barely visible. It can be seen that some are spherical, some elongated or rodlike, and some variously spiral. They commonly occur singly or heaped up more or less at random in highly populous colonies, but in a few species individuals are attached to form chains. Further study with the electron microscope has shown that bacteria have rather rigid cell walls outside the cell membranes, which is a plantlike characteristic. Many of them have at least one and usually many hair- or whiplike flagella so delicate that they are commonly not seen in classroom preparations of bacteria. The flagella provide some limited power of movement.

Bacteria are believed to have no organized nuclei or typical chromosomes, but they do contain large clumps of nucleoproteins, and their heredity is mediated by DNA as in higher organisms. Most of them reproduce by simple fission. However, as we learned in Chapter 9, some bacteria demonstrate a phenomenon resembling sexuality. In a large population of *Escherichia coli*, for example, a few organisms are capable of transmitting a chromosomelike body to other organisms through *conjugation*. This process has been observed by electron microscopy, and its genetic consequences have been traced by ingenious genetic studies of the traditional kind. Such occasional sexual exchanges of genetic material within a large population that normally reproduces by fission provide an invigorating redistribution of genetic substance that is beneficial because it tends to prevent an accumulation in the population of altered and thus debilitating genes. Thus, through the existence of "males" and "females," the advantages of sexual reproduction are available even to the lowly bacteria.

The most extraordinary thing about bacteria is that these minute and seemingly simple objects are extremely complex in molecular composition and structure. They carry out all the really basic processes of life. They do not differ fundamentally from other organisms, up to man or to a higher plant, in the complexity and general nature of their transformations of matter and energy. They testify to the unity of life and to the common basis upon which, in the course of evolution, the extraordinary diversity of living organisms has developed.

ADAPTATIONS

Bacteria are highly diverse, differing in size, shape, habitat, and reaction to the various stains used in studying them. The most significant differences among them are in their metabolism and requirements for energy and materials from their environments. As we learned in Chapter 4, some of them are fully autotrophic, needing carbon only in the form of carbon dioxide (CO_2) and synthesizing all their many and elaborate organic molecules from that simple compound, as do the green plants. We should also recall that some autotrophs are chemosynthetic, deriving their energy from inorganic chemical sources, notably sulfur or simple inorganic compounds of nitrogen. Others derive energy from light by way of pigments closely allied to the chlorophyll of green plants. Like the latter, they are photosynthetic, but unlike the latter their photosynthetic process produces no free oxygen.

Bacteria are versatile organisms. However, they are restricted by their small size and the fact that in many instances their locomotion is slight and apparently unoriented. They are further limited by the fact that they must be in a liquid medium, usually a watery one, to carry out their vital processes, including reproduction. Some can float in air or endure other dry environments for considerable periods without dying, but under such circumstances their vital processes are virtually suspended. Active life is not resumed until or unless they again reach a liquid environment. The few photosynthetic bacteria of course require light, but all bacteria are sensitive to radiation, especially ultraviolet radiation, which kills most of them on relatively brief exposure. Hospitals take advantage of this fact and kill many air-borne bacteria with ultraviolet lamps.

Within these general limitations bacteria occur in almost every place that has not been deliberately freed of them, even in places where other forms of life are rare or absent. They are particularly abundant in the waters of the earth, from the greatest depths of the ocean up to high mountain streams, and in damp soils. Unlike all animals and many plants, many (the anaerobic bacteria) can thrive without free oxygen. Indeed, some of these are actually harmed by oxygen. Others (the aerobic bacteria) tolerate or require oxygen. Different kinds of bacteria can live and reproduce at temperatures from 0 to 75°C. Quick-frozen bacteria can survive almost indefinitely at temperatures far below 0°C., and some pass through a stage that can briefly survive boiling (100°C. at sea level), but they do not reproduce in such extreme conditions. Each kind has a temperature at which it grows best. These optimal temperatures range from 12 to 60°C. for different species.

USEFUL AND HARMFUL BACTERIA

The great majority of bacteria are useful in the sense that their activities are essential for the continuous maintenance of living communities. Indeed, the whole scheme of life as it has evolved depends upon bacteria. This is true especially because bacteria are the principal (but not the only) organisms of *decay* and virtually the only organisms capable of nitrogen fixation (p. 30).[5]

When a plant or animal dies, much of the material in its body is in the form of organic

[5] Some blue-green algae can also fix nitrogen.

Hugh Spencer

19-3 Bacteria-containing nodules on the roots of a pea plant.

compounds that cannot be directly utilized by green plants. Many of these compounds are quite stable and do not tend to break down into simpler, usable materials by inorganic processes under usual conditions. Nitrogen, for instance, is mainly in the form of proteins, amino acids, and other compounds more elaborate than the nitrates required by most green plants. Carbon occurs in the same compounds and also in lipids and carbohydrates, not as the CO_2 required by green plants. If the compounds in dead plants and animals were not somehow broken down, by now some crucial materials for green plants would all be locked up in the remains of the dead, and life would have become extinct except, perhaps, for a few autotrophic bacteria.

This is where the bacteria of decay demonstrate their usefulness. Decay is simply the breaking down of organic compounds by bacteria and by some nongreen plants. An essential part of the process is the successive breaking down of proteins into amino acids and then of amino acids into ammonia. Other bacteria then oxidize the ammonia to nitrites, and still others oxidize the nitrites to nitrates,

available as the principal nitrogen source for green plants.[6]

Some bacteria, to be sure, go too far with the breaking-down process and produce free nitrogen, which cannot be used by green plants. This loss is, however, compensated for by those bacteria that can fix nitrogen. They are able to use free nitrogen, which constitutes nearly 80 per cent of the atmosphere, in the synthesis of their own proteins. By further metabolism and decay those nitrogen compounds eventually become available to other organisms. The nodules on the roots of many plants of the pea family (legumes) contain colonies of bacteria that obtain energy from the carbohydrates of the host plant and utilize part of that energy to fix nitrogen (Fig. 19–3). Soils that have become depleted of nitrates, and therefore have become infertile for green plants, can be restored by planting with nodule-bearing plants such as clover or alfalfa. (The cycles of energy and materials involving bacteria are further considered in Chapter 23.)

Although most bacteria are useful in the broad sense that they help to keep the cycles of life going, many kinds are parasites. They are responsible for most of the infectious diseases in man and other animals and for certain of the diseases of plants. Bacteria may invade any part of the body of any larger organism. If the bacteria produce substances, *toxins*, poisonous to the host, disease usually ensues. Many of the bacterial toxins are proteins. Some are excreted by the living bacteria,[7] and others are part of the bacterial cell, liberated after death of the bacterium.

Disease-producing bacteria must have arisen far back in the history of life. The survival of other organisms has depended on

[6] Plants can also use ammonia and, in some cases, amino acids as the immediate source of nitrogen.

[7] This explains the fact that there is one extremely serious disease due to bacteria that does not involve invasion of the victim by the responsible agent. Botulism is caused by eating food in which a certain bacterium (*Clostridium botulinum*) has been living. This bacterium excretes a toxin that is probably the strongest poison known in its effects on man and other mammals. Eating even a microscopic amount of this toxin is usually fatal. Yet the bacteria that produce it are common in soil, and we regularly come in contact with them without any ill effects.

their evolving one or more of a triple set of defenses: resistance to invasion by bacteria, resistance to their survival or multiplication in the host, and resistance to the effects of their toxins. The subject of parasitism and of evolutionary balance between host and parasite is discussed in Chapter 24.

Blue-green Algae

In spite of the name, only about half the *blue-green algae* (Fig. 19–4) are blue-green. The others are highly varied in color: blue, green, yellow, red, and intermediate hues. Some are unicellular and solitary, but most consist of clumps or colonies of attached cells, with little or no differentiation. Probably the most characteristic form is a filament made up of cells attached end to end. In all of them the cell or the whole colony has a mucilaginous outer wall. The cells contain nucleoproteins and their heredity involves DNA, but as in bacteria, the other main member of the group called prokaryotes, there are no chromosomes or nuclei. Reproduction is vegetative or by fission, and no evidence of a sexual or parasexual process has yet been found. Many of these characteristics are suggestive of the bacteria, especially when it is remembered that some bacteria also tend to be grouped in filamentlike strings. It has been suggested that the blue-green algae are in fact advanced, colonial bacteria or (and this may be only another way of saying the same thing) that they are extremely primitive algae, which have remained more or less at the evolutionary level at which algae arose from protists. It is clear, however, that the blue-green algae are not closely related to other (or true) present-day algae. It is possible that the simplicity of the blue-green algae is to some extent secondary. Since bacteria do have sexual processes, the absence of such a process in the blue-green algae is probably due to a secondary loss of sex in protists rather than to an original endowment. This has been rather common.

Blue-green algae are extremely abundant in the seas. The Red Sea may have been so named because of the presence of red "blue-green" algae. They also abound in lakes, streams, and the soil. Some species grow in

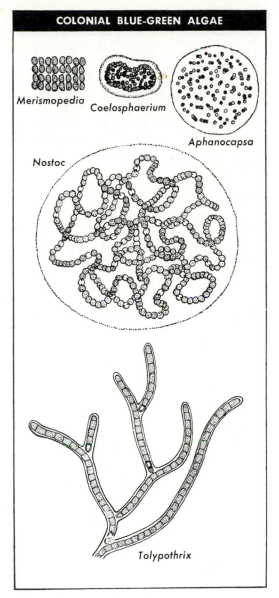

COLONIAL BLUE-GREEN ALGAE

Merismopedia   Coelosphaerium

Aphanocapsa

Nostoc

Tolypothrix

19–4   A variety of blue-green algae.

especially rigorous environments such as hot springs, heavily mineralized waters, antarctic pools, and other places where they (and often some bacteria) may be the only forms of life. How would you account for the fact that such structurally simple organisms as bacteria and blue-green algae may exist in places where complex, higher plants and animals cannot live?

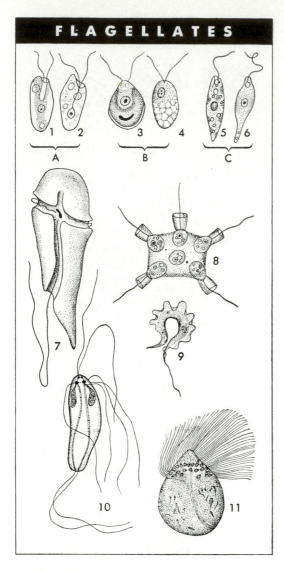

FLAGELLATES

19–5 Flagellates. *1. Cryptomonas ovata*, photosynthetic; *2. Chilomonas paramaecium*, nonphotosynthetic, closely related to *Cryptomonas*. 3. *Chlamydomonas monadina* (×610), photosynthetic; *4. Polytoma uvella* (×870), nonphotosynthetic, closely related to *Chlamydomonas*. 5. *Euglena pisciformis* (×195), photosynthetic; *6. Astasia klebsii* (×300), nonphotosynthetic, closely related to *Euglena*. (See footnote 6 in Chapter 4 for comments on these three pairs of flagellates.) 7. *Ceratodinium asymmetricum* (×600), a dinoflagellate from brackish waters. It possesses two flagella, one of which beats within the confines of a groove running around the cell wall. 8. *Protospongia haeckeli* (×320), a simple colony of eight collar-flagellate cells embedded in a common matrix. (See Chapter 21 for comment on the resemblance between *Protospongia* and the ancestors of sponges [Porifera].) A collar flagellate feeds by trapping debris and microorganisms on its collar; like perpetually moving flypaper, the collar maintains a surface streaming that brings the trapped food down to the cell surface proper, where it is engulfed in a food vacuole. This is also how sponge cells trap and ingest food. 9. *Trypanosoma giganteum* (×400), a parasitic flagellate from the blood of fishes. Other trypanosome species attack livestock and man, causing sleeping sickness. The flagellum runs alongside the cell, and an "undulating membrane" of protoplasm is stretched between the cell and the flagellum, much as a soap film can be stretched between wires. 10. *Hexamitus intestinalis* (×2000), a parasitic flagellate from the gut of frogs and other vertebrates. *Hexamitus* is one of the few flagellates having more than one nucleus per cell. The nuclei are pear-shaped bodies near the top of the cell. (The two long rodlike structures running down the cell connect the granules at the base of each flagellum; their function is unknown.) 11. *Calonympha grassii* (×630), a symbiotic flagellate from the gut of termites. (See Chapter 16 for comment on the significance of these and related flagellates to the termite hosts.)

Flagellates

The flagellates [8] (Fig. 19–5) and their allies deserve special emphasis because they are transitional between strictly plantlike and strictly animal-like organisms and because they also illustrate a possible transition from protistan to multicellular structure. It is evident that no flagellate living today can preserve in detail the characteristics of the ancient protists ancestral to multicellular organisms. It is, however, probable that the flagellates are less modified from that remote ancestry than are any other present-day protists. Sponges and numerous true plants, especially among the algae, have cells or life stages that are almost indistinguishable from flagellates. Moreover, some flagellates form colonies in which there may be some functional differentiation and also a degree of coordination among the individuals. It then becomes a mere matter of definition whether we consider such an aggregation as an advanced colonial union of individual protists or as a rudimentary grouping of cells in an individual of a higher order.

---

[8] Phylum Mastigophora (Greek for "whip-bearers"). "Flagellates" means the same thing but is derived from Latin instead of Greek.

19-6 Marine flagellates. *Top. Ceratium tripos*, a dinoflagellate. *Ceratium*, like *Ceratodinium* (Fig. 19-5, item 7) and other dinoflagellates, maintains one flagellum in motion within a groove on its body wall. (A flagellum is barely visible on the organism at the extreme upper right, waving toward 12 o'clock.) *Bottom. Noctiluca scintillans* ($\times$60), a large phosphorescent flagellate of the ocean's surface waters.

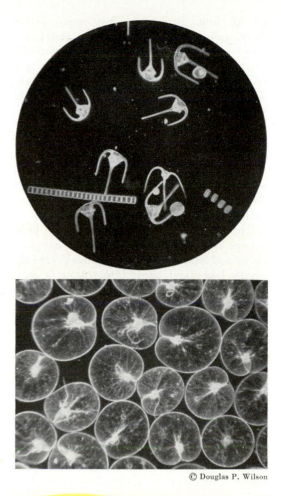

© Douglas P. Wilson

In the absence of any fossil record of the actual transition from protists to multicellular organisms, it is impossible to be sure and the subject is highly speculative. At present, however, the consensus is that multicellular plants and sponges, at least, were derived from colonies of protists somewhat like flagellates. The multicellular animals other than sponges may well have had a similar origin, but that is not the only possibility. We shall return to the problem when we discuss the multicellular animals.

Flagellates are so called because almost every one of them has one or a few long whip-like appendages, the flagella (singular, flagellum). They drive or pull the protist along by lashing. The body usually is more definite and fixed in form than in such protists as the amebas and commonly has some differentiation of organelles, although much less than in the ciliates. One of the most interesting facts about flagellates is that their various kinds have just about every sort of nutrition and associated physiology that is conceivable for organisms of their size. Some (*Euglena* and *Volvox* are common examples) contain chlorophyll, perform photosynthesis, and are generally much like plants, among which they are frequently classified. Others ingest chunks of plant or animal food and are thoroughly animal-like in nutrition. Many are organisms of decay, absorbing molecule by molecule through the cell membrane the breakdown products from the remains of other, dead organisms. (*Euglena* can do this, too, when it is not being a green "plant.") Many are parasites and cause serious diseases, although some are relatively innocuous. The trypanosomes, which cause African sleeping sickness and some other diseases, are flagellates.

Flagellates are very abundant in the sea (Fig. 19–6) and there share with the diatoms (p. 524) basic roles in the turnover of organic materials. The producers of the red tides previously mentioned are flagellates. So are the commonest of the organisms causing "phosphorescence" in sea water. One of these has the appropriate name of *Noctiluca*, the "night-shiners."

Animal-like Protists

Several major groups, phyla, of protists are particularly like animals and are often, indeed, classified as animals. They are the phyla Sarcodina, Sporozoa, and Ciliophora of the classification in the Appendix. Those who prefer to classify them as animals rather than as protists generally unite them [9] in a sin-

[9] And usually also some, at least, of the flagellates, phylum Mastigophora, of our classification.

gle phylum, Protozoa. That name means "first animals" and refers to the belief that they preserve something of the characteristics of the earliest organisms that could, by any possible definition, be considered animals. This may be true in an extremely broad sense, but again we must note that all "protozoa" now living have surely undergone profound evolutionary changes. It is unlikely that they are like the truly first animals in detail. Moreover, it is possible that the most animal-like of the "protozoa" were derived from the flagellates, which are closer to a general ancestry for animals.

The most animal-like features of these groups of protists are, first, that they eat plant or animal food, ingesting it in chunks into the protist body, and, second, that most of them are quite active. Within their small worlds they move about, scouring their surroundings for nourishment. Often there is even a sort of purposiveness in their movements, or, it would be better to say, an orientation: they tend to move toward food or better environmental conditions and away from obstacles or poor living conditions. Thus they do exhibit the rudiments of animal-like behavior. Moreover, as we shall see, the more complex among them have simple organs or organelles that are also more animal- than plantlike.

All of them are small, some as little as 2 microns in diameter, as small as many bacteria. A few, the largest of all protists, reach 3 or 4 centimeters in largest dimension. Most of them range from 100 to 300 microns, too small to be seen clearly if at all with the naked eye but a comfortable size for being studied with a light microscope. They have visible nuclei and chromosomes, and when they divide they go through essentially the same process of mitosis [10] as the cells of higher plants and animals. In these protists the usual apparatus of heredity is fully established. They reproduce by fission, and this does introduce an element in their heredity not present in the same degree, at least, in multicellular organisms. The young protist generally starts out in life not only with a full set of parental chromosomes but also with its bodily structure and material directly derived from the

[10] Some marked variations do occur; in some the spindle develops within the nuclear membrane.

parent. In most groups of these protists there also occurs from time to time an exchange of nuclear material between two individuals. This does not lead forthwith to reproduction, but over a sequence of generations it has the same genetic consequences as sexual reproduction in multicellular organisms. (See Chapter 9.)

AMEBAS AND THEIR KIN

An ameba appears as simple as a fully developed organism can be, and it is famous on that account (Fig. 19–7). It has become a living symbol of the primitive, as in the common expression of evolution "from ameba to man," although it is improbable that anything quite like an ameba ever did really figure in our ancestry.

Through a light microscope an ameba appears to be a simple lump of protoplasm, without top or bottom, front or back, and with little evidence of organelles or special differentiation except for the nucleus. Some species have within the granular cytoplasm a small, clear sphere that looks like a bubble but is really full of a watery solution. This sphere pulses and then collapses, expelling the water. This simple structure is an organ of excretion and of control of osmotic pressure. It gets rid of some waste products and also of excess water. A fresh-water ameba tends to take up more water than it loses through the cell membrane by osmosis. By forcibly expelling water, it solves the problem of how not to swell up to the bursting point.

Although we call this mechanism "simple," it is well to note that even here biology has unsolved mysteries. No one yet knows for certain how the contraction and expulsion are produced, or just what products are excreted in this way rather than through the cell membrane. It is significant that marine protists, which do not tend to take up excess water by osmosis (why not?), usually do not have this device.[11]

When an ameba is active, it moves by a streaming motion of the cytoplasm. That motion pushes out one or more irregular bulges

[11] However, a few marine protists do have it, and a few fresh-water protists do not. Can you think of a hypothesis to explain these facts? Can you design an experiment to test your hypothesis?

in the direction toward which it is moving, and pulls in bulges at the opposite end.[12] If

[12] The bulges are technically called "pseudopodia," or "pseudopods," which means "false feet."

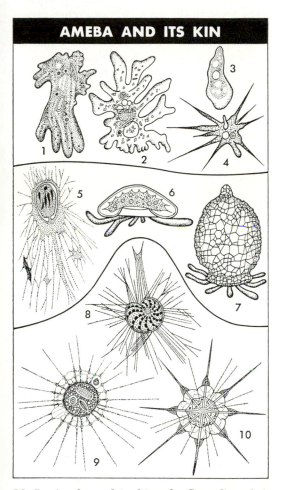

AMEBA AND ITS KIN

19–7 Ameba and its kin: the Sarcodina. *1–4.* Members of the order Amoebina, showing different forms of pseudopodia. *1. Amoeba proteus* (×40); *2. Amoeba dubia* (×40); *3. Vahlkampfia limax* (×280); *4. Amoeba radiosa* (×170). *5–7.* Members of the order Testacea, each of which possesses a shell, or test, of chitinous material, which in some species (6) is heavily thickened and in others (7) is supplemented by the addition of foreign particles. The animal can withdraw completely inside its shell or extend its pseudopodia all over it (5). *5. Gromia ovoidea* (×25); *6. Arcella discoides* (×34); *7. Difflugia urceolata* (×100). *8. Polystomella crispa* (order Foraminifera) (×15). Note the long filamentous pseudopodia. *9. Actinosphaerium eichhorni* (order Heliozoa) (×27). Note again the filamentous pseudopodia. *10. Acanthometron elasticum* (order Radiolaria). Note the siliceous spikes.

food is encountered—the food is frequently another protist—the bulges surround it, and it is taken into the body in a sac surrounded by a membrane. Digestive enzymes are secreted into the cavity thus formed, the products are absorbed through the membrane, and any undigested remnants are expelled from the protistan body.

There are a great many protists more or less similar to amebas in structure and function; tens of thousands of species of this phylum (Sarcodina), living and fossil, have been described. The most varied and perhaps the most interesting are the Foraminifera[13] (or forams, for short), most of which secrete limy shells. The shells, when suitably magnified, are often of great beauty and amazing complexity (Fig. 19–8). It is especially remarkable that such intricate structures, characteristic for each species and fixed by heredity, can be built by what look like completely structureless blobs of living material. Forams,

[13] "Hole-bearers," so called because their shells are usually pierced by many tiny holes.

19–8 The empty cases of various Foraminifera (greatly magnified).

Carl Strüwe

most of which are marine, are so abundant in the seas that much of the bottom ooze is made up of their shells. Forams are also common as fossils. They are good indexes of the ages of rocks and can readily be recovered when a well is bored in rocks originally laid down in the sea. For these reasons their study is useful in the petroleum industry, and they are the principal object of the science of micropaleontology. Forams have made an even greater contribution: much of the petroleum itself has probably been derived from their soft organic parts through the ages, although other organisms have also contributed.

The evolution of the sarcodines (amebas, forams, radiolarians) from flagellates nicely illustrates two common evolutionary principles—overlapping functions and opportunism. Pseudopodia evidently arose first as useful supplementary feeding devices in animal flagellates. Some still-living flagellates exemplify transitional stages that must have occurred in the evolution of pseudopodia. *Oikomonas* (Fig. 19–9) is an animal flagellate in which pseudopodia are developed on a localized area of the otherwise firm cell surface and engulf solid food particles. *Mastigamoeba* is a form that retains a flagellum but develops pseudopodia all over its surface. The evolution in flagellates of a highly-mobile cell surface primarily for feeding purposes created an opportunity for further evolution—of *a new mode of locomotion*. Exploitation of this evolutionary opportunity gave rise to the sarcodines, characterized by ameboid, or pseudopodial, movement. It is noteworthy that many sarcodines (forams, for instance) produce

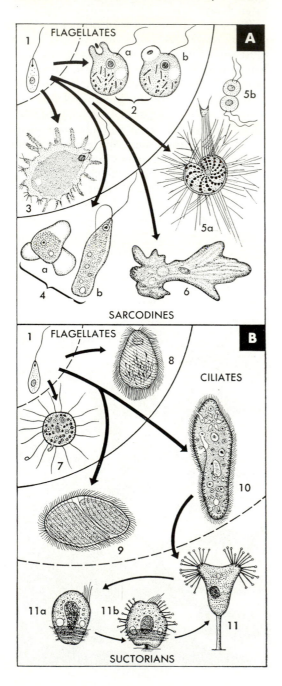

SARCODINES

SUCTORIANS

19–9 Protozoan descendants of flagellates. *A*. Flagellate–sarcodine relations. *1–3*. Flagellates: *Oikomonas* (2) and *Mastigamoeba* (3), which utilize pseudopodia for feeding. *4–6*. Sarcodines, revealing their flagellate ancestry: *Naegleria* (4) alternately pseudopodial and flagellate; *Polystomella* (5), a foram with flagellated gametes (*5b*); *Amoeba* (6), with pseudopodia (a flagellate feeding device) as organs of locomotion. *B*. Flagellate–ciliate relations. *1*. A flagellate. *7, 8*. Flagellates approaching the ciliate condition: *Multicilia* (7), with pseudopodia as a feeding device, many flagella, and more than one nucleus—the last two characters being ciliatelike; *Holomastigotoides* (8), from termite gut, with huge numbers of flagella organized in rows, like cilia. *9, 10*. Ciliates: *Opalina* (9), which lacks the differentiated macronucleus and micronucleus characteristic of true ciliates; *Paramecium* (10). *11. Acineta*, a suctorian and thus a ciliate descendant, as is evidenced by its macronuclear and micronuclear organization and the temporarily ciliated larva (*11a*), which eventually settles down (*11b*) and develops the stalk and suctorial tentacles of the adult.

19–10 Ciliates. Note the distinction between macronucleus (*MN*) and micronucleus (*mn*), which is visible in some of the drawings. The macronucleus is often a long chainlike structure, as in *1* and *4*. *1. Spirostomum ambiguum* (×40), a large freshwater ciliate. Note the oral groove curving inward on the right side; the large area shaped like an inverted hatchet and unstippled is the contractile vacuole. *2. Paramecium bursaria* (×100). The dots in the cell are symbiotic green protists. *3. Cycloposthium bipalmatum* (×210), a parasitic ciliate from the gut of horses. The cilia are compacted together into robust paddle-like processes, the cirri. The elongate structure to the right of the macronucleus is a skeletal rod. Note several contractile vacuoles lying near the macronucleus. *4. Stentor mulleri* (×48), which lives in a case it secretes. The large cilia around the "top" of the animal set up feeding currents in the adjacent water that bring microorganisms into the oral groove and mouth. *5. Ellobiophrya donacis* (×600), attached by two armlike processes to the gill bar of the mussel *Donax*. Recall how pelecypods feed (Fig. 6–2); the gill of a mussel is a rich hunting ground for a ciliate that feeds on microorganisms. As in *Stentor*, the ring of cilia sets up local currents that lead to the mouth. *6. Stylonichia mytilus* (×140), a ciliate that uses its cirri (cf. *3*) as highly coordinated leglike structures; it literally walks (*6b*) on the bottoms of ponds or on leaf surfaces under water. *7. Vorticella* sp. (×80), a ciliate with a contractile stalk that quickly pulls the main cell body away from a source of stimulation. *7b.* A free-swimming individual budded off from the colony. *7c.* Detail of the cell. The ring of cilia around the oral groove (leading deep into the cell) creates a feeding current.

---

flagellated gametes, reminders of the ancestral mode of movement.[14]

SPOROZOA

The phylum Sporozoa especially well illustrates two characteristics that are widespread among the animal-like protists: parasitism and complicated reproductive cycles.

Many of the parasitic sporozoans cause severe disease. They vie with the bacteria in the amount of misery they have caused man and other animals. The combination of parasitism and a complex reproductive cycle is well seen in the sporozoans that cause malaria. While

---

[14] In what kinds of environments would pseudopodia be more efficient organs of locomotion than flagella? How about movement through tissues of other animals? Is it possible that the evolution of ameboid movement created still further opportunities for highly successful sarcodine parasitism of other animals? Sporozoa are probably sarcodine derivatives.

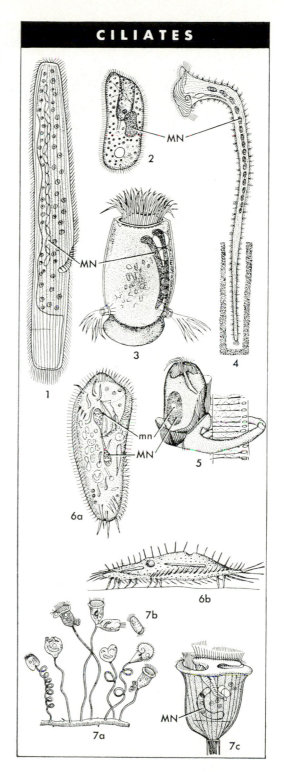

CILIATES

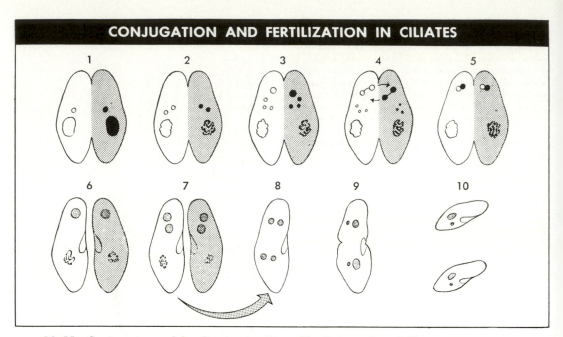

19–11 Conjugation and fertilization in ciliates like *Paramecium*. *1*. Two paramecia come in contact at the oral-groove region; technically, they are said to conjugate. *2*. The micronucleus (diploid) in each cell undergoes the first of two meiotic divisions, and the macronucleus begins the process of degeneration that continues throughout subsequent stages of conjugation and is complete by about *7*. *3*. The micronuclei undergo the second meiotic division; three of the four haploid nuclei produced by meiosis begin to degenerate, and the fourth undergoes a mitotic division to produce (in each cell) two gamete nuclei. *4*. One gamete nucleus in each cell migrates into the other cell. *5*. The nuclear migrations are complete. *6*. The two micronuclei in each cell fuse, completing fertilization; the conjugating cells fall apart. *7*. The diploid zygote nucleus divides mitotically once; the degeneration of the old macronucleus is complete. *8*. (We shall now follow only one of the two conjugating cells.) The micronuclei divide again, and the four resulting nuclei space out, two at either end of the cell. *9*. The cell divides in the oral groove region, producing (*10*) two new paramecia, each with two nuclei. One nucleus in each cell develops into a macronucleus.

living as parasites in mosquitoes, the malarial sporozoans have a sexual process, followed by multiple fission. Introduced into the human blood stream by a mosquito bite, they reproduce asexually and periodically. The cycle is completed when the biting mosquito acquires individual sporozoans from the human blood. The details are shown in Fig. 24–8.

CILIATES

The ciliates (Figs. 19–9 and 19–10) are so called because they have numerous tiny hairlike projections, or cilia (singular, cilium), which beat rhythmically and drive these protists through the water in which they live. They are by far the largest and most complexly organized protists. *Paramecium*, one of the most abundant ciliates, occurs almost

everywhere in fresh water, although, curiously enough, the way in which it colonizes isolated streams or pools is unknown. It has become a famous laboratory organism. It is easy to propagate and to study under a microscope; the range of length is about 100 to 300 microns. Its behavior is complex for a protist, and its genetics is especially enlightening, in part because it goes through the rather odd sexual process, conjugation, shown diagrammatically in Fig. 19–11.

The ciliates dramatically illustrate how much structural and physiological differentiation can occur in a small space and how far from simple even a protist may be. Although there is no division into cells, they have well-developed structures, organelles, analogous to the organ systems of multicellular animals. In

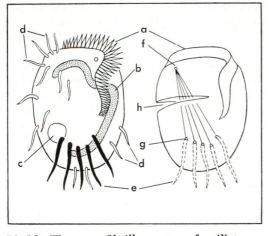

19–12 The neurofibrillar system of a ciliate, an analogue of a nervous system. *Euplotes* is another ciliate (cf. *Stylonichia*, Fig. 19–10) that uses cirri in highly coordinated locomotor movements. It can be demonstrated that the coordination of the cirri is controlled through the nervelike neurofibrils that supply them. *Left.* A general view of *Euplotes: a,* band of cilia down the oral groove to the mouth; *b,* macronucleus; *c,* contractile vacuole; *d, e,* locomotor cirri. *Right.* A view showing an experimental incision that severs the neurofibrils to five major cirri: *f,* fibrils lead out to granules (*g*) at the bases of the cirri; *h,* incision. The cut completely destroys the coordination of these cirri. Equally severe cuts elsewhere in the cell that fail to sever neurofibrils do not affect cirral coordination.

various ciliates there are locomotor systems (the cilia), the operation of which is finely coordinated; "muscular" systems of contractile fibers; reactive coordinating and conductive tracts analogous to an incipient nervous system; an alimentary canal with "mouth," "gullet," and "anus"; excretory organelles; stiffening plates analogous to a skeleton; and other organelles (Fig. 19–12). Although they are single cells, ciliates usually have two or three distinct nuclei of different sizes.

Like the sarcodines, the ciliates were probably derived from early and primitive flagellates, and there are flagellates still living that are more or less intermediate between the two phyla (see Fig. 19–9B). The ciliates illustrate interesting evolutionary phenomena. One of these is a widespread tendency to increase in size. Ciliates generally are much larger than flagellates and sarcodines, as is graphically shown in Fig. 19–13 by flagellates and sarco-

dines symbiotic and parasitic with ciliates. The differentiation of organelles and the inclusion of multiple nuclei in one cell were adaptations necessary for cells of such large size. But these constituted, in the long run, an evolutionary blind alley. Intracellular differentiation and nuclear multiplication could go just so far and no farther. The workable organization of still larger animals became possible only with multicellularity. This theme will be pursued further in Chapter 21.

Suctorians, derived from ciliates and referred to the same phylum, also exemplify the extent of complication possible at the protistan level (see Fig. 19–9B). Unlike other protists, some of them go through definite developmental stages. A young suctorian has larval characteristics and is not simply a small adult.

SLIME MOLDS

At one stage in their life history, slime molds (Fig. 19–14) are unicellular organisms with flagella and might well be classified as animal-like flagellate protists. Later, those that survive lose the flagella and become thoroughly amebalike. In both the flagellate and ameboid stages they reproduce extensively by simple fission, with mitosis of the nucleus,

19–13 The large size of ciliates. *1.* Symbiotic green flagellates in *Paramecium bursaria.* 2. Parasitic sarcodines on *Paramecium.*

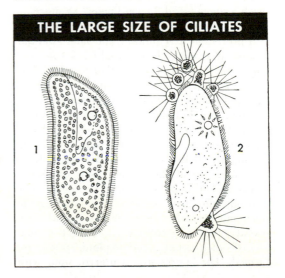

THE LARGE SIZE OF CILIATES

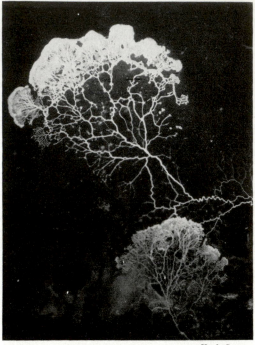

Hugh Spencer

19–14 The plasmodium of the slime mold *Physarum* **(approximately natural size).**

which is haploid in these stages. Eventually many of the ameboid individuals clump together. Then the separate cell membranes usually break down, and the result is a single mass of protoplasm with hundreds or even thousands of nuclei. This mass moves about and ingests food like a gigantic ameba; it may reach a diameter of 25 centimeters or more.

So far the slime mold's life cycle is like that of a protist, at first solitary and later colonial in a peculiar way. The colonial mass (*plasmodium*) may undergo an extraordinary differentiation that is plantlike, at least superficially. From the basal mass stalks grow upward, and bulblike expansions develop at the ends of the stalks. The remaining basal cells or nuclei, those in the stalk and those coating the bulbs, do not further reproduce. They die, just as the somatic cells of higher plants and animals die while the germ cells continue the race. In the slime molds, the inner nuclei of the bulb fuse two by two in a sexual process, resulting in diploid nuclei. Meiosis then occurs, and haploid spores are formed. The spores are scattered, and, with

luck, each may produce an individual of the flagellate stage as the cycle continues.

Are such organisms animals or plants, protistan or multicellular? The alternatives are not clear-cut, and the question hardly makes sense. Perhaps the most likely phylogenetic answer is that the slime molds evolved from protists that somehow struck off on a peculiar line of specialization of their own, not like that followed by any other organisms. This can be viewed as another independent approach toward large size and more or less multicellular organization. It was successful enough to ensure survival of some forms, but it was another blind alley that led no further.

Algae, the Grass of the Waters

We have emphasized that photosynthesis is the basic dynamic process of life. Although it may not always have been, now practically all the energy of living things is first made available to them by photosynthesis, which is also one of the key processes in building up organic compounds. We naturally think of photosynthesis in terms of green fields and forests. Yet it is true now and must always have been true that most of the photosynthesis so important in the economy of life occurs in water, especially in the vast reaches of the sea. Many protists are involved in this tremendous and ceaseless activity, and so are those betwixt-and-betweens, the blue-green algae. However, most of the photosynthesis in water —and that means most of it on earth—is performed by the many kinds of primarily aquatic true green plants, the *algae*.

Algae are symbolically the grass of the waters, not only in the sense that they are the principal energy-fixers of the meadows of the sea but also in the sense that they are the original food source for most of the untold billions of aquatic animals. This greatest of all food sources has hardly been touched by man as yet. The only significant human use—and it accounts for only a small fraction of the food consumed by humans—is indirect: the eating of fish and to still less extent of other seafood. Now that human population has outrun land-food production over large areas, experiments are under way for the cultivation and more

19–15 Diatoms. *Left.* A variety of diatom cell walls from New Zealand waters, showing the diversity of form (×67). *Right. Actinoptychus* (×130), photographed with polarized light.

direct ultilization of algae. The results are not very meaningful as yet, but these efforts may assume importance if the world population continues to increase rapidly.

Some algae are as small as bacteria. Others are more than 200 feet in length. Most of them are multicellular, but some are unicellular. The unicellular algae are not classified as protists because they seem to be related to or even derived from multicellular algae. They are, it seems, cells that get along alone but that are of essentially the same kind as those in their multicellular relatives. The vegetative parts of algae have little differentiation of tissues or organs. In many, all the vegetative cells are practically alike. Others do have some distinctions of size, shape, and function among their cells but without differentiation of such organs as roots or leaves or of such tissues as the vascular, tubelike, sap-conducting bundles of cells that occur in land plants. For this reason the algae are included among the *nonvascular* plants, as opposed to the *vascular* plants, which include most of the familiar land plants. The reproductive structures of algae,

in contrast with their vegetative parts, may be highly differentiated and complex.

Most algae are fully aquatic. Others grow in soil, where they are abundant, or in such odd places as inside or among the tissues of other plants, but always in damp or wet situations. Almost all have pigments of the chlorophyll family and are photosynthetic, but a few have lost the pigment and have become organisms of decay.

In popular language the main groups of algae are usually characterized by their predominant color: green, yellow-green, golden-brown, brown, and red. As we have already noted for the so-called blue-green algae, however, colors are not infallible means of recognizing the groups. More fundamental are details of anatomy, especially in reproductive structures, and of the life cycle. The essentials of reproduction in algae and their widely diverse life cycles have already been discussed in Chapter 9.

Although there is a tendency to lump all the algae together, botanists believe that they represent at least four main groups, each of

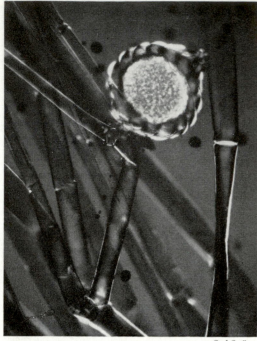

Carl Strüwe

19–16 The stonewort *Nitella*, with a fruiting body or "seed" (×23).

which may have arisen independently from protists. An evolutionary classification should, then, recognize each of these groups as a primary subdivision of the plant kingdom.[15] Algae include the familiar seaweeds and the green scum of still ponds. Among other groups peculiar in various ways are the diatoms and stoneworts. *Diatoms* (Fig. 19–15) are tiny plants, usually unicellular but occasionally in small, colonylike aggregations. They secrete unique skeletal structures made of silica (SiO_2, the same compound as that of rock crystal), sometimes complex and with a delicate beauty. Diatoms are especially important because, despite their small size, they are often so amazingly abundant as to be the main photosynthetic organisms over wide areas, especially in the sea but also in lakes and streams. The whole productivity of the oceans,

[15] See the classification in the appendix. The algae of popular speech are the phyla Chlorophyta, Chrysophyta, Phaeophyta, and Rhodophyta of our classification. The Cyanophyta, here considered protists, are also sometimes considered as algae—the blue-greens.

from the tiniest animals up to the great whales, depends largely on diatoms.

Stoneworts (Fig. 19–16) are of relatively little importance in the total economy of nature, but they are worthy of mention because they are the most complex algae. They look like an abortive attempt to evolve a higher type of plant.[16] They have "roots," branching "stems," "leaves," and strangely marked "seeds." The structures named in quotation marks are not anatomically like the true roots, stems, leaves, and seeds of higher plants, but they do serve similar purposes and show a remarkable degree of differentiation for algae. Stoneworts also secrete lime in the cell walls and may locally contribute to the formation of marl and limestone.

Fungi

The great majority of algae and of other true, multicellular plants have chlorophyll and perform photosynthesis. This is true to such an extent that "green" and "plant" are words almost automatically associated. However, the members of one large and important group of plants do not perform photosynthesis. They are the *fungi* (Fig. 19–17).[17] Everyone has seen some parts of fungi, and many fungi grow to considerable size, but they are for the most part hidden organisms. The basic unit of their structure is a threadlike element, or *hypha*. The hypha consists of an elongate cylindrical wall containing a mass of cytoplasm and hundreds of nuclei that are not separated by cross walls. The mass of hyphae that constitutes a fungal growth is collectively referred to as a *mycelium*.

The mycelium of *molds* visibly spreads on old food in damp, dark places, and sometimes on the surfaces of other plants or even of

[16] For this very reason some students do not classify them as algae but set them aside in a group of their own.

[17] They are often distinguished as "nongreen plants," although as a matter of fact there are a few green fungi, as you know if you have seen a green-molded piece of bread. The green pigment is, however, not chlorophyll. There are also a few "green plants" (members of groups mainly photosynthetic) that have no green pigment. It is so difficult to make a universally valid generalization about things as versatile as organisms or to characterize their groups by names valid for all members of the group.

19–17   A variety of fungi. *Top left.* A bracket fungus (*Fomes*) growing out from a tree trunk. *Top right. Amanita muscaria*, a poisonous mushroom. *Bottom left.* The mycelium of the mold *Penicillium chrysogenum*, showing individual hyphae. This fungus produces penicillin. *Bottom right.* Cells of brewer's yeast (*Saccharomyces cereviseae*) in the process of budding.

animals. *Yeasts* are familiar to the housewife, although what she sees is a mass of organisms rather than the small individuals. Structurally, yeasts are atypical fungi, for they consist not of hyphae but of single cells.

Mushrooms are delicacies; toadstools are a recognized danger. (This is not a scientific distinction; "toadstools" are mushrooms that are poisonous or believed to be so.) *Bracket fungi,* sometimes also called "conks," form spongy-looking but hard shelves on stumps and tree trunks. Farmers can readily see the *smut* or *rust* that may infect their grain and ruin a crop. All these easily visible organisms are fungi—masses of tightly packed hyphae —but hundreds of other fungi are not visible. Even among the readily visible ones what are seen are often only reproductive structures.

The vegetative mycelium often lies hidden in the soil or inside the tissues of a host plant or animal. It is the vegetative part of the fungus within plant and animal tissues that is harmful. When the spore masses appear, as in smuts and rusts, the damage is already done to the host. The mushrooms we eat are immature reproductive parts growing up from a mycelium buried in decaying organic matter.

All fungi are heterotrophs. Many of them require as organic food only a simple sugar and can carry on all the other necessary syntheses from there. Some have highly specific requirements for more complex organic foods. In any case all must absorb through cell membranes a carbohydrate, at least, that has been synthesized by some other organism. They are all either parasites or organisms of decay,

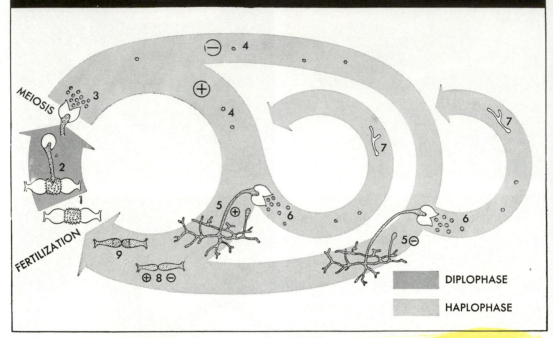

MEIOSIS

FERTILIZATION

DIPLOPHASE

HAPLOPHASE

19–18 The reproductive cycle of the bread mold *Rhizopus*. The diplophase is of minor significance, consisting only of the zygote and the sporangiophore, with its terminal sporangium that produces haploid spores (meiotic products). *1.* Zygote. *2.* Sporangiophore, grown out of the zygote, with the sporangium. *3, 4.* Haploid spores of two mating types, ⊕ and ⊖. (The term "mating type" is used where there is neither morphological nor physiological distinction between the "sexes" other than that ⊕ will mate only with ⊖ and that ⊖ will mate only with ⊕.) *5.* New mycelia (two mating types) germinated from the spores. *6.* Spores. *7.* New generation of vegetatively propagated mycelia. *8.* Two hyphae (technically progametes at this stage), one from a ⊕ mycelium and the other from a ⊖ mycelium, which come in contact and (*9*) produce gametes at their tips; fertilization consists of a fusion of the two gametes.

absorbing food from the living bodies or dead remains of other organisms.

Fungi reproduce asexually by spores, which may be as small as 1 micron in diameter, although they are usually several times that size. Such tiny objects float freely in air, and the air around us is rarely without a multitude of them. If they land on a suitable medium, the spores develop into vegetative bodies, which in turn develop anew the spore-bearing organs, often elaborate in shape and structure. Thus almost any piece of bread left exposed to air develops mold; damp shoes and luggage made of parts of dead animals become mildewed; a compost heap sprouts mushrooms —all developed from the ubiquitous air-borne spores. Most fungi also have sexual processes,

and their life cycles may be quite complex (Fig. 19–18).

The most widespread and destructive diseases of plants are caused by parasitic fungi, which do many millions of dollars of damage every year to crop plants, vegetables, and trees. Even when the plant is not killed outright, it may be stunted, its productivity reduced, or its fruit blotched or scabbed so as to be unmarketable. Diseases caused by fungi are less common in animals than in plants, but they do occur. Athlete's foot, "ringworm" (which is due to a fungus and not a worm and whose technical name is trichophytosis), and San Joaquin fever (coccidioidomycosis) are human fungal diseases.

Although some fungi do untold damage

19–19 Lichens. *Left.* A foliaceous (or leafy) lichen on a rock. *Right. Cladonia* growing tightly on the surface of a fallen tree.

from our human point of view, many others are highly useful from the same point of view. Brewing and all the industrial operations involving fermentation depend on yeasts, which have the attractive property of metabolizing sugars into alcohol and carbon dioxide. Bread raised with yeast utilizes the same reaction, but here the carbon dioxide bubbles in the dough are desired, not the alcohol. (What becomes of the alcohol?) A number of fungi, especially molds, synthesize compounds poisonous to competing organisms and particularly to bacteria. The term *antibiosis* is applied to the phenomenon in which organisms take the lives of others to preserve their own. The naturally synthesized antibacterial compounds are called *antibiotics*. They are metabolic products, formed regardless of any "need" the fungi may have for protection against invasion by or competition from bacteria. Some of them destroy bacteria that cause human diseases but are only mildly if at all toxic to the human hosts of the bacteria. The first of these life-saving products of fungi to be isolated, penicillin, was named for the mold that produces it, *Penicillium*. It has now been followed by streptomycin, aureomycin, and many others. Explorers all over the world are collecting soil samples from which fungi can be grown in laboratories. Each fungus is then tested for the antibiotic properties of its products. Some prove to be no more effective than others already known. Some turn out to be as injurious to humans as to the invading disease-producers. But one in many becomes a new weapon in the fight against disease—and our debt is increased to the fungi and to their intricate chemistry.

Lichens

Lichens (Fig. 19–19) are familiar as the scale-like, varicolored patches on rocks or tree trunks. Those who have been in the far north know another lichen as "reindeer moss," which is not really moss. A lichen is a composite of two quite distinct kinds of plants intergrown in close, obligatory association. A fungus forms a dense web of threads within which grows an alga. The fungus obtains organic food from the alga, and the alga obtains water and dissolved salts from the fungus. Neither plant can live alone under natural conditions, but together they can live on bare rocks, in arctic climates, and in other situations where few or no other plants can survive. They are pioneers that often play an important role in the first steps of breaking down rocks into soils. Their relationship is an excellent example of symbiosis, a way of living happily together discussed in Chapter 24. The alga and the fungus that together constitute a lichen never grow separately in nature. In some cases it has been possible to grow them separately in the laboratory, but then neither assumes the form characteristic of the lichen. The combined natural forms are classified in genera and species as if they were single plants, but their placement in higher categories is anomalous since the single organic system is *both* an alga and a fungus. A further complication is that the alga involved may be blue-green, hence a protist and not a true alga.

The lichens are true land plants, just as much as the higher (vascular) plants, and indeed some of them can grow in drier, more exposed environments than any other plants, and even with no soil at all. They provide stunning examples of versatility in adaptation and of alternatives for meeting adaptive requirements.

CHAPTER SUMMARY

Protists: their size, diversity, and importance for man and for life as a whole; their unicellular nature; their systematic status.

Bacteria: their extremely small size; the absence of nuclei; the great diversity in chemical and physiological respects; autotrophic and heterotrophic bacteria; the universality of bacteria in all environments; useful bacteria—their role in decay and in nitrogen fixation; harmful bacteria—their role in disease; toxins.

Blue-green algae: not true algae; absence of nuclei and of known sexual processes.

Flagellates: their importance as little-modified descendants of forms transitional between plant and animal and transitional between unicellular and multicellular life; the flagellum as a locomotory organelle; the physiological diversity of flagellate types; their ubiquity, abundance, and importance in food chains.

Animal-like protists, the Protozoa: the ameba and its kin—general form, the contractile vacuole and its functions, the food vacuole, pseudopodial locomotion, Foraminifera; the sporozoans, parasitic protists with complex life cycles; the ciliates—their form, cilia, relatively large size, complex behavior, sexual process of conjugation, complexity of structure for a unicellular organism; the slime molds—their life cycle, their fruiting bodies and spores.

Algae: photosynthetic organisms—simple plants; a primary food source for the majority of animals; the possibility of human exploitation as food; their wide range in size; unicellular and multicellular forms; their nonvascular nature; the complexity of their reproductive structures and cycles; their aquatic environments; types—seaweeds, the green scum of ponds, diatoms, stoneworts.

Fungi: nongreen (nonphotosynthetic) plants; the hypha as the unit of their structure; types—molds, yeasts, mushrooms and toadstools, bracket fungi, with the mycelium often hidden in soil or parasitized tissues; fungi as heterotrophs; their reproduction; the ubiquity of fungal spores; their possession of sex; fungi as agents of destruction and disease; useful fungi—yeast, penicillin and other antibiotics.

Lichens: tough symbionts of alga and a fungus.

20

More

Complex Plants

Flowers like those of the day lily shown here characterize the most complex of plants, the angiosperms. The angiosperms and other, more primitive land plants are the subjects of this chapter.

If you were asked for examples of plants, it is unlikely that you would come up with any of those discussed in the last chapter. Fungi are, indeed, abundant all around you, but mostly as spores too small to see or else as developed plants largely hidden in the soil or in other organisms, living or dead. Algae may be even less accessible to you unless you live near a pond or seashore. Some algae, to be sure, almost certainly are present somewhere near you, but most are so inconspicuous that you are hardly aware of them. In their active phases of growth and reproduction, at least, the algae and fungi must be in water or near it. If they are entirely immersed in water, they may grow to great size, as the giant kelps and some other seaweeds. Parts that grow out into the open air, where you live and are most likely to see them, are limited in size. A few fungi, among the mushrooms or the bracket fungi, may have exposed parts as much as 50 centimeters in diameter, but most fungi are much smaller. The combinations of algae and fungi that we call lichens are true land plants, but they are rarely very prominent. It is the higher land plants, those that push up freely from the soil into the air, that are most conspicuous in our own environment.

Conditions of Land Life

Every living cell must obtain water and other materials by absorption through the cell membrane, and such materials must be constantly available throughout the whole of an active organism. Protists are not long active unless wholly immersed in water. As long as this condition is met, there is no special further difficulty for a very small or unicellular organism. Processes of molecular diffusion suffice to distribute materials in the tiny body. The multicellular fungi and algae also get along if each cell is in contact with external water, which must of course also contain in solution the materials needed by the cells of the particular organism. Difficulties arise when some cells are internal—out of direct contact with the environment—and especially if some parts extend into the air, distant from a source of water and dissolved materials. Then in a relatively simple plant like an alga or a fungus, water must be passed along from cell to cell by osmosis and diffusion. The processes are slow and ineffective over any considerable distance inside the plant.

This is a first barrier to any great extension of a plant into the air. It has been overcome most completely by the plants that have developed vascular tissues. These tissues, in some ways analogous to the circulatory systems of animals, make possible the prompt supply of needed solutions to cells far from an external source of water.

A differentiation of organs accompanies, or indeed even preceded, development of a vascular system. A land plant needs water and minerals (in solution) from below, from the soil, and it needs sunlight and CO_2 from above. Below there develops a specialized absorptive system, the roots, and above there develops a specialized photosynthetic system, the leaves.[1] Between roots and leaves is the stem, a structural element containing the central conduits of the conductive system.

Another requirement for land life is especially associated with the stem. In water an organism may simply float, and in soil it lies among the mineral grains. There is no special requirement for support. A stem or other structure extending into the air must support itself and anything attached to it, such as fruits or a crown of leaves. Thus all the higher land plants have specialized supporting tissues. In plants growing to any considerable height, the support must be particularly strong and rigid; their stems are woody.

Still another limitation imposed by land life has to do with sexual reproduction. Gametes are, in effect, protistlike organisms that cannot long survive exposure to air. In an aquatic environment one or both of the gametes that are to unite can simply be shed into the surrounding medium. This is impossible in air. Conquest of this difficulty has involved a long series of specializations in land plants, culminating in pollen and seeds.

Bryophytes

Among living plants the *bryophytes*[2] (Fig. 20–1)—the *liverworts, hornworts,* and *mosses*—represent a group that might be called amphibious. They do not have well-developed vascular systems and are thereby restricted in size and in possible range of land environments, but they are considerably advanced over the algae in adaptation to the land. The better-developed among them do have structures superficially similar to roots, stems, and leaves.

Some bryophytes are aquatic, although none is marine. A few manage to live in arctic wastes and arid deserts. Those hardy species can survive cold and dryness by suspending vital activities until warmth and moisture come along, when they revive. Bryophytes grow most luxuriantly, however, in moist, shady places and in bogs. The low, tangled vegetation of mosses often holds water like a sponge, so that even on dry land they make for themselves what is practically an aquatic habitat. They reach a height of a few tens of centimeters at most (usually much less) and are rarely solitary. Usually they form an extensive, dense mat, a little world in itself, inhabited also by bacteria, algae, fungi, worms, insects, and snails.

We have already learned something of the reproduction and life cycle of the bryophytes (Chapter 9). The conspicuous green plant is the gametophyte. The sporophyte, with little or no chlorophyll, is in most cases a virtual parasite on the gametophyte. Only in hornworts is the sporophyte partially independent of the gametophyte. Although remarkably hardy in revival after drying, bryophytes require moisture for sexual reproduction: the sperm must swim in order to meet and fertilize the egg.

In comparison with the vascular plants, the bryophytes seem primitive and at a distinct disadvantage. Without true vascular tissue, they cannot rise high in the competition for light. They must be wet at some time if the usual life cycle is to be completed. The gametes are highly vulnerable. Yet the bryophytes have survived and are still abundant. One reason is that there are some situations for which they are really better adapted than other plants. A peat bog, characterized especially by the moss *Sphagnum,* is such a place.

---

[1] It is true that some land plants lack true roots or true leaves. These, however, have other more or less specialized organs serving the same functions.

[2] Phylum Bryophyta ("moss plants"). For the classification of these and other plants referred to in this chapter, see the appendix.

Another reason is that they are tough. If they cannot compete for the good things of life they manage, like the poor, to make do with what they have. If they cannot rise into the sunshine, they get along with more modest photosynthetic demands in shade too deep for most green plants. They are often pioneers, spreading into places not yet reached by other vegetation. When, in part through their own activities, conditions have been improved, other plants move in and the pioneers may be crowded out. Such is the usual fate of pioneers, plant or human.

Early Vascular Plants

All the groups of plants that are still to be dealt with have vascular tissue. All had the evolutionary potential to rise into the air and become upstanding growths. All have followed, to varying degrees, an evolutionary path nearly opposite to that of the bryophytes. In the bryophytes, as you know, the gametophyte became the principal or sole vegetative stage in the life cycle. That development was quite successful in a limited sort of way, but it seems to have run those plants into an evolutionary blind alley from which no radical further progress is possible. In the vascular plants the gametophyte became reduced until in the latest and most progressive groups it is microscopic and transient. The female gametophyte became a well-protected parasite on the sporophyte, and the sporophyte is in all vascular plants the principal, and usually the only, vegetative, photosynthetic phase in the life cycle.

Plants that had reached only the early phases of these important changes were at first spectacularly successful in covering the land with vegetation. Later most of them were supplanted by plants with more specialized reproduction, by means of seeds. A few relicts of earlier groups live on, however, and one such group, that of the ferns, has survived in considerable abundance.

PSILOPSIDS

The earliest known vascular plants belong to the *psilopsid* [3] group, for which there is no

[3] Subphylum Psilopsida.

Top and bottom, American Museum of Natural History; center, Hugh Spencer

20–1 Bryophytes. *Top.* A liverwort, *Conocephalum conicum. Center.* Sporophyte capsules of the moss *Pottia truncata. Bottom.* A mat of *Sphagnum* moss.

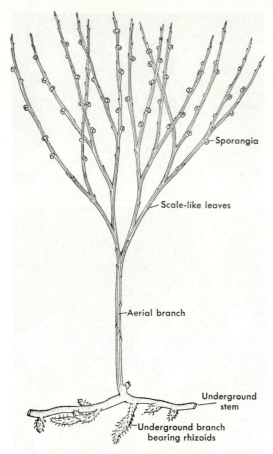

20–2 The sporophyte of *Psilotum*.

LYCOPODS (CLUB MOSSES)

The *lycopods* [4] are another group now represented only by relicts, four genera in this instance, but with fairly numerous and widespread species (Fig. 20–3). Most of the living lycopods are low herbs, although some of the vinelike species may reach as great a length as 20 meters. In their time of glory, in the coal forests of the Carboniferous, some lycopods were great forest trees, 1 to 2 meters in diameter and 30 to 40 meters high.

Lycopods have true, although sometimes poorly developed, roots as well as differentiated stems and leaves. Their anatomy in general is more complex than that of the psilopsids. They show another advance over both the psilopsids and the nonvascular plants. Some produce two distinct kinds of spores (microspores and macrospores), one of which develops into a male gametophyte and the other into a female gametophyte. The gameto-

[4] Subphylum Lycopsida.

20–3 Sporophyte of the lycopod *Lycopodium clavatum*. Note the upright branches with minute leaves (sporophylls) on which the sporangia are borne; the conelike structure formed by the sporophylls collectively is a strobilus.

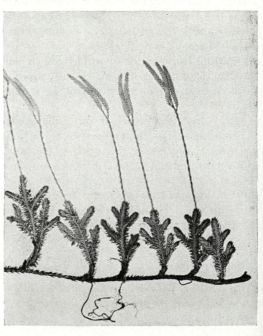

common name in English. It is nearly extinct, with only two relict genera surviving (*Psilotum*, Fig. 20–2, and *Tmesipteris*). It was a varied group early in the history of vascular plants, but was soon almost entirely replaced by more progressive forms. The whole anatomy tends to be almost diagrammatically simple. There are no true roots, a subterranean part of the stem serving the same needs. Some psilopsids have small, simple leaves, while others are leafless; in the leafless plants the aerial part of the stem carries on the necessary photosynthesis. Gametophytes are as yet unknown for the ancient psilopsids. They must have had them, but they were ill-adapted for preservation as fossils. In the surviving forms the gametophytes are small, colorless, and subterranean.

phytes are tiny but independent organisms, sometimes photosynthetic, sometimes colorless and living on organic debris in the soil.

A few fossil lycopods displayed another remarkable advance in mode of reproduction. The eggs developed in a well-protected case on the female gametophyte and were retained there even after they had been fertilized and had begun to develop into embryonic sporophytes. The results resembled seeds. (The living genus *Selaginella* exhibits the same property.) It is unlikely that the seed plants evolved from these particular lycopods, but they show clearly how such a momentous change could take place. They also illustrate the widespread occurrence of independent parallel selective processes leading to evolutionary advances.

SPHENOPSIDS (HORSETAILS)

Living *sphenopsids* [5] all belong to one genus, *Equisetum*, with only about 25 species (Fig. 20–4). They are nevertheless locally abundant and are usually easy to find around bogs or in sandy soil near streams. They are commonly known as "scouring rushes" because the stems contain gritty silica and were useful for cleaning pots and pans in the days before soap operas and the products they advertise. Our native species are rarely over a meter tall, although some tropical equisetums reach much greater heights. The equisetums, too, are relics of a group that flourished in the ancient coal forests and then included large trees.

Nothing like a seed is known in any sphenopsids, and the spores are rarely, if ever, differentiated into microspores and macrospores. Otherwise, in degree of differentiation and in life history, the sphenopsids are quite like lycopods (or ferns). They are, however, quite different in appearance and in details of anatomy. The stems are vertically ribbed and jointed, with a whorl of leaves at each joint. In the living genus the leaves are small, and photosynthesis occurs mainly in the stems, which are of course green. In some extinct species the leaves were large and were probably the main photosynthetic organs.

[5] Subphylum Sphenopsida.

American Museum of Natural History

20–4   Terminal parts of sporophyte of *Equisetum sylvaticum* (wood horsetail). Note the strobili containing sporangia.

FERNS

From primitive psilopsids three major groups of plants evolved with divergent complication in their anatomical and functional differentiation but with little change in major features of reproduction or life cycles. All three reached their climax about the time of the Carboniferous coal forests. We have seen that two of those groups, the lycopods and the sphenopsids, greatly declined thereafter but still have a few living survivors. The still more ancient and originally ancestral psilopsids also straggled along even when overshadowed by their more exuberant descendants. The third of the progressive divergent groups, the *ferns*,[6] also declined [7] in importance but is

[6] Class Filicineae.

[7] The decline of the ferns was not so great as is sometimes stated, because they were never so dominant as was once believed. Early students of paleobotany thought that ferns were the commonest plants of the Carboniferous, and you may still see that time referred to as an "Age of Ferns." It is now known, however, that many plants of that age formerly believed to be ferns because of their fernlike foliage were really seed plants.

Left, Hugh Spencer; right, New Zealand Consulate

20–5 Ferns. Most ferns are relatively small, herblike plants like the Christmas fern (*left*). A few, however, have evolved the arborescent (tree) habit like the tree ferns in New Zealand (*right*).

still so abundant and varied that its members cannot be called unsuccessful or relics (Fig. 20–5).

The reproductive and life cycle of ferns is similar to that of the vascular plants just described; it has been discussed in Chapter 9. In some respects the ferns are even less progressive than some of the ancient lycopods. Lycopods and ferns do not represent successive evolutionary steps, but are divergent groups that arose at about the same time and have remained at about the same level. The gametophyte in ferns is small, but is usually an independent, photosynthetic plant. The eggs are fertilized where they are formed, on the damp undersides of the gametophyte, but the sporophyte develops directly from the zygote, without the appearance of a seedlike, protected embryo. Sperms are still flagellated cells that must swim through an external watery medium if they are to encounter and fertilize the eggs. In this respect ferns are no better adapted to land life than are mosses. Ferns are also comparatively unprogressive in that only a few of them have differentiated microspores and macrospores. Vegetative reproduction is common in ferns.

Although few nonbotanists have ever noticed the gametophyte of a fern, the leaves, at least, of the sporophytes of various species are familiar to everyone. The leaves may be only a few centimeters in length, but in some tropical climbing ferns they may reach the astonishing length of 30 meters or more. Some ferns have simple leaves, but most have leaves divided into leaflets, which may in turn be subdivided.[8] Thus arises the lacy appearance so characteristic of the most familiar fronds of ferns. The spores are borne in sporangia on the undersides of the fronds. These are quite evident as little brown dots in many of our native ferns. In some species some of the fronds have no spores, and in some species the spores develop on specialized parts unlike the vegetative fronds.

The stems of ferns have a well-developed vascular system, and true roots are usually present. In the familiar species of our Temperate Zone the stems are usually horizontal, on or in the soil, with simple roots (occasionally absent) extending into the soil from

[8] The whole leaf is a *frond*; if it is divided into leaflets, they are called *pinnae*, and if the pinnae are further subdivided, their parts are called *pinnules*.

534   THE DIVERSITY OF LIFE

the prostrate stem. In more uniformly warm climates many ferns have large, vertical stems or trunks which may grow to great heights and develop bark: such are the *tree ferns*, common in the tropics but not extending far into regions of cold winters. The tree ferns have better-developed root systems than the herbaceous species, but even in these the roots are less extensive and complex than in most seed plants.

Ferns are most abundant in warm, moist, shady situations. Some manage to grow in cold and in arid climates, but they increase in numbers and variety toward the humid parts of the tropics. They culminate in wet tropical forests in a great profusion of forms, which include not only herbaceous species, similar to most of ours, and tree ferns, but also *creepers*, *vines*, and *epiphytes*. (Epiphytes grow on other plants, using them for attachment, but they draw no nutrition from their hosts and are not parasitic.) In rain forests such epiphytic ferns as the staghorn (Fig. 20–6) contribute to the aerial gardens that grow profusely on the upper trunks and branches of the tall forest trees. In these situations the forest floor is dark, with so little light that few photosynthetic plants can live there. The epiphytes are too small to receive the limited sunlight below, and so they utilize their tall competitors as bases and thus grow far up where they can share the light.

A few ferns are aquatic. In these species there is a differentiation of microspores and macrospores, and separate female and male gametophytes consequently develop. Can you think of any reason why this separation of the sexes is associated with aquatic life?

Early Seed Plants

Abundant as they are, the ferns still do not represent completely successful adaptation of plants to land life. The gametophytes, particularly, are highly vulnerable to adverse environmental conditions. The sperms still require environmental water, even if only a film of it, if they are to fertilize the eggs. The most sensitive and vulnerable stages in the life cycle were eliminated in the plants that evolved *seeds*. In them, as you will recall (Chapter 9),

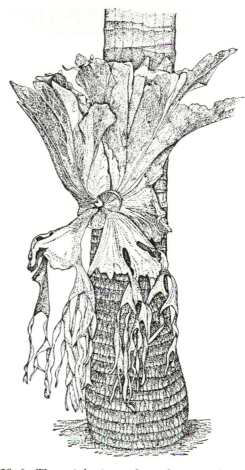

20–6 The epiphytic staghorn fern growing on a palm stem. (From W. H. Brown, *The Plant Kingdom*, Ginn, Boston, 1935.)

the partially developed male gametophyte becomes a *pollen grain*, which is often highly resistant to drying and can float for long periods in the air without dying. The female gametophyte is entirely parasitic and lives its whole life protected in the tissues of the parent sporophyte. The male gametophyte completes its development and the zygote also forms within these parental tissues. The zygote then develops further into an embryonic sporophyte, which is enclosed in protective tissue and provided with food before it is freed from the parental plant. Then it is by some agency moved off to start a new, independent sporophytic generation.

Associated with the evolution of seeds is also the most extensive and effective develop-

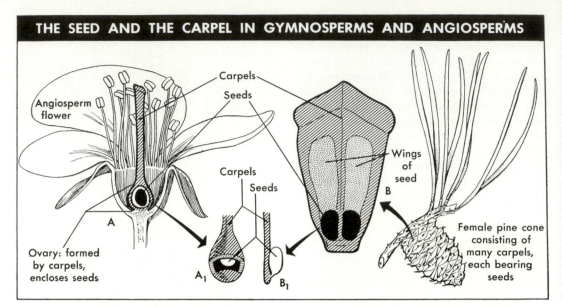

THE SEED AND THE CARPEL IN GYMNOSPERMS AND ANGIOSPERMS

20–7 The seed and the carpel in gymnosperms and angiosperms. In an angiosperm (*A*) the carpel or carpels form an ovary, which encloses the seed. In a gymnosperm (*B*) the carpel is a flat, primitively leaflike structure bearing the seeds nakedly on its surface. *A₁* and *B₁* are schematic representations of the relation of seed to carpel in angiosperm and gymnosperm, respectively; the ovary wall is cut away in *A₁* to show the seed inside.

ment of root and stem systems and tissues and of leaves. These factors have made the seed plants incomparably the most abundant and widespread of land plants. In most terrestrial environments they have nearly, although not completely, ousted and replaced all the groups of nonseed plants. Some seed plants have become aquatic, but the group is far from dominant in fresh water and is almost absent in the seas. Why has there been no significant tendency for seed plants to replace the far simpler algae in aquatic environments?

In the first seed plants to evolve, protection of the seeds was not yet perfected and flowers had not yet appeared. Spores, gametes, and seeds developed in organs sometimes of considerable complexity but without the still higher degree of organization seen in a true flower. The seeds arose on leaflike structures (technically known as *carpels* or *sporophylls*) that did not enclose them; thus the seeds were comparatively naked (although they had other protective coatings). Plants with such seeds are called *gymnosperms* (from Greek for "naked seeds"). Plants with flowers and with seeds encased in an ovary (whose tissues

are in fact the carpels, or sporophylls, which bear the seeds) are *angiosperms* ("enclosed seeds") (Fig. 20–7).

The gymnosperms, first seed plants on the scene, early (mainly in the latter part of the Paleozoic) underwent an extensive adaptive radiation. Numerous groups became divergently adapted to various environments and ways of life without, as a rule, much fundamental progressive change in basic characteristics. Some of the main branches of this basic radiation—the seed ferns, cycadeoids, and cordaites—have become extinct, probably through competition with more progressive or efficient plants. Others survive but are so diminished in numbers, diversity, and geographic distribution that they are today mere relics: the ginkgos, cycads, and joint firs.[9] Only one main branch of the early gymnosperm radiation has continued to the present in great abundance and diversity, although even it is past its heyday: the conifers.

[9] But the joint firs, most familiar to us as Mormon tea (a species of *Ephedra*), seem never to have been abundant, and they may not have arisen in the basic gymnosperm radiation.

SEED FERNS

The coal forests of the Carboniferous were dominated by shrubs and trees with fernlike leaves (Fig. 20–8). It was long assumed by paleobotanists that all these plants were, indeed, ferns. Slowly, however, the suspicion grew that some of them might be gymnosperms. At first this conclusion was based on minor anatomical peculiarities of the stems and leaves. Finally it was confirmed by finding fossils in which gymnospermous seeds were actually attached to "fern" fronds. The difficulty was one that often arises in the study of fossil plants. Roots, stems, leaves, and reproductive cells and organs may all be found, but usually not in direct association. The paleobotanists' problem then is to decide which leaf goes with what type of wood, which seed or spore with which leaves, and so on. Some of the fernlike Carboniferous vegetation has turned out really to belong to ferns, plants with spores and no seeds. Much of it, however, belonged to plants with seeds, and those plants are now generally called *seed ferns*.[10] The name is not wholly apt because the plants may better be considered gymnosperms with fernlike leaves than ferns with seeds. The group is, however, more or less intermediate between early ferns and the more advanced gymnosperms.

The seeds of the seed ferns usually grew singly and, so to speak, openly rather than in the conelike structures, or *strobili* (singular, *strobilus*), so characteristic of most gymnosperms. They were attached along a frond or leaf stalk in various ways, not unlike the varying attachments of spore cases on fern leaves. The plants were evidently primitive as gymnosperms go, and it is not surprising that the seed ferns, although early abundant, were soon replaced by more progressive groups, and are now wholly extinct.

GINKGOS

Ginkgo biloba, the maidenhair tree, is a famous living fossil (Fig. 20–9). Its foliage is much like that of a maidenhair fern, although larger, whence its popular name; but it bears true gymnospermous seeds, rather like those

[10] The accepted technical designation of the group is Pteridospermae.

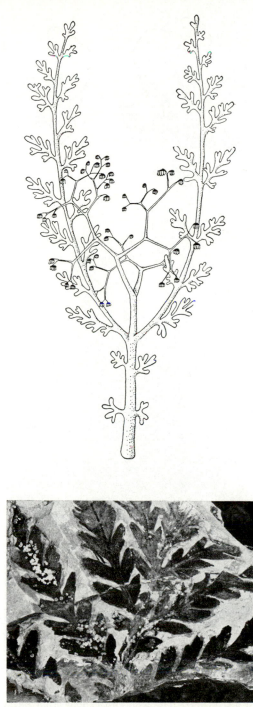

Chester A. Arnold

20–8 Seed ferns (or pteridosperms), the most primitive gymnosperms. *Top.* A seed fern's sporophyll, bearing clusters of sporangia. *Bottom.* Fossil seed-fern foliage.

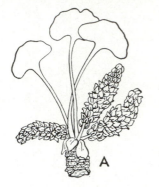
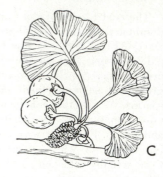

U.S. Forest Service

20–9 *Ginkgo biloba*, **the maidenhair tree.** The two sexes occur as separate trees. *Above. A.* A branch of a male tree with mature male strobili. *B.* A branch of a female tree with young ovules (female sporangia). *C.* A branch of a female tree with mature seeds. *Left.* A row of ginkgos.

plants in China.[11] The ginkgo is a striking, woody, branched deciduous tree that has recently been spread all over the world by man. It is now familiar along American streets and in parks, although few people who know it are aware of its long and romantic history. It is a tough plant, vying with the plane tree in its ability to survive in the poisonous, sooty air of large cities. How do you suppose it happens that a plant that had become virtually, if not completely, extinct in nature thrives in cultivation?

Ginkgos share with cycads a primitive feature lost in all other living gymnosperms and in angiosperms: they still have motile, flagellated sperms. The pollen, which is, as you recall, a partly developed male gametophyte, lands on the structure enclosing the egg cells. There the pollen develops further and produces sperms that swim to meet and fertilize the egg cells, much as in moss or ferns. In these primitive gymnosperms, however, the distance to be swum is very small, and the swimming is indoors; it is in a chamber enclosed by the tissue around the eggs. In other living gymnosperms and in angiosperms there are no motile sperms, and pollen tubes develop in the way already described. It is probable

of a seed fern. The wood and some other parts are more like those of a conifer. The ginkgos are an ancient main branch of gymnosperm radiation, and they were once spread over most of the earth. Only a single species has survived, mainly in the form of cultivated

[11] It used to be said that there are no living wild ginkgos, but recent reports have it that some are still growing wild in a limited area in China. It is possible, however, that these are trees that have escaped from cultivation.

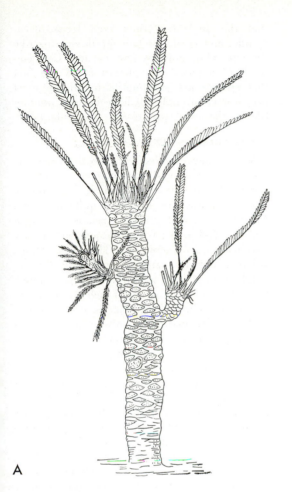

A

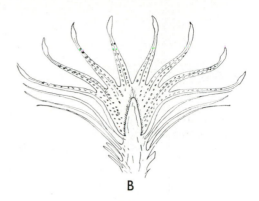

B

C

D

20–10 Cycadeoids and cycads. *A.* The extinct cycadeoid *Williamsonia sewardiana* (Jurassic, about 150 million years ago). *B.* A median section through the so-called "flower" of the cycadeoid *Cycadeoidea*; the central conical structure bore ovules, and the surrounding corolla-like structure bore "stamens" at the points marked by the circular scars. The cycadeoid "flower" is not historically related to the flowers of true flowering plants. *C.* A living cycad, *Dioon edule,* in Mexico. Note its palmlike character. *D.* A male strobilus. *E.* A single sporophyll from a male strobilus, bearing several microsporangia containing pollen. *F, G.* Sporophylls from a female strobilus, each bearing two ovules: *F,* young; *G,* older.

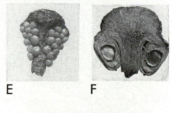

E   F

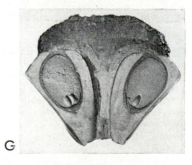

G

A, B, Chester A. Arnold; C from Chamberlain, *The Living Cycads,* Univ. of Chicago Press; D–G, from G. M. Smith *et al., A Textbook of General Botany,* 5th ed., Macmillan, New York, 1953.

MORE COMPLEX PLANTS   539

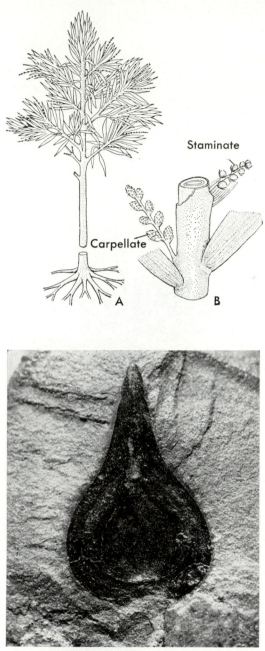

Staminate

Carpellate

A

B

Chester A. Arnold

20–11 Cordaites, extinct gymnosperms. *Top. A.*
A whole plant, with just part of the stem omitted.
Note the separate branches bearing strobili. *B.* A
portion of the main branch with two minor branches
bearing male (staminate) strobili and female (carpellate) strobili. *Bottom.* The fossil seed (×4) of
Cordianthus ampullaceus.

that the ancient, extinct seed ferns, cycadeoids, and cordaites had motile sperms like
the surviving ginkgos and cycads. Can you
think of a possible explanation for the fact
that sperms are still motile in all the most
progressive animals but have become nonmotile in the most progressive land plants?

CYCADEOIDS AND CYCADS

The extinct *cycadeoids* [12] and the relict *cycads* [13] are closely similar in general appearance and structure (Fig. 20–10). Their reproductive organs are, however, quite different.
They were probably closely related in origin
but long followed partly divergent and partly
parallel lines of evolution. They have short,
stumpy, palmlike stems, rarely reaching a
height over 2 or 3 meters and usually much
less. Most of them are unbranched; some are
simply branched. The foliage is also palmlike,
and the cultivated plants, often raised in tubs
in conservatories, are usually mistaken for
palms. (Real palms are not gymnosperms but
advanced angiosperms.)

The cycadeoids were common and widespread around the middle of the Mesozoic but
became extinct soon thereafter, geologically
speaking. In them the male (sperm-pro-
ducing) and female (egg-producing) organs
were united in a single, complex structure,
resembling a flower. Some students have
maintained that these structures were, indeed,
the forerunners of flowers. The detailed anatomy, however, is not so flowerlike, and the
consensus now is that the evolution of these
pseudo flowers and of the true flowers of the
angiosperms took place independently. This
seems to be an unusually good example of
parallel evolution.

Cycads are still rather widespread in places
with warm climates, including Florida, but
they are nowhere abundant, and they comprise relatively few genera and species. They
are at least as old as the cycadeoids and have
survived far longer, but they were probably
never so abundant as the cycadeoids in their
prime. Cycads are most readily distinguished

[12] Technically Cycadeoidales. Many works on
biology or botany call them "Bennettitales," but that
name is invalid under the technical rules of nomenclature.
[13] Cycadales.

from cycadeoids by the fact that pollen and egg cells are produced in separate strobili. In this respect the cycads resemble the conifers.

CORDAITES

The *cordaites*[14] (Fig. 20–11), now long extinct (since early Triassic), were another common group in the Carboniferous coal forests. They were tall trees with long, straplike leaves. Pollen and egg cells were produced in separate strobili. The special interest of cordaites is that they may have given rise to the conifers, although if so the separation must have occurred early in cordaitean history and among primitive members of the group.

CONIFERS

The only really common living gymnosperms are the *conifers*,[15] many of which are certainly familiar to you. They include the pines, firs, spruces, cedars, hemlocks, cypresses, redwoods, junipers, and many others; there are about 450 living species. All of them are woody perennials (living for several or many years), and most of them are trees, although some are shrubs. They include the largest and oldest (as individuals) of all living things—sequoias and redwoods (two closely related species), up to 10 meters in diameter, 100 meters in height, and 4000 years or more in age (Fig. 20–12).[16] Conifers are the dominant trees over much of the temperate zone and often form great forests made up of only one or a few species. In spite of this wide and dense distribution, it is evident that conifer forests are most common in relatively unfavorable situations: dry, cold, and windy, or with poor, sandy soil. In the most favorable situations angiosperm forests are more usual.

Most conifers have narrow, needle-like or scale-like leaves (Fig. 20–13). A few have broad leaves, and a few have reduced leaves and perform photosynthesis mostly in the stems. Most of them are *evergreen*, and the whole group is sometimes called the evergreens. Evergreens do shed their leaves, as do all trees, but they do not shed them all at once and so always have some green leaves in place. However, some *deciduous* conifers shed their leaves all at once at the onset of the cold or dry season. Among conifers familiar in this country, larches and tamaracks are deciduous. There are also some evergreen angiosperms, and the contrast evergreen–deciduous does not completely coincide with gymnosperm–angiosperm.

Conifers have separate male and female cones.[17] A pollen tube develops and the sperms are nonmotile and without flagella. As in the angiosperms, they are reduced to a minimum. Each sperm (two arise from each pollen grain or male gametophyte) consists merely of a haploid nucleus in the pollen tube. Conifer pollen usually has small winglike projections; dispersal of the pollen so that it reaches female cones and fertilizes their egg cells depends entirely on the wind. The seeds, also sometimes with a wing, may be dispersed by wind or simply by falling, rolling, and bouncing. They are frequently dispersed by rodents, which carry them away and store them but fail to eat all they take. It is interesting that the ancient cordaites had winged pollen and seeds and certainly were dependent on wind for the dispersal of both. There were no rodents, and probably no other small seed-eaters, when the cordaites and the earliest conifers lived. Some angiosperms have wind dispersal of pollen and seeds. It is, however, among the angiosperms that the most elaborate or specialized means of dispersal by animals, especially insects but also birds and mammals, occur. The spread of the angiosperms, their modernization and increasing dominance over the older gymnosperms, broadly coincides with the evolution of special groups of insects, birds, and mammals. That is surely not pure coincidence. What are some of the evolutionary concomitants of this interdependence of broadly different groups of organisms?

[14] Technically Cordaitales.

[15] Coniferales ("cone-bearers"), but, as you have seen, some other gymnosperms also have cones or something quite similar.

[16] Some pines, although not so large, may reach even greater ages.

[17] In some species, however, the cones have been secondarily modified so as to be hardly recognizable as such; the juniper has "berries," for instance, which have evolved from cones and are not true berries.

20–12 **Conifers.** *Top left.* Mountain hemlocks, *Tsuga mertensiana*, in Washington State. *Bottom left.* A monkey puzzle tree, *Araucaria araucaria*. *Right.* *Sequoia gigantea* and other conifers in Yosemite Park, California.

Conifers as a group are abundant enough, but they are past their prime, and many kinds of conifers are extinct or are relicts. Araucarias (see Fig. 20–12), for instance, were once worldwide but now occur only in a limited part of South America and in Australia and the southwestern Pacific islands. To us they seem highly exotic, but the famous petrified forest of Arizona consists largely of ancient members of the araucaria group. Even more remarkable is *Metasequoia*, known as a fossil from many parts of the world.[18] Its remains are common in the United States. It was long believed that *Metasequoia* had been extinct everywhere for tens of millions of years, but in 1944 a living *Metasequoia* forest was

[18] Paleobotanists used to confuse it with *Sequoia*. The needles of *Metasequoia* and *Sequoia* are similar in some species, but the relationship is not really very close. The needles are differently arranged, and *Metasequoia* is deciduous, whereas *Sequoia* is evergreen.

20–13   The cones of white pine. *Left.* A branch bearing female cones of three successive years, the oldest at the base and the youngest at the tip. *Right.* A branch bearing several male cones.

found in a remote part of China. This resurrected fossil is now being grown in cultivation by a number of people in the United States, but it is not yet common here.

Flowering Seed Plants

The plants with flowers, the angiosperms, are far and away the most successful of land plants today. This is true by any reasonable criterion: individual abundance, number of species (probably about 175,000), area covered, or total metabolic activity. They are also by far the most important to man. Much the greater part of our food is of angiosperm origin, directly or indirectly. Almost all of our own food plants are angiosperms: cereals, vegetables, fruits, and the rest. Likewise, almost all our food animals live mostly or entirely on angiosperms, so that when we eat them we are still eating angiosperms at one or two removes. Most of our ornamental plants are angiosperms, although here the conifers do play a considerable role. Of course all our flowers are angiosperms. Conifers are at present a more important source of wood and other forest products, but angiosperm forests are also productive. Man's existence, like that of all animals, depends on the plant kingdom, and man is one of the species that depends most heavily on the angiosperms. (It is only

fair to add that our most obnoxious weeds are also angiosperms.)

The angiosperms arose later than other groups of plants of comparable scope. They had probably evolved by the early Mesozoic, but they were not dominant until toward the end of that era. The expansion of the angiosperms coincided with the decline or extinction of most groups of gymnosperms, and undoubtedly this situation involves cause and effect. Angiosperms represent the highest level of plant evolution up to now. It would be rash to predict that no higher will occur, but it seems impossible to imagine what that higher might be. They are highest not in extending their dominance over all, for far older groups continue to be dominant in some environments, but in living in environments farthest removed from the ancestral sea and in being most successful in those particular environments. There is a parallel here to animal evolution. Among animals old groups (mollusks, fishes) continue to dominate in the water, but the highest animals in an evolutionary sense are younger groups in more recently occupied environments.

The essential features of the physiology, structure, and reproduction of flowering plants have been mentioned in previous chapters (see especially Chapter 9). In general physiology they are much like other green plants. In structure they present variations on the same themes as other vascular plants. In

20–14 **Angiosperms.** *Left.* Tulip, a monocotyledon. *Right.* Cotton plant, a dicotyledon.

reproduction, they are typical seed plants with the addition of the carpel-covered seed and the flower. The sequence in the present chapter has involved comparison with and approach to the culmination of land plants in the angiosperms. All that pertinent material should be reviewed but need not be repeated here.

The incredibly extensive array of the angiosperms runs in size from the barely visible duckweeds to great trees exceeded only by a few conifers among all living things. Comparatively few angiosperms are aquatic, but some are. On land they grow practically everywhere that any life exists. A few are parasites, although they are not disease-producers in the ordinary sense. (A hay-fever victim allergic to angiosperm pollen may quibble at that statement if he likes.) Some are organisms of decay, but the vast majority are photosynthetic. A few are in part carnivorous (Venus's flytrap and sundew, encountered in Chapter 10). Angiosperms provide most of the food for man and (nonaquatic) beast, but many are deadly poisonous to both. Their flowers may be microscopic in size or several feet across. They are of almost every imaginable shade. Some of them even have color patterns invisible to our eyes, although none of them is pure black. (Why not? Incidentally, the quest for a black

tulip or a black rose, never actually achieved, is a favorite theme in literature.) They give off scents ranging from what is to us an unbearable stench to perfumes more impelling than *Mon Nuit,* thus demonstrating that man is not the measure of all things. The "unbearable stench" is as attractive to some insects as *Mon Nuit* is supposed to be to the human male.

In somewhat more systematic vein, angiosperms are divided into *monocotyledons* and *dicotyledons* (Fig. 20–14).[19] The bulk of an angiosperm seed is made up of one or two leaflike structures (*cotyledons*) packed with food,[20] especially starch, with which the young plant begins its development. Monocots have one such organ, and dicots have two. The distinction is obvious on comparison of a grain of corn (a monocot seed) with a peanut (a dicot seed). A monocot leaf generally has numerous longitudinal veins of about equal prominence, separating at the base and converging again at the apex, with small and short cross veins more or less at right angles to the longitudinal veins; a dicot leaf usually

[19] These rather cumbersome names are usually shortened to "monocots" and "dicots."
[20] Sometimes food reserves are also present as endosperm.

has a main central vein from which others arise by successive branching.[21] In monocots bundles of vascular tissue are scattered throughout the pith; in dicots the bundles form a cylinder around the pith. Some authorities think that monocots are a more specialized group, derived from early dicots. Others think the two groups are of equal antiquity and represent a basic split that developed among the earliest angiosperms.

Familar monocots include grasses, sedges, cattails, lilies, onions, tulips, palms, and orchids. Dicots are considerably more numerous. They include magnolias, carrots, peas, mints, morning-glories, nightshades, potatoes, mustards, squashes, dandelions, sun flowers, spinach, poison ivy, and gardenias, as well as almost all the other broad-leafed shrubs and trees.

Most of our important angiosperm food plants have been cultivated since prehistoric times, although many varieties of them have been developed recently. Even early man exercised rigorous selection, so that the older cultivated plants have been changed, sometimes beyond recognition, from their wild

[21] Monocot leaves are usually described as "parallel-veined," and dicot leaves as "net-veined," but these descriptions are not literally correct.

ancestors. For instance, one of the most disputed problems in genetics and botany has to do with the wild plants from which pre-Columbian Indians evolved corn (or maize). The word "evolved" is used advisedly. The ability of man to develop new strains, even wholly new species, of cultivated plants is convincing evidence that nature has done likewise, though more slowly and less systematically from the human point of view. Much has been learned about evolutionary principles from experiments with such plants. Knowledge of evolution has, in turn, assisted efforts to develop desirable new cultivated plants. Cultivated plants are often referred to as *cultigens*.

The bulk foods of most agricultural communities have always been those rich in starch and, in different parts of the world, have usually been monocot seeds (such as corn, wheat, or rice) or dicot tubers (such as potatoes, yams, or cassava). A really fascinating exercise, unfortunately one we cannot follow up here, is to chart agricultural food supplies among primitive peoples throughout the world and to trace the history of the plants down into our modern economy.

Table 20–1 summarizes some of the characteristics of the major groups of plants.

TABLE 20–1

A summary of characteristics of the major groups of plants

(+ indicates that the stated character is the usual or the ancestral condition for the group; 0 indicates usual or ancestral absence of the stated character)

Group	General way of life	Dominance of Gametophyte	Sporophyte	Roots, stems, leaves	Vascular tissue	Seeds	Flowers
Algae	Aquatic (secondarily in other moist situations); photosynthetic	Variable		0	0	0	0
Fungi	Aquatic and moist situations; non-photosynthetic; parasites and organisms of decay	+ *	0	0	0	0	0
Bryophytes	Semiterrestrial, mostly in moist situations; photosynthetic	+	0	0†	0	0	0
Ferns (also psilopsids, lycopods, sphenopsids)	Terrestrial; photosynthetic	0	+	+	+	0	0
Gymnosperms	Terrestrial; photosynthetic	0	+	+	+	+	0
Angiosperms	Terrestrial; photosynthetic	0	+	+	+	+	+

* The terms "gametophyte" and "sporophyte" are not strictly applicable to many fungi.
† Leaves and stems of bryophytes (some authorities prefer to speak of them as scales and stalks) only superficially resemble those of vascular plants. The rootlike structures bear little resemblance to true roots. Nevertheless, there is definite organ differentiation.

CHAPTER SUMMARY

Problems raised by land life, the condition for the more complex plants: the acquisition of water and nutrients; their transportation inside the plant (the significance of vascular systems); the differentiation of root and shoot systems; the mechanical problem of support in air as against water (the significance of supporting tissues); the problem of getting the sexes together (the significance of flowers).

Bryophytes—liverworts, hornworts, and mosses; their poorly developed vascular systems; their restricted exploitation of land, predominantly moist habitats; dominance of the gametophyte; dependence on water for fertilization; modest photosynthetic demands.

Early vascular plants—the evolution of sporophyte dominance with a vascular system: psilopsids—the most primitive vascular plants—formerly abundant, two surviving genera, no roots, simple leaves, small subterranean gametophytes; lycopods—club mosses—abundant in Carboniferous, four surviving genera, true roots, microspores and macrospores, tiny gametophytes, ancient evolution of seedlike structures; sphenopsids—horsetails—abundant in Carboniferous (included tree forms), one surviving genus, leaves in whorls around the stem, no differentiation of macro- and microspores; ferns—the third main line of evolution from primitive psilopsid vascular plants—small independent gametophytes, dependence on water for fertilization, organization of the sporophyte into roots and leafy shoot systems, sporangia, abundance in warm and moist climates, diverse types—tree ferns, epiphyte ferns, and aquatic ferns. Seed plants:

fern descendants; ultimate plant adaptation to land life; the ferns' incomplete mastery of conditions of land life (vulnerability of gametophyte to desiccation); the seed as evolutionary adaptation to that condition; retention of female gametophyte in (sporangial) tissues of parent sporophyte; growth of male gametophyte and of zygote also *within* sporophyte tissue; seed as old sporangium containing embryonic sporophyte and food reserves; further evolution of root and shoot system in seed plants.

Gymnosperms—seed plants with seeds borne naked on the supporting sporophyll—their extensive radiation in the Paleozoic and their subsequent decline: seed ferns—long-extinct seed plants; gingkos—motile, flagellated sperms, sexes on separate trees; cycadeoids and cycads—palmlike appearance, cycadeoids abundant in the Mesozoic but subsequently extinct, the evolution of flowerlike structure as an example of parallel evolution, surviving cycads in warm, moist areas; cordaites—abundant in the Carboniferous, possibly the ancestors of conifers; conifers—the only common surviving gymnosperms, predominant today in unfavorable situations—needle- or scale-like leaves, evergreens, male and female cones, nonmotile sperms, wind dispersal of pollen and seeds, various modern conifers.

Angiosperms—flowering seed plants, with sporophylls encasing young seeds in an ovary; the most successful land plants today; their importance in human economy; their first appearance in the early Mesozoic and their dominance by the end of the Mesozoic; their abundance and diversity in size, appearance, and habitats; monocots and dicots.

21

Major Groups
of Animals

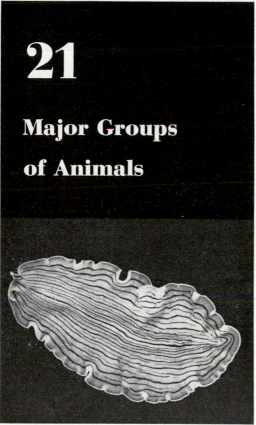

© Douglas P. Wilson

This is a flatworm, Prostheceraeus vittatus, *of the phylum Platyhelminthes. Lowly as flatworms are, they have been the subject of important discussion in attempts to elucidate evolutionary relationships among major groups of animals.*

What Is an Animal?

This chapter starts with a question that everyone thinks he can answer, but one that is often answered incorrectly. Obviously dogs, cats, and cows are animals. To many people an ant is something else, perhaps a "bug" rather than an animal. Of course, you know by now that biologists classify ants as animals, along with sponges, corals, worms, clams, and a vast array of other organisms. When it comes to the protists, as you also know, there is no complete agreement as to which are "really" animals and which are plants. The solution of calling them neither plants nor animals was found workable but rather unsatisfactory. The distinction between some of them and undoubted plants is not really clear-cut, nor is there a sharp way of excluding others from the category "animals." The best answer to "What is an animal?" is not in terms of description at all. It is this: An animal is an organism that belongs by descent and common ancestry to any of the phyla that biologists agree to call "animals." It is, however, possible to make descriptive generalizations about characteristics found more often in animals than in plants. Resemblances and differences between animals and plants have been mentioned frequently in preceding chapters. Now we need only to summarize and review some of the more important and widespread areas of contrast.

1. *Metabolism.* Most plants are photosynthetic. All those that are not were probably derived from photosynthetic ancestors. Animals probably also had photosynthetic ancestors among the protists, but no animals, as such, are photosynthetic. All derive food from other organisms.

2. *Mobility.* Most plants are attached (sessile) and nonmotile as developed organisms and are dispersed in specialized reproductive phases, such as spores or seeds. Many animals are also attached as adults, but most are mobile throughout life, and all have at least a mobile phase in development.

3. *Structure and organization.* Most animals have fixed structures to which new elements are added only at limited, usually early, phases of development. Most plants have less

fixed patterns, with new elements added at almost any time or periodically throughout life. Tissue and organ differentiation in animals is often more definite and more complex than in plants. Individual plant cells usually have rigid walls; animal cells usually do not. Plant cells usually are vacuolated when mature; animal cells usually are not.

4. *Maintenance.* In most animals the cells are either surrounded by sea water or are in a fairly constant internal environment that resembles sea water in being rich in sodium chloride (common salt). Devices for the maintenance of a constant internal environment are less evident in plants, and the internal fluids are usually low in sodium chloride.

5. *Responsiveness.* Most animals are far more responsive than any plants. With relatively few exceptions, animals have nerves and muscles, markedly unlike any plant tissues. Almost all plants lack special receptors, and the rare special receptors that do occur are few and simple. Most animals have special receptors, and these are usually numerous and complex.

6. *Reproductive cycles.* Most plants have a sexual cycle with some development of the haplophase between meiosis and fertilization. In animals development of the haplophase is highly exceptional.

Any significant contrast between two groups of organisms points up the fact that the groups have different roles in the complex interrelationships of populations in nature. Plants and animals, as a whole, clearly do have different roles, even though their characteristic roles may be modified or lost in some particular kind of plant or animal. The most basic difference is in acquisition or synthesis of foods. The other important differences are all more or less closely related to this one. The fact that animals are characteristically more mobile certainly has something to do with their getting food from other organisms rather than manufacturing it from raw materials. Differences of animals from plants in structure, maintenance, responsiveness, and even in reproduction all seem to have had some original relationships to mobility. Try to work out some of these relationships in more detail.

Major Animal Phyla

The classification in the Appendix places animals in 18 phyla, not counting the more animal-like protists, which are often considered another phylum under the name "Protozoa." Some systematists recognize more phyla, and some, fewer. For instance, the two classes of Bryozoa may have had different origins and are often called separate phyla. On the other hand, brachiopods and bryozoans have some fundamental resemblances and may have had a common origin sufficiently recently to justify uniting them in a single phylum.

The great majority of known animals, both living and extinct, belong to only 11 of the 18 phyla of our classification, those printed in heavy type. The other 7 phyla are mostly rare and of no great importance at the present time, and all but one of them are little known as fossils. *Rotifers* (phylum Trochelminthes) are now very abundant, to be sure, but they are not highly diversified, they are not key figures in the activities of life as a whole, and they have no fossil record. *Graptolites* (phylum Graptolithina) are totally extinct, known only as fossils. They were very abundant in some ancient seas, but they do not loom large in the history of life. It is also probable that we would not consider them a separate phylum if we understood their relationships, which are highly speculative at present.[1] That is also true of some or most of the smaller living phyla; if we knew their ancestries and the changes involved in their origins, we would probably place them in other, larger or better-known phyla. The *ctenophores* (phylum Ctenophora), for example, may be offshoots of the Coelenterata, and the *mesozoans* (phylum Mesozoa) may be unusually degenerated Platyhelminthes.

The best way, and indeed the only adequate way, to learn the characteristics of the various phyla of animals is not from books. As far as possible you should first see them alive in their natural surroundings or, failing that, in aquariums and zoos. You should then examine preserved specimens, models, and dissec-

[1] A current view is that they may belong in the phylum Chordata (subphylum Hemichordata; see p. 563), which would be rather exciting if true—but the evidence is poor and ambiguous.

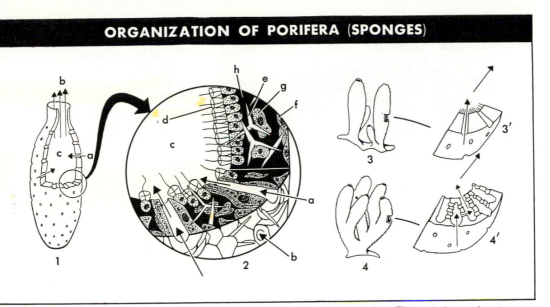

21–1 Porifera. *1.* The gross structure of a simple sponge like *Olynthus*. Water, laden with microorganisms that are captured as food, enters minute pores (*a*) all over the body surface. Passing through *c*, the body cavity, it leaves by the mouth, or osculum, at *b*. *2.* A portion of the wall of the sponge enlarged to show cellular detail: *a*, entrance pore formed by a single cell that lines the channel all the way into the body cavity (*c*); *b*, pore, surface view; *d*, collar cells that line the body cavity; *e*, skeletal spicule; *f*, external covering cell; *g*, amebocyte (an amebalike cell) embedded in the jellylike matrix (*h*) that fills most of the space between the external covering cells and the collar cells. The flow of water into the pores, through the body cavity, and out the osculum is maintained by the beating of the flagella of the collar cells, which trap and ingest microorganisms in the water (cf. Fig. 19–5). Some of the food is transported elsewhere in the sponge by the amebocytes constantly moving about the jelly matrix. Since all the water passing through the sponge enters by many pores but leaves by the one osculum, it is under considerable pressure, and its velocity carries it (depleted of food and oxygen and therefore useless to the sponge) far away from the animal. *3, 4.* Progressively more complex sponges and parts of their body walls. Sponge evolution has consisted primarily of adding complications to the body wall such that the incoming current of water passes through a succession of chambers lined with collar cells. The functional significance is that the modern sponge is able to handle more water and provide greater pressures for the exhalant current than its ancestors.

tions in museums or classroom collections and should dissect some examples yourself. Detailed descriptions and anatomical terminology can be found in works cited in the bibliography at the end of this book. This is not a textbook of zoology or anatomy, and we do not propose to describe the various phyla in detail. Concrete examples are provided by the accompanying illustrations. Beyond that, you need at this point only such brief characterization of the more important phyla as will help to fit them into the broad biological scheme of things. The next chapter will discuss somewhat more fully the three phyla that are now dominant, most varied and complex, well illustrative of certain biological principles, and in many respects most interesting to us humans.

PORIFERA (SPONGES)

Sponges (Figs. 21–1 and 21–2) resemble protistan colonies in that the separate cells seem to lead semi-independent lives within the organization as a whole. Nevertheless, the cells are of a few well-differentiated kinds with different functions, and there is some simple coordination of their activities. All sponges are aquatic, and most of them are marine. Their way of life is not very exciting; they are sessile and do little more than maintain a current of water inflowing through many small pores around their sides and outflowing

American Museum of Natural History

21–2 The sponge *Hippospongia canaliculata*. This is a colony of sponges; note the many oscula.

through a larger opening, usually at the top. They feed on microscopic particles carried along by the current. Fossil remains suggest that sponges have been numerous since early in the history of life and that they have not changed a great deal for some hundreds of millions of years. You may wonder why in so long a time sponges (and some other lowly organisms) have not evolved into anything distinctly more elaborate. The answer is similar to that of the elderly lady who was asked whether she had done much traveling, and who replied that she had not needed to travel because she had been *born* in Boston. Sponges have not gone places, evolutionarily, because they were already there. When life was young, they were already well adapted to a widely and continuously available way of life.

COELENTERATA

Coelenterates [2] are characterized by a sac-like digestive cavity with a single opening, which is surrounded by tentacles with stinging cells. All are carnivorous and snare food in their tentacles. The prey ranges from microscopic animals of many sorts to relatively large crustaceans, worms, or fishes. Like sponges, coelenterates are exclusively aquatic, and they are even more predominantly marine, although there are a few fresh-water types, such as *Hydra*.

[2] The Greek roots for Coelenterata mean literally "hollow guts."

The coelenterate's body may assume one of two basic forms: the *polyp* and the *medusa*. These are only variations on a basically similar pattern (Fig. 21–3), related to different modes of life. The polyp, an elongate cylinder, is sessile; the medusa, flatter and bell-like, is free-swimming. Polyp and medusa may alternate in the life cycle of one kind of coelenterate, but usually one form or the other is dominant or exclusively present. When the medusa is present in the life cycle, it is asexually budded off from the polyp and is the stage that effects sexual reproduction (gamete production). The free-swimming nature of the medusa also guarantees dispersal. In the evolution of coelenterates there has been a recurrent tendency for the polyp stage to become colonial. It commonly reproduces other polyps by budding; incomplete separation of the budded polyps results in a complex organism composed of many recognizably distinct but connected polyp units, all of which developed from a single zygote and can in that sense be considered parts of one individual.

Coelenterates are highly diverse. Among the different sorts are the following:

Hydroids: usually small and soft-bodied, with solitary or colonial polyps predominant but medusas often also present in the life cycle. *Hydra* and *Obelia* are both hydroids (Fig. 21–4).

Siphonophores: floating colonies with differentiated polyps; the Portuguese man-of-war (Fig. 21–5) is an example.

Jellyfishes: medusas, with a polyp generation reduced or absent (Figs. 21–4 and 21–6).

Sea anemones: [3] relatively large, solitary, soft-bodied polyps (see Fig. 21–6).

"Corals": a few solitary and a great many colonial polyps that form hard calcareous supports or skeletons. "Coral" is a popular term applied to almost all coelenterates that happen to have skeletons. They occur in a number of different groups without closer relationships than that they are all coelenterates (see Fig. 21–6).

[3] Of course, they have nothing more to do with real anemones than that someone thought they looked like flowers.

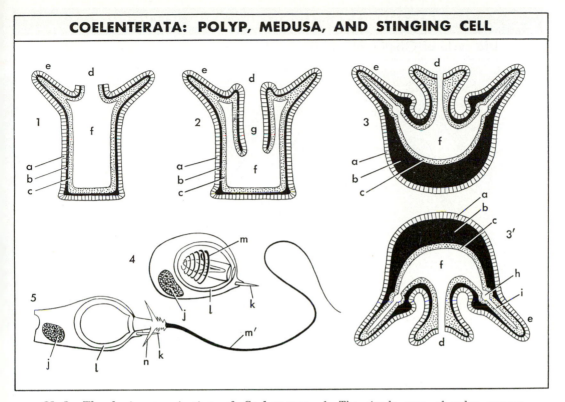

21–3   The basic organization of Coelenterata. *1.* The simple type of polyp present in hydroids: *a*, ectoderm; *b*, jellylike layer between the ectoderm and the endoderm (*c*); *d*, mouth; *e*, tentacle; *f*, digestive cavity. *2.* The more advanced type of polyp present in sea anemones, with a gullet (*g*) leading to the digestive cavity. *3.* A medusa, the predominant body form in jellyfishes and present as a motile sexual phase in hydroids: *h*, circulatory canal that passes along the perimeter; *i*, extension of the circulatory canal into the tentacle. Both *h* and *i* are continuous with the digestive cavity, whose water content is circulated through them. The medusa is essentially an inverted polyp in which the jellylike layer is extensive and the mouth is much drawn out; *3* shows it inverted to facilitate comparison with the polyp, and *3′* shows it as actually oriented. *4.* A stinging cell, or nematocyte, before discharge: *j*, nucleus; *k*, receptor bristle that, when touched, relays a stimulus to the capsule (*l*), which contracts, everting the long, coiled filament (*m*). *5.* The nematocyte after discharge: *m′*, filament; *n*, barb at the base of the filament. The nematocyte is a self-contained stimulus and response system (Chapter 13). Its function is to impale and poison prey, which is then gradually forced into the mouth by other means.

The coelenterates are another ancient phylum that has become more diversified and has changed in many minor ways but in no really fundamental way for several hundred million years. The fossil record of "corals" is particularly rich, but, oddly enough, quite a few fossil jellyfishes are also known. (Why is that odd?)

It is an interesting and disputed problem whether polyps or medusas are more primitive, that is, whether the first coelenterates fully organized as such were free-swimming or attached. The two forms are about equally old as known fossils. There are good arguments on both sides, but it seems likely that characteristic coelenterate structure arose in relationship to sessile life and that the earliest true coelenterates were therefore either polyps alone, or alternately polyps and medusas. Among reasons for this conclusion are the facts that coelenterates are radially symmetrical (p. 572) and hermaphroditic (footnote 8, p. 655). Both these features are commonly characteristic of sessile organisms.

Life cycle of *Obelia* (hydroid)

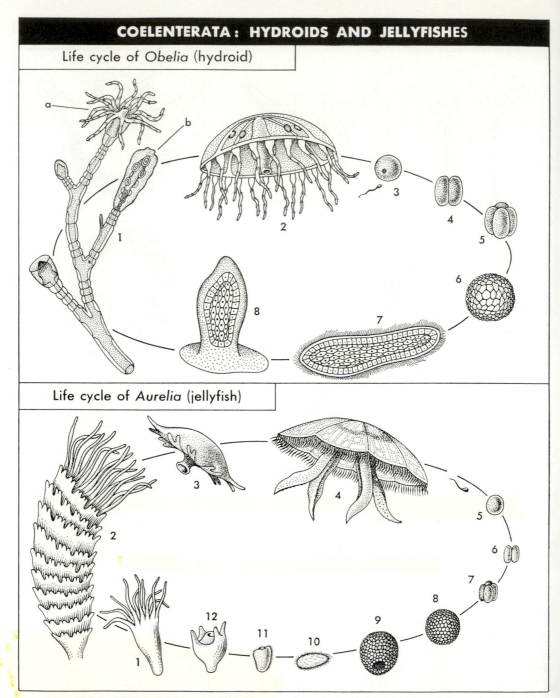

Life cycle of *Aurelia* (jellyfish)

21–4 The polyp and the medusa in the life cycles of coelenterates. In *Obelia*, a hydroid, the principal stage is the polyp, which, by asexual budding, becomes a branched colony (*1*). Some members (*a*) of the polyp colony are feeding units equipped with tentacles. Others (*b*) are reproductive units, which continuously bud off medusas (*2*). The free-swimming medusa serves to disperse the species; it is sexual, producing gametes (*3*). *4–6.* Successive stages in the development of the ciliated larva (*7*), which ultimately settles (*8*) to initiate a new polyp colony. In *Aurelia*, a jellyfish, the medusa is the dominant body form. The polyp (*1*) is a small transient stage that actively buds off (*2*) young medusas (*3*), which when fully developed (*4*) are quite large animals (6 inches in diameter). *5.* Gametes liberated by the medusa. *6–9.* Successive stages in the development of a ciliated larva (*10*), which becomes (*11–12*) a polyp.

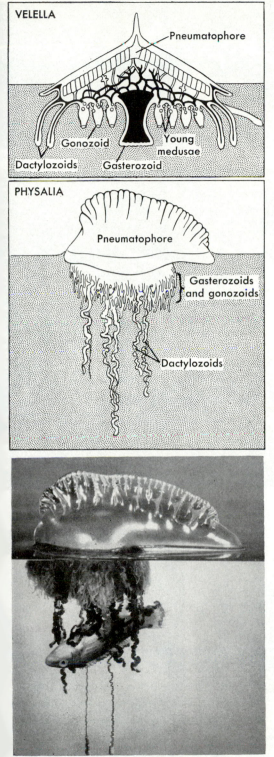

VELELLA

Pneumatophore

Gonozoid
Young medusae
Dactylozoids
Gasterozoid

PHYSALIA

Pneumatophore

Gasterozoids and gonozoids

Dactylozoids

© Douglas P. Wilson

PLATYHELMINTHES

In popular speech almost any elongated, wiggly animal without prominent legs is called a "worm." Early zoologists thought that most worms were related and put them in a phylum Vermes (Latin for "worms"). Now we know that "worm" refers properly to an *adaptive body form* that occurs in many groups of animals fundamentally different in structure and origin. Even as a body form, "worm" includes a number of different adaptive types. Of the 18 phyla of our classification, the worm body shape is a basic adaptation for eight: Platyhelminthes, Mesozoa, Nemertina, Aschelminthes, Acanthocephala, Phoronidea, Chaetognatha, and Annelida. Five other phyla, although not basically wormlike, include groups or forms that have become more or less wormlike: Ctenophora, Mollusca, Arthropoda, Echinodermata, Chordata. Many so-called worms (cutworms, inchworms, apple worms, etc.) are insect larvae. Blindworms or slow worms are legless lizards. The worm shape has evolved so often that it is clearly associated with some of the most successful and advantageous ways of life, but it seems nevertheless to be a dead end of evolution. None of the most progressive and dominant animals are wormlike (as adults, at least). The Platyhelminthes (Fig. 21–7), or flatworms, are, as the name implies, flattened from top to bottom, which makes many of them somewhat ribbon- or tapelike. Indeed, tapeworms are platyhelminths, and so are flukes (Fig. 24–7), which are also common parasites. Like many other internal parasites

---

21–5 Siphonophores. These are highly evolved coelenterates whose bodies are essentially colonies of polyps that have undergone extensive specialization or division of labor. All siphonophores are floating marine organisms. A large polyp forms a float (pneumatophore) to buoy up the rest of the colony. Another polyp (in *Velella*) or several other polyps (in *Physalia*) specialize as feeding units, a gasterozoid. Other polyps, gonozoids, are devoted entirely to reproductive activity. Finally, some are very long fingerlike processes (dactylozoids) devoted entirely to attack; they are heavily endowed with nematocysts. The two drawings represent schematically the organization in two genera of siphonophores, *Velella* and *Physalia*. The photograph shows the Portuguese man-of-war (*Physalia*) with a fish it has captured in its dactylozoids.

21–6 Jellyfishes, sea anemones, and corals.
Right. The white coral *Madrepora oculata*, from the Bay of Biscay. *Below.* The jellyfish *Gonionemus murbachi*. Note on the tentacles the knobs, which are clusters of nematocytes.

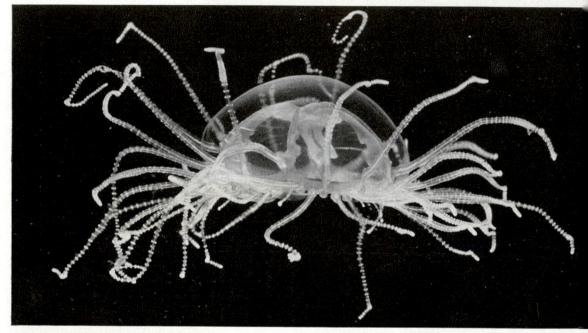

The sea anemone *Dahlia*. Note the mouth in the center of the tentacles.

The jellyfish *Chrysaora*.

The sea anemone *Metridium sessi*

(see Chapter 24), these have lost organs and tissues that occurred in their ancestry. Basic characters of the platyhelminths are better preserved in such nonparasitic forms as the planarians, already familiar to you. Look up "planarian" in the index and review what has been said about it. Most free-living flatworms resemble coelenterates in having a digestive cavity with only one opening, but in other respects they are more complex and more like higher animals; they are fully motile and bilaterally symmetrical, with sensory receptors including eyes (which do not, however, form images), specialized excretory and reproductive organs, and a fairly complex nervous system including even a brain of sorts. The group must be very old, but fossil flatworms are extremely rare and poorly preserved.

NEMATODA

Nematodes,[4] or roundworms, differ superficially from flatworms in being literally rounder and less flattened, and they differ more fundamentally in having a digestive tube with two openings—a mouth as the entrance and an anus as the exit (Fig. 21–8). Nematodes are incredibly numerous and occur practically everywhere there is any life at all, from the equator to the arctic and from mountaintops to the deep sea. An acre of soil often contains three billion or more nematodes, and you are likely to turn up a million at a time with a spade in your garden. Moreover, nematodes live as parasites in innumerable other animals and also in many plants. Fifty or so species parasitize man; you have probably been a host to them at some time and may well be right now. Some are practically harmless and others, such as hookworm and trichina, cause serious diseases. Ascaris, the commonest nematode parasitic in man, is often unpleasant but seldom disabling or fatal (see Chapter 24 on parasitism and parasites). Nematodes are among the great successes of the impersonal forces of evolution. They teach nothing about how to win friends, but they are fine examples of one way to get along in the world.

[4] Phylum Nematoda, from the Greek word for "thread." The name would be more appropriate for the horsehair worms, which do have a similar name—Nematomorpha (See Appendix).

BRYOZOA

Almost all bryozoans[5] are colonial, the colonies encrusting shells and rocks or forming mats or branching fans (Fig. 21–9). The colonies build supports of limy or horny material, and the individuals live in small cups in the skeletal framework. To that extent they resemble many corals, and they also have tentacles around the mouth, but there the resemblance to any coelenterates ends. Even the tentacles are really different because in bryozoans they have no stinging cells and do not capture food directly. The cilia create a current that carries microscopic food particles into the mouth. The internal anatomy is far more complex than in coelenterates. The digestive system is a complete tube, doubled into a U so that the anus is near the mouth, an arrangement correlated with life in a cup without a rear exit. There are nervous and muscular systems, and most bryozoans have true coeloms (see p. 571).[6] Bryozoans are another group that is exclusively aquatic and mainly marine, although fresh-water bryozoans are fairly common. They occur in all seas and are especially numerous on coral reefs, where they add materially to the stony matter. Fossil bryozoans, including many reef-dwellers, are more abundant and diverse than the living forms.

BRACHIOPODA

Brachiopods[7] ("lamp shells") are animals enclosed in two approximately equal shells. To that extent they resemble clams (phylum Mol-

[5] "Moss animals," from the fact that their colonies may look somewhat (but not very much) like moss. In some books they are called Polyzoa ("multiple animals") because they are colonial.
[6] The more abundant group (Ectoprocta) of bryozoans has a coelom and has the anus outside the ring of tentacles. A much rarer group (Endoprocta) has no coelom, and the anus is inside the ring of tentacles. These differences are considered so important by some students that they classify the two groups as distinct phyla.
[7] Phylum Brachiopoda ("arm-footed"). The name is based on an elaborate mistake. The structures carrying the tentacles were compared with arms and were supposed to have something to do with walking. They really have no functional resemblance either to arms or to feet, but the name is as good as any other or, indeed, better because all zoologists use it for this group.

lusca), but the resemblance is superficial. In clams the shells are lateral, on the right and left sides of the body. In brachiopods they are dorsoventral, on the top and bottom of the body. Thus there is a difference in symmetry.

The internal anatomy is also quite different in the two groups. Brachiopods are more complex than any phylum previously discussed, with a coelom and well-developed nervous, muscular, digestive, circulatory, excretory, and reproductive systems (Fig. 21–10). The most conspicuous peculiarity is a firm internal support, a pair of spirals or other more or less complex forms, bearing tentacles. As in bryozoans, the tentacles (or, strictly speaking, tiny hairlike cilia on the tentacles) set up feeding and respiratory currents. Brachiopods are noncolonial. An individual usually becomes attached by a fleshy stalk that protrudes through the hind end of the lower shell. They are exclusively marine and always have been. The survivors are not very abundant and di-

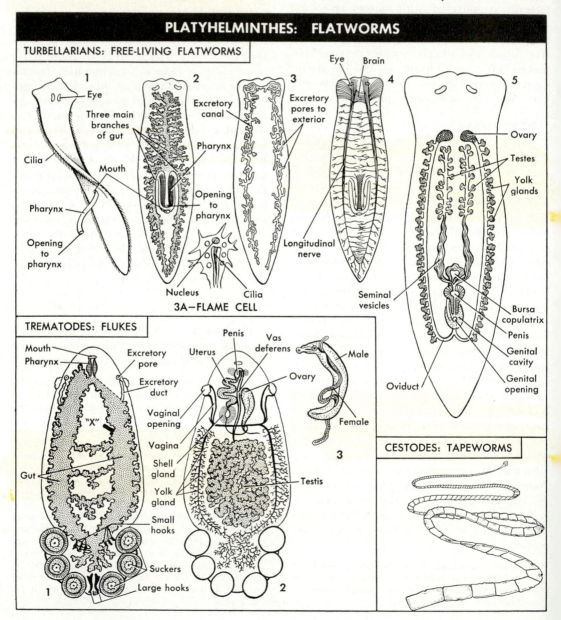

PLATYHELMINTHES: FLATWORMS

TURBELLARIANS: FREE-LIVING FLATWORMS

1 — Eye
Three main branches of gut
Cilia
Mouth
Pharynx
Opening to pharynx

2 — Excretory canal
Pharynx
Opening to pharynx
Nucleus
Cilia
3A—FLAME CELL

3 — Excretory pores to exterior
Longitudinal nerve

4 — Eye
Brain

5 — Ovary
Testes
Yolk glands
Seminal vesicles
Bursa copulatrix
Penis
Genital cavity
Genital opening
Oviduct

TREMATODES: FLUKES

Mouth
Pharynx
Excretory pore
Excretory duct
"X"
Gut
Small hooks
Large hooks
Suckers
1

Uterus
Penis
Vas deferens
Vaginal opening
Vagina
Shell gland
Yolk gland
Ovary
Testis
2

Male
Female
3

CESTODES: TAPEWORMS

verse, but they are remnants of a great population, occurring literally by the millions as fossils in rocks of many ages.

MOLLUSCA

As will be explained later in this chapter, the various phyla, especially as we see them today, do not really represent an evolutionary progression. It is, however, clear that three of the phyla, each in its own way, have gone farther than the others. Each represents a culmination of evolution in complexity of structure, in coordination of functions, and in diversity and success of adaptations. The great culminating phyla are the mollusks,[8] arthro-

[8] The name, derived through French from Latin, implies that the animals are soft-bodied. In fact, most of them have shells, and the bodies within the shells are no softer than those of most other animals, but

pods, and vertebrates. The next chapter is devoted to them. Here we need only to mention them as belonging in this conventional sequence of the major phyla and to note that the mollusks include such familiar animals as the snails, clams, and squids.

ANNELIDA

The *annelids* [9] are another group of worms (Fig. 21–11), very different indeed from the flatworms or roundworms. Annelids are also literally round worms, built on a cylindrical, tubular plan, with a mouth at one end and an anus at the other. The cylindrical surface is,

as with other systematic names the appropriateness of the derivation makes no difference now.
[9] Phylum Annelida, from a French word derived in turn from the Latin for "ring," because the body seems to be divided into a series of rings.

21–7 Platyhelminthes. The most primitive flatworms are the free-living (nonparasitic) *Turbellaria*, the planarians. *1, 2.* The planarians are flat and ribbonlike. Many still move by the primitive method of beating cilia. The alimentary system is a blind cavity, like that of coelenterates. The mouth is in the middle of the body on the under surface; a long muscular pharynx protrudes from it. *3.* A flatworm is the simplest animal with a specially developed excretory system. It consists of a series of blind ducts connecting with a pair of principal excretory canals running the length of the body. These canals vent to the outside through excretory pores. At the blind end of each excretory duct is a specialized cell (a "flame cell," *3A*), which is evidently the active secretor of nitrogenous wastes into the duct. It is ciliated, and its cilia beat actively within the cavity of the duct. *4.* The nervous system consists of a simple "brain," two longitudinal nerves (p. 376), and a network of other fibers. *5.* The reproductive system is extremely complex and one of the most interesting features of the phylum. The phylum as a whole is hermaphroditic: individual animals carry both male and female organs. This is undoubtedly a major reason why the flatworms have been so successful in exploiting the parasitic way of life (all the trematodes and cestodes are parasites [cf. p. 655]). Eggs, formed in the ovaries, pass down the long oviducts to the common genital cavity; on their way they receive yolk material (nutrients) from the extensive yolk glands. The multiple testes shed their sperm into the large seminal vesicles, where they are stored before copulation. Like the hermaphroditic flowering plants, flatworms rarely, if ever, self-fertilize; copulation involves an exchange of sperm between individuals. The muscular penis of each animal inserts sperm into the partner's bursa copulatrix. When eggs are later laid, they are fertilized with sperm released

by the bursa copulatrix—not sperm from the animal's own seminal vesicles.

Trematoda, flukes, are exclusively parasitic flatworms. *1, 2.* The anatomical organization of *Polystomum*, a trematode parasitic in the bladders of frogs. It attaches to the host by six powerful suckers and two series of hooks. The mouth, as in other trematodes, is shifted toward the anterior end. The alimentary canal is only a strong, muscular, sucking pharynx and the blind gut typical of flatworms. There are two excretory ducts. The reproductive system (*2*) is fundamentally similar to that of a planarian, but with added complications. Most significant is the extent of its development, characteristic of parasites: it comprises the larger part of the animal's body. The huge testis leads through a duct (the vas deferens) to a muscular penis that is extruded through the genital opening during copulation, entering one of the two special vaginal openings of the mate. Eggs shed by the single ovary pass along the complex oviduct. First they receive yolk from extensively branched yolk glands; next they receive sperm from the vagina; later they receive covering from the shell gland; finally, after temporary storage in the uterus, they are discharged to the exterior. A trematode has a curious duct (marked "X" in *1*) connecting the yolk glands with the gut; it probably serves to avoid waste by relaying excess yolk into the animal's gut, where it is digested. *3.* A very few trematodes, like *Schistosoma* (a parasite of human and animal bodies), have separate sexes, but they spend their lives in permanent embrace, effecting in a roundabout way the same result as hermaphroditism, so important to a parasite (cf. p. 655). *Cestoda*, tapeworms, are illustrated briefly here by *Taenia solium* (the tapeworm of man) and in more detail in Fig. 24–6.

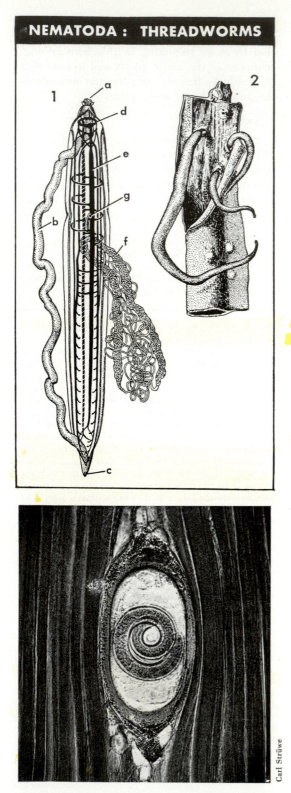

NEMATODA : THREADWORMS

however, modified by rows of bristles and often also by more complex appendages. Most strikingly, these worms are made up of numerous segments, visible externally as ringlike bulges and separated internally by partitions. The digestive tube and principal lengthwise blood vessels and nerve cords run right through many or all segments. Other organs, such as nerve ganglia, circular blood vessels, and excretory tubes, are repeated in several or many segments. Earthworms (of which there are many species) are the annelids best known to most of us. Specialized for life in moist soil, they literally eat their way through the ground, passing the dirt through the alimentary canal. Organic matter is digested from the soil, and the residue is ejected from the anus in the form of the familiar worm casting. Their reworking of the soil is so extensive that Darwin doubted whether "there are any other animals which have played such an important part in the history of the world." Do you agree? If not, what animals do you think have played a more important part?

Although earthworms are so readily obtained that they are the usual examples of annelids in biology courses, most of the annelids are marine and look and live quite differently from earthworms. They live almost everywhere in the sea, sometimes free-swimming but often in burrows or tubes. A few annelids are true parasites, and one group, the leeches (class Hirudinea), is semi-parasitic, living on the blood of vertebrates, including man. Parasitism is, however, less common among annelids than among other worms. You need not be such a worm fancier as Darwin to agree that some of the plumed,

21–8 Nematoda. *Top. 1.* Schematic representation of a dissected *Ascaris* (female), the threadworm parasitic in human intestines. The internal organs are few and simple: *a*, mouth; *b*, alimentary canal; *c*, anus; *d*, nerve ring around the pharynx; *e*, major ventral nerve cord—a dorsal nerve (crosshatched) is also visible; *f*, ovary; *g*, genital pore. *2.* The threadworm *Gigantorhynchus gigas* on the gut wall of a pig. (*Gigantorhynchus* is nowadays placed in a closely related though separate phylum, the *Acanthocephala*. However, the group is, for convenience, presented here with the nematodes, of which it was once considered a class.) *Bottom. Trichina spiralis,* the threadworm responsible for trichinosis, embedded in muscle (× approximately 100).

Carl Strüwe

© Douglas P. Wilson

21–9 Bryozoa. *Right. A.* Organization of an individual member of a bryozoan colony. Note that the whole mouth region with its tentacles can be withdrawn, by retractor muscles, within the cuticle. The avicularium is a modified individual animal that plays much the same role in the economy of the colony as does a dactylozoid in siphonophores; its "jaws" snap shut on moving particles (like small larvae) that touch it. *B.* Part of a colony of the fresh-water bryozoan *Plumatella. Above.* The marine bryozoan *Bugula turbinata* hanging from a rock (×1.2).

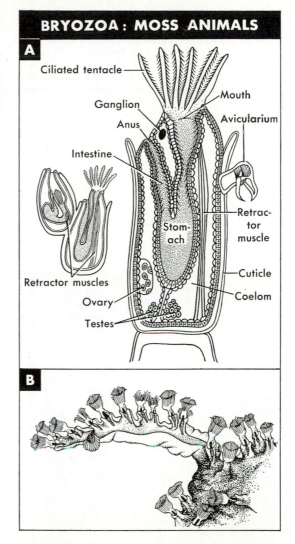

BRYOZOA: MOSS ANIMALS

brilliantly colored marine annelids (the polychaetes) are handsome and fascinating.[10]

ARTHROPODA

Arthropoda[11] is, almost beyond comparison, the largest of all phyla, plant or animal. Among other groups, it includes the crustaceans, spiders, and insects. No other phylum of organisms is anywhere near so diverse, and none except the vertebrates is so complex in structure and behavior. The arthropods will be discussed more fully in the next chapter.

ECHINODERMATA

The *echinoderms*[12] are relatively complex animals (Fig. 21–12) with a (usually) com-

[10] Admiration for these beautiful worms is reflected in the names given some of them, such as *Aphrodite* (its popular name is less enticing—"sea mouse"). The name often given to the quahog or common hard-shelled clam (mollusk), more attractive to the palate than to the eye, is *Venus mercenaria.*
[11] "Jointed feet."
[12] "Prickly skins," a name appropriate for sea urchins and extended to their relatives.

plete digestive tube, coelom, and specialized excretory, reproductive, nervous, and circulatory systems, although the last two are simpler than in most animals otherwise so complex. In spite of belonging in these respects among the "higher" phyla, they resemble the coelenterates in being radially symmetrical. At least, most adult echinoderms seem to be so. In fact, their larvae are bilaterally symmetrical, and there are traces of bilateral symmetry even in adults, but the over-all shape of most adults is radial, usually with five rays, as in common starfishes. Another unique feature of echinoderms, in addition to the true circulatory system (which is poorly developed), is a well-

21–10 Brachiopoda. *Right. 1.* A brachiopod, *Lingula*, in its burrow: *a*, with its stalk (*c*) relaxed so that the animal (*d*) can set up feeding currents; *b*, withdrawn in the burrow, with bristles (*e*) fringing the edge of the shell. *2.* A generalized view of brachiopod organization: *a*, stalk; *b*, mouth; *c*, intestine, *d*, digestive gland; *e*, ovary; *f*, coelom; *g*, dorsal shell; *h*, ventral shell; *i*, lophophore, bearing the tentacles. Note also the shell-closing muscles, which can be seen passing obliquely behind the intestine. *3.* The paths of feeding currents to the mouth (*m*). *4.* A brachiopod hanging from a rock by its stalk; this habit is commoner in the brachiopods than that of *Lingula* (*1*). *5, 6.* General views of the fossil brachiopod *Terebratula* showing the relation of the dorsal and ventral shells to the posterior and anterior ends of the animal. *Below.* A brachiopod opened to show the lophophore and tentacles.

Ralph Buchsbaum

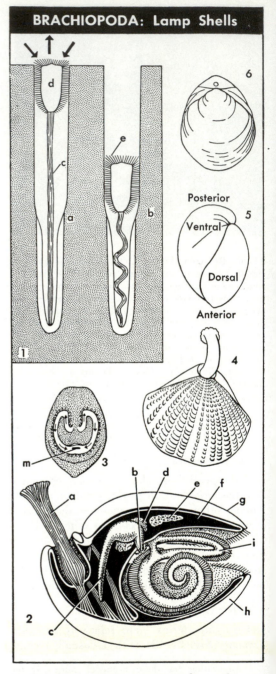

BRACHIOPODA: Lamp Shells

developed water-vascular system. This contains sea water filtered through a special sieve plate. In many echinoderms the water-circulatory system is connected to numerous "tube feet," operated by hydraulic pressure and used in slow locomotion and for grasping. Almost all echinoderms have hard, limy plates in the skin, and in some, such as the sea urchins, these may be immovably united into a hard, protective box. (Sea urchins also have movable spines jointed to the outside of the box.) All echinoderms are marine.

The echinoderms well illustrate how a single ancestral body plan may become

THE POLYCHAETE ANNELIDS

21–11 The polychaete annelids. *Left.* Although the annelids are likely to be most familiar through the common earthworm, the group to which it belongs (*Oligochaeta*) is much less abundant and typical of the phylum Annelida than the marine *Polychaeta*. ==The simplest and most typical polychaetes are free-swimming, like *Nereis*, shown here. They are carnivorous, possessing stout jaws. Their most characteristic feature is the bristle-bearing, paddle-like parapodia ("feet equivalents") borne on each segment.== Used principally for locomotion, the parapodia are also excellent respiratory structures, combining a large surface area and a rich blood supply.

Above. Many polychaetes, including the peacock worms, *Sabella favonina*, are sedentary animals that live in tubelike cases they themselves secrete. They set up currents of sea water along elaborate tentacles that lead to their mouths; and from the sea water they extract microorganisms and debris as food.

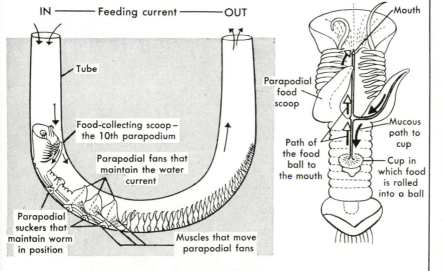

IN ——— Feeding current ——— OUT

Tube

Food-collecting scoop – the 10th parapodium

Parapodial fans that maintain the water current

Parapodial suckers that maintain worm in position

Muscles that move parapodial fans

Mouth

Parapodial food scoop

Path of the food ball to the mouth

Mucous path to cup

Cup in which food is rolled into a ball

Above. Chaetopterus pergamentaceus is a polychaete (shown out of its tube at right) that has evolved extensive specializations for its existence. ==It lives in a U-shaped tube, through which it sets up a feeding current of water propelled by many parapodia that have fused into three sets of special "fans." Parapodial suckers grip the tube, keeping the animal in position.== Two special food-collecting parapodia on the tenth segment protrude, like scoops, into the feeding current. Food particles in the current are here entangled in mucus (acting like "flypaper"), which is moved by cilia backward to a cup-shaped organ. The food particles accumulate there in a ball, which is moved forward along the groove all the way to the mouth.

adapted in radically diverse ways. Although the internal anatomy is for the most part essentially the same, the main groups of living echinoderms are extraordinarily different in appearance and habits (Fig. 21–13).

Sea lilies: body enclosed in a rigid box, with branching, flexible arms extending from around the mouth and anus; body usually sessile, attached by a stalk; no tube feet; feeding on microscopic organisms and debris.

Starfishes: body star-shaped, or five-sided, stiff but flexible; free but moving by slow crawling; tube feet strongly developed; feeding usually on clams and other

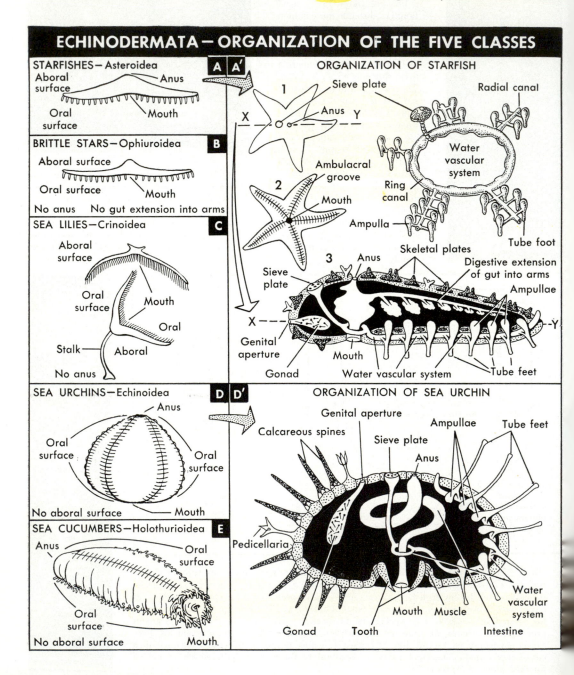

ECHINODERMATA—ORGANIZATION OF THE FIVE CLASSES

STARFISHES—Asteroidea
Aboral surface — Anus
Oral surface — Mouth

BRITTLE STARS—Ophiuroidea
Aboral surface
Oral surface — Mouth
No anus   No gut extension into arms

SEA LILIES—Crinoidea
Aboral surface
Oral surface — Mouth
Oral
Stalk — Aboral
No anus

SEA URCHINS—Echinoidea
Anus
Oral surface — Oral surface
No aboral surface — Mouth

SEA CUCUMBERS—Holothurioidea
Anus — Oral surface
Oral surface — Mouth.
No aboral surface

ORGANIZATION OF STARFISH
1 Sieve plate   Radial canal
X   Anus   Y
Ambulacral groove   Water vascular system
2   Ring canal
Mouth   Ampulla
Skeletal plates   Tube foot
3   Anus   Digestive extension of gut into arms
Sieve plate   Ampullae
X   Y
Genital aperture
Gonad   Mouth   Water vascular system   Tube feet

ORGANIZATION OF SEA URCHIN
Genital aperture
Calcareous spines   Ampullae   Tube feet
Sieve plate
Anus
Pedicellaria
Water vascular system
Gonad   Tooth   Mouth Muscle   Intestine

relatively large invertebrates, which may be digested by everting the stomach around them.

Brittle stars, serpent stars: somewhat similar to starfishes, but with long, slender arms that lash about rapidly in locomotion or in seizing prey.

Sea urchins: body enclosed in a rigid box; armless but spiny; slowly moving by means of spines and tube feet; feeding mostly on seaweeds and dead organic matter.

Sea cucumbers: body elongate (sausage-like) and leathery; no arms or spines; limy plates tiny and scattered in the skin; tentacles around the mouth; feeding on small animals or organic material in mud or sand.

The echinoderms have an exceptionally fine fossil record, extending over several hundred million years and including very numerous groups, some markedly different from any living today. It is especially noteworthy that most of the oldest forms were sessile, although among living echinoderms only crinoids are sessile, and not all of them are. What bearing may this fact have on the radial symmetry of echinoderms?

CHORDATA

Most of the chordates [13] are vertebrates— the fishes, amphibians, reptiles, birds, and mammals. They comprise another of the cul-

minating phyla to be discussed in the next chapter. Three major groups of chordates (Fig. 21–14) are not vertebrates. The Hemichordata ("half-chorded") or "acorn worms" are indeed fully wormlike in body form, and yet they have what seems to be a short equivalent of a notochord and also some other chordate characteristics (such as dorsal nerve cords and gill slits. The Urochordata ("tail-chorded," because the notochord is confined to the tail), or tunicates, look even less like vertebrates as adults, in which stage most of them are sessile, superficially spongelike creatures and some are colonial. Even the adults do have some vertebrate characteristics, however, and some of the larvae pass through a stage in which they are very much like tiny fishes. The Cephalochordata ("head-corded," because the notochord extends to the extreme tip of the head, which it does not in vertebrates), or lancelets, are quite fishlike in appearance throughout life. They have most of the basic vertebrate characters but lack vertebrae as well as a true brain and some other vertebrate features. The common form, usually available for observation or dissection, is amphioxus (the technical name of which is *Branchiostoma*). The three groups of nonvertebrate chordates are particularly mentioned and illustrated here because, as we shall

---

[13] "With a chord," because at some stage in its life a chordate has a stiffening rod, called a *notochord*, along the back.

21–12 Echinodermata. *A, A′.* A starfish, Asteroidea. The true dorsal and ventral surfaces of the bilateral larva are completely obscured in the adult, and it is convenient to recognize oral (*1*) and aboral (*2*) surfaces instead. The arms lie on five axes radiating from the mouth; these five axes are recognizable in all the echinoderms. Along the oral surface of each arm runs an ambulacral groove, in which the tube feet lie. The tube feet are extensions of the water-vascular system that connects with the external sea water via the sieve plate. The sieve plate is joined by a duct to a tube (the ring canal) encircling the alimentary canal; from this circular vessel ducts lead along each arm, with pairs of tube feet penetrating the body wall. The tube feet are effective walking organs (ambulatory, hence the name of the groove in which they lie) insofar as they are kept inflated by pressure on the sea water within; this pressure is maintained by the ampullae. *B.* Side view of

a brittle star. There is no anus, and the gut is entirely restricted to the central mass; it does not enter the five arms, as in a starfish. *C.* Side views of a free-swimming sea lily and a stalked, sedentary sea lily. *D, D′, E.* Side views of a sea urchin and a sea cucumber. As in *A′2* the ambulacral grooves bearing the tube feet are shown as crosshatches along the axes radiating from the mouth. In both these groups aboral surfaces are entirely wanting; their organization may be visualized by imagining the tips of the five arms in a starfish drawn backward (away from the mouth) and brought to a point. Note in *D′* how the arm of the water-vascular system that bears the tube feet is curled upward to the anus. Compare *D′* with *A′3.* Around the mouth a sea urchin has strong calcareous teeth moved by muscles. The pedicellaria are jointed jawlike structures. Compare them with the avicularia of Bryozoa (Fig. 21–9); the pedicellaria are not, however, degenerate individuals.

Above. The sea cucumber *Cucumaria frondosa.*

21–13 Sea lilies, starfishes, brittle stars, sea urchins, and sea cucumbers. *Right.* The starfish *Asterias.* Note the white sieve plate in the left photo and the tube feet in the right photo of the oral surface of a single arm.

Below. A fossil sea lily, *Eucalyptocrinus crassus,* from the Silurian of Indiana.

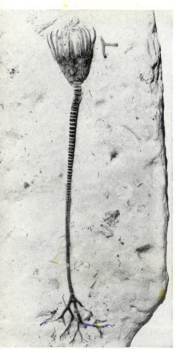

Below left. The brittle star *Ophiothrix fragilis. Below right.* The sea urchin *Echin esculentus* crawling on an aquarium wall. Note the tube feet at the upper right corr

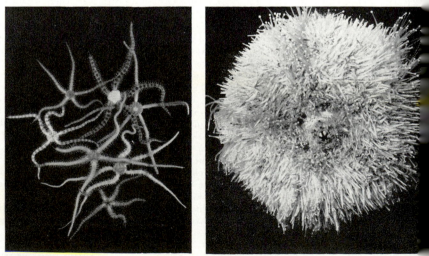

Top, F. Schensky, Helgoland; center left, Hugh Spencer; center right, © Douglas P. Wilson; bottom American Museum of Natural History; bottom right, both © Douglas P. Wilson

soon see, they cast some (rather feeble) light on the knotty problem of the origin of the vertebrates.

Adaptive Radiation Versus the Ladder of Life

The ancient Greeks were wonderful people and are rightly honored in all histories of science, and yet sometimes a horrid suspicion arises that we would understand life better today if some of their philosophical notions had been forgotten. It was apparently the Greeks who started the idea that organisms form a "ladder of life," with simplest organisms at the bottom and all others fitting in a sequence on up to the top. (Who do you suppose was put at the top? *You* were.) The ladder of life became orthodox biological theory long before evolution, as we now understand it, was even thought of. Certain early evolutionists took it over, some of them just because it was what "everybody knows" and some because they thought it was the pattern of evolution.

Many of the steps in evolution must have proceeded from the simple to the complex. That is practically a physical as well as a biological necessity for many early and some later steps. It has therefore been argued that an arrangement of living organisms from simple to complex is, or at any rate closely resembles, the actual course of evolution. But that is a non sequitur that has led to serious error and misunderstanding. Even today the old, mistaken notion of a ladder of life underlies much biological thinking and teaching. Some biologists still think of the phyla of animals as an "evolutionary series," with some dead ends and side branches, to be sure, but still on the whole one main sequence. Students are likely to study the "evolution of the vertebrates" by dissecting a dogfish, a frog, and a cat.

Some phyla must have arisen earlier than others, and some probably arose from others. It is also true that you can learn something about vertebrate evolution from dogfish, frog, and cat. The trouble is that thinking about the phyla or those vertebrates primarily as forming an evolutionary sequence gives in some essential respects a false picture of how evolution has really occurred. A truer picture is not so simple, but simplicity is no virtue unless it is also true.

Think a little more about the dogfish, the frog, and the cat. They are all living today. Therefore no one of them can be ancestral to any other one. At most the dogfish might be in some respects something like one of the ancestors of a frog, and a frog something like the ancestor of a cat. It is true that mammals (including cats) arose by way of reptiles from amphibians (and frogs are amphibians), and amphibians arose from fishes (and a dogfish is a fish). A shrewd anatomist would, however, suspect at once that a frog is not much like the amphibian ancestor of a mammal, or a dogfish much like the fish ancestor of an amphibian. Fossils show that the suspicion is certainly justified. The fishes ancestral to amphibians were radically different from a dogfish. The amphibians ancestral to reptiles and, through them, to mammals were radically different from a frog.

What, right now, is the significance of the differences between the three contemporaneous animals, dogfish, frog, and cat? It is not that they are steps in a ladder of life, each "lower" form being (or even representing or resembling) the ancestor of the next, "higher" form. The significance is that each of these animals follows a decidedly different way of life from the others. Whether we consider one lower, higher, simpler, or more complex than another (points that could be endlessly disputed and that are largely matters of verbal definition), the differences among them are plainly adaptations to living in different ways and in different environments. A cat can no more dart about chasing fishes beneath the waves of the sea than a dogfish can climb a tree and rob bird nests.

How the differences in adaptation arose, as a matter of history, is a different question entirely. It would make no difference in the facts of diverse adaptation if the cat ancestry had given rise to fishes instead of the other way around. An animal that climbs trees is not inherently "higher" than one that swims in the ocean. It happens that fishes evolved before mammals, as we know from fossils. Some fishes remained in their ancestral habitat, the

THE INVERTEBRATE CHORDATES

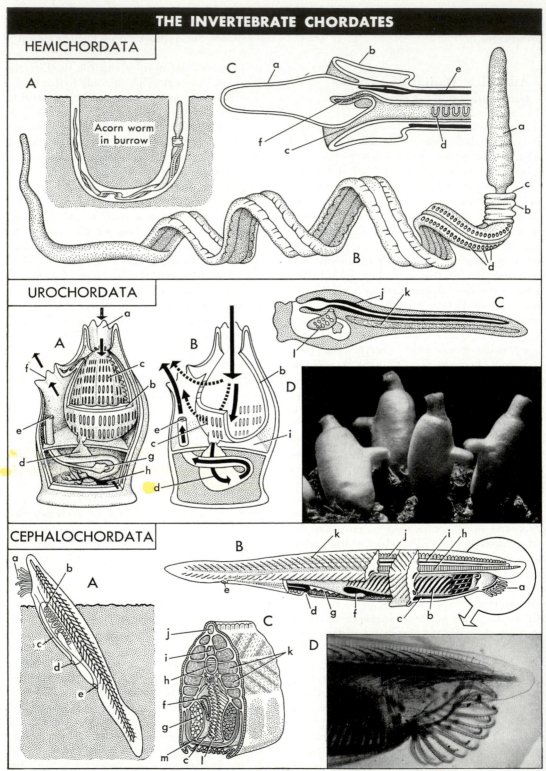

HEMICHORDATA

A — Acorn worm in burrow

UROCHORDATA

CEPHALOCHORDATA

Upper photo, © Douglas P. Wilson; lower photo, J. E. Webb

sea, and eventually evolved into dogfishes, among many others all quite different now from the ancient ancestral fishes but broadly similar in adaptive type. Other fishes acquired legs, left the ancestral habitat, and eventually evolved into cats, among many other animals now radically unfishlike. That is the history as it actually occurred. It has profound influence on just when and how the differences between dogfishes and cats arose, but it does not in the least alter the fact that the differences are adaptations to different ways of life.

Perhaps the fundamentals will be made clearer if you bring a pigeon into the series for dissection. There was no mammal stage in the ancestry of the pigeon, and no bird stage in the ancestry of the cat. The differences are purely and simply adaptive to different ways of life, which historically happened to be acquired at about the same time by different groups of organisms.

The same considerations apply to the differences between the phyla of animals—major groups that represent the most basic diversity of the animal kingdom. Some phyla are simpler than others. Sponges are certainly simpler than most other animals, but that does not necessarily mean either that sponges are older or that other animals arose from sponges. (The most competent authorities are reason-

ably sure that they did not.) In other instances, degrees of simplicity cannot be determined. Whether an earthworm or a starfish is simpler is a question that hardly makes sense when their very different anatomies are compared. The fundamental point is that the various phyla are basically different and that the differences arose as adaptations, that is, are related to different ways of life assumed by the early members of each phylum.

In great, long-lived groups like the phyla, the basic adaptive differences are obscured, and the whole situation is much complicated by diversity within each phylum. The more successful phyla have themselves become greatly diversified, and it has often happened that members of a phylum (defined in evolutionary terms) have acquired ways of life widely different from that basic for the phylum. Another problem arises from the fact that many animals, even some whole phyla, have become parasitic. Parasitism is a peculiar way of life which involves not only special new adaptations but also frequently the loss of organs and tissues of fundamental importance to nonparasites, such as the digestive system. You can seldom be sure whether a parasite is simple because it has lost complications or because it never had them. Secondary simplicity can also appear in animals that have greatly

---

21–14   The invertebrate chordates. *Top.* Hemichordata. *Balanoglossus,* the acorn—or tongue—worm. *A.* A whole animal in its U-shaped burrow in the sand. *B.* External features: *a,* burrowing proboscis; *b,* collar; *c,* mouth; *d,* gill slits. *C.* Detail of the head region: *e,* dorsal nerve cord; *f,* notochord, extending forward from the roof of the pharynx into the proboscis. The animal is a filter feeder; water and mud enter the mouth and are filtered of food at the gills; excess water passes out of the gills and oxygenates them at the same time. *Center.* Urochordata (Tunicata). *Ciona,* a sea squirt or tunicate. *A.* A whole animal with its body wall cut away to show its internal organization: *a,* mouth (or inhalant siphon), through which the feeding and respiratory current of water (see arrows) enters; *b,* wall of the much-enlarged pharynx, which is perforated by numerous gill slits (*c*); *d,* stomach; *e,* anus; *f,* exhalant siphon, by which the water current leaves the animal; *g,* gonad; *h,* heart. *B.* The pharyngeal region of the alimentary system. The inhalant and exhalant siphons are walled off by a membrane (*i*) from the body cavity proper (stippled). The cavity so formed (unstippled, above *i*) is the atrium. The feeding current entering the inhalant siphon is filtered through the

pharynx wall; the food passes on (solid arrows) into the alimentary system, and the filtered water passes (broken arrows) into the atrium and leaves by the exhalant siphon. *C.* A sea-squirt larva. It is bilaterally symmetrical, elongate, and free-swimming. It is mainly interesting because of its bearing on the origin of vertebrates (see Fig. 21–23). Note its dorsal nerve cord (*j*), notochord (*k*), and pharynx with gill slits (*l*). *D.* A photograph of sea squirts. *Bottom.* Cephalochordata. *Branchiostoma,* or amphioxus. *A.* A whole animal half-buried in sand or gravel, where it lives a semisedentary existence as a filter feeder like the other invertebrate chordates: *a,* cirri surrounding the mouth; *b,* gill slits in the pharynx; *c,* atrium; *d,* opening from the atrium to the outside; *e,* anus. *B.* The animal as though dissected from the side; *f,* liver; *g,* gonad; *h,* notochord; *i,* nerve cord; *j,* fin rays, supporting the dorsal fin; *k,* segmental muscles. The pharyngeal gill slits, as in the tunicates, open into an atrium that vents to the outside. *C.* An enlarged section through the pharyngeal region: *l,* endostyle, a gutterlike groove in the base of the pharynx along which filtered food moves into the intestine; *m,* coelom. *D.* A photograph of the head and mouth region.

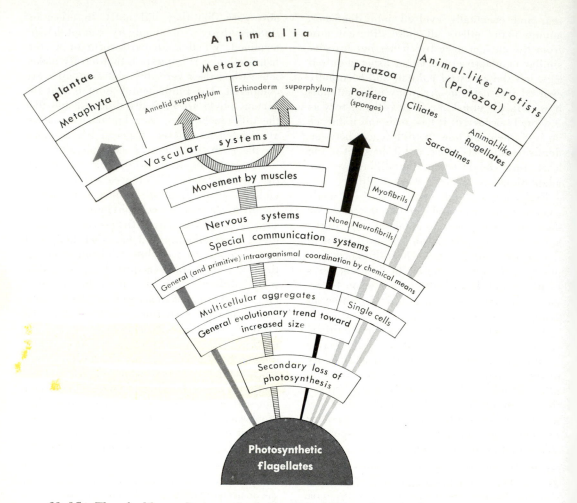

21–15 The fanlike radiation of phyla. This highly schematic diagram suggests the radiating rather than steplike pattern of the basic evolution of organisms. It also shows some of the main evolutionary advances that occurred, often independently and in different ways, in various lines. It does not show the early multiple branching of the multicellular plants (Metaphyta), and it greatly simplifies the branching of the protists and animals.

decreased in size, that have become sessile or attached, or that have made some other changes in their way of life.

It is quite clear that the animal phyla do not form a single ladder, or even two or three ladders, in which one phylum after another arose in succession. This conclusion is reinforced by historical evidence (see Chapter 29) that there is relatively little difference, geologically speaking, in the ages of most of the phyla. A picture truer than that of a ladder would be a fan (Fig. 21–15), with the phyla as ribs of the fan diverging from a remote base

as they spread by evolutionary change into different adaptive relationships and potentialities. What remains of the old idea of evolutionary sequence among the phyla (and what made it seem plausible to earlier students) is simply two points. First, some phyla did probably arise from early members (which are unknown in all cases) of other phyla, from which they inherited some complexities to begin with. Second, some phyla have changed more radically than others since they first arose, or since their common ancestry with other phyla.

Basic Characteristics of the Animal Phyla

You now have a passing acquaintance with the more important phyla, taken individually. For a broader understanding of animals and their diversity, the next step is to acquire an over-all view of the differences and resemblances among the phyla. Then, on this basis, something may be said about the origins and relationships of these major groups of animals.

Since the characteristics of the phyla are related to their ways of life, it might be supposed that they reflect adaptation to widely different physical environments. For instance, some phyla might be marine in origin, others fresh-water, and still others terrestrial. This is not the case. Most of the animal phyla certainly arose in the sea, and all of them may have. Two minor phyla (Mesozoa, Acanthocephala) are exclusively parasitic and of uncertain origin. Both include parasites of marine animals, and they may well have originated in the sea. Flatworms (Platyhelminthes) and aschelminths (especially nematodes and rotifers) are more common in fresh water or damp soil than in marine waters but do include marine species. All the other phyla seem clearly to be of marine origin.

Most phyla include both marine and fresh-water species.[14] Although both environments are aquatic, they may involve radically different physiological adaptations. Several groups, notably the nematodes and annelids, are abundant in damp soil, an environment not greatly different from fresh water in its physiological restrictions. Only three phyla include strictly terrestrial groups, animals capable of carrying on all their life activities in the open air: Mollusca (the fully terrestrial forms are the land snails), Arthropoda (insects, spiders, and some others), and Chordata (reptiles, birds, and mammals). These, too, include many marine and freshwater animals, and they are evidently of marine origin. Some of their lines of evolution became adapted to the most stringent of all physical environments, the land, where life is so difficult that no other phyla

[14] The exclusively marine nonparasitic phyla are Graptolithina, Ctenophora, Brachiopoda, Phoronidea, Chaetognatha, and Echinodermata.

have been able to cope with it. Yet the groups that did become terrestrial underwent no major changes; no new phyla emerged. It is significant that the three phyla that did conquer the terrestrial environment are also extremely successful in aquatic environments and are much the most varied and, in many respects, the most progressive of all the phyla.

The basic differences among phyla are striking anatomical features concerned with such processes as nutrition, internal transportation and maintenance, organ differentiation, coordination, and locomotion. All these features are closely interrelated. They add up to a distinct over-all structural and functional pattern for each phylum. Table 21–1 and Fig. 21–16 summarize some of the outstanding characteristics of the major phyla. The characteristics listed are not necessarily present in all the living members of each phylum but are believed to be primitive for the phylum, to have occurred in its earliest and ancestral members.

TISSUE LAYERS

The Porifera have no distinct differentiation of tissue in embryo or adult (Chapter 8). In Coelenterata there are only two reasonably distinct tissue layers (endoderm and ectoderm) in development and in the adult, although scattered cells or an irregular intermediate mass may be present between the layers. The other principal phyla have three layers (endoderm, mesoderm, ectoderm) more or less clearly involved in development, with a marked differentiation of tissues and organs within each of them.

DIGESTIVE SYSTEM

The Porifera have no special digestive organs (Chapter 6). Microscopic bits of food are carried along by currents set up by flagellated cells, and these are engulfed and digested within the cells. The Coelenterata have a central cavity with only one opening, into which food is conveyed from the tentacles. Digestion is partly in the cavity, stomachlike, and partly within cells lining it, spongelike. The digestive system of the Platyhelminthes also has only one opening, and digestion is also partly in the cavity and partly in its lining cells. The cavity is, however, much more complex than

TABLE 21–1 *Some characteristics of the major animal phyla*

Phylum	Embryonic cell layers	Digestive system	Coelom	Circulatory system	Segmentation	Symmetry	Larvae	Other features of adults probably primitive for phylum
Porifera	Indistinct	No special organ	None	Absent	Absent	Radial	Peculiar to Porifera	Sessile. Microscopic food ingested from flagella-produced currents.
Coelenterata	Two	Pouchlike; one opening	None	Absent	Absent	Radial	Peculiar to Coelenterata	Sessile. Food captured by tentacles with stinging cells.
Platy-helminthes	Three	Pouchlike; one opening	None	Absent	Absent	Bilateral	Trocho-phorelike *	Motile. Flattened, wormlike. Passive or immobilized animal food.
Nematoda	Three	Tubular; two openings	Pseudocoel	Absent	Absent	Bilateral	None	Motile. Cylindrical, wormlike. Early becoming parasitic or including some parasites.
Bryozoa	Three	Tubular; two openings	True coelom	Absent	Absent	Bilateral		Sessile but bilateral. External skeleton. Flagellated tentacles. Early becoming colonial.
Brachiopoda	Three	Tubular; two openings	True coelom	Absent	Absent	Bilateral	Trocho-phore *	Sessile but bilateral. Dorsal and ventral shells. Flagellated tentacles inside shells. Noncolonial.
Mollusca	Three	Tubular; two openings	True coelom	Present	Absent	Bilateral	Trocho-phore *	Motile, creeping on a ventral foot. Shelled (but primitive form of shell uncertain).
Annelida	Three	Tubular; two openings	True coelom	Present	Present and similar in the two phyla	Bilateral		Motile. Cylindrical, wormlike. Bristle appendages.
Arthropoda	Three	Tubular; two openings	True coelom	Present	Present and similar in the two phyla	Bilateral	None or secondary	Highly motile. Jointed legs. External skeleton.
Echino-dermata	Three	Tubular; two openings	True coelom	Present	Absent	Secondarily radial	Pluteus and pluteus-like †	Sessile or sedentary. Heavy protective skeleton. Noncolonial.
Chordata	Three	Tubular; two openings	True coelom	Present	Present, different from that in annelids and arthropods	Bilateral	Pluteus-like, † none, or secondary	Highly motile. Internal skeleton aiding propulsion

* See p. 576.   † See p. 579.

in Coelenterata, with intricate branching throughout the body. Digested products are thus available near all other tissues in spite of the absence of a special transport system. The other major phyla have a digestive tube open at both ends, so that food enters one end and residues and wastes leave the other. The tube may be almost completely simple or may be distinctly differentiated into different regions or organs. In parasitic members of these phyla, the digestive system is often reduced and may be lost.

COELOM

Porifera, Coelenterata, and Platyhelminthes have no cavity between the digestive cavity and the body wall. In Nematoda there is such a cavity, but it does not have a special lining and so is considered a pseudocoel (or "false coelom") rather than a true coelom. The other major phyla have a true coelom, with a cellular (mesodermal) lining.

SKELETAL SYSTEM

Most multicellular animals have evolved skeletal systems to support their mass of otherwise flabby tissue, to facilitate locomotion, and in some cases to serve as protection. The nature of the skeleton varies greatly in different groups, but comparison of the variations throws little light on the relationships among phyla. In many groups (coelenterates, bryozoans, brachiopods, mollusks, etc.) the skeleton is an external and usually calcareous "house" or cup. The arthropods generally have external skeletons made in part of protein, but their skeletons are elaborately jointed, permitting locomotion. Vertebrate skeletons are, of course, bony and internal. Echinoderm skeletons are also developed from internal tissue, but they are functionally external and are calcareous, with no real resemblance to bone.

CIRCULATORY SYSTEM

Porifera, Coelenterata, Platyhelminthes, Nematoda, and Bryozoa have no special circulatory systems (see Chapter 7). In them cell layers must be thin (they are usually only one or two cells in thickness) and in contact with or near body fluids or the digestive cells.

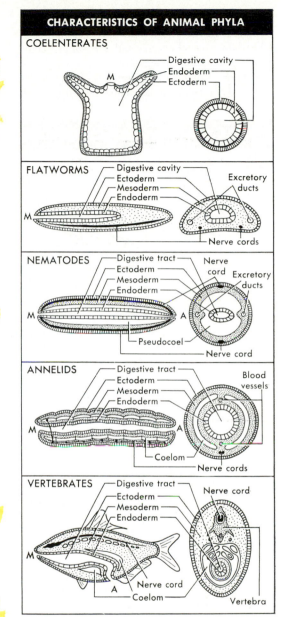

21–16 Characteristics of some animal phyla. *M*, mouth; *A*, anus.

Otherwise, the separate cells could neither obtain adequate food by diffusion nor dispose of waste products. Brachiopoda, Mollusca, Annelida, Arthropoda, Echinodermata, and Chordata have vascular circulatory systems, correlated with the development of thicker and more complex cell layers and masses.

SEGMENTATION

Three phyla, Annelida, Arthropoda, and Chordata, are characterized by some occurrence of successive segments in the length of the body during development. In annelids many of the segments are closely similar, with a repetition of various visceral organs in them. Arthropod segmentation was about as in annelids to begin with, but early became much modified by the fusion or reduction of segments, with a decreased repetition of organs. Chordate segmentation arose independently and was primarily related to locomotion. It is most evident in some muscles, nerves, and bones (notably the vertebrae) and is obscure or absent in other tissues and organs.

SYMMETRY

Most animals are more or less symmetrical in external form, at least. Among those most obviously symmetrical, some have a central axis and are radially symmetrical around this axis, much as a wheel is symmetrical around its axle (Fig. 21–17). Others (such as man) are bilaterally symmetrical: the two halves are more or less mirror images of each other on each side of a central plane. Even animals that seem to have some other kind of symmetry or none at all usually are found to have radial or bilateral symmetry when their development and ancestry are studied. Coiled snails, for instance, are not symmetrical in adult form, but their ancestors were surely bilateral, and they are still bilateral in early development. The coiling arises from the unequal growth of the two originally symmetrical sides.

The fundamental significance of symmetry is clearer if body form is also considered in terms of asymmetry, or directional differentiation. The differentiation of directions in an animal is related to locomotion or, more

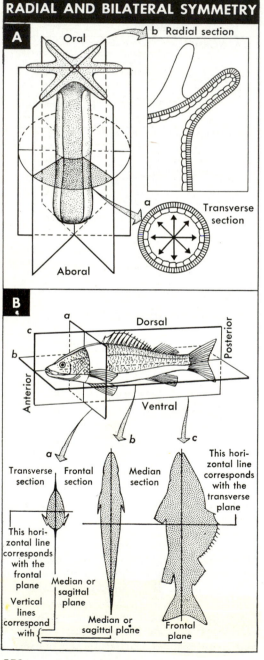

RADIAL AND BILATERAL SYMMETRY

A
Oral
b Radial section
a Transverse section
Aboral

B
Dorsal
Anterior
Posterior
Ventral

Transverse section
Frontal section
Median section
This horizontal line corresponds with the transverse plane
This horizontal line corresponds with the frontal plane
Vertical lines correspond with
Median or sagittal plane
Median or sagittal plane
Frontal plane

21–17 Animal symmetry. *A.* Radial symmetry, exemplified by a coelenterate polyp like *Hydra: a,* a transverse section, which shows perfect symmetry in all radial directions; *b,* a radial section. The significant axis is the oral–aboral (mouth-to-base) axis. *B.* Bilateral symmetry, exemplified by a fish. The upper figure shows the three planes (*a,* transverse; *b,* frontal; *c,* median or sagittal) in terms of which symmetry is discussed. A transverse section, obtained by cutting in the transverse plane, shows bilateral symmetry; it is symmetrical on the two sides of the median plane. It is not symmetrical up and down, that is, on the two sides of the frontal plane. A frontal section (cut in the frontal plane) also shows bilateral symmetry about the median plane but not about the transverse plane. A median section shows no symmetry about either the transverse or frontal planes.

broadly, to orientation with respect to the environment. In a spherical body all directions from the center are the same. Among adult organisms such a form occurs only in a few protists. They float freely in water, buoyed up so that direction of motion or of gravity has no importance in their way of life. All multicellular organisms have some degree of bodily differentiation corresponding to up (away from gravity) and down (toward gravity). If this is the only directional differentiation, body form tends to be radially symmetrical. This is especially true of sessile organisms. The direction of attachment (which usually is downward) is functionally distinct from the direction straight away from the attachment into water or air. Other directions, around and away from the up–down axis, are all about the same as far as the organisms' needs and reactions are concerned.

Animals that move under their own power usually have a habitual direction of movement, and this involves another functional differentiation of direction into forward and backward and of form into fore and aft. Fore–aft, or anterior–posterior, and up–down, or dorsal–ventral, differentiation make for symmetry only in the direction from side to side: bilateral symmetry. Since most animals are motile, or had motile ancestors at not too distant a date, most of them are bilaterally symmetrical, or they are anteroposteriorly and dorsoventrally asymmetrical.

Among the major phyla, the Porifera and Coelenterata have basic up–down differentiation and tend to be radially symmetrical. Sponges may become quite irregular in form, but in most coelenterates the radial symmetry remains beautifully apparent. The Echinodermata usually have radial symmetry, but their larvae are bilateral, and some trace of bilateral symmetry is present even in adults. It is probable that the remote ancestors of the echinoderms were motile and bilateral but that the early echinoderms themselves were sessile, their attachment leading to the development of a secondary radial symmetry; although many later echinoderms became detached and slowly motile, they retained this secondary radial symmetry. All the other animal phyla (other than Porifera, Coelenterata, and Echinodermata) were bilaterally symmetrical

originally, and most of their members still exhibit that symmetry.[15]

LARVAE

Many animals have larvae (see p. 270) that differ markedly from the adults. Some sorts of larvae are characteristic of particular phyla, and some clarify relationships among phyla, as will be discussed later in this chapter.

Origins and Relationships of the Animal Phyla

We earlier compared the animal phyla to the ribs of a fan. Of course, this is just a figure of speech, and you should guard against thinking that an analogy of this sort is precisely valid in all details. If the phyla were *exactly* like the ribs of a fan, spreading equally and all from the same point, there would be no reason for discussing the relationships between any particular phyla. (Why?) In fact, although the origins of all of them are extremely and comparably remote (all near the base of the fan), some did branch off a little later than others, and some have diverged less than others. Thus some groups of phyla seem to be more nearly related among themselves than to other phyla.

Determination of the origins and relationships of the phyla is one of the most interesting, but also one of the most difficult and at present most speculative, of phylogenetic problems. After hundreds of millions of years of separate evolution, the phyla are now so distinct that clues to their relationships are very obscure. Fossils help a little, but not much.[16] The earliest members of each phylum were extremely ancient, small, soft-bodied animals that have not in any case been preserved

[15] That the Bryozoa and Brachiopoda are bilaterally symmetrical but sessile may seem to contradict our explanation of the adaptive origin of symmetry. Their bilateralism is inherited from motile ancestors, and they retain it because of a peculiarity of their attachment, which is not at the end of an up–down axis (as in Coelenterata) but at the end of a previously developed fore–aft axis (see Fig. 21–10).

[16] Note that it is strictly the origin of phyla on which fossils throw little light. Evolution (of classes, etc.) within phyla is, in many cases, fully documented by fossil evidence. See especially Chapters 29, 30, and 31.

as fossils. At least, no fortunate paleontologist has yet found them. The earliest known (fossil) representatives of each phylum are already quite distinct from those of any other phylum. Nevertheless, they are more primitive than living forms, and so they add to the clues provided by the structure and, especially, the development of living animals.

ORIGINS OF MULTICELLULAR ANIMALS

Protistan ancestors

It is generally agreed and indeed seems to be a logical necessity that multicellular organisms—true plants and animals as we use the terms here—arose from unicellular forms, or protists according to our usage. Since the transitions occurred upwards of a billion years ago, we can be sure that no living protists are quite like the very remote ancestors of multicellular organisms. Still, it is probable that all retain some features of those ancestors and furthermore that some are more like them than others.

It was noted in Chapter 19 that the flagellates (phylum Mastigophora) are commonly considered most similar to the ancestors not only of multicellular organisms (both plants and animals) but also of animal-like protists, notably the rhizopods (amebas and relatives) and the ciliates (paramecia and their kin). These animal-like protists are frequently classified as animals in the phylum Protozoa, although it is unlikely that they particularly resemble any stage in the direct ancestry of the true animals. In fact, lowly as they are in general level of organization, most of them are manifestly too specialized in one way or another to be given a place in that ancestry. In the course of some 2 billion years, the protists must have explored practically every evolutionary development possible *without* multicellular structure. The animal-like protists, or so-called Protozoa, are especially interesting in showing how far organisms can go in the way of diversity and complexity of structure *without* becoming animals.

Flagellates as a group are not too specialized, even in the now living forms, to have given rise to multicellular organisms; they seem rather to be at just about the stage expected for that immediate ancestry. Bacteria, for instance, probably more primitive or remote in origin, lack the complex chromosomal–mitotic mechanism of inheritance present in ancestral flagellates and most or all of their descendants. The one peculiarity of flagellates, their method of locomotion, is found at least in the male gametes of almost all animals and the more primitive plants. Further evidence that the flagellates could well evolve into both plants and animals is that they include manifestly closely related species some of which are photosynthetic autotrophs, hence physiologically "plants," and others of which are non-photosynthetic heterotrophs, hence physiologically "animals." In fact, it is possible to turn some flagellates from "plants" into "animals" in a laboratory. Another item of evidence is that a number of flagellates form colonies and that the individuals in those colonies may be somewhat differentiated. This suggests a foreshadowing of multicellular organization with cell differentiation.

Size and selection pressures

Without complicated circulatory and conducting systems, such as cannot develop within one cell, the amount of cytoplasm that can be adequately supplied with needed materials and adequately coordinated in its behavior is necessarily small. There is also a low limit to the amount of cytoplasm in which chemical activities can be controlled by one nucleus. Some of the living ciliates, large in comparison with most other protists, have shown us (Chapter 19) how far strictly protistan organization can go in overcoming these difficulties. They contain special but rather simple organelles, and the problem of nuclear control is eased by a macronucleus, a sac containing many nuclei spread over the length of the cell. Since this is the extent of ciliate adaptation, the ciliates were in an evolutionary blind alley.

At the same time that there was selection pressure toward greater size, then, there was selection pressure toward greater complexity. Possibly unexploited ways of life could be followed if only other means of supply, coordination, and control could evolve. For one thing, an animal of protist size is largely at the mercy of waves and currents and cannot seek its food or a suitable environment in the active way typical of a larger animal. Better sensory and, in general, behavioral apparatus can also be packed into a larger animal. All these and other adaptive advantages could be, and were, achieved by the evolution of multicellular organization. Differentiation of the cells produced all the special tissues and organs that are present in higher animals.

Origin of sponges

It is probable that multicellular animals evolved twice independently.[17] Sponges likely had a separate origin from any other animals. For that reason they are sometimes put in a subkingdom of their own, Parazoa, while all the other multicellular animals are classified as Metazoa. Sponges almost certainly arose from flagellate colonies similar to those of the collar flagellates (Fig. 19–5), and they still contain cells practically identical with them. Although successful at their own level, the sponges were in another blind alley as regards any further advances toward active animal life. They developed no effective means of locomotion but became sessile. They feed in protistan fashion, within their various cells. Much of their internal communication is by ameboid cells slowly moving through the jelly-like matrix in which the other cells are embedded—a messenger system adequate for such inactive organisms but not for higher animals.

Origin of Metazoa

The ancestry of all the other multicellular animals, the Metazoa, evolved in a way that avoided the sponges' blind alley and led to enormously greater diversity and complication. Probably the most crucial feature of that development was the differentiation of special contractile cells, eventually becoming muscles, and special transmission cells, eventually becoming nerves. Primitive protists, probably flagellates, were subjected to various selection pressures corresponding with many different ways of life. Some conservative lines changed but little. Some became other kinds of protists, such as rhizopods and ciliates. Several became plants. One became sponges. One,

[17] Multicellular plants—algae—probably evolved from protists several times independently.

from our point of view the most promising and eventually fruitful, took this one direction that happened to have the potentialities of further animal evolution.

The exact details of the process are obscure. There are many clues, but they can be and have been interpreted in a number of different ways. The various theories are of great interest, but they are so complex and so uncertain that we shall not treat them in detail; in the absence of adequate fossil evidence, certainty is almost impossible. From the start there are two main possibilities. One, which is accepted by most biologists, is that metazoans evolved from colonial flagellates, as did sponges, but with more and dissimilar differentiation of individuals—into cells of a higher order of individual—in the colony. Perhaps the original colony was a hollow globe, something like *Volvox* (Fig. 3–20). If differential growth pushed in one side so that it came in contact with the other side, the result would be a pouchlike, radially symmetrical body with two layers of cells in its wall. This is like the basic pattern of the coelenterates. The primitive coelenterates may then have given rise to ancestral flatworms by the development of bilateral symmetry and an intermediate cell layer (mesoderm). From this stage higher animals could arise by the evolution of an anus, a coelom, and other progressive complications.

The other main possibility, which at present seems less likely, is that metazoans evolved from noncolonial protists—perhaps more like ciliates than flagellates—that already had organelles and multiple nuclei. Simple division of the whole protist by the development of membranes separating it into parts, or cells, each with one nucleus, could result immediately in a metazoan something like a flatworm. Coelenterates could then arise from ancestral proto-flatworms by divergent, partly degenerate, evolution, and other metazoans by the progressive changes suggested in the last paragraph.

Both these theories seem oversimplified, and probably neither one is quite right. The earliest metazoans need not have been and indeed probably were not much like either a modern coelenterate or a modern flatworm. In any case, we do know beyond much doubt that the ancestries of the coelenterates and of the flatworms branched off from the main line of metazoan evolution at a very early stage in that history. Within the main line three extremely important innovations then occurred.

1. A circulatory system evolved, making possible far greater bulk and complication than in the necessarily thin-walled coelenterates and flatworms. (Why must they remain thin-walled because they lack circulatory systems?)

2. An anus developed, so that food movement in the alimentary canal became one-way, and digestive and associated processes became regionally differentiated.

3. A body cavity (coelom) evolved, making possible the free and lubricated motion of internal organs suspended in it.

The Metazoa evolving these three new features seem soon to have split into two branches, the annelid and the echinoderm superphyla, within which all the higher phyla arose.

THE ANNELID SUPERPHYLUM

If you look again at Table 21–1 you will see that four of the major phyla (Bryozoa, Brachiopoda, Mollusca, and Annelida) are indicated as having larvae called *trochophores.* These larvae differ a good deal in different species and phyla, but they also have fundamental resemblances. Every one is free-living and bilateral, with a complete, regionally differentiated digestive tube and a ring of beating hairs or cilia situated anteriorly to the mouth. The ring looks somewhat like a wheel, and that is the reason for the name trochophore ("wheel-bearer"). There are other resemblances in early development in these phyla, irregularly distributed and varied in their members and yet suggesting that ancestors of the phyla did all develop in the same way. Differentiation usually starts with the first cleavage, and even in the two-celled stage each cell is destined to give rise to different tissues and organs in the course of later development. The mesoderm arises by cell cleavage between endoderm and ectoderm, and the coelom later develops as a space within the mass of mesoderm (Fig. 21–18).

The evidence is not conclusive, but it does strongly suggest that the four phyla with trochophore larvae arose from the same very remote ancestry. If so, that ancient common ancestry also had trochophore larvae. This does not tell us what the adult ancestor looked like. Presumably it did not look much like a trochophore.[18] The adult ancestor must, however, have developed from a trochophore larva and have incorporated some trochophorelike anatomical features.

Although arthropods do not have trochophore larvae, the developmental resemblances between arthropods and annelids and the strong indications of relationship even among adults leave no doubt that the two phyla had a common ancestry. Moreover, the ancestors were more annelidlike than arthropodlike; the annelids seem, on the whole, to have changed less than the arthropods. The trochophore stage in the arthropod life history disappeared early in the course of evolution. Later in some of the groups of arthropods different, newer kinds of larvae evolved, such as the wormlike larvae of many insects.

Some flatworms have larvae that resemble trochophores except that the digestive system lacks an anus, which is also true of adult flatworms. A coelom is also lacking in flatworms, but there is a mass of mesoderm that arises much as in annelids. The difference depends essentially on whether or not a space opens up in the mesoderm mass. The consensus is that flatworms probably are related to the phyla with trochophore larvae, a group of phyla sometimes called the *annelid superphylum.* Most zoologists think that the striking peculiarities of the flatworms, such as absence of anus and of coelom, are explained by their ancestry's branching off from the other phyla before these characteristics had evolved.

The annelid superphylum includes at least the major phyla Platyhelminthes, Bryozoa, Brachiopoda, Mollusca, Annelida, and Arthropoda, and probably also some of the lesser phyla. It is speculative but reasonable to imagine that there was an ancestral group, now long obscured in the mists of time, that

[18] Many zoologists used to think that it did look like a trochophore, and some charts of phylogeny still illustrate a trochophore as the ancestor of these phyla. This situation results from the wrong older interpretation of the "biogenetic law," the belief that ontogeny literally repeats phylogeny (Chapter 8).

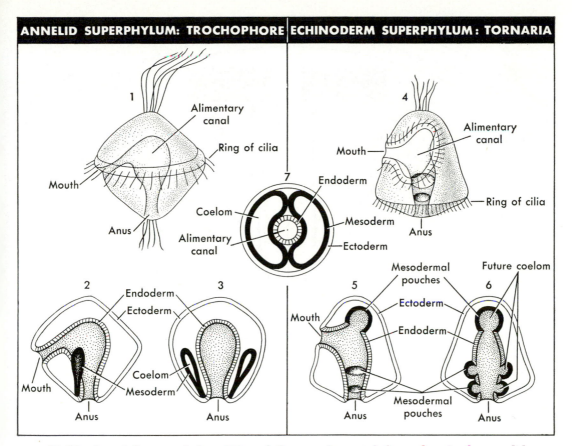

ANNELID SUPERPHYLUM: TROCHOPHORE | **ECHINODERM SUPERPHYLUM: TORNARIA**

21–18 Larval forms and the origins of the mesoderm and the coelom in the annelid and echinoderm superphyla. *1.* A side view of a trochophore. *2.* A median section of a trochophore, showing how the tissue layers are related and how the mesoderm arises as a mass (black) from special cells near the point of ectoderm–endoderm transition in the anal region. *3.* A transverse section of a trochophore, showing the coelom arising as a split within the mesodermal mass. *4.* A side view of a tornaria. *5, 6.* Median and transverse sections of a tornaria, showing how the mesoderm (black) arises as pouches from the wall of the alimentary canal. The cavities of the pouches form the coelom. *7.* A generalized transverse section of both superphyla, showing the relation of mesoderm and coelom to endoderm and ectoderm.

split or radiated into at least four basically different branches. One of the diverging, descendant groups evolved into flatworms. Another soon split again and evolved into bryozoans and brachiopods. The third evolved into mollusks. The fourth evolved into annelids, and the most primitive annelids split into two groups, one of which evolved into later annelids and one, with more radical change, into arthropods.[19] (See Fig. 21–19.)

[19] There is no fossil record of the nematodes, and the living forms are so profoundly modified that they provide no hint of broader relationships. It is possible that they are another branch of the annelid super-

THE ECHINODERM SUPERPHYLUM AND THE ORIGIN OF VERTEBRATES

We are vertebrates ourselves, and so it is not surprising that the origin of the vertebrates has long been of particular interest to zoologists. There are now no real doubts as to the origins and relationships of the various classes

phylum, affected by early adaptation to semi- or wholly parasitic habits. Rotifers (Trochelminthes) are probably members of the annelid superphylum greatly reduced in size; some adult rotifers are rather like trochophores. Several other minor phyla may also belong to the superphylum, but they are too changed or too poorly known to be classified with certainty.

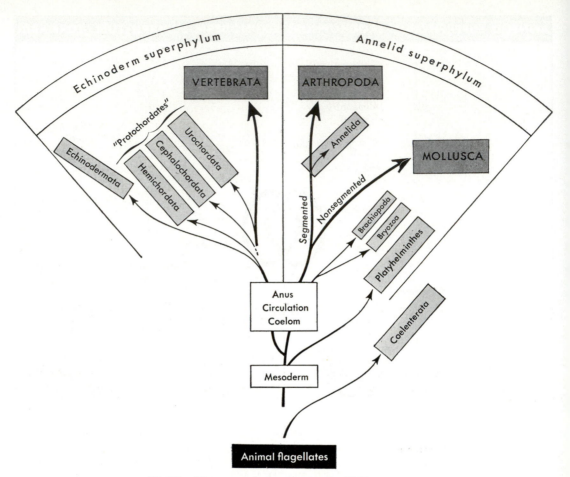

21–19 The superphyla of animals. Cf. Fig. 21–15.

within the subphylum Vertebrata (as will be shown in Chapter 22). These origins are revealed by fossils, but fossils fail to clarify the remoter origin of the very first vertebrates. The oldest known fossil vertebrates are extremely primitive, but they have no suggestive resemblances to any phylum other than the Chordata, of which they are a subphylum.

You recall (p. 566) that besides the Vertebrata there are three surviving subphyla of Chordata: Hemichordata, Urochordates, and Cephalochordata. They lack vertebrae and some other features of the earliest, as well as of recent, vertebrates, but they seem surely to have had a common origin with the vertebrates. They must have branched off from the vertebrate ancestry at a very remote time, before the vertebrates had originated as such.

It would be expected that they would by now have evolved many peculiarities of their own, which never occurred in their ancestry, and such appears to be the case. However, the mere fact that they are different, ancient branches from the same ancestry as that of vertebrates means that they might also have retained some features of that ancestry that have been lost in the vertebrates. This, too, seems to be true. There is also reason to believe that the non-vertebrate chordates have, on the whole, changed less than the vertebrates. It is the development of some members of these groups that seems at present most likely to give clues to the origin and relationships of the chordates in general, and therefore also of the vertebrates.

Vertebrates have no floating larvae. This

developmental stage probably occurred in their remote invertebrate ancestors, but as in arthropods it has been lost in the course of evolution. In the vertebrates that do have larvae (some fishes, most amphibians), the larval stage is a new evolutionary development, not inherited from the invertebrate ancestors. The hemichordates, however, have floating larvae that somewhat resemble trochophores but differ in several ways. The striking difference is that the hemichordate larva has a twisted ring of cilia that encircles the mouth, instead of a single ring anterior to the mouth as in a trochophore.[20] The hemichordate larva is extraordinarily like the larva of an echinoderm; it was, indeed, mistaken for an echinoderm when first discovered.[21] This is strong, but not conclusive, evidence that hemichordates and echinoderms have inherited their larval stages from the same very ancient marine invertebrate ancestry. Since the hemichordates seem to be an offshoot of the vertebrate ancestry, the vertebrates, too, are probably derived from the same ultimate source as the echinoderms. Because the hemichordate and echinoderm larvae are different from trochophores, the chordate–echinoderm ancestry probably early became distinct from the ancestry of the annelid superphylum. Chordates and echinoderms may, then, be considered as members of a second superphylum, sometimes called the *echinoderm superphylum* (Figs. 21–19 and 21–20).

There is some other evidence of relationships between echinoderms and chordates and of their distinction from the annelid superphylum. In both echinoderms and chordates the early cleavages of the zygote usually produce cells that are still undifferentiated, each

capable of developing into a whole adult if separated. In echinoderms and some chordates the mesoderm and coelom also develop in the same way, a way quite different from that in

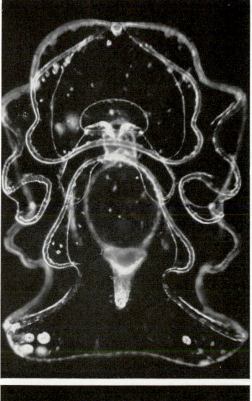

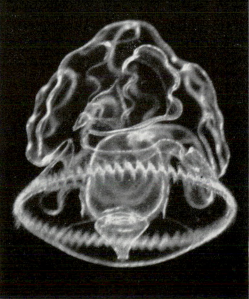

[20] The hemichordate larva is called a *tornaria*.
[21] An echinoderm larva is called a *pluteus* or *bipinnaria*.

21–20 Larvae of an echinoderm and a hemichordate. Both photographs are of living specimens, magnified. *Top.* An echinoderm (sea cucumber) larva, viewed from the front. The wavy bands of cilia are visible, as is the alimentary canal. *Bottom.* A hemichordate larva (tornaria), viewed obliquely from the right side. The mouth is obscured by the beating cilia, but the arching alimentary canal is visible, as is the anus (cf. Fig. 21–18).

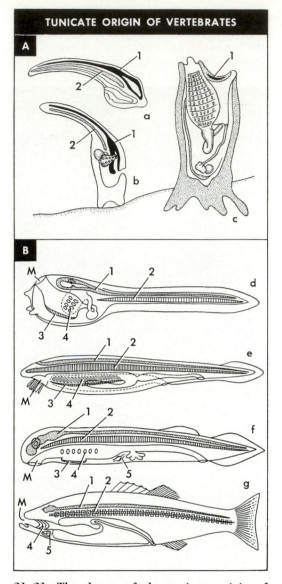

TUNICATE ORIGIN OF VERTEBRATES

21–21 The theory of the tunicate origin of vertebrates. *A.* Younger (*a*) and older (*b*) larvae of a tunicate, metamorphosing into a sedentary adult (*c*). *B.* A comparison of *d*, a tunicate larva, with *e*, a cephalochordate (amphioxus), *f*, the larva of a cyclostome (agnathous fish, p. 602), and *g*, a mature fish. Note *M*, the mouth; *1*, the dorsal nerve cord; *2*, the skeletal notochord; *3*, the endostyle, the gutter at the base of the pharynx that in higher vertebrates (including the adult cyclostome) becomes the thyroid gland; *4*, the pharyngeal gill slits; and *5*, the heart. Note also the principal point that the tunicate larva is motile and bilaterally symmetrical, thus making plausible an evolutionary connection between the sedentary tunicates and the motile vertebrates.

the annelid superphylum. Pockets or folds arise from the endoderm of the developing digestive tract. The spaces in the pockets become the coelom, and their walls become mesoderm (see Fig. 21–18). This feature of development is much modified in most vertebrates but is probably primitive for chordates as a whole. Finally, it is interesting that echinoderms and chordates share a common biochemical feature: they contain the same phosphagen (Chapter 4)—creatine. The common phosphagen in the annelid superphylum is, on the other hand, arginine.

It should be emphasized that this view does not envision the transformation of an echinoderm into a chordate—the descent of the Chordata from the Echinodermata, as such. It is believed that the two phyla arose as widely diverging lines from some group of animals now long extinct and of unknown adult structure but with early developmental stages and larvae similar to those of echinoderms. From that remote ancestry the echinoderms evolved as sessile, sedentary, or sluggish animals.

Indeed, there is some evidence that the sedentary habit may have been characteristic of the whole superphylum in some early stage of its evolution. One of the most characteristic chordate features (the pharynx perforated by gill slits) is a likely adaptation to sessile life. In fact, two of the three surviving nonvertebrate chordate groups employ the perforated pharynx as a feeding device (see Fig. 21–14). One of these groups, Urochordata (tunicates), is still predominantly sedentary. The other, Cephalochordata (amphioxus), although capable of swimming, burrows in coarse sand, tail down, and lives an effectively sedentary life, drawing in water and filtering microorganic food through its pharyngeal slits.[22] The evolution of the chordate (and therefore vertebrate) pharynx is easily understood as an adaptation to the sedentary habit. How, then, can we explain the emergence of vertebrates—the most spectacularly motile of animals—as of common origin with the sedentary tunicates (urochordates)?

Some modern tunicates give a hint as to a possible mode of vertebrate origin. Like other

[22] The chordate pharyngeal slits persist, of course, in fishes as gill slits with a respiratory function.

sessile animals, they rely heavily on motile larvae to disperse the species. The motile larvae of tunicates (Fig. 21–21) are elongate, bilaterally symmetrical forms resembling simplified caricatures of a vertebrate. Most significant is their possession of a notochord lying, as in vertebrates, dorsally to the gut. It serves the function of a skeleton for simple fishlike locomotion. Many zoologists now consider it probable that vertebrates evolved as *neotenic* tunicates (footnote 9, p. 244). According to their theory, the motile larva of some sedentary tunicate gradually assumed an increasingly larger and more important part of the whole life cycle; eventually it became capable of its own reproduction and sloughed off, so to speak, the ancestral adult stages adapted to the sedentary habit. An alternative theory, which at present seems about equally plausible, is that the common ancestor of tunicates and vertebrates was a free-swimming form somewhat like a tunicate larva. In this case the sessile adult condition is a tunicate specialization, and the ancestors of the vertebrates were not sessile.

CHAPTER SUMMARY

What is an animal? The difficulty of definition in protists; major animal characteristics —metabolism (nonphotosynthetic), mobility, definite adult form with complex tissue and organ differentiation and a general absence of rigid cell walls and cell vacuoles, maintenance of an internal environment (with high sodium chloride content), high sensitivity and responsiveness (associated with nervous tissue), an absence of development in the haplophase of the reproductive cycle.

Major animal phyla:

Porifera (sponges): colonies of semi-independent cells, aquatic and sessile; feeding-current mechanism; failure to evolve greater complexity.

Coelenterata (polyps and medusas): single opening to digestive cavity; tentacles and carnivorous nature; polyp (budding) and medusa (dispersal) in the life cycle; hydroids, siphonophores, jellyfishes, sea anemones, corals.

Platyhelminthes (flatworms): lack of exact zoological meaning for the word "worm"; flat body with one opening to digestive cavity; bilateral symmetry; receptors in the head region; nervous system; complex reproductive and excretory systems; common parasitism.

Nematoda (threadworms): anus present; parasitism and abundance.

Bryozoa ("moss animals"): colonial habit; skeleton; ciliated tentacles and feeding currents; U-shaped digestive cavity; aquatic (mainly marine) habitat; abundance as fossils.

Brachiopoda ("lamp shells"): two shells, contrasted with clams; tentacles; noncolonial habit; abundance as fossils.

Mollusca (snails, clams, squids—treated in Chapter 22).

Annelida (segmented worms): earthworms and their soil-eating habits; marine annelids; free-swimming and tube-living forms; leeches, parasitic annelids.

Arthropoda ("jointed-feet" animals— crustaceans, spiders, and insects—treated in Chapter 22).

Echinodermata ("prickly skin" animals): sea lilies, starfishes, brittle stars, sea urchins, sea cucumbers; radial symmetry; water-vascular system; tube feet; skeleton; marine habitat; abundance as fossils.

Chordata (animals "with a chord," the notochord): fishes, amphibians, reptiles, birds, mammals; Hemichordata (acorn worms); Urochordata (tunicates); Cephalochordata (amphioxus); possession of a notochord, dorsal nerve cord, and pharyngeal gill slits.

22

Mollusks,
Arthropods,
and Vertebrates

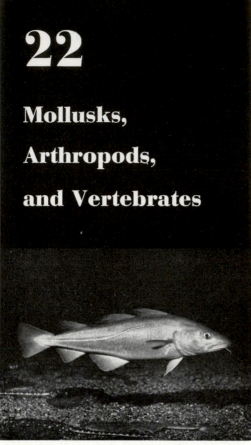

© Douglas P. Wilson

The vertebrates—exemplified here by the codfish (Gadus)—are the most complex of the three great phyla, mollusks, arthropods, and vertebrates, which represent end points of three distinct lines of evolution.

Some tourists whiz through the desert assuring each other in boredom that there is nothing there. A lounger on the beach may not spare a moment to glance at a boy toying with a crab or a shell. Yet the desert is full of life and movement for anyone willing to sit still and look, and the boy is unwittingly illustrating dramatic contrasts and resemblances in the play of life. One of the finest values of biology is that it reveals profound meanings in apparent commonplaces. The greatest poetry is not written in garrets or printed in books. It is all around us in the living world. To read it we need only to pause, to observe sympathetically, and to learn a little about the long history and the incessant activities of life.

Let us look more closely and with more understanding at the boy, the crab, and the shell. The contrasts seem obvious enough. The shell is perhaps that of a marine snail that lived in the water of the sea or burrowed in the sandy bottom. Its feeble sensory equipment could give it only a rudimentary perception of the world. Its motions were almost painfully deliberate. The crab is a species of the shore, in and out of the water but not out for long at a time. It scuttles with nervous activity constantly and seemingly with awareness, and yet it is hardly intelligent in seeking whatever animal food, living or dead, it can tear and devour. The boy was born to mastery of the land. His body is bursting with energy; his keen senses picture the fascinating world in tremendous detail; his brain is the greatest marvel in the universe.

The resemblances between the boy, the crab, and the shell require a little more thought, but there are, as you know, profound likenesses in all living things. These three are of course made up of cells with nuclei, chromosomes, and the rest. The syntheses and other metabolic processes are much the same within their cells. All are animals and must eat other organisms—all three could, indeed, live on the same food if the boy's mother were willing. In all three the food is taken in through a mouth and digested in a stomach, and wastes are similarly ejected in the three. All have nerves, muscles, glands, gonads, and other, similar organs and tissues. All three are, in short, complex and advanced animals.

They share something else which, paradoxically, depends on their differences: all of them represent culminations of the evolutionary processes. They exemplify the three phyla that have reached greatest complexity of structure, highest coordination of functions, and most manifold diversity of successful adaptions. It is the distinctive characters, the differences that so obviously exist among them, that make each of them in its own way such a high point in the expanding evolution of animal life. It will be interesting and enlightening to learn a little more about each of the major groups to which they belong: the mollusks, arthropods, and vertebrates.

Mollusks

On a smaller scale and within the limitations imposed by one ancestral anatomical plan, the *mollusks* illustrate the same sort of fanning out into adaptive diversity that is seen among the phyla of animals. As we classify them, they comprise six main subdivisions, classes, three of which are extremely diverse within themselves. The classes are so different and so old that it is hard to generalize about the phylum and impossible to say what the very first mollusks were like. The *chitons,* class Amphineura, are on the whole the simplest mollusks, and some zoologists are inclined to think that they are the most primitive. It is, however, probable that much of their simplicity is secondary, an evolutionary consequence of their becoming virtually sedentary animals. For instance, their lack of well-differentiated heads and special sense organs is more likely due to degenerative loss than to primitiveness. Nevertheless, the simplicity of the chitons does make them clear illustrations of some other basic molluscan characteristics (Fig. 22–1).[1]

[1] Chitons are not common fossils, and they appear in the known fossil record later than the apparently more specialized classes Gastropoda, Pelecypoda, and Cephalopoda. Their later appearance need not imply later evolution: it could be due to their lack of preservation and human discovery in early rocks. The evidence does, however, suggest that all classes, including the Amphineura, arose at about the same time by rather rapid divergent evolution from an ancestry not closely like any of the known classes.

More surely primitive and about equally simple are the *monoplacophorans*. These are almost bilaterally symmetrical, internally segmented mollusks with single shells. They appear in the fossil record as early as any mollusks (early Cambrian) and were supposed to have been extinct for hundreds of millions of years, but in 1952 and since some have been found living on deep ocean bottoms. The discovery of these "living fossils" was even more extraordinary than that of the more publicized *Latimeria* (p. 607).

Mollusks are especially characterized by the development of a muscular region or organ behind the mouth. This serves for crawling locomotion in many mollusks and is called the foot. Above it is the soft mass of viscera. Practically all the organ systems found in any animals are present in most mollusks: digestive (sometimes with a unique rasping device, the *radula*), circulatory (with a heart), respiratory (usually with complex gills, *ctenidia*), excretory (with "kidneys"),[2] nervous (often with brainlike ganglia and sometimes with well-developed eyes and other sense organs), muscular, and reproductive. Above and surrounding the viscera is a *mantle* of specialized tissue, usually including glands that secrete one or more shells.

The most important groups of mollusks are the classes Gastropoda (snails and their relatives), Pelecypoda (clams and relatives), and Cephalopoda (squids, octopuses, the chambered nautilus, and relatives).[3] Each of these groups has great diversity of specific adaptations evolved from a basic ancestral adaptation that is different in each class. The three other classes, equally distinctive but less di-

[2] The term "kidney" is applied rather loosely to excretory organs in different Metazoan groups. The molluscan kidney has no evolutionary relationship (no homology) with the kidney of the vertebrates (echinoderm superphylum). The molluscan kidney is clearly a much elaborated version of the simple excretory tubules (*nephridia*) found in other members of the annelid superphylum.

[3] *Gastropods* ("belly-footed") because a snail or slug seems to crawl on its belly. *Pelecypods* ("hatchet-footed") because the foot in some clams is a little like a hatchet in outline. *Cephalopods* ("head-footed") because what is the foot in other mollusks has developed into tentacles around the mouth in the head region, and these tentacles may also be used in locomotion.

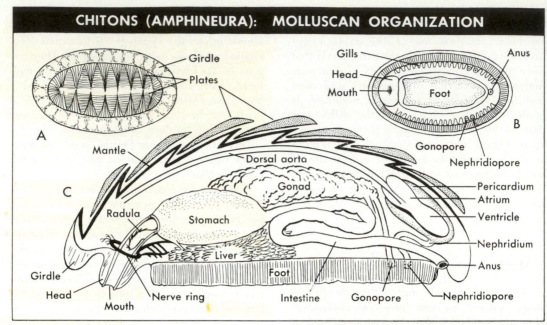

A — Girdle, Plates

B — Gills, Head, Mouth, Foot, Anus, Gonopore, Nephridiopore

C — Mantle, Dorsal aorta, Gonad, Pericardium, Atrium, Ventricle, Nephridium, Anus, Stomach, Radula, Liver, Foot, Girdle, Head, Mouth, Nerve ring, Intestine, Gonopore, Nephridiopore

22–1 Chitons (Amphineura) as an example of molluscan organization. The Amphineura, though typical of Mollusca in most respects, are atypical in the absence of special organs on the head, the multiplicity of gills (themselves atypical) in the mantle cavity, and the multiplicity of plates (shells) on the mantle. The coelom is represented by the pericardium, in which the heart lies. *Right.* A dorsal view of a chiton attached to a rock.

© Douglas P. Wilson

verse and relatively unimportant, are the Amphineura and Monoplacophora, already mentioned, and the Scaphopoda, or tusk shells (Fig. 22–2).

GASTROPODS

The deliberateness of a snail's gait is notorious. Nevertheless, snails are motile, and their slow crawling in search of food is an essential part of their way of life. Also characteristic is protection by a single shell with one opening. The early larva is bilateral, but in metamorphosis the viscera are twisted in a loop. This brings the anus from its original position behind mouth and foot to a point above the mouth. However this peculiarity may have evolved, it is highly practical for an

animal that lives in a house with only one door and that crawls about feeding on the bottom. After the twisting, the visceral organs of one side fail to develop, and further lopsided growth usually gives a spiral turn to the mantle and to the shell secreted by the mantle.

Gastropods (like all mollusks) were originally marine, and most of them still are. Some, however, have become almost fully terrestrial, although sensitive to dry air and commonly found only in moist locations. The terrestrial gastropods have lost their gills and have instead a lunglike cavity in the mantle that enables them to extract oxygen from air. Among the numerous fresh-water snails some have gills and some "lungs." Those with gills are of

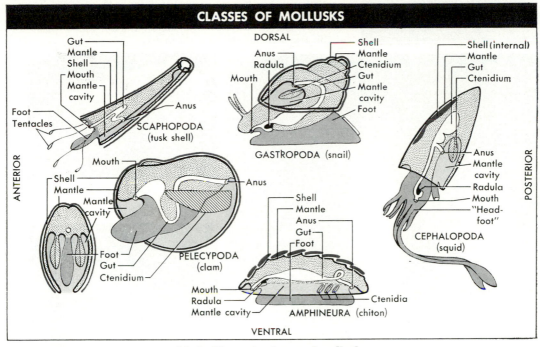

CLASSES OF MOLLUSKS

22–2 The main classes of mollusks.

marine origin, whereas those with "lungs" had terrestrial ancestors and have returned to the water. That they did not regain their long-lost gills is a good example of the irrevocability of evolution. It is paralleled by the return of air-breathing mammals, whales, to the remotely ancestral sea, where they retained lungs and did not regain the lost gills of their fish ancestors. (The gastropod "lung" is an evolutionary modification of the mantle cavity.)

Gastropods have evolved into a tremendous array of different forms, usually beautiful and often bizarre (Fig. 22–3). Wherever you live, you are familiar with some of them, especially if you have had opportunity to collect sea shells. A museum display of gastropods is one of the most impressive evidences of the diversity of life. One curiosity is that some gastropods have lost their shells. Among them are the unlovely garden slugs and the marine nudibranchs, some of which are really beautiful. These forms also illustrate the irrevocability of evolution; their viscera still show the twisting and lopsidedness that went with the development of a shell in their ancestors. It is not clear what evolutionary forces were re-

sponsible for the loss of the shell. But its loss is clearly the ultimate cause of some later evolution of protective devices substituting, so to speak, for the shell's function of protection. Some shell-less gastropods that feed on coelenterates have evolved the remarkable ability of digesting all the victim's tissues except embryonic stinging cells. These cells are borrowed, eventually appearing in the gastropods' soft skin, where they mature, conferring protection on their former predator. Such gastropods are brilliantly colored. (Might this feature have adaptive meaning?) Other shell-less gastropods, easy meat to predators, have acquired the ability to squirt out a colored solution as a sort of smoke screen when disturbed.

PELECYPODS

Pelecypods have retained bilateral symmetry and have two shells or valves, usually nearly symmetrical, one on each side of the body; hence the name "bivalve" is often applied to them. They were primitively motile, and most of them still are, but they usually lead rather sedentary lives and are even less speedy than snails. They seldom move far

22–3 Gastropods. *Left.* Garden snails, *Helix*, on a twig. The mantle cavity has become a functional lung. *Below right.* The nudibranch *Archidoris britannica*, which has no shell, no mantle cavity, and no ctenidium. The animal respires by the cluster of gills that surround the anus (at right).

Below. The whelk, *Buccinum*, a predatory marine gastropod.

Below. The shell-less garden slug, *Arion*, with freshly laid eggs.

Above. The sea hare, *Tethys*, a herbivorous tidal-zone gastropod. The foot region is developed into two large "parapodia" that enable it to swim.

Left. Limpets, *Patella*, on a soft slaty rock, which shows scars left by limpets that have died.

Top, Hugh Spencer; center left, Willis T. Hammond; center right, both ©️ Douglas P. Wilson; bottom left, ©️ Douglas P. Wilson; bottom right, American Museum of Natural History

All photos © Douglas P. Wilson except giant clam, American Museum of Natural History

Above left. The giant clam, *Tridacna*, of coral reefs, with coral growing on one of the valves. *Above right.* The queen scallop, *Chlamys*, escaping from a starfish buried in the sand. Scallops swim by opening and closing the valves, producing a jet-propulsion effect.

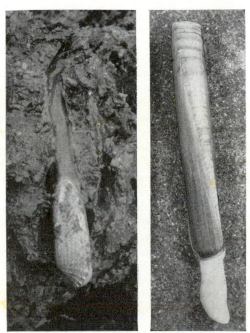

Above. Mussels, *Mytilus*, attached to rock by the byssal threads they secrete. They are being attacked by the predatory gastropod *Ocenetra*, which drills through the mussel shells with its radula. *Right. Pholas*, a pelecypod that burrows into solid rock, using the edges of its valves as a drill. *Far right. Solen*, the razor shell, with its foot extended.

Above, © Douglas P. Wilson; below,
American Museum of Natural History

22–5 Cephalopods. *Above left.* The cuttlefish, *Sepia*, a very active swimmer. Note the large eyes. *Above right.* The common octopus, *Octopus vulgaris*. Note the suction discs on its tentacles. *Right.* The chambered nautilus, *Nautilus pompilius*, with its shell cut away to show the animal in the last, largest chamber.

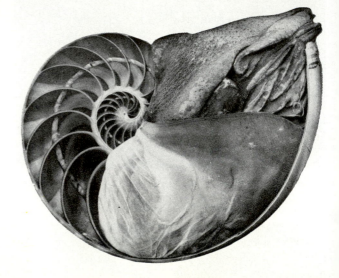

from where the larva settles down, and some, like oysters, become permanently attached. In keeping with their more sedentary lives, pelecypods have lost the differentiated head region and usually have fewer and simpler receptors than gastropods. However, many have "eye spots," simple light receptors, in the mantle along the opening between the shells.

Like so many other sedentary animals, the pelecypods feed by filtering microorganisms from their watery medium. Their mode of filtering illustrates, once again, the way the evolution of organisms exploits for new purposes structures already available, evolved initially for some other function. In the pelecypods filtering is accomplished by the typical molluscan gill, initially a respiratory structure but here much enlarged and modified as a feeding sieve.

In diversity of forms (Fig. 22–4), pelecypods are less spectacular than gastropods, but still they are extremely varied and abundant. Most are marine, although some occur in fresh water (fresh-water clams or river mussels). None are terrestrial. A few swim fairly well; scallops (*Pecten*) swim by clapping their shells together. Burrowing is a more common specialty. Many species of marine clams burrow in mud or sand, as do our common edible clams (*Venus* and *Mya*). Others (like *Teredo*) burrow in wood and cause serious damage to wooden boats and pilings. Some (like *Pholas*) burrow in solid rock. The astonishing giant clams (*Tridacna*) of the South Pacific, which may reach 6 feet in diameter, burrow in living coral reefs.

CEPHALOPODS

It is a literary tradition to play up the horror and menace of octopuses.[4] This is a bit hard on shy creatures that have seldom if ever seriously harmed a man. They, and even more particularly their relatives the squids, are also among the most complex and in almost any sense highest products of evolution (Fig. 22–5).

"Active" is the keyword for the cephalopods. They are all free-living and, on occasion, fast-moving forms. A squid has a large mantle cavity with muscular walls and a funnel-like tube (the siphon) through which water can be ejected rapidly by contraction of the walls. The animal moves swiftly by the principle of jet propulsion. If the tube is turned backward, the squid darts forward in pursuit of prey (Fig. 22–6). When it is the pursued rather than the pursuer (which seems to be the more frequent situation), it turns the tube forward and darts backward. In conjunction with its active life, the squid has a remarkably large and complex brain, as brains go among the invertebrates. It also has elaborate image-forming eyes, which work on just the same principles as our own eyes but certainly evolved entirely independently, providing a classic example of evolutionary convergence.[5] Another remarkable convergence between squids and vertebrates is that the squid, alone among invertebrates, has developed a cartilaginous internal skeleton, including a skull-like protective case around the brain. Only traces of the ancestral molluscan external shell remain.

The fact that cephalopods have tentacles with suction discs is familiar to everyone. In octopuses the tentacles serve for slower clambering about (octopuses have jet propulsion, too), and in all cephalopods the tentacles seize prey and convey it to the sharp, shearing jaws. Cephalopods are predacious; they actively pursue, kill, and devour other animals such as fishes or crabs (but not humans). Have you ever meditated on the fact that characteristics

[4] You may call them "octopodes" or "octopi" if you prefer.

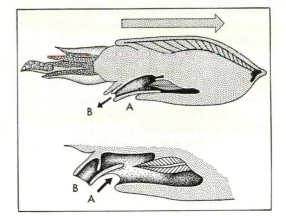

22–6   Cephalopod locomotion. *Top.* The animal moves in the direction of the stippled arrow; water is propelled as a jet (*B*, solid arrow) from the mantle cavity through the siphon. *Bottom.* The mantle cavity is recharged with water. Note the valve action. The siphon, which is muscular, can be directed backward (the reverse of the position shown) so that its jet propels the animal forward.

we admire, such as brain development, keen senses, and skillful coordination, are more likely than not to be best developed in predacious animals, whereas animals that lead quiet, respectable lives seem to have little else to recommend them?

Besides squids and octopuses (several genera and species of each), there is just one surviving genus of a markedly different group of cephalopods. This is the genus *Nautilus*. The nautilus lives in a coiled shell divided into chambers by partitions. From time to time as the animal grows, it moves to a new chamber and seals off the old one. The nautilus is a relic of the past, the remaining member of formerly very abundant groups of animals that played major roles in the history of life for tens and

[5] The convergence of their eyes with those of vertebrates has given cephalopods a curious place in the history of evolutionary thought. In the early nineteenth century the evolutionary hypothesis had not yet found a secure place in biology. In 1830 the French naturalist Geoffroy Saint-Hilaire engaged in a now-famous debate with the great French zoologist Cuvier, who was a firm opponent of the idea of evolution. Saint-Hilaire staked the case for evolution in part on the resemblance between cephalopod and vertebrate eyes. Cuvier had no trouble in demonstrating that vertebrate and mollusk eyes were quite unrelated in any sense (evolutionary or otherwise); and evolutionary thought suffered a serious setback from Saint-Hilaire's unfortunate choice of evidence.

hundreds of millions of years. The nautiloids, close relatives of the nautilus, were the first cephalopods to evolve. They swarmed in ancient seas and had many different forms (Chapter 31). Next to arise were the ammonoids, which resembled nautiloids in having chambered shells but differed in (among other things) having more complicated partitions between the chambers. The last of the ammonoids became extinct some 70 million years ago.

Arthropods

No one knows how many species of arthropods there are in the world. The number is at least a million, and estimates run as high as 10 million. Arthropods have been brought up from the deepest sea bottoms that have been dredged. They have been encountered by airplanes flying miles above the earth. They are everywhere that life exists at all. They fly, swim, hop, crawl, and just sit still. There is probably no species of organism that is not on occasion eaten by one arthropod or another. Arthropods are, in turn, eaten by many other animals. They are man's chief competitors for food and all sorts of organic materials. They include the worst of pests, but they are also essential links in maintaining the verdure-clad world as it is.

In basic structure the arthropods are somewhat like annelids, and there is no doubt that the two phyla had a common origin something over 600 million years ago. Among the more important differences are these:

The external coating or *cuticle* of arthropods is harder and serves mechanically as an external skeleton.

Arthropods have legs divided into distinct, movable segments or joints (hence the name of the phylum, which means "jointed feet").

Arthropods have muscles in definite groups mechanically related to specific movable parts. The muscles of annelids form relatively simple sheets throughout the body.

Arthropods generally have fewer segments, and there is a tendency for the segments of some regions, notably in the head,

to fuse and to become strongly differentiated in structure.

Arthropods have distinctly developed jaws. These open from side to side instead of up and down as our and other vertebrate jaws do.

Their nervous system is usually more highly developed than in annelids and is accompanied by elaborate sensory receptors, including those in the antennae and eyes.

Most of these arthropod characteristics improve or elaborate their reactions to stimuli in the environment. Their advantage over other invertebrates is largely in the efficiency and adaptability of their behavior. The vertebrates, and notably man, also owe their dominance in great part to their adaptable and efficient behavior. In arthropods the behavior is relatively inflexible in a given species, but is modified genetically in the course of evolution. Vertebrate behavior has, as a rule, a larger element of flexibility in the individual (see Chapter 14). We are likely to consider our own kind of behavioral adaptation as "better" or "higher." Arthropods are, however, from 10 to 100 times more numerous than vertebrates in species, incomparably more abundant in individuals, and divergently adapted to an even wider range of environments and habits. They easily hold their own against all the attacks of man and of other animals. Which is the more successful phylum?

Our classification (p. 837) recognizes seven classes of arthropods (Fig. 22–7). One of these (Onychophora) includes only a few rather obscure living animals, about 70 species, mostly in the Tropics.[6] The class (with the genus *Peripatus*) is of great interest because it is not fully arthropodlike and has some distinctly annelid characteristics; hence it tends to link the two phyla. Otherwise, it is unimportant. The trilobites (class Trilobita) are abundant, important fossils and interesting forerunners of the crustaceans, but they

[6] It is frequently stated that this class is known from fossils in the pre-Cambrian, which would make it among the oldest of all animals. It is, however, uncertain whether the fossils in question belong to the Onychophora or are of the age attributed to them.

have long been extinct and need not further detain us now. Centipedes (class Chilopoda), the "hundred-legs" (they may in fact have over 300 legs but 30 to 70 are more usual) and millipedes (class Diplopoda), "thousand-legs" (an exaggeration) are rather common, but still of minor importance in the phylum as a whole (Fig. 22–8). The outstanding classes are those of the crustaceans, the spiders (and relatives), and, above all, the insects.

CRUSTACEANS

All of us are familiar with some crustaceans [7] (Fig. 22–9), if only because we eat them with pleasure: lobsters, crabs, shrimps, crayfish, prawns. We have only those five common names for them, but there are literally thousands of species of these larger, free-living crustaceans, many of them edible by humans. They are the decapods or "ten-legs" among the crustaceans in general. Most of them are marine, but they are also numerous in fresh water, and some crabs can survive considerable periods in the air as long as they do not dry out. All are carnivores or scavengers, eating a wide variety of small living animals and any sort of dead animal matter.

Decapods are most familiar to us, but they are not the most numerous of crustaceans or the most important in the economy of nature. Those distinctions must be assigned to the small, even microscopic crustaceans that swarm by countless billions in all seas and most bodies of fresh water. Krill, the principal food of some of the largest whales, is composed of crustaceans (genus *Euphausia*) under an inch in length. Others, still smaller, abound in salty (example, *Artemia*) or fresh (example, *Daphnia*) ponds. More diversely specialized crustaceans well exemplify a broad adaptive radiation.

Ostracods are very small crustaceans that secrete, in addition to the usual arthropod cuticle, two protective shells, much like miniature clam shells.

Some of the *copepods* are free-living, important fish food ("brit"), but many are parasitic on or in worms, other crustaceans, echinoderms, and chordates.

[7] Class Crustacea, "with a crust or shell"; all arthropods have tough "crusts," but this feature is perhaps most evident in crustaceans.

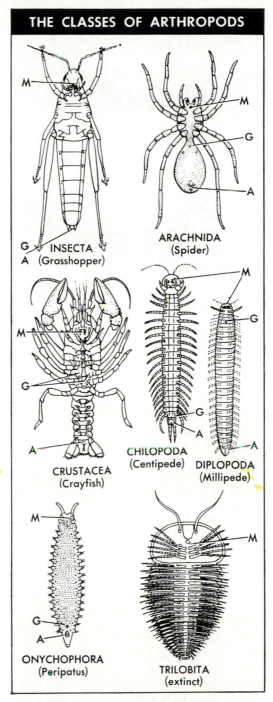

THE CLASSES OF ARTHROPODS

INSECTA (Grasshopper)

ARACHNIDA (Spider)

CRUSTACEA (Crayfish)

CHILOPODA (Centipede)

DIPLOPODA (Millipede)

ONYCHOPHORA (Peripatus)

TRILOBITA (extinct)

22–7 The classes of arthropods. *M*, mouth; *G*, genital opening; *A*, anus. Note significant differences among classes in the location of the genital opening. (The locations of the genital opening and anus in the trilobites are unknown.)

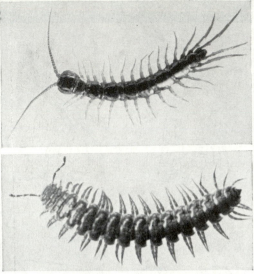

22–8 Chilopods and diplopods. *Top.* A centipede (chilopod). Note one pair of legs per segment. The centipedes are carnivores feeding mostly on other arthropods. *Bottom.* A millipede (diplopod). Note two pairs of legs per segment. The millipedes are herbivores.

Barnacles are sessile as adults, within complex limy cups secreted by the animals. A shell-less relative, *Sacculina*, parasitic on crabs, is a classic example of extreme degeneration.

Isopods and *amphipods* are small, buglike crustaceans, marine, fresh-water, or living in damp places on land: "sow bugs," "pill bugs," "wood lice," "beach fleas," and so on. They are entirely distinct from true bugs, lice, or fleas, all of which are insects.

One of the principles of evolutionary change particularly well illustrated by crustaceans is the regional specialization and differentiation of parts. Primitive crustaceans probably had many segments, most or all of them with appendages that were very similar throughout. Such a primitive condition is unknown among any true crustaceans, but it is approached in the older, extinct trilobites, closely allied to crustaceans and possibly ancestral to them. (Some authorities include trilobites in the class Crustacea.) In such a crustacean as the lobster (Fig. 22–10), no two of the 19 pairs of appendages are exactly alike, and those of different regions are highly

differentiated in form and function, as indicated in Table 22–1.

TABLE 22–1
The appendages of the lobster

Segments	Appendages and their functions
1	None. *
2–3	Antennae, different in form on the two segments. Sense organs.
4–9	Jaw parts, different on each segment. Chewing.
10	Large pincers. Offense and defense.
11–14	Leg, all different, the first two pairs with small pincers. Walking, grasping.
15–19	So-called swimmerets. Swimming forward slowly, circulating water to the gills, mating, carrying the eggs and young.
20	Broad plates. Swimming backward. (They are snapped forward under the body, an action that makes the whole animal dart backward.)
21	None. (The segment as a whole is a flat plate that assists the appendages of segment 20 in the backward-darting motion.)

* In the lobster the possible first segment is so reduced or fused with those following it that it is not apparent and its existence is sometimes doubted. However, as a first segment occurs in primitive crustaceans, there may be traces of it in the lobster.

It is an evolutionary generalization (open to exceptions, as such generalizations always are) that, when ancestral organisms have numerous parts similar in structure and function (like the appendages of trilobites), the parts in their descendants tend to be reduced in number and specialized in different places for different functions (as in the lobster appendages).

ARACHNIDS

The arachnids [8] (Fig. 22–11) are another group, like the octopuses, that have been given a largely undeserved bad name. It is true that most spiders and ticks are poisonous, but only a few are dangerous to man, and most of us are rarely under the slightest menace from them. Most spiders are beneficial to man because they prey on insects considered undesirable with better cause. Mites and ticks, also

[8] Class Arachnida, from the Greek for "spider."

22–9 Crustaceans. *Left.* A copepod, *Acartia clausi* (×88). *Right.* A decapod, the crab *Cancer magister*, from Alaska.

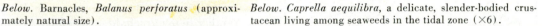

Far left. Another decapod, the common prawn *Leander serratus* (×⅔). *Left.* The sow bug or wood louse, *Armadillaria vulgare*, one of the few crustaceans that have become terrestrial.

All photos © Douglas P. Wilson except crab, U.S. Fish and Wildlife Service, and sow bug, U.S. Department of Agriculture

Below. Barnacles, *Balanus perforatus* (approximately natural size).

Below. Caprella aequilibra, a delicate, slender-bodied crustacean living among seaweeds in the tidal zone (×6).

Below. The goose barnacle, *Lepas fascicularis* (×½).

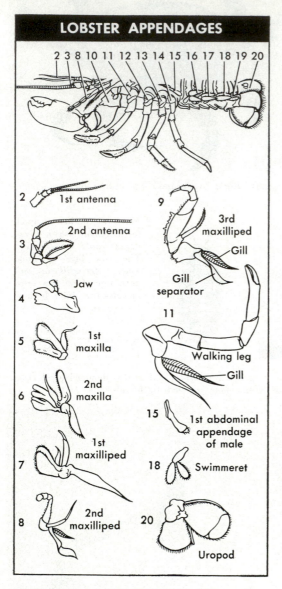

LOBSTER APPENDAGES

2 3 8 10 11 12 13 14 15 16 17 18 19 20

2 — 1st antenna

3 — 2nd antenna

4 — Jaw

5 — 1st maxilla

6 — 2nd maxilla

7 — 1st maxilliped

8 — 2nd maxilliped

9 — 3rd maxilliped / Gill / Gill separator

11 — Walking leg / Gill

15 — 1st abdominal appendage of male

18 — Swimmeret

20 — Uropod

22—10 Appendages of the lobster. The numbers are those of the segments bearing the appendages shown.

arachnids, do deserve a bad name from our point of view, for they include unpleasant and dangerous external parasites of man and domestic animals and are intermediate hosts for organisms producing some serious diseases. Daddy longlegs or harvestmen are another group of arachnids, entirely harmless and indeed helpful to man.

Most people think of arachnids as insects or bugs (which are, strictly speaking, a group of insects). Arachnids are really quite different and are easily distinguished by the fact that they have four or five [9] pairs of legs, while insects have three pairs. This is what systematists call a diagnostic or key character. It is a handy way to spot whether a particular animal is an arachnid or an insect. But an animal is not classified as an arachnid *because* it has four or five pairs of legs rather than three. It is classified in the Arachnida because it has the same ancestry as other arachnids and a different ancestry from insects over some hundreds of millions of years, as attested by all the varying characteristics of the two groups and by large numbers of fossil representatives of both. It happens that legs were early reduced to three pairs in insects and not in arachnids, so that leg number is a convenient way to tell them apart. Among some other animals, such as the centipedes, closely related forms may have quite different numbers of legs or appendages. In contrast, extremely distantly related animals may have the same number; some protists have four, as we do.

Another living arachnid looks offhand completely unlike a spider, scorpion, or tick. It is the horseshoe or king crab, genus *Limulus*. The broad protective dorsal shield (carapace) appears rather crablike, but examination of the legs (five pairs) and internal anatomy reveals indubitable phylogenetic resemblances to spiders and differences from crabs. This animal is an evolutionary relict, a famous conservative. It has changed very little in the last 200 million years or so. Still older relatives, the eurypterids (Chapter 30), are an extinct group of more scorpionlike arachnids that long played an important role in ancient seas. They included the largest of all arthropods, over 6 feet in length.

INSECTS

No one needs to be told that insects are the most successful and diverse of invertebrates. In fact, no other group of organisms of any kind, protists, plants, or animals, can begin to compare with them in diversity. At least half

[9] There are really six pairs of appendages in most arachnids, but the first pair, at least, and usually the first two pairs, are so modified that they do not look or function like legs.

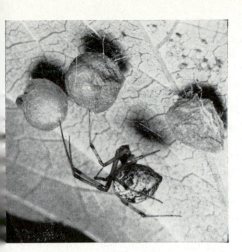

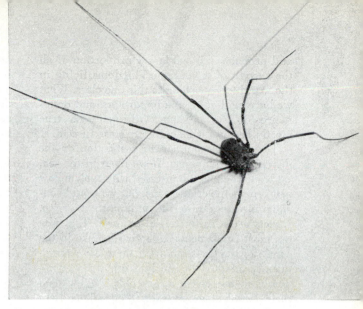

Above. The house spider, *Achaearanea*, with egg sacs on the back of a leaf. *Below.* The dog tick, *Dermacentor*, which transmits the organism causing Rocky Mountain spotted fever in man. *Bottom.* A scorpion.

Above. The harvestman, or daddy longlegs, which feeds on small insects.

22–11 Arachnids.

Below. Ventral (*left*) and dorsal (*right*) views of the king crab, *Limulus polyphemus.*

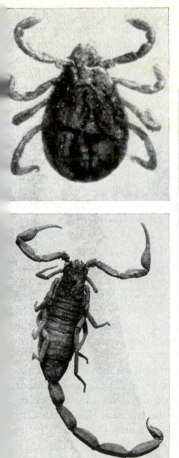

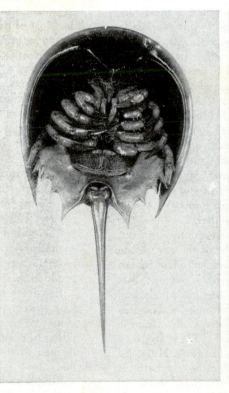

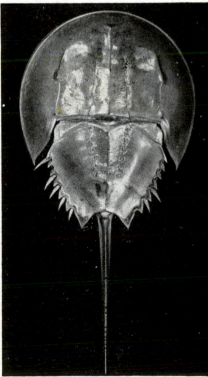

and probably a much larger proportion of all the species of all organic kingdoms living in the world today belong to this one class. What we have said about the importance and dominance of the phylum Arthropoda is true mostly because it includes the class Insecta.

There is just one strong restriction on the distribution of insects. If we lived in the sea, we would be much impressed with their fellow arthropods the crustaceans, but we would consider insects a rare and unimportant group. Insects arose as terrestrial animals. That was the most fundamental adaptive feature in the origin of the class. Many of them have subsequently adapted to life in fresh water, commonly in the larval stages only. A few pass their larval stage or even their whole lives along the shore between tides and in salty pools. There is perhaps only one (genus *Halobates*) that can be considered fully a creature of the open oceans, and it is a water strider that really lives in the air, not the water, which is the floor of its world.

One other restriction on the adaptive range of insects is a matter of size. Multicellular organisms cannot function below a minimum size which is extremely small to our eyes, to be sure, but which looks large in comparison with a bacterium or many other protists. The smallest insects (which are parasites in the eggs of other insects) are about 0.2 millimeter (that is, only about 1/125th of an inch) in length. Most insects are at least 2 millimeters long. On the other hand, there is a mechanical and physiological upper limit for the size of insects. The weight and the stress of muscles and motions are borne by an external skeleton. There is a limit to the total weight a given kind of skeleton can support. At the limit the skeleton itself either becomes too heavy to move or too light to stand up without collapsing under the weight of the rest of the body. This limit is very much higher for an internal than for an external skeleton; that is one of the advantages that mammals and other vertebrates have over insects. The respiratory and vascular systems also impose limits on the size of insects. They are very efficient for small animals but probably could not adapt to any considerable increase in bulk. Few insects are more than about 40 millimeters long, and the upper limit is

around 275 millimeters (less than a foot) for body length. Some moths have a slightly greater wingspread, up about 1 foot. A few ancient insects known as fossils had a wingspread of over 2 feet, but the body bulk was not as large in proportion.

A favorite theme of science fiction involves insects as large as men or larger. Aside from the fact that this is mechanically and biologically impossible, do you think it would make our war with the insects easier or harder? Lion hunting is more exciting and dangerous than mouse hunting, but are mice easier to exterminate? In fact, is not the small size of insects, and the accompanying small food requirement of any one of them, a reason why they are so extremely numerous and diverse and are often such pests from our viewpoint?

Most insects are small and nonaquatic. Aside from these limitations they have a bewildering variety of forms and follow almost every imaginable way of life, including many that would be inconceivable if the insects did not demonstrate them to us. They all have some features in common, which must have been involved in the evolutionary origin of the class. Most of these may be seen in a common grasshopper (Fig. 22–12). Among the most striking characteristics are these:

The head consists of six segments, thoroughly fused and practically inseparable in the adult. Appendages of one segment have become sensory antennae. Other appendages have become complex mouth parts.

The thorax, the central section of the body, is distinctly separate from the head and the abdomen and consists of three segments, each with a pair of legs.

The abdomen has 11 or fewer segments, (usually) without appendages, and the posterior segments are specialized for reproduction.

Respiration is by tracheae; the circulatory system is open, without capillaries or veins; oxygen transport by blood is unimportant because the tracheal branches carry oxygen directly to nearly every cell in the body.

Simple and compound eyes occur, as well as many other receptors in antennae and elsewhere. The nervous system is complex, with two large ganglia or "brains" in the head and a double ventral cord.

Many insects lack wings altogether, and it is probable that wings were never present in the ancestors of some of them, such as the silverfish and springtails. Thus wings were not involved in the origin of the class but certainly evolved early in its history. Most insects have wings, and most of the wingless forms (lice, fleas, wingless ants, etc.) had flying ancestors. Insects commonly have two approximately equal pairs of wings, and this condition was probably primitive. In many insects, notably the beetles, the front pair has become a protective cover. In flies (order Diptera, or "two-wings," on this account) the hind pair is greatly reduced in size and is a balancing, not flying, organ.[10] Another characteristic that is widespread but not universal in insects is the occurrence of feeding larvae, with gradual or sudden metamorphosis into adults. We have already referred to this complication in the life history (Chapter 9).

Specialists on insects (entomologists) divide the class into about 25 living orders, the number being variable because it is a matter of taste and opinion whether some related groups should be considered separate orders or suborders of one order. A dozen or more wholly extinct orders are also known. Some faint idea of the stunning diversity of insects is suggested by Fig. 22–13. Among the more numerous and familiar groups of insects are these: roaches and grasshoppers; termites (or "white ants," but they belong to a different order from true ants); dragonflies; May flies; lice; bugs[11] and aphids; caddis flies;

[10] In another group, so little known as to have no popular name (the technical name is Strepsiptera), it is the front pair of wings that is reduced. In this group wings occur only in males. The females are parasites in other insects, from which they protrude their hind ends for fertilization by the flying males. The young, both male and female, go through two distinct parasitic larval stages. Many insects have life histories even more curious and complicated than this. One could spend a lifetime (many have!) enumerating the strange and intricate lives of insects.

[11] Although in daily speech we are likely to call any insects, and even some noninsects, "bugs," bugs strictly speaking are the members of a single order of insects, the Hemiptera.

moths and butterflies; true flies; fleas; beetles; and ants, wasps, and bees.

The diversity of insects involves not only the enormous number of species and other groups but also the fact that two or more sharply distinct forms may occur within a single species or, indeed, deme. This phenomenon, which also occurs in many other groups of organisms, is called *polymorphism* ("many forms"). The different forms may characterize different stages in the life history, for instance, the wormlike larvae and flying adults of butterflies and many other insects. Polymorphism may also correspond with functional differentiation in a social organism, as in the castes of ants and other social insects. The phenomenon is most striking, however, and is most strictly defined as polymorphism when it depends neither on age nor on function. For example, among our common sulfur butterflies (genus *Colias*) most of the females are yellow, but a considerable number of them (up to 20 per cent or more in some places) are white. Why do these white sulfur butterflies not constitute a subspecies or other systematic group? What is the relationship between polymorphism and the genetic variation that occurs in all populations? Can you think of any other kinds and find other examples of polymorphism, in coelenterates, for instance? In fishes? In mammals? In mankind?

Vertebrates

You learned in the preceding chapter that the phylum Chordata is usually subdivided into four subphyla, three of which include a relatively small number of peculiar marine animals. The other subphylum, Vertebrata ("with vertebrae"), is another of the great culminations of the evolutionary processes. It is in many respects the most important and most progressive of all groups of organisms. Vertebrates are much less diverse and less abundant than insects. Their role in the total metabolic turnover of the living world is far less than the parts played by several groups of plants and also less than those of some other animals. The vertebrates are, nevertheless, highly diverse and abundant, and they do

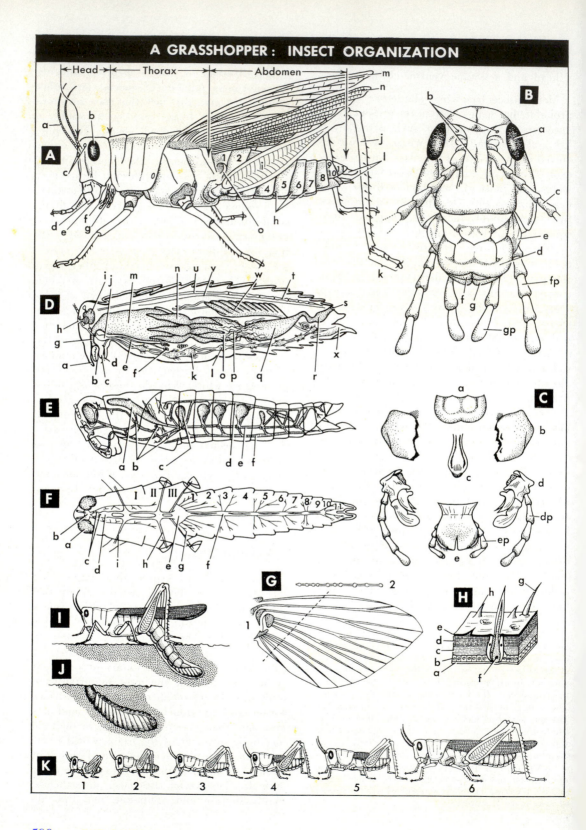

A GRASSHOPPER : INSECT ORGANIZATION

have important ecological roles. Moreover, as a whole they are characterized by the highest development of reception of environmental stimuli and the greatest flexibility and widest repertory of reactions. The vertebrates include man, who is in those respects and some others incomparably the most progressive of all organisms and who is, even from the point of view of other organisms, much the most potent force on earth today. While rightly regarding ourselves as the supreme animals, however, we should remember that the vertebrates include other groups that are also dominant and are evolutionary culminations in their own ways and in different adaptive spheres, notably the bony fishes, the perching birds, and the rodents.

Some idea of what the first vertebrates were like can be obtained by making comparisons among living vertebrates.[12] Similarities between the most diverse species, for instance,

[12] We are talking now about the first vertebrates, not the first chordates, which preceded the first vertebrates and some of which were ancestral to vertebrates. As you learned in the preceding chapter, there is some evidence of what the larvae were like in the earliest chordates, but inferences as to the adults are equivocal and disputed. For the vertebrates there is more and better evidence.

between a lamprey and a man, are almost sure to have been inherited from the earliest vertebrates. In such a comparison we need not worry much about the misleading effects of convergence and parallelism (p. 499), because the evolution of the two groups has in fact been so divergent that they are unlikely to have any similarities that are not homologies.

Still, the comparison leaves us with a very incomplete picture. For the many characteristics that are decidedly different in man and lamprey, which was the condition in the early vertebrates, or was it distinct from either? For instance, men have jaws, and lampreys do not. The human skeleton is bony, and that of a lamprey is cartilaginous. Our skin is dry and hairy; lamprey skin is smooth and slimy. What were the early vertebrates like in these respects? The evidence of early fossil vertebrates provides answers to many questions of this sort. The fossils indicate, for example, that early vertebrates were more like the lamprey in lacking jaws, more like man in having bony skeletons, and not like either in type of skin, which was covered with scales or plates. It is true that lampreys are on the whole a good deal more like the earliest vertebrates than men are. This fact is helpful in attempts

22–12 A grasshopper as an example of insect organization. *A.* The external anatomy of a female grasshopper: *a*, antenna; *b*, compound eye; *c*, ocelli (supplementary light-sensitive organs); *d*, labrum (upper lip); *e*, mandible; *f*, maxilla; *g*, labium (lower lip); *h*, spiracles (breathing apertures for the entry of air into the tracheal system); *i*, femur of the third leg; *j*, tibia of the leg; *k*, tarsal segments of the leg; *l*, ovipositor; *m*, forewing; *n*, hind wing; *o*, tympanum (organ of hearing). Note the division of the body into three major parts—head, thorax, and abdomen. The eleven abdominal segments are numbered. *B.* Details of the head: *a*, compound eye; *b*, ocelli; *c*, antenna; *d*, labrum; *e*, mandible; *f*, maxilla; *fp*, maxillary palp; *g*, labium; *gp*, labial palp. *C.* Mouth parts: *a*, labrum; *b*, mandible; *c*, hypopharynx; *d*, maxilla; *dp*, maxillary palp; *e*, labium; *ep*, labial palp. *D.* The internal anatomy (with the tracheal system omitted): *a*, labrum; *b*, mandible; *c*, hypopharynx; *d*, labium; *e*, salivary duct; *f*, salivary gland; *g*, esophagus; *h*, ocellus; *i, j*, supra- and subesophageal ganglia, forming the brain; *k*, third thoracic ganglion of the nerve cord; *l*, first abdominal ganglion of the nerve cord; *m*, crop; *n*, gastric pouch (or cecum); *o*, stomach; *p*, Malpighian (excretory) tubules; *q*, intestine; *r*, rectum; *s*, anus; *t*, heart; *u*, aorta (main blood vessel

leading to the head region); *v*, diaphragm (membrane separating the section of the body cavity containing the heart from the rest of the body cavity); *w*, ovary; *x*, vagina. *E.* The respiratory (tracheal system): *a*, thoracic air sac (reservoir of air); *b*, spiracles (air-entry apertures); *c*, first of several abdominal air sacs; *d, e, f*, dorsal, lateral, and ventral tracheal trunks. *F.* The nervous system in dorsal view: *a*, eye; *b*, supraesophageal ganglion; *c*, connective nerves (which pass around the esophagus) to *d*, subesophageal ganglion; *e*, third thoracic ganglion on the pair of ventral nerve cords; *f*, first abdominal ganglion; *g, h, i*, nerves to the third, second, and first legs, respectively. The three thoracic segments are indicated by Roman numerals, and the 11 abdominal segments by Arabic numerals. *G.* The wing of a generalized insect, showing the distribution of veins; *2* is a cross-sectional view corresponding with the dashed line on *1*. The veins carry tracheae whose walls are thickened, forming a supporting structure for the wing membrane. *H.* The body covering of a generalized insect: *a*, basement membrane; *b*, epidermis; *c, d, e*, collectively, exoskeleton; *f*, cell at the base of *g*, movable bristle; *h*, a fixed spine on the exoskeleton. *I.* A female laying eggs (ovipositing) in the ground. *J.* The egg mass. *K. 1–5*, successively older young grasshoppers or nymphs; *6*, an adult.

22–13 A variety of insects.
Above. A green and orange stinkbug (order Hemiptera, bugs) from Barro Colorado Island in the Panama Canal Zone. *Left.* An adult male Luna moth (order Lepidoptera, butterflies and moths). *Center left.* An adult dragonfly (order Odonata, dragonflies and damsel flies). *Below.* A model of the common housefly (order Diptera, flies and mosquitoes).

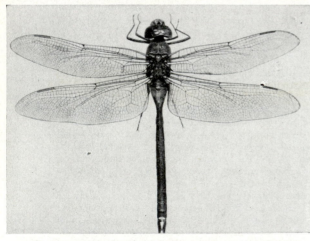

Bottom left. An adult female mosquito (order Diptera) sucking blood. *Below.* Mosquito larvae respiring at the surface of a pond.

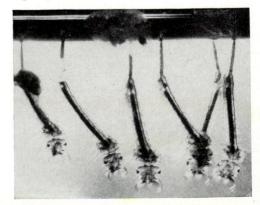

Above. A female lubber grasshopper (order Orthoptera, roaches and crickets) laying eggs.

Top left. A cicada wasp (order Hymenoptera, bees, wasps, and ants), with a cicada it has stung. *Above.* A boll weevil (order Coleoptera, beetles).

Above left. A human body louse (order Anoplura, lice), carrier of typhus fever. *Above right.* A bumblebee (order Hymenoptera), with pollen packed into special pollen baskets on its hind legs. *Below.* Adult worker ants (order Hymenoptera) and pupae.

to reconstruct the vertebrate ancestry, but plainly it does not follow that lampreys (or any other recent animals) are in all respects primitive.

Here are some of the known facts and most probable inferences about the earliest vertebrates.

They were aquatic. (Some biologists think that they originated in fresh water. This is quite uncertain, but they soon spread to both fresh and salt water.) They were motile, highly active, bilaterally symmetrical, swimming animals with fishlike bodies.

They had an internal skeleton, which included a flexible rod, the *notochord*, down the back. Around the notochord a segmented, jointed series of bones, the *vertebrae*, developed. A notochord is present in all chordates, which are named for that feature. In most later and higher vertebrates the notochord is present in the embryo only, and in later stages of life is entirely replaced by the vertebrae. There were also internal fin supports and bony plates in the skin as well as a rigid skeletal support, the *skull*, around the brain and the sense organs of the head.

The early vertebrates had powerful muscles, especially a segmented series of V-shaped muscles along the sides of the body, used in locomotion.

The mouth was a simple opening without jaws. The early vertebrates probably fed on microorganisms and on organic matter in mud and sand.

The pharynx (the part of the alimentary and respiratory canals immediately behind the mouth cavity) had a series of paired lateral openings, *gill slits*, through which water, taken in through the mouth, passed outward over the gills, which were the respiratory organs.

The rest of the alimentary canal was a relatively simple tube, perhaps somewhat twisted or looped but with little regional differentiation. There was a well-developed liver, and kidneys were also present.

Reproduction was bisexual, the sexes being different individuals. The females laid eggs, and fertilization was external.

There was a complex, closed circulatory system with capillaries. The red blood (with hemoglobin in corpuscles) was pumped by a heart with a single series of chambers.

The nervous system was already more highly developed than in any other animals. Its most striking feature was the hollow *spinal cord* above (dorsal to) the notochord and the anterior expansion of this nerve tube into a *brain*, already relatively large and well differentiated into several parts. Sense organs were also well developed, including lateral-line organs (Chapter 13), image-forming eyes essentially like those of all later vertebrates, and ears, which, however, were organs of equilibrium rather than of hearing. (In most higher vertebrates they are both.) There were only two semicircular canals (Chapter 13) in the ear, rather than the three in almost all later vertebrates.

It was from such beginnings that the whole array of vertebrates developed—lamprey, shark, trout, frog, snake, sparrow, man, and the rest. The vertebrates may be arranged in eight classes (Fig. 22–14), the first four of which are aquatic and are popularly known as fishes: Agnatha, Placodermi (the only extinct class), Chondrichthyes, and Osteichthyes. The four mainly nonaquatic classes are Amphibia, Reptilia, Aves, and Mammalia.

AGNATHS

The first vertebrates were *agnaths* (jawless fishes), which had all the ancestral characteristics just discussed but which occurred in a multitude of specific forms. They were abundant and highly diversified in the early days of vertebrate history and then dominated the realm of fishes for a relatively short time (although, even so, for some tens of millions of years; see Chapter 30). The living lampreys (Fig. 22–15) and hagfishes are relicts of this extremely ancient group. They have all the basic characteristics except bone tissue. In addition to losing bone, they have become highly specialized in some respects, notably in their feeding habits and in their elongate, eel-like bodies; their body form did not occur (as far as known) in any of the earliest verte-

THE CLASSES OF VERTEBRATES

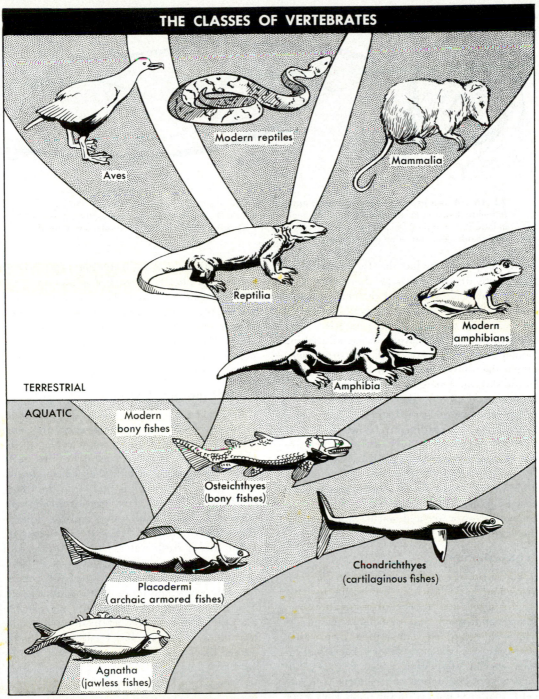

Aves

Modern reptiles

Mammalia

Reptilia

Modern amphibians

Amphibia

TERRESTRIAL

AQUATIC

Modern bony fishes

Osteichthyes (bony fishes)

Placodermi (archaic armored fishes)

Chondrichthyes (cartilaginous fishes)

Agnatha (jawless fishes)

22–14 The classes of vertebrates.

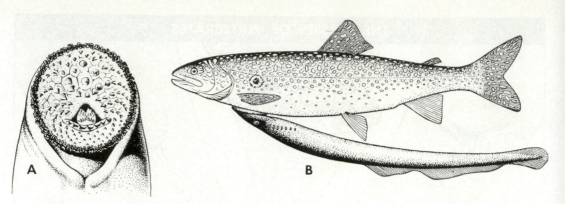

22–15 A lamprey, a contemporary agnath. *A.* The head of a lamprey, showing the sucking funnel that surrounds the mouth. Note the hooklike teeth on the sucker, and the rasping tongue in the mouth. *B.* A lamprey attached to a fish from which it is sucking blood. Note the circular mark on the fish's flank—the scar of an earlier attack.

brates in spite of the fact that they were otherwise quite diverse in shape. Lampreys have larvae that feed, as most of the early agnaths probably did, by sucking up mud containing microorganisms and organic debris. After metamorphosis they develop a round, sucking funnel lined with teeth. With this organ they attach themselves to living fishes, rasp a hole in the skin, and suck blood. (A few lampreys do not eat at all after they metamorphose into adults; they merely breed and die. This seems to be a recent evolutionary change; indeed, it may still be going on.) Hagfishes, which are not known to have larvae, also have a sucking funnel with which they eat their way right into and through the bodies of other fishes, usually attacking fish already disabled or dead. There are both fresh-water and marine lampreys. All the hagfishes are marine.

PLACODERMS

The survival of a few agnaths like the lampreys and hagfishes is probably due to their having developed a peculiar adaptation, the funnel and rasp, and associated habits in which they had no competition from other fishes. Most of the early jawless fishes had no such special adaptation and were probably mudsuckers, extracting nourishment from organic matter contained in the mud. They were soon replaced by fishes with jaws, which clearly are more effective than mere jawless openings for securing most kinds of food. Jaws evolved from a pair of hinged gill supports (Chapter 16), a remarkable example of evolutionary transformation. The earliest fishes with jaws are classified as *placoderms*.[13] The placoderms early became extinct and were replaced by more modern sorts of fishes, with more complex and still more efficient jaws. We shall have occasion to mention placoderms again in connection with the history of life in the sea (Chapter 30).

CHONDRICHTHYANS

Two groups of higher fishes evolved independently from placoderms at about the same time and, between them, soon replaced the placoderms. A possible reason for this early and basic subdivision of higher fishes into two main groups (classes) is that one—the Chondrichthyes[14]— may have been *originally* specialized for life in sea water, and the other—Osteichthyes (the next class to be discussed)—for life in fresh water.[15] In the millions of years since the groups arose, a few chondrichthyans have wandered into fresh waters, but only a few and more or less haphazardly. The group is still fundamentally marine. On the other hand, the osteichthyans, while continuing to dominate the fresh waters, early spread likewise in the sea and are now also by

[13] "Plate skins," because many of them had bony armor plates (but so did most early agnaths).

[14] "Cartilaginous fishes," because the skeletons of the living forms are entirely cartilaginous.

[15] Some books on zoology still say that the chondrichthyans were an older, more primitive group and that the osteichthyans evolved from them. This is an old theory contrary to present fossil evidence.

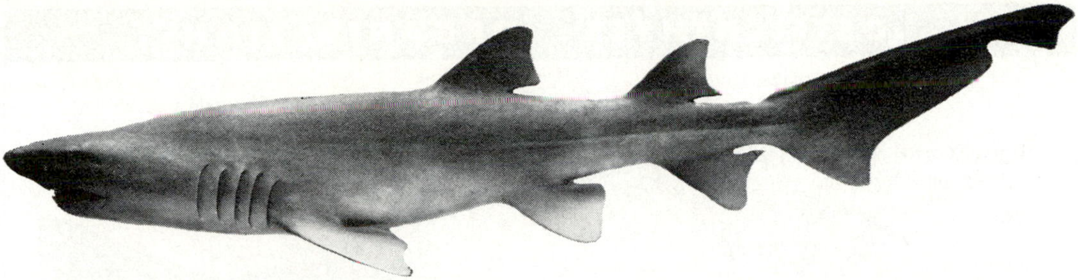

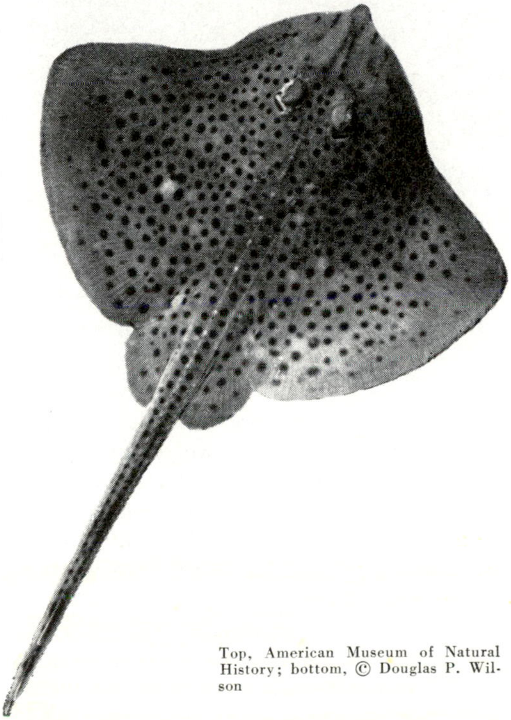

22–16 Cartilaginous fishes (Chondrichthyes). *Above.* The West Indian shark, (*Hyporion brevirostris*). Note the five conspicuous gill slits. *Left.* The spotted ray, *Raia maculata.* The functional gill slits are on the ventral surface, not visible in the photograph. Note the spiracles, which are former gill slits, immediately behind the eyes.

Top, American Museum of Natural History; bottom, © Douglas P. Wilson

far the dominant marine fishes, even though their competition has not wiped out all the chondrichthyans.

The principal chondrichthyans are the sharks and the rays (with their relatives, the skates) (Fig. 22–16). Their original physiological adaptation is seen in the way they meet the problem of osmosis in salty water—a way markedly different from and more efficient than that of the osteichthyans. In the osteichthyans the internal osmotic pressure, Δ, is less than that of sea water, and much metabolic energy is required to keep enough water in the body (Chapter 3). Chondrichthyans retain large amounts of dissolved urea in the body fluids—a peculiar specialization, for such concentrations of urea would be fatal to most animals. Together with the inorganic salts usual in such fluids, the dissolved urea raises the internal osmotic pressure to approximately that of sea water, so that marine chondrichthyans, unlike other marine fishes, are in osmotic equilibrium with their environment. In discussing the adaptations of various groups of organisms, we are likely to emphasize anatomical characteristics, because these are easy to observe. Osmotic regulation in the chondrichthyans is a good example of the fact that physiological adaptations are just as numerous and may be even more important than structural ones.

Of course, the chondrichthyans do also have characteristic anatomical features. The most obvious and universal is the one they are named for—the completely cartilaginous skeleton of recent forms. This (contrary to some earlier opinions) is a specialization. Another specialization is that the eggs have heavy, leathery shells and that fertilization is internal. The males have modified posterior paired (pelvic) fins with structures called *claspers* that aid in injecting sperms into the females. A feature that may be primitive in comparison with osteichthyans is the absence of lungs or swim bladders.

The oldest chondrichthyans are sharks, and of course sharks are still common today, changed in many details but still of the same adaptive type: elongated, streamlined, swift-swimming predators. These and other charac-

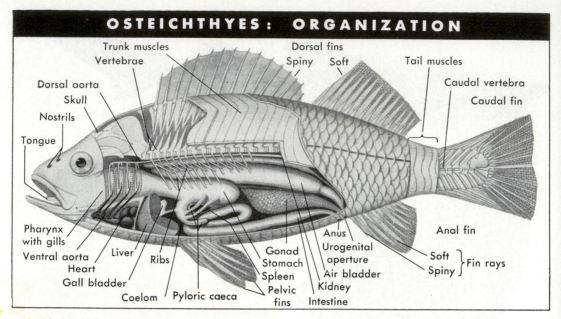

Trunk muscles
Vertebrae
Dorsal aorta
Skull
Nostrils
Tongue
Dorsal fins
Spiny Soft
Tail muscles
Caudal vertebra
Caudal fin
Pharynx with gills
Ventral aorta
Heart
Gall bladder
Liver Ribs
Coelom Pyloric caeca
Gonad
Stomach
Spleen
Pelvic fins
Anus
Urogenital aperture
Air bladder
Kidney
Intestine
Anal fin
Soft
Spiny } Fin rays

22–17 The organization of a bony fish (Osteichthyes). (The pyloric caeca are digestive pouches of the gut.)

teristics of the group are well seen in the so-called dogfishes, usual laboratory animals, which are small sharks. Later in origin and now also abundant are the skates and rays— broad, flattened forms, most of which live on the sea bottom, where they devour various invertebrates.[16]

OSTEICHTHYANS

Most osteichthyans [17] (Fig. 22–17) have retained and some have even intensified the development of bone in the skeleton. In only a few of them, represented today by the sturgeons and spoonbills, has the bone retrogressed markedly so that the skeleton is largely of cartilage. The gill slits, each of which opens separately to the exterior in the chondrichthyans, open into a chamber covered by a (usually) bony plate or flap, the *operculum*. Primitive osteichthyans had lungs as well as gills. Like the internal osmotic pressure, which is nearer that of fresh water than of salt water, the lungs were probably an adaptation to fresh-water life, providing a supplementary means of respiration when oxygen dissolved in the water became deficient. A few living osteichthyans have retained the lungs as such. These *lungfishes* (Fig. 22–18) survive as relics in Australia, South America, and Africa. In most osteichthyans, however, the former lung has lost its respiratory function and has become a swim bladder, a gas-filled sac that modifies the buoyancy of the fish as a whole and helps to maintain its position in the water. Some fishes that live on the bottom have lost the swim bladder (in the adults, at least). Can you think of a possible adaptive relationship between their habits and loss of the bladder? Do you think they live on the bottom because they have no swim bladders, or have no swim bladders because they live on the bottom? Or is neither statement correct as it stands?

The earliest osteichthyans were heavily armored, covered with bony scales coated with a hard enamel-like tissue. A few modern fishes, such as the gars, have retained the armor, but in most of them it has evolved into more flexible scales in which the harder tissues have

[16] There is also a third main group of fossil and recent chondrichthyans, the chimaeras and their relatives, rather uncommon and of minor importance. They have peculiarities related to eating hard-shelled mollusks (among other things).

[17] "Bony fishes," because in contrast with the chondrichthyans most of them have retained bony skeletons.

American Museum of Natural History

22–18 Contemporary lungfishes. *Top. Neo-ceratodus*, the Australian lungfish. *Bottom. Protopterus*, the African lungfish. The modern lungfishes are river dwellers that can surface and gulp air into their lungs; they can survive in water too foul (low in oxygen) to support other fishes, which are entirely dependent on gills for respiration.

degenerated. Some fishes (most of the eels, for example) have lost the scales altogether.

The osteichthyans, and especially the great group (usually classed as a superorder) called *teleosts*,[18] are the dominant aquatic animals today. They have become adapted to almost every aquatic environment, from the unchanging cold darkness of the deep sea to dashing mountain streams. One species or another eats practically everything that is edible by any aquatic animal. Some are sluggish, but some are the most swiftly moving of all animals. Their diversity in form is really astonishing; Fig. 22–19 gives just a hint of the extraordinary shapes among them. More than 20,000 species are known. The individual abundance of some single species is also remarkable. It is estimated that there are at least a trillion (1,000,000,000,000) herrings in the Atlantic Ocean.

The rarest of all the surviving osteichthyans are of special interest—the crossopterygians,[19] or crossopts for short. These animals were abundant in the earliest days of osteichthyan history when, with their allies the lungfishes,

[18] "Perfect bones," the implication being that they are the most perfected or progressive of the bony fishes.
[19] "Fringed fins," from the structure of the paired fins.

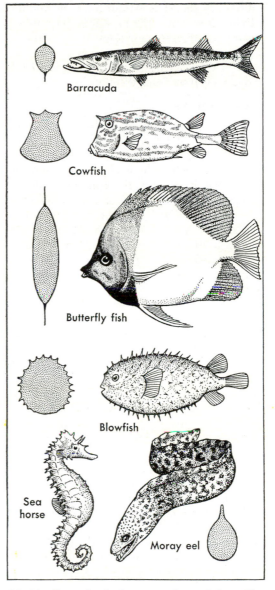

22–19 Some body forms in bony fishes. The stippled silhouettes represent transverse sections through the various fishes.

they were the dominant fishes. Thereafter they became less and less common, and until 1939 everyone thought they had been extinct for about 75 million years. Then a single specimen[20] of a living crossopt (*Latimeria*, p. 769) was caught off the east coast of South Africa,

[20] More have since been caught, in the vicinity of Madagascar.

a discovery almost as extraordinary as finding a living dinosaur.[21] The most important thing about the crossopts is not their survival, amazing as it is, but the fact that amphibians and through them all the vertebrates of the land and air evolved from them.

AMPHIBIANS

It is characteristic of evolution that the amphibians [22] did not evolve from late, specialized, progressive or perfected osteichthyans, such as the teleosts, but from primitive forms that lived near the beginning of osteichthyan history. It has usually been true that when a radical adaptive change occurs and a new major group arises, it originates from primitive and not from advanced members of the ancestral group. With progressive adaptation to any one way of life, there often comes a time when the adaptation seems to be irrevocable—a special aspect of the irrevocability of evolution in general. Then change to a markedly different way of life becomes, if not impossible, at least extremely improbable.

In the fish–amphibian transition, for instance, the presence of lungs in primitive fishes was a crucial factor. Lungs enabled the osteichthyans to live in waters that were sometimes deficient in oxygen. Later their general and increasing efficiency enabled them to spread into waters, especially those of the sea, that are continuously well oxygenated. Then the lungs lost their respiratory function and evolved other functions that made the fish still more efficient in the water but that made it practically impossible for any of their descendants to take to the land. There are few complete absolutes in nature, and even the loss

of the respiratory lungs may not have made it absolutely impossible for later fishes to have evolved some other way of obtaining oxygen from the air. In fact, there are a few teleosts that can clamber about on land for short periods. None of them has become really terrestrial, and it does seem that their commitment to the water (where, incidentally, they are getting along very well) is practically irrevocable. The historical fact is that land vertebrates did evolve on the basis of lungs and other features present in the most primitive osteichthyans but absent in the more specialized and successful of later fishes.

Since lungs were already present in fishes, it was not their appearance that marked the beginning of the vertebrate conquest of the land. It was, instead, the change of the paired fins into legs. In other respects the first amphibians were still almost completely fishlike. Probably, too, they spent most of their lives in the water. But they could, if need be, come out on land and waddle along to some other pond of water. They lived only in fresh water, and only there would the ability to go from one body of water to another be an advantage. This started their descendants on the way to becoming fully terrestrial.

Among modern amphibians (Fig. 22–20), the salamanders, although specialized or degenerate in many respects, have most nearly retained the ancestral habits, fishlike body form, and even fishlike undulatory movement of the body as a whole. They go through a larval stage in which they resemble adults rather closely except that, among a few other details, they have gills that are lost when metamorphosis occurs. The larvae of frogs and toads, familiar to everyone as tadpoles, are also fishlike, perhaps even more so than salamander larvae, because in early stages they lack legs. Their metamorphosis is more drastic, and an adult frog or toad is completely unlike a fish, and indeed unlike anything on earth, except a frog or toad. The adult anurans [23] (frogs and toads) are among the most distinctive and specialized of all the vertebrates. The most obvious specializations are that they have no tails (that is one of our most obvious specializations, too, but of course we

---

[21] But not quite. The dinosaurs have been extinct (as far as we know) for about the same length of time. But the surviving crossopt lives well down in the waters of the sea, the region least known to us of all the habitats of living things. Dinosaurs could not have become submarine forms without changing into something quite different from dinosaurs. The denizens of all recent habitats where dinosaurs could have survived are well known, and dinosaurs are not among them. Reports to the contrary have turned out to be fiction, hoaxes, or just plain mistakes.

[22] "With a double life," because some of them are aquatic as larvae and terrestrial as adults, or because some adults are literally amphibious.

[23] "Untailed."

22–20 Modern amphibians. *Above.* The common tiger salamander, *Ambystoma tigrinum.* The adult is about 8 inches long, an inhabitant of moist woodlands and streams. *Right.* The mud puppy, *Necturus maculosus,* an inhabitant of streams. The adult, attaining a length of 12 to 17 inches, retains bushy external gills.
Below. The tree frog or "spring peeper," *Hyla crucifer.* The animal is small (up to 2 inches) but possesses a loud voice amplified by the distended vocal pouch acting as a resonator.

Above, Isabelle Hunt Conant; below, Hugh Spencer

Adult (*above*) and late larva (*below*) of the green frog, *Rana clamitans.*

did not get it from the anurans) and that their hind legs are tremendous leaping mechanisms. In addition to these more obvious points, anuran anatomy, physiology, and habits are replete with other aberrant features. In particular, anurans are radically unlike the amphibians that were ancestral to the fully terrestrial vertebrates and therefore can show us little about that ancestry. On the other hand, their very aberrancy gives them a way of life in which they have few serious competitors and which makes them the only group of amphibians that was very successful in the long run.

Besides the salamanders and anurans, there is another, small and relatively unimportant group of living amphibians—the Apoda or caecilians.[24] They are blind (or practically so) and entirely limbless, and most of them burrow in wet soil. Some of them are so like earthworms in appearance and habits as to provide a remarkable example of convergence between members of two widely different phyla.

REPTILES

Even though a few kinds of amphibians came to occupy ecological positions in which they continue to have modest success, amphibians are anomalous animals. During parts of their lives they are ecologically or adaptively fishes. They are superior to true fishes only in being less rigidly and permanently confined to the water. They have succeeded and survived only because ability to leave the water when occasion demands has survival value. But as land animals they are still tied to the water. At the least (with partial exceptions of no importance in the broad picture), they must return to the water to breed, and water is still the obligatory habitat of the young.[25] On

land, then, Amphibia are usually at a disadvantage in comparison with animals able to live their whole lives there. It was inevitable that most amphibians would become extinct, given the evolutionary possibility of the rise of fully terrestrial vertebrates or, more objectively, given the fact that such vertebrates did evolve.

It was the reptiles [26] that became the first fully terrestrial vertebrates. They evolved from amphibians, or another way of putting the matter would be to say that the name "reptile" and classification in the class Reptilia are applied to a branch of early amphibians that became fully terrestrial and to the highly diversified descendants of those first fully terrestrial vertebrates. In agreement with a generalization already familiar (p. 608), the ancestors of the reptiles were early, primitive amphibians, not any later forms resembling those that now survive.

The key adaptations were in the modes of reproduction and respiration. The eggs, laid on land,[27] have leathery or limy shells that impede fatal loss of water by the embryos. Fertilization is internal (the male and female copulate). The eggs contain a large amount of food (yolk), and full development, with no larval stage, occurs before the young hatch. The newly hatched young are already essentially like adults in form and activities, although of course they later grow and, as a rule, change in proportions.

In several respects the respiratory mechanism of reptiles shows a marked advance over that of Amphibia. The lungs are more efficiently ventilated through the movement of the ribs by muscles, inflating and emptying the lungs like a bellows. Other improvements in respiration hinge on structural changes in the respiratory system, especially the heart. A partition (incomplete and still therefore "imperfect") arose in the ventricle of the heart,

[24] "Apoda" means "footless." "Caecilian" is from the generic name of one of them, ultimately from the Latin for "blind."

[25] Some modern Amphibia are also tied to the water by the ineffectiveness of their lungs as respiratory organs. Air is inadequately forced into them by the weak bellows action of the floor of the mouth cavity. The frog's skin, supplied with blood vessels, is its principal respiratory organ and, as such, must be kept moist. It is uncertain whether this dependence on the skin for respiration characterized ancestral Amphibia. It could well be that it is an evolutionary

novelty in later Amphibia, which were restricted to moist places anyway and simply exploited the skin as a potential supplement to inefficient lungs.

[26] The name is old, ultimately derived from the Latin for "crawl."

[27] In some reptiles, the eggs are retained in the mother during development and the young leave her body as they hatch. How does this process differ from the birth of the young in man and other mammals?

All photos American Museum of Natural History except garter snake, U.S. Fish and Wildlife Service, and plated lizard, Charles Halgren

22–21   Modern reptiles.
Above. The common garter snake, *Thamnophis. Right.* The tuatara of New Zealand, *Sphenodon punctatum,* a living fossil; it is the sole living representative of an ancient group of reptiles, the Rhynchocephalia.

Below. The American plated lizard, *Gerrhonotus,* with its eggs. This lizard (about a foot long) lives in fallen timber.

Above. Musk turtles, *Aromochelys odoratus* (about 5 inches long), which inhabit rivers and lakes. Note their webbed feet.
Below. The salt-marsh crocodile, *Crocodilus palustris,* of southern Asia. It reaches a length of 12 feet.

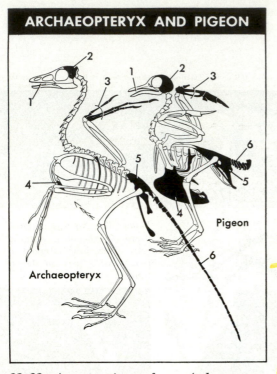

ARCHAEOPTERYX AND PIGEON

Pigeon

Archaeopteryx

22–22   A comparison of an *Archaeopteryx* and a pigeon. *1.* Teeth are present in the *Archaeopteryx* but absent in the pigeon. *2.* The brain case of the pigeon is larger than that of a reptile and the early *Archaeopteryx*. *3.* In the "hand" of the *Archaeopteryx*, three of the digits are still clearly present and separate; in the pigeon they are less distinct and partly fused. *4.* The immense development of the sternum to anchor the large flight muscles is characteristic of the pigeon but not of the *Archaeopteryx*. *5.* In the pigeon the pelvic girdle is larger than in the *Archaeopteryx* and is firmly fused to the spinal column. *6.* The tail, long and reptile-like in the *Archaeopteryx*, is much reduced in the pigeon.

separating nonoxygenated blood from other blood, freshly oxygenated, returning from the lungs.

There were of course other anatomical and functional changes throughout the body, barely perceptible in the first reptiles but becoming more pronounced in many of their descendants. Among these was a tendency for the limbs to become larger and more powerful, better able to support the body free of the ground.

The reptiles expanded and diversified into many different habitats throughout the land (except in its coldest climates), and some even

returned to the sea and competed with (and also consumed) fishes. Their dominance during the Age of Reptiles and their later decline are dramatic parts of the history of life, to which we shall later return (Chapter 30). Systematists recognize about 15 orders of the class Reptilia, living and extinct. Only four now survive (Fig. 22–21). Even among these four, one is on the verge of extinction, represented only by a few individuals of a single species, *Sphenodon punctatum,* on islands off the coast of New Zealand.[28] The other three living orders are still fairly abundant and are familiar to everyone: turtles and tortoises (Chelonia),[29] crocodiles and alligators (Crocodilia),[30] and lizards and snakes (Squamata).[31] Lizards and snakes are the most abundant living reptiles. There are about 2 thousand living species of each, and they occur in many habitats from the high seas (sea snakes) to the desert. Different as the legged, swift-running lizards and the legless, crawling snakes seem, they are rather closely related. Snakes are descendants of an early group of burrowing lizards. There are still some burrowing lizards, of later origin, that have also become legless and snakelike. Does that make them snakes? Why (or why not)?

BIRDS

The two latest and most progressive classes of land vertebrates both (but entirely separately) evolved from early reptiles. The birds, class Aves,[32] are a group in which one key characteristic opened up a whole new realm of life. This characteristic is, of course, flight. The oldest known fossil birds (*Archaeopteryx,* "ancient wing") were still almost reptilian except in one respect: they had feathered wings (Fig. 22–22). If the change had stopped there, however, birds would never have reached the great diversity and wide success that make them one of the great cli-

[28] *Sphenodon* looks just like a lizard and is, indeed, related to lizards, but its internal anatomy shows that it belongs to an order, Rhynchocephalia ("beak-headed"), that split off from the lizard ancestry before lizards, as such, had evolved.
[29] From the Greek for "tortoise."
[30] From the Latin (and earlier Greek) for "crocodile."
[31] Latin for "scaly."
[32] Latin for "birds."

maxes of evolution and that are reflected in their classification as a class. In such a case we would probably consider them simply an order of flying reptiles, which, indeed, is how we do classify another group that developed flight from reptilian ancestors—the pterosaurs ("winged lizards"), a long-continued but comparatively unsuccessful group which finally became extinct, perhaps from competition with the birds.

Along with and after their acquisition of wings, birds evolved other characteristics associated with intense and sustained activity, keen perception, and rapid and varied reactions. They have high and steady metabolic rates, with precise control of the internal environment. Along with the mammals, they are the only organisms that maintain a constant temperature (that is, are homeothermous; see Chapter 12). Feathers serve not only for flight but also as insulation (Fig. 22–23). As in mammals, the heart is four-chambered; it is completely divided into what are essentially two separate hearts. This represents the final step in efficient plumbing of the circulatory system in relation to lung respiration (Chapter 11). Bird senses, especially those of vision, equilibrium, and hearing, are particularly acute. Their brains are large and are peculiarly specialized (see Chapter 13). Their behavior, although in considerable part stereo-

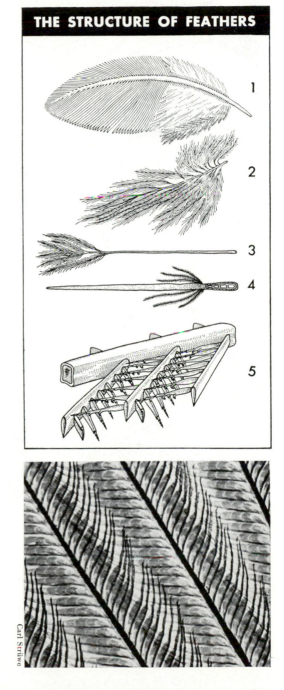

THE STRUCTURE OF FEATHERS

22–23 Feathers, one of the key structural adaptations of birds. *Top. 1.* Contour feathers cover the animal and, by establishing its smooth contours, contribute to its streamlining; they also contribute to its insulation. The much elongated contour feathers of the wings provide the effective flight surfaces of these organs. *2.* Down feathers are adaptively significant in providing insulation, a property exploited by man in the making of sleeping bags and comforters. The down feathers, like the hair of mammals, are an important part of the body temperature–control system. *3.* The small, elongate filoplumes (drawn at a much enlarged scale relative to the others) lie between the other feathers and are of uncertain function. *4.* The bristles around the mouths of birds like the whippoorwill and flycatchers are modified feathers. *5.* The structure of a contour feather. A central shaft runs along the middle (cf. *1*) and supports the wide, flat vane of the feather. The vane itself consists of parallel barbs on each side of the shaft. The barbs in their turn support barbules on their sides. The barbules have hooklets that engage the edge of barbules in the next row. Through the action of the hooklets, which can slide freely along the barbule edges they engage, the whole system of barbs and barbules is bound into the flexible wide-surfaced vane. In down feathers the barbules do not engage each other. *Bottom.* Overlapping contour feathers of a hummingbird.

Carl Strüwe

Merganser

Hummingbird

Spoonbill

Flamingo

Bittern

Crossbill

Mallard

Avocet

Nighthawk

Skimmer

Ibis

Sparrow

Puffin

Toucan

Falcon

Pelican

Kea parrot

Shoebill stork

22–24 A diversity of bird beaks.

typed or instinctive, is often very complicated.

Ancient birds, even some considerably later than *Archaeopteryx*, retained simple teeth inherited from the reptiles. Today, when there is nothing scarcer than hen's teeth, no bird has had teeth for tens of millions of years. Instead, the birds have beaks of bone covered with horn. Beaks serve not only for obtaining food, but as instruments for many purposes, from knot tying to wood boring. In their basic anatomy birds are remarkably alike. If there were no other reasons, the stringent mechanical necessities of flight would keep them so. But within the limits of the basic stereotype, they are fascinatingly diverse. The many forms of their beaks, most of which are clearly adaptive, illustrate this fact (Fig. 22–24). The feet, too, have numerous adaptive patterns, as we saw in Chapter 16. And everyone knows and enjoys something of the diversity of bird colors and patterns, which (when known) are in some cases protective, blending a bird into its background, and in others social, serving for recognition by other members of the group or as a stimulus to the opposite sex.

The systematics of recent birds has been worked out in better detail than that of any other animal group of comparable scope—not that all is yet well understood about this group that is so attractive to both amateur and professional zoologists. There are some 8600 living species, about half of them belonging to the single order of perching birds (Passeriformes, "sparrowlike"), which includes among many others the kingbirds, larks, swallows, crows, wrens, thrashers, thrushes (among which is the American robin), vireos, warblers, blackbirds, and sparrows. Other large and familiar orders are those of the herons and their allies (Ciconiiformes, "storklike"); ducks, geese, and swans (Anseriformes, "gooselike"); the predacious birds such as the hawks (Falconiformes, "falconlike"); partridges, turkeys, and their many allies (Galliformes, "chickenlike"); gulls and their kin (Charadriiformes, "ploverlike"); owls (Strigiformes, "owl-like"); and woodpeckers (Piciformes, "woodpeckerlike") (Fig. 22–25). Modern classifications recognize 25 to 30 different orders of living birds.

As you might expect, the birds most un-usual in anatomy and physiology are those that no longer have the ancestral adaptation to flight, kinds like the penguins (which still do fly, but under water) or the ostriches. Since flight is the key characteristic of the class, how would you explain the fact that some birds cannot fly? This is a very difficult question indeed, although it is possible to find a reasonable answer.

MAMMALS

The final and (to us, at least) the supreme class of vertebrates arose from the reptiles not with the appearance of any key characteristic, such as flight, but through long-continued gradual changes in many directions. On the whole, the changes made for greater mechanical, physiological, and reproductive effectiveness in ways of life similar to those of the diversified reptiles of the Age of Reptiles. Mammals [33] now occupy much of the range that was once the domain of the reptiles, although they have gone further than the reptiles ever did and although a few special kinds of reptiles still live successfully along with the mammals. Thus the mammals contrast with the birds, which by a new key adaptation entered and exploited a whole new realm of life.

The reptiles ancestral to mammals were of course very different from any living today. Here again and perhaps more than ever the comparison of recent animals is merely confusing as to the real evolutionary changes. A picture of mammalian origin from anything like a snake, lizard, crocodile, or turtle is ludicrously wrong. The reptilian ancestors of the mammals were both more primitive and more mammal-like than any later reptiles, much as the fishes ancestral to amphibians were more primitive and more amphibianlike than later fishes. In fact, the contrast is even greater because fishes somewhat amphibianlike (especially the lungfishes and *Latimeria*) have survived, but the whole great group of mammal-like reptiles became extinct soon after the evolution from it of the mammals (Chapter 30). Since the process was gradual, some characteristics of the mammals arose in animals nominally classified as reptiles, and some did not arise until after nominal mam-

[33] Class Mammalia, named for the presence in the females of milk glands or *mammae*.

22–25 A variety of modern birds. *Above.* Terns (order Charadriiformes) are marine birds that dive for their fish food. *Above right.* The blue jay (order Passeriformes), a relative of the crows, is an inverterate egg robber.

Above. The kiwi (order Apterygiformes), a flightless bird of New Zealand, has hairlike feathers that do not form a useful vane, and its sternum lacks a well-developed keel. *Below.* A female marsh hawk (order Falconiformes) is about to alight. Like all its relatives, it is a bird of prey.

Above. A female Canada goose (order Anseriformes) sits on its eggs in a nest of reeds. *Below.* The great blue heron (order Ciconiiformes), a relative of the storks and ibises, feeds on fish, frogs, and small mammals.

mals existed. The line between mammal-like reptile and mammal is arbitrary, a convenience of classification. Nevertheless, the characters that did come to stamp the mammals and that underlie their great success in the world are real enough (Fig. 22–26). Among them are these:

Warmbloodedness (homeothermy), as in birds, and external insulation, also as in birds but by hair rather than feathers. Both characters (independently developed in the two classes) are, again, related to higher metabolism and more precise and sustained internal regulation.

Complex differentiation of teeth and the development of a new joint between the lower jaw and the skull, as discussed in Chapter 16. These two characters seem to be related and together to involve greater efficiency in the utilization of food. Whatever may have been the factors in their origin, they did certainly lead to remarka-

22–26 **A comparison of a mammal and a reptile.** *1.* Skull: the mammal has a larger brain case and complex cheek teeth; its jaw consists of only one bone (the dentary), whereas the reptile's consists of several. *2.* Brain: the mammal's forebrain is relatively larger than that of the reptile. Palate: the respiratory and alimentary passages are separated by a bony palate (the heavily outlined shaded area) in the mammal. *3.* Ear ossicles: the two additional ossicles in the mammal are transformations of two jawbones in the reptile. *4.* Posture: the mammal's limbs are rotated under the body, which is thus kept off the ground—a contribution to the control of body temperature and to more rapid locomotion. *5.* Skin: the mammal's skin is covered with hair—another contribution to body-temperature control. *6.* Circulation: in the mammal the lung and body circulations are kept separate by a complete partition of the ventricle—an improvement in efficiency of respiration. *7.* Diaphragm: the thoracic cavity of the mammal is completely separated from the rest of the body cavity by a diaphragm—an improvement in efficiency of lung ventilation. *8.* Body temperature: uncontrolled in the reptile; controlled in the mammal. *9.* Growth: continuing throughout life in the reptile; limited in the mammal. *10.* Reproduction: eggs, no care of young, no milk in the reptile; placental reproduction, parental care of young, milk in the mammal. (The contrasts are here shown in the fully developed state; advanced mammal-like reptiles graded imperceptibly into primitive mammals.)

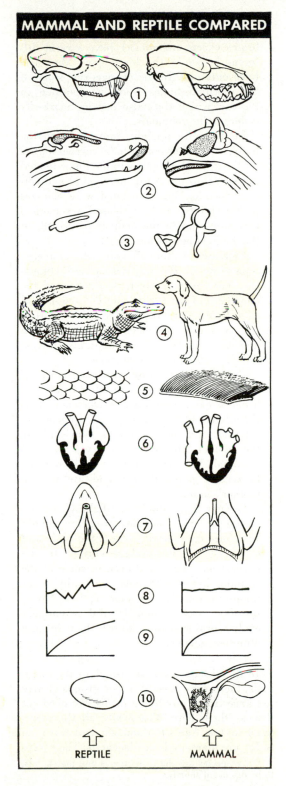

MAMMAL AND REPTILE COMPARED

REPTILE   MAMMAL

ble later diversity and efficiency of feeding in mammals. Reptiles (and indeed all vertebrates except mammals) have several different bones in the lower jaw. Mammals have only one. In the changeover, some of the reptilian bones were incorporated in the ear, which thus also became markedly different in mammals.

Limbs more upright, more beneath the body, which thus comes to be carried higher off the ground. Changes in the limbs tended in general to make for more rapid and efficient locomotion. Associated were changes in the joints, with increased mechanical precision, and in the manner of growth of the long bones (Chapter 8).

Increased (although still not complete) separation of the respiratory and alimentary passages; complete separation of the chambers of the heart; and the acquisition of a diaphragm. All three features improve the continuity and efficiency of respiration and thus are other concomitants of sustained metabolic maintenance.

Increased protection and sustenance of the young, both before and after birth, and hence greater reproductive efficiency. Fertilization and embryonic development are internal, and the embryo is nourished by the mother, through a placenta in most mammals.[34] After birth the young continue to receive parental care and are fed on mother's milk.

Greater individual modifiability of behavior, or wider behavioral reaction ranges. Increased comparative size of the brain and especially of its cerebral cortex. In the earliest mammals, and even in some of the stupider living mammals, the brain is not notably better than in some reptiles. Nevertheless, most mammals did eventually evolve brains much larger than in any other organisms.

In the obscure earliest history of the mammals, there were several major groups clearly not ancestral to any living forms. We need not discuss them here. Practically all the mammals of the Age of Mammals and down to today are marsupials or placentals. The excep-

tions are the monotremes [35] of Australia and adjacent islands, fascinating and famous because they lay eggs and yet give milk (Fig. 22–27). They are, throughout, a strange blend of characters otherwise reptilian, characters otherwise mammalian, and characters highly peculiar to themselves. They represent, not really an ancestral stage in the evolution of the mammals, but a line of descent that branched off from the mammalian ancestry at some very remote time. They have become extremely and aberrantly specialized in their own way but have also retained some ancestral characters lost in other or true mammals. Opinions differ as to when the monotremes branched off. It may well have been in a stage nominally reptilian; if so, it might be more indicative of their relationships with other creatures to consider them as surviving mammal-like reptiles rather than as mammals. The point is not very important. There is little question as to the general relationships among these groups; only the degree is in real doubt.

The marsupials and placentals are clearly groups of common origin among the mammals. It was formerly believed that the marsupials were an older group and ancestral to the placentals, but on present evidence this is highly improbable. They simply diverged from a common ancestry, and many of the peculiarities of the marsupials probably are specializations that evolved in that group and not characters formerly present in the ancestry of the placentals. The most striking of such peculiarities is that the developing embryo receives little or no nourishment from the mother while in the uterus. It is born in very immature form, crawls into a pouch on the mother's belly, hangs onto a nipple, and there completes development. You probably know that marsupials include kangaroos, wombats, and other exotic mammals of Australia, "land of marsupials." Marsupials, mostly opossums, are, however, also abundant in North and South America and were formerly even more abundant, especially in the southern continent.

Placentals are the mammals in which the fetus is nourished (through a placenta, hence

[34] Exceptions are the monotremes and marsupials, to be discussed shortly.

[35] "One-holes," because the anus and urogenital openings are not separate externally, another resemblance to reptiles.

Above. A kangaroo with young in its pouch.

Above. A wombat, with a way of life and gross appearance similar to those of the American woodchuck. *Below.* A duck-billed platypus, a monotreme.

Australian News and Information Bureau

22–27 Australian mammals: monotremes and marsupials.

the name) within the maternal uterus until development is far advanced. Although not as diverse as they were a few million years ago, placentals (Fig. 22–28) are still the dominant land vertebrates and have been throughout the Age of Mammals. A modern classification divides them into 28 orders, of which, however, only 16 have living representatives. Some species belonging to the more abundant of these orders are already quite familiar to you and well illustrate the remarkable adaptive radiation that has occurred among placentals in the last 70 million years or so: shrews and moles (Insectivora, "insect eaters"); bats (Chiroptera, "finger-wings"); armadillos (Edentata, "toothless"); rabbits (Lagomorpha, "harelike"); squirrels, porcupines, mice, and a host of other rodents (Rodentia, "gnawers"); whales (Cetacea, "whale-like"); cats and dogs (Carnivora, "meat-eaters"); elephants (Pro-

boscidea, "with a trunk"); horses (Perissodactyla, "odd-toed"); and pigs, camels, sheep, and cows (Artiodactyla, "eventoed").[36] Placentals burrow, fly, climb, run, and swim. They eat worms, fruit, insects, grass, seaweed, squids, crustaceans, bark, cocktail canapés, and each other. They live in the open sea, in tropical treetops, on Arctic ice floes, in apartment houses, and in sandy deserts.

In terms of numbers, both of species and of individuals, the outstanding placentals are the rodents. On land they swarm practically everywhere, and some are amphibious (although none are completely aquatic or

[36] Remember that the name of a systematic group does not need to be, and often is not, appropriate for all its members. Some insectivores do not eat insects; some edentates (notably the armadillos) have teeth; some carnivores do not eat meat; and so on.

Top left, Charles Halgren; top right and center left, American Museum of Natural History; center right, Ewing Galloway; bottom left to right, Sabena, Miami News Bureau, H. E. Edgerton from National Audubon Society

22–28 Some modern placental mammals. *Left.* The African leopard, *Felis pardus* (order Carnivora). *Above.* A deermouse, *Peromyscus* sp. (order Rodentia).

Below left. The mole *Scapanus latimanus* (order Insectivora), in its underground burrow. *Below right.* A South American anteater, *Myrmecophaga* (order Edentata). Note the elongate snout.

Below left. The African elephant, *Loxodonta* (order Proboscidea), a herbivorous mammal. *Below right.* The porpoise *Phocaena* (order Cetacea), a relative of the whale found in all the oceans of the world.

Left. A bat (order Chiroptera, the only true flying mammals). Flying squirrels (order Rodentia) do not really fly; they glide

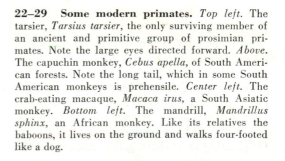

New York Zoological Society

22–29 Some modern primates. *Top left.* The tarsier, *Tarsius tarsier*, the only surviving member of an ancient and primitive group of prosimian primates. Note the large eyes directed forward. *Above*. The capuchin monkey, *Cebus apella*, of South American forests. Note the long tail, which in some South American monkeys is prehensile. *Center left*. The crab-eating macaque, *Macaca irus*, a South Asiatic monkey. *Bottom left*. The mandrill, *Mandrillus sphinx*, an African monkey. Like its relatives the baboons, it lives on the ground and walks four-footed like a dog.

marine—the only possibility they seem to have overlooked). In only one respect do they really fall short of being the dominant mammals and the climax of vertebrate evolution, but that is an important respect: they are not as smart as we are.

This is our excuse for calling the order to which we ourselves belong Primates (Latin for "the tops"). Not that all primates (Fig. 22–29) are particularly intelligent. Some living primates are below the average intelligence for mammals and so, judging by the outer form of the brain known for a few of them, were the oldest primates. The order arose among the most primitive of placentals, from early Insectivora, and seems at first to have had little to distinguish it beyond the use of the forefeet as hands and an increasing coordination of visual perception and manual response. Yet somehow these primitive creatures had the potentiality to evolve the highest intelligence ever reached by any organisms. "Potentiality" in this application is not explanatory. We only know, after the fact, that they could because they did. Just what there was about them at the time that made possible that later development is not fully clear.

The earliest primates, which were abundant and lived pretty much all over the world except in Australia and South America, were prosimians ("pre-monkeys"). Some prosimians (lemurs, bush babies, tarsiers, etc.) still survive in modified and more or less specialized form in Africa, Madagascar, and southeastern Asia. From early prosimians three other major groups evolved. The New World monkeys, Ceboidea, or ceboids in the vernacular, arose in and are still confined to Central and South America. The name "ceboid" means "cebuslike." *Cebus* is the genus of common South American capuchin monkeys. Marmosets, howlers, and others also belong to this group. Throughout the warmer parts of the Eastern Hemisphere (except Australia) lived and still live the Old World monkeys, Cercopithecoidea. This name for the whole group of Old World monkeys derives from the generic name (*Cercopithecus*) of some common African monkeys. Rhesus monkeys (much used in experimentation), baboons, mandrills, and others belong to this group. Originating somewhere in that vast area of the Old World and soon spreading throughout it was the climax group of the primates, the Hominoidea ("manlike"). Formerly much more diverse, this group now includes the gibbons, apes (orangutan, chimpanzee, and gorilla), and man.

CHAPTER SUMMARY

Mollusks, arthropods, and vertebrates: widely different animal phyla having the greatest complexity, coordination of function, and diversity of adaptation—the three great pinnacles of animal evolution.

Mollusks: the six classes, exemplified by Amphineura.
  Gastropods: slow gait; shell; marine origin; terrestrial forms with lungs; adaptive diversity.
  Pelecypods: bivalves; mostly sedentary habit and loss of sense organs; filter feeding; adaptive diversity.
  Cephalopods: organization; fast movement, with jet propulsion; complex brain and image-forming eyes; predatory nature; history and diversity.

Arthropods: immense numbers, diversity, and ubiquity; characters; the seven classes.
  Crustacea: the familiarity of the decapods; the diversity of other crustacean types; the evolutionary specialization of crustacean appendages, exemplified by the lobster.
  Arachnids: the undeserved bad reputation, since there are more useful than harmful species; the four or five pairs of appendages; diagnostic characters; *Limulus*, a living fossil.
  Insects: their success and diversity (at least one-half of all living organisms, but no marine types); primarily terrestrial habitat, with secondary invasion of fresh waters; factors responsible for size restriction; the exoskeleton and the respiratory and circulatory systems; organization, as exemplified by the grasshopper; winged and wingless insects; adaptive modifications of wings in beetles and flies; feeding larvae; polymorphism.

Vertebrates: a subphylum of the Chordata; the most progressive and important of all animal groups but less abundant and diverse than insects; their great awareness of their environment and their complexity of behavior; ancestral vertebrates; vertebrate characters, especially the aquatic (probably fresh-water) forms, notochord, vertebrae, skull, musculature, jawless mouth, gill slits in the pharynx, simple alimentary system, bisexual reproduction with external fertilization, closed circulatory system with single heart, dorsal, hollow nervous system, relatively large brain, well-developed sense organs; the eight classes.

Agnaths: the earliest vertebrates; jawless fishes, exemplified today by lampreys and hagfishes; the funnel and rasp.

Placoderms: the first vertebrates with jaws; the jaws as modified gill supports; extinction.

Chondrichthyans: the cartilaginous fishes; originally and still a marine group, exemplified by sharks and rays; urea and osmotic regulation; internal fertilization.

Osteichthyans: the bony fishes; probably originally fresh-water forms, now both fresh-water and marine; the bony skeleton; the operculum; the early acquisition of lungs; modern lungfishes; modification of the lung to a swim bladder; the bony armor of early forms and scales of modern forms; the teleosts, most successful of the bony fishes; the diversity of bony fishes; the survival of *Latimeria*.

Amphibians: their origin from early osteichthyans; the role of lungs and limbs in their origin; modern amphibians—their locomotion, larvae, metamorphosis, specialization, and divergence from early forms.

Reptiles: the restriction of amphibians to watery environments; reptiles as their descendants and the first true land vertebrates; key adaptation: shelled eggs and internal fertilization; lung ventilation and the partitioning of the ventricle; adaptive radiation; the four surviving orders.

Birds: descendants of the early reptiles; *Archaeopteryx;* bird adaptions to land life, permitting sustained high activity; the development of body-temperature control; improved respiration and circulation; the development of the brain and sense organs; the loss of teeth; the specialization of beaks and feet; the classification of modern birds; flightless birds.

Mammals: descendants of the early reptiles; the supreme class of vertebrates; their occupation of former reptilian ways of life, in contrast with the birds, which exploited different ways of life; major mammalian characters, contrasted with those of reptiles—body-temperature control, more complex teeth, more upright limbs and more rapid locomotion, improved respiration and circulation, protection and sustenance of the young (including the evolution of placenta and milk glands), the elaboration of the nervous system and behavior; monotremes and marsupials, with primitive or aberrant reproduction; placentals—represented by 28 orders, with rodents the most abundant (individuals and species) order; the primates of special interest because of their relation to man—their origin from insectivores, coordination of the hands and eyes as creating the potentiality for evolution of higher intelligence, their diversity.

Part 7

The Life of Populations and Communities

Where there is one poppy, there will be more. In fact, there will be a whole population of them, as in the photograph introducing Part 7. Associated with the poppies will be populations of other species—grasses, trees, insects, mice, hawks, and the rest—which collectively form an integrated community. One of the most striking facts of life is that individual organisms do not, because they cannot, live a wholly solitary existence. They are bound together by one cause or another—by sex, by social ties, by sharing common needs satisfied only by a particular habitat, and, above all, by dependence on one another for shelter and food. This interdependence of organisms dominates the entire subject of ecology, the study of organism-environment relationships; it raises problems and demands explanatory principles of its own, which are the topics of Part 7.

Throughout this section of the book we shall be examining communities and their constituent populations as unit living systems much as in some earlier parts we considered the individual organism as a unit living system. But this approach—of treating populations as units of life—is of course not new to you; in Part 5 on the mechanism of evolution we had to begin (Chapter 15) with a population as an entity whose genetic constitution (genetic pool) is the fundamental unit subject to evolutionary change.

Chapter 23 reviews biologically important features of the environment and some relationships to them of populations and communities. The cyclic flows of materials and the one-way transfers of energy constitute a sort of over-all metabolism of communities and their environments.

Chapter 24 considers some specific features of community organization, especially the lines of transfer of energy and materials and their consequences. Interspecific relationships within communities include consumption, competition, and the more intimately individual associations collectively known as symbiosis. Changes within communities include especially, within the human time scale, successions leading to climaxes.

Chapter 25 first deals with the growth, decline, and cyclic or periodic changes in population size and then with intraspecific interactions involved in population organization. These include cooperative aggregations culminating in societies.

In Chapter 26 the human species is treated as a biological population from a systematic and ecological point of view. The impact and dependence of this extraordinary society on natural communities are reviewed.

23

Environments, Populations, and Communities

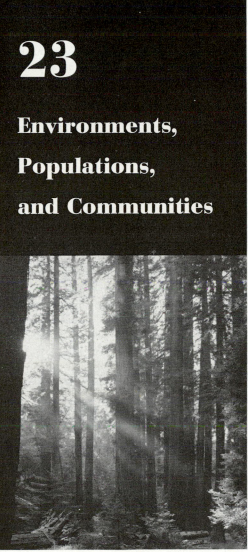

Redwood Empire Association

The sunlight that each day drenches the forest is the ultimate source of all the energy expended in the whole living community.

No organism is ever for even an instant independent of the requirements and advantages of the environment in which it lives. None lives without constantly influencing its surroundings. You cannot possibly think of an organism to which both these generalizations do not apply, and you need only to look around you to find examples of them. You would dress and feel differently if the temperature of your environment were other than it happens to be at the moment. You would not continue to live at all if the medium around you were not air but water, as it is for so many other organisms. As with every other living thing, your life depends strictly on conditions of the physical environment. Your environment also includes other people, the whole of the population of which you are a part. It, too, profoundly influences your life, and, again, interaction with the population as part of the environment normally occurs in all organisms. Moreover, you are completely dependent on other populations, other species of organisms, for many of the materials that make up your body and all of the energy that keeps you going. Like you all organisms live in communities of numerous different species and participate in the flow of materials and energy through the various specific populations.

Within a few hours or less you will be hungry, a signal that the energy and materials of your tissues need to be renewed. Perhaps you will eat a hamburger and thus transfer to yourself energy and materials previously acquired, transformed, and stored by another animal. That animal got them from plants, perhaps from grass. The grass, in its turn, acquired materials from the air and from water and solutions in the soil. The energy used in syntheses in the grass and stored in the products came from the sun. The substance of your body has come, mostly through the medium of other organisms, from the air, water, and watery solutions of the earth. All your energy was derived, through the same intermediaries, from atomic energy generated in nuclear reactions some 93 million miles away, which, incidentally, is an excellent distance at which to be from nuclear reactions. (Are there any substances in your body that were not derived from other organisms? Is there any of your energy that was not?)

Many other factors influenced the passage of materials and energy from earth and sun to you. The grass was rooted in soil. The properties of the soil that determined the possibility and the amount of growth of grass depended on many things: composition of rocks in the crust of the earth; weathering, erosion, and deposition; climate; movement of ground water; activities of earthworms, bacteria, and innumerable other organisms within the soil; effects, past and present, of plants at the given locality; and so on, through a list too long to make complete. And not only the soil affected the grass. Its growth was also directly influenced by sun, wind, rain, temperatures, insects, rodents, grazing animals, and many other things. The presence of the grass also depended on a long sequence of prior events, such as the development and growth of seeds through generation after generation, the origin of this species of grass and the long evolution of its ancestry back to the beginning of life, and indeed the still older origin of the planet on which life could and did arise.

There is nothing particularly new to you in all this, although you have perhaps not looked at it in just this way before. In turning to discussion of communities in which numerous organisms live together, we find, again, that we are not taking up a completely new and separate topic, but are simply looking more closely at a particular aspect of the living world, all aspects of which are inseparably related to all others. Everyone has some idea of the existence of communities in nature and of relationships of populations and environments, and up to this point such general knowledge has been assumed. To organize and extend such common knowledge, we shall now consider first the environment, basic relationships between populations and environments, and some consequent fundamentals of material and energy transfers in communities. These are the subjects of this chapter. Chapters 24, 25, and 26 will continue the discussion of living together in populations and communities, including the human population and its interactions with its environments. The next section of the book (Part 8) also has much to say about communities and environments but from another point of view, that of spatial or geographic distribution and association.

Environment

The word "environment" has been used repeatedly in previous chapters without being defined. As good a definition as any is that environment is the totality of extrinsic things and conditions affecting an organism. As so often happens with definitions, this one becomes blurred when it is applied to specific instances. The main trouble is that the word "extrinsic" needs defining itself. We do not went it to mean simply "external." A parasite in stomach, intestine, or in the fully internal blood stream is environmental; it is extrinsic to the essential structure and contents of the host organism. This is a special case, but it serves to stress the concept of environment construed broadly as everything that influences an organism but that is not an *intrinsic* part or condition of that organism.

In this broad sense the environment includes several distinct but simultaneous and (of course) interacting phases. It includes, first, the nonliving aspects of the place where an organism lives—its *physical environment.* It includes, too, all the living things that affect the organism—its *biotic environment.*

THE PHYSICAL ENVIRONMENT

All organisms live in water or in air. (Even those that live within the soil are effectively surrounded either by water or by air.) That is the most fundamental feature of the physical environment. The conditions of life are very different in water and in air, and most organisms are confined to one or the other. Many plants do, to be sure, live in both environments at once, with roots in water and stems and leaves in air, but the different parts are adapted to the different environments. Many animals spend part of their lives in water and part in the air, as do the numerous insects with aquatic larvae, but the organism is quite different when in one environment than when in the other. Many can temporarily leave one environment for the other, as a fish can jump into the air or a man plunge into the sea. More impressively, there are whole hosts of plants and animals, especially along the shore, that regularly undergo alternation between the two environments. Yet in almost all these plants and animals, one element is vital for the organism and the other is only briefly endured.

Radiation and climate

Organisms that live in air are profoundly affected by weather and climate, and aquatic animals are similarly affected by such factors as temperature. One of the most important factors in such conditions is *radiation*. Radiation important to living things is mainly in the form of electromagnetic waves from the sun. Radiation that we call "light," because our eyes (and those of most other animals) are sensitive to it, is most intense. (Do you think that is a coincidence?) Organisms are also sensitive in other ways to solar radiation of shorter (ultraviolet) and longer (infrared, or heat rays) wavelengths.

The importance of solar radiation cannot be overstressed because (with insignificant exceptions) it supplies all the energy available for all the processes of life in all organisms. This is the income from which all the life activity on earth must be budgeted. The influence of radiation in any particular local environment follows from this basic fact. Green plants, primary converters of solar energy into vital energy, grow only where solar radiation is received. The amount of their activity is limited by, and is roughly proportional to, the average amount of radiation in a given environment. The activities of other organisms are, in turn, limited by those of the green plants from which, directly or indirectly, practically all their energy and materials must come.

Environments with little or no solar radiation have no green plants. They are inhabited only by animals, some nongreen plants (mostly fungi), and certain protists (especially among the bacteria). These organisms in lightless environments necessarily depend on organic foods that are somehow brought in from elsewhere, from environments that do receive solar radiation. Offhand you might think that environments without solar radiation would be quite limited: caves, for instance, which do have an interesting, sparse population but are of no great importance. Actually, however, such environments are more extensive than any others. They include the soil, below its most superficial layer, and

the vast reaches of the sea below the depths to which daylight penetrates.[1] Where does the food of soil organisms come from? Of deep sea organisms?

Less solar energy is converted directly into vital energy than is expended in changing and maintaining the *temperature* of the environment. Expenditures of solar energy have been calculated (by C. Juday) for Lake Mendota in southern Wisconsin, as shown in Table 23–1.

TABLE 23–1
Expenditure of solar energy

Expenditure	Per cent of solar energy received
Reflected or otherwise lost	49.5
Absorbed in evaporation of water	25.0
Raising temperatures in the lake	21.7
Melting ice in the spring	3.0
Directly used by organisms	0.8

The figures are quite different in other environments, but everywhere only a small fraction of radiation is used directly by green plants, and a much larger part goes to warm the water or air. (In the example of the lake most of the energy tabulated as lost heats the air above the lake.) Maintenance of environmental temperature is another necessity for life as we know it, which can exist only in the range of temperatures that is, in fact, maintained on earth by solar radiation. Metabolic activities are very strongly influenced by the temperatures of organisms (Fig. 23–1), and in all plants and most animals the internal temperature depends largely on that of the environment. (What are the exceptions?) Specific adaptations of plants and animals are also related to the averages and ranges of temperatures in particular environments. You know, for instance, that orange trees require sustained warmth but apple trees thrive in

[1] Radiation is gradually absorbed and scattered as it passes through water, and there is no sharp line where radiation ceases in the sea. Its penetration varies greatly. In very clear water light may still be evident at a depth of 2000 feet. Radiation in the sea is, however, seldom significant below about 600 feet, and deeper waters have few or no living green plants. Red algae in general reach lower depths than any other plants.

regions with low winter temperatures, and that polar bears live only in the cold north and boa constrictors only in the tropics.

As the figures for Lake Mendota illustrate, enormous amounts of solar energy are also expended in the evaporation of water. This, too, is profoundly important for living things.

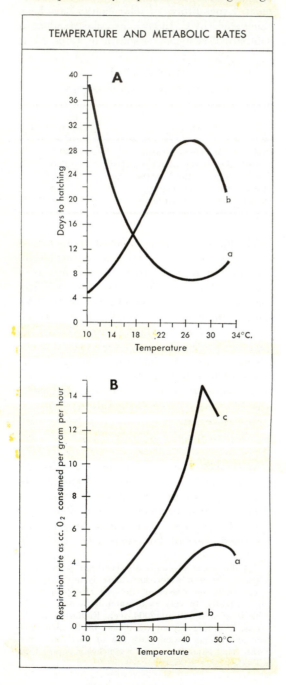

TEMPERATURE AND METABOLIC RATES

In fact, without it life would be possible only in the sea. Evaporation maintains the *humidity* of the atmosphere and is the power-input phase in the *water cycle* (Fig. 23–2). Involved are *rainfall* by condensation of evaporated water in the atmosphere and consequent maintenance of streams, lakes, and water beneath the surface of the ground (ground water). The cycle thus provides the whole of the fresh-water environment and also the enormous quantities of water required by land organisms. No farmer needs to be told that water supply is a crucial factor in the activity of plant life, and the contrast (in both plants and animals) between a well-watered New England hillside and a dry Arizona desert is well known.

Movement of water or air is another feature of the physical environment. All these factors ramify and interlock; both in water and in air movement helps to determine the distribution of temperatures, and air movement is a crucial element for rainfall. Winds and currents also influence organisms more directly in many ways. Innumerable land plants, among them the conifers and many grasses, are wind-pollinated. The animals in swift-flowing streams can remain there only by anchoring themselves somehow or by making headway against the current. Most animals in flowing fresh water have eggs that sink to the bottom, below the current, although many marine animals have floating, drifting eggs. (What

23–1 Temperature and the rate of metabolism. *A.* The effect of temperature on the rate of development in the insect *Sitona: a,* the number of days required to complete egg development at different temperatures; *b,* the reciprocal of curve *a,* representing the velocity of development. Note how the velocity increases with rising temperature up to about 29°C. and decreases at higher temperatures; 29°C. is the optimum temperature. *B.* The effect of temperature on the rate of respiration in a species of fly: *a,* in the larval stage; *b,* in the pupal stage; *c,* in the adult. The rates of metabolism, as measured by the rates of oxygen consumption, are very different for the three stages of the life cycle, but in all of them metabolic rate increases with rising temperature. The curves for the larval and adult stages again show the existence of an optimum temperature; above the optimum the rate of metabolism decreases again. How would you expect the optimum temperature to relate to the distribution of the species in nature?

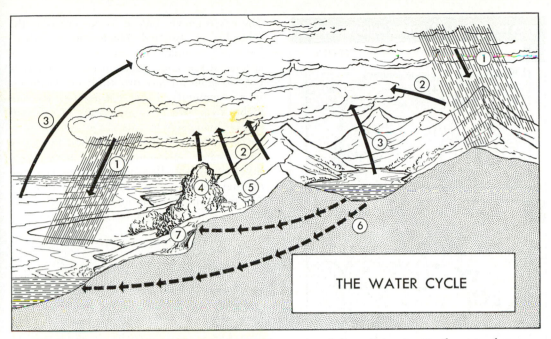

23–2   The water cycle. *1.* Precipitation as rain or snow. *2.* Immediate return to the atmosphere through evaporation. *3.* Evaporation from bodies of water—streams, lakes, oceans. *4.* Transpiration from plants. *5.* Transpiration from animals. *6, 7.* Drainage from streams and ground water, ultimately to the ocean.

would eventually happen if trout eggs, for instance, floated?) The fact that some animals have floating eggs or larvae and some plants wind-borne seeds also suggests another important effect of water and air currents: the geographic spread of organisms, a topic to which we will return (Chapter 28).

Microclimates and niches

Great care is taken in gathering the weather data on which descriptions of climates are based (climate is weather averaged over longer periods). Temperature is measured in a shady, ventilated, elevated shelter, away from heat reflections or cold pockets of the ground, pavements, or walls. Wind velocity is taken where gusts and eddies do not disturb. The recorded humidity is taken distant from the ground or any local factors that might influence it. But these are not really the conditions under which we or other organisms live. The weather, and hence the climate, of our immediate surroundings may be quite different from that recorded by the Weather Bureau. For an organism a few inches under the soil the temperature is usually different and

always less variable than for one on the surface. The climate on the floor of a forest is decidedly cooler, less sunny, more humid, and less windy than the climate in the tops of the trees (Fig. 23–3). Organisms generally have small climates, *microclimates,* of their own, which are quite diverse and more or less different from the idealized regional climate. Similarly, particular organisms do not really live in a regional environment, such as a "forest environment," but in small environments of their own. Animals burrowing in the mold, running along the ground, lurking under bark, or flitting through branches certainly have very different personal environments, both physical and biotic. The microenvironment of a particular species is one aspect of its *niche.*[2] Every regional environment has a large number of different niches.

[2] The word is French but is anglicized with the pronunciation "nitch" (not "neesh"). Some ecologists tend to confine the term to the physical environment or to think of it as the place where life is lived. However, it is also common and preferable usage to apply it to the whole of the microenvironment and further to the whole way of life of the species concerned.

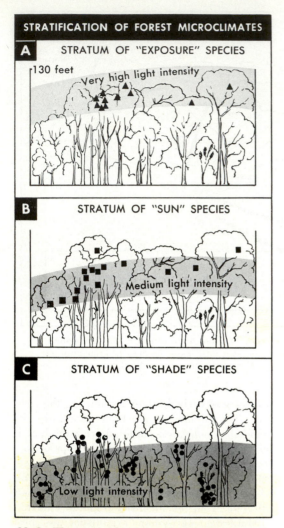

STRATIFICATION OF FOREST MICROCLIMATES

A STRATUM OF "EXPOSURE" SPECIES

130 feet
Very high light intensity

B STRATUM OF "SUN" SPECIES

Medium light intensity

C STRATUM OF "SHADE" SPECIES

Low light intensity

23–3 The stratification of forest microclimates. The existence and ecological significance of differences in microclimates is well illustrated by the vertical distribution in a forest of epiphytic ferns, orchids, and bromeliads that grow on the trees' branches. Each epiphytic species has its particular requirements for temperature, moisture, and light. The marked vertical gradients in these environmental factors result in an equally pronounced stratification in the distribution of the various epiphytic species. The figure shows the vertical distribution of three groups of bromeliad species in a South American rain forest: one of the groups (*A*) occurs only in the highest levels of the forest, where its demand for extremely high light intensities is met; another (*B*) demands moderate light intensity; and the third (*C*) requires shade and high humidity.

The environment, indeed, is not precisely the same for any two individuals, but it tends to be closely similar for members of the same species at the same stages in their lives. It is not the same for any two kinds of organisms in a community. Every species has its own niche. Even two microorganisms of different species living side by side in the soil are influenced somewhat differently by their inorganic and organic surroundings, and hence have different niches. How many niches can you distinguish in a community known to you? Are there any possible niches that are not in fact occupied by organisms?

The substratum

Some organisms, from protists to whales, spend their whole lives suspended in water, and some spend much of their lives (although none spends all of it) suspended in air. Nearly all plants and the majority of animals, however, rest on a bottom of some sort from which they project into water or air. They may be firmly attached, like a tree or a coral, or may be highly motile, like a man, but they are in contact with some surface most or all of the time. The surface to which they are attached or on which they move is their *substratum*. This, too, is a part of the physical environment that strongly influences the organisms on it. The plants of clay, sand, and rocky substrata are usually strikingly different. So, frequently, are the animals of sandy and of rocky stretches even in the same region. The shore life of mud flats, sandy beaches, and rock pools has even greater contrasts. The influence of the substratum is partly mechanical. Burrowing animals, for instance, will generally be found in a soft substratum, and sessile animals are more common on a hard substratum. The chemistry of the substratum, and of the environment in general, is also of vital importance.

The chemical environment: air and soil

Thanks to its constant and often rapid and turbulent motion, the composition of dry air is much the same almost everywhere: by volume about 78 per cent nitrogen, 21 per cent oxygen, and 1 per cent carbon dioxide.[3] Variations affecting organisms are those of water content, already mentioned, and local concentrations of gases, mostly due to human activities.

Soils and waters are markedly varied in their chemistry, and their differences profoundly affect organisms. Probably the most obvious difference is that between sea and fresh water, some of the effects of which have been discussed (Chapter 3). Fresh water does have large amounts of mineral salts in dilute solution, and sea water is not the strongest possible (saturated) solution. But some salt lakes are saturated, and have few and peculiar organisms in consequence. Every gradation between fresh water and salt lakes exists, and differences in salt content are reflected by the presence of different plants and animals and sometimes by the same organisms' assuming different forms.

Soils have a special interest because the lives of nearly all land organisms, including man, depend largely on soil. The bulk of a soil is composed of grains of minerals, especially a mixture of silica (the mineral of common sand) and clays. The mixture is formed by disintegration and decomposition of underlying rocks, or from silts and other sediments washed in by streams or blown by the wind. Among the particles of the soil are spaces, generally from a third to a half of the volume of the soil, occupied by air or water. Water in the interstices of soil is the source of most of the great amounts of water required by land plants. It is also a complex solution from which are drawn many of the other materials incorporated in all sorts of land organisms.

Soils well illustrate the fact that the contrast between physical and biotic environments is not absolute. These aspects of the environment also interact. Soil is penetrated by roots, which change it both physically (for instance, by loosening up its packed particles) and chemically (for instance, by withdrawing mineral salts from it). As plants die, parts of their organic materials are incorporated in the soil. Those organic materials, as well as parts of living plants in the soil, provide food for incredibly huge numbers of bacteria, algae, fungi, nematodes, earthworms, and other organisms. These soil organisms further modify the physical and chemical characteristics of the soil and add their excretions and dead bodies to its contents. Soil is thus itself a populous and complex community in which the inorganic and the organic interact endlessly and inextricably.

THE BIOTIC ENVIRONMENT

By definition, the biotic environment of an organism includes all the living things that affect it. The living things that affect any one organism are certain also to affect each other, directly or indirectly. They are likely, also, to exist within a circumscribed area at any one time. They are, in short, a community in themselves or a part of a larger community. Discussion of the organization of communities and of relationships in them is therefore, from another point of view, a discussion of biotic environments. They are a main subject in this and several following chapters.

Here at the outset one major distinction may be made among the various members of a community viewed as the biotic environment of any particular organism in it. There is a fundamental difference between the environmental relationships of members of different species and those among members of the same species. The members of different species have different roles; they occupy different niches. They constitute an *interspecific* ("among species") *environment*. Members of the same species have, if not quite the same, at least more closely similar roles and niches. For any individual of the species they are its *intraspecific* ("within species") *environment*. Their activities are interrelated and organized in a way unlike the relationships among different species.

Population and Environment

We have repeatedly noted that different kinds of organisms have different environmental needs or, to put it in a converse way, that different environments (or different niches) provide opportunities for different kinds of organisms. Many such relationships

---

[3] Plus relatively very small amounts of rare gases such as neon and helium, which man has learned to extract and to turn to his own uses but which are of little or no importance to any other organisms.

are indeed obvious to anyone who stops to think about them. Here we shall mention some of the principles that emerge in this new context. They arise from facts already becoming familiar to you, and some of them will be further developed in later chapters from different points of view.[4]

RANGE, LIMITS, AND OPTIMUM

There is no such thing as a constant environment. The deep sea is the most nearly constant major environment, but even there some changes occur. The deep sea is also the most difficult major environment, for it is characterized by low temperature, great pressure, no light, and hence no production of food within the community. It is perhaps the fact that these conditions, although stringent, are relatively unvarying that has made adaptation to them possible to them at all. Elsewhere change is frequent and striking: summer and winter in temperate and frigid zones; rainy and dry seasons in the tropics; daily temperature changes, storms and tides, rain and drought, strong and feeble sunlight, and many other changes everywhere.

Every individual and every species must be able to live under not one, fixed environmental circumstance but a whole *range* of environ-

[4] The subject we are starting to consider in this chapter is broadly called *ecology*. The study of the relationships of specific populations, hence of the individuals composing them, with their environments, especially physical environments, is generally included in *autecology*. The study of relationships among specifically distinct but associated and interacting populations in a community is often called *synecology*. One aspect of synecology, that of the community as a system (*ecosystem*) in which materials and energy pass from one group to another, is taken up later in this chapter, and other aspects are discussed in Chapter 24. Ecological biogeography, the subject of Chapter 27, is also sometimes included in synecology. Chapter 25 deals with interactions *within* populations and factors determining the sizes of populations. These subjects are included sometimes in autecology, sometimes in synecology, and sometimes in a subscience apart, *population ecology*. Ecologists have been particularly concerned with terminology and classification, producing a confusing array of disputed terms and definitions. We are following what seems to us a reasonable sequence in this intricately ramified science, and we shall do so with as little emphasis as possible on the complexities and conflicts of a highly technical but not yet standardized terminology.

mental circumstances. Nevertheless, the range within which life can continue indefinitely always has *limits*. No plant and only the relatively few homeothermous animals can remain active long in environmental temperatures below their own freezing points, and hence no species lives continuously in environments constantly below about 0°C. All organisms die from heat at temperatures well below 100°C., the boiling point of water. Different species differ markedly as to both the extents of their temperature ranges and the positions of those ranges on the temperature scale. Pine trees survive both greater cold and greater heat than bananas; their total range is wider.[5] The algae of warm springs cannot live at all in the range of arctic algae; the positions of the ranges are different. Similar differences in range exist for all environmental conditions—salinity of water, acidity or alkalinity of soil, light intensity, and the rest.

Within its range, each organism and each species generally has a point or a much more limited range where it does best, its *optimum*. Examples of optimum temperatures for growth in insects were given earlier in this chapter, and it is noteworthy that they are strikingly different. Similar optima exist for all environmental factors and all vital processes. This is another reason why every species has a range, and not merely an optimum, for each environmental condition under which it lives. It would be a rare coincidence if the optimum for one condition, say temperature, were invariably accompanied by the optimum for another, such as humidity. The existence of optima and the unlikelihood of different optima in a single environment stress again the importance of homeothermy and other controls of the internal environment (Chapter 12). Such controls keep the cells and organs of a body within their ranges and near their optima. Thereby the ranges of external environmental conditions within which an organism can survive are widened.

TOLERATION AND PREFERENCES

We say that a plant or animal *tolerates* the range of environments within which it can live and that it *prefers* environments in all respects

[5] Species with wide ranges are technically called *euryecious*, and those with narrow ranges *stenoecious*.

nearest its optima. Although the words "tolerance" and "preference" are scientifically acceptable in this connection, they are dangerously anthropomorphic: we can never for plants and rarely if ever for animals assume that they mean at all the same as when we apply them to ourselves. Yet we can accept them as vivid metaphors applicable to ranges and optima. It was noted in Chapter 1 that ponderosa pines are intolerant of shade, and horticulturalists often classify trees as tolerant or intolerant, accordingly. Cacti are tolerant of dry, sandy soil; irises are not. Clams are tolerant of cool, brackish, muddy water; corals are not. Additional examples are so numerous that you can supply hundreds for yourself.

It seems a truism to say that plants and animals occur only in environments they tolerate and predominantly in environments they prefer.[6] In fact, the occurrence of many species seems to be determined more by tolerance than by preference. Sagebrush (*Artemisia*), for example, grows most luxuriantly under wetter conditions than those in which it predominates. Its optimum would apparently be in regions of more abundant and stable water supply, but these are exuberantly occupied by competitors that do not tolerate drier conditions. Hence it grows not where it prefers, but where it has least effective competition. It is probably true of many other organisms, especially those in unusually rigorous environments such as deserts or the deep sea, that they live not where they would do best (as regards *physical* conditions) but where fewer competitors are able to follow them. They are, of course, still balancing optima: their physical optimum does not coincide with their biotic optimum.

ADAPTATION AND ADAPTABILITY

The means by which species meet their needs in different environments and by which environmental niches are parceled out among different species are almost incredibly varied. Some of them have already been discussed in

Chapter 16, in connection with their evolutionary explanations and implications, and others will be discussed in Chapter 24. We mention them here only to emphasize their ecological pertinence, thus tying up another of the many interconnections that unify the whole widely branching subject of biology. You will recall that the ability of a species to tolerate particular ranges of environmental conditions is achieved in the course of evolution by three processes: (1) genetic adaptation of a whole population to develop normally and reproduce in a certain environment; (2) genetic variability within a population, giving demes and individuals different tolerances and optima and thus permitting the population as a whole to maintain itself in changing environments; and (3) individual adaptation (in genetically determined reaction ranges), permitting adjustment to changed conditions even within an individual lifetime.

Still another aspect of tolerance, also already familiar in other contexts, is seen in life cycles (Chapter 9) and in the ability to suspend animation in environmental conditions that could not otherwise be tolerated. Annual plants survive the severe "temperate" winter as seeds, and annual insects as eggs or pupae. Rotifers survive the drying up of their temporary aquatic environments as inactive eggs, and various protists survive similar vicissitudes by becoming encysted. Even such high forms as mammals hibernate, with greatly lowered metabolism, when lack of food and intense cold would lead to death if their regular activities had to be continued.

COMMUNITY AND ENVIRONMENT

For the rest of this chapter and all of the next we shall consider another phase of organism–environment interaction—that involving entire, integrated communities and relationships among rather than within populations. In Chapter 1 the traffic in materials and energy in living things was introduced as a theme to be followed up later. We now turn

[6] Ecologists have coined many terms to describe tolerance and preference for particular conditions. Those with the suffix -*phile* (from the Greek for "lover") indicate preference, and those with the suffix -*phobe* (from the Greek for "fearer") indicate intolerance, thus: *hydrophile*, preferring a wet environment; *xerophobe*, intolerant of drought. Other such terms, with other roots also variously combined, are these: *dulcicole*, preferring fresh water; *calcifuge*, intolerant of limy soil.

to a detailed consideration of that traffic in communities.

CYCLES OF MATERIALS

The carbon cycle

The flow of carbon in communities is one of the most essential features of their metabolism of organic compounds. The great reservoir of carbon in nonorganic form and the source of almost all the carbon incorporated in organisms is the CO_2 in the atmosphere and dissolved in the waters of the earth. The usual first step in the utilization of carbon as a material in living things is photosynthesis by green plants (Chapter 4). The carbon thus becomes part of simple carbohydrates, and later syntheses in the same plants transfer part of it into polysaccharides, proteins, lipids, and other complex organic compounds. Animals eat plants, and the organic compounds are digested and re-synthesized (Chapter 4). Other animals eat the meat-eating animals, with still more digestions and resyntheses. Thus carbon is transferred from one organism to another through a shorter or longer, sometimes very long, sequence. In the course of the sequence, and even within any one organism of the series, the carbon atoms are constantly shifted from one kind of molecule to another. But as long as it is a vital part of an organism the carbon is in some organic compound of greater or less complexity.

Eventually most of the organic carbon becomes a part of CO_2 again and is returned to the inorganic realm of water and air. This return phase is an essential part of the cycle that has kept life going since early in its history, as the available carbon has continuously circled from air and water through plants and animals and back to air and water again. Some of the return is fairly direct. CO_2 is an end product of respiration in both plants and animals, and the respired CO_2 passes at once into the water or air of the organism's immediate environment, where it is available to start the cycle all over again.

Much carbon remains in the tissues of organisms when they die, or is eliminated by animals in waste products that are still fairly complex organic compounds, not usable as a carbon source in photosynthesis.[7] If this carbon were not somehow converted into CO_2, life would have come to an end by now. All the available carbon would be locked up in organic but nonliving form. Here is the role of the organisms of decay or putrefaction, most of which are bacteria or fungi. They attack and digest the organic materials of dead plants and animals and of excretions, reducing them to the simpler and energy-poor compounds with which the various cycles of materials and energy begin. After their work is completed, most of the carbon of organic compounds has become CO_2 again.

There are some other ways in which organic compounds can be finally broken down. Most familiar is fire or slower combustion (oxidation). When wood is burned, for instance, the carbon of its cellulose and other materials is removed from the organic compounds as CO_2. But combustion, aside from that now caused by man,[8] is not a continuous and widespread process in nature. It would not suffice to keep the carbon cycle, or other organic cycles, going. This is an extremely important principle: the continuity of organic cycles through final breakdown of organic compounds into raw materials for renewed synthesis by green plants depends on the activities of living things. That is certainly true now and has clearly been true during most of the history of life. (Can it, however, have been true at the very beginning of life?)

Some carbon is withdrawn from the cycle for long periods, if not permanently. Not all the organic compounds of dead organisms have decayed; some have been incorporated in the crust of the earth as coal, petroleum, and natural gas. By burning these, man makes their carbon again available for the organic cycle. Before man, there was little return of this stored carbon, even though some of it was liberated by combustion and by bacteria. An even larger amount of carbon is locked up in

[7] Urea, for instance, an abundant animal excretion, contains carbon. It is $CO(NH_2)_2$. Uric acid and some other excretions contain still larger percentages of carbon. Urea is utilized by some algae but not in their photosynthesis.

[8] How was coal and wood combustion initiated before the evolution of man, and how is it initiated today apart from man?

limestone, the principal mineral of which has the composition $CaCO_3$. Much, but by no means all, limestone is a result of the activities of life. Coral reefs provide a clear example of life-made limestone. The carbon of limestone may be released as CO_2 by natural processes, such as the action of weathering and natural acids, but much limestone is now deep in the earth, where its carbon will not be available for a very long time, if ever.

The most striking features of the whole carbon cycle are summed up in Fig. 23–4.

The nitrogen cycle

Nitrogen is no less essential to life than carbon. It is, as you know, part of all amino acids and proteins. Its cycle is similar to that of carbon and goes on at the same time, but there are some distinctive features.

As we have seen (Chapter 2), nitrogen makes up the greatest part of the atmosphere, but green plants can use little or no atmospheric nitrogen (N_2) in their syntheses— a marked contrast with carbon, which (as CO_2) is mainly derived from the air by green plants. Green plants require nitrogen as ammonia (NH_3) or in the form of nitrates (salts containing NO_3), and these are obtained from the nitrogen-fixing bacteria of soil, which can utilize N_2 directly in the synthesis of their own amino acids and proteins (Chapter 2). Some nitrogen-fixing bacteria live independently in the soil, and some live in nodules in roots of other plants, especially the legumes, members of the pea family. In either location, their death frees nitrogen compounds, which can be utilized by other plants, and particularly green plants.[9] The nitrogen of green plants, thus acquired either from ammonia and nitrates in the soil or from nitrogen-fixing bacteria, may be passed on to animals that eat plants, and then from animal to animal, as carbon is. As with carbon, too, nitrogen is excreted by animals (in urea, for instance) and also remains in the tissues of dead organisms. Again, it is mainly bacteria that return this locked-up nitrogen to the cycle. Bacteria of decay produce ammonia from proteins and other nitrogenous compounds. Other bacteria, nitri-

[9] There is evidence also that legume roots secrete nitrogen-containing compounds directly into the soil.

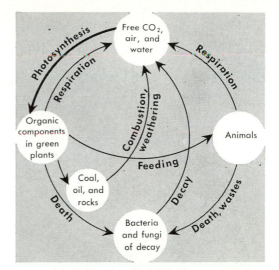

23–4 The carbon cycle.

fying bacteria, oxidize ammonia to nitrites (Chapters 4 and 19), and still others perform further oxidation to nitrates. Thus inorganic nitrogen compounds directly utilizable by green plants are restored to the soil.

Another distinction of the nitrogen cycle from the carbon cycle is that it does not necessarily or regularly involve a phase during which the nitrogen is in the atmosphere. Such a cycle as nitrates in soil → green plants → nitrifying bacteria → nitrates in soil, and so on, can continue indefinitely without the nitrogen's ever occurring as a gas, N_2, in the air. An N_2 phase may occur in a cycle, however, and when it does the return of N_2 to the atmosphere is brought about by *denitrifying bacteria.* They break down nitrates (and some other compounds) and liberate N_2. Whatever nitrogen does not simply remain in the atmosphere and does in fact continue the cycle is then returned to the soil by the action (mainly) of the nitrogen-fixing bacteria, which we met on our last time around the circle.

It may be a little misleading to speak of *the* carbon cycle and *the* nitrogen cycle. You can see that the pattern of flow, even simplified as it is in Figs. 23–4 and 23–5, does not have a single, circular course that carbon or nitrogen necessarily follows. The patterns are more complex, each with a number of different paths that the two elements may follow in a cyclic manner. The simplest course for nitro-

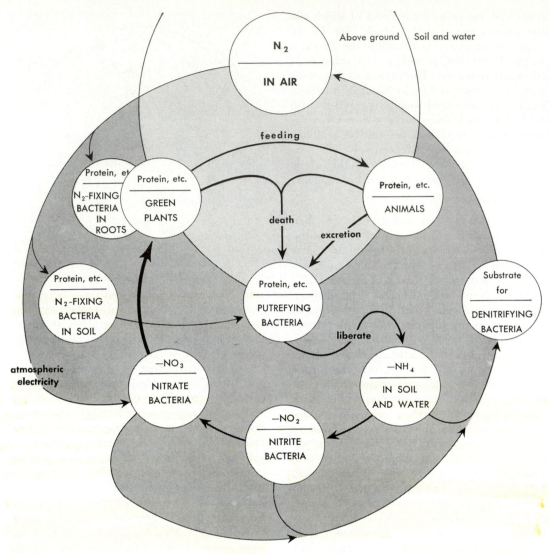

23–5   The nitrogen cycle.

gen is soil → plants → bacteria → soil, and so around again and again. What is the simplest cycle for carbon?

Consideration of all the organic cycles shows that green plants and putrefactive bacteria *must* be present if life is to continue indefinitely with a cyclic flow of materials and energy. Animals are unnecessary. In the total metabolism of life they are a side issue or an extra step that complicates the process without really contributing to it. From *this* point of view, at least, should not we animals be a little more humble?

Mineral cycles

You learned in Chapter 2 that the principal inorganic materials that become incorporated in living things are water, carbon (generally as CO_2), oxygen, nitrogen, and a variety of mineral salts. The carbon and nitrogen cycles have now been especially considered. Analogy with these cycles and what you have learned elsewhere in this and previous chapters are sufficient for you to work out the essentials of the water and oxygen cycles. (The water cycle is also summarized in Fig. 23–2.) The mineral cycles have some features of special interest and merit a diagram (Fig. 23–6) and some additional comments. Before reading these comments, review Chapter 2, on the inorganic materials of life.

With few and unimportant exceptions, the inorganic sources of all the mineral elements necessary to organisms are salts in solution in water, whether in the soil or in bodies of water. (Even so-called fresh water is in fact a dilute solution of many mineral salts, among other things.) The salt-solution phase is a part of two great cycles in nature, cycles that interlock through this phase. One, the *rock cycle*, is inorganic in essence, although it is strongly influenced by organisms, and organisms may even have direct parts in it. The mineral salts of the earth came originally, and more are still coming, from the crust of the earth. They are formed and liberated from rocks of the crust mainly by processes summed up as weathering: disintegration, and decomposition under the influence of air, water, and organisms. The soluble salts arising among the products of these processes then enter another of nature's cycles, the *water cycle*. With the water, they move through soil, streams, and lakes and eventually into the sea. As they pass through this cycle, perhaps with long stops during the trip, they are available to organisms in all the environments of life. Most of them eventually reach the oceans, where much salt remains indefinitely. Some of the mineral salts do complete a cycle and return to the crust of the earth through the processes of sedimentation.[10] They are incorporated in limestones, silts, salt beds, and other sedimentary deposits that become parts of the earth's crust. These, in turn, may weather, and their mineral salts may begin the cycle again.

In the organic cycle of the mineral salts, plants, animals, and bacteria may all acquire salts from the inorganic solutions of their environments, although this activity is most extensive in plants. All may also return salts directly to the inorganic worlds of soil and water; plants are least active in this respect. Animals also acquire salts from plants and

[10] They do not to any considerable extent go through the return phase of the water cycle, which takes place by evaporation and does not carry along appreciable amounts of minerals.

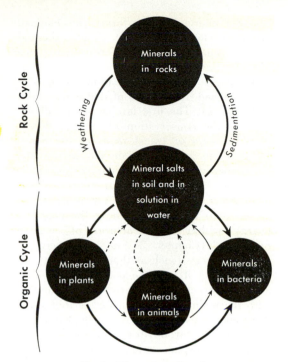

23–6 The mineral cycle.

from other animals eaten as food. Putrefactive bacteria and other organisms of decay acquire salts from dead plants and animals and, as in all organic cycles, form the last organic link in the longest, most complex sequences.

TRANSFERS OF ENERGY

We learned in Chapters 2 and 4 that all the activities of the living organism require an expenditure of energy—running a race, moving a hand, the heartbeat, even thinking. Even a plant standing motionless in a field is continually using energy in the syntheses within its living cells. The utilization and transfer of energy as fundamental features of the living world were stressed in these early chapters, where our emphasis was on the cell and the individual organism. Here we are concerned with energy transfers again—this time in relation to whole populations and communities of organisms.

Two generalizations about energy—the capacity to perform work—may be recalled from Chapter 2: (1) Energy may be potential or kinetic. A boulder on a hilltop has potential energy that becomes kinetic ("pertaining to motion") when it rolls downhill. (2) Energy assumes many different forms [mechanical, chemical, electrical, radiant (light), thermal (heat)] and may be transformed from one kind into another.

Transfers and transformations of energy are governed by the laws of thermodynamics (Chapter 2). The first law, as we recall, relates to the conservation of energy. Whenever transfers or transformations of energy occur, there is neither gain nor loss in the total energy involved in the transaction.[11]

As we have seen, the significance of the first law of thermodynamics for living things lies in their capacity to release the potential energy within the structures of organic molecules (Chapter 3) or conversely to capture energy (such as that of sunlight) by transforming it into the potential energy of chemical structure. You are already familiar with the main metabolic processes that make these energy transfers possible: the exergonic reactions of carbohydrate metabolism and oxidative phosphorylation (p. 124) and the endergonic reactions of photosynthesis (p. 113).

The second law of thermodynamics is equally important in the processes of life. We considered it in rather formal terms in Chapter 3. The second law tells us that as energy is transferred from one substance to another or transformed from one form to another, less and less of the total energy is utilizable in further transfers and transformations. Although the *total* amount cannot change, the amount that can perform work of any sort, chemical, mechanical, or other, becomes steadily smaller. The *usable* energy in a sequence of transfers tends to run down, and the whole process will come to a stop unless there is a continuing input of energy from somewhere.

We reintroduce the laws of thermodynamics simply to emphasize again their all-important significance for the activities of living things. In the cycles of *materials*, such as carbon or nitrogen, nothing is lost or

necessarily becomes unusable. All the materials that start the cycle are (or, at least, can be) returned to their original form after going through the cycle, and then they are ready to start around again. There is no reason why the process should not go on forever without addition of anything from outside. *This is not true of energy.*

In every organism, energy transfers occur through many chemical reactions, and chemical energy is also transformed into other sorts of energy, such as heat, motion, or (but this is comparatively unimportant in organisms) light. In accordance with the first law of thermodynamics, the total activity can involve only the amount of energy that the organism receives. An organism cannot generate any *new* energy. In accordance with the second law, much of the energy received is made unusable by the activities of the organism. The energy received cannot be destroyed (that would violate the first law), but it is *dissipated*, scattered in such forms (mainly as heat transferred to the environment) that it can no longer be used by that organism or by others.

Thus organisms pass on to others less energy than they received. When a herbivore eats a plant, it receives chemical energy, but the energy received is much less than the energy that the plant received from the sun. A carnivore, in turn, acquires chemical energy by eating a herbivore, but it receives from the herbivore much less energy than the latter received from the plants it ate. When, finally, the organisms of decay end the sequence, they pass on the materials of life in forms that are utilizable by other organisms, but they practically complete the dissipation of energy in the community. In fact, the flow of *energy* in a community is not a cycle at all. It is a one-way sequence in which vital energy, like all energy, follows the second law and becomes continuously less available.[12]

Since communities do keep going, and have for over a billion years, energy is obviously coming in to them continually from an outside source to compensate for what has been

[11] Everyone knows nowadays that matter can be turned into energy, as in an atomic bomb. This sort of reaction does not occur in living things and anyway is not a real exception to the law. It demonstrates only that matter itself is a form of energy. The only connection with life is that the sun's radiation involves the transformation of matter to radiant energy.

[12] Some authors of popular books claim that life does not follow the second law and draw questionable philosophical and even religious conclusions. Such authors have not thought the problem through.

lost. The source is the sun, and it, too, is subject to the second law. Some day all its energy will no longer be in usable form. Then life on earth will no longer be possible, but the event is billions of years in the future, so far away that some other catastrophe may wipe out the earth's life long before then. As to where the energy of the sun came from to begin with, the only honest *scientific* answer at present is, "We do not know."

CHAPTER SUMMARY

Environment: extrinsic things and conditions affecting an organism; physical and biotic environments; the physical environment—solar radiation, the water cycle, the movement of air and water, microclimates and niches, the substratum, the chemical environment, the air, the soil; the biotic environment—all living things affecting the organism, inter- and intraspecific environments.

Population and environment: the ranges of environmental conditions; limits at the ranges; differences in the ranges as to extent and position on the scale; optima and their relationship to range and to internal regulation; toleration and preference and their effects on the occurrence of species; adaptation and adaptability.

Community and environment: cycles of transfer of material and energy; the carbon cycle; the nitrogen cycle; mineral cycles; transfers of energy; the first law of thermodynamics—conservation of total energy; the second law of thermodynamics—dissipation of usable energy; unlike materials; the flow of energy in communities, one-way only, with continual renewal from solar energy necessary.

24

Organization
and Change
in Communities

© Douglas P. Wilson

The hermit crab makes use of an old snail shell as a house. The sea anemone feeds on leftovers from the hermit crab's meal. These are examples of the many interactions among species, as discussed in this chapter.

The Web of Life

A contemporary of Darwin's suggested that the glory of England was due to its old maids. This is the argument: The sturdy Britons were nourished by roast beef from cows, which ate clover, which was pollinated by bumblebees, which were attacked in their nests by mice, which were kept under control by cats, which were raised by old maids. The argument has become quaint with the passage of time, and perhaps it was always a little farfetched. It is, however, still valid in pointing out that the different species in a community are linked in many and curious ways. The sequence Briton ← cows ← clover is part of a chain by which materials and energy are transferred in a community. Another chain is present in bumblebees → mice → cats, which also involves the interspecific relationship of predation. The relationship between cats and old maids is a somewhat odd but legitimate example of what is called symbiosis. These and many other relationships unite all living things into a single fabric, the web of life. Since not all can be discussed at once, we shall start with the first-mentioned: food chains and some of their consequences.

Chains and Pyramids

FOOD CHAINS

In the last chapter we saw that a community as a whole is a dynamic system with flow of materials and energy, cyclic for materials and one-way for energy. This is a, or *the*, basic feature of a community as an organized association of different, interacting species. Any one sequence of species through which materials and energy pass is a *food chain*. With unimportant exceptions, all food chains start with *producers*, the photosynthetic plants that acquire energy from nonorganic forces and fix it in organic chemical form. All end with *reducers*, the organisms of decay, including fungi and especially bacteria.

The simplest chain is producer → reducer, but even so simple a sequence normally includes several species of reducers. Commonly one or many *consumers* intervene between producers and reducers. Almost all consumers are animals. (Why? What exceptions are

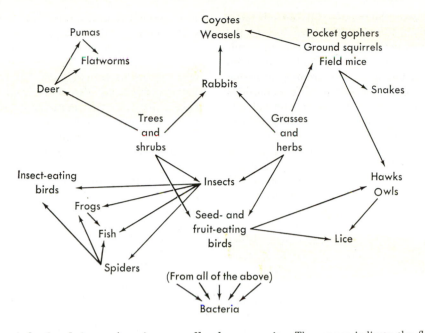

24–1 A food web in an American woodland community. The arrows indicate the flow of energy and materials (consumption of food). This diagram is suggestive only and is very incomplete. In a real community there are many more groups of organisms and many more levels and directions of consumption.

there?) There are different levels of consumers: level 1 consumers eat the producers; level 2 consumers eat level 1 consumers; and so on. Thus a usual food chain runs producer → level 1 consumer → level 2 consumer → reducer; or green plant → herbivorous animal → carnivorous animal → bacteria; or, in a still more specific example, willow → deer → puma → bacteria.

In a natural community the flow of materials and energy is much more complicated than is suggested by any one food chain. Practically all species of organisms may be consumed by more than one other species; and although some consumers and reducers obtain food from a single species, most consume numerous species. One species of plant may provide food for many species each of insects, birds, rodents, ungulates, worms, fungi, and bacteria; and one of the species of, say, rodents living on those plants may in turn be consumed by any of a dozen or more species of carnivorous snakes, birds, and mammals. Thus chains both converge and branch. There are so many branchings and cross-connections among the food chains of a whole community

that the pattern of transfer is actually a *food web* rather than a series of readily distinguishable chains (Fig. 24–1).

PYRAMIDS OF ENERGY AND MASS

In spite of the tremendous complexity of a food web, the partial chain producer → level 1 consumer → level 2 consumer, or plant → herbivore → carnivore, does sum up major aspects of food flow in any community. Reducers, although essential parts of all food chains and material cycles, complicate the picture because they derive food from *all* links in all chains and not only (in fact, least of all) from the last consumer link. They are therefore mostly omitted from this discussion of *pyramids*.

Each link in the partial chain has less available energy than the previous link, as follows from the second law of thermodynamics. The total energy of the plants in a community is greater than that of the herbivores, which in turn is greater than that of the carnivores. The total energy of the reducers is less than that of all the rest of the organisms put together, but not necessarily less than that of any one link.

The total bulk—the mass—of living substance also tends to decrease from one link to the next. This reduction is not a necessary result of physical law, as is the reduction in available energy, but it is an extremely probable result and seems to occur in all communities except those temporarily under unusual circumstances. If, for instance, you raised sheep for food and had nothing else to eat, your mass would be greater than that of the sheep if you ate the last one, but then you would starve to death. To keep going, you would have to have a flock of sheep with mass continuously much greater than yours. This is about the simplest possible illustration of the generalization that in a continuing community the total mass of later links in food chains is less than that in earlier links.

Thus each successive step in food utilization is smaller than the last, and the over-all pattern is like a pyramid, largest at the bottom (producer level) and progressively smaller toward the top (consumer levels), as shown in Fig. 24–2.

THE PYRAMID OF NUMBERS

A concept related to the pyramids of mass and energy and sometimes confused with them (although it is quite distinct) is what animal ecologists call "the pyramid of numbers." If a census is taken of animals of different sizes, smaller animals are generally found to be more numerous than larger ones (Fig. 24–2, bottom). You can probably verify this from your own experience. In any natural community known to you, what is the comparative abundance of animals of the sizes of insects, mice, and cats or dogs?

The pyramid of numbers depends in some instances and in part on the pyramid of mass. Predacious animals, those that pursue and kill prey, generally eat animals smaller or at least not much larger than themselves. Since the predators are higher in a food chain than their prey, according to the principle of the pyramid of mass the *total* bulk of the predators is considerably less than that of their prey. Therefore the predators are usually fewer than their prey. The pyramid of numbers also frequently applies, however, regardless of predator–prey relationships or of the positions of the animals in food chains. A large animal requires more food (of whatever sort it may be) than a small one, and so may have to monopolize more territory if it is to survive. Furthermore, in any given area there is food for more small than large animals. Large animals are also able to and usually do range more widely than small animals. A larger animal simply does occupy more space; think how many mosquitoes could and sometimes do occur in space the size of an elephant.[1]

Do you think that the pyramid of numbers applies to plants? If so, why? (Remember that no two green plants are successive links in a food chain.)

BUDGETS: THE LIMITATION OF LIFE

The principles of cycles of materials, of energy transfers, of food chains, and of mass and energy pyramids have important implications for the abundance and activity of the life of the earth as a whole and of living things at any one place and time. It has been shown that practically all the energy and a great part of the materials in all living things must pass through green plants. They are the basic source of foods, but they are also the bottleneck of life's activities. The total activity of life, the flow of energy through all living organisms, can proceed no faster than the fixing of that energy by photosynthesis. Photosynthesis can proceed no faster than the inflow of radiant energy from the sun. As a matter of fact, photosynthesis is very much slower. Only a small fraction of sunlight that reaches the surface of the earth is transformed into chemical energy by photosynthesis. The available data are highly inexact, but the figure must be well under 1 per cent.[2] Whatever the precise figure may be, it represents all the energy available for all the living things on earth.

[1] Another reason sometimes given for the pyramid of numbers is that smaller animals have a higher "reproductive potential," but this probably has no bearing. Larger animals do usually reproduce more slowly, but still the rate is sufficient to supply the largest population that could possibly live in any available area. As far as the reproductive potential is concerned, elephants have been in Africa long enough to be as numerous there as ants—but there is not food or room for that many elephants!

[2] See Chapter 23. The value at Lake Mendota was 0.8 per cent.

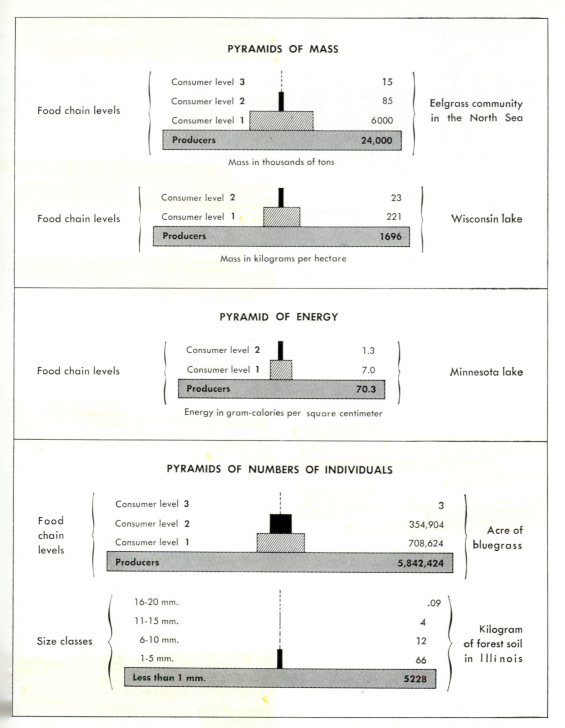

PYRAMIDS OF MASS

Food chain levels
- Consumer level 3 — 15
- Consumer level 2 — 85
- Consumer level 1 — 6000
- Producers — 24,000

Eelgrass community in the North Sea

Mass in thousands of tons

Food chain levels
- Consumer level 2 — 23
- Consumer level 1 — 221
- Producers — 1696

Wisconsin lake

Mass in kilograms per hectare

PYRAMID OF ENERGY

Food chain levels
- Consumer level 2 — 1.3
- Consumer level 1 — 7.0
- Producers — 70.3

Minnesota lake

Energy in gram-calories per square centimeter

PYRAMIDS OF NUMBERS OF INDIVIDUALS

Food chain levels
- Consumer level 3 — 3
- Consumer level 2 — 354,904
- Consumer level 1 — 708,624
- Producers — 5,842,424

Acre of bluegrass

Size classes
- 16-20 mm. — .09
- 11-15 mm. — 4
- 6-10 mm. — 12
- 1-5 mm. — 66
- Less than 1 mm. — 5228

Kilogram of forest soil in Illinois

24–2 Pyramids of mass, energy, and numbers of organisms in a community. Producers are the green plants, primary converters of environmental resources into living material. Consumers are animals. The levels represent the sequence in a food chain: consumer level 3 feeds on consumer level 2; consumer level 2 feeds on consumer level 1; and consumer level 1 feeds directly on the producers. (Reducers are omitted in these pyramids.)

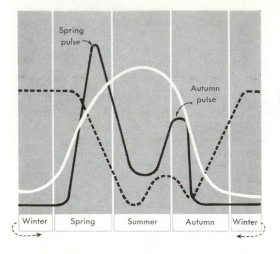

Spring
pulse

Autumn
pulse

Winter | Spring | Summer | Autumn | Winter

24–3 Seasonal fluctuations in diatom abundance in the North Atlantic. Solid black line, diatoms; dashed line, nitrates and phosphates; solid white line, light intensity. *Winter.* Surface waters are cold and poorly illuminated. Under these conditions diatom growth is inhibited, and, as a consequence, the concentration of nitrates and phosphates in the surface waters increases (during diatom growth nitrates and phosphates are incorporated into protoplasm). *Spring.* Surface waters warm up, and light intensity increases. These optimum conditions for photosynthesis, combined with a high concentration of nitrates and phosphates, lead to a spectacular spring pulse of diatom growth. *Summer.* The diatom population declines for two reasons: (1) diatoms are consumed in huge numbers by herbivorous animals; and (2) the concentration of necessary nitrates and phosphates decreases. This limitation of essential minerals itself has two causes: (1) the nitrates and phosphates have been subject to removal through the earlier diatom growth; and (2) they have not been renewed from the depths of the ocean. The main source of nitrates and phosphates is the dark, deep ocean bottom where bacteria-caused decay is high and photosynthesis is nonexistent. Actual return of the minerals to the main zone of life—the surface—depends on their upward movement from the bottom. The complex distribution of temperature conditions in the summer ocean interferes with the upward movement and thus temporarily contributes to a nitrate–phosphate deficiency that in turn contributes to a fall in the diatom population. *Autumn.* The temperature conditions of the ocean change, leading to a renewal of nitrates and phosphates at the surface and an upsurge in diatom growth. This autumn pulse in the diatom population is, however, never as great as the spring pulse because temperature and light intensity have dropped, thus limiting the rate of photosynthesis. *Winter.* Growth slackens because of the light and temperature conditions, and consequently the nitrates and phosphates accumulate to the high concentration that will again make possible a spring pulse.

Limitations of vital activity by the budget of available materials are also strong and may be more obvious in particular localities. On land the most apparent limitation is often in the water budget, as water supply is highly variable and water is the material needed in largest quantities by all living communities. The comparative scarcity of life in the sun-drenched desert is obviously not due to deficiency of solar energy but to limitation of water. The most common budget limitation in the soil is in nitrates (principal source of nitrogen) and phosphates (essential mineral salts). Agriculturalists well know that on heavily farmed land these are the principal materials that have to be renewed by fertilization. Differences in natural supplies of these materials are also reflected in different abundances of land life.

In aquatic environments there is practically no limitation of water budget, but other materials, especially, again, nitrates and phosphates, may be stringently limited. Great quantities of these materials are washed into the sea from the land, which is why marine life is particularly abundant along shores and off the mouths of rivers. Farther out to sea, nitrates and phosphates are rapidly used in upper levels of the water, where photosynthesis occurs. The return of nitrates and phosphates to the cycle by decay is more active at deeper levels, as dead organisms sink through the water and are decomposed by bacteria. They may be returned to the sunlit surface waters by rise and diffusion from deeper water or by the upwelling of deep currents along coasts. There is, for instance, such a zone of upwelling along the coast of Peru, marked by an extraordinary richness of marine life. Combined limitations of energy and material budgets are particularly well illustrated by fluctuations in the abundance of diatoms in the North Atlantic (Fig. 24–3). Since diatoms, in spite of their small size, are the most important photosynthetic organisms of the open sea, the whole life of the ocean community changes with fluctuations in their numbers.

Niches and Competition

The food web is the most pervasive feature binding a community into an organized system. There are, however, many other inter-relationships within that system. One involves the various ways of life, ecological niches, and how they are parceled out by competition. Others involve special associations, both helpful and harmful, between particular species. These further interrelationships will now be discussed in sequence.

RELATIONSHIPS AMONG NICHES

In some of the driest deserts of Arizona grows the gigantic cactus *Cereus giganteus*, commonly called "saguaro" (or "sahuaro," in either case pronounced sah-*wah*-roh). Although it is not a widespread plant and is absent from most American deserts, it is so striking and picturesque that paintings, decorations, and cartoons have made it the recognized symbol of the Southwest. There are two birds, a flicker (*Colaptes chrysoides*) and a woodpecker (*Centurus uropygialis*), that cut round openings in the spiny, ridged stems of saguaros and excavate recesses for nests in the softer internal tissues. They make more holes than they keep in use, and often an unused flicker or woodpecker hole is occupied by elf owls (*Micropallas whitneyi*), dainty little creatures no larger than sparrows. An elf owl rarely nests anywhere else.

The saguaros, living only in deserts and only in particular parts of them, have their special niche among plants. They provide, in turn, one aspect of the niches of desert flickers and woodpeckers. These birds, in their turn, excavate the homes of elf owls, and these homes are one of the defining characteristics of the elf-owl niche. The interdependence of all these organisms is evident, and so is the fact that their roles in the community are quite different; they have, in fact, different niches.

Another owl, slightly larger than an elf owl but still small as owls go, also nests in woodpecker holes in saguaros: the saguaro screech owl (*Otus asio gilmani*). A saguaro screech owl and an elf owl may be found in adjacent holes on the same cactus plant. Here, you would be inclined to say, are two species occupying the same niche, but it is not so. The elf owl feeds mainly on small insects, such as beetles, ants, or crickets. The screech owl likewise eats some insects, especially the larger

grasshoppers and locusts, as well as scorpions, but also includes in its food mice and other rodents that are seldom attacked by elf owls. There are other differences in their activities, such as the fact that the screech owls raise their young earlier in the year than do elf owls. In short, despite the identity of their homes and the similarity of their habits, these two species of owls do have distinctly different total relationships to their environment or different roles in their community, which is a way of saying that they occupy different niches.

The desert scene illustrates a principle that operates equally in any community: each species in an established community has a distinctly different niche. Known exceptions occur when one species invades the territory of another, but the duplication is then temporary. There may be other exceptions, and some naturalists think there are, but none has been definitely proved and in any case they would be exceptions to what is plainly the general rule. The reason for the rule is *competition*. Equal sharing between different species is unknown in nature. If two species really have the same niche and hence the same requirements for their continued existence, they eventually compete for those requirements and sooner or later one of them wins out. Hence the principle involved can be stated in another way: two species do not long live together if one of them can fully utilize an aspect of the environment necessary to both of them.

Competition need not extend to all environmental necessities, and it seldom does. Elf owls and screech owls do not seriously compete for woodpecker holes. Different species of fishes in the sea do not compete for water. There is plenty to go around, and there are plenty of woodpecker holes in the desert.[3] Species may compete for only one crucial thing and share everything else quite amicably. Still they cannot long exist together if they continue to compete for *anything* that is vitally essential to them.

[3] There are, in fact, enough holes so that they are regularly occupied by still another species of bird, a flycatcher (*Myiarchus tyrannulus*), and casually by many other desert animals, including lizards and snakes.

Environmental requirements of two species in the same community may be so different that there is no real overlap. Then there is no question of competition. Saguaros and woodpeckers affect each other in many ways. Aside from the fact that woodpeckers live in saguaros, both are parts of some of the same food chains: saguaros → insects → woodpeckers. Yet the niches are so different that the idea of competition as a relationship between them simply does not arise. There is, on the other hand, close similarity or wide overlap in the niches of screech owls and elf owls. A factor becomes crucial when the competition for the same thing would eliminate one or the other. It must be this factor that limits the population of each, and thus keeps them down to the point where they do not compete for those things that they do, in fact, share.

This crucial factor is to be sought among the things that are different in the overlapping niches of two species. With the two desert owls, the crucial factor is evidently food. Even in food, there is some overlap because some kinds of insects are eaten by both. In times of abundance, this does not matter. In times of scarcity, when competition is fiercest and one or the other would eventually go to the wall, they can eat different things: especially small insects for the elf owls and larger rodents for the screech owls. So the competition is relieved and does not become lethal. Among the most closely similar and most frequently competing species of a community there is always (as it seems) some such safety valve. Birds may merrily share the abundant foods of late summer, and then, when the supply dwindles and competition really begins to tell, they will take to different foods or will migrate to different regions. There are many such escapes from lethal competition. What others can you think of?

NATURE OF COMPETITION

Almost anything needed from the environment may be the object of competition. Plants compete for water or sunshine, sometimes for mineral salts—whatever is in shortest supply at a given time and place. In the desert each mesquite bush is surrounded by a zone of completely bare soil from which it has ousted competitors for water. In the forest a dense growth of seedlings thins out progressively as a few more vigorous trees preempt the sunshine and others die in their shade. In the sea one species of photosynthetic organisms may locally drive out another as it wins in competition for nutrient nitrates or phosphates.

Among animals food is the usual object of competition. This lies back of the fact that the food habits of animals are so extremely diverse and frequently so specific, characteristics that reduce competition for food between different species. Animals may, however, compete for things other than food. They may also compete for water in a dry environment, or for desirable nesting places or shelters.

We are inclined to think of competition in terms of athletic events or of struggles in which one side goes after the other and tries to beat it. Such events do occur in nature. When a coyote chases a rabbit, there is a race, with food as a prize if the coyote wins and life if the rabbit does. Ants stage epic struggles, real pitched battles between large opposing forces. Jays drive other birds away from food. But such face-to-face combat is decidedly *not* characteristic of competition in nature. The usual competition is a passive process in which each separately seeks to utilize what the other also needs. A plant in getting its water supply does not necessarily attack other plants, and a deer eating foliage may not have any contact with the other herbivores with which it is in fact competing. In this, the usual biological sense of the word, are coyote and rabbit really competing?

Competition frequently occurs among organisms similar to each other, with overlapping niches, and often phylogenetically related. It may, however, occur just as frequently between species that have some one requirement in common but that otherwise lead completely different lives. Man's most severe competitors for food are the insects. Competition between rabbits and sheep caused a crisis in wool growing in Australia.

COMPETITION AND EVOLUTION

If two species compete strongly, the frequent but not inevitable outcome is that one of them becomes extinct. This is not the only cause of extinction, but it has been a widespread cause during the long history of life. A large-scale example occurred when North and South America became reunited from one to several million years ago (see Chapter 28). Each of the previously separated continents had its own distinctive species of mammals, including rodents, carnivores, and herbivorous ungulates (hoofed mammals). North American species invaded South America, and in the ensuing competition all the ungulates, all the carnivores, and a great many of the rodents native to South America became extinct. (The native South American carnivores were marsupials.)

Competition has also played a less macabre part in evolution. As it occurs within species, that is, among the individual members of a single species, it is one of the forces inducing change through natural selection. This may be the major evolutionary role of competition, but it is not our main concern just here, where we are primarily discussing interspecific relationships. Both within species and between species, in a continuous process in which the distinction is not perfectly sharp, competition has been a force tending to develop and maintain differences among populations. It has been instrumental in the occupation of particular niches and in the multiplication of the number of niches in communities. Once a niche has been occupied, competition becomes a conservative force tending to keep the population there, impeding its change or its spread into other niches.

It is easy to see how competition works as an evolutionary force both to change species and to prevent their changing (Fig. 24–4). If two similar species compete strongly, the tendency of natural selection will be to increase the differences between them. Variants in each species least like the other species will have least competition and hence are likely to be most successful in rearing offspring. Over the generations, the species will come to occupy more distinctly separate niches. On the other hand, in a well-integrated community with numerous occupied niches, variants of any species farthest from the usual or modal adaptation to its niche are most likely to encounter competition from the occupants of other niches. The trend of selection will be adverse to these variants and will favor maintenance of the status quo.

Interspecific Interactions: Symbiosis

All the members of a community live together—this is part of the definition of a community. Two universal kinds of interactions involved in living together have how been considered: consumption (including preda-

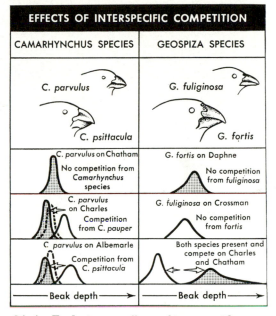

24–4 Evolutionary effects of interspecific competition. The adaptive radiation of the tree finches (*Camarhynchus* spp.) and ground finches (*Geospiza* spp.) of the Galápagos Islands has been discussed in Chapter 16. The finches also provide excellent examples of the evolutionary effects of competition between ecologically similar forms. *C. parvulus*, *C. pauper*, and *C. psittacula* are closely related insect-eating tree finches. Their beak size reflects the size of insects they eat. *C. parvulus* has the smallest beak. Variations in its size are plotted here as frequency distributions. The *parvulus* graph on all three islands is stippled; the graph for a competing species is unstippled; in the figures for the islands of Charles and Albemarle the broken-line graph is a representation of the *parvulus* graph for Chatham, given for comparison. On Chatham *C. parvulus* is the only species of the three that is present. On Charles it is in competition with *C. pauper*, which feeds on slightly larger insects than *parvulus* does. The competition has caused a specialization on still smaller insects by *parvulus*, thus minimizing the competition, and a concomitant local evolutionary decrease in the size of the *parvulus* beak. On Albemarle competition is from *C. psittacula*. The evolutionary effects of interspecific competition between *G. fuliginosa* and *G. fortis* on Charles and Chatham are clear.

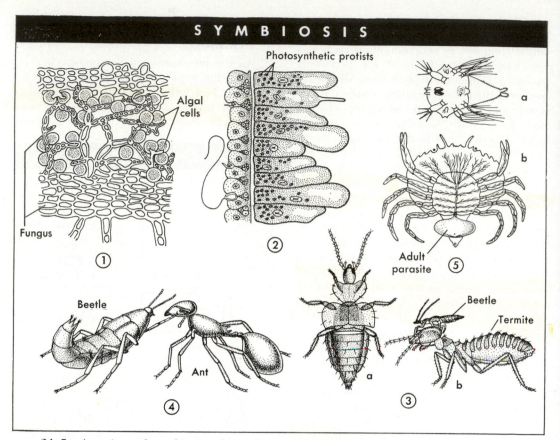

24–5   A variety of symbiotic relationships. *1. Organic mutualism:* a lichen, consisting of nonphotosynthetic fungal hyphae and photosynthetic algal cells. Neither member of this mutualistic union ever grows separately in nature. *2. Mutualism:* photosynthetic protists in cells of *Hydra;* the *Hydra* gains carbohydrates, and the protists gain water, nutrients, and shelter. *3. Commensalism: a,* a staphylinid beetle, which lives in a termite colony as a tolerated scavenger; *b,* the beetle riding on the head of a termite. It scavenges food fragments as they are passed from one worker to another. *4. Social mutualism:* a beetle that is not only tolerated but actually reared and protected by ants for the sake of its secretions. *5. Parasitism: a,* the larva of a crustacean, *Sacculina,* a relative of the barnacles; *b,* the adult *Sacculina,* a parasite of crabs, reduced to a mere pouch of reproductive tissue, which bulges out between the abdominal segments of its host, and cellular extensions throughout the body of the host, which digest the host's tissues and leave it a hollow shell.

tion) and competition. There are other relationships, more specific, less universal, but also widespread. A barnacle on a whale, a hermit crab in and a sea anemone on a snail shell, or a dog and its fleas exemplify particularly intimate ways of living together. Such special and closer associations of individuals of different species, within the looser association of the community as a whole, are given a special name, symbiosis (Fig. 24–5), which simply means "living together." [4]

The three examples given in the last paragraph embody different relationships between the symbiotic animals. Presumably a whale does not mind a few barnacles on its hide and is neither helped nor harmed by their presence. The barnacle, for its part, gets nothing out of the association except a free ride. Whale and barnacle are simply messmates who run around together, eat perhaps at the same table, but take nothing from and give nothing to each other. This is *commensalism*

[4] Many biologists apply the term "symbiosis" only to associations that are beneficial to both parties —to what we call "mutualism." The usages adopted in this book seem more convenient, accord better with the etymology of the words, and are also backed by authority.

("being at table together"). Hermit crab and sea anemone, without taking thought or having altruistic motives, are nevertheless helpful to each other. The sea anemone, with its many stinging cells, protects the crab. Sometimes it also obviates the necessity for the growing crab to leave its protective shell and seek another to fit by actually remodeling the shell. The crab seeks out food and tears it to shreds, and the sea anemone lives on such bits as come its way. The relationship is mutually beneficial, and so it is called *mutualism* (see photograph, p. 642). The dog's fleas benefit; the pup is their food as well as their home. But their unwilling host receives only annoyance or disease in return. That sort of symbiosis is *parasitism* ("eating beside [at the expense of] another").

COMMENSALISM

Distinctions between mere association in a community, commensalism, mutualism, and parasitism are not clear-cut. All sorts of intermediate relationships occur, and it is evident that one sort of relationship has often evolved into another. All the closer associations probably evolved from looser communal relationships. In commensalism one of the associates usually does derive some benefit. Whether the partnership is completely unimportant to the other (making it commensal), more or less helpful (therefore mutualistic), or harmful (parasitic) may be practically impossible to determine or may be a matter of definition or point of view.

Suppose we run briefly through a few more examples usually labeled "commensalism." You will note that one partner generally derives food, protection, transportation, or a combination of these from the other. You decide what the effect is on the other partner. Besides the barnacles that live on whales, there are barnacles that live only on those barnacles that live on whales. There are several small fishes that usually live among the tentacles of coelenterates (especially siphonophores and sea anemones), apparently immune to attack by the stinging cells, which are capable of killing other fishes and yet for these fishes provide protection and probably also some food. There are some little crustaceans (a species of isopods) that live in the mouths of fishes (menhaden), where they pick up bits of the fishes' food as it goes by. Sessile algae may grow on almost any support in the water, but at least one species grows only on the shells of certain living turtles.

MUTUALISM

Few, indeed, are the plants or animals that are not inhabited by other plants or animals or by protists. You yourself, clean and healthy as you doubtless are, are certainly a host to many bacteria and other protists on your skin and throughout your alimentary canal. Such relationships run the whole gamut, or indeed two gamuts, from being completely unimportant to the host to being quickly deadly, on one hand, or being essential to the host's life, on the other. The association is frequently beneficial to both, and often this mutualism has gone so far that neither species can survive without the other.

Many photosynthetic protists (mostly flagellates) live in the tissues of animals. The chemical exchange involved must be elaborate and is not well understood, but it is probable that in some instances the host benefits from oxygen released by photosynthesis in the guest, and the guest derives CO_2 produced by respiration in the host. In Chapter 1 you encountered an example in the protists that live in reef corals and give them their color. On the same reefs photosynthetic protists also live in and brilliantly color the mantle edges of various mollusks, notably of the giant clam *Tridacna*. This relationship is widespread. The hosts may be animal-like protists (protozoans), sponges, coelenterates (not only corals; green hydras are colored by algae in their tissues), flatworms, or mollusks.

Mutualism is common among plants. Lichens, which seem to be single plants and are even classified as if they were, are actually intimate, mutualistic organic associations of, in each case, an alga and a fungus. The fungus, nonphotosynthetic, derives food from the alga. The fungus helps to maintain the water supply necessary for growth of the alga. Most of the higher green plants have fungi that grow around or actually within their roots. Some of these fungi may be harmful to the host, but many are beneficial. It has been shown that pines and some other plants die if deprived of their root fungi, even though the exact nature of the benefit to the host is seldom clear. The benefit is obvious from nitrogen-fixing bacteria in root nodules of peas and their allies.

The wood-eating termites live on cellulose, a food that they cannot digest. It is digested by protists (species of flagellates, again) that swarm in the digestive tracts of the termites, and the termites obtain their food from the protists. These termites cannot live without their internal protists, and the protists (a distinctive species in each species of termites) cannot live without the termites, which obtain the wood on which they feed.[5] Cows and some other herbivorous mammals also acquire some food indirectly from cellulose, which they cannot themselves digest, but they are not wholly dependent on this process, as are wood-eating termites. Bacteria in the cows' alimentary canals digest some of the cellulose in grass or other forage, and the cows, in turn, later digest some of the bacteria. Most or all mammals, including man, normally have large numbers of bacteria in their intestines. Some of the bacteria seem to be helpful in synthesizing vitamins that are then absorbed by the mammalian host.

There are also many mutually beneficial associations in communities that involve no such prolonged intimacy as the examples already given. The widespread dependence between flowering plants and insects that feed on them and also pollinate them is one of the most important communal relationships in nature and is plainly mutualistic. Still more peculiar and less widely important examples include the birds that feed on ticks on mammals, and the ants that care for aphids (a group of insects) and receive a sweet secretion in return. The latter relationship suggests that of man and milk cows, and indeed the association of man with all his domestic animals and cultivated plants can be described as at least loosely mutualistic. What are the benefits given and received by man and pigs? Horses? Dogs? Wheat? Seedless oranges? Roses?

[5] See Chapter 16 for the adaptive behavior of termites related to their dependence on mutualistic flagellates.

PARASITISM

Nature and extent of parasitism

Almost everyone feels a certain repugnance for a tapeworm or a louse. Parasites are disgusting, and the fact that some of them attack us makes them all the more so. But the scientific approach does not assign praise or blame to actions that are neither good nor bad but merely natural. Nor, in fact, is it logical and sensible to do so. Why should our killing a cow for food be proper, while a tapeworm that derives food from the same cow without killing it or harming it seriously is contemptible? From the point of view of food chains and the general flow of materials and energy in a community, there is no essential difference between the herbivores and carnivores that devour plants and animals and the parasites that also derive their food from those same plants and animals. If parasitism were defined solely as living at the expense of other organisms, then the whole animal kingdom, including man, would have to be called parasitic. Even if "parasitism" is more precisely confined to particular ways in which species interact, the term is not always clearly distinguishable from other relationships. A weasel kills a rabbit and sucks its blood. A fly alights and sucks blood from a rabbit without killing it, and then the fly goes on its way. A louse spends much of its life on a rabbit, taking blood when so inclined. A protist lives in the blood stream of a rabbit, within the fluid that feeds it. Which of these are parasites?

It is worthwhile to distinguish parasitism from other relationships of foods and their consumers because it is a special case, or rather a large number of special cases with something in common. The distinguishing features are that a parasite is an organism that lives on or in another living organism for a considerable part of its life cycle, that derives its food from its host, and that is more or less harmful or, at best, not beneficial to the host.

The extent of parasitism and the diversity of parasites are astonishing. Very few living things are free of parasites. Perhaps the only organisms not subject to parasitism are a few that are, you might say, the last word in parasites themselves. Many parasites do have parasites of their own. As Swift put it, "A flea hath smaller fleas that on him prey, and these have smaller still to bite 'em, and so proceed *ad infinitum*."

From a parasite's point of view man is a habitat providing a large number of niches. The whole outside of the body is a parasite's paradise, and there are species specialized for particular niches there. Some lice, for instance, live in the head hair, and quite different lice live in pubic hair. The niches inside are more numerous. Almost every part and tissue is a potential habitat for one parasite or another. There are innumerable parasites of the alimentary canal. Others live in the lungs, blood, nerves, muscles, glands, or elsewhere.

There are insects that begin life as parasites inside the eggs of other insects. There is a species of insects [6] in which the larval females are parasites in other insects, and the larval males are parasites in the parasitic female larvae of their own species. In several species of animals, including some annelids, crustaceans, and fishes, the adult males are parasites on the adult females.

Parasites are very numerous among protists. Among plants, most of the major groups include some parasites, but parasitism is particularly common among fungi. Several comparatively small phyla of animals are wholly parasitic. The most noteworthy animal parasites are many species of flatworms, nematodes, and arthropods, including some crustaceans, innumerable insects, and many arachnoids (especially ticks and mites). Almost all other phyla have at least a few partly parasitic species. True parasitism is, however, extremely rare among vertebrates; indeed, it is practically nonexistent aside from a few partially or doubtfully parasitic fishes.[7]

Besides being of so many kinds, parasites are extremely numerous in individuals. A census is impossible, but some biologists believe

[6] *Coccophagus scutellaris*, a small wasplike insect.
[7] Lampreys and hagfishes (Chapter 22) are often called parasitic, but this is a marginal case hardly distinguishable from ordinary predation. As already mentioned, a few fishes, such as the deep-sea angler *Photocorynus*, have truly parasitic males, but the females are not parasitic. Cuckoos that impose the care of their young on other birds are parasites of a sort, but here, as in human "social parasitism," the word is really used in a different sense.

that *most* of the organisms now living are parasites. Whether we like it or not, parasitism includes many of the most successful ways of life and is one of the most important and widespread characteristics of all communities.

Conditions and adaptations of parasitism

External parasites, living on but outside organisms, exist under the peculiar condition that their substratum is alive. Yet they are in the outside world and subject to many of the same influences of the physical environment as their nonparasitic associates. Most external parasites, such as fleas, lice, or ticks, are relatively little modified structurally in adaptation for parasitism; they are usually semi-independent animals that can live on their own for short periods, at least, or can freely move from one host to another.

Internal parasitism, on the other hand, involves life in environments that are totally different from those of free and independent organisms. The peculiar conditions of life within other organisms pose problems that are met by many special and, to our eyes, strange adaptations. For one thing, the mere fact that they do live within other organisms—and must leave room for their hosts' vital activities, too—means that they must be comparatively small. Some tapeworms may be as much as 60 feet long, which is not exactly tiny, but still their bulk is far less than that of their hosts. Most parasites really are tiny, and a great many are barely visible with a microscope.

The host unwittingly provides an internal parasite with food, shelter, and a more or less stabilized environment. A tapeworm in a mammal, for instance, is bathed in pre-digested food that need only be absorbed through the tapeworm's body wall. The parasite is also sheltered from all rigors of the outer world. Some structures and processes useful in free life are of no use to such an internal parasite. It is a tendency in evolution that what is not useful is likely to disappear. Disappearance is not invariable or always prompt, and its causes are disputable, but the reality of the tendency is well established. Tapeworms do not need and do not have special sensory organs, a clearly developed brain, or a digestive system, although all these occur in their free-living relatives, and all must have occurred in ancestors of the tapeworms. Simplification and loss of some organs is almost universal in thoroughgoing internal parasites. This tendency is usually labeled "degeneration." On the other hand, it might be considered a positive adaptation to a way of life in which the lost characteristics are of no use and might be disadvantageous. Which point of view do you find more explanatory?

Parasites live under conditions that also require adaptations clearly positive and not degenerative in nature, as is well illustrated by the tapeworm (Fig. 24–6). The principal condition of an internal parasite that demands special and positive adaptation is the fact that it must be able to move from one host to another. Although it need not do so as an adult individual, sometime in its life cycle it must change hosts. In the course of the cycle it must be adapted not only to its adult environment but also to quite different environments.

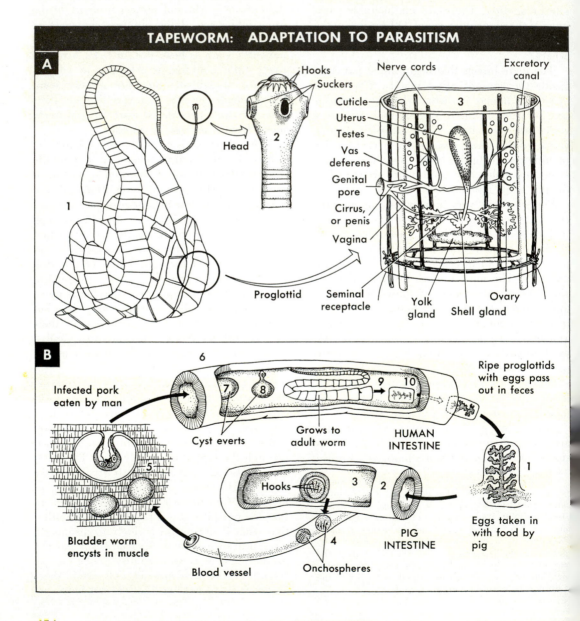

TAPEWORM: ADAPTATION TO PARASITISM

A

Hooks
Suckers
Head
Proglottid
1
2

Nerve cords
Excretory canal
Cuticle
Uterus
Testes
Vas deferens
Genital pore
Cirrus, or penis
Vagina
Seminal receptacle
Yolk gland
Shell gland
Ovary
3

B

Infected pork eaten by man

6
7   8
9   10
Ripe proglottids with eggs pass out in feces

Cyst everts
Grows to adult worm
HUMAN INTESTINE

5
Bladder worm encysts in muscle

Hooks
3   2
1
Eggs taken in with food by pig

Blood vessel
Onchospheres
4
PIG INTESTINE

Tapeworms and other internal parasites have evolved complex life cycles. Commonly the stages are a free-living, aquatic embryonic or larval stage, one or more larval stages parasitic in intermediate hosts, and a final adult stage in a host of a different species. The final host usually acquires the parasite by eating the intermediate host. The commonest tapeworms in humans are two species that have domestic livestock as intermediate hosts; one occurs in cattle, the other in hogs. They may be acquired by eating rare beef or pork. Another human tapeworm, a monster that may grow to lengths of over 60 feet, has two intermediate hosts and is acquired by man from raw or underdone fish, especially pike and pickerel but also trout and others.

Flukes (Fig. 24–7) and the protists that cause malaria (Fig. 24–8) further exemplify complex life cycles in internal parasites. The life cycle of such parasites involve very specific coordination with the habits and biochemistry of particular hosts, and often of two or more hosts in succession. One consequence is that any one egg or larva has an exceedingly small chance of encountering the right hosts at the needed times and of living to become an adult. This hazard is countered by the fact that adult animal parasites usually have extensive reproductive systems (sometimes they seem to consist of little else) and produce millions of eggs. Successful completion of a parasite's life cycle is usually so improbable that only one in millions makes it.

Indeed, in reviewing positive (rather than degenerative) adaptations to parasitic ways of life, we become impressed with the fact that the most universal adaptations relate to problems of successful reproduction. Asexual reproduction—either as simple cell fission in protists or as budding in more complex parasites like tapeworms and flukes—is more prevalent in parasites than in any other animal group. Real dangers arise if a parasite is committed to sexual reproduction while in its host. If chances are often poor that a larval parasite will encounter any suitable host, they are even poorer that it will encounter one harboring a potential mate for the parasite. It is a striking fact that the only animal phylum regularly hermaphroditic—the Platyhelminthes—has been more successful than any other in the parasitic way of life. Many Crustacea are parasites, and of these the most successful are relatives of the common barnacles. They belong to a group (the Cirripedia) that is primitively sedentary and hermaphroditic.[8] In parasitic Crustacea in which the sexes are separate, it is common for individuals of op-

[8] Note that the majority of plants are hermaphroditic (monoecious) and sedentary. Do you think the early evolution of barnacles as sedentary animals may have had any bearing on their later evolution as parasites?

24–6 A tapeworm: adaptation to parasitism.
A. The structure of *Taenia solium*, the pork tapeworm of man. *1*. The adult worm found in human intestines, with a head, or scolex, and hundreds of individual body segments, or proglottids. *2*. Detail of the head, showing a ring of hooks and three of the four suckers, all of which are used to maintain the worm within the host's intestine. *3*. An individual proglottid. Proglottids are continuously being budded off from the unsegmented region immediately behind the scolex. Each proglottid matures a complete set of both male and female reproductive organs, essentially the same as those of a trematode. (cf. Fig. 21–7). Copulation may occur between two proglottids of an individual worm (resulting in self-fertilization) or between different worms. The uterus, in which fertilized eggs are stored, has no pore to the outside. Therefore, as a proglottid matures, it becomes almost entirely filled with an ever expanding uterus full of eggs. A mature proglottid is eventually budded off and shed to the outside in the host's feces. There it bursts, liberating eggs. The adult tapeworm has these special adaptations to parasitism: (1) a heavy, enzyme-resistant cuticle; (2) absolutely no alimentary system—food is absorbed directly from the nutrient-rich environment of the host's gut. B. The life cycle of *Taenia*, which also has several features adaptive to its parasitic habit. In general all these relate to increasing its reproductive potential. First there is the continual asexual production of new young proglottids. Eggs liberated to the ground from a burst proglottid (*1*) develop into embryos (onchospheres) with six hooks. These enter a hog's intestine (*2*) with its food. Once inside the hog's intestine (*3*), the armored larvae burrow (*4*) into blood and lymph vessels, ultimately reaching voluntary muscles, where they encyst (*5*) and undergo further development into a second larval stage. This stage has a young scolex inverted within its bladderlike body; it is commonly called a "bladder worm." When infected pork is eaten by man (*6*), the cysts (*7*) germinate in the human intestine; the young scolex is everted (*8*) and develops into an adult worm (*9*), which reinitiates the cycle by liberating proglottids (*10*).

posite sex to join together permanently in their free-swimming larval stage. The female then becomes attached to a host, and the male—a midget—remains attached to the female near her genital opening.

Is parasitism an easy way of life? Parasitism seems to involve "problems" that are "solved" by such adaptations as alternation of hosts and extreme fecundity. Is that a clear, scientific way of putting the matter? Who or what posed the problems, and who or what solved them?

Parasitism is a highly successful way of life in itself, but it is a blind alley as far as any further evolutionary change is concerned. The specializations and the degenerations (if you wish to call them that) fitting a parasite for life in its peculiar environment make it completely unfit for life in any other environment. They also tend to make further adaptation to anything but a still narrower parasitic niche highly improbable if not downright impossible. It is unlikely that any free-living organisms have evolved from parasitic ancestors.

Parasitism and disease

An organism living inside another is an alien in the body. Such aliens may be helpful collaborators (mutualism), but they are likely to cause trouble. They absorb foods and other substances needed by the body of the host. Their movement or even their passive presence is likely to upset the operation of nerves, muscles, and other organs of the host. Moreover, and most important of all, they are distinct biochemical systems with proteins and other substances that are foreign to the specific chemical and regulatory system of the

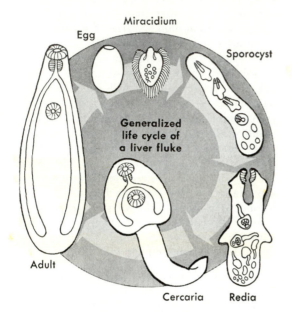

Generalized life cycle of a liver fluke

Egg — Miracidium — Sporocyst — Redia — Cercaria — Adult

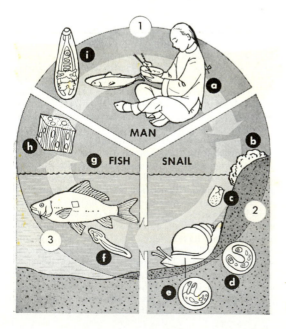

1 — MAN — SNAIL — FISH — 2 — 3
a, b, c, d, e, f, g, h, i

24–7 The life cycle of the Chinese liver fluke (*Clonorchis sinensis*). The trematode liver flukes show some adaptations to parasitism basically similar to those of the tapeworm: the life cycle involves more than one host (man, snail, and fish), and there is much asexual budding, which bolsters the total reproductive potential. *Top.* The developmental stages in the life cycle. The adult liberates eggs, which hatch as miracidium larvae. A miracidium becomes a sporocyst, which buds, internally, a large number of redia larvae. A redia in turn buds, internally, a large number of cercaria larvae. A cercaria develops into an adult liver fluke. *Bottom.* The succession of hosts. *1.* Man. Eggs are liberated in human feces (*a*, *b*), reaching fresh-water ponds, where the miracidium larvae (*c*) swim about and, on contacting a snail, enter it. *2.* Snail. Within the snail a miracidium develops into a sporocyst (*d*), which, in turn, liberates rediae (*e*). The rediae enter the snail's liver; there they themselves produce more rediae, some of which eventually produce cercariae (*f*). *3.* Fish. The cercariae leave the snail and swim free in the pond, eventually entering a fish, in whose muscles they encyst (*h*). The infective cycle is completed when the fish is eaten by man; the cysts germinate, and the adult fluke (*i*) parasitizes man.

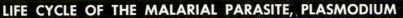

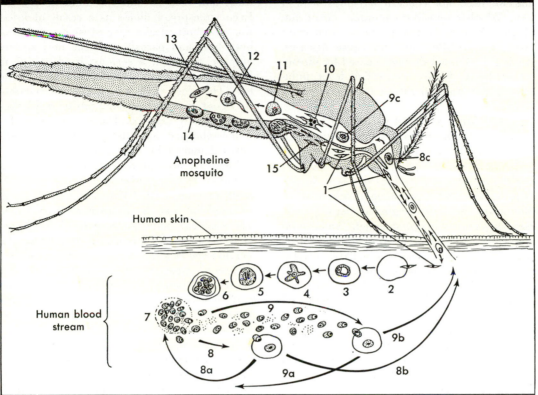

24–8   The life cycle of *Plasmodium*. Malaria is caused by a protist, the sporozoan parasite *Plasmodium*. Characteristically for a parasite, the life cycle involves more than one host and much asexual propagation. The sporozoite stage (*1*) is a convenient starting point. These elongate cells occur in the salivary glands of the mosquito. When the insect bites man, it salivates in order to lubricate the incision and introduce a chemical agent that prevents coagulation of the blood. The sporozoites infect various cells (*2*) of the victim. In a cell the parasite undergoes a development (*3–5*) before asexual reproduction (*6*), which produces large numbers of merozoites (*7*). These merozoites infect red blood cells (*8, 9*). Their subsequent history may follow either of two courses. First, they may undergo (*8a, 9a*) another asexual reproduction, producing still more merozoites. This asexual cycle takes 48 hours; it is synchronized in all the blood cells involved; at its completion the liberation of merozoites is accompanied by a liberation of toxins that cause the cyclic (every 48 hours) recurrence of fever in the host. Second, some merozoites, on infecting new blood cells, may develop into macrogametocytes (*9b*) or microgametocytes (*8b*). These cells, sucked up by a biting mosquito, enter its stomach (*8c, 9c*). A microgametocyte forms four to eight microgametes (*10*), and a macrogametocyte becomes a macrogamete (*11*). Fertilization (*12*) occurs in the stomach of the mosquito, and the motile zygote (*13*) enters the gut wall, where it becomes an oöcyst (*14*). The oöcyst eventually ruptures, liberating large numbers of sporozoites that migrate (*15*) through the mosquito's body fluids and reach the salivary glands (*1*). The cycle is then complete.

host and that may act as poisons for the host. It is small wonder, then, that parasites are the causes of a great many diseases, and especially of the *infectious diseases*. These are the diseases caused by "germs" or "microbes," names applied to all disease-producing pathogenic parasites of microscopic size. Of course

larger parasites may also cause diseases, and many diseases are not caused by parasites at all.

Most germs are viruses and parasitic protists, especially bacteria, but also many of the other protists. We have noted earlier the problem of deciding whether viruses are organisms. This is largely a matter of definition, and we need not concern ourselves with it here. Whether we call them organisms or not, viruses are organic foreign bodies that enter hosts, as parasites do, and cause diseases. Among the human diseases caused by viruses are smallpox, chicken pox, yellow fever, influenza, and infantile paralysis (poliomyelitis or "polio").

Rickettsias, a group of parasitic organisms or pseudo-organisms larger than viruses and also of disputed nature, cause typhus, spotted fever, and several other diseases. Bacteria cause most of the infectious diseases, among them scarlet fever, pneumonia, tuberculosis, diphtheria, bubonic plague, cholera, gonorrhea, syphilis, whooping cough, and undulant fever (brucellosis). Animal-like protists cause, among other diseases, malaria, several forms of trypanosomiasis,[9] including African sleeping sickness, and amebic dysentery. Fungi cause athlete's foot, "ringworm" (which is not a worm), a serious lung infection with various names including San Joaquin fever, and other diseases. Among the larger animal parasites that commonly produce human diseases in the United States are tapeworms, hookworms (which are nematodes), and trichinas or beef and pork worms (also nematodes). All these groups also have species that cause diseases in other animals and in plants. Human diseases are mentioned because they strike close to home and provide comparatively familiar examples.

In the face of this long, gruesome list, which could be greatly extended, it may seem contradictory to say that really successful, well-established parasites rarely cause serious diseases. Such is the fact, and it illustrates one aspect of the dynamic balance of nature. If a host dies quickly after infection by a parasite, the parasite also dies. Species of parasites are usually highly specific, living only in one or a few species of hosts or, what is still more restrictive, in a definite sequence of host species in the round of the life cycle. If infection regularly causes early death of the host, the

host species will almost certainly become extinct, and then so must the parasite species. Even widespread illness as a result of infection of a host species may cause extinction of both host and parasite, for the host species may be placed at a serious disadvantage in comparison with its competitors and predators. Therefore, when a host–parasite relationship continues for any long period, a balance is necessarily reached. The host species of a well-established parasite usually is not seriously damaged by the parasitism.

Balanced host–parasite species are in fact common. Most organisms, including man, are hosts to innumerable parasites that do them no good but that also do them no great harm. There are two main factors in the development of this balance. A species of parasite becoming adapted to a host tends to evolve in such a way as to cause less serious disturbance; the parasite becomes *less virulent.* How would natural selection effect such a change? At the same time the host tends toward greater success in resisting disturbance; it develops *resistance* or *immunity.* Individual resistance to disease is one of the stabilizing mechanisms of the body. The immediate reaction to infection in man and many other animals is an attack on the parasites by some of the white blood cells. A reaction of longer range is the formation of antibodies (including antitoxins), proteins that disable invading parasites or that counteract the disease-producing poisons (toxins) released by the parasites. Antibodies may remain in the host's body for a long time, even a lifetime, and while they do, they confer on the host a degree of immunity against the particular sort of infection that originally produced the antibody. There is also good evidence that host organisms may have or can build up racial resistance or immunity to an infectious disease. Unlike immunity by means of antibodies, which depends on prior infection of each individual, such resistance is hereditary. Can it be explained by natural selection? [10]

[9] From *Trypanosoma,* the genus that includes these various pathogenic species. Trypanosomiasis is common in parts of tropical America, but the disease generally called "sleeping sickness" in the United States has nothing to do with African sleeping sickness and is a virus disease.

[10] Discussion of the artificial control of human infections cannot be included in this work on general biology. Pertinent topics, on all of which you surely already have some information and can readily obtain more, would be these: prevention of infection; injection of antibodies or their stimulation by in-

If host–parasite interactions do tend to become balanced in time, why are there many serious diseases among humans and other organisms? There are several reasons. One is that all species could produce larger populations than their environments could support. Disease that weeds out individuals, especially among the very young and very old, without reducing the average numbers and vigor of the breeding adults does not necessarily endanger survival of the species as a whole and may even be beneficial to it. Nevertheless it is not beneficial to the parasites or the individual hosts killed, and in such a situation natural selection does often tend toward slow reduction in the severity of the disease. Furthermore, parasites mutate, and mutations may produce newly virulent diseases that persist until the extinction of parasite, host, or both, or until establishment of a new balance. Parasites balanced with one host may spread to others with which they are not balanced and in which they cause severe disease. This condition may persist indefinitely if parasitism in a second kind of host is only occasional and is not essential in the parasites' life cycle. Can you find examples and think of other reasons for persistence of virulent disease?

Change in Communities

Communities are dynamic systems living in and forming parts of environments that are also dynamic. They change from hour to hour, season to season, year to year, and epoch to epoch. The rhythm of day and night is reflected in the activities of all communities, and many organisms have "internal clocks" adjusted to such biologically significant environmental rhythms. In the Temperate Zone we are all familiar with the dramatic cycle of the seasons and with its effects on not only the activities but also the compositions of local communities. This cycle tends to return annually to the same condition, but it does not do so exactly. Although one rare day in June may be much like another a year later, as the years pass, they bring gradual changes.

jection of toxins and vaccines; specific drugs such as the antimalarials; antibiotics from *Penicillium* and other fungi (and their synthetic equivalents).

The gardener finds tent caterpillars or other pests more abundant some years than others. The New Englander sees the pastures of his boyhood overgrown with shrubs or merging into surrounding woods. The old swimming hole in the Midwest becomes a marsh or a prairie. The sod of the high plains, turned by courageous but injudicious pioneers, gives way to tumbleweeds and to desolate dust bowls, which yet, with renewed rain and rational treatment, may slowly recover their verdure. In the Southwest ruins of prehistoric Indian dwellings show that thousands of men once lived in wooded, fertile, watered valleys where now is a desert. The student who traces the longer history of the earth finds still greater changes. Turtles, alligators, and fishes once swarmed where now extend the dry sagebrush flats of New Mexico, and the sands of barren dunes lie buried beneath fertile fields in the Mississippi Valley.

Succession

Rhythms corresponding with daily, yearly, or other environmental cycles exemplify the dynamic aspects of communities but do not, in themselves, alter the over-all characteristics of the communities. Anyone old enough to read this book has nevertheless seen a community change in character, perhaps with breathtaking speed where man was involved, more slowly but still quite perceptibly without human intervention. A striking feature of such change is that it is not so much a modification of a single community as a *succession* of more or less different communities in one area.

EXAMPLES OF COMMUNITY SUCCESSION

Consider what is likely to happen to a lake in the northeastern United States (Fig. 24–9). The open waters of the lake have a community including protists, algae, small aquatic animals, and fishes, among other things. In shallower water near shore are pond lilies, cattails, and other comparatively large plants that grow (so to speak) with their feet in the water and their heads in the air. Silt and soil wash in and are piled up by waves, and dead vegetable matter accumulates. The lake becomes shal-

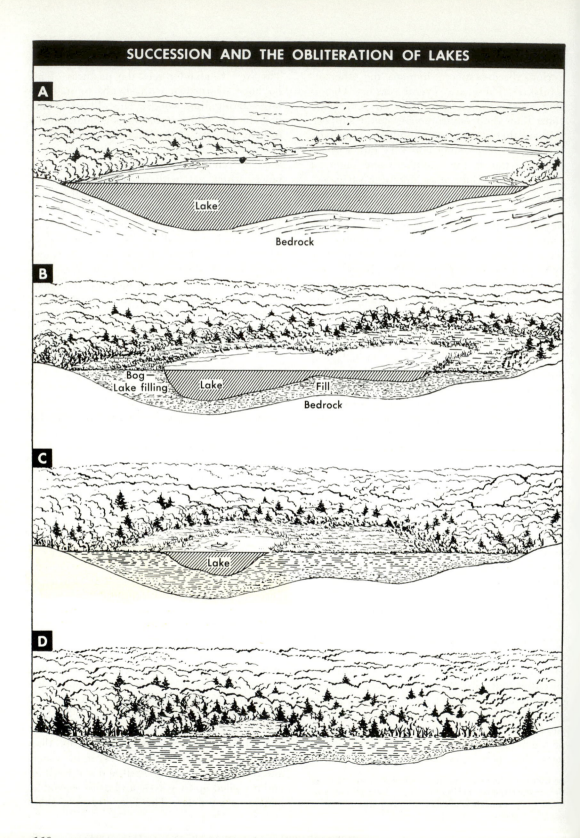

SUCCESSION AND THE OBLITERATION OF LAKES

A

Lake

Bedrock

B

Bog — Lake filling

Lake

Fill

Bedrock

C

Lake

D

lower, with marshy margins. The former shoreward rim of the lake becomes damp land on which grasses, herbs, and willows take root. The filling and the succession of new communities progress toward the center of the lake until finally open water and the aquatic community disappear. In the meantime the marshy marginal soil continues to build up as humus and silt accumulate. Eventually oaks sprout there, and finally the whole site of the vanished lake is occupied by a beech and maple forest, which follows the oaks.

Similar successions where lakes once were can often be followed over periods of thousands, even tens of thousands, of years. Changes are recorded in layers of silt, peat, and humus that fill the lake basin and in the different kinds of pollen deposited in the various layers.

Community succession is not confined to lakes or to the Temperate Zone. It occurs wherever physiographic or climatic changes take place. A coral reef grows toward the surface, its rich community changing as the species of shallower water immigrate and become more populous. Waves grind coral rock to sand and pile the sand up in shifting dunes barely above the tides. Vines take root, bind the sand in place, and contribute to it the humus of their dead tissues. Low trees, resistant to winds and spray, grow in the accumulating soil, to which they contribute in turn. Finally a copse of tall trees develops in the richer soil, and a more protected land environment is thus gradually formed. Or perhaps the climate becomes progressively drier in a region where a pine forest flourishes. The pines,

starved for water, die, and scrubby junipers grow in their place. Still drier conditions see the junipers replaced by sagebrush and greasewood. Finally there may be a cactus and thornbush desert where once a forest stood.

In all these examples, plants are stressed because they are in most instances the clearest indicators and the biological keys of the situation. In each situation, however, the association of plants is accompanied by a characteristic association of animals, and the two together make up the community.

CONVERGENCE AND CLIMAX

Let us return to northeastern United States and consider what happens to a bare rock exposed near the lake where we followed a community succession. The first living things to get a foothold on the rocks are small, scaly lichens. The lichens hasten the mechanical decomposition and chemical disintegration of the rock surface. In them dust lodges and humus begins to develop. More luxuriant lichens arise, and then moss and ferns. Soil develops along with increasing vegetation, and shrubs spring up. As the process continues, trees begin to grow, perhaps pines first while the soil is still rather rocky and comparatively poor in organic matter. Oaks follow, and in their shade appear seedlings of beech and maple, which grow up to overshade the early trees. Finally a beech and maple forest flourishes where once was bare rock, and it spreads and joins the beech and maple forest that grew up where a lake was earlier.

Different beginnings and different community successions led to the same result: the beech and maple forest community. The two successions *converged* and gave rise to the same kinds of community, or even to parts of identically the same community. Successions in the same region starting, say, on bare silt left by a flooding stream or on a cut clay bank would involve different sequences of communities, but would also tend to converge, culminating in beech and maple forest. That forest is the usual culmination or *climax* of community succession in the region. Once it is established, there is little tendency for further progressive change in the community. With the usual and incessant fluctuation of populations, the forest community persists un-

24–9 **Vegetation succession in the obliteration of lakes.** *A.* Initially a lake is surrounded by a beech and maple (hardwood) forest; a few conifers, like tamarack, stand near the edge of the lake (bog forest). *B, C.* Accumulation of humus from lake plants (water lilies, cattails, pickerelweed, etc.) forms a marginal marsh or bog invaded by sedges and sphagnum moss; then by blueberry, wintergreen, and similar plants; then by willows, poplars, and other trees; and later by tamarack, yew, dogwood, and maple. This succession, as the lake fills with humus, ends when the surrounding beech and maple forest takes over entirely. *D.* The lake is filled but still surrounded by the tamaracks and other plants of the bog forest, which are ultimately succeeded by the hardwoods.

til some further physiographic or climatic change or some interference by man starts a new community succession.

Most regions have a type of *climax community* with which all the community successions of the region usually end and which then tends to persist there. Of course, the climax is not always a beech and maple forest (although that is a common climax over much of northeastern United States), or even a forest of any kind. In many mountains the climax is a spruce and fir forest. A frequent climax on Western plateaus and other uplands is ponderosa pine forest. Mesquite communities are the climax in many more arid regions. (Mesquite is a thorny shrub belonging to the pea family.) The climax over most of the high plains east of the Rockies is grassy prairie. The nature of the climax is ultimately determined more by climate [11] than by any other factor. Local differences in climate are fairly common even within a single region, so that regions often have more than one climax in different, more limited localities. Both grassland and forest climaxes are common in some regions.

What is the usual climax community in your part of the country? Has man disturbed or destroyed that community in places? If so, has the disturbance started a new community succession? Is there a tendency for abandoned fields and clearings to follow a succession back to the original climax?

CLIMAX AND CHANGE

Here is a question that probably arose in your mind as you read about climax communities: If the community in each locality tends toward a fixed climax in time and if this has been going on for millions of years, why are there any further changes in communities? Why are not all communities everywhere set and static at their climaxes?

A climax community tends to persist, it is true, but it tends to persist only as long as no internal or external disturbance affects the community and there is no essential physiographic or climatic change in its environment. In fact, such disturbances and changes are frequent, and in the long run they are sure to

[11] The climate's ultimate effect on the community may be felt through its primary effect on soil.

happen. Inevitably, always and everywhere, the climax community does eventually change. New successions lead to new climaxes, or change is so continual that a stable climax can hardly be designated. These facts have raised some doubts as to the general validity and usefulness of the concepts of convergence and climaxes. The doubts are not very serious if we remember that convergence and climaxes are not defined as universal and eternal. Convergence is a frequent tendency, and climaxes often do plainly exist and tend to persist for periods long in terms of human life, even though they always do change finally. The concept of climax as a stable condition in nature does have validity, but only in a limited and short-range way because nature is not static.

Physiographic changes, such as the filling in of lakes, do not always tend toward climax communities. New lakes are formed, too (otherwise all would be gone by now), and in them the succession starts again. Erosion lays bare sands, clays, and harder rocks on which succession also begins from the start. Floods kill the established communities and, again, leave tracts where new communities and new successions arise. In broader view, new land rises from the sea and old land sinks into it. Mountains are worn down to plains, and, as they wear away, their communities also go through a succession, but one that goes rather from one climax to another than in sequence to a fixed climax. Not so long ago (10,000 years or so) almost the whole northern half of our continent was buried in ice, and new successions have occurred since then, not only in the glaciated areas but also in almost all others as the climates changed.

Disturbances within communities are also frequent. Forest and prairie fires (which antedate man, although he is responsible for most of them nowadays) wipe out whole communities. Population changes in climax or other well-established communities tend to fluctuate around an average, but they do not always do so. Great increases in the population of any one species may actually change the environment and the community as a whole, as will be shown in the next chapter. Decreases may go on to extinction, with a consequent long or permanent change in the composition of the community. New species frequently enter a community from elsewhere. Inevitably they change the interactions in the community, and they may profoundly change its whole nature. Moreover, communities evolve not only by the extinction of species in them and the incursions of new species, but also by evolution within their populations. The species of which a community is composed are not themselves static units. Over the generations each one of them changes, and so necessarily does the community of which it is a part.

CHAPTER SUMMARY

Food chains: producer → consumer → reducer; the pyramids of energy, of mass, of numbers; budgets of materials and energy.

Niches and competition: differences in niches; the principle that all species in any one community have different niches; the universality of competition, which is generally passive; competition leading to extinction; competition leading to the differentiation of competing forms.

Interspecific reactions: consumption, competition, symbiosis; symbiosis—commensalism, mutualism, parasitism; the great extent of parasitism; special adaptations of internal parasites especially for reproduction and host sequence; tapeworm and liver fluke as examples; diseases caused by virus, *Rickettsia*, bacteria, other protists, fungi, worms; balanced host–parasite relationships in disease; resistance and immunity.

Changes in communities: a succession of communities in one area leading to a climax; the concept of climax limited in spatial and temporal application; climaxes also going through successions in longer periods of time.

25

Organization and Change in Populations

This deer on the Kaibab Plateau is stretching to feed on pine needles because better fodder is unavailable. It is the victim of a human policy— the slaughter of its natural predators—that was founded on ignorance of the factors governing change and equilibrium in natural populations.

Communities are composed of local populations of numerous different species. The community as a whole, as we have now seen, is a dynamic system with many and diverse interrelationships among the species that compose it. Each specific population is also a system, understandable only in the context of its community, but with its own dynamics and interrelationships. Among these, population growth and decline have special importance both in themselves and in their effects on the community as a whole. In this chapter we shall first consider them. We shall then turn to some particular intraspecific interactions culminating in the evolution of societies and leading to the study of the human species and biological aspects of human society in the following chapter.

Rise and Fall of Populations

BIRTH, DEATH, AND SURVIVAL

The size of a population is apparently determined by quite simple facts of life, birth, and death. If more organisms are born than die, the net result is an increase, and if more die than are born, a net decrease occurs. The factors that determine birth and death rates are not so simple; in fact, they are very intricate. The situation is also complicated and made more interesting by the fact that the size and composition of the population is not determined by birth and death rates alone but also by how long individuals survive, that is, by *when* death occurs. A simple problem will demonstrate that fact. Suppose that in some species 1000 individuals were born and the same number died each year. Would there be a change in population from one year to another? What would be the size of the population at any one time if each individual died at the age of 1 year? If each died at the age of 10 years? In the latter population, what percentage of individuals would be 7 years old?

The problem is, of course, oversimplified by the assumption that all individuals die at the same age. This is not really true of any species. Some individuals drop out at all ages from birth to death. The percentage of individuals that die at a given age—the death rate for that age—also changes markedly through the life span. Usually the death rate is high among the very young and the very old and reaches a low point somewhere in between. This is as true of man as it is of most other organisms. The human death rate is high in the first year, drops to a low in the early teens, and then rises slowly at first (until about 60) and then with increasing rapidity. That fact of life as it is really lived contrasts with the fact that all normal organisms of the same species are usually *capable* of living for about the same length of time, a life span characteristic of the species. There is, as would be expected, some hereditary variation in potential life span. By and large, however, all members of a species born with a stock of genes usual in the species, without disabling mutations or other accidents of heredity, would tend to die of old age at about the same time. The mechanism runs down of itself after a certain length of time. Earlier death, which is more the rule than the exception in nature, is a premature failure due to an environmental incident such as competition, infection, predation, or accident.

Thus are contrasted the potential or physiological life span built into the organism and the almost always shorter actual span that it manages to achieve. Man's biblical allotment of three score and ten years is an estimate of potential span, too low an estimate for most humans. The potential span in *Homo sapiens* is the longest among mammals [1] and is longer than in the vast majority of other animals. In spite of frequent stories and news reports to the contrary, probably no birds, amphibians, or fishes have as long a potential life span as man. A few species of turtles probably have a longer span, perhaps up to about 150 years. Almost all invertebrates have comparatively short spans, from a few weeks to a few years, but some coelenterates, crustaceans, and mollusks may live almost as long as man. Most plants also have short spans; the countless annual plants are so called because their potential life is 1 year. Trees and some shrubs, how-

[1] The common belief that elephants may live longer than men is wrong. The highest authentic record for an elephant is 77 years. The maximum for man is unknown, since extreme claims are unsubstantiated by evidence and are usually tall tales. The oldest *certified* record for man is 109 years. Uncertified but probable ages may reach 120.

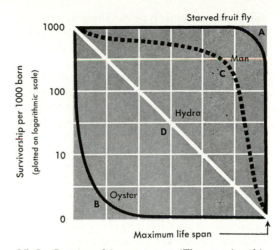

25–1 Survivorship curves. (The survivorship curve for fruit flies (*A*) applies only to their life spans under the unnatural conditions of starvation in the laboratory.)

ever, may have spans counted in centuries, and a few run into thousands of years, up to at least 4000. At the other end of the scale, the potential span for some protists may be a matter of hours or even minutes.[2]

A clear and convenient way to represent the incidence of death in a population is by a *survivorship curve* (Fig. 25–1), which shows the percentage of individuals still living at various times after birth. If most individuals live out the potential life span, the curve is nearly horizontal until that span is reached and then drops precipitously (Fig. 25–1*A*). If, on the other hand, most individuals die early in life and the survivors of that critical period have comparatively low death rates, the curve drops rapidly at first and then levels off (Fig. 25–1*B*). The first situation, practically speaking, does not occur in nature. The second seems to be fairly common. Probably more common, however, are situations intermediate between the two extremes (Fig. 25–1*C, D*).

Human survivorship curves are of the intermediate type but differ greatly in precise form in accordance with nutrition, sanitation, and medical care in a particular population. Where the level of public health is high, there is an approach to type *A*, and where it is low,

[2] Of course, the problem arises (which is only one of definition) whether a protist becomes two new individuals when it reproduces by division.

the approach is to type *B*. There is little evidence that any of the great advances in public health and medicine have increased the *potential* life span. They have greatly improved the chances that individual lives will more nearly achieve their potential span. One of the effects is to increase the percentage of older people in the population. In the United States and most other Western countries the last 50 years have been generally characterized by a declining birth rate.[3] The net result of these two factors has been that the percentage of Americans aged 20 to 45 (hence in the ages when most reproduction occurs and in the ages of most productive work) has remained fairly steady, while the percentage of younger people, under 20, has decreased and that of older people, over 45, has correspondingly increased. What influence of this change can you see in your own community? In the national political and economic scene?

MALTHUS AND THE GROWTH OF POPULATIONS

All populations, both natural and human, have tremendous capacity for increase. Elephants are notoriously slow in reproduction, but it is certainly conservative to estimate that, if all their young survived, the population would more than double every 50 years. A single couple that lived 100,000 years ago (and this is a short time in the history of natural populations) thus *could* have an astronomically enormous number of descendants today—a number represented, *at the very least*, by 4 followed by 602 zeros. That many elephants would fill the visible universe, and so we are obviously leaving something out when we say that so large a number of descendants is possible. It would be possible if all elephants reproduced at the stated rate and all offspring reached their potential life span. The result, in itself, forcefully demonstrates that such survival is absolutely impossible. Yet the example supposes that the population merely doubled in 50 years. In numerous invertebrate species one female may lay more than 2 million eggs per year. If all eggs developed and survived, the population would increase a millionfold every year—and would

[3] After World War II it climbed again, but it still has not reached its earlier height.

fill the universe during the lifetime of one generation of elephants.

If unchecked, the size of any population would increase at a fantastic rate. Its numbers would literally multiply each year, and the population growth in absolute census figures would be a curve that rapidly shot up sky-high (Fig. 25–2). The increase must be and is always checked. The environment can contain and support only so many individuals of a given species. When a population starts to grow from a few individuals in an environment favorable to it, its size shoots up at first, but then it begins to run against the barrier of the capacity of the environment. Population increase then slows down, and finally the population size is stabilized at a figure at or below environmental capacity. An increase in the carrying capacity of the environment may be quickly countered by further population growth. A decrease in the carrying capacity leads to higher death reates and so to a decline in population. The result is a tendency for populations to be as large as possible under existing conditions or, you might say, to live up to their income. Yet organisms usually keep on reproducing at rates that would rapidly raise their populations above capacity level if all the offspring survived and bred. The obvious corollary is that all offspring do not survive and breed. Only a small fraction of them do under ordinary conditions. The others necessarily are eliminated by competitors, within or outside their species, and hence by starvation, or by enemies, diseases, and other factors.

An English clergyman and economist named Thomas Robert Malthus pointed out these grim facts of population limitation in 1798. Malthus applied the principle to man and concluded that famine, pestilence, and war are inevitable brakes on the increase of human populations. Darwin noted that the Malthusian principle applies even more clearly to most organisms other than man. He saw that the inevitable decimation of offspring is a possible mechanism for evolution by natural selection. There is no serious doubt that Darwin was right on this point and that Malthusian or Darwinian selection is an important process in evolution, although we now consider that other mechanisms of selection occur and may be at least equally important, as we have stressed several times.

Malthus' conclusions as applied to man were decidedly unpleasant. We all have an ingrained human (but not scientifically sound) tendency to disbelieve what is unpleasant to us. It is not surprising, then, that it has become an article of faith for many, including some politicians, theologians, economists, and even a few biologists, that the Malthusian principle does not really apply to man. Indeed, events during the last 150 years in some progressive countries have seemed to give the lie to Malthus. In those countries, of which the United States is an outstanding example, populations have increased greatly because birth rate exceeds death rate and survival has been lengthened. Yet the standard of living, instead of dropping to minimum subsistence level, has risen notably. The explanation is of course that the extent and efficiency of production have been raised and that population growth, although large, has lagged behind increase in production. In other words, for some human populations the capacity of the environment has enlarged faster than the population. This has encouraged belief that the solution to human population problems is simply to go on increasing production as fast as possible. Unfortunately for so optimistic an outlook, it is a biological fact that that solution can work only temporarily and locally.

There are physical limits to the amount of food that the earth can possibly produce for any one species, including man. No matter how much of the earth's surface is brought into production for human use and how efficiently it is managed, the limit of environmental capacity is still there and cannot be removed. We are only temporizing with the Malthusian principle, not evading it. Eventually population must be balanced, and birth rate cannot continue to exceed death rate. That is the inevitable conclusion from biological principles. Whether the balance is to be by decrease of birth rate or increase of death rate, and at what level of subsistence and crowding the balance is to be achieved—these are economic and political problems of great magnitude and urgency. They concern the biologist not so much as a biologist but, like everyone else, as a citizen.

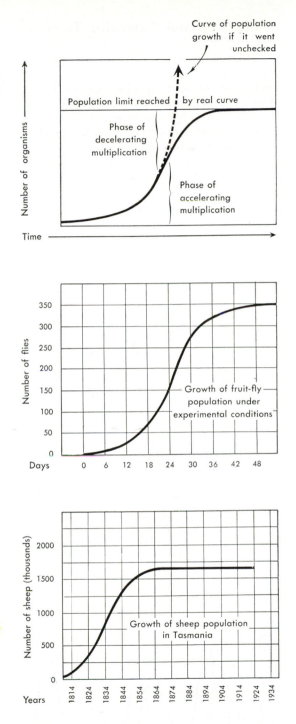

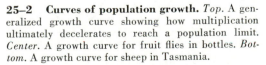

25–2 Curves of population growth. *Top.* A generalized growth curve showing how multiplication ultimately decelerates to reach a population limit. *Center.* A growth curve for fruit flies in bottles. *Bottom.* A growth curve for sheep in Tasmania.

Limiting and Balancing Factors

Apart from its controversial aspects, the Malthusian principle is certainly correct to the extent that for every species there are factors that limit population growth and determine population size by a balance of reproduction, death, and survivorship. Most of these factors are inherent in the interactions studied in the last two chapters. Here they can be reviewed briefly from this different point of view.

PREY-PREDATOR RELATIONSHIPS

All life is ultimately limited by the usable energy received from the sun. Remember that the percentage that can be directly used by life is always small. This is true not only because of the low efficiency of photosynthesis, but also because still larger amounts of energy go to keep the earth livable, especially to maintain the temperatures of its surface, water, and air and to run the water cycle on which all land organisms are dependent. If the energy now used in these ways were to decrease significantly, living populations would automatically decrease. Materials available for organisms in any community are also limited, and no population can live beyond its budget.

FOOD CHAINS

Each link in a food chain is dependent on the links before it. Population size at any point in a food chain is therefore limited by the sizes of populations in all previous links. The connection is particularly evident when, as often happens, fluctuations of population size in one species are accompanied by fluctuations in another that feeds on the first. An example will be given when we discuss periodic changes in populations later in this chapter.

Predation

Population size in food chains is influenced in two directions. The population of a food species limits the population that feeds on it, and at the same time the fact of being eaten

25–3 Prey–predator relationships. *A.* The simpler case. The prey population, *Paramecium*, grows until predation by the ciliate, *Didinium*, commences. The predator population grows until the prey is exterminated. Then the predator population—now without resources—itself declines to extinction. *B, C.* Two cases of prey–predator oscillations. The relationship here is in general the same as in *A*; however, the decline of the prey population causes a simultaneous decline of the predator population and thus relieves pressure on the prey before it reaches extinction. The prey population, under temporary reprieve, again begins to grow; but as it does so, the predator population—now with renewed resources—grows also. The system can persist for many cycles in this oscillatory state. (In *B* the prey population is protected against extinction by the regular addition ["immigrations" *1* through *6*] of small groups of individuals.) (G. F. Gause's data)

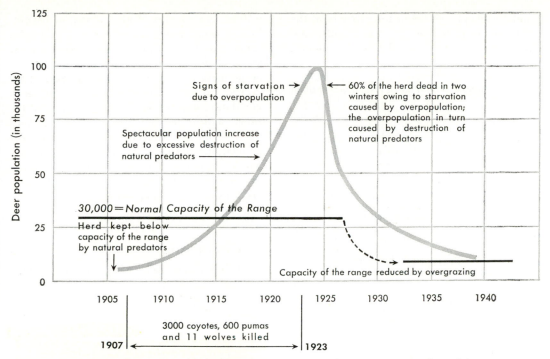

25–4 The history of the Kaibab deer. The gray line represents numbers of deer.

affects the population of the food species. In parasitism there is necessarily a balance between the numbers of parasites and the damage done to the hosts. The plant–herbivore balance is closely analogous to that of host–parasite, as the herbivore does not necessarily kill the plant on which it feeds. Reciprocal balance of populations is especially evident in the relationship called predation, in which the individual used as food is killed. The predator population necessarily has less bulk than the prey population (except in very temporary instances) and usually is much less numerous. An increase in the prey population generally increases the number of predators, but an increase in the number of predators tends in turn to decrease the prey population. The interaction can produce very complex results in the populations, but one of three reactions or a combination of them ordinarily ensues. Two of the three are illustrated in Fig. 25–3. Increasing predation may wipe out the prey population, after which the predator population either also becomes extinct or turns to other prey. This is obviously a short-range reaction that cannot persist in a balanced

community. Increased predation and decrease of prey may be followed so promptly by decrease of predators that the prey survives and becomes more abundant with lessened predation, and then predators and predation increase again, and so on. This cyclic relationship can persist indefinitely, but it is delicately balanced and may lead to extinction of the prey or to a more stable situation. In a stable balance, certainly common in nature and probably the rule in established communities, predators consume just about as many of the prey as to keep the prey population at or below the limit of environmental capacity. It can be said that predators are simply cropping excess production of the prey.

In stably balanced prey–predator populations, predation is often actually beneficial to the prey as a group, even though it destroys individuals. This fact and the unforeseen and disastrous possibilities of ignorant interference in natural communities are dramatically illustrated by the history of deer on the Kaibab Plateau in Arizona (Fig. 25–4 and the photograph on p. 664). Before 1907 the plateau had a healthy deer herd with a stable population kept well below the capacity of the vegetation (which thus also was healthy and stabilized) by heavy predation by pumas, wolves, and coyotes. With the idea of benefiting the deer by removal of their "enemies," a campaign of extermination was waged against the predators. The deer population did, indeed, increase enormously, from about 4000 in 1907 to some 100,000 in 1924. The peak population was far beyond the capacity of the range, and in the next 2 years more than half the deer starved to death. Thereafter the deer population continued to decline more slowly and by 1939 was down to 10,000, living up to the capacity of the range, now seriously damaged by overcropping. With the range still deteriorating, starvation continued to kill more deer than the predators had.

COMPETITION

The immediate cause of death of most of the Kaibab deer after the removal of predators was starvation, but the starvation resulted from excessive *intraspecific* competition for a limited food supply. The frequent result of close *interspecific* competition is the decline and ultimate extinction of one of two competing populations. Experiments have shown patterns of population change that underlie such events in nature (Fig. 25–5). Population growth of the successful competitor is slowed down, but eventually the population reaches the size it would have had without competition. The losing competitor's population starts to grow normally but soon slows down and then gradually declines to extinction.

Density

The story of the Kaibab deer points up another population factor not hitherto dis-

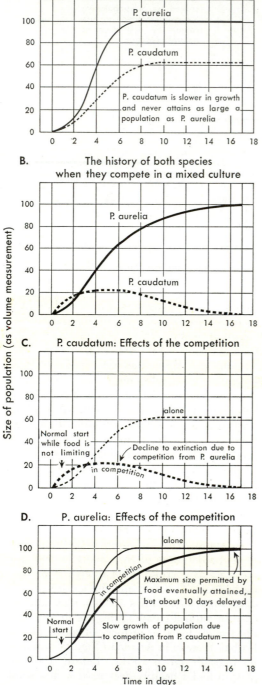

A. The history of both species when grown alone

P. aurelia

P. caudatum

P. caudatum is slower in growth and never attains as large a population as P. aurelia

B. The history of both species when they compete in a mixed culture

P. aurelia

P. caudatum

C. P. caudatum: Effects of the competition

alone

Normal start while food is not limiting

Decline to extinction due to competition from P. aurelia

in competition

D. P. aurelia: Effects of the competition

alone

in competition

Maximum size permitted by food eventually attained, but about 10 days delayed

Normal start

Slow growth of population due to competition from P. caudatum

Time in days

Size of population (as volume measurement)

25–5 An experimental study of interspecific competition. The experiments concern competition between two species of *Paramecium*—*P. aurelia* and *P. caudatum*—grown in the same culture vessels. They compete for food and other requirements. *A.* The growth of both species when grown separately under conditions similar to those in which they later compete. Note that *aurelia's* growth is more rapid; its stable population size (about 100 units) is greater than that (about 60 units) of *caudatum*. *B.* The history of the two species in competition. *C, D.* Comparison of each species, grown alone and in competition. *P. caudatum* is eliminated by the competition from *aurelia*; competition slows the population growth of *aurelia*, but it eventually attains its normal population size (about 100 units). (G. F. Gause's data) cussed—density of population, or the number of individuals concentrated in a given space. At their highest population total, the density of the deer led to intense competition for food and to starvation. There were too many deer in the available area. A little thought shows that in all such situations, environmental limitation is more a matter of density than of total population. It has repeatedly been found for a great variety of organisms that increased density slows population growth, eventually stops growth entirely, and when extreme (as for the deer) leads to reduction in the size of the population (Fig. 25–6). Usually, although not always obviously, density affects the population through the amount of food available. In experiments of growing flies under crowded conditions, it was noted that crowding reduces the number of eggs laid by a female. Other experiments demonstrated that this density effect was due mainly if not entirely to the food supply; the crowded females were undernourished and laid fewer eggs in consequence. Another frequent reason why heavy densities often react unfavorably on populations is that they promote the spread of diseases.

Other studies suggest that there is for most species a range of densities in which the population does best and that either lower or higher densities impair the maintenance and growth of the population. There are several factors involved in this phenomenon. Certainly any population can become overcrowded, and generally then limited in size. Undercrowding, which occurs in densities too thin for population maintenance, is less obvious. Widely dispersed individuals may not find mates. There is often some defensive strength in concentration of numbers. Highly social groups usually do not function well if the social unit is either too large or too small. A dense population alters the temperature, the humidity (if in the air), and the chemistry of its environment by muscular activity, respiration, excretion, and other processes. A great concentration may actually poison the environment, as when too many fish are kept in one aquarium, but there is also evidence of density ranges that condition the environment favorably for the species. Goldfishes grow better in water conditioned by other fishes than when living alone in clean water. In higher

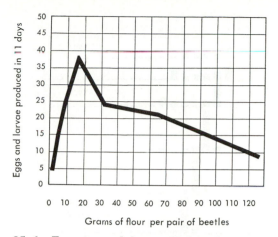

Grams of flour per pair of beetles

25–6 Experimental demonstration of an optimum population density in flour beetles. It is not surprising that egg productivity of the beetles increases as their food allotment is increased from 0 to about 18 grams of flour per pair; but note that further allotment of flour, surprisingly, leads to a decline in productivity. There is an optimum density of beetles in the flour. (Thomas Park's data)

animals there are also psychological factors, innate or learned "preferences" for lower or higher densities. What are some of the influences of density in human populations?

POPULATION PERIODICITY

Fluctuation in population size

Change in population size is rarely a simple matter of smooth increase up to the capacity of the environment or decrease to extinction. There are always fluctuations, sometimes violent ones, even when the end result is indeed either maximum or minimum (zero) population size. Many of these fluctuations are irregular and episodic, caused by environmental events (such as unusual weather conditions) so intricate or themselves so unpredictable that the population changes cannot be predicted. At present, at least, most changes in wild populations are of this type.

It is a matter of common observation that there are good years and bad years for different species of plants and animals. The fact is well known to commercial fisheries, for example, for which there are some particularly good data (Fig. 25–7). Some fluctuations seem to be completely erratic, whereas others range from vaguely to quite obviously cyclic

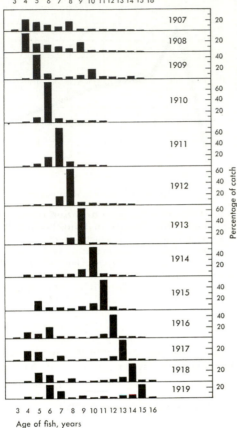

Age of fish, years
3 4 5 6 7 8 9 10 11 12 13 14 15 16

Percentage of catch

3 4 5 6 7 8 9 10 11 12 13 14 15 16
Age of fish, years

25–7 Population periodicity in the herring. 1904 was clearly a good year; it is the principal contribution to all catches through 1918, although in 1915 another good year class (1910) began to become prominent.

rhythms. Such rhythms have been especially well studied among the fur-bearers of northern Canada and Labrador, for which there

are records going as far back as 1750. The abundance of varying hares fluctuates in a somewhat irregular way but with an evident cycle of about 10 years (Fig. 25–8). Populations of lynxes have a cycle closely following that of the hares. This suggests the previously noted experimental cycle in prey-predator relationships (see Fig. 25–3), and it undoubtedly has a related cause: the lynxes prey on the hares. In this case, however, the decline in hare population is believed not to be due to predation but to overpopulation and epidemics. In the ensuing underpopulation reproduction is rapid and builds up anew to overpopulation in a few years. The lynx cycle is a secondary effect of the hare cycle, caused by changes in the food supply for the predators.

Many attempts have been made to correlate cyclic population changes with climatic or other physical cycles (of sunspots, for instance), but without success up to now. It is probable that the principal causes of regular population cycles are biological reactions within the communities themselves, as in the example of the hares and lynxes. Irregular fluctuations, on the other hand, can often be directly ascribed to environmental causes. The great reduction in the house-sparrow population of the Shetland Islands in 1926–1928 is known to have resulted from epidemic disease, and the periodic decreases in the populations of a species of British sea urchins in 1917 and 1929 followed cold winters.

Other periodic population phenomena

Numerous activities in populations are periodic and sometimes cyclic. Insofar as they are long-term characteristics of the organisms and

25–8 Population periodicity in the varying hare of Canada and Labrador. The cycle reaches its maximum about every 10 years.

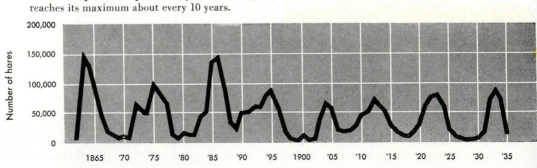

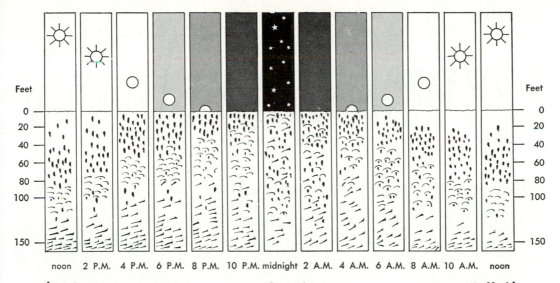

| noon | 2 P.M. | 4 P.M. | 6 P.M. | 8 P.M. | 10 P.M. | midnight | 2 A.M. | 4 A.M. | 6 A.M. | 8 A.M. | 10 A.M. | noon |

' Calanus   ⌒ Cosmetira   ⟋ Mysid

25–9   Daily periodicity in the vertical migration of marine plankters (planktonic organisms). The figure shows the cycle of up-and-down movement of three animal plankters, doubtless related to the similar daily migration of the plant plankters on which they feed. It seems likely that the migration of the plants (mostly diatoms) is aimed at attaining an optimum light intensity.

do not lead to definite population changes, they do not particularly concern us here. Nevertheless, three examples of rhythmic, environment-correlated breeding activity— daily, tidal, and seasonal—are given in Figs. 25–9, 25–10, and 25–11. Such rhythms, varying in regularity and correlated with many different environmental factors, are almost universal, and they influence abundances of species and other population phenomena markedly at given places and times. What bearing do they have, for instance, on summer and winter bird populations in your area?

Intraspecific Interactions

The dynamics of populations of course include much more than increase and decrease in population size. Each population also has some degree of internal organization, involved in interactions among its individuals. It is to some of these interactions that we now turn.

Members of the same species living together in a community necessarily affect each other. Relationships corresponding with food chains

(for example, predator–prey) and with parasite–host interactions between species are slight or absent within species.[4] Competition, aggregation (more or less analogous with commensalism between species), and cooperation (analogous with mutualism) are at least as common within as between species, but they tend to take different forms and have different evolutionary results in the two cases.

COMPETITION

Species in the same community compete if their niches overlap in some essential and exhaustible environmental requirement. The result is that each species in a stable community has its own niche, and vital overlapping tends to be eliminated. All members of a single species live (approximately) within the same niche. Obviously, intraspecific competi-

---

[4] In fact, it is not altogether uncommon in nature for animals to eat others of their own species, and we have mentioned the rare occurrence of species in which males are parasitic on females. Nevertheless, it is impossible for either predation or parasitism on other members of the species to be the usual way of life of a species as a whole. Why? Have you heard the story of the islanders all of whom made a living by doing each other's laundry?

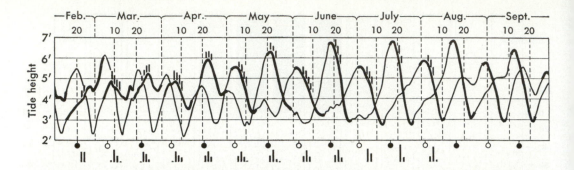

Top, Boyd Walker; bottom, Moody Institute of Science

25–10 Tidal and lunar periodicity. *Bottom.* A photograph (by flash) of a remarkable phenomenon exhibited by the grunion (*Leuresthes tenuis*) on the southern coasts of California. The grunion wriggle out of the waves at the highest points during the highest (night) high tides of the lunar cycle. Momentarily stranded on the wet beach between waves, the females wriggle down into the sand tail-first. (Note the females with only their heads protruding from the sand.) In this position the female lays her eggs while the male, curled around her body, ejaculates sperms, which travel over the female's wet flank to reach the eggs. Both male and female return to sea with the next wave that covers them. The entire grunion population's egg laying is thus synchronized, being restricted to the two periods of highest tides in the lunar cycle. Furthermore, although there are two high tides each 24 hours, the grunion limit their activity to the night high tide. The eggs, buried in the sand, hatch as young fish in time to go to sea with the next set of very high tides about 2 weeks later. *Top.* A chart showing the occurrence of grunion egg laying in relation to tidal and lunar cycles at La Jolla, California. The heights of high tides (in feet) about 24 hours apart have been connected by smooth lines. The two tides each day yield the two series of curves; the night tide is indicated by the heavier line. Grunion activity is represented by short vertical bars above the curves. Moon phases are given below (the black circles indicate new moon, and the white circles full moon). The vertical bars under each set of high tides show the relative intensities of individual runs on successive nights.

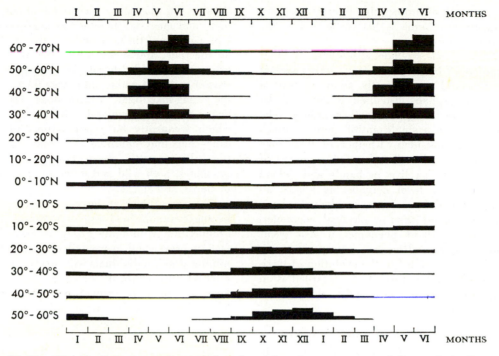

25–11 **The English sparrow's annual cycle of breeding activity at different latitudes.** Near the equator there is virtually no annual cycle; breeding is distributed nearly at random. To the north and to the south pronounced cycles develop but, of course, with a 6-month phase difference.

tion may be particularly strong, but it is equally obvious that its effect is not to move each individual into a separate niche. Each niche has a number, usually large, of places that may be occupied by individuals without lethal interference among them. Competition is for these individual places in the specific niche. The winners are, as a rule, those individuals most precisely and efficiently adapted to the niche. Competition is one of the processes that leads to natural selection. It is the one most stressed by Darwin, but in the modern concept it is only one of many selective forces. (What are some of the others?)

In its simplest form, the net effect of intra-specific competition in the community is merely to eliminate surplus population and to assign to each survivor its own place, or what the Germans call *Lebensraum* ("living space"). This minimal effect is universal. A gardener thins out seedlings as they grow. You have only to watch the annual growth of herbs in a field or the slower growth of trees in a forest to see the same process in nature. The same sort of process occurs among all groups of animals. Even fully mobile species are thinned out to the numbers that can be supported by available materials and energy.

Among animals with more complex behavior, especially arthropods and vertebrates, *territoriality* is common. Each animal has a smaller or larger territory that is somehow home to it. Often the territory centers around a literal home, such as a hive, nest, or burrow. The animal may spend all its time there; if it does stray far afield, it returns periodically to its home territory. The competitive background of such behavior is particularly evident in the many species in which the home territory is actively defended. You must have seen birds or chipmunks aggressively driving away trespassers on their preserves. Not only some birds and mammals but also some lizards, fishes, crabs, and others defend territory. This is one of the many deep-seated biological tendencies that are still present, even in their primitive forms, in man, although in him they are integrated into a far more complex network of intraspecific interactions. "An Englishman's home is his castle"—it is defended territory.

Few animals exclude absolutely all other members of their species from their personal territories. They generally admit at least a mate and perhaps also their infants and defend the territory from other groups. The many ants in a single anthill fiercely attack trespassing outsiders, whether of their own species or another. The howling monkeys of tropical America live in bands and defend their territory from neighboring bands, although the defense is more likely to be a howling contest than a physical encounter. Does this sort of phenomenon also occur among humans? How has *intra*specific competition been related to the evolution of territoriality? Can territoriality be regarded as an adaptation benefiting the individual family? The species?

AGGREGATION AND COOPERATION

Group defense of territory introduces, in relatively complicated form, other widespread kinds of relationships among members of a species. "Birds of a feather flock together." Members of a species usually tend to occur together, to be *aggregated*, and within such groups complex interactions in addition to competition may develop. In plants and in many animals, especially those with comparatively simple behavior, the aggregation is passive. Individuals do not seek out each other's company, and there is little or no differentiation of roles within the group. Even so, the aggregation as such has biological significance and survival value for the species. If nothing else, it facilitates interbreeding in biparental species.

Animals with more complex behavior are, to be sure, often solitary in habits. They live alone or briefly with a mate, and relationships within the species are dominantly competitive. What birds or mammals, for instance, do you know of which this is true? However, an opposite tendency has appeared over and over again in the course of evolution. In many species, individuals live in groups. They may actively seek each other out, and in any case they do actively maintain their aggregation. Many species of fishes gather into large schools or shoals (Fig. 25–12), which move about in a body and may stay together as a group over long periods of time. Some birds live in flocks all the time, and many others live in family groups on defended territories during nesting season and then gather into large flocks when the young take flight. "Prairie dogs" (really a genus of ground squirrels) live in "towns," which frequently have populations of hundreds and may continue over many generations. Bison, cattle, sheep, horses, and other ungulates, both wild and domestic, gather into herds and flocks.

The size of such an aggregation is limited by the available food and, sometimes, space. Competition within the group is one of the factors preventing overpopulation and determining which individuals will survive. But the relationships within the group as a whole are not predominantly competitive. They are at least tolerant, and they are usually cooperative to some degree. Membership in the group and tolerance of or cooperation with others is advantageous to each individual. Decrease of competition *within* the group is advantageous in *external* competition both with other groups of the same species and with other species. Shoaling is protective for small fishes, and it helps to promote breeding. (Herrings, for instance, shoal at mating time.) Sentinels in prairie-dog towns give warning cries at the approach of possible danger, and the signal is passed quickly to the whole population. Predators seldom successfully attacked the solid front of a bison herd.

The formation of mutually tolerant and cooperative groups has been not a universal, but nevertheless a frequent and a highly successful, trend in evolution. The widespread occurrence of the trend should be strongly emphasized. It flatly contradicts the idea that the process of evolution is one of unbounded individual competition in "nature red in tooth and claw." It is particularly important for a true understanding of the evolution of man, the biological basis of human society, and the possible future of our species.

DIVISION OF ROLES AND
THE RISE OF SOCIETIES

In most protists and plants and in a good many animals all individuals of a species are

25–12   A school of fish (*Notemigonus crysoleucas*).

so nearly alike that differences hardly matter. They all play the same role among their *conspecific* ("belonging to the same species") associates and in their community. In species with individually separate sexes (some plants and most animals), the sexes have different roles in reproduction, at least. That difference is often accompanied by marked differences in size, color, pattern, physiology, and behavior, especially in higher animals, with corresponding differences in roles even beyond the essentials of reproduction. Males and females may even belong to quite different food chains, as in mosquitoes, with males eating the juices of green plants and females eating mammalian blood. In groups that defend territory, defense is usually by the male, even though the territory may be shared with a female.

In cooperative groups there is a further evolutionary tendency for differentiation of roles, not necessarily on a sexual basis. Even in loosely organized animal aggregations, there is often an order of dominance or a "social scale" in which each individual has its place. The more dominant animals often take the lead in group activities; they are likely to have first whack at food; the more dominant among the males may monopolize the more desirable females. This dominance sequence is called the "pecking order," because it is readily seen in who pecks whom in a flock of fowls, in which dominance has been extensively studied. If you observe a flock of hens, you will almost surely find that there is one hen that pecks all the others and one hen that is pecked by all the others but never pecks back. Throughout the pecking order, each hen pecks those of lower status and is pecked by those higher in the order. Such dominance is established competitively, but once it is established it reduces competition and tension in the group. After the group is organized into a pecking order, less pecking goes on. Each hen, or each member of the sequence of whatever species, learns to know its place and gives way to its "betters" without further fuss.

Not all groups are organized into pecking orders. For instance, ants, which live in very rigidly organized groups, have none. There may, nevertheless, be a distinct division of roles. Sometimes the division is temporary. In a prairie-dog town or in a roving band of baboons, some individuals generally act as sentinels while the others go about their business, but sentinel duty rotates. Army ants on the march follow a leader, but the leadership shifts rapidly, each leader remaining at the head of the column for perhaps only an inch of the advance. On the other hand, there is often also a more or less permanent division of labor. When army ants are moving their base, the workers always carry the young, and the soldiers always protect the line. When a grazing herd is attacked, in most species it is the mature males that wheel into the first line of defense.

Differentiation of roles in its many forms makes a group more than merely an aggregation. It introduces organization within the group and makes it a *society*.

Societies

"Society" is one of many words loosely used and difficult to define. It has half a dozen different colloquial usages, which are doubtless familiar to you. Even as a technical biological term, it is used in different ways in different subdivisions of the science and by different writers. We shall confine the term to animals, although botanists sometimes apply it to plants. A society differs from other aggregations of animals in being composed mainly or entirely of members of one species, in having a degree of permanence, usually extending over the whole of a life cycle or over successive generations, and in having some measure of internal organization with differentiation of roles. It differs from a family in that the organization may include but is not confined to the roles between two parents or between parents and offspring. Still, the distinction between "aggregation" or "family" and "society" is not absolute. It is evident that various societies have gradually evolved from aggregations and from families and that the distinction must therefore be somewhat arbitrary, without a clear-cut line or moment when the group suddenly becomes a society.

In our definition of societies we include only those that have, indeed, evolved from families or other aggregations of clearly distinct individuals. A roughly, but only roughly, analogous development has also occurred rarely by evolution of colonial organisms, in which the colony arises from a single zygote and the whole colony retains organic continuity. Usually the polyps in a coelenterate colony, for instance, are virtually identical, with none of the individual differentiation of roles essential to the definition of a society. Sometimes, however, as in the coelenterate Portuguese man-of-war (*Physalia*) there is a strong differentiation of roles with polyps in the colony specialized for different functions, such as capture of prey, digestion of food, and reproduction (see Fig. 21–5). Colonies of that kind have occasionally been called "societies" ("organic societies"). Biologically, however, the whole colony is an individual organism in which differentiated organs have been evolved in an unusual way: from the colonial polyps. The difference from a society of separate individuals is radical, and the chief reason for mentioning those coelenterates here is to point out that they do *not* have a bearing on the evolution of societies as we have defined the term "society."

Groups that can loosely, at least, be called true societies have evolved independently in different kinds of organisms and in different ways. All have features in common, and something can be learned of each from any other. In some respects it is hard to distinguish them clearly. Nevertheless, they often represent separate and divergent evolutionary trends, and it is to be emphasized that what is true of one is not necessarily true, or even particularly comparable, in another. That caution applies especially to comparison between the most complex forms of animal societies: those of the social insects and of man.

INSECT SOCIETIES

Everyone knows that some insects have complex societies (Fig. 25–13). This is true at the present time of all termites and ants and of some bees and wasps. Although termites are antlike and are often called "white ants," they are not closely related to ants but more nearly to cockroaches. Societies arose separately in ancestors of the termites and of the ants and

25–13 Termites, social insects. In this royal cell of a termite colony from British Guiana, the queen (reproductive female), who has an enormously enlarged abdomen, lies at the center. The king (reproductive male) is in the left foreground. The smaller individuals are workers and soldiers.

were so successful that only social groups survive. There are still solitary, nonsocial bees and wasps, and various species even now demonstrate most of the intermediate stages between the solitary condition and such advanced societies as in the honeybees. The bees and wasps are younger groups than the termites and ants, and their still incomplete socialization began later. Societies have thus arisen repeatedly and at various times by parallel evolution among different groups of insects. All insect societies have their distinguishing features, but they also share many characteristics because they have arisen in basically similar animals and have tended to follow more or less similar lines of evolution.

All known insect societies have in common the fact that they are extended and specialized families. All have evolved from a family unit of parents and offspring. Many female insects merely lay their eggs where they have some chance of survival and give the offspring no

further care. That is doubtless the primitive condition. Some female insects make protective chambers and stock them with food on or in which the eggs are laid. No direct care is taken of the larvae hatched from the eggs, but food is there for them. Still other female insects stay with the eggs until they hatch and then protect and feed the larvae until they can shift for themselves. These varying degrees of parental care, observable especially among different species of wasps at the present time, strongly suggest that all insect societies arose through much the same stages. The decisive step in the rise of insect societies was evidently that in which the female parent and young remained together for a time as an interacting family group.

Final steps in the rise of true societies among insects occur when the young stay with each other and their mother as they mature, taking over the task of feeding and protecting further broods of young and finally also feed-

25–14 Termite castes. *Left to right.* An immature sexual form. A worker. A soldier with large mandibles.

ing the mother. In these stages there are at least three distinct social roles in the group, and the roles are generally played by individuals different in structure and appearance. The male usually has no part other than fertilizing the female and generally dies or at any rate leaves the group after that act. The reproductive female, or queen, has the primary role of laying eggs from which the other members of the social group develop. Often the queen starts a new group and rears the first brood of young, but sometimes even her first brood is reared by offspring of an older queen. The workers differ from the queen in size and structure and usually do not or cannot reproduce. Thus there are three basic *castes*, as they are called: males, queens, and workers. In ants and termites (but not in bees or wasps) there are often other castes, especially soldiers, which are large and protect the group with formidable weapons (Fig. 25–14).

Caste determination is a complex subject, not yet fully understood. In most (but apparently not all) ants, bees, and wasps the males are haploid, developing from unfertilized eggs. Queens, workers, and (in ants) soldiers, are diploid and are genetically females. The differences are usually determined by the amount or kind of food provided for the larvae, but in some species there is evidence of determination by the genetics of the zygote or by characteristics of the egg. In termites the males are diploid and the females haploid. Workers and soldiers may be either male or female genetically, although they rarely reproduce. The younger animals serve as workers and may grow up to be soldiers. Sometimes growth of certain individuals is arrested, apparently by something in the food, and such individuals remain workers throughout life. In all the social insects caste is determined by a complex interplay of genetic and developmental factors.

The important point is that the social role of any of these insects is determined either in the egg or in early developmental stages. Moreover, the behavior appropriate for each caste is *not learned* and is subject to *little modification.* Its elements are hereditary, and heredity confines its variations within quite narrow limits. The number of distinct possible roles is small, generally three and rarely over four.

In insect societies the individual has no choice of role. Changes in roles or increases in their complexity are matters of slow evolutionary change, if they occur at all. A social insect has practically no chance of surviving if separated from its own social group. The group as a whole stands or falls depending on how well it copes with environmental demands and crises. The individuals are organically separate, but they are inseparably united in the social group. Typically, they are all sisters (plus usually one mother) in any one functioning social unit. They are bound together also by deep behavioral and physiological patterns. They recognize members of their own unit (probably never as individuals) chiefly by smell, and they kill or drive out strangers. They often exchange secretions which form an actual chemical bond among them—a curious phenomenon called *trophallaxis*.

HUMAN SOCIETIES

Both human and insect societies are groups of individuals living together and reacting more or less as a unit toward other groups and with respect to many of the forces of evolution. There is some complexity of organization within each group, and different individuals have different roles. Beyond those broad resemblances, the two kinds of societies are fundamentally different, and the parallel so often drawn between them should be viewed with reserve or even suspicion. The biological basis of human society is of course to be sought among human ancestors and relatives, especially the mammals, and not among animals like the insects, which have been evolving in a strongly different direction for hundreds of millions of years.

In human societies, too, the family is basic and persistent, but not to the same extent or even in the same sense as in insects. In all mammals the mother and her suckling young, at least, remain together and so form a sort of primitive social unit. Yet a group in which differentiation of roles is no more than into father, mother, and young is not yet a society. A mammalian society is not an expanded family, but is an aggregation of adults (with their children in the persisting but subordinate family units) with some differentiation of roles other than that of male and female in their reproductive capacities. In mammals other than man the differentiation of roles seldom goes beyond dominance or leadership of some individuals over others. In civilized human societies, with their butcher, baker, and candlestick maker, the differentiation may become almost incredibly complex. An official classification of occupations in the United States lists 40,023 different occupations—and it is clearly not complete.

Heredity and development have some but comparatively little influence on determination of human social roles. Being male or female, constitutionally weaker or stronger, more or less intelligent have a bearing of course, but nevertheless leave a tremendous range of possibilities for everyone. Moreover, the precise role followed is always learned, and consequently there is a flexibility of role and choice wholly absent in insect societies. The human individual is comparatively independent and self-reliant, prospering best in an accustomed social group but able to survive (temporarily, at least) without it, able to change his role within usually broad limits, and able to transfer from one group to another. All normal members of the society are able to reproduce. The lack of differentiation in this respect, so different from the insect society's situation, takes away from the family any really *essential* biological role in the organization of society beyond the function of producing and, to some extent, training the young. This further sharp contrast with insect societies is not contradicted by the fact that with a human society, without regard for differentiation of social roles, the family remains as an essential institution with profound psychological involvements and repercussions in most aspects of life.[5]

Most mammalian social differentiation, including that of human societies, tends to override the immediate parent–child relationship of the included family units. In this and in other respects such societies are *associative* rather than strictly *familial* as in insects. They evolve from aggregations of individuals and families rather than from the single family unit. Above all, the roles are less rigidly determined and (potentially, at least) more varied and changeable.

One of the important consequences of these facts is that there is little inherent limitation in the size or the complexity of human social

[5] The clan system exemplifies a psychological ramification of family sentiment in the delimitation of a larger social unit. The clan is thought of as an expanded family, although this may be more a legal fiction than a biological fact. Even in clans and the many other complicated meshings of the family with other social units, it is obvious that the part played by the family in social structuring is very different in men and ants.

units. In insects such units can be no larger than the number of progeny produced by one female (or sometimes a few of them) and cannot achieve even remotely comparable complexity.

The flexibility of human societies in contrast with the rigidity of insect societies reflects and depends on the similar contrast in the behavioral patterns of the individuals involved (Chapter 14). This is another example of alternative solutions of evolutionary problems. Both kinds of societies are biological adaptations with high survival value, and both have been outstandingly successful in their quite different ways. Institutions and organizations within the framework of human societies may also be viewed as biological adaptations meeting, by different and generally more complex means, the same broad kinds of basic needs as are met within insect societies. Typically human organization depends on the biologically versatile, all-purpose human individual; ant organization depends on the biologically restricted, specialized, and almost depersonalized insect individual. Sociology has a biological basis and can be studied from a biological point of view, but it is a complex and highly special subject beyond the scope of general biology. Biological aspects of human populations and society are discussed in the next chapter.

CHAPTER SUMMARY

Rise and fall of populations: population size and changes in it determined by birth rate, death rate, and survivorship; the potential life span, rarely reached in nature—about a century in man, not much exceeded in any animals, from minutes to more than 4000 years in other organisms; survivorship curves.

The Malthusian principle that the potential reproductive rate of any species is enormously greater than the capacity of the environment: elephants, slow-breeding, as examples; the inevitable result that only a small fraction of offspring, or potential offspring, can possibly survive and breed in turn; the apparent exception in human populations only apparent and temporary.

Limiting and balancing factors in population growth: budgets of materials and energy; food chains, with later links on smaller budgets; predation, requiring fewer predators than prey and limiting both; mistaken interference with predation balance and the example of the Kaibab deer; competition, intraspecific and interspecific; density, with both overcrowding and undercrowding reducing populations.

Population periodicity: cyclic fluctuation, due to biological reactions within communities; the example of varying hares and lynxes; noncyclical, episodic changes from environmental incidents; rhythmic patterns, especially of breeding seasons.

Intraspecific interactions: intraspecific competition, lessened by territoriality; aggregations of organisms, passive or casual; cooperation; the importance of noncompetitive and mutually helpful aggregation in evolution; division of roles, leading to societies; pecking order; functionally differentiated colonial organisms not societies.

Insect societies: termites, ants, bees, and wasps; stages in wasps, showing origin from the association of female and young; castes, three or four in number; an insect society evolved from a family, with roles not learned and subject to little modification.

Human societies: the limited analogy with the totally distinct origin of insect societies; a human society not evolved from a family but from a mammalian aggregation of adults; the enormous number of roles, many potentially available to any individual and all learned and subject to extensive modification; the basis of a human society in a flexible determination of behavior; the importance of the individual; familial insect and associative human societies as alternative solutions of similar biological problems.

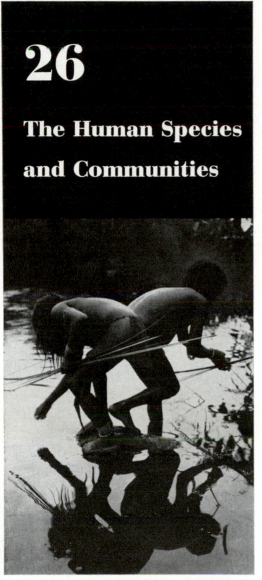

26

The Human Species and Communities

UNESCO Courier

Amazonian Indians, fishing with bow and arrow, are so clearly dependent on the immediate environment that one quickly perceives that they have—like the fish they hunt—an ecology open to analysis. And so have all the other races of man.

Man and Nature; Man in Nature

Let us take a quick tour of the world, beginning at home. You are reading these lines in a human community, probably in a town or city where little or nothing remains of the natural community that existed there before man came. Somewhere nearby are fields where the original communities have also been eradicated and replaced by rich, new communities partly (only partly) of man's choice. From those fields and from other communities, near and far, even from the opposite side of the world, ships, trains, trucks, and planes bring foods and other materials into your community. Power, fuel, and water are also poured into the community from nearer or more distant sources.

A first hop to a rain forest in the Amazon Basin offers a major contrast. Here you may find small, naked, reddish-brown people, quite different in appearance from your neighbors at home (see the photograph on this page). Some of them hack out small clearings along the edge of the forest where they plant cassava, corn, or other crops. Some gather wild fruits and vegetables and hunt the birds, mammals, and other animals of the forest. They use no power but their own; they find fuel and water where they may be. But even they are almost certain to have beads, knives, fishhooks, and other things that originally came from distant lands unknown and unimaginable to them. Another great hop may take you to central Asia, where squat, powerful, weather-beaten brown nomads live in felt houses, travel on horseback, and tend flocks of sheep. In southern Asia and its islands yellowish or tan farmers slosh through their wet rice fields. In African savannahs you may see magnificent, tall, shining black people, living in well-organized villages and owning large herds of cattle. At a village perched on the Norwegian coast you may see blond fishermen setting out to gather a crop from the sea.

Nowadays, even if we have never left home, all of us are familiar with these scenes and a thousand others over the varied surface of our planet. Taken together, they illustrate biological facts and principles of profound human significance. Human interest is likely to center on the people in the scenes and to notice their differences first. The human species is endlessly varied, and among its variations are sets of characteristics by which we distinguish local groups and races, analogous to the demes and subspecies of plants and animals. Next, however, is the fact of unity in that diversity. All men are more alike than different, biologically. All groups intergrade and all can interbreed (a fact we may not have had time to note on our hurried trip). Mankind is one species.

Each human community is found to be intimately associated with, indeed really to be a part of, a broader biological community in which there are also many other species of animals and plants. The biological communities differ in different places, and the human communities are adjusted to those differences. The human communities, too, reflect environmental conditions and have local adaptations. Yet, now that we are in the second half of the twentieth century, human communities everywhere interact in ways more extensive and intensive than was usual in earlier biological communities, and they have drawn the biological communities also into a worldwide network of interactions. Everywhere we see that man has changed the environment. He has modified all the biological communities in which he lives. The effects seem slight in the Amazonian forest and are tremendous in your home town, but they exist everywhere that man is or has been. Many of the changes are constructive, from the human point of view at least. They have oriented nature for the benefit of its dominant species. Other changes are decidedly destructive from almost any point of view. They have reduced the capacity of the environment to support our own or other species, and they constitute a serious man-made problem for man.

In taking up these themes of the present chapter, we are drawing together threads that have run through what has been said before. The science of biology is integral, and all of it has a bearing on human life. Logical compartmenting of different biological "subjects" or distinction between the biology of man and that of nature is neither possible nor desirable. The particular aspects now to be considered—man as a species of organism and as a biological element in communities—derive especially from the broader groundwork of Part 5 and of previous chapters of this Part 7. The present themes also impinge on much else, on, for instance, the political and emotional problems of racism and segregation, the economic and technological problems of food supply and industry, and problems of, in the broadest way, man's modification and utilization of his environment—problems economic, social, political, technological, and psychological, all at once. Discussion of all these problems and proposals for their solution must be left to other books. They all do have biological aspects, and it is their more strictly biological basis that is to be reviewed here.

Systematics of *Homo sapiens*

After what has been said, it is unnecessary to stress again the fact that man is an organism, subject to all the principles that apply to other living things. It is also a thoroughly familiar fact that man is an animal of the phylum Chordata, class Mammalia, and order Primates. Systematists are unanimous in considering man the only now-living member of a family Hominidae and in placing all living men in the genus *Homo*, the name of which is simply the Latin word for "man." Our present concern is with the systematics of the present human population, by itself, rather than with the phylogenetic inferences that place man among his relatives in the whole system of animal classification.[1]

The principles of systematics in general (see especially Chapter 18) apply to all organisms and therefore also to man in particular. Here, however, their application is complicated by the fact that man is unique in many respects. Man is the most widespread of all species and has an unusually high degree of what was originally local differentiation. He is the most mobile of all species, and that mobility has always blurred and has now thoroughly mixed the regional differences usual in a widespread species. On top of that, man's more strictly biological characteristics are not only overlaid by but also inextricably inter-

[1] Chapter 31 gives a fuller background to the systematic position of man among the primates.

Top left to right, Socony Mobil Oil Company, Kenya Information Office, American Museum of Natural History. Bottom left to right, American Museum of Natural History, Australian Information Office

26–1 Some races of living man. *Top left to right:* European (Portugal); African (Kenya); Asiatic (China). *Bottom left to right:* American Indian (Assiniboin); Australian aborigine.

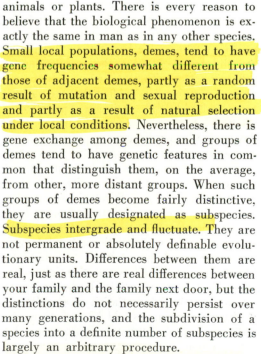

woven with cultural factors wholly absent in most species and only barely incipient in any other species of animal.

ORIGIN AND NATURE OF RACES

It is entirely obvious that mankind includes numerous groups that are physically different (Fig. 26–1). Chinese, American Indians, and Scandinavians are groups within which there is great variation, but there are also consistent differences between any two of these groups. A member of one can almost always be distinguished from a member of another. Interbreeding can and does take place between the groups, but it is far less common than breeding within one group. If, however, the whole sequence of peoples through Europe and Asia and over into the Americas is taken into consideration, it is impossible to draw a fixed line, either geographically or between populations, and say, "Here, precisely, is where the Scandinavian physical type leaves off and the Chinese begins, or the Chinese ends and the Indian starts."

That kind of situation is common in species other than *Homo sapiens*. It is, indeed, the rule for populous, widespread species of either

animals or plants. There is every reason to believe that the biological phenomenon is exactly the same in man as in any other species. Small local populations, demes, tend to have gene frequencies somewhat different from those of adjacent demes, partly as a random result of mutation and sexual reproduction and partly as a result of natural selection under local conditions. Nevertheless, there is gene exchange among demes, and groups of demes tend to have genetic features in common that distinguish them, on the average, from other, more distant groups. When such groups of demes become fairly distinctive, they are usually designated as subspecies. Subspecies intergrade and fluctuate. They are not permanent or absolutely definable evolutionary units. Differences between them are real, just as there are real differences between your family and the family next door, but the distinctions do not necessarily persist over many generations, and the subdivision of a species into a definite number of subspecies is largely an arbitrary procedure.

The major races of mankind as usually designated, such as the Mongolian or Caucasian race, are exactly like the subspecies of other species of animals. They have all the characteristics of subspecies mentioned in the last paragraph. In fact, they are subspecies, or perhaps (for reasons soon to be mentioned) it might be better to say that they *were* subspecies. The word "race" has been so abused for political and ideological reasons that some well-intentioned persons have denied the fact that such subdivisions really occur. Of course they do, and perhaps it would avoid unscientific emotionalism if we called them "subspecies" instead of "races." That would emphasize the biological facts that races or subspecies are not fixed and separate units but are arbitrary and shifting subdivisions of the species and that there is no rational way to designate them as "higher" or "lower," "primitive" or "advanced." They are simply regional populations all with the same status within the species.

Subspecies have intermediate demes that can hardly be assigned with assurance to one subspecies or another. So do human races. Subspecies include demes and groups of demes that could, as a matter of taste or preference, be distinguished by name or considered separate subspecies. So do human races. Subspecies are practically always differentiated on a geographic basis; their differences are differences between the populations of different regions. That also was surely true of races. Even now it is evident that the various races evolved by differentiation of populations in different parts of the earth. In the early days of human history populations were sparse. Family units and demes of a few interbreeding families were widely scattered and comparatively isolated. Even with occasional interbreeding among all intervening demes, passage of a gene by inheritance from a deme in, say, South Africa to one in China would take a great many generations. It might well not occur at all, especially if, as would be likely, the gene had less selective value elsewhere than in South Africa. Under such conditions genetic differentiation of populations in different regions would be sure to occur, and there is no doubt that the human races did arise in that way.

Differentiation of subspecies is sometimes (but not always or even usually) a prelude to the rise of distinct species. It becomes so if a barrier to interbreeding arises between two subspecies and persists until interbreeding ceases entirely and finally becomes impossible. If mankind had continued to live under the primitive conditions of race differentiation, it is entirely possible that speciation would have occurred and that there would have been, for instance, separate African and European species of men. That did not happen, and now it cannot happen. The partial isolation and regional divergence among groups of primitive men that *might* have made them separate species came to an end. Expanding populations reduced isolation. Human mobility, constantly increasing, practically wiped out any effectiveness of geographic barriers. There is now no place on earth where the population consists entirely of a race, or any other subdivision of the species, originally differentiated there. Effects of the prehistoric subspecies are still apparent enough, but biologically the subspecies really no longer exist as such. There is no chance at all that the same sort of geographic divergence will continue or will be resumed.

There are still barriers to interbreeding, but now they are mainly cultural or social and only secondarily or in minor degree geographic. In part they follow the lines of the old geographic subspecies, as in the case of the cultural barrier in some countries between interbreeding of descendants of the prehistoric African and European subspecies. Many of the cultural barriers, however, follow quite different lines. Just now there is little interbreeding between Communists and non-Communists, and there has long been comparatively little between members of conflicting religions. In some European countries generations passed with little interbreeding between the hereditarily (but not genetically) wealthy and the hereditarily poor, but that barrier is fast dissolving. If such cultural barriers were strong enough and continued long enough, they could lead to biological subspeciation, or even speciation, along wholly new lines. It would be entirely possible for a Communist race or a Catholic race to evolve. The chance is slim, however, for the barriers are not clear-cut, and cultural change is much more rapid than biological evolution. All we can say now is that racial divergence along the lines of the prehistoric subspecies has ceased. It seems impossible to predict whether racial differences will eventually disappear, whether renewed subspeciation on a cultural rather than geographic basis will happen, or whether there may be some other, unforeseen outcome.

RACIAL CHARACTERS

"How many races are there?" is a question that does not make biological sense. In order to answer that, races would have to be fixed or sharply definable units, and they are not. There is authority for classifying mankind into from three to 30 or more races. No two family groups are quite alike (nor yet wholly different), nor are any two demes or groups of demes. (And "family group" and "deme," although useful terms, are also definable only vaguely and not absolutely.) It is purely a matter of convenience how many and what groups of men are to be called "races." A well-considered classification [2] recognizes European (Caucasoid), African (Negroid), Asiatic (Mongoloid), American Indian, and Australoid races, and that is probably as convenient an arrangement as any. Its author recognizes that it is arbitrary and that many smaller groups cannot with assurance be placed in one or another of those races.

A concept involved in earlier attempts to specify or identify the races of mankind is that of the "pure race." That, too, is nonsense biologically, a fact more pertinent here than the further fact that it has been connected with some of the most vicious notions ever to enter the human mind. A race or subspecies is a biological, systematic group and can be defined sensibly only in terms of populations and their heredity, hence of frequencies of genes and their effects. Presumably a "pure" race would be a group of individuals all with the same genes or, at least, all free of some "impure" genes from their neighbors. No such group within a species exists in nature or can ever have existed in mankind. Adjacent demes, whether or not we classify them in different subspecies, exchange genes, and any

continuing difference is merely a matter of how frequent various genes may be. Here and there, to be sure, a deme might exist with a frequent or universal gene (more strictly allele) quite lacking in some other deme, but still over the generations genes in both would fluctuate and would be exchanged with other demes. It is not biologically possible for a deme, a race, or a subspecies to be "pure" in any meaningful sense. The idea that there were a few pure primitive races, whose mixed-up descendants we are, or that there is now such a thing as a pure Nordic, Anglo-Saxon,[3] or any other race is a fairy tale, and not a pretty one.

Adaptation is so universal in nature and is so clearly involved in differences between subspecies that it is a reasonable conclusion that many differences between human races were originally adaptive. (This does not exclude the probability that some arose from random fluctuations of gene frequencies in small groups of primitive men.) That has been hard to check, however, one way or another. There is good, but inconclusive, evidence that differences in skin color were adaptive when and where they arose. The color of the earliest men, doubtless variable as color is in all races, probably ranged through shades of brown. Darker or black skin was perhaps an adaptation in regions of damaging ultraviolet solar radiation, and lighter or white skin an adaptation to regions where that radiation (which is beneficial in *small* amounts) was deficient. Such genetic changes under the influence of selection take long periods of time. Whites have not been in Africa or Negroes in America nearly long enough for selection to have made a perceptible difference in their skin color. Is white or black skin more "primitive"? Which is "better"? Neither is more primitive than the other. Each was adaptively better under different conditions when and where it arose. Neither one seems to have a biological advantage under present conditions.

[2] By W. C. Boyd, *Genetics and the Races of Man* Little, Brown, Boston, 1950. He also lists a hypothetical "early European" race, perhaps now represented by the Basques.

[3] Anglo-Saxons, or their descendants, are not a race, anyhow, or any other even arbitrarily definable biological population. Neither, incidentally, are Jews, although some groups Jewish in religion have tended to become so, as have some other religious groups.

Skin color is so obvious that we are inclined to think of it as the main indication of race. It is, however, superficial, both literally and figuratively. It would be more interesting and more important if we could measure hereditary differences in intelligence and personality between races, but so far all attempts have failed. The difficulty is that what can be measured under the names of "intelligence" and "personality" is so strongly influenced by learning that genetic differences (although they surely exist) are obscured. Groups of American Negroes usually average lower than whites from the same region on the scoring of standard intelligence tests. It is, however, known that these tests are strongly influenced by education and cultural background and that on an average Negroes do not now have equal educational and cultural opportunities with whites of the same region anywhere in the United States. The suspicion that it is *that* difference, and not a genetic difference, that is being measured is confirmed by the fact that Negroes test higher than whites when they have a cultural advantage, for instance, in comparisons of urban northern Negroes with rural southern whites.

About all that can be said now is that genetic racial differences in average intelligence and personality are so small in comparison with cultural differences and with genetic variation *within* each race as to have doubtful importance. There is also a theoretical biological reason for expecting this result when measuring intelligence, at least. During the prehistoric period of racial differentiation, it is inconceivable that intelligence was not at a premium everywhere. It would, then, be expected that natural selection would favor intelligence in all races, although the response might not be precisely the same everywhere. (This selective trend may be another that no longer operates in some modern societies.)

Recently a more direct approach to the investigation of genetic racial differences has been begun. Study is made particularly of the blood groups, the genetics of which are known and simple and which can easily be determined in individuals. The best-known of many different blood groups (O, A, B, AB) are dependent on three alleles, the frequencies of which have been measured in many popula-

tions (Fig. 26–2). In combination with similar data on other human genes, this study provides a more rational and meaningful basis for the biological classification of mankind than older studies based on such things as skin color or head shape.

UNITY AND DIVERSITY

Man exemplifies in high degree the interplay of unity and diversity evident in the whole realm of life. Living men constitute a single species, *Homo sapiens*, by all criteria of modern definitions of species. They share a rich genetic heritage, and all are parts of an interbreeding, worldwide population. Specific unity makes it possible for desirable hereditary traits to be combined and spread, whatever their source. It makes all men fundamentally alike, with incomparably more biological resemblances than differences. The cultural apparatus has stopped or reversed primitive trends toward divergence. Mobility and communication are cultural reinforcements of the unity rooted in biological facts.

On the other hand, within the species there are innumerable differences among individuals and between groups, and both biological and cultural factors maintain diversity. Extensive variation within the bounds of one species is biologically favorable both because it extends the immediate range of adaptation and because it is the indispensable material for any progressive change. There is absolutely no evidence that any race or other group is inherently, biologically, better than another. Originally racial differences probably made *each* race better fitted biologically to the particular place where it lived, but cultural changes have largely or wholly eliminated even this local superiority of any race over any other.

Basic Ecology of *Homo sapiens*

The diversity of human demes and races was originally related to the diversity of the natural communities in which men live. It still is to some extent, but with great changes involved in man's cultural evolution, an evolution arising from and interacting closely with his biological evolution.

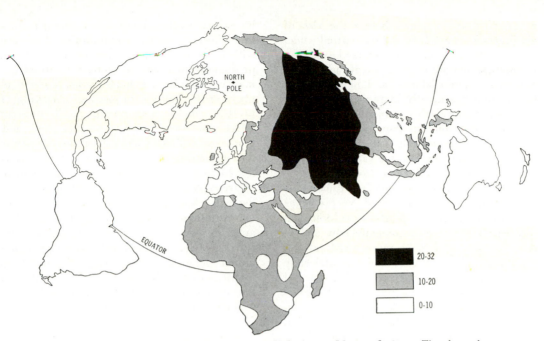

26–2   Distribution of the L^B blood group allele in world populations. The three degrees of shading indicate the percentage frequencies of the L^B allele. It is clear that the gene controlling the B blood group is not uniformly distributed among human populations; it is relatively abundant in Asiatic populations and virtually absent in American Indians. (Note that the data apply to original populations in the areas concerned; B blood groups are now common in North and South America among populations that have come from Eurasia.) Of course, the same inhomogeneous distribution applies to the other blood group alleles: L^O is relatively abundant in the Western Hemisphere and less common in Eurasia. See also Fig. 17–5, which illustrates in more detail the blood group differences between adjacent races, in that case three Indian tribes in Montana.

Man everywhere lives in interspecific animal and plant communities, and he depends on his environment no less than any other organism. It is part of his cultural development that he reacts on the environment more widely and more strongly than any other species—a fact that would require special attention to this particular species in any book on general biology, even if it were written by a squirrel or an oyster.

PRIMITIVE MAN AND ENVIRONMENTS

Man's primitive position in natural communities was that of a particularly large, vicious land animal with an unusually wide range of foods. We are *omnivorous* ("eating everything") structurally, physiologically, and psychologically. That does not mean that men do or can eat literally everything, but that they eat almost any kind of animal food large enough to be worth the trouble and also a great variety of concentrated plant foods,

especially fruits, seeds, and tubers or starchy roots. Biological consequences of this breadth of adaptation were that man could find food in almost any of the endlessly diverse natural communities of the earth and that in each community his status was complex. He became part of almost all food chains that could include a large animal.

Even before the clear establishment of civilization,[4] the human species had spread into almost all land communities. Spread into the Americas was slowest, but it occurred thou-

[4] The distinction between culture and civilization is important but not entirely clear. "Culture" includes social organization, learned activities, and constructions or manufactures of any human group. "Civilization" is simply a complex degree of culture, usually including a fixed form of government, communication beyond the face-to-face level, specialized agriculture and industry, and construction of at least semipermanent dwelling and business centers. All human groups have cultures. Nowadays most of their cultures are civilizations.

sands of years ago and before the rise of civilizations anywhere. Noncivilized man lived practically everywhere on land except on the highest mountains, on some of the smallest and most barren islands, and in Antarctica. Civilization has greatly increased the density but has not appreciably changed the extension of man's distribution. Noncivilized cultures coped with a tremendous variety of environments, and they now survive (although everywhere with some impingement of civilization) precisely in the environments most difficult for man, those that are very cold, very dry, or very warm and humid. Early, slow biological adaptation was soon accompanied by and finally almost replaced by more rapid cultural adaptation. The two were and are intricately compensatory, and degrees and forms of each have varied greatly. The Alacalufs (now-extinct natives of Tierra del Fuego) devised practically no nonbiological protection against their severe climate, while the Eskimo culture includes excellent protection by shelters and clothing.

Man's biological reaction range is exceptionally wide. In addition to that came cultural evolution which enables a man of any biological constitution to make a cultural adaptation to almost any human environment. Thus are explained why biological differentiation, speciation, did not proceed far and why man's spread over the earth did not await the slow process of biological adaptation to each radically different environment.

Primitive culture was in large part, or at least it included, adaptation to environment. Environment was thus a decisive determinant in the diverse cultures evolved among primitive men. Civilized cultures are actually no less conditioned by environments, but the conditioning is less local and there is to far greater degree a reciprocal conditioning of the environment by culture. The most primitive men were surely food-gatherers. Like an ameba, a tree, or a lizard, they accepted nature as they found it and fitted themselves into a natural community as another species among many, differing only in their increasing ability for cultural as well as biological adaptation.

The immediate factors to which primitive man became adapted in each community are,

by and large, the same environmental factors that affect all natural communities. They are still important, although the interactions are profoundly modified in modern civilized human communities. Climate, soils, water supply, and the biotic environment consisting of protists, plants, and animals still strongly affect your own life, and you can see how much more direct and decisive the effect was on primitive man. Climate and soils determined, and still do to large extent, the growth of green plants on which the whole community depends. Animals interacted, and still interact, with man as foods, as competitors, and as antagonists (parasites and predators on man, for instance). Although he is decisively a land animal, man also early developed shore communities that derived most of their food from the sea and its shore. Fishing in inland waters was also among the earliest sources of human food.

Civilization brought other environmental relationships not found in primitive cultures or in organisms other than man, and yet with an element of primitive, direct environment–organism reaction. Industry requires power and raw materials (other than food), and this has environmental relations of no significance to nonindustrial organisms. The distribution and nature of human communities is influenced by the natural distribution of water power, mineral fuels, ores, and other industrial resources. In this, however, as in all his relationships with nature, civilized man has greatly modified the directness of his interaction with the environment.

CULTURAL MODIFICATION OF HUMAN ECOLOGY

Biological adaptation to direct effects of climate involves flexibility of reaction range, evolution of a new reaction range, or both. Cultural modification of the relationship, up to the present, has been almost entirely by protection against climates: clothing and shelter. Air-conditioned homes and other buildings represent a climax of that cultural trend. It is, by the way, an interesting example of the complexity and imperfections of cultural adaptation that even now in many places men (generic, of course, including women) wear clothing demonstrably and radically inappropriate as adaptation to the existing climate. The reciprocal influence of man on climate is as yet unimportant. Men have tried to tinker with the weather, but no real effect on climate is yet evident. Other human activities, such as large drainage and irrigation projects, really have inadvertently changed local climates to some degree, but not widely or markedly.

Cultural changes have probably been greatest as regards food. The gathering of wild foods, sole resource of the earliest men and of almost all other organisms,[5] has become comparatively unimportant. It affects civilized communities on a large scale only in the form of commercial sea fishing, which is an enormous industry but yet accounts for only a small fraction of the food of mankind.

Even among wild species of aquatic foods, many are now deliberately propagated by man and not simply gathered: oysters, trout, and pond fishes, for example. Gathering of wild land foods, such as wild rice, wild berries, piñon nuts, wild rabbits, deer, or antelope, persists for variety and recreation but plays no essential part in the food supply of civilized man.

Food gathering has, as everyone knows, been almost entirely supplanted by agriculture and animal husbandry, both of which go far back in history to the dim days even before the definitive rise of early civilizations. This is one of the two most radical changes of man's relationships with environments. It is decidedly a reciprocal interaction. On the side of the organism, the food supply of *Homo sapiens* has been enormously increased and made more dependable. On the side of the environment, everywhere that cultivated plants are grown or domestic animals are raised the biotic environment has been profoundly changed. There have also been repercussions in the physical environment, for instance in the composition of soils or the localization of erosion.

The other of the two most radical cultural

[5] There is hardly any respect in which human activities are unique in kind, although most of them are unique in extent and complexity. Some ants grow cultivated plants (fungi) and do not gather wild food; some could claim to have invented animal husbandry, as they tolerate, indeed cultivate, aphids from which they obtain a sugary fluid much as we obtain milk from cows.

changes in the organism–environment relationship of *Homo sapiens* has been decreased dependence on the immediate environment of here and now. Other organisms interact with an environment with which they are in actual physiological or sensory contact at each moment. That immediate environment is of course influenced by more distant happenings, and one cannot draw a circle anywhere around any natural community and say, "Here environmental effects on the community definitely end." Nevertheless, it is true that nonhuman communities interact with their environments on a highly localized basis. That must also have been true of early *Homo sapiens*, but it decidedly is not true of modern man. He mines iron ore in Venezuela, smelts it in eastern United States, manufactures it into machinery in the Midwest, and uses the machinery in Australia. In Australia he grows sheep, which are eaten in Great Britain, where wool from the sheep is also manufactured into cloth, which is worn in South Africa. He pipes water over the mountains to Los Angeles and there lives by the millions in a desert that might support a scant dozen families living on the actual capacity of their own immediate environment.

The worldwide interlocking of human supply, commerce, and industry is thoroughly familiar to you, perhaps so familiar that you do not think about it enough. We need not belabor the point beyond emphasizing that this is a biological phenomenon. Local environmental interactions have by no means been eradicated and are as important as ever, but they are caught up into a worldwide network of interactions. It is becoming increasingly true that a reaction anywhere affects the human environment in some way and degree everywhere. *Homo sapiens* is rapidly becoming a species (the only one) in which an individual's environment is not only his own surroundings but the whole earth.

Modification of Environment

The results of man's cultural changes in his relationship to the environment have not affected man alone. They have reverberated through the whole world of life and have affected the environments of countless other organisms. If a state of nature is defined as one uninfluenced, directly or indirectly, by man, then you would search far to find a state of nature anywhere on earth today. It is certainly not to be found anywhere that man lives or travels, not in the remotest Amazonian jungle or the most barren desert. Perhaps it still exists on some unscaled Himalayan height or unplumbed oceanic deep, but even there some remote effect of human activity is likely to be felt.

All organisms influence their environments, and many have some small measure of active control over parts of their environments. The fact is obvious when a beaver builds a dam, and it is no less true when a tree sheds its leaves. Man, however, controls and modifies environments more extensively than any other organism.

LAND USE

Here in the United States and in many other of the most intensely developed parts of the world the most obvious widespread changes have been produced by man's destruction of natural communities to convert them and the space they occupied to his own uses. Indians had already considerably modified their environments in North America, but the changes they made were insignificant in comparison with what has happened since Europeans settled here. In 1492 almost the whole eastern half of North America was covered with vast forests. About 300 million acres of forests have since been destroyed. Most of the early clearing was for farmland; that is to say, biologically its purpose was to remove native communities with low supporting capacity for *Homo sapiens* and to replace them with controlled communities more productive of human food. Clearing has now become a negligible activity, simply because most of the good agricultural land was long ago cleared. Forests now remain mainly where they are themselves more productive for human use than would be farming where they stand. They are now cut for forest products, mostly lumber. Destruction of forests still goes on, although in the national forests and those controlled by the more enlightened private owners the cutting is so managed as to crop surplus growth

and keep the forest healthy. Even in that situation, the controlled crop forest is a community unlike that of virgin forest. The same is true of extensive tracts early cut for farmland and now gone back to forest by natural succession or deliberate reforestation. We are not, at this point, concerned with whether this is a "good thing" or a "bad thing" but only making the biological observation that the natural communities have been destroyed or profoundly altered. It should be added that destruction of forests still goes on extensively by fires, many of which are of human origin.

Eradication of natural communities has been at least as profound in the vast zone of former grassland between the Mississippi and the Rocky Mountains. Here almost the whole of the climax vegetation has been plowed under and all the communities associated with it profoundly affected. Even where bits of sod have been spared, spreading effects have reached. It is now literally impossible to find anywhere a natural prairie community such as those that occupied thousands of square miles only a century ago.

Besides replacing native vegetation with introduced plants, man has also drained environments too wet for his crops and put water on those too dry. In the United States enormous areas have been drained, mostly in Florida, the Mississippi Valley, and the North Central states. The mere fact of draining radically changes the environment, and of course its usual purpose is to introduce new communities. Other immense areas, mostly west of the 100th meridian, have been irrigated and their native vegetation (for none of the reclaimed "desert" land was bare) again replaced by cultivated plants and introduced weeds.

Much American farmland is used to produce food for domestic animals (of which only the turkey is native to America). Such grassland as remains has also been converted almost entirely to that human use. Introduction of exotic animals on the grasslands and man's incessant efforts to protect them from competitors and real or fancied enemies have also changed the natural communities almost beyond recognition.

Land not used to produce food or raw materials is still not free from human disturbance.

Cities, industrial plants, airfields, roads, and other such constructions wipe out the natural environment wherever they are built, and in the aggregate they cover a considerable and rapidly increasing fraction of the surface of the land. Even areas not permanently inhabited, such as the high mountain country, or those deliberately kept "natural," such as the national parks and parts of the national forests, are intensively used by man for recreation, if nothing else. No biologically minded person who joins the traffic jam in, say, Yellowstone Park can imagine for a minute that he is seeing a community natural in the sense of absence of radical human disturbance.

Even now—and intensity of use increases yearly—it is unlikely that there is an acre of land in the United States that is not sometimes and in some way used by man for his specific purposes and that has not been changed by that use. The same is true almost everywhere on earth, true to even greater extent where man's occupation has been longer, as in Europe, and true to less extent where exploitation has been less intense, as in the Amazon Basin, but nevertheless true. If you would like to return to nature, you were born too late.

DISTURBANCE OF COMMUNITIES

Let us now consider briefly but more explicitly some of the things that happen to communities as a result of human activities. The most radical thing that happens is, of course, that a community is destroyed as such and another put in its place. This is by no means a simple occurrence. When grassland is plowed up and planted in corn, the result is not simply that we have corn instead of grass. The grass and other associated green plants were, as you know, only a first link in innumerable food chains and one element in a populous community. Some of the food chains are cut off at the bottom when the grass is destroyed. The organisms in those chains die out or move elsewhere. It is man's intention that corn shall start food chains leading only to himself, but with all his controls he is unable to carry out those intentions. Some members of the destroyed community happily switch to corn, and some of the old food chains continue. Other organisms that eat corn (any part of the plant) move in, and their populations increase. New food chains develop. A whole community organization is rapidly resumed on the new basis.

Wherever man breaks the soil, drains it, waters it, or otherwise changes local conditions, he creates a new environment that is quickly exploited. Everyone who has lived in the West knows that tumbleweed (Russian thistle, *Salsola pestifer*) has great difficulty in breaking into an established community. It almost never grows anywhere but along roads, on plowed or abandoned fields, and on range so badly overgrazed that much of the native vegetation has been destroyed. Under those conditions of human disturbance it rapidly spreads everywhere. There are many other plants, including some that are natives (such as bee plant, a species of *Cleome*), that spread rapidly along the sides of roads and on other disturbed soil. It is a general observation that cleared land is rapidly invaded by some plants and not at all by others, and that there is intense selection and rapid evolution in adaptation to what is always a new biotic environment and often also a changed physical environment. Thus not only new kinds of communities arise where man has passed, but also new successions and new climaxes. Things are never again quite the same.

It seems a fairly obvious fact that interference in a food chain or in a predator–prey balance is going to react on the whole chain or on both sides of the balance. Yet men have often deliberately interfered in such situations without foreseeing the results. The destruction of predators in the Kaibab forest is a good example (p. 669). Another along the same lines happens when coyotes are eliminated because they occasionally prey on sheep. The main food supply of the coyotes is rabbits, which may increase so in numbers when the coyotes are removed that the sheep are starved by competition with the rabbits for food. Or again, widespread use of insecticides to control mosquitoes and other insect pests in some areas has led to the death of desirable fishes and birds. And that suggests still another unforeseen result of human interference which has a tinge of ironic justice: use of insecticides sometimes merely has the result of evolving, by rapid selection, a strain of insects immune to the insecticides. Experimentation with the "wonder drugs" has also developed strains of bacteria immune to the drug, and even some that require the drug in order to thrive.

To revert to the tumbleweed or Russian thistle, this is, like a great many of our weeds, an unwitting introduction by man into North America (the Hopi Indians call it "white man's plant"). All over the world man has introduced plants and animals from elsewhere, sometimes unintentionally but often deliberately. So widespread are such introductions that there is now no place on earth where all the plants and animals are natives. If the introductions become established, they are sure to disturb the native communities, and they frequently lead to the extinction of native species. A policy of bringing in plants and birds considered more desirable than the natives has decimated the native flora and fauna of the Hawaiian Islands and has effectively destroyed what was a unique biological community. Mongooses were also introduced and have had a large share in exterminating less harmful native species, as also happened in the West Indies. In Australia cacti and rabbits, also deliberately introduced, expanded explosively and almost ruined the economy of the country before they were brought under tenuous control.

Man has also wiped out numerous species single-handedly, such as the passenger pigeon and the dodo, and he reduced others, such as our bison and pronghorn antelope, to the verge of extinction before deciding to save a few for his own pleasure. On the other hand he has greatly increased the populations of some species, even aside from his domestic animals. House rats should be grateful to man who, although unwillingly, made possible their worldwide expansion. More pleasantly, some native quail find cultivated fields a fine addition to their environment and thrive in community with man.

Depletion and Conservation

All organisms derive materials and energy from their environments. All tend to increase in population and to spread in areas as greatly as is permitted by the capacity of the environment and their capacity to utilize the environment (including, of course, the biotic environment). All are checked, ultimately, by limitations of the environment and by such factors as competition and predation, which do not, as a rule, reduce the utilization of the environment but only determine the proportions of its resources that go to support each of the various species present. There is a tendency for each species to monopolize as much as possible of the supporting capacity of the environment. Man is no exception to any of these biological principles. Quite the contrary, it is precisely these principles that underlie and explain man's spread over the earth and the consequent disturbance of natural communities everywhere. That the disturbance is so incomparably greater than results of the same tendencies in any other species is merely a measure of the complexities of man's demands on the environment and of his success in enforcing them.

No organism other than man faces the environment with such a variety of needs and wants (the word is appropriate for man, but only questionably so for other organisms). Man's mental development and the cultural developments arising from it have made him eminently successful in satisfying even his extreme needs and wants. He wrests from other species what both could use and also has uses for things that have little significance for any other species. Man is in these respects the most successful of species. His capacity is greatest, but he still is acting like any species of organism in a biologically natural way. Yet it is increasingly apparent that some of man's activities in successful exploitation of his environment are reducing the capacity of that environment to support man himself. To that extent he is living beyond his income, and any species that lives beyond its biological income eventually faces decimation or extinction. Man has the advantage in that he can foresee the possibility of disaster and may find means to avoid it.

UTILIZATION VERSUS PRESERVATION

A first reaction to the widespread destruction of natural communities is often a wish that they could be preserved. Many sentimentalists understand "conservation" as opposition to any disturbance of nature. Certainly biologists, above all people, are sensitive to the interest and beauty of nature, but they must recognize that the attitude of purely sentimental conservation—that is, a desire for the preservation of untrammeled nature—is as unsound biologically as it is impossible politically or socially. As a practical matter, and regardless of whether such action would be desirable, it would now be absolutely impossible to find a natural community of any great extent completely uninfluenced by man and to keep it so. If it could be done, what benefit would arise? Any visit to such a refuge or any study of it would immediately end its pristine nature.

Man is an organism, and all organisms utilize their environments. Man will do so no matter what anyone says or does. Even his destruction of nature is biologically natural. His utilization of the whole of his environment is biologically inevitable and is desirable from the viewpoint of his own species. Surely for a human being to take any other viewpoint would be monstrous. It is, however, evident that such utilization may be either wise or foolish. Wisdom in this sense also has a biological significance. Utilization is wise if it does really, both immediately and in the long run, produce maximum benefit for our species. That is the practically and biologically sound definition of conservation.

Wise utilization for production is that which increases or at any rate does not decrease the capacity of the environment to meet human needs. Increase in capacity, although it can be spectacular, is inherently limited. It has, indeed, been found that rapid increase may actually lead to ultimate decrease, as has happened in agricultural areas too intensively farmed. The ideal would be to put production on a cyclic basis, turning to the use of man such cycles as have kept all life going continuously for some 2 billion years. That involves a thorough understanding of biological cycles, with emphasis on the facts that you cannot get more out of a cycle than goes into it and that disturbance of any part upsets the whole.

Resources that can be operated in cycles are *renewable*. The possible rate of use is limited by the rate of the cycle as a whole and by the rate of input of energy, a resource that is *not* renewable because of the second law of thermodynamics. To the extent that energy is available and that use of resources is limited to the capacity of the cycle and the cycle is kept in operation, renewable resources are inexhaustible. This *possibility* applies particularly to biological resources, to food and to raw materials derived from plants and animals. Man need never lack food or organic raw materials if his consumption of them is kept to the rate of their production by continuously balanced cycles. Unfortunately, consumption does have a tendency to run ahead of the capacity of the cycles, and some efforts to increase their capacity have actually ended in decreasing it.

Other resources are *nonrenewable*. Their production cannot be made cyclic, not, at least, as man produces them and at rates even modestly sufficient for his demands. Our present major sources of power, fuel, and light are nonrenewable. Petroleum (and its products, such as gasoline), natural gas, and coal are organic in origin, and hence theoretically renewable, but we are rapidly using up the accumulations of several hundred million years, and the rate of renewal is so extremely slow as to have no practical significance. Naturally fissionable elements, such as uranium, are a future source of power, but they are also ultimately limited in amount and nonrenewable. Most of our intensively used mineral products, notably the metals such as iron, copper, and many others, are nonrenewable. Their atoms are not destroyed by use, but they are eventually so scattered as to be unrecoverable by any process economically possible now or foreseeable. An economy as dependent on nonrenewable environmental resources as ours is cannot be permanent in the really long run. Wise utilization of nonrenewable resources is evidently the slowest possible utilization with the least possible waste. When such resources will run out is not primarily a biological problem. It does concern biology that the best chance of an indefinite future for civilization recognizably similar to ours seems to be the eventual substitution of nonrenewable materials by renewable materials, that is, in the main, of mineral production by organic production.

Here (as on p. 668) it is possible to foresee

consequences of continuous increase in the world population, consequences that few will consider pleasant. The absolute limit to otherwise unchecked population increase would be imposed (as it already has been in some countries) by food production. Since each link in a food chain dissipates energy and some materials, a maximum population would have to live as near the beginning of a food chain as possible; human food would consist solely of green plants. Because nongreen plants and all animals compete with man for food, all would be exterminated as far as possible. Pressure for food production would be such that renewable production for any other use would be minimized. Nonrenewable resources would soon be gone, and industry, along with all the products of civilization as we know it, would be reduced to a minimum. What steps will be necessary to obviate this dismal prospect?

Even on a basis of greatest production, it can be shown that wise utilization often involves as little disturbance of natural communities as is practically feasible. The natural cycles may be as productive as any that man can substitute for them. That is probably particularly true of forests, of much semiarid and arid land, and of the seas. Forests over wide areas of poor soils and hilly or mountainous country in the Temperate Zone are more productive of useful organic materials than any method yet devised for farming the same land. That is particularly true of the vast tropical forests, the productivity of which is probably the largest natural resource not yet extensively utilized by man. Cutting down those forests to clear farmland, in the way traditional to us in the Temperate Zone, has so far almost invariably failed. The hopeful new trend is the devising of techniques for cropping the exuberant production of the forests themselves. As for regions with too little rainfall for farming, irrigation is not the full answer to their utilization. There is not enough available water to irrigate them all. Many of the desert and semidesert soils do not repay irrigation even when water can be put on them. In fact, where soils were good they have sometimes been ruined by irrigation through accumulation of salts from the evaporating water. But the natural communities of such areas, adapted to their low water supply, are as productive as the

environment permits. In the sea the natural communities are extremely productive, and cropping of the natural food cycles by man is more effective than any other utilization in prospect.

Hitherto we have spoken of utilization in terms of material production. If the human population does reach the maximum capacity of the environment, no other utilization will have much significance. That point is not yet reached, and some may well hope it never is. In the meantime, we can afford and profit by nonproductive utilization of some of the surface of the earth. Space for recreation, or a silent place in which to think, or simply room to draw a clear breath—most of us think that, too, is good utilization of part of our environment. It is also of value, and practical in the fullest sense, to reserve what space we can for natural communities as little disturbed as possible, where biologists and everyone else can observe and learn more about the processes of nature by which we live and of which we are a part.

MAN AND SOIL

Conservationists are greatly concerned with soil utilization, and for good reasons. We depend on natural soils for most of our food, whether more directly as plant food or less directly as animal food. It is possible to make excellent artificial soils or soil substitutes, but so far, at least, that is so expensive that it has little promise as a possible source of the bulk of human food. It is also unlikely that the aquatic environments, rich as they are, could economically be made the main source of plants and animals usable for human food; they are not so now, at any rate. The soil, then, is the basic factor in supplying man's most basic need, a double "basic" that makes soil fundamental indeed.

It has been carefully estimated that the United States once had 1517 million acres of economically usable farm and grazing land. By the same estimate, 282 million acres had been ruined by use up to 1947 and had become essentially unusable for the predictable future. It was estimated that continued use by traditional methods would soon ruin 775 million acres more. Only 460 million acres were considered in reasonably good shape and not seriously threatened. Some parts of the world are better off, and some are worse. In North Africa, for instance, an Arab historian wrote that it was once possible to ride from Egypt to Morocco under a continuous green canopy. The statement is almost incredible, for much of the ride today is over bare sand and rock; the soil, if soil there was, is gone. That the historian may not have exaggerated widely is attested by the presence of many ruined cities, once populous and now deserted and surrounded by desolation.

Soil is theoretically a renewable resource. All soils were formed, at some time, from bare rock, sands, silts, or barren clays, and soils are continuously being formed now. The process is slow. Under the best of conditions, formation of a good agricultural soil takes centuries, and it may take thousands of years. Agricultural practices that result in loss of soil are therefore essentially mining it, using it up as a nonrenewable resource.[6]

When a farmer removes a crop from a field, he also removes material that came from the soil and that would, under natural conditions, have returned to it. Harvesting cuts off part of the cycle of materials and depletes the soil. The cycle can be kept going by interspersing harvested crops with planting that enriches the soil and by making good the losses by adding fertilizers. More serious and practically impossible to compensate for is actual physical loss of the soil by wind and water erosion after the protective natural plant cover is removed. A windstorm on May 12, 1934, swept up 300 million tons of soil from the plains east of the Rockies and began the "dust bowl" devastation that brought on an economic crisis and the migration of thousands of homeless "Okies," farmers whose farms had blown away (Fig. 26–3). In 1900 drip from a barn in Georgia started a gully in surrounding bare soil that has since spread to 3000 acres and washed away whole farms. The millions of acres of soil already lost in such ways cannot be recovered, but further

[6] These statements are only partly modified by the fact that silts spread by floods may be immediately highly fertile and have their fertility renewed each year, as in the Nile Valley. The fertile silts are soils eroded from areas upstream, and so the process robs Peter to pay Paul.

losses are being slowed down, at least, by contour plowing, terracing, and more care in maintaining a vegetation cover (Fig. 26–4). On grazing land, heavy overgrazing has also laid bare the soil and promoted erosion, besides damaging the grass community and permitting invasion by undesirable weeds. The remedy is obvious although not easy: replanting of hardy grasses and reduced grazing (Fig. 26–5).

Erosion of soil is so real and serious a problem that those combating it may be excused some exaggeration. One of them says flatly that erosion is man-made. Another publishes a picture of a scrawny cow in front of badlands, with the implication that cow and friends are responsible for the badlands. The fact is that erosion is a natural process that went on for several billion years before man appeared and that produced our badlands long before there were any cows in North America. Wherever erosion has taken the soil from farmland and grassland, it would sooner or later have done so without human help. This does not alter the fact that man has greatly hastened erosion and often localized it exactly where and when it does him the most damage. Recognition that erosion is an intermittent but inevitable and natural process everywhere on land puts the matter in truer perspective and should assist in planning to minimize the damage done by a process that cannot really be stopped.

FLOODS AND FLOOD CONTROL

Soil kept porous and covered with vegetation absorbs much rainfall. On cleared land the soil tends to pack, and more rain runs off over the surface. This is of course the immediate cause of accelerated erosion on cleared land, and it is also an interference with the water cycle that has extensive repercussions on living communities. Uplands from which forests have been cleared no longer retain water effectively, and the ground water below them is also less steadily maintained. Organic productivity is thus decreased by a lack of steady water supply. The runoff also tends to produce floods more severe than those that would occur without human disturbance. A return to natural water conditions is generally desirable and can be achieved only by

26–3 **Erosion.** *Left.* Gully erosion in Guilford County, North Carolina. *Right.* A result of wind erosion: sand blown from eroded fields and piled up on a farm in Cimarron County, Oklahoma, April, 1936.

restoring vegetation on the uplands. Even natural flooding, without intensification by human activity, is a serious problem to those who live and farm along such rivers as the Mississippi and its main tributaries.

Flood-control projects represent one of the most heroic efforts of man to control the natural environment and the intensification of natural forces resulting from man's own activities. Some attempt is made to control floods where they come from, in the runoff in the uplands, but most of the effort goes into delaying the floodwaters downstream by holding them behind storage dams. The water so impounded may also be used for generating power and for irrigation. The geologist notes that the expedient is temporary, for all the storage reservoirs will fill up with silt in a few generations at most and then be useless.

(Some have filled in a dozen years. See Fig. 26–6.) Extension of their usefulness demands that silting be reduced by the control of erosion above them. The biologist may add that the adjustment of human utilization to the natural cycles of earth and life is the only *permanent* solution.

RENEWING RENEWABLE RESOURCES

When the whole situation is reviewed, it seems that man's troubles in the utilization of his environment are rooted in biological principles. Life has kept going throughout its tremendously long history by the inpouring of energy from the sun and by interlocking inorganic and organic cycles and sequences of materials and energy on the earth. Man's manipulation of his environment has resulted in its deterioration, from his point of view.

Soil Conservation Service

26–4 Contour farming in the Elm Creek watershed, Bell County, Texas. By running his plowing and planting lines along the contours of the land, a farmer can minimize water, and therefore topsoil, runoff. The practice must often, as in this case, be a cooperative venture between adjacent landowners.

whenever he has ignored these natural processes instead of utilizing them. He accelerates the water cycle and the cycle of erosion, both of which are going to continue, man willing or not. Then he becomes alarmed, but instead of recognizing the nature of the cycles and working with them, he tries to stop them. Intensifying floods by clearing the uplands and then trying to stop them with dams is rather like whipping a horse and then shooting it when it runs away.

Similarly, man ignores food chains and

26–5 Overgrazing on a Texas range. To the left of the fence, where cattle have been kept out, there is still a good growth of grasses; to the right the land has been reduced to a minimum cover of sagebrush.

Soil Conservation Service

26–6   Mono Dam and Reservoir, Los Padres National Forest, California. Note that the reservoir is filled with silt. Dams are only temporary solutions to flood-control problems.

tries to control them by removing the one link that seems to him, often erroneously, to be responsible for trouble, as when he kills coyotes only to have the more damaging rabbits increase. He tries to get out of a cycle more than goes into it, as in the many projects that are pumping down the water table without concern for the fact that no more water is continuously available from the ground than comes into it from rainfall. Also, he nets fish without remembering that next year's population depends on the reproductive capacity of those he leaves this year.

Even more serious than man's living beyond his biological income—his depletion of natural resources—is his unchecked population growth. This has already resulted in untold misery and an unbalanced biological economy over much of the earth. If no effective corrective measures are taken, it will eventually do so everywhere. Material deficiencies are to some extent staved off by biological improvements in production, agricultural advances; and power shortages may be alleviated for a time, perhaps a long time, by the development of new sources, especially from atomic energy.[7] Nevertheless, population limitation is inevitable, and the only questions are whether it will be voluntary and at what level it will occur.

CHAPTER SUMMARY

Man in nature: the human environment in the modern Western world—mechanization, communication, the interdependence of communities—contrasted with that of primitive peoples; mankind as one species with diverse local cultures; the interaction between man and his environment.

Systematics of *Homo sapiens:* man as chordate, mammal, primate, hominid, the single species in the genus *Homo;* the unusually wide distribution of the species; local differ-

[7] It is ironic that the greatest advances in material production and in energy conversion are in countries that need them least or indeed as of today do not really need them at all. The social problems are even more pressing than those more basically biological.

entiation, of both biological and cultural characteristics; origin and nature of human races: population evolution in man as in other animals; the existence of subspecies, usually designated "races"; political and emotional overtones of the word "race"; the impermanence of subspecies as units; the breaking down of human subspecies because of human mobility, with cultural and social factors the only remaining barriers to breakdown and intermixing; racial characters: no unique answer to "How many races?" but useful to distinguish five—Australoid, American Indian, Asiatic, Negroid, Caucasoid; the concept of "pure" race as biological nonsense; skin color and other racial differences as adaptive, at least in part; no evidence for racial differences in intelligence; the use of blood-group frequencies as objective characterizations of races; no race inherently "better" in any biological sense.

Basic ecology of *Homo sapiens:* primitive man and environments; man as omnivorous and as part of all food chains open to large animals, thus permitting the success and wide dispersal of human culture; culture as, in part, adaptive; the environmental impact on human culture; the cultural modification of primitive human ecology—clothing, shelter, the human influence on environment, the transition from food gathering to agriculture, the complex interdependence of modern communities in commerce.

Modification of the environment: land use and disturbance of natural communities; in North America, the removal of forests and the development of arable land; the eradication of natural communities of animals; the introduction of foreign plants and animals; dangers in disturbing community balance (exemplified by Kaibab deer, by insecticide programs, and by introductions into Hawaii and Australia).

Depletion and conservation: the severity of man's demands on environment; problems raised and the possibility of solving them with foresight; utilization versus preservation—conservation's sentimental form and its wise form, as the planned exploitation of environment to the maximum *long-term* benefit of human species; renewable and nonrenewable resources and the problems posed by the latter (exemplified by fuel and mineral resources); erosion and flooding; the prime significance of soils in food production; the destruction of soils and factors involved in soil destruction (overfarming, wind erosion, gully erosion, etc.); erosion as a natural process seriously accelerated by man, especially by the clearing of forests and other practices that increase flooding; flood control; the need for understanding ecological processes, to live within biological income, and eventually to limit population.

Part 8

The Geography of Life

The camels (above) and the llamas (below) in our photograph are closely related animals; they have common ancestors in the not-too-distant past. We have used them to introduce Part 8 because they typify a class of problems that has long fascinated and puzzled biologists. The problems arise from present geographic distributions. The camel is Asian, and the llama is restricted to the Andean highlands of South America. How is it that such certainly related forms occur in such widely isolated areas with no contact whatsoever? Or, similarly, why is it that tapirs occur only in tropical America and Malaya? But, outstanding as these curious cases are, they do not represent the only class of problems that confronts the observer of geographic distributions. Why do cacti occur in deserts in America—and why only in American, not in African, deserts? There are striking irregularities and regularities in the distributions of organisms over the surface of the earth. The subject is a large one with distinct problems meriting special study.

The geography of life is open to two different lines of explanation, which are developed separately in Chapters 27 and 28.

First (Chapter 27), the distribution of any species of organism is ultimately limited by the distribution of suitable environments. The distribution of cacti within the Americas is limited by climate; the absence of cacti from the wet floors of American rain forests is explained by the ecological fact that they are not adapted to conditions of high moisture and low sunlight. The principle applies to all species, although the limiting environmental conditions are rarely so clearly defined as they are for cacti.

Moreover, the limiting conditions are often other organisms, which, as we stressed throughout Part 7, are major constituents of any organism's environment.

Second (Chapter 28), the distributions of organisms are only partially explained by the limited distributions of their appropriate environments. Deserts occur in Africa, but cacti do not. Full explanation of such cases demands introduction of strictly historical principles. Species arise in a particular region, and in the course of time they disperse, expanding into suitable environments elsewhere. Their dispersal is, however, contingent on some continuity of suitable habitats. The spread of cacti (originating in America) into African deserts has been blocked by the barrier of the Atlantic Ocean. The camel family originated in North America. Late in their history, some camels spread to South America and there evolved into llamas, guanacos, alpacas, and vicuñas. Others reached the Old World by way of an Alaskan–Siberian land bridge and evolved into Bactrian camels and dromedaries. With changing conditions, camels have become extinct in their original North American homeland, but they still survive in South America and in Asia

27

Ecological

Biogeography

The saguaro cactus (Cereus giganteus) *occurs only in some parts of Arizona and small adjacent areas of California and Sonora. Its strictly limited distribution within southwestern North America is determined by ecological factors, the topic of this chapter. Its presence here but absence from other regions of the world involves additional historical causes like those discussed in the next chapter.*

Bases of Biogeography

A visit to the country is a pleasure that is greatly increased by an understanding of biological principles. One thing that cannot have failed to impress you in the countryside is that different associations of plants and animals occur even quite near each other. The living things in a stream or lake are decidedly different from those on land along the shore. Farther back on a hillside or a drier meadow other local communities will appear, plainly unlike those of the shore.

Thus at one locality a walk of a few feet can take you from one community into another. If you take a longer trip you soon see that there are broader regional differences in communities. Suppose you drive from New York to Tucson. For the first couple of days you will be in regions where the predominant natural regional community, among the many different more localized communities, is usually a deciduous forest. In the flatter areas most of the forest has been artificially replaced by farmland, but it is still evident that the deciduous forest represents the usual natural climax. Farther along, west of the Mississippi, you will be in a region also extensively plowed into fields but with a natural predominance of open grassland communities. This extends onto the high plains of eastern Colorado and New Mexico, with a change in species of grass apparent if you examine the communities closely and with a botanist's eye. Then you may, depending on your route, have an interlude in mountain forests. Eventually, however, you will see the grassland grading into brush communities, dominated by such plants as sagebrush and greasewood. Finally these communities will grade into others, typical of the true desert, in which cacti, mesquite, and ocotillo are conspicuous.

If now you have the fortune to travel more widely, say to South America, you will encounter still other regional differences in natural communities. Some, such as rain forest or "jungle," live in conditions that you did not encounter on your travels in the United States. It is not surprising that all the species present are unfamiliar to you. Elsewhere, however, the communities may look almost like some you saw back home. The pampas of Argentina are like the grasslands of the United States, and much of Patagonia looks like our southwestern thornbush deserts. Here in these superficially familiar surroundings it is surprising to find that the species present are almost as distinct from ours as those in the rain forest.

We have mentioned only plants by way of example because they are fundamental in all communities and are easy to observe. The animals in the communities tend to have similar resemblances and differences on local, regional, and intercontinental scales. Many of their geographic distinctions are common knowledge. Everyone knows that you have to go up north to see a polar bear or out west to see a grizzly bear in their native haunts. You must go to Asia to see a wild tiger and to Australia to see a wild kangaroo.

These common observations show that communities are not everywhere the same. Living things have a geography of their own, which, because it relates to life, is called *biogeography*. The observations further show that biogeographic relationships require the use of an expanding scale. Some involve a few square meters, others large regions, others whole continents, and others the entire earth. At the narrow end of the scale it is obvious that the geographic positions of living things depend primarily on the environment. The fish is in the stream; the squirrel is on the hillside. As the scale broadens, that factor continues to be evident. A deciduous forest community in Pennsylvania and one in Missouri are similar because the climate, soil, and other environmental conditions are similar. They differ to the extent that the environments are not exactly the same. Farther west on the high plains the climate is different, particularly in having less average annual rainfall. In the drier climate the forest community gives way to the grasslands community, and in the still drier Southwest the grasslands give way to the desert communities.

Such resemblances and differences in communities are ecological, and their study is *ecological biogeography*. In this aspect the distribution of plants and animals and their associations in communities with different geographic positions are functional. They are adaptive. The present chapter is devoted to

that subject. Before proceeding with it, however, we should note that there is more to biogeography than that. Ecological conditions in an Australian desert can be closely matched with one in Africa, or those in a Malayan jungle with one in South America. Ecology always and everywhere affects the geography of living things, but evidently it cannot provide the whole explanation of that geography. The puzzling differences among regional *biotas* [1] that are not due primarily to ecology are due to differences in the histories of the regions. Their organisms came to them at different times and from different places; they also evolved differently once they were there. That aspect of the geography of life is *historical biogeography*, and it is the subject of the next chapter.

Some Principles of Ecological Biogeography

Ecological biogeography depends fundamentally on some broad principles with which you are already familiar. Environments differ from place to place. Every organism is adapted to the environment at the particular place where it lives. Every organism is also a member of a community and is adapted to living with other members of the community, which are, in fact, part of the organism's environment. Interrelations in the community as a whole are, further, adaptive among themselves and are also such as to adapt the whole community to the conditions prevailing in its geographic position. You have been familiarized with these principles and with examples of their operation in previous chapters.

IMPORTANCE OF PLANTS IN BIOGEOGRAPHY

Plants play a predominant role in the geography of biotas, especially of those on the land. Green plants are at the beginning of all food chains. Their nature in a given place strongly influences the nature of later links in the chains and therefore of the whole biota. Plants are also particularly sensitive to variations in the physical environment, especially

[1] A biota is the totality of organisms of a given place or region, its flora plus its fauna.

climate and soil. Although animals are also influenced by such variations, their dependence on a given set of physical factors is usually less narrowly circumscribed.

Apparent dependence of animals on climate may in reality be a dependence on a given vegetation type, which is in turn primarily dependent on climate. Grazing animals are as a rule most abundant in areas with mean annual rainfall about 12 to 30 inches, and with the precipitation irregular or tending to concentrate in a relatively short rainy season. It is unlikely that grazing animals do best in that sort of climate, in spite of the fact that they are most common there. On the contrary, they may suffer heavily from drought in such regions of moderate and unevenly distributed rainfall. But grasses, on which the grazers feed, do well in that kind of climate and are rarely the predominant vegetation in other situations. Many groups of animals are similarly restricted by the distribution of plant communities in which they find suitable food and shelter.

It is no less true in aquatic communities that plants are basic to the whole ecology. Nevertheless, as a rule and particularly in the sea, the more uniform aquatic environments have less sharply distinguished distributions of various kinds of plant communities. There other factors may be more important for biogeography. On land the kinds of plant communities are so varied, so well defined, and so basic for the whole biota that a map of vegetative provinces generally serves to indicate the ecological differentiation of biotas as a whole. Maps of broad types of plant communities [2] are given in Figs. 27–1 and 27–2. To the extent that it is ecologically determined, the distribution of animals tends, by and large, to follow the same pattern.

HORIZONTAL AND VERTICAL CONTROLS

Temperature, solar radiation, and precipitation are the main climatic controls of the plant communities. Each of these factors affects the vegetation directly or indirectly through effect on soil. Important for each of them is not only average intensity but also

[2] These major communities are sometimes called "plant formations" by botanists.

distribution through the year. Temperature and radiation have a familiar north–south gradient the effects of which are evident on the vegetation map of North America. Tropical lowlands are without frost; daily variation of temperature is greater than seasonal variation; and sunshine is intense and of about equal duration throughout the year. Northward, seasonal variation of temperature becomes greater; as we proceed northward, frost is common through a longer and longer part of the year; summers may be no cooler than in the Tropics and are often hotter, but such periods of heat decrease in length to northward; radiation is not so intense as in the Tropics, and its distribution through the year is increasingly uneven until the far north has months of continuous daylight and other months of continuous darkness.

There is also on the vegetation map of North America, especially through the Temperate Zone, an east–west differentiation. That is due mainly to changes in precipitation and subsequent evaporation. Rainfall follows a somewhat erratic pattern influenced particularly by usual air movements, by distance from the coast, and by topography. Precipitation is high on our northwest coast and in the mountains of California. Air moving eastward from there has lost moisture, and the result is an arid belt of deserts and semideserts. The Rocky Mountains again catch moisture, and the plains immediately eastward are semideserts of high, dry grassland. From there on to the Atlantic coast the general tendency, with many local irregularities, is for increase in rainfall, largely because of moist air that periodically moves northeastward from the Gulf of Mexico. Similar irregular zoning of rainfall occurs on all the continents and in all of the north–south climatic zones of temperature and radiation. Deserts as well as lush rain forests occur in the tropics, and indeed one may occur right next to the other.

Besides these horizontal zonings of climates and of biotas along with them, there is a vertical zoning. If you climb any mountain of considerable height, the climate and the life perceptibly change as you go up. The most obvious climatic change is that it gets colder. This change is similar to one that occurs if you do not climb but simply travel north

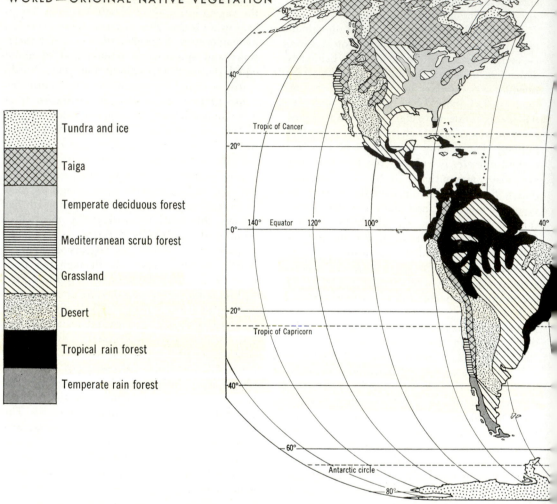

Tundra and ice

Taiga

Temperate deciduous forest

Mediterranean scrub forest

Grassland

Desert

Tropical rain forest

Temperate rain forest

27–1 The distribution of original native vegetation in the world. The categories of vegetation on a map of this scale are necessarily broadly conceived. Thus many of them—like taiga, temperate deciduous forest, and tropical rain forest—include within them a diversity of recognizably distinct subtypes. Within any one wide region there may also be quite different types of vegetation in patches too small to show on this scale.

or south from the equator. Conditions on a mountain are similar to those at sea level farther north. The tops of mountains on the equator have a considerable ecological resemblance to arctic lowlands. Timberline, above which trees do not grow, becomes lower from the southern Rockies into Alaska and finally reaches sea level at the northern tree line. These relationships between horizontal and

vertical climatic zones are shown diagrammatically in Fig. 27–3. The correspondence of an "arctic zone" on a high southern mountain and in the actual Arctic is not, however, exact. Other factors than average temperature—such as lengths of day and night, extreme temperatures, and atmospheric pressures—are quite different in the two areas and may have a distinct ecological effect.

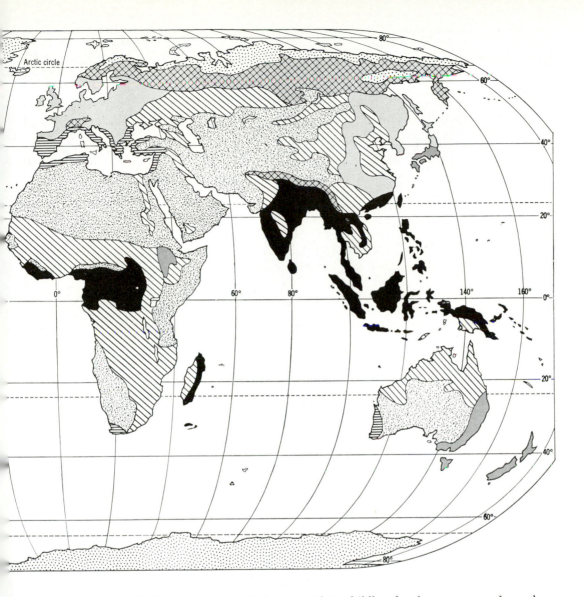

Vertical zoning of biotas is particularly clear in our Southwest, from which a brief example, Table 27–1, may be drawn. There the climatic control is precipitation more than temperature, although that is important, too. The higher the altitude, the greater the precipitation, as a rule.

Terrestrial Communities

Ecological geographic consideration of communities is like those nested Chinese boxes or wooden eggs that some of us played with in childhood; when you opened one there was always a smaller one inside. The biggest box is the whole of life on earth. Next smaller are the whole of the land communities on one hand, and of the aquatic communities on the other. So it goes on down until we are considering the life of one thicket, one meadow, or one pool. What we are now going to consider, by way of example and to bring out further biogeographic principles, is something between the extremes of the series. It concerns broadly regional kinds of communities, each of which is of course quite variable and includes many different, more local communi-

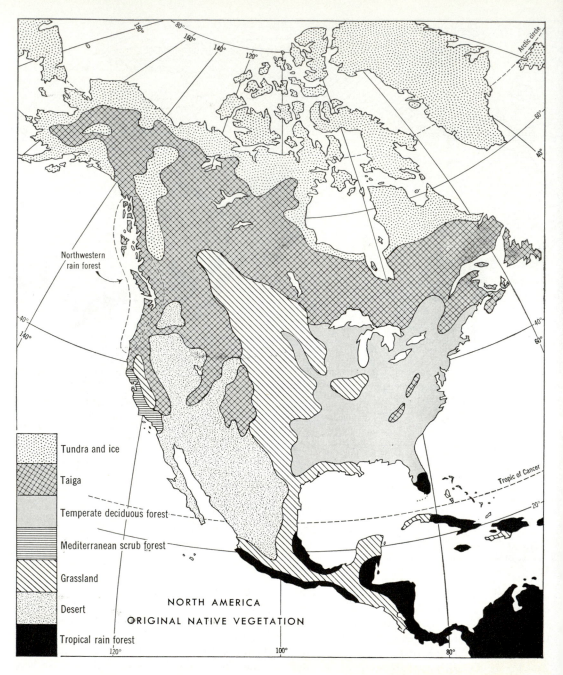

27–2 The distribution of original native vegetation in North America.

Legend:
- Tundra and ice
- Taiga
- Temperate deciduous forest
- Mediterranean scrub forest
- Grassland
- Desert
- Tropical rain forest

Northwestern rain forest

NORTH AMERICA
ORIGINAL NATIVE VEGETATION

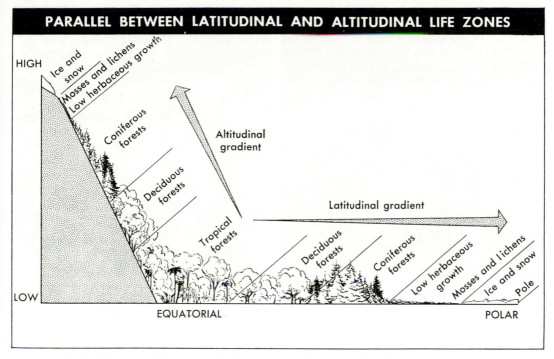

PARALLEL BETWEEN LATITUDINAL AND ALTITUDINAL LIFE ZONES

HIGH

Ice and snow

Mosses and lichens

Low herbaceous growth

Coniferous forests

Altitudinal gradient

Deciduous forests

Latitudinal gradient

Tropical forests

Deciduous forests

Coniferous forests

Low herbaceous growth

Mosses and lichens

Ice and snow

Pole

LOW

EQUATORIAL

POLAR

27–3   The parallel between the horizontal and vertical distributions of life zones.

ties.[3] The most important and widespread of them will be reviewed, but the list is not exhaustive. There are, for instance, highly distinctive kinds of geographically definable communities in salt lakes, marshes, or subterranean environments, but they are not considered here. You might consider them on

[3] Major regional community types of the sort we are going to review are often called *biomes* by ecologists.

TABLE 27–1

Altitudinal life zones in New Mexico

Altitude, feet *	Name of zone	Some characteristic plants	Some characteristic animals
2500–4500	Lower Sonoran	Creosote bush, mesquite, ocotillo	Antelope squirrel, desert fox, road runner, diamondback rattlesnake
4500–7000	Upper Sonoran	Piñon pine, tree juniper ("cedar"), sagebrush, cottonwood, cholla cactus	Prairie dog, coyote, mourning dove, meadow lark, plains rattlesnake
7000–8000	Transition	Ponderosa pine, narrow-leaf cottonwood, scrub oak, wild rose	Mule deer, Abert squirrel, mountain bobcat, black bear, wild turkey
8000–11,500	Canadian	Spruce, fir, aspen, cinquefoil, gentian	Elk, spruce squirrel, marmot, lemming mouse, Steller's jay, junco
11,500–12,500	Hudsonian	Foxtail pine	Cony, mountain jay
12,000–13,500	Arctic-Alpine	No trees. Alpine forget-me-not, spring beauty	Mountain sheep, ptarmigan

* The stated altitudes are approximate averages. In different parts of the state and on different exposures on valley and mountain slopes, the elevation of a given zone may be as much as 1000 feet above or below the stated figure, but the sequence remains the same.

27–4 **Tundra.** *Left.* A general view of the tundra at Dumb Bell Bay on Ellesmere Island, Canada. *Right.* "Reindeer moss" (a lichen) and other tundra vegetation in Finland.

your own. In each community you should especially try to answer two questions, as we shall do in the examples we discuss. The questions are: What are the special conditions of life here? How are they met by adaptations of and within the community?

TUNDRA AND ALPINE COMMUNITIES

In North America, Europe, and Asia a vast northern zone encircling the Arctic Ocean is known as the *tundra,* a word borrowed from the Russians. No similar extensive zone occurs in the Southern Hemisphere, because the south has little land in corresponding latitudes. The tundra (Fig. 27–4) has the arctic climate, cold on an average, with a long, dark winter and long or even continuous summer daylight. Frost may occur at any time of the year, and the ground is permanently frozen a few feet below the surface. During summer thaws the region is extremely wet, with saturated soils and innumerable bogs, ponds, and streams.

There are no upstanding trees in the tundra, but dwarf, shrubby alders, birches, willows, and conifers are common. Mosses, especially sphagnum, and lichens, especially "reindeer moss" (not a true moss), cover large areas. Herbs with large, brilliantly colored flowers are conspicuous and beautiful during the brief

growing season. Temperatures for growth are minimal, and surviving plants must mature without becoming large. They must resist frequent frost. Many of them can be frozen solid at any phase of life, even when in flower, and survive to resume activity when another thaw comes along. Tundra plants spend most of their lives in a state of suspended animation, active only in brief periods of warming sunshine.

Vast hordes of birds, especially waterfowl, nest in the tundra in summer, but most of them desert it in winter. Permanent residents include a few birds and mammals, warm-blooded and well protected by feathers or fur. Some of the resident birds, like the ptarmigan, and mammals, like the snowshoe hare, turn white in winter. White is protective coloration in a snowy environment and also minimizes heat loss by radiation. Musk oxen and caribou (wild reindeer) are large herbivores, dependent mainly on the abundant moss and lichens. Arctic hares and lemmings (small, ratlike rodents) are numerous and are preyed on by arctic foxes. Polar bears are amphibious; they frequent coasts and ice floes, but also wander inland on the tundra. Insects, especially flies, are so numerous as to be one of the major drawbacks of the tundra from the human point of view. Their eggs and larvae are par-

National Film Board of Canada

27–5   Taiga. A Northern spruce forest along the Alaska Highway in Canada.

ticularly cold-resistant, and the adults appear by the billions on warmer summer days. There is no lack of life on a warm day in the tundra, but the numbers of species permanently resident there are smaller than in almost any other sort of community, even the deserts.

Conditions above timberline on high mountains resemble those of the tundra, but with some differences as noted on page 708. The vegetation is similar in general appearance and in mountains of the Temperate Zone both it and much of the fauna may consist of species related to those in the Arctic. Ptarmigan (an arctic grouselike bird) and varying hares, related to arctic hares and also turning white in winter, extend far south of the tundra in alpine environments. Alpine insects are often of arctic species or closely related to them. On upper parts of very high tropical mountains, even, the general aspect of flora and fauna is like that of the arctic tundra, but the species are more distinct.

TAIGA

The *taiga* (another word we owe to the Russians) occurs in a still broader zone just south of the tundra across northern North America, Europe, and Asia (Figs. 27–1 and 27–5). Like the tundra and for the same reason, it is practically absent in the Southern Hemisphere. Winter temperatures may be as severe as in the tundra, but there is a well-defined summer growing season of 3 to 6 months. It suffices for a heavy growth of hardy trees, and the taiga as a whole is a tremendous forest. In the typical taiga the forests are coniferous, especially spruce, although several other species of conifers occur. Alder, birch, and juniper thickets are also common. Burned areas of the coniferous forest are invaded by aspens and birches, which later are succeeded by conifers again.

The moose (called an "elk" in Eurasia) occurs throughout the whole taiga, where not exterminated by man, and is its most conspicuous animal. Smaller mammals are much more varied than in the tundra. Black bears, wolves, and martens are more common in this zone than elsewhere. Fishers, wolverines, and lynxes are practically confined to it. So are some rodents, such as the northern vole, although most of the abundant rodents are races or subspecies of groups also occurring farther south. Squirrels thrive in these rich coniferous forests. So do many birds, most of which, however, are here summer breeders and migrate southward in the fall. The many insects and other invertebrates are of species that lie dormant during the severe winters.

The coniferous forests of our western

Helen Faye

27–6   Temperate deciduous forest.

mountains have some distinctive characters of their own but are essentially extensions of the taiga, occurring at increasingly high altitudes the farther south they are. Many of their species, both of plants and animals, are the same as in the typical taiga. The hemlock–hardwood forest of southern Canada and down into the Appalachians is also an extension of the taiga.

TEMPERATE DECIDUOUS FORESTS

Regions with moderate, well-distributed precipitation, with cold winters and warm summers, tend to develop communities in which deciduous trees dominate or climax the natural succession. These conditions occur in the Temperate Zones where the average annual precipitation is somewhere around 40 inches, without very well-defined dry and rainy seasons. In the United States, most of the eastern half of the country has such a climate and was formerly covered by deciduous forest (Fig. 27–6). Northward it graded into the taiga, through the hemlock–hardwood forest, and southward into the southeastern pine forests, a special local group of communities with peculiarities of soils and drainage. The British Isles and practically all of central Europe were also formerly occupied by temperate deciduous forests, and so was a large region in China and southeastern Siberia. There are similar forests in the Temperate Zone of South America, but they are not so widespread there because the precipitation is not suitable over such large areas.

The word "deciduous" implies the most obvious characteristic of this climate and the most obvious adaptation to it. Half the year or somewhat more is the growing season, when perennial plants put on their leaves and are active, while annual plants go through the whole cycle from seed to seed. The rest of the year is a period of nearly suspended animation, with trees and fields bare.

Common trees of the deciduous forest are beech, tulip, sycamore, maple, oak, hickory, elm, poplar, and birch. Chestnuts were formerly common but now have been almost eradicated in the United States by blight. The taiga and other coniferous forests include fewer species of trees, and locally a coniferous forest tends to be dominated by a single species. The deciduous forests have more varied local groupings, each of which commonly includes two or more species, as in the beech–

maple climax and oak–hickory, elm–ash–maple, or willow–cottonwood–sycamore communities. The complex distribution of these and other communities within the broader deciduous forest zone is governed by local conditions of climate, soil, and drainage.

The most striking herbivores of the deciduous forests are the browsing deer, mainly the white-tailed or Virginia deer in North America and other species in Eurasia and South America. In Eurasia wild pigs (or boars) are also characteristic of this group of communities, but they do not occur native in America. The principal predators on the larger herbivores are large cats. Our variously named puma, mountain lion, cougar, or panther (all one species, *Felis concolor*) ranges into most of the environments of North and South America. It is now extinct in the eastern forests but was originally their commonest large carnivore. Wolves, although more characteristic of the taiga, also formerly ranged widely into these forests, both in Eurasia and in North America. Foxes are still common in them. The arboreal martens are locally as common here as in the taiga, and the raccoon (absent in Eurasia) is especially abundant in our deciduous forests. These forests throughout the world are also rich in tree squirrels. Among mammals of the North American deciduous forests, over a third of the species are mainly arboreal. Tree-nesting birds are also abundant, and woodpeckers have the most obvious connection with the forest environment. The leaf- and mold-covered forest floor is a world in itself, swarming with fungi and invertebrates.

RAIN FORESTS

The lushest and most complex forest communities develop where there is an abundant and continuous water supply and a long growing season, which may be continuous through the whole year. Such *rain forests* occur in the Temperate Zone, for instance, on our northwest Pacific coast, where they have their own special characteristics and species. They are, however, most widespread and impressive in the tropics (Fig. 27–7) and subtropics. They cover most of Central America and northern South America, central Africa, southern Asia from India eastward, the East Indies and

South Pacific islands from Sumatra through New Guinea, and small parts of northeastern Australia (see Fig. 27–1).

Nowhere is life more exuberant than in the tropical rain forests. A temperate or cold climate forest frequently consists of one species of trees and rarely has a dozen. A tropical rain forest generally includes a hundred or more species of trees, and as many as 500 have been counted in one such forest. Two trees of the same species seldom stand near each other. Having noted a tree of a given species, you may have to travel for miles through the jungle before you find another. The actual species present may be totally different in different rain forests, and are sure to be if the forests compared are in widely separated regions of the earth. Always, however, there is this peculiar abundance of species in each forest, and the general aspect or structure, the ecological make-up, of a tropical rain forest is remarkably uniform wherever it may be and whatever species may compose it.

All forests and, indeed, all communities have some degree of *vertical stratification* (see Fig. 23–3). The conditions of life, the microenvironments, are different at different elevations (or depths in soil or water). Many of the organisms are adapted accordingly to those vertical differences and zones. This is particularly striking in a tropical rain forest. The main trees all grow to about the same height, generally from 20 to 40 meters in various situations. Their spreading, leafy branches there form a canopy, continuous throughout the year, which intercepts almost all the direct sunlight. Here photosynthesis is most active and flowers and fruits are abundant, but in some respects the canopy is a difficult microenvironment. Water is at a premium, for no outside supply is available except when rain is actually falling. Variation in temperature and in humidity is considerable.

Microenvironmental conditions change continuously and radically through various levels below the canopy, down to and into the forest floor. The floor is dark even at noon, and among green plants only a few with the most modest photosynthetic requirements manage to grow there. Direct rain is cut off by the umbrella of the canopy, but the lower levels have constantly high humidity and are

From H. J. Fuller and O. Tippo, *College Botany*, © Holt, Rinehart and Winston, Inc.

27–7 Tropical rain forest in Kenya, Africa. Note the epiphytes growing on the larger branches.

commonly dripping wet even when no rain is falling. The temperature is also nearly constant near the forest floor throughout the day and throughout the year. It is usually around 25°C., and in different regions seldom falls below 20 or rises above 30°C. All over the United States (except on high mountains) maximum summer temperatures are higher than in the tropical jungle. The sustained warmth of the rain-forest floor, however, and the constantly saturated atmosphere make a natural hothouse that is likely to be very trying to anyone accustomed to climates of the Temperate Zone.

Apart from the forest trees themselves, two habits of vegetation are especially characteristic of tropical rain forests: lianas and epiphytes. *Lianas* (a word of French origin) are climbing vines. Rooted in the dark forest floor, they use the standing trees as supports up which they climb toward the canopy, where they spread their leaves in the light. The rain forest is a tangle of lianas, slender or big and strong as bridge cables, looped and festooned around and among the trunks and branches of the trees.

Epiphytes are plants that grow on other plants without parasitizing them or deriving from them anything but a base on which to grow. Growing especially in the upper levels and canopy of the rain forest, they are well above the dark floor and are bathed in light, even though their own height is small. Orchids, ferns, and many other epiphytes form veritable aerial gardens among the high branches of the trees of the rain forests. Without roots reaching to a water supply in the soil, the epiphytes of a rain forest are paradoxically adapted to a dry climate. They include cacti, which store water in their pulpy tissues. The spread of cacti from desert soils to the rain forests, which are wet but not for them, is a remarkable example of what is technically called *preadaptation,* that is, an adaptation to one environment that turns out to be equally advantageous in another, apparently quite different environment. Other rain-forest epiphytes, notably the bromeliads [4]

[4] Named for a Swedish botanist, Olaf Bromel. Incidentally, and peculiarly, this family of plants includes the Spanish "moss" (not moss at all) of our South, and also the pineapple.

in South America, have their leaf bases so arranged that they catch rain and store it as in tanks against future need (Fig. 27–8). The

27–8 Epiphytic bromeliads in Trinidad, West Indies. *Top.* A large clump of *Gravisia aquilega.* A group of plants this size holds many gallons of water and supports a pond fauna high in the trees. *Bottom.* Part of the same clump brought to ground and with the leaves cut away, showing how the leaf bases overlap to form cups in which water collects (the water has spilled out from this specimen). Note the humus (dead leaves, etc.) that has accumulated; it is the plant's substitute for a soil.

little tanks form a remarkable microenvironment of their own, in which insects, frogs, and other organisms develop. Like the cacti, bromeliads have succeeded as canopy epiphytes in rain forests largely because of adaptations to a poor water supply that they acquired early in their history as desert plants.

Most of us think of the jungle, which is simply an overpopularized name for rain forest,[5] as teeming with animals. A first visit to a rain forest is disappointing in that respect, for animals are rarely seen there. Closer study reveals that animals are indeed common in those forests, although probably no more so than in our familiar Temperate Zone forests and grasslands. They are inconspicuous in the rain forest because many are nocturnal and most of those active during the day live high up in the canopy, where they are practically invisible from the ground. During the day the forest is oppressively silent, a silence likely to be broken only by the chattering or howling of monkeys (most of which are diurnal) or the squawking of parrots overhead. At dusk an ear-shattering chorus breaks out. Birds, mainly diurnal but foraging more quietly during the day, sound off as they settle for the night. Grasshoppers and allied insects make a din. They are joined most vociferously by tree frogs.

Ants, termites, flies, butterflies, beetles, and other insects are abundant in rain forest and are especially numerous in species there. Frogs also reach a sort of climax in this environment. Snakes are present but may be rather rare, contrary to accounts written more to astonish than to instruct. Mammals are less abundant in the forest than in adjacent (or, for that matter, Temperate Zone) grasslands, but they are still quite numerous. Arboreal forms include monkeys and rodents, especially squirrels. In the Old World rain forests, ground-dwelling herbivores include musk deer, small forest antelopes, and forest pigs. In South America similar ways of life are represented mainly by terrestrial rodents and peccaries. In both hemispheres the herbivores are stalked by partly arboreal carnivores, especially cats, such as the Old World leopard and the New World jaguar. Here is an illustration of the fact that the same ecological roles exist in geographically widely separated environments, but that the roles may be filled by distinct species or even by animals of different families, orders, or classes in different regions.

GRASSLANDS

In drier parts of the tropics, forests may still extend in narrow zones along watercourses where there is a good underground supply of water. These are the *gallery forests*.[6] A similar forest formation can be seen in the United States and elsewhere in the Temperate Zone, where galleries of trees may border a stream far out into a region otherwise treeless. Away from the stream galleries are vast areas in both Tropical and Temperate Zones where the water supply does not suffice for tree growth but does permit a heavy growth of grasses and other small herbs. These areas are variously called prairies, steppes, savannahs, pampas, or velds in different parts of the earth. All are ecologically similar and may be classed as *grasslands*.

The major North American grassland was the region of the high plains east of the Rocky Mountains. Most of it has now been plowed under to make way for crops. It has, however, been discovered that the drier parts of the region are more permanently productive as they were than as we have made them, and an effort is being made to return some of them to grass. Grasslands are even more extensive on other continents.

The dominant environmental restriction of the grasslands is a low, intermittent water supply. Rainfall may be only 12 to 20 inches per year. It is more than that, even up to 40 inches or thereabouts, on some grasslands but is unevenly distributed through the year. The

<hr>

[5] The word "jungle," which is of Sanskrit derivation, originally meant desert. It came to be applied to any wilderness. European travelers picked up the word in India and used it for the wilderness, as they considered it, of the rain forests there. Lately it has been so misused by explorers of the "Oh, how I suffered!" school that scientific explorers shy away from the word.

[6] The third main type of tropical forest, which we shall not further describe here, is sometimes called a *winter forest*. It occurs in areas of strongly defined rainy and dry seasons. The trees here are deciduous, shedding their leaves in the dry season and growing them again in the rainy season, the "winter" of dwellers in the tropics.

irregularity of rain, porosity and drainage of the soil, or both factors together prevent a continuous or ample supply of water to plant roots. Other environmental conditions vary greatly in grasslands and help to give each its special characteristics. A savannah in the midst of Venezuelan rain forest and a high prairie in Alberta have little in common except this: the water supply for plants is unreliable in both places. This one common feature has led to the dominance of grasses adapted to survival through unpredictable alternations of drought and downpour.

Within the grasslands different species and habits of grasses are adapted to special conditions of soil, precipitation, evaporation, and other environmental factors. The eastern, wetter parts of our North American grasslands had tall grasses, attaining heights up to 3 meters: bluestems, Indian grasses, slough grasses. The tall-grass prairie did repay plowing under; it is now our richest agricultural area, including the corn belt. In the arid western prairies short grasses predominated, especially grama and buffalo grasses, often growing among sagebrush. Between the extremes were mixed grasses, sod and bunch grasses, including needle grass, little blue stem, and wheat grass.

The grasslands swarm with animals, which are certainly more conspicuous and probably really more numerous in these communities than in any others on land. Primary consumers are the large grazing mammals. Countless millions of bisons and pronghorns[7] roamed our prairies. Even now the African grasslands support large herds of zebras and of several species of grazing antelopes. Living in open country, these grassland ungulates are all fleet of foot; they are *cursorial*. Hares and rodents are also common primary consumers in the grasslands. Some, like the hares, are likewise cursorial. Many rodents, like the prairie dogs and other ground squirrels or the pocket gophers, are *burrowing* or *fossorial* animals. Australian grasslands have herbivores very different in appearance and relationships but ecologically similar: large, grazing,

[7] Usually called "buffalo" and "antelope," respectively, but quite distinct from the Old World animals to which these names were first and are still properly applied.

U.S. Fish and Wildlife Service

27–9   Grasslands, in the National Bison Range, Moiese, Montana.

cursorial kangaroos and small, burrowing, rodentlike pouched "mice." Predators are adapted to the herbivore prey: wild dogs, lions, and the like preying on the ungulates; weasels, snakes, and others on the smaller herbivores. Herbivorous insects, such as locusts and grasshoppers, are also incredibly numerous. So are birds. What are some of our characteristic herbivorous and predacious birds in the grasslands?

DESERTS

With increasing aridity, grasslands grade into deserts (Fig. 27–10) without any sharp line of demarcation. Deserts are marked by low precipitation, generally 10 inches per year or less, which is likely to fall during a few heavy showers at erratic intervals. Deserts are also characterized by intense sunshine and very hot days, 35 or 40°C. and upward, at least during summer; and the evaporation rate is very high. Nights are generally cold,

Trans World Airlines

27–10   Desert, in the Cerbat Mountain Range, Arizona. Joshua trees dominate the scene. Note the abundance of vegetation in this true desert. Animal life is also abundant.

even in summer, and daily variations in temperature reach extremes found in no other environment.

Most annual plants in the desert are small. When a shower falls, they grow rapidly, bloom, and produce seed all within a few days. Among the most astonishing and beautiful sights on earth is one of our southwestern deserts carpeted with brilliant and many-hued flowers a few days after a spring rain. After another few days the desert is drab again, but scattered in it are millions of seeds waiting to perform the miracle anew.

Many perennial desert plants have small leaves or none at all, characters that decrease loss of water. Some have tremendously long roots, reaching deeply buried water. Others, notably the cacti in our deserts, absorb water rapidly after a rain and store it in spongy internal tissues. Can you think of any adaptive or selective explanation for the fact that most of the desert perennials are spiny or thorny?

Desert animals are also adapted to the scarcity of water and extremes of temperature. Large mammals are rare in deserts, although some Old World antelopes are adapted to extreme desert conditions. Small rodents are numerous. Almost all are burrowers, and many in different parts of the world have independ-

ently evolved bipedal, leaping locomotion. The kangaroo rat is an example in our deserts. Snakes and lizards are common in deserts, which nevertheless put sharp limitations on their activities. They are sluggish in the cold desert nights, and yet they quickly die of heat prostration in the sun. Consequently they are usually active only for short periods in the morning and evening and spend the rest of the time in burrows or crannies.

Fresh-Water Communities

Land and aquatic habitats are about as different as they can be, and yet they do intergrade along shores, in coastal lagoons, and in swamps. Some plants and animals live habitually in such transitional zones. Many plants are rooted in water but rise from it into the air. Many animals spend part of the life cycle in water and part on land, and others alternate at will between the two.

Among strictly aquatic environments, those of fresh water are less varied than those on the land but more varied than those in the sea. A primary but not sharp distinction is between the flowing water of streams and the still water of lakes. Deep lakes may have fairly complex vertical stratification, depending especially on gradients of temperature and of light penetration below the surface.

There are a good many flowering plants in fresh-water communities: water lilies, duckweeds, water hyacinths, pickerel weeds, pondweeds, and others. Almost all, however, are only semiaquatic. They float on the surface or extend above it into the air. Algae are the predominant plants of fresh water, as of all aquatic environments. Most common in fresh water are diatoms and blue-green and green algae. They are the mainstay for photosynthesis in lakes and streams and may occur in such enormous numbers near the surface, where the light is strongest, as to form a scum of pea-soup consistency.

Fresh-water faunas are rich in phyla and classes, more so than any land communities. Only three animal phyla have members completely adapted to life in the open air, but almost all phyla have fresh-water representatives. (What are the exceptions? See Chapter 21.) Among the commonest of fresh-water organisms, in addition to algae, are protists (especially bacteria and ciliates), flatworms and several other groups of worms, rotifers, arthropods such as water "fleas," crayfishes, and insects (not as adults but as aquatic larvae), snails, and, of course, fishes.

Wherever you live, there is almost sure to be a fresh-water community near you. Even a mud puddle becomes such a community if it lasts a few days. Where do its organisms come from? What food chains and other ecological factors are prominent in other fresh-water communities known to you?

Although we shall not discuss other fresh-water communities further, we shall mention one extreme as an illustration of a peculiar environmental hazard and of adaptation related to it. Torrential mountain streams would soon be swept free of organisms if their inhabitants did not have some means of staying in place in spite of the swift currents. For some of the fishes of such streams, notably trout, the situation is fairly simple. They can manage to stand still by swimming hard. Flatworms, leeches, snails, and many insect larvae in swift streams are flattened, streamlined, or limpetlike. They can cling to the bottom or to stones while the water flows by. Some caddis-

fly larvae live in cases to which they attach pebbles. The weight holds them on the bottom. Some fishes in the environment lack swim bladders, and some salamanders lack lungs; both deficiencies make the animals heavier and better able to cling to the bottom. Other fishes and many tadpoles of swift water have sucking mouths with which they cling to rocks. This adaptation has arisen several times independently in tadpoles of quite different ancestry. Some clams, snails, and insect larvae moor themselves against the current with spun fibers. Floating larvae, so common in the ocean, are unknown in fresh-water habitats.

Marine Environments and Communities

The ocean covers more than two-thirds of the face of the earth.[8] Most of the main groups of organisms among protists, plants, and animals, all three, arose in the sea and are still abundant there. Among phyla, only the Bryophyta and Tracheophyta are basically nonmarine. Even among classes, the vast majority is predominantly marine, and nearly all have at least a few representatives in the seas. The only classes of animals that probably originated on land are those of the centipedes, millepedes, insects, and the amphibians and their descendants the reptiles, birds, and mammals. There are a few secondarily marine insects, reptiles, birds, and mammals. Some of the other classes may have arisen in fresh water, but most are of marine origin and all but one (the Onychophora, Chapter 22) now have marine representatives.

The ocean, then, is the largest abode of life, and the organisms that swarm in it are more fundamentally diverse than those of fresh water or of land. The small space allotted to marine communities in this chapter is out of proportion with their overwhelming extent and richness. There are three reasons for this disproportion in emphasis. First, we have already said a great deal about marine life in other chapters. Second, for all their vastness and diversity, marine communities do not rep-

[8] The area of the ocean is about 361 million square kilometers, and that of the land about 149 million.

resent as radically divergent and sharply distinct ecological types as do those of the land. Third, the communities of the land, our own environment, are more accessible for study and are of more immediate, practical importance to humans.

MARINE ENVIRONMENTS

The part of the sea most freely open to our investigation is the shore, the *littoral* [9] zone between high and low tides. It is only a narrow band around continents and islands, and the conditions of life there are peculiar. It is, nevertheless, as crowded with living things as any part of the earth. Its outstanding environmental characteristic is of course the rhythmical ebb and flow of the tides, now covering the littoral zone completely and now leaving it exposed to air except for the many shallow tidal pools. Radiation from the sun here strikes intensely, and variations of temperature and of saltiness of the water are much more pronounced than elsewhere in the sea.

Away from the littoral zone, out beyond low tide, the most important single factor affecting marine environments is the penetration of radiation from the surface. Near the surface radiation is strongest. Here must occur most of the photosynthesis in the ocean. Temperatures here vary with the seasons, as on land. The difference between summer and winter temperature may be as much as 25°C. or even a little more, although it is less than 5°C. over most of the ocean surface. Temperatures are also zoned like the climates of the land, with a fairly constant surface temperature around 25°C. in the tropics grading off to temperatures rarely far from 0°C. near the poles. Daily, seasonal, and climatic differences of temperature in the sea are less than on land, because large bodies of water warm up and cool off more slowly than do air and land surfaces.

There is no sharp point where radiation stops with increase in depth in the sea. Light becomes dimmer and dimmer until finally there is none. The penetration of radiation varies with the clearness of the water and with latitude. It is greatest in the tropics and least in the Arctic and Antarctic. (Why?) Photosynthesis is of course confined to the illuminated zone and therefore occurs through a greater depth in tropical than in arctic regions.

Below the illuminated zone eternal darkness reigns, broken only by the glow of such organisms as create their own light. At these depths there is no daily and little seasonal or latitudinal variation in temperature, which is usually around 10 to 15°C. at the top of the dark zone and grades down to near 0°C. in the great depths of the ocean. [10]

In the deepest parts of the ocean pressure due to the overlying water reaches nearly a thousand times the atmospheric pressure at the surface, and many kinds of organisms are known to live under pressures more than 600 times that of the atmosphere. It used to be assumed that pressure made life in the greater depths difficult or impossible. Now, however, it has been discovered that pressure makes little difference to many marine organisms. How can this be so? Pressure is equal inside and out. As long as it remains constant, it is not felt any more than we feel the nearly 15 pounds per square inch of atmospheric pressure in which we normally live. Food supply and temperature are the most decisive factors in determining how many and what kinds of organisms occur at various depths below the lighted zone. The life of the oceanic deeps is still poorly known, but there are some living things at all depths, even the greatest.

The organization of communities in the ocean depends not only on depth and associated factors of light, temperature, and so on, but also on a special relationship between organisms and their surroundings. Many marine organisms simply float in the water, carried hither and yon by currents and often sinking or rising with changes in radiation and temperature. Those organisms are *planktonic* and all together make up the *plankton*. [11] Other organisms, especially fishes, swim freely in the water, actively seeking what they

[9] Some marine ecologists apply the term "littoral" to the ocean bottom as far out as light reaches it, but this is not in accord with the usual understanding of the word.

[10] Fresh water freezes at 0°C. and becomes lighter, and tends to rise as cooling drops it below 4°C. Sea water continues to become denser down to its freezing point, which is more than a degree below 0°C.

[11] From a Greek root meaning "wanderers."

may devour or evading what may devour them. They are *nektonic* and make up the *nekton*.[12] Still others live on the bottom, where they may be attached or may crawl about to a usually quite limited extent. They are *benthonic* and are the *benthos*.[13]

A combination of these ways of life and of the physical zoning of the sea determines the main life zones and kinds of communities. There is the *littoral* zone, as already noted, where most organisms are benthonic. Some plankton and nekton exist in the tidal pools or come in with the tide. Beyond the littoral zone is the *neritic* [14] zone, shallow waters in which light penetrates to the bottom. Plankton, nekton, and benthos are all abundant, and photosynthesis goes on throughout. Still farther out are the *bathyal* [15] and *abyssal* [16] zones, vertically divided into three major environments and kinds of communities. Above is the lighted open water swarming with plankton, including all the photosynthetic organisms of these zones, and with nekton. Below are the dark waters in which nearly all life is nektonic. At the bottom is the benthos, consisting of scavengers and organisms of decay, mostly bacteria.

OVER-ALL ECOLOGY OF THE OCEAN

Beyond the littoral zone, with its special features, the beginning of marine food chains is in the photosynthetic plankton. It consists mostly of microscopic diatoms and dinoflagellates, although more conspicuous green or brown algae, such as the famous Sargasso weed, also occur. Grazing, so to speak, on that grass of the sea are the many herbivores, especially various protists and crustaceans. Particularly abundant and important are the crustaceans called copepods, little more than microscopic in size but so enormously numerous that they are the chief food of the largest animals that have ever lived, the whalebone whales.

[12] Greek for "swimming."
[13] Greek for "depth," meaning by implication "living at the bottom."
[14] From a Greek root referring to the sea or to sea gods and nymphs.
[15] "Deep."
[16] "Bottomless" (which of course is not literally true).

The nektonic animals are for the most part carnivores. The whalebone whales and many fishes are among the carnivores that prey on the small herbivores. Toothed whales, sharks, squids, and many other nektonic animals prey mostly on other carnivores. There are many long food chains of the "dog eat dog" variety in the nekton.

Below the illuminated zone there are no living photosynthetic organisms. The community intake of food necessarily descends from above: sinking plankton, mostly dead when it reaches these levels, and both living and dead nektonic animals. Here, too, within the community every animal is potential prey for some other. Some of the deep sea fishes are among the most grotesque of all living things. Some can swallow whole fishes larger than themselves. Many are luminescent in this dark environment. Each light-producing species has characteristic colors and patterns of lights. The function of these animal lights is much disputed. What possibilities can you think of? It is pertinent that most deep sea fishes have functional eyes in spite of the fact that there is no light in their environment except that produced by themselves and other animals.

Finally, on the dark sea bottom the rain of dead organic matter from above comes to rest. Here are many scavengers and here, as on land, the many food chains come to an end with bacteria that break down complex organic compounds into simpler molecules. On land the products of bacterial decay are for the most part available to green plants and so start through the complex of cycles and food chains once more. On the sea bottom beyond the lighted zone, where most organic materials of the sea eventually lie, there are no photosynthetic organisms. This looks like a dead end. The total metabolism of the sea would seem to be a one-way process and not a self-renewing cycle. If that were so, life on earth would tend to run down, would indeed probably have run down by now, for even the most inland communities lose some organic material to the sea.

Fortunately the total ecology of the sea is cyclical, but the cycle is completed in an odd way. At many places near coasts, winds and currents tend to move the surface water away.

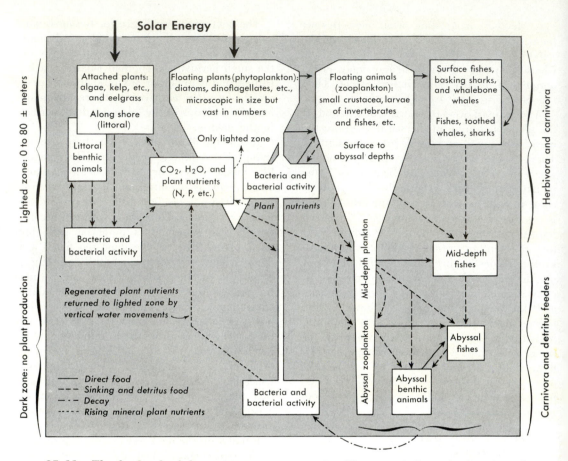

27–11   The food web of the major ocean community. The arrows indicate the directions of energy and material transfers within the community.

Its place is taken by an upwelling of water from the deeps, and with this water come the dissolved products of bacterial decay on the ocean floor. The upwelling of deep, cold water with its dissolved nutrients is periodic in some places, in others almost constant. Wherever it occurs, it has the effect of fertilizing the sea. Diatoms and other photosynthetic organisms increase enormously, and thus a cycle is completed and another begins. The main features of the entire cycle are shown in Fig. 27–11.

CHAPTER SUMMARY

Bases of biogeography: the diversity of vegetation types in the world; associated diverse faunas; regional distribution as the subject matter of biogeography; ecological versus historical biogeography.

Principles of ecological biogeography: the importance of plants; plant distribution, especially sensitive to physical environment, and its role in determining animal distributions; horizontal and vertical controls; temperature, solar radiation, and precipitation as dominant factors controlling plant distribution; the consequent parallelism between altitudinal and latitudinal distributions of plant communities (exemplified by North American communities).

Terrestrial communities: major types—tundra and alpine communities, taiga, temperate deciduous forests, rain forests with vertical stratification, grasslands and gallery forests, deserts.

Fresh-water communities: still- and flowing-water communities; examples of adaptation to fresh-water life.

Marine environments and communities: the ocean as the largest habitat of life, with marine faunas including all the animal phyla and all but two plant phyla; various marine ways of life and zones: planktonic, nektonic, and benthonic ways of life; littoral, neritic, bathyal, and abyssal zones; the over-all ecology of the ocean; the key roles of the illuminated surface zone and the accumulation of organic debris on the ocean floor; cycles in the sea.

28

Historical

Biogeography

Australian Travel Bureau

The koala "bear" (Phascolarctos) *is a marsupial. Its distribution, restricted to the Australian faunal region, is typical of many whose full explanation demands historical as well as ecological analysis.*

Ecological biogeography, studied in the preceding chapter, goes far toward explaining why plants and animals live where they do. That explanation, however, is clearly incomplete. Ecology gives a satisfactory answer to such questions as why monkeys occur in the forests of South and Central America but not in the desert and grassland regions of our West and Southwest. It gives a similarly satisfactory explanation of the distribution of monkeys through Africa and southern Asia. It does not explain why monkeys in apparently identical ecological situations in South America and Africa belong to different species, genera, and families. Still less does it explain why forests in eastern Australia, ecologically similar to those occupied by monkeys in South America, Africa, and Asia, harbor no monkeys at all. Australia does have animals similar to monkeys in habits and habitat, but they are not monkeys and the phylogenetic relationship is distant.

There are innumerable problems of that kind in all habitats. Oysters lead similar lives in the sea on the two sides of the Atlantic, but they are distinct genera. Lungfishes live in a few rivers of South America, Africa, and Australia, but they are of different genera on the three continents and they are completely absent in many other rivers apparently equally suitable for them. Large, spotted semiarboreal cats occur in both South America and Africa, but the South American jaguar is specifically distinct from the African leopard.

The explanations of all these and many similar problems of biogeography are *historical* in nature. The earth has changed during its long history, and its floras and faunas have changed with it. They have changed not only in evolving into new species, genera, and so on but also geographically, in their distribution over the face of the earth.

Many facts like those exemplified above really present a double problem and require dual historical explanations. Take the jaguar and the leopard, for instance. There is no land connection between South America and Africa today, apparently no way in which the big cats could possibly travel from one to the other. Yet the two species are closely related. At some time not long ago, geologically speaking, they had common ancestors in the same place, and those ancestors must have spread to both continents over practicable routes that no longer exist. That is one aspect of the problem. On the other hand, the jaguar and the leopard are different species, and the communities in which they live are radically different in taxonomic composition. We know that there must have been a way for land animals and plants to spread to the two continents, but how did their differences arise? That is the second aspect of the problem.

Biogeographic Regions

We have referred to the varying scale involved in biogeography. Ecological explanations of the distribution of plants and animals apply, for the most part, to the smaller end of the scale. They explain why a certain kind of community lives in one place and another kind a mile away, why one lives at the foot of a mountain and another on top, or one in northern Canada and another in southern United States. Ecological and historical aspects interact and overlap all along the scale, but on the whole the historical side becomes predominant or most evident at the larger end of the scale. The most purely historical explanations apply to the resemblances and differences of the faunas of large areas, such as whole continents or seas.

FAUNAL REGIONS ON LAND

The first approach to any scientific problem is to recognize that a problem exists. Existing facts must be observed, relationships among them must be inferred, and then an explanation must be sought. The observations from which the science of biogeography arose began in antiquity. They bore on the familiar and even obvious fact that different plants and animals live in different places. With the wide exploration of the earth from the fifteenth century onward, the nature and magnitude of the problems became more evident. Facts of the distribution of organisms over the whole earth were gathered, and broadly regional interrelationships began to appear. In the nineteenth century it became increasingly clear that there are regional patterns of floras and faunas that cannot be wholly, at least, explained by ecological factors. Such patterns occur in the sea as well as on land, but they are not, as a rule, so clear-cut in the sea, nor are the marine patterns as yet so well known. On land, regional patterns of plants are well marked, and so are those of all groups of animals that have been sufficiently studied from this point of view. Best known, in the combination of present condition and historical explanation, is the biogeography of land mammals. We shall therefore stress the mammals as an example of facts, problems, and derived principles. You should, however, bear in mind that the problems are similar and the principles the same for all groups of organisms.

By the beginning of the twentieth century the essential facts about the over-all distribution of land mammals were known. When these facts were arranged and generalized, there emerged a pattern of *land faunal regions*. Each region has some measure of general faunal resemblance throughout, and each has distinctions from any other region. The arrangement shown in Fig. 28–1 is now usual.

Faunas intergrade everywhere, and there are no such sharp lines in nature as on the map. Note, too, that this particular pattern is primarily for mammals and birds; [1] its application to other groups of animals and to plants is also generally valid but less clear. Even for mammals and birds, its application to islands other than those recently connected to continents is misleading. With these provisos the pattern has a real validity that may be briefly demonstrated.

The *Holarctic* [2] *region* has such animals as the timber wolf, hares, moose (called "elk" in Europe), and stag (called "elk" in America) that range through most of it and only marginally, if at all, elsewhere. The New World and Old World parts are distinctive in a lesser way. For instance, our commonest deer are of a genus (*Odocoileus*) absent in Eurasia, and the wild boar of Holarctic Eurasia is absent here. The Holarctic is often separated into *Nearctic* ["new (world) northern"] and the

[1] The original proposal (by Alfred Russell Wallace) in nearly this form was based on both mammals and birds.
[2] "Whole northern."

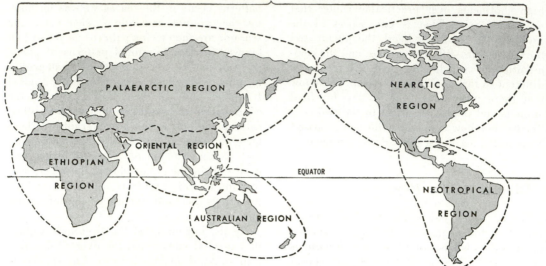

28–1   The faunal regions of the world.

Eurasian *Palaearctic* ["old (world) north-ern"] subregions.

The *Oriental region* is the haunt of the tiger, Indian elephant (a different genus from the African elephant), gibbons, and many other mammals nearly or quite confined to this region. The *Ethiopian region* is especially characterized by the giraffe, zebras, African elephant, and a great abundance of antelopes, some related to Oriental species and others sharply distinct.

The *Neotropical*[3] *region* is more distinctive than any of those already mentioned. Among the many mammals nearly or entirely confined to this region are the guinea pigs and many related rodents, New World monkeys (ceboids), sloths, true anteaters, and armadillos. The *Australian region* is even more distinctive. Its mammalian fauna consists largely of marsupials, and all belong to families that occur nowhere else.[4] The peculiar monotremes are also confined to this region. There are some native placental mammals— bats, rats, and a dog—but most of them are also of distinct species or genera.

[3] "New (world) tropical." The name is somewhat misleading. An enormous part of this region, in southern South America, is outside the tropics.

[4] There are two families of living marsupials in North and South America, one of them including the familiar opossum. but these families are not present in Australia.

SOME PROBLEMS

A biogeographic map like that of Fig. 28–1 sums up many facts, but it is still only a generalized description. A description of things is not much use and indeed is not truly a part of science unless it helps to go further, to require explanations and to find them. The descriptive data of mammalian geographic distribution do require many explanations.

There are problems here of resemblances and differences. The fauna of North America north of Mexico (that is, of the Nearctic subregion) resembles that of northern Asia much more than that of South America. Yet the Nearctic is connected to South America and not to Asia. Northern Africa, although not connected directly to Europe, has an essentially European fauna. Central and southern Africa are farther from Asia than from northern Africa, but the fauna is more like that of southern Asia.

Then there are problems of apparently conflicting resemblances and origins. One animal abundant throughout the taiga of North America, the porcupine, has its closest relatives in South America. Most of the other animals of the taiga have their closest relatives in Asia. There are a few exceptions. Mule deer and whitetail deer, which do range into the taiga but are somewhat marginal there, are

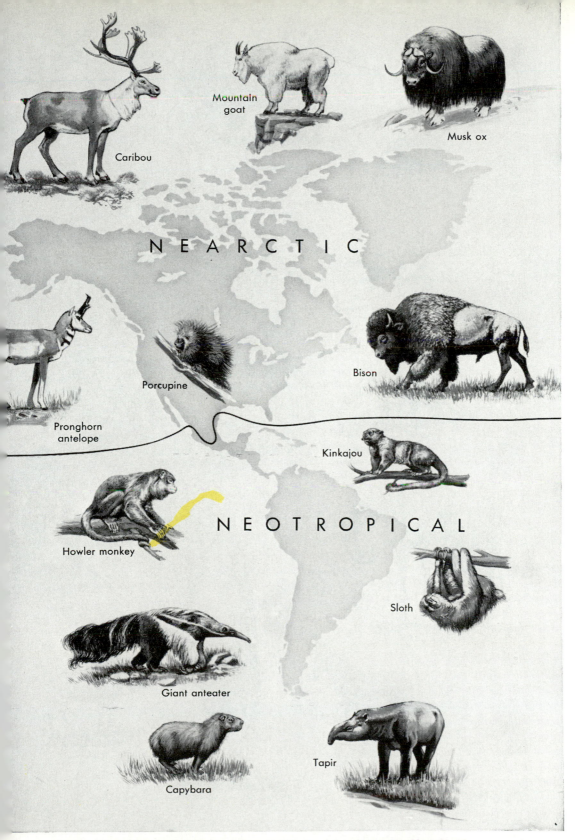

Caribou

Mountain goat

Musk ox

NEARCTIC

Pronghorn antelope

Porcupine

Bison

Kinkajou

Howler monkey

NEOTROPICAL

Sloth

Giant anteater

Capybara

Tapir

28–2 Mammals characteristic of the world's faunal regions. The New World.

Hedgehog

Bison

Reindeer

Polecat

P A L A

Bin

E T H I O P I A N

Aardvark

African elephant

Gorilla

Zebra

Giraffe

Gnu

28–2 (continued)   **Mammals characteristics of the world's faunal regions.** The Old World.

Wild ass

Marco Polo
sheep

ARCTIC

Indian
elephant

Indian tiger

Water buffalo

ENTAL

Malay
tapir

Gibbon

AUSTRALIAN

Flying
phalanger

Native cat

andicoot

Wombat

Koala

Kangaroo

more closely related to some South American deer than to any in Asia. The geographic relationships seem anomalous in themselves, and they also suggest further questions as to places of origin. Did the taiga fauna as a whole come from Asia and a few members, such as porcupines and deer, spread into South America? Or are the forms with South American affinities, the porcupine and the deer, of South American origin? (We shall give you the answer: historical evidence proves that the first explanation is correct for the deer and the second for the porcupine.)

Many of the classic problems of biogeography, including some that have never been satisfactorily solved, arise from what are called *disjunctive* (that is, unconnected) *distributions*. It rather frequently happens that two closely related groups of organisms occur in widely separated regions but that there are no equally closely related forms in between. Because the disjunctive groups are closely related, they must have had a common ancestry not long ago, geologically speaking. Therefore, ancestors of the existing groups must have spread from one region to the other, or to both regions from a third. The problem is to determine what route they followed and how. Famous examples among mammals are the tapirs, which live only in Central and South America and in southeastern Asia, and the camels, which live (as wild animals) only in South America and Asia. (Both those problems have been solved, as you will see later.) Still more puzzling examples occur among other groups of animals and among plants.

Such problems can be solved only by historical methods. Yet historical study soon raises other problems of its own. It sometimes shows that earlier faunal relationships were quite different from those of today. Thus further explanations are required. The fauna of Honduras (Central America, north of the Panama constriction) now is South American in predominant affinities. But we know from fossils that a few million years ago the mammals of Honduras had nothing to do with those of South America and were all of northern affinities. They were, in fact, more nearly related to mammals of Eurasia, even of Africa, than to those of next-door South America. At about the same time southern Europe, now part of the Holarctic region, had a fauna more closely related to that of the present Oriental region.

FAUNAL CHANGE AND EARTH CHANGE

Evidently faunas and faunal regions have not stayed put. A biogeographic map of the present world may be true enough as of now, but it is a static picture and does not convey anything of dynamic, historical processes. Those processes are the real story of biogeography. Confining attention to the static map is like looking at one frame of a motion picture instead of running through the whole film.

Organisms have developed on a constantly changing earth. Climates have changed. Mountains have arisen and been worn down. Shallow seas have advanced and retreated where now is land. Most important of all, from our present point of view, seas now separated by land have been united, lands now separated have also been united, and both seas and lands now united have at times been separated. During the latter part of geological history at least, the last hundred million years or more, major seas and lands, the oceans and continents, have had substantially their present identity.[5] Their outlines and detailed features have changed considerably, but they have existed continuously as geographic units. The connections among them have changed, however. That has had most profound effects on the distribution of organisms seen on such a large scale as in the biogeographic regions.

Historical changes in any given biotic region and indeed within any community are of four kinds: (1) Evolutionary change takes place within each of the species present in region or community. (2) The proportionate numbers of individuals of the various species change; some become more and some less abundant. (3) Some species disappear, either locally or by total extinction; this is a special case of 2, the reduction of proportionate numbers to zero. (4) New species spread into the region or community from elsewhere.

The last-mentioned kind of change is the one that is most directly geographic. Historical biogeography is concerned primarily with

[5] Some disagree with this statement, but it is the consensus.

the spread of species and of whole biotas—their dispersal. It is this that is so intimately bound up with the earth changes we have mentioned, because the earth changes open and close routes of dispersal. The geographic changes cannot, however, be wholly separate from the other kinds of change in regions and communities. Geographic spread of species is a cause of numerical changes, including extinction, in invaded communities. Such spread is generally accompanied by adaptive change in the species involved, because as they reach new environments selection tends to modify adaptation accordingly. Evolution may also be speeded up within the invaded communities, by an intensification of selection and change in its trend.

Basic to all these aspects of biogeography are the means of dispersal and the things that facilitate, hinder, or prevent it.

Dispersal and Isolation

MEANS OF DISPERSAL

All organisms have some means of dispersal. That is a necessity for living things. Can you imagine an organism that did not have any way of getting from one place to another? What would the consequences be for such a species?

The means of dispersal are most obvious in the many animals that go places under their own power. They fly, walk, crawl, or swim and so constantly change their precise geographic localities. Included in this category are most of the vertebrates, insects and other arthropods, many worms, some mollusks (such as the squids), and some coelenterates (in the medusa form, Chapter 21, although their locomotion is not so directive as in the other groups named). Most land animals (above microscopic size) and the actively swimming, nektonic aquatic animals belong to these groups.

Even among actively and directively motile organisms dispersal is not a simple matter of packing up and going somewhere else. It is to be distinguished from migrations, in which a whole population moves periodically to another region. That is a geographic movement, but only among regions already occupied on occasion and hence not an actual spread or dispersal of the species. Most animals have a strong attachment to the community into which they are born, whether that community is fixed at a single geographic locality or is mobile or migratory. Dispersal usually takes place through a sequence of generations. As the population becomes more dense, marginal individuals have a better chance if some of them move out from the center of density. Any one individual may move only a few centimeters, meters, or kilometers from where it was born. Continued over many generations, the sum of such movements may spread the species, or others derived from it, over a whole continent or ocean, or more than one.

The dispersal of protists, plants, and many animals is passive as far as the organisms themselves are concerned. Planktonic organisms are dispersed by currents in which they float. Sessile animals, such as corals, have floating larvae that are in effect temporarily planktonic and are similarly dispersed. (All sessile animals are aquatic.) Most plants have spores or seeds that are air-borne or are dispersed by animals and in other ways. Plant dispersal involves many intricate adaptations, some of which have already beeen mentioned (Chapter 9).

Insects, spiders, and other light animals are often blown for long distances by wind, and this may facilitate their dispersal. Fallen trees and mats of vegetation and debris are often floated long distances down rivers or carried for hundreds, even thousands, of kilometers by ocean currents. With them may go not only seeds but also eggs and adult animals. Violent winds, especially tornadoes, occasionally pick up salamanders, toads, frogs, and even fishes and drop them elsewhere, still living. Flying birds frequently carry live seeds for great distances. Eggs and larvae of many small aquatic animals become attached to the feet or feathers of swimming and wading birds and are carried away and deposited elsewhere in the aerial wanderings of the birds.

In the world as we see it today and not as it was in a true state of nature, man is one of the most effective agencies of dispersal. He has purposely taken domesticated animals and cultivated plants wherever he himself dispersed. He has also purposely introduced many wild animals and plants in regions where they are not native. Mice, rats, and other small animals, especially insects, have hitchhiked with man, against his own intention, in boats, wagons, automobiles, and now airplanes. As a result, there is now no region on earth where all the plants and animals are native, none introduced purposely or accidentally by man.

ROUTES OF DISPERSAL

There are so many means of dispersal for organisms that it is somewhat surprising that there are few worldwide species. For a group with completely effective dispersal the final control would, after all, be purely ecological. That is, the group would soon occur wherever the environment provided an ecological role for which it was well adapted. This is true of man and of some of his commensals and parasites, but man and the organisms most closely associated with him are special cases. Few other species are literally worldwide even in suitable environments.

Dispersal depends not only on means but also on route. An analogy is that where you drive depends not only on having an automobile, the means, but also on where a road goes, the route. Nature has many routes, from broad turnpikes to bumpy back roads. It also has many barriers, roads that are closed for one species or another and regions where, for a given species, no roads exist.

For many nektonic animals of the open sea, the whole ocean is a highway. They tend to spread widely until they encounter an environmental or ecological barrier. Distribution of plankton is most strongly affected by the great ocean currents. The over-all pattern of these currents is fairly constant now, although it must have had some radical regional changes in earlier geological times. A current from the Gulf of Mexico once flowed through what is now Central America into the Pacific, instead of doubling back into the Atlantic as the Gulf Stream. Thus routes may change in the sea as on land, but wherever they go, the ocean currents are main dispersal routes for plankton spreading downstream. Dispersal of plankton against the current is unlikely.

The winds, turbulent and erratic as they are locally, have an over-all pattern of air move-ment that has probably changed little through later geological time. For plants and animals with an air-borne dispersal phase, the zones of prevailing westerly winds, for instance, are and long have been major routes for dispersal from west to east.

Physical and climatic maps of Eurasia show that there is a pathway from western Europe clear across to northern China which could readily be traversed in all its parts by many land plants and animals. That dispersal route has been extensively followed, as is evident from the fact that some natural communities in Europe are remarkably like others in China, thousands of kilometers away. Yet they are not exactly the same in the two regions. A dispersal route, no matter how open it may be, like the Eurasian corridor or the Atlantic equatorial current, is never 100 per cent effective.

For any given group of organisms there are dispersal routes that differ in the probability of dispersal. The scale of probabilities is continuous. At one extreme are routes along which dispersal is prompt and nearly certain. At the other extreme are routes so unsuitable that dispersal along them is so unlikely as to be effectively impossible. Of course, for the group in question such an extremely low-probability route is more likely to be a barrier to spread than a dispersal route.

For whole biotas there is also a continuous scale of probabilities of dispersal or of migration. If chances are good for the spread of many or most species of a biota (although chances may still be poor or practically nil for some species), the route is a *corridor*. The Eurasian route previously mentioned has been a corridor for Holarctic floras and faunas. Other routes are more and more selective. Some species migrate readily along them, whereas others do not. The route may then be considered a *filter*, because it passes parts of biotas and holds others back. There is no sharp distinction between a corridor, which is still a filter for some individual species, and a filter, which is still a corridor for certain species. It is merely a matter of what percentage of a whole biota follows the route. A good example of a filter route is the Middle American connection between North and South America discussed later in this chapter.

BARRIERS

In a sense a barrier can be thought of as a dispersal route looked at from the other end of the scale of probability. Probabilities of dispersal for a given species along a given route may be anything from near 100 per cent to near 0. (Whether the chances are ever *exactly* 100 per cent or 0 is a moot point.) If the chances are low, the route is a barrier. In consideration of whole biotas, almost any route may be a corridor for some species and a barrier for others. A filter route is, of course, a barrier for the species that are filtered out.

A barrier is any zone physically or ecologically unsuited for the organisms impeded by it. A mountain range is a barrier to species better adapted to lowland conditions on each side of it, and the lowlands are barriers between mountain ranges. Grassland is a barrier for forest animals, and forests are barriers for plains animals. A cold ocean current is a barrier for warm-water species.

Faunal regions are delimited by major barriers. The change from the Nearctic to the Neotropical across Middle America is gradual, but it tends to center along the barrier formed by the change from temperate grassland and desert to tropical forests in Mexico. The Sahara and other deserts separate the Palaearctic and Ethiopian regions in Africa. The Himalayas and other mountains are the barrier between Palaearctic and Oriental in Asia. These barriers are not absolute. They are filters, but strong ones. They have not always been there, and faunal regions have not always been delimited as they are now.

A special kind of barrier that is one of the strongest limitations on dispersal of organisms is purely ecological. Plant seeds may be wafted for hundreds of kilometers, perhaps even across an ocean. If they land on bare soil of suitable composition and in a suitable climate, the seeds will grow. Dispersal has occurred, and the species has spread geographically to a new region. But there is a catch in the word "bare." There is practically no environment that is not already occupied by plants adapted to it. There is little chance, indeed, that the new types of seeds will land on suitable bare soil. They will land in an established community where they must compete with species already fully established there, well adapted to the community's ecology and to the environment. Exceptional invaders can overcome this tremendous handicap, but the chances are usually slim.

There is reason to believe that the spores and seeds of innumerable species of plants have crossed the South Atlantic in both directions between Africa and South America. The number of such crossings, especially by means of winds, birds, and currents, in the last few million years of geological time must have been enormous. The African and South American floras do have some related species that were probably dispersed in this way.[6] On the whole, however, the floras are very different. The great majority of migrants failed to get a foothold in the foreign communities.

Strong barriers and sweepstakes dispersal

A plains animal is not likely to cross a mountain range, but it could do so in some instances and might find food and other necessities of life on the way. The animal might be neither more nor less likely to cross an arm of the sea of equal width, but the character of the barrier is quite different. The sea is an environment in which a land animal cannot possibly carry on its normal activities. Similarly for seed plants, the sea is an area where they cannot possibly grow. As still another example, an isthmus is a barrier for marine animals where they cannot possibly colonize or sustain active life.

Barriers that represent not merely difficult but downright impossible habitats for the organisms in question are of the very strongest kind. Populations cannot spread across them by the usual processes of expansion or migration. If the barrier is crossed at all, it must be by individuals and in one jump, so to speak, not by the gradual expansion of a population. Even the strongest barriers of this kind can be crossed by many organisms and have been

---

[6] This is an expression of opinion. It is opposed by a few botanists who insist that there must have been a land connection across the South Atlantic in late geological times. The existence of such a connection late enough to be involved in the dispersal of flowering plants, at least, is denied by a strong consensus of competent geologists and biogeographers.

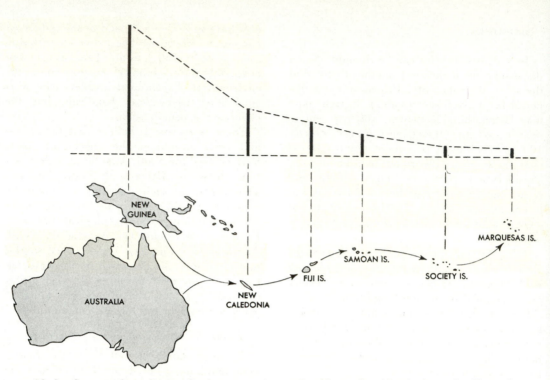

28–3 Sweepstakes dispersal. A group of weevils (Cryptorhynchinae) has island-hopped from west to east. The height of the vertical bar is proportional to the diversity (number of genera) of the group at each place. Clearly the group has been sifted out, fewer and fewer managing to follow each successive sweepstakes route. These insects are, even so, particularly good at sweepstakes dispersal.

crossed repeatedly in the long history of life. This is *waif* or *sweepstakes dispersal*, "sweepstakes" because the individual chances of dispersal over such barriers are small, as are the chances of winning a sweepstakes, and yet the event does occur (Fig. 28–3).

A great natural experiment in sweepstakes dispersal occurred when the island volcano Krakatoa, near Java in the East Indies, blew up in 1883. Every trace of life on the island was destroyed.[7] The nearest island not destroyed by the eruption is over 18 kilometers distant, and yet in only 3 years there were 11 species of ferns and 15 of flowering plants on Krakatoa. Animals soon followed, and within 25 years there were 263 species of animals resident on the island. Most of them were insects, but there were 4 species of land snails, 2 of reptiles, and 16 of birds. In 1928, 45 years after the explosion, there were 47 spe-

cies of vertebrates on the island, mostly flying forms (birds and bats), but including two kinds of rats.

The Hawaiian Islands are surrounded by a tremendous oceanic barrier and have never been connected to other land, but they had a luxuriant native land biota.[8] All the ancestors of the thousands of species of Hawaiian plants and animals reached there by sweepstakes dispersal.[9]

The possibility of sweepstakes dispersal exists for any group, but it is much higher for some than for others. It is highest for plants, especially those with wind-borne spores or seeds, and for small flying animals, especially insects but also birds and bats. This seems reasonable, and it is borne out by the data on Krakatoa and on the Pacific islands. Sweep-

[7] One species of earthworm may have pulled through the holocaust, but it probably did not.

[8] Tragically overwhelmed now by the injudicious importation of non-native plants and animals.

[9] Sweepstakes dispersal also explains the origin of the Galápagos Islands biota (Chapter 16).

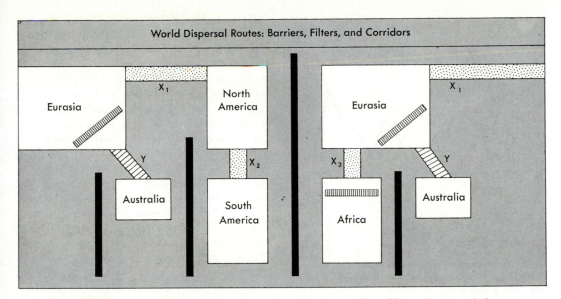

28–4 World dispersal routes: barriers, filters, and corridors. Major features of the geographic history of faunas, especially mammals, are best accounted for by considering the continental blocks and the main sea barriers as constants and the three main filter bridges or corridors and one main sweepstakes route as variables. X_1, X_2, X_3, variable filter bridges or corridors; Y, variable sweepstakes route; ||||||||||||, constant barrier during the Age of Mammals; //////, somewhat variable land barrier.

stakes dispersal of strictly marine animals across a land barrier and of strictly land (and nonflying) mammals across a sea barrier is least likely. There are no native land mammals on any of the Pacific islands beyond those immediately adjacent to Australia.[10]

One reason why the historical biogeography of land mammals has been studied more extensively than that of any other group is that they are little subject to sweepstakes dispersal. It can usually be assumed that their migration routes were on continuous land connections. Their geographic history thus is crucial in determining when and where variable earlier land connections existed. Nevertheless, it is practically certain that a few land mammals have had sweepstakes dispersal. Although exceptional, those instances have influenced mammalian biogeography markedly in some regions. Their recognition has cleared up some classic problems that seemed insoluble when it was supposed that land mammals *always* spread over continuous land.

[10] Some do have native rats, but these were introduced accidentally in Polynesian canoes.

Changing Biotas and Geography

THE WORLD CONTINENT

We have noted that the Neotropical and Australian regions are more distinctive than the Holarctic, Ethiopian, and Oriental regions. We have also mentioned that the fauna of northern North America is more like that of Asia, to which it is not connected by land, than that of South America, to which it is connected. The historical reasons for these facts may now be stated, and the general geographic background is illustrated in Fig. 28–4. Australia is now and has long been an island continent. South America is not now but was during most of the Age of Mammals, the last 70 million years or so, an island continent. Africa, Europe, Asia, and North America were separated from each other for various shorter times, but they were also connected periodically during the Age of Mammals.

As far as the mammals are concerned—and other groups of land organisms have also frequently tended to follow this pattern—Africa,

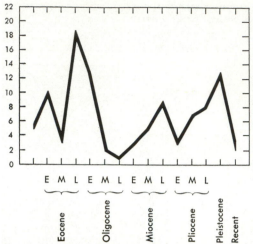

28–5 Interchange in the mammalian faunas of North America and Eurasia through the Cenozoic. The graph plots the numbers of items of evidence for the interchange. The intensity of the actual interchange doubtless followed rather closely the ups and downs shown here. Evidence indicates that the land connection, when it existed, was between Alaska and northeastern Asia. *E*, early; *M*, middle; *L*, late. For the time scale see Chapter 29.

Europe, Asia, and North America were long essentially one big land unit, a supercontinent or World Continent. There have been repeated regional isolation and differentiation from time to time, but by and large the history of land faunas has followed along the same broad lines on the World Continent. There has been frequent although always incomplete (filtered) intermigration of faunas among its different parts, the continents as we have them today.

With the aid of fossils it is possible to measure the relative intensities of intermigrations on the World Continent during the past 60 million years or so. The varying extent of dispersal of land mammals in either direction between Eurasia and North America is shown in Fig. 28–5. In general, it is reasonable to conclude that when dispersal was comparatively high there was a land connection between the continents and that when it was low there was a sea barrier. Other evidence indicates that the connection, when it existed, was between Alaska and northeastern Asia.[11]

[11] There may also have been an earlier connection across Greenland to Europe, but if so, it broke down permanently early in the Age of Mammals or before.

The connection or connections within the Old World kept the World Continent faunas sufficiently mixed so that they retained a broad similarity. The connections were filters, however, and other filters developed within the continents. As might be expected, relatively few animals especially adapted to warm climates managed to cross the northern bridge between Asia and North America, in spite of the fact that Alaska was a good deal warmer during most of the Age of Mammals than it is now. (Contrary to some popularizations of the subject, there is no good evidence that Alaska was ever tropical in climate.)

Regional differentiations of faunas thus could and did occur in spite of repeated mixing by intermigration. The World Continent did not develop a really uniform fauna, even where environmental conditions were closely similar. The alternating filter and barrier between Asia and North America gave a degree of isolation reflected now in considerable distinction of their faunas. In the Old World the east–west desert filter in Africa and the mountain filter in Asia developed during the Age of Mammals. Northern and southern faunas then became more sharply separated than they otherwise would have been, or than they were earlier in the Age of Mammals. Narrowing of the land connection between Africa and Asia and extension of the desert filter in southwestern Asia also tended to isolate what had been more nearly uniform faunas and finally resulted in the distinction we recognize now between the Ethiopian and Oriental regions.

ISLAND CONTINENTS

Similarities between the faunas of two regions naturally tend to be in proportion to the amount of intermigration that has occurred between them. Regions long connected by corridors have faunas of much the same composition, differentiated moderately on a local, largely ecological basis. Isolation, the interposition of barriers, leads to more radical differences on a regional, more historical than purely ecological, basis. The longer the period of isolation, the greater the differences. Australia and South America, long island continents, illustrate this principle, which explains the regional peculiarities of their floras and faunas.

In Australia most of the ecological roles (or ways of life) of land mammals are filled by marsupials. That is in itself a radical difference from the World Continent, where practically all such roles are filled by placentals.[12] If there had been a land migration route between the World Continent and Australia during the Age of Mammals, it seems certain that there would have been early mixture of placentals and marsupials there. Evidently Australia started the Age of Mammals with marsupials (and the now comparatively insignificant monotremes) only, and has been isolated by a strong barrier ever since.

Spreading over a whole continent with highly varied environments, the Australian marsupials early speciated profusely. Different lines rapidly became specialized in adaptation to the many possible ecological roles. They underwent, in short, an adaptive radiation on a grand scale. The roles assumed were generally similar to those of the phylogenetically distinct placentals of the World Continent. The result, as we have already mentioned in another connection (Chapter 18), was convergence between many Australian marsupials and World Continent placentals.

Later, rats, placental rodents that had evolved on the World Continent, also reached Australia. They are now numerous there and have evolved into many species and genera peculiar to the region. This is one of the facts of biogeography that can be explained only by sweepstakes dispersal. If the rats came in over a land connection, it is incredible that no other placentals accompanied them. Rats are also known to be particularly good at oversea dispersal or island-hopping. The only other native placentals of Australia are bats, dispersed by flight and winds; a wild dog, probably introduced by the aborigines; and the aborigines themselves, who came by boat.

South America must have been connected with the World Continent, undoubtedly with its North American part, early in the Age of Mammals. It started out with a far more varied stock of land mammals than did Aus-

tralia, including primitive marsupials and several groups of primitive placentals. Then the connection with North America was broken, and the mammals evolved in isolation in South America for tens of millions of years. Here, too, adaptive radiation occurred on a continental scale and here, too, there was extensive convergence toward World Continent mammals. Placentals, evolving into families and orders peculiar to South America, took over most roles. The marsupials, however, became much more diverse than they ever were on the World Continent and took over various roles. Most striking is the fact that all the predacious carnivores of island South America were marsupials. Only placental mammals evolved into predators on the World Continent.

Later on, some 30 or 35 million years ago and thus just about the middle of the Age of Mammals, two new groups appeared in South America as the rats did in Australia. The most reasonable explanation is the same: the newcomers probably got there by sweepstakes dispersal, island-hopping down from Central America, which was not then attached to South America. The newcomers were New World monkeys and rodents resembling guinea pigs; both types expanded greatly in South America and are still characteristic of that continent.

FAUNAL INTERCHANGE

In the later part of the Age of Mammals, a few million years ago, the mammalian fauna of South America was far more distinctive than it is now. It had almost nothing in common with North America or the rest of the World Continent. Then movements of the earth's crust heaved up a land connection between the two continents. The result was first a trickle and later a flood of mammals from each continent onto the other. Such mixtures of faunas after the disappearance of a barrier have often occurred, both on land and in the sea, but this is at present the clearest and most fully analyzed example (Fig. 28–6).

The filtering action of the connection was striking. Wild dogs, raccoons, cats, weasels, field mice, peccaries, deer, tapirs, and many others eventually passed the filter in great numbers from north to south. But other

[12] Opossumlike marsupials were fairly common on the World Continent early in the Age of Mammals and are still present in North America, but they never developed any considerable diversity outside the island continents.

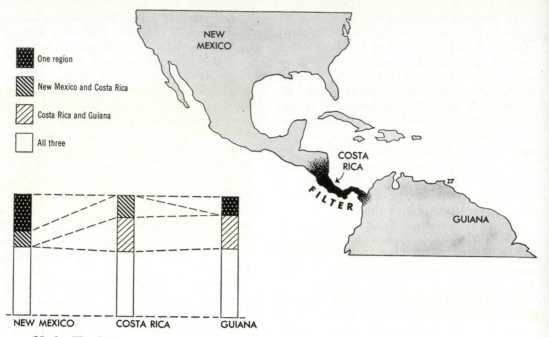

28–6   The Isthmus of Panama, a major filter route. The filter action of the isthmus is well illustrated by the graphs, which show the high proportion of families of mammals in both New Mexico and Guiana that either have entirely failed to cross the isthmus (black with white stippling) or have only gotten halfway across (diagonal lines). The total height of the column for each of the three zones represents 100 per cent of the local mammalian fauna.

common North American mammals, such as beavers, pronghorns, and bison, did not. (Why do you suppose these animals were filtered out?) From South America into North America came porcupines, capybaras (large, amphibious rodents, extinct here now but still present in South America), armadillos, glyptodonts (large extinct relatives of the armadillos), giant sloths (large extinct relatives of the living tree sloths), and perhaps opossums, although they may have been in North America all along. Most of the South American mammals, however, failed to get completely through the filter. Among others, the peculiar native ungulates, the monkeys, and the marsupial predators did not.

Animals spread in both directions between North and South America, as is usual on most dispersal routes. North American animals were, however, more successful than those of South America in making their way into the communities of the other continent. Both continents became temporarily richer in land mammals than they had been. Before the interchange North America had 27 families of

land mammals and South America 23. At the height of the interchange the figures were 34 and 36, respectively.

The increase in diversity involved some duplication of ecological roles. Animals that had evolved convergently on the two separate continents now came into direct contact and competition. Such a situation cannot last indefinitely. Ultimately one of the competing forms wins out, and one becomes extinct. The interchange was followed by the widespread extinction of species, genera, and whole families. At present North America (north of Mexico) has only 23 families of land mammals—actually fewer than before the interchange—and South America has 30. The mammals of North American origin were more successful; fewer of them became extinct. The only mammals of known or probable southern origin still present in our fauna are the porcupine, armadillo, and opossum, and they do not loom very large in the fauna. In present-day South America about half the mammals are descendants of comparatively recent invaders from North America. All its native hoofed mammals and all the marsupial predators became extinct.

The distinctiveness of the present Neotropical region involves three factors. Some old-timers that evolved in South America when it was isolated still survive there and have not crossed the filter to North America: armadillos (only one species is north of the filter), sloths, anteaters, many rodents allied to the guinea pigs, and monkeys. Some mammals really of North American origin have become extinct in North but not in South America: the camels (the llama and others) and tapirs. Other groups are still common to the two continents, but in most instances they have diverged somewhat on the two sides of the filter. For example, the deer of South America are all of distinctive species, and several belong to distinctive genera that evolved there from North American ancestors but have never spread back to the north.

FAUNAL STRATIFICATION

Wide dispersal of plants and animals has been frequent through the geological past, but it has been scattered and episodic as new dispersal routes appeared or old ones disappeared. Moreover, it has seldom if ever happened that a *whole* biota, a complete and integrated community, was dispersed all at once. There is always some filtering. Thus it happens that the regional communities we have today consist of species whose ancestors spread into that region at different times. For example, in our now decimated grasslands fauna the bison (or "buffalo") and pronghorn (or "antelope") were the prominent large herbivores, both equally at home on the high plains. But the ancestors of the bison came here from Asia quite recently, geologically speaking—a matter of some 500,000 years. The ancestors of the pronghorn, on the other hand, have been here for tens of millions of years.

The division of a fauna into different *strata*, depending on how long the various groups have been in the region, is particularly clear among the land mammals of South America. There are three readily distinguished major faunal strata there. The oldest consists of descendants of animals dispersed to South America when it was connected with the World Continent around the beginning of the Age of Mammals. Armadillos, sloths, and anteaters are prominent surviving members of that stratum. The next stratum descends from animals that reached South America by sweepstakes dispersal around the middle of the Age of Mammals: monkeys and many native rodents such as the guinea pig. The youngest stratum consists of groups that invaded from North America when the continents were reunited toward the end of the Age of Mammals: field mice, dogs, cats, deer, and many others.

As a rule, with some exceptions, older faunal strata are more distinctive and peculiar to their region than younger strata. In other words, the longer a group has been in a particular region, the more likely it is to evolve along lines different from those of its relatives elsewhere. This is eminently true of the South American strata. There is nothing like a tree sloth anywhere else on earth. South American monkeys do resemble their relatives in Africa and Asia, but they belong to a distinct superfamily. South American field mice are, on the whole, barely distinguishable from their North American relatives (Fig. 28–7).

DISJUNCTION

We are now ready to review briefly the problems of disjunctive distribution mentioned earlier in this chapter (Fig. 28–8).

Some disjunctive distributions are simply explained by the fact that a climatic or other environmental change has restricted a formerly widespread group to scattered parts of its previous range. For instance, pikas (small-eared, tailless relatives of the rabbit) occur disjunctively on various western mountain ranges and in the Yukon and adjacent parts of Alaska. They are cold-climate animals that became widely distributed during the Ice Age and now occur only where the climate is still like that of the Ice Age. *Glacial relicts*, as they are sometimes called, which are plants or animals widespread in the Ice Age and now scattered disjunctively in the far north and on mountains, are common in North America and Eurasia.

Pikas also occur on the cold northern steppes of Eurasia, which is not surprising for such a glacial relict. Disjunction between Old

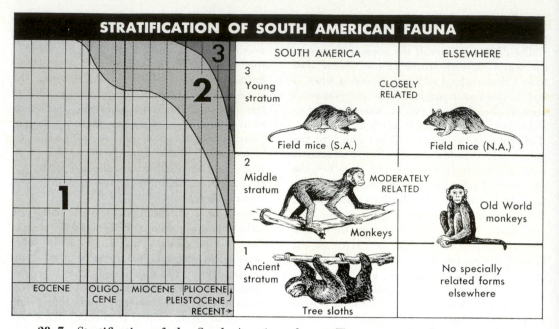

STRATIFICATION OF SOUTH AMERICAN FAUNA

	SOUTH AMERICA		ELSEWHERE
3 Young stratum	Field mice (S.A.)	CLOSELY RELATED	Field mice (N.A.)
2 Middle stratum	Monkeys	MODERATELY RELATED	Old World monkeys
1 Ancient stratum	Tree sloths		No specially related forms elsewhere

EOCENE | OLIGO-CENE | MIOCENE | PLIOCENE PLEISTOCENE RECENT→

28–7 Stratification of the South American fauna. Three strata can be recognized. *1.* Ancient, in South America since before the Eocene; no closely related forms elsewhere; sloths, anteaters, etc. *2.* Middle, in South America since the late Eocene or Oligocene; arrived by sweepstakes dispersal; New World monkeys, guinea pigs, etc. *3.* Youngest, in South America since the late Miocene and mostly after the still later reunion of North and South America; field mice, cats, dogs, etc.

World and New World pikas is clearly accounted for by the sinking of the former land bridge between Asia and Alaska. It is not unusual for a barrier to arise where a migration route used to be, producing disjunction. Many closely related pairs of marine species are disjunctively distributed on the Atlantic and Pacific coasts of Central America. This at once suggests that there was a marine migration route across the region where the isthmian land barrier now stands. The suggestion is confirmed by geological studies. What connection do these facts have with the history of the South American land fauna?

The most striking and disputed instances of disjunctive distribution involve southern land areas. The examples of the tapirs and camels have already been mentioned. Others include the marsupials in Australia and South America (absent in Eurasia), the southern beeches (*Nothofagus*) and pines (*Araucaria*) in Australia and adjacent islands and in South America (but not in Africa or the northern continents), and a group of strictly fresh-water fishes [13] in South America and Africa (and nowhere in the north). Many others exist among both plants and animals.

We know from fossil evidence that many of these now disjunctive groups formerly occurred in northern lands. There is no reasonable doubt that they spread between the Old World and the New across the Asia–Alaska land connection. Change in climate reasonably explains their present survival only in the southern parts of the two hemispheres. It is known that the northern lands now have much more severe climates than they had during most of geological time. They are only now emerging from an Ice Age. That explanation clearly applies to the tapirs, which formerly ranged all over the Holarctic region and thence spread southward into the Oriental and Neotropical regions. It also applies to the camels, which lived only in North America during most of the Age of Mammals and thence finally spread to Asia and to South America.

[13] The characins, frequently kept in tropical aquariums.

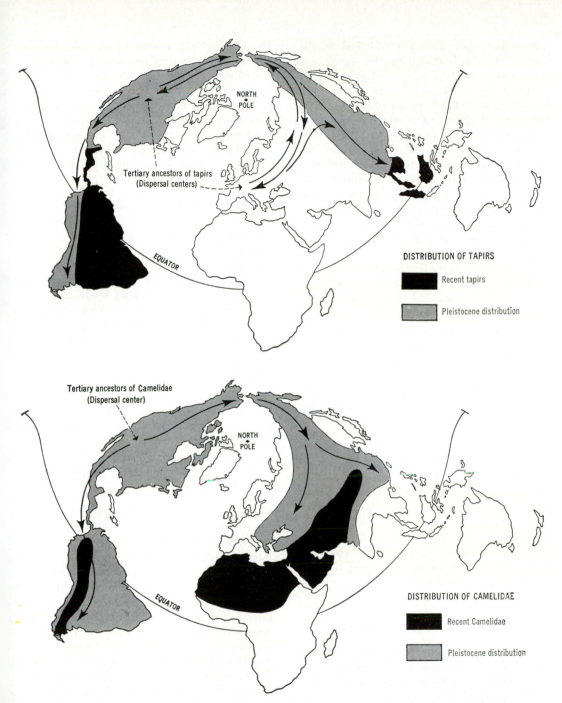

NORTH
POLE

Tertiary ancestors of tapirs
(Dispersal centers)

EQUATOR

DISTRIBUTION OF TAPIRS

■ Recent tapirs

▨ Pleistocene distribution

Tertiary ancestors of Camelidae
(Dispersal center)

NORTH
POLE

EQUATOR

DISTRIBUTION OF CAMELIDAE

■ Recent Camelidae

▨ Pleistocene distribution

28–8 Disjunctive distributions. The strange disjunction in modern distributions like those of the tapirs (*top*) and the camels (*bottom*) is explained by their earlier distributions as revealed by fossils. Present-day disjunctive distributions are relics of earlier continuous distributions.

So many examples are established by the evidence of fossils that it can be stated as a general rule that land plants and animals now disjunctively distributed in the south were formerly northern and migrated between Asia and North America.[14] For many groups there is no adequate fossil evidence, but it is usually reasonable to assume that they followed the rule. It is, however, by no means established that *all* of them did and that the rule has no exceptions. It seems probable that some southern disjunctive groups of plants, and perhaps a few animals, were really dispersed across regions now oceanic, through the tropics and farther south.

There are two ways in which such dispersal might have happened: by former land connections across what are now tropical and southern areas; or by sweepstakes dispersal across those seas. The existence of former land connections variously placed among Africa, Australia, Antarctica, and South America was formerly a popular theory. It is still sustained, in one form or another, by some biogeographers. Most, however, now believe that that theory is neither necessary nor adequate to explain why the floras and faunas of the southern continents are most decidedly distinct in spite of the presence of some disjunctive groups on two or more of them. Probably whatever migration did take place directly between tropical and southern lands was by sweepstakes dispersal over sea barriers. It may have been facilitated by island chains no longer in existence and by milder climates, so that land plants may have been able to spread along Antarctica.

CHAPTER SUMMARY

Biogeographic problems for which ecological explanation fails (exemplified by the absence of monkeys in Australia); the need for historical explanations.

Biogeographic regions: faunal regions on land—Holarctic (Palaearctic and Nearctic), Oriental, Ethiopian, Neotropical, and Australian, and typical animals; sample problems of historical biogeography, including resem-

[14] This rule was established by the great American paleontologist W. D. Matthew.

blances and differences among faunas, conflicting resemblances and origins, disjunctive distributions; faunal change and earth change—changing climates, changing connections between land masses, consequent (historical) change in faunal distributions, the four categories of historical change in a biotic region, only one of which (the regional spread or dispersal of animals) is directly geographic.

Dispersal and isolation: means of dispersal; the long-term nature of dispersal, involving successive generations; dispersal, distinguished from migrations; the passive dispersal of protists and other small organisms, the role of sea currents, winds, and severe storms, the role of birds in dispersing seeds and other organisms, the similar role of human migrations; routes of dispersal—the open sea and currents for marine forms, prevailing westerly winds for air-borne organisms, easily traversed land routes; the probability scale for dispersal of a biota; corridors, filters, and barriers; barriers as obstacles to dispersal set up by physical or ecological conditions; physical barriers—mountain ranges, deserts, oceans; ecological barriers; chance crossings of nearly absolute barriers; the idea of sweepstakes dispersal (exemplified by the histories of Krakatoa, the Hawaiian Islands, etc.).

Changing biotas and geography: the World Continent—Africa, Europe, Asia, and North America—a supercontinent with parts only intermittently isolated during Age of Mammals; relatively free dispersal within the World Continent as an explanation of broad faunal similarities, especially of the paradox provided by greater similarity between North American and Asian faunas than between North and South American faunas; South America like Australia as an island continent during most of the Age of Mammals; filters leading to the development of regional differentiation within the World Continent—Palaearctic, Nearctic, Ethiopian, and Oriental.

The island continents, South America and Australia: the marsupial fauna of Australia; the absence of placentals in Australia at the beginning of the Age of Mammals; their subsequent exclusion by a strong sea barrier; the adaptive radiation of Australian marsupials; their convergence with placental adaptive types; Australian rats' arrival by sweepstakes dispersal; marsupials and primitive placentals in the early South American fauna; their isolation on the island continent; their adaptive radiation; the late arrival of rodents and monkeys by sweepstakes dispersal; faunal interchange, exemplified by the faunas of South and North America; the development of a land connection between South and North America; interchange across it; the success of North American forms in South America; the relative failure of South American forms in North America; the analysis of fauna before and after interchange; faunal stratification, exemplified by South America; the occurrence of three strata in the fauna, each stratum entering South America at a different time; disjunction, exemplified by certain animals and plants; the two explanations— (1) some disjunctive distributions as relicts of formerly continuous distributions and (2) others as results of sweepstakes dispersal.

Part 9

The History of Life

In the seventeenth century the great French philosopher René Descartes conjectured that we would reach a much deeper understanding of the world we live in if we could only know the processes by which the world had come into being. But it was not until the early nineteenth century that his conjecture was made a reality. Before this time—in Descartes' own day, for instance—science was largely confined to observation and inference about the workings of the contemporary world. In the social sciences the introduction of historical explanations came from French and Italian writers in the eighteenth century; in geology it came from Charles Lyell and his predecessors in the late eighteenth and early nineteenth centuries; and in the life sciences it came from Darwin as late as 1859. Ever since, the scientist has perceived the world as something more than meets his eye; he has seen the present as a product of the past, and the past as something to be known and explained before the present can be understood.

In Part 9 we relate the history of life as a whole. The account begins with the young planet Earth, sterile of life, and ends with the rich diversity of living things that covers its surface today.

Chapter 29 deals mainly with the principles of historical biology, with time scales and earth history, and with the nature of the historical record of life. For the later stages of evolution—for somewhat more than the past half-billion years—life has left, in rocks, a rich record of its history in the form of fossils like that of the ancient palm leaf introducing Part 9. For still earlier stages of evolution the direct fossil record yet discovered is inadequate; but, as we shall demonstrate, the early stages are still not a book wholly closed to us.

Chapters 30 and 31 describe the history of the living world in its better-known later stages as documented in the rocks. The record reveals a long early phase in the ocean, where nearly all the major groups of organisms had their origins, and it shows how life eventually conquered the land. The history of terrestrial life focuses on the vertebrates, which have dominated the scene and have left an excellent record of their past. For a very long period, some 200 million years, reptiles were masters of most of the available ways of life for vertebrates on land. Their spectacular downfall about 70 million years ago was followed by the equally spectacular radiation of long-insignificant forms, the mammals. The final stages in the history of terrestrial life concern the progress of the mammals toward their present modern aspect, and the ultimate emergence of man.

29

Background, Beginnings, and Trends

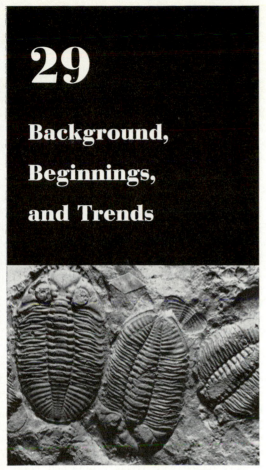

American Museum of Natural History

Fossils, such as those of trilobites shown here, are the raw material from which the paleontologist reconstructs much of life's history.

"Being" Versus "Becoming"

There are many ways of looking at a tree or at a squirrel in the tree. There is the way of a poet, who projects into the scene a complex of associations with human experiences and emotions. There is a workaday way, which also sees tree and squirrel in human terms but in terms of utility, of board feet or of squirrel pie. Then there is the scientific way, which accepts the tree and the squirrel on their own terms as things existing in the natural world. Quite independently of our reactions to them they have form, composition, and activity of their own. The purpose of science is first to ascertain but then more especially to understand, in terms deeper than mere description, those properties of existing things. Scientific understanding may be sought along many lines, but it generally follows one of two broad approaches. One is the avenue of *being*, and the other is that of *becoming*.

The law of gravitation, for instance, is something that *is*; it has not become something different since yesterday or a million years ago, and it will not become something different tomorrow or a million years hence. Most of the subject matter of the physical sciences is appropriately studied in terms of being rather than of becoming. That approach is also involved in much of biology. The processes of, say, photosynthesis in the tree or digestion in the squirrel exist and may be studied as *being* what they are and not as *becoming* something else. So may the whole tree and the whole squirrel be studied, but then it is evident that that approach alone is entirely inadequate for understanding living things. The tree and the squirrel have not always been what they are now. They have *become* so, and how they became so is essential to understanding what they are. The tree grew from a seed; the squirrel developed from a zygote in its mother's uterus. Understanding cannot stop there. It must follow the seed and the zygote back and back through the generations to times when there were no seeds or zygotes, no trees or squirrels. The long process of becoming that is evolution yields the most profound understanding of the organism that exists today.

You were introduced to that point of view

in the first chapter of this book. It has been kept constantly in mind, and much that we have said about living things has been said in terms of becoming as well as of being. Now we propose to pay even more particular attention to the fact that life as manifested in all living organisms has a *history*. We shall review that history, with special attention to the peculiarly *historical principles* involved in it.

Time and the Earth

IN THE BEGINNING

History is what happens through a span of time. To follow it, we need first of all a time scale on which to orient its events. Where should our scale start? Was there a beginning of time? This is a question that science cannot answer. It is difficult, perhaps impossible, to imagine literally infinite time, but the problem must be left to religion or philosophy. There are good scientific reasons to believe that our solar system, at least, has not always existed in anything like its present form. Since life as we know it is absolutely conditioned by the solar system, the appropriate starting point for a biological time scale might be taken as the formation of the planets more or less in their present condition.

Unfortunately there is as yet no sure and accurate way of knowing how old the solar system is. We do know beyond doubt that it is more than 3 billion years old. Certain astronomical considerations seem to place the probable upper limit at not more than 10 billion years. This is not very precise, but it is a good deal better than the old guesses, which ranged from about 6000 years to untold trillions.

We know that the solar system is more than 3 billion years old because there are rocks exposed in the earth's crust that are of about that age. The dating has been done by the study of radioactive minerals. When a mineral containing a radioactive element, uranium, for example, first crystallizes, it includes none of the products of radioactive transformation of that element. Such products then start to accumulate at a constant rate. A stable end product of the natural disintegration of uranium is one of the forms (isotopes) of lead,

and that end product usually stays in the mineral along with what is left of the uranium. The rate of formation of lead from uranium is known, and therefore the age of a mineral can be determined from the ratio between the remaining uranium and the lead produced by the disintegration of what was originally uranium. The approximate formula is

$$\frac{\text{Grams of lead}}{\text{Grams of uranium}} \times 7{,}600{,}000{,}000 = \text{Age in years}$$

Dating by this method requires good, fresh crystals of radioactive minerals that were formed at the same time as the rock containing them. To be directly useful in the study of evolution, the minerals must also be associated with rocks containing fossils. Relatively few uranium minerals have both these qualifications. In recent years another radioactive transformation, from radioactive potassium to the inert gas argon, has seemed more promising. Suitable potassium minerals are more common than uranium minerals and are more often associated with fossil-bearing rocks. The subdivision of geologic time in years is now mainly based on the potassium–argon method, especially for the last 600 million years, although the uranium–lead and several other methods are also still in use. The ages in years given in Table 29–1 have been revised in accordance with recently obtained potassium–argon dates. They are, however, still only rough approximations.

Some rocks have been dated as close to 3 billion years old. Still older rocks, without minerals now judged suitable for dating, are known to exist. It is established that approximately 3 billion years ago the earth had a solid, cool crust and that processes of rock oxidation, weathering, and erosion were already going on. This means that there was already water on the surface and that the atmosphere cannot have been extremely different from what it is now. The biological significance is that life on the earth was *possible* at that remote date and perhaps even earlier.

THE GEOLOGICAL TIME SCALE

The uranium–lead, potassium–argon, and other methods of dating in years on the basis

TABLE 29–1 *The geological time scale*

Approximate time since the beginning of the period, in millions of years	Era	Periods or epochs *	Some important events in the history of life
0.01		Recent	
2		Pleistocene	First true men. Ice Age. Mixture and later thinning out of mammalian faunas
10	CENOZOIC (The Age of Mammals)	Pliocene	Culmination of mammals. Radiation of apes
25		Miocene	
35		Oligocene	Modernization of mammalian faunas
55		Eocene	
70		Paleocene	Expansion of mammals
135	MESOZOIC	Cretaceous	Last dinosaurs. Great expansion of angiosperms
180	(The Age of Reptiles)	Jurassic	First mammals and birds
230		Triassic	First dinosaurs
280		Permian	Great expansion of primitive reptiles
345		Carboniferous †	First reptiles. Great coal forests
405	PALEOZOIC	Devonian	First amphibians. First insects
425		Silurian	First land plants
500		Ordovician	Earliest known fishes
600		Cambrian	Appearance of abundant marine invertebrates
>3000	PRE-CAMBRIAN	(Period names not well established and not needed for our purposes)	First known fossils

← YOUNGER ————— OLDER

* In technical geological use an epoch is a subdivision of a period, but the distinction is not important for our purposes. The names in this column for the Cenozoic are technically epochs, and those for the Mesozoic and Paleozoic are technically periods.
† American geologists often call the early Carboniferous "Mississippian" and the late Carboniferous "Pennsylvanian."

of radioactivity give useful approximations or orders of magnitude. In time they will doubtless improve, and it may eventually be possible to study the history of the earth and of life in terms of absolute ages in years. At present, however, the accuracy of the methods and the number of dates obtained from them are not sufficient to warrant our reliance on year dates alone in such study. We must also, and indeed principally, have recourse to a different kind of scale, one that was developed before the phenomena of radioactivity were known. The geological time scale designates the *sequence* of rocks (Fig. 29–1) and events rather than the elapsed time between them. You can see that you could follow the history

of the United States, for instance, perfectly well if you knew the order in which all events occurred even if you did not know any dates. It would also be handy to have names for successive periods. They could be named for presidents or designated arbitrarily.

The geological time scale now in general use is given in Table 29–1. Approximate times in years are also given, but in our discussion of the history of life we shall refer to the names of the eras and periods rather than the admittedly inaccurate year dates. For our purposes the whole time before the Paleozoic is lumped as Pre-Cambrian because there are few fossils in rocks of those ages and their subdivisions have little biological significance.

Union Pacific Railroad

29–1   The Grand Canyon of the Colorado River. The layers of sedimentary rocks revealed in the walls of the Grand Canyon were deposited in sequence, one above another, over spans of hundreds of millions of years. By studying the superpositions of rocks and their contained fossils in such sequences and by combining those from different regions, scientists have built up the geological time scale (Table 29–1).

We know much more about the later than about the earlier phases of the history of life, and so we use a more finely divided time scale as the Recent is approached. It is hard to visualize the vast time span involved and the increasing tempo of the history as time went on. It may help to comprehend the relative durations, at least, if we consider the history of life as if it had all occurred within the 24 hours from one midnight to the next (Fig. 29–2). Let us arbitrarily set the beginning, the first midnight, at 2 billion years ago. On that scale fossils did not become abundant until about 4:45 P.M. At 6:45 P.M. the invasion of the land by plants was under way, and by about 7:50 insects and the first amphibians had joined them. The Age of Reptiles began about 9:15. It ended, and the Age of Mammals began at about 11:10 P.M. Modern man appeared only a couple of seconds before mid-night, and the whole span of recorded human history occupies about the last ¼ second.

Origin of Life

PROBLEMS

Nothing is *directly* known about the origin of life. There could be only two kinds of direct evidence: fossils of the first organisms or the rise of similar organisms from nonliving matter today. No such fossils are known or are ever likely to be. The first organisms were almost certainly extremely small and could hardly have become fossilized or be found and recognized if they had. It was formerly believed that life is, indeed, being generated all around us anew, but we now know that the supposed evidence for this opinion was false (Chapter 3). As far as is known, no living

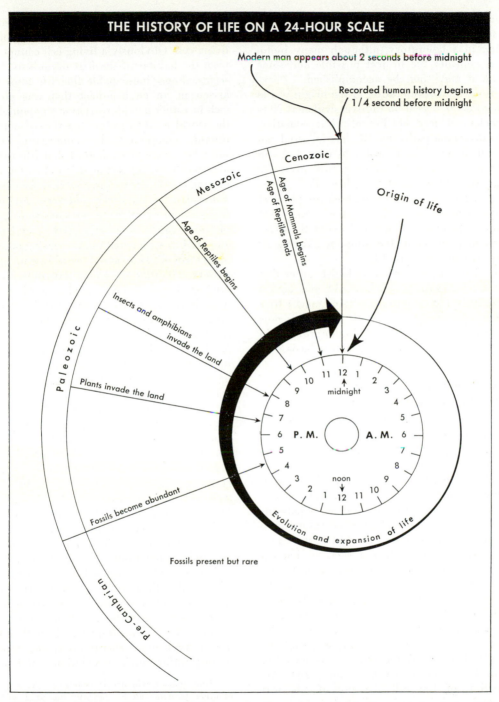

THE HISTORY OF LIFE ON A 24-HOUR SCALE

Modern man appears about 2 seconds before midnight

Recorded human history begins 1/4 second before midnight

Cenozoic

Mesozoic

Origin of life

Age of Mammals begins
Age of Reptiles ends

Age of Reptiles begins

Insects and amphibians invade the land

Paleozoic

Plants invade the land

Fossils become abundant

Evolution and expansion of life

Fossils present but rare

Pre-Cambrian

11 12 1
10   2
9   midnight   3
8   4
7   5
6 P. M.   A. M. 6
5   7
4   8
3   noon   9
2 12   11 10
1

29–2   The history of life on a 24-hour scale. For purposes of this diagram it has been arbitrarily postulated that life arose 2 billion years ago. The true figure is unknown and may be greater.

things are now being generated from the non-living. No one has yet created a truly living thing in a laboratory, although many biologists consider the project possible and some think it may soon be accomplished.

Scientific consideration of the origin of life must thus be based on indirect evidence. Nevertheless, it need not be entirely speculative. The diverging paths that life has followed can be extrapolated backward into time and can give us some idea of what the most primitive organisms of all must have been like. Moreover, a great deal is known about the physical and chemical properties of the matter that enters into living things and about the processes that go on in the simplest existing or conceivable forms of life.

In the first place, most biologists agree that the earliest forms of life could and almost certainly did arise from nonliving matter by a natural process. On the basis of what is now known there is, at least, nothing improbable in this view. We shall adopt it as the basis of a hypothesis.

A HYPOTHESIS ON THE FORERUNNERS OF LIFE

It seems to be a necessary prerequisite for the rise of life that complex organic molecules should first have arisen. Systems of such molecules may become capable of self-reproduction, at least in the sense that, if the environment supplies suitable materials and a source of energy, they can serve as patterns by which the materials are combined into likenesses of themselves. The likenesses would not always be perfect, and even at that extremely primitive level something akin to mutation would occur and variation would be present in the molecular population. Whether such molecular systems were themselves alive is a matter of definition. If the process really went along in anything like the way outlined by this hypothesis, then there was no exact point where the nonliving became alive. Acquisition of the full panoply of life was gradual.

Viruses are not recent representatives of the first organisms or of their forerunners. Viruses do nevertheless show that some organic molecules, and specifically those of DNA, can bring about the production of du-plicates of themselves in a suitable medium. The medium for viruses is itself organic, a living cell. Obviously a living cell cannot have been the habitat of the first organisms, but it seems almost inescapable that life must have arisen in an environment that was *already* rich in rather complex organic compounds. In the world as it is today, the rise of complex organic compounds by nonorganic means must be a rare event indeed, but life did not arise in the world just as it is today. On other grounds [1] it has been inferred that the original atmosphere of the earth contained little or no free oxygen or nitrogen but consisted largely of water vapor, ammonia (NH_3), and methane (CH_4), with small but increasing amounts of carbon dioxide and free nitrogen, plus a variety of other minor constituents. It has further been demonstrated experimentally that the passage of a spark (as from lightning) or of strong ultraviolet radiation (as from the sun) through such a mixture can produce complex organic compounds, including amino acids. These compounds could well have become concentrated in pools, or even in primitive seas, where they could give rise to the giant molecules hypothesized as forerunners of life and then provide a suitable medium for the reproduction and the increasing variety of such molecules. The first reproducing molecules may have been nucleic acids, and in any case nucleic acids must have participated early in the reproductive process. The basic mechanism of heredity thus arose, and natural selection could begin to act and to guide evolution into its subsequent stages.

Another step would be the aggregation of molecules that interacted favorably with each other and with their environment. If, for instance, one kind of molecule reacted with the environment in such a way as to produce oxygen as a waste product and another kind used oxygen in an energy-producing reaction in respiration, their association could be fa-

[1] One item of evidence is as follows: Atmospheres of other planets can be partially analyzed with the spectroscope. Jupiter's atmosphere consists largely of methane and ammonia, and the atmospheres of Saturn, Uranus, and Neptune are largely methane. The atmospheres of these planets have changed since the beginning of the solar system, but they have almost certainly changed much less than Earth's atmosphere.

vorable to both. The production and recombination of oxygen are used only as a simple example. It is unlikely that just these reactions were involved in the earliest stages of life. The example shows, however, that different kinds of molecules can interact in such a way that their activities are more effective when they occur together than when each occurs alone. Even at the start the interactions must have been considerably more complex than mere interchange of oxygen by only two kinds of molecules. The association of nucleic acids, proteins, and compounds required for energy transfers (Chapter 2) would be among the early steps.

Once such interacting aggregates began to form, they would tend to become more numerous and more distinctly and constantly organized in composition and pattern. A complex aggregate of favorably interacting molecules would have evident and great advantages over any remaining free-lance molecules. Given the existence of variation within molecules and within increasingly complex aggregates of molecules, natural selection would occur and would favor change in the direction of more effectively and constantly organized aggregates. The result would eventually be a true multimolecular organism, a primitive protist, alive by any definition.

EARLY ORGANISMS

At first sight it might seem logical that the earliest true organisms must have been autotrophs, performing all their necessary organic syntheses from quite simple inorganic materials and using either other inorganic materials or solar radiation as a source of energy. It might even seem absurd to think that the first organisms could have been heterotrophs when there were no other kinds of organisms to feed on. That opinion was formerly popular among students of the subject, but in recent years almost all biologists have come to an opposite conclusion.[2] Autotrophy requires more complex organization and metabolism than heterotrophy. It strains the scientific imagination

[2] It is interesting evidence of Darwin's genius that he held the view now considered most likely, although many biologists after him overlooked his opinion on this matter and the strong arguments in favor of it.

too far to think that the very first organisms can have been so complex, and no one has succeeded in visualizing in convincing detail how such organisms could originate from molecular forerunners.

The whole process becomes simpler and easier to understand on the hypothesis that the earliest organisms were, in essence, heterotrophs. Of course they could not (not all of them, anyway) feed on other organisms. They could, however, be heterotrophic in the sense that they fed on carbohydrates, amino acids, and other fairly complicated organic compounds. Then their own syntheses and energy transformations would be comparatively simple. A requirement for this hypothesis is that the earliest organisms must have lived in an environment rich in organic compounds that had not been synthesized by living organisms. We have just seen that there is good reason to believe that such environments may well have existed in the earliest waters on the earth.

Such a state of affairs could not persist indefinitely. As organisms multiplied, they would rapidly consume all the available organic materials. Synthesis by nonorganic processes would soon tend to lag behind consumption. Moreover, conditions on the earth and especially in its atmosphere were changing. The organisms themselves would cause much of the change. They would, for instance, tend to lock up much of the carbon that formerly was in methane or more complex compounds, or to convert it into CO_2, from which more complex organic molecules are less likely to be formed by nonorganic processes. Nonorganic synthesis of foods must inevitably have fallen nearly to zero, as it is today and has been through the whole history of life except in the dimmest era of beginnings. If nothing else had happened, life would have become extinct after a brief flare-up that led no further than to a few primitive protists.

Imagine what would happen if, as some needed organic compound were becoming rare in the environment, there appeared a mutant organism that could synthesize that compound from simpler and more widely available materials. The mutant organism would have a tremendous advantage over the others, and its descendants would soon become the dominant, or even the only, remaining organisms. As other compounds became rarer, additional mutations for their synthesis would be favored by selection. Step by step the trend would be toward greater self-sufficiency, toward more complete autotrophy. The culmination of this stage of evolution would be in organisms that could use CO_2 as their sole external source of carbon and solar radiation as their sole external source of energy. Such organisms would of course be photosynthetic protists and eventually plants.

In the meantime environmental conditions on earth had changed radically and could never again be those under which life originated. The surface of the earth, probably hot in still earlier times, must have cooled well below the boiling point of water when life arose. (Why?) The original ammonia and methane of the atmosphere were incorporated in the increasingly complex organic systems or were transformed (in part) into N_2 and CO_2, which probably constituted most of the atmosphere relatively soon after life arose. After the rise of photosynthesis, organisms themselves made the most important change in the atmospheric composition. They must have reduced the percentage of CO_2 in the atmosphere and also—and especially—have increased the percentage of oxygen, previously rare except in compounds. High in the atmosphere the oxygen then formed an ozone (O_3) layer, still present. That layer screens out much of the ultraviolet in the sun's radiation. The earlier influx of ultraviolet had been a possible and perhaps the most effective energy source for the nonorganic synthesis of complex molecules. As ultraviolet was cut off from the surface of the earth, visible light (not significantly absorbed by the ozone layer) became the chief source of energy through organic pigments (now mostly the chlorophylls) in living systems. These changes, largely caused by the early living systems, made further spontaneous origin of such systems impossible and gave organisms a virtual monopoly on organic syntheses.

All of the preceding is hypothetical, but it is in accordance with what we do know of the properties of living things and their environments. Even if we consider it wholly fanciful, it does illustrate possible kinds of interactions and one possibility, at least, as to what went on in the remotest, dark, unrecorded stages of evolution. When we first begin to get a little real light on the scene, the physical conditions of earth and atmosphere were much as they are today and photosynthetic organisms already existed. Their existence made possible the concomitant and subsequent evolution of other organisms that fed on them and on the materials synthesized by them: plant-eaters, parasites, and organisms of decay. Then there could also evolve still another stage: meat-eaters that feed on plant-eaters. The possible sequence in these early stages of evolution is summarized diagrammatically in Fig. 29–3.

Fossils and the Historical Record

PRINCIPLES

The history of life ceases to be hypothesis and inference and becomes direct knowledge when fossils are available. A fossil is a fact: it is a visible trace of some organism that lived in the geological past. That is the fundamental basis of *paleontology*, the study of ancient life (the Greek roots mean exactly that). Science arises from the observation of isolated facts, but it does not become truly science until those facts are seen in relationship to others and are placed in an explanatory context. The paleontological deciphering of the history of life involves, among others, these observations and inferences:

1 A fossil, an organism with definite characteristics shown by its remains. Further inference concerns characteristics and activities not directly preserved.

2 The geographic locality at which the fossil was buried and near which it must have lived, another observed fact.

3 The age of the fossil, an inference based on a wide range of geological and paleontological data.

4 Association of the fossil with others of the same species, the basis of systematic study of the fossil population.

5 Association with fossils of other species, part of the basis for the study of the community and the environment (and also of the age).

Characteristics of the rocks in which the fossil occurs and of the position and mode of burial of the fossil in those rocks—further data for the study of environment and age.

Relationships of the fossil population to others, earlier, contemporaneous, and later; comparisons and inferences leading to phylogeny and to classification.

The broader generalizations and principles of the history of life, with which we must be mainly concerned in this summary of an intricate branch of the life sciences, are derived from a vast number of detailed observations and inferences mainly of these kinds.

UNIFORMITARIANISM

There is an important principle fundamental for paleontology, geology, or any science that has historical aspects: the present is a key to the past. That principle was the subject of bitter controversy a century or two ago, when it was endowed with the formidable name of *the doctrine of uniformitarianism.* It is now accepted as true by virtually all scientists, and without it there could be no really scientific study of any kind of history.

The doctrine of uniformitarianism is that the fundamental properties of the universe, the nature and the modes of interaction of matter and energy, have not changed. They are independent of the passage of time. It is only the forms that they have taken and the status of the results of their past interactions that change through time. Liquid water on the earth always has run downhill (unless counteracted by some opposite physical force), and a certain amount of water running at a given velocity over a bed of defined coherence, grain size, and so on, always erodes that bed to a predictable degree. These facts are basic in the unchanging, uniformitarian properties of the universe. The amount of erosion that has occurred and the resultant size and shape of the eroded valley change. They have a history, although the *process* of erosion has none. Similarly we conclude that chromosomes of a given structure and composition under defined environmental conditions always did, do, and will have the same influence on a developing organism and that the same changes in the chromosomes are uniformly associated with the same changes in the organism. The changes that have, in fact, occurred in organisms have a history but, again, the processes do not.

Another way to put the principle is to recall the contrast between *being* and *becoming* made at the beginning of this chapter. The characteristics of nature that simply *are* are those inherent in the properties of matter and energy. Time is irrelevant to them; they have no history. The things that change and that *become* are configurations arising in accordance with the timelessly uniform processes. History is concerned with those "becoming" configurations.

By this principle the geologist is able to interpret past changes in the sculpturing of the face of the earth by processes that can be observed now in action and that can also be studied now experimentally. The biologist is similarly able to interpret the fossil record by processes still proceeding in living organisms and also subject to present experimentation. He feels sure that the processes were just the same in the past as they are now. Perhaps that seems obvious to you. If so, it is only because a hard-won scientific attitude now permeates much of our intellectual atmosphere. The timelessness of the properties and processes of the universe was by no means obvious to earlier thinkers. Establishment of that principle was one of the major triumphs in the history of human thought.

WHAT FOSSILS ARE

It is a wonderful thing that you can hold in your hand the remains of an organism that lived hundreds of millions of years ago. It is a common observation that dead organisms usually molder and become unrecognizable in a few years at most, sometimes even within hours. "Dust thou art, to dust returnest" is indeed the common lot of living things. The organisms of decay see to this, and it is good for the continuity of the communities of life that they do. Yet there are millions upon millions of ancient organisms that have never wholly decayed.

Fossils are being formed today, but they are only a minutely small fraction of the organisms that die.[3] It has always been true that an exceedingly small fraction of organisms has fossilized. A still smaller fraction by far has been recovered and studied. Nevertheless, the numbers of organisms have been so countless and time so long that even so minute a fraction adds up to a respectable documentation of the history of life.

The usual first condition for preservation as a fossil is burial before decay is complete.[4]

[3] Moreover, they are being formed where you are most unlikely to see them, for reasons that you will be able to discern for yourself.

[4] A few relatively young fossils have not been buried. There are exceptions to many of our statements here, but we are concerned with what is usual. Exceptional fossils may be particularly interesting,

Natural burial occurs when a dead organism sinks into mud or sand or when these and other sediments are swept over its remains by waves or streams, occasionally by winds. Organisms may be buried whole, but frequently they are already dismembered to some extent when they are buried. The same agency that buries them may break them up. Once buried, any further decay must stop short of obliteration. Fossils available to us as documents must then stay buried down to, or nearly to, the present time. An early Cambrian shell can end up in the laboratory only if it remained buried for some 600 million years and never was

but they have comparatively little importance in an over-all view of the history of life.

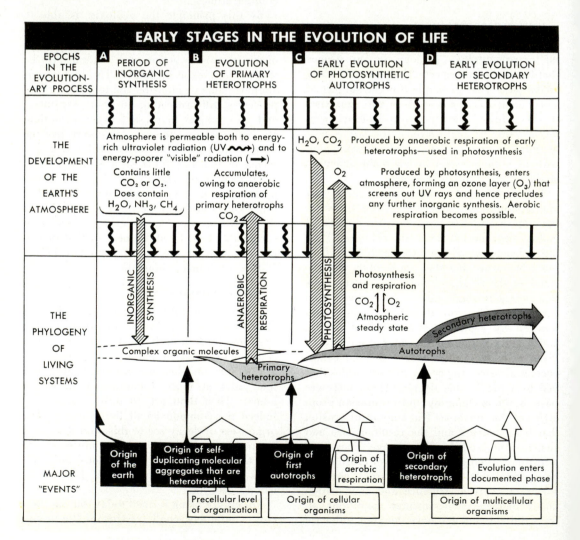

EARLY STAGES IN THE EVOLUTION OF LIFE

washed out again by erosion in all the intervening upheavals and remodelings of the earth's crust.

Erosion cannot have touched for all those years the fossil now in the laboratory, but erosion must finally have come near it. As a rule, fossils cannot be found and collected unless they are again at or near the surface of the earth where the fossil hunter can find them.[5] Fortunately, rocks of all ages are now exposed at the surface, even though they may have been deep within the crust at some time in their history. The earth's crust has been continually rising in one place, sinking in another, buckling and breaking here and there. The higher parts are constantly being worn down by processes of erosion, and thus

[5] The only important exceptions here are small fossils frequently recovered from deep drilled wells. They are useful in the petroleum industry but do not bulk large in the documentation of the history of life.

some among even the oldest rocks are now at the surface.

The processes of decay are highly effective and may continue even after burial, if burial is not too deep and not in somewhat unusual naturally antiseptic conditions. Decay usually obliterates all the soft parts of an animal, leaving only the skeletal parts, which consist mainly of resistant inorganic materials. Fossil mollusks usually consist of the shells alone. Fossil vertebrates seldom preserve anything but bones, teeth, and hard scales, if any. Fossil insects are none too common because insects are unlikely to be buried, but those that do occur are often unusually complete because of the tough, over-all external skeleton. Fossil plants also usually have lost all the soft parts, the protoplasmic cell contents. Tough leaf coatings, cell walls, spore skins, and such decay-resistant parts are, however, frequently preserved. Even the soft parts of plants and

---

29–3 Early stages in the evolution of life. This represents what we consider the most reasonable hypothesis at present. It is based on some good evidence, and there is no serious doubt that it could have happened. There is, however, no *direct* evidence that it *did* happen, and it is not presented as an established theory. Four major stages are distinguished. *A. Stage One:* nonorganic synthesis; the creation of conditions necessary for life's origin; the slow accumulation of nonorganically synthesized complex molecules such as amino acids, nucleic acids, and simple carbohydrates. Such nonorganic synthesis and accumulation of organic compounds was possible at this unique stage in evolution because of three conditions then prevailing: (1) necessary precursor compounds (NH_3, CH_4) present in the atmosphere; (2) necessary energy available in the form of ultraviolet radiation penetrating the (ozone-free) atmosphere; (3) absence of life and, therefore, exclusion of decay. The first stage ended with the origin in the accumulated organic compounds of a molecular aggregate capable of self-duplication. *B. Stage Two:* the growth of populations of these aggregates. Their expansion and evolution took place at the expense of the reservoir of complex organic molecules accumulated in Stage One; this reservoir was the source not only of chemical building blocks but also of energy, which doubtless was mobilized for metabolic use through some form of anaerobic respiration that promoted the accumulation of CO_2 in the atmosphere. The first organisms were, in fact, heterotrophs—here termed primary heterotrophs to distinguish them from now-familiar heterotrophs that arose in a different, later evolutionary context (Stage Four). Two points about Stage Two are important: (1) its extent and duration were rigidly limited by the size

of the reservoir of nonorganically synthesized resources (Stage One) on which it was dependent; (2) the consumption of this reservoir by the first heterotrophs destroyed the conditions necessary for further spontaneous origins of living systems—from then on biogenesis became the only mode of origin of organisms. *C. Stage Three:* a continuation of life, dependent on the mutational origin of the capacity to synthesize organic compounds from simple, renewable resources—dependent, in other words, on the origin of autotrophy. Some chemosynthesis may originally have played a role, but the foundation of all later life and its evolution required development of the photosynthetic ability. Photosynthesis was possible in part because of atmospheric CO_2 accumulated in Stage Two. Photosynthesis led to the oxygenation of the atmosphere, which had two principal consequences: (1) an ozone (O_3) layer formed, screening out most of the sun's ultraviolet rays and excluding them as an energy source for synthesis; (2) conditions were established for the evolution of aerobic respiration—thereafter photosynthesis and respiration maintained a more or less steady state of CO_2 and O_2 in the atmosphere. *D. Stage Four:* the probably often-repeated origin and expansion of populations of secondary heterotrophs. They are descendants of autotrophs, and they have indefinitely renewable resources of energy and materials for their growth in the form of the now well-established autotrophs that mobilize solar energy for the whole world of life. Living systems had probably attained a cellular level of organization before the origin of secondary heterotrophs. All still-surviving heterotrophs (with the exception of viruses, which are probably neither primitive nor true organisms) have a cellular organization and have almost certainly inherited it from a common ancestor.

more rarely those of animals may also be preserved as a dark film of carbonaceous material or as impressions on a rock surface.

It is a common misapprehension that fossils are petrified, that the organisms have "turned to rock." Such a thing rarely happens, at least in the usual understanding of the words. Preserved hard parts usually consist of the same material as when the organism was alive, perhaps with some recrystallization or slight chemical change. Spaces left by the decay of soft parts, for instance in the marrow of a bone or inside the cell walls of a tree trunk, are often filled with secondary deposits of a mineral, frequently silica (SiO_2 in various forms). After burial and hardening of the surrounding sediments, the hard parts of a fossil may be dissolved by percolating waters. Then a cavity, a mold, may be left, or else mineral-laden waters may precipitate silica or some other mineral in the space, producing a mineral replica of the original fossil (Figs. 29–4 and 29–5).

PRE-CAMBRIAN FOSSILS

We have already noted that there is no fossil record of the beginnings of life. From inferences as to what the earliest organisms must have been like, it seems hardly possible that they could have been preserved as fossils or recovered and recognized even if they were preserved in some way. The disappointing scarcity of the earliest fossil records unfortunately applies to a long span of geological time. Life probably existed 2 and perhaps even 3 billion years ago, but fossils become varied and abundant only with the beginning of the Cambrian, a mere 600 million years past. Thus the reasonably good fossil record as now known may not cover more than about the last fourth or fifth of the history of life. The whole span of the Pre-Cambrian is poor in fossils but is not an absolute blank. Some carbon and hydrocarbon remains of possible organic origin are on the order of 2 billion years old. At present the oldest unquestionable fossil remains that we have are microscopic algalike plants from rocks in Ontario believed to be about 1½ billion years old. The most conspicuous signs of Pre-Cambrian life are extensive limestones (some of them in the Big Belt Mountains of Montana) believed to have been deposited around primitive algae, probably blue-greens. Some of them are on the order of a billion years old, and some are younger. In later Pre-Cambrian rocks, from about 750 down to about 600 million years old, the first likely traces of animals appear—impressions that resemble worm burrows and soft-bodied coelenterates, annelids, and other animals of doubtful affinities.

Numerous and more varied animals occur in the earliest Cambrian rocks. The sudden contrast between the Pre-Cambrian rocks, in which animal fossils are so rare or dubious, and the Cambrian, in which they are abundant, poses a serious question: Why? A good scientist must be prepared to say, "I don't know," and that is at the present the correct answer. But a scientist would hardly stop there. Even though we do not know the answer, what are the possibilities, and are some more likely than others? The principal possibilities seem to be these:

1. The various groups of animals that appear in the early Cambrian did not exist earlier; they evolved then each in one enormous step (a mutation?) from protists.

2. They had existed in much the same form for a long time during the Pre-Cambrian, but earlier fossils have not been preserved or found. Perhaps they then occurred mostly under special, local conditions that made burial unlikely or the sediments of which are not preserved or have not been found and adequately searched.

3. They arose gradually but at an unusually high evolutionary rate during an immediately Pre-Cambrian span for which rocks that could contain their fossils are rare or absent on the surface at present.

4. Their Pre-Cambrian ancestors were soft-bodied and did not fossilize; their appearance in the fossil record approximately coincides with the evolution of hard parts.

Can you think of other possibilities? All of these have good arguments both pro and con, and each has been supported by competent students. To us the first seems ruled out because it is not in agreement with our understanding of later evolutionary processes. The second is a strong possibility but is unlikely to be the whole explanation. The third and fourth are also quite possible. They are not contradictory, and both could be true, as well as the second.

OVER-ALL RECORD

With the Cambrian begins a rich, continuous record of the history of life in the form of fossils. There are some groups that we infer must have been present and important but that are nevertheless rare, although not entirely absent, as fossils. Among these are bacteria and worms. (What basis is there for inferring that they were present and important even when not represented by fossils? Why would they tend to be underrepresented in the fossil record?) On the whole the record is amazingly good when we consider the infinitesimal chance that any one organism, or even a representative of any one species, has had of winding up on a paleontologist's desk.

Some Major Tendencies

In the next two chapters some of the events in the history of life since the beginning of the Cambrian will be examined in logical and chronological sequence. Attention will of course be given not merely to the events but also to the principles and processes underlying those events and illustrated by them. First we propose to take a quick look at the history as a whole, as it is schematically represented in Figs. 29–6 and 29–7. In this scheme and in the ever-changing panorama of plants and animals that it symbolizes, there are certain broad tendencies. Those tendencies tie in with much that you have already learned; they are understandable in the light of previous chapters. They also provide a background against which the drama of Chapters 30 and 31 will be played.

EXPANSION

The first striking generalization about the over-all tendency of evolution is that life has expanded. The total number of living things has increased. The bulk of matter in living form has increased. The number of different kinds of organisms has increased. Their diversity, in the sense of the extent of differences among them and not only in the number of species, has increased. All of these increases may be viewed as aspects of the expansion of life, although they are quite distinct from each other.

If you think of the time when life was just getting under way and compare it with the present day, it is quite obvious that life must have expanded tremendously in every possible way. The record of the first and most radical phases of that expansion is unfortunately poor or wholly lost. Yet expansion is still clearly evident in the last quarter or so of the time, when we do have a fairly good record. Expansion is seen in Figs. 29–6 and 29–7, where it reflects especially the increase in number of kinds of organisms. Fig. 29–7 also shows increase in basic diversity with the increase in numbers of phyla, and this is still more evident when classes, for instance, are counted against a time scale. New structural extremes, new levels of organization, have continually arisen without an equal loss of the old. Divergence has continued. There is far more difference among an ameba, a maple tree, and a man living today than among any organisms that were alive in the Cambrian.

It is noteworthy that expansion of this sort, at least, has not been constant. There have been times of especially great outbreaks, one might say, of multiplicity and diversity: one occurred in the Devonian for both plants and animals. There have even been times when in this respect evolution seemed to be losing ground: both plants and animals seem on the whole to have decreased in multiplicity and diversity between the Permian and the Triassic. Nevertheless, the general tendency for increase in this respect is plain.

Increase in numbers of individuals and in total bulk of living matter cannot be read directly from the fossil record. (Why not?) That there must have been tremendous expansion in these respects, too, since life began is obvious. There is much uncertainty and room for differences of opinion as to the rates of increase at various times and as to whether these kinds of expansion continued into later phases of the history. Numbers of individuals tend to be in inverse ratio to their size (Chapter 24). The evolution of large plants may have involved a distinct decrease in the total

29–4 Fossil hunting and preparation in the field and laboratory. *Left.* A paleontologist uncovering a fossil. *Below.* A fossil exposed in the field. It is a giant bony fish, *Portheus*, which lived more than 70 million years ago. *Below left.* A fossil being covered with plaster of Paris for shipment to the laboratory. *Below right.* A fossil being carefully removed from the plaster casing.

American Museum of Natural History

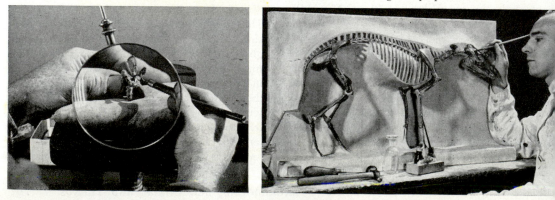

Below left. Cleaning a vertebra of the fossil reptile *Hesperosuchus*.
Below right. A mounted skeleton of eohippus (*Hyracotherium*) in the final stages of preparation.

29–5 A diversity of fossils. *Left.* Fragments of the plant *Otozamites* from the Petrified Forest of Arizona. This cycad lived about 200 million years ago. *Below.* An ammonite from the Jurassic, about 150 million years old.

Below. Fossil dinosaur tracks in Texas.

Below. Detail of the dinosaur *Corythosaurus*, showing muscle and skin on the tail vertebrae.

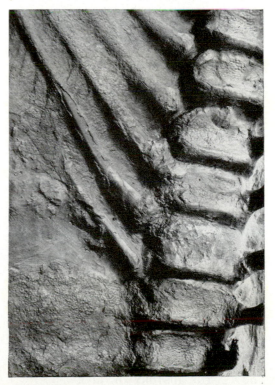

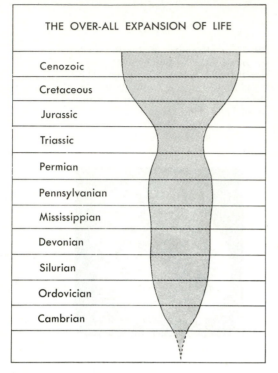

THE OVER-ALL EXPANSION OF LIFE
Cenozoic
Cretaceous
Jurassic
Triassic
Permian
Pennsylvanian
Mississippian
Devonian
Silurian
Ordovician
Cambrian

29–6 The expansion of life. The width of the pathway is approximately in proportion to the known diversity of organisms (plant and animal) at various times in the past. Note the constriction of the pathway during the late Permian and Triassic; this reflects the Permo-Triassic crisis (Chapter 30).

number of photosynthetic organisms, at least.

Among the factors limiting the bulk of living matter, or its total turnover or metabolic activity, are input of solar energy and efficiency of photosynthesis. There is no reason to believe that either radiation input or photosynthetic efficiency has increased significantly since well back in the Pre-Cambrian. Even a few primitive species in the Pre-Cambrian might have filled their environment with as great a bulk of living matter and might have had as much metabolic turnover as was possible—and the possibility may not have increased much or at all since then in a particular environment, such as the sea. However, two other factors tend to increase both numbers and total bulk of organisms: invasion of new environments and lengthening of food chains. Both those tendencies were plainly still active long after the Cambrian.

OCCUPATION OF ENVIRONMENTS

The expansion of life has been accompanied by both more extensive and more intensive occupation of the possible environments for life on the earth. In fact the spread in environments has been, in a sense, one of the main reasons for the general expansion. The most spectacular example of extension in environment, spread into great areas hitherto unoccupied, was the invasion of the land by plants and animals. Until Silurian times life was confined to the water, as far as we know.[6] Then started a movement into land environments not completed until some hundreds of millions of years later.

Increasingly intensive occupation of environments is also evident in the fossil record. This takes the form of an increasingly fine subdivision of ecological niches. Along with it goes increasing specialization of the organisms in the niches. As a simple example, instead of one species of animal eating ten kinds of plants there may be ten species each eating one kind of plant.

Another factor in environmental expansion is the fact that the expansion in itself creates new environments that can be occupied in their turn. For instance the high "gardens" of epiphytes in the crowns of forest trees (Chapter 27) occupy an environment that did not exist until the forest trees evolved. Lengthening of food chains also interlocks with expansion in general and occupation of environments and subdivisions of niches in them. Increasingly large plants and animals give increasing opportunities for evolution of new links in food chains leading from them. Each new species of, say, mammals is a possible environment for a new species of parasite, which may itself link up a new food chain.

CHANGE: PERSISTENCE AND REPLACEMENT

Another evident over-all aspect of the whole of evolution is its constant state of flux. Nowhere in the fossil record is there any time of static equilibrium. Change has been slower at

[6] There may have been some protists, simple plants, and perhaps a few wormlike animals earlier in moist soils. If so, they have not been certainly identified as fossils.

some times than at others and faster in some groups of organisms than in others, but some change has always been going on. There is not even any evidence that evolution tends to approach an equilibrium or that it has, on an average, slowed down as if it might eventually reach completion. There are, it is true, some biologists who think that such a final state has now been essentially reached. To them we can only say that change has *always* occurred through the past and we see no reason why it should stop now. The next 10 million years should settle the point, but there may be no biologists around when it is decided.

Some other biologists hold that the rate of evolutionary change has tended to accelerate ever since life began and that flux is now at its greatest so far. That is a tricky point that we cannot discuss adequately here. It depends largely on just what is meant by the rate of evolution, which is not so simple to define as you might think. Part of the argument has to do with the expansion of life, which is clearly a fact but which has not shown constant acceleration and may not be going on at all right now. Unlike change in a more general sense, expansion has not been constant in the past and cannot continue indefinitely into the future. There are limits to the amounts and kinds of organisms that earth can accommodate at any one time.

How can there be change without expansion? The fossil record is replete with examples. Some groups of organisms have persisted for long times without notable change. They get into a comfortable rut, a persisting environment to which they are well adapted, and they remain there. However, there are more examples of the replacement of now extinct organisms by others better adapted to what are essentially the same environments. Environments suitable for seed ferns, cycadeoids, cordaites, and other early land plants exist today, but those plants do not. The flowering plants (and to less extent the conifers) have replaced them.

Expansion and replacement are two of the main factors in the constant change among living things, and, if expansion is limited, replacement has not so far appeared to be.

COMPLICATION AND IMPROVEMENT

Increasing complication and improvement of biological functioning are the changes most often mentioned in discussions of the overall tendency of evolution. You may be surprised that we have left them till the last. We have done so purposely with the intention of de-emphasizing them, because we believe that they are overemphasized and somewhat misunderstood in much popular thought and teaching. Increasing complication has certainly been an important factor in various phases of the history of life, but it has been far from universal as an evolutionary tendency, and it has not been particularly noticeable in the last few hundred million years of history. The most important point to stress about improvement, change for the better, or progress in evolution is that it is not inherent.[7] It is not built into the nature of the universe or of life. Improvement has occurred in the course of evolution, not because it is an inherent and general tendency, but where and when it had a natural, immediate cause: selection. This statement could fall into the fallacy of circular reasoning if it were taken for granted that what selection favors is improvement.

There can be many definitions of improvement, and selection does not necessarily favor improvement by a particular definition. Nor does evolution inevitably tend toward improvement by any definition. Nevertheless, what may reasonably be called improvement has occurred so frequently that it may be considered a common, not universal, tendency in the history of life. It has taken many forms, but especially these three: (1) increasingly precise and effective adaptation to a particular way of life; (2) marked change in structure and function, making possible ways of life, occupation of environments, and so on, increasingly far removed from the ancestral ways and environments; (3) increasing per-

---

[7] A few biologists think that it is. This is of course an inference, not an observable fact, and therefore it is open to dispute. In such matters we have tried to follow the consensus of competent modern biologists. When the point is important and there is strong disagreement, we have played fair with you and said so. Biology, like all sciences, is not a finished structure with all doubts settled.

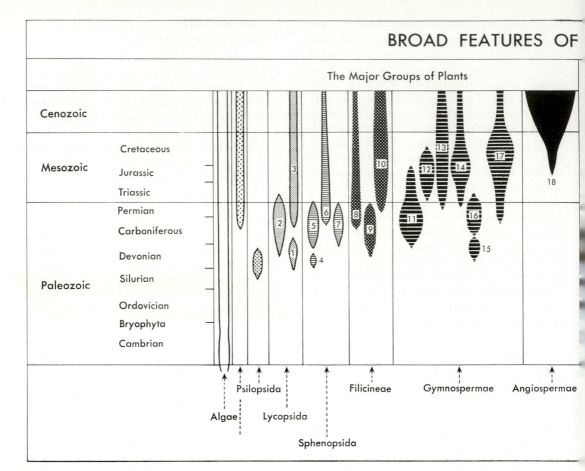

The Major Groups of Plants

29–7   Broad features of the fossil record of life. The width of each pathway is approximately in proportion to the known variety of organisms in the phylum through time. Note these general points: (1) Diversity increases through time. (2) Evidence of the Permo-Triassic crisis is striking: there is a constriction in virtually every animal phylum during the Triassic. (3) Within the plants, which are given here in greater detail than the animals, the generalization is clear that the more complex groups (for example, gymnosperms and angiosperms) arise progressively later in the fossil record. (The same generalization is clear for vertebrate animals, for example, in Figs. 31–4 and

ception of the environment and increasing complexity, flexibility, and appropriateness of reactions to stimuli.

Tendency 1 is most widespread and least important in this over-all view of the history of life, although of great importance among the details of the history. You can supply plenty of examples from your own observations or reading, in previous chapters of this book or elsewhere. A good example of tendency 2 is the rise of root–stem–leaf differentiation and of vascular tissue in plants, making possible true land life, or of the egg in reptiles, realizing the same possibility.

Tendency 3 is peculiar to animals and is especially pertinent to the rise of man. In spite of considerable evidence to the contrary, we do have a right to consider ourselves an improvement over other organisms. To a lesser degree, improvement of a generally similar kind is evident among many other evolutionary lines of animals.

CHAPTER SUMMARY

Being versus becoming, two broad approaches to scientific explanation: the explanation of tree and squirrel as entities in being and as

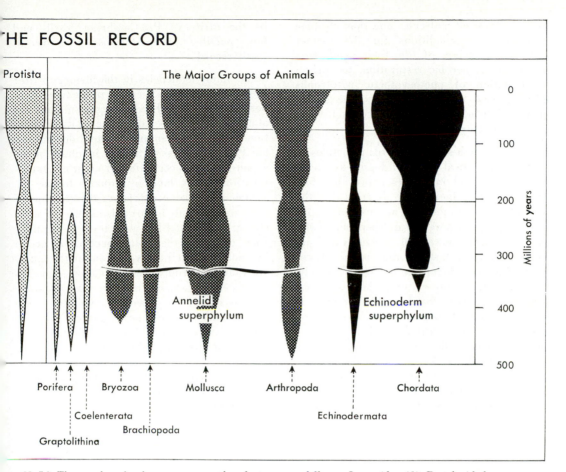

Protista — The Major Groups of Animals

Annelid superphylum

Echinoderm superphylum

Millions of years

0 — 100 — 200 — 300 — 400 — 500

Porifera — Bryozoa — Mollusca — Arthropoda — Chordata

Coelenterata — Echinodermata

Brachiopoda

Graptolithina

31–5.) The numbered subgroups among the plants are as follows: *Lycopsida:* (*1*) Protolepidodendrales, (*2*) Lepidodendrales, (*3*) Lycopodiales; *Sphenopsida:* (*4*) Hyeniales, (*5*) Calamitales, (*6*) Equisetales, (*7*) Sphenophyllales; *Filicineae:* (*8*) Marattiales, (*9*) Coenopteridales, (*10*) Filicales; *Gymnospermae:* (*11*) Pteridospermae, (*12*) Cycadeoidales, (*13*) Cycadales, (*14*) Ginkgoales, (*15*) Pityeae, (*16*) Cordaiteae, (*17*) Coniferales; *Angiospermae:* (*18*) monocotyledons and dicotyledons. (Plant data from C. I. Arnold, *An Introduction to Paleobotany*, McGraw-Hill, New York, 1947; animal data from G. G. Simpson, *Life of the Past*, Yale Univ. Press, New Haven, Conn., 1953.)

entities in the process of becoming; historical principles in the understanding of becoming.

Time and the earth: beginnings—estimates of the ages of planets (3 to 10 billion years), radioactivity as a tool for dating, the possibility that life is more than 3 billion years old; the geological time scale, primarily a scale of sequence only—its approximation to an absolute time scale by radioactivity datings, analogy of the history of life with a 24-hour day.

Origin of life: problems—fossil evidence unavailable, the difficulty of experimental re-

creation of processes involved, restriction of discussion largely to indirect evidence, the scientific belief in the natural origin of life from nonliving world; a hypothesis on the forerunners of life—the nonorganic synthesis of complex organic molecules from atmospheric constituents energized by lightning and ultraviolet radiation, the slow accumulation of such molecules on sterile earth, the ultimate origin of aggregates of such molecules capable of self-duplication and, therefore, alive; early organisms—the heterotrophic status of the first living systems, their utilization of the accumulation of nonorganically synthesized molecules and thus ultimate destruction of conditions for the further spontaneous origin of life, their production of CO_2 by anaerobic respiration, the origin by mutation of photosynthetic autotrophs utilizing CO_2 and solar energy, the subsequent origin of modern heterotrophs feeding on the autotrophs.

Fossils and the historical record: principles of the paleontological method; fossils as facts; observations and inferences concerning fossils; uniformitarianism, the doctrine that the present is a key to the past, and its wide application—physical laws are independent of the passage of time, but actual structures and organisms have a history, that is, are time-dependent; the definition of fossils; conditions for their preservation and ultimate recovery by man; fossils predominantly of former hard parts of organisms; fossils not usually petrified; Pre-Cambrian fossils— their scarcity although both plants and (later) animals are known; life's beginnings not recorded in the rocks; an adequate fossil record spanning only perhaps the last quarter of life's total history; the question posed by the rarity of Pre-Cambrian fossils and four possible interpretations; the over-all fossil record.

Some major tendencies in the history of life: life's expansion—an increase in the total number and diversity of living organisms but not at a constant rate, periods of rapid diversification in the Devonian and of rapid decrease in the Permian-Triassic, factors involved in the expansion; the occupation of environments— invasion of the land by aquatic forms, the further subdivision of niches, the creation of new environments by organisms; evolution as change—continual flux with no static equilibrium, change as the replacement of one group by another as well as expansion, or the multiplication of groups; complication and improvement of organic structure (usually over-emphasized as an aspect of evolution since evolutionary change is never guaranteed to be progressive)—the problem of defining evolutionary progress and three interpretations of the term—(1) increasingly precise adaptations, (2) marked structural changes, making possible new ways of life, and (3) increasing perception.

30

Ancient Seas and Conquest of the Land

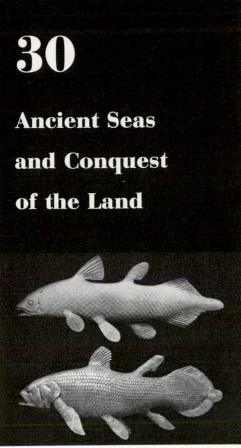

American Museum of Natural History

Diplurus (top) lived more than 200 million years ago; Latimeria (bottom) is still in existence. Both are members of the crossopterygians, the group whose descendants conquered the land and gave rise to all land vertebrates.

If you had a time machine, set it for 600,000,-000 B.C., and took off, you would have a rude shock when you arrived. Even though you started from a mountaintop, you might find yourself floundering in a sea when you landed. That would be an unpleasant demonstration of the fact that the face of the earth has changed radically. The main ocean basins and continental masses probably already existed in the Cambrian,[1] but their outlines were different. There were mountains, but not where our mountains now stand. Shallow seas, arms of the oceans, flooded far into the interior of the continental blocks, across what is now dry land.

If you had the luck to arrive on land, all would seem well at first. The air would be breathable. Clouds, winds, and rain would be familiar. The climate would probably be better than where, or we should say *when*, you now live—better, at least, if you like it to be rather warm throughout the year, without sharp alternation of hot summers and cold winters. At second glance, the land would seem completely alien to you. It would be alien, not so much because of the unfamiliar topography as because it would be completely bare. No grass, no shrubs, no trees, no buzzing insects, singing birds, or scurrying rodents— no life at all. You would quickly starve unless you could reach the seashore, and even there everything would be unfamiliar. There would be seaweeds, shellfishes, and other marine life in some abundance, but all of kinds you never saw before. There would be no fishes. You might eke out a dreary existence by eating shellfishes, if they proved not to be poisonous to you, but at best you would be deeply impressed with the fact that the earth has not

[1] This statement is disputed among geologists and is still far from certain. Practically all agree that the Pacific basin already existed. One school thinks that the North Atlantic existed but that a continental mass extended from South America across where the South Atlantic now is to Africa and thence across where the Indian Ocean now is to southern Asia and Australia. Still another school holds that all the continental masses were united or nearly so and that the only major ocean basin was the Pacific, then vastly larger than it is now. We are not concerned with the paleogeographers' disputes. That lands and seas were then quite different from what they are now is certain, regardless of their exact shapes.

always been the pleasant world you live in now.

Cambrian and Ordovician Seas

APPEARANCE OF THE ANIMAL PHYLA

Fossil animals suddenly become abundant with the beginning of the Cambrian, and almost all the phyla appear as fossils during that period, a remarkable phenomenon that we have already discussed. The suddenness of the change from the almost barren Pre-Cambrian rocks is real enough, but it is not true that all the phyla mysteriously show up at precisely the same time. The Cambrian was a very long period, on the order of 100 million years, and the various major groups of animals straggle into the fossil record throughout that long span. Some, notably the vertebrates, do not appear until some time in the Ordovician, also a long period of some 75 million years. Many major groups probably were actually originating during the long time represented by the Cambrian and Ordovician. We need not look for the origins of all of them in the Pre-Cambrian.

The straggling into the record of major groups of animals and the great expansion of life in the Cambrian and Ordovician can be told better in figures than in words. Table 30–1 shows the numbers of phyla and classes of animals (including animal-like protists) definitely known as fossils at the stated times or earlier.

Before the end of the Ordovician all the

TABLE 30–1

Numbers of classes of animals surely known to have existed in subdivisions of Cambrian and Ordovician time

Time	Number of phyla	Number of classes
Cambrian		
Early	8	12
Middle	10	20
Late	11	22
Ordovician		
Early	11	27
Middle	12	32
Late	12	33

protistan and animal phyla that are at all likely to be preserved as fossils were definitely present. Most of them had already become highly diversified, as shown in classification by their subdivision into classes and lesser groups, including a great number of species. The land was still barren and there was probably little life in lakes and rivers, but the seas were full of organisms in great abundance and variety. Life in the sea has changed greatly since then, but its over-all ecology was already well established at that time, over 425 million years ago. Later changes within the marine environments have been replacements rather than expansion.

Since the Ordovician innumerable groups have died out, but as they disappeared their places were simply taken over by other groups, generally of more recent origin. Among animals and animal-like protists that are at all likely to leave a fossil record, there are only 12 phyla and 31 classes in the present seas. That actually represents a slight decrease from the 33 classes known for late Ordovician seas. The recent phyla are the same as those of the Ordovician. Several of the classes are of later origin and have replaced extinct classes present in the Ordovician. Replacement has been more and more complete at lower levels of the hierarchy of classification. No species have survived from Ordovician to Recent, and perhaps only one genus has: *Lingula,* a primitive brachiopod, has hung on, little changed, like an animal Methuselah. In terms of lack of essential difference from its present form, *Lingula* may be the oldest living thing on earth. Some of the protists and algae have probably changed as little since times still more remote, but their claims are not supported by the clear evidence of fossils.

LIFE OF THE LATE ORDOVICIAN

We cannot follow in even a summary way the life of the seas through all of the geological periods. A quick review of Ordovician life (Fig. 30–1) will, however, show characteristic ancient marine floras and faunas. Some of the most striking subsequent changes will be mentioned later. For characteristics of the various groups refer to Chapters 21 and 22, and for their classification see the Appendix.

Flagellates, forams, and radiolarians are

protists known from fossils to have existed in late Ordovician. They seem to have been rather similar to some forms living today. No other protists are likely to fossilize, but probably most or all of the other main protist groups were also in existence at that time. It is a certain inference that bacteria were abundant. Algae have left a poor record, but they are positively known to have existed in the Ordovician, and they were probably highly abundant and quite varied. Diatoms, which carry out such a large proportion of the photosynthesis in present-day waters, are not known from the Ordovician or for long after. Since they are fairly common fossils later (in the Jurassic), it is likely that they had not yet evolved in earlier times and that the lack of their fossils in the Ordovician is not just a failure of preservation or discovery. Fungi and bryophytes are unknown from the Ordovician, but fungi, at least, may well have existed then. Their tissues are so soft that their fossil records consist of only a few chance discoveries. Vascular plants probably had not yet evolved in the Ordovician.[2]

Sponges were present in Ordovician seas and do not seem to have undergone any really striking changes since then. Coelenterates were abundant, and some of them built large reefs, as they do today. Ordovician reefs were, however, formed mostly by groups more primitive than the true corals, now usually dominant on reefs. The true corals had just appeared (middle Ordovician). Floating graptolites occurred in enormous numbers and would have been the most unfamiliar of Ordovician organisms to a time-traveling zoologist. There is nothing like them today. Bryozoans were generally similar to those still living. Brachiopods were much more numerous than they are now, when they are almost down to relict status. Most of them had by late Ordovician developed heavy, ribbed, calcareous shells, but the really fancy ones were to come later in the Paleozoic.

The clamlike pelecypods, principal rivals of the brachiopods, which they have now almost entirely replaced, had appeared only in the early Ordovician and were just beginning to be common in the late Ordovician. The sea

---

[2] There is some evidence for the existence of earlier vascular plants, but it is inconclusive at present.

snails (gastropods) are an older group of mollusks (since early Cambrian) and were abundant in Ordovician seas. They looked much like some modern forms, but they have since become considerably more diverse, and there has been almost complete replacement of earlier groups of low taxonomic scope (families, genera, and species). The most striking mollusks of the Ordovician, and perhaps of all time, were not clams or snails but cephalopods, nautiloids related to the now relict chambered nautilus. Some were coiled and somewhat resembled their surviving relative. Others were straight and reached great lengths, up to 5 meters or so. They were much the largest animals of their time. The ammonites, relatives of the nautiloids with more complex partitions between the chambers, had not yet appeared in the Ordovician. The squids and octopuses, now the only abundant cephalopods, were not to evolve until much later.

The dominant arthropods during most of the Paleozoic were the bizarre, now extinct trilobites, which reached their climax in the Ordovician. There were a few true crustaceans, but nothing like the crabs or lobsters, which represent a group that later completely replaced the trilobites, had yet appeared. Strangest of Ordovician arthropods, to our eyes, were the eurypterids, which look something like crustaceans but are really more nearly related to spiders and scorpions. Their remains are rare in Ordovician rocks and usually much broken. It has been suggested that they were originally a fresh-water group and that their Ordovician fossils had been washed down by rivers into the sea. No fossil insects are known from the Ordovician, and they almost certainly had not yet evolved. That event has to await the covering of the land with vegetation.

Echinoderms were another abundant group in late Ordovician seas. They included ancestors of all the living groups and were not very strikingly different from these. They also included a greater variety of extinct groups, some primitive and some peculiar divergent lines.

It is in the Ordovician that the first fossil vertebrates appear, completing the roster of the animal phyla as far as these can readily be preserved as fossils. Strange as it seems to

30–1 Some faunas of early Paleozoic seas.
Middle Cambrian, western North America: *1*, a jelly-fish, (scyphozoan) ; *2*, the spongelike *Archeocyathus*; *3*, a trilobite, *Ogygopsis*; *4*, an arachnid, *Sidneyia*; *5*, a crustacean, *Barrella*; *6*, an annelid worm; *7*, a holothurian (echinoderm) ; *8*, a crustacean, *Hymeno-caris*; *9*, a trilobite, *Neolenus.*

Middle Ordovician, central North America: *1*, a straight-shelled (orthoconic) nautiloid cephalopod; *2*, a gastropod; *3*, a small trilobite, *Calymene*; *4*, a large trilobite, *Isotelus*; *5*, massive coral; *6*, branching coral; *7*, two solitary corals.

Middle Silurian, Illinois—a coral-reef community: *1*, a stalked (sessile) cystoid echinoderm; *2*, a cephalopod mollusk, *Phragmoceras*; *3*, honeycomb coral, *Favosites*; *4*, tube coral, *Syringopora*; *5*, chain coral, *Halysites*; *6*, a solitary coral; *7*, a nautiloid cephalopod; *8*, a trilobite, *Isotelus*; *9*, another trilobite, *Actinurus*; *10*, brachiopods, *Pentamerus*; *11*, brachiopods, *Leptaena*; *12*, a cephalopod mollusk, *Cyrtorizoceras*.

Late Silurian, New York: *1*, a eurypterid, *Pterygotus*; *2*, a snail, *Pycnomphalus*; *3*, a eurypterid, *Carcinosoma*; *4*, a eurypterid, *Hughmilleria*.

think of a sea without fishes, that was the state of the seas before this time. It may, indeed, still have been true at this time. Few exposed rocks of Ordovician age were laid down in streams or lakes, but the scattered and fragmentary remains of the earliest known fishes are found in situations where they may have been washed down by streams to a beach or sea. Some authorities also believe on theoretical grounds that the first fishes probably evolved in running fresh water. If so, and especially if the eurypterids also arose in fresh water, plants (algae) must already have been abundant in those waters. Then there were probably also numerous fresh-water protists and a variety of invertebrates. Nevertheless, life in fresh water must have been far less common or diversified than in the sea then, or than in fresh waters today.

The first fishes were the jawless agnaths, primitive, few, and unimpressive. Their appearance is, however, one of the most dramatic events in the history of life. It marked the rise of the phylum that was to become dominant in every sphere that it invaded and that was eventually to produce the writers and the readers of this book.

The Age of Fishes

The Silurian, next period after the Ordovician, was no exception to the rule of ceaseless change in the history of life. Graptolites declined, corals expanded, species and genera arose and died out. It was, nevertheless, a period without any striking innovations among aquatic organisms. Its most important biological event was the still feeble beginning of the occupation of the land, an event to which we will return.

The next period, the Devonian, was a period of accelerated evolutionary activity in many groups of organisms. Most important (from the human point of view, at any rate) is the fact that fishes first became common in the Devonian and that their most basic differentiation occurred mainly in that period. For these reasons the Devonian is often called the Age of Fishes.

The scanty Ordovician and Silurian fossil vertebrates are all, or nearly all,[3] jawless fishes, agnaths. They were well represented in the early Devonian, including bottom-living forms with broad, flattened head shields (cephalaspids) as well as more active swimmers (such as pteraspids) (Fig. 30–2). They declined rapidly thereafter, and by the end of the Devonian they were all extinct except for the unknown lines that led to the living lampreys and hagfishes.

This is one of the most striking examples of the historical principle of *replacement*. The Devonian agnaths were plainly being erased from the ecological picture by another, more efficient group of fishes, the placoderms. The placoderms arose from agnaths, but not from those they replaced in the Devonian. They evolved from primitive Silurian agnaths by the transformation of a set of gill arches into movable jaws, among other changes. Throughout most of the Devonian the placoderms expanded as the agnaths contracted. Toward the end of the Devonian the placoderms themselves were being replaced by still other groups of fishes, the Chondrichthyes and Osteichthyes. Most placoderms became extinct at the close of the Devonian, although a few struggled on into the Permian.

In their heyday the placoderms were highly diversified. Most normal in appearance, to eyes used to recent fishes, was a group of little, sharklike forms (acanthodians, Fig. 30–3). This group also happens to be the one that survived into the early Permian before it became extinct. Another abundant group included bottom-livers and mud-feeders (antiarchs), evidently ecologically similar to the flattened agnaths and competing successfully with them. A third group (arthrodirans) must have been violently predacious. Most members of the group had powerful, gaping jaws with sharp shearing blades. They must have been the scourge of Devonian waters. They came in all sizes, suitable to prey on almost anything that moved. The largest of them were monsters 10 meters or so in length, quite as fearsome as any sharks today.

The fishes destined to replace the later placoderms arose from early placoderms, just as the placoderms arose from early agnaths

[3] There is some not completely satisfactory evidence of placoderms in the late Silurian.

and then replaced the later agnaths. The replacing groups, finally successful in that they are the dominant fishes today, are the Chondrichthyes and the Osteichthyes. Both arose in the Devonian and were becoming abundant by the end of that period. The Devonian chondrichthyans included sharks not too unlike those of today and several other divergent branches, most of which have become extinct.[4] The flattened skates and rays, now quite common, evolved from sharks at a much later date, in the Jurassic. Their strong dorsoventral flattening is a distinctive adaptation that has evolved several times in different groups. There were placoderms quite skatelike in body form and probably in habits.

It has previously been mentioned that the early chondrichthyans seem to have been adapted mainly to salt water and the osteichthyans to fresh water (Chapter 22). In the Devonian the osteichthyans were still mostly fresh-water fishes, although their eventually highly successful invasion of salt water was probably already under way. Even in the Devonian the osteichthyans were differentiated into three basic groups (see Fig. 30–3), not then very obviously or fundamentally distinct but with decidedly different prospects in the history of life. One group, that of the crossopterygians or crossopts (Fig 30–4), gave rise in the later Devonian to the amphibians and, through them, to all the vertebrates of the land and air. After the amphibians had appeared, other lines of crossopts lingered on in diminishing numbers. They are represented today by the single relict *Latimeria*.[5]

Another basic group of Devonian osteichthyans, that of the paleoniscoids, gave rise to a great radiation of fishes in both salt and fresh waters. That radiation, beginning in the Permian and continuing through the Mesozoic, produced nearly all our present-day fishes plus a number of extinct Mesozoic groups later replaced by the teleosts, which have been the dominant fishes since the Cretaceous. (The teleosts were also products of

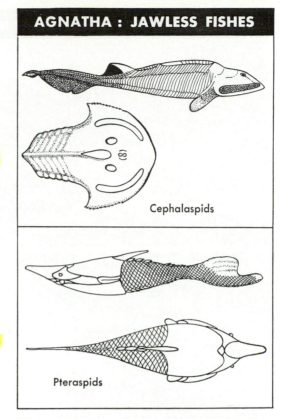

AGNATHA : JAWLESS FISHES

Cephalaspids

Pteraspids

30–2 Agnatha, earliest and jawless fishes. *Top.* A side view of *Cephalaspis* and a dorsal view of the head and trunk region of *Kieraspis*, showing the paired eye sockets (round), the median nostril, and the pair of crescent-shaped apertures supposed to be electric organs like those of the modern electric eel. *Bottom.* Side and dorsal views of *Pteraspis*.

the Mesozoic radiation originally rooted in the paleoniscoids.)

The third basic group of Devonian osteichthyans was that of the lungfishes. They were among the most common fishes in the later Devonian, but since then they have dwindled steadily until now they are represented by only three relict genera. The lungfishes illustrate particularly well two phenomena that are widespread in the history of life: change of pace and stagnation. They were rather slowly differentiated from other osteichthyans during the early Devonian. Once the group was well established, it evolved with increasing rapidity, culminating around the transition from middle to late Devonian. This rapid change was, however, abortive. The lungfishes

[4] One branch does live on in the comparatively uncommon chimaeras.

[5] A second living genus was named, but it is now considered synonymous with *Latimeria*.

30–3 Some Devonian fishes. *Top.* A Devonian "aquarium": *1, Osteolepis,* a crossopterygian (Osteichthyes); *2, Holoptychius,* another crossopt; *3, 6, Climatius,* an acanthodian (Placodermi); *4, Diplacanthus,* another acanthodian; *5, Coccosteus,* a primitive arthrodiran (Placodermi); *7, Cheirolepis,* a paleoniscoid (Osteichthyes) (see also center left); *8, 9, Dipterus,* a primitive lungfish (Osteichthyes). *Left center. Cheirolepis,* a paleoniscoid. The paleoniscoids gave rise to nearly all our present-day fishes, including the dominant teleosts. *Bottom. Dinichthys,* an arthrodire (Placodermi), which reached 30 feet in length.

seem to have been squeezed, so to speak, between the two other groups that preempted the main possibilities for expansion: descendants of the paleoniscoids occupied most of the aquatic vertebrate niches, and descendants of the crossopts moved onto the land and eventually occupied its vertebrate niches. By the end of the Paleozoic, lungfish evolution had slowed almost to a standstill. The species became stereotyped and confined essentially to one narrow niche in which they have remained ever since without any further important changes.

Apart from the fishes, the most important event in Devonian seas was probably the rise of the ammonites, cephalopods that were to be the dominant marine mollusks of the Mesozoic. Their ancestral group, that of the nautiloids, continued but was dwindling rapidly in the Devonian: another of the innumer-

30–4   The vertebrates conquer the land. *Top left.* Devonian lobe-fin fish, *Eusthenopteron*, crawling out of the water. *Bottom left.* The skeleton of *Eusthenopteron. Top right.* Primitive Carboniferous-Permian labyrinthodont amphibians, *Diplovertebron. Bottom right.* The skeleton of *Diplovertebron.* Note that the limb bones of *Diplovertebron* show the same pattern, now familiar, as those of all higher vertebrates; this pattern is not clear in *Eusthenopteron.*

able instances of replacement. Most important of all, however, is the fact that the movement of life onto the land was in full swing during the Devonian and was an accomplished fact by the end of the period.

Occupation of the Land

DIFFICULTIES

We have previously had occasion to refer to the difficulties of life on land and to some of the adaptations of plants and animals to terrestrial environments. The outstanding difficulty is that land organisms are no longer surrounded by water, as their ancestors all were. Water must still be obtained from somewhere: by root absorption from the soil, by breathing water vapor from air (a scanty resource in most situations), by drinking liquid water on the surface, by eating plants and animals (which contain water), or by metabolism of carbohydrates and lipids in such a way as to release water. Once acquired, water must be conserved against evaporation. All the fully terrestrial plants and animals have external coverings of one sort or another by which evaporation is controlled and limited. In spite of this needed protection against desiccating air, it is necessary to obtain CO_2 (in the case of plants) and O_2 from the air, and also to discharge waste gases into the air. Land plants have stomata and associated structures. Land animals have lungs, tracheae, or pouches or modified gills that function as lungs. Gravity, no special problem to an organism buoyed up by water, becomes serious for all but the smallest land organisms. Most land animals, excluding only small and wormlike forms, have strong skeletons, external or internal. All true land plants have supportive tissues, especially in the stems, and all of any great size are woody. Extremes of temperature from season to season outside the tropics and from day to night everywhere (including the tropics) are far greater on land than in the water. Land plants and animals have numerous adaptations to these fluctuations. Some have been noted in previous pages, and you can probably think of others.

Only four phyla include organisms that can be considered fully and progressively adapted to land life: Tracheophyta among plants; Mollusca, Arthropoda, and Chordata (Vertebrata) among animals.

EARLIEST LAND PLANTS AND ANIMALS

In the historical sequence of occupation of the land, it seems necessary that plants should have led the way. Animals require accessible plant material at the beginning of their food chains. This is borne out by the fossil record, in which land plants begin to appear before any land animals. The first certain and fairly common fossils of land plants appear in the middle Silurian, and they are for the most part small psilopsids (Chapter 20). At least one may be a forerunner of the lycopods or transitional between the psilopsids and that group. Basic divergence of the subphyla of vascular plants was probably already under way, but it had not yet gone far.

The earliest known possible land animal is a scorpion from the late Silurian. It is enough like the modern scorpions, which are fully terrestrial, to suggest that it had the same habits. It is, however, also similar to its aquatic forerunners, and so the question is not entirely settled.

During the Devonian, land plants became common. Some of them reached the size of trees, and the first forests grew on the earth. Psilopsids continued, but they had become scarce by late Devonian. The incoming, replacing groups were the lycopsids, sphenopsids, and ferns (Chapter 20), all of which were abundant by late Devonian. A few primitive gymnosperms (cordaites) had also appeared. These groups continued to expand greatly in the Carboniferous, when they formed the great coal forests along with the newly evolved seed ferns.[6]

In the Devonian are found a few animals that were certainly terrestrial. All are arthropods: a mite, several forerunners of the spiders probably not yet advanced enough to be definitely classified as spiders, and a creature that may similarly have been a forerunner of the insects.

By the end of the Devonian some of the crossopts had developed legs and had become amphibians.[7] They still had fishlike tails and were probably still almost as aquatic as fishes. Nevertheless, they were the first vertebrates to walk on land, and their descendants were to rise to dominance in the new environment (see Fig. 30–4).

The Permo-Carboniferous

The Carboniferous and Permian, the last two periods of the Paleozoic following the Devonian, had much in common and may be considered together.

COAL FORESTS

The Carboniferous, "carbon-bearing," is so called because many extensive coal deposits are of that age, including our Appalachian coal fields in Pennsylvania and adjacent states.[8] The coal is the compressed remains of plants that grew in widespread, swampy forests. Deep burial for long periods of time has driven out the more volatile constituents of the plant tissues, and the compacted residue has a higher percentage of carbon than the original plants. It still retains traces of the original structure, however, and associated

---

[6] Some textbooks say that seed ferns appeared in the middle Devonian, on the evidence of a fossil forest found near Gilboa, New York. It has, however, been established that the supposed "seeds" of the trees in the forest are really spore cases. This does not exclude the possibility or even the probability that seed ferns did occur in the Devonian.

[7] There is some dispute whether the earliest surely identified amphibians are from the latest Devonian or earliest Carboniferous. In geological perspective the difference is not great and does not matter to us.

[8] There are, however, many rich coal deposits of younger ages in the United States and elsewhere.

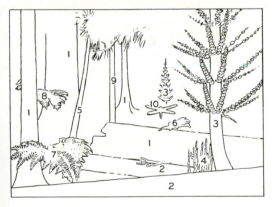

30–5 A Carboniferous forest. Lycopsids: *1*, various species of *Sigillaria*; *2*, two species of *Lepidodendron*. Sphenopsids: *3*, *Calamites*; *4*, *Sphenophyllum*. Ferns: *5, Caulopteris*; *6, Mariopteris*. Gymonosperms: *7, 8*, two species of *Neuropteris*; *9, Cordaites*. Insects: *10*, a primitive dragonfly, *Meganeura monyi*.

shales and sandstones are often rich in well-preserved fossil plants. Thus we are well acquainted with the compositions of the forests and the structures of their plants (Fig. 30–5). You are already familiar with the main groups represented: lycopsids such as *Lepidodendron* and *Sigillaria*; seed ferns, including *Neuropteris*; *Cordaites* and its relatives; and the conifers, which did not, however, become distinct until the late Carboniferous, when *Walchia* and other genera appeared. In the late Permian and early Triassic, cycadeoids, cycads, and ginkgos appeared, but they had no part in the earlier coal forest.

LAND INVERTEBRATES

Animal life, so rare hitherto, swarmed in Carboniferous forests. Here appeared the first land snails, the only mollusks to complete the great transition from water to air. Undoubted terrestrial scorpions were common, and so were relatives and ancestors of the spiders. Centipedes were present. Most striking and

ANCIENT SEAS AND CONQUEST OF THE LAND 779

Peabody Museum of Natural History

30–6 A small mayfly-like insect, *Dunbaria*, from the Permian in England.

most important of the land invertebrates were insects. The majority of them in the Carboniferous belonged to orders unfamiliar in aspect and now extinct, but cockroaches were already there, and so were forerunners of the dragonflies. Several other recent orders appeared in the Permian, among them the mayflies (Fig. 30–6), thrips, bugs, lacewings, and beetles. Most of the modern orders (16 out of 24) are known from the Jurassic or earlier. Most Carboniferous insects were small, but some were remarkable for their large size. One dragonflylike giant is believed to have had a wingspread of nearly 75 centimeters. No known later insect reaches such dimensions. It is noticeable that the groups of insects now particularly associated with flowers were absent in these early faunas. They begin to appear in the fossil record in the Jurassic and spread greatly during the Cretaceous and early Cenozoic, in close parallel with the fossil record of the flowering plants.

AMPHIBIANS

Amphibians were particularly common during the Carboniferous and scarcely less so in the Permian. The Permo-Carboniferous is sometimes called the Age of Amphibians, although, as a matter of fact, amphibians were already much outnumbered by reptiles in the Permian. None of the modern groups had yet evolved. Most Permo-Carboniferous amphibians were labyrinthodonts.[9] Typically they were clumsy brutes with four short, sprawling legs, big, flattened heads, and stubby tails.

[9] "Labyrinth-toothed," so called because the tissues of the teeth seen in cross section have an intricate, mazelike pattern.

Eryops (Fig. 30–7), an early Permian labyrinthodont about 1½ meters long, is an example of that type. There were, however, many other kinds, sizes, and shapes. Among the most bizarre were some that were long and snakelike, or others with extraordinary triangular heads.

REPTILES

The transition of the vertebrates to land life was completed when the reptiles arose. Since the change was gradual, it is difficult to point to an exact time and say, "Here the reptiles appear," but some of the latest Carboniferous fossils were already true reptiles, and reptiles were abundant in the Permian. Most of the Permian reptiles belonged to only two main groups, neither of which was at all like any reptiles living today. The *root reptiles*,[10] cotylosaurs, were most primitive. They intergraded with the earlier labyrinthodont amphibians and often looked not unlike those ancestors, although they did develop some more unusual later forms. Even more common were the *mammal-like reptiles*. The earliest of them, the pelycosaurs,[11] included the ancestry of the whole group (hence also of the mammals) and also some divergent, extinct lines of creatures like *Dimetrodon* (Fig. 30–8), with a fantastic fin on the back, supported by the elongated spines of the vertebral column. The somewhat later and eventually more varied therapsids[12] were the dominant land animals of the late Permian and early Triassic and were the immediate ancestors of the mammals. They are best known from South Africa, where they have been discovered in almost incredible numbers, but some have been found in Russia, China, Brazil, the United States, and elsewhere. They doubtless occurred on all the continental land areas of the Permian and Triassic.

One of the oddest features of the history of life is that reptiles had hardly completed the

[10] So called because they are the stock from which most or all of the higher land vertebrates were derived.

[11] "Basin reptiles," so called because of the basin-like pelvis.

[12] "Mammal-arched," because the cheekbone or zygomatic arch was constructed as in the mammals and not as in other reptiles.

30-7 *Eryops*, an early Permian labyrinthodont amphibian.

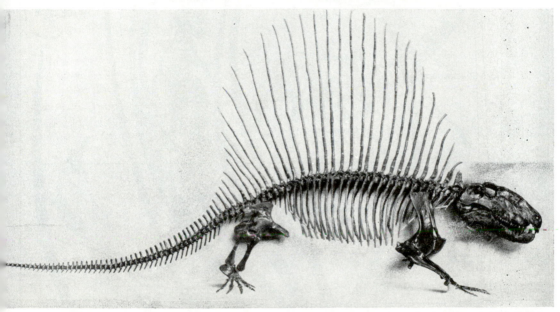

30-8 *Dimetrodon*, an early mammal-like reptile.

transition from water to land before some of them took to the water again. Among the earliest known reptiles is a little group of aquatic (probably fresh-water) fish-eaters. The group became extinct almost at once, but several other, later groups of reptiles also became aquatic. The main lines of reptilian evolution are shown in Fig. 30–9.

The Age of Reptiles

The Permian was already an age of reptiles, but the term is usually applied to the next three periods, the Triassic, Jurassic, and Cretaceous, composing the Mesozoic.

PERMO-TRIASSIC CRISIS

The later Permian and early Triassic were times of crisis in the history of life. Evolution in many groups was exceptionally rapid. Decline or extinction of once-flourishing kinds of organisms was accompanied by sudden expansion of others. The ammonites (cephalopod mollusks) were one of the rapidly changing, expanding groups of the Permian, and they became extremely abundant and varied in the Triassic. On the other hand, the brachiopods

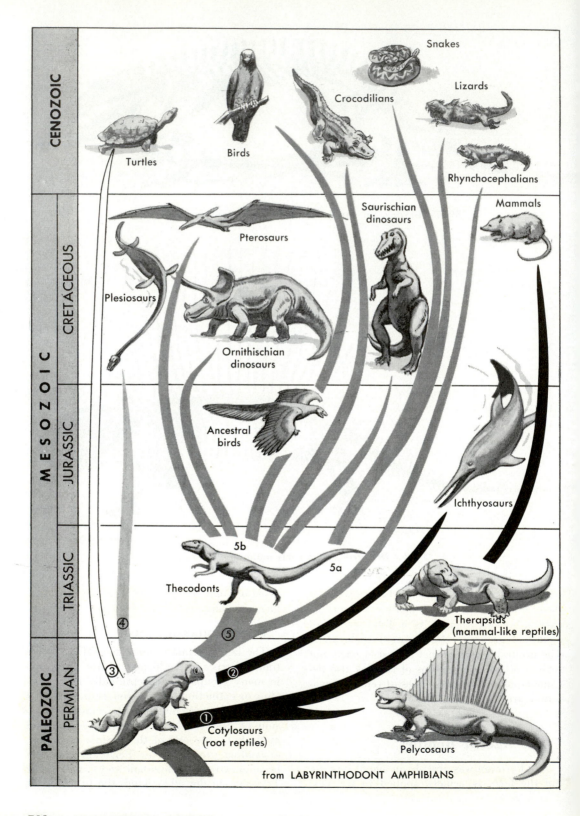

Snakes

Crocodilians

Lizards

Turtles

Birds

Rhynchocephalians

Saurischian dinosaurs

Mammals

Pterosaurs

Plesiosaurs

Ornithischian dinosaurs

Ancestral birds

Ichthyosaurs

5b

Thecodonts

5a

Therapsids (mammal-like reptiles)

④

③

⑤

②

① Cotylosaurs (root reptiles)

Pelycosaurs

from LABYRINTHODONT AMPHIBIANS

CENOZOIC

M E S O Z O I C

CRETACEOUS

JURASSIC

TRIASSIC

PALEOZOIC

PERMIAN

declined greatly, and many of their groups became extinct in the Permian. They were never again the common marine shells that they had been through most of the Paleozoic. Trilobites, which had been slowly declining since the Ordovician, finally became extinct in the Permian. So did several old groups of echinoderms and of corals. More modern groups of mollusks, echinoderms, and crustaceans appeared in the Triassic and expanded rather steadily thereafter.

The fishes, too, changed markedly. The last placoderms disappeared. Most of the archaic chondrichthyans and osteichthyans declined, their places taken by more progressive offshoots that evolved and expanded steadily through the Mesozoic and on to now.

On land most of the great coal-forest trees became extinct, while the cycadeoids, cycads, ginkgos, and conifers expanded. The flowering plants probably originated in the Triassic, although the early fossil evidence is obscure. A likely hypothesis is that they originated in upland environments, poorly represented in the fossil record. Flowering plants were still rare, if they existed at all, in lowlands in the Jurassic. In the Cretaceous, however, they began their remarkable expansion. By the end of that period they had replaced the great majority of other land plants and already had the dominance that they retain today.

Most of the Paleozoic groups of amphibians were victims of the Permo-Triassic crisis. One circumscribed group of labyrinthodonts was fairly common (highly so in some particularly favorable localities) in the Triassic, but even it disappeared at the end of that period. Thereafter, as far as is known, only the immediate ancestors of the comparatively few modern amphibians continued.

30–9 The radiation of reptiles. The root reptiles, or cotylosaurs, were derived from the labyrinthodont amphibians. From the cotylosaurs five major lines of reptilian evolution can be traced: (*1*) to the mammal-like reptiles and mammals; (*2*) to the ichthyosaurs; (*3*) to the turtles; (*4*) to the plesiosaurs; and (*5*) to the thecodonts (*5b*) and the snakes, lizards, and rhynchocephalians (*5a*). From the Triassic thecodonts another major radiation took place: to the crocodiles, birds, and flying reptiles (pterosaurs), as well as the great dinosaur orders, the Ornithischia and the Saurischia.

Changes among the reptiles were not less dramatic. The two main groups of late Permian reptiles did survive and, for a time, thrive in the Triassic, but they were nearly extinct at the end of that period. Only a few very advanced mammal-like reptiles continued into the Jurassic, although others of their lineages were then represented by the mammals themselves.

In the Triassic there was a great radiation of new reptilian orders. It set the pattern for the great diversity and dominance of reptiles that continued throughout the Mesozoic. Most impressive and famous of the innumerable Mesozoic reptiles were the many kinds of dinosaurs.

DINOSAURS

The animals popularly called *dinosaurs* [13] technically comprise two different orders of reptiles, the Ornithischia and Saurischia. [14] During their heyday, in the Jurassic and Cretaceous, they were extremely numerous and varied. They are known for their huge size, and some of them, *Brontosaurus* and its relatives, were indeed enormous, reaching about 20 meters in length and perhaps 25 metric tons in weight. No larger animals ever walked on land. (But some larger ones still swim the seas.) However, not all dinosaurs were large. They came in many shapes and sizes, some not much bigger than chickens.

In the early Triassic dinosaurs did not exist as such. Their ancestry was then represented by a primitive group, the thecodonts, that radiated widely. They gave rise not only to the two orders of dinosaurs but also to the crocodiles, the flying reptiles (pterosaurs), and the birds. The earliest true dinosaurs, in the late Triassic, were rather slender, mainly bipedal forms such as *Coelophysis*. The later Saurischia include not only the large, herbivorous, quadrupedal, long-necked and long-tailed allies of *Brontosaurus* but also a host of bipedal, mostly carnivorous forms. Some of the latter were tiny, as dinosaurs go, but some, like *Tyrannosaurus*, were very large.

[13] From the Greek for "terrible reptiles." The first syllable should be pronounced *dye-*, not *din-*.

[14] From Greek roots for "birdlike pelvis" and "lizardlike pelvis."

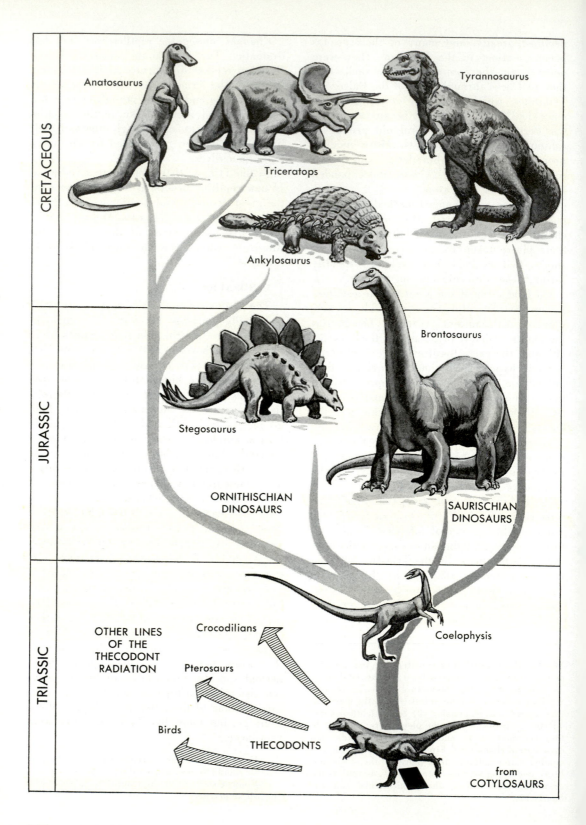

CRETACEOUS

Anatosaurus

Triceratops

Tyrannosaurus

Ankylosaurus

JURASSIC

Stegosaurus

Brontosaurus

ORNITHISCHIAN
DINOSAURS

SAURISCHIAN
DINOSAURS

TRIASSIC

OTHER LINES
OF THE
THECODONT
RADIATION

Crocodilians

Pterosaurs

Coelophysis

Birds

THECODONTS

from
COTYLOSAURS

All the Ornithischia were primarily herbivorous, but in other respects they became even more diverse than the Saurischia. There were four main groups: the plated dinosaurs, *Stegosaurus* and its relatives, quadrupeds with rows of bony plates down the back; the armored dinosaurs such as *Ankylosaurus*, animated four-footed tanks almost completely enclosed in a bony carapace; the duck-billed dinosaurs, including *Anatosaurus*, mostly bipedal, many of them strong swimmers and more or less amphibious in habits; and the frilled dinosaurs, typified by *Triceratops*, quadrupeds with a bony, frill-like extension of skull bones over the neck and in the more advanced forms usually also with horns. Restorations of all the genera of dinosaurs named as examples are pictured in Fig. 30–10, and the skeleton of *Tyrannosaurus* is shown in Fig. 30–11. About 250 genera in all are known, and they have been found on every continent except Antarctica. The richest known deposits are, however, in North America, and our seven examples are all American forms.

SOME OTHER MESOZOIC REPTILES

The dinosaurs were the dominant medium-sized to large terrestrial vertebrates of the Mesozoic, but they did not occupy all possible reptilian niches. Along with them lived a great variety of other reptiles with different habits and habitats. Land reptiles mostly below dinosaur size included, even in the Triassic, the *rhynchocephalians* (Chapter 22). *Lizards* appeared in the Jurassic, and *snakes* in the Cretaceous. The amphibious to fully aquatic *crocodiles* evolved at the end of the Triassic and have been quite abundant ever since, although recently somewhat reduced in range and numbers. *Turtles*, apparently at first amphibious in fresh water but later spreading widely from the high seas to the deserts, appeared in the Triassic. They have been expanding quite steadily every since. The groups mentioned in this paragraph are the only ones of all the Mesozoic reptilian hordes that still survive. They have undergone no really essential changes since the Cretaceous, although of course many new species have arisen and there has been much change in detail.

Several groups of reptiles became fully aquatic and marine in the Mesozoic (Fig. 30–12). The *plesiosaurs*[15] had a broad, flattened body, four paddles, and a long neck or tail, or both. Someone has likened them to a snake threaded through a turtle. (But they had no turtle-like external armor.) They appeared in the Triassic and were abundant in the Jurassic and through the Cretaceous. The *ichthyosaurs*[16] looked something like sharks and even more like porpoises or dolphins. They also appeared in the Triassic and were most numerous in the Jurassic. Only a few survived into the Cretaceous, and they died out well before the end of that period. Both these groups of marine reptiles ate fish, and the ichthyosaurs, at least, also relished active cephalopods. Food habits of extinct animals usually have to be inferred from the tooth and jaw apparatus, but stomach contents have been found in the fossil skeletons of ichthyosaurs.

In the late Cretaceous, another group of large, marine, fish-eating reptiles evolved and had a brief but lively career: the *mosasaurs*.[17] They were overgrown lizards that went to a marine existence. There are no fully aquatic lizards today, but the Galápagos lizard swims out to sea in search of food. With appropriate variation, that is a situation in which natural selection could produce a new race of mosasaurlike lizards.

INTO THE AIR

Once the land was occupied, there remained only one great unoccupied sphere accessible to life on the earth: the atmosphere. Even now there are no completely aerial organisms;

[15] "Nearly reptiles." We are not sure what the namer had in mind when he applied that name to a group completely reptilian, however queer in appearance and habits.
[16] "Fish reptiles," from the fishlike external appearance.
[17] "Reptiles of the River Meuse" (in Belgium). They were first found in that region, although now best known from Kansas.

30–10 The history of the dinosaurs. The earliest (late Triassic) dinosaurs were saurischian and are represented by the small *Coelophysis*; they were part of the radiation of thecodonts. The ornithischian dinosaurs early diverged as a distinct order.

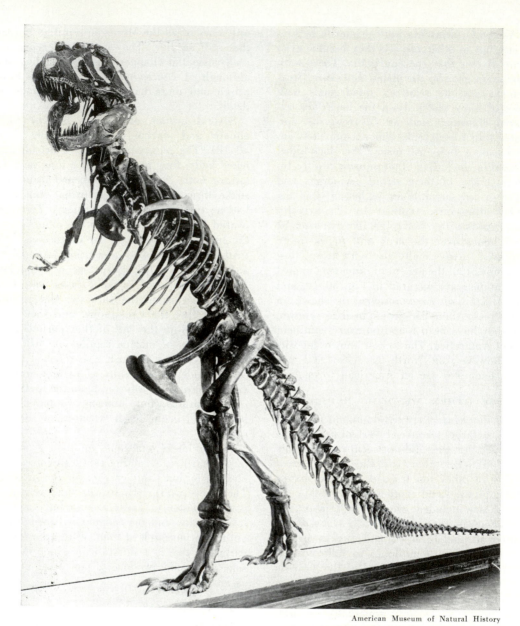

30–11   *Tyrannosaurus rex.*

none live their whole lives suspended in air. Fully terrestrial animals are, however, already aerial in physiological adaptation. They do live surrounded by air, derive oxygen directly from that medium, and are adequately protected from the perils that air has for animals of aquatic ancestry. To be launched fully into the atmosphere for longer or shorter periods, they require only an additional mechanical adaptation, some structure that can sustain them in air against the pull of gravity.

Flight was already completely achieved by insects in the Carboniferous and has remained an insect specialty ever since, although there are, of course, a good many nonflying insects. The only other organisms that evolved sus-

30–12 Aquatic reptiles of the Mesozoic. *Top.* Plesiosaurs (one of which holds a fish in its mouth) and ichthyosaurs (jumping). *Bottom.* Fossil skeletons of *Ichthyosaurus quadricissus* and young, which were evidently born alive in a manner analogous to that of mammals.

tained, directed flight [18] were two related lineages of late Triassic or early Jurassic reptiles (Fig. 30–13). They arose, independently of each other, from the ancestral group that also independently gave rise to the dinosaurs. One enjoyed considerable but transient success: the pterosaurs ("wing reptiles," Figs. 30–13 and 30–14). The other, after a slow start, went on to become so distinctive and so widespread and diverse that it is no longer considered reptilian but a new, higher class: the birds.

The pterosaurs flew by means of a thin membrane of skin, stretched between body, forearm, and the enormously elongated fourth finger. (The fifth finger was lost.) The mechanism was similar to a bat's wing but apparently less efficient. In bats several fingers sup-

[18] This excludes the many organisms that float passively in air at some time in their lives: plant spores and seeds, spiders with silken "parachutes," and so on. It also excludes a number of animals that make directed glides but not sustained flights: "flying" fishes, "flying" squirrels, and a few others.

port the wing. Some pterosaurs were as small as sparrows. Others were giants with wingspreads up to 8 meters, the largest animals that ever flew. Most of the known pterosaurs are believed to have skimmed over the Mesozoic seas catching fish, much as terns do today. Whether there were inland species with different habits is a moot point.

The first known bird is *Archaeopteryx* ("ancient wing") from the middle Jurassic of Germany (Figs. 30–13 and 30–15). It had teeth, a long, jointed tail, and so many other reptilian characters that it might well be considered a reptile if identified only by its bones. By an extremely rare and fortunate chance, however, impressions of feathers, which generally do not fossilize, were preserved. They show that *Archaeopteryx* had a feathered wing attached to its otherwise reptilian forearm and hand.

This rare discovery has a bearing on general theories of evolution. It has been maintained that really radical adaptive changes

30–13 A Jurassic scene. On the ground is the small dinosaur *Ornitholestes*; in the air and on the trees are pterosaurs and the ancestral birds *Archaeopteryx*. The trees are cycadeoids.

30–14 A pterosaur, or flying reptile. The fossil skeleton of *Nyctosaurus*.

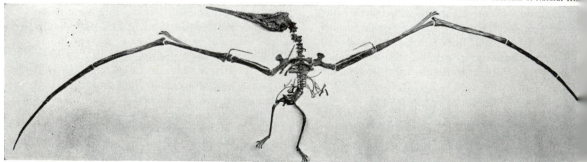

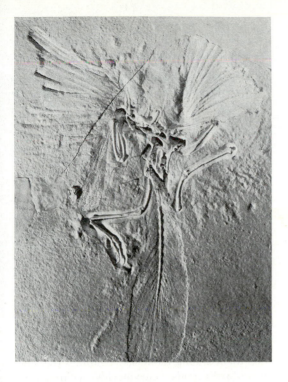

30–15 *Archaeopteryx. Left.* A photograph of an actual fossil. *Right.* A reconstruction of the animal. The feathers (birdlike feature) are clear in the fossil. The animal also possessed teeth and an incompletely specialized forelimb (reptile-like features).

must have taken place all at once, in one mutational leap or saltation. A living, workable intermediate between, say, a reptile and a bird, a fish and an amphibian, or a land mammal and a whale is considered unthinkable by followers of this school. But *Archaeopteryx* is about as completely intermediate between a reptile and a bird as one can imagine. Furthermore, some likewise rare latest Devonian or earliest Carboniferous animals have turned out to be almost perfectly midway between fishes and amphibians. The intermediates between land mammals and whales have not yet been found, but in view of these other discoveries, who can reasonably claim that they never existed?

MAMMALS

The once great group of the mammal-like reptiles became steadily more and more mammal-like through the Permian and the Triassic. In the latest Triassic and early Jurassic a few fragmentary fossils suggest that some of them may then have become mammals by definition. The process was gradual, and the distinction is necessarily arbitrary at first. Remains of unquestionable mammals appear in the middle Jurassic and at intervals thereafter. The mammal-like reptiles, those that did not make the grade and become mammals, declined greatly toward the end of the Triassic and became extinct during the Jurassic.

The history of mammals will be summarized in Chapter 31. Here it suffices to record that they did appear during the Mesozoic, but that they remained obscure, overshadowed by the reptiles, until the end of that era.

THE GREAT DYING

The late Cretaceous was a time of widespread extinction. Among the many invertebrate groups that declined or died then, the ammonites are most striking. They are extremely abundant in most Cretaceous marine rocks, but near the end of that period they disappeared completely and for good. It is interesting that they had *almost* become extinct long before, at the end of the Triassic, when only one small group, probably only a single genus, survived. From it, however, a tremendous new radiation occurred in the Jurassic and into the Cretaceous. The failure of even one genus to pull through the similar crisis at the end of the Cretaceous and to restock Cenozoic seas may have been due to competition from then abundant squidlike cephalopods—but that is speculative.

The great marine reptiles, plesiosaurs, ichthyosaurs, and mosasaurs, also became extinct in the Cretaceous. Yet, oddly enough, nothing radical happened to the fishes living along with them and on which they fed. There was some extinction and replacement, but the dominant modern teleost families were becoming established in the late Cretaceous, and for the most part they continued their expansion through what was for many animals the time of the great dying.

On land, all the dinosaurs became extinct. Of all the hordes of Mesozoic reptiles, only four orders survived (Chapter 22), and one of them (Rhynchocephalia) dwindled to relict status.

Innumerable attempts have been made to explain the great dying, and especially the extinction of the dinosaurs, but none is satisfactory. It is agreed that there was some widespread environmental change, but no one has come up with a really plausible and well-supported theory as to just what the change was. In any attempt to explain this mystery, certain facts should be kept in mind:

Groups that became extinct were of many wholly different kinds living in entirely different environments.

Other groups living along with them and in the same general environments did not become extinct, and some did not even undergo evident change.

It is not logically necessary or probable that any *one* factor caused the extinction.

The extinction was not really sudden, although it has been said to be. It went on over millions and tens of millions of years. Many kinds of dinosaurs gradually disappeared, and only a few were left at

the very end. Ichthyosaurs were already rare in the early Cretaceous, and they disappeared long before the end of the period.

Whatever may have been its exact causes and sequences, the great dying did occur. It closed an era, and it opened the world to a new era, the Age of Mammals.

CHAPTER SUMMARY

Cambrian and Ordovician seas: the appearance of some major groups and the great expansion of life generally in the Cambrian and Ordovician; life still marine and the land barren; few fresh-water faunas; replacements in the post-Ordovician marine faunas; life of the late Ordovician—most protist groups present; no diatoms, bryophytes, or vascular plants; sponges, coelenterates, graptolites, bryozoans, and brachiopods, the latter much more abundant than later; pelecypods and gastropods common; nautiloid cephalopods dominant among mollusks; trilobites the dominant arthropods; eurypterids; echinoderms abundant; the first appearance of fishes, the agnaths.

The Age of Fishes: faunal changes in the Silurian: the Devonian as the Age of Fishes; replacement of agnaths by placoderms; placoderm types—acanthodians, antiarchs, and arthrodires; the replacement of placoderms by Chondrichthyes and Osteichthyes; the three basic osteichthyan groups—crossopterygians, paleoniscoids, and lungfishes; evolutionary derivatives of the three groups.

Occupation of the land: difficulties facing organisms on land, fully overcome by only four plant and animal phyla—Tracheophyta, Mollusca, Arthropoda, and Chordata; earliest land plants and animals—Silurian psilopsids and scorpions; Devonian land plants abundant—psilopsids, lycopsids, sphenopsids, ferns, and some primitive gymnosperms; Devonian land animals—arthropods; the first appearance of amphibians.

The Permo-Carboniferous, close of the Paleozoic: coal forests and the principal genera—*Lepidodendron, Sigillaria, Neuropteris, Cordaites, Walchia;* land invertebrates—land snails, scorpions, centipedes, cockroaches, and dragonflies; amphibians—labyrinthodonts, like *Eryops;* reptiles abundant by the Permian—cotylosaurs, mammal-like reptiles, and others.

The Age of Reptiles, the Mesozoic: the Permo-Triassic crisis; widespread extinctions; replacements in marine faunas; replacement of coal forests by cycadeoids, cycads, ginkgos, and conifers; the emergence of angiosperms; the decline of amphibians; major changes in the reptilian faunas; two dinosaur groups derived from thecodonts, examples being *Coelophysis, Tyrannosaurus, Brontosaurus, Anatosaurus, Stegosaurus, Ankylosaurus, Tri-*ceratops; a few still-surviving Mesozoic reptiles—rhynchocephalians, lizards, snakes, crocodiles, turtles; extinct aquatic groups—plesiosaurs, ichthyosaurs, mosasaurs; into the air, with pterosaurs, the flying reptiles; birds; the bearing of *Archaeopteryx* on theories of evolution; rise of mammals; mammal-like reptiles, evolving slowly through Permo-Triassic times and becoming full mammals by late Triassic and early Jurassic; the great dying—widespread extinctions in the Cretaceous (ammonites as a spectacular case; also plesiosaurs, ichthyosaurs, and mosasaurs) but survival of all major fish groups; extinction of all the dinosaurs; the problem of explaining the great dying; cautions in its interpretation.

31

Modernization of the Living World

Massie from Missouri Resources Division

Highly evolved mammals like the horse exemplify the modernization of the living world that occurred during the Cenozoic.

By the end of the Cretaceous we are far along in the history of life, with a mere 70 million years or so to go. What remains chronologically is the Cenozoic era, often popularly designated as the Age of Mammals. Its brief last part is the Age of Man, who is a mammal, too. From our vantage point at the time called "now," the greatest interest of the last 70 million years is that they did eventuate in the modern world. The origin of what is now, the modernization of the earth and its life, thus becomes a main theme in this chapter. Before we concentrate on the mammals and man, the theme involves some considerations of principle and of the more wide-ranging modernization of other groups of organisms.

The history of life is not the sort of thing we could imagine if there were no fossil record to set us right. On the evidence of present-day organisms alone, attempted reconstruction of the history would probably postulate a relatively simple divergence and successive rise of the various living phyla, classes, orders, and so on, and the steadily progressive advance of each toward its present condition. Such was the postulate, expressed or unexpressed, of most of the early thinkers about evolution, before a substantial historical record had been recovered.

A sequence of origins of progressively "higher" groups and the subsequent modernization of each have, indeed, occurred in the course of evolution. The actual history, however, is both more complicated and more fitful and seemingly erratic. How could one possibly infer from present conditions that the environments now dominated by mammals were long occupied not by earlier mammals or their reptilian ancestors but by hordes of other reptiles with neither descendants nor similar substitutes in the present world?

It is still more strange that, after mammals had arisen, a very long time passed before they began to assume their present roles. The modernization of floras and faunas has been neither steady nor straightforward. It has often advanced in, geologically speaking, short spurts, now in one group, now in another. It has also involved innumerable detours when dominance and seeming progress were by groups of organisms that were, in fact, to have little or no part in the modern world of life.

Modernization in Aquatic Environments

Aquatic environments were the first to be occupied by living things. It might be expected and is on the whole correct that they would have approached their present aspect at earlier times than the land environments. You have seen that even in the late Ordovician, more than 425 million years ago, the seas, at least, swarmed with plants and animals about as diverse as those now living in the same environments. The aquatic phyla were the same then as now. Yet in detail the modernization of these biotas has not involved the progressive change of most of the lesser groups, the classes, orders, families, and so on, then dominant. Instead, most of these groups have become extinct without issue and have been replaced by others whose direct ancestors were few and obscure in the Ordovician.

AQUATIC PLANTS

The fossil record of aquatic plants is inadequate. Most of them are soft-bodied and usually fossilize as a structureless smear of carbon or not at all. Nevertheless, we do know that algae were already common far back in the Pre-Cambrian. It is probable that most of the main groups of algae were present in the Paleozoic. They have never lost their dominance in all aquatic environments. Certainly many new species have evolved more recently, but on the whole the aspect and general composition of aquatic floras seems already to have been modern several hundred million years ago. The most recent important change for which there is good evidence was the appearance of the diatoms, first found as fossils in the Jurassic and increasingly common since then. Now they account for a good fraction of the photosynthesis in most aquatic environments.

INVERTEBRATES

A few modern groups of aquatic invertebrates date back without really profound change from the early Paleozoic. Most of them, however, have had extensive replacement since then within groups and adaptive niches. Corals, bryozoans, and clams, for instance, had much the same roles in Paleozoic seas as they do now and did not look very different superficially. Nevertheless, there has been almost complete replacement in these groups, not only of species and genera but of families and superfamilies, even, in some instances, of orders. Without going into detail, it may be said that for these groups and for most other aquatic invertebrates the beginning of the definitive replacement tended to be most evident during the Permo-Triassic crisis.

Modernization of aquatic invertebrates proceeded apace during the Mesozoic, and in most respects it was essentially complete in the late Cretaceous. In only one respect would the faunas of late Cretaceous seas have seemed particularly strange to anyone familiar with modern seas: the now extinct ammonites were still abundant (Fig. 31–1). When they disappeared at the beginning of the Cenozoic, the aquatic invertebrate faunas were almost thoroughly modern. Changes during the Cenozoic have their own fascination, but they were matters of detail, of not very distinctive changes among species and genera and their precise geographic distributions.

FISHES

Although they got started so much later, the fishes rather closely paralleled the aquatic invertebrates in their modernization. As already noted, the archaic groups of fishes dominant in the Paleozoic dwindled during or before the Permo-Triassic crisis. The beginning of dominance of higher bony fishes was evident in the Triassic. The most progressive main group, that of the teleosts, had appeared and was spreading in the Jurassic. By the end of the Cretaceous it was fully dominant and modern in aspect. Changes during the Age of Mammals have, again, been matters of detail.

OTHER AQUATIC AND AMPHIBIOUS VERTEBRATES

The archaic amphibians also mostly disappeared in the Permo-Triassic crisis. None are known after the Triassic. The rise of the modern groups is poorly documented by fossils, for some unknown reason, but frogs almost modern in appearance are known from the Jurassic. The salamanders may be equally old, but none is yet known before the Cretaceous.

SOME CRETACEOUS AMMONITES

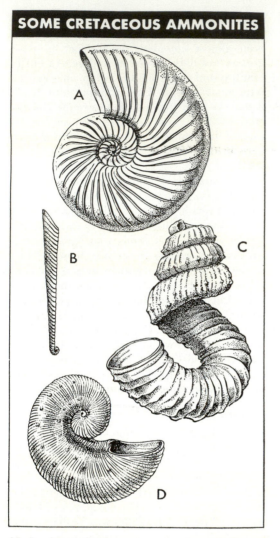

31–1   Some Cretaceous ammonites. *A. Oxytropidoceras acutocarinatum* ($\times \frac{2}{3}$). *B. Baculites compressus* ($\times \frac{5}{6}$). *C. Heteroceras sp.* ($\times \frac{2}{3}$). *D. Scaphites nodosus* ($\times \frac{2}{3}$).

The rise of several groups of aquatic reptiles in the latest Paleozoic and during the Mesozoic introduced a bizarre and unmodern note in the aquatic faunas of those times. The extinction of most of them in the Cretaceous left the turtles and crocodiles (with their allies the alligators) as the common amphibious or aquatic reptiles. Neither group has changed essentially during the Cenozoic. The most striking change in aquatic environments during the Cenozoic, practically the only change of really deep significance, was

the rise and spread of aquatic mammals, whales, seals, sea cows, and their relatives. They began to appear in the Eocene and were common and essentially modernized in the Miocene.

Modernization in Land Environments

PLANTS

As you already know, the dominant, archaic vegetation of the Carboniferous vanished, for the most part, during the Permo-Triassic crisis. In the Triassic and Jurassic there was a sort of interim dominant floral type composed mainly of ferns, cycadeoids, cycads, ginkgos, and conifers. Most of those groups do survive now, but in greatly diminished numbers. Their dominance and the scarcity of flowering plants gave the plant life of the Triassic and Jurassic a decidedly nonmodern aspect. By late Cretaceous time, however, the aspect was fully modern. No distinctive groups of plants are known from the late Cretaceous that do not survive today. The flowering plants were then already decidedly dominant, and almost all their modern groups date from then or from quite early in the Cenozoic. Changes during the middle and late Cenozoic involved little more than the shifting of established floras as climates changed.

INSECTS

Modernization of insects also began during the Permo-Triassic crisis, when most of the archaic groups that had appeared in the Carboniferous became extinct. No important groups (for instance, no orders) that are now extinct evolved after the Permian. The whole picture of insect evolution during the Mesozoic was one of steady expansion by the evolution of new groups that were to continue onward into the modern world. Expansion in detail went on also through the Cenozoic and is perhaps still going on, but most if not all of the main groups of insects were already present and modern in form in the late Cretaceous. Furthermore, no important known groups of Cretaceous insects have become extinct. As far as we can judge from a spotty fossil record, the insects come closer than any

31–2   The great toothed diver (*Hesperornis regalis*). A large, wingless, swimming bird of the Cretaceous. Note that it possessed teeth.

other group to the postulate of simple expansion mentioned at the beginning of this chapter.

REPTILES

As we said earlier, the modernization of land reptiles was essentially only a matter of extinction of the dinosaurs. That left the lizards, snakes, and tortoises, all of which had nearly reached their modern form while the dinosaurs still lived. A good deal of expansion in detail and sometimes intricate speciation have occurred among lizards and snakes during the Cenozoic.

BIRDS

Birds rarely fossilize except under unusual circumstances, so that their fossil record is spotty. Patient accumulation has, however, revealed the essentials of their history. They arose in the Jurassic and became fully bird-like, essentially modern in structure, by the end of the Cretaceous. Some of them then still

had teeth, but their scanty remains suffice to show that their skeletons were no longer semi-reptilian but completely avian. They had, moreover, undergone some sharp divergence or adaptive radiation, for, along with normal flying birds, there were large, wingless, swimming birds at that time (Fig. 31–2).

Expansion and subdivision of the birds, their main adaptive radiation, apparently occurred early in the Cenozoic. Most of the orders of birds were already present in the Eocene. Miocene birds generally differ hardly at all from their recent descendants. Despite the comparatively thin fossil record, birds surely flitted throughout the Cenozoic in great numbers, and it would be as true to call that the Age of Birds as to call it the Age of Mammals.

Large flightless birds evolved on all the larger land areas of the Cenozoic. Three groups have survived: ostriches in Africa and Arabia (and formerly over most of Eurasia); rheas in South America; and cassowaries and

American Museum of Natural History

31–3 *Diatryma,* **a large flightless bird of the Eocene.**

the closely related emus in Australia and New Guinea. More have become extinct: *Diatryma* in North America (Fig. 31–3), with relatives

in Europe; moas in New Zealand; "elephant birds" (*Aepyornis* and relatives) in Madagascar; and several kinds in South America. It is an oddity of evolution that the birds, after acquiring flight and dominating a new environment, repeatedly gave rise to lines in which flight was lost. The large running birds have competed, on the whole quite successfully, with mammals.[1]

History of the Mammals

MESOZOIC MAMMALS

You recall that the mammals evolved from mammal-like reptiles in the late Triassic or earliest Jurassic (Figs. 31–4 and 31–5). They

[1] An earlier theory had it that the running birds held their own because they evolved where mammalian carnivores were absent (New Zealand) or supposedly ineffective (Madagascar, Australia, ancient but not modern South America). But they also evolved in North America, Eurasia, and Africa along with admittedly highly effective mammalian carnivores. Here is an evolutionary problem not fully solved.

31–4   The historical record of the vertebrates. The widths of the pathways roughly approximate the relative numbers of known genera in the classes.

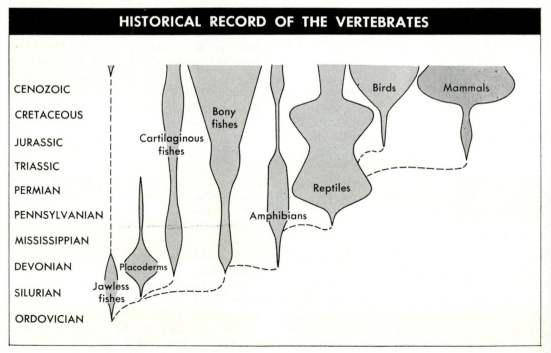

HISTORICAL RECORD OF THE VERTEBRATES

CENOZOIC
CRETACEOUS
JURASSIC
TRIASSIC
PERMIAN
PENNSYLVANIAN
MISSISSIPPIAN
DEVONIAN
SILURIAN
ORDOVICIAN

Cartilaginous fishes
Bony fishes
Birds
Mammals
Reptiles
Amphibians
Placoderms
Jawless fishes

are about as old as flowering plants, turtles, or crocodiles, and probably older than teleost fishes, lizards, snakes, or birds. They are by and large the most progressive, by most definitions the very highest, of all organisms. Yet their modernization and their rise to dominance in their own environments was slower than for most other groups dominant in one environment or another at various times in the history of life. For more than 100 million years after they first appeared, mammals cut a small figure in the world. That duration is considerably longer than the whole era of their later dominance, the Age of Mammals. By the late Jurassic they had some diversity, but even up to the end of the Cretaceous they remained small, mostly about mouse-sized, and rare.[2] (See Fig. 31–6).

The mammals of the Jurassic fall into only three quite circumscribed adaptive types: one group probably ate seeds and small fruits; another comprised carnivores predatory in a small way; most of them (the third adaptive type) were what is commonly called "insectivorous," which means that they ate almost any small food bits, mostly of animal origin. By the end of the Cretaceous, about 65 million years later, their evolutionary change was considerable. Yet, as far as the record shows, mammals were if anything even more circumscribed ecologically. The seedeaters were still there, relatively little changed, and there were somewhat more progressive and varied insectivorous to partly carnivorous forms.

Why were the mammals so obscure for so long? We do not know, but we can offer a hypothesis. When amphibians arose, they had a new way of life with no competitors. Expansion was rapid, geologically speaking, as you would expect. Early reptiles had to still greater extent access to new ways of life and new environments empty of any possible competitors. They not only expanded rapidly but also eventually all but wiped out the amphibians, whose environments overlapped theirs. When mammals arose, the situation was quite different. Niches or adaptive types accessible to the mammals, where they would much later become dominant, were already extensively occupied by well-adapted reptiles.

In the over-all picture of the ecology of the Mesozoic lands, mammals appear as if they were merely a few specialized reptilian offshoots, occupying a few narrow niches to which they were confined by the pressure of other, temporarily successful specialized reptilian lineages. That mammals did survive through the Jurassic and Cretaceous and managed to hang onto some, at least, of their small niches is a tribute to their inherently more efficient physiology and reproduction. Brains probably did not help at this stage of the game, for what evidence we have suggests that Mesozoic mammals were not significantly brighter than their reptilian neighbors.

BEGINNING OF THE AGE OF MAMMALS

The Cenozoic began at an unfortunate time from the fossil-hunters' point of view. In the long seesaw between the uplift and the wearing down of the continents it was a time of predominant, widespread uplift. A result was that erosion was likewise widespread. Most of the eroded material was reworked and eventually deposited beyond the margins of what is now dry land. Few deposits of sediments were made and left where we can now get at them. Consequently, there are few available deposits of the beginning of the Cenozoic, the early Paleocene, where fossils of land animals can be found. In spite of intensive search, earliest Paleocene mammals have been found in only one part of the earth: in and near the Rocky Mountains in the United States. Later Paleocene mammals are known from Europe, Asia, and South America, and probably early Paleocene mammals will be found elsewhere eventually. In the meantime we can follow the detailed change from Age of Reptiles to Age of Mammals only in the Rocky Mountain region. Obviously a good deal that we do not know about must have been going on elsewhere.

[2] The Mesozoic mammals were not mice, which evolved much later, but most of them were of about mouse size. The largest may have been about the size of a house cat. It is of course possible that larger and more varied mammals evolved during the Mesozoic in places where no fossils have been preserved or found. Yet it seems improbable that larger mammals can have existed for very long before the Paleocene or can have become very diversified and abundant without leaving any trace at all in the record we do have.

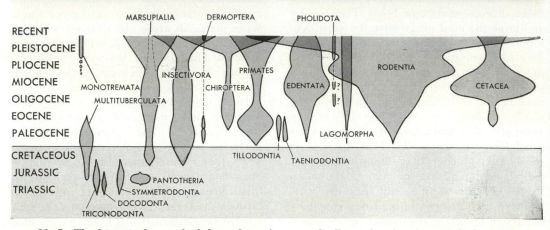

31–5 The historical record of the orders of mammals. For each order of mammals the width of the pathway is proportional to its known diversity in each of the geological periods in which it lived. The Mesozoic-Cenozoic boundary (Cretaceous-Paleocene) is indicated. The mammals as a class had emerged from their reptilian ancestry by the Jurassic, in which period five distinct orders were present. Of these the Pantotheria probably gave rise to all later orders except the monotremes. Note the great variety in the histories of the different orders: some reach great diversity and maintain it for a long time before slowly declining to extinction (Notungulata) or to near extinction (Proboscidea); some rapidly reached a maximum and subsequently decline rapidly (Condylarthra) or slowly (Perissodactyla). A striking generalization is that no order expands slowly to a climax and

In the Rocky Mountain region the latest Cretaceous rocks contain dinosaurs in some numbers and also tiny mammals. The mammals included one holdover group from the Jurassic, the old seedeaters,[3] which were to become extinct during the Eocene. The "insectivorous" forms were mostly marsupials, closely similar to our modern opossums, which are still thriving in spite of the fact that they have hardly changed at all since that remote date. With them were a few true insectivores, that is, placental mammals classified in the order Insectivora.

In the earliest Paleocene rocks the dinosaurs are completely absent; they seem to have disappeared with startling suddenness. The fauna is already dominated by mammals, although as yet these are of only a few kinds. Opossumlike marsupials and placental insectivores are still present and will continue to be, in one place or another, throughout the Cenozoic, but they are in a small minority. The common mammals are somewhat larger.

[3] *Multituberculates,* so called from the many tubercles or cusps on their grinding teeth.

The largest in the early Paleocene was about the size of a not-too-robust collie. Most of these moderate-sized early mammals were much alike. Strong differentiation among placental mammals had not yet occurred. They were long, rather short-legged, running (probably not very swiftly) on feet with five toes. Tails were long, heavy at the base, and heads were small in proportion. The teeth suggest that most of these animals were omnivorous, some with a tendency to rely more on plant food and some with a tendency to rely more on animal food (Fig. 31–7).

Later faunas in the American Paleocene show marked and relatively rapid expansion and divergence of the placental mammals. By late Paleocene there are many kinds: numerous hoofed herbivores, some of them now well over a meter in height; fairly specialized predacious carnivores of many sizes and kinds; small forerunners of the monkeys; the first, still very rare, rodents; and others. The later Paleocene faunas from Europe and Asia suggest that much the same sort of expansion had been going on there. It was probably shared

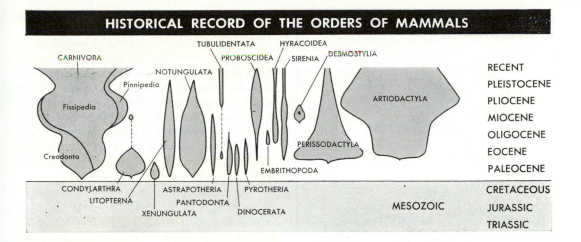

HISTORICAL RECORD OF THE ORDERS OF MAMMALS

then declines rapidly. Only the rodents seem to have maintained anything like a steady increase throughout their history. Some internal detail is given for the carnivores to show what is generally typical of the history of an order. The early expansion of the carnivores was almost entirely due to the now long-extinct Creodonta, which underwent a quite wide adaptive radiation. One product of this radiation was a basic new carnivore type, the Fissipedia (dogs, raccoons, bears, cats, etc.). Fissipedes themselves underwent a strong adaptive radiation and in so doing gradually replaced the creodonts. Still later the Pinnipedia (seals, walruses) arose as a product of fissipede radiation. Pinnipedes, invading a wholly new adaptive zone (the sea), have never threatened to replace the fissipedes from which they evolved.

by the whole World Continent, the intermittently united land masses of Africa, Eurasia, and North America. In South America a great expansion was at least well under way in the late Paleocene, although it was already peculiar to that island continent.

What had happened is fairly clear: the mammals were finally inheriting the earth. Extinction of most of the Mesozoic land reptiles left empty environments which were occupied by the mammals in a worldwide adaptive radiation most active during the Paleocene. The mammals, more efficient and more capable of divergent adaptation, eventually went further than the reptiles ever had in occupation and subdivision of the environmental niches. All this only makes more mysterious and more exasperating the problem of *why* the Mesozoic reptiles became extinct and left the environmental opportunity to the mammals. The record seems to make it clear that the dinosaurs did not become extinct because the mammals had expanded, but on the contrary that the mammals expanded because the dinosaurs previously had become extinct.

MODERNIZATION OF MAMMALS

Eocene and Oligocene mammals

For reasons with which you are already familiar (Chapter 28), modernization of the South American mammalian faunas was long delayed. It did not occur until the late Pliocene and Pleistocene. Elsewhere, on the World Continent, basic modernization occurred most prominently in the Eocene, with a filling in of detail in the Oligocene and later. Probably all the main groups, the orders, of living placental mammals were already in existence in the Eocene.[4] Most of the living families were present by the end of the Oligocene, at latest. It is significant that no family of placental mammals still living dates from before the

[4] There are 16 living orders of placental mammals, of which 13 are definitely known from fossils in the Eocene. The other three (scaly anteaters, aardvarks, and hyraxes) do not appear in the fossil record until later. They are, however, all small groups that originated in the Old World tropics, from which few Eocene fossils are known. Their apparent absence in the Eocene is almost surely due only to lack of discovery.

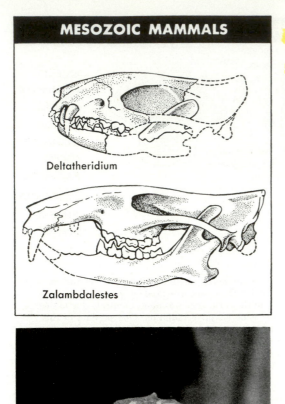

MESOZOIC MAMMALS

Deltatheridium

Zalambdalestes

American Museum of Natural History

31–6 Mesozoic mammals. *Top.* Skulls (partially restored) of two small Cretaceous mammals, *Deltatheridium* (skull 1¾ inches long) and *Zalambdalestes* (skull 2 inches long). *Bottom.* A photograph of the skull of a Cretaceous mammal.

31–7 *Ectoconus*, a Paleocene mammal.

American Museum of Natural History

Eocene, but that most of them do date from before the Miocene.

There were more orders of placental mammals in the early Eocene than there have ever been since. The modern orders were arising, and the older, Paleocene orders were not yet extinct. Expansion of the modern groups at the expense of the older ones, which were for the most part entirely replaced, is very evident through the Eocene (Fig. 31–8). By the middle of the Oligocene the number of orders was almost down to the present level.

Eocene and Oligocene faunas look strange to us in spite of the rapid modernization that was going on. There are three main reasons for their exotic look: some of the striking and peculiar ancient groups were not yet extinct; some modern groups had different geographic distributions then and occurred in unexpected places; and, as a point of detail, the direct ancestors of our present mammals had not yet reached just their present forms.

Ancient and now extinct groups of mammals prominent in the Eocene and Oligocene include, among the herbivores, the uintatheres, the titanotheres, and, in North America only, the oreodonts (Fig. 31–9). Among

31–8 The modernization of the mammalian fauna of North America. The diagram shows the percentages of North American mammals belonging to different groups as they changed through the Age of Mammals. Extensive turnover from archaic to modern groups is especially evident in the Eocene and Oligocene. (The early native primates in North America were all of archaic, prosimian groups; modernization of primates occurred, but in the Old World.)

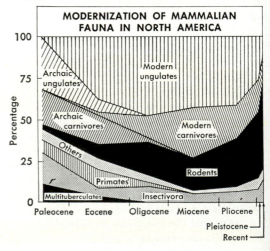

MODERNIZATION OF MAMMALIAN FAUNA IN NORTH AMERICA

Eohippus

Uintathere

Titanothere

Camel

Oreodont

Oligocene
sabertooth

31–9 Some Eocene and Oligocene mammals.

Moeritherium

Gomphotherium

Woolly
mammoth

striking extinct carnivores, the sabertooths [5] appeared in the earliest Oligocene and were prominent throughout the World Continent until the Pleistocene, when they became extinct. Startling later changes in geographic distribution are illustrated by the presence of abundant camels and horses in North America through much of the Cenozoic, in spite of the fact that there are no native camels or horses here now.

Equally striking, especially at later dates, are the changes in areas occupied by the bulky proboscidians, the elephants and their allies (Fig. 31–10). Mastodonts, the more primitive members of the group, enter the fossil record in Egypt in the Oligocene. In the Miocene they spread everywhere on the World Continent and were, for instance, common in North America from the late Miocene through the Pleistocene. In the Pleistocene the mammoths or elephants (mammoths are simply the extinct species of elephants) arose and also spread abundantly everywhere on the World Continent, including North America. Now they occur, as natives, only in southeastern Asia and central and southern Africa.

The horse family

Changes in modern groups from the Eocene onward are well illustrated by that classic example of evolution, the horse family (Figs. 31–11 and 31–12). Its earliest known member is eohippus (*Hyracotherium*), in the early Eocene (see Fig. 31–9). "Little eohippus, no bigger than a fox," generally stood about 45 to 60 centimeters high. It had four toes on

[5] Frequently called "saber-toothed tigers," but they were not tigers.

31–10 Some proboscidians. *Moeritherium*, the earliest known proboscidian (from the Eocene of Egypt) was apparently amphibious in habit and about the size of a pig. There was little in its structure to suggest the striking morphological change in the faces of its later relatives. *Gomphotherium*, a mastodont from the Miocene of North America, looked somewhat more familiar to us as a member of the elephant lineage; it had short tusks (enlarged incisor teeth) in both upper and lower jaws. The lower jaw was elongated, and an elephantlike trunk was already present. In later mastodonts and in the elephants the lower tusks were lost, and the lower jaw became shortened. *Mammuthus primigenius*, the woolly mammoth of the Pleistocene, was a fully evolved elephant.

the front feet and three on the hind, each toe ending in a tiny hoof. The small head lacked the heavy muzzle of modern horses and had comparatively large eyes set near the middle, not far back as in our horses. The teeth were simple, not fit for grazing but only for browsing on soft vegetation.

The contrast between little eohippus and the large horses of today is great. Yet almost all the intermediate stages are known, a powerful demonstration not only that evolution is a fact but also of how it has occurred. This group, at least, convincingly demonstrates that progressive change is a process of spread of small mutations and new combinations of genes and chromosomes in variable populations. Other examples, and there are now many about as good as that of the horse family, suggest that this is the rule in evolutionary sequences, although not necessarily a rule with no exceptions.

It is to be emphasized that the picture of steady, gradual change from eohippus to *Equus*, the modern horses, still commonly given in popular discussions, is quite incorrect. The true history of the horse family does not show a lineage that gradually increased in size, reduced the number of toes, and developed higher, more complex teeth from eohippus to *Equus*. In the first place, there was not one lineage but, at times, dozens of them. The phylogeny is intricately branched, although all but a few of the branches have now become extinct. Increase in size and change in the feet were not constant but sporadic. There was no significant increase in average size for the first 20 million years or the last 5 or more. Between times, increase was usual but not constant. After the Eocene (when there was no noticeable change in the feet), horses developed three mechanically and functionally different kinds of feet. Each type arose comparatively rapidly, and once it had arisen each tended to remain constant in most lineages, not to change steadily toward another type. All three were common in different groups of late Cenozoic horses, although only one type has survived today. The change from eohippus to *Equus* did occur, of course, but it occurred irregularly through the complex phylogeny shown in greatly oversimplified fashion in Fig. 31–11.

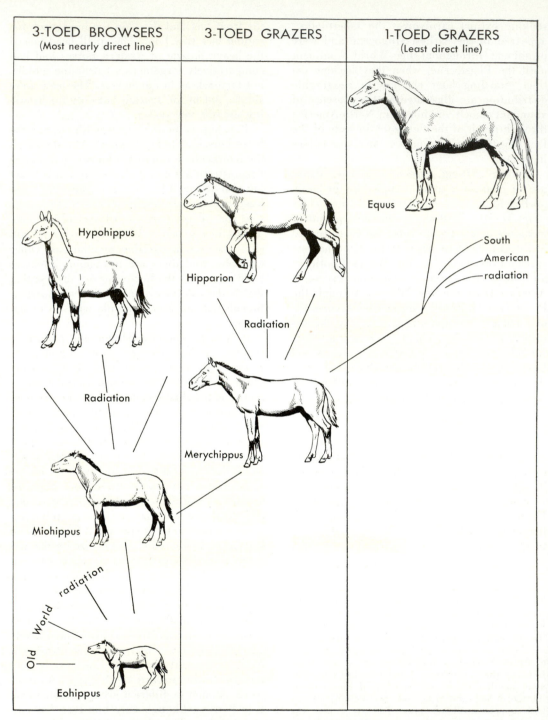

3-TOED BROWSERS (Most nearly direct line)	3-TOED GRAZERS	1-TOED GRAZERS (Least direct line)

Equus

South American radiation

Hypohippus

Hipparion

Radiation

Radiation

Merychippus

Miohippus

Old World radiation

Eohippus

31–11 History of the horses. *Miohippus* was only one product of an early radiation from eohippus. It was still three-toed and browsed on trees and bushes. From *Miohippus* many forms arose, some of which remained three-toed browsers; but one of them, *Merychippus*, made the major step to the grazing habit; it began exploiting the extensive grasslands of the Miocene period. The line eventually leading to the one-toed *Equus* from *Merychippus* was, again, only one of many in the adaptive radiation of the three-toed grazers. Thus the lineage from eohippus to *Equus* is the least direct of those that might be traced through the succession of adaptive radiations. This diagram is greatly simplified and shows only a few of the many genera known.

Forerunners of Man

We may forgive ourselves for being more interested in the evolution of man and his relatives than in that of horses or other non-human groups. The fossil record of human origin and of the order Primates in general is now quite extensive. Although it still has gaps, many will doubtless be filled as larger gaps in the past have been filled by new finds of fos-

31–12 The evolution of the forefoot in the horse family. The single toe (stippled in the figure) of the modern horse (*Equus*) is the sole survivor of four toes present in the ancestral form eohippus (technically *Hyracotherium*). *Hypohippus* and *Hipparion* show stabilization of the mechanism, without advance toward the condition of *Equus*. The feet, of different sizes, have been reduced to the same length for comparison.

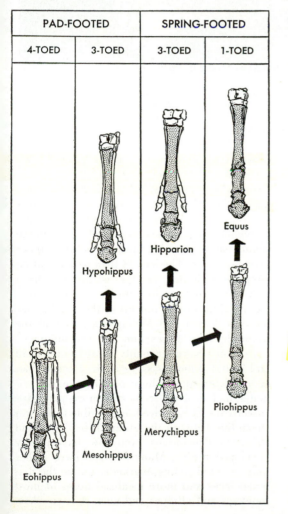

PAD-FOOTED		SPRING-FOOTED	
4-TOED	3-TOED	3-TOED	1-TOED

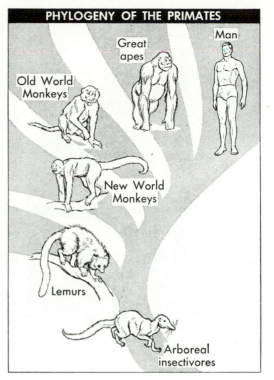

PHYLOGENY OF THE PRIMATES

31–13 The phylogeny of the primates. The figure is not scaled either to size of animal or to time.

sils. Some gaps may always remain, because the fossils are simply not preserved or where we can find them, but the main features of the history are already clear (Fig. 31–13).

That the primates have not left as complete a fossil record as, for instance, the horses is readily understandable. The greater part of the history of the horse family—its most central and progressive part, at least—took place in one region, a region rich in fossils and well explored by bone diggers: central North America. The history of the primates was more far-flung, with crucial episodes in several different regions. Moreover, central parts of the history occurred in tropical areas, which are, as a rule, neither very rich in fossil deposits nor as yet thoroughly explored for them. The primate way of life also militates against a good fossil record. Most primates are and have been arboreal, and tree dwellers are comparatively unlikely to be buried and to fossilize. It has further been suggested that higher primates were too bright to be fossilized with any great frequency. A shrewd

31–14 **Prosimians.** *Left.* An Eocene prosimian, *Notharctus.* Note the grasping hands and feet; the thumb closes over the branch in the opposite direction to the fingers. *Right.* A living prosimian, *Loris tardigradus* (the slender loris). Note the thumb and the manner in which both eyes are directed forward, permitting stereoscopic vision; the latter is an advanced character in which this particular prosimian has paralleled the higher primates.

animal has a better chance to avoid being mired down or swept away in a flood, or suffering other accidents that could readily lead to burial and fossilization.

EARLY PROSIMIANS

The main groups of primates were briefly mentioned in Chapter 22. The history starts with the most primitive of these groups, that of the *prosimians* or premonkeys. Their oldest known fossils are found in the middle Paleocene of the Rocky Mountain region. Prosimians (Fig. 31–14) were abundant through the Eocene, not only in North America but also in Eurasia and probably in Africa, where the Eocene fossil record is extremely poor as far as yet discovered. They are absent from the known North American record from the Miocene onward, but they lived on in parts of the Old World. They now live only in tropical Asia and adjacent islands, in Africa, and in the large island of Madagascar. The continental forms are mostly rare and comprise only a few species. One group of

prosimians, that of the lemurs, is, however, common and diversified in Madagascar, where they are the only primates (except man). This abundant survival of so primitive a group is an interesting evolutionary phenomenon. Their ancestors gained access to the island early in the Cenozoic, before higher primates arose. The lemurs then underwent an adaptive radiation on the island and have been protected from effective competition. Later primates did not cross the sea barrier between Africa and Madagascar, so that the island became an asylum for prosimians.

Primates were probably more widespread and included more species in the Eocene than at any later time. Those primitive early primates had evolved from still more primitive Insectivora. It is, indeed, impossible to draw a sharp line between the orders Insectivora and Primates. Prosimians, even now, have poorly developed brains. Many of them have rather long, pointed snouts, markedly unlike the flattened faces and more localized noses of most monkeys and all apes and men. A basic characteristic of primates present in early prosimians is the retention and further evolution of grasping hands and, often, feet, with apposable thumbs and big toes. In connection with this adaptation, the insectivore claws early began to evolve into nails.

There were innumerable divergent lines among Paleocene and Eocene primates and primatelike insectivores. Many of them became specialized in peculiar ways and have become extinct. The ones most significant for further evolution retained simple, primitive teeth and had a tendency toward enlargement of the cranium, reduction of the snout, and enlargement of the eyes, which also tended toward a more anterior than lateral outlook.

MONKEYS

The Old World monkeys, *cercopithecoids*, and New World monkeys, *ceboids* (Chapter 22), are fully distinct phylogenetic units. They have always had approximately the same geographic distribution that they have now: cercopithecoids in the warmer parts of Africa, Asia, and Europe,[6] and ceboids in South and Central America. Both groups evolved from prosimians, probably from closely related prosimians, but they arose separately, on different continents. (Why did not these warm-climate animals migrate in either direction between Eurasia and North America?)

Both groups first appeared in the Oligocene. They differ from prosimians in similar ways. They have larger, more effective brains, with expansion especially of the upper part of the cortex. The brain case correspondingly forms a larger proportion of the head. The snout is reduced, and the flattened monkey face is characteristic. The eyes are pointed more nearly forward, and the fields of vision of the two eyes overlap widely. The shortened muzzle involves a shortening of upper and lower jaws, and the number of teeth is reduced. The molars are squared, but remain simple. Hands and feet are grasping, but no more so than in some prosimians.

Differences between cercopithecoids and ceboids are not profound or important. They are merely indications of the fact that the two were of somewhat different ancestry. Ceboids have retained some primitive features lost in cercopithecoids, but they are more advanced in some respects. They are not a lower group, but just a different one. They have no direct bearing on the ancestry of the cercopithecoids, apes, or man.

APES

The name "ape" is sometimes applied to a monkey, but strictly it means a member of one particular family of the Primates, the Pongidae (Fig. 31–15). Sometimes they are called *anthropoid*[7] *apes* to make the distinction clear. The living apes are the *gorilla* and *chimpanzee* of central Africa, the *orangutan*[8] of Sumatra and Borneo, and the *gibbons* of southeastern Asia. All are above average size for primates, and gorillas are big brutes, the males becoming heavier and stronger than any men. All lack tails.[9] They have relatively larger brains and brain cases than the monkeys and are in general more intelligent.

All recent apes except the gorilla are highly arboreal, spending most of their lives in trees. Gibbons, smallest of the apes, have tremendously long arms and are astonishing acrobats. They swing (*brachiate*) and jump for great distances between branches. When they do come to the ground, they walk on their hind legs, holding up the long arms as balancers. The other apes habitually walk on all fours unless they have been taught to walk clumsily on the hind legs as a circus or vaudeville trick. Gorillas also climb trees, and their immediate ancestors were evidently arboreal, but now they spend much of their time on the ground. Chimpanzees and orangutans do not swing from limb to limb with the abandon of gibbons, but they are agile four-handed climbers and spend most of their time in trees.

Apes have always been confined to the warmer parts of the Old World. The earliest

---

[6] The only European monkeys now are the famous "apes" (really monkeys, not apes) of the Rock of Gibraltar, but monkeys used to occur more widely in southern Europe.

[7] "Manlike." Some of the confusion may have arisen from German usage, in which *Affen* correctly applies to what we call "monkeys," and our "apes," strictly speaking, are *Menschenaffen*, "men-monkeys."

[8] A Malay name meaning "forest man."

[9] A few monkeys are also tail-less. Incidentally, ability to hang by the tail, sometimes considered typical of all monkeys, is confined to a few South American monkeys.

Gibbon

Chimpanzee

Gorilla

Orangutan

31–15 **The great apes, Pongidae.** Note the hands and eyes. The gibbon is brachiating.

known fossils occur in the Oligocene, along with the first Old World monkeys. Apes and cercopithecoids share some characters present neither in the ceboids nor in the prosimians. The earliest forms are hard to tell apart. It is apparent that apes either were derived from a branch of the earliest cercopithecoids or that they and the cercopithecoids had an immediate common ancestry among the prosimians. The distinction is more one of terminology than of significant difference in the phylogeny.

In the Miocene, apes became more varied than they are today. Their remains are not common, but diligent search has turned up fragments of many species in the Miocene of eastern Africa and the Pliocene of India and a few in the Miocene and Pliocene of Europe. There was a great expansion of the family starting at about the beginning of the Miocene. Some of the members of that late Cenozoic complex were aberrant and became extinct. Others were ancestral to the modern apes and were already becoming specialized in similar ways. Still others, especially in the Miocene, were comparatively light, agile forms not yet strongly specialized for arboreal life and lacking other characters peculiar to the surviving apes. *Proconsul* (Fig. 31–16), from the Miocene of Kenya, is the most completely known and is among the less specialized.

AUSTRALOPITHECINES

In 1925 was discovered the first of a remarkable group of fossil primates known as the *australopithecines* ("southern monkeys [or apes]"). Since then, several nearly complete skulls and numerous teeth, jaw fragments, and other parts of skeletons have been collected. They are definitely known only from Africa, although some fragments from southern Asia and Europe may belong to the same group (Figs. 31–17 and 31–18).

The australopithecine brain was comparable in size and complexity with that of the larger living apes. It was smaller than in any normal modern human. The skull as a whole is apelike in appearance, but it has a number of more manlike anatomical details. Among these is evidence that it was set squarely on top of the backbone, not thrust forward as in the apes.

31–16 *Proconsul*, fossil ape from the Miocene of Kenya, Africa.

The teeth are more manlike than apelike. The pelvis and limb and vertebral parts suggest that the australopithecines walked upright, or nearly so, a posture decidedly human in contrast with that of the apes.

The australopithecines now known were quite varied and covered a considerable span of time, from about 2 million to about ½ million years ago. Some of them were seemingly too aberrant (for instance, in great enlargement of the molar teeth and development of a crest on the skull) to figure in the direct ancestry of modern man. There is also evidence that when the later, at least, of them lived, more advanced members of the human family had already evolved. It is, however, now the consensus of competent specialists that early and comparatively unspecialized australopithecines did represent the human ancestry around the beginning of the Pleistocene. The combination of a brain still practically on the ape level with nearly human dentition and posture is remarkable and was not anticipated in earlier speculation about intermediaries between apes and men.

Whether to call the australopithecines "apes," "men," or something else is really a semantic and not a scientific question. Scientifically the consensus now places them as pre-*Homo* hominids, that is, as members of the human family that had not attained the status of the living genus of humans. An important

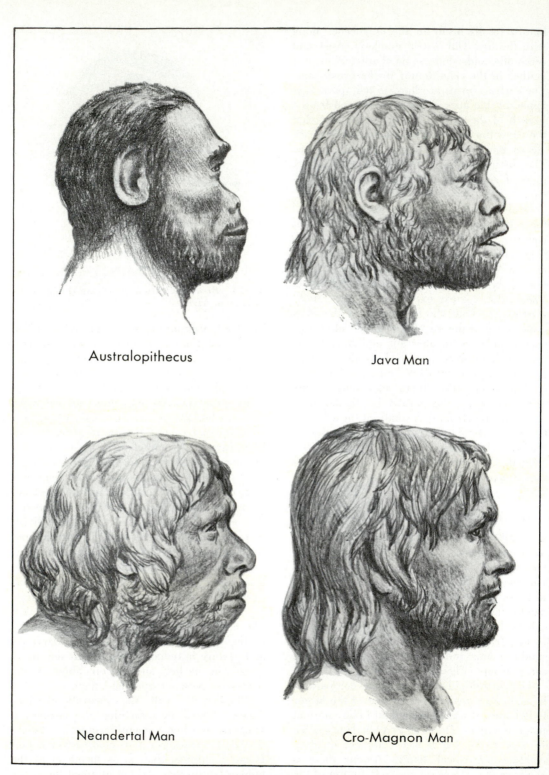

Australopithecus

Java Man

Neandertal Man

Cro-Magnon Man

31–17 *Australopithecus* **and fossil men.**

but not necessarily decisive point is whether they made tools. Living chimpanzees frequently use and occasionally make very simple tools, usually sticks or straws, but most anthropologists consider the regular making of special tools a definitely human trait beyond the capacity of the true ape (pongid) brain. Although stone tools certainly and perhaps others of bones, teeth, and horns have been found associated with australopithecines, it is not yet certain that they manufactured them.[10]

Man's Place in Nature

HUMAN ORIGINS

Even the pre-evolutionary biologists recognized that man is an animal, fundamentally like other animals but distinguished as a species by higher and different development of intelligence. In the eighteenth century Linnaeus, an antievolutionist, classified man, *Homo sapiens*, in the order Primates with the prosimians, monkeys, and apes. That classification is still accepted by biologists. Knowledge of evolution makes it evident that man's distinctive characteristics arose in the same way as those of other species of animals, by more or less gradual change in varying populations. That man's material being evolved

[10] In any event, whether the australopithecines did or did not make tools would not in itself establish their phylogenetic relationship to *Homo*. It could have an indirect bearing on that question, but it would directly indicate only the degree of advance toward or in parallel with *Homo* in one evolutionary direction.

31–18   Skulls of an ape and of members of the human family. In comparing apes and men as a group, it is useful to recognize two sets of skull features: one set (called paleanthropic) is more apelike and characterizes earlier members of the human family; the other set (called neanthropic) characterizes modern men. Paleanthropic skull characters include (1) small brain-case volume; (2) brain case shallow relative to length; (3) heavy bony brow ridges over the eyes; (4) jaws protruding forward from a vertical dropped from the eyes; (5) receding lower jaw—no chin. Neanthropic skull characters are the opposite: (1) large brain-case volume; (2) brain case deep relative to length; (3) no brow ridges; (4) jaws not protruding in front of the eyes; (5) chin protruding forward.

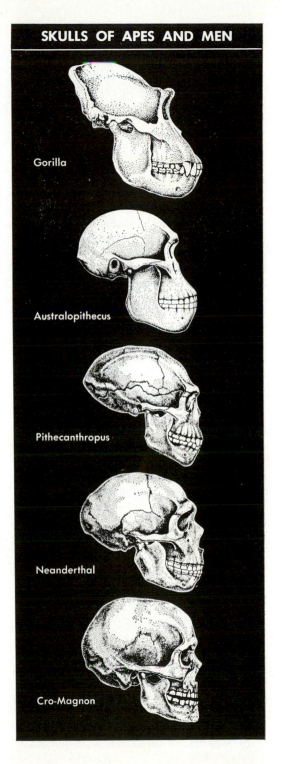

SKULLS OF APES AND MEN

Gorilla

Australopithecus

Pithecanthropus

Neanderthal

Cro-Magnon

from other and (in intelligence, at least) lower animals is about as certain as a scientific conclusion can be. That man has an immortal soul and that other animals do not is a proposition that scientific investigation can neither affirm nor deny. You are free to reach your own conclusions about that or to accept whatever religious tenet seems worthy of your faith.

Man's classification as a primate is now accepted as valid on a phylogenetic basis. His physical ancestry was, at some time in the past, the same as that of all the other primates. His precise relationships among the primates are still disputed in detail. Even scientists cannot always avoid being swayed by emotion and prejudice when they deal with their own origins. There is, nevertheless, a clear consensus as to the most important points in the light of modern knowledge.

It is obvious to the most casual visitor to zoo or circus that the apes are the most manlike of living nonhumans. This has been confirmed by intensive investigation of structure, physiology, and behavior. It is, however, impossible that any living ape represents the human ancestry or is even very closely similar to it. Obviously no animal now living can be literally ancestral to animals of any other living species. Moreover, all the living apes have specializations that surely did not occur in our ancestry. The apes undoubtedly had a common ancestry with us, but since then they have diverged definitely from the line of our later ancestors. It is not impossible and is in fact probable that some of the primitive Miocene members of the ape family were our ancestors. The australopithecines add to the weight of evidence favoring that conclusion.

Modern man's most distinctively human characteristic is his brain. But the origin of man and of his brain was dependent on older features, reflecting earlier primate adaptations. Three of the most important of those prehuman acquisitions, in the order in which they evolved, were grasping hands, binocular, stereoscopic vision, and upright posture.

The earliest primates were arboreal, as so many still are. A fundamental adaptation to arboreal life was the apposable thumb, already present in the oldest known fossil prosimians. With it, the hand was able to grasp the branches of trees. That ancient adaptation now makes possible for us all our most complex manipulations and hence all the tools we make and use as *our* most powerful adaptive equipment.

In the earliest primates and the most primitive of those still surviving, the eyes are directed more laterally, somewhat as in a long-nosed dog. The two fields of vision overlap only slightly, and an object is ordinarily seen with only one eye at a time. In various lines of descent there was an early trend for the eyes to point more forward, so that the fields overlapped more widely. Then an object was generally seen with both eyes at once, with *binocular* ("two-eyed") vision. This, in turn, provided the basis for an actual fusion of the images from both eyes as perceived in the brain, giving *stereoscopic* ("solid [three-dimensional] viewing") vision. These developments were also probably adaptations to arboreal life, in which acute vision and depth perception have special importance. On the ground there is always something under foot, but an animal moving from branch to branch must know precisely where the next branch is. In man, more than in most other vertebrates, vision has become the dominant sense, the source of our most valuable information about the world around us. And this priceless stereoscopic vision has combined with the earlier grasping hand in *hand–eye coordination*, the basis of tool use.

Primates are basically four-footed in locomotion and posture. In the larger arboreal monkeys and in the apes there was a trend toward brachiation, swinging from tree branches by the arms instead of walking along the branches on four feet. This involved a straightening of the trunk of the body and an adaptive elongation of the arms. Our own ancestors never became extremely specialized brachiators like the modern apes (especially the gibbons), but some advance along this trend may have facilitated the upright posture when our ancestors came down to walk and live on the ground. That posture freed the hands from use in locomotion and permitted efficient specialization of fore and hind limbs—fore (now upper) limbs for manipulation only, hind (now lower) limbs for locomotion only.

Thus our ancient arboreal primate ancestors supplied the basis for becoming human. The australopithecines had grasping hands, stereoscopic vision, and erect posture. After that stage came the final great expansion of the brain—and we were fully human.

FOSSIL MEN

Just when man arose and what fossils should be classed as human are matters of definition. Not one but many "missing links" are known, and it is arbitrary which of them we choose to call "men." Some would confine the designation to populations of *Homo sapiens* just like those now living. Others insist that even the australopithecines should be called "men." In the present section we are concerned with the members of the human family more advanced than the australopithecines. All are limited to the Pleistocene, and not the earliest Pleistocene as far as is known. None of them is likely to be more than a million years old, and most of them must be under 500,000. Dating is still inexact in this range. In any event, man has a respectable antiquity from the viewpoint of human history, but he is a newcomer, a Johnny-come-lately, in comparison with most other species of animals.

Except for the australopithecines, the most primitive fossil men are represented especially by a group of skulls, jaws, and other fragments found in Java and another found near Peking in China. The most primitive Java men (see Figs. 31–17 and 31–18) have been called *Pithecanthropus* ("monkey [or ape] men"), and the Peking men *Sinanthropus* ("China men"). They are, however, so similar that they probably represent merely different demes, or at most different subspecies, of a single specific population. The present consensus is that the species was not generically distinct from modern man, and the name *Homo erectus* is applied to it. Recently fossil remains of the same general type have been found in northern and central Africa, and members of the species probably occurred throughout most of the Old World. Apparently none of them ever discovered America.

Skulls of this archaic group retained apelike characters that are reduced or lost in all living men. The brain case was small. Brain size varied greatly, from about 775 cubic centimeters to about 1300. The average was about 1000, which is neatly intermediate between living apes and men.[11] There were heavy ridges above the eyes. The jaws and teeth protruded in front, and the chin retreated as in apes.

NEANDERTAL MEN

Later than the group just discussed there lived another large population, distinctly more advanced than *Homo erectus* but still more primitive than modern man (see Figs. 31–17 and 31–18). The scattered finds at many places in Europe, Asia, and Africa show individual and probably racial differences, but they are enough alike for all to be considered as members of one variable population, the *Neandertals*[12] or Neandertaloids (that is, Neandertal-like).

The Neandertals were short, stocky, powerful people with large, heavy-boned heads. Brow ridges were present and the chin was retreating, but neither character was as extreme as in the older Java and Peking men. The forehead was retreating and the brain low, but surprisingly enough the total size of the brain was about as great as in modern man. We do not know how we would stack up with Neandertals in an intelligence test, but we do know that their intelligence was considerable in amount and human in quality. They constructed a large variety of beautifully made tools and were successful hunters. They sometimes buried their dead and put offerings or sacrifices in the graves, which implies that they had a religion and rituals.

EARLY HOMO SAPIENS

Several fossil skulls have been discovered that are somewhat older than the typical Neandertals but that show greater resemblance to modern man. Among these are the Swanscombe (England) skull, unfortunately only a few fragments, and the more nearly complete

[11] The average cranial capacity is about 500 cubic centimeters for recent gorillas and about 1350 cubic centimeters for modern man. Both vary considerably.
[12] From the Neander Valley ("Tal" in German) near Düsseldorf, Germany, where the first described remains were found. An older spelling, "Neanderthal," is still in common use, but it is antiquated and encourages mispronunciation.

Steinheim and Ehringsdorf (Germany) skulls.[13] They are not fully modern in appearance and do have some primitive and Neandertaloid traits, but they are more like *Homo sapiens* than are the most fully developed later Neandertals. They suggest that the history of man in the Pleistocene was not a simple line that could be symbolized by Java man → Neandertal man → modern man.

There are two main possibilities. One is that Pleistocene man constituted a single species but was split up into highly varying demes, some more *Homo sapiens*-like, even at an early date, and others retaining or even accentuating more archaic characters. Selection within the species eventually eliminated archaic variants. This idea is supported by the fact that some fossil men do seem to represent contemporaneous intermediates between the Neandertaloid and the more extreme modern types. The other main possibility is that the Neandertals were a separate branch of mankind, a species that became extinct and was replaced in competition with forerunners of *Homo sapiens* The contrast between these two theories is not absolutely clear-cut.

True *Homo sapiens*, indistinguishable from modern man, is first known in the last glacial stage of the Pleistocene. Cro-Magnon men, first found in France, lived at least 20,000 and probably not over 50,000 years ago (see Figs. 31–17 and 31–18). All discoveries of younger fossil men are also true *Homo sapiens*, with no more variation or regional differentiation than still occurs.

In spite of inevitable doubts as to certain details, the increasing modernization of human populations during the Pleistocene is a fact, and it bears eloquent witness to the rise of *Homo sapiens* by natural evolutionary processes.

EARLY MAN IN AMERICA

No men truly primitive in a biological sense ever reached the Americas as far as is yet known. America was peopled by repeated invasions from Asia, principally although perhaps not exclusively by way of Alaska. The oldest surely dated traces of man in America are only about 10,000 years old, although

[13] All these discoveries are named for the places where they were found.

some finds may be twice that age or more. In any case, it seems fairly certain that man did not reach America until the closing stage of the Pleistocene. By then *Homo sapiens*, with various racial and lesser distinctions, was universal in the Old World. Only *Homo sapiens*, mainly or entirely mongoloid in origin, ever reached the New World before the European discovery.

And What of the Future?

This brings the biological aspects of the history of life up to date. *Homo sapiens* has evolved and has expanded to nearly all the lands of the earth. He dominates the land environments as no other species or larger group of organisms ever has. He is, so far, the culmination of the whole incredibly long and complex evolutionary process.

It is natural to sit back at this point and to speculate on the future history of life. The simplest assumption would be that past trends will continue indefinitely. It is, however, one of the lessons of history that trends do not continue indefinitely. They eventually change direction or stop. There is reason to believe that the trend of physical evolution by which man arose has now stopped. Under present conditions man's future biological evolution is more likely to be degenerative than progressive. But man himself is a new factor in the history of life and one that seems just now to be highly unpredictable.

The being and the becoming of the universe, its rules and its history, will always limit and condition what can occur. Man, for the first time ever, has conscious knowledge of many of the rules and much of the history. He can use that knowledge to modify and guide his own destiny and the destinies of other organisms. How he will in fact use that awesome power is hidden in the darkness of the future. Perhaps the only safe prediction is that you and your descendants, whether you want to or not, whether indeed you know it or not, will have the most decisive influence on the future history of life.

CHAPTER SUMMARY

The Cenozoic, Age of Mammals: 70 million years duration, with man's appearance toward the end; the vicissitudes and complexity of life's history as revealed by fossils.

Modernization of aquatic environments—a process of replacement of lesser groups (classes, orders, families) within the same ancient phyla:
Aquatic plants: very little change; the Jurassic appearance of diatoms the only major event.
Invertebrates: the extinction of ammonites by the early Cenozoic the outstanding event.
Fishes: by the end of the Cretaceous, the full emergence of teleosts as the dominant fish group.
Other aquatic vertebrates: the extinction of archaic amphibians (Triassic); the rise of aquatic reptiles in the Mesozoic and of aquatic mammals in the Cenozoic.

Modernization of land environments:
Plants: the disappearance of the predominantly lycopsid-sphenopsid forests during the Permo-Triassic; forests of the Triassic-Jurassic mostly ferns and gymnosperms (cycadeoids, cycads, ginkgos, and conifers); by the late Cretaceous forests of modern aspect—predominantly angiosperms.
Insects: their modernization during the Mesozoic, essentially complete by the late Cretaceous.
Reptiles: the extinction of dinosaurs, leaving rhynchocephalians, lizards, snakes, and tortoises as the only surviving reptiles.
Birds: their poor fossil record; their Jurassic origin; their modernization nearly complete by the end of the Cretaceous; adaptive radiation in the Cenozoic, including the origin of large flightless forms.

History of the mammals:
Mesozoic mammals: the Triassic-Jurassic origin of mammals; their slow early evolution; their small size and scarcity up to the end of the Cretaceous; the three adaptive types of Jurassic mammals; the effect of reptilian competition on the history of Mesozoic mammals.
Beginning of the Age of Mammals: the geological factors responsible for the poor early Cenozoic fossil record; faunal changes from the Cretaceous to the Paleocene; the archaic character of early Paleocene mammals; the diversification of mammals by the late Paleocene; their adaptive radiation, made possible by the evacuation of niches by reptilian extinctions.
Modernization of mammals: their post-Paleocene history; the modernization of South American faunas delayed to the Pliocene; on the World Continent most of the modernization occurring in the Eocene and Oligocene; extinct Eocene and Oligocene groups—uintatheres, titanotheres, oreodonts, sabertooths; the histories of elephants and horses; the complexity of the eohippus–*Equus* lineage.

Forerunners of man: the fossil record of primates; the effect of their arboreal way of life on the quality of the record.
Early prosimians: lemurs; their early distribution and more recent radiation in Madagascar; insectivore-primate relationships; primate characteristics.
Monkeys: Old World and New World types.
Apes: anthropoids; the four living types; their large size and predominantly arboreal habit; their restrictions to warm climates and other characteristics; their relationship to Old World monkeys, the Miocene expansion; *Proconsul*.
Australopithecines, a late group of fossil anthropoids from Africa; their manlike features; their position in the human family.

Man's place in nature:
Human origins: man as a primate; evolution from lower forms; the probability of ancestors among Miocene apes; distinctive biological attributes of man— brain, grasping hand, binocular-stereoscopic vision, upright posture; the early evolution of these characters as adaptations to aboreal life.
Fossil men: most fossil hominids less than a million years old; *Homo erectus* group and its skull characters; Nean-

dertal men—wide distribution, size, skull characters, tools, and other cultural relicts.

Early *Homo sapiens:* the Swanscombe, Steinheim, and Ehringsdorf skulls; the relation of the Neandertals to modern men; the Cro-Magnon men, the first completely modern type of about 20,000 to 50,000 years ago.

Early man in America: the mongoloid immigration from Asia; the oldest remains at least 10,000 years old but all fully modern.

The future: man's current dominant biological role; the human evolutionary future: evolution, now including social and cultural evolution, potentially in human power to control.

Part 10

The History of Biology

On a time scale that measures in the billions of years man's time on earth is as nothing. On a 24-hour scale for the whole history of life we found *Homo sapiens* emerging just about 2 seconds ago. The whole of recorded human history fills only the final quarter-second before the midnight that is now.

Capacity for thought and analysis is the key adaptation that accounts for the incredible rate of human progress in this last quarter-second of life's history. Man uses his mind to greater biological effect than any organism ever used its bodily adaptation; his unparalleled mastery over the physical environment is due to the technology and social systems produced by the evolution of knowledge. The advent of thinking man was, in fact, a great turning point in life's history: it set off an entirely new trend and tempo in evolutionary progress; it initiated *cultural* evolution. The *biological* evolution of man is, to be sure, still proceeding, and perhaps as rapidly as ever; but the rate of change in the human gene pool is dwarfed by the pace at which our social and intellectual adaptation—our culture—is evolving.

Our task as biologists in this book is now essentially complete because the study of cultural evolution is the traditional domain of the human historian. Knowledge of its historical growth is, however, essential to the proper understanding of a science, and accordingly we have devoted Part 10 to an outline of the history of biology. This history, like that of all culture, has been a complex evolution subject to many influences. Its full analysis would show how the progress of biological insight has interacted with other aspects of cultural evolution; how it has usually depended on, and sometimes led, the growth of the other sciences; how the invention of tools and techniques has played a role; how social forces in general have shaped its course, and how society in turn has been profoundly affected by that ultimate self-consciousness of life which biology represents. The culture of man, no less than his body, is an integrated whole; it has evolved as a whole and should be studied as a whole.

Part 10 is too brief to allow a full study of the history of biology; it is a summary of some major trends, events, and persons that are central to the fuller story. In making this summary, we have a secondary purpose: to specify and characterize the numerous sciences into which the broad field of biology has now become split.

However one tells the story—in full or, as here, in brief—two features always stand out: one is the astonishing youth of the science of life; and the other is the towering influence of Charles Darwin, whose portrait introduces Part 10. If we must count the whole of recorded human history as a quarter of a second, how shall we count the century since publication (1859) of *The Origin of Species?* Darwin's book marks the entry of biology into the mainstream of scientific thought, which until then had been largely shaped by the growth of physics. Darwin's concepts of evolution and natural selection are still the general principles that bring order into the diverse facts of life.

32

The Long Search

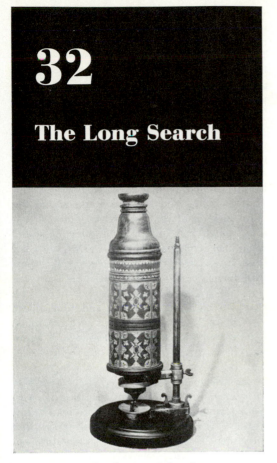

The invention and utilization of the microscope marked a great turning point in the history of biology. This instrument belonged to Robert Hooke, one of the leading figures of early microscopy.

Man has been trying to learn the secrets of life for centuries and millennia. The methods and knowledge that constitute biology today are an accumulation that began among our prehistoric ancestors. The search has continued with sometimes wavering but generally accelerating pace and success up to the present minute. It still has far to go, and it continues unabated. Already it is one of the great achievements of mankind. The history of the search is something we can all be proud of, unlike the shabby histories of dynasties and conquests. Even though summarized in irreducibly small compass, that history is also one of the best ways to review what the science of biology is—its scope, aims, and subdivisions. We have chosen the historical approach to give a final over-all view of the science that you have now studied.

One point is likely to be obscured in so brief a review of the history of a science and must be strongly emphasized now. The acquisition of biological knowledge has been the work of tens of thousands of devoted students. Some, of course, have been more brilliant or (what is not necessarily at all the same thing) have turned out to be historically more important than others. We shall name a few. The men we name were all great, although the list cannot begin to include all of the great nor can it be claimed to designate the very greatest of them. The men named typify an epoch, mark a culmination of achievement, or were involved in a turning point in biological research. Even in these respects they did not stand alone, but were surrounded and upheld by innumerable others who cannot be named here.

No scientist ever made a discovery of his own. Every one had many predecessors and contemporaries whose work was just as essential to his discovery as anything that he did. The idea of evolution and also that of natural selection were already old when Darwin wrote *The Origin of Species* (1859). That does not belittle Darwin's position as the greatest of all biologists. It does put the matter in better perspective. Science is a cumulative, social product. It includes but does not wholly consist of the work of individuals who can be labeled "great."

We will not name any living biologists in

this historical review. There are now more biologists than ever before. Many of them are certainly as brilliant as any in the past, and some of them will surely be adjudged great by future historians. So many are brilliant that to select a few in this brief space would be unfair and misleading. Selection of the greatest must be left to posterity.

Natural History

Biology began as a branch of natural history. Natural history began as an attempt to observe and describe the physical universe. The expression "natural history" is still used in almost that sense as a broad and popular term to refer to all the natural sciences collectively, as in the names of our many museums of natural history. Research and study in the natural sciences are, however, now carried on under the names of the innumerable special sciences into which the old, almost all-embracing subject of natural history has inevitably been split.

In the ancient world, natural history, and particularly that part of it dealing with animals, reached a culmination among the Greeks and in the person of Aristotle (384–322 B.C.). He compiled descriptions, excellent on the whole, of most of the phenomena of animal life known in his day and culture. Neither he nor any of the other ancients advanced deeply and correctly into an *understanding* of the principles beneath the observed phenomena. With a few lucky exceptions they were unable to formulate the right questions because they held philosophies that were fundamentally not scientific, in the modern definition of the word. Beyond the level of simple observation, and sometimes even at that level, they tended to ask and to answer questions about nature in terms of their non-scientific philosophies rather than in terms of nature itself.

Many centuries passed before it was usual for the students of nature to approach the subject in an actually naturalistic way, to rely on nature itself to suggest the questions and to supply the answers. Among the landmarks in this change of attitude were the discoveries that the earth is not the center of the universe

(Copernicus, 1473–1543) and that hitherto mysterious activities in this universe can be reduced to mathematical law (Isaac Newton, 1642–1727). The gradual and often painful liberation of inquiry from philosophical, authoritarian, and dogmatic preconceptions gave rise to science in the full, modern sense.

CLASSIFICATION AND SYSTEMATICS

The period in which scientific attitude and method were coming into form was also the great period of navigations and explorations. Europeans discovered the rest of the world. Strange plants and animals were among the most striking of the discoveries. Natural history, which had degenerated after Aristotle, sometimes into sheer fantasy, was revived on a broader and more objective basis. Collections were formed and compendia written. Georgius Agricola (1490–1555), although more famous for his work on mining and mineralogy, also made systematic biological collections. Konrad von Gesner (1516–1565) and Leonhard Fuchs (1501–1566), among the first who may be called biologists in a nearly modern sense, also collected. They used the new arts of printing from type and illustrating with woodcuts to publish huge works on natural history. Andrea Cesalpino (1519–1603) first attempted a really serious classification, in the modern spirit, of the whole plant kingdom.

The more orderly approach to natural history and the flood of new discoveries increased the desire and the need to classify organisms. Efforts were made to establish classifications on a natural system, rather than on superficial resemblances or fancied properties. John Ray (1627–1705), an English botanist, classified both plants and animals and was one of the greatest forces in production of a truly systematic science of systematics. The preeminent place of the Swedish botanist Linnaeus (1707–1778) has been mentioned elsewhere (Chapter 18). He formalized the consistent use of the hierarchy and system of nomenclature still in use. Modern classification is conveniently dated from him, although it has changed much since his day, and in some ways the earlier Ray seems more modern in retrospect.

The climax of description and classification for their own sake was reached by Linnaeus, his contemporaries, and his early successors. Buffon (1707–1788) produced a voluminous natural history so complete and so well written that it is still a household work in France. Georges Cuvier (1769–1832) also advanced the systematics of animals and was among the first to insert extinct animals into the system.

The pre-Darwinian systematists, those we have named and many others, were largely concerned with developing a "natural classification." From our vantage point, with all their work and all that done since then available, it appears that they never achieved a *natural* definition of a "natural classification." Their criteria as to what was natural in classification was merely intuitive, occasionally acute and occasionally absurd, or it was based on philosophical considerations with no evident connection with the material facts of nature. A workable connection appeared only when the truth of evolution was generally recognized. Phylogeny is a material fact of nature (even when, like many other such facts, it is incompletely known to us),[1] and it does provide an evolutionary basis for truly natural classification.

Evolutionary classification is mainly post-Darwinian. Its great expansion, even in Darwin's lifetime, is exemplified in the overoptimistic work of the German zoologist Haeckel (1834–1919). He produced elaborate phylogenetic trees and corresponding classifications. They have not stood up well in detail, but they pointed to one of the directions in which the science of systematics was to develop.

To this day the identification and classification of organisms, one of the earliest activities in the field of natural history, is a main preoccupation of most botanists and zoologists. After the first excitement of applying evolutionary concepts to classification had worn off, it seemed for a time to be settling into dull routine. Recently, however, this science of *systematics* has been rejuvenated, and it is now unusually active and interesting. The revival is due largely to two related factors. First is the change in point of view from classifying individuals to classifying populations. Second is a broader concept of the basis and significance of systematics. The science is no longer merely classification in the sense of pigeonholing and labeling. It is the study of the diversity of organisms and of their relationships among themselves. As such it ramifies into most other branches of biology, notably biogeography, ecology, and genetics.

The science of systematics has become so vast and so complex that no one now attempts to do original research in more than a small part of the whole field. The primary division is into *botany* and *zoology*, with *protistology* (study of protists) often lately added as a third. Even a systematic zoologist, for instance, does not work on the whole animal kingdom. Among many other specialties, he may be an *entomologist* (student of insects), a *malacologist* (student of mollusks), an *ichthyologist* (student of fishes), a *herpetologist* (student of amphibians and reptiles), an *ornithologist* (student of birds), or a *mammalogist* (student of mammals).

LIFE SCIENCES AND THE SCOPE OF BIOLOGY

The fractionation of systematics into many different specialties illustrates a trend in all the sciences, necessary as the bulk of knowledge has increased and the skills required have become so different. What our forebears were content to call "natural history" has become a score of major sciences comprising hundreds of specialties. Among the sciences having to do with life or, briefly, the *life sciences*, systematics probably most nearly retains the approach and interests of the earlier natural historians. It is, of course, part of the broader science or superscience of *biology*.

Biology, literally "the study of life," should logically include any study of living (or formerly living) things. In usual practice it does not. Several sciences just as much concerned with living things are not generally considered parts of biology and have not been so considered in this book. They have become so complex in themselves and have developed such different materials, methods, and aims that they are most conveniently, if not logically, excluded from the general study of life.

[1] The reconstructed phylogenies in books are theoretical and may possibly be proved wrong by later discoveries, but phylogeny as it actually occurred by descent from parent to offspring was real and factual.

Among the more important of these life sciences apart from biology are *agriculture*, the study of cultivated plants and domesticated animals; *medicine*, the study of disturbances of functions and their correction; *anthropology*, the study of man; *psychology*, the study of behavior; and *sociology*, the study of human communities.

It is of course impossible to draw a sharp line between biology and the other life sciences. General biology is a necessary basis for the understanding not only of the various sciences, such as systematics, included in biology itself, but also of all the other life sciences. There is overlap all along the line. The student of animal behavior is simultaneously biologist and psychologist. Bacteriology, physiology, and anatomy, among other subjects, belong equally to biology and to medicine. Subdivision is necessary, but it must be remembered that any subdivision of the whole subject "natural history" or "science" or, grandly, "knowledge" is arbitrary.

The Organism in Its Environment

The great explorations of the fifteenth, sixteenth, and seventeenth centuries were seldom carried out by scientists. The explorers sought booty, empire, trade, and also, to be sure, geographic information but with aims seldom directly scientific. They incidentally revealed new worlds for science, as well, and we have seen that they stimulated the study of natural history. Later on, expeditions were sent out with the primary purpose of obtaining scientific data. Some scientific exploration was done in the eighteenth century or even earlier, and innumerable expeditions are now sent out as a matter of routine by museums and other scientific institutions. The golden age of scientific exploration was, however, the nineteenth century.

Many of the great naturalists of the nineteenth century served an apprenticeship in science on long voyages, and many based their major contributions on that experience. Darwin (1809–1882) went around the world on the *Beagle* as a young man and made observations that were to lead to his theory of evolution. T. H. Huxley (1825–1895), champion of Darwin and a great biologist in his own right, cruised on the *Rattlesnake*. Many oceanographic expeditions studied marine life, and brought back specimens for the work of innumerable specialists. The British *Challenger* expedition (1872–1876) is among the most famous.

As a result of expeditions and of individual travels by naturalists, organisms were being studied more intensively and over wider areas in the field, out where they live all over the earth. Attention was turning increasingly not only to the classification of animals and their anatomy and physiology but also to their distribution and to their lives in their natural environments. Thus from the broad stream of natural history two more special sciences came to be distinguished: *biogeography* and *ecology*.

BIOGEOGRAPHY

Some ideas about the distribution of plants and animals date back to antiquity, indeed to prehistory. The compendious natural histories of the seventeenth and later centuries customarily designated the habitat of each organism described. Such designation has always been an accepted and necessary part of systematics. The clear development of a scientific biogeography, involving enlightening generalization and explanatory theory as well as flat descriptive statement, was nevertheless a rather late development in biology.

The earlier status of attempts at scientific biogeography is amusingly illustrated by the polemic between the eminent French naturalist Buffon and the eminent American president and amateur naturalist Jefferson (1743–1826). Buffon stated that American animals are smaller than their European relatives because the climate is wetter. Jefferson reasonably pointed out, with evidence, first, that American animals are not smaller and, second, that the American climate is not wetter. There that particular hypothesis rests up to now. Alexander von Humboldt (1769–1859) spent 5 years in South America in his thirties and then returned to Germany to make fundamental contributions not only to biology but also to geology and meteorology. When he died he was well along with a work that was, quite simply, to include everything known about the universe, scientific, historical, and artistic! Before that, he produced a study of the geography of plants that was the modern starting point for both biogeography and ecology.

The English naturalist Alfred Russell Wallace (1823–1913) was another of the great field biologists, with a year in Brazil and several in the East Indies. He is most famous because he worked out the theory of natural selection independently of Darwin and, thanks to Darwin's integrity, published it at the same time. His greatest contribution, however, was a book on the geographical distribution of animals (1876) in which he firmly established biogeography as a science. In fact, very little of fundamental importance has been since added to his main subject, the present regional distribution of land birds and mammals.

Wallace saw and stressed the fact that regional distributions must be explained on a historical basis. He tried to develop historical principles in the light of knowledge of his day, but the knowledge was insufficient for more than a bare start. Historical biogeography must necessarily rest on a synthesis of information from fossils and from living organisms. The most fundamental contribution in this respect was probably that by the American paleontologist W. D. Matthew (1871–1930). In 1915 he published an unpretentious paper on the subject which has stimulated and oriented most of the progress made since then.

Today many biologists, especially ecologists, systematists, and paleontologists, are actively working on problems of biogeography, but there are few biogeographers as such. The division of the subject into ecological and historical aspects has been exemplified in this book (Chapters 27 and 28). Ecological marine biogeography is just now coming into its own as one of the most promising fields for research.

ECOLOGY

Humboldt's pioneering work was, in modern terms, as much ecological as biogeographical. Much earlier natural history presaged the science of ecology, and so did much of the biogeographic and marine biological work that grew out of it in the nineteenth century. It was not, however, until the last quarter of that century that more explicitly ecological concepts came to the fore and that a distinct science of ecology began to develop. The first general works specifically devoted to this subject were by the Danish biologist J. E. B. Warming (1841–1924) in 1895 and the Swiss A. F. W. Schimper (1856–1901) in 1898. Both were botanists writing on plant ecology.

Main themes of animal ecology also emerged from the earlier studies of less specialized naturalists. Such themes are prominent in Darwin's works. Animal ecology, as such, nevertheless was slow to develop, slower than plant ecology. That side of the science is mainly a product of the twentieth century.

At present ecology is among the most active and most fascinating of the biological sciences. It seems that a really general ecology is still in the formative stage, and here is one of the enticing fronts of advancing knowledge. Most field studies of living organisms now have an ecological orientation. Basic attitudes and principles of ecology have also permeated most of the biological sciences, and some other sciences as well. Considerable sections of this book have been devoted to exposition of those principles, and we now need only recall a few of the most essential of them: the universality and nature of adaptation; the concept of populations as dynamic groups (remember, too, what a revolution this concept has worked in systematics and genetics as well as in ecology proper); the principle that all organisms are parts of multispecific communities, which are organized and interacting units.

Related to these ecological principles is the broader biological principle of levels of organization: molecular, cellular, individual, population (or intraspecific society), and community (or interspecific aggregation), each level including and to some extent analogous with all lower levels but each with phenomena and principles peculiar to it. Ecology also illustrates the broad contacts of biology with other sciences, in this case particularly with sociology.

The Organism
in the Laboratory

What may still be called the naturalist's approach to biology is particularly evident in the biological sciences whose rise has now been briefly scanned: systematics, biogeography, and ecology. There were other approaches, even in the earliest formative times for science. One of these was what we would now call the laboratory approach. Organisms were dissected or experiments were made with them. The distinction is far from sharp; all biological sciences are so interlocked and overlapping that none can be absolutely distinguished from the others. Experiments are an integral part of ecology, for instance, and animal behavior is frequently studied outside the laboratory. Nevertheless, in origin and in dominant techniques ecology has been a field science, and animal behavior a laboratory science.

HUMAN AND
COMPARATIVE ANATOMY

The study of human anatomy is and has always been more an adjunct to medicine than a part of biology in the usual sense, but of course the connection between medicine and biology is intimate, and the separation more or less arbitrary. In ancient and medieval times dissection of human cadavers was generally considered impious and was often illegal. Such anatomical knowledge as existed came mostly from dissection of monkeys, pigs, and a few other animals. Yet that dissection was done as an adjunct to human medicine. Doctors who would have been shocked by any suggestion of evolutionary relationships between men and monkeys nevertheless treated their patients as if they were monkeys! In the fourteenth century dissection of humans was legally authorized in Bologna, then one of the few world centers of learning. Slowly through the next two or three centuries dissection became a recognized part of the medical curriculum.[2]

The modern study of human anatomy dates

[2] Even so, dissection was rarely performed either by students or professors. A flunky did the dirty work while, from his podium, the professor read to the students what Galen had written about the various organs more than a thousand years earlier.

from 1543, as nearly as a fixed date can be assigned. In that year Andreas Vesalius (1514–1564) published a book based on his own dissections and describing objectively what he had seen in the human body. The book was not popular at the time, because most doctors still preferred the confused and often absurd anatomical notions inherited from the Romans (especially Galen, about 130–200 A.D.). Nevertheless, Vesalius' work did slowly persuade anatomists that the way to learn about the human body was to look at it, rather than to read Galen—just as the naturalists were beginning to look at nature instead of reading Aristotle.

More definitely biological is the study of comparative anatomy. The dissection of a pig in order to find out what the human body is like was, in a sense, comparative anatomy—except that comparison was rarely made. By the late sixteenth century followers of Vesalius were extending their observations to various animals and actually making comparisons. As early as 1555 the French zoologist Pierre Belon (1517–1564) [3] pointed out correspondences—what we now call "homologies"—between the bones of a bird and of a man.

Human anatomy as usually studied has always been mainly descriptive, and so has been much comparative anatomy. To attain stature as a true science, interpretive principle and explanatory theory must be added. There was already a hint of these in Belon, but a truly theoretical comparative anatomy was first consistently developed by Goethe (1749–1832), Lorenz Oken (1779–1851), and others in their period. Its theoretical basis was typological, embodying the principle of the archetype (Chapter 18). Most modern biologists reject that principle, and yet it laid a firm basis for the evolutionary comparative anatomy that was to follow. It was the English zoologist Richard Owen (1804–1892) who anachronistically carried typological (but not evolutionary) comparative anatomy to its highest point, although he lived well into the Darwinian period.

[3] You have perhaps noticed that most great biologists had longer lives than was usual in their times. Belon's 47 years of life were not short for the sixteenth century, but they would have been longer if he had not been murdered. Most biologists, even the explorers, have died in bed.

Since 1859 (the date of publication of *The Origin of Species*), the theoretical basis of comparative anatomy has become evolutionary. Resemblances and differences are traced in terms of ancestry and changing adaptation. The fundamental principle of homology, intuitively recognized by Belon and other pre-evolutionary biologists, has become evolutionary. No one landmark stands out clearly in this general change, but perhaps the work of Robert Wiedersheim (1848–1923) is as good an example as any. His great handbook of comparative vertebrate anatomy is a painstaking compilation of facts presented in the new spirit.

Human anatomy and comparative anatomy today are generally taught as descriptive subjects technologically prerequisite for a medical career. As such they have little scientific biological interest. Interpretive, evolutionary comparative anatomy has passed largely into the hands of the paleontologists, among whom it continues to be a lively and progressive subject for research. Another sort of interpretive anatomy is functional, and that too is lively and is undergoing a renaissance. It connects anatomy with physiology, from which, indeed, anatomy has never been and could not be wholly divorced.

PHYSIOLOGY

Up until the seventeenth century, physiology, philosophy, and theology assigned the greatest importance to the circulatory system, without any idea that it is a circulatory system. The Greeks thought that the arteries contained air. Galen corrected that but taught that blood ebbs and flows in the veins. Centuries later it was still believed that the heart, liver, and blood were the seats and nutrients of the soul and of the various "spirits" of the body. Among the theologians who made mystical studies of the circulatory system was Michael Servetus (1511–1553), whose views on this and other subjects so annoyed the Calvinists that he was burned at the stake.

It was William Harvey (1578–1657) who demonstrated in a small work published in 1628 that the blood does circulate and that its movement is purely mechanical, produced by the heart, which is simply a pump. The discovery was so revolutionary in bringing the very seat of the soul into the realm of material science that the origin of scientific biology is commonly dated from 1628. Yet Harvey never freed himself from other mistakes of the Aristotelian system and never fully envisioned a purely scientific approach to biology. That the old concepts died hard is illustrated by the fact that Emanuel Swedenborg (1688–1772) made really important advances in physiology but did so in terms of a mystical theology so that he is remembered now as a theologian rather than a physiologist.

In the meantime, workers who were primarily anatomists and physicians made strides toward a purely naturalistic physiology. It can, however, hardly be said that a physiology entirely scientific, as we now consider science, was achieved until the nineteenth century. Here the name of Claude Bernard (1813–1878) stands out. In a series of brilliantly planned and executed experiments, he demonstrated most of the basic features of animal metabolism and showed how its processes tend to maintain equilibrium in a constantly changing system.

Since Bernard fundamental progress has been made in physiology, for the most part along special lines and often in relationship to other branches of the life sciences. A *general* or *cellular physiology* was developed, a study not of organ systems in the complex vertebrate body but of physical and chemical conditions and events within single cells. This is, indeed, in the last analysis the basis of functioning of the organ systems as well. It led in turn to the sciences of *biophysics* and *biochemistry*, so characteristic of present-day biology and to be mentioned again later in this chapter. Research in *medical physiology*, including such topics as the action of hormones and of drugs, continues to be extremely active and also to follow, in part, biochemical lines. *Sensory and nerve physiology* is basic to the study of behavior and through the latter to psychology. This general field of biological study deserves brief separate notice.

Nerve and sensory physiology

The early physiological anatomists were frequently concerned with the nervous system. That was the main physiological preoccupation of Swedenborg, who localized the activities of the "soul" (consciousness and related phenomena) in the cerebral cortex and believed that various parts of the cortex were connected by nerves to various other parts of the body. A contemporaneous universal genius, Albrecht von Haller (1708–1777), studied the "irritability" and "sensibility" of tissues and organs and connected these phenomena with the nervous system. The works of Swedenborg and Haller now seem primitive in many respects, but between them they pointed the way to most of the fundamentals of nerve and sensory physiology.

It is in this sphere that the long and still not entirely conclusive philosophical struggle between the vitalists and the materialists in biology comes especially to the fore. The vitalists maintain that life is or involves something nonmaterial, forever outside possible observation or scientific explanation. The materialists believe that the phenomena peculiar to living things arise from the nature and complexity of their organization and do not involve either materials or processes absent in the nonliving world. One of the important lessons of the history of science is that the materialistic approach is a *necessary* part of scientific method. Most vitalists accept that as an essential restriction in research, but they insist that when all the resources of science have been exhausted there still will remain a vital element in life beyond the reach of scientific method.

It is impossible, at least at present, to *prove* either the vitalist or the materialist position. It is, nevertheless, a fact that a main trend in the history of biology has been that more and more phenomena earlier ascribed to the mysterious "vital force" have been explained by purely material processes. The crucial and as yet not completely conclusive test comes in dealing with the incomparably complex phenomena of the nervous system, with its mysteries of sensation, perception, consciousness, and mind.

The point is well illustrated by two of the founders of modern nerve physiology. J. P. Müller (1801–1858) was one of the last investigators who successfully ranged over almost the whole field of biology. He studied, among many things, the actions of sensory and motor nerves in animals. In his own work

and that of his many famous students he laid a firm foundation for modern nerve physiology, even though his own philosophical conclusions were vitalistic. In 1840 Müller assigned to one of his students, Emil Du Bois-Reymond (1818–1896), the study of electrical phenomena in nerves and muscles. Du Bois-Reymond soon saw that these material phenomena provide a possible explanation, at least, for the supposedly vitalistic operation of the nervous system. He refuted the whole vitalistic philosophy in a way thoroughly convincing to most, but not all, later biologists.

Most subsequent work on nerve physiology, still a highly active field, has followed and expanded the sort of experimental approach taken by Müller and Du Bois-Reymond. Even more important, they were pioneers in the application of the physical to the biological sciences, a movement that is now at a most active and productive phase.[4]

Animal behavior

Approaching a related field from a different direction, the work of nineteenth-century and earlier naturalists abounds in descriptions of animal behavior. With few exceptions the observations were anecdotal, and their interpretation was anthropomorphic (Chapter 14). They smacked as much of folklore as of science. However, there were exceptions, as, for example, the work of Jean Henri Fabre (1823–1915). He was not a rigorous scientist and is sometimes dismissed as a mere popularizer (his name does not even appear in some voluminous histories of biology). Nevertheless, he made extensive firsthand observa-

[4] The importance of Müller and more especially of Du Bois-Reymond in the history of biology has generally been underestimated. Du Bois-Reymond is barely mentioned in some books on the subject, probably because his original research was confined to a single and narrow field. Nevertheless, his methods and views mark a turning point in biology. It is worth noting that he did not subscribe to the extreme mechanism that became popular among nineteenth-century scientists, the view that all the phenomena of the universe are in principle mechanical and completely predictable. He held that, although the phenomena of life are nonvitalistic and can be reduced to those of the physical sciences, the basic concepts of the physical sciences are themselves abstractions not open to ultimate explanation. That reservation makes his work seem all the more modern.

tions on insect behavior, reported his observations accurately, and strenuously rejected anthropomorphic interpretations.[5]

We have noted (Chapter 14) the influence of Lloyd Morgan (1852–1936) and Jacques Loeb (1859–1924) on the rise of animal behavior as a distinct science. Its growth was marked by the introduction of controlled experimental methods and by the correlation of behavior with nerve physiology. Here the work of Ivan Pavlov (1849–1936) is also of major historical importance. Its scope and applicability are decidedly limited in the field of animal behavior as a whole, but his work marked and accelerated the rise of careful experimentation and physiological interpretation.

The ramifications of the science of animal behavior into other life sciences are many and crucial. Relationships with anatomy and physiology are obviously particularly intimate. The ecology of animals cannot be disentangled from their behavior. Comparative psychology is simply animal behavior as studied by psychologists. Experimental psychology is largely concerned with the physiological basis of animal behavior, especially of the species *Homo sapiens* and *Rattus rattus*. All such studies are now very active, and a more meaningful, evolutionary science of comparative behavior is beginning to emerge.

Within the Organism

SCIENCE AND TECHNIQUE

The history of a science is greatly influenced by the development of new instruments and techniques. Major advances in science often depend on technological improvements: the invention of the telescope in astronomy, the invention of air pumps in the physics of an earlier day or of particle accelerators in our own time, the manufacture of delicate scales in chemistry, and many others. This aspect of history is important, but it can be overemphasized. Scientific advances do not necessarily demand new apparatus, nor does technological progress inevitably advance science. It is noteworthy that recognition of evolution, the most revolutionary single event in the history of biology, involved no instrumentation and no techniques that had not been used for centuries. It is a grave and common mistake to confuse gadgetry with science.

The biological sciences we have so far considered started out as naked-eye sciences and are so to a large extent even today. It is true that elaborate apparatus is now sometimes used in them, in physiological and behavioral experiments, for example. Much of their content is, however, accessible to anyone with his unaided senses, his hands, and such simple instruments as have been available since early civilizations: a few knives, dishes, and the like. We now turn to several biological sciences that have more decisively depended on special apparatus and advanced laboratory techniques. We can more definitely say that progress in them would not have gotten beyond a primitive level if it had not been accompanied by invention.

In the first rank among inventions important in the history of biology are instruments for enlarging visible things: simple lenses, then compound microscopes of increasing complexity in structure and operation, and finally the electron microscope. They have other uses as well, but their greatest impact in biology has come from seeing more deeply into the fine structure within the organism. Then there has been increasing application to biology of apparatus, methods, and concepts primarily developed in the sciences of physics and chemistry. This is a motley group, from the simplest application of pressure or of a chemical reagent to an organism on to the use of electrophoresis, tracer isotopes, and many other methods in modern biology. With these methods, too, one of the most important of their many results is the ability to get down farther and farther into the organism, down to the processes that go on at the molecular level.

MICROORGANISMS

In the seventeenth century simple lenses with great magnifying power (superior to the power of the early compound microscopes) began to be constructed. Biologists soon ap-

[5] In one of the fascinating passages of his *Entomological Memoirs* he gave Erasmus Darwin (Charles' grandfather) a lambasting for daring to suggest that wasps reason.

plied the lenses to everything they could think of. Among the most enthusiastic were two Dutch naturalists, Jan Swammerdam (1637–1680), who first saw red blood corpuscles and made anatomical studies of small insects, and Anton Van Leeuwenhoek (1632–1723), who first saw bacteria, other protists, and sperms. Their observations and those of other early enthusiasts were unsystematic but so wide-ranging that most of what can be seen under moderate magnifications had been described well before the nineteenth century.

The most stunning of the early discoveries under the microscope was the whole world of otherwise invisible microorganisms, mostly protists, the existence of which had hardly been suspected. It was found, too, that many visible and known organisms had unsuspected microscopic phases in their life cycles, such as spores or small eggs. These discoveries had a bearing on the then generally accepted doctrine of spontaneous generation. We have already mentioned the parts played at an early date by Redi (1627–1697) and much later by Pasteur (1822–1895) in disproving that doctrine.

Even before microorganisms had been seen it had been speculated that infection was caused by invisible "seeds of disease." With increased acquaintance with bacteria and other protists, the speculation became a conviction. Robert Koch (1843–1910) was among those who demonstrated beyond question that some, at least, of the infectious diseases are caused by "germs." Koch also invented a technique for obtaining and growing cultures of microorganisms in the laboratory. That invention, extremely simple once it was thought of, made possible the science of *bacteriology*, which is fundamental to modern medicine. More broadly, with modifications and additions Koch's technique has made possible physiological and other experiments on all kinds of living microorganisms and has been instrumental in the rise of the more strictly biological science of *microbiology*. In this one case a whole science has been born of a technique.

Just to show how broadly such inventions can ramify, the techniques of culturing microorganisms are now extensively used in popula-tion and physiological genetics and also in the commercial growing of orchids (which have microscopic seeds) for corsages.

CELLS AND TISSUES

The microscope made it possible to carry anatomy to greater depths, to study not only the organs and tissues visible to the naked eye but also their microscopic make-up. Here, too, a great surprise lay in wait, for it soon appeared that skin, muscles, and so on are not continuous structures but are made up of minute, discrete units, the cells.

Cells, at first the hollow spaces among the walls of dead plant tissues (Hooke, 1635–1703) and later living protoplasmic cells, were among the first things seen by the earliest microscopists. Tissues in the modern sense, as distinct from the organs made up various tissues, were first clearly described and named at the end of the eighteenth century by Xavier Bichat (1771–1802). It took longer to bring the two observations together and to realize that tissues are made up of cells. Bichat failed to do so because from some quirk of human nature he did not take microscopic observations seriously.[6]

The idea that all tissues develop from and are composed of cells and cell secretions was first clearly expressed by Schwann (1810–1882). Although Schwann himself did not successfully carry the idea much further, that is the basis for one of the most important of all biological principles: the fundamental similarity of developmental units and of intracellular processes in all living organisms. The modern concept of the cell as the structural unit of life was essentially reached in the 1860's by Max Schultze (1825–1874). By 1858 Virchow had realized that all cells are the offspring of other cells, and paved the way for union of the cell theory with that of evolution.

The study of tissues, frequently from the medical point of view, has become the science of *histology*. The study of cells is *cytology*. The discovery that heredity is carried and development controlled (for the most part, at least) by structures that can be made visible

[6] Bichat was one of the shortest-lived of great biologists, and a genius of his caliber might well have taken the next step if he had lived longer.

within the cells has greatly stimulated twentieth-century research in cytology. This field is sometimes now considered a separate science, *cytogenetics.*

EMBRYOLOGY

The development of some plants and animals from seeds and eggs was known to primitive man and was considered by philosophers and scientists from the Greeks on. The part of this process most easily visible to the naked eye is, however, simple growth. The far more fundamental processes of differentiation were the subject of speculation by Aristotle and many later students, but a scientific approach to them was hardly possible until the invention of the microscope.

Swammerdam and other early microscopists started embryology off on a false track and began a long controversy that seems, in retrospect, to have been futile and unnecessary. It is, however, a futile thing in itself to apply that adjective to mistakes that were inevitable in the historical setting of their own times. Such faltering has occurred in all of man's intellectual pursuits, not to mention his social or political ones. Other important examples in biology are: the doctrine of special creation; belief in spontaneous generation; and the theory of the inheritance of acquired characters.

The false track in embryology was *preformationism,* the theory that the egg or sperm contains a miniature of the adult and all its organs so that embryological development is considered as nothing but growth. The preformationists did not boggle at the idea that Adam's body contained in miniature the bodies of all the humans that ever have lived or ever will live on earth. They produced drawings of human sperm with a "homunculus," a little man, crouching inside.

There were opponents of preformationism from the start, but before the nineteenth century most of them were, in their own ways, equally wide of the mark. Modern embryology may be said to take form in the work of Von Baer (1792–1876). He discovered the microscopic mammalian egg and traced it through fertilization to final form. He showed that differentiation is gradual and occurs in the course of repeated cellular division. He

systematized the principle of embryonic cell layers (which had been seen before). He also made comparative studies, showed the value of embryology in the study of homologies and relationships among mammals, and noted that the embryos of different vertebrates resemble each other more than do the adults.

Since Von Baer an enormous amount of descriptive work on embryology has been done. Von Baer's theoretical interpretations have in large part stood up and have been put on an evolutionary basis. Haeckel's reinterpretation in terms of "recapitulation" (Chapter 8) went too far and has required modification, returning actually more nearly to Von Baer's original statements. A brilliant statement of evolutionary comparative embryology was produced by Francis M. Balfour (1851–1882) as a young man (an accident ended his life in its prime).

In the present century embryology has become almost entirely experimental and is mainly focused on seeking biochemical and biophysical explanations for organization and differentiation in the developing individual. Much progress has been made, but the connection between the genetic system of the zygote and the structure of the developed organism is still full of profound mysteries. Here are some of the most important unsolved problems of biology.

BIOCHEMISTRY AND BIOPHYSICS

The modern sciences of biochemistry and biophysics have developed from the field of physiology in general and are hardly distinguishable from some phases of physiology. The tendency now is for physiology to be restricted to studies with a more directly biological approach, in terms of organs and organisms. Biochemistry and biophysics study related phenomena less in terms of the organism than in terms of specific chemical reactions and physical changes. The physiologist may study chemical input and output in the anatomical system of a leaf or the speed, force, and continuity of contraction in a given muscle. The biochemist may study the sequence of special reactions involved in photosynthesis, wherever that occurs, and the biophysicist may concern himself with internal changes of state in muscle fibers generally.

The instrumentation of biochemistry and biophysics is particularly elaborate and is adapted, for the most part, from chemistry and physics.

A leader in the application of physical and chemical methods to physiology was J. P. Müller, who has already been mentioned. Besides his own important work, he turned into these fields a host of his students, including Schwann and Du Bois-Reymond, also previously mentioned, as well as Hermann von Helmholtz and other eminent biologists. Helmholtz (1821–1894) was a surgeon who did important work in both pure physics and pure biology and, along with Emil Du Bois-Reymond, was a pioneer in combining the two. He was ingenious in devising apparatus and he used physical methods in the study of vision and hearing.

Justus von Liebig (1803–1873) was at first a straight chemist and then turned to the chemical aspects of physiology. Through his writing and teaching he led a group of nineteenth-century biochemists who attacked, and in their major aspects solved, such basic problems as the oxygen, carbon, and nitrogen cycles in plants and animals and the chemistry of digestion. Among the numerous later biologists who ushered in the modern science of biochemistry Emil Fischer (1852–1919) may be mentioned. With new reagents and methods, he made great strides in deciphering the molecular structures of complex organic compounds, still an active field of research.

A still more specialized modern development of biochemistry is the field of *molecular biology*, which at present is particularly concerned with the structure, properties, and functions of DNA. As DNA is the (or the principal) medium by which heredity is carried, molecular biology has led to a reorientation or a new outgrowth in the field of genetics, *cell heredity*, which seeks to explicate the genetic behavior of single-celled organisms and individual cells of multicellular organisms and individual cells of multicellular organisms in biochemical or molecular terms.

Great stress on biochemistry and only slightly less on biophysics is characteristic of biology at the present time. These have recently been the most intensely cultivated and rapidly advancing of the biological sciences. Although other fields have not been neglected, some biologists feel that these may even have been unduly stressed. After all, understanding of the *organism* as such is the primary object of the biological sciences; many of the answers are within the scope of biochemistry or biophysics, but many are not. At present a return to a more balanced view is evident.

Organisms Through Time

Naturalists at first had a nearly static view of nature. They studied it as it is, and few of them saw any reason to envision any radical changes in the past or future. In fact, until well into the nineteenth century many biologists accepted the then current theological opinion that the world is less than 6000 years old, a time so short as practically to exclude a historical view of nature. All the biological sciences reviewed up to this point were originally unconcerned with periods of time longer than the life cycles of individual organisms. Some of them have since been revolutionized by the introduction of historical concepts, notably systematics, biogeography, ecology, and comparative anatomy, and all have been influenced by those concepts, but that was not and in some instances still is not their primary concern.

The principal biological sciences that are necessarily, inherently concerned with changes through time longer than a generation are paleontology and genetics.

PALEONTOLOGY

Paleontology is one of the sciences that belongs equally to the physical and the biological sciences. It is as much a part of geology as it is of biology, just as biochemistry is as much chemistry as biology. Fossils, the principal objects of paleontological study, occur in sedimentary rocks, the study of which is part of the science of geology. Many of the principles basic for paleontology are more geological than biological.

Fossils were known to the ancient philosophers, some of whom recognized them as the remains of ancient organisms. Later that correct opinion was often discarded, and it was not until around the beginning of the nineteenth century that the basic principles necessary for a science of paleontology were generally accepted. The most essential of these principles are: fossils are traces of once living organisms and can be studied and classified as such; most of them belong to species and larger groups no longer living; they occur in a definite historical sequence of extinct floras and faunas, living organisms being simply the latest members of that sequence.

The final essential step for the foundation of the science of paleontology was recognition of rock sequences containing characteristic sequences of extinct organisms. As usual, many students contributed to that result and several reached it about the same time, but among them the English professional civil engineer and amateur geologist W. Smith (1769–1839) is worthy of special mention.

It was the French zoologist Georges Cuvier, as much as any one man, who made paleontology a distinct biological science. He published systematic descriptions of many extinct species and recognized major features of their sequence, although necessarily in a crude and incomplete way. His accomplishment was brilliant, but it must be added that there were then in geology and paleontology two of the important wrong turns that checker the history of science, and Cuvier took them both. He believed in catastrophism rather than uniformitarianism (Chapter 29) and in special creation rather than evolution. So great was Cuvier's authority that most paleontologists followed his false lead. Their progress was greatly impeded by the fact that they had to be slowly persuaded to uniformitarianism by the geologists and to evolution by the biologists, although, now that they have been persuaded, it is obvious that their science preeminently reflects both principles. Uniformitarianism was established especially by the British geologists James Hutton (1726–1797) and Charles Lyell (1797–1875). Some remarks on the rise of evolution are made later in this chapter.

Modern paleontology, uniformitarian and evolutionary, was a product of the late nineteenth century. The paleobotanists, micropaleontologists (students of microscopic fossils, mostly protists), and invertebrate pa-

leontologists have contributed greatly. The more biological approach and integration with other biological sciences have, however, been due in greater measure to the vertebrate paleontologists. Among the most important of these are the Russian Waldemar Kovalevsky [7] (1842–1883) and the Americans E. D. Cope (1840–1897) and H. F. Osborn (1857–1935).

The discovery of new species of fossils continues today at an accelerated pace, and descriptive, systematic paleontology has never been more active. Most significant in paleontology at present, however, is a broadening of the base and aims of the science. Its data, and also its methods, are being combined with those of all the other biological sciences to produce a historical view of all aspects of biology. Paleontology, strictly speaking, is the principal parent of what seems to be emerging as a new biological science, *historical biology*.

GENETICS

Genetics was not, in origin, historical to the same extent or in the same sense as paleontology. Some aspects of it have, however, become so, and genetics has always and necessarily been concerned with time, with changes through generations and less with events in single life cycles.

Notions of heredity, some correct and some fantastic, have been current since man became articulate. As a distinct science worthy of the name, however, genetics is the youngest of all the main biological sciences. Attempts to study heredity before 1900 were marked by a whole series of wrong turnings. You are already familiar with some of these: inheritance by blood and the more sophisticated but equally wrong idea of inheritance by the assemblage of particles from all over the body; blending inheritance; the inheritance (as such) of acquired characters. Variations of these incorrect theories dominated biological thought about heredity right up until the twentieth century.

You are also familiar (from Chapter 6) with the importance of the methods and re-

[7] As with many Russian names, this one has been transliterated into our alphabet in several different ways.

sults of Mendel (1822–1884) and with their rediscovery by De Vries (1848–1935) and others. The elaboration of formal or, strictly speaking, Mendelian genetics was largely due to Morgan (1867–1945) and his students and followers. That branch of genetics is concerned principally with tracing the heredity of single, readily separable gene mutations, the recombinations of such mutations, and their association and sequence in the chromosomes.

Such work is still basic for genetics, but at present the main interests of geneticists have followed other lines of what has become a vast and increasingly ramifying science. Interactions between genes and effects of the whole genetic apparatus have assumed greater importance than the study of single genes. Joint study of chromosomes and genes—cytogenetics—has become a subscience of its own, with essential contributions to systematics and other biological sciences. Direct studies of the physiological effects of genes are made with the mold *Neurospora* and an increasing number of other organisms. The nature and action of genes and chromosomes are being intensively investigated with biochemical and biophysical methods. There is increasing interest in extranuclear inheritance, through the cytoplasm or particles in it, and its relationship to the primary, chromosomal genetic mechanism. The relationship between the genetic apparatus and differentiation of the developing organism is a vital and still poorly understood subject, as was noted in the discussion of embryology.

Among all these divergent developments of genetics, few have provoked greater interest or wider implications than *population genetics*. Random and selective changes in the genetic compositions of populations, isolating mechanisms, breeding patterns, and many other population phenomena are studied in the field, in the laboratory, and by mathematical models. The importance of these studies is their bearing on the universal and intricate fact of evolution.

EVOLUTION

We have been following the subdivision of biology into the various biological sciences in the framework of the history of the subject.

Evolution is not a science, nor is it a discovery or a principle that has emerged from one of the biological sciences. It is a fact that is true of the subject matter of all of biology in general, and of all of the life sciences. Also, our knowledge and theories of how it operates do not come from any one of the life sciences but in varying degrees from all of them. Thus evolution requires separate mention here, and, since it is the most pervasive of all the principles you have learned about life, this final mention of it may fittingly close your introduction to biology.

The ancients and the theological philosophers who were the principal persons to carry Western intellectual thought up to the dawning of the scientific age had little or no inkling of evolution in the modern sense. They saw a certain unity in nature; they spoke of affinities or relationships among organisms; and they discerned a (highly irregular and imperfect) sequence of living things, the "ladder of life" (Chapter 21). It is easy to read our evolutionary interpretation into what they said, and what they said was capable of logical development into evolutionary theory. But they never clearly said and evidently did not believe that the unity, affinities, and sequence of living things are material facts, due to modifications of physical descent, and not only philosophical or metaphysical ideas.

The concept of evolution as a material fact of life began to appear, at first hesitantly and vaguely, when biology was just beginning to be a science, back in the sixteenth and seventeenth centuries. By the end of the eighteenth century evolution was a definitely stated and widely known concept, but a concept that was rejected by the majority of the biologists of that period. Its eventual acceptance depended in part on further accumulation of facts that finally could not reasonably be interpreted in any other way. General acceptance also awaited the development of underlying and still more basic principles about nature. There had to be recognition that the age of the earth is to be counted not in thousands of years but in millions, at least. There also had to be a choice between catastrophism and uniformitarianism, for although catastrophism does not logically exclude evolution it produced a cast of mind far less favorable to the idea of evolution than did uniformitarianism. Those preliminaries for acceptance of evolution were gifts from geology to biology. It is highly significant that Darwin was strongly influenced by Lyell, who assigned great antiquity to the earth and who figured largely in the triumph of uniformitarianism.

On the biological side, acceptance of evolution awaited the slowly growing conviction that there are similarities throughout the whole realm of life and that they have a physical and not only a metaphysical basis. The cell theory was developing along with the concept of evolution, and it was no coincidence that the universality of cells and of evolution reached full recognition at almost the same time. Finally, acceptance of evolution had to await the proposal of some plausible theory as to how it operates.

He had many more hesitant and speculative predecessors, but the French zoologist Lamarck (1744–1829) was the first really thoroughgoing evolutionist. He proclaimed that all species are the related products of evolution, and he proposed a theory that was plausible, in the state of biology at the time, even though we now consider it incorrect. He believed that organisms had followed a pattern of increasing perfection (the old "ladder of nature") and had also become divergent from that pattern through local adaptation. Adaptation was claimed to be the cumulative, inherited effect of responses to environmental stimuli.

After Lamarck evolutionary views were increasingly held by other biologists, but, as everyone knows, their final acceptance came after the publication in 1859 of *The Origin of Species* by Darwin. Darwin, a great and wide-ranging naturalist even aside from his work on evolution, marshaled evidence from the whole field of the biology of his day. He demonstrated that the present state of the living world can be reasonably explained only as the result of evolution. As the major, but in his opinion not the only, cause of evolution he proposed natural selection. Darwin's explanation of evolution, truly brilliant as far as it went, was inevitably incomplete, principally because so little was then known about heredity.

After Darwin there was long dispute between the adherents of natural selection and those, called Neo-Lamarckians, who believed in more direct environmental influences on heredity. The dispute was settled as far as most biologists are concerned by the study of genetics, especially population genetics, in the twentieth century. Genetics has rehabilitated Darwin's natural selection as the cause of adaptation, giving it a broader, somewhat different, but not contradictory meaning. Genetics has also supplied the principal elements lacking in Darwin's own theory.

We do not need to discuss how the modern theory of evolution has permeated all the biological sciences, nor how all of them in synthesis have supported the truth of evolution and deepened our knowledge of its processes. This whole book is a demonstration of those results.

A book on biology might close with a statement of the importance and fascination of the subject and of how it enriches your understanding of life. For this, too, we let the whole book speak. If we have not conveyed this message by now, no final word will do it.

CHAPTER SUMMARY

The long search for the secrets of life: the work of many; science as a historical growth of ideas and insights.

Natural history: the original broad meaning of the term; the transition from a philosophical, authoritarian, dogmatic approach to a naturalistic attitude through the works of Copernicus and Newton; classification and systematics—the collections and compendia of Agricola, Gesner, Fuchs; the first serious classifications of Cesalpino, Ray, Linnaeus, Buffon, Cuvier; post-Darwinian evolutionary classification; the systematic subsciences; life sciences and the scope of biology—the growth of biology from natural history into many and diverse special sciences.

The organism in its environment: the great naturalist explorers of the nineteenth century, especially Darwin and Huxley; biogeography—Humboldt, Wallace, Matthew; ecology—the pioneers Warming and Schimper, full development only in the twentieth century.

The organism in the laboratory: the laboratory, as against the naturalist's, approach to biology; human and comparative anatomy—the history of human dissection, Vesalius and the birth of modern anatomy, Belon and the beginnings of comparative anatomy, Goethe, Oken, Owen, typological anatomy, the evolutionary basis to modern comparative anatomy as expressed by Wiedersheim; physiology—ancient ideas on the workings of the body from Galen and Servetus, Harvey and the beginnings of modern physiology, Claude Bernard, the emergence of general and cellular physiology and of biophysics and biochemistry; nerve and sensory physiology—Swedenborg's mystical views, Haller, Müller and Du Bois-Reymond as cofounders of modern nerve physiology, the refutation of vitalism, the electrical phenomena of nerves as the basis of the nervous system's activity; animal behavior—the nineteenth-century work anecdotal and anthropomorphic except for that of Fabre, Morgan and Loeb and the origins of the modern work, Pavlov's correlation of behavior and nerve physiology.

Within the organism: science and technique—the influence of techniques and instruments on scientific advance, the importance of the microscope in biology, modern techniques of physics and chemistry; microorganisms—the influence of the microscope, Swammerdam and Leeuwenhoek, the spontaneous-generation problem of Redi and Pasteur, bacteria and disease as related by Koch, modern microbiology; cells and tissues—the impact of the microscope again, the beginnings of histology with Hooke and Bichat, the cell theory as developed by Schwann, Schultze, Virchow, cytology and cytogenetics; embryology—preformationism as a blind alley, the beginnings of modern embryology, Von Baer, Balfour and evolutionary comparative embryology, modern experimental embryology and its relation to genetics; biochemistry and biophysics as derivative studies from physiology—Müller's students Du Bois-Reymond, Schwann, and Helmholtz (the latter on the physics of vision and hearing), Liebig, Fischer, modern biochemistry and biophysics.

Organisms through time: paleontology as both geology and biology—beginnings of the modern science in the early nineteenth century, paleontological principles, W. Smith and the sequence of strata, Cuvier's great contributions and two false steps, Hutton, Lyell, and uniformitarianism, the beginnings of vertebrate paleontology, Kovalevsky, Cope, Osborn, emergence of a historical biology; genetics—early ideas and the beginnings of modern formal genetics, Mendel, De Vries, Morgan, the recent trend to study of gene action, gene and cytoplasm, gene and embryo, population genetics; evolution—the first glimmerings in the post-Renaissance period, the statement as a hypothesis by the close of the eighteenth century, the factors influencing its ultimate scientific acceptance, Lamarck as the first thoroughgoing evolutionist, Darwin and natural selection, post-Darwinian debate between Neo-Lamarckians and selectionists, the recent influence of genetics, especially in settling the Neo-Lamarckian–Darwinian debate in favor of Darwin.

APPENDIX

A Classification of Organisms

As explained in Chapter 18, no two authorities agree on every detail in the classification of organisms. The following arrangement seems reasonable to us. It generally represents either the consensus or a compromise when significant conflict occurs. Most of the names in current use for major groups in protistology, botany, and zoology appear here. The classification may thus serve as a glossary of systematic names as well as an indication of the tremendous basic diversity of living things.

All phyla recognized by us are listed, although some systematists would add several more. A few phyla and subphyla that are *relatively* unimportant in the modern world are named in lightface type. Most of the classes are listed, and these include the groups sometimes considered as phyla. When classes are not designated within a phylum, either the phylum is small and has a single class, or a proposed division into classes is not satisfactory. The symbol † indicates that a group is totally extinct. Of course, virtually all the phyla and classes include known extinct smaller groups.

For each phylum or class at least one genus is named by way of example. As far as possible, the genera given as examples are available for demonstration or dissection as well as being mentioned elsewhere in this book.

Kingdom Protista

PHYLUM SCHIZOPHYTA or **SCHIZO-MYCETES**, bacteria: *Bacillus, Escherichia, Azotobacter, Clostridium, Pneumococcus*

PHYLUM CYANOPHYTA, "blue-green algae" (many are not blue-green in color, and none seems to be closely allied to other or true algae): *Anabaena, Nostoc, Oscillatoria*

PHYLUM MASTIGOPHORA, flagellates:

> Class Phytomastigina, plantlike flagellates: *Euglena, Volvox, Chlamydomonas, Cryptomonas*

> Class Dinoflagellata, dinoflagellates: *Ceratium, Peridinium, Gymnodinium, Gonyaulax, Ceratodinium, Noctiluca*

> Class Zoomastigina, animal-like flagellates: *Trypanosoma, Polytoma, Chilomonas, Astasia, Oikomonas, Mastigamoeba, Hexamitus, Calonympha*

PHYLUM SARCODINA or **RHIZO-PODA:** rhizopods, *Amoeba;* forams, *Globigerina;* radiolarians, *Lychnaspis*

PHYLUM SPOROZOA: *Plasmodium*

PHYLUM CILIOPHORA

> Class Ciliata, ciliates: *Paramecium, Stentor, Stylonichia, Euplotes, Epidinium*

PHYLUM MYXOMYCETES, slime molds: *Lycogala, Physarum*

Kingdom Plantae (Plants)

Subkingdom Algae (sometimes considered protists and sometimes excluding fungi)

PHYLUM CHLOROPHYTA, green algae

Class Chlorophyceae, grass-green algae: *Ulothrix, Oedogonium, Spirogyra, Closterium*

Class Charophyceae, stoneworts: *Chara, Nitella*

PHYLUM CHRYSOPHYTA

Class Xanthophyceae, yellow-green algae: *Vaucheria, Tribonema, Botrydium*

Class Chrysophyceae, golden-brown algae: *Synura, Dinobryon*

Class Bacillariophyceae, diatoms: *Navicula, Pinnularia, Tabellaria, Actinoptychus*

PHYLUM PHAEOPHYTA, brown algae: *Ectocarpus, Laminaria, Fucus*

PHYLUM RHODOPHYTA, red algae: *Porphyra, Batrachospermum, Nemalion, Lithophyllum, Thamnion, Polysiphonia*

PHYLUM MYCOPHYTA, fungi

Class Phycomycetes, tube fungi: bread molds, *Rhizopus;* water molds, *Saprolegnia;* white rusts and downy mildews, *Albugo;* chytrids, *Chytridium*

Class Ascomycetes, sac fungi: bread molds, *Neurospora;* yeasts, *Saccharomyces;* blue and green molds, *Aspergillus, Penicillium;* powdery mildews, *Microsphaera;* cup fungi, *Sclerotinia;* morels, *Morchella*

Class Basidiomycetes, club fungi: mushrooms, *Psalliota;* toadstools, *Amanita;* bracket fungi, *Fomes;* smuts and rusts, *Puccinia*

(*Fungi imperfecti:* Under this name botanists place a large number of disease-producing fungi, such as those causing athlete's foot or beet leaf spot. They are probably ascomycetes and basidiomycetes in which the sexual cycle has been lost or is unknown. This is not, strictly, a unit of classification but a catchall for fungi that have not been classified.)

(**Lichens:** These composite organisms are obligatory symbiotic associations of a fungus and an alga; in *ascolichens* the fungus is an ascomycete, and in *basidiolichens* it is a basidiomycete.)

Subkingdom Metaphyta

PHYLUM BRYOPHYTA

Class Anthoceropsida or Anthocerotae, hornworts: *Anthoceros*

Class Hepaticopsida or Hepaticae, liverworts: *Marchantia, Riccia, Conocephalum*

Class Bryopsida or Musci, mosses: *Sphagnum, Andreaea, Mnium, Funaria, Pottia*

PHYLUM TRACHEOPHYTA, vascular plants

Subphylum Psilopsida

Class † Psilophytales: † *Rhynia,* † *Psilophyton*

Class Psilotales: *Psilotum, Tmesipteris*

Subphylum Lycopsida: lycopods
† *Lepidodendron,* † *Sigillaria, Lycopodium, Selaginella*

Subphylum Sphenopsida: † *Calamites, Equisetum*

Subphylum Pteropsida, ferns and seed plants

Class Filicineae, ferns: *Ophioglossum, Cyathea, Polypodium, Aspidium, Azolla*

Class Gymnospermae, gymnosperms: † seed ferns, † *Neuropteris;* † cycadeoids, † *Cycadeoidea,* † *Williamsonia;* cycads, *Zamia, Dioon;* ginkgos, *Ginkgo;* † cordaites, † *Cordaites;* conifers, *Pinus, Abies, Tsuga, Taxus, Sequoia, Metasequoia, Araucaria;* joint firs, *Ephedra*

Class Angiospermae, angiosperms, flowering plants

DICOTS: magnolias, *Magnolia;* snake root, *Aristolochia;* eucalypts, *Eucalyptus;* oaks, *Quercus;* elms, *Ulmus;* maples, *Acer;* beeches, *Fagus, Nothofagus;* peaches, *Amygdalus;* cacti, *Cereus;* blackberries, *Rubus;* peas, *Pisum;* nightshades, *Solanum;* sages, *Salvia;* mustards, *Brassica;* dandelions, *Taraxacum;* ragweeds, *Ambrosia*

MONOCOTS: grasses, *Panicum, Stipa;* sedges, *Cyperus;* lilies, *Lilium;* tulips, *Tulipa;* yuccas, *Yucca;* palms, *Sabal;* orchids, *Cypripedium, Ophrys, Cryptostylis*

Kingdom Animalia (Animals)

Subkingdom Parazoa

PHYLUM PORIFERA, sponges

Class † Pleospongiae: † *Archeocyathus*

Class Calcarea or Calcispongiae, chalky sponges: *Scypha*

Class Hexactinellida or Hyalospongiae, glass sponges: *Hyalonema*

Class Demospongiae, horny sponges; bath sponges, *Spongia*

Subkingdon Metazoa

PHYLUM COELENTERATA or CNIDARIA, coelenterates

Class Hydrozoa: *Hydra; Obelia; Gonionemus;* Portuguese man-of-war, *Physalia, Velella*

Class † Stromatoporoidea: † *Clathrodictyon*

Class Scyphozoa, jellyfishes: *Aurelia, Chrysaora*

Class Anthozoa: corals, *Astrangia, Madrepora;* sea anemones, *Metridium, Dahlia*

PHYLUM † GRAPTOLITHINA, † graptolites: † *Didymograptus*

PHYLUM CTENOPHORA, comb jellies: *Cestum*

PHYLUM PLATYHELMINTHES, flatworms

Class Turbellaria, planarians: *Planaria, Dugesia*

Class Trematoda, flukes: *Fasciola, Polystomum, Schistosoma*

Class Cestoda, tapeworms: *Taenia*

PHYLUM MESOZOA: *Rhopalura*

PHYLUM NEMERTINA or NEMERTEA, ribbon worms: *Lineus*

PHYLUM ASCHELMINTHES (classes highly diverse; each is sometimes considered a phylum)

Class Rotifera or Trochelminthes, rotifers or wheel animalcules: *Rotaria, Asplanchna*

Class Gastrotricha: *Chaetonotus*

Class Kinorhyncha or Echinoderida: *Echinoderes*

Class Nematomorpha or Gordiacea, horsehair worms: *Paragordius, Nectonema*

Class Nematoda, roundworms (much the most important group of Aschelminthes and taken to represent the whole phylum in the discussion of principal phyla in the text): *Ascaris, Trichinella, Turbatrix, Ancylostoma*

PHYLUM ACANTHOCEPHALA, spiny-headed worms: *Gigantorhynchus*

PHYLUM BRYOZOA or **POLYZOA,** bryozoans, sea mosses, or moss animals (each class is often considered a phylum):

Class Endoprocta: *Urnatella*

Class Ectoprocta: *Plumatella, Bugula*

PHYLUM BRACHIOPODA, brachiopods or lampshells

Class Inarticulata: *Lingula*

Class Articulata: *Laqueus, Terebratulina*

PHYLUM PHORONIDEA: *Phoronis*

PHYLUM CHAETOGNATHA, arrow worms: *Sagitta*

PHYLUM MOLLUSCA, mollusks

Class Amphineura, chitons: *Chiton, Neomenia*

Class Monoplacophora: *Neopilina*

Class Gastropoda, gastropods: snails, *Helix, Lymnaea, Planorbis;* whelks, *Buccinum, Ocenebra;* slugs, *Arion;* limpets, *Patella;* nudibranchs, *Archidoris;* sea hares, *Tethys*

Class Scaphopoda, tooth-shells: *Dentalium*

Class Pelecypoda or Lamellibranchia or Bivalvia, pelecypods: clams, mussels, *Venus, Anodonta, Mya, Pecten, Chlamys, Tridacna, Pholas, Teredo, Solen, Mytilus*

Class Cephalopoda, cephalopods: squids, *Loligo;* octopuses, *Octopus;* nautilus, *Nautilus*

PHYLUM ANNELIDA, annelids, segmented worms

Class Polychaeta, polychaetes, sandworms: *Neanthes, Nereis, Aphrodite, Chaetopterus*

Class Oligochaeta, oligochaetes: earthworms, *Lumbricus*

Class Archiannelida: *Polygordius*

Class Hirudinea, leeches: *Hirudo*

Class Gephyrea, sipunculid, echiuroid, and priapulid worms (often placed in one, two, or three separate phyla): *Sipunculus, Echiurus, Priapulus*

PHYLUM ARTHROPODA, arthropods (several other small classes are often recognized)

Class Onychophora: *Peripatus*

Class † Trilobita, † trilobites: † *Triarthrus,* † *Neolenus,* † *Ogygopsis,* † *Isotelus,* † *Calymene*

Class Crustacea, crustaceans: brine shrimps, *Artemia;* † *Barrella;* † *Hymenocaris;* water fleas, *Daphnia;* copepods, *Cyclops;* cirripeds (barnacles), *Balanus, Lepas, Sacculina;* wood lice, *Armadillaria, Caprella;* euphausids, *Euphausia;* prawns, *Leander;* lobsters, *Homarus;* crabs, *Cancer*

Class Arachnida: † eurypterids, † *Pterygotus,* † *Carcinosoma,* † *Hughmilleria;* spiders, *Eurypelma, Theridion;* scorpions, *Vejovis;* king crabs, *Limulus;* ticks, *Dermacentor;* sea spiders, *Pycnogonum*

Class Chilopoda, centipedes: *Lithobius*

Class Diplopoda, millipedes: *Julus*

Class Insecta, insects: cockroaches, *Periplaneta;* grasshoppers, *Melanoplus;* dragonflies, † *Dunbaria, Libellula;* bugs, *Cimex, Halobates;* butterflies, *Papilio, Colias;* flies, *Musca, Drosophila;* beetles, *Calosoma;* ants,

Pogonomyrmex; bees, wasps, and allies, *Bombus, Vespa, Coccophagus*

PHYLUM ECHINODERMATA, echinoderms

Class † Cystoidea, † cystoids: † *Caryocrinites*

Class † Edrioasteroidea, † edrioasteroids: † *Edrioaster*

Class † Blastoidea, † blastoids † *Pentremites*

Class Crinoidea, crinoids, sea lilies: *Antedon*

Class Asteroidea, starfishes: *Asterias*

Class Ophiuroidea, serpent stars, brittle stars: *Ophiura*

Class Echinoidea, sea urchins: *Strongylocentrotus, Cidaris*

Class Holothuroidea, sea cucumbers: *Cucumaria*

PHYLUM CHORDATA, chordates

Subphylum Hemichordata, tongue worms (acorn worms): *Balanoglossus*

Subphylum Urochordata, or **Tunicata,** tunicates: ascidians, *Ciona*

Subphylum Cephalochordata, lancelets: *Branchiostoma* (amphioxus)

Subphylum Vertebrata, vertebrates

SUPERCLASS PISCES, aquatic vertebrates, fishes

Class Agnatha, agnaths, jawless fishes: † *Cephalaspis,* † *Kieraspis,* † *Pteraspis;* lampreys, *Petromyzon*

Class † Placodermi, † placoderms: † *Climatius,* † *Diplacanthus,* † *Coccosteus,* † *Dinichthys,* † *Pterichthyodes*

Class Chrondrichthyes or Elasmobranchii: sharks, *Squalus;* rays, *Raja; Chimaera*

Class Osteichthyes, bony fishes: † *Cheirolepis;* sturgeon, *Acipenser;* trout,

Salmo; perch, *Perca;* anglerfish, *Photocorynus;* crossopterygians: † *Eusthenopteron, Latimeria;* lungfishes: † *Dipterus, Epiceratodus, Protopterus*

SUPERCLASS TETRAPODA, land vertebrates, tetrapods

Class Amphibia, amphibians: † labyrinthodonts, † *Diplovertebron,* † *Eryops;* salamanders, *Ambystoma, Necturus;* frogs, *Rana;* toads, *Bufo;* tree toads, *Hyla;* caecilians (Apoda), *Gymnophis*

Class Reptilia, reptiles: † cotylosaurs; turtles and tortoises, *Testudo, Aromochelys;* † ichthyosaurs † *Ichthyosaurus;* † plesiosaurs; rhynchocephalians, *Sphenodon;* lizards, *Gerrhonotus, Crotophytus;* snakes, *Thamnophis;* † thecodonts; alligators, *Alligator,* and crocodiles, *Crocodilus;* † pterosaurs, † *Nyctosaurus,* † *Pteranodon;* † dinosaurs, † *Coelophysis,* † *Ornitholestes,* † *Tyrannosaurus,* † *Brontosaurus,* † *Anatosaurus,* † *Stegosaurus,* † *Ankylosaurus,* † *Triceratops;* † mammallike reptiles, † *Dimetrodon,* † *Cynognathus*

Class Aves, birds: † *Archeopteryx;* † *Aepyornis;* † *Hesperornis;* kiwis, *Apteryx;* † *Diatryma;* pigeons, *Columba;* chickens, *Gallus;* owls, *Micropallus, Otus;* woodpeckers, *Centurus;* flickers, *Colaptes;* flycatchers, *Myiarchus;* finches, *Geospiza, Camarrhynchus, Certhidia*

Class Mammalia, mammals: Monotremes, *Ornithorhynchus.* Marsupials, including opossums, *Didelphis,* and others (see Figs. 19–7 and 23–27). † Mesozoic placentals, † *Deltatheridium,* † *Zalambdalestes.* Primates, including: † *Notharctus; Loris;* monkeys, *Cebus, Cercopithecus, Macaca;* apes, *Gorilla* and others; † *Australopithecus;* men, *Homo.* Anteaters, *Myrmecophagus.* Rodents: rats, *Rattus;* mice, *Mus, Peromyscus;* hamsters, *Cricetus.* Carnivores: † Creodonts, † sabertooths; dogs, *Canis;* cats, *Felis;* seals, *Phoca.* Cetaceans: whales, *Orcinus;* porpoises, *Phocaena.* Ungulates: † condylarths, † *Phenacodus;* † uintatheres, † *Uintatherium;* † titanotheres, † *Bron-* *tops;* elephants, *Loxodonta;* horses, † *Hyracotherium* ("Eohippus"), † *Miohippus, Equus;* deer, *Cervus, Odocoileus,* † *Megaloceros;* pigs, *Sus;* sheep, *Ovis;* cows, *Bos.*

SUGGESTIONS FOR FURTHER READING

Part 1. Introduction

A rich source of collateral reading is the monthly magazine *Scientific American*. Its articles are authoritative and (as of the date of publication) up to the minute accounts of specific scientific subjects, often biological. Many of them are reprinted in pamphlet form by Freeman, San Francisco, from whom a list may be obtained. They are too numerous and in part too specialized for separate citation here.

The following are a few books on the biological sciences in general, on the nature of science, and (Roe) on the characteristics of scientists. Most of the botanies and zoologies cited for Part 6 and the histories cited for Part 10 also contain material of interest for the whole subject of biology.

Bates, M., *The Nature of Natural History*, Scribner's, 1950; Scribner's, 1961 (pap.).
Beck, W. S., *Modern Science and the Nature of Life*, Harcourt, Brace & World, 1957; Natural History Library, Doubleday, 1961 (pap.).
Gabriel, M. L., and S. Fogel, eds., *Great Experiments in Biology*, Prentice-Hall, 1955 (pap.).
Gray, P., *The Encyclopedia of the Biological Sciences*, Reinhold, 1961.
Hall, T. S., *A Source Book in Animal Biology*, McGraw-Hill, 1951.
Roe, A., *The Making of a Scientist*, Dodd, Mead, 1952; Apollo Editions, 1961 (pap.).
Simpson, G. G., *This View of Life: The World of an Evolutionist*, Harcourt, Brace & World, 1964.
Waddington, C. H., *The Nature of Life*, Atheneum, 1962.

Part 2. The Basis of Life

In addition to the periodicals *Science, Nature*, and *Scientific American*, many excellent annual publications regularly report on recent developments in cell physiology and biochemistry. They include the following: the various *Annual Reviews* series published by Annual Reviews, Inc., Palo Alto—particularly the *Annual Review of Biochemistry*, the *Annual Review of Plant Physiology*, and the *Annual Review of Microbiology*; the series published by Academic Press, New York, including *Advances in Carbohydrate Chemistry, Advances in Protein Chemistry, Advances in Virus Research, Progress in Nucleic Acid Research*, and the *International Review of Cytology*; the series *Advances in Enzymology and Related Subjects of Biochemistry*, published by Interscience Publishers, New York, and the volumes *Symposia on Quantitative Biology*, published by the Biological Laboratory at Cold Spring Harbor. Books of interest are as follows:

Adams, M. H. *Bacteriophages*, Interscience, 1959.
Allen, J. M., ed., *The Molecular Control of Cellular Activity*, McGraw-Hill, 1962.
Anfinsen, C. B., *The Molecular Basis of Evolution*, Wiley, 1959; Science Editions, Wiley, 1963 (pap.).
Baldwin, E., *The Nature of Biochemistry*, Cambridge Univ. Press, 1962 (also pap.).
Baldwin, E., *An Introduction to Comparative Biochemistry*, 4th ed., Cambridge Univ. Press, 1964 (also pap.).
Baldwin, E., *Dynamic Aspects of Biochemistry*, 4th ed., Cambridge Univ. Press, 1964.
Bourne, G. H., *Division of Labor in Cells*, Academic Press, 1962 (pap.).
Boyer, P. D., H. Lardy, and K. Myrbäck, eds., *The Enzymes*, 2nd ed., Vols. 1–8, Academic Press, 1959–63.
Brachet, J., and A. E. Mirsky, eds., *The Cell: Biochemistry, Physiology, Morphology*, Vols. 1–6, Academic Press, 1959–64.
Burnet, F. M., and W. M. Stanley, eds., *The Viruses: Biochemical, Biological, and Biophysical Properties*, Vols. 1–3, Academic Press, 1959.
Chargaff, E., and J. N. Davidson, eds., *The Nucleic Acids*, Vols. 1–3, Academic Press, 1955–60.
Cohn, N. S., *Elements of Cytology*, Harcourt, Brace & World, 1964.
Colowick, S. P., and N. O. Kaplan, eds., *Methods in Enzymology*, Vols. 1–7, Academic Press, 1955–64.
Deuel, H. J., Jr., *The Lipids: Their Chemistry and Biochemistry*, Vols. 1–3, Interscience, 1951–57.
Dixon, M., and E. C. Webb, *Enzymes*, 2nd. ed., Academic Press, 1964.
Esau, K., *Plant Anatomy*, Wiley, 1953.
Fieser, L. F., and M. Fieser, *Organic Chemistry*, 3rd ed., Heath, 1956.
Florkin, M., and H. S. Mason, eds., *Comparative Biochemistry*, Vols. 1–6, Academic Press, 1960–64.
Fruton, J. S., and S. Simmonds, *General Biochemistry*, 2nd ed., Wiley, 1958.
Galston, A. W., *The Life of the Green Plant*, 2nd ed., Prentice-Hall, 1964 (also pap.).

Gunsalus, I. C., and R. Y. Stanier, eds., *The Bacteria: A Treatise on Structure and Function*, Vols. 1–5, Academic Press, 1960–64.

Henderson, L. J., *The Fitness of the Environment: An Inquiry into the Biological Significance of the Properties of Matter*, Peter Smith, 1959; Beacon, 1958 (pap.).

Ingram, V. M., *The Hemoglobins in Genetics and Evolution*, Columbia Univ. Press, 1963.

Kasha, M., and B. Pullman, eds., *Horizons in Biochemistry*, Academic Press, 1962.

Kluyver, A. J., and C. B. van Niel, *The Microbe's Contribution to Biology*, Harvard Univ. Press, 1956.

Kurtz, S. M. ed., *Electron Microscopic Anatomy*, Academic Press, 1964.

Luria, S. E., *General Virology*, Wiley, 1953.

McElroy, W. D., *Cell Physiology and Biochemistry*, 2nd ed., Prentice-Hall, 1964 (also pap.).

Meyer, B. S., D. B. Anderson, and R. H. Böhning, *Introduction to Plant Physiology*, Van Nostrand, 1960.

Michelson, A. M., *The Chemistry of Nucleosides and Nucleotides*, Academic Press, 1963.

Neurath, H., and K. Bailey, *The Proteins: Chemistry, Biological Activity and Methods*, Vol. 1, Part A–Vol. 2, Part B, Academic Press, 1953–54.

Pauling, L., *General Chemistry: An Introduction to Descriptive Chemistry and Modern Chemical Theory*, 2nd ed., Freeman, 1958.

Pimentel, G. C., and A. L. McClellan, *The Hydrogen Bond*, Freeman, 1960.

Porter, K. R., and M. A. Bonneville, *An Introduction to the Fine Structure of Cells and Tissues*, Lea & Febiger, 1963.

Ray, P. M., *The Living Plant*, Holt, Rinehart and Winston, 1963 (pap.).

Sinnott, E. W., *Plant Morphogenesis*, McGraw-Hill, 1960.

Stanier, R. Y., M. Doudoroff, and E. A. Adelberg, *The Microbial World*, Prentice-Hall, 1957.

Stent, G. S., *Molecular Biology of Bacterial Viruses*, Freeman, 1963.

Swanson, C. P., *Cytology and Cytogenetics*, Prentice-Hall, 1957.

White, A., P. Handler, and E. L. Smith, *Principles of Biochemistry*, 3rd ed., McGraw-Hill, 1964.

Wilson, E. B., *The Cell in Development and Heredity*, 3rd ed., Macmillan, 1925.

Part 3. Reproduction: The Continuity of Life

Allen, J. M., ed., *The Nature of Biological Diversity*, McGraw-Hill, 1963.

Arey, L. B., *Developmental Anatomy*, 6th ed., Saunders, 1954.

Barth, L. G., *Embryology*, rev. ed., Holt, Rinehart and Winston, 1953.

Berrill, N. J., *Sex and the Nature of Things*, Dodd, Mead, 1953; Apollo Editions, 1964 (pap.).

Bonner, D. M., *Heredity*, 2nd ed., Prentice-Hall, 1964 (also pap.).

Bonner, J. T., *Morphogenesis*, Princeton Univ. Press, 1952; Atheneum, 1963 (pap.).

Brachet, J., *The Biochemistry of Development*, Pergamon Press, 1960.

Corner, G. W., *Ourselves Unborn*, Yale Univ. Press, 1944.

Jacob, F., and E. L. Wollman, *Sexuality and the Genetics of Bacteria*, Academic Press, 1961.

Levine, R. P., *Genetics*, Holt, Rinehart and Winston, 1962 (pap.).

Mazia, D., and A. Tyler, eds., *The General Physiology of Cell Specialization*, McGraw-Hill, 1963.

Patten, B. M., *Human Embryology*, 2nd ed., McGraw-Hill, 1953.

Rugh, R., *Vertebrate Embryology: The Dynamics of Development*, Harcourt, Brace & World, 1964.

Sager, R., and F. H. Ryan, *Cell Heredity*, Wiley, 1961.

Singleton, W. R., *Elementary Genetics*, Van Nostrand, 1962.

Srb, A. M., and R. D. Owen, *General Genetics*, Freeman, 1952.

Stern, C., *Principles of Human Genetics*, Freeman, 1960.

Sussman, M., *Animal Growth and Development*, Prentice-Hall, 1960 (also pap.).

Taylor, J., ed., *Molecular Genetics*, Academic Press, 1963.

Thompson, D'A., *On Growth and Form*, Putnam, 1917.

Waddington, C. H., *Principles of Embryology*, Macmillan, 1956.

Waddington, C. H., *New Patterns in Genetics and Development*, Columbia Univ. Press, 1962.

Part 4. The Maintenance and Integration of the Organism

Progress in certain aspects of maintenance and integration may be conveniently followed in the *Annual Review of Psychology* and the *Annual Review of Physiology* (both published by Annual Reviews) and *Recent Progress in Hormone Research* and *Vitamins and Hormones* (both published by Academic Press).

Ariëns Kappers, C. U., *The Evolution of the Nervous System*, Bohn, 1929.

Beach, F. A., *Hormones and Behavior*, Hoeber, 1948.

Bonner, J., and A. W. Galston, *Principles of Plant Physiology*, Freeman, 1952.

Carlson, A. J., and V. Johnson, *The Machinery of the Body*, Univ. of Chicago Press, 1948.

Davson, H., *A Textbook of General Physiology*, 3rd ed., Little, Brown, 1964.

Dethier, V. G., and E. Stellar, *Animal Behavior: Its Evolutionary and Neurological Basis*, 2nd ed., Prentice-Hall 1964 (also pap.).

Lorenz, K. Z., *King Solomon's Ring: New Light on Animal Ways*, Crowell, 1952; Time Reading Program, 1964 (pap.).

Prosser, C. L., ed., *Comparative Animal Physiology*, Saunders, 1950.

Roe, A., and G. G. Simpson, eds., *Behavior and Evolution*, Yale Univ. Press, 1958.

Scheer, B. T., *Comparative Physiology*, Wiley, 1948.

Schmidt-Nielsen, K., *Animal Physiology*, 2nd ed., Prentice-Hall, 1964 (also pap.).

Scott, J. P., *Animal Behavior*, Univ. of Chicago Press, 1958; Natural History Press, Doubleday, 1963 (pap.).

Thorpe, W. H., *Learning and Instinct in Animals*, Harvard Univ. Press, 1956.

Tinbergen, N., *Social Behavior in Animals*, Wiley, 1953.

Von Frisch, K., *The Dancing Bees*, Harcourt, Brace & World, 1955; Harvest Books, Harcourt, Brace & World, 1961 (pap.).

Waters, R. H., D. A. Rethlingshafer, and W. E. Caldwell, *Principles of Comparative Psychology*, McGraw-Hill, 1960.

Part 5. The Mechanism of Evolution

Darwin, C., *The Origin of Species; and the Descent of Man*, Modern Library, 1948 (and many other editions).

Dobzhansky, Th., *Genetics and the Origin of Species*, Columbia Univ. Press, 1951.

Dobzhansky, Th., *Evolution, Genetics, and Man*, Wiley, 1955.

Grant, V., *The Origin of Adaptations*, Columbia Univ. Press, 1963.

Huxley, J. S., *Evolution: The Modern Synthesis*, Harper, 1943.

Lack, D., *Darwin's Finches*, Cambridge Univ. Press, 1947.

Mayr, E., *Animal Species and Evolution*, Harvard Univ. Press, 1963.

Moody, P. A., *Introduction to Evolution*, Harper, 1962.

Rensch, B., *Evolution Above the Species Level*, Columbia Univ. Press, 1960.

Simpson, G. G., *The Meaning of Evolution*, Yale Univ. Press, 1949; Yale Univ. Press, 1960 (pap.).

Simpson, G. G., *The Major Features of Evolution*, Columbia Univ. Press, 1953.

Simpson, G. G., *This View of Life; The World of an Evolutionist*, Harcourt, Brace & World, 1964.

Stebbins, G. L., *Variation and Evolution in Plants*, Columbia Univ. Press, 1950.

Tax, S., ed., *Evolution After Darwin* (3 vols.), Univ. of Chicago Press, 1960.

Part 6. The Diversity of Life

The number of works descriptive of particular groups of organisms is enormous. The following list includes a few of the more general books and a very small sprinkling of special studies. Any reader of this book should also be interested in observing and identifying organisms in the open, and for that purpose field guides are indispensable. We have not listed these individually but point out that two excellent series are published by Doubleday, New York, and by Houghton Mifflin, Boston, the latter under the editorship of R. T. Peterson. We also call attention to the multivolume Cambridge Natural History, which is now out of date and yet not superseded by any similar work of equal merit. *The New Naturalist* series published by Collins, London, is devoted to British fauna and flora. It is, however, of outstanding interest to any student of biology and has no parallel in the United States, where many popularizations of natural history are either pictorially magnificent but intellectually barren or, at another extreme, literary works with little scientific content.

Anderson, E. A., *Plants, Life and Man*, Little, Brown, 1952.

Blair, W. F., A. P. Blair, P. Brodkorb, F. R. Cagle, and G. A. Moore, *Vertebrates of the United States*, McGraw-Hill, 1957.

Bonner, J., and A. W. Galston, *Principles of Plant Physiology*, Freeman, 1959.

Borradaile, L. A., and F. A. Potts, *The Invertebrata*, Macmillan, 1935.

Boulière, F., *The Natural History of Mammals*, Knopf, 1954.

Brown, W. H., *The Plant Kingdom*, Ginn, 1935.

Clark, W. E. Le G., *History of the Primates*, British Museum (Natural History), 1956.

Clark, W. E. Le G., *The Antecedents of Man*, Harper & Row, 1963.

Corliss, J. O., *The Ciliated Protozoa*, Pergamon Press, 1961.

Fisher, J., *Birds as Animals*, Heinemann, 1939.

Frost, S. W., *General Entomology*, McGraw-Hill, 1942.

Gertsch, W. J., *American Spiders*, Van Nostrand, 1949.

Hyman, L. H. *The Invertebrates* (5 vols.), McGraw-Hill, 1940–59.

Mayr, E., E. B. Lindsley, and R. L. Usinger, *Methods and Principles of Systematic Zoology*, McGraw-Hill, 1953.

Moore, J. A., *Principles of Zoology*, Oxford Univ. Press, 1957.

Oginsky, E. L., and W. W. Umbreit, *An Introduction to Bacterial Physiology*, Freeman, 1954.

Romer, A. S., *Man and the Vertebrates*, Univ. of Chicago Press, 1941.

Rothschild, Lord, *A Classification of Living Animals*, Longmans, Green, 1961.

Schery, R. W., *Plants for Man*, Prentice-Hall, 1952.

Simpson, G. G., *Principles of Animal Taxonomy*, Columbia Univ. Press, 1961.

Sinnot, E. W., and K. S. Wilson, *Botany: Principles and Problems*, McGraw-Hill, 1955.

Smith, G. M., *Cryptogamic Botany*, McGraw-Hill, 1955.

Storer, T. I., and R. L. Usinger, *General Zoology*, McGraw-Hill, 1957.

Young, J. Z., *The Life of Vertebrates*, Oxford Univ. Press, 1950.

Young, J. Z., *The Life of Mammals*, Oxford Univ. Press, 1957.

Part 7. The Life of Populations and Communities

Allee, W. C., *The Social Life of Animals*, Norton, 1951.

Allee, W. C., A. E. Emerson, O. Park, T. Park, and K. P. Schmidt, *Principles of Animal Ecology*, Saunders, 1949.

Andrewartha, H. G., and L. C. Birch, *The Distribution and Abundance of Animals*, Univ. of Chicago Press, 1954.

Benedict, R., *Patterns of Culture*, Houghton Mifflin, 1934; Mentor Books, New American Library, 1946 (pap.).

Boyd, W. C., *Genetics and the Races of Man*, Heath, 1950.

Clark, G. L., *Elements of Ecology*, Wiley, 1954.

Clements, F. E., and V. E. Shelford, *Bio-ecology*, Wiley, 1939.

Coon, C. S., *The Story of Man*, Knopf, 1954.

Daubenmire, R. F., *Plants and Environment*, Wiley, 1947.

Davis, D. H., *The Earth and Man*, Macmillan, 1942.

Dice, L. R., *Natural Communities*, Univ. of Michigan Press, 1952.

Dobzhansky, Th., *Mankind Evolving*, Yale Univ. Press, 1962.

Elton, C. S., *The Ecology of Animals*, Wiley, 1946.

Graham, E. H., *Natural Principles of Land Use*, Oxford Univ. Press, 1944.

Hardin, G., *Nature and Man's Fate*, Rinehart, 1959.

Haskins, C. P., *Of Societies and Man*, Norton, 1951.

Kluckhohn, C., *Mirror for Man*, McGraw-Hill, 1949.

Kroeber, A. L., *Anthropology*, Harcourt, Brace & World, 1948.

Michener, C. D., and M. H. Michener, *American Social Insects*, Van Nostrand, 1951.

Odum, E. P., *Fundamentals of Ecology*, Saunders, 1959.

Oosting, H. J., *Plant Communities*, Freeman, 1956.

Pearl, R., *The Natural History of Populations*, Oxford Univ. Press, 1939.

Wheeler, W. M., *Social Life Among the Insects*, Harcourt, Brace & World, 1923.

Part 8. The Geography of Life

Cain, S. A., *Foundations of Plant Geography*, Harper, 1944.

Darlington, P. J., Jr., *Zoogeography: The Geographical Distribution of Animals*, Wiley, 1957.

Ekman, S., *Zoogeography of the Sea*, Sidgwick and Jackson, 1953.

Hesse, R., W. C. Allee, and V. P. Schmidt, *Ecological Animal Geography*, Wiley, 1951.

Matthew, W. D., *Climate and Evolution*, New York Academy of Science, 1939.

Newbigin, M. I., *Plant and Animal Geography*, Methuen, 1936.

Schimper, A. F. W., *Plant Geography upon an Ecological Basis*, Oxford Univ. Press, 1903.

Sclater, W. L., and P. L. Sclater, *The Geography of Mammals*, Kegan Paul, Trench, Trubner, 1899.

Simpson, G. G., *Evolution and Geography*, Oregon State Board of Education, 1953.

Wallace, A. R., *The Geographical Distribution of Animals* (2 vols.), Cambridge Univ. Press, 1876.

Wallace, A. R., *Island Life*, Macmillan, 1911.

Part 9. The History of Life

A basic reference work in this field is the monumental *Treatise on Invertebrate Paleontology*, edited by R. C. Moore, in the process of publication in many volumes by the Geological Society of America, New York, and the University of Kansas Press, Lawrence, Kansas.

Andrews, H. N., *Ancient Plants and the World They Lived in*, Comstock, 1947.

Arnold, C. A., *An Introduction to Paleobotany*, McGraw-Hill, 1947.

Beerbower, J. R., *Search for the Past*, Prentice-Hall, 1960.

Easton, W. H., *Invertebrate Paleontology*, Harper, 1960.

Gregory, W. K., *Evolution Emerging* (2 vols.), Macmillan, 1951.

Howells, W., *Mankind in the Making*, Doubleday, 1959.

Howells, W., ed., *Ideas on Human Evolution*, 1949–61, Harvard Univ. Press, 1962.

Kummel, B., *History of the Earth*, Freeman, 1961.

Moore, R. C., C. G. Lalicker, and A. G. Fischer, *Invertebrate Fossils*, McGraw-Hill, 1952.

Romer, A. S., *Vertebrate Paleontology*, Univ. of Chicago Press, 1945.

Simpson, G. G., *Horses*, Oxford Univ. Press, 1951.

Simpson, G. G., *Life of the Past*, Yale Univ. Press, 1953.

Part 10. The History of Biology

No history of biology is entirely satisfactory in our opinion, but the most important facts are given in the books listed. General works on the history of science almost invariably slight biology in comparison with the physical sciences. There are many biographies of individual biologists, often of the dull official kind but occasionally more sprightly; we can cite only a few. Some of the most fascinating historical readings in biology are accounts of explorations, especially in the nineteenth century; again, we can cite only a few. The books edited by Gabriel and Fogel and by Hall and cited in the reading list for Part I are also historical in nature.

Asimov, I., *A Short History of Biology*, Doubleday, 1964; American Museum Science Books, Doubleday, 1964 (pap.).

Bates, W., *The Naturalist on the River Amazon*, Dent, 1940.

Dampier, W. C., *A History of Science*, Macmillan, 1942.

Darwin, C., *The Voyage of the Beagle*, Dent, 1950.

De Beer, G., *Charles Darwin*, Doubleday, 1964.

Dubos, R. C., *Louis Pasteur, Free Lance of Science*, Little, Brown, 1950.

Dupree, A. H., ed., *Asa Gray 1810–1888*, Harvard Univ. Press, 1959.

Huxley, T. H., *Diary of the Voyage of H. M. S. Rattlesnake*, Chatto, 1935.

Irvine, W., *Apes, Angels and Victorians*, McGraw-Hill, 1955.

Locy, W. A., *Biology and Its Makers*, Holt, 1910.

Lurie, E., *Louis Agassiz, a Life in Science*, Univ. of Chicago Press, 1960.

Sarton, G., *A Guide to the History of Science*, Chronica Botanica, 1952.

Schuchert, C., and C. M. Levene, *O. C. Marsh, Pioneer in Paleontology*, Yale Univ. Press, 1940.

Singer, C., *A History of Biology*, rev. ed., Abelard-Schuman, 1959.

Wallace, A. R., *Travels on the Amazon and Rio Negro*, Ward, Lock, 1889.

Yonge, C. M., *A Year on the Great Barrier Reef*, Putnam, 1930.

ACKNOWLEDGMENTS FOR ILLUSTRATION SOURCES

Drawings by CARU Studios and others.

Opening to Part 1: photo by Hugh Spencer.

1–3 Adapted from R. Buchsbaum, *Animals Without Backbones,* rev. ed., Univ. of Chicago Press, 1948.
1–4 Adapted from Buchsbaum.

Opening to Part 2: photo by Carl Strüwe.

2–12 Adapted from H. Neurath and K. Bailey, eds., *Proteins: Chemistry, Biological Activity, and Methods,* Vol. 1, Part A, Academic Press, 1953.
3–2 Adapted from E. B. Wilson, *The Cell in Development and Heredity,* 3rd ed., Macmillan, 1925.
3–3 Adapted from figures in R. O. Greep, ed., *Histology,* Blakiston, 1954.
3–6 Reprinted with permission. © 1961 by Scientific American, Inc. All rights reserved.
3–10 From J. M. Allen, *The Nature of Biological Diversity,* McGraw-Hill, 1963.
3–11 From K. R. Porter and M. A. Bonneville, *An Introduction to the Fine Structure of Cells and Tissues,* Lea & Febiger, 1963.
3–12 From Allen.
3–13 From A. Lehninger, *Sci. Am.,* Vol. 205 (Sept., 1961).
3–14 From Dr. Wilhelm Bernhard, Institute for Cancer Research, Villejuif (Seine).
3–16 Adapted from C. L. Prosser, ed., *Comparative Animal Physiology,* Saunders, 1950.
3–19 Adapted from T. I. Storer, *General Zoology,* McGraw-Hill, 1943.
3–20 Adapted from L. H. Hyman, *The Invertebrates: Protozoa Through Ctenophora,* McGraw-Hill, 1940; L. A. Kenoyer, H. N. Goddard, and D. D. Miller, *General Biology,* 3rd ed., Harper, 1953.
3–21 Adapted with permission from G. M. Smith, E. M. Gilbert, G. S. Bryan, R. J. Evans, and J. T. Stauffer, *A Textbook of General Botany,* 5th ed., Macmillan, 1953.
3–22 Adapted from Smith *et al.;* in part from E. W. Sinnott and K. S. Wilson, *Botany: Principles and Problems,* 5th ed., McGraw-Hill, 1953.
3–23 Adapted from W. H. Brown, *The Plant Kingdom: A Textbook of General Botany,* Ginn, 1935.
3–24 Adapted from Smith *et al.*
3–27 From A. Lehninger, *Sci. Am.,* Vol. 205 (Sept., 1961).
3–35 Adapted from H. V. Neal and H. W. Rand, *Chordate Anatomy,* Blakiston, 1939; after Bremer.
3–36 Adapted from Greep; Storer; Neal and Rand; P. E. Smith, ed., *Bailey's Textbook of Histology,* 10th ed., Williams & Wilkins, 1940.
4–10 From A. Lehninger, *Sci. Am.,* Vol. 205 (Sept., 1961).

Opening to Part 3: photo by Dr. L. B. Shettles.

Opening to Chapter 5: photo from J. Brachet and A. E. Mirsky, eds., *The Cell,* Vol. 3, Academic Press, 1961.

5–1 Top photos by Chester F. Reather, FBPA, Johns Hopkins Univ. School of Medicine; lower photo by Richard D. Grill, Carnegie Institution of Washington.
5–5 From Dr. A. H. Sparrow, Brookhaven National Laboratory.
5–8 From D. M. Bonner, *Heredity,* Prentice-Hall, 1958.
7–6 A from 14th Symposium on Fundamental Cancer Research, *Cell Physiology of Neoplasia,* Univ. of Texas Press, 1960; B from J. H. Taylor, P. S. Woods, and W. L. Hughes, *Proc. Natl. Acad. Sci. U.S.,* Vol. 43 (1957).
7–8 Redrawn with permission. © 1963 by Scientific American, Inc. All rights reserved.
7–9 Redrawn with permission. © 1963 by Scientific American, Inc. All rights reserved.
7–15 Adapted from E. W. Sinnott, L. C. Dunn, and Th. Dobzhansky, *Principles of Genetics,* 4th ed., McGraw-Hill, 1950.
7–16 Adapted with permission from H. Reidel, *Arch. Entwicklungsmech. Organ.,* Vol. 132 (1935) (Springer-Verlag).
8–1 Adapted from A. F. Huettner, *Fundamentals of Comparative Embryology of the Vertebrates,* rev. ed., Macmillan, 1949.
8–2 Adapted from Wilson.
8–3 Adapted from L. G. Barth, *Embryology,* rev. ed., Holt, Rinehart and Winston, 1953.
8–4 Adapted from Huettner.
8–5 Adapted from J. Z. Young, *The Life of Vertebrates,* Oxford Univ. Press, 1950.
8–6 Adapted from Huettner; Storer.
8–7 Adapted from Barth.
8–8 Adapted from Barth.
8–10 Adapted from V. B. Wigglesworth, *The Principles of Insect Physiology,* 3rd ed., Methuen, London, 1947.
8–11 Adapted from C. S. Minot, *The Problem of Age, Growth, and Death,* Putnam, 1908.
8–12 Adapted from W. Etkin, *College Biology,* Crowell, 1950.
8–13 Adapted from T. J. Parker and W. A. Haswell, *A Textbook of Zoology,* 6th ed., Macmillan, 1940; in part from Storer.
8–14 Adapted from D'A. Thompson, *On Growth and Form,* Cambridge Univ. Press, 1942.
9–1 Adapted from E. Strasburger, *Textbook of Botany,* 6th ed., Macmillan, 1930.
9–4, 9–5, 9–6, 9–7, 9–8, 9–9, 9–11 Details of plant structure adapted from Smith *et al.*
9–12 Adapted from Sinnott and Wilson.
9–13 Adapted from P. Knuth, *Handbook of Flower Pollination,* Oxford Univ. Press, 1906; B and C adapted from Strasburger; D and E adapted with permission from O. Ames, *Botan. Museum Leaflets,* Vol. 5, No. 1 (1937) (Harvard Univ. Press).
9–17 Adapted from Buchsbaum.
9–18 Adapted from Storer.
9–19 Adapted from Young, *The Life of Vertebrates;* after J. S. Huxley, *Proc. Zool. Soc. London* (1914), with permission.
9–23 Adapted from A. S. Romer, *Man and the Vertebrates,* 3rd ed., Univ. of Chicago Press, 1941.

9–25 From B. M. Patten, *Human Embryology,* 2nd ed., McGraw-Hill, 1953; after Schroeder.
9–26 Adapted from W. F. Pauli, *The World of Life,* Houghton Mifflin, 1949; Storer.

Opening to Part 4: photo by Carl Strüwe.

10–2 Adapted in part from Storer.
10–4 Adapted from M. Thomas, *Plant Physiology,* Churchill, London, 1940.
10–5 Adapted in part from G. Hardin, *Biology: Its Human Implications,* 2nd ed., Freeman, 1952.
10–6 Adapted from Buchsbaum.
10–10, 10–12, 10–13, 10–14 Adapted from Storer.
11–1 Adapted from P. B. Weisz, *Biology,* McGraw-Hill, 1954.
11–2 Adapted from Thomas.
11–5 Adapted from Hardin; L. A. Borradaile and F. A. Potts, *The Invertebrata,* Cambridge Univ. Press, 1938; T. S. Hall and F. Moog, *Life Science,* Wiley, 1955.
11–7 Adapted from Pauli; in part from Hall and Moog.
11–10 Adapted from many sources.
11–11 Adapted from Storer.
11–13 Adapted from R. R. Kudo, *Protozoology,* 3rd ed., Thomas, 1946.
11–14 Adapted from E. Baldwin, *An Introduction to Comparative Biochemistry,* Cambridge Univ. Press, 1948.
12–1 Adapted from G. S. Carter, *A General Zoology of the Invertebrates,* 3rd ed., Sidgwick & Jackson, London, 1948.
12–2 Adapted from Hall and Moog.
12–3 Adapted from C. A. Villee, *Biology,* 2nd ed., Saunders, 1954.
12–6 Adapted from photos in Villee.
12–7 Adapted from C. D. Turner, *General Endocrinology,* Saunders, 1948; A. J. Carlson and V. Johnson, *The Machinery of the Body,* 3rd ed., Univ. of Chicago Press; Patten.
12–8 Adapted from photos in Turner.

Opening to Chapter 13: photo with permission from J. Z. Young, *Doubt and Certainty in Science,* Oxford Univ. Press, 1956.

13–1 Adapted from H. S. Jennings, *Behavior of the Lower Organisms,* Columbia Univ. Press, 1906.
13–3 Adapted from D. M. Pace and B. W. McCashland, *College Physiology,* Crowell, 1955; A. S. Romer, *The Vertebrate Body,* Saunders, 1950.
13–4 Adapted from L. Plate, *Allgemeine Zoologie und Abstammungslehre,* Fischer, Jena, 1924.
13–5 Adapted from J. S. Rogers, T. H. Hubbell, and C. F. Byers, *Man and the Biological World,* 2nd ed., McGraw-Hill, 1952; in part from Pace and McCashland; in part from K. Von Frisch, *The Dancing Bees,* Harcourt, Brace & World, 1955.
13–6 Adapted from Von Frisch; in part from Prosser.
13–7 Adapted from Parker and Haswell.
13–8 Adapted from Carter.
13–9 Adapted from Storer; Weisz; Pace and McCashland.
13–10 Adapted from Storer.
13–12 Adapted from Parker and Haswell; Storer.
13–13 Adapted from M. Demerec, ed., *Biology of Drosophila,* Wiley, 1950.
13–14, 13–15, 13–16 Adapted from Romer, *The Vertebrate Body*.
13–18 Adapted with permission from W. Penfield and T. Rasmussen, *The Cerebral Cortex of Man,* Macmillan, 1950.
13–19 Adapted from T. I. Storer and R. L. Usinger, *Elements of Zoology,* McGraw-Hill, 1955.
14–2 Adapted from Prosser.
14–3 B adapted with permission from J. Gray and H. W. Lissmann, *J. Exptl. Biol.,* Vol. 15 (1938); after Carter.

14–4 Adapted from Storer.
14–6, 14–7, 14–8 Adapted from Jennings.
14–9 Adapted from Hall and Moog.
14–10 Adapted from N. Tinbergen, *Social Behavior in Animals,* Methuen, London, 1953.
14–11 Adapted from Young, *Doubt and Certainty in Science*.
14–14 Adapted from K. D. Roeder, ed., *Insect Physiology,* Wiley, 1953; after Schneirla.
14–15 Adapted from Roeder.
14–16 Adapted from Von Frisch.

Opening to Part 5: photo © Fouke Fur Company, 1946.

15–4 Adapted with permission from S. C. Reed and E. W. Reed, *Evolution,* Vol. 4 (1950).
16–1 Adapted from Parker and Haswell; Storer; Borradaile and Potts; A. D. Imms, *Insect Natural History,* Blakiston, 1951; R. Hesse and F. Doflein, *Tierbau und Tierleben,* Teubner, Leipzig and Berlin, 1914.
16–5 Adapted from J. J. S. Cornes, *Nature,* Vol. 140 (1937).
16–6 Adapted from Hesse and Doflein.
16–9, 16–10, 16–12 Adapted with permission from D. Lack, *Darwin's Finches,* Cambridge Univ. Press, 1947.
16–13 Adapted from many sources, including Pauli; Romer, *Man and the Vertebrates; Columbia University Laboratory Manual in General Zoology,* 1945.
17–5 Based with permission on data from W. C. Boyd, *Genetics and the Race of Man,* Little, Brown, 1950.
17–6, 17–7 Adapted with permission from E. Mayr, *Systematics and the Origin of Species,* Columbia Univ. Press, 1942.
17–8 Adapted with permission from A. H. Miller, *Univ. Calif. (Berkeley) Publ. Zool.,* Vol. 44, No. 3 (1941).
17–9 Adapted with permission from J. Clausen, *Stages in the Evolution of Plant Species,* Cornell Univ. Press, 1951.

Opening to Part 6: photo © Douglas P. Wilson.

Opening to Chapter 18: adapted with permission from W. B. Scott, *A History of Land Mammals in the Western Hemisphere,* rev. ed., Macmillan, 1937.

18–3 Adapted from Storer; Romer, *Man and the Vertebrates*.
18–4 Adapted with permission from H. M. Kyle, *The Biology of Fishes,* Macmillan, 1937.
18–5 Adapted from Storer.
18–7 Adapted from A. E. Brehms, *Brehms Tierleben, Allgemeine Kunde Des Tierreichs,* 4th ed., Bibliographisches Institut, Leipzig and Vienna, 1912.
19–2, 19–4 Adapted from Smith *et al*.
19–5 Adapted from Kudo; G. M. Smith, *Cryptogamic Botany,* Vol. I., McGraw-Hill, 1938.
19–7, 19–9, 19–10 Adapted from Kudo; Borradaile and Potts.
19–12 Adapted from Buchsbaum.
19–13 Adapted from Hyman.
20–2, 20–3, 20–8 (top), 20–9 (A, B, C), 20–10 (D-G), 20–11 (A, B), 20–13 Adapted with permission from Smith *et al*.
21–1 Adapted from Hall and Moog.
21–3 Adapted from Parker and Haswell.
21–4 Adapted from Buchsbaum.
21–5 Adapted from Borradaile and Potts.
21–7 Adapted from Parker and Haswell; in part after W. S. Bullough, *Practical Invertebrate Anatomy,* Macmillan, 1950, with permission.
21–8 Adapted with permission from R. W. Hegner and K. A. Stiles, *College Zoology,* 6th ed., Macmillan, 1951.

21–9 Adapted from Buchsbaum; in part with permission from R. W. Pennak, *Fresh-Water Invertebrates of the United States,* © 1953, The Ronald Press Company.
21–10 Adapted from Buchsbaum; Borradaile and Potts.
21–11 Drawing adapted from Borradaile and Potts.
21–12 Adapted from Borradaile and Potts; Buchsbaum.
21–14 Drawings adapted from Storer; in part from Hall and Moog; lower photo by Dr. J. E. Webb from *Proc. Zool. Soc. London,* Vol. 125 (1955).
21–16, 21–17 Adapted from Storer.
21–21 A adapted from Hardin; B adapted from Storer; Hesse and Doflein.
22–1 Adapted from Storer.
22–6 Adapted from Borradaile and Potts.
22–7, 22–12 Adapted from Storer.
22–13 Luna moth and ants, Hugh Spencer; stinkbug and housefly, American Museum of Natural History; dragonfly, boll weevil, and louse, U.S. Department of Agriculture; mosquitoes, wasp, and bumblebee, Lynwood Chace from National Audubon Society; grasshopper, John R. Clawson from National Audubon Society.
22–14 Adapted from Romer, *Man and the Vertebrates.*
22–15 Adapted from Young, *The Life of Vertebrates.*
22–17 Adapted from Storer.
22–19 Adapted from J. R. Norman, *A History of Fishes,* Wyn, 1951.
22–22 Adapted with permission from E. H. Colbert, *Evolution of the Vertebrates,* Wiley, 1955.
22–23 Adapted from Storer.
22–25 Terns, Australian Information Service; jay and heron, American Museum of Natural History; kiwi, New Zealand Consulate; goose, U.S. Fish and Wildlife Service; hawk, Charles Halgren.
22–26 Adapted from Colbert.

Opening to Part 7: photo by Chamber of Commerce, San Jose, Calif.

23–1 Adapted from Wigglesworth.
23–3 Adapted from C. S. Pittendrigh, *Evolution,* Vol. 2 (1948).
24–2 Adapted from E. P. Odum, *Fundamentals of Ecology,* Saunders, 1949.
24–3 Adapted from W. C. Allee, A. E. Emerson, O. Park, T. Park, and K. P. Schmidt, *Principles of Animal Ecology,* Saunders, 1949; after O. Park, W. C. Allee, and V. E. Shelford, *A Laboratory Introduction to Animal Ecology and Taxonomy,* Univ. of Chicago Press, 1939, with permission.
24–4 Adapted with permission from Lack.
24–5 Parts 1 and 2 adapted from Turtox Key Card, courtesy, General Biological Supply House, Inc., Chicago; part 3 from Allee *et al.;* after Emerson, with permission.
24–6 Adapted from Storer; Parker and Haswell.
24–7 Adapted from Buchsbaum.
24–9 Adapted from A. K. Lubeck, *Geomorphology,* McGraw-Hill, 1939; after Dachnowski.
25–1 Adapted from Odum.
25–2 Adapted from Allee *et al.;* after R. Pearl, *The Biology of Population Growth,* Knopf, 1930 with permission; J. Davidson, *Trans. Roy. Soc. S. Australia,* Vol. 62 (1938).
25–3 Adapted from Allee *et al.;* after Gause.
25–4 Adapted from Allee *et al.;* after A. Leopold,

Wisconsin Conserv. Dept. Publ., No. 321 (1943), with permission.
25–5 Adapted from Allee *et al.;* after Gause.
25–6 Adapted from Allee *et al.;* after T. Park, *J. Exptl. Zool.,* Vol. 65 (1933), with permission.
25–7 Adapted from Young, *The Life of Vertebrates.*
25–8 Adapted from Allee *et al.;* after D. A. MacLulich, *Univ. Toronto Biol. Ser.,* No. 43 (1937), with permission of Univ. of Toronto Press.
25–9 Adapted from Allee *et al.;* after E. S. Russell and C. M. Yonge, *The Seas,* Warne, 1928, with permission.
25–10 Drawing reproduced with permission from Dr. Boyd W. Walker, Univ. of California at Los Angeles.
25–11 Adapted from G. L. Clarke, *Elements of Ecology,* Wiley, 1954; after J. R. Baker, *Proc. Zool. Soc. London* (1938), with permission.
26–2 Based on data by Sinnott, Dunn, and Dobzhansky.

Opening to Part 8: photos by Standard Oil Company of New Jersey.

27–3 Adapted from Allee *et al.;* after Wolcott.
27–8 After Pittendrigh.
27–11 Adapted from Allee *et al.;* after H. U. Sverdrup, M. W. Johnson, and R. H. Fleming, *The Oceans,* Prentice-Hall, 1942, with permission.
28–4, 28–5, 28–6, 28–7 Adapted from G. G. Simpson, *Evolution and Geography,* Oregon State Board of Education, 1953.
28–8 Adapted with permission from W. D. Matthew, *Climate and Evolution* (Spec. Publ., Vol. 1), N.Y. Academy of Sciences, 1950.

Opening to Part 9: photo from American Museum of Natural History.

29–6 Adapted from Simpson, *Evolution and Geography.*
30–2 Adapted from A. S. Romer, *Vertebrate Paleontology,* 2nd ed., Univ. of Chicago Press, 1945; Young, *The Life of Vertebrates.*
30–9, 30–10 Adapted with permission from E. H. Colbert, *Dinosaur Book,* McGraw-Hill, © 1954, American Museum of Natural History.
31–1 Adapted from C. O. Dunbar, *Historical Geology,* Wiley, 1949.
31–4 Adapted from G. G. Simpson, *The Meaning of Evolution,* Yale Univ. Press, 1949.
31–5 Adapted from Simpson, *The Meaning of Evolution.*
31–6 Adapted from Romer, *Vertebrate Paleontology;* after Simpson.
31–8 Adapted from Simpson, *Evolution and Geography.*
31–9 Adapted from Scott.
31–11, 31–12 Adapted from G. G. Simpson, *Horses,* Oxford Univ. Press, 1951.
31–13 Adapted from Romer, *Man and the Vertebrates.*
31–15 Gibbon and gorilla, New York Zoological Society; chimpanzee, American Museum of Natural History; orangutan, San Diego Zoological Garden.
31–18 Adapted from Romer, *Vertebrate Paleontology.*

Opening to Part 10: photo from American Museum of Natural History.

INDEX

(Page numbers in italics refer to illustrations.)

227; closed, 225; differentiation as process of, see Differentiation; evolution of, 240–43; and genes, 217; and genetic control, problem of, 234–36; growth as process of, 224, 225; morphogenesis as process of, 224, 225–32; open, 225

Deviation, in developmental mechanism, 241, 242

Devonian period, 463, 751 (table), 761, *764, 766, 774, 775, 776, 778* and n., 789

De Vries, Hugo, 166, 173 and n., 211 n., 480, 481, 834

Diabetes mellitus, 219, 220, 315, 342 (table), 343 n., 344

Dialysis, 61, *61,* 62

Diaphragm, 334, *335, 336*

Diastole, 312, 314, *314,* 315

Diatoms, 249 n., *486,* 515, *523,* 524, 721, 723, 724, 771, 793; fluctuations in numbers of, 646, *646*

Diatryma, 796, *796*

Dicotyledons, 96, 544, *544,* 545

Didinium, in prey–predator relationships, *668*

Dielectric constant, defined, 34

Differentiation, 225, 227, 228, *228,* 235, 236; and organ formation, 232

Difflugia urceolata, 517

Diffusion, and control of internal environment, 331; of molecules, 83–84

Digestion, 287–92; chemical reactions of, 288 (table); extracellular, 290, *290;* intracellular, 289, 290

Digestive systems, 289–92, *290, 291*

Dihybrid, 172

Dihydroxyacetone, 41, 42

Dimetrodon, 780, *781*

Dinichthys, 776

Dinoflagellates, 723

Dinosaurs, 608 n., *782, 783, 784, 785, 786, 787, 788,* 790, 795, 798, 799

Dionaea, 285–86, *286, 389*

Dioon edule, 539

Dipeptide, defined, 48

Diphtheria, 658

Diplacanthus, 776

Diploidy, 155–56, *156,* 160, 161

Diplophase, 249, 264–65, 266

Diplopoda, 591, *591, 592*

Diplovertebron, 777

Diplurus, 769

Diptera, 597

Dipterus, 776

Disaccharides, 41, 289 (table)

Dispersal of organisms, 733–37; barriers to, 735–37, *737;* routes of, 734, 735, *737;* sweepstakes, 736, *736,* 737, 739

Distribution, frequency, 469, 470, *470*

Distributions, disjunctive, 732, 741–42, *743,* 744

Divergent evolution, 501–02, 599

DNA, 15, 50, 130 and n., 131, 181, 197, 220, 221, 425 n., 493 n., 754, 832; and bacteria, 75, 197, 247, 510; in chromosomes, 75, 197; composition of, 50 (table), 200; double-stranded structure of, 51, *51,* 200, 201, 204; in gene mutation, 214, 215; genetic significance of, 197–98, 200–01; and heavy nitrogen, 201 and n., 202, *202;* in kappa particles, 222; pro-

tein synthesis controlled by, 206–11; replication of, 201–04, *202,* 211; single-stranded, 204 and n.; structure of, *52,* 200–01; synthesized by Kornberg, 203; see also Chromosomes; Genes; Genetics; Heredity; Reproduction

Dogs, 619; Pavlov's experiments with, 407, *408;* sense of smell in, 371

Dollo, Louis, 498 n.

Dominant allele, 169, 172, 215

Donax, 519

Dragonfly, *600*

Drive, biological, defined, 411

Drosera, 285, *286*

Drosophila melanogaster, 162, 175, 177, 425; chromosomes of, 156, *162,* 177, *177,* 180, 182, 185, *190,* 195, 196, *196;* eye color of, genes for, 180, 182, 196, *198, 199,* 216; genes of, 180, 182, 185, 189, 190, 195, 196, *198, 199,* 213, 214, 216, 218, 427; gynandromorphism in, chromosomal basis of, *184,* 185; interaction of genes in, 218; and interbreeding, 482; linkage maps for chromosomes of, *190;* mutations in, 195, 214; nervous system of, *378;* nondisjunction of X-chromosomes in, *183,* 184; non-random mating in, 434, 434–35; sex-linked heredity in, 180, *181,* 182; vestigial gene in, action of, 218, *218*

Dry ice, 29

Du Bois-Reymond, Emil, 828 and n., 832

Duck, canvasback, adaptive specialization of feet of, *446*

Dulcicole, 635 n.

Dunbaria, 780

Duodenum, *288,* 289, *341*

Dwarfism, 342 (table), 346 and n.

Dysentery, amebic, 658

Eagle, adaptive specialization of feet of, *446*

Ear (phonoreceptor), 362, 367, 368, 369, *370,* 463, *464,* 465

Eardrum, in mammals, 369

Earthworms, 284, *307,* 558; brains of, rudiments of, 402; digestion in, 290, *291;* hearts of, 306, *308;* locomotion of, muscles in, 391, *391;* reflexes of, 406–07; training of, 403

East, E. M., 191

Echinoderm superphylum, *577, 577–81, 578*

Echinoderms, 321, 402, 403, *486,* 553, 559–60, *562,* 562–63, 569 n., 571, 573, *578,* 771, *772, 773,* 783; characteristics of, 570 (table); larvae of, 579, *579;* phosphagen contained in, 580

Echinus, 147, 149, 564

Ecological opportunity, 457, 461

Ecology, 634 n., 824, 825; defined, 489; of *Homo sapiens,* 688–91; of ocean, 723–24; see also Stratification

Ecosystem, 634 n.

Ectocarpus, 251, 252, *252*

Ectoderm, 230, 232, 233, *577*

Ectoponus, 800

Edentata, 619

Eel, moray, *607*

Effectors, 354, 359, *359,* 367, 388–94; muscular, 390–92; in plants, 388–89, *389;* in protists, 389–90

Efferent nerves, 377, 384

Eggs (egg cells), 70, 630, 631; of *Aepyornis,* 70; cleavage of fertilized, *227,* 227–29, *228, 229;* of coral animal, 10; differentiation of, 228, *228;* fertilization of, 142, 143, 227, 246 and n., 248–49, 266–70, 267, *268, 269, 276;* of frog, 228, 229, *229;* human, *138,* 139, 142, 143; maturation of, 225–26, *226,* 227; of sea urchin, 88, 147, *149,* 228, *228,* 349; yolk of, 227, 229, *229,* 270; see also Germ cells

Ehringsdorf skull, 814

Ejaculation, 274

Electric current, as kinetic energy, 52–53

Electrolyte (ionic compound), 27 and n., 31, 35, 36

Electromagnetic radiations, spectrum of, *356;* and climate, 629

Electron acceptor, 116

Electron donor, 115, 116

Electron-dot symbols, 25–26, *26*

Electron micrograph, 71 and n., 111

Electron microscope, 71 and n., 73, 75, 76, 77, 78, 510, 829

Electron transfer, 27, 31

Electrons, addition of, as definition of reduction, 57; mass of, 24; negative electric charge of, 24; removal of, as definition of oxidation, 57; sharing, 26, *26;* and shells, 24–25, *25, 26, 26, 27*

Elementary particles, in atoms, 23 ff.; see also Electrons; Neutrons; Protons

Elements, 23; isotopes of, 116 n.; major, of living organism, 28–31; number of, 23; symbols for, 23; trace, 28, 31, 63

Elephant, 619, *620,* 644 n., 665 n., 728, *730, 731,* 803; heart rate in, 332

Elk, 242, 727

Ellobiophyra donacis, 519

Embolism, 309

Embolus, 309

Embryo, amphibian, 233–34, *234;* human, 143, *143,* 144, *240,* 240 and n., 274, *276;* shark, *240*

Embryo sac, 257 n.

Embryology, 831

Emotion, study of, 416–17

Empirical formula, in chemistry, defined, 37

Emu, 796

Emulsion, defined, 82

Endergonic reaction, 56, 57, 106, 118

Endocrine glands, 135, 320, 340; of man, 340–41, *341,* 342, 343 n., 344, *345,* 346–47, 349; of other animals, 348, 349

Endoderm, 230, 232, *577,* 580

Endometrium, *276*

Endoplasmic reticulum, 72, 75, 76, *76, 77, 78, 79,* 111

Endosperm, 258

Energy, of activation, 55, *55,* 118; atomic, 700; budget limitations of, 644, 646; defined, 52; free, of reaction, *55,* 56–57 kinetic, 52, 53, 54, 55, 639; potential, 52, 54, *54,* 639; pyramid of, 643, 644, *645;* transformations of, 53 and n., 639–41

Entelurus aequoreus, 453

Entomological Memoirs, 829 n.

Entomologists, 597, 823

Entropy, 56 and n.

Liver, bile from, 288; excretion by, 322–23, 331; glucose produced by, 130 *n.*, 331; glucose transported to, 295; plasma solutions regulated by, 331
Liverworts, 530, *531*
Lizard, *611*, 612, 720, *782*, 785, 795
Llama, *702*, 703
Lobelia, 263
Lobster, appendages of, 592 (table), *594;* elements in plasma of, 329 (table), 330
Locomotion, 388, 390; of earthworm, muscles in, 391, *391;* of flagellates, 518; of *Hydra*, 402, *402;* of mollusks, 583, *589*
Locusts, temperature regulation in, 337–38
Loeb, Jacques, 397, 400, 401, 829
Loris tardigradus, 806
Louse, *601*, 652, 653
Loxodonta, 620
Lumbricus, 290, *291*
Lungfish, 606, *607,* 608, 615, 726, 775, 776, *776*
Lungs, 463; and respiration in man, 300, *301,* 334, *335, 336;* and water loss, 322
Lupine plants, growth curves of, *237*
Lycopods, *532,* 532–33, 545 (table), 778
Lycopsids, 778, 779, *779*
Lyell, Charles, 747, 833, 835
Lymph, 104, 308, 317, 331
Lymph nodes, 317, *318*
Lymphatic system, 317, *318*
Lymphocytes, *311,* 317
Lynx, 713; population periodicity in, 672
Lyon, Mary, 181
Lyrebird, Australian, *387*
Lysine, 46 (table), 209 (table)
Lysosomes, 78, 111

Macaca irus, 621
Macallum, A. B., 329
Macaque, *621*
McLeod, J. J. R., 344
Macromolecule, 39
Macronucleus, 574
Macronutrients, 31
Maculae of saccule and utricle, *370*
Madrepora oculata, 554
Magnesium, as activator for enzymes, 63; amount in human body, 28 (table); in blood plasma, 329 (table); ionic bonds formed by, 27; as nutrient, 31; in sea water, 329 (table)
Malacologist, defined, 823
Malaria, 519, 655, 658
Malleus, *370*, 463, *464*
Malonic acid, 60
Malonyl coenzyme A, 133
Maltase, 289 (table)
Malthus, Thomas Robert, 666, 667
Maltose, 41, 289 (table)
Mammalogist, defined, 823
Mammals, and adaptive radiation, 565; Age of, 618, 619, 737, 738 and *n.*, 739 and *n.*, 741, 742, 752, 790, 792, 793, 797; brain function in, 381–83; compared with reptiles, *617*, 617–18; convergent evolution of placental and marsupial, 499, *500;* ears of, 361, 368–69, 463, *464*, 465; Eocene, 799 and *n.*, 800, *800, 801,* 803; evolution of, 615, 616–19, 622;

eyes of, *364;* fertilization in, 269; growth curve of, 237; heart of, *313;* historical record of orders of, *798, 799;* internal environment of, control of, 330–31, 337, 338, 339; marsupial, *see* Marsupials; Mesozoic, 796–97, *800;* modernization of, 797, 799–800, *800,* 803; Oligocene, 799, 800, *800, 801,* 803; Paleocene, 797, 798, 806; placental, *see* Placentals; reproduction in, man as example of, 273–76, *274, 275, 276*
Mammary glands, 273, *345*
Mammoth, woolly, *802,* 803
Man, Age of, 792; chromosomes of, 155, *179,* 181–82, 195; circulatory mechanism in, 312, 314, *314, 315, 316;* concept formation in, 415; Cro-Magnon, 814; early, in America, 814; ears of, 361, 369, *370;* ecology of, 688–91; elements in plasma of, 329 (table); endocrine glands of, 340–41, *341, 342,* 343 *n.*, 344, *345,* 346–47, 349; evolutionary advance represented by, 463; eyes of, 361, *364,* 367; fossil, 813; growth curves of, 237, 238, *238,* 239, *239,* 239–40; heart of, 312, 314, *314,* 315, *326;* hormones of, 340–47, 349; interbreeding by, 686; internal environment of, control of, 330–31, 332–36, *333, 335, 336,* 337, 338, 339; Java, 813, 814; kidney of, *324,* 324–25; land use by, 692–93; life span of, 665 and *n.*, 666; lung respiration in, 300, *301,* 334, *335, 336;* lymphatic system in, 317, *318;* modification of environment by, 691–94; Neandertal, 813, 814; nervous system of, *378;* nose of, 361, 371, *372;* origins of, 811–14; Peking, 813; primitive, 689, 690; races of, *685,* 685–88; reproduction in, 273–76, *274, 275, 276;* sensitivity of, to radiations of different wavelengths, *356;* societies formed by, 681–82; soil utilized by, 696–97; systematics of, 684–88; urinary system in, *323,* 324–25
Mandrill, *621,* 622
Mandrillus sphinx, 621
Manganese, 31, 115
Mantle, in mollusks, 583
Mariopteris, 779
Marmoset, 622
Marsupials, 242, 618, 728, 739, 742, 798; and placentals, convergent evolution of, 499, *500*
Marten, 713, 715
Mastigamoeba, 518, *518*
Mastigophora, *see* Flagellates
Mastodont, 803
Maternal effect, in genetics, 223
Mathematics, meaning of proof in, 16 *n.*
Matthew, W. D., 744 *n.*, 825
Maturation, and behavior, 417; *see also* Growth
Maze learning, 410, *410, 412*
Mechanical energy, defined, 52
Medawar, P. B., 239
Medical physiology, 827
Medicine, as life science, 824, 826
Medulla, adrenal, 135, 342 (table), 343 *n.*, 346, 385

Medulla oblongata, *333,* 334, *336, 379, 379,* 380
Medusas, in life cycle of coelenterates, 265, 266, *267,* 550, 551, *551, 552*
Meganeura monyi, 779
Megaspores, 257
Meiosis, 156–57, *157,* 158, 160, 171, 174, *174,* 212, 213, 227, 247, 248, 249; compared with mitosis, *159;* crossover at, 187, *188;* significance of, 160–61
Melanin, 218
Mendel, Gregor, 166, 173, 174, 175, 180, 189, 213, 473, 834; experiments of, on pea plants, 166–72, *167, 171, 172, 173,* 177, 430; and rule of independent assortment, 172, 174, 185; and segregation, genetic, 171–72, *173,* 174, 185; tabulation of results obtained by, 168, 169; tests of hypothesis by, 170–71, *171*
Mendelian (genetic) population, 423–31, *424,* 439, 834
Merismopedia, 513
Meristematic tissue, 91, *93,* 94, 225
Merychippus, 804, 805
Meselson, M., 201
Mesoderm, 232, 233 *n.*, 235, 576, *577,* 579, 580
Mesohippus, 805
Mesozoans, 548, 553, 569
Meoszoic era, 751 (table), *766,* 775, 776, 781, *782,* 783, 785, 789, 790, 793, 794, 796, 799
Mesquite communities, 662
Metabolism, 30, 106–12, 117, 547; and cell structure, 111–12; cellular, control of, 133–37; defined, 106; glucose, 120, *121;* intermediary, central position of acetyl CoA in, 123, *123;* rate of, and temperature, 630; universality of, 106–07
Metamorphosis, 272, 348, 349, 597, 608
Metaphase, *152,* 153, 158, *159,* 160, 174, *174,* 176
Metasequoia, 542 and *n.*
Metazoa, 575; origin of, 575–76
Methane, 36, 37, 39, 754 and *n.*, 756
Methionine, 46 (table), 209 (table)
Methyl ether, 37
Methyl group, 40 (table)
Metridium sessile, 554
Microbiology, 830
Microchemical analysis, of chromosomes, 197
Microclimates, 631, *632*
Micrograph, electron, 71 and *n.*, 111
Micron, defined, 508 *n.*
Micronutrients, 31
Micropallas whitneyi, 647
Microscelis leucocephalus, 476
Microscope, electron, 71 and *n.*, 73, 75, 76, 77, 78, 510, 829; Hooke's, *821;* invention of, *821;* light, 71, 510
Microspores, 256
Midbrain, 378, 379, *379,* 380, 381, 382, 385
Middle ear, 369, *370*
Millipedes, 591, *592*
Mimicry, 455
Mimosa, 389
Mineral cycles, 638–39, *639*
Minerals, essential to life, 31; in human body, 28

Minot, C., 239
Miocene epoch, 751 (table), *798, 799*, 800, 803, 806, 809, 812
Miohippus, 804
Mites, 592
Mitochondria, *72, 76, 77,* 77–78, *78, 79, 80,* 111, 221 and *n.,* 222, 359; elementary particles of, *127,* 128; role of, in critic acid cycle. 125–26, *126*
Mitosis, 89, 149 and *n.,* 150 and *n., 150, 151, 152,* 153–55, 157, 212, 234; compared with meiosis, *159;* significance of, 160
Moa, 796
Mode, and frequency distribution, 469
Modifying genes, action of, 217, *217*
Moeritherium, 802
Molar solution, defined, 33 *n.*
Molds, 509; antibacterial compounds synthesized by, 527; life cycle of, 526, *526;* mycelium of, 524, *525;* slime, 509, 521–22, *522;* spores of, 526, *526*
Mole, 499, *500,* 619, *620*
Mole (gram molecule), defined, 33 *n.*
Molecular biology, 22–63, 832
Molecules, 22, 23; diffusion of, 83–84, glucose, model of, *22, 43;* hydrogen fluoride, 27, *27;* motion of, as heat, 53, 83; protein, size of, 48 *n.;* three-dimensional structure of, 36–37, *37*
Mollusks, 553, 557, 569, 576, 577, *578,* 583–90, *584, 585, 586, 587, 589,* 778, 783; characteristics of, 570 (table); chemoreceptors of, 369; circulatory system of, 571; eyes of, *364,* 365; foot of, 583; hormones of, 348, 349; locomotion in, 583, *589;* of Ordovician period, 771
Molybdenum, 31
Monarcha castaneoventris, 476
Monas, flagellary motion in, 390, *390*
Mongoloid race, 687
Mongoose, 694
Monkeys, *621,* 622, 728, 740, 741, 807; howler, 622, 676, *729;* rhesus, 622; rudimentary concept formation in, 415
Monocotyledons, 96, 544, *544,* 545
Monoecious plants, 260
Monoglycerides, 289 (table)
Monomer, defined, 39
Monoplacophorans, 583, 584
Monosaccharides, 41, 42, 129, 289 (table)
Monotremes, 618 and *n., 619,* 728
Moose, 727
Morgan, Lloyd, 395, 396, 397, 400, 401, 408, 415, 829
Morgan, T. H., 177 *n.,* 180, 187
Morphogenesis, 224, 225–32
Morphology, defined, 22
Morula, 227
Mosasaur, 785 and *n.,* 790
Mosquito, *600*
Mosses, 530, *531,* 712; reproductive cycle in, 253–54, *255*
Moth, Luna, *600;* Yucca, *261,* 262
Motion receptors, 362, 369
Motivation, in learning, 410–12
Motor neurons, *359,* 360
Mouse, 217, *217,* 499, *500,* 619
Mouth, in early vertebrates, 602; taste organs in, 361, 372, *373*

Muller, H. J., 214
Müller, J. P., 828 and *n.,* 832
Multicilia, 518
Multiple-factor inheritance, 189, 191, *192*
Multituberculates, 798 and *n.*
Muscle(s), 99–101, *103,* 390, 392; ciliary, *364;* heart, 99, 100, *101,* 332, 357; intracellular respiration of, 295, *296,* 297; skeletal, 99, 100, *101,* 357, 391; smooth, 99, 100, *101;* striated, 99, 100, *101,* 357, 391; tonus of, 392; visceral, 391
Muscle-skeleton mechanisms, 393
Mushrooms, 525, *525*
Musk ox, *729*
Mussels, 587
Mutagenic agents, 214, 215
Mutants, auxotrophic, 205
Mutational equilibrium, 426–27, 430
Mutations, *211,* 211–15, 419, 426, 430, 440, 480; auxotrophic, 205; back, 426, 427; chromosome, 212; forward, 426, 427; frequency of, 213–14; gene, 212, 213–15, 221, 481; harmful, 440; induced in *Neurospora,* 204–05, *205;* induced by x-ray, 204, 205, 214; molecular mechanisms of, 214–15; as raw materials for evolution, 430
Muton, 221
Mutualism, *650,* 651–52
Mya, 588
Mycelium, 524, 525, *525*
Myelin, 101, *102;* from frog nerve, 64
Myiarchus tyrannulus, 647 *n.*
Myopia, *364*
Myrmecophaga, 620
Mytilus, 587
Myxedema, 342 (table), 343

NAD, 62, 123, 125 and *n.,* 126, 293
NADH, 62
NADP, 62, 116, 129, 133, 134, 293
NADPH₂, 116, 117, 123, 129, 130 *n.*
Naegleria, 518
Natural gas, 695
Natural selection, 17, 18 *n.,* 419, 426, 431–41, 765, 825; and adaptive coloration, 452–56; changing concepts of, 438; creative, 438–39; as historical process, 456–65; and inherited information, 466; limitations of, 440; nature of, 437–41; stabilization of, 439–40; *see also* Evolution; Nonrandom reproduction
Nautiloids, 590, 771, 776
Natuilus, 588, 589, 771
Neandertal man, 813, 814
Neanthropic skull, *811*
Nearctic subregion, 727, *728, 729,* 735
Nectar of flowers, 262
Necturus maculosus, 609
Negroid race, 687
Nekton, 723 and *n.,* 734
Nematocyte, *551*
Nematodes, 555, *558,* 569, *569,* 577 *n.,* 658; characteristics of, 570 (table)
Nemertina, 553
Neoceratodus, 607
Neo-Darwinism, 451 and *n.*
Neo-Lamarckism, 451 and *n.,* 452, 835

Neolenus, 772
Neon, 25, 633 *n.*
Neopallium, *379,* 380, *380,* 381, 382
Neoteny, 242 *n.*
Neotropical region, 728, *728,* 735, 737, 741, 742
Nephridia, 583 *n.*
Nereis, 561
Neritic zone, 723 and *n.*
Nerve cells, 101, *102,* 357–58, 359, *359,* 360, 405
Nerve fibers, 101, *102,* 357, 358, 359, 361, 377
Nerve impulses, 99, 101, 357, 358–59, 360, 377, 392 *n.*
Nerve-net behavior, 375, *376,* 401–02, *402*
Nerve physiology, 827, 828, 829
Nervous system, 339, 357, 361, 828; autonomic, 384, *384;* and behavior, 401–03; central, 377, 416; in early vertebrates, 602; and endocrine system, 340, 349, 350; history of, 374–83; and internal environment, 383–85; peripheral, 377; visceral, 383 and *n.*
Neural folds, 233
Neural plate, 233, *233*
Neurofibrillar system, of ciliate, 521, *521*
Neurohumors, 385
Neurohypophysis, 341, 342 (table), 343 *n.*
Neurons (nerve cells), 101, *102,* 357–58, 359, *359,* 360, 405
Neuropteris, 779, 779
Neurospora crassa, in gene-enzyme experiments, 204, *205,* 205 *n.,* 834
Neutrons, 24
Newton, Isaac, 820
Niacin, 294 (table)
Niches, 631 and *n.,* 632; and competition, 646–49; relationships among, 647–48
Nicotinamide, 62, 125 *n.*
Nicotinic acid, 293, 294 (table)
Night blindness, 294 (table)
Nitella, 524, *524*
Nitrates, 30, 128, 512, 637, 646
Nitrites, 30, 512, 637
Nitrogen, 30–31, 39, 131, 754; amount in dry air, 632; amount in human body, 28 (table); atomic weight of, 201 *n.;* as constituent of cell, 128; covalence of, 36; in end products of protein metabolism, 323; heavy, and experiments in DNA replication, 201 and *n.,* 202, *202;* in proteins, 47, 323; symbol for, 23
Nitrogen cycle, 637–38, *638*
Nitrogen fixation, 30, 109, 511 and *n.,* 512, 637, 652
Nitrous acid, 214
Noctiluca, 515, *515*
Node(s), atrioventricular, *333;* lymph, 317, *318;* sinoatrial, *333*
Nomenclature, 504–05; rule of priority for codes of, 505
Nonelectrolyte, defined, 31
Nonose, 41
Nonvascular plants, 303 and *n.,* 328
Nose, 371, *372;* human, 361, 371, *372*
Nostoc, 513
Notemigonus crysoleucas, 677
Notharctus, 806
Nothofagus, 742

Rhynchocephalians, *782,* 785, 790
Riboflavin, 62, 125, 294 (table)
Ribonucleic acid, *see* RNA
Ribonucleosides, 130 *n.,* 131
Ribonucleotides, 50, 129, 130
Ribose, 41, 42, 50, 62, 203 *n.;* synthesis of, 129–30
Ribose 5-phosphate, 130
Ribosomes, 75, *76, 77, 78,* 111, 206, *210*
Ribulose, 41, 42,
Rickets, 294 (table)
Rickettsias, 509, 658
Ring structure, 41, 42
Ringworm, 526, 658
RNA, 50, 51, 129, 130 and *n.,* 131, 134, 197, 198 *n.,* 203 *n.,* 405; composition of, 50 (table); messenger, 206, 207 and *n.,* 208, *208,* 209, 221; in nucleolus, 75; in protein synthesis, 51, 206, 207 and *n.,* 208, *208,* 209; transfer, 206, 207
Rock cycle, 639
Rocky Mountain spotted fever, 595
Rodentia, 619
Rods of retina, *364,* 367
Root, 91; tissue organization in, 97–98, *98*
Root cap, *98*
Root hairs, 98, *98,* 286, *287*
Root pressure, 286, 304
Rotifers, 548, 569, 577 *n.*
Round window, *370*
Roundworms, *see* Nematodes
Rust, grain infected by, 525

Sabella favonina, 561
Sabertooth, *801,* 803
Saccharomyces cereviseae, 525
Saccharomyces exiguus, in prey–predator relationship, *668*
Saccule of ear, *370*
Sacculina, 592, *650*
Saguaro cactus, 647, 648, *705*
Saint-Hilaire, Geoffrey, 589 *n.*
Salamander, 227, 242, 608, *609,* 793; gastrula in, 233; organizers in, 233; regeneration in, 235
Salivary glands, 320, *320*
Salmon, homing of, 373
Salsola pestifer, 693
Salt, defined, 27; dissociated in water, 34–35
Saltation, 480, 481
Salvia, 261, 262
Sample, in statistics, 175, 430, 503; biased, 433
Sampling errors, 430–31
San Joaquin fever, 526, 658
Sap, 304, 305, 330
Sarcodina, 515, 517, *517,* 518, *518*
Sarracenia, 285, *286*
Saturated fatty acids, 45
Saurischia, *782,* 783 and *n., 784,* 785
Scala media, *370*
Scala tympani, *370*
Scala vestibuli, *370*
Scallop, *587,* 588
Scapanus latimanus, 620
Scaphites nodosus, 794
Scaphopoda, 584, *585*
Scarlet fever, 658
Scenedesmus, 116
Schimper, A. F. W., 825
Schleiden, Matthias, 147
Schultze, Max, 830
Schwann, Theodor, 64, 65, 68, 106, 147, 830, 832
Schwann cells, 101, *102*

Science (scientific inquiry), conceptual scheme of, 16–18; as cumulative product, 821; explanation in, 14, 15, 16; generalization in, 14, 15, 16; hypotheses formed in, 15–16; method used in, 14, 828; motivations for, 13–14; natural, defined, 16 *n.;* nature of, 13–16; observation in, 14, 15; organization of, 12–13; origins of, 16–17; probability in, 16; proof in, meaning of, 16 and *n.;* questions and answers in, 14, 443; and technology, 14, 829; testability in, 14; theory in, 16; usefulness of, 14
Sclerenchyma, 92, *93*
Sclerotic coat, *364*
Scorpion, *595,* 778, 779
Scurvy, 293, 294 (table)
Sea anemone, 401, 550, *554,* 651
Sea cucumber, *9,* 563, *564;* respiratory trees in, 299, *299*
Sea hare, *586*
Sea horse, *607*
Sea lily, 562, *564*
Sea urchin, 560, *562,* 563; eggs of, *88,* 147, *149,* 228, *228,* 349
Seal, *418,* 419; temperature regulation in, 337
Seaweed, 529
Secondary protein structure, 48
Secondary spermatocytes, 157
Second-order reactions, 53 *n.,* 58
Secretin, 340 *n.,* 341
Secretion, 319–20; as cell extrusion process, 76; internal, 340 *ff.*
Secretory granules, in specialized cells, 77
Seed coat, 258
Seed ferns, 537, *537*
Seed plants, 255–58, *256,* 260; early, 535–43, *536, 537, 538, 539, 540, 542;* flowering, *see* Flowering plants
Seeds, dispersal of, 263–64, *264,* 541
Segmentation, 572
Segregation, in genetics, 171–72, *173, 174,* 174–75, *176,* 185
Selaginella, 533
Selection, interspecific, 483–84; natural, *see* Natural selection, sexual, 435
Selection pressure, 437
Selenium, 31
Self-fertilization, 277
Self-pollination, 260
Semen, 142, 274
Semicircular canals, 369, *370,* 602
Semipermeability, of cell membrane, 84–85, *85*
Sense organs, *see* Receptors
Sensory neurons, *359,* 360
Sensory physiology, 827
Sepal, *256, 258*
Sepia, 588
Sequoia gigantea, 542
Serine, 46 (table); 60, 209 (table)
Serum, blood, 216, 308, 309
Servetus, Michael, 827
Sexual behavior, 411
Sexual reproduction, 144, 160, 245, 246, 247, 248; advantages of, 247; nuclear cycle in, 157, *157;* plus vegetative reproduction, 265–66; variability promoted by, 247; *see also* Chromosomes; DNA; Genes; Genetics; Heredity
Sexual selection, 435

Shark, 102, *313,* 322, 605, *605,* 606, 723
Sheep, 619; Marco Polo, *731*
Shells, electron, 24–25, *25,* 26, *26, 27*
Shoot system, 91 *ff.*
Shrew, 619; brain of, *379*
Shrimp, 447
Shunt, oxidative, 129, 130
Sibling species, 501 *n.*
Sickle-cell anemia, 208
Sidneyia, 772
Sieve plates, 92, *94*
Sight, sense of, *see* Eye; Photoreceptors
Sigillaria, 779, *779*
Silicon, 39
Silurian period, 751, (table), 764, *766, 773,* 774, 778
Sinanthropus, 813
Sinuses, 307 and *n.,* 308, *372*
Siphonophores, 550, *553*
Sitona, 630
Skate (chondrichthyan), 605, 606
Skeletal muscle, 99, 100, *101,* 357, 391
Skeletal system, 74, 102, 391, 392, 571; *see also* Bones; Cartilage
Skin, receptors in, 361; tissue organization of, 99, *100*
Skull, 602, *811;* Ehringsdorf, 814; neanthropic, *811;* paleanthropic, *811;* Steinheim, 814; Swanscombe, 813
Sleeping sickness, African, 515, 658 and *n.*
Slime molds, 509, 521–22, *522*
Sloth, 728, *729,* 741
Small intestine, 288, 289 (table)
Smallpox, 658
Smell receptors, 361, 362, 369, 371, *372*
Smith, W., 833
Smooth muscle, 99, 100, *101*
Smut, grain infected by, 525
Snail, 584, *584, 585, 586,* 771, *773,* 779; circulatory system in, *308;* and symmetry, 572; *see also* Gastropoda
Snake, *611,* 612, 720, *782,* 785, 795
Social behavior, 417
"Social Darwinism," 438
Societies, defined, 678; human, 681–82; insect, 678–81, *679, 680*
Sociology, as life science, 824
Sodium, amount in human body, 28 (table); atomic structure of, *25,* 27; in blood plasma, 329 (table); ionic bond with chlorine, 27; as nutrient, 31; in sea water, 329 (table)
Sodium-24, half-life of, 116 *n.*
Sodium bicarbonate, 36
Sodium chloride, formula for, 27; freezing point related to concentrated solution of, *86;* as ionic compound, 27
Sodium hydroxide, 35, 36
Soil, composition of, 633; erosion of, 697, *698;* man's utilization of, 696–97; organisms in, 633
Sol state of colloid, 82 and *n.*
Solar system, age of, 750
Solen, 587.
Solution, defined, 81; molar, defined, 33 *n.*
Soma, in one-way relation to germ cells, 165–66, *166*
Sonneborn, Tracy, 221
Sound receptors, 362, 367, 368, 369, *370,* 463, *464,* 465